HANDBOOK OF HYDROXYBENZOPHENONES

Handbook of Hydroxybenzophenones

by

Robert Martin

*formerly of the Institut Curie,
Paris, France*

SPRINGER-SCIENCE+BUSINESS MEDIA, B.V.

A C.I.P. Catalogue record for this book is available from the Library of Congress.

ISBN 978-94-010-5872-8   ISBN 978-94-011-4347-9 (eBook)
DOI 10.1007/978-94-011-4347-9

Printed on acid-free paper

To my dear Angèle, always so touchingly devoted to me throughout these many years. She reminds me every day of the sweet title of this song by Jean Ferrat: "What would I be without you?".

Robert MARTIN

CONTENTS

Chapter 2   Substituted Hydroxybenzophenones   *(Class of METHANONES).*

PREFACE

Two centuries of organic chemistry have already yielded millions of molecules, either synyhesized or isolated as natural products created by biosynthesis, but much still remains to be done. Therefore, from time to time, it is probably useful to gather and classify the scattered data concerning this or that class of compounds in order to save time for chemists planning new syntheses or natural products isolations.

As a continuation of his previous Handbook of Hydroxyacetophenones, R. MARTIN now tackles the Hydroxybenzophenones and related aromatic ketones in a new Handbook including circa 1900 molecules and 1500 references up to June 1999.

The data concern the syntheses routes or natural origin, physicochemical and spectroscopical characteristics available in the literature and, should the occasion arise a criticism of dubious structures or constants.

Since hydroxyketones are widely utilized, themselves or as intermediates, in numerous syntheses interesting, among others, pharmaceuticals, dyes, agrochemicals, perfumes and plastic preservatives, this handbook will be of great value for both academic and industrial research chemists. The multiple ways of hydroxybenzophenones syntheses herein described will certainly help chemists, as most of these methods can be applied to prepare analogues in aromatic and even in some heterocyclic series.

The three exhaustive indexes enable the reader to rapidly find a compound and its isomers if previously published or patented and to select the best way for preparation.

Considering the large amount of information included in this Handbook, I guess it will be entirely successful.

P. Demerseman

ACKNOWLEDGEMENTS

I wish to express my heartily thanks to Dr. Pierre Demerseman who accepted me in his Laboratory at Institut Curie in 1987, and kindly revised my manuscript. The foreword of this Handbook was also redacted by Dr. Pierre Demerseman. I most appreciate this mark of kindness.

I am also grateful to Dr. J.-P. Buisson, always so amiable and efficient, whose knowledge of word-processing largely contributed to the final page-setting of this work.

My thanks are also directed to Prof. Claude Monneret, Head of the Chemical Department at Institut Curie, who has always been so benevolent to me, and all his collaborators for their warm welcome at each of my visits.

I thank my son Serge Martin for friendly advice on the english redaction of this book. Moreover, Mr. Serge Martin was a constant aid to me as regards data processing.

The author also thanks Prof. Jean Paul Guetté for all his good advices.

Various friends who readily accepted to translate foreign publications are also to be acknowledged here, in particular Dr. Jean Burkhard who has been of unvaluable help for translating german papers over the last 30 years. The diverse abbreviations used in ancient reviews - particularly *Chemisches Zentralblatt* - had no secrets for him.
In this connection, thanks are due to Mrs. Feiga Weisbuch for her precious assistance as regards rumanian and russian texts, as well as to Miss Marie-Françoise Liachenko and Dr. Daniel Dauzonne. I wish to express my thanks to Mrs. Mireille Guyonneau, Mrs. Elisabeth Matarasso, Mrs. Françoise Rémy and Mrs. Simonne Rissé for their keen contribution to my bibliographic research, as well as the Orsay University Library for their helpful kindness towards me for 30 years now.
I am also grateful to Mrs. Colette Ledoux for judicious advice in the field of scientific edition.

Before closing, I would like to remember my dear departed. My affectionate thoughts are turned towards Prof. Léon Denivelle who transmitted to me his passion for aromatic organic Chemistry in 1945, and Prof. Albert Kirrmann who accepted me among his students in 1961 and was always so amiable and well-disposed whenever I went to him. I cannot mention without emotion Prof. Albert Saint-Maixen who largely communicated to me his knowledge on analytical Chemistry.
I also have a personal thought towards my friends from the industry who left us too soon. I am particularly thankful to Drs. Henri Barbier, Félix Lepors and Henri Ruelleux (SPCA, Ltd.) who gave me the practical means to carry about my work on aromatic hydroxyketones. In this firm, I started my research on the Fries reaction. I also wish to acknowledge the late Dr. François Krausz who, at that time, made me benefit from his precious advice.

Robert MARTIN

INTRODUCTION

Acylphenols are used as starting material for an extremely large number of syntheses in organic chemistry, leading to a wide range of applications. For this reason, it seemed interesting to offer a dictionary of Benzoylphenols or Hydroxybenzophenones, since they constitute the most numerous and currently used representatives of this series of compounds. Moreover, most syntheses of Hydroxybenzophenones can be extended to their other homologs not described here.

The dictionary covers over 1,900 Hydroxybenzophenones methodically classified under the official nomenclature of **"Methanones"** according to the International System (I.U.P.A.C.) and the recommendations given in the Chemical Abstracts "Collective Index" (9 CI) since 1972.

About 1,500 bibliographic references are compiled in this work. Names of periodicals are abbreviated according to the Chemical Abstracts Service Source Index (C.A.S.S.I.). Whenever Hydroxybenzophenones can be obtained from plants, sources and corresponding references are given.

For each compound described, the different protocols of synthesis are presented as well as the main physico-chemical characteristics and references of spectroscopic data. Besides, the usual abbreviations are also indicated at the end of this Dictionary.

For precise and quick research of an Hydroxybenzophenone, you can refer either to the classification by molecular formula (**Molecular Formula Index**) or to the **Chemical Abstracts Registry Numbers** table. An entry **Usual Names Index** including the current names of some Hydroxyacetophenones and their precursors is also available.

Finally, a glance through any chapter of this Dictionary will inform the reader on the diverse ways of synthesizing Hydroxybenzophenones. These methods can be used to obtain hitherto unknown analogs in the related series.

PART 1 - MONOAROYLPHENOLS

Chapter 1. Unsubstituted Hydroxybenzophenones *(Class of METHANONES)*

1. MONOHYDROXYBENZOPHENONES

(2-Hydroxyphenyl)phenylmethanone

[117-99-7] $C_{13}H_{10}O_2$ mol.wt. 198.22

Syntheses

-Preparation by Friedel-Crafts acylation of benzene in the
   presence of aluminium chloride,
   *with 2-hydroxybenzoyl chloride (salicylic acid chloride)
   [546], (52%) [265], (39%) [346], at temperature <60° [244];
*with o-anisoyl chloride [21, 385, 498, 554, 775, 946, 1306, 1425], (65%) [1397], (57%) [188],
   (53%) [1111], (37%) [179], (30%) [345]. Demethylation occurred during the Friedel-Crafts
   acylation, especially in the presence of ferric chloride at 130-140° [21];
*with 2-ethoxybenzoyl chloride in a boiling water bath (42%) [1397].
-Preparation by dealkylation,
*of 2-methoxybenzophenone,
   with aluminium bromide in refluxing benzene for 4 h (96%) [1059];
   with 48% hydrobromic acid in boiling acetic acid for 2 h [1309];
   with boron tribromide in methylene chloride at r.t. for 12 h [766], according to [891];
*of 5-tert-butyl-2-methoxybenzophenone with aluminium chloride in benzene at 65-70° for 45 h
   (79%) [774, 775]. The starting keto ether was obtained by reaction of benzoyl chloride with
   4-tert-butylanisole in the presence of zinc chloride in refluxing tetrachloroethane for 40 h [774];
*of 5-tert-butyl-2-hydroxybenzophenone with aluminium chloride in benzene at 65-70° for 45 h
   (63%) [775].
-Preparation by Fries rearrangement of phenyl benzoate,
*with aluminium chloride [692],
without solvent
   at 200° for 20 min [766] or for 15 min (26%) [864], at 190° for 7 h (26%) [312], at 140° (24%)
   [490] or at 140° for 15 min (27%) [311], at 130° for 1 h, then at 160° for 1 h (30%) [1447],
   between 100 to 150° for 3 h (minor product) [1363], or at 63° for 4 h (<2%) [1449],
with solvent
   in nitrobenzene at 98° for 5 h (<9%) or at 120° for 2 h (15%) [126];
*with an aluminium chloride and sodium chloride mixture at 140-200° (37%) [490];
*with aluminium bromide in chlorobenzene at 110° for 15 min (2%) or for 4 h (20%) [325];
*with aluminium iodide in refluxing acetonitrile (82°) for 10 h (23%) [844];
*with boron trifluoride-etherate in the presence of phenol in refluxing benzene for 3 h (68%) [725];
*with ferric chloride in a boiling water bath for 6 h [611];
*with titanium tetrachloride in nitrobenzene at 60° for 18 h (7%) [311] or in refluxing nitromethane
   for 30 min (15%) [1446];
*with TFMS (trifluoromethanesulfonic acid) in tetrachloroethane at 170° for 24 h in a sealed tube
   (39%) [386];
*with Nafion-H, a polymeric perfluorinated resin sulfonic acid, in refluxing nitrobenzene for 12 h
   (24%) [1004];
*with Nafion-XR, a H+-form ion exchange resin, at 150° for 4 h (23%) [1231];
*with polyphosphoric acid at 100° (1%) [960];

*with K10 montmorillonite, in refluxing N,N-dimethylformamide for 8 h (17%) [661] or in the
presence of Na-exchanged montmorillonite clay as catalyst for 5 h at 140° (24%) [1409].
-Preparation by saponification of 2-(benzoyloxy)benzophenone [65], with sodium hydroxide in
refluxing dilute ethanol for 1 h [1052] or for 10-15 min (45%) [64] or with potassium hydroxide
[1055] in methanol at 15° for 24 h (70%) [1129]. The starting keto ester was obtained by oxidation
of 2-benzylphenyl benzoate with chromium trioxide in boiling acetic acid [65], of 2,3-diphenyl-
benzofuran [64, 1052] and of 2-(4-methylphenyl)-3-phenylbenzofuran or 2-(4-methoxyphenyl)-
3-phenylbenzofuran [1055]. The starting keto ester can be also obtained by photooxidation of
2,3-diphenylbenzofuran in chloroform in the presence of methylene blue as initiator [1129].
-Also obtained by reaction of benzotrichloride with phenol in the presence of aqueous sodium
hydroxide in a water bath (14%) [527], (1%) [561].
-Preparation by reaction between phenyloxymagnesium bromide complexed with HMPT and
benzaldehyde in refluxing benzene for 48 h (30%) [250].
-Also obtained by reaction of salicylaldehyde with iodobenzene by using a catalyst system of
palladium chloride/lithium chloride in the presence of sodium carbonate in N,N-dimethyl-
formamide at 100° for 3.5 h (91%) [1204].
-Also obtained by reduction of o-hydroxybenzophenone 2,4-dinitrophenylhydrazone with stannous
chloride dihydrate in the presence of concentrated hydrochloric acid in boiling dilute acetic acid for
1 h (73%) [310].
-Also obtained (poor yields) by action of benzoic acid with phenol,
*in the presence of polyphosphoric acid at 100° (1%) [960];
*in the presence of Amberlyst-15 in refluxing chlorobenzene for 48 h (10%) [582].
-Also obtained by action of benzoyl chloride,
*with phenol in the presence of aluminium chloride or titanium tetrachloride in nitrobenzene at 60°
for 18 h (6-7%) [308];
*with phenyl borate in the presence of aluminium chloride in refluxing carbon disulfide for 2 h
(6%) [1373].
-Also obtained (poor yield) by diazotization of 2-aminobenzophenone, followed by hydrolysis of
the diazonium salt formed* [497, 498], (6%) [499] or by thermal decomposition of 2-benzoyl-
benzenediazonium fluoborate in dilute sulfuric acid between 25 to 50° (37%) [345]; *In these
conditions, the fluorenone was the major compound formed [498].
-Also obtained (poor yield) from o-bromophenyl benzoate on treatment with n-butyllithium in a
mixture of ethyl ether, hexane and tetrahydrofuran at -70° for 2 h, followed by treatment with
saturated aqueous ammonium chloride (7%) [598, 916].
-Also obtained (poor yield) by pyrolysis of o-(allyloxy)benzophenone at 750°, under pressure of
1x10^{-3} Torr, for 2 h (8%) [243].
-Also obtained by cleavage of phenylindoxazene with fuming hydriodic acid (d = 1.70) in the
presence of red phosphorous at 140-160° for 6-7 h. The intermediate compound formed
(2-hydroxybenzophenone imine), unstable, gave by hydrolysis the expected ketone [283].
-Preparation from thioxanthen-9-one 10,10-dioxide (SM) by a three-steps synthesis: refluxing SM
in 2% sodium hydroxide-65% dioxane-water solution for 4 h gave the 2-(2-hydroxybenzoyl)-
phenylsulfinic acid (25%). The former, by reaction with mercuric chloride in refluxing aqueous
acetic acid for 4 h, led to the 2-chloromercuri-2'-hydroxybenzophenone (69%). Removal of the
chloromercury group was achieved with concentrated hydrochloric acid in refluxing ethanol for
2 h (91%) [160].
-Also obtained by heating xanthone with lead monoxide [498].
-Also obtained by action of an alkaline solution of sulfur on 4-nitrodiphenylmethane [1394].
-Also obtained by photo-Fries rearrangement of phenyl benzoate,
*in methanol (3%) [102], (48%) [1335], using 3000 Å lamp (35%) or using 2537 Å lamp (41%)
[1070];
*in isopropanol (35%) [1070], (<4%) [102];
*in butanol (48%) [1070];
*in ethanol at 30° for 3 days (20%) [52] or at r.t. for 70 h (18%) [600];
*in water (50%) [268]. Adding amylose, α-cyclodextrin or β-cyclodextrin does not improve the
yield (45%, 34% and 18% yields, respectively) [268], although in the presence of β-cyclodextrin, a
99% yield has been obtained [1335]; in water in the presence of soluble starch (67%) [268] or in
the presence of SDS (sodium dodecyl sulfate), a micelle, was obtained a 74% yield [1276];

*in hexane in the presence of zeolites for 1 h (92-98%) [1069];
*in benzene for 12-24 h (55%) [1335] or at 52° for 8 h (13%) [419];
*in cyclohexane at 55° for 24 h (11%) [419];
*in pentane (14%) [102]. Adding silica gel does not improve the yield (12%);
*in dioxane at 61° for 24 h (13%) [419];
*in ethyl ether (11%) [1070];
*without solvent on K10 montmorillonite using microwave radiations (640 W, 10 min) (25%) [661].
-Also obtained by photo-Fries rearrangement of phenyl salicylate in ethanol for 76 h (19%) [600].
-Also refer to: [200, 202, 219, 243, 317, 348, 384, 421, 438, 598, 642, 665, 666, 810, 1068, 1123, 1303].
N.B.: Na salt [1331].

oil [1059];
m.p.   41° [312, 561, 1052, 1309], 40-41° [188, 499], 39-41° [265],
       39-40° [498, 946, 1129], 39° [67, 310, 311, 385, 1111, 1397],
       38°8 [1437], 38°5-39° [345, 354], 38-39° [160, 661, 774, 775, 1276],
       38° [250, 490, 1055], 37°5-38° [346], 37-39° [400], 37° [606, 864, 1178, 1179],
       36°5-37° [419], 36°5 [608], 36-38° [244], 36° [283], 35-36° [754];
b.p.$_{0.2}$ 124-126° [243], b.p.$_{1.5}$ 127-133° [1425], b.p.$_{1.9}$ 136-138° [160],
b.p.$_{12}$ 170-185° [775], b.p.$_{14}$ 175° [179], b.p.$_{15}$ 177° [527],
b.p.$_{10}$ 200° [611], b.p.$_{560}$ 250° [283];
^{1}H NMR [243, 250, 267, 610, 661, 754, 996, 1129, 1177, 1178, 1179, 1181];
^{13}C NMR [243, 754, 1177];
IR [68, 250, 354, 661, 754, 792, 946, 996, 1052, 1129, 1178, 1179, 1268, 1313, 1315, 1316];
UV [354, 606, 607, 610, 793, 794, 946, 1392, 1406]; MS [125, 250];
pK_a [606, 608, 791, 792];   TLC [377];
polarographic study [1223];   thermal behaviour [1178, 1179].

(3-Hydroxyphenyl)phenylmethanone

[13020-57-0] $C_{13}H_{10}O_2$ mol.wt. 198.22

Syntheses

-Preparation by aromatization of 5-benzoyl-2-cyclohexenone,
*in the presence of 10% Pd/C in refluxing xylene for 6 h (60%) [158];
*in the presence of lithium bromide and cupric bromide in refluxing acetonitrile for 1 h (75%) [158].
-Preparation by reaction of m-anisoyl chloride with benzene in the presence of excess aluminium chloride first at 20°, then at reflux for 2 h (95%) [82].
-Preparation by demethylation of m-methoxybenzophenone (SM),
*with boron tribromide in methylene chloride (55-60%). SM was obtained by condensation of m-methoxybenzoyl chloride with benzene in the presence of aluminium chloride [346];
*with aluminium bromide in refluxing benzene for 2 h (89%) [1059];
*with 48% hydrobromic acid [1397] in refluxing acetic acid [1309], for 6 h (60%) [251]. SM was prepared by a two-steps synthesis: first, formation of 3-methoxybenzhydrol (SM1) by condensation of a Grignard reagent (prepared from m-iodoanisole) with benzaldehyde. Then, oxidation of SM1 by adding a solution of potassium dichromate in dilute sulfuric acid in an hot solution of SM1 in acetic acid and heating for 15 min in a water bath (60%) [251].
-Preparation by reductive deamination of 2-amino-5-hydroxybenzophenone [478].
-Preparation by diazotization of m-aminobenzophenone, itself obtained by reduction of m-nitrobenzophenone [1289, 1306].
-Also obtained by treatment of m-bromophenol in tetrahydrofuran with tert-butyllithium in pentane for 15 min at -78° under argon, after which a solution of phenyl-N,O-dimethylhydroxamide in tetrahydrofuran was slowly added (68%). -Refer to: Chem. Abstr., 119, 49010t (1993)*.

-Also refer to: [103].

m.p. 118° [82], 117° [251], 116° [478, 1059, 1289, 1309, 1397], 115° [346], 114-116°*;
b.p.$_{20-30}$ 285-295° [251];
IR [180].

(4-Hydroxyphenyl)phenylmethanone

[1137-42-4] $C_{13}H_{10}O_2$ mol.wt. 198.22

Syntheses

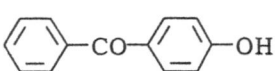

-Preparation by condensation of benzotrichloride,
*with phenol in the presence of aluminium chloride in carbon
  disulfide at 0° (90%) [973] or in the presence of
zinc oxide in a water bath [365];
*with p-bromophenol in 30% sodium hydroxide solution at 80-85° (81%) [1009]. There is
  substitution of the bromine atom by the benzoyl group.
-Preparation by Fries rearrangement of phenyl benzoate,
*with aluminium chloride [920], (55%) [220],
without solvent
  at 140° (usually for 15 min), (quantitative yield) [187], 88-91% [470, 1143], (69%) [490], (65%)
  [864], at 63° for 4 h (19%) [1449], at 130° for 1 h and at 160° for 1 h (9%) [1447],
with solvent
  in nitrobenzene at 98° for 5 h (64%) [126], at 120° for 2 h (54%) [126] or at 45° for 3 h (4%)
  [877];
  in refluxing ethylene dichloride (84°) for 1.5 h (37%) [548];
*with an aluminium chloride and sodium chloride mixture, at 140-200° (50%) [490];
*with aluminium iodide in refluxing acetonitrile (82°) for 10 h (46%) [844];
*with aluminium bromide in chlorobenzene at 110° for 4 h (17%) [325];
*with ferric chloride in a boiling water bath for 6 h (28%) [611];
*with titanium tetrachloride at 60° for 18 h (83%) [311] or in refluxing nitromethane for 30 min
  (36%) [1446];
*with stannic chloride at 120° for 18 h (7%) [188];
*with hydrofluoric acid at 55° for 4 h (70%) [986];
*with polyphosphoric acid at 80° for 2.5 h (25%) [459], at 100° (6%) [960] or in a boiling water
  bath for 30 min (6%) [961];
*with Nafion-H, a polymeric perfluorinated resin sulfonic acid, in refluxing nitrobenzene for 12 h
  (48%) [1004];
*with Nafion-XR, a H$^+$-form ion exchange resin, at 150° for 4 h under nitrogen (33%) [1231];
*with ion-exchanged stratified clay catalyst, in the presence of phenol, at 180° for 4 h (45%) [441];
*with K 10 Montmorillonite as catalyst for 5 h at 140° (80%) [1409].
-Preparation by dealkylation,
*of p-methoxybenzophenone,
  with aluminium bromide in refluxing benzene for 4 h (95%) [1059];
  with aluminium chloride at 200-210° for 1.5 h [547];
  with 48% hydrobromic acid in refluxing acetic acid (good yield) [1309], for 12 h [182];
  with hydrochloric acid by heating between 145 and 150° during 3-4 h [1289]. There is elimination
  of methyl chloride;
*of p-ethoxybenzophenone,
  with aluminium bromide in refluxing benzene for 4 h [1059];
  with aluminium chloride [93]. The starting keto ether [93] was obtained by Friedel-Crafts acylation
  of phenetole with benzoyl chloride in the presence of aluminium chloride according to [547];
  with hydrobromic acid in boiling acetic acid for 2 days [941];
*of 3-tert-butyl-4-hydroxybenzophenone with aluminium chloride in refluxing benzene [880].
-Preparation by reaction of benzoyl chloride,

*with phenol,
  in the presence of aluminium chloride at 100-130° (77%) [490] or at 75° (19%) [562];
  in the presence of aluminium chloride in nitrobenzene at 60° (92%) [490], at 60° for 18 h (88%)
  [308] or at 45° for 3 h (22%) [877];
  in the presence of aluminium chloride in ethylene dichloride at 85° (5%) [490];
  in the presence of aluminium chloride in boiling carbon disulfide [1066], (10%) [1289];
  in the presence of titanium tetrachloride in nitrobenzene at 60° (76%) [490] or at 60° for 18 h
  (87%) [308];
  in the presence of zinc chloride (10%) [1289] according to [364];
*with phenyl borate in the presence of aluminium chloride in refluxing carbon disulfide for 2 h
  (45%) [1373];
*with p-nonylphenol in the presence of aluminium chloride at 160° for 30 min (57%) [1108].
-Also obtained by reaction of benzoic acid with phenol,
  in the presence of polyphosphoric acid at 75° (91%) [490], at 75° for 3 h (51%) [1291] or at 75°
  for 20 h (46%) [1291]; in a boiling water bath for 10 min (<4%) [961]; at 100° (9%) [960] or at
  100° for 20 min (16%) [959]; at 160° for 30 min (30%) [1291];
  in the presence of boron trifluoride at 100° for 1 h (52%) [730];
  in the presence of stannic chloride at 120° for 18 h (20%) [188];
  in the presence of trifluoromethanesulfonic acid, in a tube stoppered, at r.t. for 1 h [1138];
  in the presence of sulfophenyl group-containing polysiloxanes at 180° for 20 h (52%) [1351];
  in the presence of Amberlyst-15 in refluxing chlorobenzene for 48 h (9%) [582].
-Also obtained by reaction of p-(benzoyloxy)benzophenone with phenol in the presence of
  aluminium chloride in refluxing ethylene dichloride for 4 h (91%) [548].
-Preparation by diazotization of p-aminobenzophenone, followed by hydrolysis of the diazonium
  salt obtained [283, 385, 941, 1289], (54%) [1111].
-Also obtained by oxidation of p-hydroxydiphenylcarbinol with DDQ (2,3-dichloro-5,6-dicyano-
  benzoquinone) in dioxane at r.t. for 15 min (82%) [149].
-Also obtained by reduction of p-hydroxybenzophenone 2,4-dinitrophenylhydrazone with stannous
  chloride dihydrate in the presence of concentrated hydrochloric acid in boiling acetic acid for 1 h
  (93%) [310].
-Also obtained during the particular Fries rearrangements of some aryl esters during which there is
  usually an elimination of alkyl group,
*Also obtained by Fries rearrangement of p-nonylphenyl benzoate with aluminium chloride at 130-
  135° for 4 h (51%) [1108];
*Also obtained by Fries rearrangement of p-tert-butylphenyl benzoate with hydrofluoric acid at 55°
  for 1 h (26%) and at 25° for 2 h (15%) [986] or with aluminium chloride at 140° (major product)
  [743];
*Also obtained (by-product) by Fries rearrangement of phenyl o-benzoyloxybenzoate with
  aluminium chloride at 180° for 3 h (26%) [171].
-Preparation by heating at 120-130° during 12 h a mixture of phenol, benzanilide imidochloride and
  aluminium chloride. The keto anil obtained was hydrolyzed with ice-cold hydrochloric acid (21%)
  [1061]. The same reaction using anisole gave a 43% yield [1061].
-Preparation by saponification,
*of p-(benzoyloxy)benzophenone,
  with 10% sodium hydroxide at reflux for 3 h (good yield) [548];
  with potassium hydroxide in ethanol [364, 366, 508], (51%) [920, 921].
  The starting keto ester was obtained by reaction of benzoyl chloride with phenyl benzoate in the
  presence of zinc chloride at 175-180° [366, 920, 921] or by reaction of benzotrichloride with the
  same ester in the presence of zinc oxide [366]. This same keto ester (m.p. 112°5) can be also
  prepared by heating a benzoyl chloride/phenol mixture in the presence of zinc powder [508].
*of p-(carbomethoxyoxy)benzophenone with N sodium hydroxide solution at reflux for 15 min
  [424].
-Also obtained by hydrolysis of,
*p-(chloroacetoxy)benzophenone with fuming nitric acid at r.t. for several days (27%) [1090];
*p-benzoylphenyl 5-benzoylsalicylate [171].
-Also obtained by cleavage of benzaurine on heating its dilute aqueous sodium hydroxide solution
  (1%) in a water bath with bubbling air (44%) [622].

-Also obtained by photo-Fries rearrangement of phenyl benzoate,
*in isopropanol (41-42%) [102, 1070];
*in methanol (40%) [1070], irradiation for 1 h (37-43%) [1069] or for 12-24 h (30%) [1335];
*in butanol (36%) [1070];
*in ethanol, at 30° for 3 days (28%) [52] or at r.t. for 70 h (16%) [600];
*in water for 24 h (16%) [268]; adding soluble starch, amylose or α or β-cyclodextrin does not
 improve the yield (18-20%) [268]; in water, in the presence of SDS (sodium dodecyl sulfate), a
 micelle, one obtains a 23% yield [1276];
*in benzene for 12-24 h (30%) [1335], 1 h (37-43%) [1069] or at 52° for 8 h (20%) [419];
*in hexane for 1 h (37-43%) [1069];
*in pentane (13%) [102]; adding silica gel does not improve the yield (10%) [102];
*in cyclohexane at 55° for 24 h (29%) [419];
*in dioxane at 61° for 24 h (23%) [419];
*in ethyl ether (22%) [1070].
-Also refer to: [249, 292, 296, 421, 705, 810, 1100, 1432].
N.B.: K salt [534, 1121].

Isolation from natural sources

-From leaves of *Talauma mexicana* (Magnoliceae) [831];
-From leaves of *Yoloxochitl* (Talauma mexicana) [1033].

m.p.   135-136° [187, 188, 877], 135° [182, 310, 311, 424, 606, 1309],
       134-135° [622, 1061, 1291],
       134° [346, 364, 365, 366, 490, 547, 611, 730, 864, 1033, 1406],
       133°8-133°9 [921], 133°7 [941], 133°5-134°5 [973],
       133-134° [171, 1276], 133-133°5 [1111],
       133° [385, 961, 1059, 1066, 1121], 132°5 [1060], 132-135° [530, 531],
       132-134° [459], 132-133°5 [548, 986], 132-132°5 [354],
       132° [419, 959], 130-132° [52, 1090], 115-120° [470];
b.p.$_{24}$ 261° [941], b.p.$_{0.45}$ 168-170° [1121];
^{1}H NMR [1004, 1108, 1138], ^{13}C NMR [1004],
IR [180, 354, 530, 531, 792, 1108],   UV [354, 606, 793, 1391, 1406], MS [125, 234];
pK_a [606, 791, 792];   TLC [377];   cryoscopic study [93].

2. DIHYDROXYBENZOPHENONES

2.1. *Hydroxy groups located on one ring*

(2,3-Dihydroxyphenyl)phenylmethanone

[52870-68-5]   C$_{13}$H$_{10}$O$_3$ mol.wt. 214.22

Syntheses

-Preparation by hydrolytic rearrangement of 2-benzoylacetyl-
 2,5-dimethoxytetrahydrofuran with refluxing 0.1 N
 hydrochloric acid in aqueous dioxane (95%) [190].
-Preparation by reaction of benzoic acid with pyrocatechol in the presence of Amberlyst-15 in
 refluxing chlorobenzene for 9 h (68%) [582].
-Preparation by refluxing 2,3-dimethoxybenzophenone with hydrobromic acid (d = 1.5) and acetic
 acid for 8 h [120].
-Also refer to: [32, 253, 261].

   m.p.   65° [120], 64-65° [190];   Spectra (NA).

(2,4-Dihydroxyphenyl)phenylmethanone
(Resbenzophenone, Benzoresorcinol, Uvinul 400)

[131-56-6] $C_{13}H_{10}O_3$ mol.wt. 214.22

Syntheses

-Obtained by condensation of benzanilide with resorcinol in the presence of zinc chloride and phosphorous oxychloride at 130-140° for 1 h (25%) [1244].
-Also obtained by condensation of benzamide with resorcinol in the presence of zinc chloride and phosphorous oxychloride at 100-140° for 1 h (22%) [1244].
-The condensation of benzanilide imidochloride with resorcinol in the presence of aluminium chloride in ethyl ether gave a keto anil (92%). Which, by hydrolysis with concentrated hydrochloric acid in refluxing ethanol for 3 h yielded the expected ketone (89%) [1061].
-Also obtained by condensation of benzoic acid with resorcinol,
*in the presence of hydrofluoric acid at 75° for 4 h in an autoclave (99%) [363];
*in the presence of boron trifluoride at 160° for 2 h in a sealed tube (69%) [993] or in nitrobenzene at 80° for 30 min (96%) [390];
*in the presence of stannic chloride at 140-150° for 30 min (30%) [1244];
*in the presence of zinc chloride (Nencki reaction), between 150° and 170° (for usually 15-30 min) [113, 748, 805, 1358, 1415], (20%) [822], (5%) [1244];
*in the presence of zinc chloride and phosphorous oxychloride at 65° for 3 h [982];
*in the presence of zinc chloride and a mixture of polyphosphoric acid/85% phosphoric acid (60:40) at 40°. Then, during 1.5 h, phosphorous trichloride was added and the mixture heated at 60° for 16 h (91%) [1302];
*in the presence of phosphorous pentoxide at 120-130° for 15 min (poor yield) [1244];
*in the presence of Amberlite IR-120 at 160° for 2-3 h (25%) [1088];
Always by condensation of benzoic acid with resorcinol, but with stirring and azeotropic removal of water, in refluxing chlorobenzene (130°);
*in the presence of ion exchange resins,
-Dowex-50W-X-8 for 53 h (69%) [582];
-Amberlyst-15 for 6 h (67%) [582], for 24 h (62%) [584]; the same reaction carried out in refluxing p-chlorotoluene (162°) for 2 h gave 59% yield [582];
-Nafion-117 for 6 h (46%) [582] or for 24 h (45%) [584];
*in the presence of Zeolite-H-beta for 72 h (42%) [582] or for 24 h (20%) [584]; the same reaction carried out in refluxing p-chlorotoluene (162°) for 16-18 h gave 70% yield [582] or in refluxing n-butylbenzene for 3 h under nitrogen gave 88% yield [582] and 40% yield [584];
*in the presence of polyphosphoric acid or silicotungstic acid for 23 h (19%) [582];
*in the presence of methanesulfonic acid for 23 h (17%) [582, 584];
*in the presence of phosphoric acid for 24 h (9%) [582, 584].
-Also obtained by reaction of benzoyl chloride with resorcinol [851],
*in the presence of zinc chloride and hydrochloric acid gas in chlorobenzene at 120° for 7 h (86%) [988] or without hydrochloric acid (72%) [988];
*in the presence of aluminium chloride in ethylene dichloride at 65° for 5 h (75%) [189];
*in the presence of aluminium chloride in nitrobenzene at r.t. for 48 h (70%) [333] or first at r.t., then at 70° for 5 h (83%) [1259, 1260];
*in the presence of aluminium chloride in refluxing carbon disulfide [1066], for 4 h [356];
*in the presence of Nafion-XR, a H+-form ion exchange resin, for 4 h under nitrogen at 160° [1230] or at 150° (70%) [1231].
-Preparation by demethylation of 2,4-dimethoxybenzophenone with aluminium bromide in refluxing benzene for 4 h (98%) [1059]. Also obtained by using aluminium chloride in refluxing carbon disulfide for 30 min (3%) [718].
-Preparation by Fries rearrangement of,
*resorcinol monobenzoate,
with zinc chloride in the presence of hydrochloric acid gas in chlorobenzene at 120° for 7 h (88%) [988];

with aluminium chloride at 180° for 2 h (68%) [171];
with Nafion-XR, a H⁺-form ion exchange resin, at 150-160° for 4 h under nitrogen (72%) [1230, 1231];
*resorcinol dibenzoate with aluminium chloride between 100 to 150° [1066].
-Preparation by condensation of benzoic anhydride with resorcinol,
*in the presence of a few drops of concentrated sulfuric acid at 130° for some min (60%) [633];
*in the presence of Amberlite IR-120, a cation exchange resin (sulfonic acid type), at 160° for 2-3 h (30%) [1088];
*in the presence of Zeokarb 225 at 160° for 2-3 h (30%) [1088].
-Also obtained by decarboxylation,
*of 5-benzoyl-2,4-dihydroxybenzoic acid with dilute hydrochloric acid at 160-170° in a sealed tube [49] or by heating in a sealed tube [339];
*of 3-benzoyl-2,6-dihydroxybenzoic acid with a few drops of concentrated hydrochloric acid in boiling dilute acetic acid for 15-18 h [1238].
-Also obtained by saponification of 2,4-di(benzoyloxy)benzophenone with 10% sodium hydroxide in refluxing ethanol for 2 h [356] or with potassium hydroxide in ethanol [367]. The starting diester was prepared by Fries rearrangement of resorcinol dibenzoate with zinc chloride at 100-120° [367].
-Also obtained by condensation of benzotrichloride with resorcinol,
*in hot water [748] or in water at 65° for 4 h (42%) [1260];
*in 50% aqueous methanol at 70-80° (90%) [1460] or in dilute methanol at 50° for 3 h (87%) [1259, 1260];
*in 50% aqueous isopropanol at 70-80° (90%) [1460] or in dilute isopropanol between 30 to 80° for 2-4 h (80-84%) [1260];
*in 50% aqueous acetic acid solution (89%) [1460] or in dilute acetic acid between 30 to 80° for 2-4 h (80-84%) [1260];
*in 50% aqueous dioxane solution (82%) [1460] or in dilute dioxane between 30 to 80° for 2-4 h (80-84%) [1260];
*in ethanol [1058] or in dilute ethanol between 30 to 80° for 2-4 h (80-84%) [1260];
*in dilute ethylcellosolve between 30 to 80° for 2-4 h (80-84%) [1260];
*in hydrofluoric acid in the presence of water at -10° for 4 h, then at r.t. overnight (96%) [393]. The same reaction carried out in the presence of methanol or octanol at -10°, then at r.t. for 4 h yielded 96% and 93%, respectively [393].
-Preparation by reaction of benzonitrile with resorcinol (Hoesch reaction) in the presence of zinc chloride and hydrochloric gas at 50° for 8 h (93%) [1430] or during 12 days (89%) [1079, 1470, 1471], or in ethyl ether at <5° for 3 h, then at 5° for 20 h (35%) [1259, 1260].
-Also obtained by treatment of resorcinol with phenyl benzoate in the presence of boron trifluoride-etherate in refluxing benzene for 3 h (40%) [725].
-Also obtained from β-(4-benzoyl-3-hydroxyphenoxy)propionic acid by boiling for 5 min with 10% aqueous sodium hydroxide [1274].
-Also obtained by photo-Fries rearrangement of resorcinol monobenzoate [970].
-Preparation by reaction of di(carbomethoxy)-β-resorcylic acid chloride with benzene in the presence of aluminium chloride at 60-70° for 1 h, then at 80° for 1 h (70%) [425].
-Also refer to: [23, 43, 53, 271, 285, 364, 384, 483, 565, 615, 707, 732, 810, 856, 1120, 1411, 1432].

m.p.   148° [171], 146° [356], 145° [333, 340, 1067, 1217, 1358, 1359],
       144-146° [558, 1088], 144-145° [582, 584, 822, 1238, 1244],
       144° [367, 393, 608, 633, 718, 1079], 143°5-144° [354],
       143-144° [49, 748, 988, 993, 1061, 1430, 1460], 143° [1059, 1066, 1274],
       142°6-144°2 [1302], 142°5-143° [501], 142-144° [1058], 141-143° [1259, 1260];
¹H NMR [267], IR [354, 1259, 1260], UV [271, 354, 474, 475, 515, 607, 793, 794, 970, 1067, 1079, 1156, 1217, 1259, 1260, 1406, 1411];
TLC [377, 1396];   HPLC [270, 390];   pK_a [608, 791];
polarographic study [1223];   gel permeation chromatography [301, 1145];
vapour pressure [500, 1217].

(2,5-Dihydroxyphenyl)phenylmethanone *(Quinbenzophenone)*

[2050-37-5] $C_{13}H_{10}O_3$ mol.wt. 214.22

Syntheses

-Preparation by Fries rearrangement of hydroquinone
  dibenzoate with aluminium chloride [337, 1247], (good
  yield) [334], at 140° for 1 h (33%) [47], at 190-200° for 1 h
  30 min (50%) [338] or at 200° for 20 min [766].
-Preparation by oxidation of 5-(benzoyloxy)-2,3-diphenyl-
benzofuran with chromium trioxide in boiling acetic acid for 2 h, followed by saponification of the
keto ester formed with boiling 8% sodium hydroxide in ethanol [357].
-Also obtained by hydrolysis of quinbenzophenone monobenzoate — 5-(benzoyloxy)-2-hydroxy-
benzophenone — (easily obtained by Fries rearrangement of hydroquinone mono or dibenzoate),
with concentrated sulfuric acid at r.t. for overnight [47] or with 85% sulfuric acid [338].
-Also obtained by saponification of 2,5-di(benzoyloxy)benzophenone with alcoholic potassium
hydroxide [736].
-Also obtained by reaction of benzoic acid with hydroquinone in the presence of boron trifluoride at
160° for 1 h in a sealed tube (61%) [993].
-Preparation by demethylation,
*of 2-hydroxy-5-methoxybenzophenone with hydriodic acid in refluxing acetic acid and acetic
  anhydride mixture for 1.5 h (87%) [191];
*of 2,5-dimethoxybenzophenone (SM),
  with boiling hydriodic acid [568];
  with an excess of hydrobromic acid [1060];
  with boron tribromide in methylene chloride at 22° (78%) [1048] or at r.t. for 12 h [766],
  according to [891]. SM was obtained by acylation of hydroquinone dimethyl ether with benzoyl
  chloride in methylene chloride in the presence of aluminium chloride at 0° (85%) [1048].
-Preparation by reaction of phenyl benzoate with hydroquinone in the presence of boron trifluoride-
etherate in refluxing benzene for 3 h (60%) [725].
-Also obtained by daylight irradiation of a 1,4-benzoquinone/benzaldehyde mixture [736] in
benzene under nitrogen for 5 days (60%) [762].
-Also obtained by irradiation of α-hydroxybenzyl-1,4-benzoquinone in benzene for 4 days (65%)
[1350].
-Also refer to: [33, 222, 341, 763, 791, 871, 1277, 1333, 1350].

  m.p. 125-126°1 [191], 125° [337, 338, 736, 1247], 124-126° [568],
        124-125° [47, 357], 124° [993], 123-124° [1350], 122° [1060], 121-123° [762];
  ^{1}H NMR (Sadtler: standard n° 30288 M) [1350],
  IR (Sadtler: standard n° 57333) [1350], UV [607, 793, 794];
  TLC [377, 762]; pK_a [791].

(2,6-Dihydroxyphenyl)phenylmethanone

[63411-81-4] $C_{13}H_{10}O_3$ mol.wt. 214.22

Synthesis

-Preparation by treatment of 8-benzoyl-7-hydroxy-4-methyl-
  coumarin (8-benzoyl-4-methylumbelliferone) [821, 1283] or
  of 8-benzoyl-7-hydroxy-4-phenylcoumarin (8-benzoyl-
  4-phenylumbelliferone) [822] with refluxing aqueous
  sodium hydroxide (64-78%) [821], (48%) [822].

-Also refer to: [644].

  m.p. 135° [821, 822, 1406]; ^{1}H NMR [267], UV [1406].

(3,4-Dihydroxyphenyl)phenylmethanone

[10425-11-3]   C13H10O3   mol.wt. 214.22

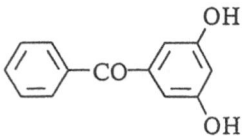

Syntheses

-Preparation by Fries rearrangement of pyrocatechol dibenzoate,
*in the presence of aluminium chloride in nitrobenzene at 100° for 4 h (quantitative yield) [1142] or heated on a steam bath for 6 h [1139];
*in the presence of Nafion-XR, a H+-form ion exchange resin, at 175° for 4 h under nitrogen (38%) [1231].
-Preparation in two steps by condensation of pyrocatechol with benzanilide imidochloride after acidic hydrolysis of the intermediate keto anil (30%) [1061].
-Also obtained by demethylation,
*of 3-hydroxy-4-methoxybenzophenone with boiling hydriodic acid (90%) [138];
*of 3,4-dimethoxybenzophenone with boiling hydriodic acid [138] or with aluminium chloride in refluxing benzene for 6 h (64%) [306].
-Also obtained by reaction of benzoyl chloride with pyrocatechol dibenzoate in the presence of zinc chloride, followed by saponification of the 3,4-di(benzoyloxy)benzophenone formed with a boiling ethanolic sodium hydroxide solution [364].
-Also obtained by reaction of benzoyl chloride with pyrocatechol in the presence of phosphorous trichloride [956].
-Preparation by condensation of benzoic anhydride with pyrocatechol in the presence of zinc chloride at 180° for 5-6 h [1136].
-Preparation by adding 30% hydrogen peroxide solution to the solution of 3-formyl-4-hydroxy-benzophenone in 10% sodium hydroxide and stirring the mixture at r.t. for 2 h (Dakin oxidation) (42%) [121].
-Also obtained by hydrolysis of 3-(benzoyloxy)-4-hydroxybenzophenone by heating gently with 50% sulfuric acid for 2 h (74%) [1211].
-Also refer to: [305, 465, 466, 714, 879, 1120].

m.p.   148-149° [306], 147-148° [1139], 145° [364],
        134° [138, 1061, 1136, 1142, 1211], 133-134° [121];
hemihydrate [364], monohydrate [138, 1211];
Spectra (NA);   pK_a [1139];   molecular orbital studies [1258].

(3,5-Dihydroxyphenyl)phenylmethanone

                                        C13H10O3   mol.wt. 214.22

Syntheses

-Preparation by reaction of diphenylcadmium with 3,5-di-acetoxybenzoyl chloride in refluxing benzene for 1 h. The keto ester obtained (66%) by saponification with refluxing 5% sodium hydroxide for 4-5 h gave the expected ketone (53%) [612].
-Also obtained by heating a mixture of 3,5-dicarbomethoxybenzoyl chloride and benzene in the presence of aluminium chloride, first at 70-75° for 1 h and at 75-80° for 45 min (37%). -Refer to: Chem. Abstr., 7, 2563^2 (1913)*.

m.p.   160-162°*, 148° (anhydrous) [612], 84° (monohydrate) [612];
Spectra (NA).

2.2. *Hydroxy groups located on both rings*

Symmetrical ketones

Bis(2-hydroxyphenyl)methanone

[835-11-0] $C_{13}H_{10}O_3$ mol.wt. 214.22

Syntheses

-Preparation by diazotization of 4,4'-diamino-2,2'-di-
   hydroxybenzophenone in diluted hydrochloric acid, followed
   by treatment with 50% phosphorous acid at 0° for 1 h, then
   at r.t. for 24 h (34%) [662].
-Also obtained by Fries rearrangement of phenyl o-hydroxybenzoate with aluminium chloride at
   200° for 20 min [766].
-Also obtained from xanthone,
*by heating with potassium hydroxide in ethanol at 180° for 4 h in a sealed tube [496, 1072];
*by fusion with anhydrous potassium hydroxide [1060, 1118], at 205-210° for 20 min [202],
   (64%) [109] or by heating at 200° (45%) [817].
-Also obtained by photo-Fries rearrangement of phenyl salicylate (salol) [970] or of phenyl
   carbonate in ethanol for 150 h (4%) [600].
-Also obtained by demethylation of 2,2'-dimethoxybenzophenone,
*with boron trichloride in methylene chloride first at -70°, then at r.t. for 8 h (97%) [330];
*with boron tribromide in methylene chloride at r.t. for 12 h [766], according to [891].
-Also obtained by hydrolysis of 2,2'-dihydroxybenzophenone ditosylate by treatment with 0.5 N
   sodium hydroxide in dilute methanol at r.t. for 8 days (98%) [815].
-Also refer to: [41, 594, 606, 810, 814, 859, 958, 1290, 1296, 1297, 1457].
N.B.: Ba, NH_4 [1118] and K salts [1072, 1118].

   m.p. 62-63° [109], 59°5 [1060], 59-60° [496, 1118, 1296, 1297],
         55-57° [817], 54-56° [662];   b.p.$_{0.15}$ 128-132° [817];
   ^{1}H NMR [177, 267], EPR [267], IR [817], UV [817, 970], MS [125];
   pK_a [791];   polarographic analysis [677];   HPLC [270].

Bis(3-hydroxyphenyl)methanone

[611-80-3] $C_{13}H_{10}O_3$ mol.wt. 214.22

Synthesis

-Preparation from 3,3'-dinitrobenzophenone by reduction,
   diazotization of 3,3'-diaminobenzophenone obtained
   (m.p. 173-174°) [462], followed by hydrolysis of the
   diazonium salt obtained [230, 462, 1072, 1297, 1403].
-Also refer to: [599, 752].

   m.p. 163-164° [462], 162-163° [1296, 1297];   IR [180].

Bis(4-hydroxyphenyl)methanone

[611-99-4] $C_{13}H_{10}O_3$ mol.wt. 214.22

Syntheses

-Preparation by demethylation of 4,4'-dimethoxy-
   benzophenone with aluminium bromide in refluxing
   benzene for 4 h (88%) [1059].

-Also obtained by complete dealkylation of 4,4'-diethoxybenzophenone with hydrobromic acid in boiling acetic acid for 2 days [941].

-Preparation by hydrolysis of 4,4'-dichlorobenzophenone with aqueous sodium hydroxide in the presence of cuprous chloride or cupric chloride in an autoclave at 230-240° for 2 h (95-98%). The same reaction using cuprous oxide at 200° for 2 h gave a 82% yield [1026, 1027]. Other methods using copper compounds for hydrolysis of the 4,4'-dichlorobenzophenone in the presence of sodium hydroxide [214, 813], for 1 h at 270° in a steel autoclave (98-99%) [926].

-Preparation by Fries rearrangement of phenyl p-anisate (phenyl p-methoxybenzoate) with aluminium chloride at 140° (80%) [1143] or at 160° [514].

-Also obtained by reaction of p-hydroxybenzoic acid with phenol,

*in the presence of zinc chloride and a mixture of polyphosphoric acid/85% phosphoric acid (60:40) at 40°. Then, during 1.5 h, phosphorous trichloride was added and the mixture heated at 60° for 16 h (90%) [1302];

*in the presence of hydrofluoric acid at 75° in an autoclave (88%) [363];

*in the presence of boron trifluoride in hydrofluoric acid (84%) [318] or in nitrobenzene at 80° for 30 min (67%) [390];

*in the presence of polyphosphoric acid in a boiling water bath for 30 min (20%) [961] or at 100° for 20 min (47%) [959];

*in the presence of trifluoromethanesulfonic acid at r.t. for overnight (93%) [1138];

*by heating in the presence of stannic chloride for 10 h [905].

-Also obtained by treatment of 4,4'-diphenoxybenzophenone with 16.7% sodium hydroxide at 300° for 30 min (57%) [724].

-Also obtained (poor yield) by Fries rearrangement of oxalic acid diphenyl ester with hydrofluoric acid in carbon tetrachloride at 80° for 4 h in an autoclave (<7%) [322].

-Also obtained by reaction of salicylic acid with phenol in the presence of stannic chloride (by-product) [112, 905].

N.B.: According to Michael [905], the formation of this ketone was due to the presence of a small amount of p-hydroxybenzoic in the commercial salicylic acid used [1022]. In contrast, Baeyer [112] claimed that this reaction proceeds *via* a transposition.

-Also obtained by oxidation of 4,4'-di-(benzoyloxy)diphenylmethane with chromium trioxide in refluxing acetic acid for 8 h, followed by saponification of the 4,4'-di-(benzoyloxy)benzophenone formed with potassium hydroxide in refluxing ethanol [92, 1297, 1298, 1299].

-Preparation by acylation of phenol with p-(trichloromethyl)phenyl p-(trichloromethyl)benzoate in methylene chloride in the presence of aluminium chloride at 0-5° over 1 h, then at r.t. for 1 h (yield 83%), followed by alkaline hydrolysis of the resulting keto ester (93-95%) [581].

-Also obtained by melting phenolphthalein,

*with potassium hydroxide (75%) [1473], according to [115];

*with sodium hydroxide [232, 1358], (quantitative yield) [114].

-Also obtained by reduction of 4-hydroxy-4'-nitrobenzophenone with stannous chloride in the presence of hydrochloric acid, followed by diazotization of the 4-amino-4'-hydroxybenzophenone formed and hydrolysis of the diazonium salt obtained [92].

-Also obtained on treatment of aurin (p-rosolic acid) with water at high temperature, between 220 to 250° [249] or at >270° [820].

-Also obtained by treatment of rosaniline with water at high temperature (>270°) [819, 820].

-Also obtained by reaction of carbon tetrachloride with phenol in the presence of zinc chloride at 120° (36%) [485].

-Also obtained by hydrogenation of 4,4'-dihydroxy-3,3',5,5'-tetraiodobenzophenone in ethanol in the presence of 10% Pd/C and sodium acetate (66%) [979].

-Also refer to: [284, 319, 810, 906].

N.B.: Na [926, 1331] and K salts [570].

m.p.   218° [1138], 213°5 [941], 213-214° [979, 1059], 210°4-212° [1302],
       210-213° [318], 210-212° [1308],
       210° [249, 1027, 1296, 1297, 1298, 1299, 1358, 1359], 209-214° [363],
       208-210° [92], 207° [959, 961], 206° [114, 232, 322, 514, 1406],
       205°5-206°5 [905], 205-207° [820], 200° [819];
¹H NMR [318, 1138], IR [180, 318, 1138, 1281], UV [979, 1406];

GLC [318];   HPLC [270, 390];   polarographic study [677];
gel permeation chromatography [1145].

Asymmetric ketones

(2-Hydroxyphenyl)(3-hydroxyphenyl)methanone

$C_{13}H_{10}O_3$ mol.wt. 214.22

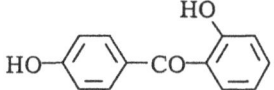

Syntheses

-Preparation by diazotization of 2,3'-diaminobenzophenone
  followed by hydrolysis of the obtained diazonium salt
  [1297].
-Also obtained by Fries rearrangement of phenyl m-methoxy-
benzoate (phenyl m-anisate) with aluminium chloride at 120° or 160° for 2 h [1168].
-Also obtained (poor yield) by reaction of m-methoxybenzoyl chloride with phenol in the presence
  of aluminium chloride in nitrobenzene at 160° for 45 min (4%) [87].
-Preparation by demethylation of 2,3'-dimethoxybenzophenone in the presence of aluminium
  chloride in refluxing chlorobenzene for 1 h (86%) [87].
-Also refer to: [86, 415].

m.p.   127° [1168], 126° [1297], 124-127° [87], 121-122° [1296];
^{1}H NMR [87], IR [87], UV [87].

(2-Hydroxyphenyl)(4-hydroxyphenyl)methanone

[606-12-2] $C_{13}H_{10}O_3$ mol.wt. 214.22

Syntheses

-Preparation by diazotization of 2,4'-diamino-2',4-di-
  hydroxybenzophenone, followed by decomposition of the
  diazonium salt obtained in the presence of 50% hypophosphorous acid (57%) [662].
-Preparation from 2-hydroxy-4'-nitrobenzophenone by reduction with stannous chloride and
  hydrochloric acid, followed by diazotization of the 4-amino-2'-hydroxybenzophenone formed and
  hydrolysis of the diazonium salt obtained [92].
-Preparation by hydrolysis of 2,4'-dichlorobenzophenone with sodium hydroxide for 1 h at 270° in
  a steel autoclave (98-99%) [926].
-Preparation by diazotization of 2,4'-diaminobenzophenone, followed by hydrolysis of the
  diazonium salt obtained [1297].
-Also obtained by condensation of salicylic acid with phenol,
*in the presence of stannic chloride at 115-125° (major product) [92, 112, 495, 775, 905, 906, 1058,
  1358];
*in the presence of zinc chloride (poor yield) [905];
*in the presence of hydrofluoric acid at 75° in an autoclave [363];
*in the presence of boron trifluoride in nitrobenzene at 80° for 30 min (83%) [390];
*in the presence of polyphosphoric acid at 100° for 20 min (2%) [959].
-Also obtained by Fries rearrangement,
*of phenyl o-methoxybenzoate with aluminium chloride at 120-160° for 2 h [1168], with aluminium
  chloride in the presence of phenol at 180° for 3 h (44%) [171];
*of phenyl salicylate (salol),
  with aluminium chloride at 120-160° for 2 h [1168], at 140° for 3 h (75%) [50] or at 180-182° for
  3 h (53%) [171];
  with stannic chloride at 115-120° for 18 h (54%) [188];

*of phenyl o-(benzoyloxy)benzoate with aluminium chloride in the presence of phenol at 180° for
  3 h (36%) [171];
*of o-(nicotinyloxy)phenyl benzoate with aluminium chloride at 140-145° for 2 h (by-product)
  (9%) [170].
-Preparation by dealkylation of 5-tert-butyl-2,4'-dimethoxybenzophenone with aluminium chloride
  in benzene at 65-70° for 45 h (75%) [775].
-Preparation by reaction of hydrobromic acid (d = 1.49) with 2,4'-di(4-methoxybenzoyloxy)-
  benzophenone in refluxing acetic acid for 2 h (86%) [1309].
-Also obtained by hydrolysis of p-(salicyloyl)phenyl salicylate [170].
-Also obtained by treatment of 4-acetoxy-2'-methoxybenzophenone with aluminium chloride at 153-
  155° for 2 h (19%) [171]. There is simultaneously demethylation and hydrolysis of the acetoxy
  group.
-Also obtained by decarboxylation of 2-hydroxy-5-(salicyloyl)benzophenone by melting with
  potassium hydroxide (63%) [188].
-Also obtained by photo-Fries rearrangement,
*of phenyl salicylate (salol) [970], in ethanol for 76 h (22%) [600];
*of phenyl carbonate in ethanol for 150 h (11%) [600].
-Also refer to: [247, 719, 720, 810].
N.B.: Na salt [926, 1331].

        m.p.   150-151° [112], 150° [1168], 148° [170, 171, 495], 147-149° [188],
               147-148° [1058], 146-147° [363, 775], 145° [50], 144° [92],
               143-145° [662], 143-144° [905, 906],
               142° [1296, 1297, 1358, 1359], 141° [959], 138-141° [1309];
        ^{1}H NMR (Sadtler: standard n° 38497 M),
        IR (Sadtler: standard n° 65535 K), UV [970], MS [125];
        gel permeation chromatography [1457];   HPLC [390].

(3-Hydroxyphenyl)(4-hydroxyphenyl)methanone

[611-81-4] $C_{13}H_{10}O_3$ mol.wt. 214.22

Syntheses

-Obtained by Fries rearrangement of phenyl m-anisate
 without solvent in the presence of aluminium chloride for
 2 h between 120 to 160° [1168].
-Also obtained by reaction of m-hydroxybenzoic acid with
 phenol,
*in the presence of polyphosphoric acid for 20 min at 100° (7%) [959];
*in hydrofluoric acid at 30° under a pressure of 30 psi boron trifluoride in an autoclave for 4 h
  [1365], (89%) [626].
-Also obtained by reaction of m-methoxybenzoyl chloride with phenol in the presence of aluminium
  chloride in nitrobenzene at 160° for 45 min (12%) [87].
-Preparation by diazotization of 3,4'-diaminobenzophenone (m.p. 121-122°) [462], followed by
  hydrolysis of the diazonium salt obtained [462, 1297].
-Preparation by demethylation of 3,4'-dimethoxybenzophenone with 48% hydrobromic acid in a
  acetic anhydride/acetic acid mixture (1:1) for 15 h at reflux (quantitative yield) [624].
-Also refer to: [344, 623, 625].

        m.p.   203° [626], 200° [462], 198° [1296], 197° [1297], 196-197° [1168],
               195-200° [87], 195° [959], 193-194° [624];
        Spectra (NA).

3. TRIHYDROXYBENZOPHENONES

3.1. *Hydroxy groups located on one ring*

Phenyl(2,3,4-trihydroxyphenyl)methanone *(Alizarine Yellow A)*

[1143-72-2] $C_{13}H_{10}O_4$ mol.wt. 230.22

Syntheses

-Preparation by reaction of benzoic acid with pyrogallol,
*in the presence of zinc chloride during 3 h at 145° [105, 495, 1358];

*in the presence of Amberlyst-15 (a strongly acid ion exchanger) in chlorobenzene during 10 h at 131-132° (60%) [5];
*in the presence of Amberlite IR-120 or Zeokarb 225 (cation exchange resins, sulfonic acid type) during 3 h at 160° (14%) [1088];
*in the presence of boron trifluoride in ethyl ether at 0° (44%) [245, 246].
-Preparation by reaction of benzoic anhydride with pyrogallol [105, 495, 1088],
*in the presence of concentrated sulfuric acid or polyphosphoric acid during 15 min at reflux (14%) [1088];
*in the presence of zinc chloride at 145° [105, 495];
*in the presence of Amberlite IR 120 or Zeokarb 225 during 3 h at 160° (16%) [1088].
-Preparation by reaction of benzoyl chloride with pyrogallol [105, 495, 1231],
*in the presence of Nafion-XR, a H+-form ion exchange resin, during 4 h at 150° under nitrogen (78%) [1231];
*in the presence of zinc chloride at 145° [105, 495].
-Preparation by reaction of benzotrichloride with pyrogallol [1020], in the presence of zinc chloride at 145° [106, 495].
-Preparation by reaction of 2,3,4-triacetoxybenzoyl chloride with benzene in the presence of aluminium chloride at 40° for 4 h (55%) [426].
-Preparation in two steps by condensation of pyrogallol with benzanilide imidochloride after acidic hydrolysis of the intermediate keto anil (31%) [1061].
-Also obtained by hydrolysis of gallacetophenone monobenzoate in the presence of concentrated sulfuric acid during 4 h into a cooling system (10%) [335].
-Preparation by Fries rearrangement of,
*2,3-dihydroxyphenyl benzoate with Nafion-XR during 4 h at 150° under nitrogen (81%) [1231];
*pyrogallol tribenzoate with aluminium chloride during 2 h at 160-170° (15%) [335].
-Also refer to: [465, 466, 516, 801, 1182, 1300, 1333, 1334].
N.B.: K, Pb and Na salts are obtained according to [495].

   m.p.   146° [245, 246], 141-143° [1088], 140-141° [495, 1061], 140° [335],
          139-140° [1358, 1359], 138-139° [426];
   UV [246, 802, 1358]; pK_a [802].

Phenyl(2,3,5-trihydroxyphenyl)methanone

                                     $C_{13}H_{10}O_4$ mol.wt. 230.22

Synthesis

-Refer to: [425].

m.p. and Spectra (NA).

Phenyl(2,4,5-trihydroxyphenyl)methanone

[14894-91-8] $C_{13}H_{10}O_4$ mol.wt. 230.22

Synthesis

-Preparation by hydrolysis of 2,4,5-triacetoxybenzophenone
   with hydrochloric acid in refluxing ethanol for 2 h (77%)
   [191].

m.p. 129°1 [191];   Spectra (NA).

Phenyl(2,4,6-trihydroxyphenyl)methanone *(Phlorbenzophenone)*

[3555-86-0] $C_{13}H_{10}O_4$ mol.wt. 230.22

Syntheses

-Preparation by Fries rearrangement of phloroglucinol
   tribenzoate with aluminium chloride at 160-170° for 2 h
   (95%) [335].
-Preparation by reaction of benzonitrile with phloroglucinol in
   the presence of zinc chloride and hydrochloric acid in
ethyl ether at r.t. for overnight, followed by hydrolysis of the ketimine salt formed with dilute
sulfuric acid [577], (65%) [585] (Hoesch reaction).
-Preparation in two steps by condensation of phloroglucinol with benzanilide imidochloride after
   acidic hydrolysis of the intermediate keto anil (39%) [1061].
-Preparation by reaction of benzoyl chloride with phloroglucinol in the presence of aluminium
   chloride in a ethyl ether/nitrobenzene mixture [1144].
-Preparation by condensation of benzoyl chloride with 1,3,5-trimethoxybenzene, followed by
   demethylation [1043].
-Also obtained by reaction of phenyl benzoate with phloroglucinol in the presence of boron
   trifluoride-etherate in refluxing benzene for 3 h (25%) [725].
-Preparation by demethylation of 2,4,6-trimethoxybenzophenone with hydriodic acid (d = 1.7) in
   acetic anhydride at 115-120° for 4 h (83%) [1044].
-Also refer to: [25, 702, 705, 1100].

Isolation from natural sources

-From *Helichrysum triplinerve* (Asteraceae) [1099];
-From *genus Leontonyx* [192].

m.p. 170-171° [1044], 165° [335, 585, 725, 1061], 164-165° [1144], 164° [577];
^{1}H NMR [1044], UV [577].

Phenyl(3,4,5-trihydroxyphenyl)methanone

[60487-86-7] $C_{13}H_{10}O_4$ mol.wt. 230.22

Synthesis

-Preparation by reaction of tri(carbomethoxy)galloyl chloride
   with benzene in the presence of aluminium chloride at
   70-80° for 1.5 h, followed by saponification of the keto ester
   formed with N sodium hydroxide in a water bath for 15 min
   (65%) [424].

-Also refer to: [425, 950].

     m.p.   177-178° [424];   Spectra (NA).

3.2. *Hydroxy groups located on both rings*

(2,3-Dihydroxyphenyl)(4-hydroxyphenyl)methanone

[129726-78-9] $C_{13}H_{10}O_4$ mol.wt. 230.22

Synthesis

-Preparation by total demethylation of 2,3,4'-trimethoxy-benzophenone with hydrobromic acid (d = 1.5) in refluxing acetic acid [120].

     m.p.   169° [120];   Spectra (NA).

(2,4-Dihydroxyphenyl)(2-hydroxyphenyl)methanone

[13087-18-8] $C_{13}H_{10}O_4$ mol.wt. 230.22

Syntheses

-Preparation by diazotization of 4,4'-diamino-2,2'-di-hydroxybenzophenone, followed by treatment of the diazonium salt formed with 50% hypophosphorous acid (26%) [662].
-Also obtained (poor yield) by action of salicylic acid on resorcinol at 195-200° for 15 h, without a dehydrating agent [495, 905, 906, 1358] or in the presence of zinc chloride and phosphorous oxychloride (40%) [233].
-Also obtained (poor yield) by condensation of o-acetoxybenzonitrile with resorcinol in the presence of zinc chloride and hydrochloric acid in ethyl ether at 0° (Hoesch reaction), followed by hydrolysis of the intermediate compound obtained with boiling 0.5 N sodium hydroxide (<5%) [83]. In this reaction, the major compound was 3-hydroxyxanthone (18%).
-Also refer to: [483, 810] and [1073] (Japanese patent).

     m.p.   134-135° [233], 133-134° [83, 905, 906], 132-133° [662], 130-132° [1358, 1359];
     UV [177, 233, 1358], MS [125];   TLC [377];
     paper chromatography [911].

(2,4-Dihydroxyphenyl)(3-hydroxyphenyl)methanone

[837-60-5] $C_{13}H_{10}O_4$ mol.wt. 230.22

Synthesis

-Preparation by total demethylation of 2,3',4-trimethoxy-benzophenone,
*with 60% hydrobromic acid in refluxing acetic acid for 6 h [87, 415], (87%) [817];
*with aluminium chloride in boiling chloroform for 2.5 h [817].
-Also refer to: [815, 886, 887].

     m.p.   178-182° [817], 178-180° [887];
     ^{1}H NMR [87], IR [817, 887], UV [817, 887], MS [887].

(2,4-Dihydroxyphenyl)(4-hydroxyphenyl)methanone

[1470-79-7] $C_{13}H_{10}O_4$ mol.wt. 230.22

Syntheses

-Preparation by reaction of 4-hydroxybenzoic acid with resorcinol,
*in the presence of zinc chloride at 160° (Nencki reaction) [748, 1358];
*in the presence of zinc chloride and phosphorous oxychloride for 4 days at r.t. (78%) [1415];
*in the presence of zinc chloride and a mixture of polyphosphoric acid/85% phosphoric acid (60:40) at 27°. Then, during 2 h, phosphorous trichloride was added between 27 to 37° and the mixture heated at 60° for 16 h (quantitative yield) [1302];
*in the presence of hydrofluoric acid at 75° in an autoclave [363].
-Preparation by reaction of β-resorcylic acid with phenol in the presence of zinc chloride and a mixture of polyphosphoric acid/85% phosphoric acid (60:40) at 27° [1302], (40%) [233]. Then, during 2 h, phosphorous trichloride was added between 27 to 37° and the mixture heated at 60° for 16 h [1302].
-Also refer to: [59, 225, 259, 260, 395, 713, 958, 1198, 1410].

   m.p. 203°4-204° [1302], 200-201° [748, 1415], 200° [1358, 1359], 199-200° [233];
   dihydrate [748];   IR [991], UV [233, 991, 1358], MS [125];
   TLC [377], paper chromatography [911].

(2,5-Dihydroxyphenyl)(2-hydroxyphenyl)methanone

[183106-13-0] $C_{13}H_{10}O_4$ mol.wt. 230.22

Syntheses

-Preparation by total demethylation of 2,2',5-trimethoxy-benzophenone with boron tribromide in methylene chloride at 0° under nitrogen atmosphere for 3 h (85%) [1250].
-Preparation by condensation of salicylic acid with hydroquinone in the presence of zinc chloride for 45 min at 125-140° (Nencki reaction). -Refer to: Chem. Abstr., **44**, 1271f (1950)*.

   m.p. 149-150° [1250], 98°*. One of the reported melting points is obviously wrong.
   ^{1}H NMR [1250], IR [1250], UV [1250], MS [1250];   TLC [1250].

(2,5-Dihydroxyphenyl)(4-hydroxyphenyl)methanone

[120506-56-1] $C_{13}H_{10}O_4$ mol.wt. 230.22

Syntheses

-Preparation by heating a mixture of 4-hydroxybenzoic acid and hydroquinone in nitrobenzene in the presence of boron trifluoride at 80° for 30 min (64%) [390]. **N.B.**: In the patent, this compound was erroneously called 3',4,5'-tri-hydroxybenzophenone (assay 5, table page 3) [390].
-Preparation by Friedel-Crafts acylation of hydroquinone dimethyl ether with p-anisoyl chloride (AlCl₃/CS₂/ 3 h at r.t.), followed by demethylation (AlCl₃/Toluene/ 1 h at 120°) (65%). -Refer to: Chem. Abstr., **44**, 1271f (1950)*.

   m.p. 162°*;   Spectra (NA).

(2,6-Dihydroxyphenyl)(2-hydroxyphenyl)methanone

[82-69-9] $C_{13}H_{10}O_4$ mol.wt. 230.22

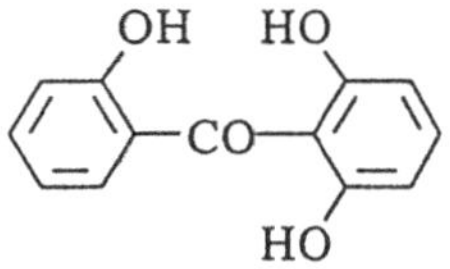

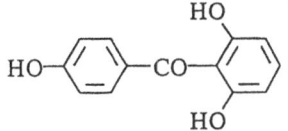

m.p. 155°*; IR*.

Synthesis

-Preparation by hydrolysis of 7-hydroxy-8-(o-hydroxy-
benzoyl)-4-methylcoumarin with sodium hydroxide in dilute
ethanol (52%). -Refer to: Chem. Abstr., **114**, 42490n
(1991)*.

(2,6-Dihydroxyphenyl)(3-hydroxyphenyl)methanone

[21554-76-7] $C_{13}H_{10}O_4$ mol.wt. 230.22

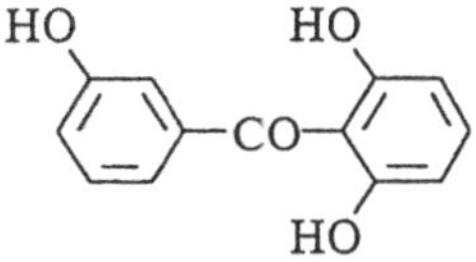

Synthesis

-Preparation by demethylation of 2,3',6-trimethoxy-
benzophenone in the presence of aluminium chloride in
refluxing chlorobenzene for 1 h (47%) [87].
-Also refer to: [86, 415].

m.p. 124-125° [87]; IR [87], UV [87].

(2,6-Dihydroxyphenyl)(4-hydroxyphenyl)methanone

[131425-90-6] $C_{13}H_{10}O_4$ mol.wt. 230.22

Synthesis

-Preparation by hydrolysis of 7-hydroxy-8-(p-hydroxybenzoyl)-4-methylcoumarin (SM) with sodium hydroxide in
refluxing dilute ethanol (57%). SM was obtained by Fries
rearrangement of 7-(p-methoxybenzoyloxy)-4-methyl-
coumarin with aluminium chloride, first at 190°, then
at 200° for 1.5 h (52%, m.p. 181°). -Refer to: Chem. Abstr., **114**, 42490n (11991)*.

m.p. 170°*; IR*.

(3,4-Dihydroxyphenyl)(4-hydroxyphenyl)methanone

$C_{13}H_{10}O_4$ mol.wt. 230.22

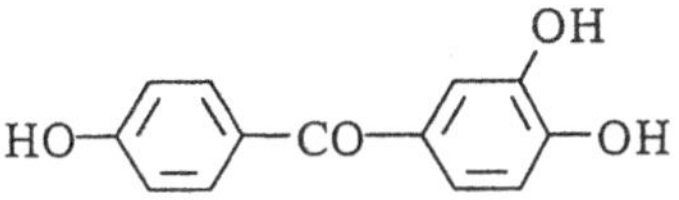

Synthesis

-Preparation by Friedel-Crafts acylation of pyrocatechol
dimethyl ether with p-anisoyl chloride ($AlCl_3/CS_2$/ 3 h at
r.t.), followed by demethylation ($AlCl_3$/toluene/ 1 h at
120°) (65%). -Refer to: Chem. Abstr., **44**, 1271f
(1950)*.

m.p. 205°*; Spectra (NA).

(3,5-Dihydroxyphenyl)(4-hydroxyphenyl)methanone

[129020-58-2] $C_{13}H_{10}O_4$ mol.wt. 230.22

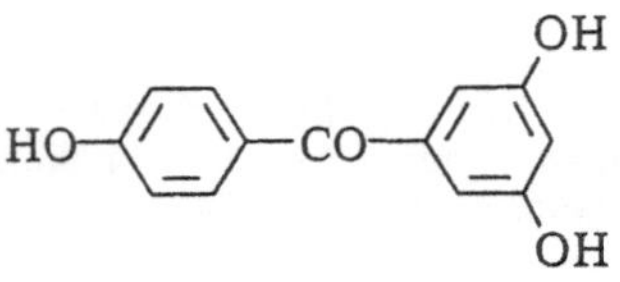

Synthesis

-Refer to: [828].

m.p. and Spectra (NA).

4. TETRAHYDROXYBENZOPHENONES

4.1. *Hydroxy groups located on one ring*

Phenyl(2,3,4,6-tetrahydroxyphenyl)methanone

[198879-06-0] $C_{13}H_{10}O_5$ mol.wt. 246.22

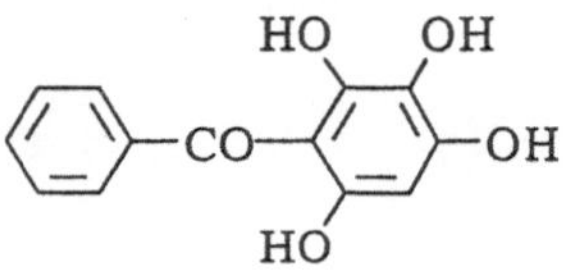

Synthesis

-Preparation by Friedel-Crafts acylation of 1,2,3,5-tetra-
hydroxybenzene with benzoyl chloride (46%). -Refer to:
Chem. Abstr., **128**, 3540y (1998)*.

m.p. (NA); ^{1}H NMR*, ^{13}C NMR*, MS*.

4.2. *Hydroxy groups located on both rings*

Symmetrical ketones

Bis(2,3-dihydroxyphenyl)methanone

[35042-50-3] $C_{13}H_{10}O_5$ mol.wt. 246.22

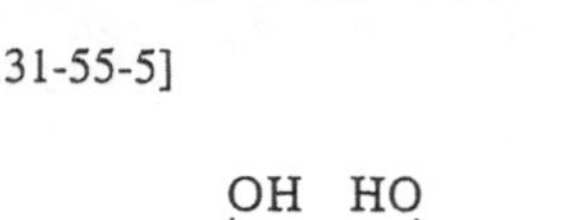

Synthesis

-Preparation by demethylation of 2'-hydroxy-2,3,3'-tri-
methoxybenzophenone with 48% hydrobromic acid in
refluxing acetic acid for 5 h (78%) [420].

m.p.   121-122° [420, 422];   IR [420], UV [420].

Bis(2,4-dihydroxyphenyl)methanone *(Uvinul-D-50)*

[131-55-5] $C_{13}H_{10}O_5$ mol.wt. 246.22

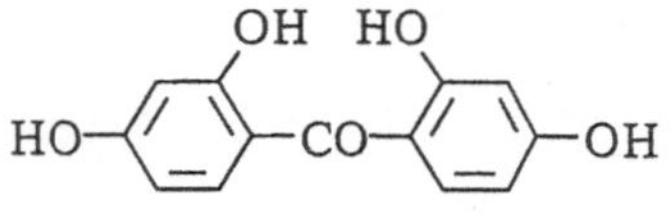

Syntheses

-Obtained (small amount) by melting a fluorescin
chloride/sodium hydroxide mixture in the presence of a
very few water at 230-240° for 2-3 h [903, 904].
-Preparation by reaction of phosgene with resorcinol in the
presence of zirconium chloride in nitrobenzene at 70° (74%) [620].

-Preparation by Fries rearrangement of 3-hydroxyphenyl 2,4-dihydroxybenzoate with zirconium chloride in nitrobenzene at 70° (94%) [620].

-Preparation by reaction of β-resorcylic acid with resorcinol,

*in the presence of zinc chloride and phosphorous oxychloride at 80-90° for 45 min, or at r.t. for 48 h (75%) [507], in sulfolane for 2 h at 50° (92%) [8];

*in the presence of zinc chloride and a mixture of polyphosphoric acid/85% phosphoric acid (60:40) at 60° for 1 h. Then, during 1.5 h, phosphorous trichloride was added and the mixture heated at 60° for 8.5 h (85%) [1302];

*in the presence of boron trifluoride in nitrobenzene at 80° for 30 min (59%) [390].

-Preparation by condensation of 2,4-diacetoxybenzonitrile with resorcinol in the presence of zinc chloride and hydrochloric acid in ethyl ether (Hoesch reaction). The 2,4-diacetoxy-2',4'-dihydroxybenzophenone imine hydrochloride thus formed (m.p. 195° (d)) was hydrolyzed with boiling 25% aqueous sulfuric acid for 15 min [83], (33%) [1263].

-Also refer to: [61, 225, 285, 384, 421, 455, 565, 603, 615, 810, 1067, 1212].

m.p. 201° [1067], 200-201° [1302], 198° [1217], 196-198° [507, 1263], 193-195° [903, 1406];
UV [474, 475, 515, 1067, 1217, 1406];
TLC [377];   HPLC [390];   vapour pressure [1217];
gel permeation chromatography [1145].

Bis(2,5-dihydroxyphenyl)methanone

$C_{13}H_{10}O_5$ mol.wt. 246.22

Syntheses

-Obtained by gently heating b-isoeuxanthone (2,7-dihydroxyxanthone) with anhydrous potassium hydroxide [496].

-This benzophenone can be obtained by total dealkylation of its tetramethyl ether by the usual methods. This one (m.p. 109°) was prepared by reaction of dimethylgentisic acid chloride (2,5-dimethoxybenzoyl chloride) with hydroquinone dimethyl ether (2,5-dimethoxybenzene) in the presence of aluminium chloride in carbon disulfide [717].

N.B.: Pb salt [496].

m.p. and Spectra (NA).

Bis(3,4-dihydroxyphenyl)methanone

[61445-49-6] $C_{13}H_{10}O_5$ mol.wt. 246.22

Synthesis

-Refer to: [399, 465, 466, 752].

m.p. and Spectra (NA).

Asymmetric ketones

(2,3-Dihydroxyphenyl)(2,4-dihydroxyphenyl)methanone

[37728-10-2] $C_{13}H_{10}O_5$ mol.wt. 246.22

Synthesis

-Preparation by reaction of 2,3-dihydroxybenzoic acid with
  resorcinol in the presence of freshly fused zinc chloride and
  phosphorous oxychloride at 65-70° for 3 h (31%) [422].
-Also refer to: [421, 465, 1038].

m.p. 160-161° [422];   IR [422], UV [422].

(2,3-Dihydroxyphenyl)(2,5-dihydroxyphenyl)methanone

[35040-37-0] $C_{13}H_{10}O_5$ mol.wt. 246.22

Synthesis

-Preparation by demethylation of 2-hydroxy-2',3',5'-tri-
  methoxybenzophenone with 48% hydrobromic acid in
  refluxing acetic acid for 4 h (64%) [420].

m.p. 190-191° [420]; ^{1}H NMR [420], IR [420].

(2,3-Dihydroxyphenyl)(2,6-dihydroxyphenyl)methanone

[25577-01-9] $C_{13}H_{10}O_5$ mol.wt. 246.22

Syntheses

-Preparation by reaction of boron tribromide with
  2,2',3'-tris[(ethoxycarbonyl)oxy]-6-methoxybenzophenone
  (SM) in methylene chloride at r.t. for 5 h under nitrogen
  (61%) [830]. SM was obtained by reaction of ethyl
  chloroformate with 2,2',3'-trihydroxy-6-methoxybenzo-
  phenone in the presence of potassium carbonate in
refluxing acetone for 5-6 h (91%).
-Also obtained (by-product) by demethylation of 2,2',3',6-tetramethoxybenzophenone with boron
tribromide in benzene at r.t. for 5 h (16%) [830].

m.p. 173-174° [830]; ^{1}H NMR [830], IR [830], UV [830], MS [830].

(2,3-Dihydroxyphenyl)(3,4-dihydroxyphenyl)methanone

[37728-15-7] $C_{13}H_{10}O_5$ mol.wt. 246.22

Syntheses

-Preparation by reaction of 2,3-dihydroxybenzoic acid with
  pyrocatechol in the presence of freshly fused zinc chloride
  and phosphorous oxychloride at 65-70° for 3 h (32%) [422].
-Preparation by Friedel-Crafts acylation of veratrole with 2,3-dimethoxybenzoyl chloride in the
presence of aluminium chloride at 30-40° for 16 h. The 2,3,3',4'-tetramethoxybenzophenone
formed was demethylated by heating with pyridinium chloride [466].

-Also refer to: [421, 465, 1038].

   m.p.   200° [466], 141-142° [422]. One of the reported melting points is obviously wrong.
   UV [422].

(2,4-Dihydroxyphenyl)(2,5-dihydroxyphenyl)methanone

[61234-44-4] $C_{13}H_{10}O_5$ mol.wt. 246.22

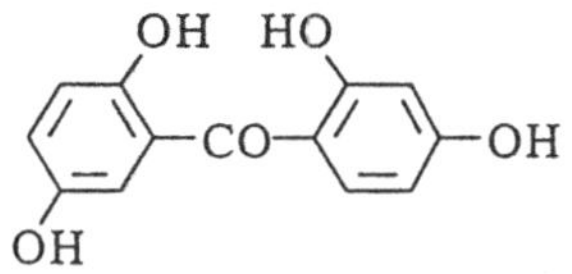

Syntheses

-Preparation by demethylation of 2,4,5'-trihydroxy-
 2'-methoxybenzophenone with hydrobromic acid in boiling
 acetic acid for 2 h (79%) [398].
-Preparation [397] according to [1402].
-Also refer to: [415].

   m.p.   237° [398];   Spectra (NA).

(2,4-Dihydroxyphenyl)(3,4-dihydroxyphenyl)methanone

[61445-50-9] $C_{13}H_{10}O_5$ mol.wt. 246.22

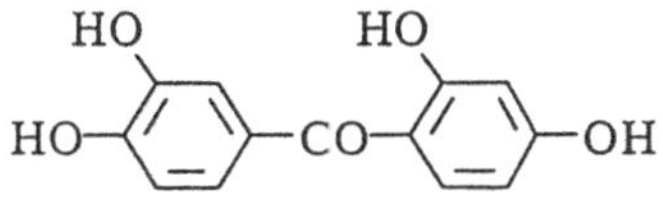

Synthesis

-Preparation by reaction of protocatechuic acid with
 resorcinol,
*in the presence of zinc chloride at 160° [748, 1358]
 (Nencki reaction);
*in the presence of zinc chloride and phosphorous oxychloride for 4 days at r.t. (69%) [1415].
-Also refer to: [465, 466].

   m.p.   201-202° [748, 1415], 199-200° [1358, 1359];   dihydrate [748];   UV [1358].

(2,5-Dihydroxyphenyl)(2,6-dihydroxyphenyl)methanone

[88331-62-8] $C_{13}H_{10}O_5$ mol.wt. 246.22

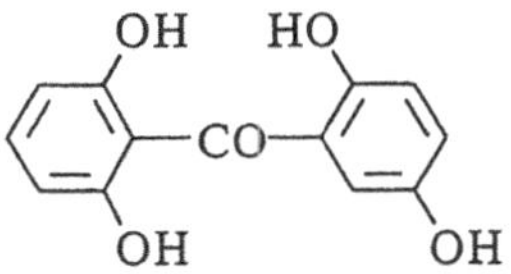

Synthesis

-Refer to: [1333, 1334].

m.p. and Spectra (NA).

(2,6-Dihydroxyphenyl)(3,4-dihydroxyphenyl)methanone

[25576-99-2] $C_{13}H_{10}O_5$ mol.wt. 246.22

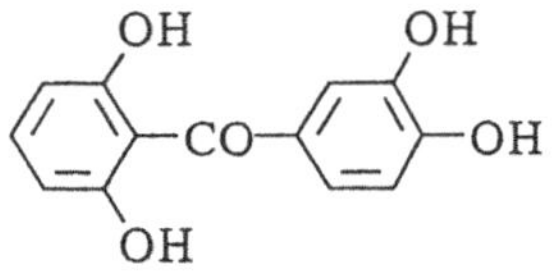

Synthesis

-Preparation by demethylation of 2,3',4',6-tetramethoxy-
 benzophenone with boron tribromide in methylene chloride
 at r.t. for 24 h (51%) [830].
-Also refer to: [1097].

m.p.   113-119° (d) [830];
^{1}H NMR [830], IR [830], UV [830], MS [830].

(2-Hydroxyphenyl)(2,3,4-trihydroxyphenyl)methanone

[42204-63-7] $C_{13}H_{10}O_5$ mol.wt. 246.22

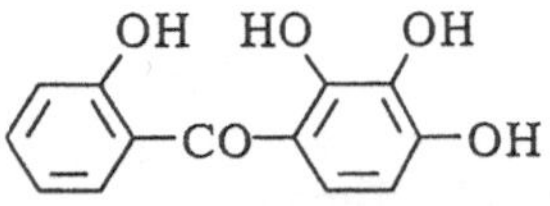

Syntheses

-Preparation by condensation of salicylic acid with pyrogallol,
*in the presence of zinc chloride at 145° for 3 h (Nencki reaction) [495, 1358];

*in the presence of zinc chloride and phosphorous oxychloride at 60-70° for 2 h (50%) [507].
-Also obtained by condensation of o-acetoxybenzonitrile with pyrogallol in the presence of zinc chloride and hydrochloric acid in ethyl ether at 0° (Hoesch reaction). The intermediate compound obtained was hydrolyzed in boiling water for 1 h (19%) [83].
-Also refer to: [421, 1038].
N.B.: Na salt [495].

m.p.   (sesquihydrate) 103-104° [507], 102° [495], 100° [83];
          (anhydrous) 149° [83, 495, 507], 145-147° [1358];
UV [1358].

(2-Hydroxyphenyl)(2,4,6-trihydroxyphenyl)methanone

$C_{13}H_{10}O_5$ mol.wt. 246.22

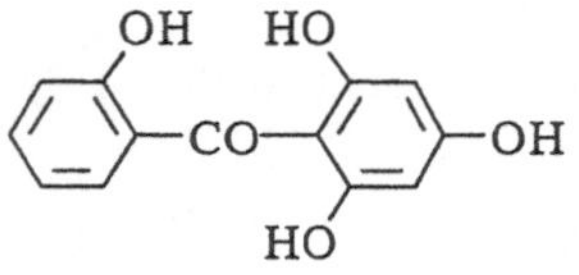

Synthesis

-Obtained by reaction of salicylonitrile with phloroglucinol (Hoesch reaction) Karrer [703]. N.B.: Nevertheless, Nishikawa and Robinson [978] were unable to confirm Karrer's view of the structure of the isolated product.
-Also refer to: [1303].

m.p. and Spectra (NA).

(3-Hydroxyphenyl)(2,3,4-trihydroxyphenyl)methanone

[105443-53-6] $C_{13}H_{10}O_5$ mol.wt. 246.22

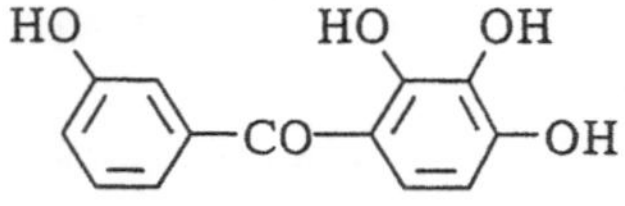

Synthesis

-Preparation by benzoylation of pyrogallol with m-hydroxybenzoic acid,
*in the presence of Amberlyst-15 in refluxing toluene under azeotropical removal of water [1400];

*in the presence of boron trifluoride or its complexes [1443].
-Also refer to: [1257, 1400, 1443] (Japanese patents).

m.p. and Spectra (NA).

(3-Hydroxyphenyl)(2,4,6-trihydroxyphenyl)methanone

[26271-33-0] $C_{13}H_{10}O_5$ mol.wt. 246.22

Syntheses

-Preparation by demethylation of 2,4,6-trihydroxy-
   3'-methoxybenzophenone with aluminium chloride in
   refluxing chlorobenzene for 1 h (95%) [85, 87, 415].
-Preparation from phloroglucinol [397] according to [1402].
-Also obtained (small quantities) by degradation of Gentisein
in the presence of sodium hydroxide and potassium hydroxide (1 g each) at 300-310° under
nitrogen for 3 h [814].
-Preparation by reaction of m-hydroxybenzonitrile with phloroglucinol (38%) (Hoesch reaction)
[978].
-Also refer to: [86, 398].

Isolation from natural source

-From fresh *Gentiana lutea* rhizome (Gentianaceae) [84, 85, 814].

   m.p.   246° (d) [978], 235-238° [85, 87, 814];
   ^{1}H NMR [85, 87], IR [85, 87], UV [85, 87];
   TLC [85];   GC [85].

(4-Hydroxyphenyl)(2,3,4-trihydroxyphenyl)methanone

[31127-54-5] $C_{13}H_{10}O_5$ mol.wt. 246.22

Syntheses

-Preparation by benzoylation of pyrogallol with
   p-hydroxybenzoic acid in the presence of Amberlyst-15
   in refluxing toluene for 21 h under azeotropical removal
   of water (86%) [1400].
-Preparation by Friedel-Crafts acylation of pyrogallol trimethyl ether with p-anisoyl chloride
   (AlCl$_3$/CS$_2$/ 3 h at r.t.), followed by demethylation (AlCl$_3$/toluene/ 1 h at 120°) (65%). -Refer to:
   Chem. Abstr., **44**, 1271f (1950)*.
-Also refer to: [7, 734, 849, 1000] and Chem. Abstr., **129**, 88025c, 129007u, 142608k, 237674t,
   237675u, 237676v (1998).

   m.p.   219°*;   Spectra (NA).

(4-Hydroxyphenyl)(2,4,5-trihydroxyphenyl)methanone

[58115-12-1] $C_{13}H_{10}O_5$ mol.wt. 246.22

Synthesis

-Obtained by total demethylation of 2-hydroxy-4,4',5-tri-
   methoxybenzophenone with hydriodic acid in acetic
   anhydride [371].

   unstable compound [371];   m.p. and Spectra (NA).
   m.p. (of tetraacetate) 132-134° [371].

(4-Hydroxyphenyl)(2,4,6-trihydroxyphenyl)methanone *(Iriflophenone)*

[52591-10-3] $C_{13}H_{10}O_5$ mol.wt. 246.22

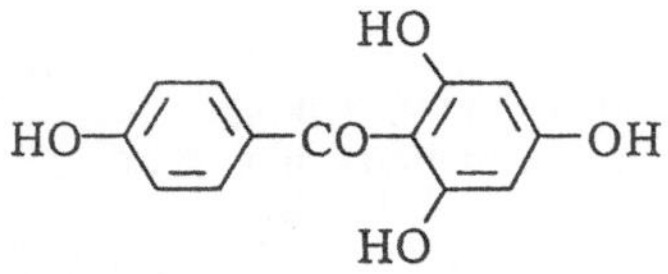

Synthesis

-Preparation by reaction of p-hydroxybenzonitrile with
 phloroglucinol [593], (15%) [978] (Hoesch reaction).

Isolation from natural sources

-From branchwood of *Morus alba* (Moraceae) [797, 831, 1294];
-From rhizomes of *Iris Germanica Linn.* (white flowered variety) [350];
-From rhizoma of *Iris florentina* (Iridaceae) [62].
N.B.: Dihydrate [978, 1294].

    m.p. 210° [978, 1294], 208-210° [350], 207-208° [62];
    ^{1}H NMR [62], IR [62], UV [62], MS [62, 350].

5. PENTAHYDROXYBENZOPHENONES

5.1. *Hydroxy groups located on one ring*

Only one ketone possible, not yet described.

5.2. *Hydroxy groups located on both rings*

(2,3-Dihydroxyphenyl)(3,4,5-trihydroxyphenyl)methanone

[114415-01-9] $C_{13}H_{10}O_6$ mol.wt. 262.22

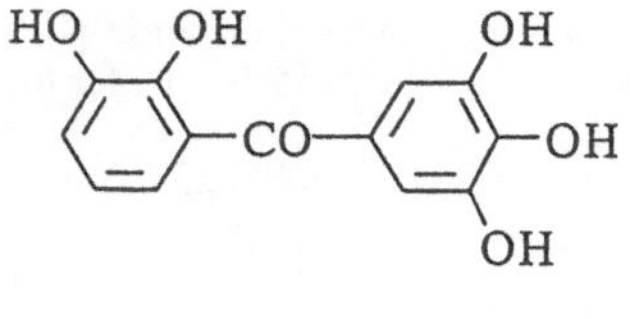

Synthesis

-Preparation by reaction of gallic acid with pyrocatechol in
 chlorobenzene in the presence of methanesulfonic acid for
 6 h between 65 to 75° (90%) or by using boron trifluoride-
 etherate instead of methanesulfonic acid as catalyst (10%).
-Refer to: Chem. Abstr., **108**, 204327v (1988).

    m.p. and Spectra (NA).

(2,4-Dihydroxyphenyl)(2,3,4-trihydroxyphenyl)methanone

[92379-42-5] $C_{13}H_{10}O_6$ mol.wt. 262.22

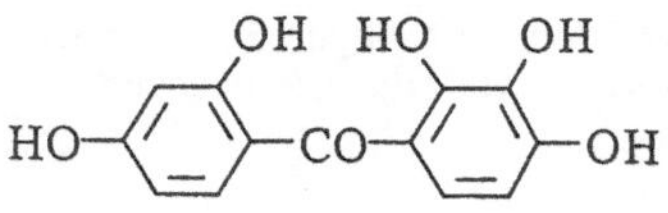

Syntheses

-Preparation by reaction of 2,3,4-trihydroxybenzoic acid
 with resorcinol,
*in the presence of zinc chloride and phosphorous
 oxychloride at 70-80° for 1.5 h (65%) [507];

*in the presence of zinc chloride and a mixture of polyphosphoric acid/85% phosphoric acid
  (60:40) at 27°. Then, during 2 h, phosphorous trichloride was added between 27 to 37° and the
  mixture heated at 60° for 16 h [1302].
-Preparation by reaction of β-resorcylic acid with pyrogallol,
*in the presence of zinc chloride and phosphorous oxychloride at 70-80° for 1 h [507], in sulfolane
  at 50° for 2 h [8];
*in the presence of zinc chloride and a mixture of polyphosphoric acid/85% phosphoric acid
  (60:40) at 27°. Then, during 2 h, phosphorous trichloride was added between 27 to 37° and the
  mixture heated at 60° for 16 h [1302];
*in the presence of boron trifluoride in tetrachloroethane at 110° for 1 h [1443].
-Also obtained by condensation of 2,4-diacetoxybenzonitrile with pyrogallol in the presence of zinc
  chloride and hydrochloric acid in ethyl ether (Hoesch reaction). The intermediate compound
  obtained was hydrolyzed in boiling water for 2 h (22%) [83].
-Also refer to: [712, 713, 734, 737, 761, 849, 958, 1015].

     m.p. 187-188° [507], 187° [83]; dihydrate [83];
     Spectra (NA); TLC [377].

(2,4-Dihydroxyphenyl)(3,4,5-trihydroxyphenyl)methanone

[10425-09-9] $C_{13}H_{10}O_6$ mol.wt. 262.22

Syntheses

-Preparation by reaction of gallic acid with resorcinol in
  the presence of zinc chloride at 145° for 3 h [1358]
  (Nencki reaction).
-Preparation by condensation of 2,4-dihydroxybenzoic
  acid with pyrogallol in the presence of zinc chloride for
45 min at 125-140° (Nencki reaction). -Refer to: Chem. Abstr., **44**, 1271f (1950)*.
-Also refer to: [465, 466].

     m.p. 253°*, 243-245° [1358, 1359]; UV [1358]; TLC [377].

(2,5-Dihydroxyphenyl)(3,4,5-trihydroxyphenyl)methanone

$C_{13}H_{10}O_6$ mol.wt. 262.22

Synthesis

-Preparation by the Friedel-Crafts reaction. -Refer to: Chem.
  Abstr., **65**, 18519h (1966)*.

m.p. (NA); UV*.

(2,6-Dihydroxyphenyl)(2,3,4-trihydroxyphenyl)methanone

[112232-16-3] $C_{13}H_{10}O_6$ mol.wt. 262.22

Synthesis

-Refer to: [601] (Japanese patent).

m.p. and Spectra (NA).

(2,6-Dihydroxyphenyl)(2,4,6-trihydroxyphenyl)methanone

$C_{13}H_{10}O_6$ mol.wt. 262.22

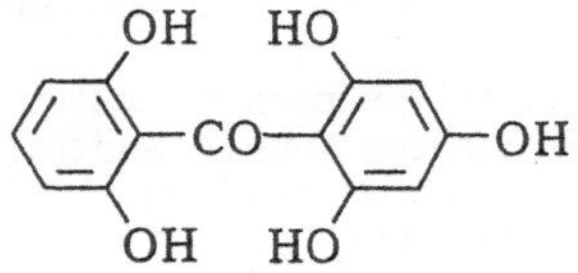

Synthesis

-Obtained (poor yield) by treatment of 1,3,8-trihydroxy-
xanthone (m.p. 265°) with aqueous potassium hydroxide for
30 min at 240-250° (6%). -Refer to: Chem. Abstr., **50**,
15523[b] (1956)*.

m.p. 162°*;   Spectra (NA).

(3,4-Dihydroxyphenyl)(2,3,4-trihydroxyphenyl)methanone

[61445-51-0] $C_{13}H_{10}O_6$ mol.wt. 262.22

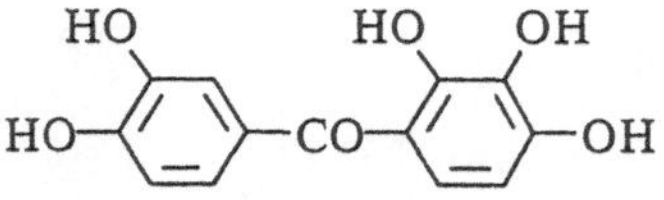

Synthesis

-Refer to: [465, 466].

m.p. and Spectra (NA).

(3,4-Dihydroxyphenyl)(2,3,6-trihydroxyphenyl)methanone

[25577-03-1] $C_{13}H_{10}O_6$ mol.wt. 262.22

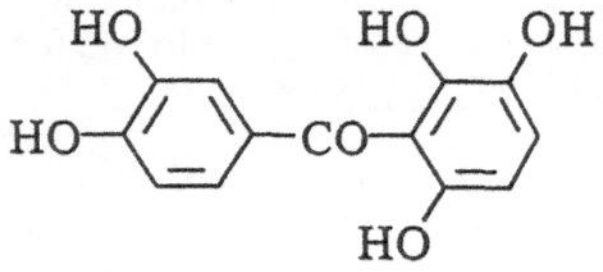

Synthesis

-Preparation by demethylation of 2,3,3',4',6-pentamethoxy-
benzophenone with boron tribromide in methylene chloride
at r.t. for 8 h (70%) [830].

m.p. 131-135° [830];
^{1}H NMR [830], IR [830], UV [830], MS [830].

(3,4-Dihydroxyphenyl)(2,4,5-trihydroxyphenyl)methanone

[61445-52-1] $C_{13}H_{10}O_6$ mol.wt. 262.22

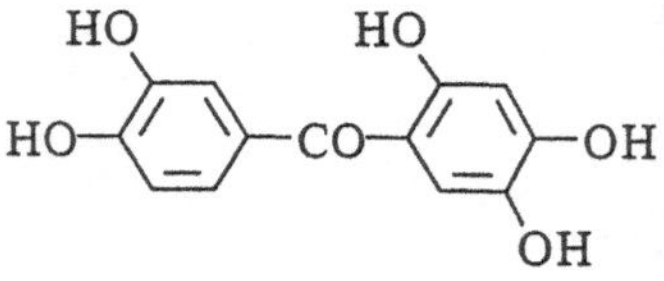

Synthesis

-Preparation by reaction of 3,4-(diacetoxy)benzonitrile with
hydroxyhydroquinone in the presence of zinc chloride and
hydrochloric acid in a chloroform and ethyl ether mixture,
treatment with 10% sulfuric acid, followed by hydrolysis
of the ketimine sulfate formed with boiling
water for 1 h (21%) [755] and Chem. Abstr., **22**, 4519[7] (1928).
-Also refer to: [465, 466].

m.p. 242° [755] and Chem. Abstr., **22**, 4519[7] (1928);
Spectra (NA).

(3,4-Dihydroxyphenyl)(2,4,6-trihydroxyphenyl)methanone *(Maclurin, Morin)*

[519-34-6] $C_{13}H_{10}O_6$ mol.wt. 262.22

Syntheses

-Preparation by reaction of 3,4-dihydroxybenzonitrile with phloroglucinol in the presence of zinc chloride and hydrochloric acid in ethyl ether (Hoesch reaction), first at r.t. for 1 h, then at 50-60° for 4 h (37%) [586].

-Also obtained by condensation of protocatechuic acid with phloroglucinol [750].
-Also refer to: [84, 86, 87, 276, 415, 700, 705, 836, 1100].

Isolation from natural sources

-From heartwood of *Symphonia globulifera* L. (Guttiferae) [829];
-From yellow wood of *Maclura tinctoria* L. D. Don (Moraceae) [580, 831], so called *Morus tinctoria* L. *(Maclura aurantiaca Nutt.)* [278, 1417] or *Chlorophora tinctoria* L. Gaud [797], (major product) [278, 1294];
-From bark of *Laguncularia racemosa* Garten (Moraceae) [831];
-From yellow sapwood of *Acacia catechu* and *Acacia catechuoides* (Moraceae) [831, 976];
-From yellow sapwood of *Acacia sundra* (Moraceae) [831, 976];
-From branches [831] or sawdust (small amounts) [1294] of *Morus Alba* Linn. (Moraceae);
-From bark of *Coto* (Lauraceae) (main component) [277].
N.B.: Ba [157, 580], Ca [580] and Pb salts [580, 836].

monohydrate [276, 586], (Wagner) [580]; sesquihydrate (Delffs) [580];
m.p. 226-230° [87], 222-224° [797], 220-222° [976], 220-221° [1294], 220° [586];
^{1}H NMR [87], IR [87], UV [87];   TLC [87].

(3,4-Dihydroxyphenyl)(2,4,6-trihydroxyphenyl-*1,3,5-* $^{14}C_3$)methanone
(Maclurin-1,3,5- $^{14}C_3$)

[75629-21-9] $C_{13}H_{10}O_6$ mol.wt. 268.22

Synthesis

-Preparation by reaction of protocatechuonitrile (3,4-di-hydroxybenzonitrile) with phloroglucinol-2,4,6-^{14}C (Hoesch reaction). -Refer to: Chem. Abstr., **93**, 217975b (1980)*.

m.p. 199-200°*;   Sp. act. 3.05 x 10^7 dpm/mM*.

(3,4-Dihydroxyphenyl)(3,4,5-trihydroxyphenyl)methanone

[56609-45-1] $C_{13}H_{10}O_6$ mol.wt. 262.22

Synthesis

-Preparation by reaction of gallic acid with pyrocatechol in the presence of zinc chloride for 1 h at 140-145° [1078] (Nencki reaction).
-Also refer to: [242, 376, 465, 466].

m.p. 266° [1078];   UV [1078].

(3,5-Dihydroxyphenyl)(2,4,6-trihydroxyphenyl)methanone

[53250-52-5] $C_{13}H_{10}O_6$ mol.wt. 262.22

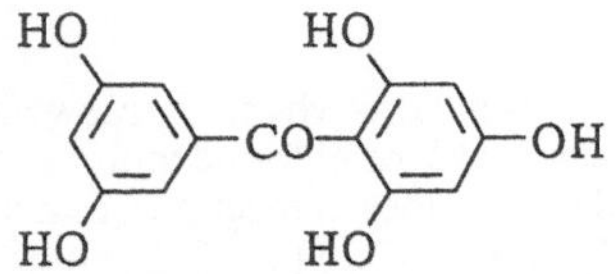

Synthesis

-Not yet described.

Isolation from natural source

-From the heartwood of *Garcinia pedunculata* (Guttiferae) [700, 1100].

m.p. 258-260° [1100];   MS [1100].

(2-Hydroxyphenyl)(2,3,5,6-tetrahydroxyphenyl)methanone

$C_{13}H_{10}O_6$ mol.wt. 262.22

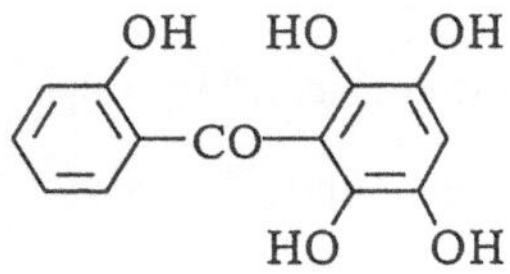

Synthesis

-Refer to: [1303].

m.p. and Spectra (NA).

(4-Hydroxyphenyl)(2,3,4,5-tetrahydroxyphenyl)methanone

[112232-17-4] $C_{13}H_{10}O_6$ mol.wt. 262.22

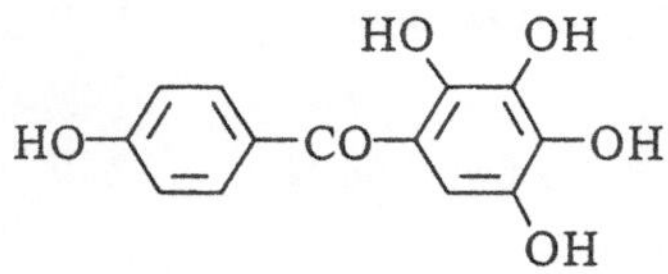

Synthesis

-Refer to: [601] (Japanese patent).

m.p. and Spectra (NA).

6. HEXAHYDROXYBENZOPHENONES

Symmetrical ketones

Bis(2,3,4-trihydroxyphenyl)methanone

[75440-84-5] $C_{13}H_{10}O_7$ mol.wt. 278.22

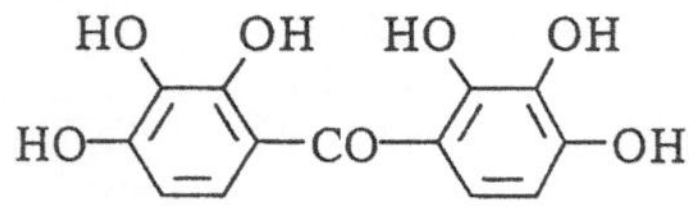

Syntheses

-Preparation by reaction of 2,3,4-trihydroxybenzoic acid with pyrogallol in the presence of zinc chloride for 45 min at 125-140° (Nencki reaction). -Refer to: Chem. Abstr., **44**, 1271f (1950)*.

-Preparation by heating a mixture of 2,3,4-trihydroxybenzoic acid, pyrogallol and excess of phosphorous oxychloride with zinc chloride for 2 h at 70-80° (73%). -Refer to: Chem. Abstr., **50**, 1787g (1956); **51**, 8736a (1957)**.

-Also refer to: Chem. Abstr., **102**, 73168v (1985) and **118**, 90911e (1993).

m.p.   244-245°**, 240°*;   Spectra (NA).

Bis(3,4,5-trihydroxyphenyl)methanone

[111621-53-5] $C_{13}H_{10}O_7$ mol.wt. 278.22

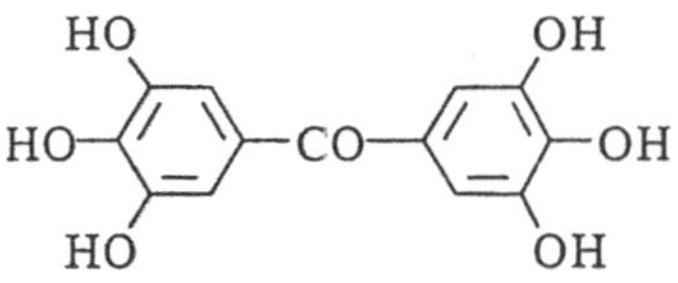

Synthesis

-One of metabolites of tannic acid formed on *in vitro* incubation with rat liver microsomes. -Refer to: Chem. Abstr., **108**, 1953y (1988)*.

m.p. (NA); ^{1}H NMR*, IR*, UV*, MS*.

Asymmetric ketones

(2,3,4-Trihydroxyphenyl)(2,4,5-trihydroxyphenyl)methanone

[153812-71-6] $C_{13}H_{10}O_7$ mol.wt. 278.22

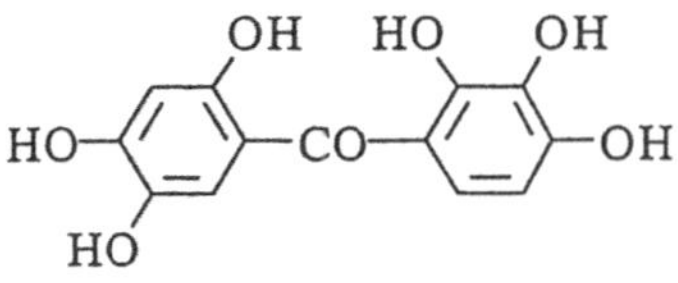

Synthesis

-Refer to: Chem. Abstr., **120**, 204680c (1994).

m.p. and Spectra (NA).

(2,3,4-Trihydroxyphenyl)(3,4,5-trihydroxyphenyl)methanone *(Exifone, Adlone)*

[52479-85-3] $C_{13}H_{10}O_7$ mol.wt. 278.22

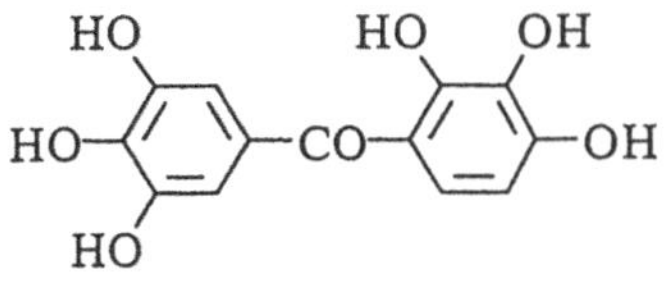

Synthesis

-Preparation by condensation of gallic acid with pyrogallol, *in the presence of zinc chloride at 120° (good yield) [185] or at 145° for 3 h [1358] (Nencki reaction);
*in the presence of zinc chloride and phosphorous oxychloride at 80° for 2 h, *via* Fries rearrangement [466].

-Also refer to: [104, 376, 465, 801] and Chem. Abstr., **80**, 103486u (1974); **89**, 186079y (1978); **106**, 207622g (1987); **107**, 109260p (1987); **109**, 86223d (1988); **112**, 104863f (1990); **125**, 5333v (1996); **129**, 88025c (1998).

m.p.   275-280° [1358, 1359], 272-273° [185], 270° [466];
^{1}H NMR -Refer to: Chem. Abstr., **115**, 196535p (1991), UV [1358, 1359];
pK_a [802];   HPLC -Refer to: Chem. Abstr., **112**, 48178x (1990).

(2,4,5-Trihydroxyphenyl)(3,4,5-trihydroxyphenyl)methanone

[119427-61-1] $C_{13}H_{10}O_7$ mol.wt. 278.22

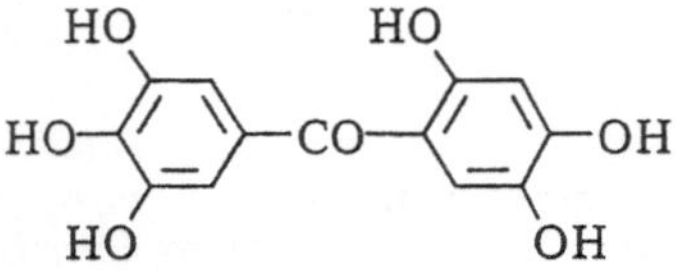

Synthesis

-Refer to: [466, 737, 958].

m.p. and Spectra (NA).

(2,4,6-Trihydroxyphenyl)(3,4,5-trihydroxyphenyl)methanone

[112005-19-3] $C_{13}H_{10}O_7$ mol.wt. 278.22

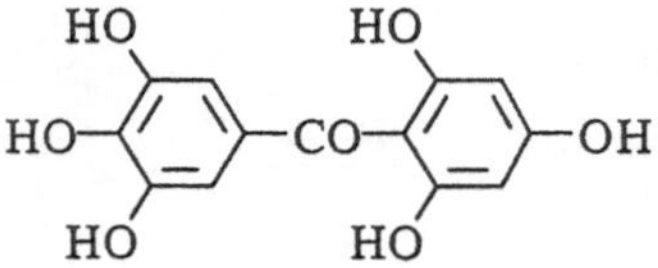

Synthesis

-Refer to: [934] and [601] (Japanese patent).

m.p. and Spectra (NA).

Chapter 2. Substituted Hydroxybenzophenones *(Class of METHANONES)*

1. MONOHYDROXYBENZOPHENONES

1.1. *Substituents located on the hydroxylated ring*

Phenyl(2,3,5-trichloro-6-hydroxyphenyl)methanone

[7396-96-5] $C_{13}H_7Cl_3O_2$ mol.wt. 301.56

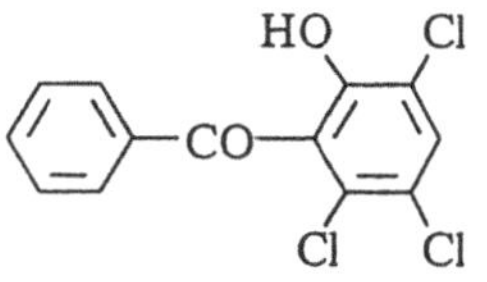

Synthesis

-Preparation by Fries rearrangement of 2,4,5-trichloro-
phenyl benzoate with aluminium chloride for 30 min at
150-160° [375].

m.p. 143-144° [375];   Spectra (NA).

(3-Hydroxy-2,4,6-triiodophenyl)phenylmethanone

[91692-34-1] $C_{13}H_7I_3O_2$ mol.wt. 575.91

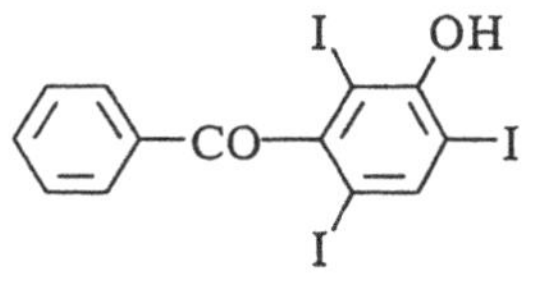

Syntheses

-Preparation by treating a solution of 3-hydroxybenzo-
phenone in methanolic sodium hydroxide with iodine monochloride in aqueous sodium chloride (75%) [251, 252].
-Also obtained by iodination of 3-hydroxybenzophenone
with iodine monochloride [122].

m.p. 190° [251, 252];   Spectra (NA).

(3-Bromo-5-fluoro-4-hydroxyphenyl)phenylmethanone

[579-15-7] $C_{13}H_8BrFO_2$ mol.wt. 295.11

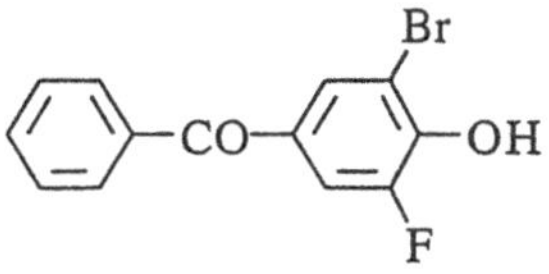

Synthesis

-Preparation by reaction of bromine with 3-fluoro-
4-hydroxybenzophenone in acetic acid [241].

m.p. 187° [241];   Spectra (NA).

(3,5-Dibromo-2-hydroxyphenyl)phenylmethanone

[111277-24-8] $C_{13}H_8Br_2O_2$ mol.wt. 356.01

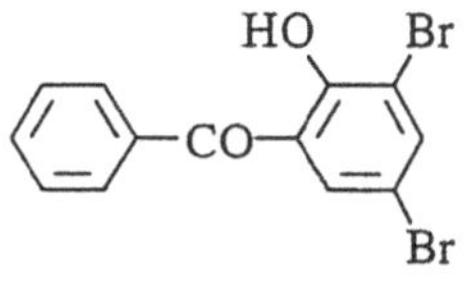

Syntheses

-Preparation by adding a solution of bromine in chloroform
to a solution of 2-hydroxybenzophenone in ethanol [183,
283, 611, 1009].
-Preparation by Friedel-Crafts acylation of benzene with
3,5-dibromo-2-hydroxybenzoyl chloride in the presence of
aluminium chloride in a steam bath for 1 h (70%) [183] or at 80° (>80%) [838].

-Also obtained (poor yield) by alkaline condensation of benzotrichloride with 2,4-dibromophenol in
  30% aqueous sodium hydroxide solution at 80-85° (8%) [1009].
-Also refer to: [54, 200, 837].

   m.p.   129-130° [838], 128-129° [1009], 126° [283, 611], 121-122° [183];   Spectra (NA).

(3,5-Dibromo-4-hydroxyphenyl)phenylmethanone

[26733-16-4] $C_{13}H_8Br_2O_2$ mol.wt. 356.01

Syntheses

-Preparation by reaction of bromine with 4-hydroxybenzo-
  phenone in acetic acid [182, 534, 756], in chloroform [1009]
  or in water [182].
-Preparation by adding a 5% solution of bromine in
  potassium bromide to a solution of 4-hydroxybenzophenone
  in dilute aqueous potassium hydroxide at 5° [182].
-Also obtained by treatment of a mixture of sodium acetate and 4-methoxybenzophenone with
  bromine in a sealed tube at 140° [182].
-Preparation by Fries rearrangement of 2,6-dibromophenyl benzoate with aluminium chloride
  without solvent at 120° for 3 h (40%) [655].
-Also obtained (poor yield) by alkaline condensation of benzotrichloride with 2,4,6-tribromophenol
  in 30% aqueous sodium hydroxide solution at 90-95° (2%) [1009].
-Also refer to: [1353].
N.B.: K salt [534].

   m.p.   178° [655], 155° [182], 152-153° [756], 151-152° [534], 150-152° [1009];
   MS [234];   voltammetric studies [1286].

(2-Chloro-4-fluoro-6-hydroxyphenyl)phenylmethanone

[169781-84-4] $C_{13}H_8ClFO_2$ mol.wt. 250.66

Syntheses

-Obtained by Fries rearrangement of 3-chloro-5-fluorophenyl
  benzoate with aluminium chloride at 200° for 20 min [766].
-Also obtained by demethylation of 2-chloro-4-fluoro-
  6-methoxybenzophenone with boron tribromide in methylene
  chloride at r.t. for 12 h [766], according to [891].

m.p. and Spectra (NA).

(5-Chloro-2-hydroxy-3-nitrophenyl)phenylmethanone

[85052-26-2] $C_{13}H_8ClNO_4$ mol.wt. 277.66

Synthesis

-Preparation by nitration of 5-chloro-2-hydroxybenzo-
  phenone,
*with 26% nitric acid in acetic acid at r.t. for 12 h (50%)
  [566];
*with nitric acid (d = 1.38) in acetic acid and methylene
  chloride at r.t. for overnight [291];
*with 60% nitric acid in the presence of one drop of concentrated sulfuric acid at r.t. for 25 min
  (89%) [932].

-Also refer to: [1040].

  m.p. 70°5-71° [566], 68-70° [291], 56-58° [932];
  ^{1}H NMR [932], IR [566, 932], MS [932].

(5-Chloro-2-hydroxy-4-nitrophenyl)phenylmethanone

[66306-91-0] $C_{13}H_8ClNO_4$ mol.wt. 277.66

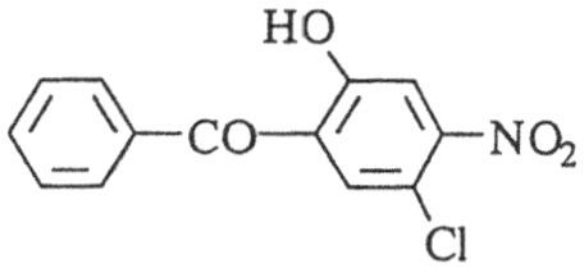

Synthesis

-Refer to: [669].

m.p. and Spectra (NA).

(2,3-Dichloro-4-hydroxyphenyl)phenylmethanone

[62967-12-8] $C_{13}H_8Cl_2O_2$ mol.wt. 267.11

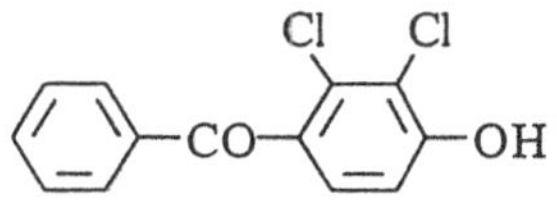

Synthesis

-Preparation by demethylation of 2,3-dichloro-4-methoxy-
 benzophenone (SM),
*with aluminium chloride, in refluxing methylene chloride
 overnight [653] or in refluxing benzene for 5 h [1270];
*with pyridinium chloride at 180° for 2 h (99%) [618].
SM was obtained by Friedel-Crafts acylation of 2,3-dichloroanisole with benzoyl chloride in the
presence of aluminium chloride in methylene chloride [653], first at 0° for 30 min, then at r.t. for
6 h (42%) [618] or in ethylene dichloride for 2 h [1270].

  m.p. 123-126° [653]; Spectra (NA).

(2,4-Dichloro-6-hydroxyphenyl)phenylmethanone

[34199-75-2] $C_{13}H_8Cl_2O_2$ mol.wt. 267.11

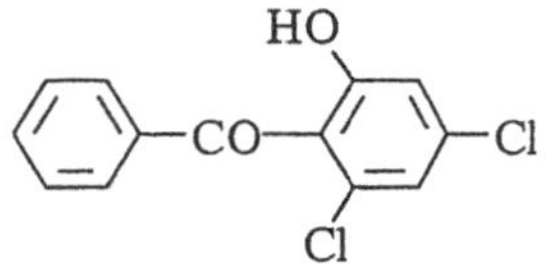

Synthesis

-Preparation by Fries rearrangement of 3,5-dichlorophenyl
 benzoate with aluminium chloride for 30 min at 150-160°
 [375].

m.p. 156-157° [375]; Spectra (NA).

(2,5-Dichloro-4-hydroxyphenyl)phenylmethanone

[123574-94-7] $C_{13}H_8Cl_2O_2$ mol.wt. 267.11

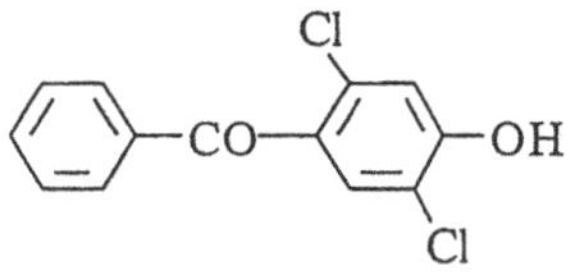

Synthesis

-Preparation by Friedel-Crafts acylation of 2,5-dichloro-
 phenol with benzoyl chloride in the presence of aluminium
 chloride in refluxing ethylene dichloride for 33 h (61%)
 [1195].
-Also refer to: [744].

m.p.   161-163° [1195];
^{1}H NMR [1195], IR [1195], MS [1195].

(2,6-Dichloro-4-hydroxyphenyl)phenylmethanone

[34183-13-6] $C_{13}H_8Cl_2O_2$ mol.wt. 267.11

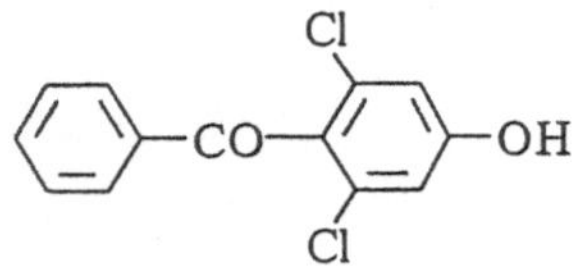

Synthesis

-Obtained by Fries rearrangement of 3,5-dichlorophenyl
  benzoate with aluminium chloride in chlorobenzene at
  140-150° for 20 min or in nitrobenzene at 75° for 24 h
  [1214].

m.p.   135-136° [1214];   Spectra (NA).

(3,4-Dichloro-2-hydroxyphenyl)phenylmethanone

[54923-64-7] $C_{13}H_8Cl_2O_2$ mol.wt. 267.11

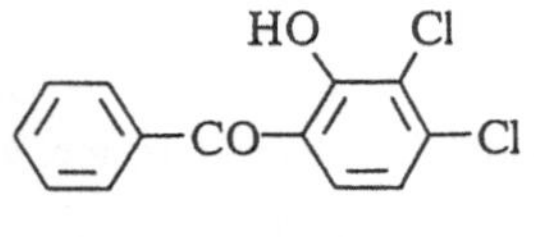

Synthesis

-Obtained by chromic oxidation (CrO$_3$) of 6,7-dichloro-
  3-phenylbenzofuran, followed by saponification of obtained
  keto ester [1133].

m.p.   141°5 [1133];   IR [1133].

(3,4-Dichloro-5-hydroxyphenyl)phenylmethanone

[113730-38-4] $C_{13}H_8Cl_2O_2$ mol.wt. 267.11

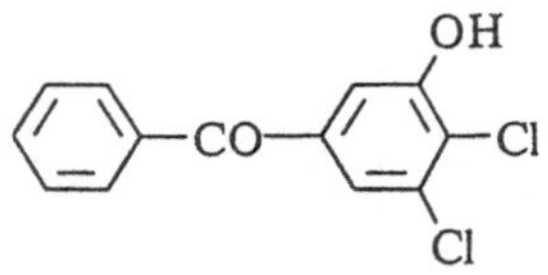

Synthesis

-Preparation by demethylation of 3,4-dichloro-5-methoxy-
  benzophenone with boron tribromide in methylene chloride
  at 5° for 1 h, then at r.t. for 20 min (98%) [535].

m.p.   177° [535]; hemihydrate [535];
^{1}H NMR [535].

(3,5-Dichloro-2-hydroxyphenyl)phenylmethanone

[7396-92-1] $C_{13}H_8Cl_2O_2$ mol.wt. 267.11

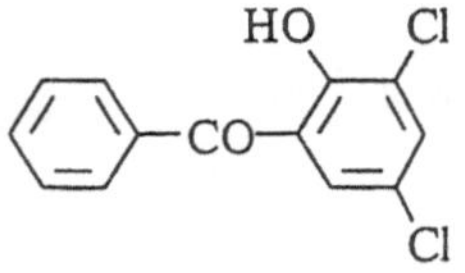

Syntheses

-Preparation by adding aluminium chloride into a solution of
  3,5-dichloro-2-hydroxybenzoyl chloride in benzene and
  heating at 60° for 3-4 h (60-70%) [1266].
-Also obtained by chromic oxidation (CrO$_3$) of 5,7-dichloro-
  3-phenylbenzofuran, followed by saponification of the
obtained keto ester [1133].
-Preparation by reaction of benzoyl chloride with 2,4-dichlorophenol in the presence of aluminium
chloride at 180° for 23 min (46%) [605].

-Preparation by Fries rearrangement of 2,4-dichlorophenyl benzoate (SM) with aluminium chloride at 140° for 30 min (quantitative yield). SM was obtained by heating benzoyl chloride with aluminium tris(2,4-dichlorophenoxide) in a water bath for 30 min [759].
-Also refer to: [54, 1042, 1265, 1312].

   m.p. 116° [1266], 115° [1133], 114-115° [759], 113-114° [605];
   IR [1133], UV [475].

(3,5-Dichloro-4-hydroxyphenyl)phenylmethanone

[34183-06-7] $C_{13}H_8Cl_2O_2$ mol.wt. 267.11

Syntheses

-Preparation by Fries rearrangement of 2,6-dichlorophenyl benzoate with aluminium chloride,
*without solvent at 154° for 2.5 h (71%) [1357];
*in chlorobenzene at 140-150° for 20 min [1214];
*in nitrobenzene at 75° for 24 h [1214].
-Preparation by demethylation of 3,5-dichloro-4-methoxybenzophenone with 48% hydrobromic acid in refluxing acetic acid [182].
-Preparation by passing chlorine into a solution of p-hydroxybenzophenone and sodium acetate in acetic acid during some hours [182].
-Also obtained by Fries rearrangement of 4-tert-butyl-2,6-dichlorophenyl benzoate with aluminium chloride at 140° (sole isolated reaction product). There is a total dealkylation [743].
-Also refer to: [958].

   m.p. 149-150° [1214], 148° [182], 145-146° [1357];   Spectra (NA).

(4,5-Dichloro-2-hydroxyphenyl)phenylmethanone

[58430-25-4] $C_{13}H_8Cl_2O_2$ mol.wt. 267.11

Synthesis

-Preparation by Friedel-Crafts acylation of 3,4-dichloro-phenol with benzoyl chloride in the presence of aluminium chloride at 180° for 35 min [219].

   m.p. 115-116° [219];   Spectra (NA).

(2,5-Difluoro-4-hydroxyphenyl)phenylmethanone

[179018-49-6] $C_{13}H_8F_2O_2$ mol.wt. 234.20

Synthesis

-Preparation by Fries rearrangement of 2,5-difluorophenyl benzoate with aluminium chloride at 160° for 2 h [256, 257].

   m.p. (NA);   MS [256, 257].

(3,5-Difluoro-2-hydroxyphenyl)phenylmethanone

[183280-20-8] $C_{13}H_8F_2O_2$ mol.wt. 234.20

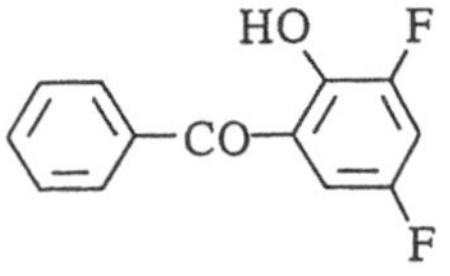

Synthesis

-Preparation by Fries rearrangement of 2,4-difluorophenyl
 benzoate with aluminium chloride at 150-180° for 20 min
 (35%) [765].

m.p.   83°5 [765];
  ^{1}H NMR [765], UV [765], MS [765].

(3,5-Difluoro-4-hydroxyphenyl)phenylmethanone

[179018-48-5] $C_{13}H_8F_2O_2$ mol.wt. 234.20

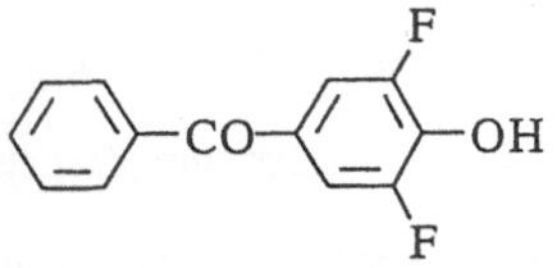

Synthesis

-Preparation by Fries rearrangement of 2,6-difluorophenyl
 benzoate with aluminium chloride at 160° for 2 h [256, 257].

m.p. (NA);   MS [256, 257].

(2-Hydroxy-3,5-diiodophenyl)phenylmethanone

$C_{13}H_8I_2O_2$ mol.wt. 450.01

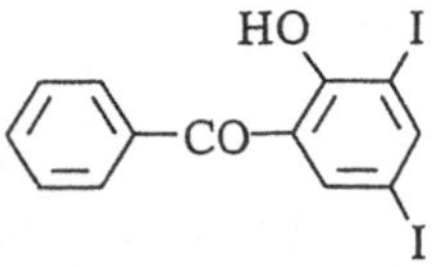

Synthesis

-Preparation by reaction of 3,5-diiodosalicylic acid chloride
 with benzene in the presence of aluminium chloride in
 carbon disulfide at 50° [1220].
-Also refer to: [54, 1219].

m.p.   116° [1220];   Spectra (NA).

(4-Hydroxy-3,5-diiodophenyl)phenylmethanone

[70036-74-7] $C_{13}H_8I_2O_2$ mol.wt. 450.01

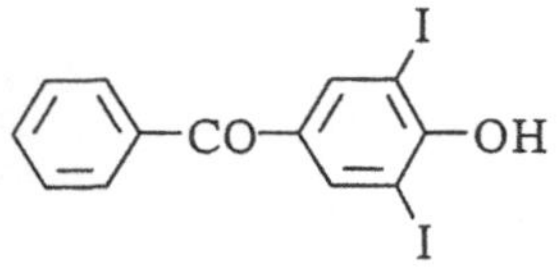

Synthesis

-Preparation by treatment of 4-hydroxybenzophenone with
 iodine monochloride,
*in the presence of sodium acetate in acetic acid [182];
*at steam bath temperature [122, 897].

m.p.   145° [182];   Spectra (NA).

(2-Hydroxy-3,5-dinitrophenyl)phenylmethanone

$C_{13}H_8N_2O_6$ mol.wt. 288.22

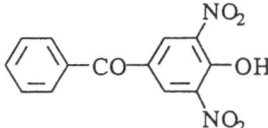

Synthesis

-Obtained by action of potassium hydroxide with 2-bromo-3,5-dinitrobenzophenone (SM) in ethanol. SM was prepared by Friedel-Crafts acylation of benzene with 2-bromo-3,5-dinitrobenzoyl chloride in the presence of aluminium chloride (97%, m.p. 153-154°). -Refer to: Chem. Abstr., **20**, 1229[8] (1926)*.

N.B.: K salt, m.p. 180-200° (d)*.

m.p. and Spectra (NA).

(4-Hydroxy-3,5-dinitrophenyl)phenylmethanone

$C_{13}H_8N_2O_6$ mol.wt. 288.22

Syntheses

-Preparation by heating a solution of 4-hydroxybenzophenone in nitric acid (d = 1.4) at 40° for 30 min [182].
-Preparation by reaction of 10% aqueous sodium hydroxide with 4-chloro-3,5-dinitrobenzophenone (SM) in refluxing ethanol for 20 min (75%). SM was prepared from p-chlorobenzoic acid by a three-step synthesis: at first, nitration of this acid with nitric acid/concentrated sulfuric acid at 140° for 1.5 h. The 4-chloro-3,5-dinitrobenzoic acid formed (95%, m.p. 159°), on treatment with phosphorous pentachloride in refluxing benzene gave the 4-chloro-3,5-dinitrobenzoyl chloride. Then, by adding aluminium chloride to the reaction mixture and heating this one in a water bath for 30 min, SM was obtained (90%, m.p. 118°) [1399].

m.p. 138° [182], 136° [1399]; Spectra (NA).

(3-Bromo-2-hydroxyphenyl)phenylmethanone

[147321-82-2] $C_{13}H_9BrO_2$ mol.wt. 277.11

Synthesis

-Refer to: [1161].

m.p. and Spectra (NA); TLC [1161]; HPLC [1161].

(3-Bromo-4-hydroxyphenyl)phenylmethanone

[89899-44-5] $C_{13}H_9BrO_2$ mol.wt. 277.11

Syntheses

-Preparation by bromination of p-hydroxybenzophenone with bromine [756], in acetic acid [182, 534] or in chloroform [1009].
-Preparation by demethylation of 3-bromo-4-methoxybenzophenone (SM) with 48% hydrobromic acid in refluxing acetic acid. SM was prepared by reaction of bromine with 4-methoxybenzophenone in the presence of sodium acetate in acetic acid

at 100° for 6 h [182].
-Also obtained (poor yield) by aqueous alkaline condensation of benzotrichloride with 4-bromo-
phenol or 2,4-dibromophenol in 30% sodium hydroxide solution at 80-85° (<3%) [1009].
-Also obtained (poor yield) by reaction of benzoyl chloride with o-bromophenol in the presence of
ferric chloride [756].
-Also refer to: [593, 1285].
N.B.: K salt [534].

m.p. 183° [182], 182-183° [1009], 180-181° [534, 756];   Spectra (NA);
polarographic study [1284]; voltammetric study [1287]; TLC [1161]; HPLC [1161].

(4-Bromo-2-hydroxyphenyl)phenylmethanone

[6723-04-2]   C13H9BrO2   mol.wt. 277.11

Syntheses

-Preparation by oxidation of 6-bromo-2,3-diphenylbenzo-
furan with chromium trioxide in boiling acetic acid, followed
by saponification of the obtained keto ester —
(2-benzoyloxy)-4-bromobenzophenone — with sodium
hydroxide in refluxing ethanol [998, 1052], (70%) [995].
-Also obtained by Fries rearrangement of m-bromophenyl benzoate with aluminium chloride [692],
at 200° for 20 min [766].
-Also obtained by demethylation of 4-bromo-2-methoxybenzophenone with boron tribromide in
methylene chloride at r.t. for 12 h [766], according to [891].

m.p. 83° [995, 998], 79° [1052];   IR [68, 998, 1052], UV [72].

(5-Bromo-2-hydroxyphenyl)phenylmethanone

[55082-33-2]   C13H9BrO2   mol.wt. 277.11

Syntheses

-Preparation by Fries rearrangement of p-bromophenyl
benzoate with aluminium chloride without solvent at 150° for
15 min (42%) [183].
-Preparation by oxidation of 5-bromo-2,3-diphenylbenzo-
furan with chromium trioxide in boiling acetic acid,
followed by saponification of the obtained keto ester — 2-(benzoyloxy)-5-bromobenzophenone —
with sodium hydroxide [1052] with potassium hydroxide in refluxing ethanol (65%) [575].
-Also obtained (poor yield) by alkaline condensation of benzotrichloride with p-bromophenol or
2,4-dibromophenol in 30% aqueous sodium hydroxide solution at 80-85° (2%) [1009].
-Also refer to: [1441].

m.p. 111-112° [183], 110° [575], 109° [1052];   IR [1052].

(2-Chloro-4-hydroxyphenyl)phenylmethanone

[81375-00-0]   C13H9ClO2   mol.wt. 232.67

Syntheses

-Preparation by Fries rearrangement of 3-chlorophenyl
benzoate without solvent,
*with aluminium chloride for 2 h at 150° (25%) [172] or at
160° [256, 257] or for 15 min at 175° (8%) [1446];

*with titanium tetrachloride for 15 min at 175° (17%) [1446].
-Preparation by diazotization of 4-amino-2-chlorobenzophenone, followed by hydrolysis of the
 diazonium salt so obtained (66%) [385].

     m.p. 118-119° [172], 115° [385];   MS [256, 257].

(2-Chloro-6-hydroxyphenyl)phenylmethanone

[81374-99-4]   C13H9ClO2   mol.wt. 232.67

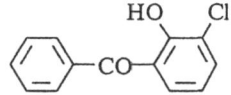

Synthesis

-Obtained by reaction of sodium iodide with 2-chloro-
 6-methoxybenzophenone (SM) in the presence of trimethylsilyl chloride in acetonitrile into an autoclave at 130° for 24 h
 (36%). SM was prepared from 2-amino-6-chlorotoluene by
 a five-step synthesis [385].

    m.p. 105° [385];   13C NMR [385].

(3-Chloro-2-hydroxyphenyl)phenylmethanone

[35582-86-6]   C13H9ClO2   mol.wt. 232.67

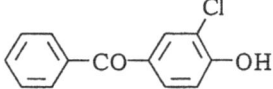

Syntheses

-Preparation by decarboxylation of 2-(3-chloro-2-hydroxy-
 benzoyl)benzoic acid in quinoline at 250° during 15 min in
 the presence of silver carbonate (80%) [506, 555].

-Also obtained from o-chlorophenol,
*by reaction with benzoyl chloride in the presence of aluminium chloride in tetrachloroethane for
 3 h at 120-130° (17%) [555];
*by reaction with benzoic acid in the presence of Amberlyst-15 in refluxing chlorobenzene during
 70 h (18%) [582].
-Also obtained by chromic oxidation (CrO3) of 7-chloro-3-phenylbenzofuran, followed by
 saponification of the obtained keto ester [1133].
-Also obtained (by-product) by Fries rearrangement of 2-chlorophenyl benzoate with aluminium
 chloride without solvent at 152-155° for 2 h (6%) [172].
-Also refer to: [665, 666, 999, 1215].

    m.p. 95° [172], 93° [1133], 92°5-93° [555], 91°7-92°5 [506];
    IR [506, 1133], UV [506].

(3-Chloro-4-hydroxyphenyl)phenylmethanone

[55191-20-3]   C13H9ClO2   mol.wt. 232.67

Syntheses

-Preparation by Fries rearrangement of o-chlorophenyl
 benzoate with aluminium chloride without solvent for 2 h at
 152-155° (88%) [172] or at 160° [256, 257].
 -Also obtained by Fries rearrangement of 4-tert-butyl-
 2-chlorophenyl benzoate with aluminium chloride without
 solvent for 10 min at 140° (28%) [743].
-Preparation by condensation of benzoyl chloride,

*with 2-chlorophenol in the presence of aluminium chloride in tetrachloroethane for 3 h at 120-
130° [506, 555], (46%) [555] or by heating in the presence of ferric chloride without solvent [967];
*with 2-chloroanisole in the presence of aluminium chloride in tetrachloroethane for 3 h at 120-
130° (48%) [555].
-Preparation by heating 3-chloro-4-methoxybenzophenone in the presence of aluminium chloride in
tetrachloroethane for 1 h at 100-110° (98%) [555].
-Also obtained by reaction of benzotrichloride (1 mole) with o-chlorophenol (1 mole) in the
presence of water (1 mole) in solution of hydrofluoric acid (85%) [393].
-Also refer to: [15, 511].

m.p.   182° [172], 181° [393], 180-181° [555], 179°6-181°3 [506],
        179-180° [743], 176° [967];
IR [506], MS [256, 257, 1463].

(4-Chloro-2-hydroxyphenyl)phenylmethanone

[2985-80-0] $C_{13}H_9ClO_2$ mol.wt. 232.67

Syntheses

-Preparation by Fries rearrangement of m-chlorophenyl
benzoate,
*with aluminium chloride [692, 1223], without solvent,
at 140° for 1 h (54%) [608], at 150° for 2 h (58%) [172], at 175° for 15 min (80%) [1446] or at
200° for 20 min [766];
*with titanium tetrachloride for 15 min at 175° (67%) [1446].
-Also obtained by degradation of 6-chloro-2,3-diphenylbenzofuran with chromium trioxide in
boiling acetic acid during 40 min or 2 h, followed by saponification of the 2-benzoyloxy-4-chloro-
benzophenone so formed with 2 N or 4 N sodium hydroxide in refluxing aqueous ethanol for 1 h
or 15 min [997, 1052].
-Preparation by reaction of 4-chloro-2-methoxybenzoyl chloride (SM) with benzene in the presence
of aluminium chloride at r.t. [385] according to [188]. SM was obtained from 2-amino-
4-chlorotoluene by a four-step synthesis.
-Also obtained by demethylation of 4-chloro-2-methoxybenzophenone with boron tribromide in
methylene chloride at r.t. for 12 h [766], according to [891].
-Also refer to: [810, 1016, 1133].

m.p.   77° [172], 76-77° [997], 74°5 [1052], 74° [385, 608];
^{1}H NMR [610], IR [997, 1052, 1313, 1315], UV [607, 610, 793, 794];
pK_a [608, 791];   polarographic study [1223];   TLC [377].

(5-Chloro-2-hydroxyphenyl)phenylmethanone (Light Absorber HCB)

[85-19-8] $C_{13}H_9ClO_2$ mol.wt. 232.67

Syntheses

-Preparation by Fries rearrangement of p-chlorophenyl
benzoate with aluminium chloride [692],
*without solvent [1438], at 180° for 10 min (85%) [944], at
155-157° for 2 h (90%) [172], at 130-140° for 1 h [554],
(72%) [254] or at 100-150° for 0.5-3 h (68%) [1363];
*in refluxing chlorobenzene for 4 h (14%) [1450].
-Preparation by photo-Fries rearrangement of p-chlorophenyl benzoate in benzene, in cyclohexane
or in isopropanol at 53-54° for 19 h (51%, 37% and 29% yields, respectively) [419].
-Preparation from p-chlorophenol,

*by reaction with benzoyl chloride in the presence of aluminium chloride without solvent at 180-195° for 35 min (87%) [605], in boiling carbon disulfide [1066] or in tetrachloroethane at 120-130° for 3 h (10%) [555];
*by reaction with benzoic acid in the presence of boron trifluoride at 180° for 4 h, in a sealed tube (70%) [729].
-Preparation by reaction of benzoyl chloride with p-chloroanisole in the presence of aluminium chloride in tetrachloroethane at 120-130° for 3 h (60%) [555].
-Preparation by reaction of chromium trioxide with 5-chloro-2,3-diphenylbenzofuran in refluxing acetic acid, followed by alkaline hydrolysis of the keto ester so obtained [1052], (quantitative yield) [64, 71].
-Preparation from 2-chlorothioxanthen-9-one 10,10-dioxide (SM) by a three-step synthesis: SM by refluxing in a solution of 2% sodium hydroxide in dioxane/water mixture (65:35) for 4 h gave the 2-(2-hydroxy-5-chlorobenzoyl)phenylsulfinic acid (37%). The former, by reaction with mercuric chloride in refluxing aqueous acetic acid for 4 h, led to the 2-chloromercuri-2'-hydroxy-5'-chloro-benzophenone (74%). Removal of the chloromercury group was achieved with concentrated hydrochloric acid in refluxing ethanol for 2 h (84%) [160].
-Also obtained by reaction of 5-chloro-2-hydroxybenzaldehyde with iodobenzene by using a catalyst system of palladium chloride/lithium chloride in the presence of sodium carbonate in N,N-dimethylformamide at 100° for 6 h (51%) [1204].
-Also refer to: [67, 200, 219, 226, 285, 348, 522, 554, 613, 665, 666, 932, 969, 1117, 1147, 1241, 1324].

m.p. 97° [172], 96-97° [254], 96° [944], 95°5-96° [608], 95-95°5 [555],
95° [1052], 94-95° [64, 71, 1450], 94° [729, 1066, 1217],
93°7-95° [419], 93-94° [160, 1438], 93° [67], 82-85° [1363];
b.p.$_{0.3}$ 147-149° [729];
^{1}H NMR [610, 1204], EPR [267], IR [67, 160, 419, 1052, 1313, 1315],
UV [474, 475, 607, 610, 793, 794, 1217], MS [1204]; pK_a [608, 768, 791];
TLC [377]; polarographic study [1223]; vapour pressure [1217].

(2-Fluoro-4-hydroxyphenyl)phenylmethanone

[179018-47-4]   C13H9FO2   mol.wt. 216.21

Synthesis

-Preparation by Fries rearrangement of m-fluorophenyl benzoate with aluminium chloride at 160° for 2 h [256, 257].

m.p. (NA); MS [256, 257].

(2-Fluoro-5-hydroxyphenyl)phenylmethanone

[145300-05-6]   C13H9FO2   mol.wt. 216.21

Synthesis

-Preparation by aromatization of 5-benzoyl-4-fluoro-3-cyclo-hexenone in the presence of cupric bromide and lithium bromide in refluxing acetonitrile during 1 h (70%) [158].

m.p. 102° [158]; ^{1}H NMR [158], IR [158].

(3-Fluoro-2-hydroxyphenyl)phenylmethanone

[183280-19-5] $C_{13}H_9FO_2$ mol.wt. 216.21

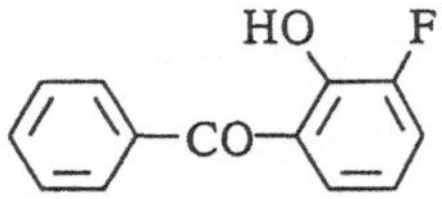

Synthesis

-Preparation by Fries rearrangement of o-fluorophenyl
   benzoate with aluminium chloride at 150-180° for 20 min
   (25%) [765].

m.p. 56° [765];
^{1}H NMR [765], UV [765], MS [765].

(3-Fluoro-4-hydroxyphenyl)phenylmethanone

[365-14-0] $C_{13}H_9FO_2$ mol.wt. 216.21

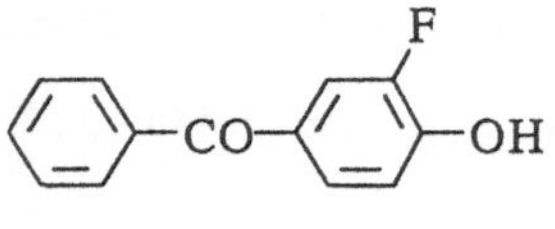

Syntheses

-Preparation by demethylation of 3-fluoro-4-methoxy-
   benzophenone with refluxing pyridinium chloride for
   30 min [241].
-Preparation by Fries rearrangement of o-fluorophenyl
benzoate with aluminium chloride at 160° for 2 h [256, 257].

m.p. 123° [241];   MS [256, 257].

(4-Fluoro-2-hydroxyphenyl)phenylmethanone

[169781-83-3] $C_{13}H_9FO_2$ mol.wt. 216.21

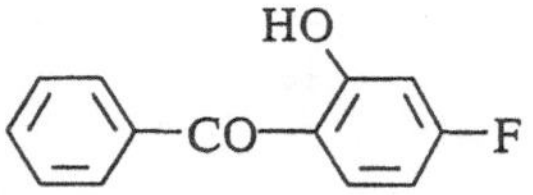

Syntheses

-Obtained by Fries rearrangement of m-fluorophenyl
   benzoate with aluminium chloride at 200° for 20 min [766].
-Also obtained by demethylation of 4-fluoro-2-methoxy-
benzophenone with boron tribromide in methylene chloride at r.t. for 12 h [766], according to
[891].

m.p. and Spectra (NA).

(5-Fluoro-2-hydroxyphenyl)phenylmethanone

[362-47-0] $C_{13}H_9FO_2$ mol.wt. 216.21

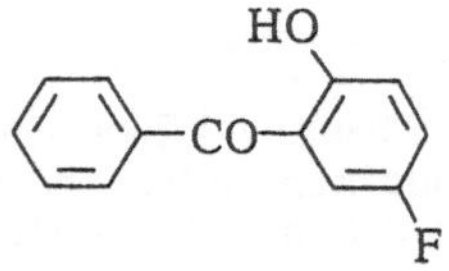

Syntheses

-Preparation by demethylation of 5-fluoro-2-methoxy-
   benzophenone with refluxing pyridinium chloride [236].
-Preparation by Fries rearrangement of p-fluorophenyl
   benzoate with aluminium chloride [692].
-Also refer to: [613].

m.p. 77° [236];   Spectra (NA).

(4-Hydroxy-3-iodophenyl)phenylmethanone

[170744-87-3] $C_{13}H_9IO_2$ mol.wt. 324.12

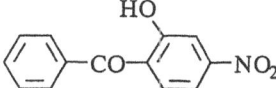

Syntheses

-Preparation by demethylation of 3-iodo-4-methoxybenzo-
phenone with 48% hydrobromic acid in refluxing acetic acid
[182].
-Preparation by treating an alkaline solution of 4-hydroxy-
benzophenone with 5% iodine in aqueous solution of potassium bromide [182].
-Preparation by reaction of iodine monochloride with 4-hydroxybenzophenone in acetic acid
solution at r.t. [897].
-Also refer to: [60].

m.p. 184° [182]; Spectra (NA).

(2-Hydroxy-3-nitrophenyl)phenylmethanone

[182499-95-2] $C_{13}H_9NO_4$ mol.wt. 243.22

Synthesis

-Obtained by reaction of a 68% nitric acid/96% sulfuric acid
mixture with o-hydroxybenzophenone first at <10°, then at
r.t. for 2 h (32%) [765].
-Also refer to: [1433].

m.p. 88°5 [765];
^{1}H NMR [765], UV [765], MS [765].

(2-Hydroxy-4-nitrophenyl)phenylmethanone

[1834-88-4] $C_{13}H_9NO_4$ mol.wt. 243.22

Syntheses

-Preparation by oxidation of 6-nitro-2,3-diphenylbenzofuran
with chromium trioxide in acetic acid, followed by
saponification of the keto ester so obtained (2-benzoyloxy-
4-nitrobenzophenone) in the presence of 2 N sodium
hydroxide in refluxing ethanol for 15 min (60%) [69].
-Preparation by reaction of 2-hydroxy-4-nitrobenzoyl chloride with benzene in the presence of
aluminium chloride in nitrobenzene at 100-110° for 2.5 h (60%) [451].
-Also obtained (poor yield) by Fries rearrangement of m-nitrophenyl benzoate with aluminium
chloride at 170° for 2 h (11%) [451].
-Also obtained (poor yield) by reaction of benzoyl chloride with m-nitrophenol in the presence of
aluminium chloride at 170° for 2 h (10%) [1338].

m.p. 115° [67], 114-115° [69], 112-113° [451, 608], 108° [1338];
^{1}H NMR [610], IR [68, 792], UV [607, 610, 793, 794];
pK_a [608, 790, 791, 792]; TLC [377].

(2-Hydroxy-5-nitrophenyl)phenylmethanone

[18803-19-5] $C_{13}H_9NO_4$ mol.wt. 243.22

Syntheses

-Preparation by nitration of o-hydroxybenzophenone,
*with 65% nitric acid in acetic acid at r.t. for 24 h (32%) [608];
*with a 68% nitric acid/96% sulfuric acid mixture first at <10°, then at r.t. for 2 h (32%) [765].

-Preparation by reaction of potassium hydroxide with 2-chloro-5-nitrobenzophenone,
*in the presence of a few water at 150-160° for 5-6 h (quantitative yield) [1398], (56%) [554];
*in refluxing aqueous ethylene glycol (135°) for 6 h (17%) (by-product) [226].

-Preparation by oxidation of 5-nitro-2,3-diphenylbenzofuran with chromium trioxide in acetic acid, followed by saponification of the keto ester obtained — 2-(benzoyloxy)-5-nitrobenzophenone — with 2 N sodium hydroxide in refluxing ethanol [69].

-Also obtained (by-product) by diazotization of 2-amino-5-nitrobenzophenone (4%). The 2-nitrofluorenone was the major compound obtained (75%) [1398].

-Also obtained (poor yield) by Fries rearrangement of p-nitrophenyl benzoate with aluminium chloride in nitrobenzene at 170° for 1 h (4%) [450].

-Preparation from 2-nitrothioxanthen-9-one 10,10-dioxide (SM) by a three-step synthesis: SM by refluxing in a solution of 2% sodium hydroxide in dioxane/water mixture (65:35) for 2 h, gave the 2-(2-hydroxy-5-nitrobenzoyl)phenylsulfinic acid (89%). The former by reaction with mercuric chloride in refluxing aqueous acetic acid for 4 h led to the 2-chloromercuri-2'-hydroxy-5'-nitro-benzophenone (82%). Removal of the chloromercury group was achieved with concentrated hydrochloric acid in refluxing ethanol for 2 h (84%) [160].

-Also obtained by reaction of 2-hydroxy-5-nitrobenzaldehyde with iodobenzene by using a catalyst system of palladium chloride/lithium chloride in the presence of sodium carbonate in N,N-di-methylformamide at 120° for 22 h (58%) [1204].

-Preparation by dealkylation of 2-(2-hydroxyethoxy)-5-nitrobenzophenone (SM) in methylene chloride with boron tribromide at r.t. for 18 h (94%) [226]. SM was obtained by reaction of potassium hydroxide with 2-chloro-5-nitrobenzophenone in refluxing aqueous ethylene glycol (135°) for 6 h (65%).

-Also refer to: [217, 532, 553, 593, 708, 709, 776, 777].

m.p. 128-129° [226], 126°5 [765], 124-124°5 [1398], 124° [69, 608], 123-124° [160], 122-124° [450], 122-123° [554];
^{1}H NMR [267, 610, 1204], UV [607, 610, 793, 794], MS [1204];
pK_a [608, 790, 791]; TLC [377].

(3-Hydroxy-4-nitrophenyl)phenylmethanone

[182499-94-1] $C_{13}H_9NO_4$ mol.wt. 243.22

Synthesis

-Refer to: [1433].

m.p. and Spectra (NA).

(4-Hydroxy-2-nitrophenyl)phenylmethanone

$C_{13}H_9NO_4$ mol.wt. 243.22

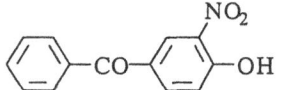

Synthesis

-Obtained (poor yield) by Fries rearrangement of m-nitrophenyl benzoate with titanium tetrachloride in benzene or in
   nitromethane at 40° for 25 h (3%) [1446].

m.p. and Spectra (NA).

(4-Hydroxy-3-nitrophenyl)phenylmethanone

[5464-98-2] $C_{13}H_9NO_4$ mol.wt. 243.22

Syntheses

-Preparation by treatment of 4-methylamino-3-nitrobenzo-
   phenone with boiling 10% aqueous sodium hydroxide for
   48 h (52%) [11].
-Preparation by demethylation of 4-methoxy-3-nitrobenzo-
phenone with 48% hydrobromic acid in refluxing acetic acid [182].
-Preparation by nitration of 4-hydroxybenzophenone with nitric acid (d = 1.5) in an acetic acid and
anhydride acetic mixture at 50-60° [182] or with nitric acid (d = 1.4) at 0° for 2 h [182].
-Also obtained (poor yield) by Fries rearrangement of o-nitrophenyl benzoate with aluminium
chloride in nitrobenzene at 165-175° for 1 h (<4%) [450].
-Also obtained from 4-bromo-3-nitrobenzophenone (m.p. 124°) by heating with a mixture of
sodium acetate and acetamide at 175-200° for 2 h (good yield) [198].
-Also refer to: [533, 1087, 1446].

   m.p. 120-121° [198], 94° [182], 91-92° [450], 90° [11]. One of the reported melting points
   is obviously wrong.   Spectra (NA).

(2-Amino-4-chloro-5-hydroxyphenyl)phenylmethanone

[62492-59-5] $C_{13}H_{10}ClNO_2$ mol.wt. 247.68

Synthesis

-Preparation by demethylation of 2-amino-4-chloro-
   5-methoxybenzophenone with boron tribromide in
   methylene chloride at r.t. for 4 h (86%) [746].

   m.p. 165-167° [746]; ^{1}H NMR [746].

(2-Amino-5-chloro-3-hydroxyphenyl)phenylmethanone

[28363-58-8] $C_{13}H_{10}ClNO_2$ mol.wt. 247.68

Synthesis

-Obtained by hydrolysis of 2-benzoyl-4-chloro-6-hydroxy-
   benzanilide with refluxing 2 N aqueous sodium hydroxide
   for 5 h (80-82%) [457, 746].
-Also refer to: [1209].

m.p.   166-168° [457], 166-167° [746];   ¹H NMR [457].

(2-Amino-5-chloro-4-hydroxyphenyl)phenylmethanone

[62492-60-8] $C_{13}H_{10}ClNO_2$ mol.wt. 247.68

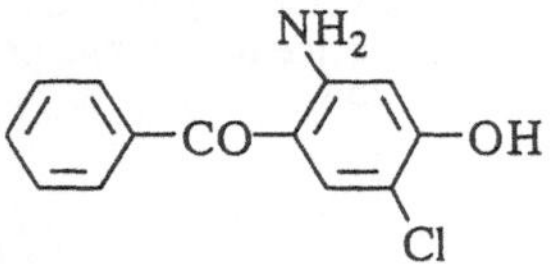

Synthesis

-Preparation by adding acetic anhydride to an ice cooled
solution of 2-amino-5-chloro-4-methoxybenzophenone in
57% aqueous hydriodic acid and heating the resulting
mixture at reflux for 12 h (75%) [746].

m.p.   151-153° [746];   ¹H NMR [746];   TLC [746].

(3-Amino-5-chloro-2-hydroxyphenyl)phenylmethanone

[85052-43-3] $C_{13}H_{10}ClNO_2$ mol.wt. 247.68

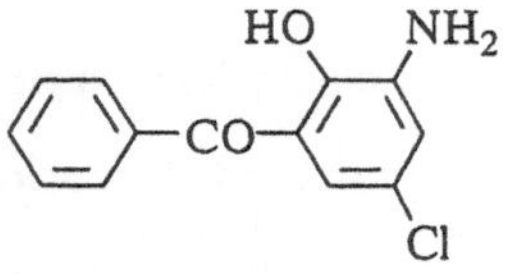

Synthesis

-Obtained from 5-chloro-2-hydroxy-3-nitrobenzophenone,
*by reduction with stannous chloride and hydrochloric acid
in refluxing methanol for 6 h [291];
*by hydrogenation in the presence of 10% Pd/C in a
chloroform/ethanol mixture for 2 h [932].

m.p.   94-95° [291], 93-95° [932];
¹H NMR [932], IR [932], MS [932];   TLC [932].

(3-Amino-5-chloro-2-hydroxyphenyl)phenylmethanone *(Hydrochloride)*

[85052-44-4] $C_{13}H_{10}ClNO_2$,HCl mol.wt. 284.15

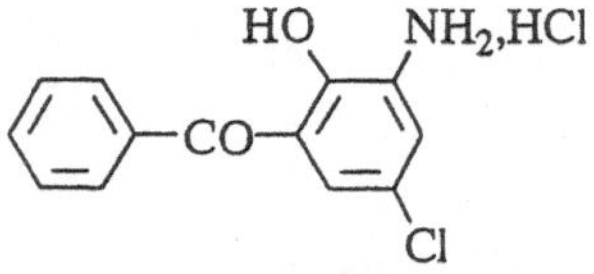

Synthesis

-Obtained by treatment of 3-amino-5-chloro-2-hydroxy-
benzophenone with 2 N hydrochloric acid (27%) [932].

m.p. and Spectra (NA).

(3-Amino-5-fluoro-2-hydroxyphenyl)phenylmethanone

                                                $C_{13}H_{10}FNO_2$ mol.wt. 231.23

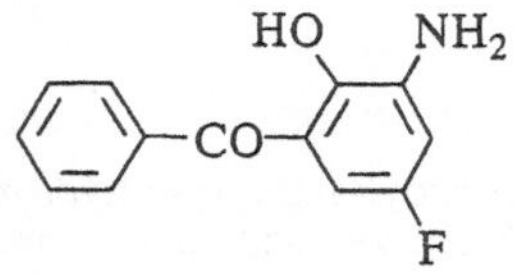

Synthesis

-Refer to: [932].

m.p. and Spectra (NA).

(2-Amino-3-hydroxy-5-nitrophenyl)phenylmethanone

[60302-91-2] $C_{13}H_{10}N_2O_4$ mol.wt. 258.23

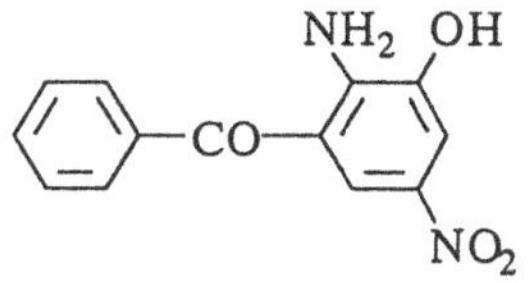

Synthesis

-Not yet described.

Isolation from natural source

-Major metabolite appearing mainly as conjugated compound, in the urine of rabbits fed nitrazepam. It was isolated after enzymatic hydrolysis with β-glucuronidase. -Refer to: Chem. Abstr., **85**, 116484v (1976)*; **90**, 197343b (1979)*.

m.p. (NA); ^{1}H NMR*, IR*, UV*, MS*.

(2-Amino-3-hydroxyphenyl)phenylmethanone

$C_{13}H_{11}NO_2$ mol.wt. 213.24

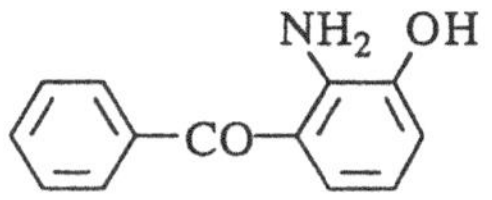

Synthesis

-Preparation by treatment of 4-benzoyl-2,3-dihydro-benzoxazol-2-one (SM) with refluxing 20% hydrochloric acid for 10 h (68%). SM was obtained by UV light irradiation of N-benzoyl-2,3-dihydrobenzoxazol-2-one in acetonitrile for 48 h (21%) [627].

m.p.   129°5-130°5 [627];   Spectra (NA).

(2-Amino-4-hydroxyphenyl)phenylmethanone

$C_{13}H_{11}NO_2$ mol.wt. 213.24

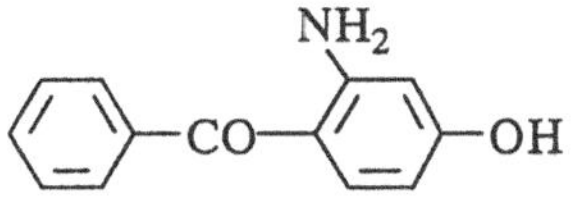

Synthesis

-Refer to: [828].

m.p. and Spectra (NA).

(2-Amino-5-hydroxyphenyl)phenylmethanone

[17562-32-2] $C_{13}H_{11}NO_2$ mol.wt. 213.24

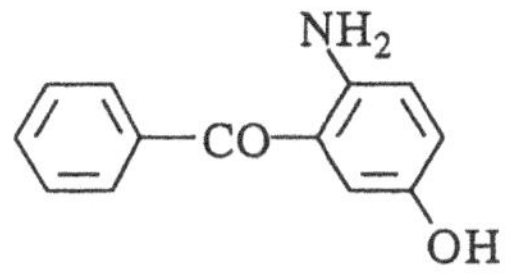

Syntheses

-Preparation by demethylation of 2-amino-5-methoxy-benzophenone with boiling 48% hydrobromic acid [590], for 8 h (85%) [1307].
-Also obtained by UV light irradiation of 3-phenyl-2,1-benz-isoxazole in 66% sulfuric acid at 80-90° (88-95%) [478].
-Also obtained by heating N-benzoyl-p-methoxyaniline with bismuth chloride (5 mol excess) at 200-230° for 3 min (72%) [145].

m.p.   127-128° [590, 1307], 127° [478];   UV [145], MS [145];   TLC [145].

(3-Amino-4-hydroxyphenyl)phenylmethanone

[42404-41-1]   C13H11NO2   mol.wt. 213.24

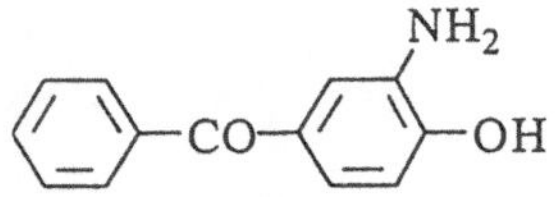

Syntheses

-Preparation by hydrolysis of 3-(acetylamino)-4-hydroxy-
benzophenone with refluxing 10 N hydrochloric acid for
30 min (80%) [31].
-Preparation by hydrolysis of 3-benzamido-4-hydroxy-
benzophenone (SM) with concentrated hydrochloric acid in refluxing acetic acid for 16 h (50%).
SM was obtained by Fries rearrangement of 2-benzamidophenyl benzoate with aluminium chloride
for 3 h at 160° under nitrogen (20%, m.p. 210°) [174].

    m.p.   164-165° [31], 154° [174]; ^{1}H NMR [31], IR [31].

(3-Amino-4-hydroxyphenyl)phenylmethanone *(Hydrochloride)*

[87855-75-2]   C13H11NO2,HCl   mol.wt. 249.70

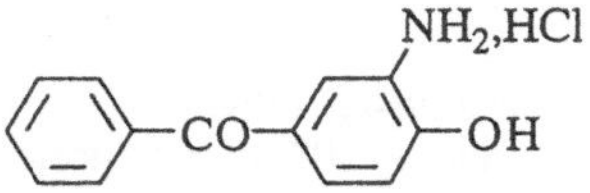

Synthesis

-Refer to: [627, 1013] (Japanese patents).

    m.p. and Spectra (NA).

(4-Amino-2-hydroxyphenyl)phenylmethanone

[3333-96-8]   C13H11NO2   mol.wt. 213.24

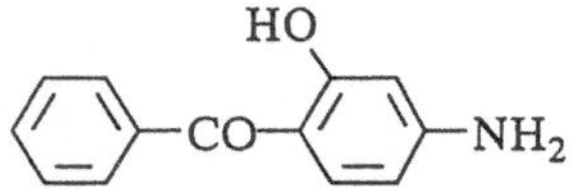

Syntheses

-Obtained by hydrolysis of 4-acetamido-2-hydroxybenzo-
phenone,
*with boiling 5 N hydrochloric acid [658];
*with refluxing 48% hydrobromic acid [1307].
-Preparation by hydrolysis of 4-benzamido-2-hydroxybenzophenone (SM) with concentrated
hydrochloric acid in refluxing acetic acid for 16 h (61%). SM was obtained by Fries rearrangement
of 3-benzamidophenyl benzoate with aluminium chloride for 3 h at 160° (27%, m.p. 155°) [174].
-Also refer to: [1135].

    m.p.   127-128° [1307], 125° [174, 658];
    Spectra (NA).

(4-Amino-3-hydroxyphenyl)phenylmethanone

[31684-63-6]   C13H11NO2   mol.wt. 213.24

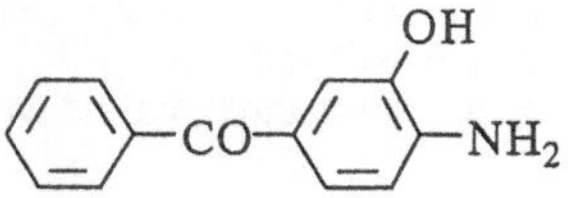

Synthesis

-Preparation from 6-benzoyl-2(3*H*)-benzoxazolinone,
*by alkaline hydrolysis with boiling 10% aqueous sodium
hydroxide solution for 4 h (90-100%) [31, 148, 196], (80%)
[952];
*by treatment with refluxing 20% hydrochloric acid during 40 h (88%) [627].

m.p.   164° [31, 196, 952], 134-135° [627]. One of the reported melting points is obviously wrong.
^{1}H NMR [31], IR [31], MS [148].

(5-Amino-2-hydroxyphenyl)phenylmethanone

[119798-76-4] $C_{13}H_{11}NO_2$ mol.wt. 213.24

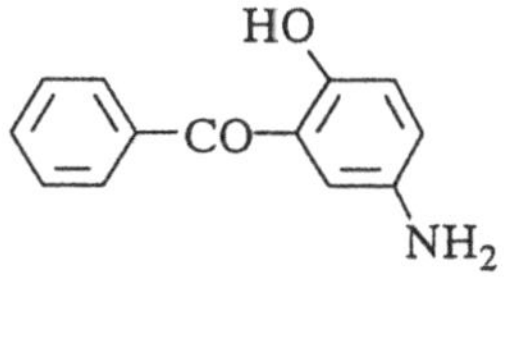

Syntheses

-Preparation by hydrolysis of 5-benzamido-2-hydroxy-benzophenone (SM) with concentrated hydrochloric acid in refluxing acetic acid for 16 h (71%). SM was obtained by Fries rearrangement of 4-benzamidophenyl benzoate with aluminium chloride for 3 h at 160° under nitrogen (40%, m.p. 168°) [174].

-Also obtained from the corresponding hydrochloride by treatment with sodium carbonate in aqueous solution [461].

m.p.   107° [174, 461];   Spectra (NA).

(5-Amino-2-hydroxyphenyl)phenylmethanone *(Hydrochloride)*

$C_{13}H_{11}NO_2,HCl$ mol.wt. 249.70

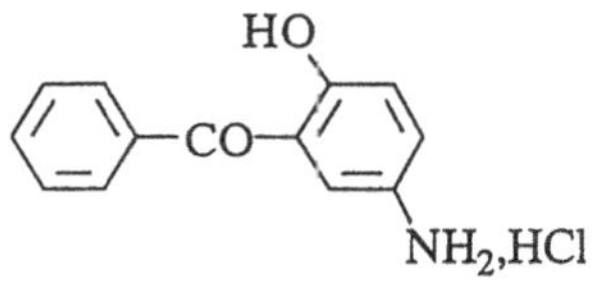

Synthesis

-Obtained by electrolytic reduction of m-nitrobenzophenone in concentrated sulfuric acid for 30 h, followed by action of hydrochloric acid gas in ethyl ether on the amino ketone so obtained [461].

m.p. and Spectra (NA).

[2-Hydroxy-5-(trifluoromethyl)phenyl]phenylmethanone

[72083-16-0] $C_{14}H_9F_3O_2$ mol.wt. 266.22

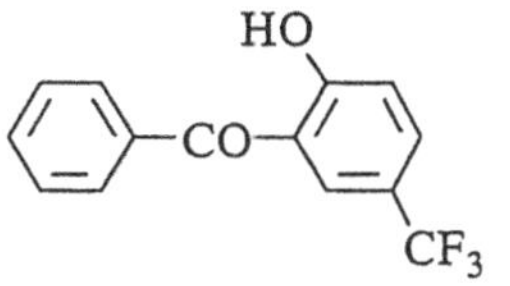

Syntheses

-Preparation by adding phenylmagnesium bromide to 2-methoxy-5-(trifluoromethyl)benzonitrile, followed by hydrolysis of the intermediate imino compound formed, then demethylation of 2-methoxy-5-(trifluoromethyl)-benzophenone so obtained [1336].

-Preparation by demethylation of 2-methoxy-5-(trifluoromethyl)benzophenone with boron trichloride in methylene chloride at -60° for 1 h, then at r.t. [693].

m.p.   84-85° [693];   Spectra (NA).

Phenyl(3,5,6-trifluoro-2-hydroxy-4-methoxyphenyl)methanone

[32541-21-2] $C_{14}H_9F_3O_3$ mol.wt. 282.22

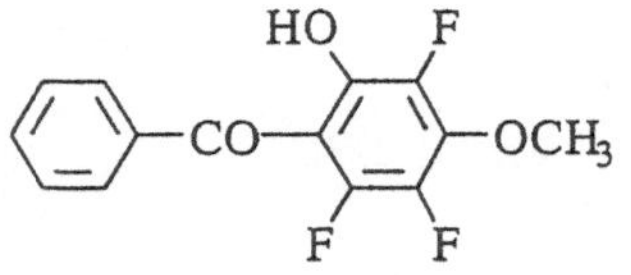

Synthesis

-Preparation by partial demethylation of 2,4-dimethoxy-3,5,6-trifluorobenzophenone in methylene chloride in the presence of aluminium chloride at 20° for 3-6 h (94%) [839].

m.p. 56-58° [839];
^{1}H NMR [839], IR [839], UV [839].

(2-Bromo-6-hydroxy-3-methoxy-4-nitrophenyl)phenylmethanone

[40990-74-7] $C_{14}H_{10}BrNO_5$ mol.wt. 352.14

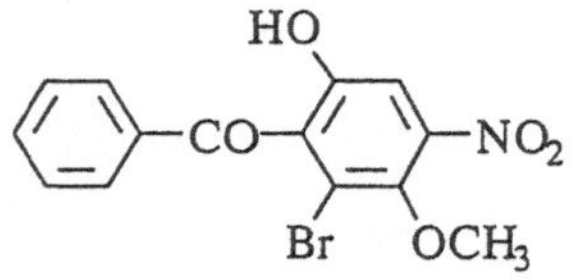

Syntheses

-Preparation by saponification of 2-(benzoyloxy)-6-bromo-5-methoxy-4-nitrobenzophenone (SM) with potassium hydroxide in ethanol (71%). SM was obtained by oxidation of 4-bromo-5-methoxy-6-nitro-2,3-diphenylbenzofuran with chromium trioxide in boiling acetic acid for 30 min [574].
-Also obtained by hydrolysis of 6-bromo-5-methoxy-4-nitro-2-(4-nitrobenzoyloxy)benzophenone [574].

m.p. 126° [574];   Spectra (NA).

(2,4-Dibromo-6-hydroxy-3-methoxyphenyl)phenylmethanone

[40990-66-7] $C_{14}H_{10}Br_2O_3$ mol.wt. 386.04

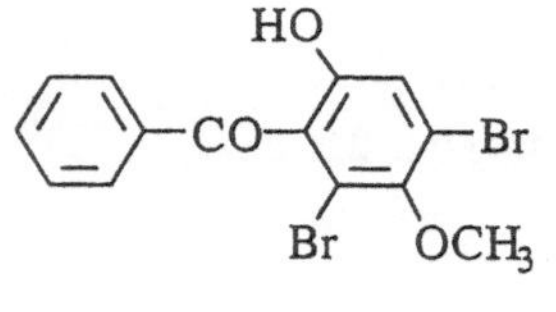

Synthesis

-Preparation by saponification of 2-(benzoyloxy)-4,6-dibromo-5-methoxybenzophenone (SM) with potassium hydroxide in ethanol (65%). SM was obtained by oxidation of 4,6-dibromo-5-methoxy-2,3-diphenylbenzofuran with chromium trioxide in refluxing acetic acid for 30 min [574].

m.p. 138° [574];   Spectra (NA).

(3,5-Dichloro-2-hydroxy-4-methoxyphenyl)phenylmethanone

[158547-83-2] $C_{14}H_{10}Cl_2O_3$ mol.wt. 297.14

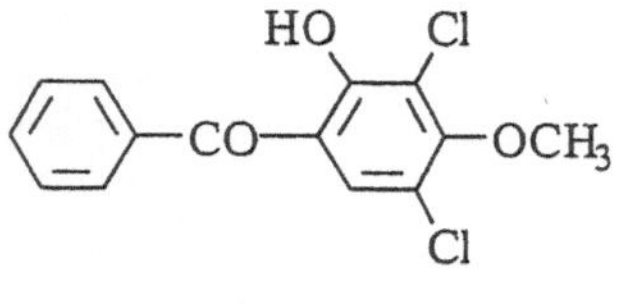

Synthesis

-Formation (trace) by 30 min UV light irradiation of oxybenzone in chlorinated water containing 5 ppm Cl [1348].

m.p. (NA);   MS [1348];   GC [1348].

(3-Bromo-2-hydroxy-5-methylphenyl)phenylmethanone

[6723-09-7] $C_{14}H_{11}BrO_2$ mol.wt. 291.14

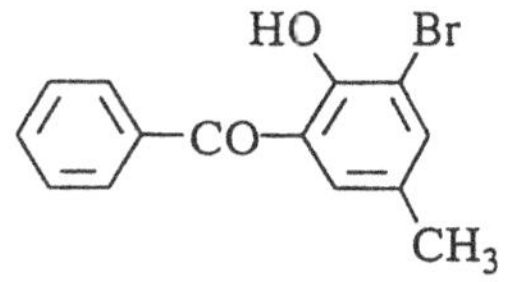

Synthesis

-Preparation by bromination of 2-hydroxy-5-methyl-
  benzophenone [995], in aqueous acetic acid (86%) [1440].
-Also refer to: [258, 1039, 1245, 1246].

m.p.   82° [998], 80-82° [995], 77-78° [1440];   IR [68].

(3-Bromo-4-hydroxy-5-methylphenyl)phenylmethanone

$C_{14}H_{11}BrO_2$ mol.wt. 291.14

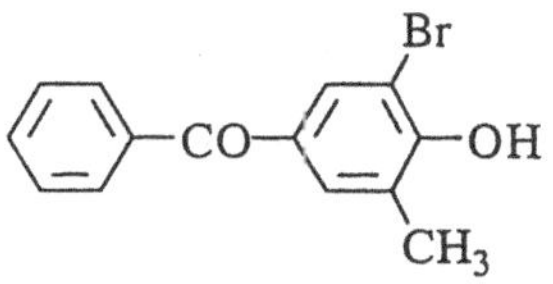

Synthesis

-Preparation by action of bromine on 4-hydroxy-3-methyl-
  benzophenone in acetic acid [142].

m.p. 130-131° [142];   Spectra (NA).

(4-Bromo-2-hydroxy-3-methylphenyl)phenylmethanone

[6758-89-0] $C_{14}H_{11}BrO_2$ mol.wt. 291.14

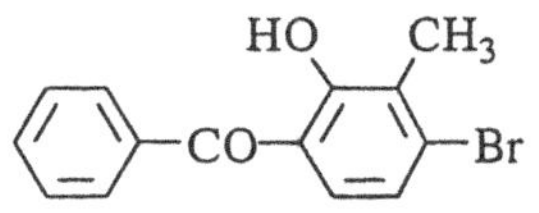

Synthesis

-Obtained by chromic oxidation (CrO_3) of 6-bromo-
  7-methyl-2,3-diphenylbenzofuran, followed by alkaline
  hydrolysis of keto ester so obtained [995, 998].

m.p.   96° [995, 998];   IR [68, 998], UV [72].

(4-Bromo-2-hydroxy-5-methylphenyl)phenylmethanone

[6723-07-5] $C_{14}H_{11}BrO_2$ mol.wt. 291.14

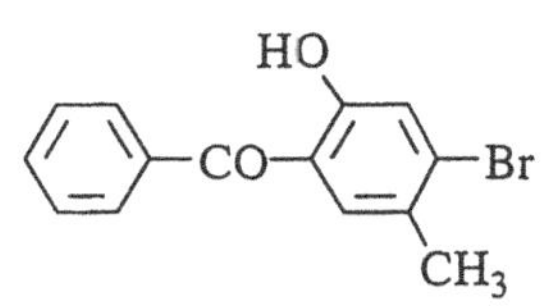

Synthesis

-Preparation by reaction of chromium trioxide with 6-bromo-
  5-methyl-2,3-diphenylbenzofuran in acetic acid, followed by
  saponification of the keto ester formed (2-benzoyloxy-
  4-bromo-5-methylbenzophenone) with sodium hydroxide in
  ethanol [998], (73%) [995].

m.p.   122° [995, 998];
^{1}H NMR [996], IR [996], UV [72].

(5-Bromo-2-hydroxy-3-methylphenyl)phenylmethanone

[6723-13-3] $C_{14}H_{11}BrO_2$ mol.wt. 291.14

Synthesis

-Preparation by bromination of 2-hydroxy-3-methyl-
benzophenone [998], in chloroform [995].

m.p. 83° [995, 998];   IR [68, 998, 1313, 1315].

(4-Bromo-2-hydroxy-3-methoxyphenyl)phenylmethanone

[65202-49-5] $C_{14}H_{11}BrO_3$ mol.wt. 307.14

Synthesis

-Preparation by saponification of 2-(benzoyloxy)-4-bromo-
3-methoxybenzophenone (SM) with potassium hydroxide in
boiling ethanol (70%). SM was obtained by oxidation of
6-bromo-7-methoxy-2,3-diphenylbenzofuran with
chromium trioxide in boiling acetic acid (65%) [1].

m.p.   240-243° [1];   Spectra (NA).

(5-Bromo-2-hydroxy-4-methoxyphenyl)phenylmethanone

[3286-93-9] $C_{14}H_{11}BrO_3$ mol.wt. 307.14

Synthesis

-Preparation from 2-(benzoyloxy)-5-bromo-4-methoxy-
benzophenone (SM) by treatment with potassium
hydroxide in refluxing ethanol for 1 h (68%). SM was
obtained by oxidation of 5-bromo-6-methoxy-2,3-di-
phenylbenzofuran with chromium trioxide in refluxing
acetic acid for 30 min [574].

m.p.   125° [574];   Spectra (NA).

(3-Chloro-2-hydroxy-5-methylphenyl)phenylmethanone

$C_{14}H_{11}ClO_2$ mol.wt. 246.69

Syntheses

-Preparation by Fries rearrangement of 2-chloro-4-methyl-
phenyl benzoate with aluminium chloride without solvent,
*at 140° for 10 min (92%) [1143];
*at 120-130° for 1 h (67%) [254];
*at 160° for 15 min (14%) [98] (here, by using an old
aluminium chloride).
-Also obtained by action of aluminium chloride with esters mixtures* without solvent at 150° for
15 min [98],
*2-chloro-4-methylphenyl benzoate and p-tolyl acetate (50% yield);
*p-tolyl benzoate and 2-chloro-4-methylphenyl acetate (40% yield);
*2-chloro-4-methylphenyl benzoate and mesityl acetate (14% yield).

m.p. 81° [254], 71° [98, 1143]; Spectra (NA).

(3-Chloro-4-hydroxy-5-methylphenyl)phenylmethanone

[34183-08-9] $C_{14}H_{11}ClO_2$ mol.wt. 246.69

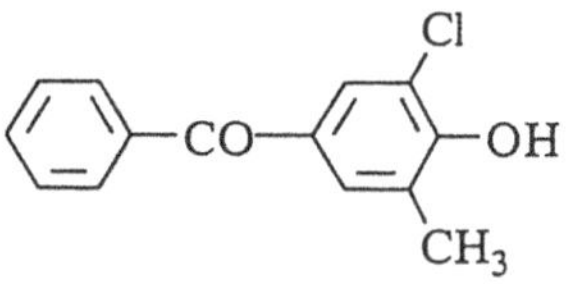

Synthesis

-Preparation by Fries rearrangement of 2-chloro-6-methyl-
  phenyl benzoate with aluminium chloride,
*in chlorobenzene at 140-150° for 20 min or in nitrobenzene
  at 75° for 24 h [1214];
*without solvent at 160° for 45 min (45%) [1221].

m.p. 126-127° [1214]; Spectra (NA).

(4-Chloro-2-hydroxy-5-methylphenyl)phenylmethanone

[33561-92-1] $C_{14}H_{11}ClO_2$ mol.wt. 246.69

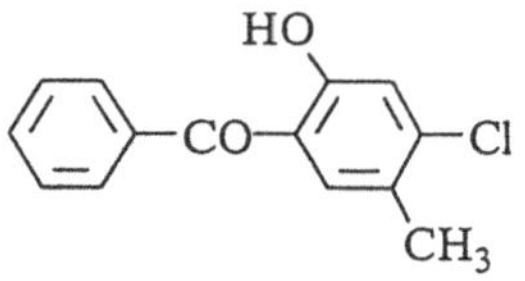

Syntheses

-Obtained by Fries rearrangement of 3-chloro-4-methyl-
  phenyl benzoate with aluminium chloride at 200° for 20 min
  [766].
-Also obtained by demethylation of 4-chloro-2-methoxy-
  5-methylbenzophenone with boron tribromide in methylene
chloride at r.t. for 12 h [766], according to [891].
-Also obtained by oxidation of 6-chloro-5-methyl-2,3-diphenylbenzofuran with chromium trioxide
 in boiling acetic acid for 2 h, followed by saponification of the 2-(benzoyloxy)-4-chloro-5-methyl-
 benzophenone formed with 2 N sodium hydroxide in boiling ethanol for 15 min [997].
-Also refer to: [1314, 1315].

m.p. 108-109° [997]; IR [997, 1314, 1315].

(5-Chloro-2-hydroxy-3-methylphenyl)phenylmethanone

[53347-30-1] $C_{14}H_{11}ClO_2$ mol.wt. 246.69

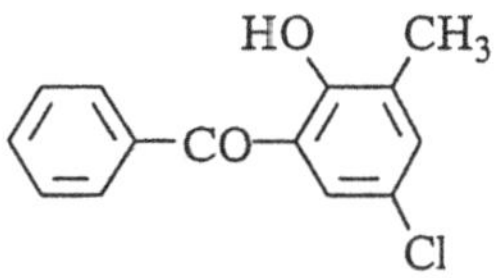

Synthesis

-Refer to: [666] (compound 7e) and [665, 1222].

m.p. and Spectra (NA).

(5-Chloro-2-hydroxy-4-methylphenyl)phenylmethanone

[68751-90-6] $C_{14}H_{11}ClO_2$ mol.wt. 246.69

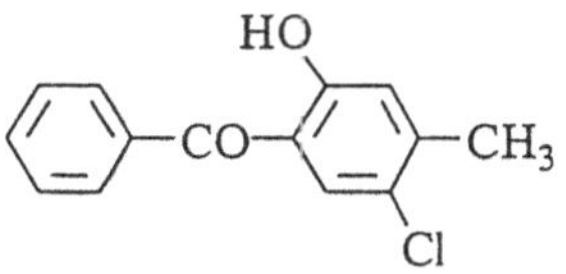

Syntheses

-Preparation by Fries rearrangement of 4-chloro-3-methyl-
  phenyl benzoate in the presence of,
*aluminium chloride [692] without solvent at 140° for
  10 min (quantitative yield) [1143];

*Nafion-XR, a H+-form ion exchange resin, at 175° for 4 h (37%) [1230, 1231].
-Preparation by oxidation of 5-chloro-6-methyl-2,3-diphenylbenzofuran with chromium trioxide in
  boiling acetic acid for 40 min, followed by saponification of the keto ester so obtained with 4 N
  sodium hydroxide in refluxing ethanol for 1 h [1052].
-Also obtained by reaction of benzoyl chloride with 4-chloro-3-methylphenol in the presence of
  Nafion-XR at 175° for 4 h (28%) [1231].
-Also refer to: [224, 348, 560].

     m.p.   142° [1143], 140° [1052];   IR [1052].

[3-(Chloromethyl)-4-hydroxyphenyl]phenylmethanone

[14898-76-1] $C_{14}H_{11}ClO_2$ mol.wt. 246.69

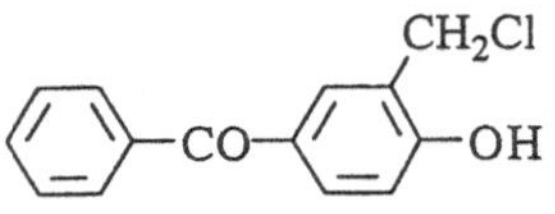

Syntheses

-Preparation by passing hydrogen chloride through a mixture
of 40% formaldehyde solution, concentrated hydrochloric
acid and p-hydroxybenzophenone in acetic acid at r.t. for
2 h (64%) [121].
-Other chloromethylation process [572], (73%) [292].
-Also refer to: [1464].

     m.p.   156° (d) [121];   Spectra (NA).

[5-(Chloromethyl)-2-hydroxyphenyl]phenylmethanone

[120973-82-2] $C_{14}H_{11}ClO_2$ mol.wt. 246.69

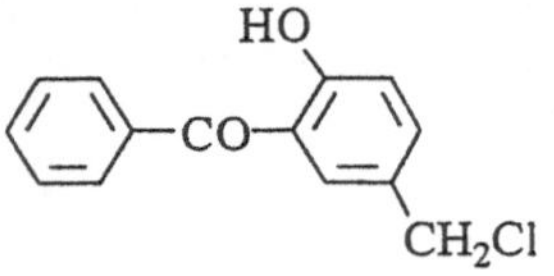

Synthesis

-Refer to: [525].

m.p. and Spectra (NA).

(2-Chloro-6-hydroxy-4-methoxyphenyl)phenylmethanone

[136741-50-9] $C_{14}H_{11}ClO_3$ mol.wt. 262.69

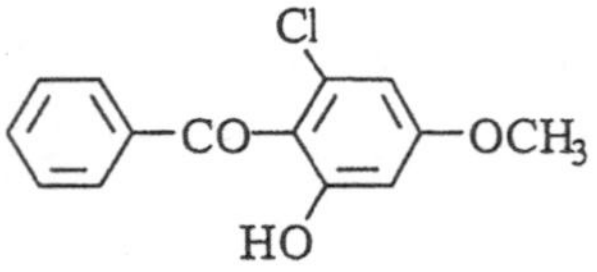

Synthesis

-Preparation by reaction of benzoyl chloride with
5-chlororesorcinol dimethyl ether in the presence of
aluminium chloride in ethylene dichloride (72%) [1193].

m.p. 96-98° [1193];   Spectra (NA).

(3-Chloro-2-hydroxy-4-methoxyphenyl)phenylmethanone

[158547-82-1] $C_{14}H_{11}ClO_3$ mol.wt. 262.69

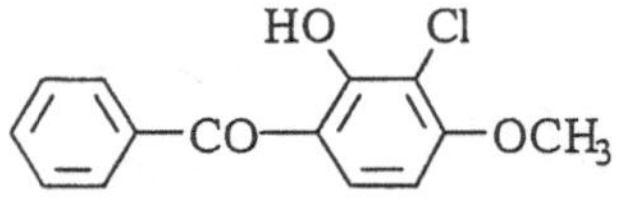

Synthesis

-Formation (trace) by 30 min UV light irradiation of
oxybenzone in chlorinated water containing 5 ppm Cl
[1348].

m.p. (NA);   MS [1348];   GC [1348].

(5-Chloro-2-hydroxy-4-methoxyphenyl)phenylmethanone

[3286-91-7]   C14H11ClO3   mol.wt. 262.69

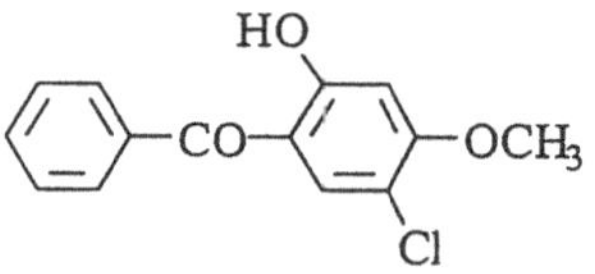

Syntheses

-Preparation by reaction of benzoyl chloride with
4-chlororesorcinol dimethyl ether in the presence of
aluminium chloride in ethylene dichloride (66%) [1193].
-Formation (trace) by 30 min UV light irradiation of
oxybenzone in chlorinated water containing 5 ppm Cl
[1348].

m.p. 118-119° [1193];
^{1}H NMR [1193], IR [1193], MS [1193, 1348];   GC [1348].

(2-Hydroxy-3-methyl-4-nitrophenyl)phenylmethanone

[4072-22-4]   C14H11NO4   mol.wt. 257.25

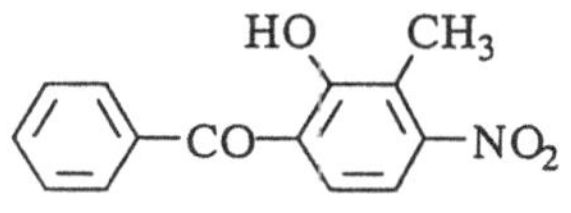

Synthesis

-Preparation by oxidation of 7-methyl-6-nitro-2,3-diphenyl-
benzofuran with chromium trioxide in acetic acid, followed
by saponification of the keto ester so formed — the
2-(benzoyloxy)-3-methyl-4-nitrobenzophenone — with
sodium hydroxide in boiling dilute ethanol (70%) [70].

m.p. 98° [70], 97° [67];
^{1}H NMR [996], IR [68, 996, 1314, 1315], UV [72].

(2-Hydroxy-3-methyl-5-nitrophenyl)phenylmethanone

[18619-93-7]   C14H11NO4   mol.wt. 257.25

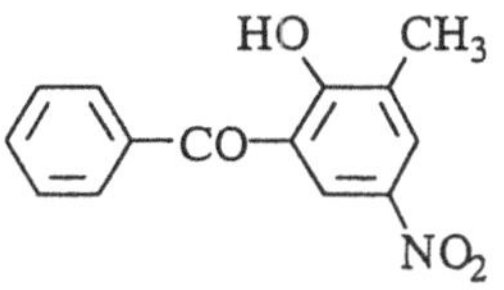

Synthesis

-Preparation by reaction of nitric acid (d = 1.42) with
2-hydroxy-3-methylbenzophenone in acetic acid under
stirring overnight at r.t. (70%) [70].

m.p. 122° [70];   IR [68].

(2-Hydroxy-4-methyl-5-nitrophenyl)phenylmethanone

[68430-99-9]   C14H11NO4   mol.wt. 257.25

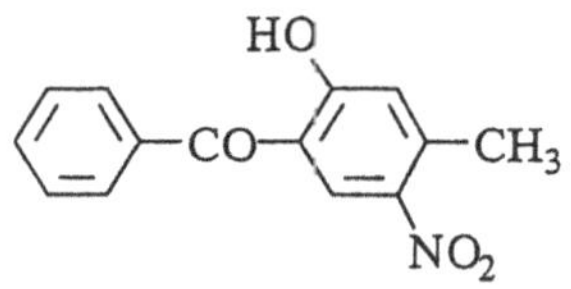

Synthesis

-Refer to: [471].

m.p. and Spectra (NA).

(2-Hydroxy-5-methyl-3-nitrophenyl)phenylmethanone

[4072-26-8] $C_{14}H_{11}NO_4$ mol.wt. 257.25

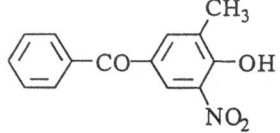

Synthesis

-Preparation by nitration of 2-hydroxy-5-methylbenzo-
   phenone with nitric acid (d = 1.42) in acetic acid overnight at
   r.t. (82%) [70] or at 32° [1083].
-Also refer to: [258, 1041].

   m.p.   68° [67], 67-68° [70];   IR [68].

(2-Hydroxy-5-methyl-4-nitrophenyl)phenylmethanone

[4072-24-6] $C_{14}H_{11}NO_4$ mol.wt. 257.25

Synthesis

-Preparation by oxidation of 5-methyl-6-nitro-2,3-diphenyl-
   benzofuran with chromium trioxide in acetic acid, followed
   by saponification of the keto ester so formed — the
   2-(benzoyloxy)-5-methyl-4-nitrobenzophenone — with
   sodium hydroxide in boiling dilute ethanol (86%) [70].

   m.p.   92° [67, 70];   IR [68, 1314, 1315].

(4-Hydroxy-3-methyl-5-nitrophenyl)phenylmethanone

[103555-87-9] $C_{14}H_{11}NO_4$ mol.wt. 257.25

Synthesis

-Preparation from o-methylanisole in three steps: at first,
   acylation of o-methylanisole with benzoyl chloride in the
   presence of ferric chloride during 3 h at 20°. Then,
   demethylation of the 4-methoxy-3-methylbenzophenone so
   formed by treatment with refluxing 55% hydriodic acid for
10 h and subsequent nitration of the 4-hydroxy-3-methylbenzophenone so obtained with a
concentrated nitric acid/concentrated sulfuric acid mixture with ice cooling (57%) [872].

   m.p. and Spectra (NA).

(2-Hydroxy-3-methoxy-6-nitrophenyl)phenylmethanone

[65202-46-2] $C_{14}H_{11}NO_5$ mol.wt. 273.25

Synthesis

-Preparation by saponification of 2-(benzoyloxy)-
   3-methoxy-6-nitrobenzophenone (SM) with potassium
   hydroxide in boiling ethanol (70%). SM was obtained by
   oxidation of 7-methoxy-4-nitro-2,3-diphenylbenzofuran with
   chromium trioxide in boiling acetic acid (80%) [1].

   m.p.   72-73° [1];   Spectra (NA).

(2-Hydroxy-4-methoxy-5-nitrophenyl)phenylmethanone

[41123-21-1]   C$_{14}$H$_{11}$NO$_5$ mol.wt. 273.25

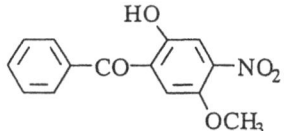

Synthesis

-Preparation by saponification of 2-(benzoyloxy)-
 4-methoxy-5-nitrobenzophenone (SM) with potassium
 hydroxide in ethanol (67%). SM was obtained by oxidation
 of 6-methoxy-5-nitro-2,3-diphenylbenzofuran with
 chromium trioxide in boiling acetic acid for 30 min [574].

m.p. 132° [574]; ^{1}H NMR [267].

(2-Hydroxy-5-methoxy-4-nitrophenyl)phenylmethanone

[40990-72-5]   C$_{14}$H$_{11}$NO$_5$ mol.wt. 273.25

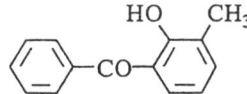

Synthesis

-Preparation by saponification of 2-(benzoyloxy)-5-methoxy-
 4-nitrobenzophenone (SM) with potassium hydroxide in
 ethanol (78%). SM was obtained by oxidation of
 5-methoxy-6-nitro-2,3-diphenylbenzofuran with chromium
 trioxide in boiling acetic acid for 30 min [574].

m.p. 102° [574];   Spectra (NA).

(2-Hydroxy-3-methylphenyl)phenylmethanone

[4072-08-6]   C$_{14}$H$_{12}$O$_2$ mol.wt. 212.25

Syntheses

-Preparation by reaction of 2-hydroxy-3-methylbenzoyl
 chloride with benzene in the presence of aluminium chloride
 at 40-50° for 4 h (72%) [1109], (23%) [400].
-Preparation by oxidation of 7-methyl-2,3-diphenylbenzofuran with chromium trioxide in refluxing
 acetic acid for 40 min, followed by hydrolysis of the keto ester so obtained (2-benzoyloxy-
 3-methylbenzophenone) with 4 N sodium hydroxide in refluxing ethanol for 1 h [1052], (50%)
 [70].
-Preparation by reaction of benzoic acid with o-cresol in the presence of Amberlyst-15 in refluxing
 chlorobenzene for 47 h (50%) [582].
-Also obtained (by-product) by Fries rearrangement of o-tolyl benzoate in the presence of
 aluminium chloride in refluxing chlorobenzene for 4 h (<6%) [1448].
-Also obtained (poor yield) by reaction of benzoyl chloride with o-tolyl borate in the presence of
 aluminium chloride in refluxing carbon disulfide for 2 h (2%) [1373].
-Preparation by reaction between (o-tolyloxy)magnesium bromide complexed with HMPT and
 benzaldehyde in refluxing benzene for 48 h (54%) [250].
-Also refer to: [917].

yellow oil [250];
m.p. 48° [67], 47-48° [70], 47° [1052];
b.p.$_2$ 152-155° [1109], b.p.$_{0.1}$ 114-117° [400];
^{1}H NMR (Sadtler: standard n° 57894 M) [250, 996],
IR (Sadtler: standard n° 84942 K) [67, 68, 250, 996], UV [917], MS [250].

(2-Hydroxy-4-methylphenyl)phenylmethanone

[3098-18-8] $C_{14}H_{12}O_2$ mol.wt. 212.25

Syntheses

-Preparation by Friedel-Crafts acylation of m-cresol,
*with benzoic acid in the presence of boron trifluoride at 160°
  for 2 h (94%) [730] or in the presence of Amberlyst-15 in
  refluxing chlorobenzene for 65 h (37%) [582];
*with benzoyl chloride in the presence of aluminium chloride at 75° [562], in nitrobenzene at 60°
  for 18 h (8%) [308] or with titanium tetrachloride in the same conditions (17%) [308].
-Preparation by Fries rearrangement of m-tolyl benzoate,
*in the presence of aluminium chloride,
without solvent
  at 200° for 5 min (81%) [213] or 20 min [766], at 175° for 15 min (95%) [1143], (82%) [1446],
  (31%) [721], at 140-150° [373], at 140° for 15 min (89%) [311], at 130-134° for 24 h (18%) [244]
  at 90° for a short time (50%) [302], at 60-70° for 1 h (40%) [923],
with solvent
  in nitrobenzene, at 60-63° for 13-18 h (15%) [126, 309, 311];
*in the presence of titanium tetrachloride,
without solvent
  at 175° for 15 min (75%) [1446], at 140° for 15 min (82%) [311],
with solvent
  in nitrobenzene at 60° for 18 h (27%) [311].
-Preparation from 2-methoxy-4-methylbenzophenone by demethylation,
*with refluxing pyridinium chloride for 1 h (74%) [237];
*with boron tribromide in methylene chloride at r.t. for 12 h [766], according to [891].
-Preparation by dehydrogenation of 6-benzoyl-3-methyl-2-cyclohexen-1-one by heating at reflux
  for 30 min in the presence of 5% Pd/BaSO$_4$ catalyst [796].
-Preparation by oxidation of various substituted benzofurans* with chromium trioxide in boiling
  acetic acid, followed by saponification of the keto ester so formed (good yields),
  * 6-methyl-2,3-diphenylbenzofuran [78, 1052], 6-methyl-3-phenyl-2-(4-methoxyphenyl)-
  benzofuran or 6-methyl-3-phenyl-2-(4-methylphenyl)benzofuran [1055].
-Also obtained by isomerization of 4-hydroxy-2-methylbenzophenone with aluminium chloride at
  180-190° for 20 min (92%) [1143].
-Also obtained by condensation of m-cresol with benzotrichloride in the presence of aqueous
  sodium hydroxide in a water bath for 4 h (7%) [527].
-Also obtained (poor yield) by diazotization of 2-amino-4-methylbenzophenone, followed by
  hydrolysis of the diazonium salt so obtained (18%) [1131].
-Also obtained by reaction of benzoyl chloride with m-tolyl borate in the presence of aluminium
  chloride in refluxing carbon disulfide for 2 h (19%) [1373].
-Also obtained by photo-Fries rearrangement of m-tolyl benzoate,
*in water in the presence of sodium dodecyl sulfate (a micelle) for 7 h under nitrogen (73%)
  [1276];
*in ethanol or in the presence of β-cyclodextrin solid for 12-24 h (29% and 96% yields,
  respectively) [1335].
-Preparation by reaction of 2-methoxy-4-methylbenzoyl chloride with benzene in the presence of
  aluminium chloride in tetrachloroethane [385], according to [188].
-Also refer to: [317, 483, 1326].

yellow oil [213];
m.p. 65° [67, 73], 64° [373], 63° [78, 302, 309, 311, 562, 721], 62°5 [730],
      61-63° [1276], 61-62° [237], 61° [385, 923], 60-61° [608],
      60° [527, 1055, 1131], 59-60° [796], 59° [1052], 58° [244];
b.p.$_{15}$ 233-234° [923], b.p.$_{17}$ 230-240° [237], b.p.$_{14}$ 195-215° [302, 721];
^{1}H NMR (Sadtler: standard n° 20031 M) [213, 267, 610],

IR (Sadtler: standard n° 47040) [68, 73, 1052, 1276], UV [73, 126, 607, 610, 793, 794];
TLC [377]; pK_a [608, 791]; polarographic study [1223].

(2-Hydroxy-5-methylphenyl)phenylmethanone

[1470-57-1] $C_{14}H_{12}O_2$ mol.wt. 212.25

Syntheses

-Preparation by Fries rearrangement of p-tolyl benzoate,
*with aluminium chloride [692],
without solvent
at 180° for 10 min (80%) [944], at <150° for 8 min (94%)
[1143], at 140° (71%) [754], according to [183], at 140°
for 10-15 min (quantitative yield) [1143], (88%) [311], (80%) [721], at 140° for 20 min, then for a
short time at 200° (94%) [302, 1213], at 140° for 1 h [608], at 130-140° for 1 h (88%) [254] or at
130° for 30 min (55%) [1185],
with solvent
in refluxing o-dichlorobenzene for 3 h or in refluxing p-chlorotoluene for 1 h (89%) [1450];
in refluxing 1,2,4-trichlorobenzene for 1 h (85%) [1450];
in refluxing chlorobenzene for 1 h (63%) [1450] or for 4 h (82%) [1450], (37%) [1448];
in refluxing nitromethane for 30 min (12%) [1446];
*with titanium tetrachloride,
without solvent
at 140° for 15 min (78%) [311],
with solvent
in refluxing nitromethane for 10 min (41%) [1446];
*with beryllium chloride at 145-150° for 30 min, then at 160° for a short time (69%) [208, 209];
*with Nafion-XR, a H+-form ion exchange resin, at 150° for 4 h under nitrogen (52%) [1231];
*with Nafion-H, a polymeric perfluorinated resin sulfonic acid in refluxing nitrobenzene for 12 h
(70%) [1004];
*with hydrofluoric acid at 55° for 4 h or 6 h (52% and 76% yields, respectively) [986].
-Preparation by photo-Fries rearrangement of p-tolyl benzoate,
*in water in the presence of sodium dodecyl sulfate (a micelle) for 7 h under nitrogen (95%)
[1276];
*in benzene at 55° for 19 h (53%) [419];
*in pentane (37%) or in the presence of silica gel (42%) [102];
*in isopropanol at 57° for 20 h (34%) [419];
*in cyclohexane at 52° for 23 h (25%) [419];
*in dioxane at 63° for 24 h (20%) [419];
*in ethanol in the presence of sulfuric acid for 70 h [641].
-Preparation by condensation of benzotrichloride with p-cresol,
*in the presence of aluminium chloride in carbon disulfide for 2 h (75%) [973]. The 6,12-diphenyl-
2,8-dimethyl-6,12-epoxy-*6H,12H*-dibenzo[b,f][1,5]dioxocin, an intermediate compound, was also
formed under these conditions (29%). This "dioxocin", by hydrolysis with concentrated sulfuric
acid at r.t., gave the expected ketone (91%) [973];
*in the presence of aqueous sodium hydroxide in a water bath for 4 h (33%) [527] or in 32-40%
aqueous sodium hydroxide at 70-80° (67%) [1339].
-Preparation by Friedel-Crafts acylation of p-cresol with benzoyl chloride,
*in the presence of aluminium chloride,
without solvent
at 180° for 30 min (64-72%) [758],
with solvent
in tetrachloroethane at 100-110° for 4 h (80%) [1110, 1111];
in nitrobenzene at 60° for 18 h (33%) [308];
in carbon disulfide, in addition with ethyl iodide, by heating in a water bath [1358]. An intermediate
compound develops at the start of reaction (4-methylphenetole);

*in the presence of titanium tetrachloride, in nitrobenzene at 60° for 18 h (51%) [308].
-Preparation by reaction of chromium trioxide with 2,3-diphenyl-5-methylbenzofuran in acetic acid
    at 60° for 2.5 h [65], at 80° for 3 h [78] or at reflux for 40 min [1052], followed by saponification
    of the keto ester formed with sodium hydroxide [65, 1052], (74%) [78].
    The same ketone was also obtained by chromic oxidation of 5-methyl-2-(4-methylphenyl)-
    3-phenylbenzofuran or of 5-methyl-2-(4-methoxyphenyl)-3-phenylbenzofuran, followed by
    saponification of the keto ester so formed with potassium hydroxide [1055].
-Preparation by diazotization of 2-amino-5-methylbenzophenone, followed by hydrolysis of the
    diazonium salt so obtained (25%) [1131].
-Also obtained by reaction of benzoyl chloride,
*with p-methylanisole in the presence of aluminium chloride in carbon disulfide [94];
*with p-methylphenetole in the presence of aluminium chloride, followed by treatment of the
    2-ethoxy-5-methylbenzophenone so formed with aluminium chloride in carbon disulfide at 60-70°
    for 8 h [91, 93], according to [547];
*with p-tolyl borate in the presence of aluminium chloride in refluxing carbon disulfide for 2 h
    (15%) [1373].
-Also obtained (poor yield) by photo-Fries rearrangement of 2-methoxy-4-methylphenyl benzoate
    (creosol benzoate) in benzene or ethanol for 4 h (6-7%) [206].
-As an historical curiosity, this compound has also been obtained by Fries rearrangement of p-tolyl
    benzoate with aluminium chloride in the presence of another aromatic compounds, at 150° for
    15 min [98]: mesitol (83%); 2,6-dimethylphenyl acetate (53%); 2-chloro-4-methylphenyl acetate
    (20-34%); mesityl acetate (18%).
-Also refer to: [200, 258, 348, 349, 384, 563, 665, 666, 770, 963, 965, 1004, 1083, 1104, 1252, 1324,
    1339, 1446, 1472].

    m.p.   87° [254, 302], 85° [311, 534], 84°5 [1110], 84-85° [1276],
            84° [67, 73, 91, 98, 209, 527, 608, 721, 772, 1111, 1131, 1143],
            83°6-84° [973], 83°5-85° [754, 986], 83°5-84° [1185, 1450], 83-83°5 [94],
            83° [944], 82-83° [65], 82-82°5 [1339], 82° [70, 78, 1358, 1359],
            81-84° [400], 81° [1052, 1055], 79-82° [419], 70-72° [1213];
    b.p.$_{1.5}$ 153-156° [754], b.p.$_{1-2}$ 119-122° [973];
    ^{1}H NMR [610, 754, 772, 996, 1213, 1339], ^{13}C NMR [754],
    IR [68, 754, 772, 792, 996, 1052, 1213, 1313, 1315, 1339],
    UV [607, 610, 772, 793, 794, 948, 996, 1339], MS [1339];
    pK_a [608, 768, 772, 791, 792];
    TLC [377, 1211];   gas chromatography study [1343];
    polarographic study [1223];   cryoscopic study [93].

(2-Hydroxy-6-methylphenyl)phenylmethanone

[50597-28-9]   C$_{14}$H$_{12}$O$_2$ mol.wt. 212.25

Syntheses

-Preparation from 4-methyl-2,3-diphenylbenzofuran by
    oxidation with chromium trioxide in acetic acid at 60° for
    3 h, followed by saponification of the keto ester so formed
    with 10% sodium hydroxide in boiling ethanol [74, 75].
-Preparation by Fries rearrangement of m-cresyl benzoate
with trifluoromethanesulfonic acid in a sealed tube at 170° for 24 h (49%) [386].
-Also obtained from 2-methyl-6-nitroaniline by a eight-step synthesis. The final step consists in
    making to react sodium iodide and trimethylsilyl chloride with 2-methoxy-6-methylbenzophenone
    in acetonitrile at 130° for 24 h into an autoclave (15%) [385].
-Also obtained by photo-Fries rearrangement of m-cresyl benzoate in ethanol for 24 h (25%) and in
    the presence of β-cyclodextrin (34%) [1335].

    m.p.   73° [385], 57-63° [74]; ^{13}C NMR [385], MS [385].

(3-Hydroxy-2-methylphenyl)phenylmethanone

[74167-87-6] $C_{14}H_{12}O_2$ mol.wt. 212.25

Synthesis

-Preparation by diazotization of 3-amino-2-methyl-
benzophenone, followed by hydrolysis of the diazonium salt
obtained (56%) [82], according to [731].

m.p. 123° [82];   Spectra (NA).

(3-Hydroxy-4-methylphenyl)phenylmethanone

$C_{14}H_{12}O_2$ mol.wt. 212.25

Synthesis

-Preparation by diazotization of 3-amino-4-methylbenzo-
phenone, followed by hydrolysis of the diazonium salt
obtained (76%) [263].

m.p. 132-133° [263];   Spectra (NA).

(4-Hydroxy-2-methylphenyl)phenylmethanone

[10425-07-7] $C_{14}H_{12}O_2$ mol.wt. 212.25

Syntheses

-Preparation by Fries rearrangement of m-tolyl benzoate,
*in the presence of aluminium chloride,
without solvent
at 60-70° for 1 h, then at r.t. for 24 h (62%) [923]; at 90°
for a short time (32%) [302]; at 120° for 80 min (40%) [1143]; at 130-134° for 24 h (21%) [244]
or at 160° for 2 h [256, 257],
with solvent
in nitrobenzene at 25-30° for 90 h (45%) [126]; at 60° for 5 h (60%) [1143], (47%) [309], for
13 h (51%) [309] or for 18 h (67%) [311]; or at 62-63° for 18 h (42%) [126];
in chlorobenzene at reflux for 4 h (46%) [1448];
in nitromethane at reflux for 30 min (33%) [1446];
*in the presence of titanium tetrachloride, in refluxing nitromethane for 30 min (44%) [1446];
*in the presence of Nafion-XR, a H+-form ion exchange resin, at 150° for 4 h (67%) [1231].
-Also obtained from p-benzoylthymol by elimination of isopropyl group with aluminium chloride in
chlorobenzene at r.t. for 20 h, then at 50° for 4 h (80%) [648].
-Preparation by demethylation of 4-methoxy-2-methylbenzophenone with refluxing pyridinium
chloride for 1 h (74%) [237].
-Also obtained by condensation of benzotrichloride with m-cresol in the presence of aqueous
sodium hydroxide in a water bath for 4 h (6%) [527].
-Preparation by reaction of benzoyl chloride with m-cresol in the presence of aluminium chloride or
titanium tetrachloride in nitrobenzene at 60° for 18 h (67% and 61% yields, respectively) [308].
The same reaction with aluminium chloride without solvent at 75° gave a 10% yield [562].
-Also obtained (poor yield) by reaction of benzoyl chloride with m-tolyl borate in the presence of
aluminium chloride in refluxing carbon disulfide for 2 h (5%) [1373].

-Preparation by acylation of m-tolyl benzoate with benzoyl chloride in the presence of zinc chloride at 130° for 1 h, followed by saponification of the p-keto ester so formed with sodium hydroxide in boiling ethanol [140].
-Also obtained by photo-Fries rearrangement of m-tolyl benzoate in ethanol during 24 h (30%) [1335].

> m.p. 135-136° [244], 130° [385], 129° [237, 302, 309, 311, 562, 923],
>   128° [140, 527, 648];
> b.p.$_{15}$ 235-240° [923], b.p.$_{13}$ 220-240° [302], b.p.$_{17}$ 230-240° [237];
> ^{1}H NMR (Sadtler: standard n° 20030 M), ^{13}C NMR [385],
> IR (Sadtler: standard n° 47039), UV [126, 923], MS [256, 257].

(4-Hydroxy-3-methylphenyl)phenylmethanone

[5326-42-1]   C$_{14}$H$_{12}$O$_2$ mol.wt. 212.25

Syntheses

-Preparation by Fries rearrangement of o-tolyl benzoate,
*with aluminium chloride,
without solvent
        at 160° for 1 h [292], (90-91%) [302, 567] or
2 h [256, 257]; at 140° for 15 min (quantitative yield) [1143], (86%) [311], (70%) [51], (10%) [721] and for 4 h [1440],
with solvent
   in nitrobenzene at 60° for 18 h (91%) [311], in phenyl ether at 170° for 30 min (68%) [303], in refluxing nitromethane for 30 min (60%) [1446] or in refluxing chlorobenzene for 4 h (45%) [1448];
*with titanium tetrachloride, in nitrobenzene at 60° for 18 h (89%) [311] or in refluxing nitromethane for 30 min (86%) [1446];
*with Nafion-XR, a H$^+$-form ion exchange resin, at 175° for 4 h under nitrogen (45%) [1231].
-Also obtained by photo-Fries rearrangement of o-tolyl benzoate in ethanol between 60 to 75 h (20%) [1034].
-Preparation by Friedel-Crafts acylation of o-cresol with benzoyl chloride,
*in the presence of aluminium chloride in nitrobenzene at 60° for 18 h (90%) [308] or without solvent [484], at 75° (31%) [562];
*in the presence of titanium tetrachloride in nitrobenzene at 60° for 18 h (87%) [308].
-Also obtained by reaction of benzoyl chloride,
*with o-bromotoluene in the presence of aluminium chloride at 75° for 5 h (38%) [562];
*with o-tolyl borate in the presence of aluminium chloride in refluxing carbon disulfide for 2 h (51%) [1373];
*with o-tolyl benzoate in the presence of zinc chloride at 130° for 1 h, followed by saponification of the p-keto ester formed with sodium hydroxide in boiling ethanol [141].
-Also obtained by treatment of p-benzoylcarvacrol with aluminium chloride in chlorobenzene, first at r.t. for 20 h, then at 50° for 4 h (68%) [648].
-Also obtained by reaction of benzotrichloride with o-cresol in the presence of aqueous sodium hydroxide at 80° several hours [1019] or in a water bath for 4 h (22%) [527].
-Also obtained by heating a mixture of benzoic acid and o-cresol in the presence of Tonsil [1216].
-Preparation by diazotization of 4-amino-3-methylbenzophenone, followed by hydrolysis of the resulting diazonium salt (76%) [263].
-Also refer to: [782, 783].

> m.p. 174-175° [263], 173-175° [256, 257], 173-174° [302, 721, 1034],
>   173° [303, 311, 562, 648], 172-173° [527], 172-172°5 [141],
>   172° [1143, 1216, 1440], 170-171° [1019], 169° [51], 163° [484];
> b.p.$_{12-15}$ 240-260° [302, 567, 721];
> Spectra (NA).

(2-Hydroxy-3-methoxyphenyl)phenylmethanone

[65202-31-5] $C_{14}H_{12}O_3$ mol.wt. 228.25

Syntheses

-Preparation by saponification of 2-benzoyloxy-3-methoxy-
benzophenone (SM) with potassium hydroxide in refluxing
ethanol for 1 h (73%) [1]. SM was obtained by oxidation of
7-methoxy-2,3-diphenylbenzofuran with chromium
trioxide in refluxing acetic acid for 30 min (78%).
-Preparation by reaction of phenylmagnesium bromide with 2-hydroxy-3-methoxybenzonitrile in
ethyl ether in a water bath for 2.5 h (88%) [199].
-Also obtained by reaction of 2-hydroxy-3-methoxybenzaldehyde with iodobenzene by using a
catalyst system of palladium chloride/lithium chloride in the presence of sodium carbonate in
N,N-dimethylformamide at 100° for 8 h (70%) [1204].

  b.p._3 220-225° [199];   m.p. 59° [199];
  ^{1}H NMR [1204], MS [1204].

(2-Hydroxy-4-methoxyphenyl)phenylmethanone *(Cyasorb UV-9, Oxybenzone, Sumisorb 110)*

[131-57-7] $C_{14}H_{12}O_3$ mol.wt. 228.25

Syntheses

-Preparation by partial methylation of 2,4-dihydroxy-
benzophenone [757, 785],
*with methyl iodide in the presence of sodium hydroxide
[805];
*with a methyl halide [42];
*with dimethyl sulfate in the presence of sodium hydroxide [805] or alkaline solution [751, 1058].
-Also obtained by partial demethylation of 2,4-dimethoxybenzophenone (SM),
*with aluminium chloride in refluxing carbon disulfide for 30 min [29, 364, 748, 1068], (42%)
  [718]. SM was prepared by reaction of benzoyl chloride with resorcinol dimethyl ether in the
  presence of aluminium chloride (good yield) [718];
*with aluminium chloride in nitrobenzene [536, 1068];
*with aluminium chloride or aluminium bromide in chlorobenzene at 90-95° (good yield) [602];
*with excess beryllium chloride in refluxing toluene for 3 h (90%) [1250];
*in refluxing 40% hydrobromic acid for 5 h [1068].
-Preparation by Friedel-Crafts acylation of resorcinol dimethyl ether with benzoyl chloride,
*in the presence of aluminium chloride in ethylene dichloride (85%) [1193] or in ethyl ether [807];
*in the presence of aluminium chloride in a chlorobenzene/N,N-dimethylformamide mixture (22:1)
  at 115° [539, 1067], (74%) [1079];
*in the presence of a zinc chloride/aluminium chloride mixture in ethylene dichloride, first between
  0 to 5°, then at 10° for 1 h and at 65° for 6 h (71%) [479].
-Preparation by saponification of 2-(benzoyloxy)-4-methoxybenzophenone (SM) with sodium
  hydroxide [1052] or potassium hydroxide [558] in refluxing ethanol for 1 h. SM was obtained by
  oxidation of 6 methoxy-2,3-diphenylbenzofuran with chromium trioxide in boiling acetic acid for
  30-40 min [1052].
-Preparation by Fries rearrangement of m-methoxyphenyl benzoate,
*in the presence of aluminium chloride [692];
*in the presence of hydrochloric acid and a small amount of ferric chloride (60%) [1469].
-Preparation by reaction of benzotrichloride with resorcinol monomethyl ether in hydrofluoric acid
  in the presence of water at 0°, then at r.t. for 7 h (55%) [393].
-Preparation by Friedel-Crafts acylation of resorcinol monomethyl ether with benzoyl chloride,
*in the presence of boron trichloride in benzene, first at -10°, then at reflux for 10 h (85%) [1064];

*in the presence of titanium tetrachloride in benzene, first at -10°, then at reflux for 14 h (77%) [1064];
*in the presence of iron powder at 240-260° for 2 h (50%) [1468];
*in the presence of ferric chloride-nitromethane complex at 185-195° for 20 min (51%) [1468].
-Preparation by reaction of benzoyl chloride with resorcinol dimethyl ether,
*in chlorobenzene in the presence of titanium tetrachloride for 1 h at 120° (78%) [1140];
*without solvent or in o-dichlorobenzene, in the presence of ferric chloride, at 160 to 200° for 7-11 h (by-product) [1467].
-Also obtained from β-(2-benzoyl-5-methoxyphenoxy)propionic acid by heating on a steam bath with a 10% aqueous sodium hydroxide solution for some minutes [1274].
-Preparation from 2-iodo-5-methoxyphenyl benzoate by rearrangement on treatment with n-butyl-lithium in a mixture of ethyl ether, hexane and tetrahydrofuran at -70° for 2 h, then treatment with saturated aqueous ammonium chloride solution (<16%) [598].
-Also obtained by reaction of benzoic acid with resorcinol monomethyl ether,
*in the presence of polyphosphoric acid on a water bath for 20 min (27%) [1104];
*in the presence of boron trifluoride at 80° for 30 min (59%) [390].
-Also refer to: [3, 23, 43, 131, 200, 213, 219, 285, 348, 383, 483, 615, 810, 856, 913, 965, 968, 1089, 1107, 1112, 1302, 1303, 1323, 1324, 1342, 1348, 1411].
N.B.: Na [630, 1193], Sn salts [1058].

oil [1104];
m.p.   69° [1140], 66° [751, 1406], 65° [1064, 1079, 1468], 64-66° [1193],
          64° [785, 1058, 1217], 63-65° [1250], 63-64° [479, 558], 63° [1274], 62-63° [501],
          62° [608, 718], 61° [606, 1052], 55° [393];
b.p. 359-361° [501];
^{1}H NMR [177, 267, 403, 598, 610, 1250], IR [785, 991, 1052, 1068, 1250],
UV [474, 475, 515, 606, 607, 610, 785, 793, 794, 991, 1067, 1156, 1217, 1250, 1406, 1411];
GC [1343];   HPLC [270];   TLC [377, 1250, 1396];
polarographic study [1223];   pK_a [606, 608, 768, 791];
vapour pressure [500, 1217];   gel permeation chromatography [1145].

(2-Hydroxy-4-methoxyphenyl)phenylmethanone-[14]C

[17655-53-7] $C_{14}H_{12}O_3$ mol.wt. 230.25

Synthesis

-Refer to: [374].

m.p. and Spectra (NA).

(2-Hydroxy-5-methoxyphenyl)phenylmethanone (UV 9)

[14770-96-8] $C_{14}H_{12}O_3$ mol.wt. 228.25

Syntheses

-Preparation by Friedel-Crafts acylation of hydroquinone dimethyl ether with benzoyl chloride in the presence of aluminium chloride [1060], in carbon disulfide first at 25° for 48 h, then at 50° for 30 min (55%) [191] or at r.t. for 48 h (by-product) [717], (10%) [716].
-Also obtained by selective demethylation of 2,5-dimethoxybenzophenone (SM),
*with hydriodic acid [568, 717];

*with excess beryllium chloride in refluxing toluene for 3 h (90%) [1250];
*with aluminium chloride [809], in nitromethane at 20° for 24 h (65%) [863] or in benzene under nitrogen at 80° for 12 h (90%) [1047, 1048].

N.B.: SM was prepared by reaction of benzoyl chloride with hydroquinone dimethyl ether, either in the presence of stannic chloride in nitromethane at 20° for 1 h (78-94%) [863] or in the presence of aluminium chloride in carbon disulfide at r.t. for 48 h (78%) [716], (74-82%) [717].

-Preparation by partial methylation of 2,5-dihydroxybenzophenone with dimethyl sulfate in the presence of sodium hydroxide in dilute ethanol at 80-90° for 30 min (33%) [804].

-Also obtained by Fries rearrangement of p-methoxyphenyl benzoate with titanium tetrachloride without solvent at 120° for 1 h (20-35%) [863].

-Preparation by oxidation of 5-methoxy-2,3-diphenylbenzofuran with chromium trioxide in boiling acetic acid for 30-40 min, followed by saponification of the keto ester formed, the 2-(benzoyloxy)-5-methoxybenzophenone with sodium hydroxide in refluxing ethanol for 1 h [1052, 1054]. The same result was obtained by using 5-methoxy-2-(4-methoxyphenyl)-3-phenylbenzofuran or 5-methoxy-2-(4-methylphenyl)-3-phenylbenzofuran [1055].

-Also obtained (poor yield) by UV light irradiation of 2,5-dimethoxybenzophenone in carbon tetrachloride for 400 h (3%) [809].

-Also refer to: [99, 541, 766, 1112, 1324].

m.p.   84-85°5 [191], 84° [863, 1060], 83°5 [1047], 83-84° [718],
82-85° [568], 82-84° [1250], 82-83° [608], 81°5-82° [804],
81-82° [809], 81° [1052, 1054, 1055], 78° [716, 717];
^{1}H NMR (Sadtler: standard n° 30289 M) [267, 610, 809, 1250],
EPR [267], IR (Sadtler: standard n° 57334) [809, 863, 1052, 1250],
UV [607, 610, 793, 794, 863, 1250];
pK_a [608, 791];   TLC [377, 809, 1250];   polarographic study [1223].

(2-Hydroxy-6-methoxyphenyl)phenylmethanone

[20034-63-3] $C_{14}H_{12}O_3$ mol.wt. 228.25

Syntheses

-Preparation from 2-iodo-3-methoxyphenyl benzoate by rearrangement on treatment with n-butyllithium in a mixture of ethyl ether, hexane and tetrahydrofuran at -70° for 2 h, followed by treatment with saturated aqueous ammonium chloride (51%) [598].

-Also obtained by Fries rearrangement of m-methoxyphenyl benzoate in the presence of trifluoro-methanesulfonic acid in tetrachloroethane at 170° for 24 h in a sealed tube (78%) [386].

-Preparation by partial methylation of 2,6-dihydroxybenzophenone with dimethyl sulfate,
*in the presence of potassium carbonate in refluxing acetone for 7 h [644];
*in the presence of potassium hydroxide in benzene in a water bath for 1 h (61%) [1283].

-Preparation by partial demethylation of 2,6-dimethoxybenzophenone with aluminium chloride in benzene at 0° for 2 h (43%) [1283].

-Also refer to: [24, 831].

m.p.   140-141° [1283], 140° [644], 106-107° [598]. There is discrepancy between the two melting points.

^{1}H NMR [598], IR [598], MS [598];   TLC [1283].

(3-Hydroxy-4-methoxyphenyl)phenylmethanone

[66476-03-7]   C$_{14}$H$_{12}$O$_3$ mol.wt. 228.25

Syntheses

-Preparation from 3-(benzoyloxy)-4-methoxybenzophenone (SM) (m.p. 95°5-96°5) [137] by saponification [622] with sodium hydroxide in refluxing ethanol (good yield) [137, 138] or with potassium hydroxide in refluxing methanol (98%) [862]. SM was obtained by Friedel-Crafts acylation of guaiacol benzoate with benzoyl chloride in the presence of zinc chloride [137, 138, 622] or in the presence of stannic chloride in nitromethane for 1 h at 20° (82%) [862].
-Preparation by partial methylation of 3,4-dihydroxybenzophenone with dimethyl sulfate in the presence of potassium carbonate in refluxing acetone for 90 min (70%) [306].
-Also refer to: [1318, 1319].

m.p.   132°5-133° [306], 131-132° [137, 138, 622], 117° [862];
^{1}H NMR (Sadtler: standard n° 28215 M) [306],
IR (Sadtler: standard n° 55287) [862], UV [306, 862].

(4-Hydroxy-2-methoxyphenyl)phenylmethanone

[21112-64-1]   C$_{14}$H$_{12}$O$_3$ mol.wt. 228.25

Syntheses

-Preparation by saponification of 4-(benzoyloxy)-2-methoxybenzophenone (SM) with potassium hydroxide in refluxing aqueous ethanol for 1 h (95%) [558]. SM was obtained from resbenzophenone by a two-step synthesis (4-O-benzoylation and 2-O-methylation).
-Also obtained from β-(4-benzoyl-3-methoxyphenoxy)propionic acid by heating with 10% aqueous sodium hydroxide [805, 1274].

m.p.   124° [805, 1274], 123°5-124°5 [558]; ^{1}H NMR (Sadtler: standard n° 28221 M),
IR (Sadtler: standard n° 55293) [558], UV [558].

(4-Hydroxy-3-methoxyphenyl)phenylmethanone

[51439-89-5]   C$_{14}$H$_{12}$O$_3$ mol.wt. 228.25

Syntheses

-Preparation by decarboxylation of 2-(4-hydroxy-3-methoxybenzoyl)benzoic acid in the presence of cupric acetate monohydrate in quinoline at 250-254° for 35 min (72%) [505].
-Also obtained by Friedel-Crafts acylation,
*of veratrole with benzoyl chloride in the presence of aluminium chloride in carbon disulfide overnight at r.t. [1119];
*of guaiacol with benzoic acid in the presence of polyphosphoric acid at 120° for 1 h [256, 257].
-Also obtained by cleavage of 3,3'-dimethoxybenzaurine* on heating its dilute aqueous sodium hydroxide solution (1%) in a water bath with air bubbling [622].
N.B.: *Synonyms: 4-hydroxy-3,3'-dimethoxyfuchsone and 4-[(4-hydroxy-3-methoxyphenyl)-phenylmethylene]-3-methoxy-2,5-cyclohexadien-1-one.
-Also refer to: [1318, 1319].

m.p.   100-101°5 [505], 97-98° [622, 1119];   IR [505, 1282], UV [505], MS [256, 257].

(5-Hydroxy-2-methoxyphenyl)phenylmethanone

[80427-34-5] $C_{14}H_{12}O_3$ mol.wt. 228.25

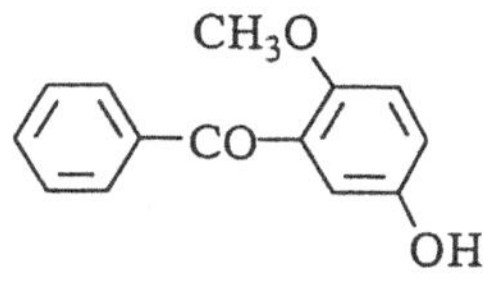

Synthesis

-Preparation by acylation of p-methoxyphenyl benzoate with benzoyl chloride in the presence of stannic chloride in nitromethane at 20° for 2 days, followed by saponification of the resulting keto ester — 5-(benzoyloxy)-2-methoxy-benzophenone — with sodium hydroxide in refluxing methanol for 1 h (90%) [863].

-Also claimed to be obtained by partial demethylation of 2,5-dimethoxybenzophenone with hydriodic acid [717]. Nevertheless, the structure of this compound was erroneous. In that case, it probably was its isomer the 2-hydroxy-5-methoxybenzophenone [568].

m.p. 111° [863], 78° [717]. One of the reported melting points is obviously wrong.
^{1}H NMR (Sadtler: standard n° 30290 M),
IR (Sadtler: standard n° 57335) [863], UV [863].

(2-Amino-4-hydroxy-6-methylphenyl)phenylmethanone

[54439-89-3] $C_{14}H_{13}NO_2$ mol.wt. 227.26

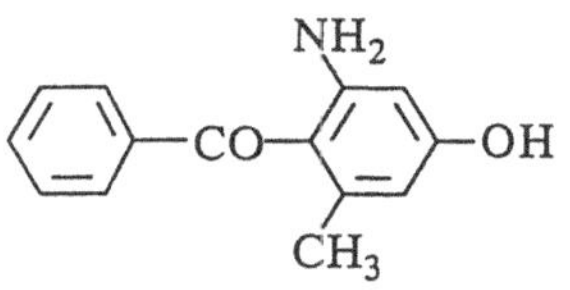

Synthesis

-Preparation from 3'-methyl-5-phenyl-3,5'-diisoxazolyl-methane by performing hydrogenolysis and subsequent hydrolysis with hydrochloric acid (86%) [90].

m.p. 162° [90]; ^{1}H NMR [90], MS [90].

(3-Amino-2-hydroxy-5-methylphenyl)phenylmethanone

$C_{14}H_{13}NO_2$ mol.wt. 227.26

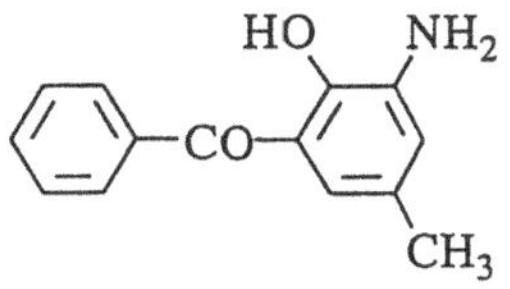

Synthesis

-Refer to: [932].

m.p. and Spectra (NA).

(4-Amino-2-hydroxy-6-methylphenyl)phenylmethanone

[54439-92-8] $C_{14}H_{13}NO_2$ mol.wt. 227.26

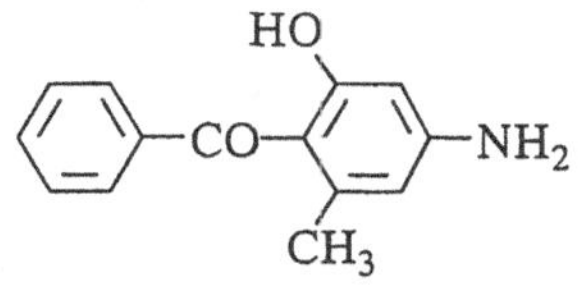

Synthesis

-Obtained (trace) from 5-methyl-3'-phenyl-3,5'-diisoxazolyl-methane by performing hydrogenolysis and subsequent hydrolysis with hydrochloric acid [90].

m.p. 125° [90]; ^{1}H NMR [90], MS [90].

[3-Hydroxy-4-(methylamino)phenyl]phenylmethanone

[54903-59-2] $C_{14}H_{13}NO_2$ mol.wt. 227.26

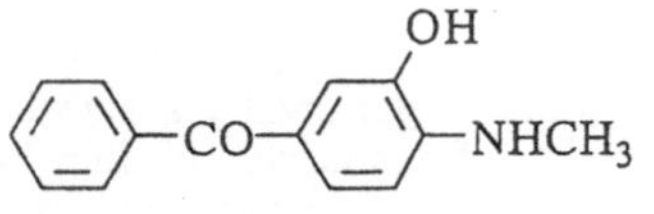

Synthesis

-Preparation from 6-benzoyl-3-methylbenzoxazolinone by alkaline hydrolysis with boiling 10% aqueous sodium hydroxide solution [148], (90-100%) [196], 85% [952].

m.p. 165° [196, 952];   Spectra (NA).

(3-Ethynyl-4-hydroxyphenyl)phenylmethanone

[183589-15-3] $C_{15}H_{10}O_2$ mol.wt. 222.24

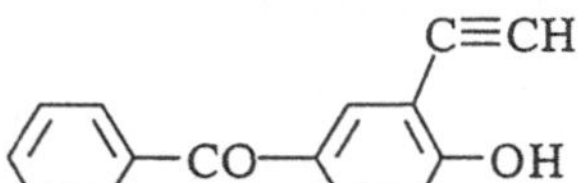

Synthesis

-Obtained by treatment of p-benzoyl-o-[(trimethylsilyl)-ethynyl]phenol in methanol in the presence of potassium fluoride at 25° for 1 h (75%) [60].

m.p. 145-146° [60]; ^{1}H NMR [60], ^{13}C NMR [60], IR [60], MS [60].

(3,5-Dibromo-2-hydroxy-4,6-dimethoxyphenyl)phenylmethanone

$C_{15}H_{12}Br_2O_4$ mol.wt. 416.07

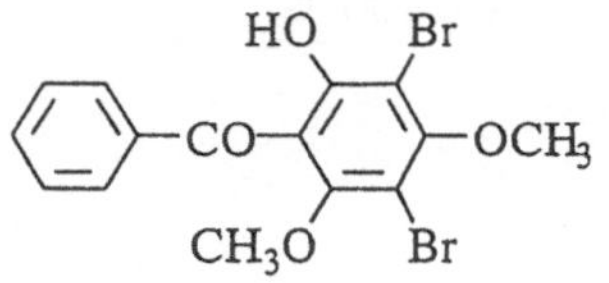

Synthesis

-Preparation by reaction of bromine with monobromo-hydrocotoin in chloroform [645].

m.p. 95° [645];   Spectra (NA).

[2-(Acetyloxy)-4-hydroxyphenyl]phenylmethanone

[145747-24-6] $C_{15}H_{12}O_4$ mol.wt. 256.26

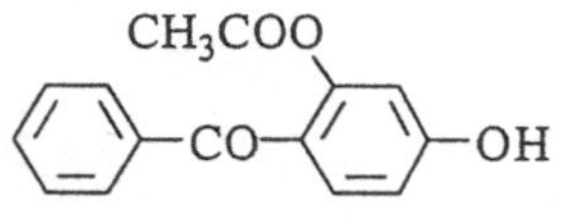

Synthesis

-Obtained by enzymatic deacetylation of 2,4-diacetoxy-benzophenone in the presence of porcine pancreas lipase in a tetrahydrofuran/n-butanol mixture at 40-45° (70%) [1036].

m.p. and Spectra (NA).

[4-(Acetyloxy)-2-hydroxyphenyl]phenylmethanone

[18803-24-2] $C_{15}H_{12}O_4$ mol.wt. 256.26

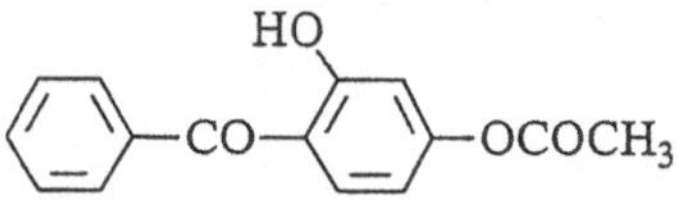

Syntheses

-Preparation by partial hydrolysis of 2,4-diacetoxy-benzophenone in the presence of trifluoroacetic acid containing 5% of water at 65° for 5 min (98%) [545].

-Preparation by reaction of acetyl chloride (1 equiv) with resbenzophenone [281].
-Also refer to: [377, 607, 608, 610].

m.p. 93°6-94° [281], 92-93° [608];
^{1}H NMR [610], UV [607, 610];   TLC [377];   pK_a [608].

(3-Bromo-2-hydroxy-4,5-dimethylphenyl)phenylmethanone

[143815-12-7]   C$_{15}$H$_{13}$BrO$_2$ mol.wt. 305.17

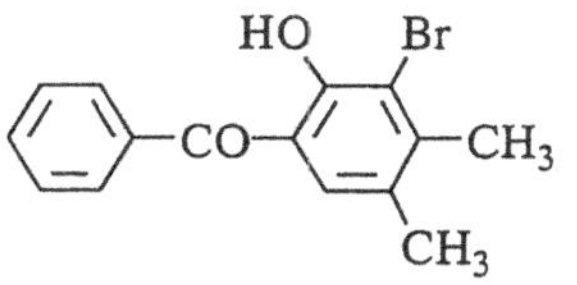

Syntheses

-Preparation by Fries rearrangement of 2-bromo-4,5-di-
 methylphenyl benzoate with aluminium chloride without
 solvent at 140° for 2 h (24%) [866].
-Preparation by reaction of bromine with 2-hydroxy-4,5-di-
 methylbenzophenone in boiling acetic acid [142].

m.p. 134-135° [142, 866];
^{1}H NMR (Sadtler: standard n° 59390 M) [866],
IR (Sadtler: standard n° 86546 K) [866], UV [866], MS [866].

(3-Bromo-2-hydroxy-5,6-dimethylphenyl)phenylmethanone

[143815-11-6]   C$_{15}$H$_{13}$BrO$_2$ mol.wt. 305.17

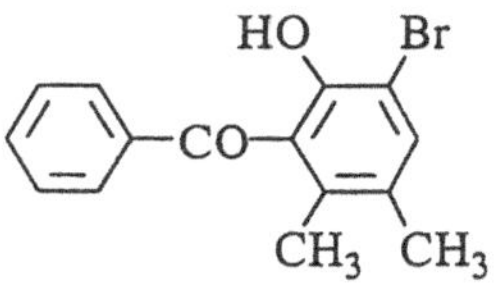

Syntheses

-Preparation by Fries rearrangement of 2-bromo-4,5-di-
 methylphenyl benzoate with titanium tetrachloride without
 solvent at 140° for 2 h (60%) [866].
-Also obtained (by-product) by Fries rearrangement of
 2-bromo-4,5-dimethylphenyl benzoate with aluminium
chloride without solvent at 140° for 2 h (10%) [866].

m.p. 157-158° [866];
^{1}H NMR (Sadtler: standard n° 59392 M) [866],
IR (Sadtler: standard n° 86548 K) [866], UV [866], MS [866].

(3-Bromo-4-hydroxy-2,5-dimethylphenyl)phenylmethanone

                                      C$_{15}$H$_{13}$BrO$_2$ mol.wt. 305.17

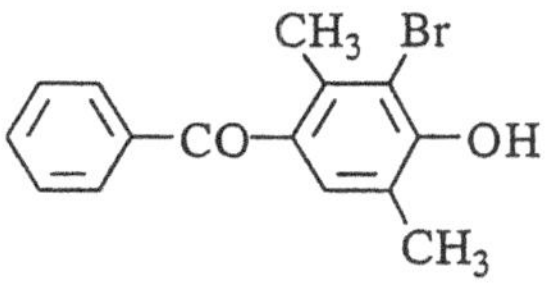

Synthesis

-Preparation by reaction of bromine with 4-hydroxy-2,5-di-
 methylbenzophenone in acetic acid [142].

m.p. 115-116° [142];   Spectra (NA).

(3-Bromo-6-hydroxy-2,5-dimethylphenyl)phenylmethanone

$C_{15}H_{13}BrO_2$ mol.wt. 305.17

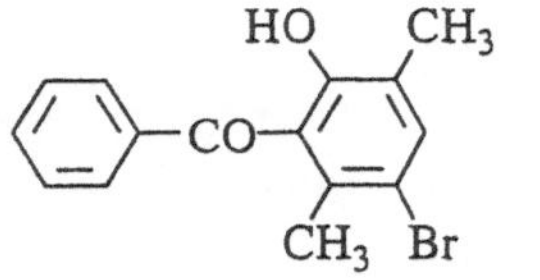

Synthesis

-Preparation by bromination of 2-hydroxy-3,6-dimethyl-
benzophenone in chloroform [995, 998].

m.p.   133° [995, 998];   IR [68, 998].

(4-Bromo-2-hydroxy-3,6-dimethylphenyl)phenylmethanone

[6721-06-8] $C_{15}H_{13}BrO_2$ mol.wt. 305.17

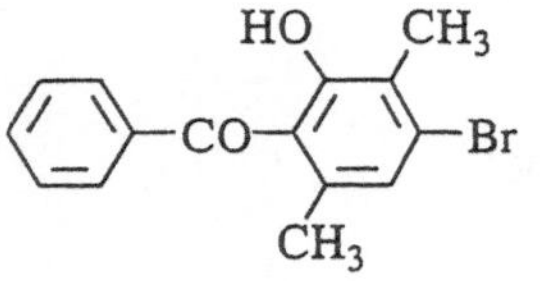

Synthesis

-Obtained by chromic oxidation (CrO_3) of 6-bromo-4,7-di-
methyl-2,3-diphenylbenzofuran, followed by saponification
of the keto ester so obtained [995, 998].

m.p.   117° [995, 998];   IR [68, 998], UV [72].

(4-Bromo-6-hydroxy-2,3-dimethylphenyl)phenylmethanone

[143815-13-8] $C_{15}H_{13}BrO_2$ mol.wt. 305.17

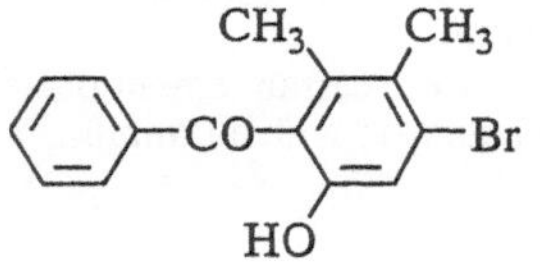

Synthesis

-Obtained (by-product) by Fries rearrangement of 2-bromo-
4,5-dimethylphenyl benzoate with aluminium chloride
without solvent at 140° for 2 h (22%) [866].

m.p.   194-195° [866];
^{1}H NMR (Sadtler: standard n° 59394 M) [866],
IR (Sadtler: standard n° 86550 K) [866], UV [866], MS [866].

[4-(2-Bromoethoxy)-2-hydroxyphenyl]phenylmethanone

[18902-63-1] $C_{15}H_{13}BrO_3$ mol.wt. 321.17

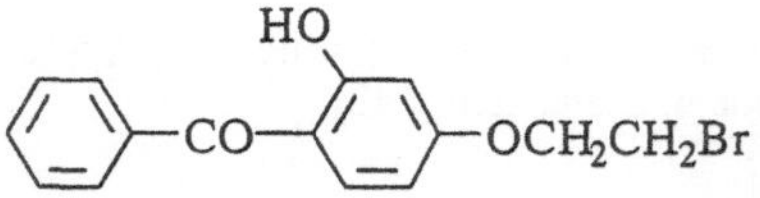

Synthesis

-Preparation by reaction of ethylene dibromide with
resbenzophenone,
*in the presence of sodium hydroxide in refluxing
dilute ethanol for 15 h (47%) [557, 558];
*in the presence of sodium methoxide in diisobutylketone between 130-140° (78%) [388].
-Also refer to: [119, 1423, 1424].

m.p.   90-95° [388], 87-88° and 97-98° [557, 558]; double melting point (two allotropic
forms);
Spectra (NA).

(3-Bromo-2-hydroxy-4,6-dimethoxyphenyl)phenylmethanone

$C_{15}H_{13}BrO_4$ mol.wt. 337.17

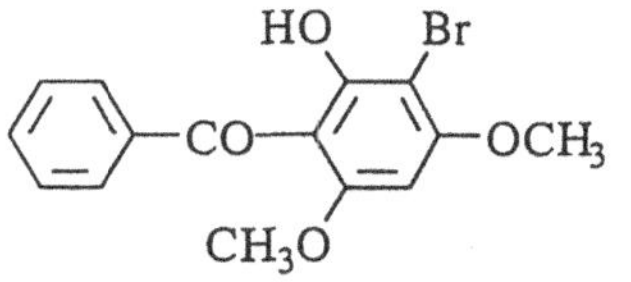

Synthesis

-Preparation by reaction of bromine with hydrocotoin in chloroform [645].

m.p. 147° [645];   Spectra (NA).

(3-Chloro-6-hydroxy-2,4-dimethylphenyl)phenylmethanone

[34174-02-2] $C_{15}H_{13}ClO_2$ mol.wt. 260.72

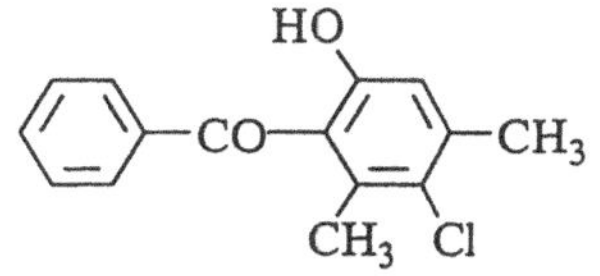

Synthesis

-Preparation by Fries rearrangement of 4-chloro-3,5-dimethylphenyl benzoate with aluminium chloride for 30 min at 150-160° [375].

m.p. 203-204° [375];   Spectra (NA).

(4-Chloro-2-hydroxy-3,6-dimethylphenyl)phenylmethanone

[33561-94-3] $C_{15}H_{13}ClO_2$ mol.wt. 260.72

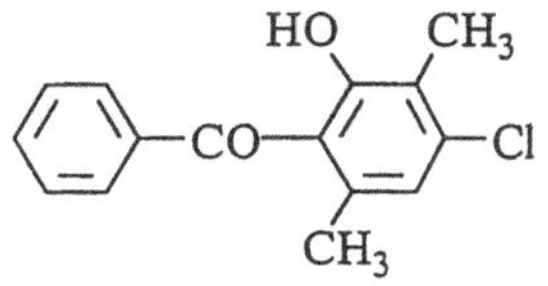

Synthesis

-Obtained by oxidation of 6-chloro-4,7-dimethyl-2,3-diphenylbenzofuran with chromium trioxide in boiling acetic acid for 2 h, followed by saponification of the resulting 2-(benzoyloxy)-4-chloro-3,6-dimethylbenzophenone with 2 N sodium hydroxide in boiling ethanol for 15 min [997].

m.p. 102° [997];   IR [997].

(5-Chloro-2-hydroxy-3,4-dimethoxyphenyl)phenylmethanone

[140665-35-6] $C_{15}H_{13}ClO_4$ mol.wt. 292.72

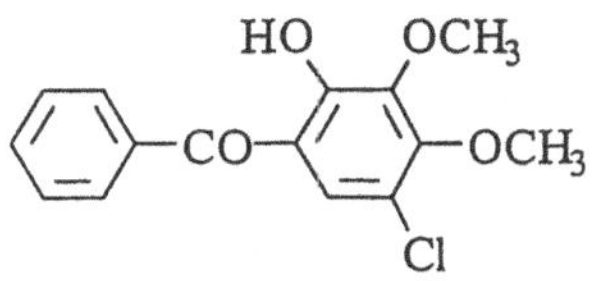

Synthesis

-Preparation by reaction of sulfuryl chloride with 2-hydroxy-3,4-dimethoxybenzophenone in methylene chloride overnight at r.t. (70%) [1194].

m.p. 84-85° [1194];
1H NMR [1194], IR [1194], MS [1194].

(2-Hydroxy-3,6-dimethyl-4-nitrophenyl)phenylmethanone

[18619-94-8]   C₁₅H₁₃NO₄   mol.wt. 271.27

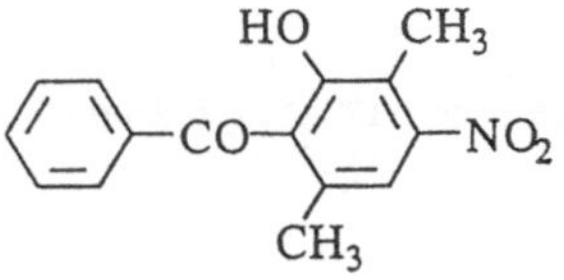

Synthesis

-Refer to: [68, 1314, 1315, 1316].

m.p. (NA);   IR [68, 1314, 1315, 1316].

(2-Hydroxy-3,6-dimethyl-5-nitrophenyl)phenylmethanone

[18619-95-9]   C₁₅H₁₃NO₄   mol.wt. 271.27

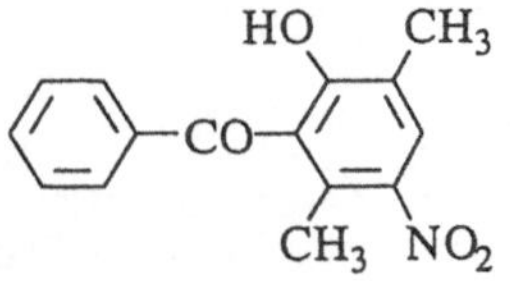

Synthesis

-Refer to: [68].

m.p. (NA);   IR [68].

(3-Hydroxy-4,6-dimethyl-2-nitrophenyl)phenylmethanone

[20010-69-9]   C₁₅H₁₃NO₄   mol.wt. 271.27

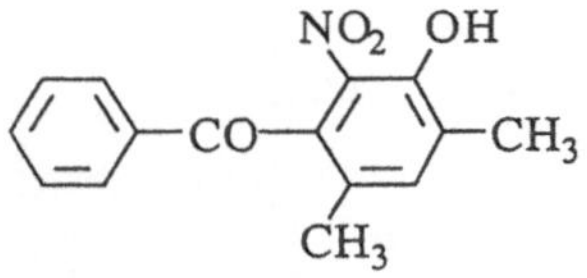

Synthesis

-Obtained by treatment of 5-amino-2,4-dimethylbenzo-
phenone (SM) in nitric acid (d = 1.34) with an aqueous
sodium nitrite solution between 0 to 10°, then at r.t. for 2 h
and at 50° for 3 h (43%). SM was prepared by acylation of
2,4-dimethylacetamide with benzoyl chloride in carbon
disulfide in the presence of aluminium chloride (62%, m.p. 101°) [382].

      m.p.   119°5-120° [382];   Spectra (NA).

(4-Hydroxy-2,6-dimethyl-3-nitrophenyl)phenylmethanone

[100923-75-9]   C₁₅H₁₃NO₄   mol.wt. 271.27

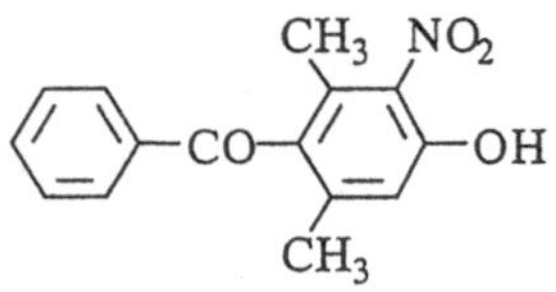

Synthesis

-Refer to: [1256].

m.p. (NA);
¹H NMR [676, 1256], ¹³C NMR [1180], IR [1256].

(3-Ethyl-2-hydroxyphenyl)phenylmethanone

[56394-91-3]   C₁₅H₁₄O₂   mol.wt. 226.27

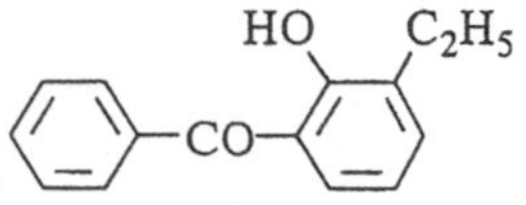

Synthesis

-Preparation by reaction of 3-ethylsalicylic acid chloride
(3-ethyl-2-hydroxybenzoyl chloride) with benzene in the
presence of aluminium chloride overnight at 100° (58%)
[400, 1205, 1207, 1208].

-Preparation by Fries rearrangement of o-ethylphenyl benzoate (SM) with aluminium chloride at 160-170° for 30 min (61%). SM was obtained by reaction of benzoyl chloride with aluminium tris(o-ethylphenoxide) [759].

b.p.$_{0.14}$ 123-126° [1205, 1207], b.p.$_{0.4}$ 123-126° [400, 1208],
b.p.$_3$ 165-167 [759]; n_D^{22} = 1.6081 [1205, 1207];
IR [1205, 1207].

(3-Ethyl-4-hydroxyphenyl)phenylmethanone

[67217-94-1]   C$_{15}$H$_{14}$O$_2$ mol.wt. 226.27

Syntheses

-Preparation by decarboxylation of 2-(3'-ethyl-4'-hydroxy-benzoyl)benzoic acid in the presence of cupric acetate in refluxing quinoline (254°) for 40 min (98%) [1162].
-Also obtained (by-product) by Fries rearrangement of o-ethylphenyl benzoate with aluminium chloride at 160-170° for 30 min (12%) [759].

m.p. 138-140°2 [1162], 136° [759];   Spectra (NA).

(4-Ethyl-2-hydroxyphenyl)phenylmethanone

[78473-50-4]   C$_{15}$H$_{14}$O$_2$ mol.wt. 226.27

Synthesis

-Preparation by Fries rearrangement of m-ethylphenyl benzoate in the presence of aluminium chloride without solvent at 140° for 4 h [266].

-Also refer to: [401].

oil [266];   b.p. and Spectra (NA).

(5-Ethyl-2-hydroxyphenyl)phenylmethanone

[3132-42-1]   C$_{15}$H$_{14}$O$_2$ mol.wt. 226.27

Syntheses

-Preparation by Fries rearrangement of p-ethylphenyl benzoate with aluminium chloride at 140° for 4 h [266].
-Preparation by reaction of benzoyl chloride with p-ethyl-phenol in the presence of aluminium chloride in tetra-chloroethane at 105° for 22 h (60%) [400].
-Preparation by condensation of benzotrichloride with p-ethylphenol in the presence of 32-40% aqueous sodium hydroxide at 70-80° (70%) [1339].
-Also refer to: [1340, 1341].

m.p. 74° [266], 69°5-70° [1339], 69-72° [400];
^{1}H NMR [1339], IR [1339], UV [1339], MS [1339];
gas chromatography study [1343].

(2-Hydroxy-3,4-dimethylphenyl)phenylmethanone

[14770-98-0] $C_{15}H_{14}O_2$ mol.wt. 226.27

Syntheses

-Preparation by Fries rearrangement of 2,3-dimethylphenyl
    benzoate in the presence of aluminium chloride at 180° for
    10 min (60%) [944].

-Preparation by oxidation of 6,7-dimethyl-2,3-diphenylbenzofuran with chromium trioxide in
  boiling acetic acid, followed by saponification of the keto ester so obtained (2-benzoyloxy-3,4-di-
  methylbenzophenone) [76, 1052, 1054].
-Also refer to: [317].

m.p.   45° [76], 42° [944, 1051, 1052, 1054];   IR [76, 1052].

(2-Hydroxy-3,5-dimethylphenyl)phenylmethanone

[16762-34-8] $C_{15}H_{14}O_2$ mol.wt. 226.27

Syntheses

-Preparation by Fries rearrangement of 2,4-dimethylphenyl
    benzoate in the presence of aluminium chloride (major
    product) [895], at 130-140° for 4 h (50%) [97], at 140-160°
    for 20 min (83%) [331] or at 180° for 10 min (75%) [944].

-Preparation by chromic oxidation of various substituted 5,7-dimethylbenzofurans (5,7-dimethyl-
  2,3-diphenylbenzofuran [1052]; 5,7-dimethyl-2-(4-methylphenyl)-3-phenylbenzofuran [1055] and
  2-(4-methoxyphenyl)-5,7-dimethyl-3-phenylbenzofuran [1055]), followed by saponification of the
  resulting keto esters.
-Preparation by Friedel-Crafts acylation of 2,4-dimethylanisole or of 2,4-dimethylphenetole with
  benzoyl chloride in the presence of aluminium chloride in refluxing carbon disulfide (10% and
  30% yields, respectively) [895].
-Preparation by reaction of benzoyl chloride with 2,4-dimethylphenol in the presence of aluminium
  chloride in nitrobenzene at 50-60° for 18 h (27%) [331].
-Also refer to: [244].

oil [1052, 1055];   m.p. 40-41° [97, 184];
b.p.$_{20}$ 202° [895], b.p.$_{12}$ 198-200° [944], b.p.$_{17}$ 192-194° [97];   IR [1052].

(2-Hydroxy-3,6-dimethylphenyl)phenylmethanone

[4072-17-7] $C_{15}H_{14}O_2$ mol.wt. 226.27

Synthesis

-Preparation by oxidation of 4,7-dimethyl-2,3-diphenyl-
    benzofuran with chromium trioxide in acetic acid at 60°,
    followed by hydrolysis of the keto ester so obtained
    (2-benzoyloxy-3,6-dimethylbenzophenone) with 10%
    sodium hydroxide [74, 1429].

-Also refer to: [75, 317].

m.p.   104° [67, 73, 1429], 103-104° [74];
IR [67, 68, 1429], UV [73].

(2-Hydroxy-4,5-dimethylphenyl)phenylmethanone

[4072-14-4] $C_{15}H_{14}O_2$ mol.wt. 226.27

Syntheses

-Preparation by reaction of chromium trioxide with 5,6-dimethyl-2,3-diphenylbenzofuran [78, 1052] or 5,6-dimethyl-2-(4-methylphenyl)-3-phenylbenzofuran [1055] in acetic acid, followed by saponification of the keto ester so formed (2-benzoyloxy-4,5-dimethylbenzophenone).

-Preparation by Fries rearrangement of 3,4-dimethylphenyl benzoate without solvent, in the presence of,
*aluminium chloride between 120 to 150° (quantitative yield) [110] or at 140° for 20 min (66%) [183];
*titanium tetrachloride at 140° for 2 h (75%) [866].

-Also obtained by isomerization of 2-hydroxy-4,6-dimethylbenzophenone in the presence of aluminium chloride between 140 to 180° for several hours (quantitative yield) [110]. There is a methyl group migration.

-Preparation by reaction of benzoyl chloride with 3,4-dimethylphenyl benzoate in the presence of zinc chloride at 110°, followed by saponification of the keto ester formed (2-benzoyloxy-4,5-dimethylbenzophenone) with sodium hydroxide in boiling ethanol [143].

-Preparation by reaction of pyridinium chloride with 2-methoxy-4,5-dimethylbenzophenone by boiling for 50 min [1151].

-Also obtained by reaction of benzoyl chloride with 4-methylthymol methyl ether (4,5-dimethyl-2-isopropylanisole) in the presence of aluminium chloride in carbon disulfide at r.t. for 23 h [1151].

-Preparation by saponification of 2-benzoyloxy-4,5-dimethylbenzophenone with potassium hydroxide in refluxing ethanol for 1 h (64%) [575].

-Also obtained from 3,4-dimethylbenzophenone by nitration with nitric acid in acetic anhydride at -70°, then at 5° overnight. The diene obtained (2-benzoyl-4,5-dimethyl-4-nitro-1,4-dihydrophenyl acetate as cis and trans mixture) was added to a concentrated solution of sodium methoxide in methanol [423].

-Also obtained (by-product) by reaction of aluminium chloride with 2-bromo-4,5-dimethylphenyl benzoate without solvent at 140° for 2 h (7%) [866].

-Also obtained (by-product) by treatment of 2-bromo-4,5-dimethylbenzophenone with 25% aqueous ammonium hydroxide in ethanol into an autoclave at 180-190° for 5 h under 40 atmospheres (15%) [262].

-Preparation by reaction of 3,4-xylenol with benzoic acid in the presence of boron trifluoride-ethyl ether complex, at 128-129° for 7 min, followed by treatment of the difluoroboroxy chelate formed with boiling aqueous ethanol for 15 min (63%) [1305].

-Also refer to: [575, 1154, 1155].

m.p. 112-113°5 [423], 111° [67, 110], 110-111° [78, 143, 183, 262, 866, 1305], 110° [1151], 108° [1052, 1055];
^{1}H NMR (Sadtler: standard n° 59389 M) [423, 866, 1305],
IR (Sadtler: standard n° 86545 K) [67, 68, 423, 866, 1052, 1055, 1151, 1305],
UV [866, 1305], MS [423, 866].

(2-Hydroxy-4,6-dimethylphenyl)phenylmethanone

[2929-45-5] $C_{15}H_{14}O_2$ mol.wt. 226.27

HO

Syntheses

-Preparation by reaction of benzoyl chloride with 3,5-di-
methylphenol in the presence of aluminium chloride in
nitromethane at 25° for 3 h, then at reflux for 1 h [1181],
(45%) [35].
-Preparation by Fries rearrangement of 3,5-dimethylphenyl
benzoate with aluminium chloride [74, 385, 1439], without solvent between 120-150° (quantitative
yield) [110], (55%) [1439] or in nitrobenzene at 62-63° for 18 h (80%) [126].
-Preparation by oxidation of 4,6-dimethyl-2,3-diphenylbenzofuran with chromium trioxide in acetic
acid at 60°, followed by saponification of the keto ester so obtained (2-(benzoyloxy)-4,6-di-
methylbenzophenone) with 10% sodium hydroxide in boiling ethanol [74].
-Also obtained by reaction of benzoic acid with 3,5-dimethylphenol in the presence of boron
trifluoride in a sealed tube at 160° for 2 h (58%) [730].
-Also obtained (by-product) by treatment of 2-bromo-4,5-dimethylphenyl benzoate with aluminium
chloride at 140° for 2 h (8%) [866].
-Also refer to: [75, 147, 244, 317, 386].

m.p.   143-143°5 [126], 143° [110], 142-143° [35], 142° [74, 730], 141-142° [866],
        140° [67, 73, 385, 609], 139-140° [1439], 134° [1178, 1179];
^{1}H NMR [609, 676, 866, 1177, 1178, 1179, 1181],
^{13}C NMR [1177], IR [66, 68, 609, 866, 1178, 1179],
UV [609, 866, 907], MS [866];
pK_a [609];   polarographic study [1223];   thermal behaviour [1178, 1179].

(4-Hydroxy-2,3-dimethylphenyl)phenylmethanone

[107931-09-9] $C_{15}H_{14}O_2$ mol.wt. 226.27

CH₃ CH₃

Synthesis

-Preparation by demethylation of 2,3-dimethyl-4-methoxy-
benzophenone (SM) with pyridinium chloride at 180° for
2 h. SM was prepared by reaction of phenylmagnesium
bromide with 2,3-dimethyl-4-methoxybenzaldehyde in ethyl
ether at 0° for 30 min [618].

m.p. and Spectra (NA).

(4-Hydroxy-2,5-dimethylphenyl)phenylmethanone

[62262-03-7] $C_{15}H_{14}O_2$ mol.wt. 226.27

CH₃

Syntheses

-Preparation by Friedel-Crafts acylation of 2,5-dimethyl-
phenyl benzoate with benzoyl chloride in the presence of
zinc chloride, followed by saponification of the p-keto ester
so obtained with sodium hydroxide in boiling ethanol
(quantitative yield) [143].
-Preparation by reaction of hydriodic acid with 4-methoxy-2,5-dimethylbenzophenone in boiling
acetic acid (56%) [895].

-Also obtained (by-product) by Fries rearrangement of 2,5-dimethylphenyl benzoate with
aluminium chloride at 130-140° for 4 h [895].
-Also refer to: [147].

   m.p. 166-167° [150, 895];   Spectra (NA).

(4-Hydroxy-2,6-dimethylphenyl)phenylmethanone

[81375-01-1] $C_{15}H_{14}O_2$ mol.wt. 226.27

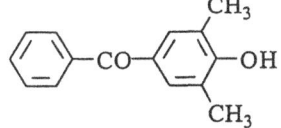

Syntheses

-Preparation by reaction of sodium iodide with 4-benzyloxy-
2,6-dimethylbenzophenone (SM) in the presence of
trimethylsilyl chloride in acetonitrile into an autoclave at 130°
for 24 h (60%). SM was prepared from 4-bromo-3,5-di-
methylphenol by a four-step synthesis [385].

-Preparation by diazotization of 4-amino-2,6-dimethylbenzophenone, followed by hydrolysis of the
diazonium salt obtained (34%) [35].
-Also refer to [34, 1256].

   m.p. 115° [35, 385];   b.p.$_{0.01}$ 153-155° [385];
      ^{1}H NMR [35, 676, 1256], ^{13}C NMR [385, 1180], IR [35, 1256].

(4-Hydroxy-3,5-dimethylphenyl)phenylmethanone

[5336-56-1] $C_{15}H_{14}O_2$ mol.wt. 226.27

Syntheses

-Preparation by decarboxylation of 2-(4-hydroxy-3,5-di-
methylbenzoyl)benzoic acid in the presence of cupric acetate
monohydrate in refluxing quinoline for 30 min (98%)
[1160].

-Preparation by demethylation of 4-methoxy-3,5-dimethyl-
benzophenone (SM) with aluminium chloride without solvent at 100-110° for 3 h [96]. SM was
obtained by Friedel-Crafts acylation of 2,6-dimethylanisole with benzoyl chloride according to
usual method.
-Preparation by Fries rearrangement of 2,6-dimethylphenyl benzoate with aluminium chloride
without solvent [440], at 130-140° (46%) [95].
-Also obtained by photo-Fries rearrangement of 2,6-dimethylphenyl benzoate in isopropanol at 26°
for 24 h (33%) [102, 419] or in pentane in the presence of silica gel (23%) [102].
-Also obtained (poor yield) by reaction of cumene hydroperoxide with 4-benzyl-2,6-dimethyl-
phenol and air bubbling in the presence of cobalt phthalate in cumene at 80-100° for 100 h (6%)
[1113].
-Also refer to: [244, 452, 502, 593, 760].

   m.p. 145° [440], 143°3-144°5 [1160], 143° [1113], 142-142°5 [95],
      141-142° [96], 139-141°6 [419];
   ^{1}H NMR [440].

(5-Hydroxy-2,4-dimethylphenyl)phenylmethanone

$C_{15}H_{14}O_2$ mol.wt. 226.27

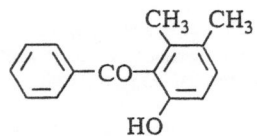

Syntheses

-Preparation by Friedel-Crafts acylation of 2,4-dimethyl-
phenyl benzoate with benzoyl chloride in the presence of
zinc chloride in boiling chloroform, followed by
saponification of the 5-(benzoyloxy)-2,4-dimethyl-
benzophenone so obtained with sodium hydroxide in
refluxing ethanol (45%) [827].
-Also obtained (poor yield) by Fries rearrangement of 2,4-dimethylphenyl benzoate with aluminium
chloride without solvent at 130-140° for 4 h (4%) [97].
-Obtained (by-product) by Friedel-Crafts acylation,
*of 2,4-dimethylanisole with benzoyl chloride in the presence of aluminium chloride in refluxing
carbon disulfide for 3-5 h (8%) [895];
*of 2,4-dimethylphenetole with benzoyl chloride in the presence of aluminium chloride in boiling
carbon disulfide [895].
-Preparation by demethylation of 2,4-dimethyl-5-methoxybenzophenone,
*with hydriodic acid (d = 1.7) in boiling acetic acid at 130-140° for 2 h (70%) [895];
*with aluminium chloride in boiling carbon disulfide for 8 h (32%) [895].

m.p.   145-146° [827], 140-141° [97, 895];   Spectra (NA).

(6-Hydroxy-2,3-dimethylphenyl)phenylmethanone

[108478-10-0] $C_{15}H_{14}O_2$ mol.wt. 226.27

Syntheses

-Preparation by oxidation of 4,5-dimethyl-2,3-diphenyl-
benzofuran with chromium trioxide in acetic acid, followed
by alkaline hydrolysis of the resulting keto ester with
refluxing 10% sodium hydroxide in ethanol for 2 h [74].
-Also obtained by saponification of 2,3-dimethyl-6-benzoyl-
oxybenzophenone with refluxing 10% sodium hydroxide for 30 min [74].
-Preparation by debromination of 3-bromo-2-hydroxy-5,6-dimethylbenzophenone in the presence
of copper powder in caproic acid at 220° for 15 min (82%) [866].
-Also obtained (poor yield) by Fries rearrangement of 3,4-dimethylphenyl benzoate with titanium
tetrachloride without solvent at 140° for 2 h (5%) [866].
-Also refer to: [75].

m.p.   124-125° [866], 114-115° [74];
^{1}H NMR (Sadtler: standard n° 59393 M) [866],
IR (Sadtler: standard n° 86549 K) [866], UV [866], MS [866].

(4-Ethoxy-2-hydroxyphenyl)phenylmethanone

[15889-70-0] $C_{15}H_{14}O_3$ mol.wt. 242.27

Syntheses

-Preparation by reaction of ethyl bromide with resbenzophenone, in the presence of sodium hydroxide in dilute
ethanol, first at 30-40° for 1 h, then at 65-75° for 16 h
(35%) [805].

-Preparation by reaction of ethyl p-toluenesulfonate with resbenzophenone in the presence of potassium carbonate in boiling water for 8 h (75%) [1259, 1260].
-Preparation by Friedel-Crafts acylation of 1,3-diethoxybenzene with benzoyl chloride in the presence of a zinc chloride/aluminium chloride mixture, first between 0 to 5°, then at 10° for 1 h and finally at 65° for 6 h [479].
-Also refer to: [43, 602, 913].

m.p. 54°5 [805], 50-52° [1259, 1260];   Spectra (NA).

(2-Hydroxy-3-methoxy-5-methylphenyl)phenylmethanone

[17603-92-8] $C_{15}H_{14}O_3$ mol.wt. 242.27

Synthesis

-Preparation by photo-Fries rearrangement of 2-methoxy-4-methylphenyl benzoate in benzene or in ethanol during 4 h (40% and 63% yields, respectively) [206].

m.p. (NA); ^{1}H NMR [206], IR [206].

(2-Hydroxy-3-methoxy-6-methylphenyl)phenylmethanone

[129103-91-9] $C_{15}H_{14}O_3$ mol.wt. 242.27

Synthesis

-Preparation from 2-iodo-6-methoxy-3-methylphenyl benzoate by rearrangement on treatment with n-butyllithium in a mixture of ethyl ether, hexane and tetrahydrofuran at -70° for 2 h, then treatment with saturated aqueous ammonium chloride (55%) [598].

m.p. 128-129°5 [598]; ^{1}H NMR [598], IR [598].

(2-Hydroxy-4-methoxy-3-methylphenyl)phenylmethanone

[83803-88-7] $C_{15}H_{14}O_3$ mol.wt. 242.27

Syntheses

-Preparation by methylation of 2,4-dihydroxy-3-methyl-benzophenone with methyl iodide in the presence of potassium carbonate in refluxing acetone for 2 h [652].
-Also obtained by methylation of resbenzophenone with methyl iodide in the presence of potassium hydroxide in methanol at 0° (26%) [1242]. There is an introduction of one methyl group on the nucleus.
-Also obtained by methylation of resbenzophenone [757] according to the method [1347].
-Also refer to: [23].

m.p. 125° [652, 757, 1242], 124-125° [1406];
UV [1406].

84   HYDROXYBENZOPHENONES

(2-Hydroxy-4-methoxy-5-methylphenyl)phenylmethanone

[59954-97-1]   C15H14O3   mol.wt. 242.27

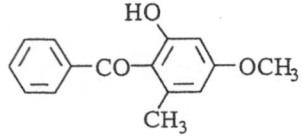

Syntheses

-Preparation by reaction of benzoyl chloride with 4-methyl-
resorcinol dimethyl ether in the presence of aluminium
chloride in ethylene dichloride (97%) [1193].
-Preparation by oxidation of 6-methoxy-5-methyl-2,3-di-
phenylbenzofuran with chromium trioxide in refluxing

acetic acid, followed by alkaline hydrolysis of the keto ester so obtained (2-benzoyloxy-
4-methoxy-5-methylbenzophenone) with potassium hydroxide in refluxing ethanol (73%) [2].

m.p. 188° [2], 88-89° [1193]. One of the reported melting points is obviously
wrong. Spectra (NA).

(2-Hydroxy-4-methoxy-6-methylphenyl)phenylmethanone

[23573-43-5]   C15H14O3   mol.wt. 242.27

Syntheses

-Preparation by partial methylation of 2,4-dihydroxy-
6-methylbenzophenone with dimethyl sulfate in the presence
of potassium carbonate in refluxing acetone for 4 h (74%)
[26].
-Also obtained by UV light irradiation of 3-methoxy-5-methyl-
phenyl benzoate in ethanol at 20° for 70 h (21%) [16].

m.p. 93-94° [26], 87-88° [16];   1H NMR [16, 141], IR [16], UV [16].

(2-Hydroxy-5-methoxy-4-methylphenyl)phenylmethanone

[59954-92-6]   C15H14O3   mol.wt. 242.27

Synthesis

-Preparation by oxidation of 5-methoxy-6-methyl-2,3-di-
phenylbenzofuran with chromium trioxide in refluxing acetic
acid for 30 min, followed by alkaline hydrolysis of the
resulting keto ester (2-benzoyloxy-5-methoxy-4-methyl-
benzophenone) with potassium hydroxide in refluxing
ethanol (75%) [2].

m.p. 114° [2];   Spectra (NA).

(2-Hydroxy-6-methoxy-4-methylphenyl)phenylmethanone

[23565-66-4]   C15H14O3   mol.wt. 242.27

Synthesis

-Obtained by UV light irradiation of 3-methoxy-5-methyl-
phenyl benzoate in ethanol at 20° for 70 h (13%) [16].

m.p. 70-71° [16];   1H NMR [16], IR [16], UV [16].

(4-Hydroxy-2-methoxy-6-methylphenyl)phenylmethanone

[23565-67-5] $C_{15}H_{14}O_3$ mol.wt. 242.27

Synthesis

-Obtained by UV light irradiation of 3-methoxy-5-methyl-
 phenyl benzoate in ethanol at 20° for 70 h (8%) [16].

m.p. 123-124° [16];   ¹H NMR [16], IR [16], UV [16].

(5-Hydroxy-4-methoxy-2-methylphenyl)phenylmethanone

                                              $C_{15}H_{14}O_3$ mol.wt. 242.27

Synthesis

-Preparation by saponification of 5-(benzoyloxy)-
 4-methoxy-2-methylbenzophenone (SM) (m.p. 95-96°)
 with sodium hydroxide in boiling dilute ethanol (quantitative
 yield). SM was obtained by condensation of benzoyl
 chloride with creosol benzoate in the presence of zinc
 chloride [139].

m.p. 150° [139];   Spectra (NA).

(2-Hydroxy-3,4-dimethoxyphenyl)phenylmethanone

[7508-32-9] $C_{15}H_{14}O_4$ mol.wt. 258.27

Syntheses

-Preparation by reaction of benzoyl chloride with pyrogallol
 trimethyl ether,
*in the presence of aluminium chloride in ethylene
  dichloride between 0 to 5°, then at r.t. for 3 h and finally at
  reflux for 1 h (96%) [1194] or for 2 h (47%) [213] and
also in refluxing methylene chloride for 2 h (22%) [1407].
N.B.: This benzophenone is, by mistake, named, 2-hydroxy-3,4-dimethoxybenzaldehyde (synthetic
  example 20, compound a, page 39), in the patent PCT Int. Appl. WO 96 05,188 of 22 February
  1996 [213].
*in the presence of zinc chloride (good yield) [137], at 100° for 1 h [132];
*in the presence of mercuric chloride instead of aluminium chloride without solvent at 115° (reflux)
  for 1-3 h (40%) [857, 858].
-Preparation by reaction of methyl iodide with 2,3,4-trihydroxybenzophenone monosodium salt in
 the presence of sodium carbonate at 160° for 6 h [495]. By using lead salt instead of sodium salt in
 the same conditions, the yield was decreased.
-Also obtained by reaction of methyl iodide with 2,3,4-trihydroxybenzophenone in the presence of
 lithium carbonate in N,N-dimethylformamide at 30° for 15 h under nitrogen (17%) [802].
-Also obtained by reaction of dimethyl sulfate with 2,3-dihydroxy-4-methoxybenzophenone or with
 2,4-dihydroxy-3-methoxybenzophenone in the presence of alkali [950].

m.p. 185-187° [802]. This melting point is obviously wrong. 132-134° [1194],
      131° [495, 1407], 130-131° [132, 137], 127-128° [213], 120-121° [950];
¹H NMR [213, 802, 1194], ¹³C NMR [802], IR [1194], UV [802],
MS [802, 1194]; pK_a [802].

(2-Hydroxy-4,5-dimethoxyphenyl)phenylmethanone

[36896-99-8] $C_{15}H_{14}O_4$ mol.wt. 258.27

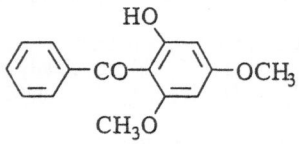

Syntheses

-Preparation by partial methylation of 2,5-dihydroxy-4-methoxybenzophenone with dimethyl sulfate in the presence of potassium carbonate in refluxing acetone [512, 807].
-Preparation by partial demethylation of 2,4,5-trimethoxy-benzophenone with aluminium chloride in boiling carbon disulfide for 1 h [134, 135].
-Also obtained by Friedel-Crafts acylation of 1,2,4-trimethoxybenzene with benzoyl chloride in the presence of aluminium chloride,
*in carbon disulfide (by-product) [778], (15%) [134, 135];
*in ethylene dichloride between 0 to 5°, then at r.t. for 4 h and finally at reflux for 2 h (97%) [1194].
-Preparation by reaction of benzoyl chloride with 3,4-dimethoxyphenol in the presence of boron trichloride in methylene chloride at r.t. for 2.5 h (51%) [213].
N.B.: This benzophenone is, by mistake named, 2-hydroxy-4,5-dimethoxybenzaldehyde (synthetic example 1, compound h, page 21), in the patent PCT Int. Appl. WO 96 05,188 of 22 February 1996 [213].
-Also obtained from O-methyldalbergin (6,7-dimethoxy-4-phenylcoumarin) (SM) by oxidation with neutral permanganate. SM was prepared by methylation of dalbergin (6-hydroxy-7-methoxy-4-phenylcoumarin), itself isolated from *Dalbergia sissoo* [1234].
-Also refer to: [1107, 1324].

m.p.   109-110° [807, 1194], 106-107° [134, 135, 512], 104-105° [213];
^{1}H NMR [213, 512, 1194], IR [1194], MS [1194].

(2-Hydroxy-4,6-dimethoxyphenyl)phenylmethanone *(Hydrocotoin)*

[34425-64-4] $C_{15}H_{14}O_4$ mol.wt. 258.27

Syntheses

-Preparation from 2-bromo-3,5-dimethoxyphenyl benzoate on treatment with sec-butyllithium in a mixture of ethyl ether, hexane and tetrahydrofuran at -70° for 2 h, followed by treatment with saturated aqueous ammonium chloride (47%) [598].
-Preparation by Friedel-Crafts acylation of phloroglucinol trimethyl ether with benzoyl chloride in the presence of aluminium chloride [26], in ethyl ether for 8 h [1097].
-Preparation by action of boron trichloride with 2,4,6-trimethoxybenzophenone in methylene chloride for 30 min at r.t. under nitrogen (92%) [831].
-Preparation by heating a mixture of benzoyl chloride and 3,5-dimethoxyphenyl benzoate in the presence of zinc chloride in benzene, followed by saponification of the keto ester so formed [425, 1074].
-Also obtained (poor yield) by reaction of benzonitrile with phloroglucinol dimethyl ether in the presence of zinc chloride and hydrochloric acid in ethyl ether (Hoesch reaction) [29], (3%) [704].
-Also obtained from 5,7-dimethoxy-4-phenylcoumarin,
*by oxidation with potassium permanganate in acetone for 30 min (<5%) [1076];
*by ozonolysis in a mixture of carbon tetrachloride and chloroform (25%) [1076].
-Also obtained (poor yield) by UV light irradiation of 2,4,6-trimethoxybenzophenone in carbon tetrachloride for 500 h (3%) [809].
-Also refer to: [28, 30, 137, 950, 1416].

Isolation from natural sources

-From the heartwood of *Allanblackia fluoribunda* Oliver (Guttiferae; subfamily Clusioideae) [831];
-From the Paracotobark or *Aniba pseudocoto* Rusby (Kostermans) (Lauraceae) [396, 645];
-From the *Coto* bark (Lauraceae) [275].

    m.p. 99-100° [26], 98° [273, 278, 645, 704, 900], 97-98° [809], 96-98° [831],
        96-97° [598], 95-96° [1076], 93-95° [1074];
    ^{1}H NMR [598, 809, 831, 1097], IR [831], UV [831, 1076], MS [598, 831];
    GLC [809].

(4-Hydroxy-2,6-dimethoxyphenyl)phenylmethanone

$C_{15}H_{14}O_4$ mol.wt. 258.27

Syntheses

-Preparation by saponification of 4-(benzoyloxy)-2,6-di-
  methoxybenzophenone with 10% potassium hydroxide in
  methanol at r.t. for 2 h [248].
-Also obtained by reaction of benzonitrile with phloroglucinol
  dimethyl ether in the presence of zinc chloride and
hydrochloric acid (14%) (Hoesch reaction) [704].
-Also refer to: [950].

    m.p. 178-179° [248], 177° [704]; Spectra (NA).

[2-Hydroxy-4-(2-hydroxyethoxy)phenyl]phenylmethanone

[16909-78-7] $C_{15}H_{14}O_4$ mol.wt. 258.27

Syntheses

-Preparation by reaction of ethylene chlorohydrin
  (2-chloroethanol) with resbenzophenone,
*in the presence of sodium hydroxide, in boiling
  water for 6 h (40%) [558] or in water at 70° for
18 h (91%) [528];
*in the presence of sodium carbonate in refluxing dilute ethanol for 16 h (58%) [706].
-Preparation by reaction of ethylene carbonate with resbenzophenone,
*in the presence of a quaternary ammonium salt (for example benzyltriethylammonium chloride) at
140-150° [524], for 11 h (93%) [523];
*in the presence of an alkaline metal or alkaline earth metal salt as catalyst (sodium ethylenediamine
tetraacetate (0.02 mol); calcium citrate monohydrate (0.02 mol); sodium nitriloacetate (0.02 mol);
sodium oleate (0.05 mol); sodium stearate (0.05 mol)), for 7-8 h at 155° (92-93% yields) [89];
*in the presence of sodium methoxide in diisobutylketone for 50 min at 140° [388].
-Preparation by treatment of resbenzophenone with ethylene oxide,
*in methanol in the presence of sodium methoxide in autoclave at 150°, under 16.4 atmospheres, for
6 h (54%) [706];
*in the presence of aqueous potassium hydroxide at 105 to 110° into an autoclave (<90 psig) under
nitrogen (99%) [727].
-Also refer to: [36, 55, 56, 464, 1229, 1422, 1466] and also [779, 918, 937, 1014, 1352] (Japanese
patents).

    m.p. 93°5-95°5 [488], 92-93° [706], 91°5-97°5, this gap of 6° C appears in the patent
    U.S. US 4,978,797 [727], 91-92° [558], 89-90° [606];
    UV [606]; gel chromatography [1145]; pK_a [606]; TLC [377].

[2-Hydroxy-3-(hydroxymethyl)-4-methoxyphenyl]phenylmethanone

[80604-76-8] $C_{15}H_{14}O_4$ mol.wt. 258.27

HO   CH₂OH

(structure: phenyl-CO-benzene ring with HO, CH₂OH, OCH₃ substituents)

Synthesis

-Obtained by adding a 27% solution of formaldehyde to a
solution of 2-hydroxy-4-methoxybenzophenone in aqueous
sodium hydroxide and tetrahydrofuran then stirring the
mixture at r.t. for 5.5 h (31%) [129, 130].

-Also refer to: [131].

m.p. and Spectra (NA).

[2-Hydroxy-5-(hydroxymethyl)-4-methoxyphenyl]phenylmethanone

[80501-48-0] $C_{15}H_{14}O_4$ mol.wt. 258.27

HO

(structure: phenyl-CO-benzene ring with HO, OCH₃, CH₂OH substituents)

CH₂OH

Synthesis

-Preparation by adding a 37% solution of formaldehyde to a
solution of 2-hydroxy-4-methoxybenzophenone in aqueous
sodium hydroxide and tetrahydrofuran then stirring the
mixture at r.t. for 24 h (75%) [129]. In the same conditions,
but using a 27% solution of formaldehyde and stirring the
mixture at r.t. for 5.5 h, the keto alcohol was obtained in a yield of only 47% [129, 130].

-Also refer to: [128, 131].

m.p. and Spectra (NA).

(2-Amino-3-hydroxy-4,6-dimethylphenyl)phenylmethanone

$C_{15}H_{15}NO_2$ mol.wt. 241.29

NH₂ OH

(structure: phenyl-CO-benzene ring with NH₂, OH, CH₃, CH₃ substituents)

CH₃

Synthesis

-Preparation by reduction of 2-nitro-3-hydroxy-4,6-dimethyl-
benzophenone with sodium hydrosulfite in dilute ethanol
(67%) [382].

m.p. 125° [382];   Spectra (NA).

[4,5-Dichloro-3-hydroxy-2-(2-propenyl)phenyl]phenylmethanone

[113730-42-0] $C_{16}H_{12}Cl_2O_2$ mol.wt. 307.18

CH₂=CHCH₂   OH

(structure: phenyl-CO-benzene ring with CH₂=CHCH₂, OH, Cl, Cl substituents)

Cl

Synthesis

-Obtained by heating 5-(allyloxy)-3,4-dichlorobenzo-
phenone at 235° for 8 min (67%) (Claisen rearrangement)
[535].

m.p. 105° [535];   ¹H NMR [535].

[4-Hydroxy-3-(2-propenyl)phenyl]phenylmethanone

[73720-75-9] $C_{16}H_{14}O_2$ mol.wt. 238.29

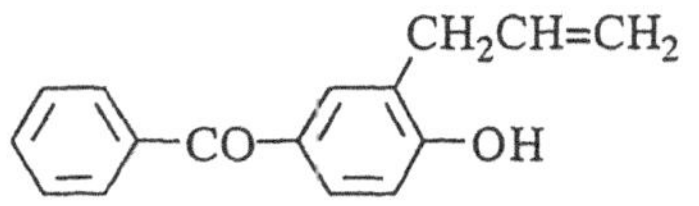

Synthesis

-Preparation by Claisen rearrangement of 4-(allyloxy)-
benzophenone in refluxing phenyl ether for 1 h (73%)
[1431] or in diethylaniline at 207-218° for 3.5 h (47%)
[935].
-Also refer to: [593].

m.p. 129-130°6 [1431];   UV [1431].

[2-Hydroxy-4-(1-propenyloxy)phenyl]phenylmethanone

$C_{16}H_{14}O_3$ mol.wt. 254.29

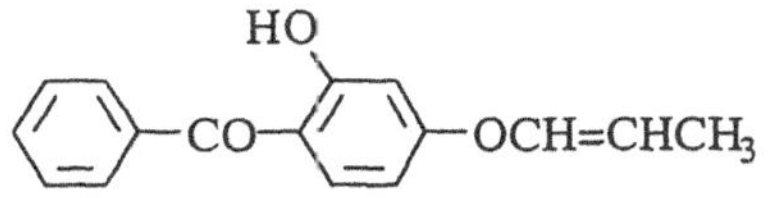

Synthesis

-Obtained (poor yield) by reaction of
epichlorohydrin with resbenzophenone in the
presence of aqueous potassium hydroxide at r.t.
for 18 h (<6%) [803].

m.p. 97-98°5 [803];   Spectra (NA).

[2-Hydroxy-4-(2-propenyloxy)phenyl]phenylmethanone

[2549-87-3] $C_{16}H_{14}O_3$ mol.wt. 254.29

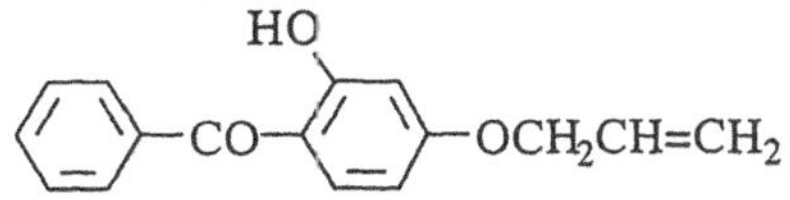

Synthesis

-Preparation by reaction of allyl chloride (3-chloro-
propene) with resbenzophenone (90%), according
to [387].

m.p. 65-68° [387];   Spectra (NA);   TLC [377].

[2-Hydroxy-4-(oxiranylmethoxy)phenyl]phenylmethanone

[19389-82-3] $C_{16}H_{14}O_4$ mol.wt. 270.28

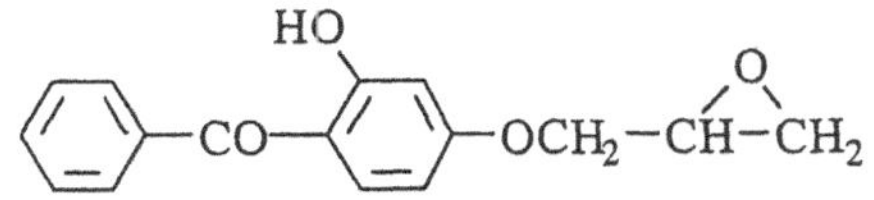

Synthesis

-Preparation by reaction of epichlorohydrin
with 2,4-dihydroxybenzophenone in the
presence of an aqueous potassium hydroxide
solution at 88° for 2 h [840].

m.p. 99-100° [489], 98°5-99° [840];
IR [840], UV [840];   TLC [377].

[2-Hydroxy-3-(1-methylethyl)phenyl]phenylmethanone

[33621-54-4]   C16H16O2   mol.wt. 240.30

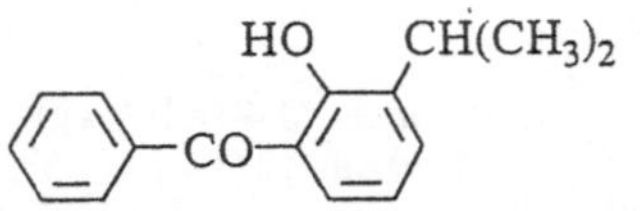

Synthesis

-Preparation by oxidation of 7-isopropyl-2,3-diphenyl-
    benzofuran with chromium trioxide in acetic acid at 70° for
    2 h, followed by alkaline hydrolysis of the resulting
keto ester (2-benzoyloxy-3-isopropylbenzophenone) in boiling dilute ethanol for 15 min [76].

Yellow liquid [76];   b.p.3.5 170-175° [76];   IR [76].

[2-Hydroxy-5-(1-methylethyl)phenyl]phenylmethanone

[20401-89-2]   C16H16O2   mol.wt. 240.30

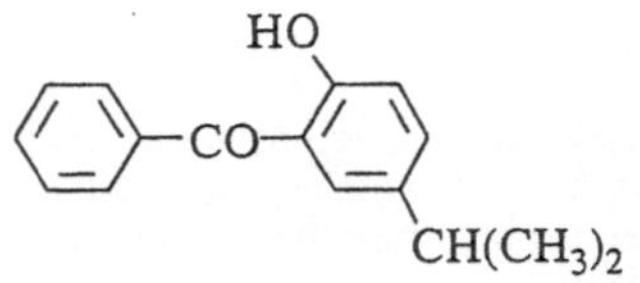

Syntheses

-Preparation by Fries rearrangement of p-isopropylphenyl
    benzoate with aluminium chloride [1223].
-Also obtained by photo-Fries rearrangement of p-isopropyl-
    phenyl benzoate in pentane, with or without silica gel (32-
    35%) [102].

m.p. (NA);
1H NMR [610], IR [610];
TLC [377];   polarographic study [1223].

[4-Hydroxy-3-(1-methylethyl)phenyl]phenylmethanone

[83938-73-2]   C16H16O2   mol.wt. 240.30

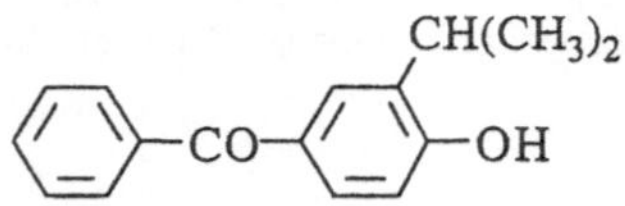

Synthesis

-Preparation by decarboxylation of 2-(4-hydroxy-3-iso-
    propylbenzoyl)benzoic acid in the presence of cupric
    acetate monohydrate in refluxing quinoline for 45 min
    (97%) [1164].

m.p.   134°1-135°2 [1164];   1H NMR [1164], IR [1164], UV [1164].

(2-Hydroxy-3-propylphenyl)phenylmethanone

[108294-70-8]   C16H16O2   mol.wt. 240.30

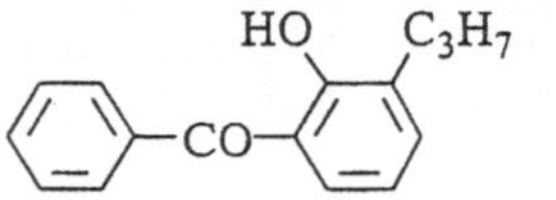

Synthesis

-Preparation by reaction of 2-hydroxy-3-propylbenzoyl
    chloride with benzene in the presence of aluminium chloride
    in refluxing carbon disulfide (42%) [400].

m.p.   34-35° [400];   Spectra (NA).

(4-Hydroxy-3-propylphenyl)phenylmethanone

[183013-50-5]   C16H16O2   mol.wt. 240.30

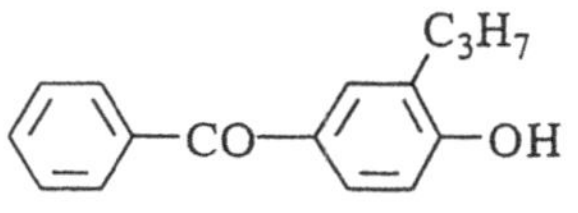

Synthesis

-Preparation by catalytic hydrogenation of 3-allyl-4-hydroxy-
benzophenone (SM). SM was obtained by reaction of allyl
bromide with 4-hydroxybenzophenone in the presence of
potassium carbonate, followed by Claisen rearrangement of
the 4-(allyloxy)benzophenone so formed [1432].

m.p. and Spectra (NA).

(2-Hydroxy-3,4,6-trimethylphenyl)phenylmethanone

[33621-48-6]   C16H16O2   mol.wt. 240.30

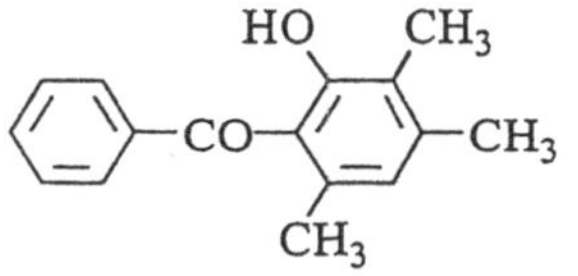

Synthesis

-Obtained by oxidation of 2,3-diphenyl-4,6,7-trimethyl-
benzofuran with chromium trioxide in acetic acid at 65° for
1.5 h, then saponification of the resulting keto ester, the
2-(benzoyloxy)-3,4,6-trimethylbenzophenone, with 10%
sodium hydroxide in boiling dilute ethanol for 30 min [1429].
-Also refer to: [317].

m.p. 102° [1429];   IR [1429].

(2-Hydroxy-3,5,6-trimethylphenyl)phenylmethanone

[33634-16-1]   C16H16O2   mol.wt. 240.30

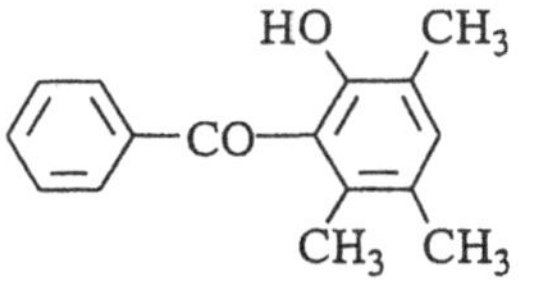

Synthesis

-Obtained by oxidation of 2,3-diphenyl-4,5,7-trimethyl-
benzofuran with chromium trioxide in acetic acid at 65° for
1.5 h, then saponification of the resulting keto ester, the
2-(benzoyloxy)-3,5,6-trimethylbenzophenone, with 10%
sodium hydroxide in boiling dilute ethanol for 30 min [1429].

m.p. 127° [1429];   IR [1429].

(4-Ethyl-2-hydroxy-5-methoxyphenyl)phenylmethanone

[59623-15-3]   C16H16O3   mol.wt. 256.30

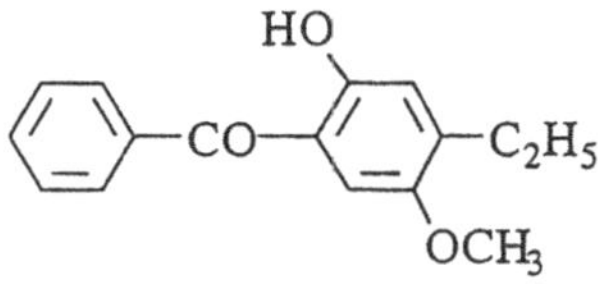

Synthesis

-Preparation by saponification of 2-(benzoyloxy)-4-ethyl-
5-methoxybenzophenone (SM) with potassium hydroxide
in refluxing ethanol for 1 h (73%). SM was obtained by
oxidation of 6-ethyl-5-methoxy-2,3-diphenylbenzofuran
with chromium trioxide in refluxing acetic acid for 30 min
(80%, m.p. 140°) [577].

m.p. 124° [577];   Spectra (NA).

(5-Ethyl-2-hydroxy-4-methoxyphenyl)phenylmethanone

[59623-21-1] $C_{16}H_{16}O_3$ mol.wt. 256.30

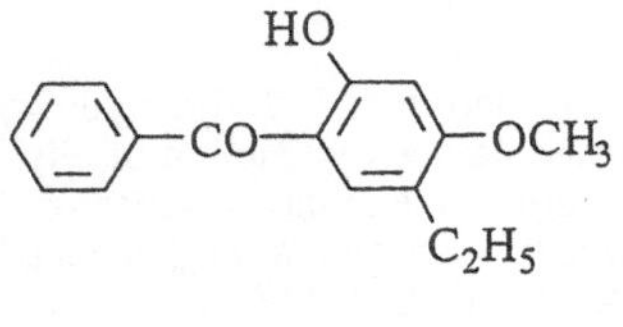

Synthesis

-Preparation by saponification of 2-(benzoyloxy)-5-ethyl-
4-methoxybenzophenone (SM) with potassium hydroxide
in refluxing ethanol for 1 h (67%). SM was obtained by
oxidation of 5-ethyl-6-methoxy-2,3-diphenylbenzofuran
with chromium trioxide in refluxing acetic acid for 30 min
(64%, m.p. 181°) [577].

    m.p. 170° [577];   Spectra (NA).

[2-Hydroxy-4-(1-methylethoxy)phenyl]phenylmethanone

                                   $C_{16}H_{16}O_3$ mol.wt. 256.30

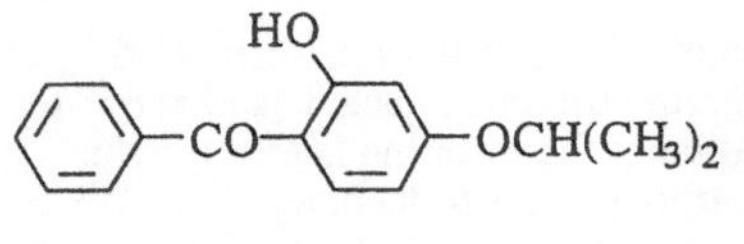

Syntheses

-Preparation by reaction of isopropyl bromide with
resbenzophenone in the presence of potassium
hydroxide in dilute ethanol at 75-80° for 15 h (38%)
[803].
-Preparation by Friedel-Crafts acylation of 1,3-diisopropoxybenzene with benzoyl chloride in the
presence of a zinc chloride/aluminium chloride mixture, first between 0 to 5°, then at 10° for 1 h
and finally at 65° for 6 h [479].

    m.p. 42-42°5 [803];   Spectra (NA).

(2-Hydroxy-4-propoxyphenyl)phenylmethanone

[3088-11-7] $C_{16}H_{16}O_3$ mol.wt. 256.30

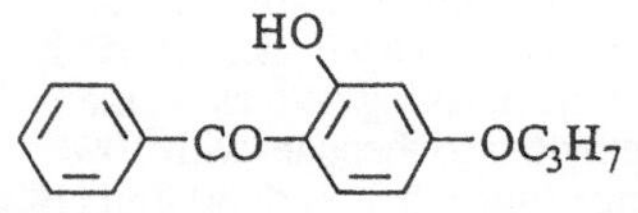

Synthesis

-Preparation by reaction of n-propyl bromide with
resbenzophenone in the presence of sodium hydroxide in
dilute ethanol at 70-80° for 13 h (30%) [805].
-Also refer to: [43].

    m.p. 67° [805];   Spectra (NA).

(2-Hydroxy-5-propoxyphenyl)phenylmethanone

                                   $C_{16}H_{16}O_3$ mol.wt. 256.30

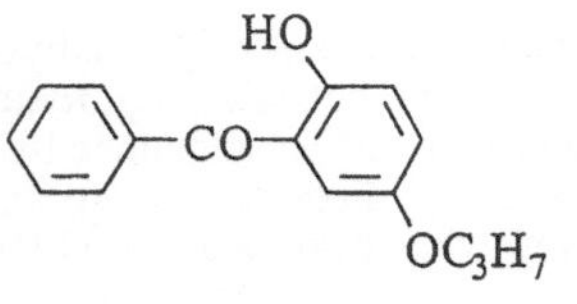

Synthesis

-Preparation by reaction of n-propyl bromide with 2,5-di-
hydroxybenzophenone in the presence of sodium hydroxide
in dilute ethanol at 70-80° for 28 h (25%) [804].

    m.p. and Spectra (NA).

(5-Ethoxy-2-hydroxy-4-methoxyphenyl)phenylmethanone

[52811-38-8] $C_{16}H_{16}O_4$ mol.wt. 272.30

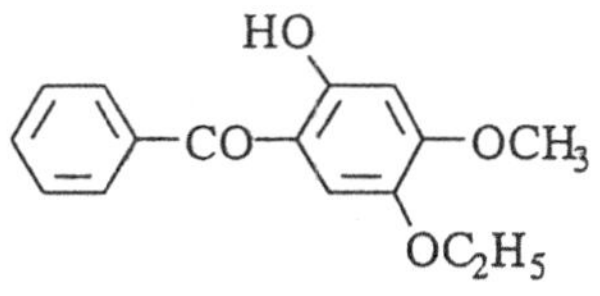

Synthesis

-Preparation by partial ethylation of 2,5-dihydroxy-
4-methoxybenzophenone with diethyl sulfate in the
presence of potassium carbonate in refluxing acetone for
5 h (75%) [778].

   m.p. 100-102° [778];
   ^{1}H NMR [778], IR [778], UV [778].

(2-Hydroxy-4,6-dimethoxy-3-methylphenyl)phenylmethanone

                                               $C_{16}H_{16}O_4$ mol.wt. 272.30

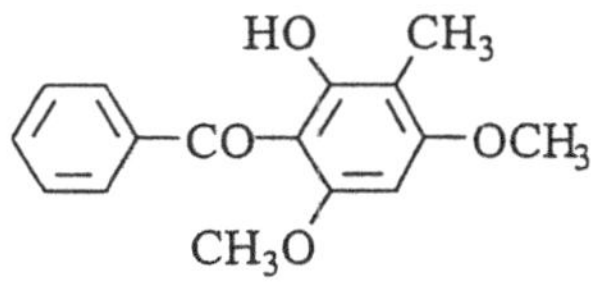

Syntheses

-Preparation by reaction of methyl iodide with 2,4,6-tri-
hydroxy-3-methylbenzophenone or with 2,6-dihydroxy-
4-methoxy-3-methylbenzophenone in the presence of
potassium carbonate in boiling acetone for 8 h [889, 1081].

-Preparation by reaction of diazomethane with 2,4,6-trihydroxy-3-methylbenzophenone in ethyl
ether (86%) [1081].
-Also refer to: [274, 1075].

Isolation from natural source

-From the leaves of *Leptospermum luehmannii* (F. M. Bailey) (Myrtaceae) (minor product) [654,
831, 976, 1081].

   m.p. 137° [654, 1081], 136-137° [889];
   ^{1}H NMR [1081], UV [1081].

(2-Hydroxy-4,6-dimethoxy-5-methylphenyl)phenylmethanone

                                               $C_{16}H_{16}O_4$ mol.wt. 272.30

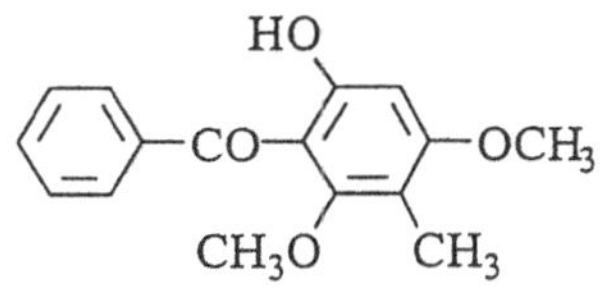

Synthesis

-Not yet described.

Isolation from natural source

-From the leaves or rhizomes of *Agonis luehmannii* (now
Leptospermum luehmannii) (F. M. Bailey) (Myrtaceae) (major product) [654, 831, 976, 1081].

   m.p. 110° [1081], 104° [654];
   ^{1}H NMR [1081], IR [1081], UV [1081];   GC [1081].

[2-Hydroxy-4-(2-hydroxypropoxy)phenyl]phenylmethanone

[22546-86-7]   C$_{16}$H$_{16}$O$_4$ mol.wt. 272.30

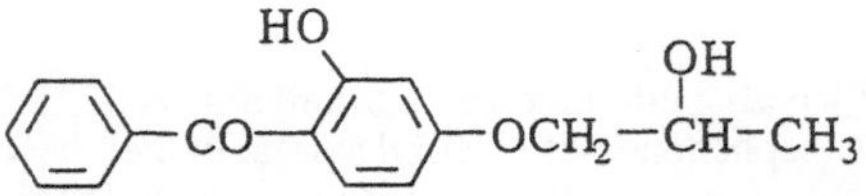

Synthesis

-Preparation by reaction of propylene oxide with resbenzophenone in the presence of aqueous potassium hydroxide into an autoclave (140°) under nitrogen (66%) [727].

m.p. 78-80° [727];   Spectra (NA).

[2-Hydroxy-4-(2-methoxyethoxy)phenyl]phenylmethanone

[27992-95-6]   C$_{16}$H$_{16}$O$_4$ mol.wt. 272.30

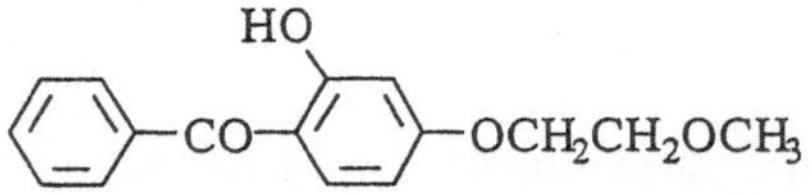

Syntheses

-Preparation by reaction of 2-methoxyethyl chloride (2-chloroethyl methyl ether) with resbenzophenone (91%), according to [387].
-Preparation by reaction of 2-methoxyethylene bromide with 2,4-dihydroxybenzophenone in the presence of sodium carbonate in acetone at 50-55° (74%) [443].

m.p. 40-41° [387];   Spectra (NA).

[2-Hydroxy-3,5-di(hydroxymethyl)-4-methoxyphenyl]phenylmethanone

[80501-47-9]   C$_{16}$H$_{16}$O$_5$ mol.wt. 288.30

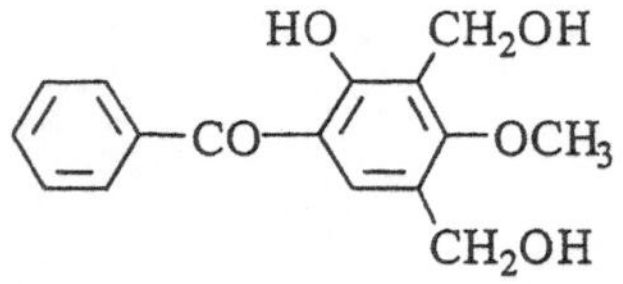

Synthesis

-Obtained (poor yield) by adding a 27% solution of formaldehyde to a solution of 2-hydroxy-4-methoxy-benzophenone in aqueous sodium hydroxide and tetrahydrofuran then stirring the mixture at r.t. for 5.5 h (3%) [129, 130].

m.p. and Spectra (NA).

(2-Hydroxy-3,4,5-trimethoxyphenyl)phenylmethanone

[42833-88-5]   C$_{16}$H$_{16}$O$_5$ mol.wt. 288.30

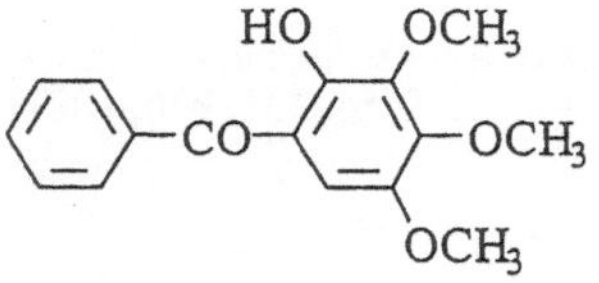

Synthesis

-Preparation by reaction of benzoyl chloride with 1,2,3,4-tetramethoxybenzene in the presence of aluminium chloride in ethyl ether for 22 h [1097].

m.p. (NA); ^{1}H NMR [1097].

[3-[(Dimethylamino)methyl]-4-hydroxyphenyl]phenylmethanone *(Hydrochloride)*

[82506-20-5] $C_{16}H_{17}NO_2$,HCl   mol.wt. 291.78

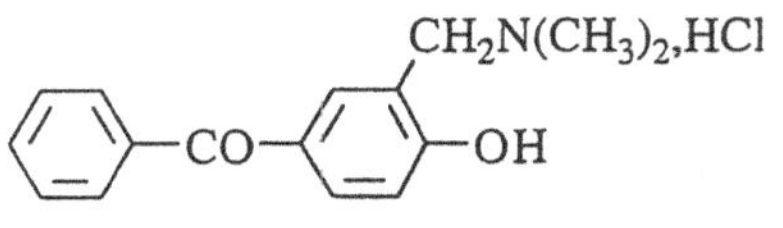

CH$_2$N(CH$_3$)$_2$,HCl

〈_〉—CO—〈_〉—OH

Synthesis

-Preparation by aminomethylation of p-hydroxybenzophenone with dimethylamine and formaline
in water at 35-40° for 4 h (52%) [773].

m.p. 77° [773]; ^{1}H NMR [773], IR [773].

[2,3-Bis(acetyloxy)-4-hydroxyphenyl]phenylmethanone

[177703-36-5] $C_{17}H_{14}O_6$ mol.wt. 314.29

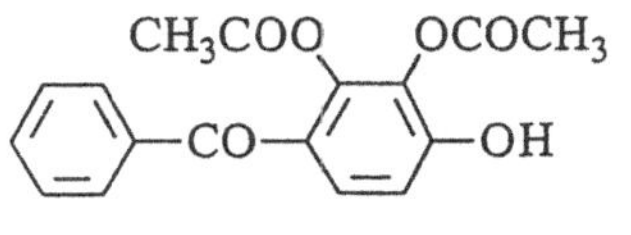

CH$_3$COO   OCOCH$_3$

〈_〉—CO—〈_〉—OH

Synthesis

-Obtained by action of acetic anhydride with 2,3,4-tri-
hydroxybenzophenone in the presence of lithium carbonate
in N,N-dimethylformamide at r.t. for 15 h under nitrogen
(12%) [802].

m.p. (NA); ^{1}H NMR [802].

[3,4-Bis(acetyloxy)-2-hydroxyphenyl]phenylmethanone

[177703-35-4] $C_{17}H_{14}O_6$ mol.wt. 314.29

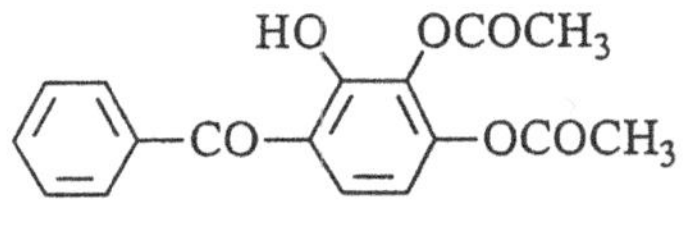

HO   OCOCH$_3$

〈_〉—CO—〈_〉—OCOCH$_3$

Synthesis

-Obtained by action of acetic anhydride with 2,3,4-tri-
hydroxybenzophenone in the presence of lithium
carbonate in N,N-dimethylformamide at r.t. for 15 h
under nitrogen (28%) [802].

m.p. (NA); ^{1}H NMR [802].

[3-(2-Butenyl)-4-hydroxyphenyl]phenylmethanone

[96825-03-5] $C_{17}H_{16}O_2$ mol.wt. 252.31

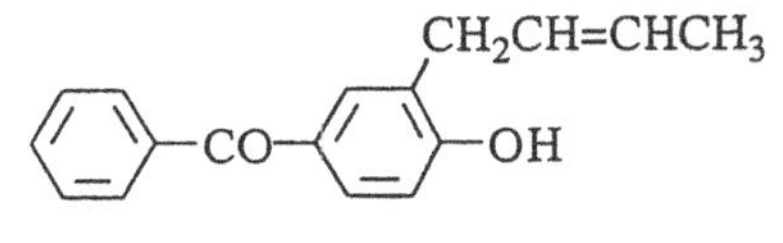

CH$_2$CH=CHCH$_3$

〈_〉—CO—〈_〉—OH

Synthesis

-Preparation by condensation of p-hydroxybenzophenone with 1,3-butadiene in the presence of
orthophosphoric acid in petroleum ether at 30-35°
for 24 h (77%) [30].

m.p. 104-106° [30]; ^{1}H NMR [30].

[4-Hydroxy-3-(1-methyl-2-propenyl)phenyl]phenylmethanone

[73720-57-7] $C_{17}H_{16}O_2$ mol.wt. 252.31

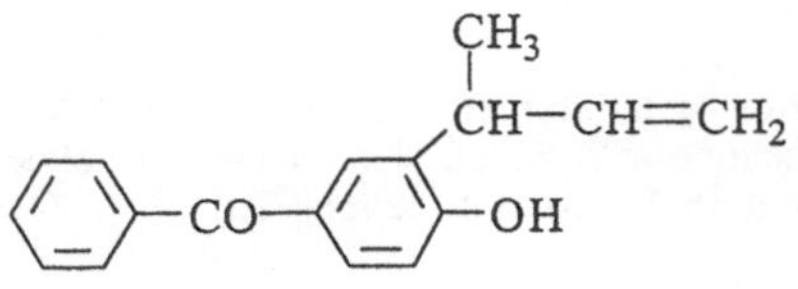

Synthesis

-Preparation by Claisen rearrangement of
  4-(γ-methallyloxy)benzophenone [935].
-Also refer to: [934].

m.p. and Spectra (NA).

[4-Hydroxy-3-(2-methyl-2-propenyl)phenyl]phenylmethanone

[112005-09-1] $C_{17}H_{16}O_2$ mol.wt. 252.31

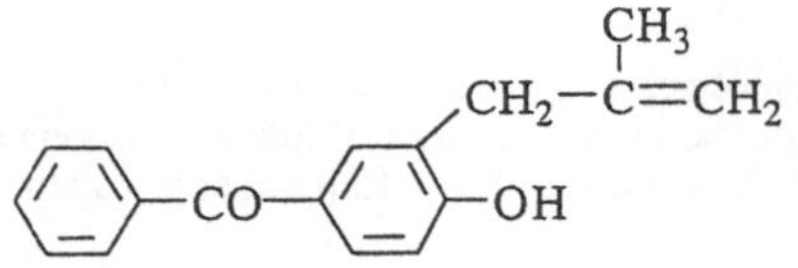

Synthesis

-Preparation by heating 4-(β-methallyloxy)-
  benzophenone in diethylaniline for 3.5 h between
  207 to 218° (73%) (Claisen rearrangement) [935].
-Also refer to: [934].

m.p.   131-133° [935]; ^{1}H NMR [935].

[2-Hydroxy-6-methoxy-3-(2-propenyl)phenyl]phenylmethanone

$C_{17}H_{16}O_3$ mol.wt. 268.31

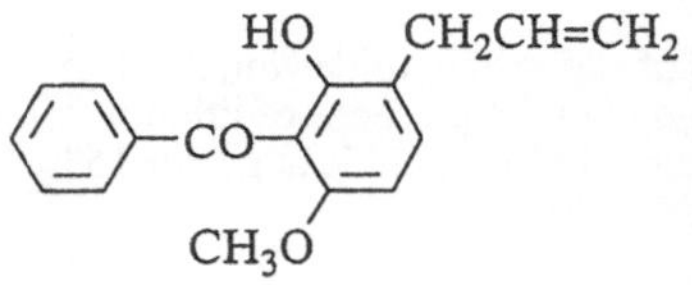

Syntheses

-Preparation by Claisen rearrangement of 2-allyloxy-
  6-methoxybenzophenone in boiling diethylaniline for
  40 min (40%) [1283].
-Also obtained (by-product) by partial methylation of
  3-allyl-2,6-dihydroxybenzophenone with dimethyl
sulfate in boiling benzene in the presence of potassium hydroxide for 10 h (16%) [1283].

b.p.$_3$ 192-194° [1283]; d_4^{20} = 0.888 [1283]; n_D^{20} = 1.5960 [1283];
TLC [1283];   Spectra (NA).

[6-Hydroxy-2-methoxy-3-(2-propenyl)phenyl]phenylmethanone

$C_{17}H_{16}O_3$ mol.wt. 268.31

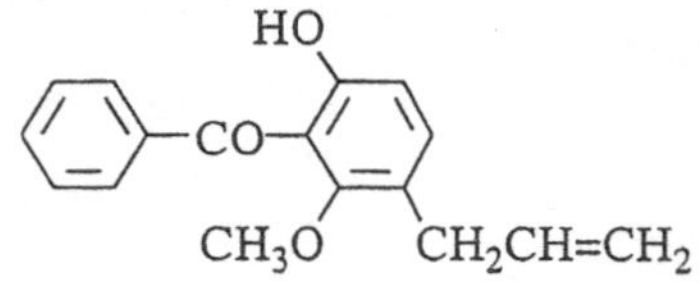

Syntheses

-Preparation by partial methylation of 3-allyl-2,6-di-
  hydroxybenzophenone with dimethyl sulfate in boiling
  benzene in the presence of potassium hydroxide for
  10 h (50%) [1283].
-Obtained (by-product) by Claisen rearrangement of
2-allyloxy-6-methoxybenzophenone in boiling diethylaniline for 40 min (10%) [1283].

yellow oil [1283];   b.p. and Spectra (NA);   TLC [1283].

[3-Bromo-2-hydroxy-6-methyl-5-(1-methylethyl)phenyl]phenylmethanone

[143815-17-2] $C_{17}H_{17}BrO_2$ mol.wt. 333.22

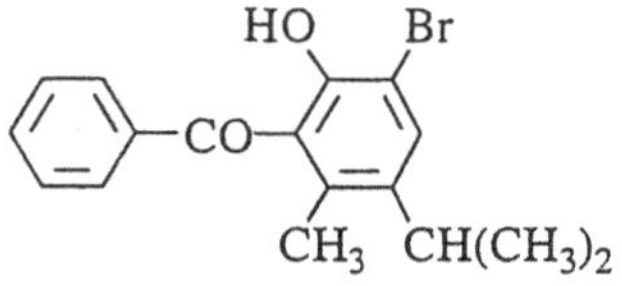

Synthesis

-Preparation by Fries rearrangement of 2-bromo-5-methyl-
4-isopropylphenyl benzoate with titanium tetrachloride
without solvent at 140° for 2 h (62%) [866].

m.p. 119-120° [866];
^{1}H NMR (Sadtler: standard n° 59400 M) [866],
IR (Sadtler: standard n° 86556 K) [866], UV [866], MS [866].

[3-Bromo-6-hydroxy-2-methyl-5-(1-methylethyl)phenyl]phenylmethanone

[16846-17-6] $C_{17}H_{17}BrO_2$ mol.wt. 333.22

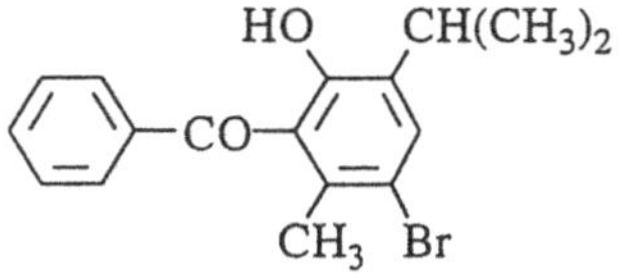

Synthesis

-Preparation by bromination of 2-hydroxy-6-methyl-3-iso-
propylbenzophenone [998], with bromine in chloroform at
r.t. for 6 h [360].

m.p. 129° [998], 128-129° [360], 74° [609]. One of the reported melting points is
obviously wrong.
^{1}H NMR [609], IR [68, 609, 998], UV [609]; pK_a [609]; polarographic study [1223].

[4-Bromo-2-hydroxy-6-methyl-3-(1-methylethyl)phenyl]phenylmethanone

[16846-13-2] $C_{17}H_{17}BrO_2$ mol.wt. 333.22

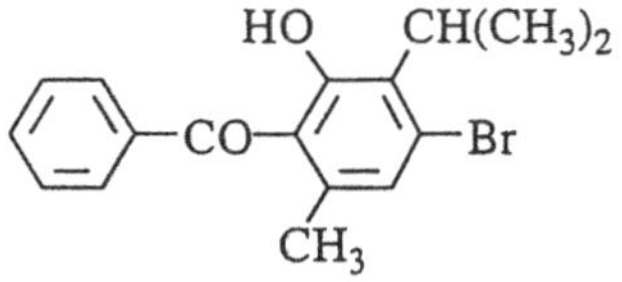

Synthesis

-Preparation by reaction of chromium trioxide with
6-bromo-4-methyl-2,3-diphenyl-7-isopropylbenzofuran
in boiling acetic acid, followed by saponification of the
keto ester so formed [998], by action of potassium
hydroxide in ethanol in a water bath for 1 h [360].

m.p. 91° [360, 998];   IR [68, 77].

[3-Chloro-5-(1,1-dimethylethyl)-2-hydroxyphenyl]phenylmethanone

$C_{17}H_{17}ClO_2$ mol.wt. 288.77

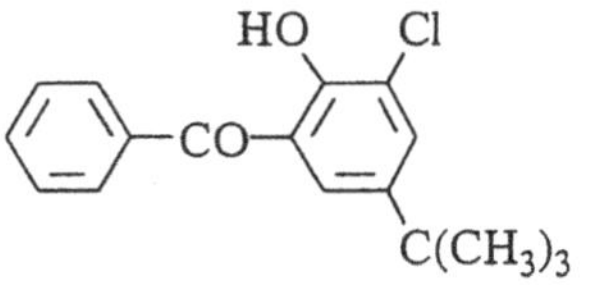

Synthesis

-Obtained (poor yields) by photo-Fries rearrangement of two
substituted phenyl esters* in benzene [743],
*4-tert-butyl-2-chlorophenyl benzoate (7%);
*4-tert-butyl-2,6-dichlorophenyl benzoate (11%). In this case,
one chlorine atom was eliminated.

m.p. 55-59° [743];   b.p.$_{0.2}$ 150° [743];   Spectra (NA).

[5-Chloro-3-(1,1-dimethylethyl)-2-hydroxyphenyl]phenylmethanone

[52196-47-1] $C_{17}H_{17}ClO_2$ mol.wt. 288.77

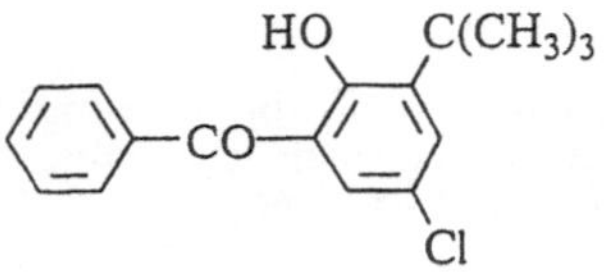
HO
C(CH3)3
CO
Cl

Synthesis

-Refer to: [768].

m.p. and Spectra (NA); pK_a [768].

[4-(Chloromethyl)-2-hydroxy-3-propylphenyl]phenylmethanone

[97582-40-6] $C_{17}H_{17}ClO_2$ mol.wt. 288.77

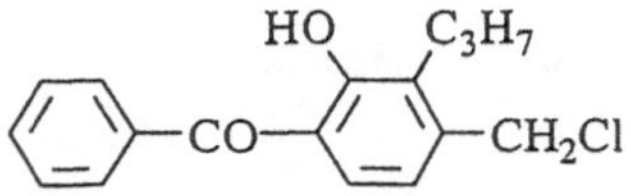
HO
C3H7
CO
CH2Cl

Synthesis

-Preparation by reaction of ethyl chloroformate with 4-(dimethylamino)methyl-2-hydroxy-3-propylbenzophenone in toluene cooled with an ice water bath for 2 h, then at r.t. for 16 h (55%) [353].

-Also refer to: [352, 1432].

m.p. 47-49° [353]; Spectra (NA); HPLC [353].

[5-(1,1-Dimethylethyl)-2-hydroxy-3-nitrophenyl]phenylmethanone

[85052-33-1] $C_{17}H_{17}NO_4$ mol.wt. 299.33

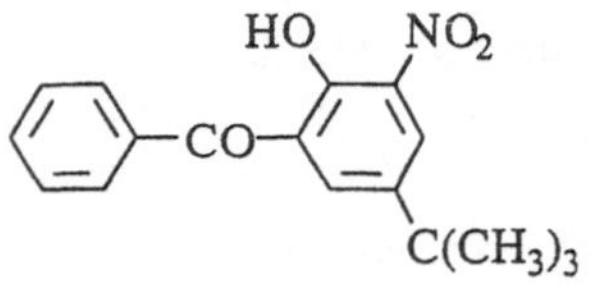
HO
NO2
CO
C(CH3)3

Synthesis

-Preparation by reaction of a concentrated nitric acid/ concentrated sulfuric acid mixture with 5-tert-butyl-2-hydroxybenzophenone in methylene chloride at 5° for 30 min (42%) [932].

m.p. (NA); ^{1}H NMR [932], MS [932].

[2-Hydroxy-6-methyl-3-(1-methylethyl)-4-nitrophenyl]phenylmethanone

$C_{17}H_{17}NO_4$ mol.wt. 299.33

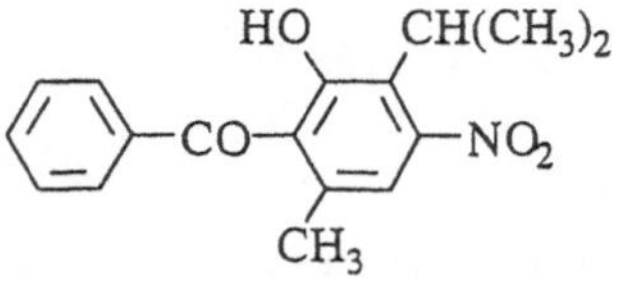
HO
CH(CH3)2
CO
NO2
CH3

Synthesis

-Refer to: [68, 72].

m.p. (NA); IR [68], UV [72].

[2-Hydroxy-6-methyl-3-(1-methylethyl)-5-nitrophenyl]phenylmethanone

$C_{17}H_{17}NO_4$ mol.wt. 299.33

Synthesis

-Refer to: [68].

m.p. (NA);   IR [68].

[3-(1,1-Dimethylethyl)-2-hydroxyphenyl]phenylmethanone

[24248-99-5] $C_{17}H_{18}O_2$ mol.wt. 254.33

Syntheses

-Preparation by oxidation of 7-tert-butyl-2,3-diphenyl-
benzofuran with chromium trioxide in acetic acid at 70° for
2 h, followed by alkaline hydrolysis of the keto ester
obtained (2-benzoyloxy-3-tert-butylbenzophenone) in boiling diluted ethanol for 15 min [76].
-Also obtained by photo-Fries rearrangement of 2-tert-butylphenyl benzoate in benzene (35%) [880].
-Also obtained by reaction between (o-tert-butylphenoxy)magnesium bromide complexed with
HMPT and benzaldehyde in refluxing benzene for 48 h (18%) [250].
-Also refer to: [917].

pale yellow viscous oil [880], m.p. 55° [76], 53° [250];   b.p.$_{0.01}$ 92° [880];
^{1}H NMR [250, 880], IR [76, 250, 880], MS [250].

[3-(1,1-Dimethylethyl)-4-hydroxyphenyl]phenylmethanone

[16928-03-3] $C_{17}H_{18}O_2$ mol.wt. 254.33

Syntheses

-Obtained by partial dealkylation of 3,5-di-tert-butyl-
4-hydroxybenzophenone by UV light irradiation in
cyclohexane (50%) [880].
                -Also obtained by photo-Fries rearrangement of 2-tert-
butylphenyl benzoate in benzene (20%) [880].
-Preparation by Friedel-Crafts acylation of o-tert-butylphenol with benzoyl chloride in ethylene
dichloride in the presence of titanium tetrachloride, first with ice cooling, then for 5 h at r.t. (44%)
[733].

m.p. 179-180° [880]; ^{1}H NMR [880], IR [880].

[4-(1,1-Dimethylethyl)-2-hydroxyphenyl]phenylmethanone

[39000-51-6] $C_{17}H_{18}O_2$ mol.wt. 254.33

Syntheses

-Obtained by Fries rearrangement in hydrofluoric acid,
*of m-tert-butylphenyl benzoate at 25° for 2 h (37%) [986];
*of 3,5-di-tert-butylphenyl benzoate at 55° for 4 h (10%)
[986].

m.p. 81-82° [986];   Spectra (NA).

[5-(1,1-Dimethylethyl)-2-hydroxyphenyl]phenylmethanone

[10425-05-5] $C_{17}H_{18}O_2$ mol.wt. 254.33

Syntheses

-Preparation by Fries rearrangement of p-tert-butylphenyl
  benzoate with,
*titanium tetrachloride without solvent at 140° for 15 min
  (47%) [864] or in nitromethane at r.t. for 7 days (29%)
  [864];
*aluminium chloride without solvent at 180° for 15 min (35%) [864].
-Preparation by demethylation of 5-tert-butyl-2-methoxybenzophenone in the presence of,
*in a refluxing mixture of 47% hydrobromic acid (2 vol) and 57% hydriodic acid (1 vol) in acetic
  acid (10 vol) (89%) [636];
*aluminium chloride in benzene at 50-60° for 5 h (63%) [775].
-Also obtained by UV light irradiation of two substituted phenyl esters in benzene: p-tert-
  butylphenyl benzoate (45%) [743], 2-chloro-4-tert-butylphenyl benzoate. In this last case, there is
  an elimination of chlorine atom (21%) [743].
-Also obtained by photo-Fries rearrangement of p-tert-butylphenyl benzoate in ethanol for 10 h or
  in ethyl ether [418].
-Preparation by reaction of benzotrichloride with p-tert-butylphenol in the presence of 30% aqueous
  sodium hydroxide at 75-80° for 30 min, and hydrolysis of the resulting ester as a side product by
  steam distillation (56-65%) [754, 1339].
-Preparation by treatment of o-hydroxybenzophenone at 120° with a mixture of isobutylene/
  nitrogen (1:1) or with tert-butyl chloride in the presence a macroreticular acid ion exchanger as
  catalyst (Wofatit OK 80) for 1 h (52%) [111].
-Also refer to: [421, 917, 932, 1303, 1442].

   m.p. 67-68° [775], 67° [743, 864], 63-65° [754], 52-53° [1339];
   b.p.$_{12}$ 195-197° [775], b.p.$_2$ 164-166° [754], b.p.$_{0.15}$ 125-130° [111];
   ^{1}H NMR [754, 1339], ^{13}C NMR [754], IR [743, 754, 864, 1339], UV [864, 1339],
   MS [1339];   gas chromatography study [1343];   TLC [377].

[2-Hydroxy-3-methyl-6-(1-methylethyl)phenyl]phenylmethanone

[33829-50-4] $C_{17}H_{18}O_2$ mol.wt. 254.33

Synthesis

-Refer to: [996].

m.p. (NA); ^{1}H NMR [996], IR [996], UV [996].

[2-Hydroxy-4-methyl-5-(1-methylethyl)phenyl]phenylmethanone

[108974-20-5] $C_{17}H_{18}O_2$ mol.wt. 254.33

Syntheses

-Preparation by Fries rearrangement of p-thymyl benzoate
  with titanium tetrachloride at 140° for 2 h (82%) [866].
-Claimed to be prepared by two methods:
*by reaction of 2-methoxy-4-methyl-5-isopropyl-

benzoyl chloride (SM1) with benzene in the presence of aluminium chloride at 80° for 3 h, then at r.t. for 20 h (33%) [1152];
*from 2-methoxy-4-methyl-5-isopropylbenzophenone (SM2) by heating with pyridinium chloride for 1 h [1152].
However, the structures of SM1 and SM2 were incorrect.
N.B.: All the results of reference [1152] were erroneous. Only the first route was correct. The ^{1}H NMR spectra confirms the above structure [866].

> m.p. 152°5 [1152] (this melting point is incompatible with a non-vicinal ortho-acylphenol structure), 48-49° [866]; b.p.$_{20}$ 282-287° [1152];
> ^{1}H NMR (Sadtler: standard n° 59398 M) [866],
> IR (Sadtler: standard n° 86554 K) [866], UV [866], MS [866].

[2-Hydroxy-6-methyl-3-(1-methylethyl)phenyl]phenylmethanone

[4072-16-6] $C_{17}H_{18}O_2$ mol.wt. 254.33

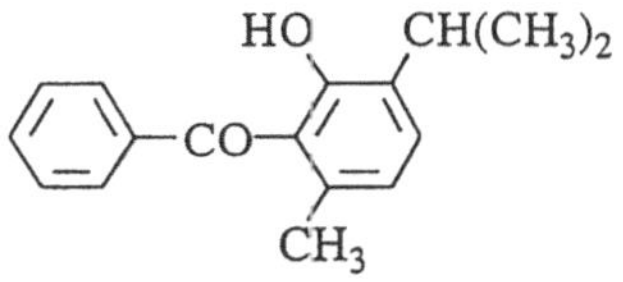

Syntheses

-Obtained by oxidation of 4-methyl-7-isopropyl-2,3-diphenylbenzofuran with chromium trioxide in refluxing acetic acid (82%) [65, 360], followed by saponification of the keto ester so formed [65, 74, 360].

-Obtained by refluxing mixture of benzoic acid/thymol/aluminium chloride for 12 h (76%) [1317] (no reproductive reaction).
-Also refer to: [75].

> m.p. 104° [609], 97° [67, 73, 74, 360, 998], 44° [1317]. One of the reported melting points is obviously wrong.
> ^{1}H NMR [609, 996], IR [68, 77, 609, 996, 1314, 1315, 1316], UV [609, 996]; polarographic study [1223]; pK_a [609].

[2-Hydroxy-6-methyl-4-(1-methylethyl)phenyl]phenylmethanone

$C_{17}H_{18}O_2$ mol.wt. 254.33

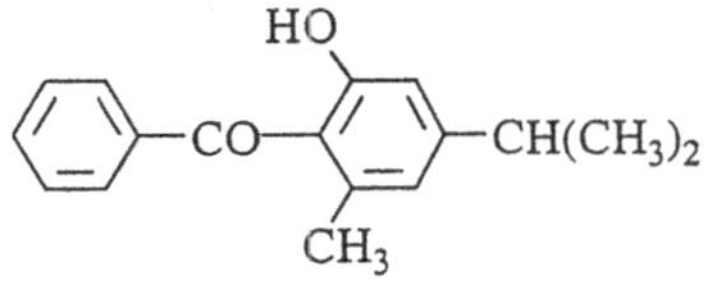

Synthesis

-Preparation (compound II-f) by saponification of 2-(benzoyloxy)-6-methyl-4-isopropylbenzophenone, itself obtained by chromic oxidation of 4-methyl-6-isopropyl-2,3-diphenylbenzofuran [75].

> m.p. 80-83° [75]; IR [75].

[4-Hydroxy-2-methyl-5-(1-methylethyl)phenyl]phenylmethanone

[28178-94-1] $C_{17}H_{18}O_2$ mol.wt. 254.33

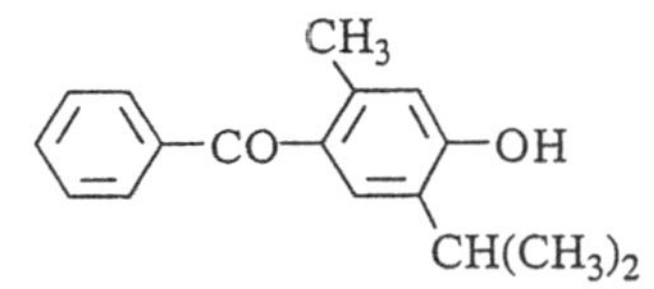

Syntheses

-Preparation by reaction of benzoyl chloride with thymol in the presence of aluminium chloride [360], (98%) [1018], in nitrobenzene at r.t. overnight (good yield) [1144].
-Preparation by demethylation of 4-methoxy-2-methyl-

5-isopropylbenzophenone with pyridinium chloride at 205-215° for 1.5 h (95%) [1148] or 15 min [239].
-Preparation by Fries rearrangement of thymyl benzoate with aluminium chloride in nitrobenzene at 60° for 5 h (70%) [1143] or at 50° in the same conditions (21%) [470].
-Also obtained by reaction of benzotrichloride with thymol in the presence of stannic chloride at 60-65° for 15 h (20%) [1018].
-Also obtained by saponification of 4-(benzoyloxy)-2-methyl-5-isopropylbenzophenone (SM) with sodium hydroxide in boiling ethanol [360]. SM was obtained by oxidation of 4-desylthymyl benzoate with chromium trioxide in boiling acetic acid.
-Also obtained (by-product) from p-thymyl benzoate by heating with aluminium chloride [1150].
-Also refer to: [795], [1230] (Japanese patent).

m.p.   154° [1148], 153° [239, 360, 1143, 1144], 152°5 [1150],
150-150°5 [1018], 138-144° [470];
Spectra (NA).

[4-Hydroxy-5-methyl-2-(1-methylethyl)phenyl]phenylmethanone

[99821-75-7]   C$_{17}$H$_{18}$O$_2$ mol.wt. 254.33

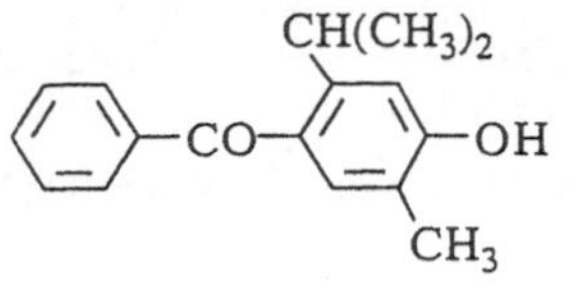

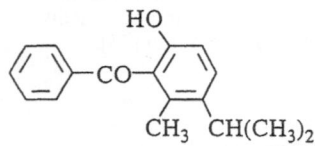

Syntheses

-Preparation by Fries rearrangement of 2-methyl-5-isopropyl-phenyl benzoate,
*in the presence of aluminium chloride in nitrobenzene at 60° for 5 h (60%) [1143];
*in the presence of Nafion-XR, a H$^+$-form ion exchange resin, at 150° for 4 h under nitrogen (31%) [1231].
-Also obtained by reaction of benzoyl chloride with carvacrol,
*in the presence of Nafion-XR at 150° for 4 h under nitrogen (20%) [1231];
*in the presence of aluminium chloride in nitrobenzene at r.t. for 48 h (4%) [647].
-Also refer to: [1230] (Japanese patent).

m.p.   172-173° [1143], 126° [647]. One of the reported melting points is obviously wrong.   Spectra (NA).

[6-Hydroxy-2-methyl-3-(1-methylethyl)phenyl]phenylmethanone

[108974-21-6]   C$_{17}$H$_{18}$O$_2$ mol.wt. 254.33

Syntheses

-Preparation from 3-bromo-2-hydroxy-6-methyl-5-iso-propylbenzophenone (SM1) by reductive removal of bromine with copper powder in caproic acid at 220° for 15 min (83%) [866]. SM1 was obtained from p-thymol by a three-step synthesis (bromination, esterification and Fries rearrangement with titanium tetrachloride).
-Also obtained (by-product) by Fries rearrangement of p-thymyl benzoate with titanium tetrachloride at 140° for 2 h (10%) [866].
-Claimed to be prepared by heating 6-methoxy-2-methyl-3-isopropylbenzophenone (SM2) with pyridinium chloride at reflux for 1.5 h (85%) [1152]. SM2 was obtained by Friedel-Crafts acylation of 3-methyl-4-isopropylanisole with benzoyl chloride in the presence of aluminium chloride in carbon disulfide at r.t. The structure of SM2 was erroneous. The true structure of SM2 must be the 2-methoxy-4-methyl-5-isopropylbenzophenone.
N.B.: All the results of reference [1152] were erroneous. Only the first route was correct. The ^{1}H NMR spectra confirms the above structure [866].

m.p. 142-143° [866];
yellow viscous oil [1152];   b.p.$_{14}$ 207° [1152]; $n_D^{25.5}$ = 1.5950 [1152];
^{1}H NMR (Sadtler: standard n° 59401 M) [866],
IR (Sadtler: standard n° 86557 K) [866], UV [866], MS [866].

[2-Hydroxy-5-(1-methylpropyl)phenyl]phenylmethanone

[59746-97-3]   C$_{17}$H$_{18}$O$_2$ mol.wt. 254.33

Synthesis

-Preparation by reaction of benzotrichloride with
4-sec-butylphenol in hydrofluoric acid in the presence
of water at -10°, then between 0 to -10° for 2 h, at r.t.
for 7 h and at 80° for 30 min into an autoclave (60%)
[393].

yellow oil [393];   b.p.$_1$ 150-155° [393];   Spectra (NA).

[4-Hydroxy-3-(1-methylpropyl)phenyl]phenylmethanone

[124979-07-3]   C$_{17}$H$_{18}$O$_2$ mol.wt. 254.33

Synthesis

-Preparation by Friedel-Crafts acylation of 2-sec-
butylphenol — 2-(1-methylpropyl)phenol — with
benzoyl chloride in ethylene dichloride in the presence
of titanium tetrachloride, first at 0°, then at r.t. [733].

m.p. 95-96° [733];   Spectra (NA).

(4-Butoxy-2-hydroxyphenyl)phenylmethanone

[15131-43-8]   C$_{17}$H$_{18}$O$_3$ mol.wt. 270.33

Syntheses

-Preparation by reaction of n-butyl bromide with
resbenzophenone,
*in the presence of sodium hydroxide in dilute ethanol in a
water bath (40%) [805];
*in the presence of potassium carbonate in cyclohexanone at 145° for 6 h (72%) [1260], (48%)
[1259].
-Preparation by alkylation of resbenzophenone with a butyl halide [42].
-Preparation by Friedel-Crafts acylation of 1,3-di-n-butoxybenzene with benzoyl chloride,
*in the presence of a zinc chloride/aluminium chloride mixture, first between 0 to 5°, then at 10° for
1 h and at 65° for 6 h [479];
*in the presence of aluminium chloride in chlorobenzene at 90° [43].
-Preparation by partial dealkylation of 4-butoxy-2-methoxybenzophenone or 2,4-dibutoxy-
benzophenone with aluminium chloride in chlorobenzene at 80-100° [602].
-Also refer to: [381, 384, 856, 913, 1067].

m.p. 56-57° [805], 52-53° [1259], 50-53° [1260];
UV [1067];   TLC [377];   pK_a [768].

(5-Butoxy-2-hydroxyphenyl)phenylmethanone

$C_{17}H_{18}O_3$ mol.wt. 270.33

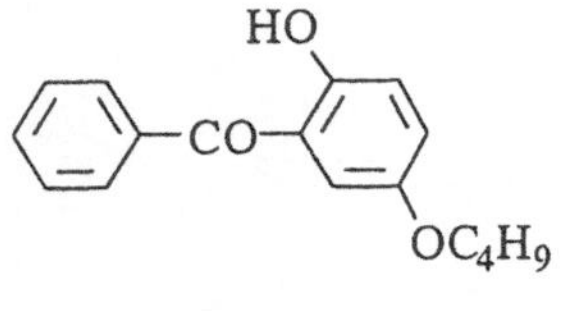

Synthesis

-Obtained by reaction of n-butyl bromide with 2,5-di-
hydroxybenzophenone in the presence of sodium hydroxide
in dilute ethanol at 80-85° for 35 h (21%) [804].

m.p. 42° [804];   Spectra (NA).

(2-Hydroxy-3-methyl-4-propoxyphenyl)phenylmethanone

[172479-20-8] $C_{17}H_{18}O_3$ mol.wt. 270.33

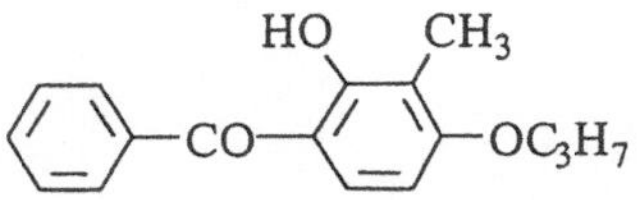

Synthesis

-Prepared by standard techniques [1065].

m.p. (NA);   UV [1065].

(2-Hydroxy-6-methyl-4-propoxyphenyl)phenylmethanone

[172479-19-5] $C_{17}H_{18}O_3$ mol.wt. 270.33

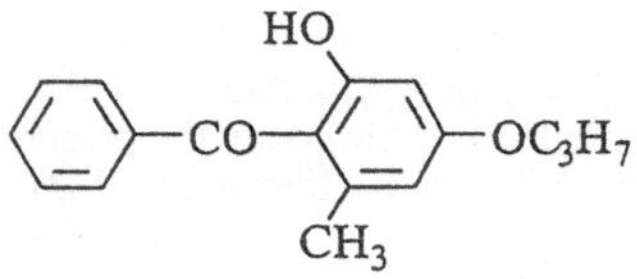

Synthesis

-Prepared by standard techniques [1065].

m.p. (NA);   UV [1065].

[2-Hydroxy-4-(1-methylpropoxy)phenyl]phenylmethanone

$C_{17}H_{18}O_3$ mol.wt. 270.33

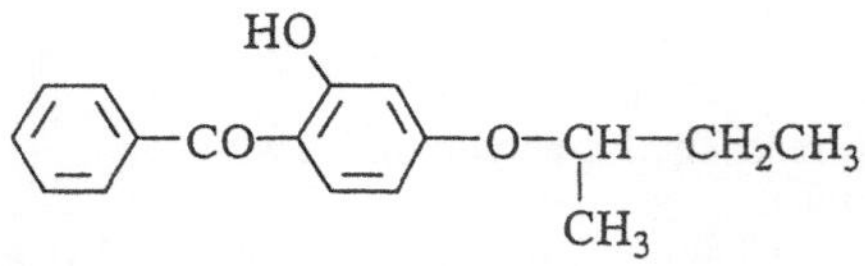

Synthesis

-Preparation by reaction of sec-butyl bromide
with resbenzophenone in the presence of
potassium hydroxide in dilute ethanol at 80-
90° for 20 h (29%) [803].

m.p. 41° [803];   Spectra (NA).

[2-Hydroxy-4-(2-methylpropoxy)phenyl]phenylmethanone

$C_{17}H_{18}O_3$ mol.wt. 270.33

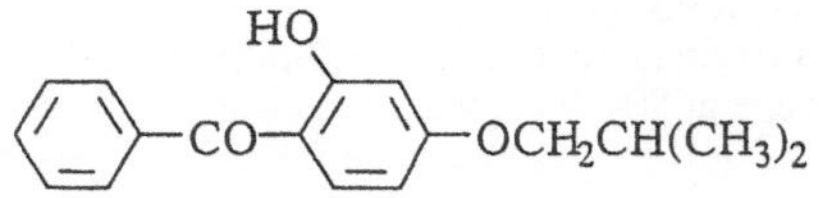

Synthesis

-Preparation by reaction of isobutyl bromide with
resbenzophenone in the presence of potassium
hydroxide in dilute ethanol at 80-92° for 18 h
(23%) [803].

m.p. 100-100°5 [803];   Spectra (NA).

[3-Amino-5-(1,1-dimethylethyl)-2-hydroxyphenyl]phenylmethanone

[85052-51-3] $C_{17}H_{19}NO_2$ mol.wt. 269.34

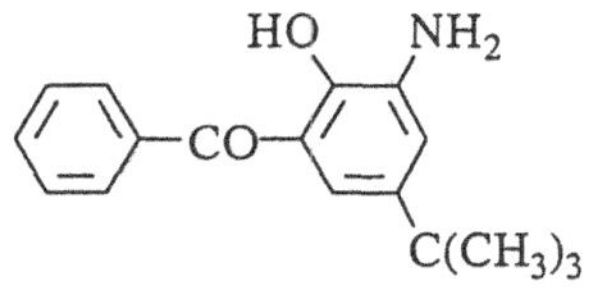

Synthesis

-Preparation by reduction of 5-tert-butyl-2-hydroxy-3-nitro-benzophenone with titanium trichloride in a benzene/tetrahydrofuran mixture for 2 h at r.t. (34%) [932].

m.p. 123-124° [932];
^{1}H NMR [932], IR [932], MS [932]; TLC [932].

[3-Amino-5-(1,1-dimethylethyl)-2-hydroxyphenyl]phenylmethanone *(Hydrochloride)*

[85069-31-4] $C_{17}H_{19}NO_2,HCl$ mol.wt. 305.81

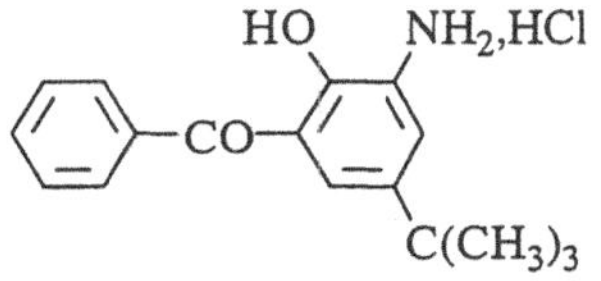

Synthesis

-Obtained by treatment of 3-amino-5-tert-butyl-2-hydroxy-benzophenone with 2 N hydrochloric acid [932].

m.p. and Spectra (NA).

[4-Hydroxy-3-(3-methyl-2-butenyl)phenyl]phenylmethanone

[63565-02-6] $C_{18}H_{18}O_2$ mol.wt. 266.34

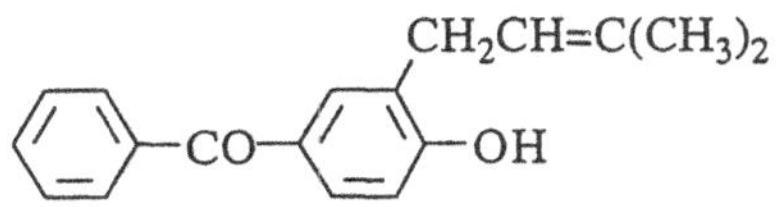

Synthesis

-Obtained (poor yield) by reaction of prenyl bromide with p-hydroxybenzophenone,
*in the presence of sodium methoxide in refluxing methanol for 4 h (8%) [875];
*in the presence of silver oxide in dioxane at r.t. for 2 h (13%) [874].

m.p. 77° [875], 76-77° [874];
^{1}H NMR [875], IR [875], UV [875].

[2-Hydroxy-4-[(3-methyl-2-butenyl)oxy]phenyl]phenylmethanone

[63564-99-8] $C_{18}H_{18}O_3$ mol.wt. 282.34

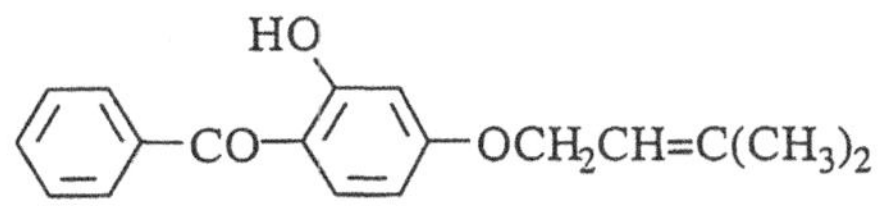

Synthesis

-Obtained by reaction of prenyl bromide with resbenzophenone,
*in the presence of potassium carbonate in refluxing acetone for 4 h (30%) [875];
*in the presence of sodium methoxide in refluxing methanol for 4 h [875].

m.p. 82° [875]; ^{1}H NMR [875], IR [875], UV [875].

[2-(1,1-Dimethylethyl)-4-hydroxy-6-methylphenyl]phenylmethanone

[133721-73-0] $C_{18}H_{20}O_2$ mol.wt. 268.36

Synthesis

-Obtained by treating a solution of 2-(tert-butyl)-4-[[(tert-butyl)dimethylsilyl]oxy]-6-methylbenzophenone in ethanol with 2 N aqueous hydrochloric acid for 1 h at 75° in a sealed tube (94%) [992].

m.p. 150-152° [992]; ^{1}H NMR [992], IR [992], MS [992].

[3-(1,1-Dimethylethyl)-2-hydroxy-5-methylphenyl]phenylmethanone

[52196-46-0] $C_{18}H_{20}O_2$ mol.wt. 268.36

Synthesis

-The reaction of benzoyl chloride with a pentane solution of [AlMe(dbmp)$_2$] leads to acylation of one of the dbmp ligands and affords [AlMe(dbmp)(bhmbp)]. Hydrolysis of this complex with a saturated solution of ammonium chloride gave the attempted ketone (75%) [1082].

N.B.: Hdbmp = 2,6-di-tert-butyl-4-methylphenol;
Hbhmbp = 3-tert-butyl-2-hydroxy-5-methylbenzophenone.

m.p. >240° [1082];
^{1}H NMR [1082], ^{13}C NMR [1082]; IR [1082], MS [1082];   pK_a [768].

[3-(1,1-Dimethylethyl)-2-hydroxy-6-methylphenyl]phenylmethanone

[14963-84-9] $C_{18}H_{20}O_2$ mol.wt. 268.36

Synthesis

-Preparation by saponification of 2-(benzoyloxy)-3-tert-butyl-6-methylbenzophenone (SM) with sodium hydroxide in boiling ethanol for 2 h. SM was obtained by oxidation of 7-tert-butyl-4-methyl-2,3-diphenylbenzofuran with chromium trioxide in acetic acid at 60° (m.p. 127°) [75].
-Also refer to: [73, 1223].

m.p. 125-126° [609], 77° [75]. One of the reported melting points is obviously wrong.
^{1}H NMR [609], IR [77, 609], UV [609];   pK_a [609];   polarographic study [1223].

[5-(1,1-dimethylpropyl)-2-hydroxyphenyl]phenylmethanone

[110701-33-2] $C_{18}H_{20}O_2$ mol.wt. 268.36

Synthesis

-Preparation by reaction of benzotrichloride with p-tert-pentylphenol in the presence of 30% aqueous sodium hydroxide at 75-80° for 30 min, and hydrolysis of the resulting ester as a side product (43%) [754].

m.p. 37-38°5 [754]; b.p.$_2$ 164-167° [754];
^{1}H NMR [754], ^{13}C NMR [754], IR [754].

[2-Hydroxy-5,6-dimethyl-3-(1-methylethyl)phenyl]phenylmethanone

[109252-33-7] C$_{18}$H$_{20}$O$_2$ mol.wt. 268.36

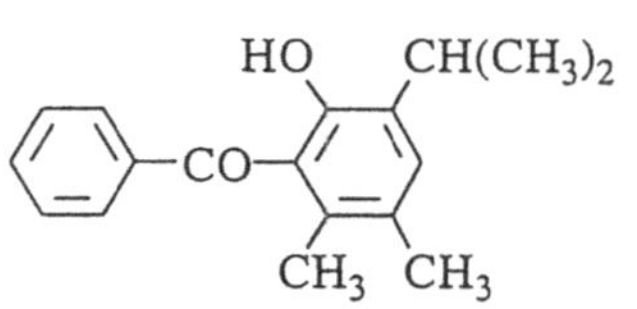

Synthesis

-Obtained by Friedel-Crafts acylation of 4-methylthymol
(4,5-dimethyl-2-isopropylphenol) with benzoyl chloride in
carbon disulfide in the presence of aluminium chloride at r.t.
for 2 days (16%) [1151].

b.p.$_{0.15}$ 148-149° [1151]; m.p. 89-90° [609];
^{1}H NMR [609], IR [609], UV [609];
polarographic study [1223]; pK_a [609].

[5-(1,1-Dimethylethyl)-2-hydroxy-4-methoxyphenyl]phenylmethanone

C$_{18}$H$_{20}$O$_3$ mol.wt. 284.36

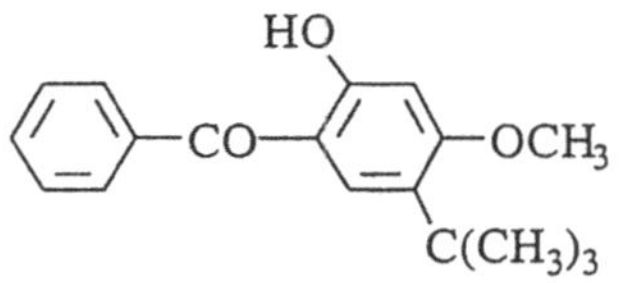

Synthesis

-Preparation by selective methylation of 5-tert-butyl-2,4-di-
hydroxybenzophenone with dimethyl sulfate in refluxing
methyl ethyl ketone for 7 h in the presence of potassium
carbonate (81%) [587].

m.p. 93° [587, 588]; Spectra (NA).

[2-Hydroxy-4-(1-methylbutoxy)phenyl]phenylmethanone

C$_{18}$H$_{20}$O$_3$ mol.wt. 284.36

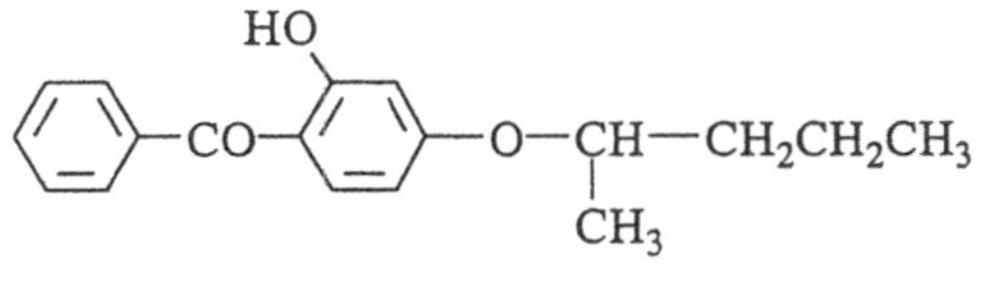

Synthesis

-Obtained by reaction of sec-amyl
bromide with resbenzophenone in the
presence of potassium hydroxide in
dilute ethanol at 80-90° for 16 h (17%)
[803].

m.p. 70-72° [803]; Spectra (NA).

[2-Hydroxy-4-(3-methylbutoxy)phenyl]phenylmethanone

[36130-62-8] C$_{18}$H$_{20}$O$_3$ mol.wt. 284.36

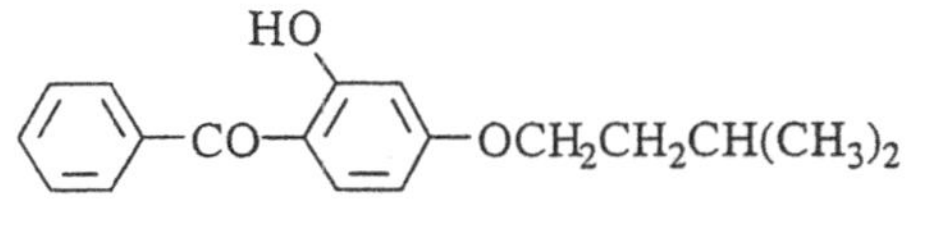

Syntheses

-Preparation by Friedel-Crafts acylation of
1,3-diisoamyloxybenzene with benzoyl
chloride in the presence of a zinc chloride/
aluminium chloride mixture, first between

0 to 5°, then at 10° for 1 h and at 65° for 6 h [479].

-Preparation by reaction of isoamyl chloride (1-chloro-3-methylbutane) with resbenzophenone (90%) [387].

     m.p. 38-40° [387];   Spectra (NA).

[2-Hydroxy-4-(pentyloxy)phenyl]phenylmethanone

[83937-21-7] $C_{18}H_{20}O_3$ mol.wt. 284.36

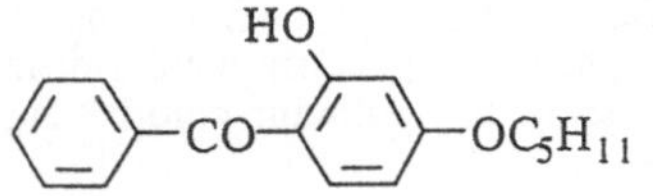

Synthesis

-Preparation by reaction of n-amyl bromide with resbenzophenone,
*in the presence of potassium hydroxide in dilute ethanol at 80-90° for 18 h (25%) [803];
*in the presence of potassium carbonate in refluxing acetone for 20 h [150].
-Also refer to: [79] (Japanese patent).

     m.p. 49° [803], 45° [150];   Spectra (NA).

[2-Hydroxy-4,6-bis(methoxymethoxy)-3-methylphenyl]phenylmethanone

[74627-90-0] $C_{18}H_{20}O_6$ mol.wt. 332.35

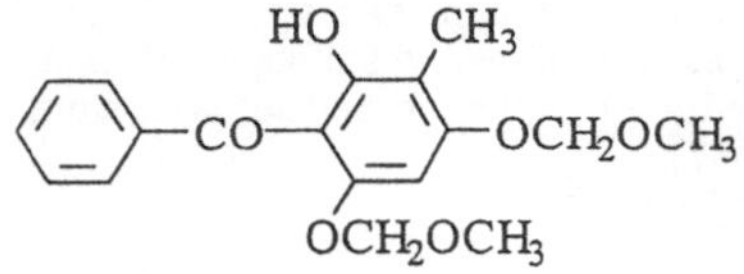

Synthesis

-Preparation by alkylation of 2,4,6-trihydroxy-3-methylbenzophenone with chloromethyl methyl ether in acetone in the presence of potassium carbonate for 1 h (55%) [57].

     yellow oil [57];   b.p. (NA); ^{1}H NMR [57], MS [57];   TLC [57].

[3-(Aminomethyl)-5-(1,1-dimethylethyl)-2-hydroxyphenyl]phenylmethanone

[75060-99-0] $C_{18}H_{21}NO_2$ mol.wt. 283.37

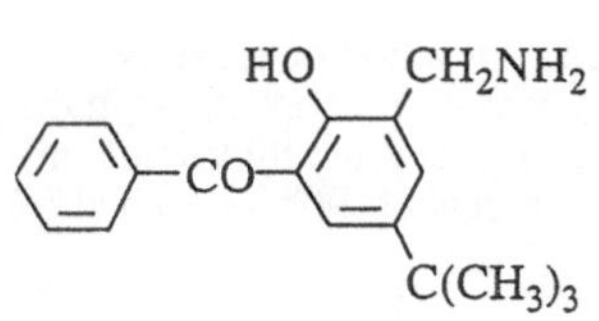

Synthesis

-Refer to: [636]; look at the hydrochloride below.

m.p. and Spectra (NA).

[3-(Aminomethyl)-5-(1,1-dimethylethyl)-2-hydroxyphenyl]phenylmethanone
(Hydrochloride)

[75060-64-9] $C_{18}H_{21}NO_2,HCl$ mol.wt. 319.83

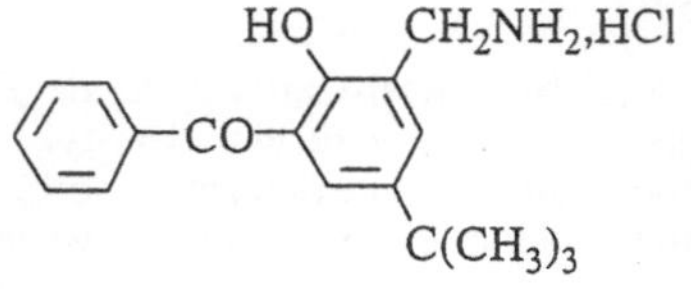

Synthesis

-Preparation by reaction of concentrated hydrochloric acid with 2-benzoyl-4-tert-butyl-6-(N-chloroacetyl-aminomethyl)phenol in refluxing ethanol for 20 h (92%) [636].

     m.p. 227-228° [636];   Spectra (NA).

[4-(3-Bromophenoxy)-2-hydroxyphenyl]phenylmethanone

[35698-51-2] $C_{19}H_{13}BrO_3$ mol.wt. 369.21

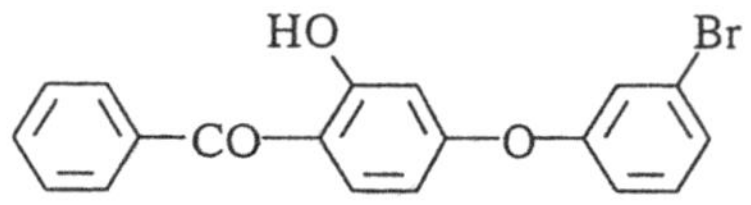

Synthesis

-Refer to: [101] (compound 9).

m.p. (NA);   UV [101].

(2-Hydroxy-4-phenoxyphenyl)phenylmethanone

[35698-39-6] $C_{19}H_{14}O_3$ mol.wt. 290.32

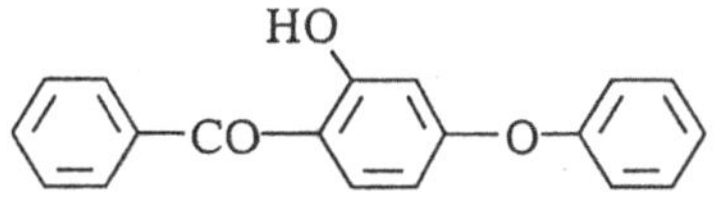

Synthesis

-Preparation by reaction of benzoyl chloride with
3-methoxydiphenyl ether in chlorobenzene in the
presence of aluminium chloride, first at r.t., then at
90-95° for 4 h (75%) [101].

m.p. 64-65° [101];   UV [101].

[3-(1-Hexynyl)-4-hydroxyphenyl]phenylmethanone

[183589-20-0] $C_{19}H_{18}O_2$ mol.wt. 278.35

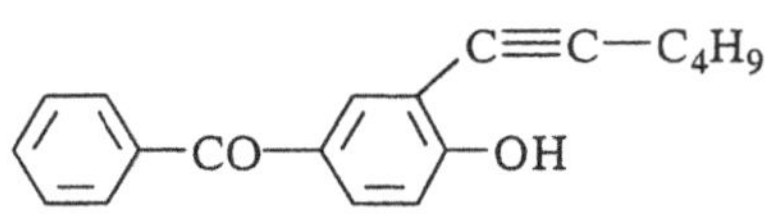

Synthesis

-Obtained by treatment of p-benzoyl-o-(hex-1-
ynyl)phenyl acetate in acetone in the presence of
2 N hydrochloric acid at 60° for 20 h (30%) [60].

oil [60];   b.p. (NA);   Spectra (NA).

(3-Cyclohexyl-4-hydroxyphenyl)phenylmethanone

[23299-02-7] $C_{19}H_{20}O_2$ mol.wt. 280.37

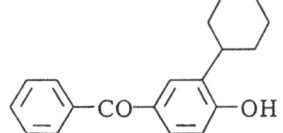

Synthesis

-Preparation by Friedel-Crafts acylation of 2-cyclohexyl-
phenol with benzoyl chloride in ethylene dichloride in the
presence of titanium tetrachloride, first at 0°, then at r.t.
[733].

m.p. 199-200° [733];   Spectra (NA).

(5-Cyclohexyl-2-hydroxyphenyl)phenylmethanone

[3097-56-1] $C_{19}H_{20}O_2$ mol.wt. 280.37

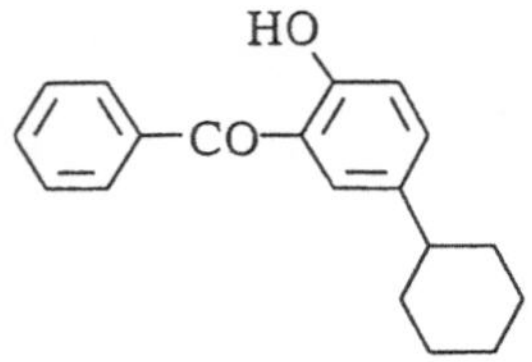

Synthesis

-Preparation by reaction of benzotrichloride with 4-cyclo-hexylphenol in hydrofluoric acid in the presence of water at -10°, then between 0 to -10° for 2 h, at r.t. for 7 h and at 80° for 30 min into an autoclave (50%) [393].

yellow oil [393];   b.p._1 175-180° [393];   Spectra (NA).

[3-(2-Butenyl)-2-hydroxy-4,6-dimethoxyphenyl]phenylmethanone

[96836-14-5] $C_{19}H_{20}O_4$ mol.wt. 312.37

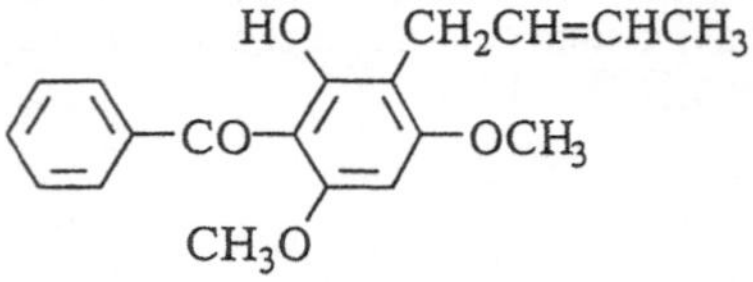

Synthesis

-Preparation by reaction of 2-hydroxy-4,6-di-methoxybenzophenone with 1,3-butadiene in the presence of orthophosphoric acid (70%) [30].

m.p.   93-95° [30]; ^{1}H NMR [30], IR [30].

[3-(1,1-Dimethylethyl)-2-hydroxy-5,6-dimethylphenyl]phenylmethanone

[14963-88-3] $C_{19}H_{22}O_2$ mol.wt. 282.38

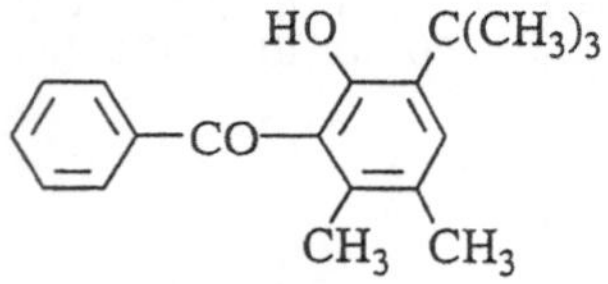

Synthesis

-Preparation by saponification of 2-(benzoyloxy)-3-tert-butyl-5,6-dimethylbenzophenone (SM) with sodium hydroxide in boiling ethanol for 2 h. SM was obtained by oxidation of 7-tert-butyl-4,5-dimethyl-2,3-diphenyl-benzofuran with chromium trioxide in acetic acid

at 60° (m.p. 156-157°) [75].
-Also refer to: [77].

m.p.   91-92° [75], 89-90° [609];
^{1}H NMR [609], IR [77], UV [609];
pK_a [609];   polarographic study [1223].

(2-Hydroxy-6-methyl-3-pentylphenyl)phenylmethanone

[26940-71-6] $C_{19}H_{22}O_2$ mol.wt. 282.38

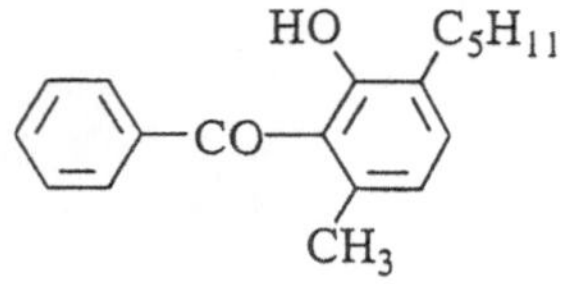

Synthesis

-Refer to: [73, 1223].

m.p. 74° [609];
^{1}H NMR [609], IR [609], UV [609];
pK_a [609]; polarographic study [1223].

[4-Hydroxy-3,5-bis(1-methylethyl)phenyl]phenylmethanone

[738-15-8] $C_{19}H_{22}O_2$ mol.wt. 282.38

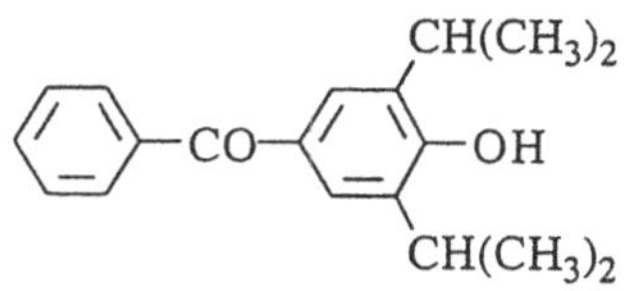

CH(CH$_3$)$_2$
OH
CH(CH$_3$)$_2$

Synthesis

-Obtained by photo-Fries rearrangement of 2,6-di-
isopropylphenyl benzoate (SM) in isopropyl alcohol
(56%) [102, 419], of SM absorbed on a silica gel-pentane
(40%) or on dry silica gel (70%) [102].

m.p. 113°5-114° [419] (a phase change occurred at 98°5); Spectra (NA).

[4-(Hexyloxy)-2-hydroxyphenyl]phenylmethanone

[3293-97-8] $C_{19}H_{22}O_3$ mol.wt. 298.38

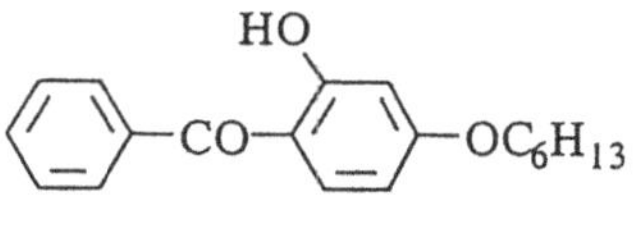

HO
CO OC$_6$H$_{13}$

Synthesis

-Obtained by reaction of n-hexyl bromide with resbenzophenone,
*in the presence of sodium hydroxide in dilute ethanol in a
water bath for 18 to 23 h (15%) [805];
*in the presence of potassium carbonate in refluxing acetone for 20 h [150].

m.p. 55°5 [805], 52° [150]; UV [1156].

[5-(Benzoyloxy)-3,4-dichloro-2-hydroxyphenyl]phenylmethanone

$C_{20}H_{12}Cl_2O_4$ mol.wt. 387.22

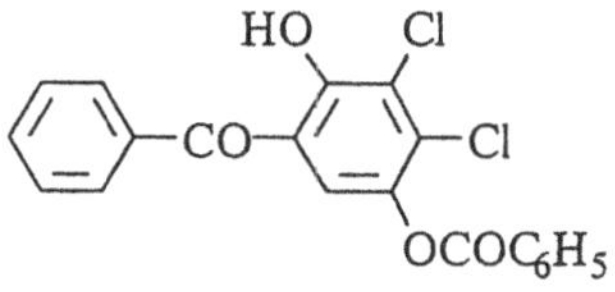

HO Cl
CO Cl
OCOC$_6$H$_5$

Synthesis

-Obtained by reaction of 2,3-dichloro-1,4-benzoquinone with
benzaldehyde in the presence of benzoyl peroxide at 80° or
in the absence of benzoyl peroxide at 155° [745].

m.p. and Spectra (NA).

[2,3-Dichloro-4-hydroxy-5-(phenylmethoxy)phenyl]phenylmethanone

[103843-60-3] $C_{20}H_{14}Cl_2O_3$ mol.wt. 373.23

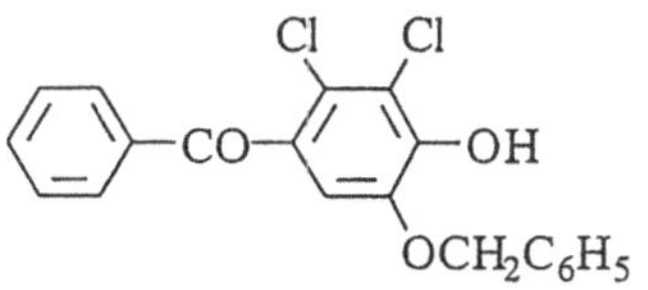

Cl Cl
CO OH
OCH$_2$C$_6$H$_5$

Synthesis

-Preparation by reaction of benzyl bromide with 2,3-di-
chloro-4,5-dihydroxybenzophenone in the presence of
sodium hydride in N,N-dimethylformamide at r.t. for
10-15 min (77%) [634, 635].

m.p. 171-173° [634, 635];
^{1}H NMR [634, 635], IR [634, 635].

[2,3-Dichloro-5-hydroxy-4-(phenylmethoxy)phenyl]phenylmethanone

[103843-65-8] $C_{20}H_{14}Cl_2O_3$ mol.wt. 373.23

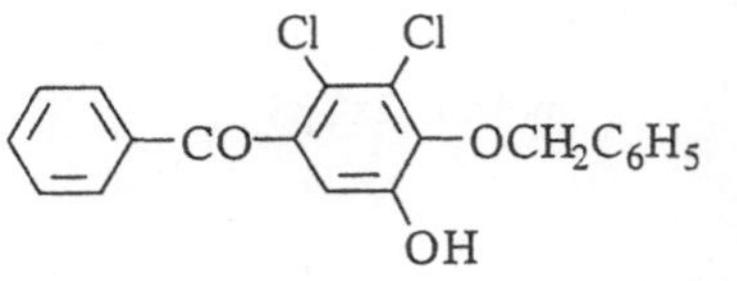

Synthesis

-Obtained by reaction of benzyl bromide with 2,3-di-
chloro-4,5-dihydroxybenzophenone in the presence of
sodium hydride in N,N-dimethylformamide at 100°
for 2 h (34%) [634].
-Also refer to: [635].

m.p. 111-112° [634]; ^{1}H NMR [634], IR [634].

[2-(Benzoyloxy)-4-hydroxyphenyl]phenylmethanone

$C_{20}H_{14}O_4$ mol.wt. 318.33

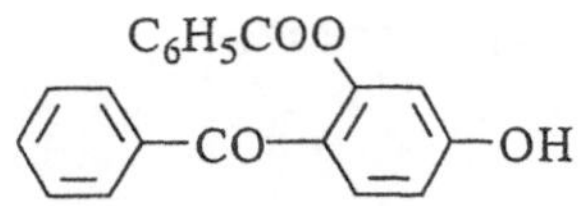

Synthesis

-Preparation from 2,4-dihydroxybenzophenone as starting
material *via* the 2-hydroxy-4-methoxymethoxy-
benzophenone and 2-benzoyloxy-4-methoxymethoxy-
benzophenone [593].

m.p. and Spectra (NA).

[3-(Benzoyloxy)-2-hydroxyphenyl]phenylmethanone

[97971-72-7] $C_{20}H_{14}O_4$ mol.wt. 318.33

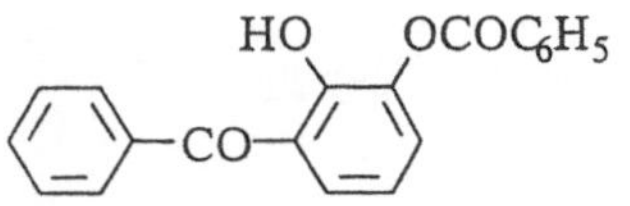

Synthesis

-Obtained by photo-Fries rearrangement of 1,2-di-
(benzoyloxy)benzene in benzene for 8 h under nitrogen
(15%) [1254].

m.p. 74° [1254]; ^{1}H NMR [1254].

[3-(Benzoyloxy)-4-hydroxyphenyl]phenylmethanone

[76346-15-1] $C_{20}H_{14}O_4$ mol.wt. 318.33

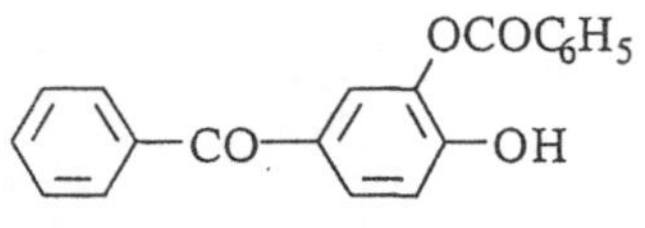

Synthesis

-Obtained (poor yield) by reaction of benzoyl peroxide
with p-hydroxybenzophenone [562] in refluxing
chloroform for 16 h (9%) [1211].

m.p. 109-110° [1211];   Spectra (NA);   TLC [1211].

[4-(Benzoyloxy)-2-hydroxyphenyl]phenylmethanone

[18803-25-3]   C$_{20}$H$_{14}$O$_4$ mol.wt. 318.33

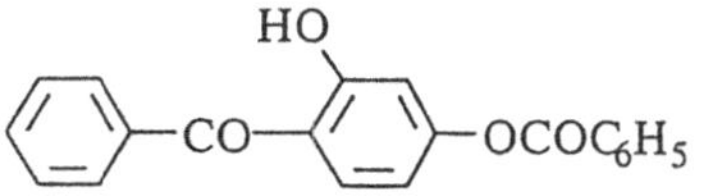

Syntheses

-Preparation by photo-Fries rearrangement of resorcinol dibenzoate in benzene for 8 h under nitrogen (35%) [1254].
-Also obtained by reaction of benzoyl chloride,
*with resbenzophenone [281], in the presence of sodium hydroxide in refluxing ethanol/ethyl ether mixture for 6 h (22%) [558];
*with resorcinol [851].

m.p. 94°5-95°5 [549], 93°5-94°5 [558], 90° [281], 84° [608], 77-80° [1254];
^{1}H NMR [610], UV [610]; pK_a [608]; TLC [377].

[5-(Benzoyloxy)-2-hydroxyphenyl]phenylmethanone

[97971-74-9]   C$_{20}$H$_{14}$O$_4$ mol.wt. 318.33

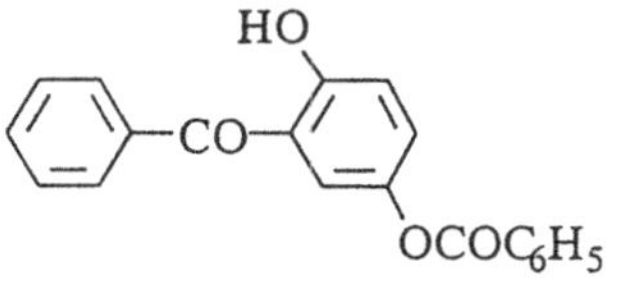

Syntheses

-Obtained by Fries rearrangement,
*of quinol monobenzoate in the presence of aluminium chloride (3.3 mol) at 140° for 1 h (14%) [47];
*of quinol dibenzoate in the presence of aluminium chloride (2 mol) at 190-200° for 1.5 h [338] or (3.3 mol or 5.5 mol of catalyst) at 140° for 1 h (33%) [47];
*of p-methoxyphenyl benzoate in the presence of aluminium chloride (3.3 mol) at 130-132° for 1 h or at 150-155° for 1 h (<9%) [48].
-Preparation by photo-Fries rearrangement of quinol dibenzoate in benzene for 8 h under nitrogen (30%) [1254].

m.p. 213-214° [1254], 96° [338], 94-95° [47]. There is a discrepancy between the different melting points indicated in literature. 94-96° are more likely, due to chelation.
^{1}H NMR [1254].

Phenyl 5-benzoyl-2-hydroxybenzoate *5-Benzoyl-2-hydroxybenzoic acid phenyl ester*

[124208-60-2]   C$_{20}$H$_{14}$O$_4$ mol.wt. 318.33

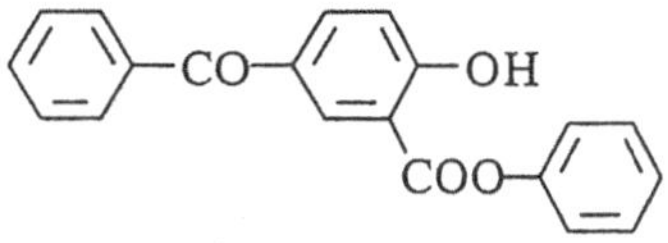

Synthesis

-Obtained by Fries rearrangement of phenyl 2-benzoyl-oxybenzoate with aluminium chloride without solvent at 142-145° for 2 h (31%) or in nitrobenzene as a solvent at r.t. for 2 days (17%) [171].

m.p. 82-84° [171]; Spectra (NA).

[2-Hydroxy-4-[(4-nitrophenyl)methoxy]phenyl]phenylmethanone

[36419-36-0] $C_{20}H_{15}NO_5$ mol.wt. 349.34

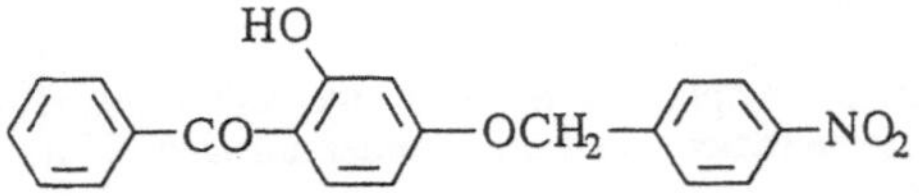

Synthesis

-Refer to: [1067].

m.p. 172° [1067];   UV [1067].

[2-Hydroxy-4-(methylphenoxy)phenyl]phenylmethanone

[35698-46-5] $C_{20}H_{16}O_3$ mol.wt. 304.35

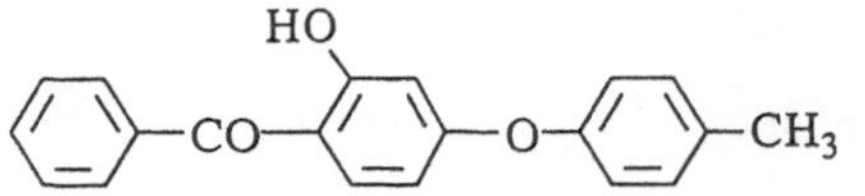

Synthesis

-Preparation by reaction of benzotrichloride with
 3-hydroxy-4'-methyldiphenyl ether in solution
 of 2.5 N sodium hydroxide in the presence of
 potassium iodide at 80° [101].

m.p.   82-85° [101];   UV [101].

[2-Hydroxy-4-(phenylmethoxy)phenyl]phenylmethanone

[6079-76-1] $C_{20}H_{16}O_3$ mol.wt. 304.35

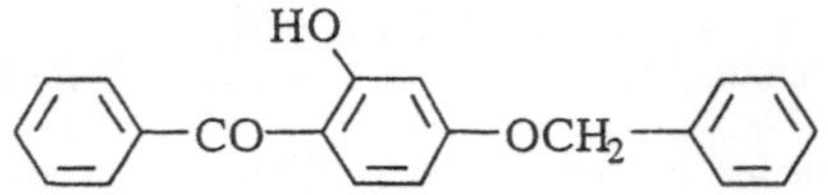

Synthesis

-Preparation by reaction of benzyl chloride with
 2,4-dihydroxybenzophenone
 (resbenzophenone),
*in the presence of potassium carbonate in
refluxing acetone [150], (34%) [954];
*in the presence of potassium carbonate and potassium iodide in N,N-dimethylformamide at 100°
for 1.5 h (95%) [387].
-Also refer to: [615, 631, 682].

m.p.   121-122° [387], 120-121° [150, 800, 954];   Spectra (NA).

(2-Hydroxy-5,6-dimethyl-3-pentylphenyl)phenylmethanone

[26881-03-8] $C_{20}H_{24}O_2$ mol.wt. 296.41

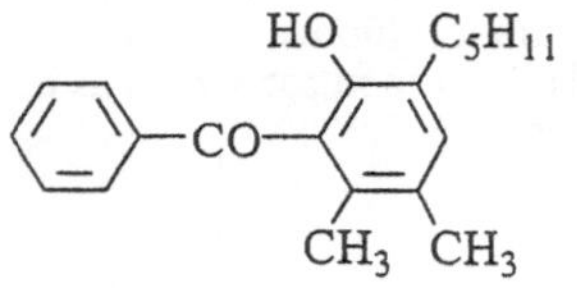

Synthesis

-Refer to: [73, 1223].

m.p.   85° [609];
^1H NMR [609], IR [609], UV [609];
pK_a [609];   polarographic study [1223].

[4-(Heptyloxy)-2-hydroxyphenyl]phenylmethanone *(Uvistat 247)*

[3550-43-4] $C_{20}H_{24}O_3$ mol.wt. 312.41

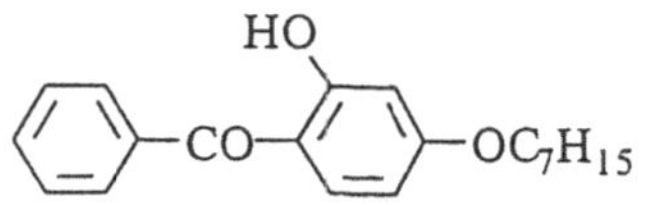

Synthesis

-Obtained by reaction of n-heptyl bromide with
 resbenzophenone,
*in the presence of potassium hydroxide in dilute ethanol
 at 85-95° for 22 h (13%) [803];
*in the presence of potassium carbonate in refluxing acetone for 20 h [150].
-Also refer to: [123, 124, 617, 1262].

   m.p. 40° [150, 803];
   Spectra (NA);   gel permeation chromatography [1145].

[4-Hydroxy-3-(phenylethynyl)phenyl]phenylmethanone

[183589-17-5] $C_{21}H_{14}O_2$ mol.wt. 298.34

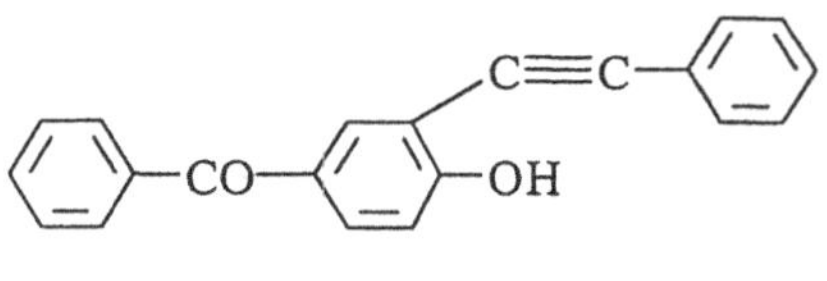

Synthesis

-Obtained by treatment of p-benzoyl-o-(phenylethynyl)phenyl acetate in acetone in the
 presence of 2 N hydrochloric acid at 60° for
 16 h (40%) [60].

   m.p. 139-140° [60];
   ${}^1H$ NMR [60], ${}^{13}C$ NMR [60], IR [60], MS [60].

[2-Hydroxy-5-(1-phenylethyl)phenyl]phenylmethanone

[125182-23-2] $C_{21}H_{18}O_2$ mol.wt. 302.37

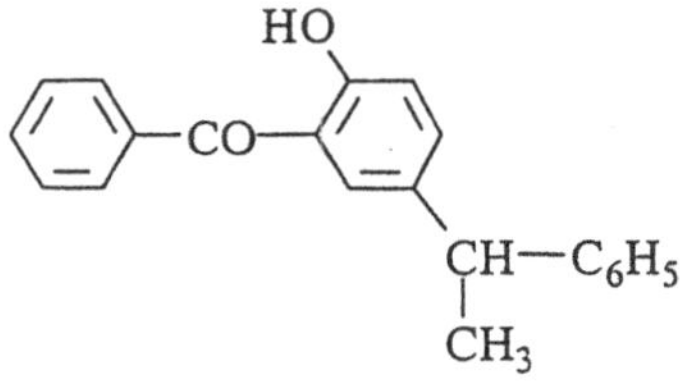

Synthesis

-Preparation by treatment of o-hydroxybenzophenone at
 120° with styrene under nitrogen in the presence of a
 macroreticular acid ion exchanger as catalyst
 (Wofatit OK 80) for 1-2 h (90%) [111].

   b.p.$_{0.15}$ 200-210° [111];   Spectra (NA).

[2-Hydroxy-3-nitro-5-(1,1,3,3-tetramethylbutyl)phenyl]phenylmethanone

                               $C_{21}H_{25}NO_4$ mol.wt. 355.44

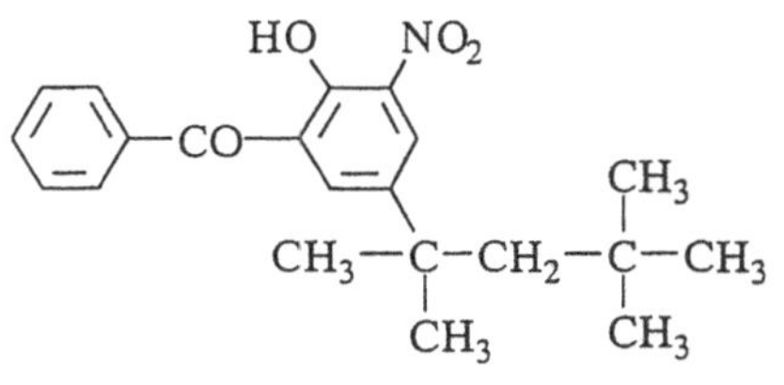

Synthesis

-Obtained by reaction of 65% nitric acid with
 2-hydroxy-5-tert-octylbenzophenone in acetic
 acid at 20° [1037].

oil [1037];   b.p. and Spectra (NA).

[2,4-Bis(1,1-dimethylethyl)-6-hydroxyphenyl]phenylmethanone

[13113-73-0] $C_{21}H_{26}O_2$ mol.wt. 310.44

Synthesis

-Obtained by photo-Fries rearrangement of 3,5-di-tert-
  butylphenyl benzoate,
*in benzene [1268], (29%) [417]; after 27.5 h, it had
  decreased to 8% [417];
*in ethanol for 3 h (48%) [418];
*in isopropanol for 3 h (37%) and 52% after 6 h; after 17 h, it had decreased to 16% [418];
*in a mixture of ethanol/ethyl ether (9:1) for 3 h (34%) [418];
*in N,N-dimethylformamide for 3 h (9%); (16%) after 1 h [418];
*in n-hexane and tetrahydrofuran for 3 h (8% and 12% yields, respectively) [418];
*in dioxane and glyme (1,2-dimethoxyethane) for 3 h (2% and 5% yields, respectively) [418].
-Also refer to: [421].

  m.p. 202°5 [417]; ^{1}H NMR [1268], IR [1268].

[3,5-Bis(1,1-dimethylethyl)-2-hydroxyphenyl]phenylmethanone

[24242-58-8] $C_{21}H_{26}O_2$ mol.wt. 310.44

Syntheses

-Obtained from (3,5-di-tert-butyl-2-hydroxyphenyl)-
  phenylcarbinol by oxidation with DDQ (2,3-dichloro-
  5,6-dicyanobenzoquinone) in dioxane at r.t. for 16 h
  (71%) [149].
-Also obtained by photo-Fries rearrangement of 2,4-di-
tert-butylphenyl benzoate in benzene under nitrogen (32%) [880].
-Preparation by treatment of o-hydroxybenzophenone at 120° with a mixture of isobutylene/
  nitrogen (1:1) or with tert-butyl chloride in the presence of a macroreticular acid ion exchanger as
  catalyst (Wofatit OK 80) for 10 h (80%) [111].

  b.p.$_{0.15}$ 140-145° [111];   m.p. 61-62° [880], 60-62° [149];
  ^{1}H NMR [880], IR [880].

[3,5-Bis(1,1-dimethylethyl)-4-hydroxyphenyl]phenylmethanone

[7175-89-5] $C_{21}H_{26}O_2$ mol.wt. 310.44

Syntheses

-Preparation by reaction of benzoyl chloride with 2,6-di-tert-
  butylphenol in the presence of aluminium chloride [767],
  (93%) [1080], 70% [1321], (50-55%) [293, 1233].
-Preparation by reaction of benzoic acid with 2,6-di-tert-
  butylphenol in the presence of trifluoroacetic anhydride at
  r.t. [980], for 3 h (65%) [981].
-Preparation by oxidation of (3,5-di-tert-butyl-4-hydroxyphenyl)phenylcarbinol with 2,3-dichloro-
  5,6-dicyanobenzoquinone (DDQ) in dioxane at r.t. for 15 min (95%) [149].
-Preparation by reaction of benzoyl chloride with sodium 2,6-di-tert-butylphenoxide (SM) in
  dioxane between 60 to 80° (24%) [282]. SM was obtained by reaction of sodium with 2,6-di-tert-
  butylphenol in methanol, then solvent elimination at 80°.
-Also obtained by basic hydrolysis of α(3,5-di-tert-butyl-4-hydroxyphenyl)benzyl benzoate with
  potassium hydroxide in refluxing dilute ethanol for 30 min (90%) [282].
-Also refer to: [880].

m.p. 132-134° [980, 981], 125°5-126°5 [1321], 125-126° [282],
124-125° [293, 1233], 123-124° [1080];
^{1}H NMR [981, 1321], IR [1233, 1321].

[2-Hydroxy-5-(1,1,3,3-tetramethylbutyl)phenyl]phenylmethanone

[4090-99-7] $\qquad$ C$_{21}$H$_{26}$O$_2$ $\qquad$ mol.wt. 310.44

Syntheses

-Preparation by photo-Fries rearrangement of p-tert-octyl-
phenyl benzoate (45%) [1037] according to [743].
-Preparation by reaction of benzotrichloride with p-tert-
octylphenol in the presence of 30% aqueous sodium
hydroxide at 75-80° for 30 min, and hydrolysis of the
resulting ester as a side product (61%) [1339], (44%) [754]. The same reaction was carried out in
the presence of sodium iodide, during 2.5 h at 80° (39%) [873].
-Preparation by reaction between (p-octylphenoxy)magnesium bromide complexed with HMPT and
benzaldehyde in refluxing benzene for 48 h (36%) [250].

yellow oil [250]; m.p. 34-35° [1339];
b.p.$_{2.5}$ 176-179° [754], b.p.$_1$ 140° [873];
^{1}H NMR [250, 754, 1339], ^{13}C NMR [754],
IR [250, 754, 1339], UV [873, 1339], MS [250, 1339];
gas chromatography study [1343].

[2-Hydroxy-4-(isooctyloxy)phenyl]phenylmethanone

[33059-05-1] $\qquad$ C$_{21}$H$_{26}$O$_3$ $\qquad$ mol.wt. 326.44

Synthesis

-Preparation by reaction of isooctyl chloride
with resbenzophenone in the presence of
sodium carbonate and sodium iodide in
acetone for 4 h at 150° (97%) [559].

m.p. and Spectra (NA).

[2-Hydroxy-4-(octyloxy)phenyl]phenylmethanone *(Octabenzone, Cyasorb UV 531)*

[1843-05-6] $\qquad$ C$_{21}$H$_{26}$O$_3$ $\qquad$ mol.wt. 326.44

Syntheses

-Preparation by reaction of n-octyl chloride with 2,4-di-
hydroxybenzophenone,
*in the presence of a mixture of sodium carbonate,
triethylamine and potassium iodide in refluxing butanol
for 15 h (90%) [738];
*in the presence of potassium carbonate in cyclohexanone at 145° for 5 h (66%) [1259, 1260];
*in the presence of potassium hydroxide and antimony triiodide in diethylene glycol at 150° for 1 h
(93%) [387];
*in the presence of sodium bicarbonate and potassium iodide in 1-methylpyrrolidone for 2 h at
150° (96%) [521].
-Preparation by reaction of n-octyl bromide with 2,4-dihydroxybenzophenone,

*in the presence of potassium carbonate,
 in water at 110-115° for 4 h (84%) [628];
 in cyclohexanone at 110° for 6 h (75%) [1260];
 in methyl isobutyl ketone at 110° for 10 h (70%) [1260];
 in methyl n-hexyl ketone at 161° for 4 h (71%) [1260];
 in acetone at 58° for 20 h (54%) [1260] or in refluxing acetone for 16 h [63].
*in the presence of sodium hydroxide,
 in dilute ethanol in a water bath for 18-23 h (13-15%) [805];
 in cyclohexanone at 145° for 5 h (51%) [1260].
-Preparation by alkylation of resbenzophenone with an octyl halide [42].
-Preparation by reaction of n-octyl p-toluenesulfonate with 2,4-dihydroxybenzophenone,
*in the presence of potassium carbonate in boiling water for 5 h (74%) [1259, 1260];
*in the presence of sodium carbonate in boiling water for 8 h (63%) [1260];
*in the presence of potassium hydroxide in boiling water for 8 h (61%) [1260].
-Preparation by reaction of n-octyl benzenesulfonate with 2,4-dihydroxybenzophenone in the
 presence of potassium carbonate in boiling water for 8 h (66%) [1260].
-Preparation by reaction of benzoyl chloride with resorcinol dioctyl ether in chlorobenzene in the
 presence of titanium tetrachloride for 1 h at 120° (69%) [1140].
-Also refer to: [40, 79, 271, 348, 483, 509, 595, 615, 617, 840, 856, 896, 910, 913, 968, 1067, 1089,
 1165, 1269, 1275, 1325, 1367, 1465].

  m.p. 50° [1067], 48-50° [387], 48-49° [1140], 47-48° [1259, 1260],
    45-46° [63, 805, 1370];
  ^{1}H NMR [610], IR [1259, 1260], UV [610, 1067, 1156, 1259, 1260];
  TLC [377, 1396]; gel permeation chromatography [1145].

[2-Hydroxy-5-(octyloxy)phenyl]phenylmethanone

[4998-51-0] $C_{21}H_{26}O_3$ mol.wt. 326.44

Synthesis

-Obtained (poor yield) by reaction of n-octyl bromide with
2,5-dihydroxybenzophenone in the presence of sodium
hydroxide in dilute ethanol at 90-95° for 50 h (7%) [804].

  m.p. 26-27° [804]; Spectra (NA).

[4-[(2-Ethylhexyl)oxy]-2-hydroxyphenyl]phenylmethanone (Dastib 242)

[2549-90-8] $C_{21}H_{26}O_3$ mol.wt. 326.44

Synthesis

-Preparation by reaction of 1-chloro-
2-ethylhexane with resbenzophenone in
the presence of barium hydroxide and
arsenic tribromide in dimethyl
sulfoxide at 80° for 3 h (93%) [387].

  b.p.$_1$ 213-216° [387]; ^{1}H NMR [610], UV [610];
  gel permeation chromatography [301, 1145].

[6-Hydroxy-4-methoxy-3-methyl-2-(phenylmethoxy)phenyl]phenylmethanone

[74627-93-3] $C_{22}H_{20}O_4$ mol.wt. 348.40

Synthesis

-Preparation by partial methylation of 2-benzyloxy-
3-methyl-4,6-dihydroxybenzophenone with dimethyl
sulfate in the presence of potassium carbonate in refluxing
acetone for 10 min (96%) [57].

colourless oil [57];   b.p. (NA); ^{1}H NMR [57];   TLC [57].

(2-Hydroxy-5-isononylphenyl)phenylmethanone

[59802-03-8] $C_{22}H_{28}O_2$ mol.wt. 324.46

Synthesis

-Preparation by reaction of benzotrichloride with 4-iso-
nonylphenol in hydrofluoric acid in the presence of water
at -10°, then between -10 to 0° for 2 h, at r.t. for 7 h and at
80° for 30 min into an autoclave [393].

yellow oil [393];   b.p.$_1$ 190° [393];   Spectra (NA).

(2-Hydroxy-5-nonylphenyl)phenylmethanone

[58085-73-7] $C_{22}H_{28}O_2$ mol.wt. 324.46

Syntheses

-Preparation by Fries rearrangement of p-nonylphenyl
benzoate with aluminium chloride [212], without solvent at
140-145° for 30-50 min (52%) [1108].
-Preparation by Friedel-Crafts acylation of p-nonylphenol
with benzoyl chloride in the presence of aluminium chloride
in ethylene dichloride at 160° for 30-60 min (37%) [1108].
-Preparation by reaction of benzotrichloride with p-nonylphenol in the presence of 30% aqueous
sodium hydroxide (60-70%) [1010], at 75-80° for 30 min, and hydrolysis of the ester formed as a
side product (42%) [754].
-Also refer to: [491, 966].

m.p. 35-36° [754];   b.p.$_5$ 205-208° [754];   TLC [80], GC [80], HPLC [80].
^{1}H NMR [754, 1108], ^{13}C NMR [754], IR [754, 1108].

(2-Hydroxy-5-tert-nonylphenyl)phenylmethanone

[111547-84-3] $C_{22}H_{28}O_2$ mol.wt. 324.46

Synthesis

-Refer to: [1343].

m.p. and Spectra (NA);
gas chromatography study [1343].

[2-Hydroxy-3-methyl-4-(octyloxy)phenyl]phenylmethanone

[52220-72-1] $C_{22}H_{28}O_3$ mol.wt. 340.46

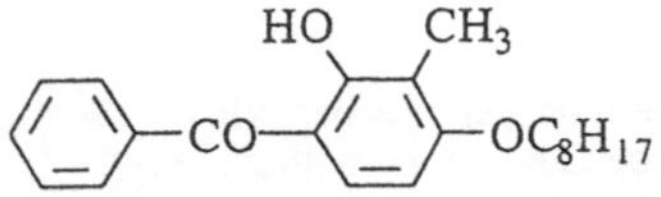

Synthesis

-Preparation by partial alkylation of 2,4-dihydroxy-3-methylbenzophenone with octyl chloride in the presence of sodium bicarbonate and potassium iodide in 1-methyl-pyrrolidone for 2 h at 150° (92%) [521].

m.p. and Spectra (NA).

[2-Hydroxy-5-methyl-4-(octyloxy)phenyl]phenylmethanone

[52220-73-2] $C_{22}H_{28}O_3$ mol.wt. 340.46

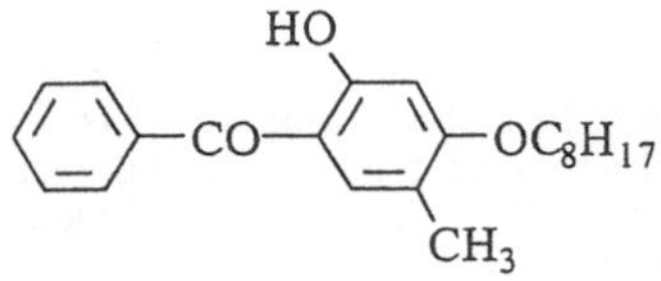

Synthesis

-Preparation by partial alkylation of 2,4-dihydroxy-5-methylbenzophenone with octyl chloride in the presence of sodium bicarbonate and potassium iodide in 1-methyl-pyrrolidone for 2 h at 150° (91%) [521] or in refluxing methyl cellosolve [520].

m.p. and Spectra (NA).

[2-Hydroxy-4-(nonyloxy)phenyl]phenylmethanone

$C_{22}H_{28}O_3$ mol.wt. 340.46

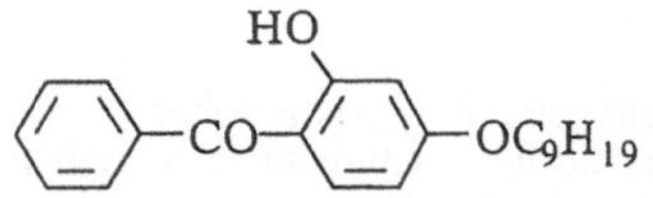

Synthesis

-Obtained by reaction of n-nonyl bromide with resbenzo-phenone in the presence of sodium hydroxide in dilute ethanol in a water bath for 18-23 h (28-30%) [805].
-Also refer to: [617].

m.p.   50°5 [805];   Spectra (NA).

[4-(4-Butylphenoxy)-2-hydroxyphenyl]phenylmethanone

[35698-49-8] $C_{23}H_{22}O_3$ mol.wt. 346.94

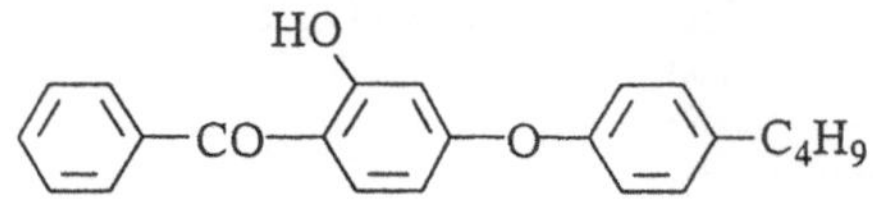

Synthesis

-Refer to: [101] (compound 7).

m.p. (NA);   UV [101].

[4-(Decyloxy)-2-hydroxyphenyl]phenylmethanone

[2162-63-2] $C_{23}H_{30}O_3$ mol.wt. 354.49

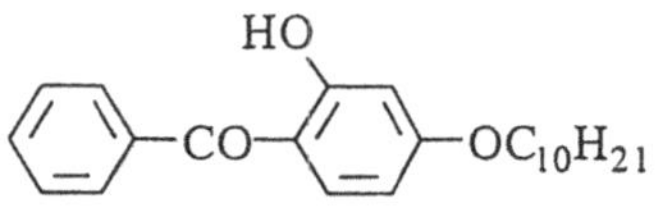

Synthesis

-Obtained by reaction of n-decyl bromide with resbenzophenone,
*in the presence of sodium hydroxide in dilute ethanol at
110-115° for 15 h (18%) [805];
*in the presence of potassium carbonate in refluxing acetone for 20 h [150].
-Also refer to: [483, 615].

   m.p.   50° [805], 47-50° [150];   IR [991], UV [991].

[2-Hydroxy-4-(isodecyloxy)phenyl]phenylmethanone

[55909-78-9] $C_{23}H_{30}O_3$ mol.wt. 354.49

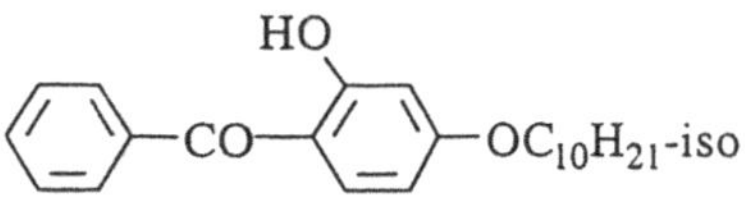

Synthesis

-Refer to: [1145].

m.p. and Spectra (NA);
gel permeation chromatography [1145].

[2-Hydroxy-4-[2-(octylthio)ethoxy]phenyl]phenylmethanone

[36130-66-2] $C_{23}H_{30}O_3S$ mol.wt. 386.56

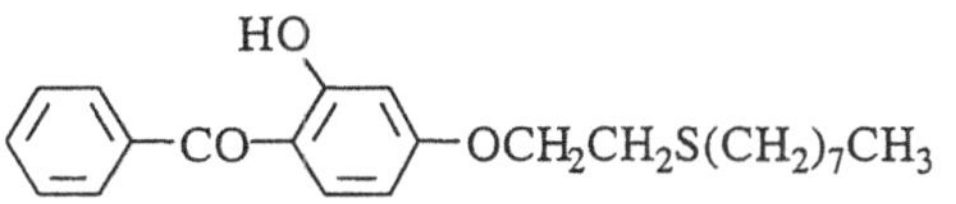

Synthesis

-Preparation by reaction of 2-chloroethyl
octyl sulfide (1-(2-chloroethylsulfanyl)-
octane) with resbenzophenone (82%),
according to [387].

   b.p.$_{0.2}$ 253-258° [387];   Spectra (NA).

[4-[(3,7-Dimethyl-2,6-octadienyl)oxy]-2-hydroxy-6-methoxyphenyl]phenyl-methanone (E)

[140158-57-2] $C_{24}H_{28}O_4$ mol.wt. 380.48

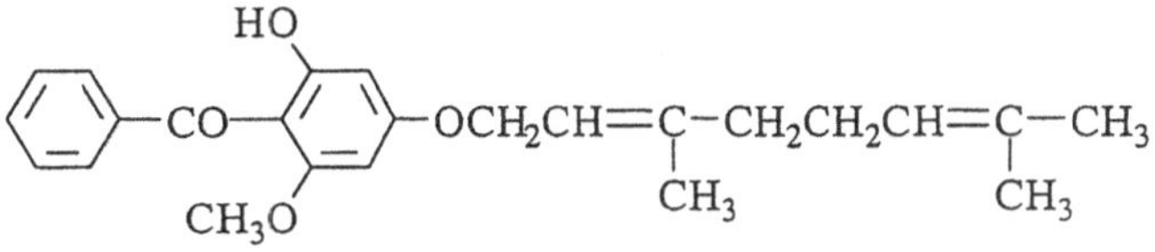

Synthesis

-Not yet described.

Isolation from natural sources

-From *Helichrysum triplinerve* (Asteraceae) [1099];
-From *genus Leontonyx* [192].

   m.p. (NA); ^{1}H NMR [1099], MS [1099].

[2-Hydroxy-4-(undecyloxy)phenyl]phenylmethanone

$C_{24}H_{32}O_3$ mol.wt. 368.52

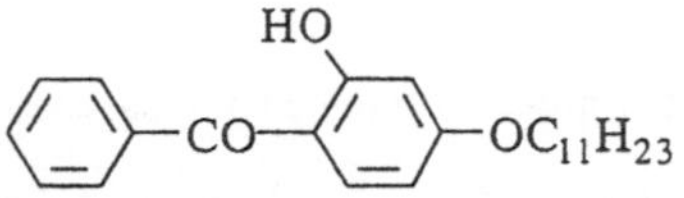

Synthesis

-Refer to: [228].

m.p. and Spectra (NA).

(5-Dodecyl-2-hydroxy-3-nitrophenyl)phenylmethanone

[35698-17-0] $C_{25}H_{33}NO_4$ mol.wt. 411.54

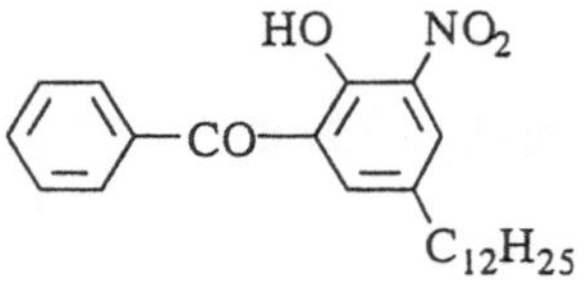

Synthesis

-Preparation by reaction of fuming nitric acid with 5-dodecyl-
 2-hydroxybenzophenone in acetic acid/acetic anhydride
 solution for 40 min at 15-22° (92%) [469, 881, 882].

b.p.$_{0.8}$ 210° [469, 882];   Spectra (NA).

(5-Dodecyl-2-hydroxyphenyl)phenylmethanone

[35698-16-9] $C_{25}H_{34}O_2$ mol.wt. 366.54

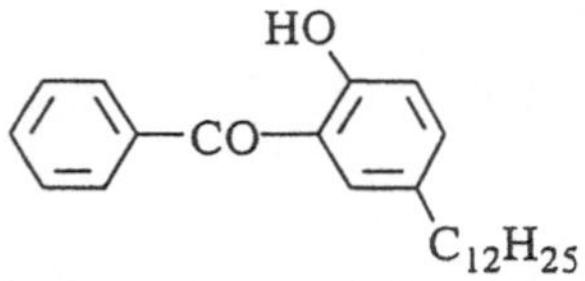

Synthesis

-Preparation by reaction of benzenyl trichloride (benzotrichloride) with p-dodecylphenol in carbon disulfide in
 the presence of aluminium chloride, first between -20 to
 -15°, then at 0 to 5° for 1 h (38%) [469, 881, 882].

m.p. and Spectra (NA).

[5-(1,1-Dimethylethyl)-2-hydroxy-4-(octyloxy)phenyl]phenylmethanone

[55913-02-5] $C_{25}H_{34}O_3$ mol.wt. 382.54

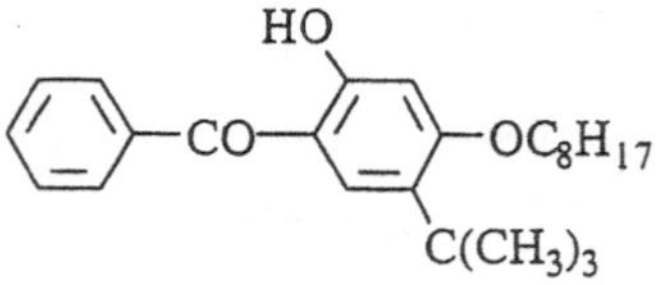

Synthesis

-Refer to: [1145].

m.p. and Spectra (NA);
gel permeation chromatography [1145].

[4-(Dodecyloxy)-2-hydroxyphenyl]phenylmethanone

[2985-59-3] $C_{25}H_{34}O_3$ mol.wt. 382.54

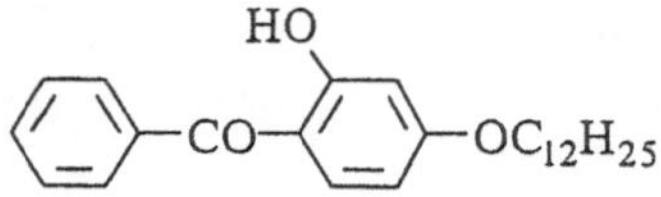

Syntheses

-Preparation by reaction of 1-bromododecane (dodecyl
 bromide or lauryl bromide) with resbenzophenone in the
 presence of potassium carbonate in refluxing acetone for
 16 h [63] or for 20 h [150].

-Preparation by alkylation of resbenzophenone with a dodecyl halide [42].
-Also refer to: [229, 381, 384, 615, 646, 913, 1389].

m.p. 49-50° [63] 48° [150];
EPR [267], IR [991], UV [991];
TLC [377]; gel permeation chromatography [1145].

[2-Hydroxy-4-(phenylmethoxy)-3-(phenylmethyl)phenyl]phenylmethanone

$C_{27}H_{22}O_3$ mol.wt. 394.47

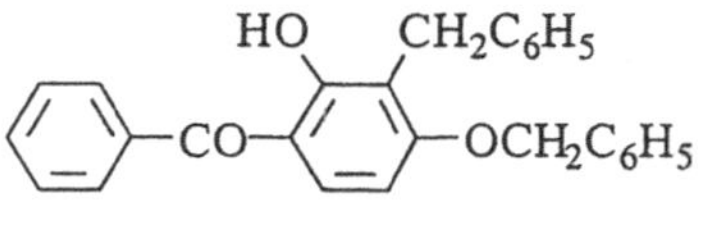

Synthesis

-Preparation by reaction of benzyl chloride with
3-benzyl-2,4-dihydroxybenzophenone in the presence
of potassium carbonate in refluxing acetone for 8 h
(31%) [954].

m.p. 92-93° [954]; Spectra (NA).

[2-Hydroxy-4-(4-nonylphenoxy)phenyl]phenylmethanone

[35698-50-1] $C_{28}H_{32}O_3$ mol.wt. 416.56

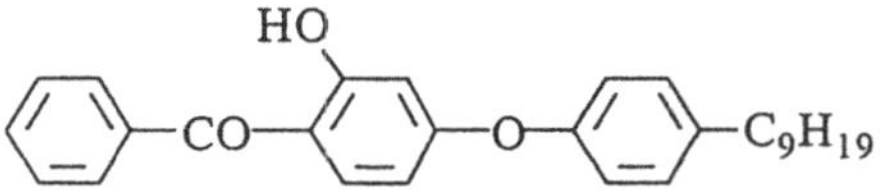

Synthesis

-Refer to: [101] (compound 8).

m.p. (NA);   UV [101].

[2-Hydroxy-3,5-bis(3-methyl-2-butenyl)-4-[(3-methyl-2-butenyl)oxy]phenyl]-phenylmethanone

[63565-03-7] $C_{28}H_{34}O_3$ mol.wt. 418.58

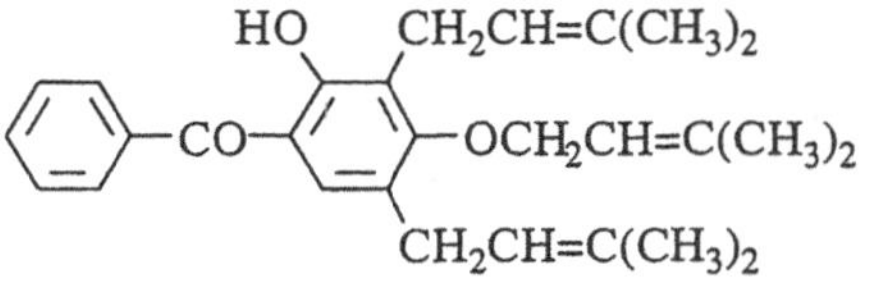

Synthesis

-Obtained (poor yield) by reaction of prenyl
bromide with resbenzophenone in the
presence of sodium methoxide in refluxing
methanol for 4 h (<2%) [875].

m.p. 98° [875]; ^{1}H NMR [875], IR [875], UV [875].

[2-Hydroxy-3-(3-methyl-2-butenyl)-4,6-bis[(3-methyl-2-butenyl)oxy]phenyl]-phenylmethanone

[63565-08-2] $C_{28}H_{34}O_4$ mol.wt. 434.58

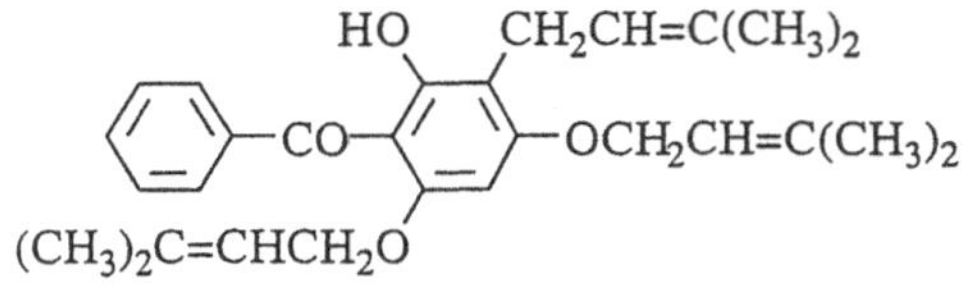

Synthesis

-Obtained by reaction of prenyl bromide
with 2,4,6-trihydroxybenzophenone in the
presence of potassium carbonate in
refluxing acetone for 4 h (21%) [875].

m.p.   112° [875];
^{1}H NMR [875], IR [875], UV [875].

[2-Hydroxy-3,5-bis(1-phenylethyl)phenyl]phenylmethanone

[125182-24-3] $C_{29}H_{26}O_2$ mol.wt. 406.52

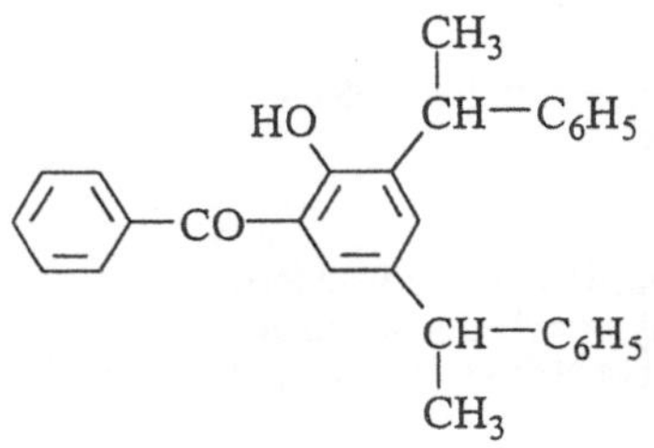

Synthesis

-Preparation by treatment of o-hydroxybenzophenone at
  120° with styrene under nitrogen in the presence of a
  macroreticular acid ion exchanger as catalyst
  (Wofatit OK 80) for 3-4 h (80%) [111].

b.p.$_{0.15}$ 220-225° [111];   Spectra (NA).

[5-(1,1-Dimethylethyl)-4-(dodecyloxy)-2-hydroxyphenyl]phenylmethanone

$C_{29}H_{42}O_3$ mol.wt. 438.65

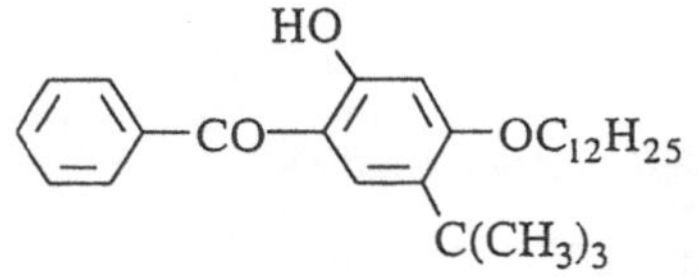

Synthesis

-Preparation by selective alkylation of 5-tert-butyl-
  2,4-dihydroxybenzophenone with dodecyl bromide in
  refluxing methyl ethyl ketone in the presence of
  potassium carbonate [587].

m.p.   74°5-75°5 [587, 588];   Spectra (NA).

[4-(Hexadecyloxy)-2-hydroxyphenyl]phenylmethanone

[3457-17-8] $C_{29}H_{42}O_3$ mol.wt. 438.65

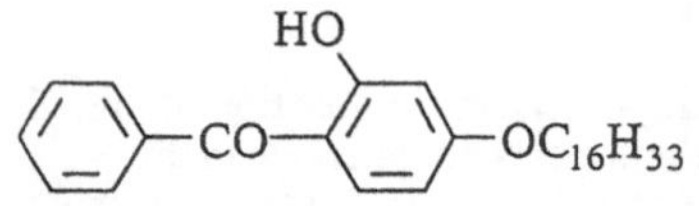

Synthesis

-Obtained by partial alkylation of resbenzophenone with
  hexadecyl bromide (cetyl bromide) in the presence of
  potassium carbonate in refluxing acetone for 20 h [150].
-Also refer to: [61] and [771] (Japanese patent).

m.p.   50° [150];   Spectra (NA).

[5-(Hexadecyloxy)-2-hydroxyphenyl]phenylmethanone

[131664-12-5] $C_{29}H_{42}O_3$ mol.wt. 438.65

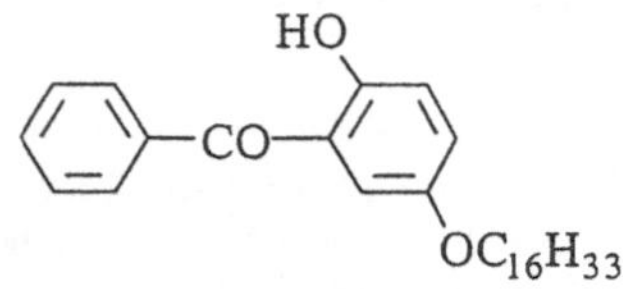

Synthesis

-Refer to: [1172] (Japanese patent).

m.p. and Spectra (NA).

[2-Hydroxy-4-(octadecanoyloxy)phenyl]phenylmethanone
Octadecanoic acid, 4-benzoyl-3-hydroxyphenyl ester

[65953-50-6] $C_{31}H_{44}O_4$ mol.wt. 480.69

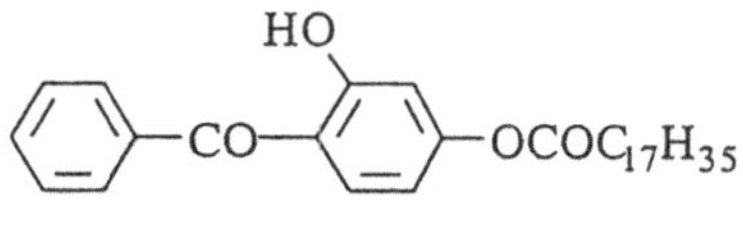

Synthesis

-Refer to: [1156].

m.p. (NA);   UV [1156].

[2-Hydroxy-4-(octadecyloxy)phenyl]phenylmethanone

[3457-13-4] $C_{31}H_{46}O_3$ mol.wt. 466.70

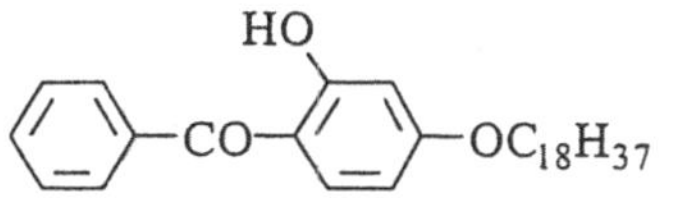

Syntheses

-Preparation by reaction of stearyl chloride (octadecyl
 chloride or 1-chlorooctadecane) with resbenzophenone in
 the presence of sodium hydroxide and phosphorous
 triiodide in aqueous diethylene glycol at 175° for 1.5 h
 (94%) [387].
-Preparation by alkylation of resbenzophenone with an octadecyl halide [42].
-Also refer to: [913].

m.p. 55-56° [387];   Spectra (NA);   TLC [377].

[2-Hydroxy-4-(nonadecyloxy)phenyl]phenylmethanone

$C_{32}H_{48}O_3$ mol.wt. 480.73

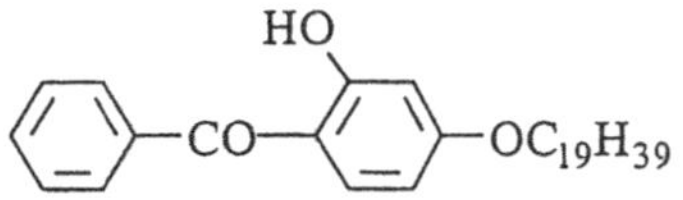

Synthesis

-Obtained by partial alkylation of resbenzophenone with
 nonadecyl bromide in the presence of potassium
 carbonate in refluxing acetone for 20 h [150].

m.p. 53° [150];   Spectra (NA).

**[2-Hydroxy-3,5-bis(3-methyl-2-butenyl)-4,6-bis[[(4-methylphenyl)sulfonyl]oxy]phenyl]-
phenylmethanone**

[83611-03-4] $C_{37}H_{38}O_8S_2$ mol.wt. 674.84

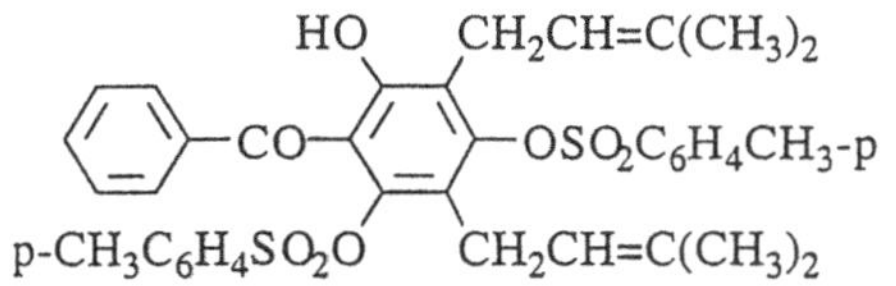

Synthesis

-Obtained by ditosylation of 2,4,6-trihydroxy-
 3,5-diprenylbenzophenone with p-toluene-
 sulfonyl chloride in the presence of potassium
 carbonate in refluxing acetone for 8 h (39%)
 [1043].

m.p. 57° [1043]; ^{1}H NMR [1043].

1.2. *Substituents located on the other ring*

(2-Hydroxyphenyl)(2,3,4,5,6-pentafluorophenyl)methanone

[32541-24-5] $C_{13}H_5F_5O_2$ mol.wt. 288.17

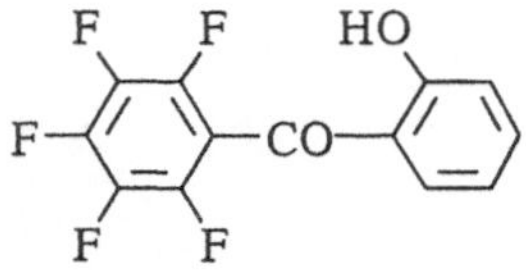

Synthesis

-Preparation by demethylation of 2,3,4,5,6-pentafluoro-
2'-methoxybenzophenone (SM) in methylene chloride in the
presence of aluminium chloride at 20° for 3-6 h (68%)
[839]. SM was obtained in two steps: first, preparation of
2'-methoxy-2,3,4,5,6-pentafluorobenzhydrol by
condensation of o-methoxybenzaldehyde with pentafluorophenylmagnesium bromide in boiling
ethyl ether for 2 h (96%). Then, this "benzhydrol" was oxidized with chromium trioxide in acetic
acid at 20° for 20 h (95%) [839].

m.p. 78-79° [839]; ^{1}H NMR [839], IR [839], UV [839].

(3-Bromo-4-chlorophenyl)(4-hydroxyphenyl)methanone

[78930-23-1] $C_{13}H_8BrClO_2$ mol.wt. 311.56

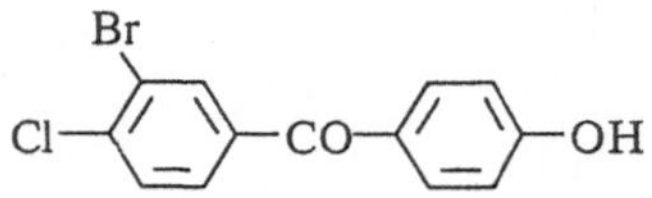

Synthesis

-Preparation by reaction of 3-bromo-4-chlorobenzoyl
chloride with phenol in the presence of aluminium
chloride [296].

m.p. 193-194° [296];   Spectra (NA).

(4-Bromo-2-fluorophenyl)(4-hydroxyphenyl)methanone

[192443-11-1] $C_{13}H_8BrFO_2$ mol.wt. 295.11

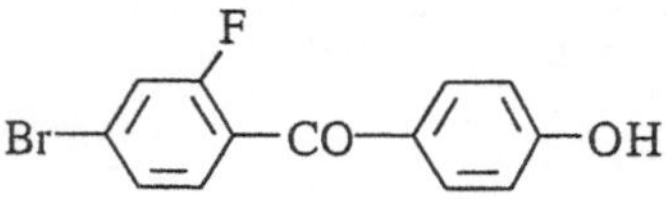

Synthesis

-Preparation by demethylation of 4-bromo-2-fluoro-
4'-methoxybenzophenone (SM) with 62% aqueous
hydrobromic acid in refluxing acetic acid (98%). SM was
obtained by Friedel-Crafts acylation of anisole with
4-bromo-2-fluorobenzoyl chloride in nitrobenzene in the presence of aluminium chloride, first at
temperature <6°, then at r.t. overnight (88%, m.p. 93-94°). -Refer to: Chem. Abstr., **127**, 108921f
(1997)*.

m.p. (NA);   MS*.

(2,4-Dibromophenyl)(2-hydroxyphenyl)methanone

                                                $C_{13}H_8Br_2O_2$ mol.wt. 356.01

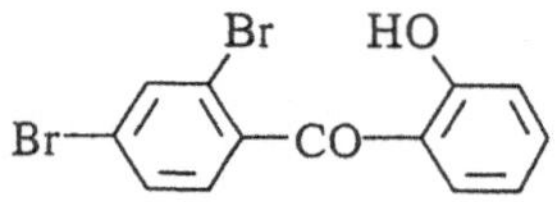

Synthesis

-Refer to: [1303].

m.p. and Spectra (NA).

(4-Chloro-3-iodophenyl)(4-hydroxyphenyl)methanone

[83888-75-9] $C_{13}H_8ClIO_2$ mol.wt. 358.56

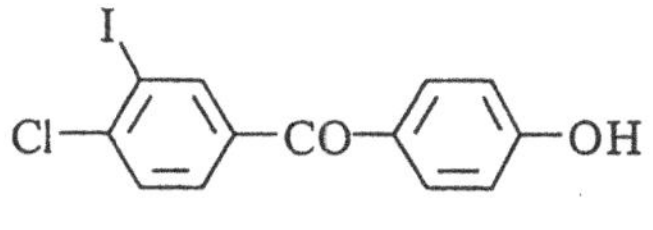

Synthesis

-Preparation by reaction of 4-chloro-3-iodobenzoyl
chloride with phenol in the presence of aluminium
chloride [296].

m.p. 191-192° [296]; Spectra (NA).

(2-Chloro-4-nitrophenyl)(2-hydroxyphenyl)methanone

[72090-64-3] $C_{13}H_8ClNO_4$ mol.wt. 277.66

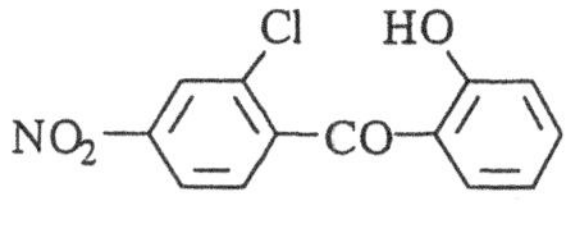

Synthesis

-Obtained by Fries rearrangement of phenyl 2-chloro-4-nitro-
benzoate with aluminium chloride without solvent at 120° or
at 160° [1169].

m.p. and Spectra (NA).

(2-Chloro-4-nitrophenyl)(4-hydroxyphenyl)methanone

[72103-42-5] $C_{13}H_8ClNO_4$ mol.wt. 277.66

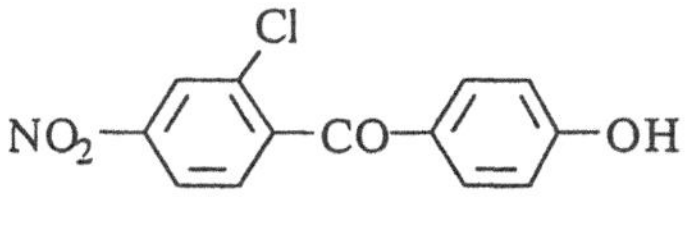

Synthesis

-Obtained by Fries rearrangement of phenyl 2-chloro-
4-nitrobenzoate with aluminium chloride without solvent
at 120° or at 160° [1169].

m.p. and Spectra (NA).

(2-Chloro-5-nitrophenyl)(2-hydroxyphenyl)methanone

[72090-65-4] $C_{13}H_8ClNO_4$ mol.wt. 277.66

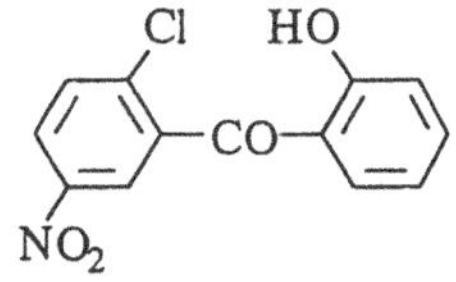

Synthesis

-Obtained by Fries rearrangement of phenyl 2-chloro-5-nitro-
benzoate with aluminium chloride without solvent at 120° or
at 160° [1169].

m.p. and Spectra (NA).

(2-Chloro-5-nitrophenyl)(4-hydroxyphenyl)methanone

[72090-66-5] $C_{13}H_8ClNO_4$ mol.wt. 277.66

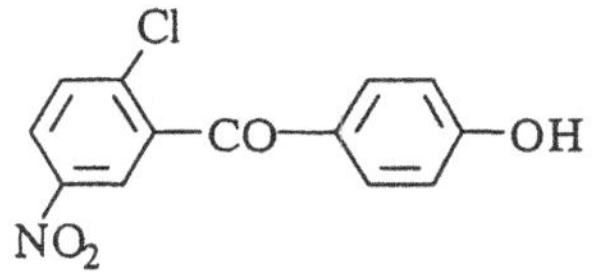

Synthesis

-Obtained by Fries rearrangement of phenyl 2-chloro-5-nitro-
benzoate with aluminium chloride without solvent at 120° or
at 160° [1169].

m.p. and Spectra (NA).

(4-Chloro-3-nitrophenyl)(4-hydroxyphenyl)methanone

[93958-85-1] $C_{13}H_8ClNO_4$ mol.wt. 277.66

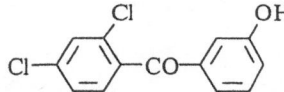

Synthesis

-Preparation by demethylation of 4-chloro-4'-methoxy-
   3-nitrobenzophenone with pyridinium chloride at 220-230°
   for 30 min (55%) [20].

m.p. 185° [20];   IR [20].

(2,4-Dichlorophenyl)(2-hydroxyphenyl)methanone

[46795-43-1] $C_{13}H_8Cl_2O_2$ mol.wt. 267.11

Synthesis

-Refer to: [1303] (compound VI-A).

m.p. and Spectra (NA).

(2,4-Dichlorophenyl)(3-hydroxyphenyl)methanone

[62810-56-4] $C_{13}H_8Cl_2O_2$ mol.wt. 267.11

Syntheses

-Preparation by reaction of m-anisoyl chloride with m-di-
   chlorobenzene in the presence of excess aluminium chloride
   first at 20°, then at reflux for 2 h (71%) [82, 1292].
-Preparation by demethylation of 2,4-dichloro-3'-methoxybenzophenone with aluminium chloride in
   refluxing chlorobenzene [847].
-Also obtained by diazotization of 3'-amino-2,4-dichlorobenzophenone followed by hydrolysis of
   the diazonium salt so obtained [847].

m.p. 135° [82, 1292], 70° [847]. One of the reported melting points is obviously wrong.
Spectra (NA).

(2,4-Dichlorophenyl)(4-hydroxyphenyl)methanone

[34183-01-2] $C_{13}H_8Cl_2O_2$ mol.wt. 267.11

Syntheses

-Preparation by reaction of 2,4-dichlorobenzoyl chloride
   with phenol in the presence of aluminium chloride [296].
-Preparation by reaction of 2,4-dichlorobenzotrichloride
with phenol,
*in hydrofluoric acid in the presence of water at -10°, then at 15° overnight and at 80° for 30 min
[393];
*in methylene chloride in the presence of aluminium chloride, first at 0 to 3° for 30 min, then at 20°
for 3 h (52%) [486].
-Preparation by Fries rearrangement of phenyl 2,4-dichlorobenzoate with aluminium chloride in
chlorobenzene at 140-150° for 20 min or in nitrobenzene at 75° for 24 h [1214].

m.p.   140-141° [1214], 134-135° [296], 132-135°5 [486];   Spectra (NA).

(2,6-Dichlorophenyl)(4-hydroxyphenyl)methanone

[61002-53-7]　　　　　　　　　　$C_{13}H_8Cl_2O_2$　　mol.wt. 267.11

Synthesis

-Refer to: [909].

m.p. and Spectra (NA).

(3,4-Dichlorophenyl)(3-hydroxyphenyl)methanone

[62810-54-2]　　　　　　　　　　$C_{13}H_8Cl_2O_2$　　mol.wt. 267.11

Syntheses

-Preparation by reaction of m-anisoyl chloride with o-dichlorobenzene in the presence of excess aluminium chloride
first at 20°, then at reflux for 2 h (41%) [82, 1292].

-Preparation by demethylation of 3,4-dichloro-3'-methoxybenzophenone with aluminium chloride in
refluxing chlorobenzene [847].

m.p.　140° [847], 130° [82, 1292];　Spectra (NA).

(3,4-Dichlorophenyl)(4-hydroxyphenyl)methanone

[60013-02-7]　　　　　　　　　　$C_{13}H_8Cl_2O_2$　　mol.wt. 267.11

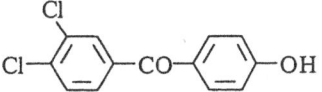

Syntheses

-Preparation by reaction of 3,4-dichlorobenzoyl chloride,
　*with phenol in the presence of aluminium chloride [296];
　*with phenetole in the presence of aluminium chloride

in carbon disulfide at r.t. overnight (40%) [207].
*with anisole in the presence of aluminium chloride at 70° for 4 h, followed by demethylation of the
obtained ketone with 48% hydrobromic acid in refluxing acetic acid for 47 h [1293].
-Preparation by dealkylation of 3,4-dichloro-4'-ethoxybenzophenone with aluminium chloride in
refluxing carbon disulfide [207].

m.p.　172-174° [207], 158°5-170° [296]. A typing error probably occurred in the published
data.
Spectra (NA).

(3,5-Dichlorophenyl)(4-hydroxyphenyl)methanone

[119427-60-0]　　　　　　　　　　$C_{13}H_8Cl_2O_2$　　mol.wt. 267.11

Synthesis

-Refer to: [511, 958] and [737] (compound 155).

m.p. and Spectra (NA).

(2,4-Difluorophenyl)(2-hydroxyphenyl)methanone

[46795-44-2] $C_{13}H_8F_2O_2$ mol.wt. 234.20

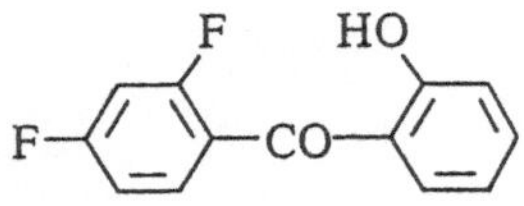

Synthesis

-Refer to: [1303] (compound V-A).

m.p. and Spectra (NA).

(3,5-Difluorophenyl)(4-hydroxyphenyl)methanone

[148253-49-0] $C_{13}H_8F_2O_2$ mol.wt. 234.20

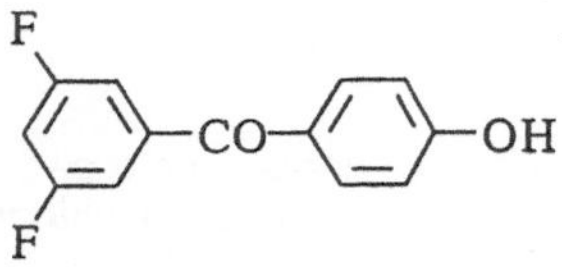

Synthesis

-Preparation by demethylation of 3,5-difluoro-4'-methoxy-
benzophenone (SM) with 48% hydrobromic acid in
refluxing acetic acid for 4 h (91%) [272, 550]. SM was
obtained by Friedel-Crafts acylation of anisole with 3,5-di-
fluorobenzoyl chloride in ethylene dichloride in
the presence of aluminium chloride at r.t. under nitrogen for 3 h (88%) [272].

m.p. 134-135° [272, 550];
^{1}H NMR [272, 550], ^{13}C NMR [272, 550], IR [272, 550], MS [272, 550].

(3,5-Dinitrophenyl)(4-hydroxyphenyl)methanone

[51339-44-7] $C_{13}H_8N_2O_6$ mol.wt. 288.22

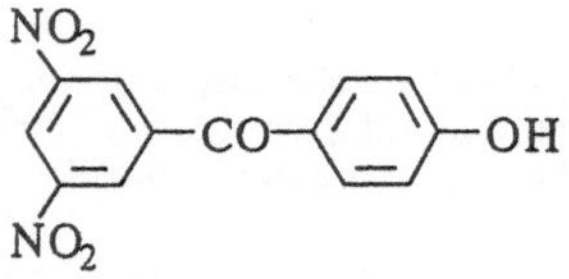

Syntheses

-Preparation by reaction of 3,5-dinitrobenzoyl chloride with
phenetole in the presence of aluminium chloride in carbon
disulfide, first between 0 to 5° for 8.5 h, then at r.t. overnight
(60%) [207].
-Obtained (by-product) by reaction of 3,5-dinitrobenzoyl
chloride with anisole in the presence of aluminium chloride [922].

m.p. 196-197° (d) [207], 176-177° [922];   Spectra (NA).

(2-Bromophenyl)(2-hydroxyphenyl)methanone

[99515-47-6] $C_{13}H_9BrO_2$ mol.wt. 277.11

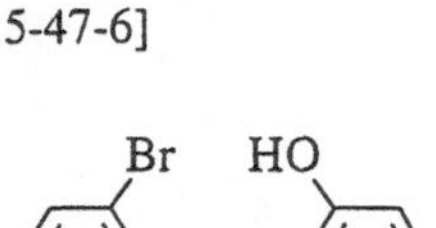

Synthesis

-Preparation from 2'-bromo-5-tert-butyl-2-methoxy-
benzophenone by total dealkylation with aluminium chloride
in benzene at 65-70° for 45 h (75%) [775], (60-80%) [774].
-Also refer to: [1183].

m.p. 76-77° [774, 775];   Spectra (NA).

(2-Bromophenyl)(4-hydroxyphenyl)methanone

$C_{13}H_9BrO_2$ mol.wt. 277.11

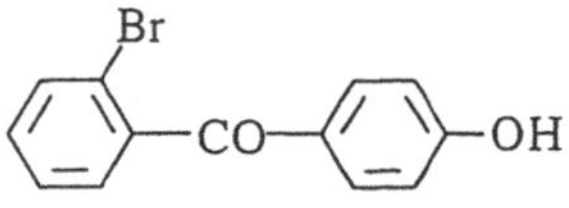

Synthesis

-Preparation by dealkylation of 2-bromo-4'-ethoxy-
benzophenone with hydrobromic acid (d = 1.49) in boiling
acetic acid for 2 days. -Refer to: Chem. Abstr., **17**, 3497[5]
(1923)*.

b.p.$_{10}$ 260°*;   m.p. 114°*;   Spectra (NA).

(3-Bromophenyl)(2-hydroxyphenyl)methanone

$C_{13}H_9BrO_2$ mol.wt. 277.11

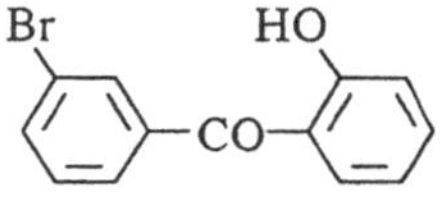

Synthesis

-Preparation by total dealkylation of 5-tert-butyl-3'-bromo-
2-methoxybenzophenone with aluminium chloride in
benzene at 65-70° for 45 h (60%) [775], 60-80% [774].

m.p. 77-78° [774, 775];   Spectra (NA).

(3-Bromophenyl)(3-hydroxyphenyl)methanone

[62810-50-8] $C_{13}H_9BrO_2$ mol.wt. 277.11

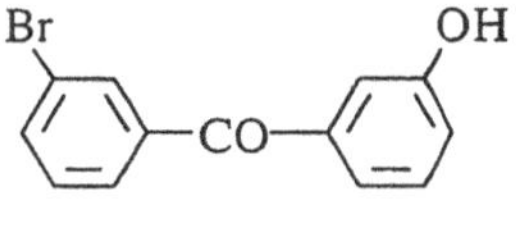

Synthesis

-Preparation by demethylation of 3-bromo-3'-methoxy-
benzophenone with aluminium chloride in refluxing
chlorobenzene [847].

m.p. 110° [847];   Spectra (NA).

(3-Bromophenyl)(4-hydroxyphenyl)methanone

$C_{13}H_9BrO_2$ mol.wt. 277.11

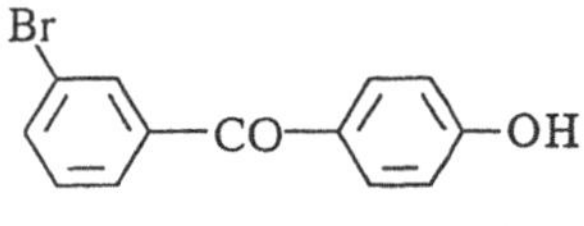

Synthesis

-Preparation by dealkylation of 3-bromo-4'-ethoxy-
benzophenone (SM) with hydrobromic acid (d = 1.49) in
boiling acetic acid for 2 days. SM was obtained by action of
m-bromobenzoyl chloride with phenetole in carbon disulfide
in the presence of aluminium chloride at 55° (80%, m.p. 79°5). -Refer to: Chem. Abstr., **17**, 3497[5]
(1923)*.

m.p. 171°*;   Spectra (NA).

(4-Bromophenyl)(2-hydroxyphenyl)methanone

[2038-92-8]   C13H9BrO2   mol.wt. 277.11

Br—⟨ ⟩—CO—⟨ ⟩ (HO)

Syntheses

-Preparation by Fries rearrangement of phenyl p-bromobenzoate [447], with aluminium chloride at 140° for 30 min
  (23%) [759].
-Preparation by reaction of o-methoxybenzoyl chloride with
bromobenzene in the presence of aluminium chloride (Friedel-Crafts) [37].
-Preparation by diazotization of 2-amino-4'-bromobenzophenone, followed by hydrolysis of the
diazonium salt so obtained (28%) [1332]. The 3-bromofluorenone was the major compound
obtained.

m.p. 98° [37, 1332], 90-92° [759];
^{1}H NMR [37], IR [37].

(4-Bromophenyl)(3-hydroxyphenyl)methanone

[62810-46-2]   C13H9BrO2   mol.wt. 277.11

Br—⟨ ⟩—CO—⟨ ⟩ (OH)

Syntheses

-Preparation by reaction of m-anisoyl chloride with bromobenzene in the presence of an excess of aluminium chloride:
  first at 20°, then at reflux for 2 h (65%) [82, 1292].
-Preparation by Friedel-Crafts acylation of bromobenzene
with m-nitrobenzoyl chloride, reduction of the obtained 4-bromo-3'-nitrobenzophenone and
diazotization of the resulting 3-amino-4'-bromobenzophenone, followed by hydrolysis of the
diazonium salt [847].
-Preparation by demethylation of 4-bromo-3'-methoxybenzophenone with aluminium chloride in
refluxing chlorobenzene [847].
-Also refer to: [436].

m.p. 170° [847], 167° [82, 1292];   Spectra (NA).

(4-Bromophenyl)(4-hydroxyphenyl)methanone

[4369-50-0]   C13H9BrO2   mol.wt. 277.11

Syntheses

Br—⟨ ⟩—CO—⟨ ⟩—OH   -Preparation by reaction of p-bromobenzoyl chloride with
                          phenol in the presence of aluminium chloride [296].
-Preparation by isomerization of 2-hydroxy-4'-bromobenzophenone with trifluoromethanesulfonic
acid at 110° for 3 h (42%) [447].
-Preparation by Fries rearrangement of phenyl p-bromobenzoate with aluminium chloride without
solvent at 140° for 30 min (47%) [759] or in nitrobenzene at 60° for 40 h (67%) [591, 592, 1157].
-Also obtained by reaction of hydrobromic acid with 4-bromo-4'-ethoxybenzophenone in refluxing
acetic acid [941].
-Also obtained by diazotization of 4-amino-4'-bromobenzophenone, followed by hydrolysis of the
obtained diazonium salt [942].
-Preparation by demethylation of 4-bromo-4'-methoxybenzophenone (SM) with aluminium chloride
in refluxing benzene for 8 h. SM was obtained by Friedel-Crafts acylation of anisole with
p-bromobenzoyl chloride [1159].
-Also refer to: [15, 476, 511, 909].

m.p. 192-193° [591, 592, 1157], 191° [759, 941, 942], 187°5-191° [296]; IR [1159]; TLC [1159].

(2-Chlorophenyl)(2-hydroxyphenyl)methanone

[70288-96-9] $C_{13}H_9ClO_2$ mol.wt. 232.67

Syntheses

-Obtained by total dealkylation of 5-tert-butyl-2'-chloro-2-methoxybenzophenone with aluminium chloride in benzene at 65-70° for 45 h (60-80%) [774], (73%) [775].

-Also obtained by Fries rearrangement of phenyl o-chlorobenzoate (SM) with aluminium chloride [692, 1169], at 140° for 30 min (32%) [759]. SM was prepared by heating o-chlorobenzoyl chloride with aluminium tris(phenoxide) in a water bath for 30 min [759].

-Also obtained (poor yield) by reaction of o-chlorobenzoyl chloride with phenyl borate in the presence of aluminium chloride in tetrachloroethane at 100° (3%) [1373].

-Also refer to: [123, 1363].

m.p. 92° [759], 58-59° [774, 775]. There is a discrepancy between the two melting points. Spectra (NA).

(2-Chlorophenyl)(3-hydroxyphenyl)methanone

[62810-53-1] $C_{13}H_9ClO_2$ mol.wt. 232.67

Synthesis

-Preparation by demethylation of 2-chloro-3'-methoxy-benzophenone with aluminium chloride in refluxing chlorobenzene [847].

m.p. 126° [847]; Spectra (NA).

(2-Chlorophenyl)(4-hydroxyphenyl)methanone

[55270-71-8] $C_{13}H_9ClO_2$ mol.wt. 232.67

Syntheses

-Preparation by Fries rearrangement of phenyl o-chloro-benzoate in the presence of aluminium chloride [472, 931, 1169], without solvent for 2 h at 160° [931] or in nitrobenzene at 60° (31%) [472].

-Preparation by Fries rearrangement of phenyl o-chlorobenzoate (SM) with aluminium chloride at 140° for 30 min (56%). SM was obtained by heating o-chlorobenzoyl chloride with aluminium tris(phenoxide) in a water bath for 30 min [759].

-Preparation by reaction of o-chlorobenzoyl chloride,

*with phenyl borate in the presence of aluminium chloride in tetrachloroethane at 100° (57%) [1373];

*with phenol trimethylsilyl ether in the presence of aluminium chloride in refluxing methylene chloride for 2 h (54%) [649].

-Also obtained by reaction of o-chlorobenzoic acid with phenol in the presence of polyphosphoric acid for 20 min at 100° (13%) [959].

-Also obtained (trace) by reaction of EKONOL(TM), an aromatic polyester, behaves as a Friedel-Crafts acylating reagent, with chlorobenzene in triflic acid solution at 25° for 18 h (1%) [287].

-Also refer to: [532, 909].

m.p.   165° [931], 128° [1373], 119-121° [472], 118° [959], 112° [759], 102-104° [649];
There is a discrepancy between the various melting points.
Spectra (NA);   HPLC [287].

(3-Chlorophenyl)(2-hydroxyphenyl)methanone

[72090-60-9] $C_{13}H_9ClO_2$ mol.wt. 232.67

Syntheses

-Preparation by heating a mixture of 3-(3-chlorobenzoyl)-
4-methoxybenzoic acid and pyridinium chloride at 200° for
20 h (65%) [144]. There are simultaneous demethylation
and decarboxylation.
-Also obtained by Fries rearrangement of phenyl m-chlorobenzoate,
*with aluminium chloride between 120 and 160° [18];
*in the presence of Nafion-H, a polymeric perfluorinated resin sulfonic acid, in refluxing
 nitrobenzene for 12 h (21%) [1004].

m.p.   89° [18], 87° [144];
^{1}H NMR [1004], ^{13}C NMR [1004], IR [144].

(3-Chlorophenyl)(3-hydroxyphenyl)methanone

[62810-42-8] $C_{13}H_9ClO_2$ mol.wt. 232.67

Synthesis

-Preparation by demethylation of 3-chloro-3'-methoxy-
benzophenone with aluminium chloride in refluxing
chlorobenzene (81%) [847].

m.p.   104° [847];   Spectra (NA).

(3-Chlorophenyl)(4-hydroxyphenyl)methanone

[61002-52-6] $C_{13}H_9ClO_2$ mol.wt. 232.67

Syntheses

-Preparation by Fries rearrangement of phenyl m-chlorobenzoate,
*in the presence of aluminium chloride without solvent at
 120° or at 160° [18];
*in the presence of Nafion-H, a polymeric perfluorinated resin sulfonic acid, in refluxing
 nitrobenzene for 12 h (54%) [1004].
-Also obtained (poor yield) by reaction of m-chlorobenzoic acid with phenol in the presence of
 polyphosphoric acid at 100° for 20 min (5%) [959].
-Also obtained by reaction of ethyl nitrite with 2-amino-5-chloro-4'-hydroxybenzophenone in
 refluxing ethanol (elimination of amino group) [1474].
-Also refer to: [532, 909, 1169].

m.p.   172° [18], 169° [959], 161° [1474];
^{1}H NMR [1004], ^{13}C NMR [1004].

(4-Chlorophenyl)(2-hydroxyphenyl)methanone

[2985-79-7]   C13H9ClO2   mol.wt. 232.67

Syntheses

-Preparation by Fries rearrangement of phenyl p-chlorobenzoate,
*with aluminium chloride without solvent between 120 and
160° [850, 1447], (29%) [1447], at 200° for 20 min
[766] or in refluxing chlorobenzene for 10 h (53%) [1447];
*with trifluoromethanesulfonic acid at 45-55° (6%) [1426].
-Preparation by reaction of 2-hydroxybenzoyl chloride with chlorobenzene in the presence of
aluminium chloride in refluxing carbon disulfide (72%) [400].
-Also obtained by photo-Fries rearrangement of phenyl p-chlorobenzoate in cyclohexane or in
benzene between 46-52° (42-49%) [419].
-Also obtained by reaction of p-chlorobenzoyl chloride with phenyl borate in the presence of
aluminium chloride in tetrachloroethane at 100° (18%) [1373].
-Preparation by treatment of 5-tert-butyl-4'-chloro-2-methoxybenzophenone with aluminium
chloride in benzene at 65-70° during 45 h (60-80%) [775, 1426].
-Also obtained by reaction of salicylaldehyde with p-iodochlorobenzene by using a catalyst system
of palladium chloride/lithium chloride in the presence of sodium carbonate in N,N-dimethyl-
formamide at 100° for 2 h (57%) [1204].
-Also obtained by demethylation of 4-chloro-2'-methoxybenzophenone with boron tribromide in
methylene chloride at r.t. for 12 h [766], according to [891].
-Also refer to: [348].

m.p.   77-78° [850], 75-76° [419], 74-75° [774, 775], 70-72° [400],
68-71°5 [1447];   1H NMR [1204], MS [1204].

(4-Chlorophenyl)(3-hydroxyphenyl)methanone

[62810-39-3]   C13H9ClO2   mol.wt. 232.67

Synthesis

-Preparation by diazotization of 3-amino-4'-chlorobenzo-
phenone (SM), followed by hydrolysis of the diazonium salt
obtained (81%) [847], (84%) [731]. SM was obtained by
Friedel-Crafts acylation of chlorobenzene with m-nitro-
benzoyl chloride, followed by reduction of the resulting 4-chloro-3'-nitrobenzophenone [847].
-Also refer to: [1371].

m.p.   154-155° [731], 154° [847];   Spectra (NA).

(4-Chlorophenyl)(4-hydroxyphenyl)methanone

[42019-78-3]   C13H9ClO2   mol.wt. 232.67

Syntheses

-Preparation by Fries rearrangement of phenyl p-chlorobenzoate,
*in the presence of aluminium chloride, without solvent, at 160° for 5 min [256, 257], between 120
to 160° (good yield) [850] or at 130° for 1 h, then at 160° for 1 h (10%) [1447] or in refluxing
chlorobenzene for 10 h (52%) [1447];
*in the presence of trifluoromethanesulfonic acid at 45-55° (94%) [1426].

-Preparation by reaction of p-chlorobenzotrichloride with phenol in hydrofluoric acid in the presence of water at 0°, then at r.t. overnight (91%) [393].
-Also obtained by reaction of p-chlorobenzoyl chloride,
*with phenol in the presence of aluminium chloride [296];
*with anisole in the presence of aluminium chloride [510], at 70° for 4 h [1293]. The 4'-chloro-4-methoxybenzophenone so formed [510], (67%) [1293], gave the expected ketone by demethylation with 48% hydrobromic acid in refluxing acetic acid for 47 h [1293] or with aluminium chloride in refluxing chlorobenzene for 1.5 h [510];
*with phenyl borate in the presence of aluminium chloride in tetrachloroethane at 100° (32%) [1373].
-Also obtained by reaction of p-chlorobenzoic acid with phenol,
*in the presence of hydrofluoric acid for 6 h at 75° in an autoclave under pressure (74%) [1137];
*in the presence of polyphosphoric acid at 100° for 20 min (2%) [959];
*in the presence of a trifluoromethanesulfonic acid at r.t. for one day [1138].
-Also obtained by reaction of p-hydroxybenzoic acid with chlorobenzene in the presence of trifluoromethanesulfonic acid at 100° for 5 days (50%) [1138].
-Preparation by Friedel-Crafts acylation of chlorobenzene with p-anisoyl chloride in the presence of aluminium chloride at 120° for 3 h (50-70%) [1330].
-Preparation by diazotization of 4-amino-4'-chlorobenzophenone [940], followed by hydrolysis of the diazonium salt so obtained [941].
-Also obtained by reaction of EKONOL$^{(TM)}$, an aromatic polyester as Friedel-Crafts reagent, with chlorobenzene in triflic acid solution at 25° for 18 h (28%) [287]. Similar results can be obtained using hydrofluoric acid/boron trifluoride or aluminium chloride in place of triflic acid [287].
-Also obtained (poor yield) by photo-Fries rearrangement of phenyl p-chlorobenzoate in benzene at 52° for 19 h (14%) or in cyclohexane at 46° for 19 h (6%) [419].
-Also obtained from 4-chloro-4'-fluorobenzophenone by reaction under nitrogen with potassium hydroxide in aqueous dimethyl sulfoxide at 60° for 18 h [1121].
-Also refer to: [294, 298, 314, 327, 878, 908, 909, 990, 1077, 1182, 1264, 1349, 1458].
N.B.: K salt [1121].

m.p.   179-181° [296], 179°5 [1121], 179°2 [941], 179° [393, 959], 178°5-180°5 [1137], 175-176° [850], 173-175° [1330], 172°5-173°8 [419], 170-171° [1447];
b.p.$_{13}$ 257° [941];   TLC [1330];   HPLC [287];
^{1}H NMR [1138, 1330], ^{13}C NMR [287], MS [287, 1330].

(4-Chlorophenyl)(4-hydroxyphenyl)methanone-^{14}C

[60044-21-5]   C$_{13}$H$_9$ClO$_2$ mol.wt. 234.67

Synthesis

-Preparation by demethylation of ^{14}C-4'-chloro-4-methoxybenzophenone (SM) with 48% hydrobromic acid in refluxing acetic acid for 48 h (74%). SM was obtained by reaction of ^{14}C-p-chlorobenzoyl chloride with anisole in the presence of aluminium chloride at 50° for 3 h (82%) [841].

m.p. and Spectra (NA).

(2-Fluorophenyl)(2-hydroxyphenyl)methanone

C$_{13}$H$_9$FO$_2$ mol.wt. 216.21

Synthesis

-Preparation by Fries rearrangement of phenyl o-fluorobenzoate with aluminium chloride [692].

m.p. and Spectra (NA).

(2-Fluorophenyl)(4-hydroxyphenyl)methanone

$C_{13}H_9FO_2$ mol.wt. 216.21

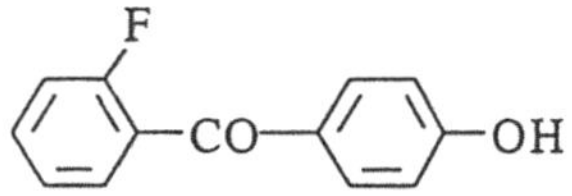

Synthesis

-Obtained (trace) by reaction of EKONOL(TM), an aromatic
polyester as Friedel-Crafts reagent, with fluorobenzene in
triflic acid solution at 75° for 2 h or at 25° for 18 h (1%)
[287].

m.p. and Spectra (NA);   HPLC [287].

(3-Fluorophenyl)(3-hydroxyphenyl)methanone

[62810-55-3] $C_{13}H_9FO_2$ mol.wt. 216.21

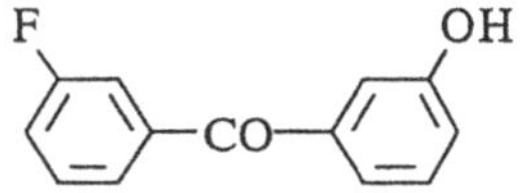

Synthesis

-Preparation by demethylation of 3-fluoro-3'-methoxy-
benzophenone with aluminium chloride in refluxing
chlorobenzene [847].

m.p. 64° [847];   Spectra (NA).

(3-Fluorophenyl)(4-hydroxyphenyl)methanone

[190728-34-8] $C_{13}H_9FO_2$ mol.wt. 216.21

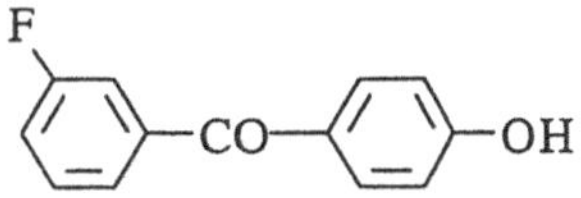

Synthesis

-Refer to: Chem. Abstr., **127**, 34137f (1997).

m.p. and Spectra (NA).

(4-Fluorophenyl)(2-hydroxyphenyl)methanone

[62666-37-9] $C_{13}H_9FO_2$ mol.wt. 216.21

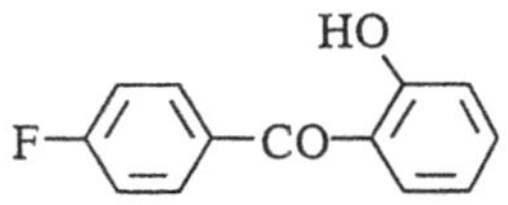

Syntheses

-Preparation by Fries rearrangement of phenyl p-fluorobenzoate with aluminium chloride,
*without solvent at 200° for 20 min [766];
*in nitrobenzene at 140-145° for 3 h (35%) [447].
-Also obtained by demethylation of 4-fluoro-2'-methoxybenzophenone with boron tribromide in
methylene chloride at r.t. for 12 h [766], according to [891].
-Also obtained (by-product) by reaction of p-fluorobenzoic acid with phenol in the presence of
hydrofluoric acid for 6 h at 75° in an autoclave under pressure (5%) [1137].
-Also refer to: [448].

m.p. and Spectra (NA).

(4-Fluorophenyl)(3-hydroxyphenyl)methanone

[62810-47-3] $C_{13}H_9FO_2$ mol.wt. 216.21

Syntheses

-Preparation by reaction of m-anisoyl chloride with fluorobenzene in the presence of an excess of aluminium chloride:
  first at 20°, then at reflux for 2 h (43%) [82].
-Preparation by reaction of m-hydroxybenzoic acid with
fluorobenzene in the presence of a hydrofluoric acid/boron trifluoride mixture at r.t. for 6 h into an
autoclave under 30 psig of boron trifluoride (61%) [1365].
-Preparation by diazotization of 3-amino-4'-fluorobenzophenone followed by hydrolysis of the
obtained diazonium salt [847].
-Preparation by demethylation of 4-fluoro-3'-methoxybenzophenone with aluminium chloride in
refluxing chlorobenzene [847].
-Also refer to: [437, 848, 1292].

m.p.   105° [847, 848], 103° [82, 1292], 102° [437], 99-99°5 [1365];
    ^{1}H NMR [1365], ^{13}C NMR [1365], ^{19}F NMR [1365].

(4-Fluorophenyl)(4-hydroxyphenyl)methanone

[25913-05-7] $C_{13}H_9FO_2$ mol.wt. 216.21

Syntheses

-Preparation by reaction of p-fluorobenzoyl chloride with
  phenol,
*in the presence of aluminium chloride [296];
*in the presence of boron trifluoride in hydrofluoric acid (83%) [318].
-Preparation by isomerization of 4-fluoro-2'-hydroxybenzophenone,
*with trifluoromethanesulfonic acid at 120° for 5 h (75%) or with perfluoroethanesulfonic acid at
  120° for 3 h (53%) [447];
*by dissolution in toluene at 110° and the resultant solution cooled to 3° (97%) [448].
-Preparation by Fries rearrangement of phenyl p-fluorobenzoate,
*with hydrofluoric acid between -10 to 0° (63%) [446];
*with aluminium chloride without solvent at 160° for 5 min [256, 257] or in nitrobenzene at 140-
  145° for 3 h (65%) [447].
-Preparation by reaction of p-fluorobenzoic acid with phenol,
*in the presence of hydrofluoric acid for 6 h at 75° in an autoclave under pressure (90%) [1137];
*in the presence of boron trifluoride in hydrofluoric acid (64%) [318];
*in the presence of trifluoromethanesulfonic acid overnight at r.t. (77%) [1138].
-Also obtained by reaction of EKONOL(TM), an aromatic polyester, behaves as a Friedel-Crafts
  acylating reagent, with fluorobenzene in triflic acid solution at 75° for 2 h (74%) or at 25° for 18 h
  (67%) [287]. Similar results can be obtained using hydrofluoric acid/boron trifluoride or
  aluminium chloride in place of triflic acid [287].
-Preparation by reaction of 4,4'-difluorobenzophenone (1 mol) with potassium hydroxide (2 mol) in
  aqueous dimethyl sulfoxide at 60° for 18 h (81%) [1121].
-Also refer to: [15, 178, 201, 295, 314, 448, 529, 678, 1225, 1304, 1365, 1421, 1458].
N.B.: Na [178] and K salts [570, 1121].

m.p.   170°6-172° [318], 169°5-171°5 [296], 169-171° [88], 168° [1121],
    166°5-168°5 [1138];
^{1}H NMR [318, 1138], ^{13}C NMR [287], IR [318, 1138], MS [287, 1137];
GLC [318];   HPLC [287].

(4-Hydroxyphenyl)(4-iodophenyl)methanone

[113275-52-8] $C_{13}H_9IO_2$ mol.wt. 324.12

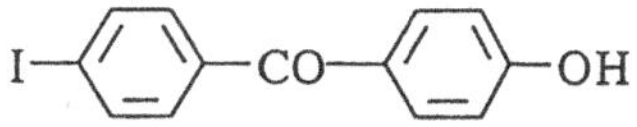

Synthesis

I—⟨ ⟩—CO—⟨ ⟩—OH   -Refer to: [15].

m.p. and Spectra (NA).

(2-Hydroxyphenyl)(2-nitrophenyl)methanone

[22293-32-9] $C_{13}H_9NO_4$ mol.wt. 243.22

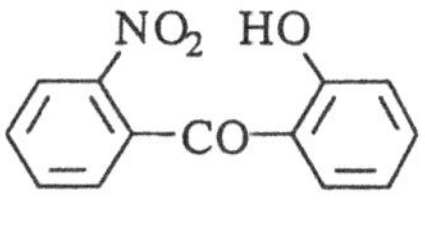

Synthesis

-Obtained by photo-Fries rearrangement of phenyl o-nitrobenzoate in ethanol during 60-75 h (23%) [1034].
-Also refer to: [1169].

m.p. 104° [1034];   IR [1034], UV [1034].

(2-Hydroxyphenyl)(3-nitrophenyl)methanone

[36412-61-0] $C_{13}H_9NO_4$ mol.wt. 243.22

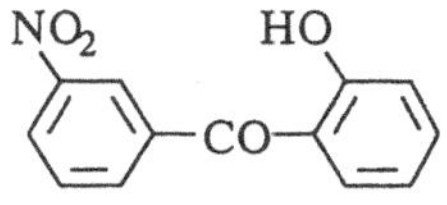

Syntheses

-Preparation by diazotization of 2-amino-3'-nitrobenzo-
phenone followed by thermal decomposition of the
2-(3'-nitrobenzoyl)benzenediazonium fluoborate formed
with 0.05 M sulfuric acid at 45 or 65° (67%) [345].
-Also obtained (poor yield) by Fries rearrangement of phenyl m-nitrobenzoate with aluminium
chloride at 120 or 160° for 2 h (5%) [1166, 1169].
-Also obtained by reaction of m-nitrobenzoyl chloride with anisole in the presence of aluminium
chloride in refluxing carbon disulfide for 2 h (16%) [205]. In this reaction, 4-methoxy-
3'-nitrobenzophenone was the major product.
-Also refer to: [404].

m.p. 101° [1166], 96-97° [345], 93°5-94°5 [205];
IR [205], UV [205].

(2-Hydroxyphenyl)(4-nitrophenyl)methanone

[68223-20-1] $C_{13}H_9NO_4$ mol.wt. 243.22

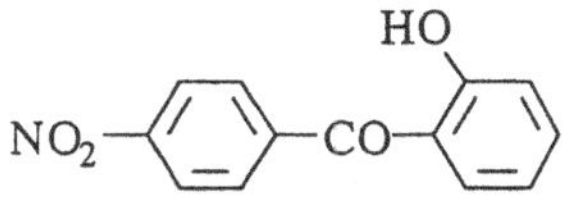

Syntheses

-Obtained by Fries rearrangement of phenyl p-nitrobenzoate
with aluminium chloride [1169],
*without solvent at 160° (26%) [1447], (15%) [1167];
*in refluxing chlorobenzene (8%) [1447].
-Also obtained (by-product) by Fries rearrangement of phenyl p-nitrobenzoate (SM) with
aluminium chloride at 140° for 30 min (16%). SM was obtained by heating p-nitrobenzoyl
chloride with aluminium tris(phenoxide) in a water bath for 30 min [759].

-Also obtained by photo-Fries rearrangement of phenyl p-nitrobenzoate in ethanol during 60-75 h (19%) [1034].
-Also obtained (by-product) by reaction of p-nitrobenzoyl chloride with phenetole in the presence of aluminium chloride in carbon disulfide [92, 99, 1456].
-Also refer to: [1063].

m.p.   114° [1167], 112° [1034], 111-113° [92], 111° [759], 108-110° [1447];
IR [1034], UV [1034].

(3-Hydroxyphenyl)(4-nitrophenyl)methanone

[147029-77-4] $C_{13}H_9NO_4$ mol.wt. 243.22

Synthesis

-Preparation by demethylation of 3-methoxy-4'-nitro-benzophenone with 62% hydrobromic acid in refluxing acetic acid for 4 h (70%) [152].

m.p.   117° [152];   Spectra (NA).

(4-Hydroxyphenyl)(2-nitrophenyl)methanone

[61101-88-0] $C_{13}H_9NO_4$ mol.wt. 243.22

Syntheses

-Preparation by reaction of o-nitrobenzoyl chloride with (trimethylsilyl)phenol in the presence of stannic chloride in refluxing methylene chloride for 2 h (56%) [649].
-Also obtained (trace) by reaction of o-nitrobenzoic acid in the presence of polyphosphoric acid at 100° for 20 min (0.1%) [959].
-Also refer to: [1169].

m.p.   165-167° [649], 122° [959]. There is a discrepancy between the two melting points.
Spectra (NA).

(4-Hydroxyphenyl)(3-nitrophenyl)methanone

[72090-63-2] $C_{13}H_9NO_4$ mol.wt. 243.22

Syntheses

-Preparation by acylation of phenetole with m-nitrobenzoyl chloride in ethyl ether in the presence of aluminium chloride, then dealkylation of the 4-ethoxy-3'-nitrobenzophenone so formed with the same catalyst [93], in boiling carbon disulfide (60-70°) for 8 h [91] according to [547].
-Preparation by Fries rearrangement of phenyl m-nitrobenzoate with aluminium chloride [1169] without solvent at 120° or at 160° for 2 h (32%) [1166].
-Also obtained (poor yield) by reaction of m-nitrobenzoyl chloride with phenyl borate in the presence of aluminium chloride in tetrachloroethane at 100° (9%) [1373].
-Also obtained (trace) by reaction of m-nitrobenzoic acid with phenol in the presence of polyphosphoric acid at 100° for 20 min (1%) [959].
-Also refer to: [680, 681, 1182].

m.p.   173° [91, 1166], 171° [959];   Spectra (NA);   cryoscopic study [93].

(4-Hydroxyphenyl)(4-nitrophenyl)methanone

[18920-70-2] $C_{13}H_9NO_4$ mol.wt. 243.22

Syntheses

NO_2—⟨_⟩—CO—⟨_⟩—OH   -Obtained by Fries rearrangement of phenyl p-nitrobenzoate with aluminium chloride [1169],
*without solvent at 140° for 30 min (52%) [759], at 120° for 2 h (21%) [1167] or first at 130° for
1 h, then at 160° for 1 h (2%) [1447];
*in refluxing chlorobenzene for 5 h (24%) [1447].
N.B.: Phenyl p-nitrobenzoate failed to undergo the Fries rearrangement in the presence of
aluminium chloride [202].
-Preparation by dealkylation,
*of 4-methoxy-4'-nitrobenzophenone with aluminium chloride in boiling carbon disulfide or
without solvent at 100-120° [92];
*of 4-ethoxy-4'-nitrobenzophenone with aluminium chloride [93], in boiling carbon disulfide or
without solvent at 100-120° [92].
-Preparation by reaction of p-nitrobenzoyl chloride with phenetole in the presence of aluminium
chloride in ethyl ether [93] or in carbon disulfide. Simultaneous deethylation take place during the
reaction [92].
-Also obtained (trace) by reaction of p-nitrobenzoic acid with phenol in the presence of
polyphosphoric acid at 100° for 20 min [959].
-Also refer to: [909, 1025, 1182].
N.B.: Na salt [1063].

m.p. 193-195° [1447], 192° [1167], 190-192° [92], 190° [759, 959];
IR [180]; cryoscopic study [93].

(2-Amino-5-chlorophenyl)(2-hydroxyphenyl)methanone

[62492-57-3] $C_{13}H_{10}ClNO_2$ mol.wt. 247.68

Synthesis

-Preparation by demethylation of 2-amino-5-chloro-
2'-methoxybenzophenone with boron tribromide in
methylene chloride at r.t. for 4 h (88%) [746].

m.p. 74-76° [746]; ^1H NMR [746].

(2-Amino-5-chlorophenyl)(3-hydroxyphenyl)methanone

[62492-58-4] $C_{13}H_{10}ClNO_2$ mol.wt. 247.68

Synthesis

-Preparation by demethylation of 2-amino-5-chloro-
3'-methoxybenzophenone with boron tribromide in
methylene chloride at r.t. for 4 h (84%) [746].

m.p. 190-192° [746]; ^1H NMR [746].

(2-Amino-5-chlorophenyl)(4-hydroxyphenyl)methanone

[784-41-8] $C_{13}H_{10}ClNO_2$ mol.wt. 247.68

Syntheses

-Preparation by cleavage of 5-chloro-3-(p-hydroxyphenyl)-2,1-benzisoxazole (other name: 5-chloro-3-(p-hydroxy-phenyl)anthranil) (SM),
*by heating at reflux with aluminium iodide for 50 min (87%) [753];

*by reaction with concentrated hydrochloric acid and an excess of tin in boiling ethanol or acetic acid. SM (m.p. 241°) was prepared by condensation of o-nitrobenzaldehyde with phenol in the presence of hydrogen chloride or phosphorous oxychloride in cold acetic acid [1474];
*by hydrogenation in the presence of 10% Pd/C in ethyl acetate between 40-60° at a pressure of 3 atmospheres (quantitative yield) [1419].
-Preparation by demethylation of 2-amino-5-chloro-4'-methoxybenzophenone with boron tribromide in methylene chloride at r.t. for 4 h (80%) [746].

m.p.   174° [1474], 173-175° [1419], 172-173° [753], 166-168° [746]; ^1H NMR [746].

(4-Amino-3-nitrophenyl)(4-hydroxyphenyl)methanone

[60014-09-7] $C_{13}H_{10}N_2O_4$ mol.wt. 258.23

Synthesis

-Preparation by reaction of ammonia with 4-chloro-4'-hydroxy-3-nitrobenzophenone in dimethyl sulfoxide at 100° for 6 h (72%) [20].
-Also refer to: [833].

m.p.   218° [20];   IR [20], MS [20].

(2-Aminophenyl)(2-hydroxyphenyl)methanone

[13134-93-5] $C_{13}H_{11}NO_2$ mol.wt. 213.24

Syntheses

-Preparation by demethylation of 2-amino-2'-methoxy-benzophenone (SM),
*with concentrated hydrobromic acid in refluxing acetic acid for 24 h (85%) [155];

*with aluminium chloride in refluxing benzene for 1 h (94%) [1445]. SM was obtained according to [799].
-Also obtained by action of an excess ammonia on the 2,2'-dihydroxybenzophenone in ethanol [495].

m.p.   222° [495];   red oil [155], yellow [1445];
^1H NMR [1445], IR [1445], MS [1445];   TLC [1445].

(2-Aminophenyl)(2-hydroxyphenyl)methanone *(Hydrochloride)*

$C_{13}H_{11}NO_2$,HCl   mol.wt. 249.70

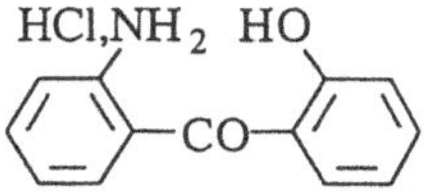

Synthesis

-Obtained from the corresponding amine [495].

m.p. 242° [495];   Spectra (NA).

(2-Aminophenyl)(3-hydroxyphenyl)methanone

[38824-12-3] $C_{13}H_{11}NO_2$ mol.wt. 213.24

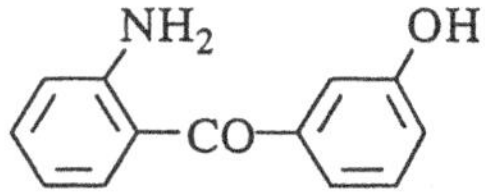

Synthesis

-Refer to: [204].

m.p. and Spectra (NA).

(3-Aminophenyl)(2-hydroxyphenyl)methanone

[35486-64-7] $C_{13}H_{11}NO_2$ mol.wt. 213.24

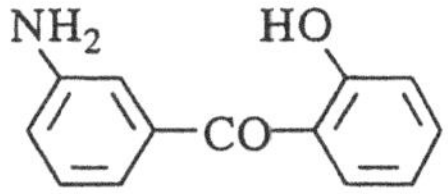

Synthesis

-Preparation by reduction of 2-hydroxy-3'-nitrobenzo-
  phenone with ammonium ferrous sulfate (86%) [205].
-Also refer to: [203, 204].

m.p. 119-120° [205];   IR [205], UV [205].

(4-Aminophenyl)(2-hydroxyphenyl)methanone

[13134-94-6] $C_{13}H_{11}NO_2$ mol.wt. 213.24

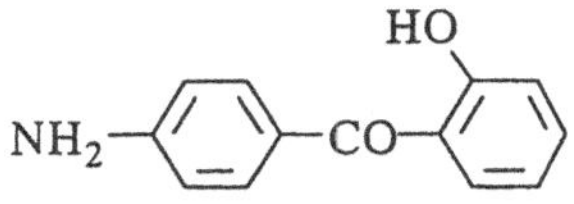

Synthesis

-Preparation by hydrogenation of 2-hydroxy-4'-nitro-
  benzophenone in the presence of Raney nickel in methanol
  (89%) [1456].
-Also refer to: [1135].

m.p. 138-139° [1456];   Spectra (NA).

(4-Aminophenyl)(4-hydroxyphenyl)methanone

[14963-34-9] $C_{13}H_{11}NO_2$ mol.wt. 213.24

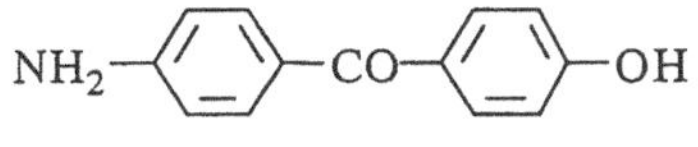

Synthesis

-Obtained by heating 4-hydroxy-4'-nitrodiphenylmethane
  (SM), sulfur and sodium hydroxide in 50% ethanol in a
  boiling water bath for 7 h (60%) [1392]. In this
reaction, oxidation of the methylene group to a carbonyl group occurred together with reduction of

the nitro group. The oxidizing agent for the methylene group was the tetrasulfide [1394]. SM was obtained by diazotization of 4-amino-4'-nitrodiphenylmethane (83%), according to [1393].
-Also refer to: [1391].

m.p. 184° [1392];   UV [1391, 1392].

(3,4-Diaminophenyl)(4-hydroxyphenyl)methanone

[93958-45-3]   C13H12N2O2   mol.wt. 228.25

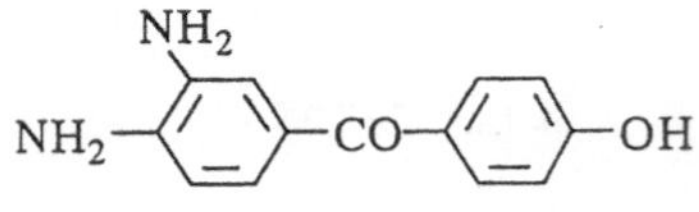

Synthesis

-Preparation by catalytic hydrogenation of 4-amino-4'-hydroxy-3-nitrobenzophenone in the presence of Raney nickel in ethanol in a Paar hydrogenator at 3 kg/cm2 pressure for 4 h [20].

solid mass [20];   m.p. and Spectra (NA).

(2-Hydroxyphenyl)[2-(trifluoromethyl)phenyl]methanone

[205319-41-1]   C14H9F3O2   mol.wt. 266.22

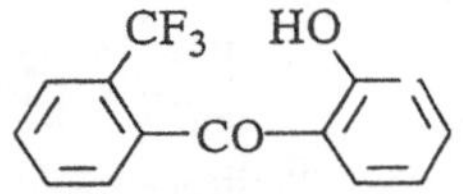

Synthesis

-Refer to: Chem. Abstr., **128**, 257232e (1998).

m.p. and Spectra (NA).

(3-Hydroxyphenyl)[3-(trifluoromethyl)phenyl]methanone

[62810-48-4]   C14H9F3O2   mol.wt. 266.22

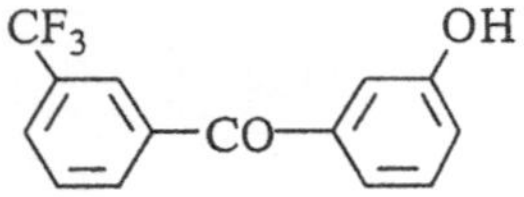

Synthesis

-Preparation by demethylation of 3-methoxy-3'-(trifluoromethyl)benzophenone with refluxing pyridinium chloride or with 48% hydrobromic acid [847].

m.p. 78° [847];   Spectra (NA).

(3-Hydroxyphenyl)[4-(trifluoromethyl)phenyl]methanone

[21084-29-7]   C14H9F3O2   mol.wt. 266.22

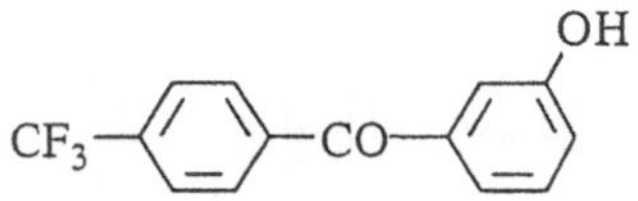

Syntheses

-Preparation by demethylation of 3-methoxy-4'-(trifluoromethyl)benzophenone with refluxing pyridinium chloride for 90 min (72%) or with 48% hydrobromic acid [847].
-Preparation by dealkylation of 4-(trifluoromethyl)-3'-ethoxybenzophenone with pyridinium bromide at 210° for 0.5 h [362].

m.p. 130° [847], 127-128° [362];   Spectra (NA).

(4-Hydroxyphenyl)[2-(trifluoromethyl)phenyl]methanone

[190728-32-6] $C_{14}H_9F_3O_2$ mol.wt. 266.22

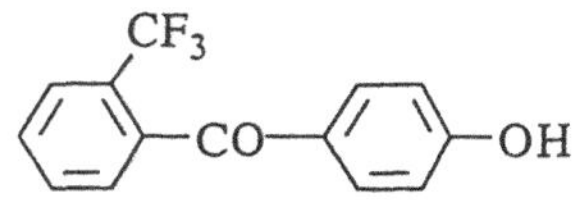

Synthesis

-Refer to: Chem. Abstr., **127**, 34137f (1997).

m.p. and Spectra (NA).

(4-Hydroxyphenyl)[3-(trifluoromethyl)phenyl]methanone

[732-55-8] $C_{14}H_9F_3O_2$ mol.wt. 266.22

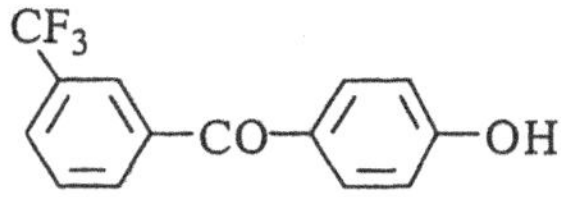

Syntheses

-Preparation by reaction of m-(trifluoromethyl)benzoyl
  fluoride with phenol in hydrofluoric acid at 100° for 6 h
  under 5 atmospheres (92%) [902].
-Preparation by reaction of m-(trifluoromethyl)benzoyl
chloride with phenetole in the presence of aluminium chloride in carbon disulfide between 0 to 5°,
then at r.t. for overnight (21%) [207].

m.p.   144-145° [207];   Spectra (NA).

(4-Hydroxyphenyl)[4-(trifluoromethyl)phenyl]methanone

[21084-27-5] $C_{14}H_9F_3O_2$ mol.wt. 266.22

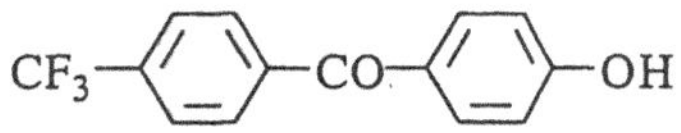

Syntheses

-Preparation by reaction of p-(trifluoromethyl)benzoyl
  chloride with phenol in the presence of aluminium
  chloride [296].
-Also obtained from 4-fluoro-4'-(trifluoromethyl)benzophenone by reaction under nitrogen with
  potassium hydroxide in aqueous dimethyl sulfoxide at 60° for 18 h [1121].
-Preparation by dealkylation of 4-(trifluoromethyl)-4'-ethoxybenzophenone with pyridinium
  bromide at 210° for 0.5 h (77%) [362].
-Also refer to: [15, 1458].
N.B.: K salt [1121].

m.p.   147° [1121], 144-145° [362], 142-143° [296];   Spectra (NA).

(2,3-Dichloro-4-methoxyphenyl)(4-hydroxyphenyl)methanone

[92285-27-3] $C_{14}H_{10}Cl_2O_3$ mol.wt. 297.14

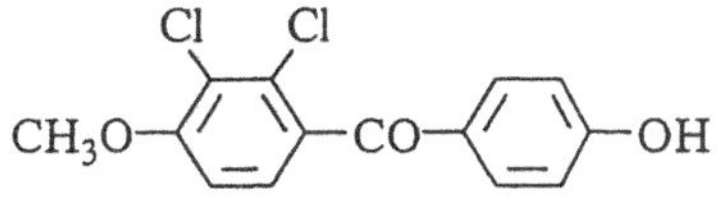

Synthesis

-Preparation by adding 2,3-dichloro-4-methoxy-
  4'-nitrobenzophenone in a mixture of acetaldoxime and
  sodium hydroxide in N,N-dimethylformamide cooled
  in an ice bath, and the mixture stirred overnight
at r.t. (75%) [1071].

m.p.   215-217° [1071];   Spectra (NA).

(3,5-Dichloro-2-methoxyphenyl)(2-hydroxyphenyl)methanone

$C_{14}H_{10}Cl_2O_3$ mol.wt. 297.14

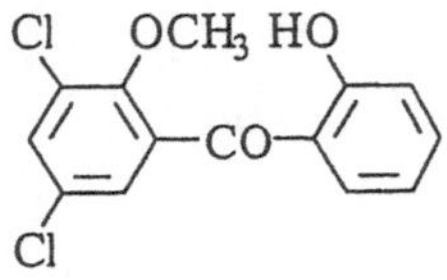

Synthesis

-Preparation from 2-iodophenyl 3,5-dichloro-2-methoxy-
  benzoate on treatment with n-butyllithium in a mixture of
  ethyl ether, hexane and tetrahydrofuran at -70° for 2 h,
  followed by treatment with saturated aqueous ammonium
  chloride (68%) [598].

pale yellow oil [598];   b.p. (NA);
^{1}H NMR [598], IR [598], MS [598].

(3,5-Dichloro-4-methoxyphenyl)(2-hydroxyphenyl)methanone

[129103-88-4] $C_{14}H_{10}Cl_2O_3$ mol.wt. 297.14

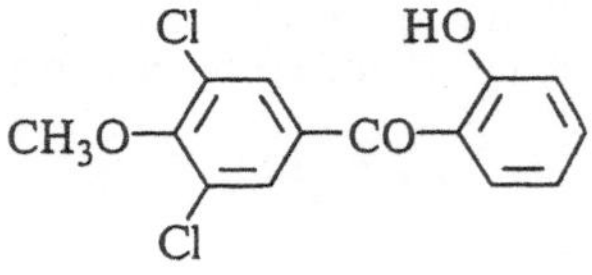

Synthesis

N.B.: This benzophenone was mentioned in the [Chem.
  Abstr., **113**, 131832x (1990)]. It has never been prepared by
  authors [598]. The paper actually concerns the 3,5-di-
  chloro-2'-hydroxy-2-methoxybenzophenone or
  (3,5-dichloro-2-methoxyphenyl)(2-hydroxyphenyl)-
methanone. In entry 6 (table 1) of the paper [598], R$_3$ = H, but this information was not indicated
in this one [1134].

m.p. and Spectra (NA).

(2-Chloro-4-methylphenyl)(4-hydroxyphenyl)methanone

[98155-82-9] $C_{14}H_{11}ClO_2$ mol.wt. 246.69

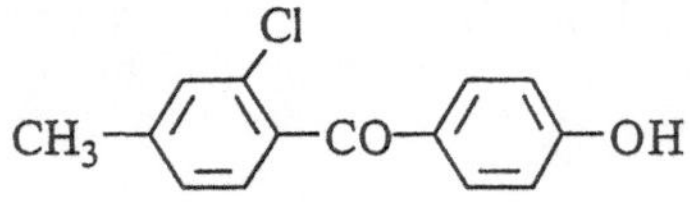

Synthesis

-Refer to: [486] (compound 33).

m.p. and Spectra (NA).

(3-Chloro-4-methylphenyl)(4-hydroxyphenyl)methanone

[83885-15-8] $C_{14}H_{11}ClO_2$ mol.wt. 246.69

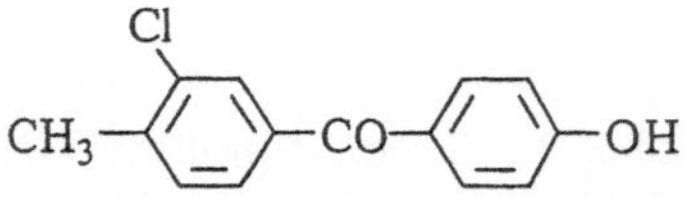

Synthesis

-Preparation by reaction of 3-chloro-4-methylbenzoyl
  chloride with phenol in the presence of aluminium
  chloride [296].

m.p.   153-154° [296];   Spectra (NA).

(4-Chloro-2-methylphenyl)(4-hydroxyphenyl)methanone

[98155-76-1] $C_{14}H_{11}ClO_2$ mol.wt. 246.69

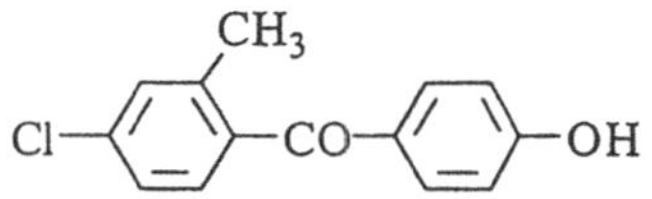

Synthesis

-Refer to: [486] (compound 9).

m.p. and Spectra (NA).

(4-Chloro-3-methylphenyl)(4-hydroxyphenyl)methanone

[83885-20-5] $C_{14}H_{11}ClO_2$ mol.wt. 246.69

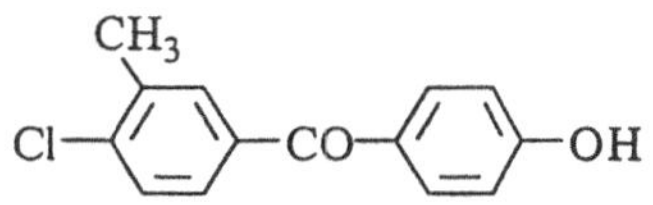

Synthesis

-Preparation by reaction of 4-chloro-3-methylbenzoyl chloride with phenol in the presence of aluminium chloride [296].

m.p.   154-156° [296];   Spectra (NA).

(3-Chloro-4-methoxyphenyl)(4-hydroxyphenyl)methanone

[83885-14-7] $C_{14}H_{11}ClO_3$ mol.wt. 262.69

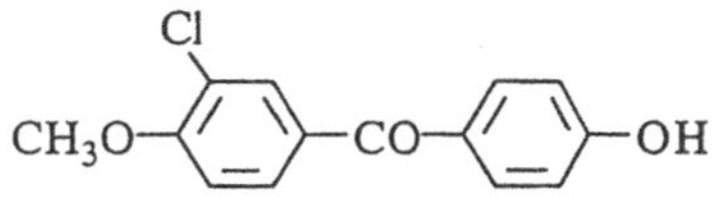

Synthesis

-Preparation by reaction of 3-chloro-4-methoxybenzoyl chloride with phenol in the presence of aluminium chloride [296].

m.p.   167-168° [296];   Spectra (NA).

(3-Fluoro-4-methylphenyl)(3-hydroxyphenyl)methanone

[62810-52-0] $C_{14}H_{11}FO_2$ mol.wt. 230.24

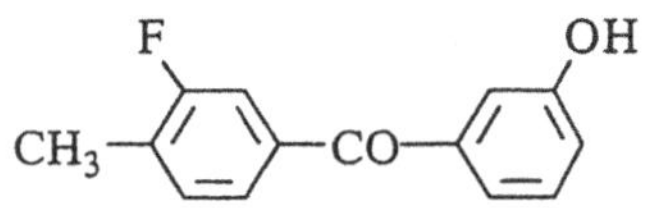

Synthesis

-Preparation by diazotization of 3'-amino-3-fluoro-4-methyl-benzophenone followed by hydrolysis of the diazonium salt obtained [847].

m.p.   120° [847];   Spectra (NA).

(2-Hydroxyphenyl)(2-methylphenyl)methanone

[51974-19-7] $C_{14}H_{12}O_2$ mol.wt. 212.25

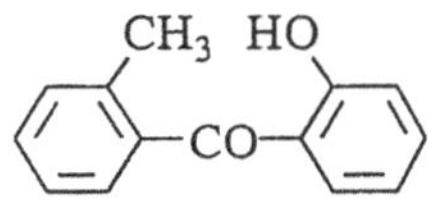

Syntheses

-Preparation by hydrogenation of 5-chloro-2-hydroxy-2'-methylbenzophenone in ethanolic solution in the presence of 10% Pd/C and potassium acetate at r.t. under atmosphere pressure (98%) [1363].

-Also obtained by photo-Fries rearrangement of phenyl o-toluate in methanol or isopropanol (36-39%) and in benzene or ethyl ether (22%) [1070].
-Also obtained (by-product) by diazotization of 2-amino-2'-methylbenzophenone with sodium nitrite in 5 N hydrochloric acid (3%). 1-methylfluorenone was the major product obtained in this reaction [834].

　　　oil [1363];　　m.p. 65-67° [834];　　^{1}H NMR [1363].

(2-Hydroxyphenyl)(3-methylphenyl)methanone

[33785-66-9] $C_{14}H_{12}O_2$ mol.wt. 212.25

Syntheses

-Preparation from 2-methylthioxanthen-9-one 10,10-dioxide (SM) by a three-step synthesis: SM, by refluxing in 2% sodium hydroxide-65% dioxane-water solution for 18 h gave the 2-(2-hydroxybenzoyl)-4-methylphenylsulfinic acid (72%). The former, by reaction with mercuric chloride in refluxing acetic acid for 4 h led to the 2-chloromercuri-2'-hydroxy-5-methylbenzophenone (74%). Removal of the chloromercury group was achieved with concentrated hydrochloric acid in refluxing ethanol for 2 h (82%) [160].
-Also obtained by Fries rearrangement of phenyl m-toluate in the presence of aluminium chloride in refluxing carbon disulfide, then elimination of the solvent and heating at 150° for 3 h (20%) [160].
-Also obtained by reaction of m-toluoyl chloride with phenyl borate in the presence of aluminium chloride in tetrachloroethane at 100° (13%) [1373].

　　　oil [160];　　b.p.$_{0.9}$ 140-141° [160];　　^{1}H NMR [160], IR [160].

(2-Hydroxyphenyl)(4-methylphenyl)methanone

[19434-30-1] $C_{14}H_{12}O_2$ mol.wt. 212.25

Syntheses

-Preparation by reaction of pyridinium chloride on the 2-methoxy-4'-methylbenzophenone (SM) at reflux (33%). SM was obtained by reaction of p-tolunitrile on the o-methoxyphenylmagnesium bromide in ethyl ether (44%) [146].
-Preparation by Fries rearrangement of phenyl p-methylbenzoate,
*with aluminium chloride in refluxing chlorobenzene for 10 h (40%) or without solvent, at 160° (22%) [1447] or at 180° for 10 min (15%) [944];
*with Nafion-H, a polymeric perfluorinated resin sulfonic acid, in refluxing nitrobenzene for 12 h (18%) [1004].
-Also obtained by reaction of aluminium chloride with 5-tert-butyl-2-methoxy-4'-methyl-benzophenone in benzene at 65-70° for 45 h (60 to 80%) [774, 775].
-Also obtained by reaction of 2-methoxybenzoyl chloride with toluene in the presence of aluminium chloride [21, 775], (72%) [1397], (47%) [345].
-Demethylation occurred during the Friedel-Crafts acylation, especially in the presence of ferric chloride at 130-140° [21].
-Also obtained by thermal decomposition of 2-(4'-methylbenzoyl)benzenediazonium fluoborate in 0.05 M sulfuric acid at 25° (36%) [345].
-Also refer to: [243].

　　　m.p.　　61°5 [1397], 61-63° [146], 61-62° [345], 58-60° [1447], 40° [944],
　　　　　　　39-40° [774, 775]. There is a discrepancy between the various melting points.

　　^{1}H NMR [1004], ^{13}C NMR [1004].

(3-Hydroxyphenyl)(2-methylphenyl)methanone

[147029-78-5] $C_{14}H_{12}O_2$ mol.wt. 212.25

Synthesis

-Preparation by condensation of the Grignard reagent of anisole with o-toluoyl chloride, followed by demethylation of the resulting methyl ether (excellent yield) [152].

m.p. 112° [152];   Spectra (NA).

(3-Hydroxyphenyl)(4-methylphenyl)methanone

[62810-49-5] $C_{14}H_{12}O_2$ mol.wt. 212.25

Syntheses

-Preparation by reaction of m-anisoyl chloride with toluene in the presence of an excess of aluminium chloride: first at 20°, then at reflux for 2 h (78%) [82, 1292].
-Preparation by diazotization of 3-amino-4'-methyl-benzophenone followed by hydrolysis of the diazonium salt obtained [847].
-Also obtained by demethylation of 3-methoxy-4'-methylbenzophenone with aluminium chloride in refluxing chlorobenzene [847].

m.p. 126° [82, 1292], 121° [847];   Spectra (NA).

(4-Hydroxyphenyl)(2-methylphenyl)methanone

[52981-01-8] $C_{14}H_{12}O_2$ mol.wt. 212.25

Syntheses

-Preparation by reaction of o-toluic acid with phenol in the presence of polyphosphoric acid at 100° for 20 min (47%) [959].
-Also obtained by photo-Fries rearrangement of phenyl o-toluate in methanol or isopropanol (21%) and in ethyl ether or benzene (6-8%) [1070].

m.p. 96° [959];   Spectra (NA).

(4-Hydroxyphenyl)(3-methylphenyl)methanone

[71372-37-7] $C_{14}H_{12}O_2$ mol.wt. 212.25

Syntheses

-Preparation by Fries rearrangement of phenyl m-toluate in the presence of aluminium chloride in refluxing carbon disulfide for 2 h, then elimination of the solvent and heating at 150° for 3 h (major product) [160].
-Preparation by reaction of m-toluoyl chloride with phenetole in the presence of aluminium chloride in carbon disulfide between 0 to 5° for 8 h, then at r.t. overnight (50%) [207].
-Also obtained by reaction of m-toluic acid with phenol in the presence of polyphosphoric acid at 100° for 20 min (19%) [959].

-Also obtained (poor yield) by reaction of m-toluoyl chloride with phenyl borate in the presence of aluminium chloride in tetrachloroethane at 100° (5%) [1373].

m.p. 166° [959], 165-166° [160], 163-164° [207]; Spectra (NA).

(4-Hydroxyphenyl)(4-methylphenyl)methanone

[134-92-9] $C_{14}H_{12}O_2$ mol.wt. 212.25

Syntheses

CH_3—⟨_⟩—CO—⟨_⟩—OH

-Preparation by reaction of p-methylbenzoyl chloride with phenol in the presence of aluminium chloride [296].

-Preparation by reaction of p-toluic acid with phenol in the presence of polyphosphoric acid at 100° for 20 min (24%) [959].
-Preparation by Fries rearrangement of phenyl p-toluate,
*with aluminium chloride without solvent at 130° for 1 h, then 160° for 1 h (20%) or in refluxing chlorobenzene for 7.5 h (57%) [1447];
*with Nafion-H, a polymeric perfluorinated resin sulfonic acid, in refluxing nitrobenzene for 12 h (45%) [1004].
-Preparation by diazotization of 4-amino-4'-methylbenzophenone, followed by hydrolysis of the diazonium salt obtained (50%) [1475].
-Also refer to: [481, 482, 589, 1114, 1462].

m.p. 171-173° [296], 170° [959], 166-167° [1475], 161-162° [1447];
^{1}H NMR [1004], ^{13}C NMR [1004].

(4-Hydroxyphenyl)[4-(methylthio)phenyl]methanone

[83888-61-3] $C_{14}H_{12}O_2S$ mol.wt. 244.31

Synthesis

CH_3S—⟨_⟩—CO—⟨_⟩—OH

-Preparation by reaction of p-(methylthio)benzoyl chloride with phenol in the presence of aluminium chloride [296].

m.p. 133-134° [296]; Spectra (NA).

(2-Hydroxyphenyl)(2-methoxyphenyl)methanone

[21147-18-2] $C_{14}H_{12}O_3$ mol.wt. 228.25

Syntheses

OCH₃ HO
⟨_⟩—CO—⟨_⟩

-Preparation from 2-iodophenyl 2-methoxybenzoate by treatment with n-butyllithium in a mixture of ethyl ether, hexane and tetrahydrofuran at -70° for 2 h, followed by treatment with saturated aqueous ammonium chloride (93%) [598].
-Preparation by hydrogenolysis of 2-(benzyloxy)-2'-methoxybenzophenone (SM) in ethyl acetate in the presence of 10% Pd/C at r.t. under atmospheric pressure (91%). SM was obtained by reaction of 2-(benzyloxy)benzaldehyde with 2-methoxyphenylmagnesium bromide in tetrahydrofuran [1363].

-Also obtained by partial demethylation of 2,2'-dimethoxybenzophenone with boron trichloride in methylene chloride: first at -70°, then at r.t. for 30 min. From 2,2'-dimethoxybenzophenone, one methyl group is lost rapidly and a second somewhat more slowly [330].
-Also obtained (poor yield) by Fries rearrangement of phenyl o-anisate with aluminium chloride at 160° [1169].
-Preparation by partial methylation of 2,2'-dihydroxybenzophenone with dimethyl sulfate in the presence of potassium carbonate in refluxing acetone for 5 h [657].
-Also refer to: [496].

oil [598, 1363];   m.p. 77° [657]; ^{1}H NMR [598, 1363], IR [598], MS [125, 598].

(2-Hydroxyphenyl)(3-methoxyphenyl)methanone

[21554-73-4] $C_{14}H_{12}O_3$ mol.wt. 228.25

Syntheses

-Preparation from 2-methoxythioxanthen-9-one 10,10-dioxide (SM) by a three-step synthesis: SM by refluxing in 2% sodium hydroxide-65% dioxane-water solution for 18 h gave the 2-(2-hydroxybenzoyl)-4-methoxyphenylsulfinic acid (68%). The former by reaction with mercuric chloride in refluxing aqueous acetic acid for 4 h led to the 2-chloromercuri-2'-hydroxy-5-methoxybenzophenone (68%). Removal of the chloromercury group was achieved with concentrated hydrochloric acid in refluxing ethanol for 2 h (83%) [160].
-Also obtained by Fries rearrangement of phenyl m-anisate in the presence of aluminium chloride in refluxing carbon disulfide for 2 h, then elimination of the solvent and heating at 150° for 3 h (34%) [160].
-Also obtained by selective demethylation of 2,3'-dimethoxybenzophenone in the presence of boron trichloride in methylene chloride at r.t. for 30 min [87, 415].
-Also obtained (poor yield) by reaction of 3-methoxybenzoyl chloride with anisole in the presence of aluminium chloride in refluxing carbon disulfide for 2 h [87].
-Also refer to: [1169].

light yellow oil [160];   m.p. 40-42° [87];   b.p.$_{0.5}$ 148-149° [160];
^{1}H NMR [87, 160], IR [87, 160], UV [87].

(2-Hydroxyphenyl)(4-methoxyphenyl)methanone

[18733-07-8] $C_{14}H_{12}O_3$ mol.wt. 228.25

Syntheses

-Preparation from 2-iodophenyl p-anisate on treatment with n-butyllithium in a mixture of ethyl ether, hexane and tetrahydrofuran at -70° for 2 h, followed by treatment with saturated aqueous ammonium chloride (<18%) [598].
-Preparation by Fries rearrangement of phenyl p-anisate in the presence of aluminium chloride [1169].
-Also obtained by reaction of salicylaldehyde with p-iodoanisole by using a catalyst system of palladium chloride/lithium chloride in the presence of sodium carbonate in N,N-dimethyl-formamide at 100° for 6 h (81%) [1204].
-Preparation by partial demethylation of 2,4'-dimethoxybenzophenone with aluminium chloride in chlorobenzene at 80-100° [602].
-Also refer to: [1104].

m.p. (NA); ^{1}H NMR [1204], UV [1067], MS [1204].

(3-Hydroxyphenyl)(4-methoxyphenyl)methanone

[103203-53-8] $C_{14}H_{12}O_3$ mol.wt. 228.25

Syntheses

-Obtained by saponification of 3-acetoxy-4'-methoxy-
benzophenone with 2 N sodium hydroxide in refluxing
dilute ethanol for 30 min (98%). The starting material was
obtained by Friedel-Crafts acylation of anisole with
m-acetoxybenzoyl chloride in the presence of aluminium chloride in carbon disulfide at 5° for 20 h
(15%) [1149].
-Also obtained by selective demethylation of 3,4'-dimethoxybenzophenone in the presence of boron
trichloride in methylene chloride at r.t. for 30 min [87].

m.p. 133° [1149], 130-133° [87]; Spectra (NA).

(4-Hydroxyphenyl)(2-methoxyphenyl)methanone

[72090-61-0] $C_{14}H_{12}O_3$ mol.wt. 228.25

Syntheses

-Preparation by reaction of o-methoxybenzoic acid with
phenol in the presence of polyphosphoric acid at 100° for
20 min (61%) [959] or at 75-85° for 3 h (56%) [170].
-Preparation by Fries rearrangement of phenyl o-methoxybenzoate with polyphosphoric acid at 100° for 20 min (43%) [867].
-Also refer to: [171, 173, 175, 1169].
N.B.: Na salt [1331].

m.p. 152-153° [867], 149° [959], 147-149° [170];
^1H NMR (Sadtler: standard n° 38490 M),
IR (Sadtler: standard n° 65528 K), UV [867].

(4-Hydroxyphenyl)(3-methoxyphenyl)methanone

[72090-62-1] $C_{14}H_{12}O_3$ mol.wt. 228.25

Syntheses

-Preparation by Fries rearrangement of phenyl m-anisate
in the presence of aluminium chloride in refluxing carbon
disulfide for 2 h, then elimination of the solvent and
heating at 150° for 3 h (major product) [160].
-Preparation by reaction of m-methoxybenzoyl chloride with phenetole in the presence of
aluminium chloride in carbon disulfide between 0 to 5°, then at r.t. overnight (51%) [207].
-Also obtained by reaction of m-methoxybenzoic acid with phenol in the presence of
polyphosphoric acid at 100° for 20 min (15%) [959].
-Also refer to: [1169].

m.p. 141-142° [207], 138° [959], 137-138° [160]; Spectra (NA).

(4-Hydroxyphenyl)(4-methoxyphenyl)methanone

[61002-54-8] $C_{14}H_{12}O_3$ mol.wt. 228.25

Syntheses

CH_3O—⬡—CO—⬡—OH

-Preparation by reduction of 4-methoxy-4'-nitro-
  benzophenone with stannous chloride and
  hydrochloric acid, followed by diazotization of the
resulting 4-amino-4'-methoxybenzophenone and hydrolysis of the diazonium salt [92].
-Preparation by Fries rearrangement of phenyl p-anisate,
*with aluminium chloride,
  without solvent at 120° or at 160° [514];
  in nitromethane at 20° for 170 h (35%) [867];
  in nitrobenzene at 75° for 6 h (49%) [867] or at 80° for 1 h (21%) [156];
*with titanium tetrachloride,
  without solvent at 95-100° for 30 min (11%) [867];
  in nitromethane at 20° for 170 h (76%) [867];
*with stannic chloride in nitromethane at 20° for 170 h (14%) [867];
*with polyphosphoric acid in a water bath for 30 min (43%) [961].
-Also obtained by reaction of p-anisic acid with phenol in the presence of polyphosphoric acid by
  heating in a water bath for 30 min (34%) [961] or at 100° for 20 min (75%) [959].
-Also obtained by reaction of p-hydroxybenzoic acid with anisole,
*in the presence of polyphosphoric acid in a boiling water bath for 20 min (25%) [961];
*in the presence of zinc chloride and a mixture of polyphosphoric acid/85% phosphoric acid
  (60:40) at 40°. Then, during 1.5 h, phosphorous trichloride was added and the mixture heated at
  60° for 16 h (91%) [1302];
*in the presence of boron trifluoride in nitrobenzene at 80° for 30 min (74%) [390].
-Preparation by Friedel-Crafts acylation,
*of phenol with p-methoxybenzoyl chloride in the presence of aluminium chloride [296];
*of anisole with p-(acetoxy)benzoyl chloride in the presence of aluminium chloride in carbon
  disulfide at 15° for 1 h. The saponification of the 4-(acetoxy)-4'-methoxybenzophenone formed
  with sodium hydroxide in refluxing dilute ethanol for 30 min gave the expected ketone (95%)
  [1149].
-Preparation by hydrolysis of 4-(4-anisoyloxy)-4'-methoxybenzophenone with concentrated
  sulfuric acid on standing for 10 min (79%) [961].
-Preparation from 4-fluoro-4'-methoxybenzophenone by reaction under nitrogen with potassium
  hydroxide in aqueous dimethyl sulfoxide at 60° for 18 h [1121].
-Also refer to: [173, 175, 210, 211, 442, 476, 749, 752, 909, 1169, 1320, 1458].
N.B.: Na [1331] and K salts [1121].

m.p.   155° [1121], 154° [1149], 153-154° [156], 151-152° [92],
        151° [867, 959, 961], 150°6-151°8 [1302], 150-151° [514],
        114°5-145°5 [296]. A typing error probably occurred in the published data.
^{1}H NMR (Sadtler: standard n° 38498 M), IR (Sadtler: standard n° 65536 K), UV [867].

[2-(Acetyloxy)phenyl](4-hydroxyphenyl)methanone

[145723-29-1] $C_{15}H_{12}O_4$ mol.wt. 256.26

Synthesis

⬡ OCOCH$_3$ —CO—⬡—OH

-Obtained by photooxygenation of 3-(4-hydroxyphenyl)-
  2-methylbenzofuran in methylene chloride at 5° (60%) [12].

m.p.   125-126° [12]; ^{1}H NMR [12], ^{13}C NMR [12], IR [12], UV [12].

[4-(2-Bromoethoxy)phenyl](4-hydroxyphenyl)methanone

[79578-62-4] $C_{15}H_{13}BrO_3$ mol.wt. 321.17

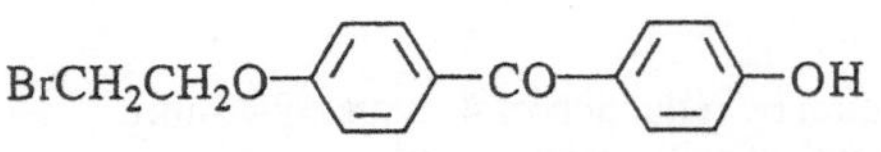

Synthesis

$BrCH_2CH_2O$—⟨phenyl⟩—CO—⟨phenyl⟩—OH -Preparation by reaction of 4-hydroxybenzoic
                                                acid with β-bromophenetole [1158], in solution
                                                of a polyphosphoric acid/85% phosphoric
acid/zinc chloride mixture. The solution was heated to 50-60°, phosphorous trichloride was added
during 1 h and the mixture heated for 20 h at 70° (74%) [715].

m.p. 139-142° [715], 136-138° [1158]; ^{1}H NMR [715], IR [715], MS [715].

(2,3-Dimethyl-5-nitrophenyl)(2-hydroxyphenyl)methanone

[110969-51-2] $C_{15}H_{13}NO_4$ mol.wt. 271.27

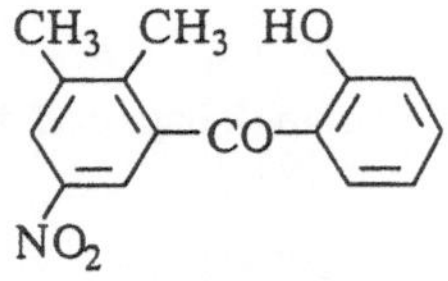

Synthesis

-Obtained by Fries rearrangement of phenyl 2,3-dimethyl-
5-nitrobenzoate with aluminium chloride without solvent at
160° for 2 h (17%) [843].

m.p. 126° [843]; ^{1}H NMR [843], IR [843], UV [843].

(2,3-Dimethyl-5-nitrophenyl)(4-hydroxyphenyl)methanone

[110969-52-3] $C_{15}H_{13}NO_4$ mol.wt. 271.27

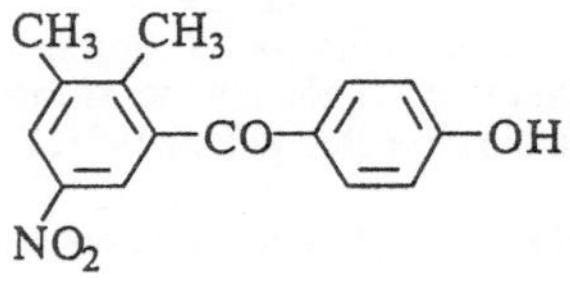

Synthesis

-Preparation by Fries rearrangement of phenyl 2,3-dimethyl-
5-nitrobenzoate with aluminium chloride without solvent at
160° for 2 h (55%) [843].

m.p. 212° [843]; ^{1}H NMR [843], IR [843], UV [843].

(2,3-Dimethylphenyl)(4-hydroxyphenyl)methanone

[134994-27-7] $C_{15}H_{14}O_2$ mol.wt. 226.27

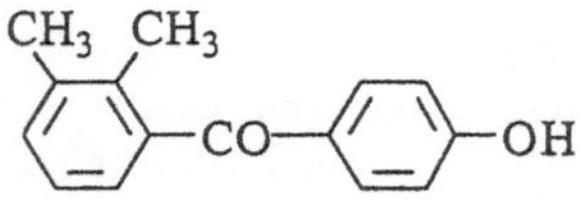

Synthesis

-Refer to: [511].

m.p. and Spectra (NA).

(2,4-Dimethylphenyl)(2-hydroxyphenyl)methanone

[143824-87-7] $C_{15}H_{14}O_2$ mol.wt. 226.27

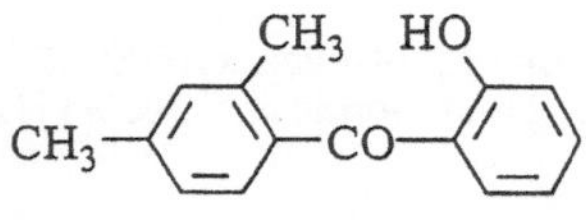

Synthesis

-Preparation by reaction of salicylic acid with m-xylene in
hydrofluoric acid at 60° for 4 h in an autoclave (79%) [81].

m.p. and Spectra (NA).

(2,4-Dimethylphenyl)(3-hydroxyphenyl)methanone

[74167-90-1] $C_{15}H_{14}O_2$ mol.wt. 226.27

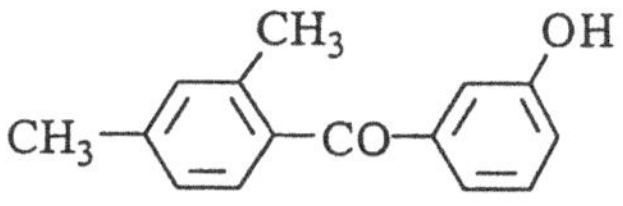

Syntheses

-Preparation by reaction of m-anisoyl chloride with
m-xylene in the presence of excess aluminium chloride at
first at 20°, then at reflux for 2 h (79%) [82, 1292].
-Preparation by Friedel-Crafts acylation of m-xylene with
m-nitrobenzoyl chloride, reduction of the 2,4-dimethyl-3'-nitrobenzophenone so obtained and
diazotization of the 3'-amino-2,4-dimethylbenzophenone formed, followed by hydrolysis of the
diazonium salt [847].
-Also refer to: [436].

m.p. 116° [82, 1292];   Spectra (NA).

(2,4-Dimethylphenyl)(4-hydroxyphenyl)methanone

[116173-30-9] $C_{15}H_{14}O_2$ mol.wt. 226.27

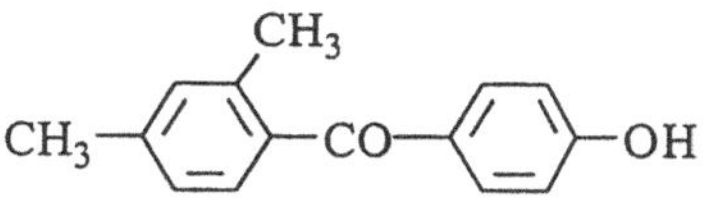

Synthesis

-Obtained by reaction of EKONOL(TM), an aromatic
polyester as Friedel-Crafts reagent, with m-xylene in
triflic acid solution at 25° for 18 h (90%) [287]. Similar
results can be obtained using hydrofluoric
acid/boron trifluoride or aluminium chloride in place of triflic acid [287].
-Also refer to: [442] (Japanese patent).

m.p. (NA); ^{13}C NMR [287], MS [287];   HPLC [287].

(2,5-Dimethylphenyl)(2-hydroxyphenyl)methanone

$C_{15}H_{14}O_2$ mol.wt. 226.27

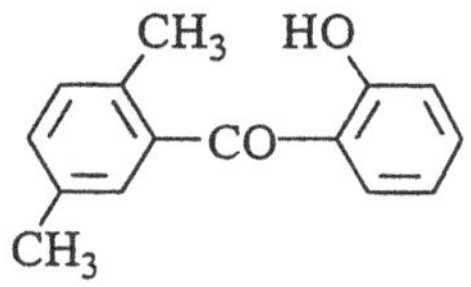

Synthesis

-Obtained by reaction of salicylic acid with p-xylene in
hydrofluoric acid at 60° for 4 h in an autoclave (18%) [81].

m.p. and Spectra (NA).

(2,6-Dimethylphenyl)(4-hydroxyphenyl)methanone

[61002-55-9] $C_{15}H_{14}O_2$ mol.wt. 226.27

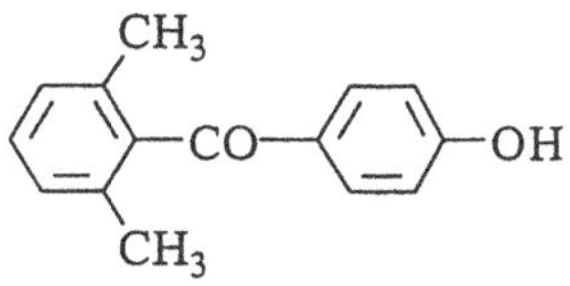

Synthesis

-Preparation by Friedel-Crafts acylation of anisole with
2,6-dimethylbenzoyl chloride in the presence of aluminium
chloride [909].

m.p. 155° [909];   Spectra (NA).

(3,4-Dimethylphenyl)(3-hydroxyphenyl)methanone

[62810-57-5]   C$_{15}$H$_{14}$O$_2$ mol.wt. 226.27

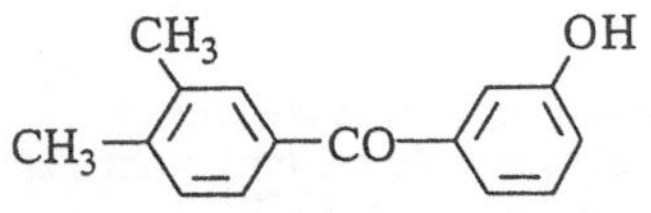

Synthesis

-Preparation by diazotization of 3'-amino-3,4-dimethyl-
 benzophenone followed by hydrolysis of the diazonium salt
 so obtained [847].

      m.p. 116° [847];   Spectra (NA).

(3,4-Dimethylphenyl)(4-hydroxyphenyl)methanone

                                      C$_{15}$H$_{14}$O$_2$ mol.wt. 226.27

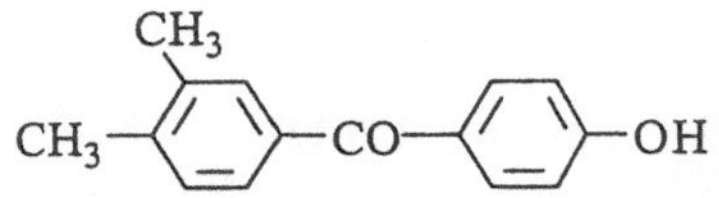

Syntheses

-Preparation by reaction of 3,4-dimethylbenzoyl chloride
 with phenol in the presence of aluminium chloride [296].
-Preparation by Friedel-Crafts acylation of o-xylene with
p-anisoyl chloride in the presence of aluminium chloride in a boiling water bath for 1 h (major
product, 40% yield) [235].
-Preparation by demethylation of 3,4-dimethyl-4'-methoxybenzophenone with boiling pyridinium
chloride (80%) [235].

      m.p. 131° [235];   b.p.$_{17}$ 265-266° [235];   Spectra (NA).

(3,5-Dimethylphenyl)(4-hydroxyphenyl)methanone

                                        C$_{15}$H$_{14}$O$_2$ mol.wt. 226.27

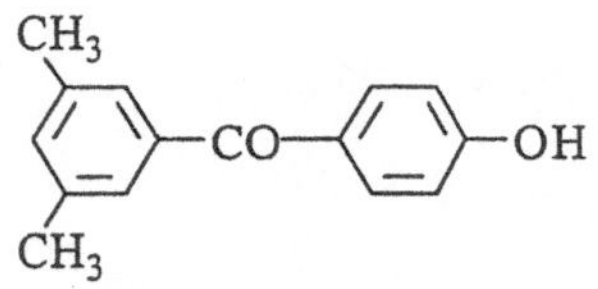

Synthesis

-Preparation by Friedel-Crafts acylation of anisole with
 3,5-dimethylbenzoyl chloride, followed by demethylation of
 the resulting 4'-methoxy-3,5-dimethylbenzophenone with
 pyridinium bromide [593].

      m.p. and Spectra (NA).

(4-Ethylphenyl)(2-hydroxyphenyl)methanone

[82520-51-2]   C$_{15}$H$_{14}$O$_2$ mol.wt. 226.27

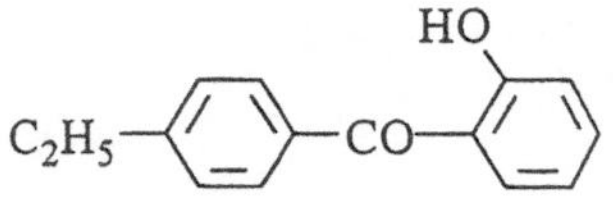

Synthesis

-Obtained (by-product) by action of o-anisoyl chloride with
 ethylbenzene [21],
*in the presence of aluminium chloride between 25 to 60°
 for 2 h;
*in the presence of ferric chloride at 130-140° for 5 h.

      b.p.$_7$ 165-167° [21];   d$_4^{20}$ = 1.1203 [21];   n$_D^{20}$ = 1.6072 [21];
      Spectra (NA).

(4-Ethylphenyl)(4-hydroxyphenyl)methanone

[83888-76-0] $C_{15}H_{14}O_2$ mol.wt. 226.27

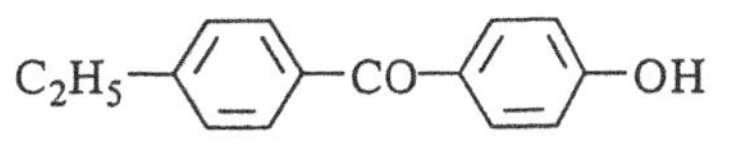

Synthesis

C_2H_5—⟨ ⟩—CO—⟨ ⟩—OH

-Preparation by reaction of p-ethylbenzoyl chloride with phenol in the presence of aluminium chloride [296].

m.p. 99-100° [296];   Spectra (NA).

(4-Ethoxyphenyl)(4-hydroxyphenyl)methanone

[13380-65-9] $C_{15}H_{14}O_3$ mol.wt. 242.27

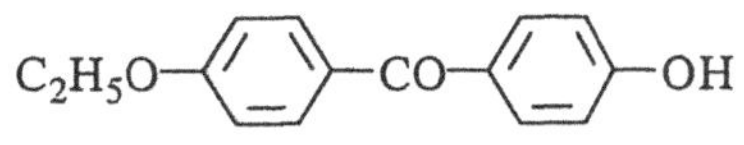

Synthesis

C_2H_5O—⟨ ⟩—CO—⟨ ⟩—OH

-Obtained (by-product) by action of ethyl iodide with 4,4'-dihydroxybenzophenone in the presence of potassium hydroxide in refluxing ethanol for 3 h [1298].

-Also refer to: [407, 596, 597, 1035].

m.p. 146-147° [1298];   Spectra (NA).

(4-Hydroxyphenyl)(2-methoxy-5-methylphenyl)methanone

$C_{15}H_{14}O_3$ mol.wt. 242.27

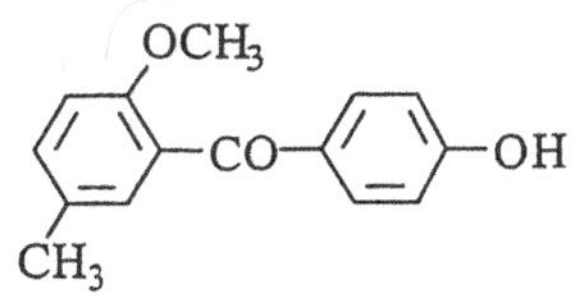

Synthesis

-Refer to: [99].

m.p. 160° [99];   Spectra (NA).

(2,3-Dimethoxyphenyl)(2-hydroxyphenyl)methanone

[129103-87-3] $C_{15}H_{14}O_4$ mol.wt. 258.27

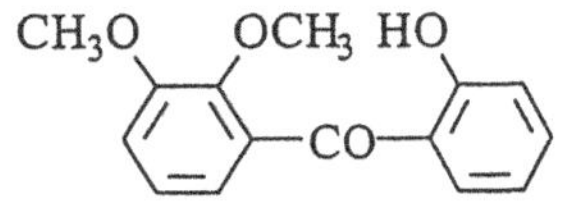

Synthesis

-Preparation from 2-iodophenyl 2,3-dimethoxybenzoate by rearrangement with n-butyllithium in a mixture of ethyl ether, hexane and tetrahydrofuran at -70° for 2 h, then treatment with saturated aqueous ammonium chloride (quantitative yield) [598].

m.p. 75-77° [598]; ^{1}H NMR [598], MS [598].

(2,4-Dimethoxyphenyl)(2-hydroxyphenyl)methanone

[108475-95-2] $C_{15}H_{14}O_4$ mol.wt. 258.27

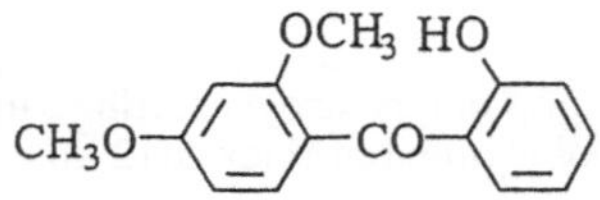

Synthesis

-Obtained, in mixture with 2-hydroxy-2',4-dimethoxy-
 benzophenone, by reaction of 2-methoxybenzoyl chloride
 with resorcinol dimethyl ether in ethyl ether in the presence
 of aluminium chloride for 8 h at r.t. (total yield: 65%) [825,
 826].

-Also refer to: [541].

m.p. (NA); ^{1}H NMR [825, 826].

(2,4-Dimethoxyphenyl)(3-hydroxyphenyl)methanone

$C_{15}H_{14}O_4$ mol.wt. 258.27

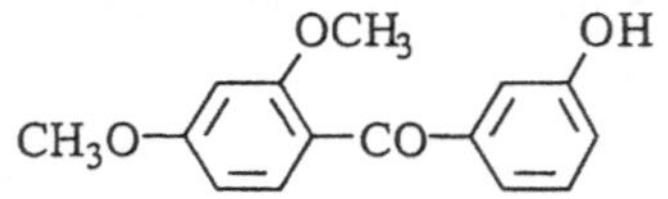

Synthesis

-Preparation by saponification of 3'-(acetyloxy)-2,4-di-
 methoxybenzophenone (SM) with sodium hydroxide in
 refluxing dilute ethanol for 30 min [1149]. SM was
 obtained by Friedel-Crafts acylation of resorcinol
dimethyl ether with m-acetoxybenzoyl chloride in the presence of aluminium chloride in carbon
disulfide at 0° for 20 h (10%).

m.p. 163° [1149];   Spectra (NA).

(2,4-Dimethoxyphenyl)(4-hydroxyphenyl)methanone

[41351-30-8] $C_{15}H_{14}O_4$ mol.wt. 258.27

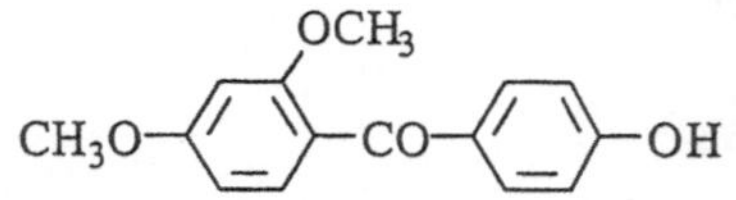

Syntheses

-Preparation by Friedel-Crafts acylation of resorcinol
 dimethyl ether with p-hydroxybenzoic acid,
*in the presence of zinc chloride and phosphorous
 oxychloride at 60-65° for 1.5 h (71%) [1408] or in
nitrobenzene at 60° for 2-3 h [1128], (64%) [1210], according to the method [1104];
*in the presence of polyphosphoric acid [593].
-Also obtained [1213] according to the method of [216].
-Also refer to: [965, 1083, 1084, 1127].
N.B.: Cs salt [1127, 1128].

m.p.   138-139° [1210], 135-137° [1128], 134-136° [1408];
 ^{1}H NMR [1128, 1210, 1408], ^{13}C NMR [1210],
 IR [1128, 1210, 1408];   TLC [1408].

(2,5-Dimethoxyphenyl)(2-hydroxyphenyl)methanone

[183106-14-1]   C15H14O4   mol.wt. 258.27

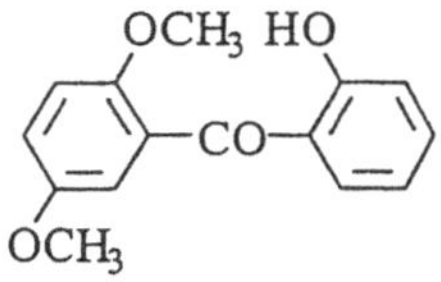

Synthesis

-Obtained by partial demethylation of 2,2',5-trimethoxy-
 benzophenone,
*with boron tribromide in methylene chloride at 0° for 1.5 h
 (50%) [1250];
*with boron trichloride in methylene chloride at 0° for 3 h
 (50%) [1250];
*with boron trifluoride-etherate in refluxing benzene for 6 h (40%) or in refluxing toluene for 4 h
 (40%) [1250];
*with beryllium chloride in refluxing benzene for 8-10 h (60%) or in refluxing toluene for 5 h
 (60%) [1250].
N.B.: In these experiments, only the reactions using boron halides were carried out under nitrogen
 atmosphere.

m.p. 98-100° [1250];   1H NMR [1250], IR [1250], UV [1250], MS [1250].

(2,6-Dimethoxyphenyl)(2-hydroxyphenyl)methanone

[129103-86-2]   C15H14O4   mol.wt. 258.27

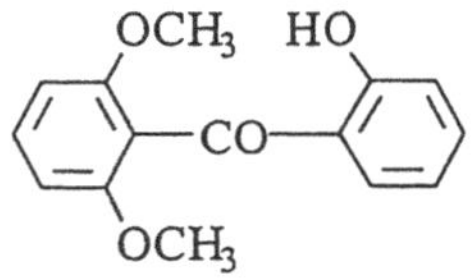

Synthesis

-Preparation from 2-iodophenyl 2,6-dimethoxybenzoate by
 rearrangement with n-butyllithium in a mixture of ethyl ether,
 hexane and tetrahydrofuran at -70° for 2 h, then treatment
 with saturated aqueous ammonium chloride (quantitative
 yield) [598].

m.p. 120°5-121°5 [598];   1H NMR [598], IR [598], MS [598].

(3,4-Dimethoxyphenyl)(2-hydroxyphenyl)methanone

[183106-12-9]   C15H14O4   mol.wt. 258.27

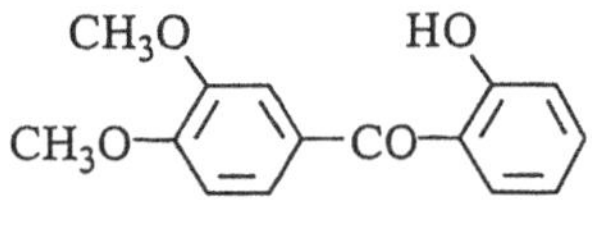

Synthesis

-Preparation by selective demethylation of 2',3,4-tri-
 methoxybenzophenone with excess beryllium chloride in
 refluxing toluene for 3.5 h (92%) [1250].

m.p. 76-78° [1250];
1H NMR [1250], IR [1250], UV [1250], MS [1250];   TLC [1250].

(3,4-Dimethoxyphenyl)(4-hydroxyphenyl)methanone

[26955-00-0]   C15H14O4   mol.wt. 258.27

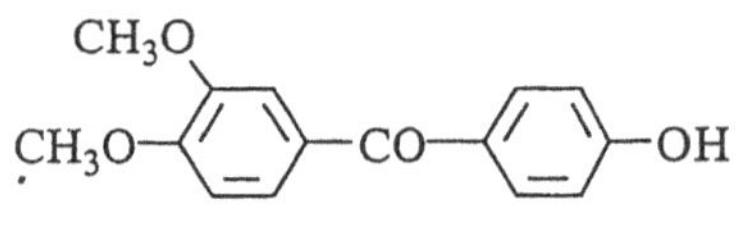

Synthesis

-Obtained by condensation of veratric acid with phenol
 [218].
-Also refer to: [164].

m.p. 166-167° [218];   Spectra (NA).

(3,5-Dimethoxyphenyl)(4-hydroxyphenyl)methanone

C15H14O4   mol.wt. 258.27

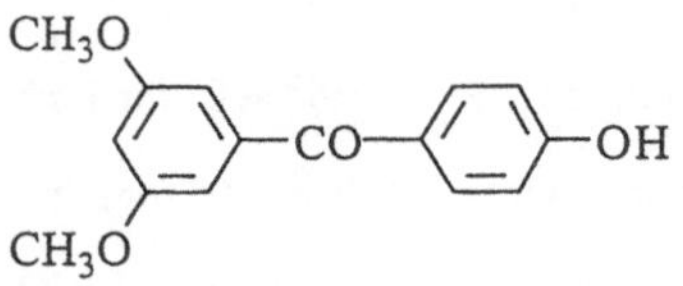

CH3O

CH3O ⟨ ⟩—CO—⟨ ⟩—OH

Synthesis

-Preparation by Friedel-Crafts acylation of phenol with
 3,5-dimethoxybenzoyl chloride [593].

m.p. and Spectra (NA).

(2-Hydroxyphenyl)[2-(methoxymethoxy)phenyl]methanone

[59410-99-0]   C15H14O4   mol.wt. 258.27

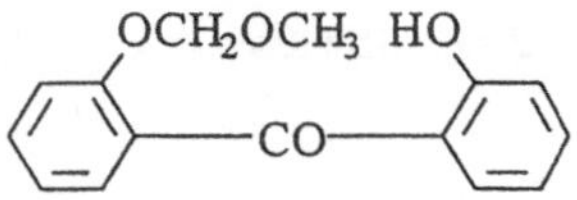

OCH2OCH3 HO

⟨ ⟩—CO—⟨ ⟩

Synthesis

-Preparation by reaction of dimethoxymethane with 2,2'-di-
 hydroxybenzophenone in the presence of p-toluenesulfonic
 acid and a "Linde" type 3 Å molecular sieve in refluxing
 methylene chloride overnight under nitrogen (69%) [1455].

m.p. 43°5 [1455]; ^{1}H NMR [1455].

(2-Hydroxyphenyl)[4-(methoxymethoxy)phenyl]methanone

[31772-30-2]   C15H14O4   mol.wt. 258.27

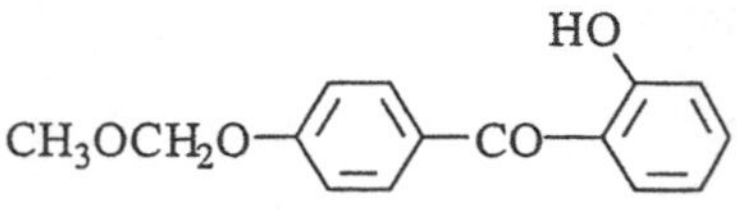

HO

CH3OCH2O—⟨ ⟩—CO—⟨ ⟩

Synthesis

-Preparation by reaction of chloromethyl methyl ether
 with 2,4'-dihydroxybenzophenone monosodium salt
 (4') in toluene at r.t. overnight (64%) [710].

oil [710];   b.p. (NA);   IR [710].

[4-(Dimethylamino)phenyl](3-hydroxyphenyl)methanone

C15H15NO2   mol.wt. 241.29

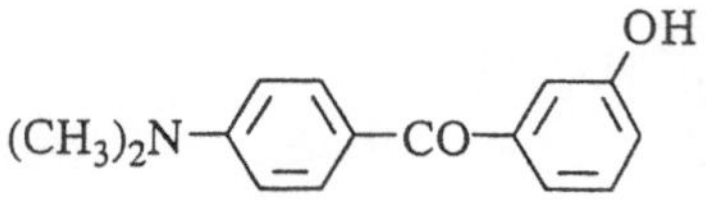

OH

(CH3)2N—⟨ ⟩—CO—⟨ ⟩

Synthesis

-Preparation by heating 3-methoxycarbonyloxybenzoic
 acid anilide and N,N-dimethylaniline in the presence of
 phosphorous oxychloride on a water bath for 4 h (50%)
 [1243].

m.p. 185-187° [1243];   Spectra (NA).

[4-(Dimethylamino)phenyl](4-hydroxyphenyl)methanone

[102827-03-2] $C_{15}H_{15}NO_2$ mol.wt. 241.29

Syntheses

$(CH_3)_2N$—⟨◯⟩—CO—⟨◯⟩—OH

-Preparation by demethylation of 4-(dimethylamino)-
4'-methoxybenzophenone (SM) with aluminium
bromide in refluxing benzene for 4 h (95%)
[1059]. SM was obtained by condensation of N-(p-anisoyl)aniline with dimethylaniline [315].
-Preparation by heating p-methylcarbonatobenzanilide and N,N-dimethylaniline in the presence of
phosphorous oxychloride on a water bath for 4 h (60%) [1243].
-Also obtained by treating benzophenone-4,4'-bis(trimethylammonium chloride) with sodium
methoxide in refluxing methanol for 3 h [1346].

m.p. 200° [1059], 199-200° [315, 1243], 198-200° [1346];
Spectra (NA).

[2-(Acetyloxy)-5-methoxyphenyl](2-hydroxyphenyl)methanone

[83570-59-6] $C_{16}H_{14}O_5$ mol.wt. 286.28

Synthesis

-Obtained by adding silica gel to a solution of 2'-acetoxy-
2-hydroxy-5-methoxybenzophenone in ethyl ether, then
elimination of the solvent, and the resulting powder allowed
to stand at r.t. during 42 h (75%) [351]. There is a
transacylation on silica gel.

m.p. 70-73° [351]; ^{1}H NMR [351], IR [351].

(2-Hydroxyphenyl)[4-(1-methylethyl)phenyl]methanone

[35839-45-3] $C_{16}H_{16}O_2$ mol.wt. 240.30

Synthesis

$(CH_3)_2CH$—⟨◯⟩—CO—

-Preparation by reaction of salicyloyl chloride with
cumene in nitrobenzene in the presence of aluminium
chloride, first between 4 to 8° for 50 min, then at 40°
for 3 h and at r.t. overnight (60%) [469, 882].

-Also refer to: [881].

b.p.$_{2.25}$ 165-167° [469, 882];   Spectra (NA).

(4-Hydroxyphenyl)(4-propylphenyl)methanone

[64357-90-0] $C_{16}H_{16}O_2$ mol.wt. 240.30

Synthesis

C_3H_7—⟨◯⟩—CO—⟨◯⟩—OH

-Preparation by Friedel-Crafts acylation of phenol with
p-propylbenzoyl chloride [531].

m.p. (NA);   IR [530, 531].

(2-Hydroxyphenyl)(2,4,6-trimethylphenyl)methanone

[46863-20-1]   C$_{16}$H$_{16}$O$_2$ mol.wt. 240.30

Syntheses

-Obtained by photo-Fries rearrangement of phenyl mesitoate
  (phenyl 2,4,6-trimethylbenzoate),
*in methanol (39%) or in methanol in the presence of
  β-cyclodextrin (99%) [1335];
*in pentane (19%) [102];
*in benzene at 40° for 332 h (9%) or in hexane at 40° for 109 h (6%) [419].
-Also obtained from mesityl o-methoxyphenyl ketone (2'-methoxy-2,4,6-trimethylbenzophenone)
  by cleavage of methoxy group with hydriodic acid or with a binary mixture prepared from
  magnesium and iodine in refluxing toluene/butyl ether solution (poor yield) [453].
-Preparation by reaction of 2-methoxybenzoyl chloride with mesitylene in the presence of
  aluminium chloride in benzene (65%) [1181].

m.p.   94-95° [1181], 82° [1178, 1179], 81-83° [419], 81-82° [453];
^{1}H NMR [1177, 1178, 1179, 1181], ^{13}C NMR [1177],
IR [453, 1178, 1179, 1181];   thermal behaviour [1178, 1179].

(3-Hydroxyphenyl)(2,4,6-trimethylphenyl)methanone

[76981-50-5]   C$_{16}$H$_{16}$O$_2$ mol.wt. 240.30

Syntheses

-Preparation by reaction of m-anisoyl chloride with
  mesitylene (1,3,5-trimethylbenzene) in the presence of an
  excess of aluminium chloride: first at 20°, then at reflux for
  2 h (82%) [82].
-Preparation by Friedel-Crafts acylation of mesitylene with
m-anisoyl chloride, followed by demethylation of the 3'-methoxy-2,4,6-trimethylbenzophenone so
obtained with pyridinium bromide [593].

m.p.   130° [82];   Spectra (NA).

(4-Hydroxyphenyl)(2,4,6-trimethylphenyl)methanone

[2004-55-9]   C$_{16}$H$_{16}$O$_2$ mol.wt. 240.30

Syntheses

-Preparation by Friedel-Crafts acylation of phenol with
  mesitoyl chloride in the presence of aluminium
  chloride in carbon disulfide [530, 531].
-Also obtained by photo-Fries rearrangement of phenyl
  mesitoate (phenyl 2,4,6-trimethylbenzoate),
*in methanol (46%) [1335];
*in pentane (24%) [102];
*in benzene at 40° for 332 h (12%) [419];
*in hexane at 40° for 109 h (11%) [419].

m.p.   167-168° [419]; ^{1}H NMR [419], IR [419, 530, 531], UV [419].

(2-Hydroxyphenyl)(2,3,4-trimethoxyphenyl)methanone

[147188-09-8] $C_{16}H_{16}O_5$ mol.wt. 288.30

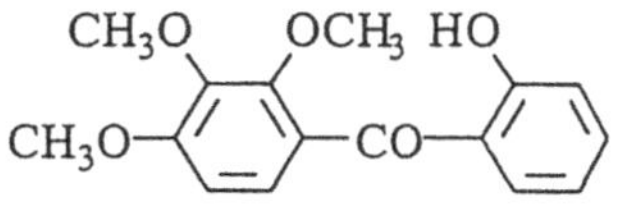

CH$_3$O   OCH$_3$ HO

CH$_3$O—⟨ ⟩—CO—⟨ ⟩

-Also refer to: [556].

Synthesis

-Obtained, in mixture with 2-hydroxy-2',3,4-trimethoxy-benzophenone, by reaction of 2-methoxybenzoyl chloride with pyrogallol trimethyl ether in ethyl ether in the presence of aluminium chloride for 8 h at r.t. (total yield: 56-59%) [825, 826].

m.p. (NA); ^{1}H NMR [825].

(2-Hydroxyphenyl)(2,4,5-trimethoxyphenyl)methanone

[147188-07-6] $C_{16}H_{16}O_5$ mol.wt. 288.30

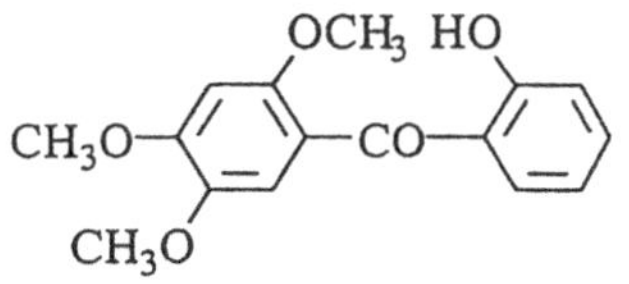

OCH$_3$ HO

CH$_3$O—⟨ ⟩—CO—⟨ ⟩

CH$_3$O

-Also refer to: [826].

Synthesis

-Obtained, in mixture with 2-hydroxy-2',4,5-trimethoxy-benzophenone, by reaction of 2-methoxybenzoyl chloride with 1,2,4-trimethoxybenzene in ethyl ether in the presence of aluminium chloride for 8 h at r.t. (total yield: 56%) [825].

m.p. (NA); ^{1}H NMR [825].

(2-Hydroxyphenyl)(2,4,6-trimethoxyphenyl)methanone

[147188-05-4] $C_{16}H_{16}O_5$ mol.wt. 288.30

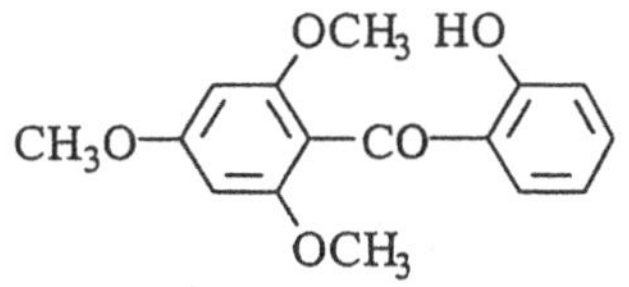

OCH$_3$ HO

CH$_3$O—⟨ ⟩—CO—⟨ ⟩

OCH$_3$

-Also refer to: [826].

Synthesis

-Obtained, in mixture with 2-hydroxy-2',4,6-trimethoxy-benzophenone, by reaction of 2-methoxybenzoyl chloride with phloroglucinol trimethyl ether in ethyl ether in the presence of aluminium chloride for 8 h at r.t. (total yield: 59%) [825].

m.p. (NA); ^{1}H NMR [825].

(3-Hydroxyphenyl)(2,4,6-trimethoxyphenyl)methanone

 $C_{16}H_{16}O_5$ mol.wt. 288.30

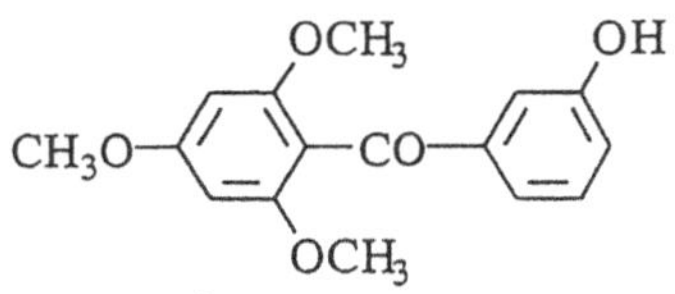

OCH$_3$ OH

CH$_3$O—⟨ ⟩—CO—⟨ ⟩

OCH$_3$

Synthesis

-Preparation by reaction of m-hydroxybenzonitrile with phloroglucinol trimethyl ether (Hoesch reaction) [593].

m.p. and Spectra (NA).

(4-Hydroxyphenyl)(2,4,6-trimethoxyphenyl)methanone

[41351-32-0] $C_{16}H_{16}O_5$ mol.wt. 288.30

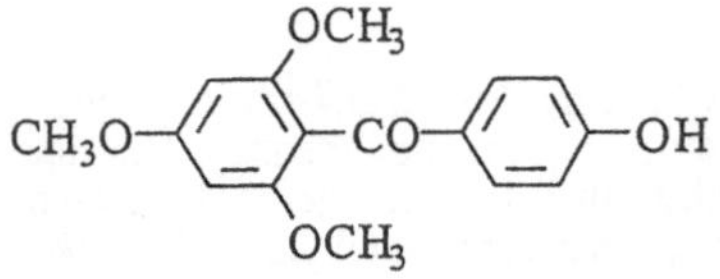

Synthesis

-Refer to: [593].

m.p. and Spectra (NA).

(4-Hydroxyphenyl)(3,4,5-trimethoxyphenyl)methanone

[14938-63-7] $C_{16}H_{16}O_5$ mol.wt. 288.30

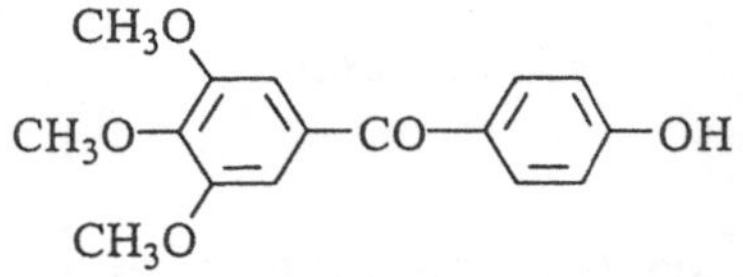

Synthesis

-Preparation by Friedel-Crafts acylation of phenol with 3,4,5-trimethoxybenzoyl chloride in nitrobenzene in the presence of aluminium chloride, first at 10°, then for 24 h at r.t. (45%) [504].

m.p.   120° [504];   Spectra (NA).

(4-Butylphenyl)(4-hydroxyphenyl)methanone

[190728-33-7] $C_{17}H_{18}O_2$ mol.wt. 254.33

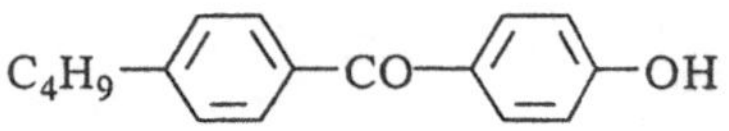

Synthesis

-Refer to: Chem. Abstr., **127**, 34137f (1997).

m.p. and Spectra (NA).

[4-(1,1-Dimethylethyl)phenyl](2-hydroxyphenyl)methanone

$C_{17}H_{18}O_2$ mol.wt. 254.33

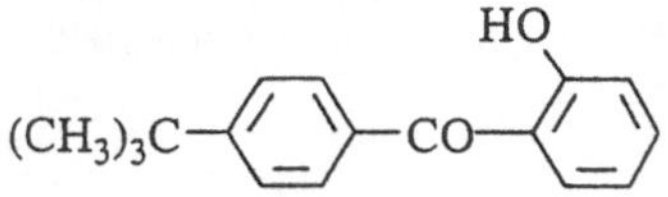

Synthesis

-Refer to: [810].

m.p. and Spectra (NA).

[4-(1,1-Dimethylethyl)phenyl](4-hydroxyphenyl)methanone

[55044-96-7] $C_{17}H_{18}O_2$ mol.wt. 254.33

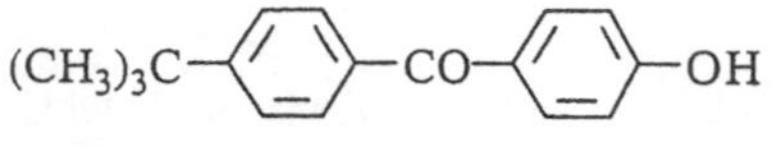

Synthesis

-Preparation by reaction of p-tert-butylbenzoyl chloride with phenol in the presence of aluminium chloride [296].

m.p.   130-131° [296];   Spectra (NA).

[4-(1,1-Dimethylethoxy)phenyl](4-hydroxyphenyl)methanone

$C_{17}H_{18}O_3$ mol.wt. 270.33

Synthesis

$(CH_3)_3C-O-\langle\rangle-CO-\langle\rangle-OH$

-Obtained by decomposition of the benzophenone-4,4'-bis(trimethylammonium chloride) with sodium tert-butoxide in refluxing alcohol for 3 h [1346].

m.p. 85° [1346]; Spectra (NA).

(4-Hydroxyphenyl)(4-pentylphenyl)methanone

[64357-91-1] $C_{18}H_{20}O_2$ mol.wt. 268.36

Synthesis

$C_5H_{11}-\langle\rangle-CO-\langle\rangle-OH$

-Preparation by reaction of 4-pentylbenzoyl chloride with phenol in the presence of aluminium chloride in carbon disulfide [530, 531].

m.p. (NA); IR [530, 531].

(3-Hydroxyphenyl)[4-methoxy-2-methyl-5-(1-methylethyl)phenyl]methanone

[109250-48-8] $C_{18}H_{20}O_3$ mol.wt. 284.36

Synthesis

$(CH_3)_2CH$ ... OH
$CH_3O-\langle\rangle-CO-\langle\rangle$
CH_3

-Obtained by saponification of 3'-acetoxy-4-methoxy-2-methyl-5-isopropylbenzophenone (SM) with sodium hydroxide in dilute ethanol at reflux for 30 min (73%). SM was prepared by Friedel-Crafts acylation of thymol methyl ether with m-acetoxybenzoyl chloride in the presence of aluminium chloride in carbon disulfide at 0° for 20 h (12%) [1149].

m.p. 106° [1149]; Spectra (NA).

(4-Hydroxyphenyl)[4-methoxy-2-methyl-5-(1-methylethyl)phenyl]methanone

[109250-49-9] $C_{18}H_{20}O_3$ mol.wt. 284.36

Synthesis

$(CH_3)_2CH$
$CH_3O-\langle\rangle-CO-\langle\rangle-OH$
CH_3

-Obtained by saponification of 4'-acetoxy-4-methoxy-2-methyl-5-isopropylbenzophenone (SM) with sodium hydroxide in dilute ethanol at reflux for 30 min (80%). SM was prepared by Friedel-Crafts acylation of thymol methyl ether with p-acetoxy-benzoyl chloride in the presence of aluminium chloride in carbon disulfide at 15° for 3 h (17%) [1149].

b.p.$_{16}$ 265-268° [1149]; m.p. 65° [1149]; Spectra (NA).

(4-Hydroxyphenyl)[6-methoxy-2-methyl-3-(1-methylethyl)phenyl]methanone

[109250-50-2] $C_{18}H_{20}O_3$ mol.wt. 284.36

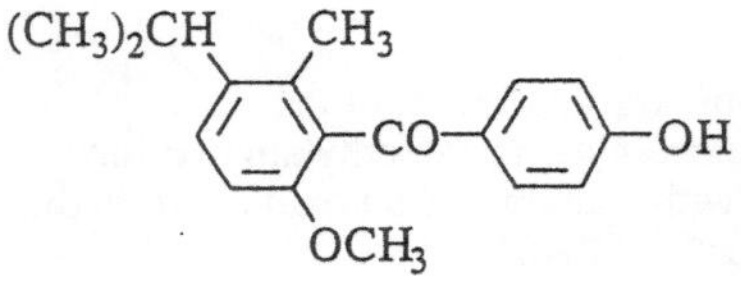

Synthesis

-Obtained by saponification of 4'-acetoxy-6-methoxy-2-methyl-3-isopropylbenzophenone (SM) with sodium hydroxide in dilute ethanol at reflux for 30 min (79%). SM was prepared by Friedel-Crafts acylation of p-thymol methyl ether (3-methyl-4-isopropylanisole) with p-acetoxybenzoyl chloride in the presence of aluminium chloride in carbon disulfide at 15° for 3 h (7%) [1149].

b.p.$_{0.3}$ 212-213° [1149];   Spectra (NA).

(2,4-Dimethoxy-3-propylphenyl)(2-hydroxyphenyl)methanone

[115296-03-2] $C_{18}H_{20}O_4$ mol.wt. 300.35

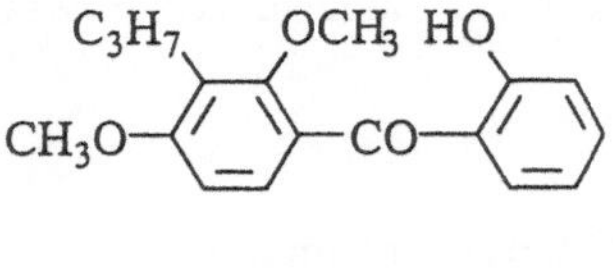

Synthesis

-Obtained (poor yield) by reaction of o-anisoyl chloride with 1,3-dimethoxy-2-propylbenzene in methylene chloride in the presence of aluminium chloride, first at 0° for 2 h then at r.t. for 1 h (13%) [456].

oil [456];   b.p. and Spectra (NA).

(2-Hydroxyphenyl)(2-phenoxyphenyl)methanone

[194548-68-0] $C_{19}H_{14}O_3$ mol.wt. 290.32

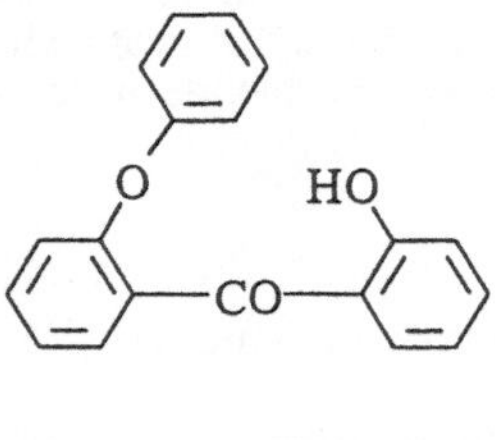

Synthesis

-Refer to: Chem. Abstr., **127**, 205428h (1997).

m.p. and Spectra (NA).

(3-Hydroxyphenyl)(4-phenoxyphenyl)methanone

[76981-53-8] $C_{19}H_{14}O_3$ mol.wt. 290.32

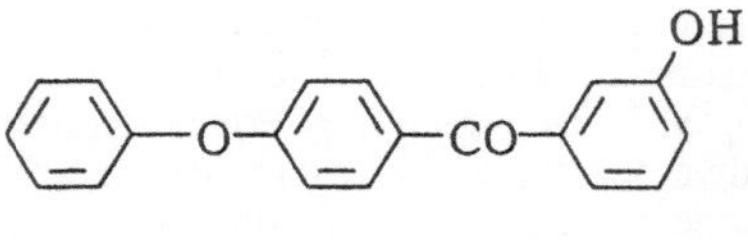

Synthesis

-Preparation by reaction of m-anisoyl chloride with diphenyl oxide in the presence of an excess of aluminium chloride: first at 20°, then at reflux for 2 h (46%) [82].

m.p. 142° [82];   Spectra (NA).

(4-Hydroxyphenyl)(4-phenoxyphenyl)methanone

[78930-16-2] $C_{19}H_{14}O_3$ mol.wt. 290.32

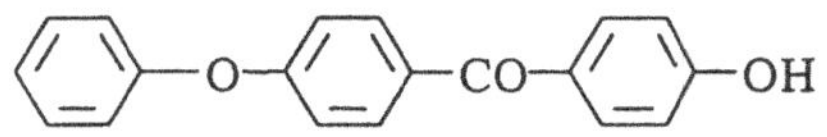

Synthesis

-Preparation from 4-fluoro-4'-phenoxy-
benzophenone by reaction under nitrogen with
potassium hydroxide in aqueous dimethyl
sulfoxide at 60° for 18 h [1121].

-Also refer to: [1458].
N.B.: K salt [1121].

m.p. 143° [1121]; Spectra (NA).

[2-[(6-Bromohexyl)oxy]phenyl](4-hydroxyphenyl)methanone

[31772-32-4] $C_{19}H_{21}BrO_3$ mol.wt. 377.28

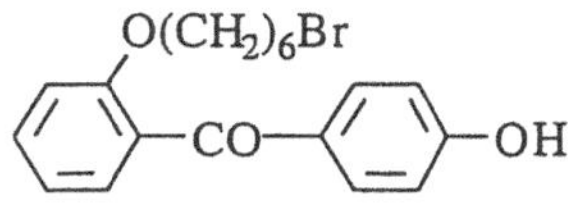

Synthesis

-Preparation by heating 2-(6-bromohexyloxy)-4'-methoxy-
benzophenone in refluxing aqueous acetic acid in the
presence of a few concentrated sulfuric acid for 15 min
(95%) [710].

oil [710]; b.p. (NA); IR [710].

[4-[(6-Bromohexyl)oxy]phenyl](2-hydroxyphenyl)methanone

$C_{19}H_{21}BrO_3$ mol.wt. 377.28

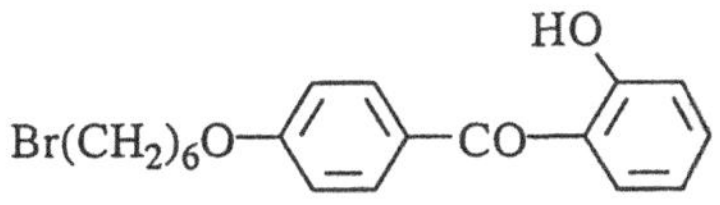

Synthesis

-Preparation by reaction of 1,6-dibromohexane with
2,4'-dihydroxybenzophenone in ethanol in the presence
of potassium hydroxide, first at r.t. for 1 h then at
reflux for 40 min (63%) [710].

yellow oil [710]; b.p. (NA); IR [710].

[4-[(7-Bromoheptyl)oxy]phenyl](2-hydroxyphenyl)methanone

$C_{20}H_{23}BrO_3$ mol.wt. 391.30

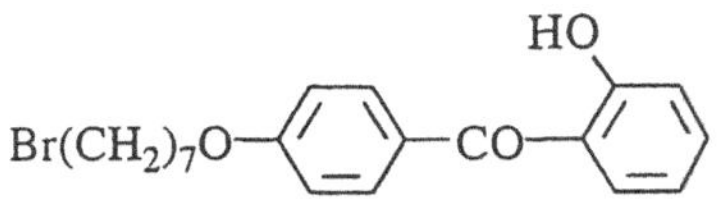

Synthesis

-Preparation by reaction of 1,7-dibromoheptane with
2,4'-dihydroxybenzophenone in ethanol in the presence
of potassium hydroxide, first at r.t. for 1 h then at
reflux for 40 min [710].

yellow oil [710]; b.p. (NA); IR [710].

[4-(Heptyloxy)phenyl](2-hydroxyphenyl)methanone

[33213-89-7] $C_{20}H_{24}O_3$ mol.wt. 312.41

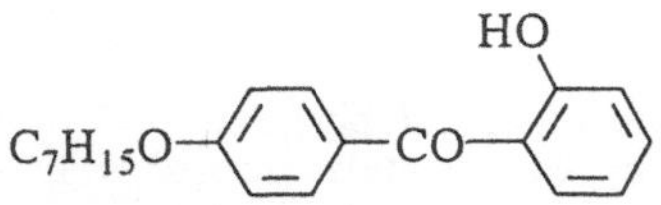

HO

$C_7H_{15}O$—⟨ ⟩—CO—⟨ ⟩

Synthesis

-Refer to: [617].

m.p. and Spectra (NA).

[4-[(8-Bromooctyl)oxy]phenyl](2-hydroxyphenyl)methanone

$C_{21}H_{25}BrO_3$ mol.wt. 405.33

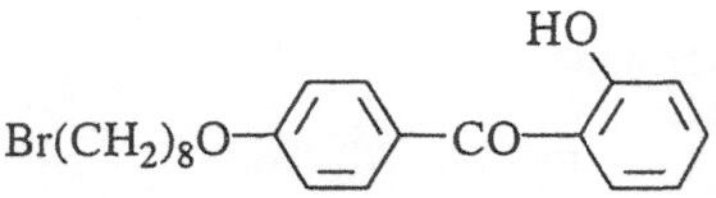

HO

$Br(CH_2)_8O$—⟨ ⟩—CO—⟨ ⟩

Synthesis

-Preparation by reaction of 1,8-dibromooctane with 2,4'-dihydroxybenzophenone in ethanol in the presence of potassium hydroxide, first at r.t. for 1 h, then at reflux for 40 min (57%) [710].

yellow oil [710];   b.p. (NA);   IR [710].

[4-[(9-Bromononyl)oxy]phenyl](2-hydroxyphenyl)methanone

$C_{22}H_{27}BrO_3$ mol.wt. 419.36

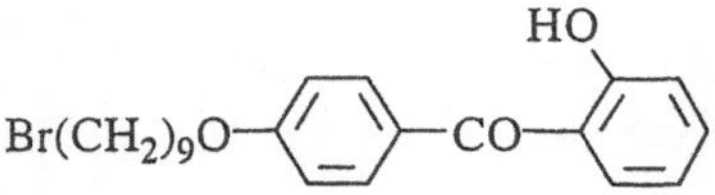

HO

$Br(CH_2)_9O$—⟨ ⟩—CO—⟨ ⟩

Synthesis

-Preparation by reaction of 1,9-dibromononane with 2,4'-dihydroxybenzophenone in ethanol in the presence of potassium hydroxide, first at r.t. for 1 h, then at reflux for 40 min (74%) [710].

yellow oil [710];   b.p. (NA);   IR [710].

[4-[(10-Bromodecyl)oxy]phenyl](2-hydroxyphenyl)methanone

$C_{23}H_{29}BrO_3$ mol.wt. 433.39

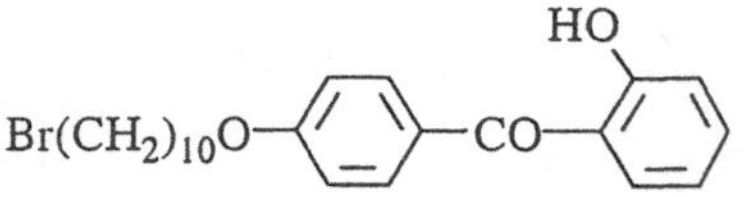

HO

$Br(CH_2)_{10}O$—⟨ ⟩—CO—⟨ ⟩

Synthesis

-Preparation by reaction of 1,10-dibromodecane with 2,4'-dihydroxybenzophenone in ethanol in the presence of potassium hydroxide, first at r.t. for 1 h, then at reflux for 40 min (68%) [710].

yellow oil [710];   b.p.$_5$ 138-141° [710];   IR [710].

[4-[(11-Bromoundecyl)oxy]phenyl](2-hydroxyphenyl)methanone

$C_{24}H_{31}BrO_3$ mol.wt. 447.41

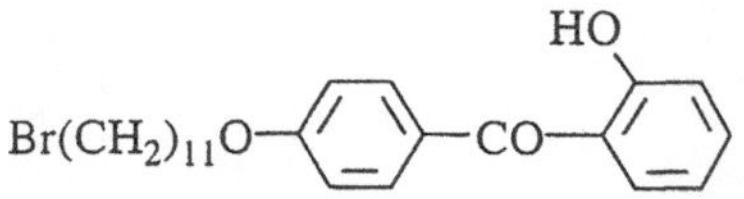

HO

$Br(CH_2)_{11}O$—⟨ ⟩—CO—⟨ ⟩

Synthesis

-Preparation by reaction of 1,11-dibromoundecane with 2,4'-dihydroxybenzophenone in ethanol in the presence of potassium hydroxide, first at r.t. for 1 h, then at reflux for 40 min (quantitative yield) [710].

yellow oil [710];   b.p. (NA);   IR [710].

[4-[(12-Bromododecyl)oxy]phenyl](2-hydroxyphenyl)methanone

C$_{25}$H$_{33}$BrO$_3$ mol.wt. 461.44

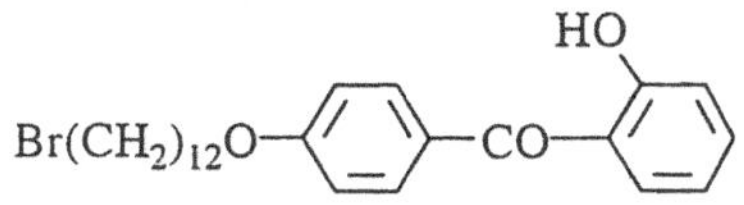

Synthesis

Br(CH$_2$)$_{12}$O—〈 〉—CO—〈 〉 HO

-Preparation by reaction of 1,12-dibromododecane with 2,4'-dihydroxybenzophenone in ethanol in the presence of potassium hydroxide, first at r.t. for 1 h, then at reflux for 40 min (46%) [710].

yellow crystals [710];   m.p. (NA);   IR [710].

(4-Dodecylphenyl)(2-hydroxyphenyl)methanone

[35698-22-7]   C$_{25}$H$_{34}$O$_2$ mol.wt. 366.54

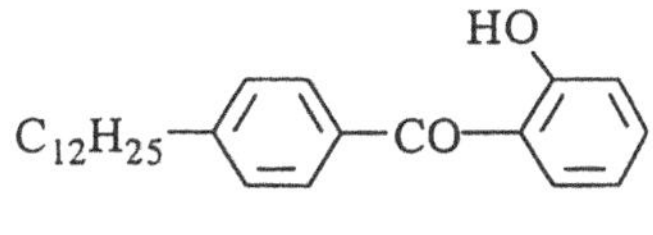

Synthesis

C$_{12}$H$_{25}$—〈 〉—CO—〈 〉 HO

-Obtained by reaction of 2-hydroxybenzoyl chloride with dodecylbenzene in nitrobenzene in the presence of aluminium chloride for 4 h at 40° then 16 h at r.t. (16%) [469, 882].

b.p.$_{0.1}$ 175-195° [469, 882];   Spectra (NA).

[4-(Hexadecyloxy)phenyl](4-hydroxyphenyl)methanone

[129727-61-3]   C$_{29}$H$_{42}$O$_3$ mol.wt. 438.65

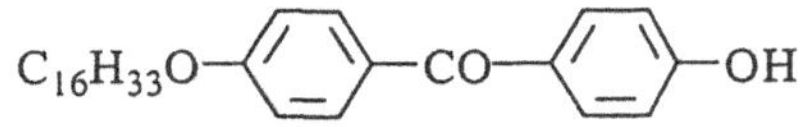

Synthesis

C$_{16}$H$_{33}$O—〈 〉—CO—〈 〉—OH

-Refer to: [1197] (Japanese patent).

m.p. and Spectra (NA).

1.3. *Substituents located on both rings*

(2,4-Dichlorophenyl)(2,3,5-trichloro-6-hydroxyphenyl)methanone

[34171-61-4]   C$_{13}$H$_5$Cl$_5$O$_2$ mol.wt. 370.45

Synthesis

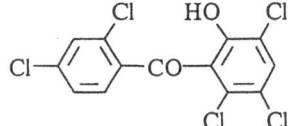

-Preparation by Fries rearrangement of 2,4,5-trichlorophenyl 2,4-dichlorobenzoate with aluminium chloride for 30 min at 150-160° [375].

m.p. 81-82° [375];   Spectra (NA).

(3,4-Dichlorophenyl)(2,3,5-trichloro-6-hydroxyphenyl)methanone

[34171-60-3] $C_{13}H_5Cl_5O_2$ mol.wt. 370.45

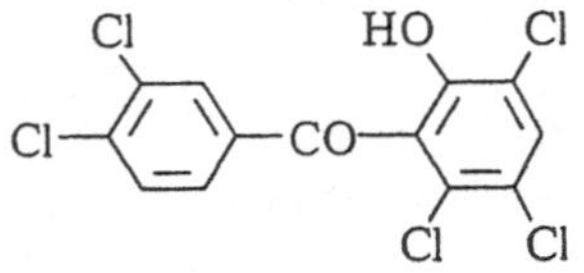

Synthesis

-Preparation by Fries rearrangement of 2,4,5-trichlorophenyl
 3,4-dichlorobenzoate with aluminium chloride for
 30 min at 150-160° [375].

m.p. 217-218° [375];   Spectra (NA).

(4-Bromophenyl)(2,3,5-trichloro-6-hydroxyphenyl)methanone

[34171-63-6] $C_{13}H_6BrCl_3O_2$ mol.wt. 380.45

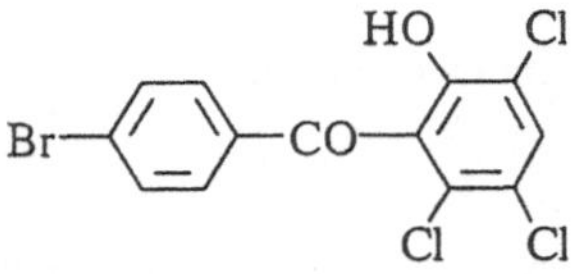

Synthesis

-Preparation by Fries rearrangement of 2,4,5-trichlorophenyl
 p-bromobenzoate with aluminium chloride for 30 min at
 150-160° [375].

m.p. 185-186° [375];   Spectra (NA).

(2-Chlorophenyl)(2,3,5-trichloro-6-hydroxyphenyl)methanone

[34174-12-4] $C_{13}H_6Cl_4O_2$ mol.wt. 336.00

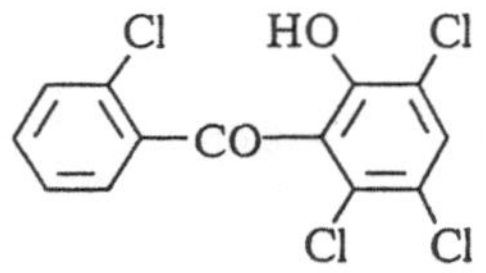

Synthesis

-Preparation by Fries rearrangement of 2,4,5-trichlorophenyl
 o-chlorobenzoate with aluminium chloride for 30 min at
 150-160° [375].

m.p. 94-95° [375];   Spectra (NA).

(4-Chlorophenyl)(2,3,5-trichloro-6-hydroxyphenyl)methanone

[34171-59-0] $C_{13}H_6Cl_4O_2$ mol.wt. 336.00

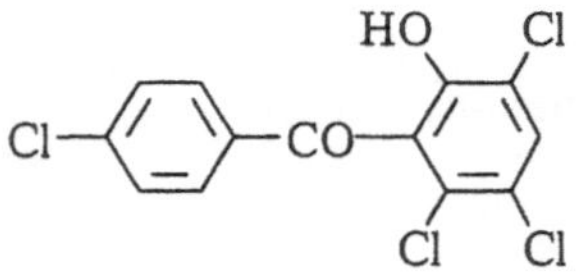

Synthesis

-Preparation by Fries rearrangement of 2,4,5-trichlorophenyl
 p-chlorobenzoate with aluminium chloride for 30 min at
 150-160° [375].

m.p. 176-177° [375];   Spectra (NA).

(2,3-Dichloro-4-hydroxyphenyl)(2,4-dichlorophenyl)methanone

[34183-05-6] $C_{13}H_6Cl_4O_2$ mol.wt. 336.00

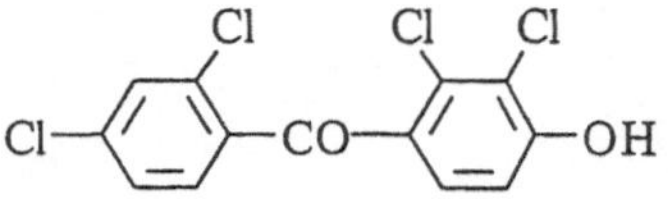

Synthesis

-Obtained by Fries rearrangement of 2,3-dichlorophenyl
 2,4-dichlorobenzoate with aluminium chloride in
 chlorobenzene for 20 min at 140-150° or in nitrobenzene
 for 24 h at 75° [1214].

m.p. 164-165° [1214];   Spectra (NA).

(2,3-Dichloro-4-hydroxyphenyl)(3,4-dichlorophenyl)methanone

[72482-75-8] $C_{13}H_6Cl_4O_2$ mol.wt. 336.00

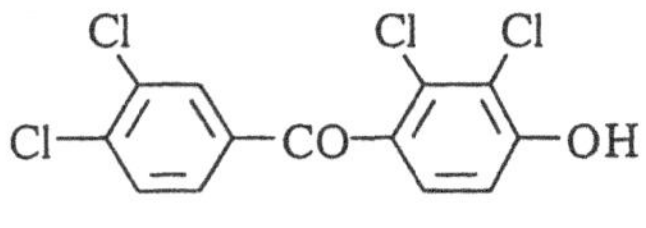

Synthesis

-Preparation by demethylation of 2,3,3',4'-tetrachloro-
  4-methoxybenzophenone (SM) with aluminium chloride,
  *in refluxing methylene chloride for overnight [653];
  *in refluxing benzene for 5 h, then at r.t. for 18 h [1270].
SM was obtained by Friedel-Crafts acylation of 2,3-dichloroanisole with 3,4-dichlorobenzoyl
chloride in the presence of aluminium chloride in refluxing methylene chloride [653] or in ethylene
dichloride at 60° for 1 h [1270].

m.p. 179-180° [1270];   Spectra (NA).

(2,4-Dichloro-6-hydroxyphenyl)(2,4-dichlorophenyl)methanone

[34174-05-5] $C_{13}H_6Cl_4O_2$ mol.wt. 336.00

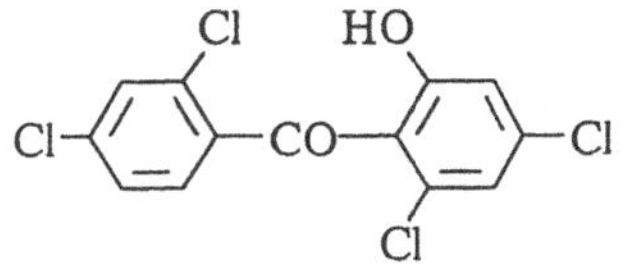

Synthesis

-Preparation by Fries rearrangement of 3,5-dichlorophenyl
  2,4-dichlorobenzoate with aluminium chloride for 30 min at
  150-160° [375].

m.p. 105-106° [375];   Spectra (NA).

(2,4-Dichloro-6-hydroxyphenyl)(2,5-dichlorophenyl)methanone

[34174-09-9] $C_{13}H_6Cl_4O_2$ mol.wt. 336.00

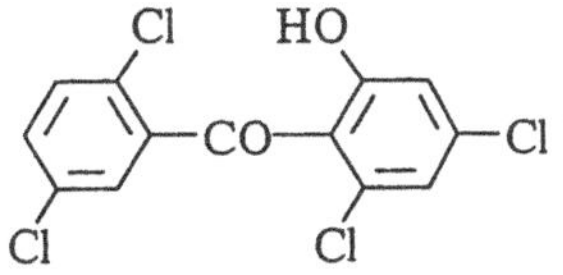

Synthesis

-Preparation by Fries rearrangement of 3,5-dichlorophenyl
  2,5-dichlorobenzoate with aluminium chloride for 30 min at
  150-160° [375].

m.p. 107-108° [375];   Spectra (NA).

(2,4-Dichloro-6-hydroxyphenyl)(2,6-dichlorophenyl)methanone

[34786-96-4] $C_{13}H_6Cl_4O_2$ mol.wt. 336.00

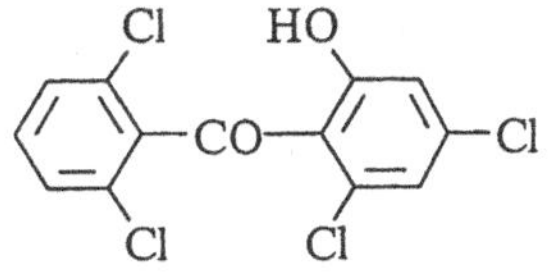

Synthesis

-Preparation by Fries rearrangement of 3,5-dichlorophenyl
  2,6-dichlorobenzoate with aluminium chloride for 30 min at
  150-160° [375].

m.p. 110-111° [375];   Spectra (NA).

(2,4-Dichloro-6-hydroxyphenyl)(3,4-dichlorophenyl)methanone

[31656-23-2] $C_{13}H_6Cl_4O_2$ mol.wt. 336.00

Synthesis

-Preparation by Fries rearrangement of 3,5-dichlorophenyl 3,4-dichlorobenzoate with aluminium chloride first 10 min at 140-150°, then for 30 min at 150-160° (80%) [375].

m.p. 138-139° [375];   Spectra (NA).

(2,5-Dichloro-4-hydroxyphenyl)(2,4-dichlorophenyl)methanone

[34183-04-5] $C_{13}H_6Cl_4O_2$ mol.wt. 336.00

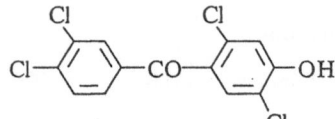

Synthesis

-Obtained by Fries rearrangement of 2,5-dichlorophenyl 2,4-dichlorobenzoate with aluminium chloride in chlorobenzene for 20 min at 140-150° or in nitrobenzene for 24 h at 75° [1214].

m.p. 147-148° [1214];   Spectra (NA).

(2,5-Dichloro-4-hydroxyphenyl)(3,4-dichlorophenyl)methanone

[34183-03-4] $C_{13}H_6Cl_4O_2$ mol.wt. 336.00

Synthesis

-Obtained by Fries rearrangement of 2,5-dichlorophenyl 3,4-dichlorobenzoate with aluminium chloride for 20 min at 140-150° or in nitrobenzene for 24 h at 75° [1214].

m.p. 160-161° [1214];   Spectra (NA).

(2,6-Dichloro-4-hydroxyphenyl)(3,4-dichlorophenyl)methanone

[34183-00-1] $C_{13}H_6Cl_4O_2$ mol.wt. 336.00

Synthesis

-Obtained by Fries rearrangement of 3,5-dichlorophenyl 3,4-dichlorobenzoate with aluminium chloride,
*in chlorobenzene for 20 min at 140-150° or in nitrobenzene for 24 h at 75° [1214];
*without solvent, first for 10 min at 140-150°, then for 30 min at 150-160° (by-product) [375].

m.p. 180-181° [375, 1214];   Spectra (NA).

(3,5-Dichloro-4-hydroxyphenyl)(2,4-dichlorophenyl)methanone

[34182-98-4] $C_{13}H_6Cl_4O_2$ mol.wt. 336.00

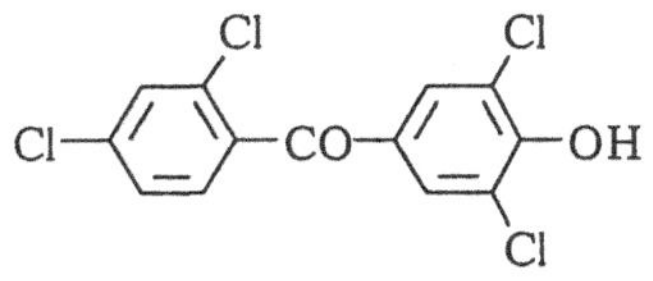

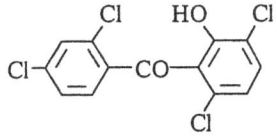

Synthesis

-Preparation by Fries rearrangement of 2,6-dichlorophenyl 2,4-dichlorobenzoate with aluminium chloride in chlorobenzene at 140-150° for 20 min or in nitrobenzene at 75° for 24 h [1214].
-Also refer to: [1353] (Japanese patent).

m.p. 183°5-184°5 [1214];   Spectra (NA).

(3,5-Dichloro-4-hydroxyphenyl)(3,4-dichlorophenyl)methanone

[34189-57-6] $C_{13}H_6Cl_4O_2$ mol.wt. 336.00

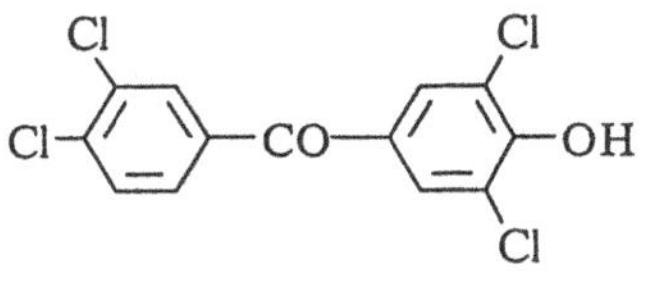

Synthesis

-Preparation by Fries rearrangement of 2,6-dichlorophenyl 3,4-dichlorobenzoate with aluminium chloride,
*in chlorobenzene for 20 min at 140-150° (89%) [1214];
*in nitrobenzene for 24 h at 75° (89%) [1214].

m.p. 202-203° [1214];   Spectra (NA).

(3,6-Dichloro-2-hydroxyphenyl)(2,4-dichlorophenyl)methanone

[34171-57-8] $C_{13}H_6Cl_4O_2$ mol.wt. 336.00

Synthesis

-Preparation by Fries rearrangement of 2,5-dichlorophenyl 2,4-dichlorobenzoate with aluminium chloride for 30 min at 150-160° [375].

m.p. 115-116° [375];   Spectra (NA).

(4-Bromophenyl)(2,4-dichloro-6-hydroxyphenyl)methanone

[34174-06-6] $C_{13}H_7BrCl_2O_2$ mol.wt. 346.01

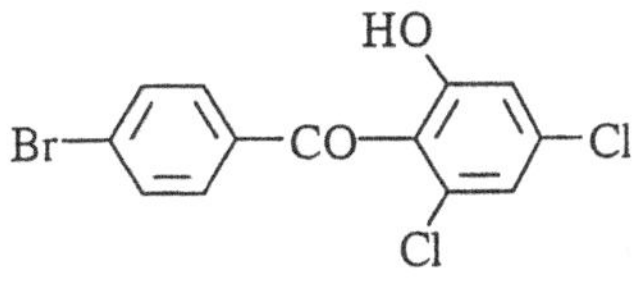

Synthesis

-Preparation by Fries rearrangement of 3,5-dichlorophenyl p-bromobenzoate with aluminium chloride for 30 min at 150-160° [375].

m.p. 157-158° [375];   Spectra (NA).

(4-Bromophenyl)(2,6-dichloro-4-hydroxyphenyl)methanone

[34183-12-5] $C_{13}H_7BrCl_2O_2$ mol.wt. 346.01

Synthesis

-Obtained by Fries rearrangement of 3,5-dichlorophenyl p-bromobenzoate with aluminium chloride in chlorobenzene for 20 min at 140-150° or in nitrobenzene for 24 h at 75° [1214].

      m.p. 147-148° [1214];   Spectra (NA).

(4-Bromophenyl)(3,5-dichloro-4-hydroxyphenyl)methanone

[34183-07-8] $C_{13}H_7BrCl_2O_2$ mol.wt. 346.01

Synthesis

-Preparation by Fries rearrangement of 2,6-dichlorophenyl p-bromobenzoate with aluminium chloride in chlorobenzene for 20 min at 140-150° or in nitrobenzene for 24 h at 75° [1214].

      m.p. 173-174° [1214];   Spectra (NA).

(4-Bromophenyl)(2,5-difluoro-4-hydroxyphenyl)methanone

[192437-36-8] $C_{13}H_7BrF_2O_2$ mol.wt. 313.10

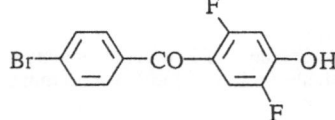

Synthesis

-Preparation by demethylation of 4'-bromo-2,5-difluoro-4-methoxybenzophenone (SM) with 62% aqueous hydrobromic acid in refluxing acetic acid for 13 h (95%). SM was obtained by Friedel-Crafts acylation of 2,5-difluoroanisole with 4-bromobenzoyl chloride in nitrobenzene in the presence of aluminium chloride, first at temperature <6°, then at r.t. overnight (80%). -Refer to: Chem. Abstr., **127**, 108766j (1997).
-Also refer to: Chem. Abstr., **127**, 108921f (1997).

      m.p. and Spectra (NA).

(4-Chlorophenyl)(5-fluoro-2-hydroxy-3-nitrophenyl)methanone

[85052-27-3] $C_{13}H_7ClFNO_4$ mol.wt. 295.65

Synthesis

-Preparation by reaction of 60% nitric acid with 4'-chloro-5-fluoro-2-hydroxybenzophenone in acetic acid at r.t. for 30 min (84%) [932].

      m.p. (NA);   yellow crystals [932];
      ^{1}H NMR [932], IR [932], MS [932];   TLC [932].

(2,3-Dichloro-4-hydroxyphenyl)(2-fluorophenyl)methanone

[72498-54-5] $C_{13}H_7Cl_2FO_2$ mol.wt. 285.10

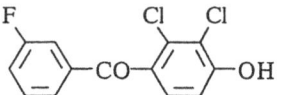

Synthesis

-Preparation by demethylation of 2,3-dichloro-2'-fluoro-4-methoxybenzophenone (SM) with aluminium chloride in refluxing benzene for 5 h [1270], (89%) [1272] or in refluxing methylene chloride overnight [653]. SM was obtained by Friedel-Crafts acylation of 2,3-dichloroanisole with o-fluorobenzoyl chloride in the presence of aluminium chloride in refluxing methylene chloride [653] or in ethylene dichloride for 2 h [1270].
-Also refer to: [1271].

m.p.   128-131° [1270, 1272];   Spectra (NA).

(2,3-Dichloro-4-hydroxyphenyl)(3-fluorophenyl)methanone

[72482-40-7] $C_{13}H_7Cl_2FO_2$ mol.wt. 285.10

Syntheses

-Preparation by demethylation of 2,3-dichloro-3'-fluoro-4-methoxybenzophenone (SM) with aluminium chloride in refluxing methylene chloride overnight. SM was obtained by Friedel-Crafts acylation of 2,3-dichloroanisole with m-fluorobenzoyl chloride in the presence of aluminium chloride in refluxing methylene chloride [653].
-Preparation by reaction of m-fluorobenzoyl chloride with 2,3-dichloroanisole in the presence of aluminium chloride [1270]. Here, in this process, their is simultaneously acylation and demethylation in one step.

m.p. and Spectra (NA).

(2,3-Dichloro-4-hydroxyphenyl)(4-fluorophenyl)methanone

[62967-10-6] $C_{13}H_7Cl_2FO_2$ mol.wt. 285.10

Synthesis

-Preparation by demethylation of 2,3-dichloro-4'-fluoro-4-methoxybenzophenone (SM),
*with aluminium chloride in refluxing methylene chloride overnight (72%) [653];
*with pyridinium chloride at 200° for 1 h [1270].
SM was obtained by Friedel-Crafts acylation of 2,3-dichloroanisole with p-fluorobenzoyl chloride in the presence of aluminium chloride in methylene chloride first at 5°, then at reflux for 2 h (50%) [653] or in ethylene dichloride at 60° for 1 h [1270].
-Also refer to: [1272].

m.p.   155-159° [653];   Spectra (NA).

(5-Chloro-2-hydroxy-3-nitrophenyl)(4-chlorophenyl)methanone

[85052-24-0] $C_{13}H_7Cl_2NO_4$ mol.wt. 312.11

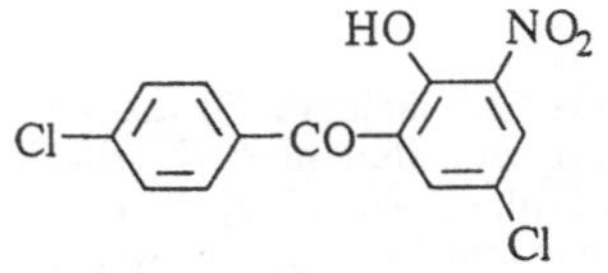

Synthesis

-Preparation by reaction of 60% nitric acid with 4',5-di-
chloro-2-hydroxybenzophenone in the presence of one drop
of concentrated sulfuric acid at r.t. for 25 min (92%) [932].

m.p. (NA);   crystals [932];
^{1}H NMR [932], IR [932], MS [932];   TLC [932].

(2,3-Dichloro-4-hydroxyphenyl)(4-nitrophenyl)methanone

[92285-28-4] $C_{13}H_7Cl_2NO_4$ mol.wt. 312.11

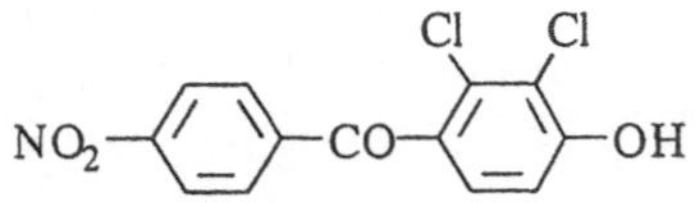

Synthesis

-Preparation by demethylation of 2,3-dichloro-
4-methoxy-4'-nitrobenzophenone (SM) with
aluminium chloride in refluxing ethylene dichloride
for 1.5 h (95%) [1071] or in refluxing methylene
chloride overnight [653]. SM was obtained by Friedel-Crafts acylation of 2,3-dichloroanisole with
p-nitrobenzoyl chloride in the presence of aluminium chloride in refluxing methylene chloride
[653].

m.p.   201-204° [1071];   Spectra (NA).

(2-Chloro-4-hydroxyphenyl)(2,4-dichlorophenyl)methanone

[34294-62-7] $C_{13}H_7Cl_3O_2$ mol.wt. 301.56

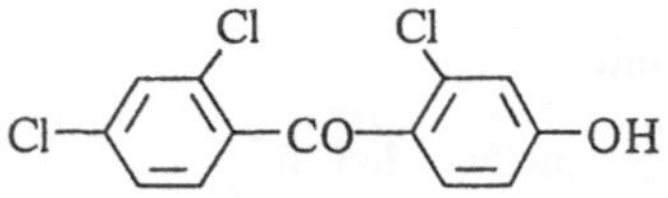

Synthesis

-Obtained by Fries rearrangement of m-chlorophenyl
2,4-dichlorobenzoate with aluminium chloride in chlorobenzene for 20 min at 140-150° or in nitrobenzene for
24 h at 75° [1214].

m.p.   163-164° [1214];   Spectra (NA).

(3-Chloro-4-hydroxyphenyl)(2,4-dichlorophenyl)methanone

[34182-96-2] $C_{13}H_7Cl_3O_2$ mol.wt. 301.56

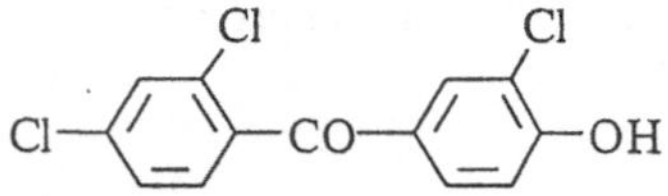

Synthesis

-Preparation by Fries rearrangement of o-chlorophenyl
2,4-dichlorobenzoate with aluminium chloride in chlorobenzene for 20 min at 140-150° or in nitrobenzene for
24 h at 75° [1214].

m.p.   144-145° [1214];   Spectra (NA).

(5-Chloro-2-hydroxyphenyl)(2,4-dichlorophenyl)methanone

[72089-86-2] $C_{13}H_7Cl_3O_2$ mol.wt. 301.56

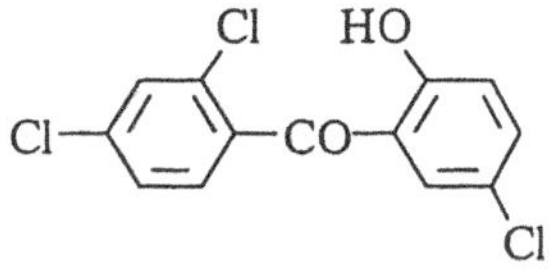

Synthesis

-Preparation by Fries rearrangement of p-chlorophenyl 2,4-dichlorobenzoate with aluminium chloride at 190° for 15 min [693].

m.p. 96-97° [693];   Spectra (NA).

(5-Chloro-2-hydroxyphenyl)(3,4-dichlorophenyl)methanone

$C_{13}H_7Cl_3O_2$ mol.wt. 301.56

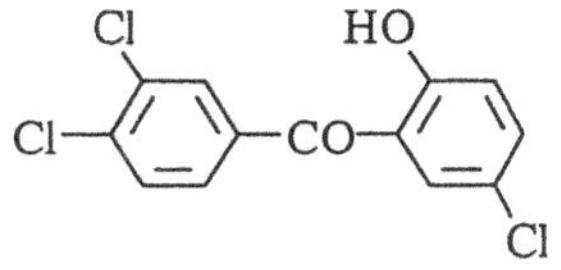

Synthesis

-Preparation by reaction of 3,4-dichlorobenzoyl chloride with p-chlorophenol in the presence of aluminium chloride, first at 100° for 3 min, then at 178-180° for 23 min (53%) [605].

m.p. 92-92°5 [605];   Spectra (NA).

(2-Chlorophenyl)(2,3-dichloro-4-hydroxyphenyl)methanone

[72482-80-5] $C_{13}H_7Cl_3O_2$ mol.wt. 301.56

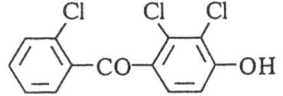

Synthesis

-Preparation by demethylation of 2,2',3-trichloro-4-methoxybenzophenone (SM) with aluminium chloride in refluxing benzene for 5 h, then at r.t. for 18 h. SM was obtained by reaction of o-chlorobenzoyl chloride with 2,3-dichloroanisole in ethylene dichloride in the presence of aluminium chloride at 60° for 1 h [1270].

m.p. 74-77° [1270];   Spectra (NA).

(2-Chlorophenyl)(2,4-dichloro-6-hydroxyphenyl)methanone

[34174-11-3] $C_{13}H_7Cl_3O_2$ mol.wt. 301.56

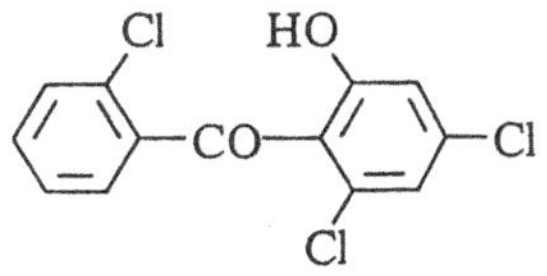

Synthesis

-Preparation by Fries rearrangement of 3,5-dichlorophenyl o-chlorobenzoate with aluminium chloride for 30 min at 150-160° [375].

m.p. 72-73° [375];   Spectra (NA).

(2-Chlorophenyl)(3,5-dichloro-2-hydroxyphenyl)methanone

$C_{13}H_7Cl_3O_2$ mol.wt. 301.56

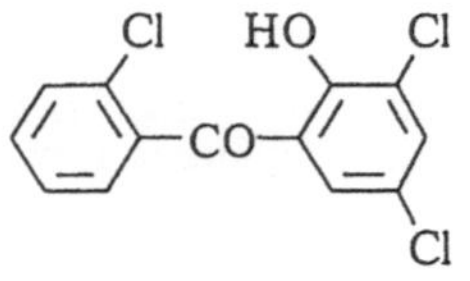

Synthesis

-Preparation by reaction of o-chlorobenzoyl chloride with
2,4-dichlorophenol in the presence of aluminium chloride at
180° [605].

m.p. 92° [605];   Spectra (NA).

(2-Chlorophenyl)(3,5-dichloro-4-hydroxyphenyl)methanone

[34183-18-1] $C_{13}H_7Cl_3O_2$ mol.wt. 301.56

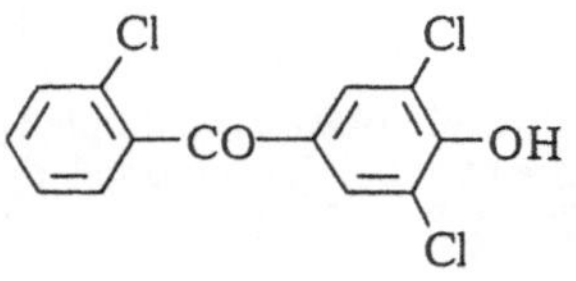

Synthesis

-Preparation by Fries rearrangement of 2,6-dichlorophenyl
o-chlorobenzoate with aluminium chloride in chlorobenzene
for 20 min at 140-150° or in nitrobenzene for 24 h at 75°
[1214].

m.p. 162-163° [1214];   Spectra (NA).

(3-Chlorophenyl)(2,3-dichloro-4-hydroxyphenyl)methanone

$C_{13}H_7Cl_3O_2$ mol.wt. 301.56

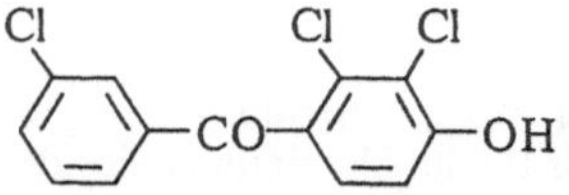

Synthesis

-Preparation by demethylation of 2,3,3'-trichloro-4-methoxy-
benzophenone (SM) with aluminium chloride in refluxing
methylene chloride overnight. SM was obtained by Friedel-
Crafts acylation of 2,3-dichloroanisole with m-chloro-
benzoyl chloride in the presence of aluminium chloride in refluxing methylene chloride [653].

m.p. and Spectra (NA).

(4-Chlorophenyl)(2,3-dichloro-4-hydroxyphenyl)methanone

[72498-76-1] $C_{13}H_7Cl_3O_2$ mol.wt. 301.56

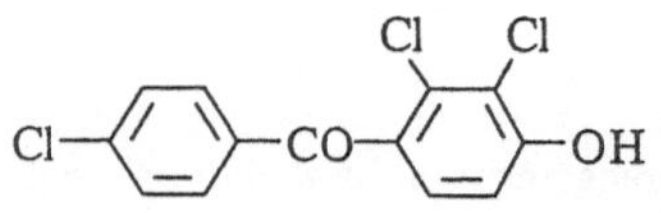

Synthesis

-Preparation by demethylation of 2,3,4'-trichloro-
4-methoxybenzophenone (SM) with aluminium chloride in
refluxing methylene chloride overnight. SM was obtained by
Friedel-Crafts acylation of 2,3-dichloroanisole with
p-chlorobenzoyl chloride in the presence of aluminium chloride in refluxing methylene chloride
[653].
-Also refer to: [1270].

m.p. and Spectra (NA).

(4-Chlorophenyl)(2,4-dichloro-6-hydroxyphenyl)methanone

[34171-58-9] $C_{13}H_7Cl_3O_2$ mol.wt. 301.56

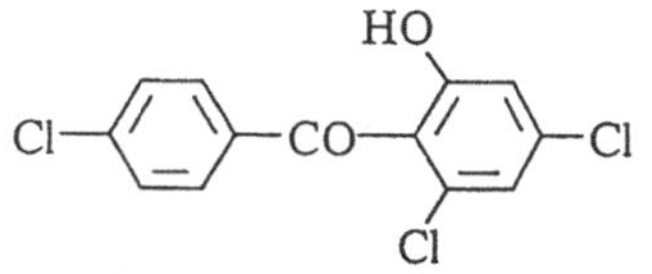

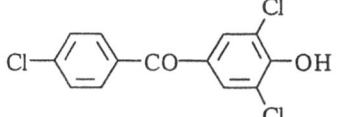

Synthesis

-Preparation by Fries rearrangement of 3,5-dichlorophenyl p-chlorobenzoate with aluminium chloride for 30 min at 150-160° [375].

m.p. 161-162° [375]; Spectra (NA).

(4-Chlorophenyl)(2,6-dichloro-4-hydroxyphenyl)methanone

[34183-11-4] $C_{13}H_7Cl_3O_2$ mol.wt. 301.56

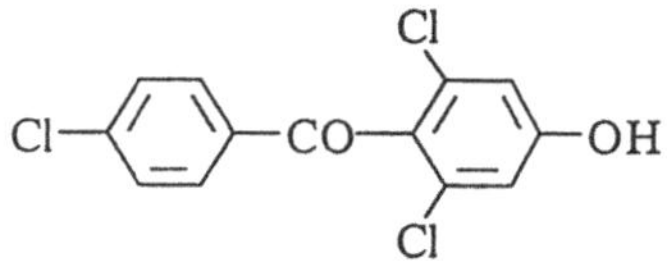

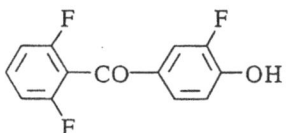

Synthesis

-Obtained by Fries rearrangement of 3,5-dichlorophenyl p-chlorobenzoate with aluminium chloride in chloro-benzene for 20 min at 140-150° or in nitrobenzene for 24 h at 75° [1214].

m.p. 137°5-138° [1214]; Spectra (NA).

(4-Chlorophenyl)(3,5-dichloro-4-hydroxyphenyl)methanone

[34182-97-3] $C_{13}H_7Cl_3O_2$ mol.wt. 301.56

Synthesis

-Preparation by Fries rearrangement of 2,6-dichlorophenyl p-chlorobenzoate with aluminium chloride in chloro-benzene for 20 min at 140-150° or in nitrobenzene for 24 h at 75° [1214].

m.p. 179-180° [1214]; Spectra (NA).

(2,6-Difluorophenyl)(3-fluoro-4-hydroxyphenyl)methanone

[161581-97-1] $C_{13}H_7F_3O_2$ mol.wt. 252.19

Synthesis

-Preparation by demethylation of 4-methoxy-2,3',6-tri-fluorobenzophenone (SM) with 62% aqueous hydrobromic acid in acetic acid at 125° (97%). SM (m.p. 79-83°) was obtained by Friedel-Crafts acylation of o-fluoroanisole with 2,6-difluorobenzoyl chloride in nitrobenzene in the presence of aluminium chloride [15].

m.p. (NA); MS [15].

(2-Bromophenyl)(5-chloro-2-hydroxyphenyl)methanone

[92739-90-7] $C_{13}H_8BrClO_2$ mol.wt. 311.56

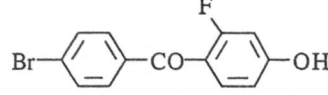

Synthesis

-Refer to: [686].

m.p. and Spectra (NA).

(4-Bromophenyl)(3-chloro-4-hydroxyphenyl)methanone

[161582-04-3] $C_{13}H_8BrClO_2$ mol.wt. 311.56

Synthesis

-Refer to: [15].

m.p. and Spectra (NA).

(3-Bromophenyl)(5-fluoro-2-hydroxyphenyl)methanone

[62433-28-7] $C_{13}H_8BrFO_2$ mol.wt. 295.11

Synthesis

-Preparation by Fries rearrangement of p-fluorophenyl
   m-bromobenzoate with aluminium chloride at 130° for 2 h
   (91%) [1295].

m.p. 159-160° [1295];   Spectra (NA).

(4-Bromophenyl)(2-fluoro-4-hydroxyphenyl)methanone

[161581-99-3] $C_{13}H_8BrFO_2$ mol.wt. 295.11

Synthesis

-Preparation by demethylation of 4'-bromo-2-fluoro-
   4-methoxybenzophenone (SM) with 62% aqueous
   hydrobromic acid in acetic acid at 125°. SM was obtained
   by Friedel-Crafts acylation of m-fluoroanisole
with p-bromobenzoyl chloride in nitrobenzene in the presence of aluminium chloride [15].
-Also refer to: [945] and Chem. Abstr., **127**, 108766j (1997).

m.p. (NA);   MS [15].

(4-Bromophenyl)(3-fluoro-4-hydroxyphenyl)methanone

[161581-98-2] $C_{13}H_8BrFO_2$ mol.wt. 295.11

Synthesis

-Preparation by demethylation of 4'-bromo-3-fluoro-
   4-methoxybenzophenone (SM) with 62% aqueous
   hydrobromic acid in acetic acid at 125°. SM was obtained
   by Friedel-Crafts acylation of o-fluoroanisole

with p-bromobenzoyl chloride in nitrobenzene in the presence of aluminium chloride [15].
-Also refer to: Chem. Abstr., **127**, 108766j (1997).

m.p. 183-184° [15];   Spectra (NA).

(3-Bromo-4-hydroxyphenyl)(4-bromophenyl)methanone

[161582-02-1]   C13H8Br2O2   mol.wt. 356.01

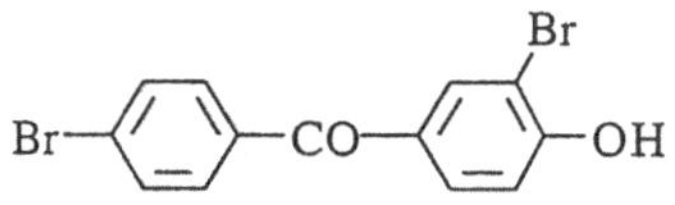

Synthesis

-Refer to: [15].

m.p. and Spectra (NA).

(4-Chloro-2-hydroxyphenyl)(4-fluorophenyl)methanone

[169781-85-5]   C13H8ClFO2   mol.wt. 250.66

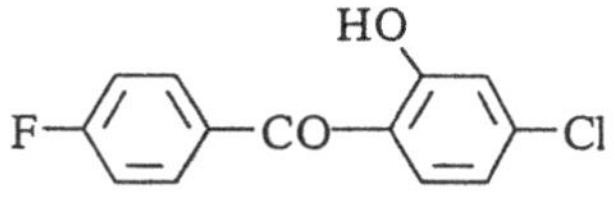

Syntheses

-Obtained by Fries rearrangement of m-chlorophenyl
p-fluorobenzoate with aluminium chloride at 200° for
20 min [766].
-Also obtained by demethylation of 4-chloro-4'-fluoro-
2-methoxybenzophenone with boron tribromide in methylene chloride at r.t. for 12 h [766],
according to [891].

m.p. and Spectra (NA).

(5-Chloro-2-hydroxyphenyl)(2-fluorophenyl)methanone

[65185-33-3]   C13H8ClFO2   mol.wt. 250.66

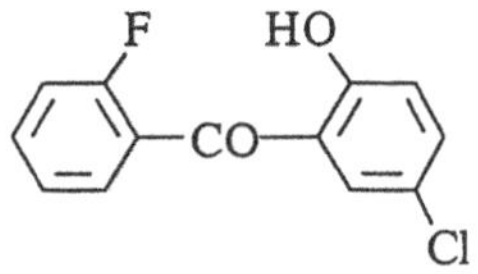

Syntheses

-Preparation by Friedel-Crafts acylation of p-chlorophenol
with o-fluorobenzoyl chloride in the presence of aluminium
chloride at 195° for 25 min (70%) [226].
-Preparation by Fries rearrangement of p-chlorophenyl
o-fluorobenzoate with aluminium chloride [692].
-Also refer to: [290].

m.p. 76-80°5 [226];   Spectra (NA).

(5-Chloro-2-hydroxyphenyl)(3-fluorophenyl)methanone

[62433-31-2]   C13H8ClFO2   mol.wt. 250.66

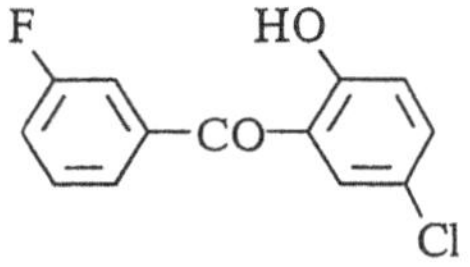

Synthesis

-Preparation by Fries rearrangement of p-chlorophenyl
m-fluorobenzoate with aluminium chloride at 130° for 2 h
(88%) [1295].

m.p. 153° [1295];   Spectra (NA).

(2-Chlorophenyl)(2-fluoro-4-hydroxyphenyl)methanone

[87750-64-9] $C_{13}H_8ClFO_2$ mol.wt. 250.66

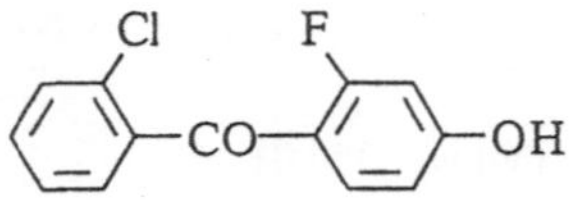

Synthesis

-Preparation by reaction of o-chlorobenzoyl chloride with
m-fluorophenol in an hydrofluoric acid solution in the
presence of boron trifluoride at 0° for 1 h [343].

m.p. and Spectra (NA).

(2-Chlorophenyl)(4-fluoro-2-hydroxyphenyl)methanone

[87750-63-8] $C_{13}H_8ClFO_2$ mol.wt. 250.66

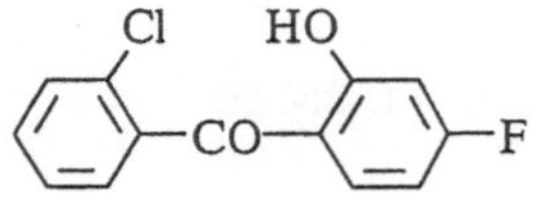

Synthesis

-Preparation by reaction of o-chlorobenzoyl chloride with
m-fluorophenol in an hydrofluoric acid solution in the
presence of boron trifluoride at 0° for 1 h [343].

m.p. and Spectra (NA).

(2-Chlorophenyl)(5-fluoro-2-hydroxyphenyl)methanone

[2341-94-8] $C_{13}H_8ClFO_2$ mol.wt. 250.66

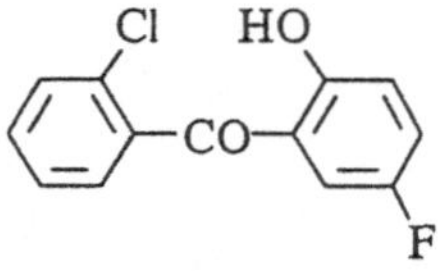

Synthesis

-Preparation by Fries rearrangement of p-fluorophenyl
o-chlorobenzoate with aluminium chloride [692], at 130° for
2 h (95%) [656].
-Also refer to: [165].

m.p. 65° [656];   Spectra (NA).

(3-Chlorophenyl)(5-fluoro-2-hydroxyphenyl)methanone

[62666-38-0] $C_{13}H_8ClFO_2$ mol.wt. 250.66

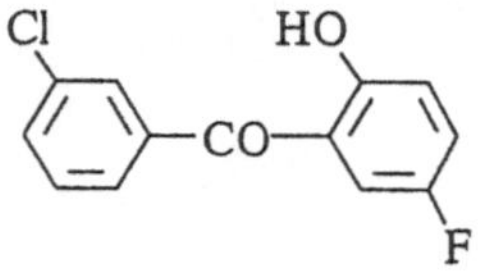

Synthesis

-Preparation by Fries rearrangement of p-fluorophenyl
m-chlorobenzoate with aluminium chloride [692].
-Also refer to: [165].

m.p. and Spectra (NA).

(4-Chlorophenyl)(3-fluoro-4-hydroxyphenyl)methanone

[83885-18-1] $C_{13}H_8ClFO_2$ mol.wt. 250.66

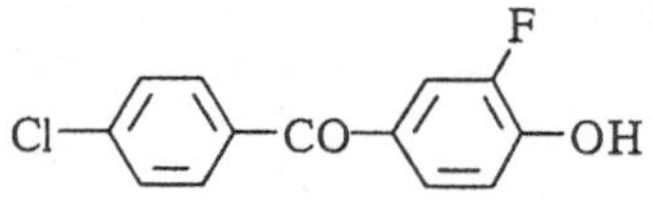

Synthesis

-Preparation by reaction of p-chlorobenzoyl chloride with
o-fluorophenol in the presence of aluminium chloride
[296].

m.p. 173-176°5 [296];   Spectra (NA).

(4-Chlorophenyl)(4-fluoro-2-hydroxyphenyl)methanone

[169781-86-6] $C_{13}H_8ClFO_2$ mol.wt. 250.66

Syntheses

-Obtained by Fries rearrangement of m-fluorophenyl
 p-chlorobenzoate with aluminium chloride at 200° for
 20 min [766].
-Also obtained by demethylation of 4'-chloro-4-fluoro-.
2-methoxybenzophenone with boron tribromide in methylene chloride at r.t. for 12 h [766],
according to [891].

m.p. and Spectra (NA).

(4-Chlorophenyl)(5-fluoro-2-hydroxyphenyl)methanone

[62433-26-5] $C_{13}H_8ClFO_2$ mol.wt. 250.66

Syntheses

-Preparation by Fries rearrangement of p-fluorophenyl
 p-chlorobenzoate,
*with titanium tetrachloride at 150° for 18 h (81%) [932];
*with aluminium chloride [117, 692], at 130° for 2 h (98%)
 [1295] or at 200° for 5 min [39] or for 15 min (65%) [697].
-Preparation by reaction of p-chlorobenzoyl chloride with p-fluorophenol in hydrofluoric acid in the
 presence of boron trifluoride at 80° for 20 h (74%) [343]. **N.B.**: In the patent, this compound was
 erroneously named 3-fluoro-4'-chloro-2-hydroxybenzophenone (assay 21, page 13) [343].
-Also refer to: [165, 231, 324, 408, 409, 410, 411, 412, 477, 686, 687, 939, 947, 1031, 1459].

m.p. 174° [1295], 67°6 [697]. One of the reported melting points is obviously wrong.
^{1}H NMR [932], ^{13}C NMR [697], IR [697, 932], MS [932];   TLC [932];
HPLC [1030]; pK_a [409].

(4-Chlorophenyl)(5-fluoro-2-hydroxyphenyl)methanone-^{14}C

[82589-26-2] $C_{13}H_8ClFO_2$ mol.wt. 252.66

Synthesis

-Preparation by Fries rearrangement of p-fluorophenyl
 p-chloro-[carboxyl-^{14}C]benzoate with aluminium chloride at
 200° for 5 min (76%) (51.8 mCi/mmol) [39].

m.p. and Spectra (NA).

(3-Chloro-4-hydroxyphenyl)(4-iodophenyl)methanone

[161582-03-2] $C_{13}H_8ClIO_2$ mol.wt. 358.56

Synthesis

-Refer to: [15].

m.p. and Spectra (NA).

(3-Chloro-4-hydroxyphenyl)(4-nitrophenyl)methanone

$C_{13}H_8ClNO_4$ mol.wt. 277.66

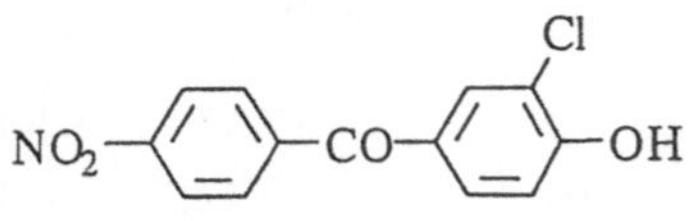

Synthesis

-Obtained by Fries rearrangement of o-chlorophenyl p-nitrobenzoate with aluminium chloride at 120° for 2 h [1345].

m.p. 196° [1345];   Spectra (NA).

(4-Chloro-2-hydroxyphenyl)(3-nitrophenyl)methanone

[22293-33-0] $C_{13}H_8ClNO_4$ mol.wt. 277.66

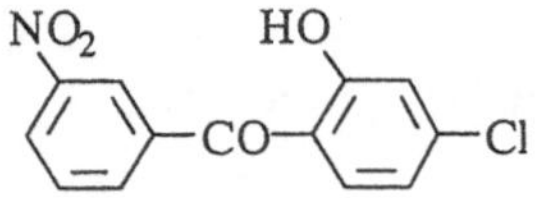

Synthesis

-Obtained by photo-Fries rearrangement of m-chlorophenyl m-nitrobenzoate in ethanol for 60-75 h (35%) [1034].

m.p. 126° [1034];   IR [1034], UV [1034].

(4-Chloro-2-hydroxyphenyl)(4-nitrophenyl)methanone

[22359-51-9] $C_{13}H_8ClNO_4$ mol.wt. 277.66

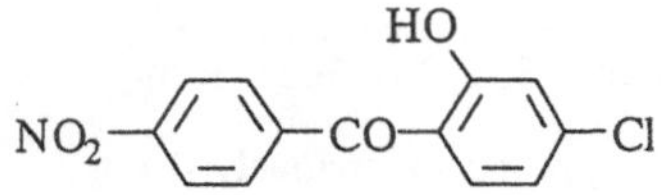

Syntheses

-Preparation by Fries rearrangement of m-chlorophenyl p-nitrobenzoate with aluminium chloride at 120° for 2 h [1345].

-Also obtained by photo-Fries rearrangement of m-chlorophenyl p-nitrobenzoate in ethanol for 60-75 h (13%) [1034].

m.p. 160° [1345], 117° [1034];   IR [1034], UV [1034].

(5-Chloro-2-hydroxyphenyl)(3-nitrophenyl)methanone

[126260-47-7] $C_{13}H_8ClNO_4$ mol.wt. 277.66

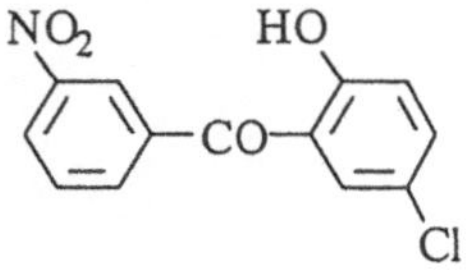

Syntheses

-Obtained (poor yield) by Fries rearrangement of p-chloro-phenyl m-nitrobenzoate with aluminium chloride in refluxing chlorobenzene for 4 h (11%) [1450].

-Obtained by photo-Fries rearrangement of p-chlorophenyl m-nitrobenzoate in ethanol for 60-75 h (35%) [1034].

m.p. 126° [1034], 125-126° [1450];   IR [1034], UV [1034].

(5-Chloro-2-hydroxyphenyl)(4-nitrophenyl)methanone

[84443-36-7] $C_{13}H_8ClNO_4$ mol.wt. 277.66

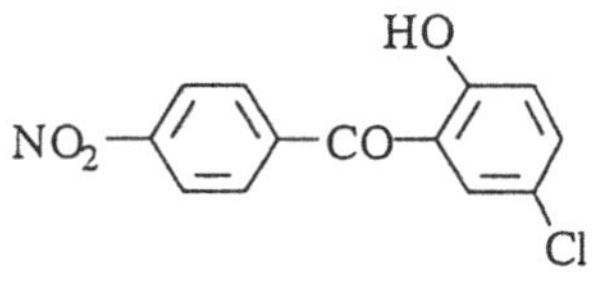

Syntheses

-Obtained by Fries rearrangement of p-chlorophenyl
 p-nitrobenzoate,
*with aluminium chloride at 120° for 2 h [1345];
*with UV light irradiation in ethanol during 60-75 h (13%)
 [1034].

-Also refer to: [670].

m.p.   120° [1345], 117° [1034];   Spectra (NA).

(2-Chlorophenyl)(2-hydroxy-5-nitrophenyl)methanone

[95263-98-2] $C_{13}H_8ClNO_4$ mol.wt. 277.66

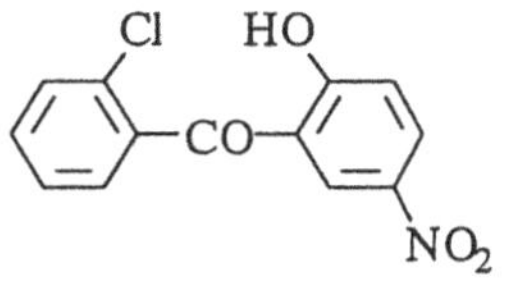

Synthesis

-Obtained by alkaline degradation of nizofenone fumarate
 [1171].
-Also refer to: [709].

m.p. and Spectra (NA).

(4-Chlorophenyl)(2-hydroxy-5-nitrophenyl)methanone

[124071-26-7] $C_{13}H_8ClNO_4$ mol.wt. 277.66

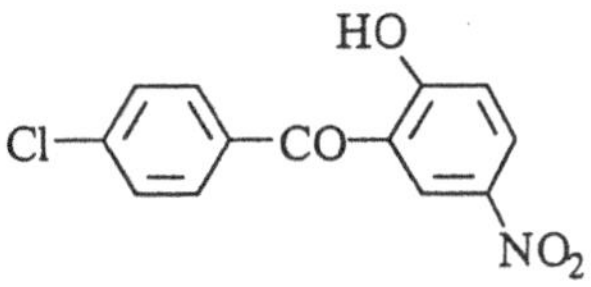

Synthesis

-Preparation according to the method [1398], (23%) [1232].

m.p.   170-172° [1232];   Spectra (NA).

(2-Chloro-5-hydroxyphenyl)(4-chlorophenyl)methanone

[62810-45-1] $C_{13}H_8Cl_2O_2$ mol.wt. 267.11

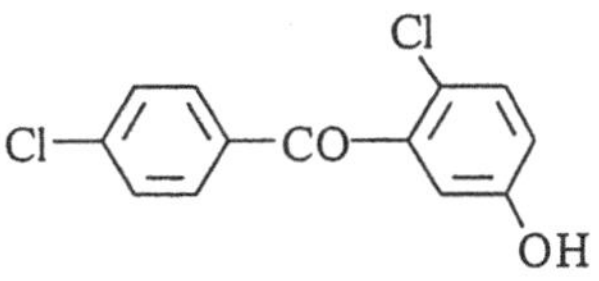

Syntheses

-Preparation by diazotization of 5-amino-2,4'-dichloro-
 benzophenone followed by hydrolysis of the diazonium salt
 so obtained [847].
-Also obtained by demethylation of 2,4'-dichloro-5-methoxy-
 benzophenone with aluminium chloride in refluxing
 chlorobenzene [847].

m.p.   172° [847];   Spectra (NA).

(3-Chloro-2-hydroxyphenyl)(3-chlorophenyl)methanone

[41796-26-3] $C_{13}H_8Cl_2O_2$ mol.wt. 267.11

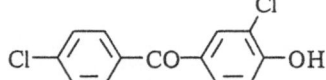

Synthesis

-Refer to: [1005].

m.p. and Spectra (NA).

(3-Chloro-4-hydroxyphenyl)(2-chlorophenyl)methanone

$C_{13}H_8Cl_2O_2$ mol.wt. 267.11

Synthesis

-Preparation by demethylation of 2',3-dichloro-4-methoxy-
  benzophenone with pyridinium chloride at reflux for 30 min
  [241].

m.p. 156° [241];   Spectra (NA).

(3-Chloro-4-hydroxyphenyl)(4-chlorophenyl)methanone

[34189-58-7] $C_{13}H_8Cl_2O_2$ mol.wt. 267.11

Syntheses

-Preparation by reaction of p-chlorobenzoyl chloride with
  o-chlorophenol in the presence of aluminium chloride
  [296].
-Preparation by demethylation of 3,4'-dichloro-
4-methoxybenzophenone in refluxing pyridinium chloride for 30 min [241].
-Preparation by Friedel-Crafts acylation of chlorobenzene with 3-chloro-4-methoxybenzoyl chloride
in the presence of aluminium chloride at 120° for 3 h (50-70%) [1330].
-Preparation by Fries rearrangement of o-chlorophenyl p-chlorobenzoate with aluminium chloride
in chlorobenzene at 140-150° for 20 min or in nitrobenzene at 75° for 24 h [1214].

m.p. 178-179° [1330], 176-177°5 [1214], 168° [241], 52°5-53°5 [296]. One of the
reported melting points is obviously wrong.
^{1}H NMR [1330], MS [1330];   TLC [1330].

(4-Chloro-2-hydroxyphenyl)(4-chlorophenyl)methanone

[60805-31-4] $C_{13}H_8Cl_2O_2$ mol.wt. 267.11

Syntheses

-Preparation by Friedel-Crafts acylation of chlorobenzene
  with 4-chloro-2-methoxybenzoyl chloride in the presence of
  aluminium chloride at 120° for 3 h (50-70%) [1330].
-Also obtained by Fries rearrangement of m-chlorophenyl
p-chlorobenzoate with aluminium chloride at 200° for 20 min [766].
-Also obtained by demethylation of 4,4'-dichloro-2-methoxybenzophenone with boron tribromide in
methylene chloride at r.t. for 12 h [766], according to [891].

m.p. 78-79° [1330]; ^{1}H NMR [1330], MS [1330];   TLC [1330].

(4-Chloro-3-hydroxyphenyl)(4-chlorophenyl)methanone

[60805-30-3] $C_{13}H_8Cl_2O_2$ mol.wt. 267.11

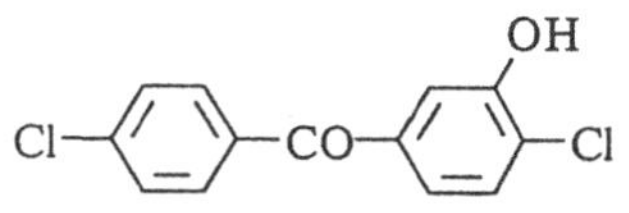

Synthesis

-Preparation by Friedel-Crafts acylation of chlorobenzene with 4-chloro-3-methoxybenzoyl chloride in the presence of aluminium chloride at 120° for 3 h (50-70%) [1330].

m.p. 160-162° [1330];
^{1}H NMR [1330], MS [1330]; TLC [1330].

(5-Chloro-2-hydroxyphenyl)(2-chlorophenyl)methanone

[61785-35-1] $C_{13}H_8Cl_2O_2$ mol.wt. 267.11

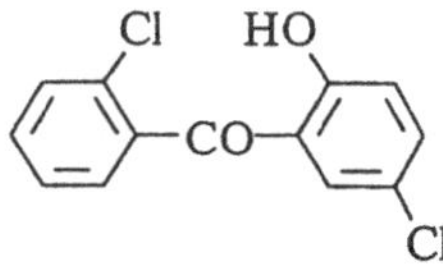

Syntheses

-Preparation by reaction of o-chlorobenzoyl chloride with p-chlorophenol in the presence of aluminium chloride at 180° [605].
-Preparation by Fries rearrangement of p-chlorophenyl o-chlorobenzoate with aluminium chloride [694].
-Also refer to: [289, 290, 686, 690, 695].

m.p. 107°9 [694], 106°5-107°5 [605]; Spectra (NA).

(5-Chloro-2-hydroxyphenyl)(3-chlorophenyl)methanone

[61785-36-2] $C_{13}H_8Cl_2O_2$ mol.wt. 267.11

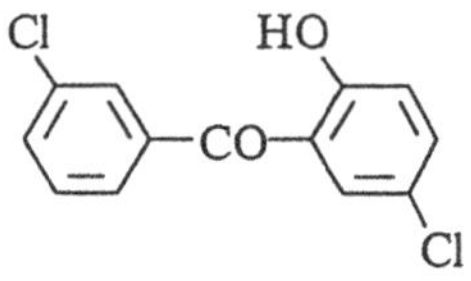

Synthesis

-Preparation by Fries rearrangement of p-chlorophenyl m-chlorobenzoate with aluminium chloride for 30 min at 160° (64%) [1147].

m.p. 72° [1147]; Spectra (NA).

(5-Chloro-2-hydroxyphenyl)(4-chlorophenyl)methanone

[61785-37-3] $C_{13}H_8Cl_2O_2$ mol.wt. 267.11

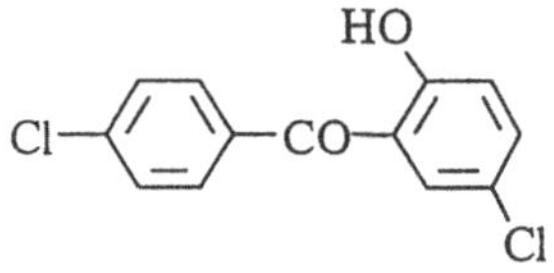

Synthesis

-Obtained by Fries rearrangement of p-chlorophenyl p-chlorobenzoate with titanium tetrachloride at 150° for 18 h (23%) [932].
-Also refer to: [226, 240, 605].

m.p. (NA); yellow crystals [932];
^{1}H NMR [932], IR [932], MS [932];
TLC [932]; HPLC [1030].

(2-Fluoro-4-hydroxyphenyl)(3-nitrophenyl)methanone

[194290-75-0] $C_{13}H_8FNO_4$ mol.wt. 247.20

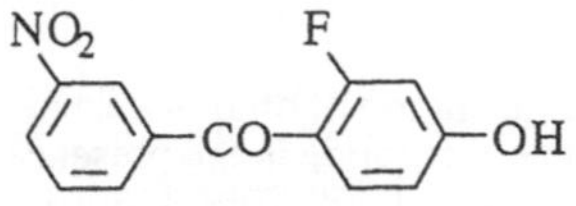

Synthesis

-Obtained by Friedel-Crafts acylation of m-fluorophenol with
m-nitrobenzoyl chloride in ethylene dichloride in the
presence of aluminium chloride at r.t. for 4 h (general
procedure C; compound 33 i). -Refer to: Chem. Abstr., **127**,
190681j (1997).

m.p. and Spectra (NA).

(2-Fluoro-5-hydroxyphenyl)(3-nitrophenyl)methanone

[194290-73-8] $C_{13}H_8FNO_4$ mol.wt. 247.20

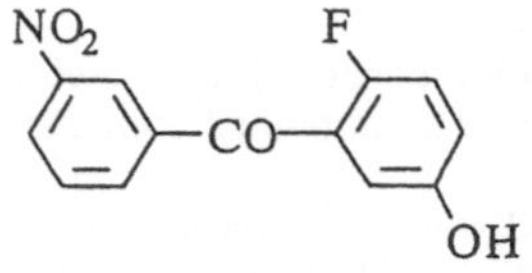

Synthesis

-Obtained by Friedel-Crafts acylation of p-fluorophenol with
m-nitrobenzoyl chloride in ethylene dichloride in the
presence of aluminium chloride at r.t. for 4 h (General
Procedure C; compound 33 h). -Refer to: Chem. Abstr., **127**,
190681j (1997).

m.p. and Spectra (NA).

(5-Fluoro-2-hydroxyphenyl)(2-nitrophenyl)methanone

[62433-27-6] $C_{13}H_8FNO_4$ mol.wt. 247.20

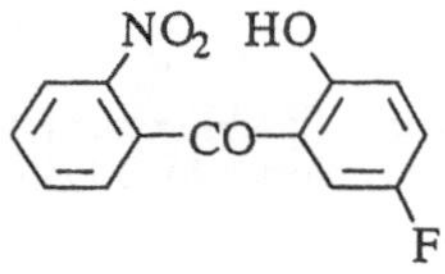

Synthesis

-Preparation by Fries rearrangement of p-fluorophenyl
o-nitrobenzoate with aluminium chloride at 130° for 2 h
(86%) [1295].

m.p.   123° [1295];   Spectra (NA).

(5-Fluoro-2-hydroxyphenyl)(4-nitrophenyl)methanone

[62433-30-1] $C_{13}H_8FNO_4$ mol.wt. 247.20

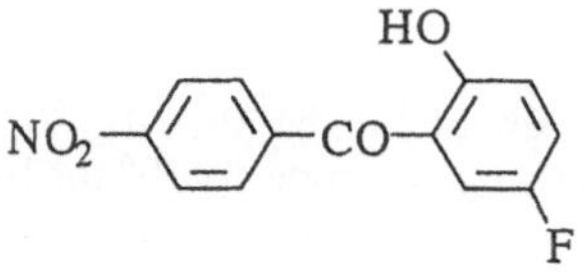

Synthesis

-Preparation by Fries rearrangement of p-fluorophenyl
p-nitrobenzoate with aluminium chloride at 130° for 2 h
(93%) [1295].

m.p.   141° [1295];   IR [1295].

(4-Fluoro-2-hydroxyphenyl)(4-fluorophenyl)methanone

[153411-29-1] $C_{13}H_8F_2O_2$ mol.wt. 234.20

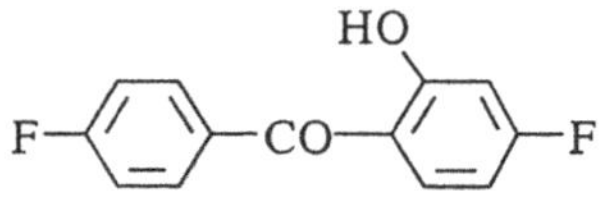

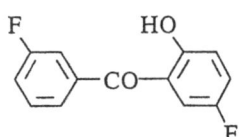

Syntheses

-Obtained by Fries rearrangement of m-fluorophenyl
 p-fluorobenzoate with aluminium chloride at 200° for
 20 min [766].
-Also obtained by demethylation of 4,4'-difluoro-2-methoxy
benzophenone with boron tribromide in methylene chloride at r.t. for 12 h [766], according to
[891].
-Also refer to: [1427].

    m.p. and Spectra (NA).

(5-Fluoro-2-hydroxyphenyl)(2-fluorophenyl)methanone

$C_{13}H_8F_2O_2$ mol.wt. 234.20

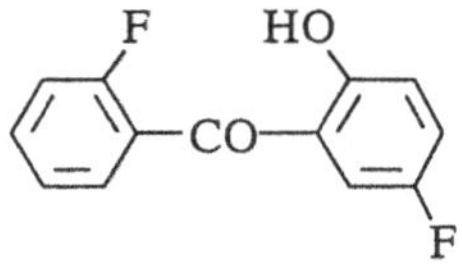

Synthesis

-Preparation by Fries rearrangement of p-fluorophenyl
 o-fluorobenzoate with aluminium chloride [692].

    m.p. and Spectra (NA).

(5-Fluoro-2-hydroxyphenyl)(3-fluorophenyl)methanone

$C_{13}H_8F_2O_2$ mol.wt. 234.20

Synthesis

-Preparation by Fries rearrangement of p-fluorophenyl
 m-fluorobenzoate with aluminium chloride [692].

    m.p. and Spectra (NA).

(5-Fluoro-2-hydroxyphenyl)(4-fluorophenyl)methanone

[2559-64-0] $C_{13}H_8F_2O_2$ mol.wt. 234.20

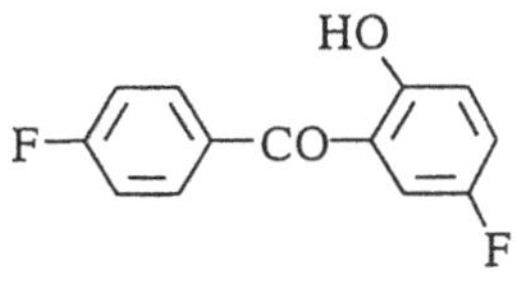

Synthesis

-Preparation by Fries rearrangement of p-fluorophenyl
 p-fluorobenzoate with aluminium chloride [692], without
 solvent at 130° for 2 h (83%) [656].
-Also refer to: [1404, 1405].

    b.p.$_{30}$ 205° [656];   Spectra (NA).

(2-Hydroxy-4-nitrophenyl)(3-nitrophenyl)methanone

[1834-89-5] $C_{13}H_8N_2O_6$ mol.wt. 288.22

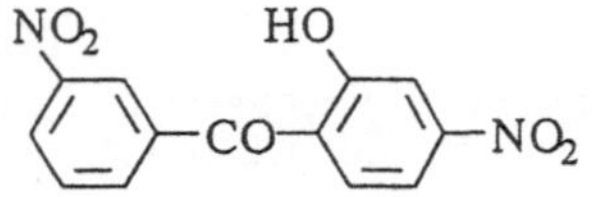

Syntheses

-Obtained (trace) by reaction of m-nitrobenzoyl chloride with m-nitrophenol in the presence of aluminium chloride without solvent at 175-180° for 2.5 h (0.5%) [1338].

-Also obtained (poor yield) by Fries rearrangement of m-nitrophenyl m-nitrobenzoate with aluminium chloride without solvent at 175° for 2 h (2%) [389].

m.p. 158-159° [389], 157-158° [1338]; Spectra (NA).

(2-Hydroxy-5-nitrophenyl)(4-nitrophenyl)methanone

$C_{13}H_8N_2O_6$ mol.wt. 288.22

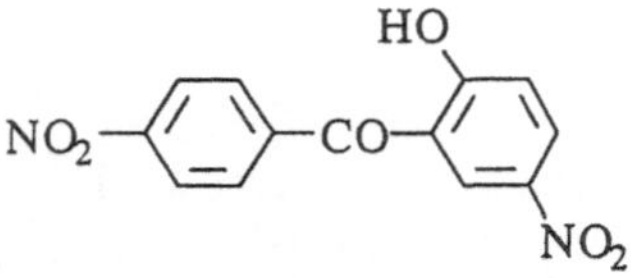

Synthesis

-Obtained by photo-Fries rearrangement of p-nitrophenyl p-nitrobenzoate in benzene at 50° for 35 h (11%) [419].

m.p. 189° [419]; Spectra (NA).

(4-Hydroxy-3-nitrophenyl)(3-nitrophenyl)methanone

[37567-45-6] $C_{13}H_8N_2O_6$ mol.wt. 288.22

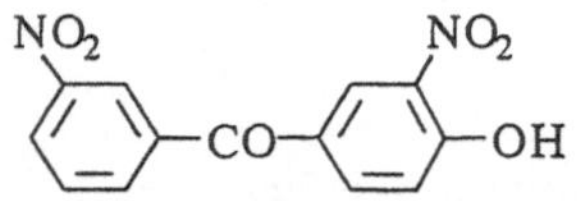

Syntheses

-Preparation by reaction of 48% hydrobromic acid with 3,3'-dinitro-4-methoxybenzophenone (SM) in refluxing acetic acid. SM was obtained from 3'-nitro-4-methoxy-benzophenone by a two-step synthesis [182].

-Preparation by hydrolysis of 4-chloro-3,3'-dinitrobenzophenone with 5-15% sodium hydroxide at 155-160° for 1-1.5 h (97-98%) [927].

m.p. 165° [182], 149°8-150°6 [927]; Spectra (NA).

(4-Hydroxy-3-nitrophenyl)(4-nitrophenyl)methanone

[37567-41-2] $C_{13}H_8N_2O_6$ mol.wt. 288.22

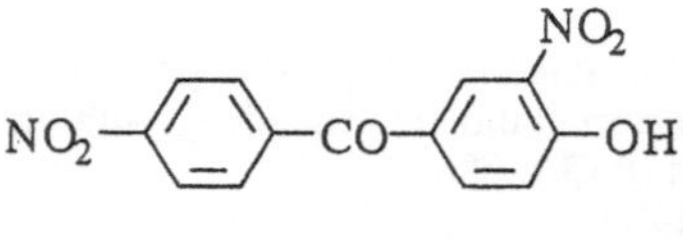

Synthesis

-Preparation by hydrolysis of 4-chloro-3,4'-dinitro-benzophenone with 5-15% aqueous sodium hydroxide at 155-160° for 1-1.5 h (97-98%) [927].

m.p. 154°5-154°8 [927]; Spectra (NA).

(3-Amino-5-chloro-2-hydroxyphenyl)(4-fluorophenyl)methanone

$C_{13}H_9ClFNO_2$ mol.wt. 265.67

HO   NH$_2$

F—⟨ ⟩—CO—⟨ ⟩

Cl

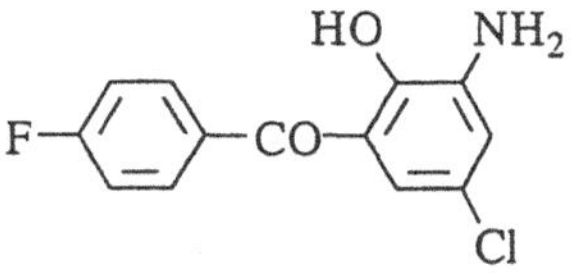

Synthesis

-Refer to: [932].

m.p. and Spectra (NA).

(3-Amino-5-fluoro-2-hydroxyphenyl)(4-chlorophenyl)methanone

[85052-42-2] $C_{13}H_9ClFNO_2$ mol.wt. 265.67

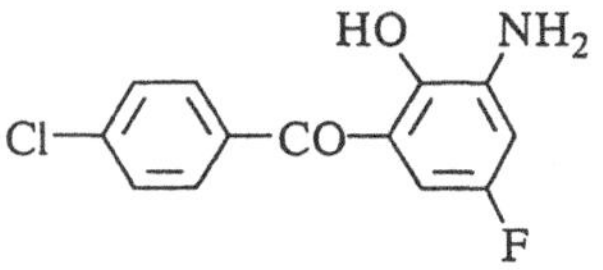

HO   NH$_2$

Cl—⟨ ⟩—CO—⟨ ⟩

F

Synthesis

-Preparation by hydrogenation of 4'-chloro-5-fluoro-
2-hydroxy-3-nitrobenzophenone in the presence of 10%
Pd/C in a chloroform/ethanol mixture for 2 h (82%) [932].

m.p. 124-127° [932];
^{1}H NMR [932], IR [932], MS [932];   TLC [932].

(3-Amino-5-fluoro-2-hydroxyphenyl)(4-chlorophenyl)methanone *(Hydrochloride)*

[85052-69-3] $C_{13}H_9ClFNO_2,HCl$ mol.wt. 302.14

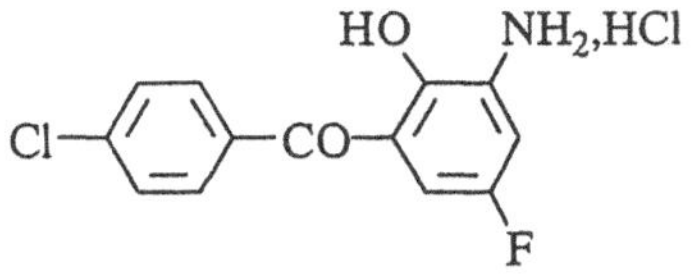

HO   NH$_2$,HCl

Cl—⟨ ⟩—CO—⟨ ⟩

F

Synthesis

-Obtained by reaction of hydrochloric acid with
3-amino-4'-chloro-5-fluoro-2-hydroxybenzophenone
[932].

m.p. and Spectra (NA).

(3-Amino-5-chloro-2-hydroxyphenyl)(4-chlorophenyl)methanone

[85052-41-1] $C_{13}H_9Cl_2NO_2$ mol.wt. 282.13

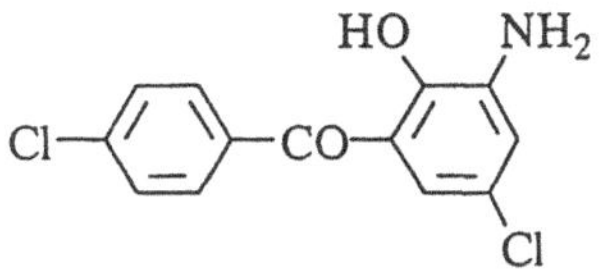

HO   NH$_2$

Cl—⟨ ⟩—CO—⟨ ⟩

Cl

Synthesis

-Preparation by hydrogenation of 4',5-dichloro-2-hydroxy-
3-nitrobenzophenone in the presence of 10% Pd/C in a
chloroform/ethanol mixture for 2 h (21%) [932].

m.p. 91-94° (d) [932];
^{1}H NMR [932], IR [932], MS [932];   TLC [932].

(3-Amino-5-chloro-2-hydroxyphenyl)(4-chlorophenyl)methanone *(Hydrochloride)*

[85052-68-2] $C_{13}H_9Cl_2NO_2$,HCl mol.wt. 318.60

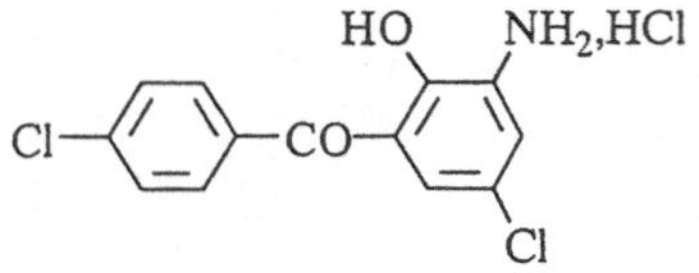

Synthesis

-Obtained by reaction of hydrochloric acid with
  3-amino-4',5-dichloro-2-hydroxybenzophenone [932].

m.p. and Spectra (NA).

(3-Amino-5-fluoro-2-hydroxyphenyl)(4-fluorophenyl)methanone

$C_{13}H_9F_2NO_2$ mol.wt. 249.22

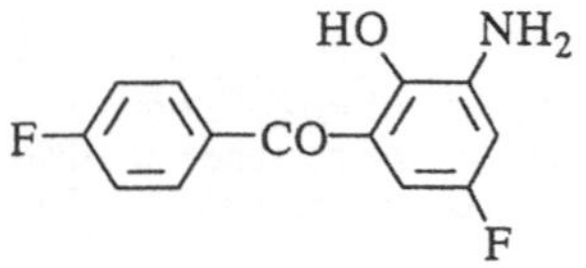

Synthesis

-Refer to: [932].

m.p. and Spectra (NA).

(2-Amino-5-hydroxyphenyl)(2-chlorophenyl)methanone

[61871-78-1] $C_{13}H_{10}ClNO_2$ mol.wt. 247.68

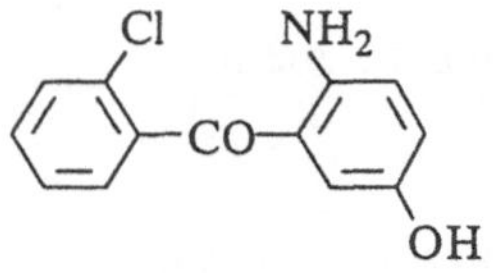

Synthesis

-Obtained by heating N-(o-chlorobenzoyl)-p-methoxyaniline
  with bismuth chloride (5 mol excess) at 180-200° for 3 min
  (80%) [145].

m.p. (NA);   UV [145], MS [145];   TLC [145].

(4-Amino-3-hydroxyphenyl)(4-chlorophenyl)methanone

[123172-45-2] $C_{13}H_{10}ClNO_2$ mol.wt. 247.68

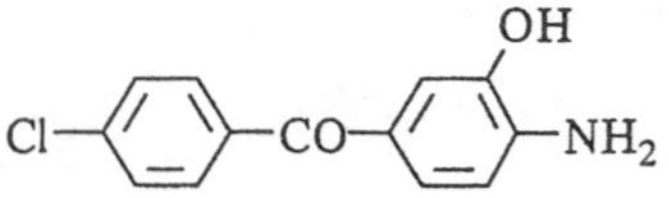

Synthesis

-Preparation from 6-(p-chlorobenzoyl)benzoxazolinone by
  heating with sodium hydroxide (80%) [952].

m.p.   196-198° [952];   Spectra (NA).

(4-Amino-2-hydroxyphenyl)(4-nitrophenyl)methanone

$C_{13}H_{10}N_2O_4$ mol.wt. 258.23

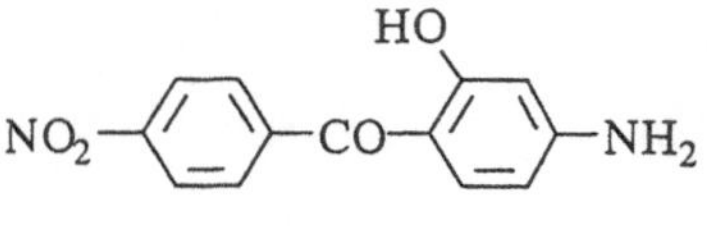

Synthesis

-Preparation by hydrolysis of 4-acetamido-2-hydroxy-
  4'-nitrobenzophenone with boiling 50% hydrochloric
  acid [658].

m.p.   228° [658];   Spectra (NA).

(3-Amino-4-hydroxyphenyl)(3-aminophenyl)methanone

[37567-47-8]   C13H12N2O2   mol.wt. 228.25

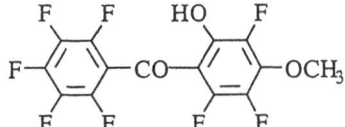

Synthesis

-Preparation by hydrogenation of 4-hydroxy-3,3'-dinitro-
benzophenone in the presence of Raney nickel in water at
90° for 1 h under 90 atmospheres (70%) [927].

   m.p. >178° (d) [927];   Spectra (NA).

(3-Amino-4-hydroxyphenyl)(4-aminophenyl)methanone

[37567-42-3]   C13H12N2O2   mol.wt. 228.25

Synthesis

-Preparation by hydrogenation of 4-hydroxy-3,4'-di-
nitrobenzophenone in the presence of Raney nickel in
water at 90° for 1 h under 100 atmospheres (71%)
[927].

   m.p. 191°5-191°8 [927];   Spectra (NA).

(2,3,4,5,6-Pentafluorophenyl)(2,3,5-trifluoro-6-hydroxy-4-methoxyphenyl)methanone

[32541-22-3]   C14H4F8O3   mol.wt. 372.17

Syntheses

-Preparation by partial demethylation of 2,4-dimethoxy-
2',3,3',4',5,5',6,6'-octafluorobenzophenone (SM) in
methylene chloride in the presence of aluminium
chloride at 0-5° for 3-6 h (63%) [839]. SM was obtained
by condensation of methyl 2,4-dimethoxy-
3,5,6-trifluorobenzoate with pentafluorophenylmagnesium bromide in ethyl ether at 20° for 2 h
(28%).
-Preparation by partial methylation of 2,4-dihydroxy-2',3,3',4',5,5',6,6'-octafluorobenzophenone with
diazomethane in ethyl ether between -10 to 0° (70%) [839].

   m.p. 68-69° [839];   1H NMR [839], IR [839], UV [839].

(2-Hydroxy-4-methoxyphenyl)(2,3,4,5,6-pentafluorophenyl)methanone

[32541-23-4]   C14H7F5O3   mol.wt. 318.20

Synthesis

-Preparation by partial demethylation of 2',4'-dimethoxy-
2,3,4,5,6-pentafluorobenzophenone (SM) in methylene
chloride in the presence of aluminium chloride at 0-5°
for 3-6 h (69%) [839]. SM was obtained in two steps:
first, preparation of 2,4-dimethoxy-
phenylbis(pentafluorophenyl)carbinol by condensation of methyl 2,4-dimethoxybenzoate with

pentafluorophenylmagnesium bromide in ethyl ether at 20° for 2 h (37%). Then, this carbinol was treated with potassium fluoride in boiling acetone for 4 h (98%) [839].

m.p.   152°5-154° [839]; ^{1}H NMR [839], IR [839], UV [839].

(5-Chloro-2-hydroxyphenyl)[2-(trifluoromethyl)phenyl]methanone

$C_{14}H_8ClF_3O_2$ mol.wt. 300.66

Synthesis

-Preparation from 2-bromo-4-chloro-(2-methoxyethoxy)-methoxybenzene and o-(trifluoromethyl)benzaldehyde as the starting materials [963].

m.p.   71-72° [963];   Spectra (NA).

(5-Fluoro-2-hydroxyphenyl)[4-(trifluoromethyl)phenyl]methanone

[183280-21-9] $C_{14}H_8F_4O_2$ mol.wt. 284.21

Synthesis

-Preparation by demethylation of 5-fluoro-2-methoxy-4'-(trifluoromethyl)benzophenone with boron tribromide in methylene chloride under argon: first at 0°, then at r.t. for 12 h (98%) [765].

m.p.   58°5 [765]; ^{1}H NMR [765], UV [765], MS [765].

(4-Bromophenyl)[2-(dibromomethyl)-4-hydroxyphenyl]methanone

[192443-53-1] $C_{14}H_9Br_3O_2$ mol.wt. 448.94

Synthesis

-Preparation by demethylation of 4'-bromo-2-dibromo-methyl-4-methoxybenzophenone (SM) with boron tribromide in methylene chloride at -78° (91%). SM was obtained by bromination of 4'-bromo-4-methoxy-2-methylbenzophenone with N-bromosuccinimide in carbon tetrachloride in the presence of few dibenzoyl peroxide under irradiation at r.t. (90%; MS). -Refer to: Chem. Abstr., **127**, 108921f (1997)*.

m.p. (NA);   MS*.

(3-Chloro-2-hydroxy-4-methoxyphenyl)(2,3-difluorophenyl)methanone

[72482-00-9] $C_{14}H_9ClF_2O_3$ mol.wt. 298.67

Synthesis

-Preparation by reaction of 2,3-difluorobenzoyl chloride with 2-chlororesorcinol dimethyl ether in ethylene dichloride in the presence of aluminium chloride: first at 5°, then at r.t. and at reflux for 30 min [1270].

m.p.   161-162° [1270];   Spectra (NA).

(3-Chloro-2-hydroxy-4-methoxyphenyl)(2,5-difluorophenyl)methanone

[72482-10-1] $C_{14}H_9ClF_2O_3$ mol.wt. 298.67

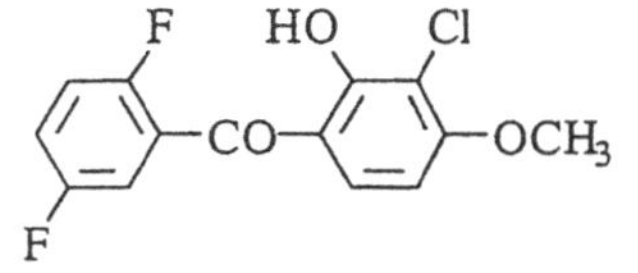

Synthesis

-Preparation by reaction of 2,5-difluorobenzoyl chloride with 2-chlororesorcinol dimethyl ether in ethylene dichloride in the presence of aluminium chloride: first at 5-10°, then at r.t. and at reflux for 30 min [1270].

m.p. 178-180° [1270]; Spectra (NA).

(5-Chloro-2-hydroxy-4-methoxyphenyl)(2,4-difluorophenyl)methanone

[136741-46-3] $C_{14}H_9ClF_2O_3$ mol.wt. 298.67

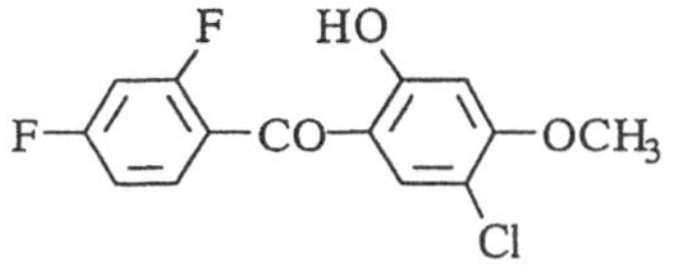

Synthesis

-Preparation by reaction of 2,4-difluorobenzoyl chloride with 4-chlororesorcinol dimethyl ether in the presence of aluminium chloride in ethylene dichloride (99%) [1193].

m.p. 136-137° [1193]; Spectra (NA).

(5-Chloro-2-hydroxy-4-methoxyphenyl)(2,6-difluorophenyl)methanone

[136741-45-2] $C_{14}H_9ClF_2O_3$ mol.wt. 298.67

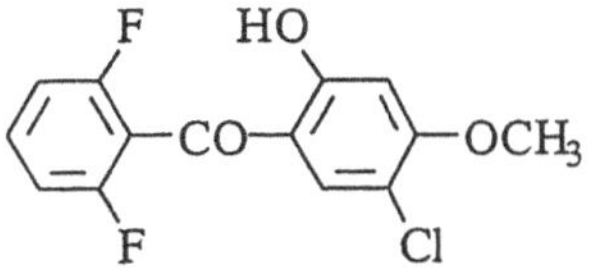

Synthesis

-Preparation by reaction of 2,6-difluorobenzoyl chloride with 4-chlororesorcinol dimethyl ether in the presence of aluminium chloride in ethylene dichloride (92%) [1193].

m.p. 132-133° [1193]; Spectra (NA).

(2,6-Dichlorophenyl)(4-hydroxy-2-methyl-5-nitrophenyl)methanone

[183725-86-2] $C_{14}H_9Cl_2NO_4$ mol.wt. 326.14

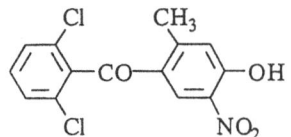

Synthesis

-Preparation by demethylation of 2',6'-dichloro-4-methoxy-2-methyl-5-nitrobenzophenone with aluminium chloride in methylene chloride, first for 30 min at 20° and then for 1 h at 45° (73%) [313].

m.p. 170° [313]; Spectra (NA).

(3-Chloro-4-hydroxy-5-methylphenyl)(2,4-dichlorophenyl)methanone

[34182-99-5] $C_{14}H_9Cl_3O_2$ mol.wt. 315.58

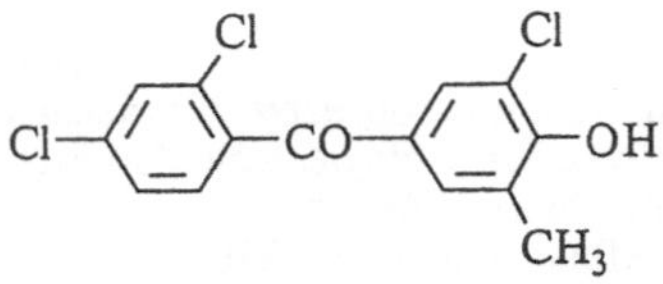

Synthesis

-Preparation by Fries rearrangement of 2-chloro-6-methyl-
phenyl 2,4-dichlorobenzoate with aluminium chloride in
chlorobenzene for 20 min at 140-150° or in nitrobenzene
for 24 h at 75° [1214].

m.p.   168-170° [1214];   Spectra (NA).

(3-Chloro-6-hydroxy-2-methylphenyl)(2,5-dichlorophenyl)methanone

[34174-13-5] $C_{14}H_9Cl_3O_2$ mol.wt. 315.58

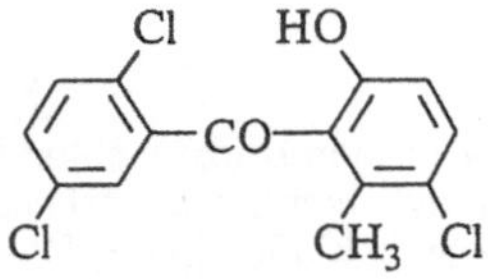

Synthesis

-Preparation by Fries rearrangement of 4-chloro-3-methyl-
phenyl 2,5-dichlorobenzoate with aluminium chloride for
30 min at 150-160° [375].

m.p.   122-123° [375];   Spectra (NA).

(2-Methylphenyl)(2,3,5-trichloro-6-hydroxyphenyl)methanone

[34171-64-7] $C_{14}H_9Cl_3O_2$ mol.wt. 315.58

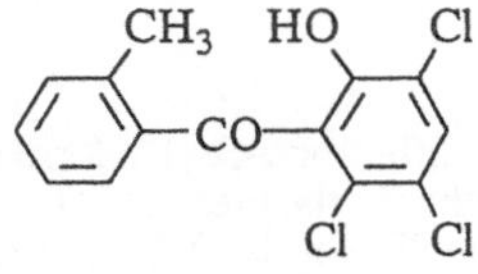

Synthesis

-Preparation by Fries rearrangement of 2,4,5-trichlorophenyl
o-toluate with aluminium chloride for 30 min at 150-160°
[375].

m.p.   125-126° [375];   Spectra (NA).

(4-Methylphenyl)(2,3,5-trichloro-6-hydroxyphenyl)methanone

[34171-62-5] $C_{14}H_9Cl_3O_2$ mol.wt. 315.58

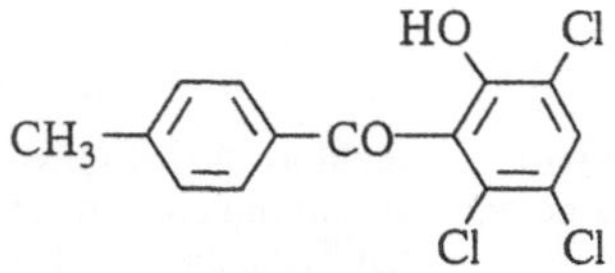

Synthesis

-Preparation by Fries rearrangement of 2,4,5-trichlorophenyl
p-toluate with aluminium chloride for 30 min at 150-160°
[375].

m.p.   141-142°5 [375];   Spectra (NA).

(2,6-Difluorophenyl)(4-hydroxy-3-methoxy-5-nitrophenyl)methanone

[134612-75-2] $C_{14}H_9F_2NO_5$ mol.wt. 309.23

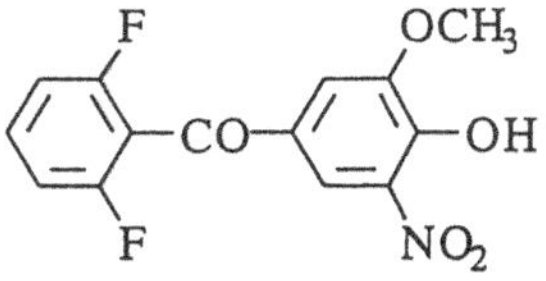

Synthesis

-Preparation by nitration of 2',6'-difluoro-4-hydroxy-
3-methoxybenzophenone with 65% nitric acid in acetic acid
at 20° [164].

m.p. 147-149° [164];   Spectra (NA).

(3-Bromo-2-hydroxy-5-methylphenyl)(3-fluorophenyl)methanone

$C_{14}H_{10}BrFO_2$ mol.wt. 309.13

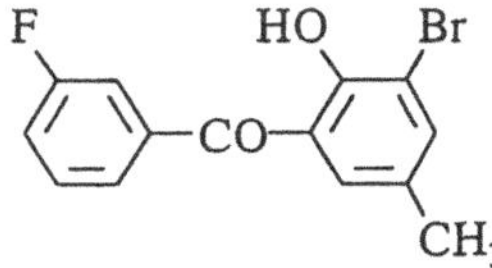

Synthesis

-Preparation by treatment of 3'-fluoro-2-hydroxy-5-methyl-
benzophenone sodium salt with bromine in aqueous
potassium bromide solution [472].

m.p. 130-131° [472];   Spectra (NA).

(3-Bromophenyl)(3-fluoro-2-hydroxy-5-methylphenyl)methanone

[55270-81-0] $C_{14}H_{10}BrFO_2$ mol.wt. 309.13

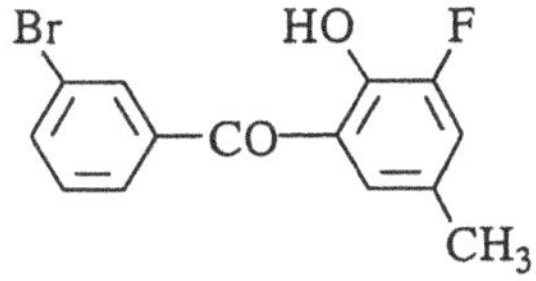

Synthesis

-Not yet described. This compound was mentioned in the
Chem. Abstr., **82**, 170585e (1975). Nevertheless, it is not
described in the original paper [472].

m.p. and Spectra (NA).

(2-Bromo-5-nitrophenyl)(2-hydroxy-5-methylphenyl)methanone

$C_{14}H_{10}BrNO_4$ mol.wt. 336.14

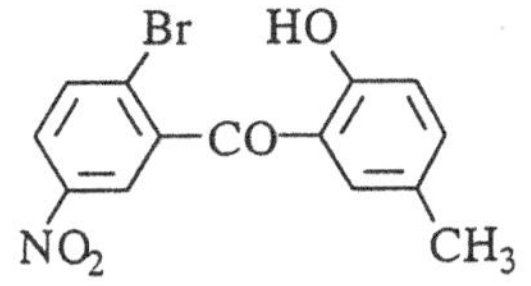

Synthesis

-Preparation by reaction of 2-bromo-5-nitrobenzoyl chloride
with p-cresol methyl ether in the presence of aluminium
chloride in carbon disulfide at r.t. for 1 h, then in a boiling
water bath for 2-4 h (78%) [895].

m.p. 151-152° [895];   Spectra (NA).

(3,5-Dibromo-2-hydroxyphenyl)(4-methylphenyl)methanone

$C_{14}H_{10}Br_2O_2$ mol.wt. 370.04

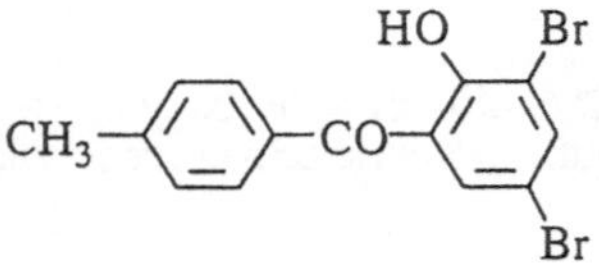

Synthesis

-Preparation by reaction of bromine with 2-hydroxy-4'-methylbenzophenone in chloroform [1397].

m.p. 132°5 [1397];   Spectra (NA).

(2-Chloro-6-hydroxy-4-methoxyphenyl)(2-fluorophenyl)methanone

[72482-27-0] $C_{14}H_{10}ClFO_3$ mol.wt. 280.68

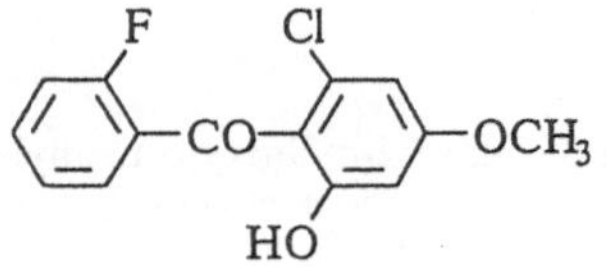

Syntheses

-Preparation by reaction of 2-fluorobenzoyl chloride with 5-chlororesorcinol dimethyl ether in the presence of aluminium chloride in refluxing ethylene dichloride (70%) [1193], 96% [1272].
-Preparation by partial demethylation of 2-chloro-2'-fluoro-4,6-dimethoxybenzophenone (SM) in ethylene dichloride in the presence of aluminium chloride at reflux (90°) for 3 h [1270]. SM was obtained by reaction of o-fluorobenzoyl chloride with 5-chlororesorcinol dimethyl ether in methylene chloride in the presence of aluminium chloride at r.t. for 4 h [1270].

m.p.   111-113° [1193], 108-110° [1272], 85-90° [1270];
Spectra (NA).

(3-Chloro-2-hydroxy-4-methoxyphenyl)(2-fluorophenyl)methanone

[72482-07-6] $C_{14}H_{10}ClFO_3$ mol.wt. 280.68

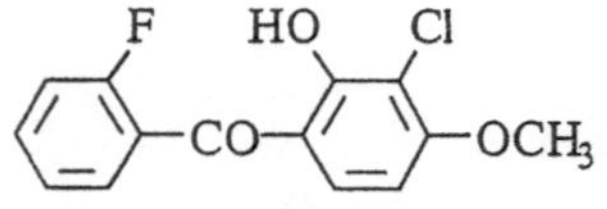

Synthesis

-Preparation by reaction of o-fluorobenzoyl chloride with 2-chlororesorcinol dimethyl ether in the presence of aluminium chloride in refluxing ethylene dichloride for 30 min [1270], (67%) [1272].

m.p.   132-133° [1270, 1272];   Spectra (NA).

(5-Chloro-2-hydroxy-4-methoxyphenyl)(2-fluorophenyl)methanone

[130556-06-8] $C_{14}H_{10}ClFO_3$ mol.wt. 280.68

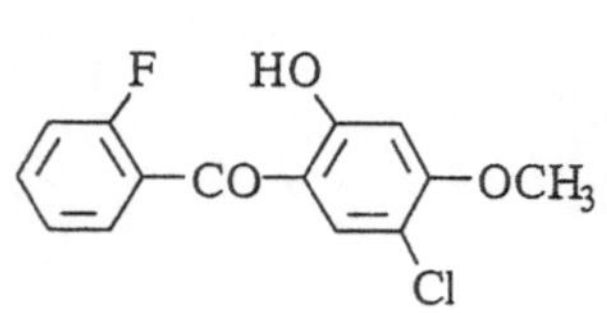

Synthesis

-Preparation by reaction of 2-fluorobenzoyl chloride with the 4-chlororesorcinol dimethyl ether in the presence of aluminium chloride in ethylene dichloride (64%) [1193].

m.p.   133-134° [1193];   Spectra (NA).

(2-Chloro-4-nitrophenyl)(2-hydroxy-5-methylphenyl)methanone

$C_{14}H_{10}ClNO_4$ mol.wt. 291.69

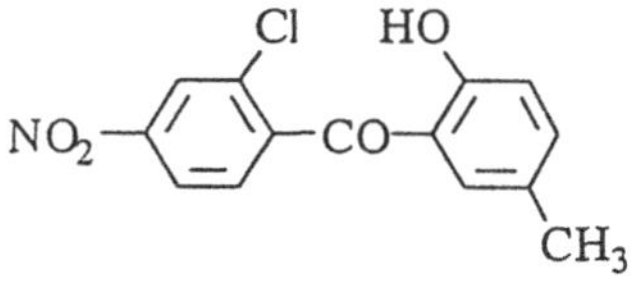

Synthesis

-Preparation by demethylation of 2-chloro-2'-methoxy-
  5'-methyl-4-nitrobenzophenone with excess boiling
  pyridinium chloride for 30 min (57%) [1153].

m.p. 130° [1153];   Spectra (NA).

(2-Chloro-5-nitrophenyl)(2-hydroxy-5-methylphenyl)methanone

[37883-98-0] $C_{14}H_{10}ClNO_4$ mol.wt. 291.69

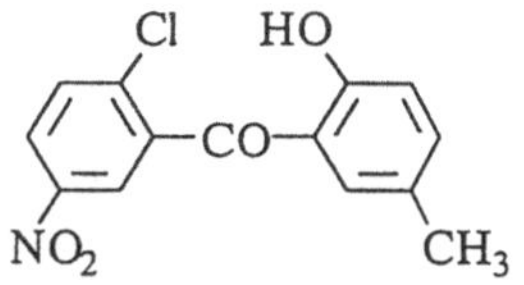

Synthesis

-Refer to: [1153].

m.p. and Spectra (NA).

(5-Chloro-2-hydroxy-3-nitrophenyl)(4-methoxyphenyl)methanone

[85052-28-4] $C_{14}H_{10}ClNO_5$ mol.wt. 307.69

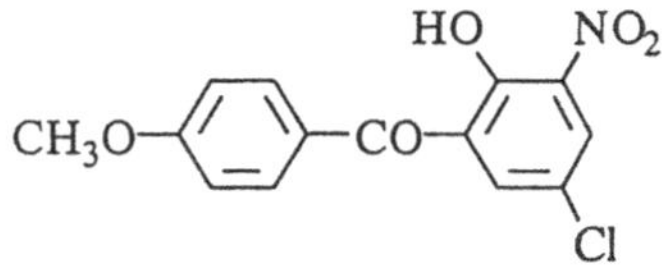

Synthesis

-Preparation by reaction of 60% nitric acid with 5-chloro-
  2-hydroxy-4'-methoxybenzophenone in acetic acid at r.t.
  for 30 min (79%) [932].

m.p. (NA); ^{1}H NMR [932], IR [932], MS [932];   TLC [932].

(2-Chlorophenyl)(4-hydroxy-3-methoxy-5-nitrophenyl)methanone

[134612-76-3] $C_{14}H_{10}ClNO_5$ mol.wt. 307.69

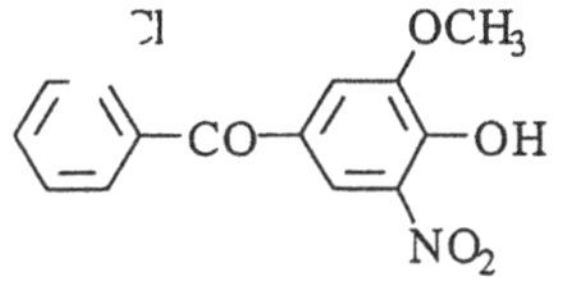

Synthesis

-Preparation by nitration of 2'-chloro-4-hydroxy-3-methoxy-
  benzophenone with 65% nitric acid in acetic acid at 20°
  [164].

m.p. 123-125° [164];   Spectra (NA).

(3-Chlorophenyl)(4-hydroxy-3-methoxy-5-nitrophenyl)methanone

[134612-77-4] $C_{14}H_{10}ClNO_5$ mol.wt. 307.69

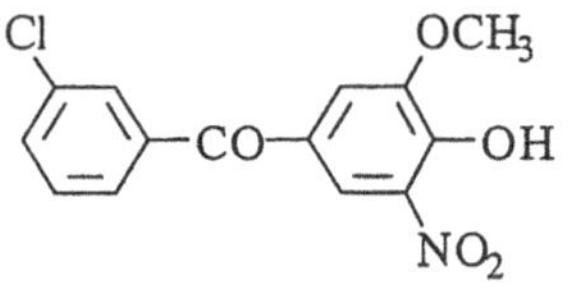

Synthesis

-Preparation by nitration of 3'-chloro-4-hydroxy-3-methoxy-
  benzophenone with 65% nitric acid in acetic acid at 20°
  [164].

m.p. 152-154° [164];   Spectra (NA).

(4-Chlorophenyl)(4-hydroxy-3-methoxy-5-nitrophenyl)methanone

[134612-78-5] $C_{14}H_{10}ClNO_5$ mol.wt. 307.69

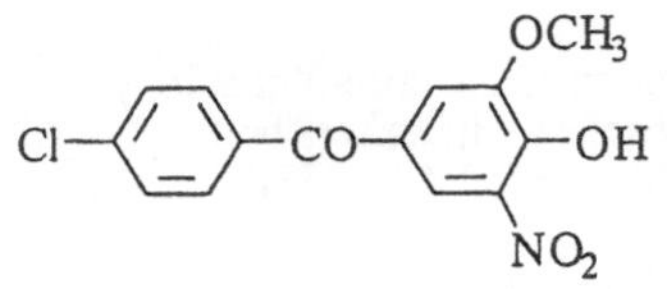

Synthesis

-Preparation by nitration of 4'-chloro-4-hydroxy-
3-methoxybenzophenone with 65% nitric acid in acetic acid
at 20° [164].

m.p. 129-131° [164];   Spectra (NA).

(3-Chloro-4-hydroxy-5-methylphenyl)(2-chlorophenyl)methanone

[34183-19-2] $C_{14}H_{10}Cl_2O_2$ mol.wt. 281.14

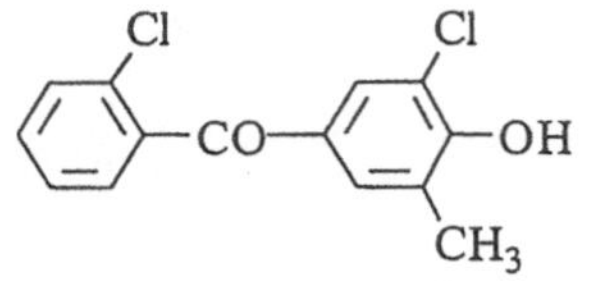

Synthesis

-Preparation by Fries rearrangement of 2-chloro-6-methyl-
phenyl o-chlorobenzoate with aluminium chloride in chlorobenzene for 20 min at 140-150° or in nitrobenzene for 24 h
at 75° [1214].

m.p. 164-165° [1214];   Spectra (NA).

(3-Chloro-4-hydroxy-5-methylphenyl)(4-chlorophenyl)methanone

[34183-09-0] $C_{14}H_{10}Cl_2O_2$ mol.wt. 281.14

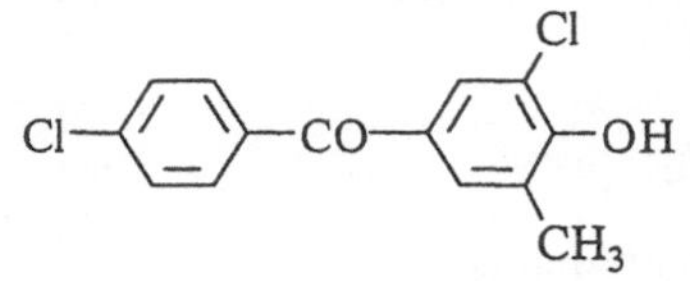

Synthesis

-Preparation by Fries rearrangement of 2-chloro-6-methyl-
phenyl p-chlorobenzoate with aluminium chloride in
chlorobenzene for 20 min at 140-150° or in nitrobenzene
for 24 h at 75° [1214].

m.p. 138-139° [1214];   Spectra (NA).

(5-Chloro-2-hydroxy-3-methylphenyl)(2-chlorophenyl)methanone

[86914-82-1] $C_{14}H_{10}Cl_2O_2$ mol.wt. 281.14

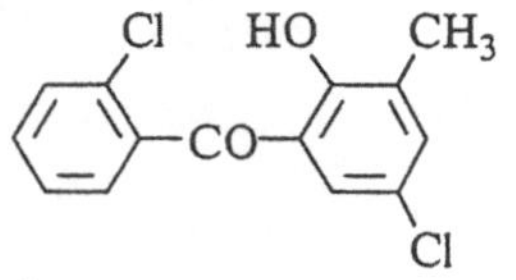

Synthesis

-Preparation by Fries rearrangement of 4-chloro-2-methyl-
phenyl o-chlorobenzoate with aluminium chloride at 180° for
10 min [688].

m.p. 87-88° [688];   Spectra (NA).

(5-Chloro-2-hydroxy-3-methylphenyl)(3-chlorophenyl)methanone

[86914-87-6]   C$_{14}$H$_{10}$Cl$_2$O$_2$ mol.wt. 281.14

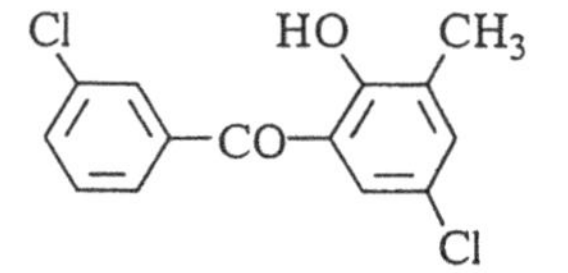

Synthesis

-Preparation by Fries rearrangement of 4-chloro-2-methyl-
  phenyl m-chlorobenzoate with aluminium chloride at 180°
  for 10 min [688].

m.p. 130-131° [688];   Spectra (NA).

(5-Chloro-2-hydroxy-3-methylphenyl)(4-chlorophenyl)methanone

[86914-72-9]   C$_{14}$H$_{10}$Cl$_2$O$_2$ mol.wt. 281.14

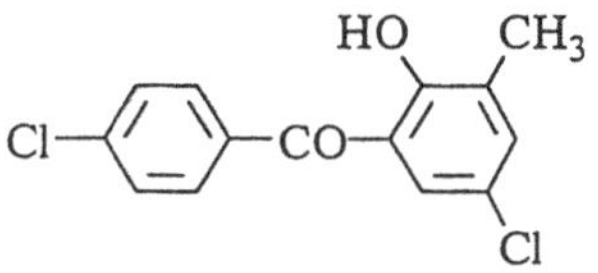

Synthesis

-Preparation by Fries rearrangement of 4-chloro-2-methyl-
  phenyl p-chlorobenzoate with aluminium chloride at 160°
  [688].
-Also refer to: [685, 686, 689, 696].

m.p. 41-42° [688];   Spectra (NA).

(5-Chloro-2-hydroxy-4-methylphenyl)(2-chlorophenyl)methanone

                                                C$_{14}$H$_{10}$Cl$_2$O$_2$ mol.wt. 281.14

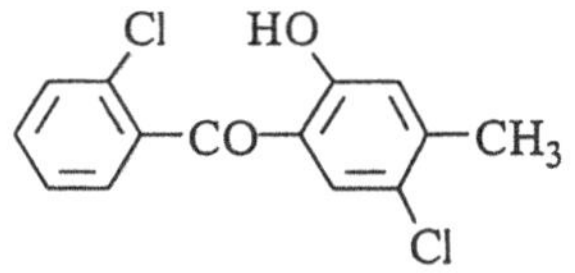

Synthesis

-Refer to: [241].

m.p. 105° [241];   Spectra (NA).

(2,3-Dichloro-4-hydroxyphenyl)(2-methylphenyl)methanone

[72482-84-9]   C$_{14}$H$_{10}$Cl$_2$O$_2$ mol.wt. 281.14

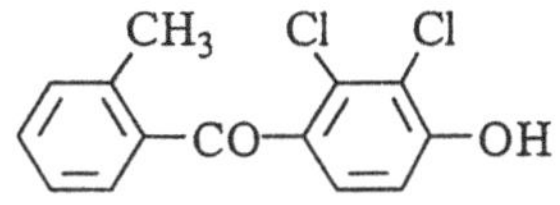

Synthesis

-Preparation by demethylation of 2,3-dichloro-4-methoxy-
  2'-methylbenzophenone (SM) with aluminium chloride in
  refluxing benzene for 5 h, then at r.t. for 18 h. SM was
  obtained by reaction of o-toluoyl chloride with 2,3-di-
chloroanisole in ethylene dichloride in the presence of aluminium chloride at 60° for 1 h [1270].

m.p. and Spectra (NA).

(2,3-Dichloro-4-hydroxyphenyl)(3-methylphenyl)methanone

                                                C$_{14}$H$_{10}$Cl$_2$O$_2$ mol.wt. 281.14

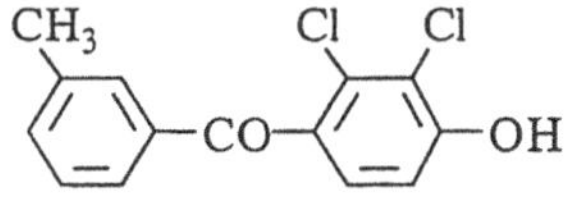

Synthesis

-Preparation by demethylation of 2,3-dichloro-4-methoxy-
  3'-methylbenzophenone (SM) with aluminium chloride in
  refluxing methylene chloride overnight. SM was obtained by
  Friedel-Crafts acylation of 2,3-dichloroanisole with

m-toluoyl chloride in the presence of aluminium chloride in refluxing methylene chloride [653].

m.p. and Spectra (NA).

(2,3-Dichloro-4-hydroxyphenyl)(4-methylphenyl)methanone

[72498-72-7]   C$_{14}$H$_{10}$Cl$_2$O$_2$ mol.wt. 281.14

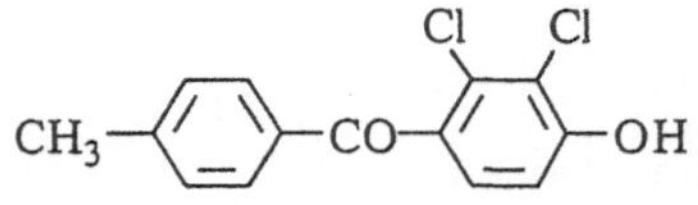

Synthesis

-Preparation by demethylation of 2,3-dichloro-
4-methoxy-4'-methylbenzophenone (SM) with
aluminium chloride in refluxing methylene chloride
overnight. SM was obtained by Friedel-Crafts
acylation of 2,3-dichloroanisole with p-methylbenzoyl chloride in the presence of aluminium
chloride in refluxing methylene chloride [653].
-Also refer to: [1270].

m.p. and Spectra (NA).

(2,4-Dichloro-6-hydroxyphenyl)(2-methylphenyl)methanone

[34174-08-8]   C$_{14}$H$_{10}$Cl$_2$O$_2$ mol.wt. 281.14

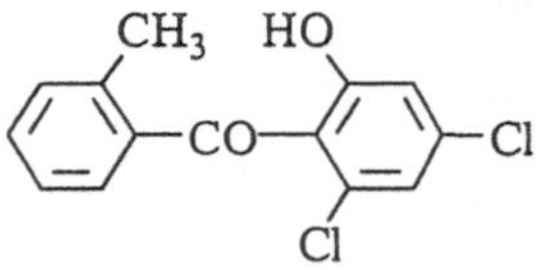

Synthesis

-Preparation by Fries rearrangement of 3,5-dichlorophenyl
o-toluate with aluminium chloride for 30 min at 150-160°
[375].

m.p.   117-118° [375];   Spectra (NA).

(2,4-Dichloro-6-hydroxyphenyl)(4-methylphenyl)methanone

[34174-07-7]   C$_{14}$H$_{10}$Cl$_2$O$_2$ mol.wt. 281.14

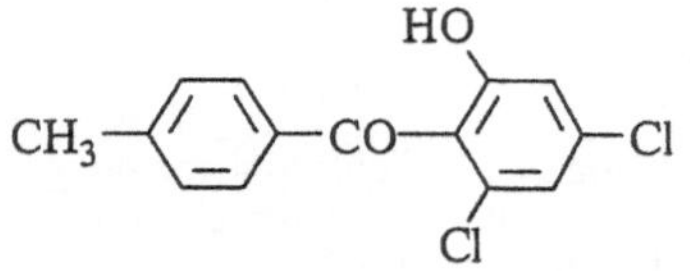

Synthesis

-Preparation by Fries rearrangement of 3,5-dichloro-
phenyl p-toluate with aluminium chloride for 30 min at
150-160° [375].

m.p.   177-178° [375];   Spectra (NA).

(2,6-Dichloro-4-hydroxyphenyl)(4-methylphenyl)methanone

[34183-14-7]   C$_{14}$H$_{10}$Cl$_2$O$_2$ mol.wt. 281.14

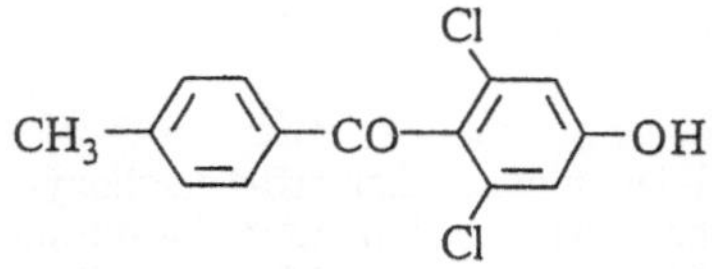

Synthesis

-Obtained by Fries rearrangement of 3,5-dichlorophenyl
p-toluate with aluminium chloride in chlorobenzene for
20 min at 140-150° or in nitrobenzene for 24 h at 75°
[1214].

m.p.   165-166° [1214];   Spectra (NA).

(3,5-Dichloro-4-hydroxyphenyl)(2-methylphenyl)methanone

[34183-17-0] $C_{14}H_{10}Cl_2O_2$ mol.wt. 281.14

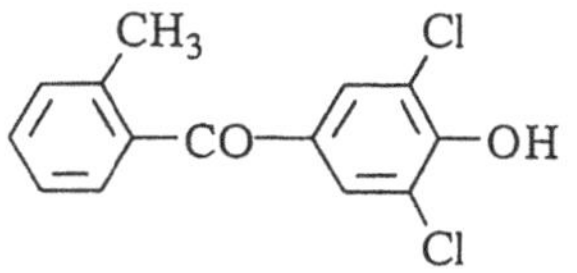

Synthesis

-Preparation by Fries rearrangement of 2,6-dichlorophenyl o-toluate with aluminium chloride in chlorobenzene at 140-150° for 20 min or in nitrobenzene at 75° for 24 h [1214].

m.p. 163-164° [1214];   Spectra (NA).

(3,5-Dichloro-4-hydroxyphenyl)(3-methylphenyl)methanone

[70036-75-8] $C_{14}H_{10}Cl_2O_2$ mol.wt. 281.14

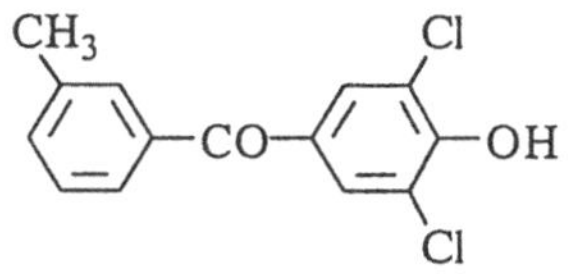

Synthesis

-Refer to: [1353] (Japanese patent).

m.p. and Spectra (NA).

(3,5-Dichloro-4-hydroxyphenyl)(4-methylphenyl)methanone

[34183-10-3] $C_{14}H_{10}Cl_2O_2$ mol.wt. 281.14

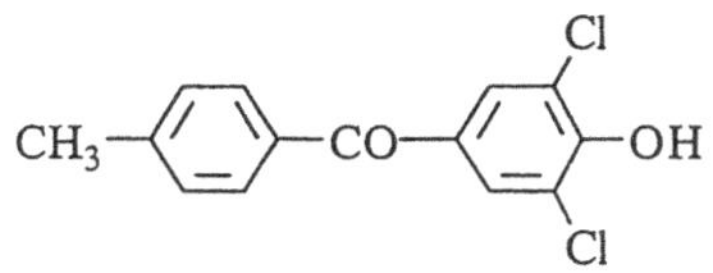

Synthesis

-Preparation by Fries rearrangement of 2,6-dichloro-phenyl p-toluate with aluminium chloride in chloro-benzene at 140-150° for 20 min or in nitrobenzene at 75° for 24 h [1214].

m.p.   163-164° [1214];   Spectra (NA).

(2,3-Dichlorophenyl)(2-hydroxy-5-methylphenyl)methanone

[77151-84-9] $C_{14}H_{10}Cl_2O_2$ mol.wt. 281.14

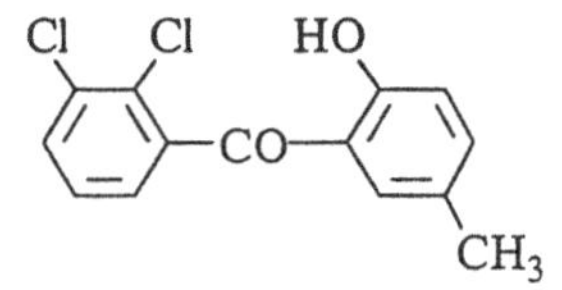

Synthesis

-Preparation by reaction of 2,3-dichlorobenzoyl chloride with p-cresol in the presence of aluminium chloride for 8 h at 190° [348].

m.p. and Spectra (NA).

(2,4-Dichlorophenyl)(2-hydroxy-4-methylphenyl)methanone

[59746-93-9] $C_{14}H_{10}Cl_2O_2$ mol.wt. 281.14

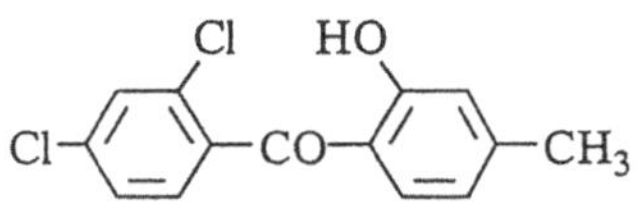

Synthesis

-Preparation by reaction of 2,4-dichlorobenzotrichloride with m-cresol in hydrofluoric acid in the presence of water at -10°, then at 15° overnight and at 80° for 30 min (87%) [393].

m.p.   83° [393];   Spectra (NA).

(2,4-Dichlorophenyl)(2-hydroxy-5-methylphenyl)methanone

$C_{14}H_{10}Cl_2O_2$ mol.wt. 281.14

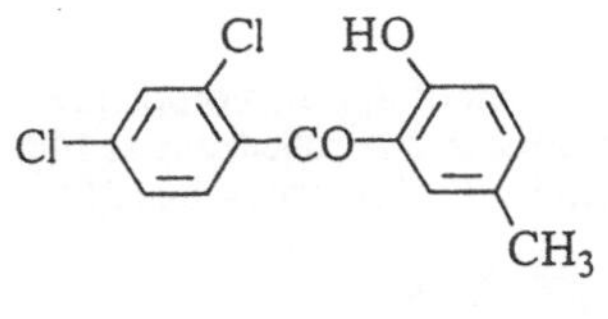

Syntheses

-Preparation by demethylation of 2,4-dichloro-2'-methoxy-
5'-methylbenzophenone with excess boiling pyridinium
chloride for 30 min (72%) [1153].

-Preparation by Fries rearrangement of p-cresyl 2,4-dichloro-
benzoate with aluminium chloride without solvent at 120° for
2 h [651].

-Preparation by reaction of 2,4-dichlorobenzotrichloride with p-cresol in the presence of aluminium
chloride in carbon disulfide at 0° (22% to 32% yields) [975].

-Also obtained by sulfuric acid hydrolysis of 6,12-di(2,4-dichlorophenyl)-2,8-dimethyl-
6,12-epoxy-*6H,12H*-dibenzo[b,f][1,5]dioxocin (56%) [975].

     m.p.   92°1-92°8 [975], 92-93° [651], 91-92° [1153];
     Spectra (NA).

(2,4-Dichlorophenyl)(3-hydroxy-2-methylphenyl)methanone

[74167-88-7] $C_{14}H_{10}Cl_2O_2$ mol.wt. 281.14

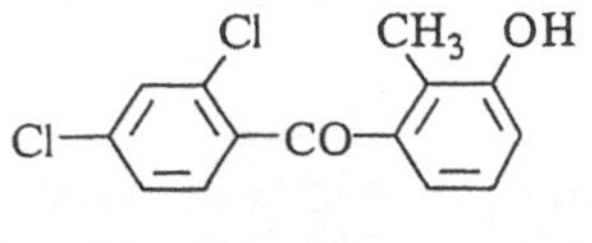

Synthesis

-Preparation by diazotization of 3-amino-2',4'-dichloro-
2-methylbenzophenone, followed by hydrolysis of the
resulting diazonium salt (70%) [82], according to [731].

     m.p.   98° [82];   Spectra (NA).

(2,4-Dichlorophenyl)(4-hydroxy-2-methylphenyl)methanone

$C_{14}H_{10}Cl_2O_2$ mol.wt. 281.14

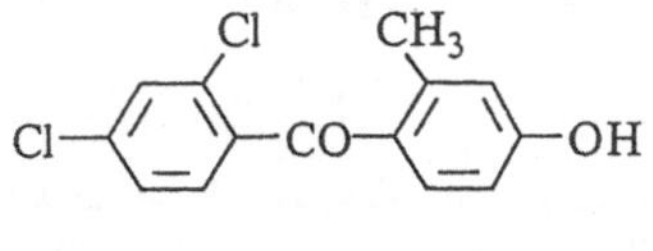

Synthesis

-Obtained by Fries rearrangement of m-cresyl 2,4-di-
chlorobenzoate with aluminium chloride without solvent at
120° for 2 h [651].

-Also refer to: [486] (compound 11).

     m.p.   155° [651];   Spectra (NA).

(2,4-Dichlorophenyl)(4-hydroxy-3-methylphenyl)methanone

$C_{14}H_{10}Cl_2O_2$ mol.wt. 281.14

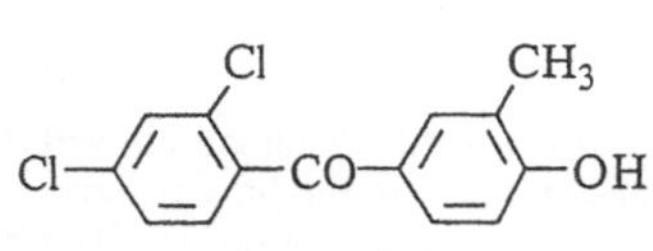

Synthesis

-Preparation by Fries rearrangement of o-cresyl 2,4-di-
chlorobenzoate with aluminium chloride without solvent at
120° for 2 h [651].

     m.p.   165° [651];   Spectra (NA).

(2,6-Dichlorophenyl)(2-hydroxy-5-methylphenyl)methanone

[174186-21-1] $C_{14}H_{10}Cl_2O_2$ mol.wt. 281.14

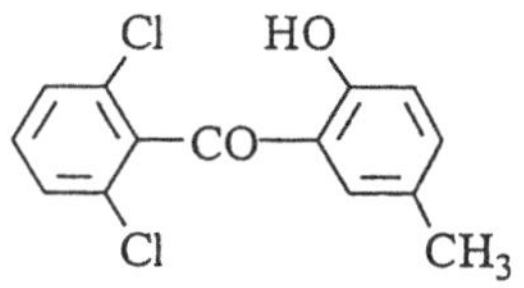

Syntheses

-Preparation by reaction of 2,6-dichlorobenzoyl chloride with p-cresol in the presence of aluminium chloride for 8 h at 190° [348].

-Preparation by Fries rearrangement of 4-methylphenyl 2,6-dichlorobenzoate with aluminium chloride in refluxing ethylene dichloride for 3 h (68%) [722].

m.p. 139-141° [722]; ^{1}H NMR [722].

(3,4-Dichlorophenyl)(2-hydroxy-5-methylphenyl)methanone

$C_{14}H_{10}Cl_2O_2$ mol.wt. 281.14

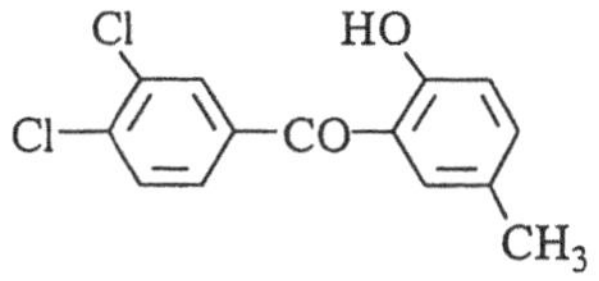

Syntheses

-Preparation by Fries rearrangement of p-cresyl 3,4-dichloro-benzoate with aluminium chloride without solvent at 120° for 2 h [651].

-Preparation by reaction of 3,4-dichlorobenzotrichloride in the presence of aluminium chloride in carbon disulfide at 0° (56%) [975].

-Also obtained by sulfuric acid hydrolysis of the 6,12-di(3,4-dichlorophenyl)-2,8-dimethyl-6,12-epoxy-*6H,12H*-dibenzo[b,f][1,5]dioxocin (76%) [975].

m.p. 90°2-90°6 [975], 90-91° [651]; Spectra (NA).

(3,4-Dichlorophenyl)(3-hydroxy-2-methylphenyl)methanone

[76981-65-2] $C_{14}H_{10}Cl_2O_2$ mol.wt. 281.14

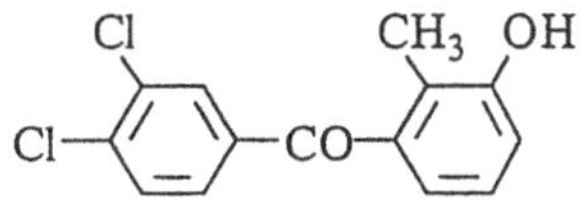

Synthesis

-Preparation by diazotization of 3-amino-3',4'-dichloro-2-methylbenzophenone, followed by hydrolysis of the resulting diazonium salt (48%) [82], according to [731].

m.p. 140° [82]; Spectra (NA).

(3,4-Dichlorophenyl)(4-hydroxy-2-methylphenyl)methanone

$C_{14}H_{10}Cl_2O_2$ mol.wt. 281.14

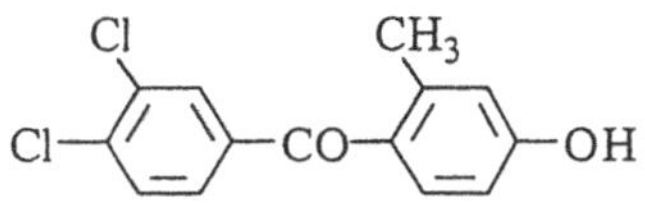

Synthesis

-Obtained by Fries rearrangement of m-cresyl 3,4-dichloro-benzoate with aluminium chloride without solvent at 120° for 2 h [651].

m.p. 95° [651]; Spectra (NA).

(3,4-Dichlorophenyl)(4-hydroxy-3-methylphenyl)methanone

$C_{14}H_{10}Cl_2O_2$ mol.wt. 281.14

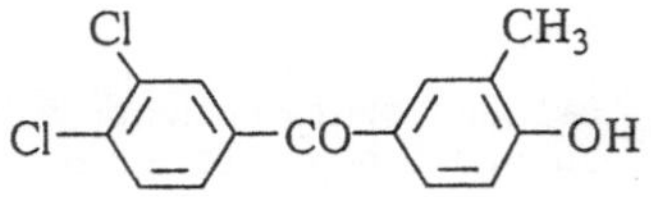

Synthesis

-Preparation by Fries rearrangement of o-cresyl 3,4-dichlorobenzoate with aluminium chloride without solvent at 120° for 2 h [651].

m.p. 209° [651];   Spectra (NA).

[5-Chloro-2-hydroxy-3-(hydroxymethyl)phenyl](4-chlorophenyl)methanone

[95304-56-6] $C_{14}H_{10}Cl_2O_3$ mol.wt. 297.14

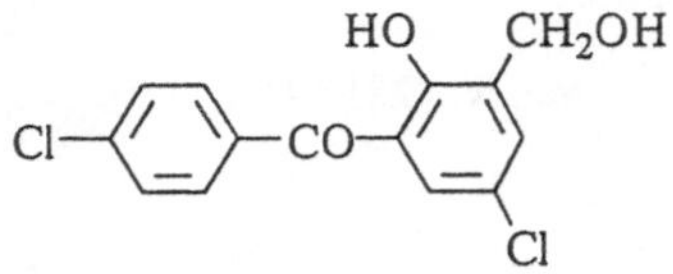

Synthesis

-Refer to: [691].

m.p. and Spectra (NA).

(5-Chloro-2-hydroxy-3-methoxyphenyl)(4-chlorophenyl)methanone

[95304-54-4] $C_{14}H_{10}Cl_2O_3$ mol.wt. 297.14

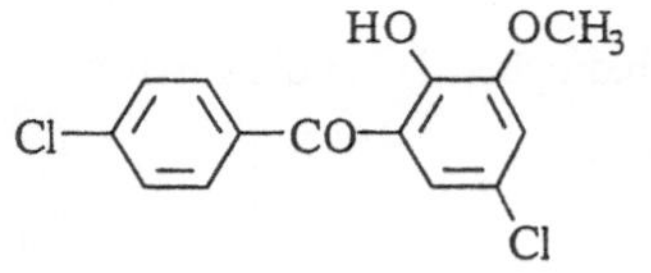

Synthesis

-Refer to: [691].

m.p. and Spectra (NA).

(5-Chloro-2-hydroxy-4-methoxyphenyl)(2-chlorophenyl)methanone

[136741-43-0] $C_{14}H_{10}Cl_2O_3$ mol.wt. 297.14

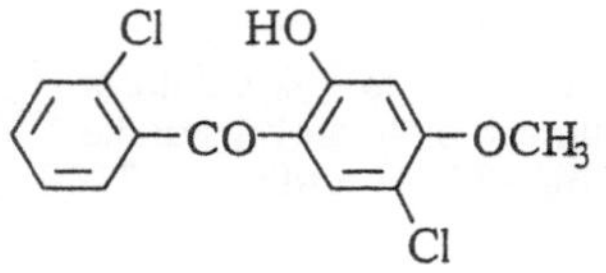

Synthesis

-Preparation by reaction of 2-chlorobenzoyl chloride with 1-chloro-2,4-dimethoxybenzene in the presence of aluminium chloride in ethylene dichloride (77%) [1193].

m.p. 98-99° [1193];   Spectra (NA).

(5-Chloro-2-hydroxy-4-methoxyphenyl)(4-chlorophenyl)methanone

[87118-99-8] $C_{14}H_{10}Cl_2O_3$ mol.wt. 297.14

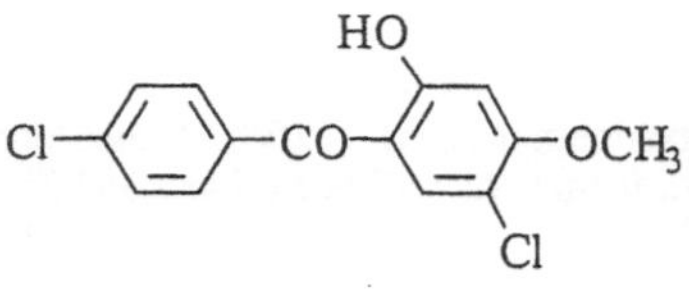

Synthesis

-Preparation by reaction of p-chlorobenzoyl chloride with 1-chloro-2,4-dimethoxybenzene in the presence of aluminium chloride in methylene chloride at r.t. for 20 h (66%) [933].

crystals [933];   m.p. (NA);   ¹H NMR [933], IR [933], MS [933];   TLC [933].

(2,3-Dichloro-4-hydroxyphenyl)(4-methoxyphenyl)methanone

[78235-'8-4] $C_{14}H_{10}Cl_2O_3$ mol.wt. 297.14

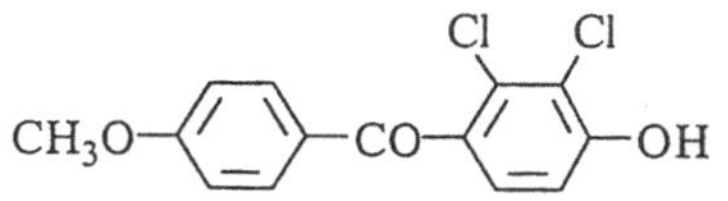

Synthesis

-Preparation by selective demethylation of 2,3-di-
chloro-4,4'-dimethoxybenzophenone (SM) with
aluminium chloride in refluxing methylene chloride
overnight. SM was obtained by Friedel-Crafts
acylation of 2,3-dichloroanisole with p-methoxybenzoyl chloride in the presence of aluminium
chloride in refluxing methylene chloride [653, 1071].
-Also refer to: [1024, 1025].

m.p. and Spectra (NA).

(3,5-Dichloro-4-hydroxyphenyl)(4-methoxyphenyl)methanone

[34183-20-5] $C_{14}H_{10}Cl_2O_3$ mol.wt. 297.14

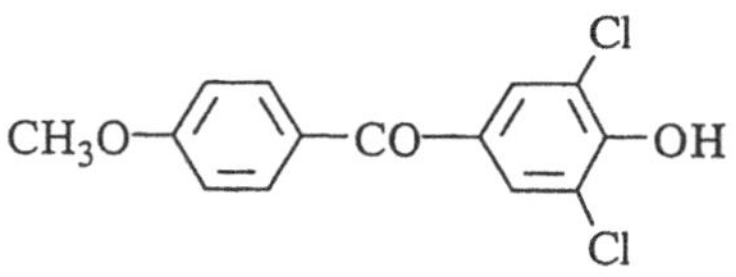

Synthesis

-Preparation by Fries rearrangement of 2,6-dichloro-
phenyl p-anisate with aluminium chloride in chlorobenzene at 140-150° for 20 min or in nitrobenzene at
75° for 24 h [1214].

m.p. 239-240° [1214];   Spectra (NA).

(2,3-Dichlorophenyl)(2-hydroxy-4-methoxyphenyl)methanone

$C_{14}H_{10}Cl_2O_3$ mol.wt. 297.14

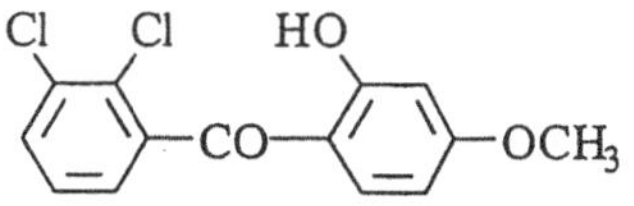

Synthesis

-Preparation by reaction of 2,3-dichlorobenzoyl chloride with
m-methoxyphenol in the presence of aluminium chloride for
8 h at 190° [348].

m.p. and Spectra (NA).

(2,4-Dichlorophenyl)(2-hydroxy-4-methoxyphenyl)methanone

$C_{14}H_{10}Cl_2O_3$ mol.wt. 297.14

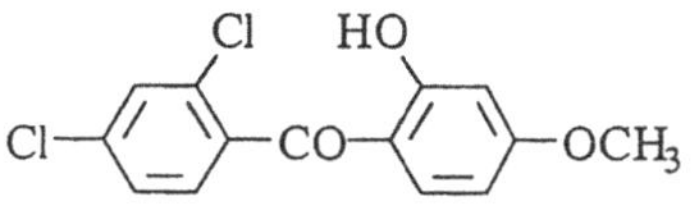

Synthesis

-Preparation by partial demethylation of 2',4'-dichloro-
2,4-dimethoxybenzophenone with aluminium chloride
(or aluminium bromide) in chlorobenzene at 90-95°
(good yields) [602].

m.p. and Spectra (NA).

(2,6-Dichlorophenyl)(2-hydroxy-4-methoxyphenyl)methanone

[77156-44-6] $C_{14}H_{10}Cl_2O_3$ mol.wt. 297.14

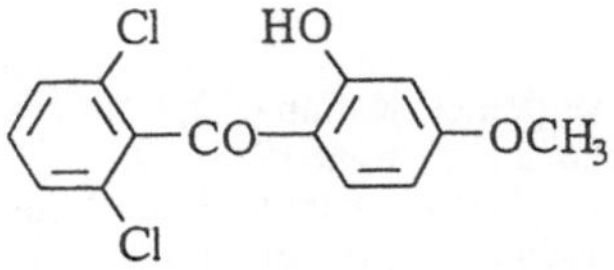

Synthesis

-Preparation by reaction of 2,6-dichlorobenzoyl chloride with m-methoxyphenol in the presence of aluminium chloride for 8 h at 190° [348].

m.p. and Spectra (NA).

(2-Fluorophenyl)(4-hydroxy-3-methoxy-5-nitrophenyl)methanone

[125629-31-4] $C_{14}H_{10}FNO_5$ mol.wt. 291.24

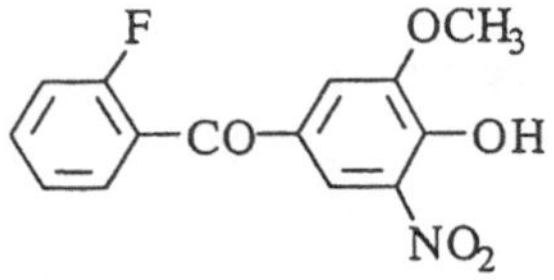

Synthesis

-Preparation by nitration of 2'-fluoro-4-hydroxy-3-methoxy-benzophenone with 65% nitric acid at 20° for 90 min (94%) [164, 197].

m.p.   150-152° [197], 127-129° [164];
^{1}H NMR [197], MS [197].

[2-(Fluoro-^{18}F)phenyl](4-hydroxy-3-methoxy-5-nitrophenyl)methanone

[190585-66-1] $C_{14}H_{10}FNO_5$ mol.wt. 290.24

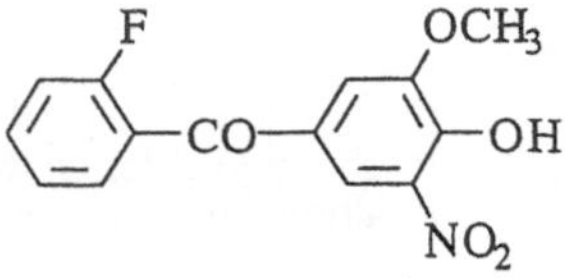

Synthesis

-Obtained by adding methyl iodide to a solution of 2'-[^{18}F] fluoro-3,4-dihydroxy-5-nitrobenzophenone in N,N-di-methylformamide first treated with sodium hydride at 0° (90%). -Refer to: Chem. Abstr., **127**, 17465u (1997).

m.p. and Spectra (NA).

(3-Fluorophenyl)(4-hydroxy-3-methoxy-5-nitrophenyl)methanone

[134612-73-0] $C_{14}H_{10}FNO_5$ mol.wt. 291.24

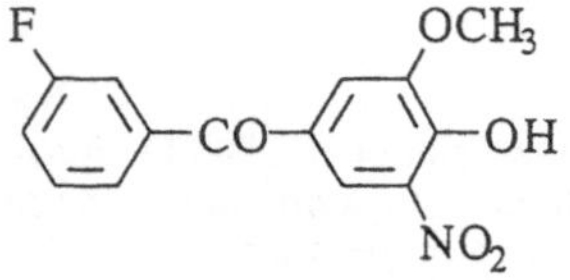

Synthesis

-Preparation by nitration of 3'-fluoro-4-hydroxy-3-methoxy-benzophenone with 65% nitric acid in acetic acid at 20° [164].

m.p.   168-170° [164];   Spectra (NA).

(4-Fluorophenyl)(4-hydroxy-3-methoxy-5-nitrophenyl)methanone

[134612-74-1] $C_{14}H_{10}FNO_5$ mol.wt. 291.24

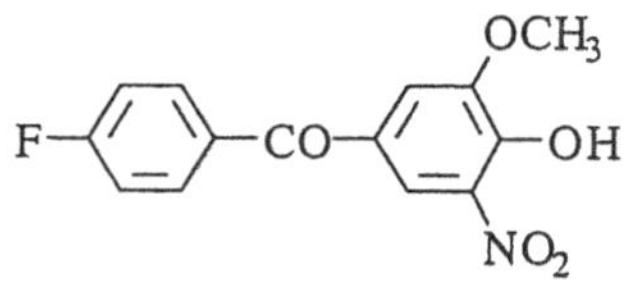

Synthesis

-Preparation by nitration of 4'-fluoro-4-hydroxy-
3-methoxybenzophenone with 65% nitric acid in acetic acid
at 20° [164].

m.p. 126-128° [164];   Spectra (NA).

(2,6-Difluorophenyl)(4-hydroxy-3-methoxyphenyl)methanone

[134612-34-3] $C_{14}H_{10}F_2O_3$ mol.wt. 264.23

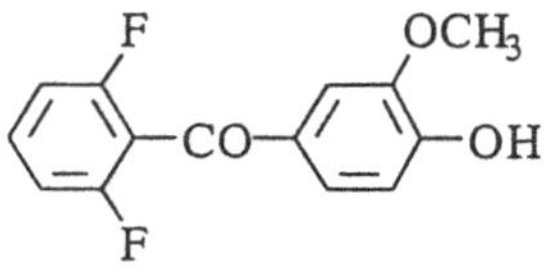

Synthesis

-Preparation by reaction of 33% hydrobromic acid in acetic
acid with 4-(benzyloxy)-2',6'-difluoro-3-methoxy-
benzophenone in methylene chloride at 20-25° [164].

m.p. 130-132° [164];   Spectra (NA).

(2-Hydroxy-5-methyl-3-nitrophenyl)(3-nitrophenyl)methanone

[67246-06-4] $C_{14}H_{10}N_2O_6$ mol.wt. 302.24

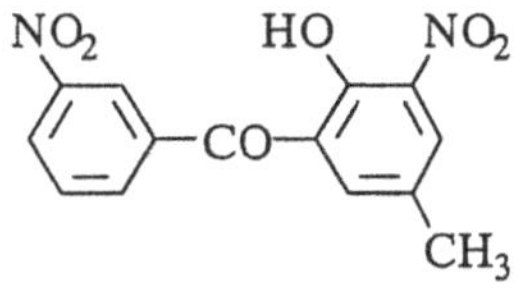

Synthesis

-Obtained by reaction of nitric acid (d = 1.42) with
2-hydroxy-5-methylbenzophenone in acetic acid at 100°
[1083].

m.p. 118° [1083]; ^{1}H NMR [1083].

(4-Hydroxy-3-methoxy-5-nitrophenyl)(2-nitrophenyl)methanone

[190522-98-6] $C_{14}H_{10}N_2O_7$ mol.wt. 318.24

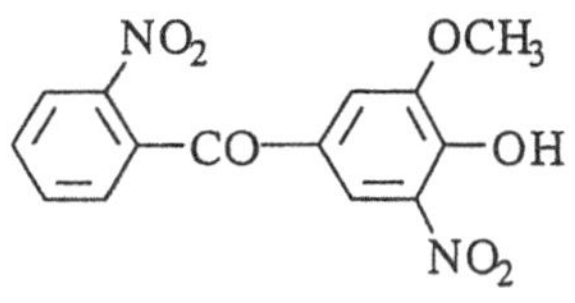

Synthesis

-Preparation by reaction of 85% nitric acid with 4-hydroxy-
3-methoxy-2'-nitrobenzophenone in acetic acid, first
15 min at 0°, then 1.5 h at 20° (77%). -Refer to: Chem.
Abstr., **127**, 17465u (1997)*.

m.p. (NA); ^{1}H NMR*, MS*.

(2-Bromophenyl)(2-hydroxy-5-methylphenyl)methanone

[55270-73-0] $C_{14}H_{11}BrO_2$ mol.wt. 291.14

Syntheses

-Preparation by reaction of 2-bromobenzoyl chloride with
 4-methoxytoluene in the presence of aluminium chloride,
*in refluxing carbon disulfide for 3 h (80%) [895];
*without solvent for 1 h at 150° (68%) [472].
-Preparation by Fries rearrangement of 4-methylphenyl
2-bromobenzoate with aluminium chloride without solvent for 10 min at 140° (quantitative yield)
[1143].
-Preparation by demethylation of 2'-bromo-2-methoxy-5-methylbenzophenone with excess boiling
pyridinium chloride for 1 h (87%) [1153].

   m.p. 78°5 [895], 76-77° [1143], 76° [1153], 75°5 [472];   Spectra (NA).

(3-Bromophenyl)(2-hydroxy-4-methylphenyl)methanone

                                       $C_{14}H_{11}BrO_2$ mol.wt. 291.14

Syntheses

-Preparation by reaction of m-bromobenzoic acid with
 m-cresol in the presence of alumina in methanesulfonic acid
 for 1 h at 140° (85%).
-Preparation by Fries rearrangement of m-cresyl m-bromo-
benzoate with alumina in methanesulfonic acid for 2 h at 160° (70%). -Refer to: Chem. Abstr., 130,
81248q (1999)*.

   m.p. 88°*; ^{1}H NMR*, IR*, UV*, MS*.

(3-Bromophenyl)(2-hydroxy-5-methylphenyl)methanone

[55270-77-4] $C_{14}H_{11}BrO_2$ mol.wt. 291.14

Synthesis

-Preparation by reaction of 3-bromobenzoyl chloride with
 4-methoxytoluene in the presence of aluminium chloride
 without solvent for 1 h at 150° (20%) [472].

   m.p. 88-89° [472];   Spectra (NA).

(3-Bromophenyl)(4-hydroxy-2-methylphenyl)methanone

                                       $C_{14}H_{11}BrO_2$ mol.wt. 291.14

Synthesis

-Obtained by reaction of m-bromobenzoic acid with m-cresol
 in methanesulfonic acid with or without phosphorous
 pentoxide, for 12 h at 100° (30%). -Refer to: Chem. Abstr.,
 130, 81248q (1999).

   m.p. and Spectra (NA).

(4-Bromophenyl)(2-hydroxy-4-methylphenyl)methanone

$C_{14}H_{11}BrO_2$ mol.wt. 291.14

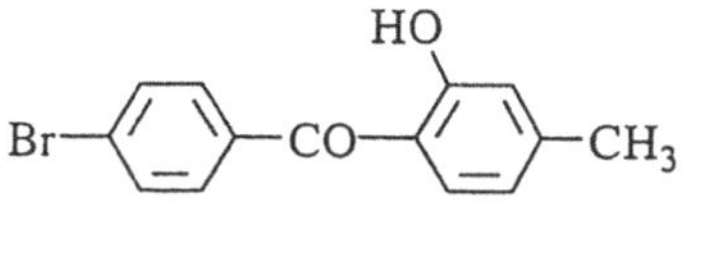

Syntheses

-Preparation by acylation of m-cresol with p-bromobenzoic acid in methanesulfonic acid in the presence of
alumina for 40 min at 140° (80%).
-Preparation by Fries rearrangement of m-cresyl
p-bromobenzoate with alumina in methanesulfonic acid for 2 h at 160° (83%). -Refer to: Chem.
Abstr., **130**, 81248q (1999)*.

   m.p. 66°*;
   ^{1}H NMR*, IR*, UV*, MS*.

(4-Bromophenyl)(2-hydroxy-5-methylphenyl)methanone

$C_{14}H_{11}BrO_2$ mol.wt. 291.14

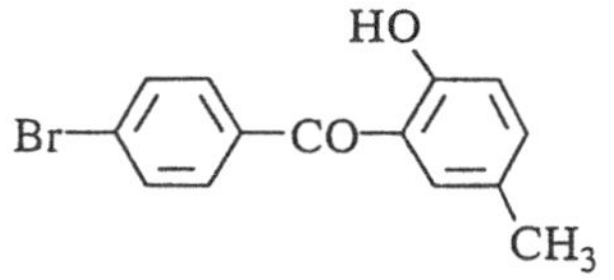

Synthesis

-Preparation by Fries rearrangement of p-cresyl p-bromobenzoate with aluminium chloride without solvent at 140° for
10 min (90%) [1143].

   m.p. 79° [1143];   b.p.$_{11}$ 210° [1143];   Spectra (NA).

(2-Bromophenyl)(2-hydroxy-4-methoxyphenyl)methanone

[183106-15-2] $C_{14}H_{11}BrO_3$ mol.wt. 307.14

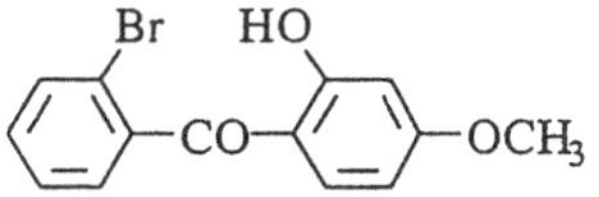

Synthesis

-Preparation by selective demethylation of 2'-bromo-2,4-di-
methoxybenzophenone with excess beryllium chloride in
refluxing toluene for 3.5 h (90%) [1250].

   m.p. 96-98° [1250];
   ^{1}H NMR [1250], IR [1250], UV [1250], MS [1250];   TLC [1250].

(2-Bromophenyl)(2-hydroxy-5-methoxyphenyl)methanone

[183106-23-2] $C_{14}H_{11}BrO_3$ mol.wt. 307.14

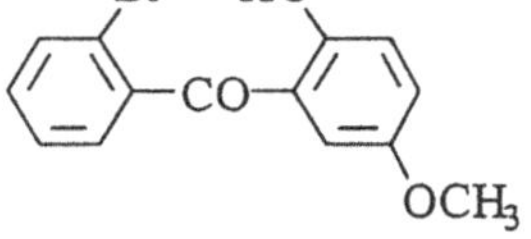

Synthesis

-Preparation by selective demethylation of 2'-bromo-2,5-di-
methoxybenzophenone with excess beryllium chloride in
refluxing toluene for 3.5 h (90%) [1250].

   pale yellow oil [1250];   b.p. (NA);
   ^{1}H NMR [1250], IR [1250], UV [1250], MS [1250];   TLC [1250].

(4-Bromophenyl)(2-hydroxy-4-methoxyphenyl)methanone

$C_{14}H_{11}BrO_3$ mol.wt. 307.14

Synthesis

-Refer to: [43, 602].

m.p. and Spectra (NA).

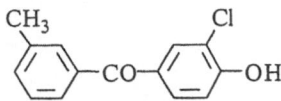

(3-Chloro-4-hydroxyphenyl)(2-methylphenyl)methanone

$C_{14}H_{11}ClO_2$ mol.wt. 246.69

Synthesis

-Obtained by Friedel-Crafts acylation of o-chloroanisole with
o-toluoyl chloride in the presence of aluminium chloride in
tetrachloroethane at 120-130° for 3 h (25%) [555].

m.p.   128-129° [555];   Spectra (NA).

(3-Chloro-4-hydroxyphenyl)(3-methylphenyl)methanone

$C_{14}H_{11}ClO_2$ mol.wt. 246.69

Syntheses

-Obtained (poor yield) by Friedel-Crafts acylation of
o-chloroanisole with m-toluoyl chloride in the presence of
aluminium chloride in tetrachloroethane at 120-130° for 1 h
(14%) [555].
-Preparation by demethylation of 3-chloro-4-methoxy-3'-methylbenzophenone with aluminium
chloride in tetrachloroethane at 100-110° for 1 h (quantitative yield) [555].

m.p.   145-146° [555];   Spectra (NA).

(5-Chloro-2-hydroxyphenyl)(2-methylphenyl)methanone

[52980-94-6] $C_{14}H_{11}ClO_2$ mol.wt. 246.69

Syntheses

-Preparation by Friedel-Crafts acylation of p-chloroanisole
with o-toluoyl chloride in the presence of aluminium
chloride in tetrachloroethane at 120-130° for 1 h (49%)
[555] or at 150° for 7 h [963].
-Also obtained by reaction of o-toluoyl chloride with
p-chlorophenol in the presence of aluminium chloride in tetrachloroethane at 120-130° for 2 h
(19%) [555].
-Preparation by Fries rearrangement of p-chlorophenyl o-toluate with aluminium chloride at 100-
150° for 0.5-3 h (71%) [1363].
-Also refer to: [893, 1444].

m.p.   67°5-68° [555], 67-68° [1363], 65-66° [963];   Spectra (NA).

(5-Chloro-2-hydroxyphenyl)(3-methylphenyl)methanone

[52980-95-7] $C_{14}H_{11}ClO_2$ mol.wt. 246.69

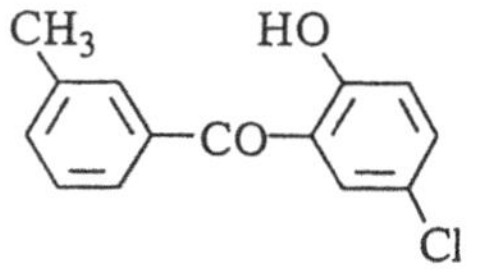

Synthesis

-Preparation by Friedel-Crafts acylation of p-chloroanisole with m-toluoyl chloride in the presence of aluminium chloride in tetrachloroethane at 120-130° for 1 h (63%) [555].

-Also refer to: [1444].

m.p. 106-106°5 [555]; Spectra (NA).

(5-Chloro-2-hydroxyphenyl)(4-methylphenyl)methanone

[116544-78-6] $C_{14}H_{11}ClO_2$ mol.wt. 246.69

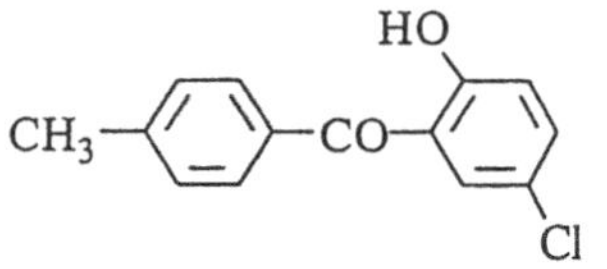

Synthesis

-Preparation by Fries rearrangement of p-chlorophenyl p-toluate with aluminium chloride at 180° for 10 min (75%) [944].

m.p. 90° [944]; ^{1}H NMR [944], IR [944].

(2-Chlorophenyl)(2-hydroxy-3-methylphenyl)methanone

$C_{14}H_{11}ClO_2$ mol.wt. 246.69

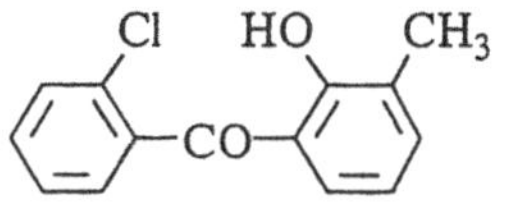

Synthesis

-Obtained (by-product) by Fries rearrangement of o-cresyl o-chlorobenzoate in the presence of aluminium chloride without solvent according to [184], (12%) [616].

m.p. 72°3-72°8 [616]; Spectra (NA).

(2-Chlorophenyl)(2-hydroxy-4-methylphenyl)methanone

[107623-97-2] $C_{14}H_{11}ClO_2$ mol.wt. 246.69

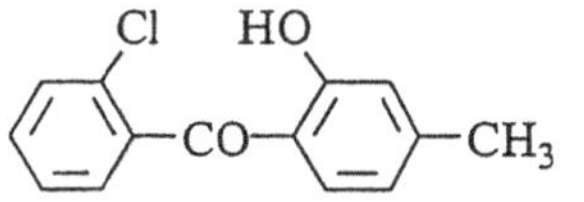

Syntheses

-Preparation by Fries rearrangement of m-cresyl o-chloro-benzoate,
*with polyphosphoric acid at 130° for 40 min (87%) [167];
*with aluminium chloride between 120 to 160° for 2 h [931].
-Preparation by Friedel-Crafts acylation of m-cresol with o-chlorobenzoic acid in the presence of polyphosphoric acid at 110° for 4 h (77%) [167].
-Also obtained by isomerization of 2'-chloro-4-hydroxy-2-methylbenzophenone (para isomer) with polyphosphoric acid at 130° for 4 h (60%) [167].

m.p. 106° [167, 931]; ^{1}H NMR [167], IR [167], UV [167].

(2-Chlorophenyl)(2-hydroxy-5-methylphenyl)methanone

[6280-52-0]   C₁₄H₁₁ClO₂   mol.wt. 246.69

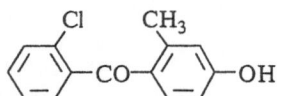

Syntheses

-Preparation by Fries rearrangement of p-cresyl o-chlorobenzoate in the presence of aluminium chloride [692] without
   solvent at 120 and 160° [472, 931, 1143], (71%) [616].
-Preparation by reaction of 2-chlorobenzotrichloride with
p-cresol in the presence of aluminium chloride in carbon disulfide at 0° (57%) [975].
-Also obtained by sulfuric acid hydrolysis of 6,12-di(2-chlorophenyl)-2,8-dimethyl-
   6,12-epoxy-*6H,12H*-dibenzo[b,f][1,5]dioxocin (74%) [975].
-Preparation by reaction of o-chlorobenzoyl chloride with p-cresol in the presence of aluminium
   chloride at 100° for 4 min, then at 175° for 6 min (61%) [605].
-Preparation by reaction of o-chlorobenzoic acid with p-cresol in the presence of 80%
   polyphosphoric acid at 190° for 3 h (46%) [144].
-Preparation by demethylation of 2'-chloro-2-methoxy-5-methylbenzophenone with excess boiling
   pyridinium chloride for 1 h (91%) [1153].
-Also refer to: [289].

  m.p. 80° [931], 78° [144, 1143], 77-78° [605], 77° [1153], 76°3-77°2 [616],
    76-77° [975], 75-77° [472];
  b.p.₁₅ 195° [1143], b.p.₁₋₂ 141-145° [975]; IR [144].

(2-Chlorophenyl)(4-hydroxy-2-methylphenyl)methanone

[92103-15-6]   C₁₄H₁₁ClO₂   mol.wt. 246.69

Syntheses

-Preparation by Fries rearrangement of m-cresyl o-chlorobenzoate without solvent in the presence of,
*aluminium chloride at 160° for 2 h [931];
*polyphosphoric acid at 70° for 6 h (42%) [1249].
-Also obtained by Friedel-Crafts acylation of m-cresol with o-chlorobenzoic acid in the presence of
   polyphosphoric acid at 90° for 1 h (31%) [1249].
-Preparation by reaction of o-chlorobenzoyl chloride with m-tolyl borate in the presence of
   aluminium chloride in tetrachloroethane at 100° (31%) [1373].

  m.p. 159-160° [931], 153° [1249, 1373]; ¹H NMR [1249], IR [1249], UV [1249].

(2-Chlorophenyl)(4-hydroxy-3-methylphenyl)methanone

C₁₄H₁₁ClO₂   mol.wt. 246.69

Syntheses

-Preparation by action of o-chlorobenzoyl chloride with
   o-tolyl borate in the presence of aluminium chloride in
   tetrachloroethane at 100° (72%) [1373].
-Preparation by Fries rearrangement of o-cresyl o-chloro-
benzoate in the presence of aluminium chloride without solvent (35%) [616], at 120 or 160° for
2 h [931].

  m.p. 167°9-168°6 [616], 167° [1373], 162° [931]; Spectra (NA).

(3-Chlorophenyl)(2-hydroxy-3-methylphenyl)methanone

$C_{14}H_{11}ClO_2$ mol.wt. 246.69

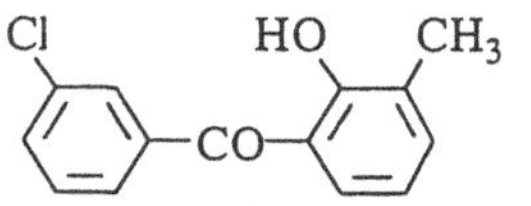

Synthesis

-Obtained (by-product) by Fries rearrangement of o-cresyl
 m-chlorobenzoate in the presence of aluminium chloride
 without solvent according to [184], (17%) [616].

m.p. 69°5-70°3 [616];   Spectra (NA).

(3-Chlorophenyl)(2-hydroxy-4-methylphenyl)methanone

[67548-59-8] $C_{14}H_{11}ClO_2$ mol.wt. 246.69

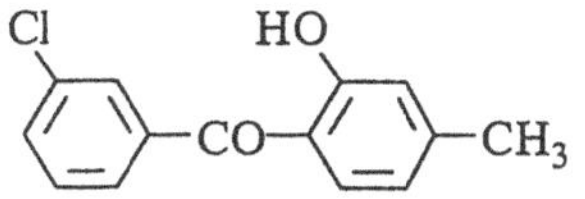

Synthesis

-Preparation by Fries rearrangement of m-cresyl m-chlorobenzoate in the presence of aluminium chloride,
 *without solvent at 120 and 160° [18];
 *in nitrobenzene at 62-63° for 72 h (28%) [126].

-Also refer to: [786].

m.p. 89°5-90°5 [126], 89° [18];   Spectra (NA).

(3-Chlorophenyl)(2-hydroxy-5-methylphenyl)methanone

[6280-54-2] $C_{14}H_{11}ClO_2$ mol.wt. 246.69

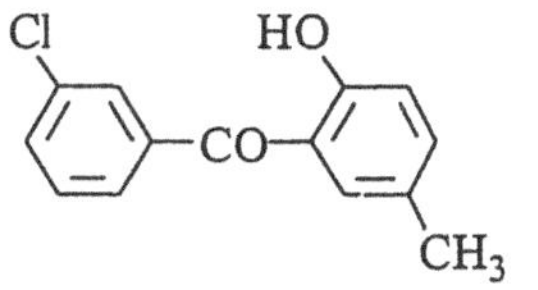

Synthesis

-Preparation by Fries rearrangement of p-cresyl m-chlorobenzoate in the presence of,
 *aluminium chloride without solvent at 120 and 160° [18,
 472], (45%) [616];
 *Nafion-H, a polymeric perfluorinated resin sulfonic acid, in
 refluxing nitrobenzene for 12 h (71%) [1004].

m.p. 72° [18], 70°5-71°5 [616], 70-71°5 [472], 69-70° [772];
^{1}H NMR [772], IR [772], UV [772];
pK_a [772, 792];   polarographic study [1223].

(3-Chlorophenyl)(4-hydroxy-2-methylphenyl)methanone

$C_{14}H_{11}ClO_2$ mol.wt. 246.69

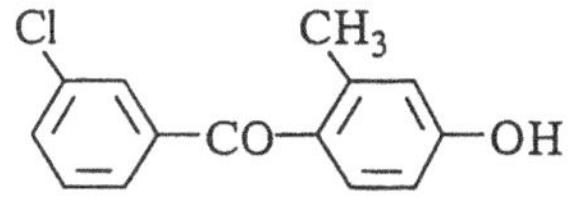

Synthesis

-Obtained by Fries rearrangement of m-cresyl m-chlorobenzoate with aluminium chloride in nitrobenzene at 62-63°
 for 72 h (40%) [126] or without solvent at 120 or 160° [18].

m.p. 125-125°5 [126], 108° [18]. One of the reported melting points is obviously
wrong.
Spectra (NA).

(3-Chlorophenyl)(4-hydroxy-3-methylphenyl)methanone

$C_{14}H_{11}ClO_2$ mol.wt. 246.69

Synthesis

-Preparation by Fries rearrangement of o-cresyl m-chlorobenzoate with aluminium chloride without solvent according
   to [184], (55%) [616] and at 120 or 160° [18].

   m.p.   151°2-151°8 [616], 149-150° [18];   Spectra (NA).

(4-Chlorophenyl)(2-hydroxy-3-methylphenyl)methanone

[6279-04-5] $C_{14}H_{11}ClO_2$ mol.wt. 246.69

Syntheses

-Obtained (by-product) by Fries rearrangement of o-cresyl
   p-chlorobenzoate in the presence of aluminium chloride
   without solvent (12%) [616].

-Preparation by reaction of 3-methylsalicylic acid chloride (2-hydroxy-3-methylbenzoyl chloride)
   with chlorobenzene in the presence of aluminium chloride at 100° overnight (47%) [400, 1205,
   1206, 1207, 1208].
-Also refer to: [1146, 1436].

   m.p.   61°5-62° [616], 61-63° [400, 1208], 55-58° [1205, 1206, 1207];
   b.p.$_{0.5}$ 148-152° [1205, 1206, 1207, 1208];   Spectra (NA).

(4-Chlorophenyl)(2-hydroxy-4-methylphenyl)methanone

[107622-28-6] $C_{14}H_{11}ClO_2$ mol.wt. 246.69

Syntheses

-Preparation by Fries rearrangement of m-cresyl p-chlorobenzoate with aluminium chloride between 120 to 160°
   (good yield) [850].

-Preparation by reaction of p-chlorobenzoic acid with m-cresol in the presence of boron trifluoride
   at 160° for 2 h (70%) [730].
-Also obtained by reaction of p-chlorobenzoyl chloride,
*with m-tolyl borate in the presence of aluminium chloride in tetrachloroethane at 100° (26%) [1373];
*with m-cresol in the presence of aluminium chloride at 105° for 22 h (22%) [400].

   m.p.   83-84° [850], 81°5 [730], 80-81° [400];   Spectra (NA).

(4-Chlorophenyl)(2-hydroxy-5-methylphenyl)methanone

[6279-05-6] $C_{14}H_{11}ClO_2$ mol.wt. 246.69

Syntheses

-Preparation by Friedel-Crafts acylation of p-cresol,
*with p-chlorobenzoic acid in the presence of boron
   trifluoride at 160° for 5 h (89%) [730];
*with p-chlorobenzoyl chloride in the presence of aluminium
   chloride without solvent (78%) [400, 1208], or in

o-dichlorobenzene at 135° for 2 h (quantitative yield) [637].
-Preparation by Fries rearrangement of p-cresyl p-chlorobenzoate with aluminium chloride without
   solvent at 120 and 160° [850], (51%) [616].
-Also obtained by reaction of 4-chlorobenzotrichloride with p-cresol in the presence of aluminium
   chloride in carbon disulfide at 0° (28%) [975].
-Also obtained by photo-Fries rearrangement of p-cresyl p-chlorobenzoate in methylene chloride
   (major product) [1418].
-Also obtained by sulfuric acid hydrolysis of 6,12-di(4-chlorophenyl)-2,8-dimethyl-
   6,12-epoxy-*6H,12H*-dibenzo[b,f][1,5]dioxocin (93%) [975].
-Also obtained (poor yield) by diazotization of 2-amino-4'-chloro-5-methylbenzophenone and
   decomposition of the resulting diazonium salt in 0.1 M sulfuric acid at 60° (10%) [1390].
-Also refer to: [643, 1205, 1206, 1207, 1412, 1436].

m.p. 71° [850], 69° [772], 68° [400], 66°9-67°6 [616], 66°4-67°2 [975], 66-67° [730];
b.p.$_{0.1}$ 168-174° [1208], b.p.$_{0.25}$ 160-170° [730], b.p.$_{1-2}$ 143-145° [975];
^{1}H NMR [637, 772], IR [637, 772, 792], UV [772];
pK_a [772, 792];   polarographic study [1223].

(4-Chlorophenyl)(3-hydroxy-2-methylphenyl)methanone

[74167-86-5] $C_{14}H_{11}ClO_2$ mol.wt. 246.69

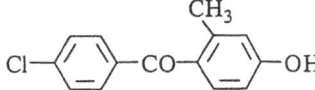

Synthesis

-Preparation by diazotization of 3-amino-4'-chloro-
   2-methylbenzophenone, followed by hydrolysis of the
   diazonium salt so obtained (66%) [82], according to [731].

-Also refer to: [103, 436].

m.p. 100° [82];   Spectra (NA).

(4-Chlorophenyl)(3-hydroxy-4-methylphenyl)methanone

[74177-55-2] $C_{14}H_{11}ClO_2$ mol.wt. 246.69

Synthesis

-Refer to: [82, 436].

m.p. and Spectra (NA).

(4-Chlorophenyl)(4-hydroxy-2-methylphenyl)methanone

[61002-51-5] $C_{14}H_{11}ClO_2$ mol.wt. 246.69

Syntheses

-Preparation by Fries rearrangement of m-cresyl p-chlorobenzoate in the presence of aluminium chloride without
   solvent at 120 and 160° (good yield) [850].
-Also obtained by reaction of p-chlorobenzoyl chloride,
*with m-tolyl borate in the presence of aluminium chloride in tetrachloroethane at 100° (11%)
   [1373];
*with m-cresol in the presence of aluminium chloride in nitrobenzene, first at 15° and at r.t. for
   28 h [1293].
-Also refer to: [6, 909].

m.p.   116-117° [850];   Spectra (NA).

(4-Chlorophenyl)(4-hydroxy-3-methylphenyl)methanone

[6279-06-7] $C_{14}H_{11}ClO_2$ mol.wt. 246.69

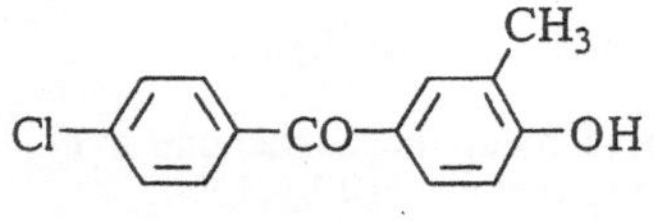

Syntheses

-Preparation by Fries rearrangement of o-cresyl p-chlorobenzoate with aluminium chloride without solvent at 120
and 160° (good yield) [850], (52%)[616].

-Preparation by reaction of p-chlorobenzoyl chloride with o-tolyl borate in the presence of
aluminium chloride in tetrachloroethane at 100° (47%) [1373].

-Preparation by reaction of p-chlorobenzoyl chloride with o-cresol in the presence of aluminium
chloride [296].

m.p.   211-211°5 [296], 210°5-211°5 [616], 209-210° [850];   Spectra (NA).

(4-Chlorophenyl)(5-hydroxy-2-methylphenyl)methanone

$C_{14}H_{11}ClO_2$ mol.wt. 246.69

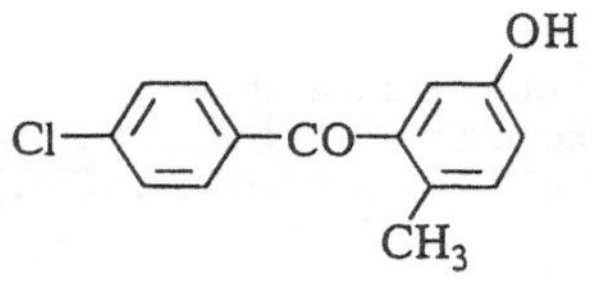

Synthesis

-Preparation by diazotization of 5-amino-4'-chloro-2-methyl-
benzophenone, followed by hydrolysis of the resulting
diazonium salt (82%) [82], according to [731].

m.p.   188° [82];   Spectra (NA).

(5-Chloro-2-hydroxyphenyl)(2-methoxyphenyl)methanone

[159819-70-2] $C_{14}H_{11}ClO_3$ mol.wt. 262.69

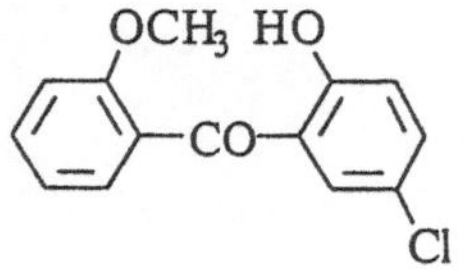

Synthesis

-Preparation from 2-bromo-4-chloro-(2-methoxyethoxy)-
methoxybenzene and o-anisaldehyde as the starting materials
[963].

m.p.   94-95° [963];   Spectra (NA).

(5-Chloro-2-hydroxyphenyl)(4-methoxyphenyl)methanone

[85052-20-6] $C_{14}H_{11}ClO_3$ mol.wt. 262.69

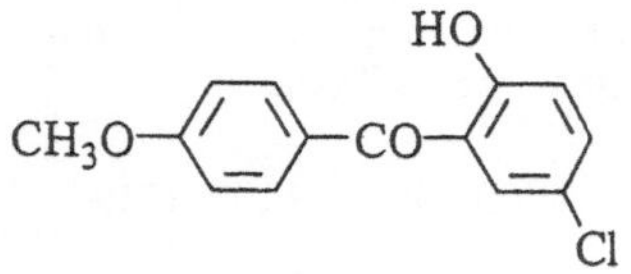

Synthesis

-Preparation by reaction of p-anisoyl chloride with p-chloroanisole in the presence of aluminium chloride in methylene
chloride at r.t. for 18 h under nitrogen (41%) [932].

m.p. (NA);   crystals [932];
^{1}H NMR [932], IR [932], MS [932];   TLC [932].

(2-Chlorophenyl)(2-hydroxy-4-methoxyphenyl)methanone

[107517-49-7]   C$_{14}$H$_{11}$ClO$_3$ mol.wt. 262.69

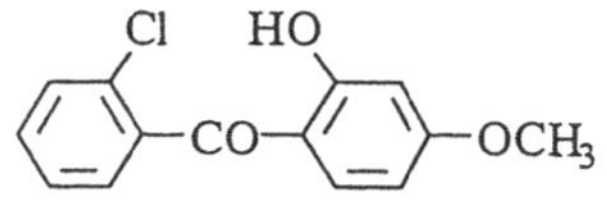

Syntheses

-Preparation by selective demethylation of 2'-chloro-2,4-di-
methoxybenzophenone with excess beryllium chloride in
refluxing toluene for 3 h (90%) [1250].
-Preparation by reaction of o-chlorobenzoyl chloride with
resorcinol dimethyl ether in tetrachloroethane in the presence of aluminium chloride at 90° [43].
-Also refer to: [383, 384].

m.p. 85-88° [43], 74-75° [1250];
^{1}H NMR [1250], IR [1250], UV [1067, 1250];   TLC [1250].

(2-Chlorophenyl)(2-hydroxy-5-methoxyphenyl)methanone

[183106-21-0]   C$_{14}$H$_{11}$ClO$_3$ mol.wt. 262.69

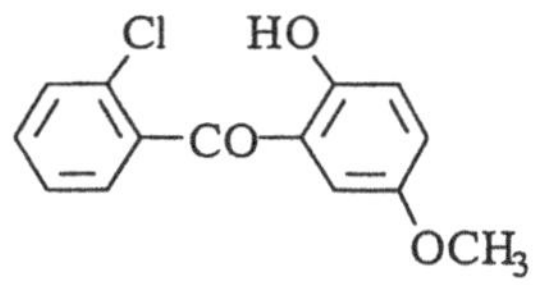

Synthesis

-Preparation by selective demethylation of 2'-chloro-2,5-di-
methoxybenzophenone with excess beryllium chloride in
refluxing toluene for 3 h (90%) [1250].

pale yellow oil [1250];   b.p. (NA);
^{1}H NMR [1250], IR [1250], UV [1250], MS [1250];   TLC [1250].

(2-Chlorophenyl)(4-hydroxy-3-methoxyphenyl)methanone

[134612-35-4]   C$_{14}$H$_{11}$ClO$_3$ mol.wt. 262.69

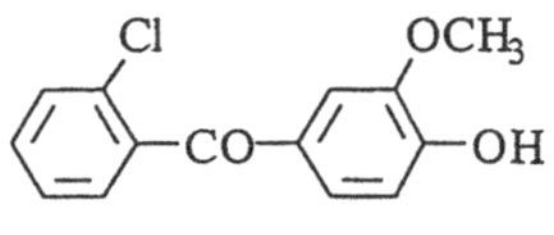

Synthesis

-Preparation by debenzylation of 4-(benzyloxy)-2'-chloro-
3-methoxybenzophenone with 33% hydrobromic acid/acetic
acid in methylene chloride at 20-25° [164].

m.p. and Spectra (NA).

(3-Chlorophenyl)(2-hydroxy-4-methoxyphenyl)methanone

[96410-70-7]   C$_{14}$H$_{11}$ClO$_3$ mol.wt. 262.69

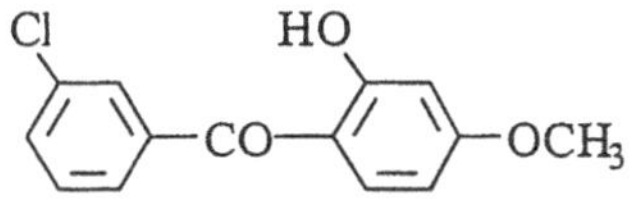

Synthesis

-Preparation by reaction of m-chlorobenzonitrile with
m-methoxyphenol in the presence of aluminium chloride for
8 h at 190°, followed by hydrolysis of the ketimine formed
[348].
-Also refer to: [1380, 1381, 1382, 1383] (Japanese patents).

m.p. and Spectra (NA).

(3-Chlorophenyl)(4-hydroxy-3-methoxyphenyl)methanone

[134612-36-5]   C14H11ClO3   mol.wt. 262.69

Synthesis

-Preparation by debenzylation of 4-(benzyloxy)-3'-chloro-
3-methoxybenzophenone with 33% hydrobromic acid/acetic
acid in methylene chloride at 20-25° [164].

m.p.   136-138° [164];   Spectra (NA).

(4-Chlorophenyl)(2-hydroxy-4-methoxyphenyl)methanone

[85-28-9]   C14H11ClO3   mol.wt. 262.69

Syntheses

-Preparation by oxidation of 6-methoxy-2-phenyl-
3-(4-chlorophenyl)benzofuran with chromium trioxide
in boiling acetic acid for 40 min, followed by
saponification of the resulting keto ester
— 2-(benzoyloxy)-4'-chloro-4-methoxybenzophenone — with 4 N sodium hydroxide in refluxing
ethanol for 1 h [1052].
-Preparation by reaction of p-chlorobenzoyl chloride with resorcinol dimethyl ether,
*in the presence of aluminium chloride;
  in a chlorobenzene/N,N-dimethylformamide mixture (22:1) at 115° [539, 1067];
  in tetrachloroethane first at r.t., then at 90° [43];
*in the presence of titanium tetrachloride in chlorobenzene for 1 h at 120° (74%) [1140];
*in the presence of ferric chloride, without solvent or in o-dichlorobenzene, at 180-200° for 6-7 h
  (by-product) [1467].
-Preparation by partial demethylation of 4'-chloro-2,4-dimethoxybenzophenone with aluminium
  chloride or aluminium bromide in chlorobenzene at 90-95° (good yield) [602].
-Preparation by Fries rearrangement of m-methoxyphenyl p-chlorobenzoate with aluminium
  chloride [692].
-Also refer to: [79] (Japanese patent) and [219, 348, 383, 855, 968, 1089].

     m.p.   115° [1140], 113° [1467], 111° [1052], 109-112° [43];
     IR [1052], UV [1067].

(4-Chlorophenyl)(4-hydroxy-3-methoxyphenyl)methanone

[134612-37-6]   C14H11ClO3   mol.wt. 262.69

Synthesis

-Preparation by debenzylation of 4-(benzyloxy)-4'-chloro-
3-methoxybenzophenone with 33% hydrobromic
acid/acetic acid in methylene chloride at 20-25° [164].

m.p.   114-116° [164];   Spectra (NA).

(5-Fluoro-2-hydroxyphenyl)(2-methylphenyl)methanone

$C_{14}H_{11}FO_2$ mol.wt. 230.24

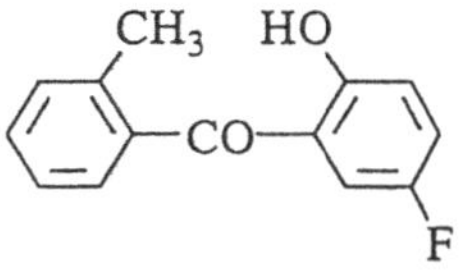

Synthesis

-Preparation by Fries rearrangement of p-fluorophenyl
o-toluate with aluminium chloride [692].

m.p. and Spectra (NA).

(5-Fluoro-2-hydroxyphenyl)(3-methylphenyl)methanone

[342-18-7] $C_{14}H_{11}FO_2$ mol.wt. 230.24

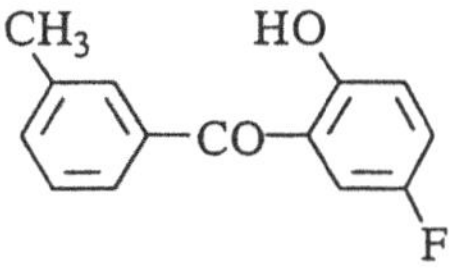

Synthesis

-Preparation by Fries rearrangement of p-fluorophenyl
m-toluate with aluminium chloride [692] without solvent at
130° for 2 h (92%) [656].

b.p.$_{15}$ 185° [656];   Spectra (NA).

(5-Fluoro-2-hydroxyphenyl)(4-methylphenyl)methanone

[62433-29-8] $C_{14}H_{11}FO_2$ mol.wt. 230.24

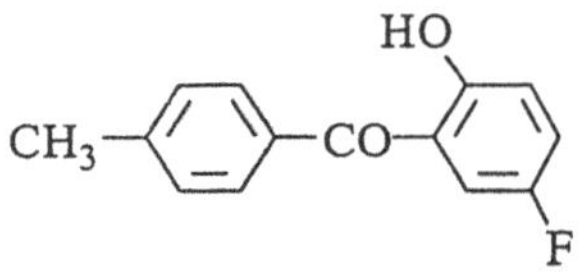

Synthesis

-Preparation by Fries rearrangement of p-fluorophenyl
p-toluate with aluminium chloride at 150-180° for 20 min
(92%) [765] or at 130° for 2 h (82%) [1295].

m.p. 89° [1295], 75-76° [765];   Spectra (NA).

(2-Fluorophenyl)(2-hydroxy-5-methylphenyl)methanone

[55270-76-3] $C_{14}H_{11}FO_2$ mol.wt. 230.24

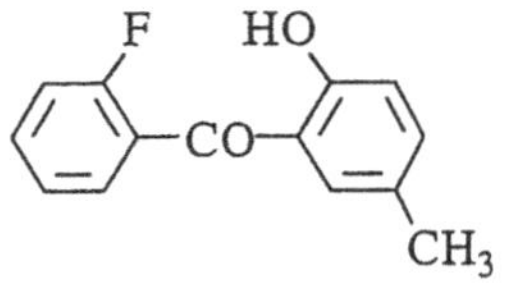

Synthesis

-Preparation by Fries rearrangement of p-cresyl 2-fluoro-
benzoate with aluminium chloride [692] without solvent at
160° (36%) [472].

m.p. 72-73° [472];   Spectra (NA).

(3-Fluorophenyl)(2-hydroxy-5-methylphenyl)methanone

[55270-80-9] $C_{14}H_{11}FO_2$ mol.wt. 230.24

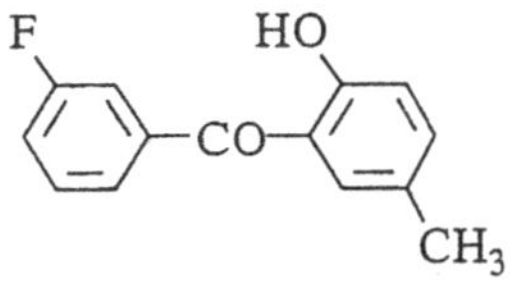

Synthesis

-Preparation by reaction of 3-fluorobenzoyl chloride with
4-methoxytoluene in the presence of aluminium chloride at
150° for 1.5 h (21%) [472].

m.p. 31-34° [472];   b.p.$_{15}$ 70-74° [472]; $n_D^{27.4}$ = 1.5946 [472];   Spectra (NA).

(4-Fluorophenyl)(2-hydroxy-4-methylphenyl)methanone

[108294-71-9]   C$_{14}$H$_{11}$FO$_2$ mol.wt. 230.24

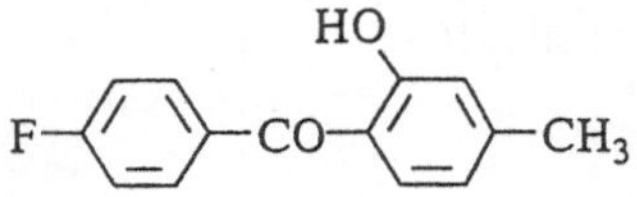

Synthesis

-Obtained (poor yield) by reaction of p-fluorobenzoyl
chloride with m-cresol in the presence of aluminium
chloride in tetrachloroethane at 105° for 22 h (9%) [400].

m.p. 78° [400];   Spectra (NA).

(4-Fluorophenyl)(4-hydroxy-2-methylphenyl)methanone

[32192-52-2]   C$_{14}$H$_{11}$FO$_2$ mol.wt. 230.24

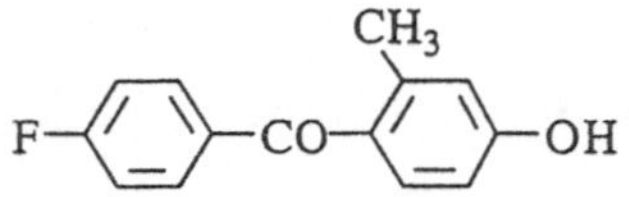

Synthesis

-Preparation by reaction of p-fluorobenzoyl chloride with
m-cresol in nitrobenzene in the presence of aluminium
chloride first at 0°, then at 60° for 20 h (33%) [769].

m.p. 110-112° [769]; ^{1}H NMR [769].

(5-Fluoro-2-hydroxyphenyl)(4-methoxyphenyl)methanone

[727-93-5]   C$_{14}$H$_{11}$FO$_3$ mol.wt. 246.24

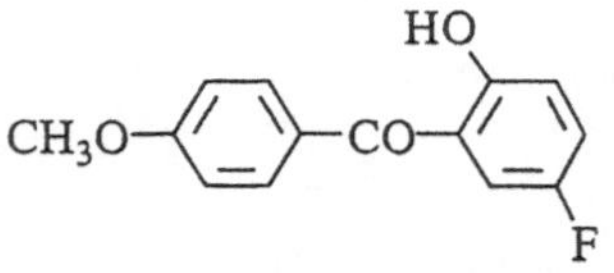

Synthesis

-Preparation by Fries rearrangement of p-fluorophenyl
p-anisate without solvent,
*with aluminium chloride at 130° for 2 h (60%) [656];
*with titanium tetrachloride at 160° for 20 min (49%) [329],
according to [698].

m.p. 79°5 [329];   b.p.$_{30}$ 260° [656];
^{1}H NMR [328, 329], MS [328, 329];   HPLC [328].

(5-Fluoro-2-hydroxyphenyl)[4-(methoxy- ^{11}C)phenyl]methanone

[161585-22-4]   C$_{14}$H$_{11}$FO$_3$ mol.wt. 245.24

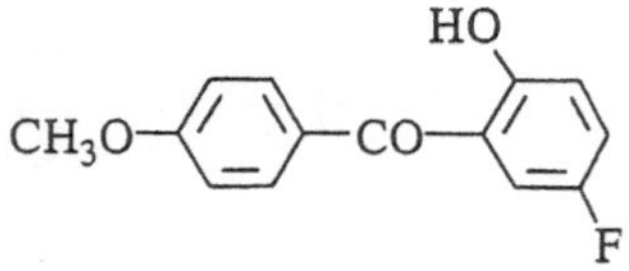

Synthesis

-Preparation by partial methylation of 2,4'-dihydroxy-
5-fluorobenzophenone with [^{11}C] methyl iodide in
N,N-dimethylformamide at -45° [329].

m.p. (NA); ^{1}H NMR [328, 329], MS [328, 329];
HPLC [328].

(2-Fluorophenyl)(2-hydroxy-4-methoxyphenyl)methanone

[3119-88-8] $C_{14}H_{11}FO_3$ mol.wt. 246.24

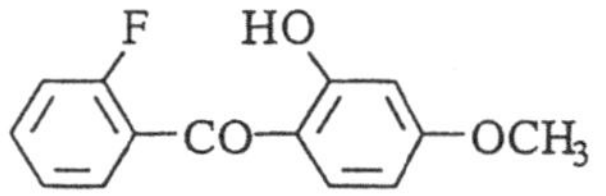

Syntheses

-Preparation by reaction of 2-fluorobenzoyl chloride with resorcinol dimethyl ether in the presence of aluminium chloride in refluxing ethylene dichloride for 30 min (78%) [1272] or refluxing hexane for 8 h [501]. Titanium tetrachloride can also be used instead of aluminium chloride [501].

-Preparation by selective demethylation of 2'-fluoro-2,4-dimethoxybenzophenone with excess beryllium chloride in refluxing toluene for 3 h (92%) [1250].

m.p. 149-150° (d) [1250], 53-54° [1272], 49-50° [501]. One of the reported melting points is obviously wrong.

^{1}H NMR [1250], IR [1250], UV [1250];   TLC [1250];   vapour pressure [500].

(2-Fluorophenyl)(2-hydroxy-5-methoxyphenyl)methanone

[183106-19-6] $C_{14}H_{11}FO_3$ mol.wt. 246.24

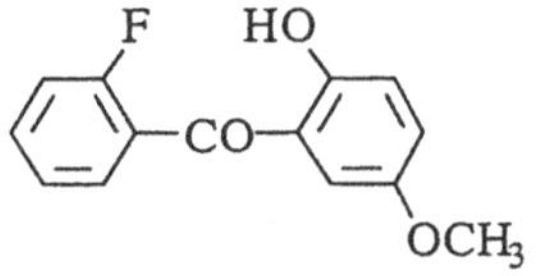

Synthesis

-Preparation by selective demethylation of 2'-fluoro-2,5-dimethoxybenzophenone with excess beryllium chloride in refluxing toluene for 3.5 h (90%) [1250].

pale yellow oil [1250];   b.p. (NA);

^{1}H NMR [1250], IR [1250], UV [1250], MS [1250];   TLC [1250].

(2-Fluorophenyl)(4-hydroxy-3-methoxyphenyl)methanone

[125629-30-3] $C_{14}H_{11}FO_3$ mol.wt. 246.24

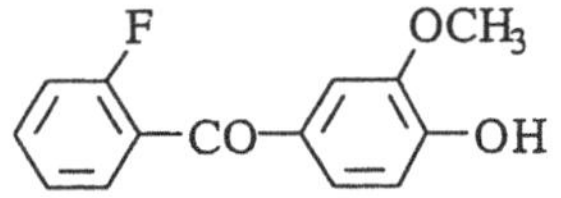

Synthesis

-Preparation by debenzylation of 4-(benzyloxy)-2'-fluoro-3-methoxybenzophenone with 33% hydrobromic acid/acetic acid in methylene chloride at r.t. for 2 h [164], (92%) [197].

m.p. 84-86° [197]; ^{1}H NMR [197], MS [197].

(3-Fluorophenyl)(2-hydroxy-4-methoxyphenyl)methanone

$C_{14}H_{11}FO_3$ mol.wt. 246.24

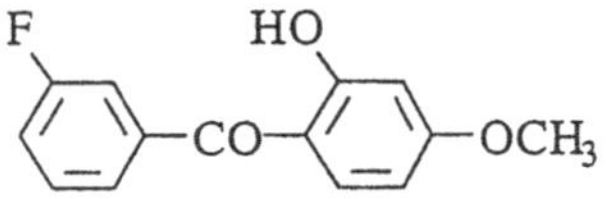

Synthesis

-Preparation by reaction of m-fluorobenzoyl chloride with resorcinol dimethyl ether in the presence of aluminium chloride or titanium tetrachloride in refluxing n-hexane for 8 h [501].

m.p. 88°5-89°5 [501];   Spectra (NA);   vapour pressure [500].

(3-Fluorophenyl)(2-hydroxy-5-methoxyphenyl)methanone

$C_{14}H_{11}FO_3$ mol.wt. 246.24

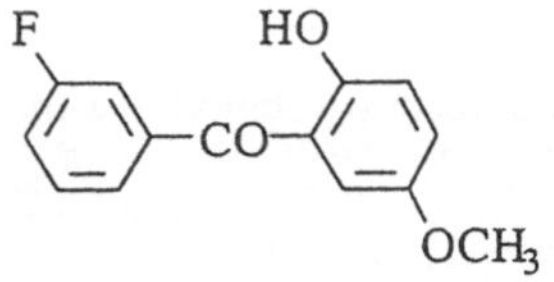

Synthesis

-Obtained by partial demethylation of 3'-fluoro-2,5-di-
  methoxybenzophenone with aluminium chloride in benzene
  at 80° [1048].

m.p. and Spectra (NA).

(3-Fluorophenyl)(4-hydroxy-3-methoxyphenyl)methanone

[134612-32-1] $C_{14}H_{11}FO_3$ mol.wt. 246.24

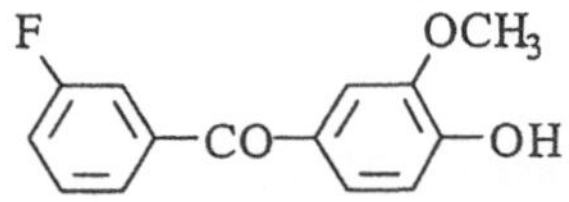

Synthesis

-Preparation by debenzylation of 4-(benzyloxy)-3'-fluoro-
  3-methoxybenzophenone with 33% hydrobromic acid/acetic
  acid in methylene chloride at r.t. [164].

m.p. 133-135° [164];   Spectra (NA).

(4-Fluorophenyl)(2-hydroxy-4-methoxyphenyl)methanone

[3602-47-9] $C_{14}H_{11}FO_3$ mol.wt. 246.24

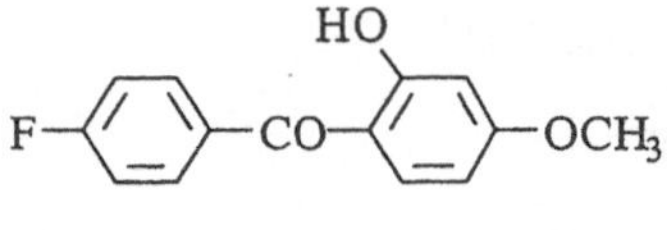

Syntheses

-Preparation by reaction of p-fluorobenzoyl chloride with
  resorcinol dimethyl ether in the presence of aluminium
  chloride or titanium tetrachloride in refluxing hexane for
  8 h [501].
-Preparation by reaction of dimethyl sulfate with 4'-fluoro-2,4-dihydroxybenzophenone in the
  presence of 10% potassium hydroxide first at 35°, then at reflux for 30 min (79%) [394].

m.p. 88-89° [501], 86-88° [394];
Spectra (NA);   vapour pressure [500].

(4-Fluorophenyl)(2-hydroxy-5-methoxyphenyl)methanone

[162657-93-4] $C_{14}H_{11}FO_3$ mol.wt. 246.24

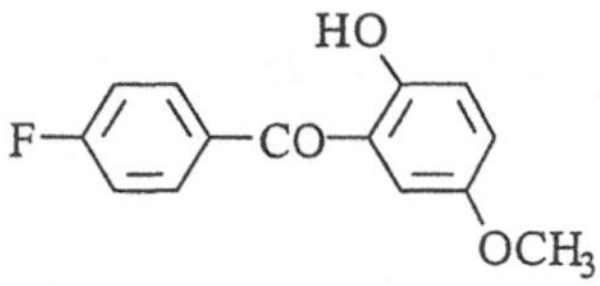

Synthesis

-Preparation by partial demethylation of 4'-fluoro-2,5-di-
  methoxybenzophenone with aluminium chloride in benzene
  under nitrogen at 80° for 12 h (66%) [1047, 1048].

m.p. 93° [1047];
^{1}H NMR [1047], ^{13}C NMR [1047], MS [1047].

(4-Fluorophenyl)(4-hydroxy-3-methoxyphenyl)methanone

[134612-33-2] $C_{14}H_{11}FO_3$ mol.wt. 246.24

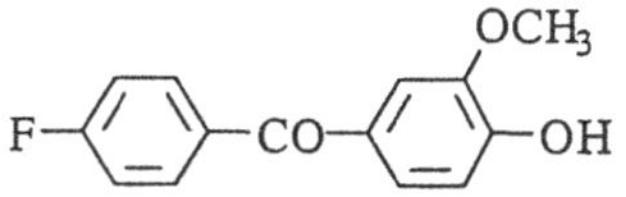

Synthesis

-Preparation by debenzylation of 4-(benzyloxy)-4'-fluoro-
3-methoxybenzophenone with 33% hydrobromic acid/
acetic acid in methylene chloride at r.t. [164].

m.p. 139-141° [164];   Spectra (NA).

(2-Hydroxy-3-methylphenyl)(3-nitrophenyl)methanone

$C_{14}H_{11}NO_4$ mol.wt. 257.25

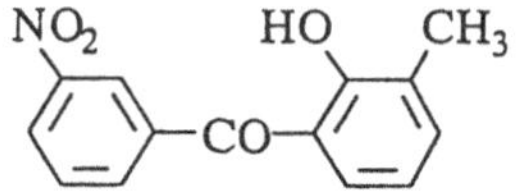

Syntheses

-Obtained (poor yield) by Fries rearrangement of o-tolyl
m-nitrobenzoate with aluminium chloride at 120 or at 160°
for 2 h (<8%) [1166].
-Also obtained (poor yield) by reaction of m-nitrobenzoyl
chloride with o-tolyl borate in the presence of aluminium chloride in tetrachloroethane at 100° (3%)
[1373].

m.p. 115° [1166];   Spectra (NA).

(2-Hydroxy-3-methylphenyl)(4-nitrophenyl)methanone

[65611-78-1] $C_{14}H_{11}NO_4$ mol.wt. 257.25

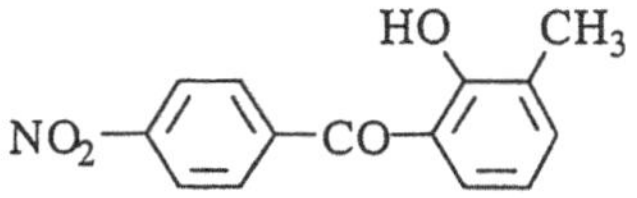

Syntheses

-Preparation by Fries rearrangement of o-tolyl p-nitrobenzoate with aluminium chloride at 120° [1167].
-Preparation by reaction between (o-tolyloxy)magnesium
bromide complexed with HMPT and p-nitrobenzaldehyde
in refluxing benzene for 48 h (68%) [250].

m.p. 118° [1167], 113° [250];
^{1}H NMR [250], IR [250], MS [250].

(2-Hydroxy-4-methylphenyl)(3-nitrophenyl)methanone

$C_{14}H_{11}NO_4$ mol.wt. 257.25

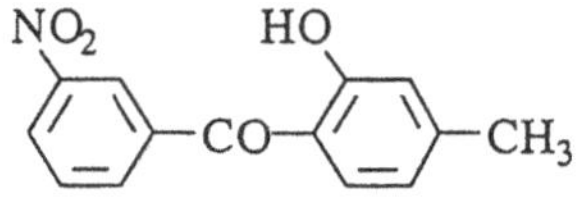

Syntheses

-Obtained by reaction of m-nitrobenzoyl chloride with m-tolyl
borate in the presence of aluminium chloride in tetrachloroethane at 100° (25%) [1373].
-Also obtained by Fries rearrangement of m-tolyl m-nitrobenzoate with aluminium chloride at 120 or at 160° for 2 h (15%) [1166].

m.p. 132° [1166];   Spectra (NA).

(2-Hydroxy-4-methylphenyl)(4-nitrophenyl)methanone

$C_{14}H_{11}NO_4$ mol.wt. 257.25

Synthesis

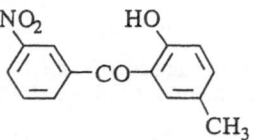

-Preparation by Fries rearrangement of m-tolyl p-nitrobenzoate with aluminium chloride at 120° for 2 h
(major product) (<26%) [1167].

m.p.   134° [1167];   Spectra (NA).

(2-Hydroxy-5-methylphenyl)(3-nitrophenyl)methanone

[53669-31-1] $C_{14}H_{11}NO_4$ mol.wt. 257.25

Syntheses

-Preparation by Fries rearrangement of p-cresyl m-nitrobenzoate with aluminium chloride without solvent at 120-
160° for 2 h [772], (62%) [1166] or in refluxing
o-dichlorobenzene for 2 h (30%) [1450] or in refluxing
chlorobenzene for 3 h (22%) [1450].
-Also obtained (by-product) by reaction of m-nitrobenzoyl chloride with p-methylanisole in the
presence of aluminium chloride without solvent at 140° for 30 min (11%) [472] or in refluxing
carbon disulfide [205].

m.p.   104-105° [1166], 102-103° [772], 102° [205], 99-100° [1450], 98-100° [472];
^{1}H NMR [772], IR [205, 772, 792], UV [205, 772];
pK_a [772, 792].

(2-Hydroxy-5-methylphenyl)(4-nitrophenyl)methanone

[53669-32-2] $C_{14}H_{11}NO_4$ mol.wt. 257.25

Syntheses

-Preparation by deethylation of 2-ethoxy-5-methyl-
4'-nitrobenzophenone (SM) in the presence of aluminium
chloride in carbon disulfide at 60-70° for 8 h [91, 93]
according to [547]. SM was obtained by reaction of
p-nitrobenzoyl chloride with p-methylphenetole in the
presence of aluminium chloride.
-Preparation by Fries rearrangement of p-cresyl p-nitrobenzoate with aluminium chloride at 160° for
2 h (60%) [1167] or at 140° for 30 min (52%) [759]. In this case, the starting ester was obtained
by heating p-nitrobenzoyl chloride with aluminium tris(p-methylphenoxide) in a boiling water bath
for 30 min [759].
-Also obtained by reaction of p-nitrobenzoyl chloride with p-methylanisole in the presence of
aluminium chloride in boiling carbon disulfide during several hours (by-product) [99] or in
ethylene dichloride first at 0-5°, then at 50° for 1 h [291].
-Also refer to: [291].

m.p.   149-150° [291], 143° [1167], 142-143° [91, 99], 142° [759];
IR [792]; pK_a [792];   cryoscopic study [93].

(2-Hydroxy-6-methylphenyl)(4-nitrophenyl)methanone

$C_{14}H_{11}NO_4$ mol.wt. 257.25

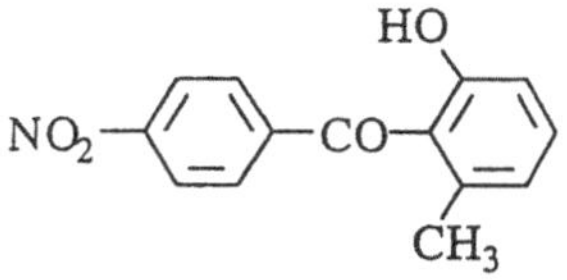

Synthesis

-Obtained (by-product) by Fries rearrangement of m-tolyl
p-nitrobenzoate with aluminium chloride at 120° for 2 h
(poor yield) [1167].

m.p. 143° [1167]; Spectra (NA).

(4-Hydroxy-2-methylphenyl)(3-nitrophenyl)methanone

[107558-23-6] $C_{14}H_{11}NO_4$ mol.wt. 257.25

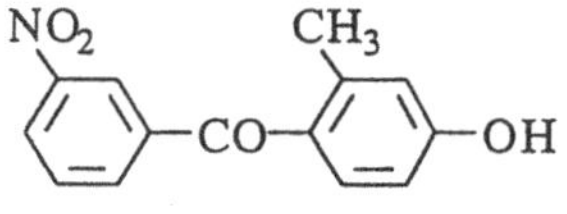

Syntheses

-Obtained by reaction of m-nitrobenzoyl chloride with
m-tolyl borate in the presence of aluminium chloride in
tetrachloroethane at 100° (17%) [1373].
-Also obtained by Fries rearrangement of m-tolyl m-nitrobenzoate with aluminium chloride at 120 or at 160° for 2 h (10%) [1166].

m.p. 200° [1166]; Spectra (NA).

(4-Hydroxy-2-methylphenyl)(4-nitrophenyl)methanone

[203060-34-8] $C_{14}H_{11}NO_4$ mol.wt. 257.25

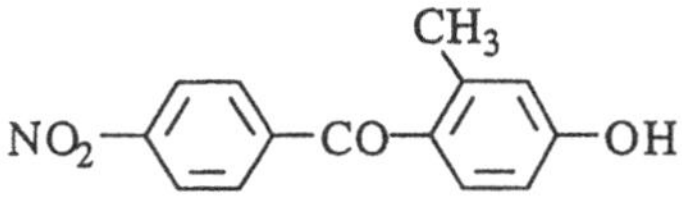

Syntheses

-Preparation by Fries rearrangement of m-tolyl p-nitrobenzoate with aluminium chloride at 120 or at 160° for
2 h (21%) [1167].
-Preparation by reaction of p-nitrobenzoyl chloride with
m-methylanisole in the presence of aluminium chloride in carbon disulfide at 25° for 4 h, followed
by demethylation of the keto ether so formed, that is to say 4-methoxy-2-methyl-4'-nitro-
benzophenone (68%) [1344].

m.p. 194° [1167], 191-192° [1344]; Spectra (NA).

(4-Hydroxy-3-methylphenyl)(3-nitrophenyl)methanone

$C_{14}H_{11}NO_4$ mol.wt. 257.25

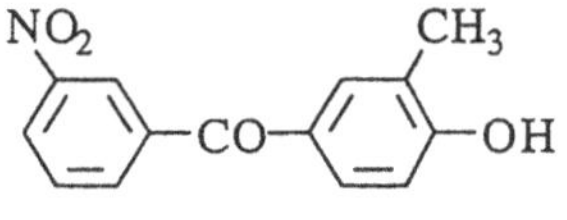

Syntheses

-Preparation by reaction of m-nitrobenzoyl chloride with
o-tolyl borate in the presence of aluminium chloride in
tetrachloroethane at 100° (41%) [1373].
-Also obtained by Fries rearrangement of o-tolyl m-nitrobenzoate with aluminium chloride at 120° or at 160° for 2 h (15%) [1166].

m.p. 182-183° [1166]; Spectra (NA).

(4-Hydroxy-3-methylphenyl)(4-nitrophenyl)methanone

$C_{14}H_{11}NO_4$ mol.wt. 257.25

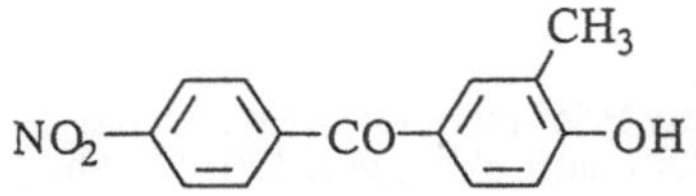

Syntheses

-Preparation by Fries rearrangement of o-tolyl p-nitrobenzoate with aluminium chloride at 120° (25%)
[1167].
-Preparation by reaction of p-nitrobenzoyl chloride with
o-methylanisole in the presence of aluminium chloride in carbon disulfide at 25° for 4 h, followed
by demethylation of the resulting keto ether, that is to say 4-methoxy-3-methyl-4'-nitro-
benzophenone (66%) [1344].

m.p. 215-216° [1344], 215° [1167]; Spectra (NA).

(2-Hydroxy-4-methoxyphenyl)(3-nitrophenyl)methanone

[126077-53-0] $C_{14}H_{11}NO_5$ mol.wt. 273.25

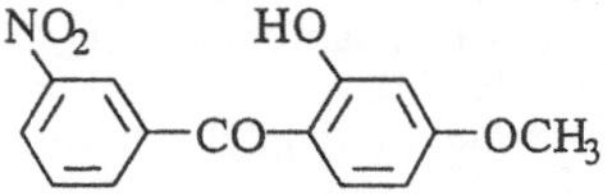

Synthesis

-Refer to: [432].
-Also refer to: [433, 434].

m.p. and Spectra (NA).

(2-Hydroxy-4-methoxyphenyl)(4-nitrophenyl)methanone

[6994-36-1] $C_{14}H_{11}NO_5$ mol.wt. 273.25

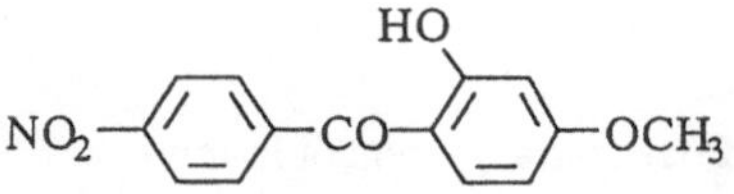

Synthesis

-Preparation by reaction of p-nitrobenzoyl chloride with
m-methoxyphenol in the presence of aluminium
chloride in carbon disulfide at 25° for 4 h (55%)
[1344].

-Also refer to: [1252].

m.p. 149° [1344]; Spectra (NA).

(2-Hydroxy-5-methoxyphenyl)(4-nitrophenyl)methanone

[80427-39-0] $C_{14}H_{11}NO_5$ mol.wt. 273.25

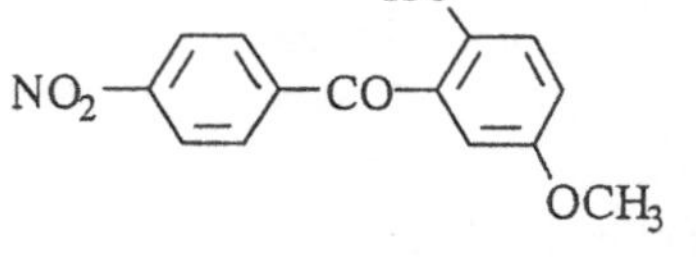

Syntheses

-Preparation by Friedel-Crafts acylation of hydroquinone
dimethyl ether with p-nitrobenzoyl chloride in the
presence of stannic chloride in nitromethane at 20° for
1 h, followed by demethylation of the resulting ketone
(75%) with aluminium chloride in nitromethane at 20° for
24 h (64%) [863].
-Also obtained by Fries rearrangement of p-methoxyphenyl p-nitrobenzoate with titanium
tetrachloride without solvent at 120° for 1 h (20 to 35%) [863].

m.p. 127° [863];
^{1}H NMR (Sadtler: standard n° 35276 M) [863],
IR (Sadtler: standard n° 62644 K) [863], UV [863].

(4-Hydroxy-3-methoxyphenyl)(2-nitrophenyl)methanone

[190522-97-5] $C_{14}H_{11}NO_5$ mol.wt. 273.25

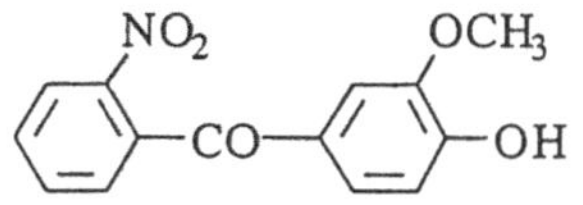

Synthesis

-Preparation by adding a solution of 30% hydrobromic acid to a solution of 4-(benzyloxy)-3-methoxy-2'-nitro-benzophenone in acetic acid, within 10 min at 0° and stirring for 1 h at 20° (98%). -Refer to: Chem. Abstr., **127**, 17465u (1997)*.

m.p. (NA); ^1H NMR*, MS*.

(5-Hydroxy-2-methoxyphenyl)(4-nitrophenyl)methanone

[80427-35-6] $C_{14}H_{11}NO_5$ mol.wt. 273.25

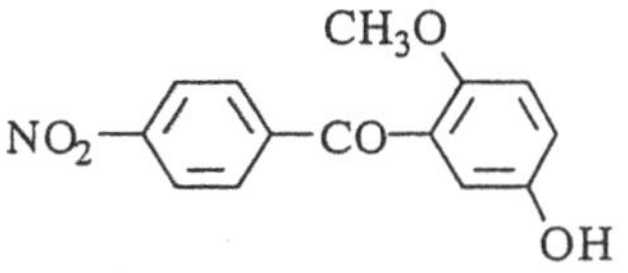

Syntheses

-Preparation by reaction of p-nitrobenzoyl chloride with hydroquinone monomethyl ether in the presence of aluminium chloride in carbon disulfide at 25° for 4 h (50%) [1344].
-Preparation by reaction of p-nitrobenzoyl chloride with p-methoxyphenyl p-nitrobenzoate in the presence of stannic chloride in nitromethane at 20° for 2 days (49%). The m-keto ester formed, the 4-methoxy-3-(4-nitrobenzoyl)phenyl 4-nitrobenzoate, (49%), gave the expected ketone by saponification with sodium hydroxide in refluxing methanol for 1 h (quantitative yield) [863].

m.p.   129° [863], 117° [1344];
^1H NMR (Sadtler: standard n° 35277 M) [863],
IR (Sadtler: standard n° 62645 K) [863], UV [863].

(3-Amino-5-chloro-2-hydroxyphenyl)(4-methylphenyl)methanone

$C_{14}H_{12}ClNO_2$ mol.wt. 261.71

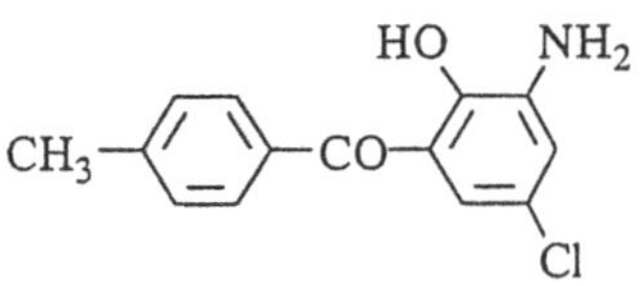

Synthesis

-Refer to: [932].

m.p. and Spectra (NA).

(2-Amino-5-chlorophenyl)(2-hydroxy-5-methylphenyl)methanone

$C_{14}H_{12}ClNO_2$ mol.wt. 261.71

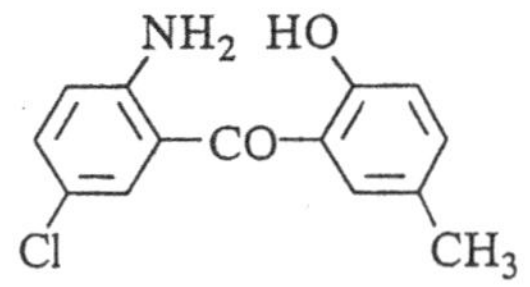

Synthesis

-Obtained from 5-chloro-3-(2-hydroxy-5-methylphenyl)-anthranil (other name: 5-chloro-3-(2-hydroxy-5-methyl-phenyl)-2,1-benzisoxazole) (SM) by reaction with concentrated hydrochloric acid and an excess of tin in boiling acetic acid. SM (m.p. 210°) was prepared by

condensation of o-nitrobenzaldehyde with p-cresol in the presence of hydrogen chloride and phosphorous oxychloride in acetic acid [1474].
N.B.: Na salt [1474].

m.p.   115° [1474];   Spectra (NA).

(3-Amino-2-hydroxy-5-methylphenyl)(4-chlorophenyl)methanone

$C_{14}H_{12}ClNO_2$ mol.wt. 261.71

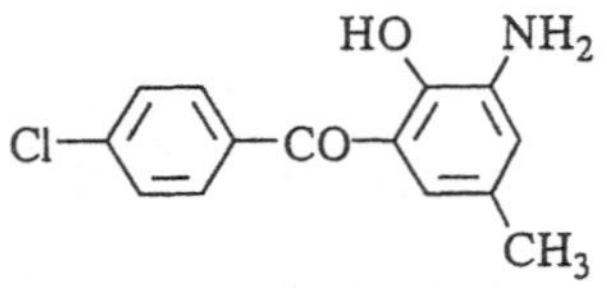

Synthesis

-Refer to: [932].

m.p. and Spectra (NA).

(4-Chlorophenyl)[3-hydroxy-4-(methylamino)phenyl]methanone

[123172-46-3] $C_{14}H_{12}ClNO_2$ mol.wt. 261.71

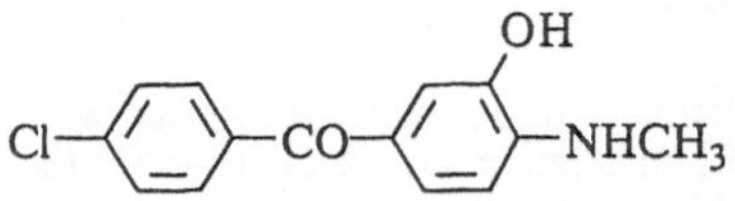

Synthesis

-Preparation by alkaline degradation of 6-(4-chloro-benzoyl)-3-methylbenzoxazolinone (85%) [952].

m.p.   144° [952];   Spectra (NA).

(3-Amino-5-chloro-2-hydroxyphenyl)(4-methoxyphenyl)methanone

[85052-70-6] $C_{14}H_{12}ClNO_3$ mol.wt. 277.71

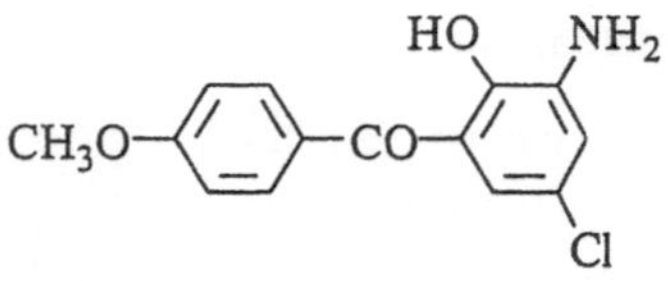

Synthesis

-Preparation by reduction of 5-chloro-2-hydroxy-4'-methoxy-3-nitrobenzophenone with sodium hydrosulfite in a aqueous ammonia/tetrahydrofuran mixture for 15 min (71%) [932].

m.p.   125-135° [932];
^{1}H NMR [932], IR [932], MS [932];   TLC [932].

(3-Amino-5-chloro-2-hydroxyphenyl)(4-methoxyphenyl)methanone *(Hydrochloride)*

[85052-38-6] $C_{14}H_{12}ClNO_3,HCl$ mol.wt. 314.18

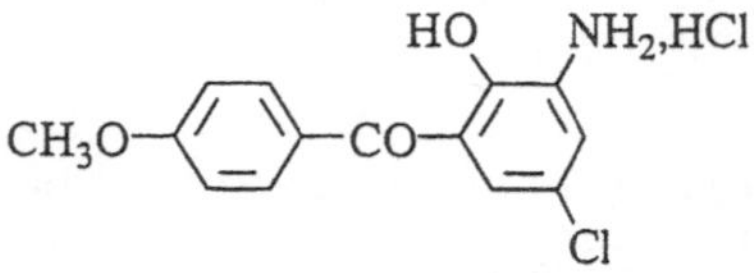

Synthesis

-Obtained by treatment of 3-amino-5-chloro-2-hydroxy-4'-methoxybenzophenone with concentrated hydrochloric acid [932].

m.p. and Spectra (NA).

(3-Amino-5-fluoro-2-hydroxyphenyl)(4-methylphenyl)methanone

$C_{14}H_{12}FNO_2$ mol.wt. 245.25

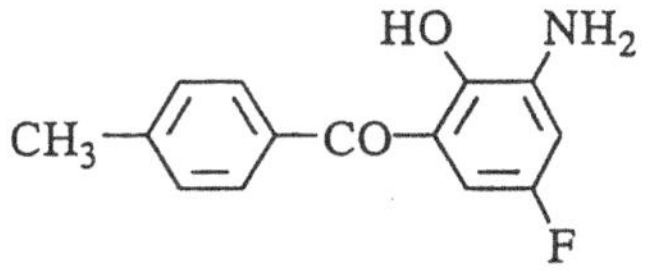

Synthesis

-Refer to: [932].

m.p. and Spectra (NA).

(3-Amino-2-hydroxy-5-methylphenyl)(4-fluorophenyl)methanone

$C_{14}H_{12}FNO_2$ mol.wt. 245.25

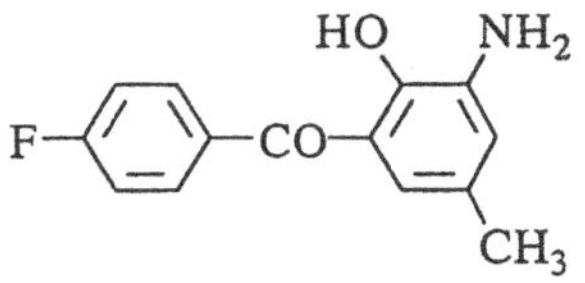

Synthesis

-Refer to: [932].

m.p. and Spectra (NA).

(3-Amino-5-fluoro-2-hydroxyphenyl)(4-methoxyphenyl)methanone

$C_{14}H_{12}FNO_3$ mol.wt. 261.25

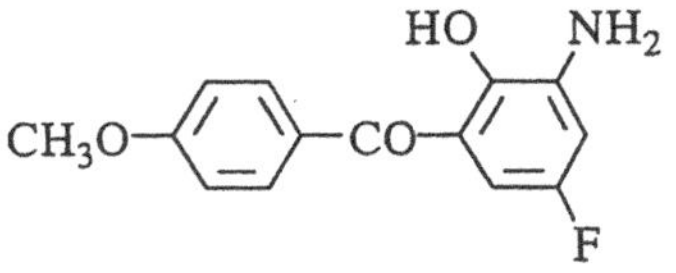

Synthesis

-Refer to: [932].

m.p. and Spectra (NA).

(2-Amino-4-methoxyphenyl)(2-hydroxy-4-nitrophenyl)methanone

$C_{14}H_{12}N_2O_5$ mol.wt. 288.26

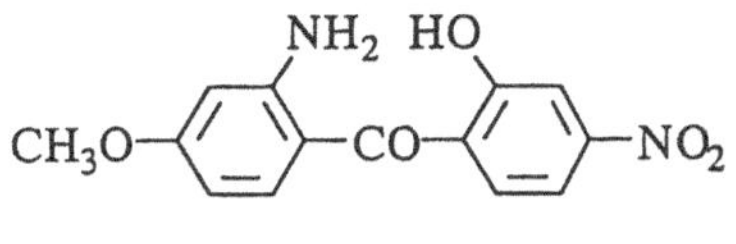

Synthesis

-Preparation by treatment of 2-acetamido-2'-hydroxy-4-methoxy-4'-nitrobenzophenone with refluxing 20% hydrochloric acid for 3 h (91%) [662].

m.p. 224-226° [662]; Spectra (NA).

(4-Amino-2-methoxyphenyl)(2-hydroxy-4-nitrophenyl)methanone

$C_{14}H_{12}N_2O_5$ mol.wt. 288.26

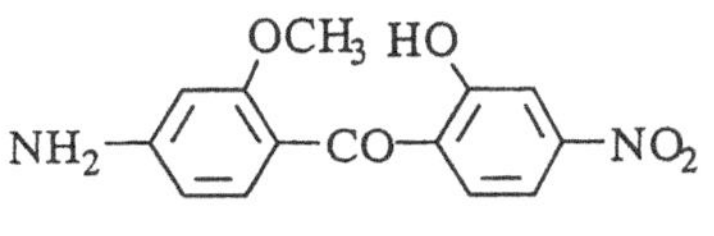

Synthesis

-Preparation by treatment of 4-acetamido-2'-hydroxy-2-methoxy-4'-nitrobenzophenone with refluxing 20% hydrochloric acid for 3 h (78%) [662].

m.p. 176-178° [662]; Spectra (NA).

(2-Hydroxy-5-methylphenyl)(2-mercaptophenyl)methanone

[127024-47-9] $C_{14}H_{12}O_2S$ mol.wt. 244.31

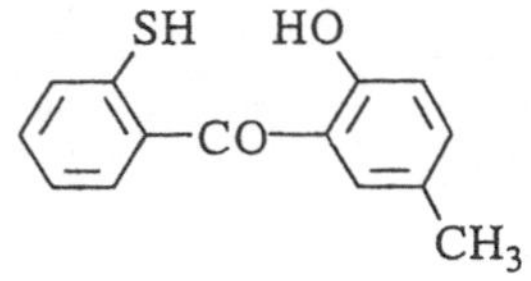

Synthesis

-Obtained by reduction of 2-(2-hydroxy-5-methylbenzoyl)-
phenyl disulfide (SM) in the presence of zinc powder in
refluxing acetic acid for 2 h (40%) [154]. SM was obtained
(poor yields) by UV light irradiation of p-cresyl
2-mercaptobenzoate or of p-cresyl 2-(acetylthio)benzoate in
benzene for 1 h (14% and 7% yields, respectively).

    m.p.   164-167° [154];
    ^{1}H NMR [154], IR [154], MS [154].

(2-Hydroxy-5-methoxyphenyl)(2-mercaptophenyl)methanone

[127024-46-8] $C_{14}H_{12}O_3S$ mol.wt. 260.31

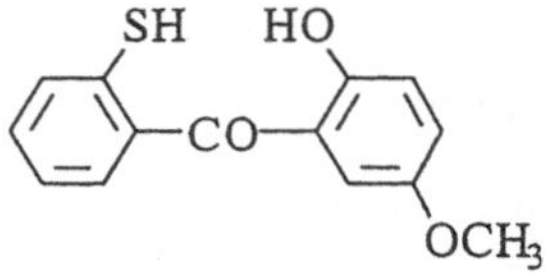

Synthesis

-Obtained by treatment of 2'-(ethoxycarbonylthio)-
2-hydroxy-5-methoxybenzophenone (SM) in methanol with
potassium carbonate at r.t. for 70 min (33%) [154]. SM was
prepared by UV light irradiation of p-methoxyphenyl
2-(ethoxycarbonylthio)benzoate in benzene for 2 h (39%).

    m.p. (NA);
    ^{1}H NMR [154], IR [154], UV [154], MS [154].

(5-Amino-2-hydroxyphenyl)(4-methylphenyl)methanone

$C_{14}H_{13}NO_2$ mol.wt. 227.26

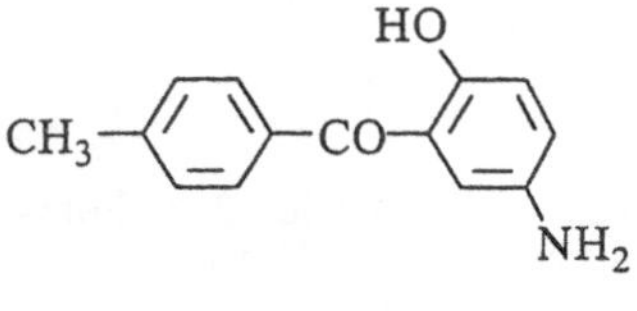

Synthesis

-Obtained by electrolytic reduction of 4-methyl-3'-nitro-
benzophenone in concentrated sulfuric acid [461].

m.p. 93° [461];   Spectra (NA).

(5-Amino-2-hydroxyphenyl)(4-methylphenyl)methanone *(Hydrochloride)*

$C_{14}H_{13}NO_2,HCl$ mol.wt. 263.72

Synthesis

-Preparation from the corresponding amino ketone (see
above) [461].

m.p. and Spectra (NA).

(2-Aminophenyl)(2-hydroxy-5-methylphenyl)methanone

[131946-77-5] $C_{14}H_{13}NO_2$ mol.wt. 227.26

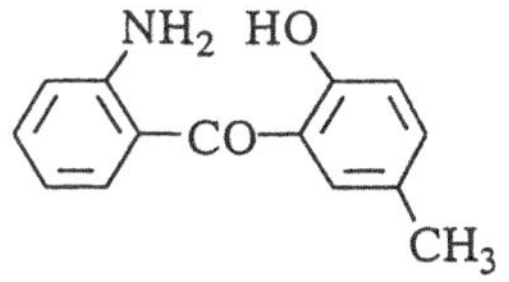

Synthesis

-Obtained (poor yield) by photo-Fries rearrangement of
p-cresyl anthranilate (p-cresyl 2-aminobenzoate) in benzene
for 10 h (4%) [153].

m.p. and Spectra (NA).

(2-Aminophenyl)(2-hydroxy-5-methylphenyl)methanone *(Hydrochloride)*

[55270-74-1] $C_{14}H_{13}NO_2,HCl$ mol.wt. 263.72

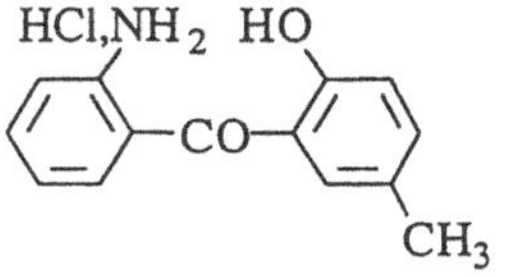

Synthesis

-Obtained by adding 2'-bromo-2-hydroxy-5-methyl-
benzophenone or 2'-chloro-2-hydroxy-5-methyl-
benzophenone to a solution of potassium amide in liquid
ammonia, isolation of the amino compound, then treatment
with 2 N hydrochloric acid (19% and 12% yields,
respectively) [472].

m.p.   175-179° [472];   Spectra (NA).

(3-Aminophenyl)(2-hydroxy-5-methylphenyl)methanone

[35486-63-6] $C_{14}H_{13}NO_2$ mol.wt. 227.26

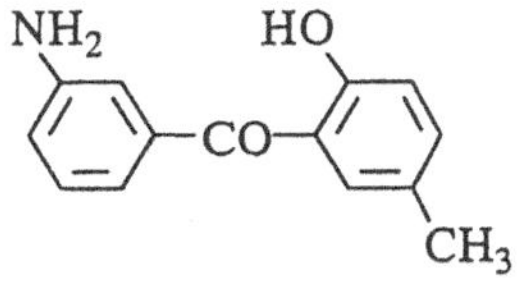

Synthesis

-Preparation by reduction of 2-hydroxy-5-methyl-3'-nitro-
benzophenone with ammonium ferrous sulfate (84%) [205].
-Also refer to: [204].

m.p.   115-120° [205];   IR [205], UV [205].

(3-Aminophenyl)(2-hydroxy-5-methylphenyl)methanone *(Hydrochloride)*

[55270-78-5] $C_{14}H_{13}NO_2,HCl$ mol.wt. 263.72

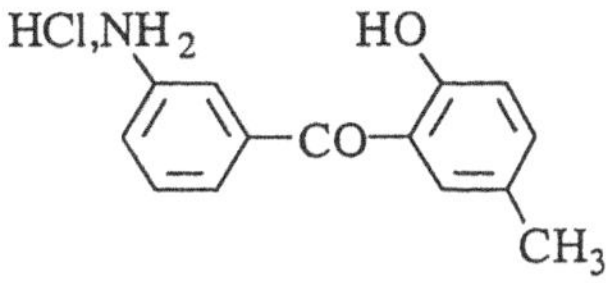

Syntheses

-Obtained by adding 3'-chloro-2-hydroxy-5-methyl-
benzophenone to a solution of potassium amide in liquid
ammonia, isolation of the amino compound, then treatment
with 2 N hydrochloric acid (20%) [472].
-Also obtained by adding aqueous ammonia to a mixture of
2-hydroxy-5-methyl-3'-nitrobenzophenone and ferrous sulfate in aqueous ethanol at 80-85°,
isolation of the amino compound, then treatment with 2 N hydrochloric acid (33%) [472].

m.p.   165-167° [472];   Spectra (NA).

(4-Aminophenyl)(2-hydroxy-5-methylphenyl)methanone

[106612-60-6] $C_{14}H_{13}NO_2$ mol.wt. 227.26

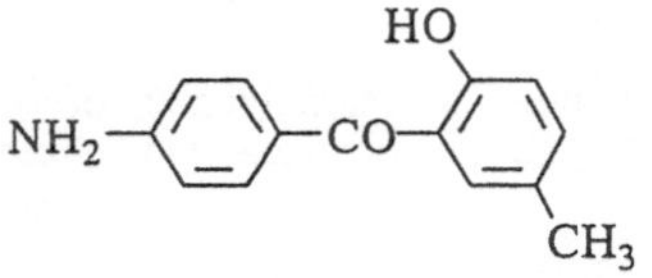

Syntheses

-Preparation by hydrogenation of 2-hydroxy-5-methyl-
  4'-nitrobenzophenone in the presence of 10% Pd/C as a
  catalyst in ethanol under pressure (0.14 MPa) for 1 h
  [291].
-Also obtained by reduction of 2-methoxy-5-methyl-
4'-nitrobenzophenone (m.p. 101-102°) [99],
*with stannous chloride/hydrochloric acid (poor yield);
*with ammonium sulfide in refluxing ethanol (good yield).

   m.p.   137-138° [291], 137° [99];   Spectra (NA).

(4-Aminophenyl)(2-hydroxy-5-methylphenyl)methanone *(Hydrochloride)*

                                               $C_{14}H_{13}NO_2,HCl$ mol.wt. 263.72

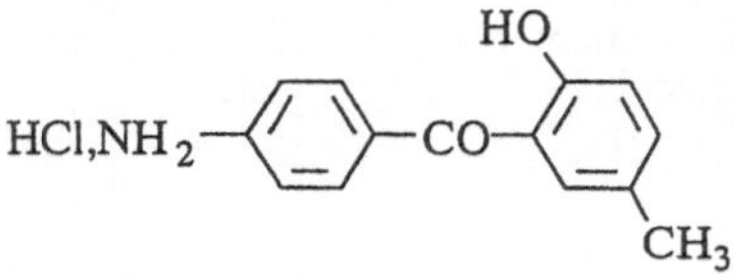

Synthesis

-Preparation by action of hydrochloric acid with the
  corresponding amino ketone (see above) in ethyl ether
  [99].

      m.p. and Spectra (NA).

(2-Aminophenyl)(2-hydroxy-5-methoxyphenyl)methanone

[131946-76-4] $C_{14}H_{13}NO_3$ mol.wt. 243.26

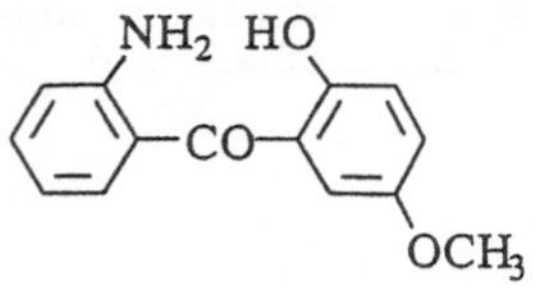

Syntheses

-Preparation by saponification of 2'-acetylamino-2-hydroxy-
  5-methoxybenzophenone with 10% sodium hydroxide
  (25%) [153].
-Also obtained (poor yield) by photo-Fries rearrangement of
  p-methoxyphenyl o-aminobenzoate in benzene for 5 h (7%)
  [153].

   m.p. (NA);   ^{1}H NMR [153], IR [153], UV [153], MS [153].

(4-Aminophenyl)(2-hydroxy-4-methoxyphenyl)methanone

[6994-37-2] $C_{14}H_{13}NO_3$ mol.wt. 243.26

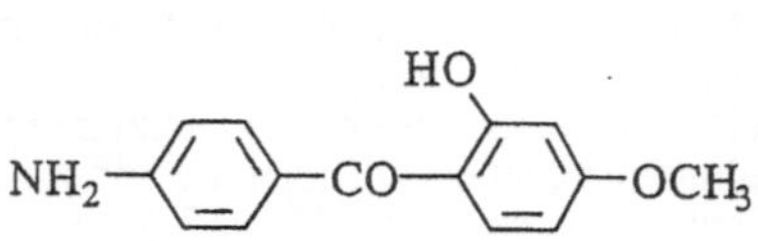

Synthesis

-Refer to: [615].

      m.p. and Spectra (NA).

(2,4-Diaminophenyl)(2-hydroxy-6-methylphenyl)methanone

$C_{14}H_{14}N_2O_2$ mol.wt. 242.28

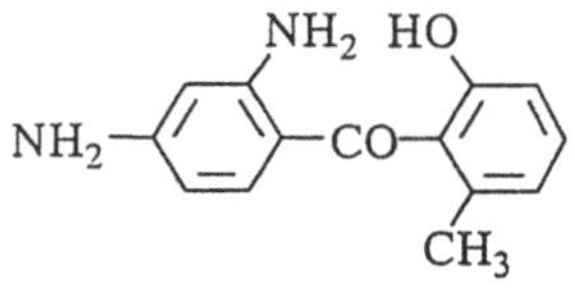

Synthesis

-Preparation by reaction of 2,4-diaminobenzoyl chloride with trimethylsilyl derivative of m-cresol in the presence of stannic chloride (or titanium tetrachloride or aluminium chloride) in refluxing methylene chloride for 2 h [649].

m.p. and Spectra (NA).

(3-Chloro-2-hydroxy-4-methoxyphenyl)[2-(trifluoromethyl)phenyl]methanone

[72482-16-7] $C_{15}H_{10}ClF_3O_3$ mol.wt. 330.69

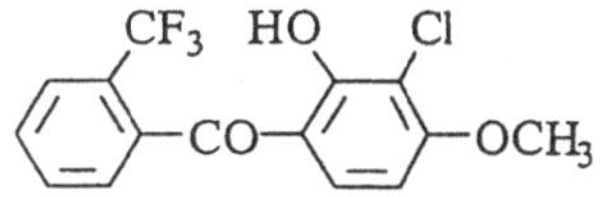

Synthesis

-Preparation by reaction of o-(trifluoromethyl)benzoyl chloride with 2-chlororesorcinol dimethyl ether in ethylene dichloride in the presence of ferric chloride first at 5-7°, then at r.t. for 18 h and at reflux for 30 min [1270].

m.p. 101-102° [1270];   Spectra (NA).

(3,5-Dichloro-2-hydroxy-4,6-dimethylphenyl)(2,4-dichlorophenyl)methanone

[34174-15-7] $C_{15}H_{10}Cl_4O_2$ mol.wt. 364.05

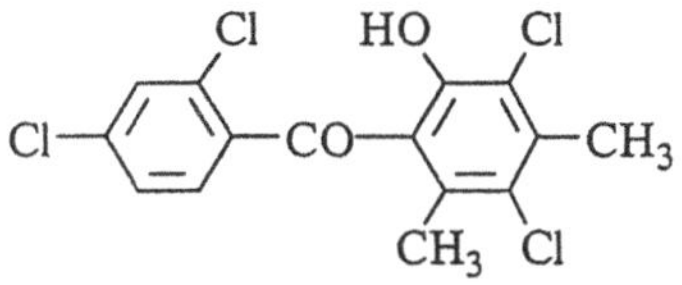

Synthesis

-Preparation by Fries rearrangement of 2,4-dichloro-3,5-dimethylphenyl 2,4-dichlorobenzoate with aluminium chloride for 30 min at 150-160° [375].

m.p. 129-130° [375];   Spectra (NA).

(3,5-Dichloro-2-hydroxy-4,6-dimethylphenyl)(3,4-dichlorophenyl)methanone

[34174-14-6] $C_{15}H_{10}Cl_4O_2$ mol.wt. 364.05

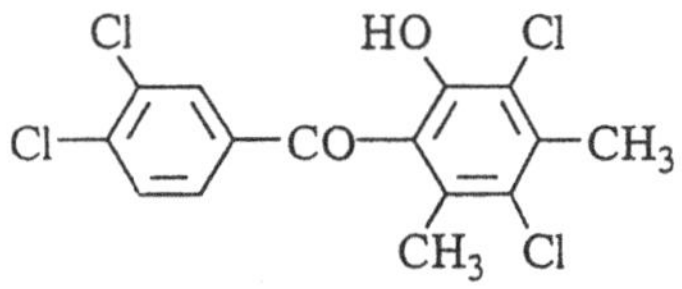

Synthesis

-Preparation by Fries rearrangement of 2,4-dichloro-3,5-dimethylphenyl 3,4-dichlorobenzoate with aluminium chloride for 30 min at 150-160° [375].

m.p. 136-137° [375];   Spectra (NA).

(4-Hydroxy-3-methoxy-5-nitrophenyl)[2-(trifluoromethyl)phenyl]methanone

[134612-82-1] $C_{15}H_{10}F_3NO_5$ mol.wt. 341.24

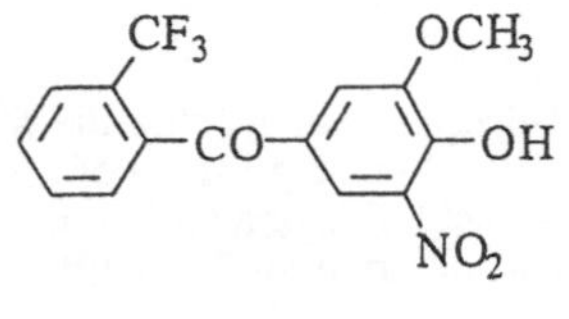

Synthesis

-Preparation by nitration of 4-hydroxy-3-methoxy-2'-(trifluoromethyl)benzophenone with 65% nitric acid in acetic
acid at 20° [164].

m.p. 138-140° [164];   Spectra (NA).

(4-Hydroxy-3-methoxy-5-nitrophenyl)[4-(trifluoromethyl)phenyl]methanone

[134611-75-9] $C_{15}H_{10}F_3NO_5$ mol.wt. 341.24

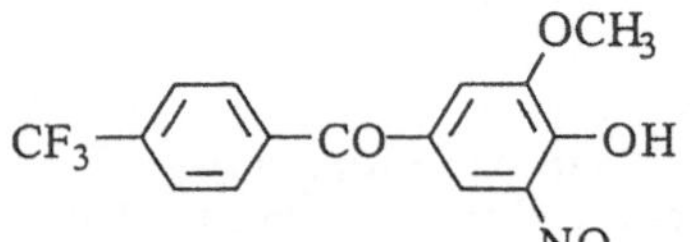

Synthesis

-Preparation by nitration of 4-hydroxy-3-methoxy-
4'-(trifluoromethyl)benzophenone with 65% nitric acid in
acetic acid at 20° for 90 min [164].

m.p. 172° [164];   Spectra (NA).

(3-Chloro-6-hydroxy-2,4-dimethylphenyl)(2,4-dichlorophenyl)methanone

[34171-56-7] $C_{15}H_{11}Cl_3O_2$ mol.wt. 329.61

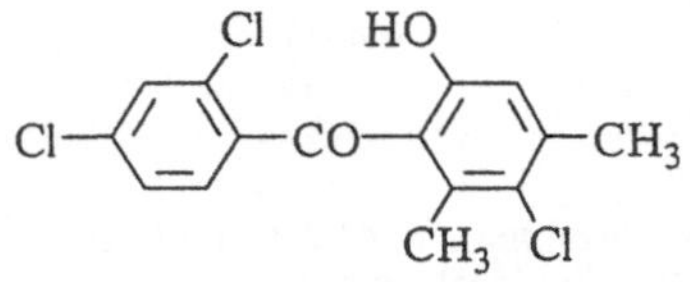

Synthesis

-Preparation by Fries rearrangement of 4-chloro-3,5-di-
methylphenyl 2,4-dichlorobenzoate with aluminium
chloride for 30 min at 150-160° [375].

m.p. 103-104° [375];   Spectra (NA).

(2-Chlorophenyl)(3,5-dichloro-2-hydroxy-4,6-dimethylphenyl)methanone

[34174-16-8] $C_{15}H_{11}Cl_3O_2$ mol.wt. 329.61

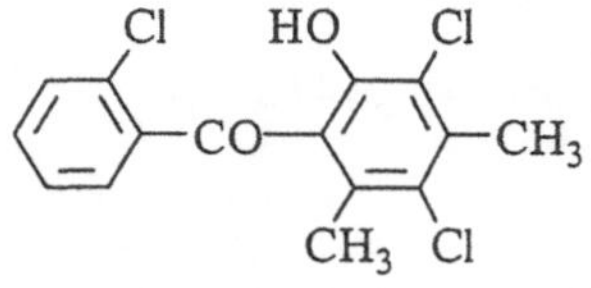

Synthesis

-Preparation by Fries rearrangement of 2,4-dichloro-3,5-di-
methylphenyl o-chlorobenzoate with aluminium chloride for
30 min at 150-160° [375].

m.p. 96-97° [375];   Spectra (NA).

(2,4-Dimethylphenyl)(2,3,5-trichloro-6-hydroxyphenyl)methanone

[34174-01-1] $C_{15}H_{11}Cl_3O_2$ mol.wt. 329.61

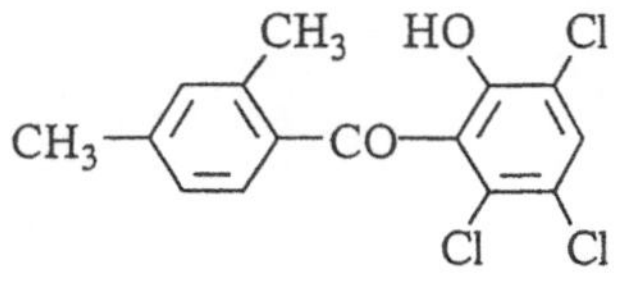

Synthesis

-Preparation by Fries rearrangement of 2,4,5-trichlorophenyl
2,4-dimethylbenzoate with aluminium chloride for 30 min at
150-160° [375].

m.p. 103-104° [375];   Spectra (NA).

(3,4-Dimethylphenyl)(2,3,5-trichloro-6-hydroxyphenyl)methanone

[34174-00-0] $C_{15}H_{11}Cl_3O_2$ mol.wt. 329.61

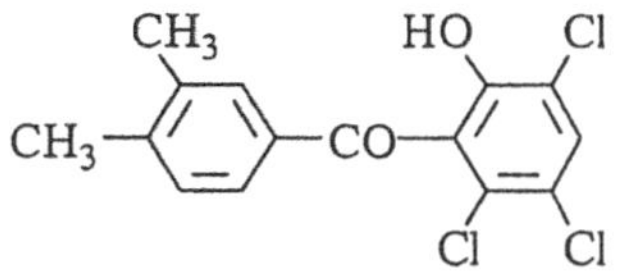

Synthesis

-Preparation by Fries rearrangement of 2,4,5-trichloro-
phenyl 3,4-dimethylbenzoate with aluminium chloride for
30 min at 150-160° [375].

m.p. 163-164° [375];   Spectra (NA).

(2-Hydroxy-5-methylphenyl)[3-(trifluoromethyl)phenyl]methanone

                                       $C_{15}H_{11}F_3O_2$ mol.wt. 280.25

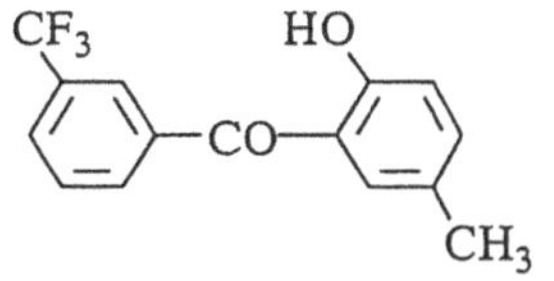

Synthesis

-Preparation by reaction of m-(trifluoromethyl)benzoyl
chloride with p-cresol in the presence of aluminium chloride
for 8 h at 190° [348].

m.p. and Spectra (NA).

(2-Hydroxy-4-methoxyphenyl)[2-(trifluoromethyl)phenyl]methanone

[3119-86-6] $C_{15}H_{11}F_3O_3$ mol.wt. 296.21

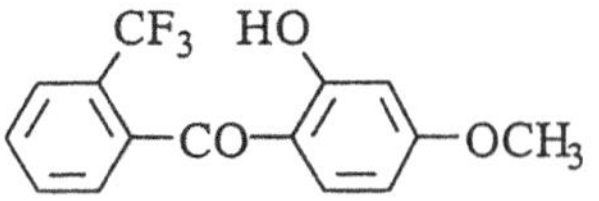

Synthesis

-Preparation by reaction of o-(trifluoromethyl)benzoyl
chloride with resorcinol dimethyl ether in the presence of
aluminium chloride or titanium tetrachloride in refluxing
n-hexane for 8 h [501].
-Also refer to: [500].

m.p. 95-95°5 [501];   b.p. 358-362° [501];
Spectra (NA);   vapour pressure [500].

(2-Hydroxy-4-methoxyphenyl)[3-(trifluoromethyl)phenyl]methanone

[7396-89-6] $C_{15}H_{11}F_3O_3$ mol.wt. 296.21

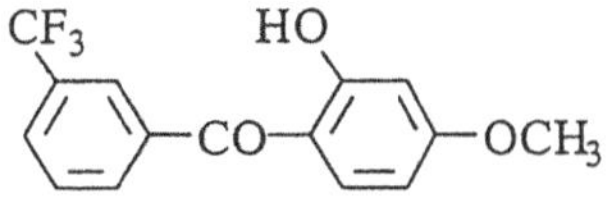

Syntheses

-Preparation by reaction of m-(trifluoromethyl)benzoyl
chloride with resorcinol dimethyl ether in the presence of
aluminium chloride or titanium tetrachloride in refluxing
n-hexane for 8 h [501].
-Preparation by reaction of m-(trifluoromethyl)benzonitrile with m-methoxyphenol in the presence
of aluminium chloride for 8 h at 190°, followed by hydrolysis of the ketimine so formed [348].

m.p. 65°5-66° [501];   b.p. 360-362° [501];
Spectra (NA);   vapour pressure [500].

(2-Hydroxy-4-methoxyphenyl)[4-(trifluoromethyl)phenyl]methanone

[7396-90-9] $C_{15}H_{11}F_3O_3$ mol.wt. 296.21

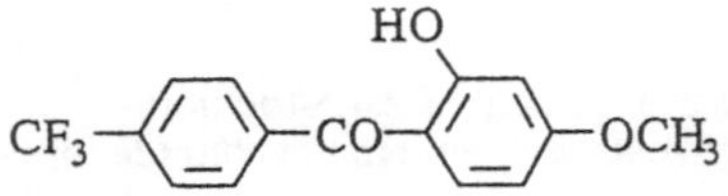

Synthesis

-Preparation by reaction of p-(trifluoromethyl)benzoyl
chloride with resorcinol dimethyl ether in the presence
of aluminium chloride or titanium tetrachloride in
refluxing n-hexane for 8 h [501].
-Also refer to: [500].

m.p. 66°5-67° [501];   b.p. 380-385° [501];
Spectra (NA);   vapour pressure [500].

(4-Hydroxy-3-methoxyphenyl)[2-(trifluoromethyl)phenyl]methanone

[134612-41-2] $C_{15}H_{11}F_3O_3$ mol.wt. 296.21

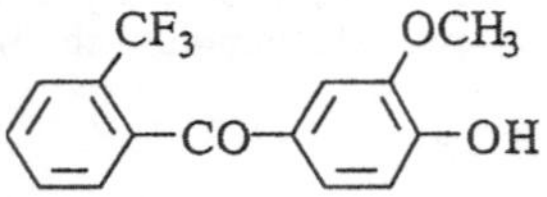

Synthesis

-Preparation by reaction of 33% hydrobromic acid in acetic
acid with 4-(benzyloxy)-3-methoxy-2'-(trifluoromethyl)-
benzophenone in methylene chloride at 20-25° [164].

m.p. 115-117° [164];   Spectra (NA).

(4-Hydroxy-3-methoxyphenyl)[4-(trifluoromethyl)phenyl]methanone

[134611-74-8] $C_{15}H_{11}F_3O_3$ mol.wt. 296.21

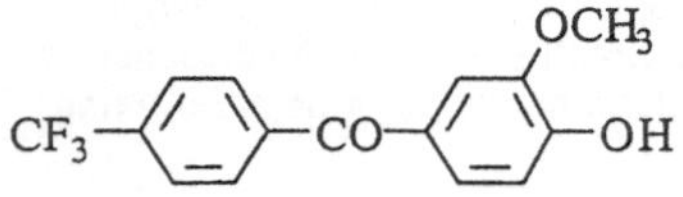

Synthesis

-Preparation by reaction of 33% hydrobromic acid in acetic
acid with 4-(benzyloxy)-3-methoxy-4'-(trifluoromethyl)-
benzophenone in methylene chloride at 20° for 90 min
[164].

m.p. 97° [164];   Spectra (NA).

[4-(2-Bromoethoxy)phenyl](4-hydroxy-3-iodophenyl)methanone

[79578-67-9] $C_{15}H_{12}BrIO_3$ mol.wt. 447.07

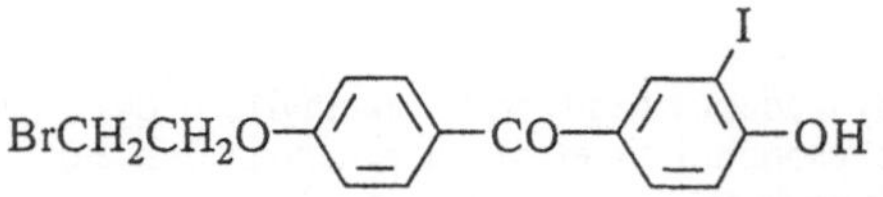

Synthesis

-Obtained (poor yield) by reaction of 3-iodo-
4-hydroxybenzoic acid with β-bromophenetole
in solution of a polyphosphoric acid/85%
phosphoric acid/zinc chloride mixture. The
solution was heated at 50°, phosphorous trichloride was added during 1 h and the mixture was
heated for 2.5 h at 70° (11%) [715].

m.p. 175-176°5 [715];
^{1}H NMR [715], IR [715], MS [715].

(3,5-Dibromo-2-hydroxy-4-methoxyphenyl)(4-methoxyphenyl)methanone

[66666-25-9] $C_{15}H_{12}Br_2O_4$ mol.wt. 416.07

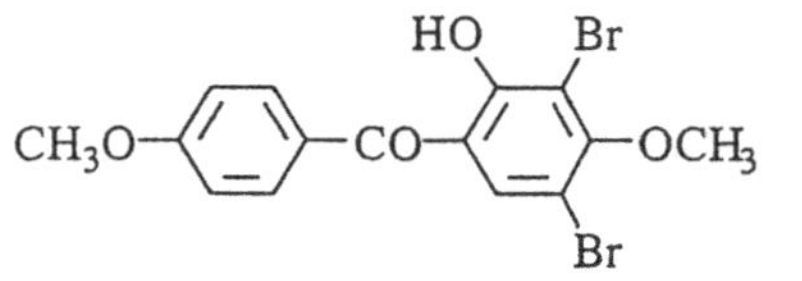

Synthesis

-Preparation by saponification of 2-(p-anisoyloxy)-3,5-dibromo-4,4'-dimethoxybenzophenone (SM) with potassium hydroxide in refluxing ethanol for 1 h (66%). SM was obtained by oxidation of 5,7-dibromo-6-methoxy-2,3-bis(p-methoxyphenyl)-benzofuran with chromium trioxide in refluxing acetic acid for 45 min (70%) [578].

m.p. 200° [578]; Spectra (NA).

(5-Chloro-2-hydroxy-3,4-dimethoxyphenyl)(4-fluorophenyl)methanone

[140665-40-3] $C_{15}H_{12}ClFO_4$ mol.wt. 310.71

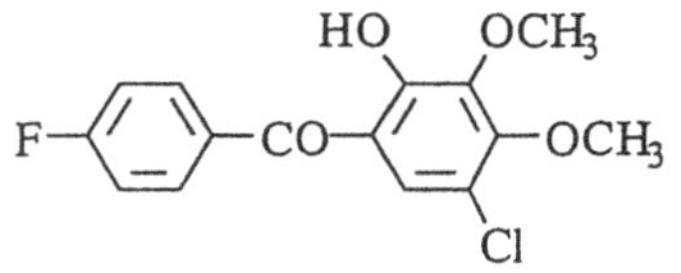

Synthesis

-Preparation by reaction of sulfuryl chloride with 4'-fluoro-2-hydroxy-3,4-dimethoxybenzophenone in methylene chloride at r.t. overnight (69%) [1194].

m.p. 130-131° [1194]; Spectra (NA).

(2-Chloro-4-hydroxyphenyl)(2,3-dimethyl-5-nitrophenyl)methanone

[110969-66-9] $C_{15}H_{12}ClNO_4$ mol.wt. 305.72

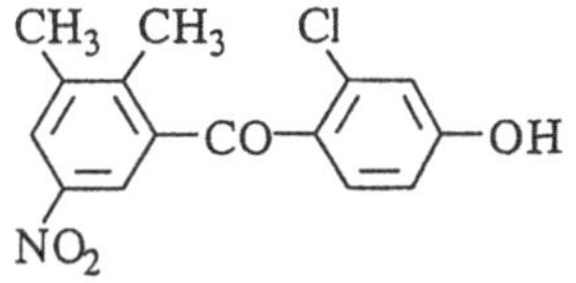

Synthesis

-Obtained (by-product) by Fries rearrangement of m-chlorophenyl 2,3-dimethyl-5-nitrobenzoate with aluminium chloride at 160° for 2 h (11%) [843].

m.p. 214° [843]; ^{1}H NMR [843], IR [843], UV [843].

(3-Chloro-2-hydroxyphenyl)(2,3-dimethyl-5-nitrophenyl)methanone

[110969-62-5] $C_{15}H_{12}ClNO_4$ mol.wt. 305.72

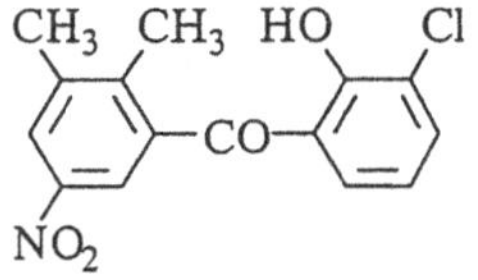

Synthesis

-Obtained (by-product) by Fries rearrangement of o-chlorophenyl 2,3-dimethyl-5-nitrobenzoate with aluminium chloride at 160° for 2 h (6%) [843].

m.p. 130° [843]; ^{1}H NMR [843], IR [843], UV [843].

(3-Chloro-4-hydroxyphenyl)(2,3-dimethyl-5-nitrophenyl)methanone

[110969-63-6] $C_{15}H_{12}ClNO_4$ mol.wt. 305.72

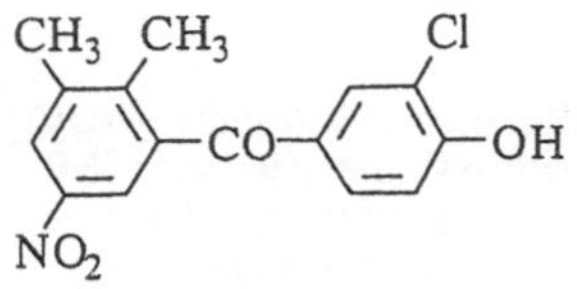

Synthesis

-Preparation by Fries rearrangement of o-chlorophenyl
 2,3-dimethyl-5-nitrobenzoate with aluminium chloride at
 160° for 2 h (60%) [843].

m.p. 195° [843]; ^{1}H NMR [843], IR [843], UV [843].

(4-Chloro-2-hydroxyphenyl)(2,3-dimethyl-5-nitrophenyl)methanone

[110969-65-8] $C_{15}H_{12}ClNO_4$ mol.wt. 305.72

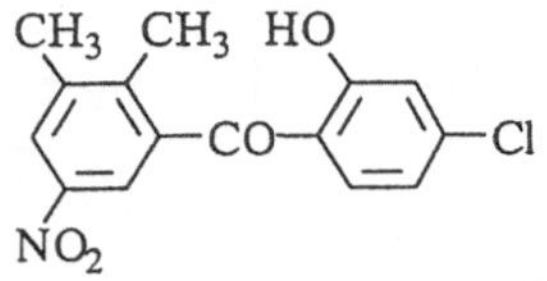

Synthesis

-Preparation by Fries rearrangement of m-chlorophenyl
 2,3-dimethyl-5-nitrobenzoate with aluminium chloride at
 160° for 2 h (40%) [843].

m.p. 154° [843]; ^{1}H NMR [843], IR [843], UV[843].

(5-Chloro-2-hydroxyphenyl)(2,3-dimethyl-5-nitrophenyl)methanone

[110969-68-1] $C_{15}H_{12}ClNO_4$ mol.wt. 305.72

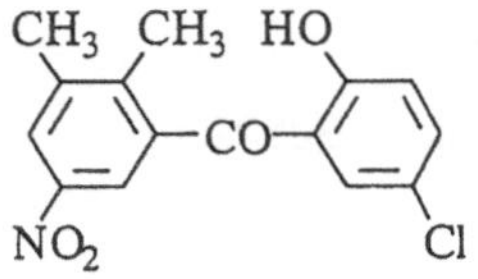

Synthesis

-Preparation by Fries rearrangement of p-chlorophenyl
 2,3-dimethyl-5-nitrobenzoate with aluminium chloride at
 160° for 2 h (70%) [843].

m.p. 147° [843]; ^{1}H NMR [843], IR [843], UV [843].

(5-Chloro-3-ethyl-2-hydroxyphenyl)(4-chlorophenyl)methanone

[93575-71-4] $C_{15}H_{12}Cl_2O_2$ mol.wt. 295.16

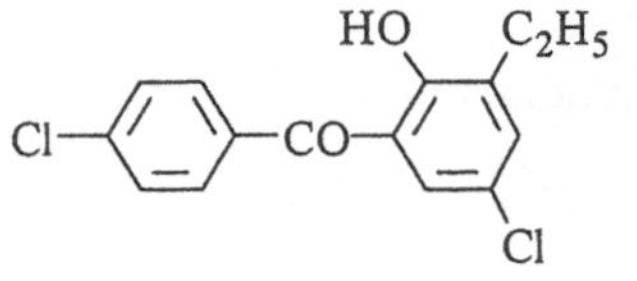

Synthesis

-Preparation by Fries rearrangement of 4-chloro-2-ethyl-
 phenyl p-chlorobenzoate with aluminium chloride at 160°
 for 15 min [698].

m.p. 47-49° [698];   Spectra (NA).

(3-Chloro-6-hydroxy-2,4-dimethylphenyl)(2-chlorophenyl)methanone

[34174-03-3] $C_{15}H_{12}Cl_2O_2$ mol.wt. 295.16

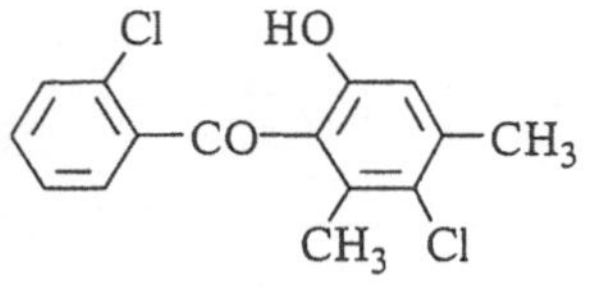

Synthesis

-Preparation by Fries rearrangement of 4-chloro-3,5-di-
 methylphenyl o-chlorobenzoate with aluminium chloride for
 30 min at 150-160° [375].

m.p. 91-92° [375];   Spectra (NA).

(5-Chloro-2-hydroxy-3-methylphenyl)(4-chloro-2-methylphenyl)methanone

[86914-83-2] $C_{15}H_{12}Cl_2O_2$ mol.wt. 295.16

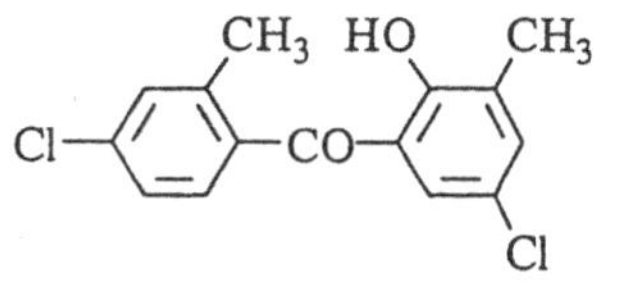

Synthesis

-Preparation by Fries rearrangement of 4-chloro-2-methyl-
phenyl 4-chloro-2-methylbenzoate with aluminium chloride
at 180° for 10 min [688].

m.p. 53-54° [688];   Spectra (NA).

(2,3-Dichloro-4-hydroxyphenyl)(2,3-dimethylphenyl)methanone

[72482-89-4] $C_{15}H_{12}Cl_2O_2$ mol.wt. 295.16

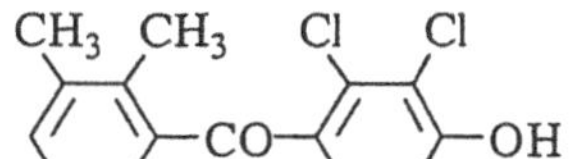

Synthesis

-Preparation by demethylation of 2,3-dichloro-4-methoxy-
2',3'-dimethylbenzophenone (SM) with aluminium chloride
in refluxing benzene for 5 h, then at r.t. for 18 h. SM was
obtained by reaction of 2,3-dimethylbenzoyl chloride with
2,3-dichloroanisole in ethylene dichloride in the presence of aluminium chloride at 60° for 1 h
[1270].

m.p. and Spectra (NA).

(2,4-Dichlorophenyl)(3-ethyl-2-hydroxyphenyl)methanone

[61466-78-2] $C_{15}H_{12}Cl_2O_2$ mol.wt. 295.16

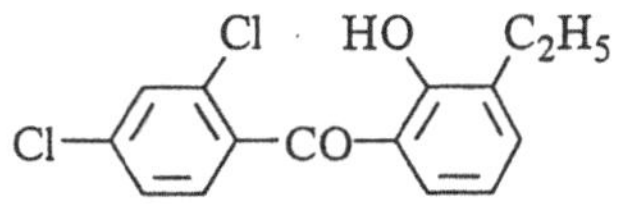

Synthesis

-Obtained by reaction of 2,4-dichlorobenzoyl chloride with
o-ethylphenol in the presence of aluminium chloride in
tetrachloroethane at 105° for 21-22 h (86-91%) [1206,
1207], (6%) [400].

oil [400];   b.p.$_{0.06}$ 156-172° [1206, 1207]. A typing error probably occurred in the
published data.
$n_D^{22} = 1.6163$ [1206, 1207];   Spectra (NA).

(2,4-Dichlorophenyl)(5-ethyl-2-hydroxyphenyl)methanone

[61466-83-9] $C_{15}H_{12}Cl_2O_2$ mol.wt. 295.16

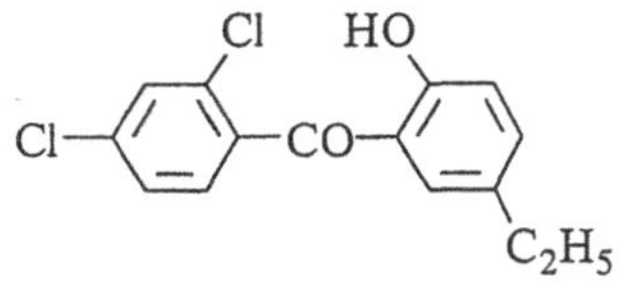

Synthesis

-Preparation by reaction of 2,4-dichlorobenzoyl chloride
with p-ethylphenol in the presence of aluminium chloride in
tetrachloroethane at 105° for 22 h (34%) [400, 1206].

m.p. 44°6 [1206], 44-46° [400];   b.p.$_{0.45}$ 148-151° [1206];
Spectra (NA).

(3,4-Dichlorophenyl)(5-ethyl-2-hydroxyphenyl)methanone

[61466-87-3] $C_{15}H_{12}Cl_2O_2$ mol.wt. 295.16

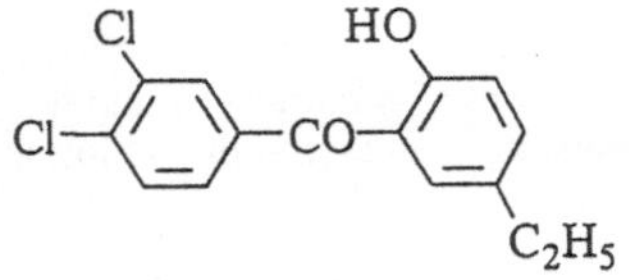

Synthesis

-Preparation by reaction of 3,4-dichlorobenzoyl chloride with
p-ethylphenol in the presence of aluminium chloride in
tetrachloroethane at 105° for 22 h (30%) [400], (17%)
[1206].

m.p.   50° [1206];   b.p.$_{0.35}$ 175-177° [400, 1206];   Spectra (NA).

(2,4-Dichlorophenyl)(2-hydroxy-4,6-dimethylphenyl)methanone

[34203-52-6] $C_{15}H_{12}Cl_2O_2$ mol.wt. 295.16

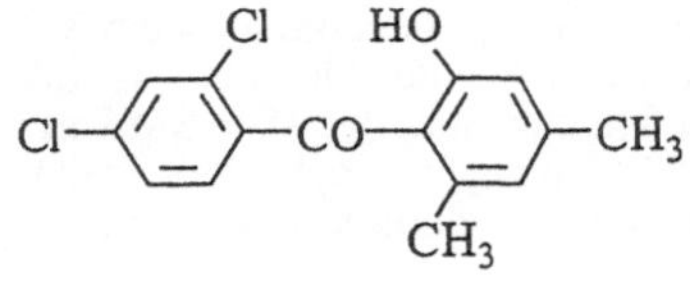

Synthesis

-Preparation by Fries rearrangement of 3,5-dimethyl-
phenyl 2,4-dichlorobenzoate with aluminium chloride for
30 min at 150-160° [375].

m.p.   94-95° [375];   Spectra (NA).

(3,4-Dichlorophenyl)(4-hydroxy-2,6-dimethylphenyl)methanone

[34183-02-3] $C_{15}H_{12}Cl_2O_2$ mol.wt. 295.16

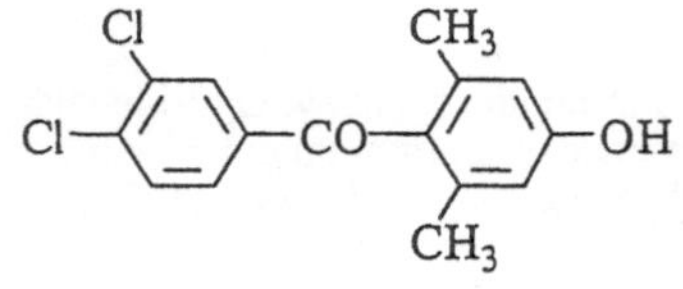

Synthesis

-Obtained by Fries rearrangement of 3,5-dimethylphenyl
3,4-dichlorobenzoate with aluminium chloride in chlorobenzene for 20 min at 140-150° or in nitrobenzene for
24 h at 75° [1214].

m.p.   138-139° [1214];   Spectra (NA).

(2,6-Dichlorophenyl)(2-hydroxy-4-methoxy-6-methylphenyl)methanone

[183726-73-0] $C_{15}H_{12}Cl_2O_3$ mol.wt. 311.16

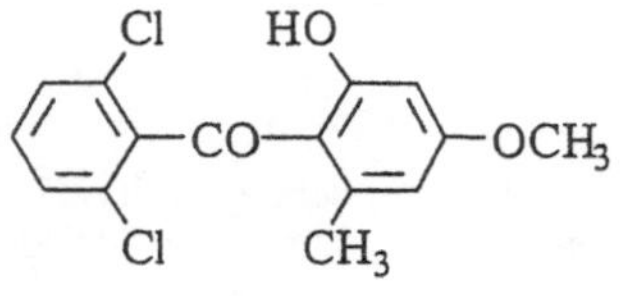

Synthesis

-Refer to: [313].

m.p. and Spectra (NA).

(2,6-Dichlorophenyl)(5-hydroxy-4-methoxy-2-methylphenyl)methanone

[183724-10-9] $C_{15}H_{12}Cl_2O_3$ mol.wt. 311.16

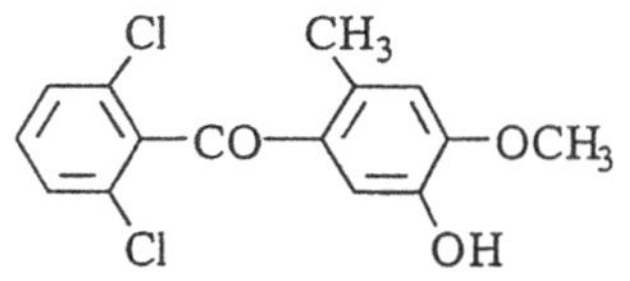

Synthesis

-Preparation by partial demethylation of 2',6'-dichloro-
4,5-dimethoxy-2-methylbenzophenone with 33%
hydrobromic acid in acetic acid for 1.5 h at 75° (45%)
[313].

m.p. 152° [313];   Spectra (NA).

(2,6-Dichlorophenyl)(2-hydroxy-4,5-dimethoxyphenyl)methanone

[183725-20-4] $C_{15}H_{12}Cl_2O_4$ mol.wt. 327.16

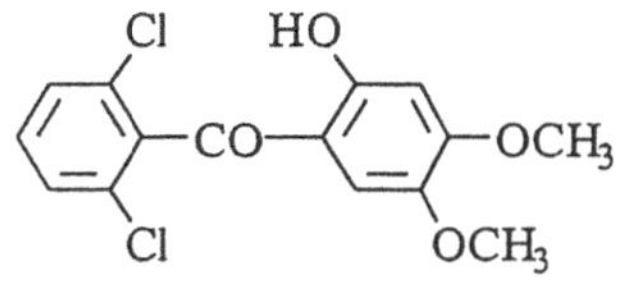

Syntheses

-Preparation by Friedel-Crafts acylation of 3,4-dimethoxy-
phenol with 2,6-dichlorobenzoyl chloride [313].
-Preparation by partial demethylation of 2',6'-dichloro-
2,4,5-trimethoxybenzophenone [313].
-Preparation by diazotization of 2-amino-2',6'-dichloro-
4,5-dimethoxybenzophenone (compound 3) [313].

m.p. 80° [313];   Spectra (NA).

[4-Hydroxy-3-(methoxymethyl)-5-nitrophenyl](2-nitrophenyl)methanone

$C_{15}H_{12}N_2O_7$ mol.wt. 332.27

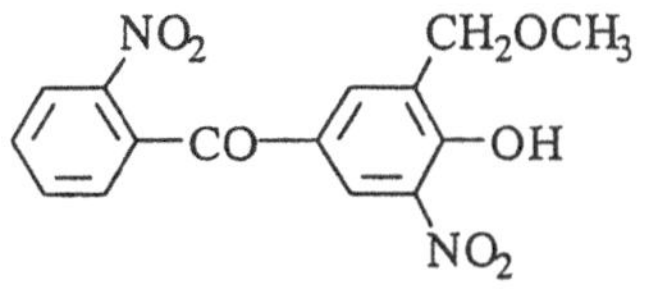

Synthesis

-Obtained by reaction of chloromethyl methyl ether in
methylene chloride with 3,4-dihydroxy-2',5-dinitro-
benzophenone in tetrahydrofuran in the presence of
N,N-diisopropylethylamine (Huenig's base)
for 40 min at 0° (41%). -Refer to: Chem. Abstr., 127,
17465u (1997)*.

m.p. (NA);   ¹H NMR*.

(2,4-Dimethoxyphenyl)(4-hydroxy-3,5-dinitrophenyl)methanone

[67246-02-0] $C_{15}H_{12}N_2O_8$ mol.wt. 348.27

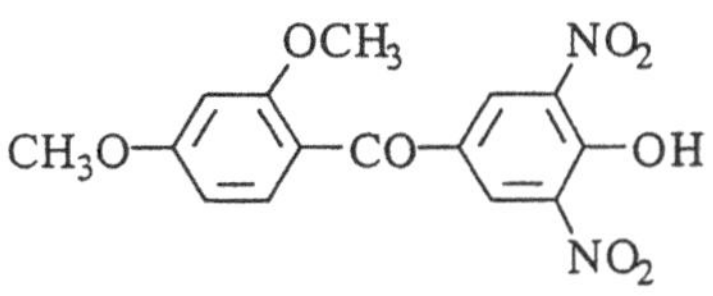

Synthesis

-Obtained by reaction of nitric acid (d = 1.42) with
4-hydroxy-2',4'-dimethoxybenzophenone in acetic acid
at 32° [1083].

m.p. 157° [1083];   ¹H NMR [1083].

(2-Hydroxy-4-methoxy-3,5-dinitrophenyl)(2-methoxyphenyl)methanone

[79204-71-0] $C_{15}H_{12}N_2O_8$ mol.wt. 348.27

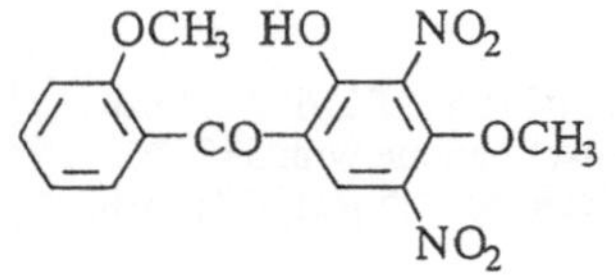

Synthesis

-Preparation by oxidation of 2,3-bis(2-methoxyphenyl)-
5,7-dinitro-6-methoxybenzofuran with chromium trioxide
in acetic acid, followed by saponification of the keto ester
so formed with potassium hydroxide in ethanol (60%)
[579].

m.p. 120° [579];   IR [579].

(2-Hydroxy-4-methoxy-3,5-dinitrophenyl)(4-methoxyphenyl)methanone

[66666-08-8] $C_{15}H_{12}N_2O_8$ mol.wt. 348.27

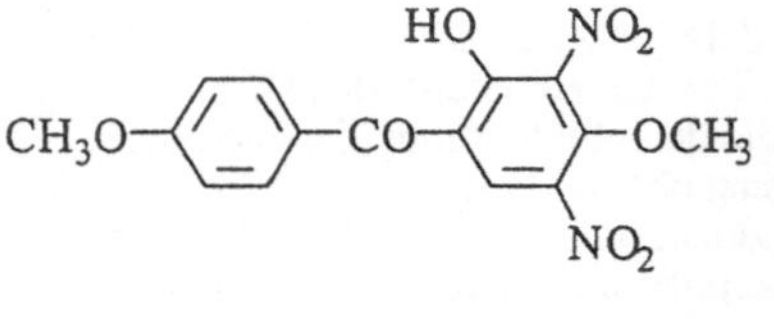

Synthesis

-Preparation by saponification of 2-(p-anisoyloxy)-
4,4'-dimethoxy-3,5-dinitrobenzophenone (SM) with
potassium hydroxide in refluxing ethanol for 1 h
(70%). SM was obtained by oxidation of 5,7-di-
nitro-6-methoxy-2,3-bis(p-methoxyphenyl)-
benzofuran with chromium trioxide in refluxing acetic acid for 45 min (72%) [578].

m.p. 140° [578];   Spectra (NA).

(4-Bromophenyl)(5-ethyl-2-hydroxyphenyl)methanone

[108294-74-2] $C_{15}H_{13}BrO_2$ mol.wt. 305.17

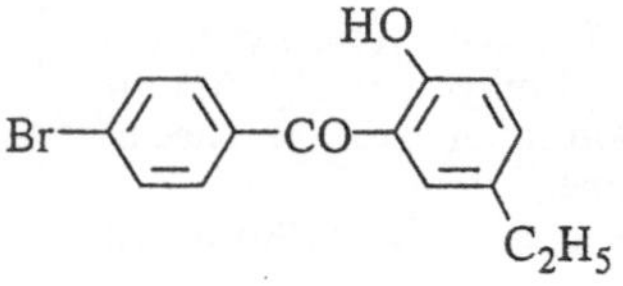

Synthesis

-Preparation by reaction of p-bromobenzoyl chloride with
p-ethylphenol in the presence of aluminium chloride in
tetrachloroethane at 105° for 22 h (45%) [400].

b.p.$_{0.95}$ 186° [400];   Spectra (NA).

(2-Bromophenyl)(2-hydroxy-3,5-dimethylphenyl)methanone

[86914-81-0] $C_{15}H_{13}BrO_2$ mol.wt. 305.17

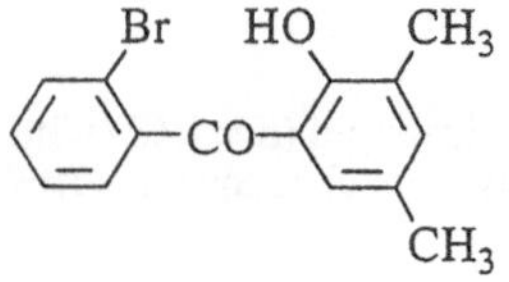

Synthesis

-Preparation by Fries rearrangement of 2,4-dimethylphenyl
o-bromobenzoate with aluminium chloride at 180° for
10 min [688].

m.p. 86-89° [688];   Spectra (NA).

(4-Bromophenyl)(2-hydroxy-4,6-dimethylphenyl)methanone

[34174-04-4] $C_{15}H_{13}BrO_2$ mol.wt. 305.17

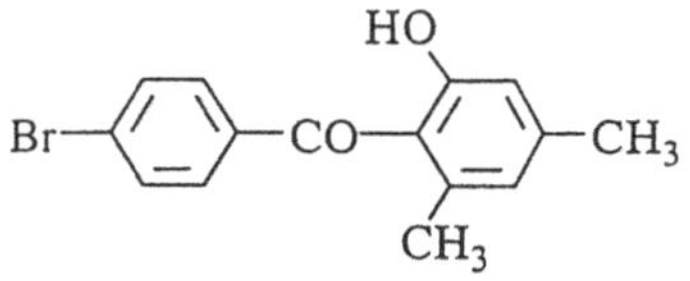

Synthesis

-Preparation by Fries rearrangement of 3,5-dimethyl-
phenyl p-bromobenzoate with aluminium chloride for
30 min at 150-160° [375].

m.p. 147-148° [375];   Spectra (NA).

(4-Bromophenyl)(4-hydroxy-2,6-dimethylphenyl)methanone

[34183-16-9] $C_{15}H_{13}BrO_2$ mol.wt. 305.17

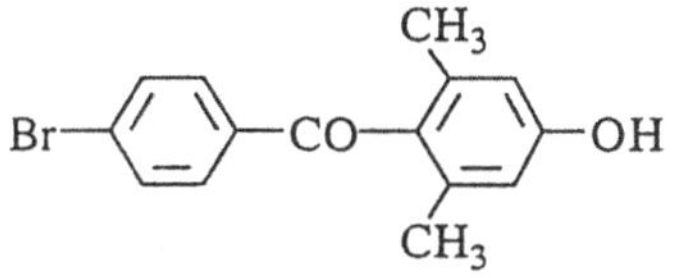

Synthesis

-Obtained by Fries rearrangement of 3,5-dimethylphenyl
p-bromobenzoate with aluminium chloride in chlorobenzene at 140-150° for 20 min or in nitrobenzene at 75°
for 24 h [1214].

m.p.   125-126° [1214];   Spectra (NA).

(4-Chloro-2-hydroxy-5-methylphenyl)(2-methylphenyl)methanone

[170799-18-5] $C_{15}H_{13}ClO_2$ mol.wt. 260.72

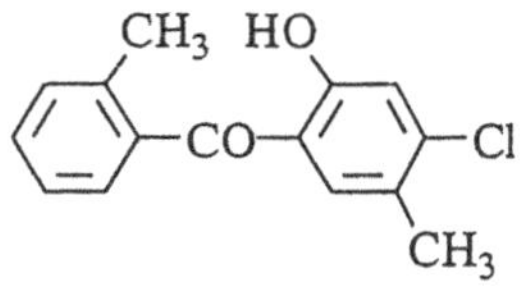

Synthesis

-Preparation by Fries rearrangement of 3-chloro-4-methyl-
phenyl o-toluate with aluminium chloride at 100-150° for
0.5-3 h (31%) [1363].

m.p. 72-73° [1363];   Spectra (NA).

(5-Chloro-2-hydroxy-3-methylphenyl)(2-methylphenyl)methanone

[86914-77-4] $C_{15}H_{13}ClO_2$ mol.wt. 260.72

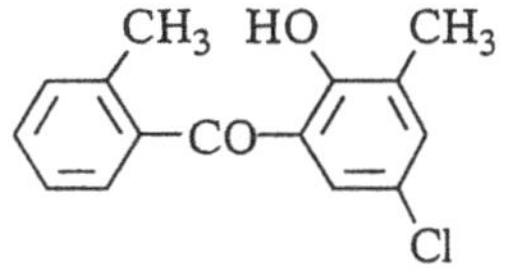

Synthesis

-Preparation by Fries rearrangement of 4-chloro-2-methyl-
phenyl o-toluate with aluminium chloride at 180° for 10 min
[688].

m.p. 38° [688];   Spectra (NA).

(5-Chloro-2-hydroxy-3-methylphenyl)(3-methylphenyl)methanone

[86914-90-1] $C_{15}H_{13}ClO_2$ mol.wt. 260.72

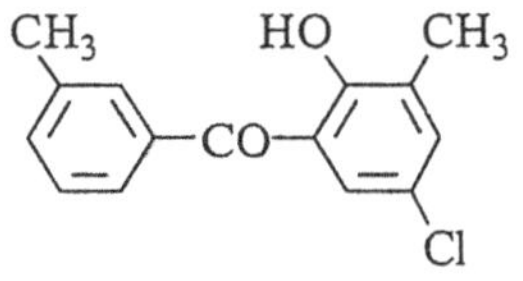

Synthesis

-Preparation by Fries rearrangement of 4-chloro-2-methyl-
phenyl m-toluate with aluminium chloride at 180° for 10 min
[688].

m.p. 108-109° [688];   Spectra (NA).

(5-Chloro-2-hydroxy-3-methylphenyl)(4-methylphenyl)methanone

[86914-86-5] $C_{15}H_{13}ClO_2$ mol.wt. 260.72

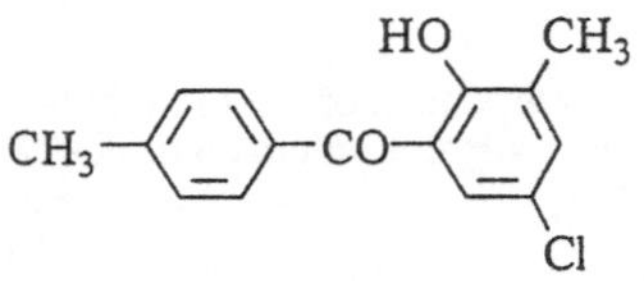

Synthesis

-Preparation by Fries rearrangement of 4-chloro-2-methyl-
   phenyl p-toluate with aluminium chloride at 180° for
   10 min [688].

m.p. 76-77° [688];   Spectra (NA).

(5-Chloro-2-hydroxy-4-methylphenyl)(2-methylphenyl)methanone

[170799-04-9] $C_{15}H_{13}ClO_2$ mol.wt. 260.72

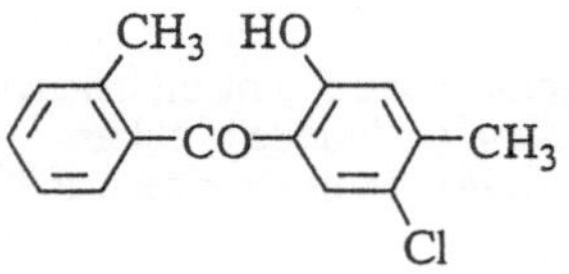

Synthesis

-Preparation by Fries rearrangement of 4-chloro-3-methyl-
   phenyl o-toluate with aluminium chloride at 100-150° for
   0.5-3 h (76%) [1363].

m.p. 57-58° [1363];   Spectra (NA).

(3-Chloro-2-methylphenyl)(2-hydroxy-5-methylphenyl)methanone

$C_{15}H_{13}ClO_2$ mol.wt. 260.72

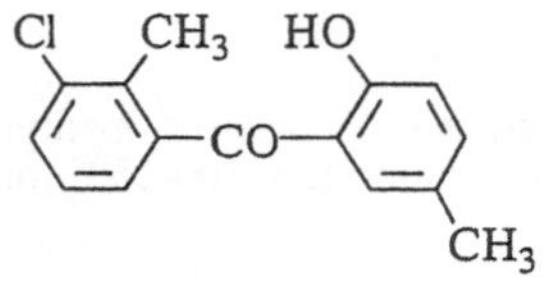

Synthesis

-Preparation by reaction of 3-chloro-2-methylbenzoyl
   chloride with p-cresol in the presence of aluminium chloride
   for 8 h at 190° [348].

m.p. and Spectra (NA).

(2-Chlorophenyl)(5-ethyl-2-hydroxyphenyl)methanone

[108294-72-0] $C_{15}H_{13}ClO_2$ mol.wt. 260.72

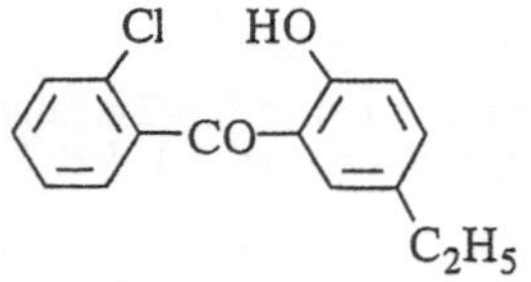

Synthesis

-Preparation by reaction of o-chlorobenzoyl chloride with
   p-ethylphenol in the presence of aluminium chloride in
   tetrachloroethane at 105° for 22 h (60%) [400].

b.p.$_{0.45}$ 138-140° [400];   Spectra (NA).

(3-Chlorophenyl)(5-ethyl-2-hydroxyphenyl)methanone

[61466-85-1] $C_{15}H_{13}ClO_2$ mol.wt. 260.72

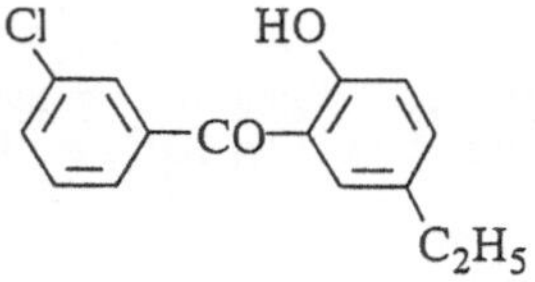

Synthesis

-Preparation by reaction of m-chlorobenzoyl chloride with
   p-ethylphenol in the presence of aluminium chloride in
   tetrachloroethane at 105° for 22 h (66%) [400, 1206].

b.p.$_{0.25}$ 153-156° [400, 1206];   Spectra (NA).

(4-Chlorophenyl)(2-ethyl-4-hydroxyphenyl)methanone

[61466-73-7] $C_{15}H_{13}ClO_2$ mol.wt. 260.72

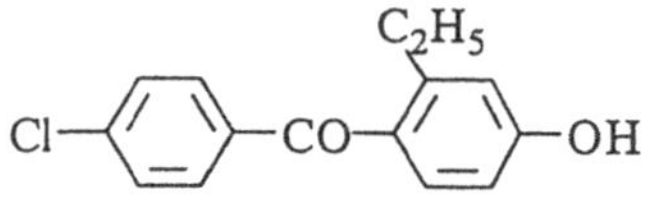

Synthesis

-Obtained (by-product) by reaction of p-chlorobenzoyl chloride with m-ethylphenol in the presence of aluminium chloride in tetrachloroethane at 105° for 22 h (12%) [1206].

-Also refer to: [1436].

  m.p. and Spectra (NA).

(4-Chlorophenyl)(3-ethyl-2-hydroxyphenyl)methanone

[61466-80-6] $C_{15}H_{13}ClO_2$ mol.wt. 260.72

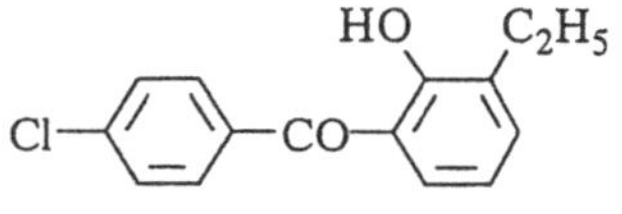

Synthesis

-Preparation by reaction of 3-ethylsalicylic acid chloride (3-ethyl-2-hydroxybenzoyl chloride) with chlorobenzene in the presence of aluminium chloride at 100° overnight (31%) [400, 1208], (23%) [1206].

  m.p. 72-73° [400, 1206, 1208];   Spectra (NA).

(4-Chlorophenyl)(4-ethyl-2-hydroxyphenyl)methanone

[56394-72-0] $C_{15}H_{13}ClO_2$ mol.wt. 260.72

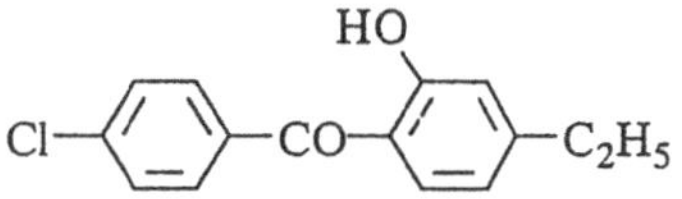

Syntheses

-Preparation by reaction of p-chlorobenzoyl chloride with m-ethylphenol in the presence of aluminium chloride in tetrachloroethane at 105° for 22 h (83%) [400, 1208], (48%) [1205, 1206].

-Obtained from the corresponding oxime (m.p. 159-161°) by treatment with sodium metabisulfite in refluxing ethanol for 40 h [1206].

  m.p. 52° [400, 1206, 1208];   b.p.$_{0.09}$ 155-160° [1205];   Spectra (NA).

(4-Chlorophenyl)(5-ethyl-2-hydroxyphenyl)methanone

[56394-67-3] $C_{15}H_{13}ClO_2$ mol.wt. 260.72

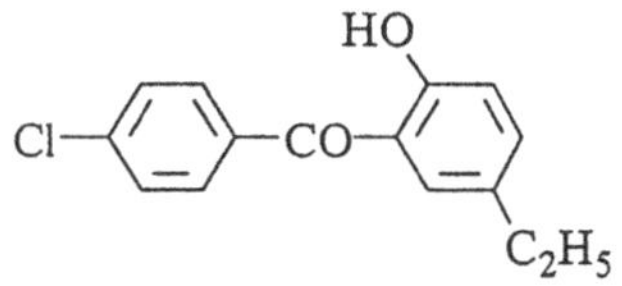

Syntheses

-Preparation by reaction of p-chlorobenzoyl chloride with p-ethylphenol in the presence of aluminium chloride in tetrachloroethane at 105° for 22 h (72%) [400, 1208], (43-49%) [1205, 1206, 1207].

-Preparation by Fries rearrangement of p-ethylphenyl p-chlorobenzoate with aluminium chloride in tetrachloroethane at 125° for 6 h [1207].

-Also refer to: [401].

  m.p. 41°5-43°5 [400, 1208], 35-38° [1206, 1207];

b.p.$_{0.09}$ 147-150° [1208], b.p.$_3$ 150-160° [400], b.p.$_{0.3}$ 160-168° [1205, 1206, 1207]; Spectra (NA).

(2-Chlorophenyl)(2-hydroxy-4,5-dimethylphenyl)methanone

[170799-17-4] $C_{15}H_{13}ClO_2$ mol.wt. 260.72

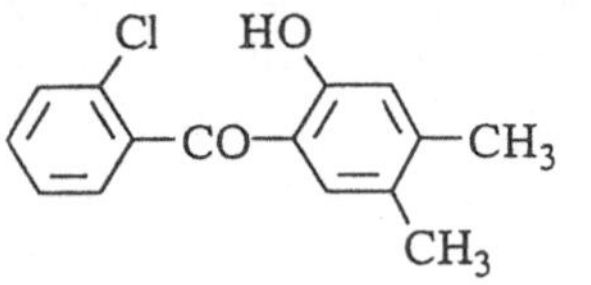

Synthesis

-Preparation by Fries rearrangement of 3,4-dimethylphenyl o-chlorobenzoate with aluminium chloride at 100-150° for 0.5-3 h (58%) [1363].

m.p. 81-82° [1363];   Spectra (NA).

(4-Chlorophenyl)(2-hydroxy-3,5-dimethylphenyl)methanone

[86914-84-3] $C_{15}H_{13}ClO_2$ mol.wt. 260.72

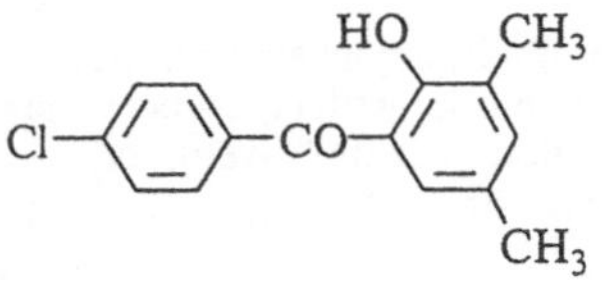

Synthesis

-Preparation by Fries rearrangement of 2,4-dimethylphenyl p-chlorobenzoate with aluminium chloride at 180° for 10 min [688].

m.p. 45-46° [688];   Spectra (NA).

(4-Chlorophenyl)(2-hydroxy-4,6-dimethylphenyl)methanone

[34199-74-1] $C_{15}H_{13}ClO_2$ mol.wt. 260.72

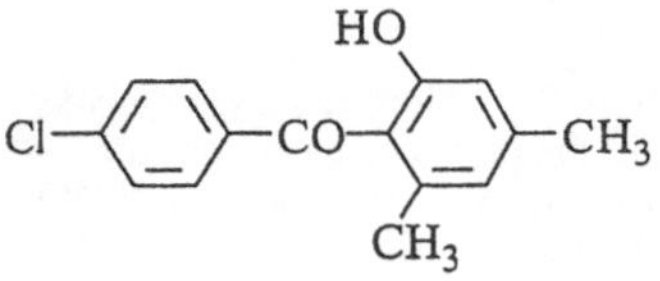

Syntheses

-Preparation by reaction of p-chlorobenzoyl chloride with 3,5-dimethylanisole in the presence of stannic chloride in benzene, first between 0 to 5°, then at r.t. for 3 h. The 4'-chloro-4,6-dimethyl-2-methoxybenzophenone obtained (27%) by demethylation with 48% hydrobromic acid gave the expected ketone [1293].
-Preparation by Fries rearrangement of 3,5-dimethylphenyl p-chlorobenzoate with aluminium chloride for 30 min at 150-160° [375].

m.p. 132-133° [375];   Spectra (NA).

(4-Chlorophenyl)(4-hydroxy-2,6-dimethylphenyl)methanone

[34183-15-8] $C_{15}H_{13}ClO_2$ mol.wt. 260.72

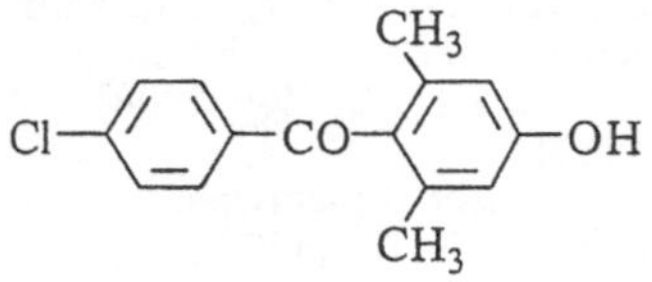

Syntheses

-Preparation by reaction of p-chlorobenzoyl chloride with 3,5-dimethylanisole in the presence of stannic chloride in benzene, first between 0 to -5°, then at r.t. for 3 h. The 4-(p-chlorobenzoyl)-3,5-dimethylanisole obtained (40%) by demethylation with 48% hydrobromic acid gave the expected ketone [1293].

-Also obtained by Fries rearrangement of 3,5-dimethylphenyl p-chlorobenzoate with aluminium chloride in chlorobenzene at 140-150° for 20 min or in nitrobenzene at 75° for 24 h [1214].

m.p. 139-140° [1214]; Spectra (NA).

(4-Chlorophenyl)(4-hydroxy-3,5-dimethylphenyl)methanone

[61002-59-3] $C_{15}H_{13}ClO_2$ mol.wt. 260.72

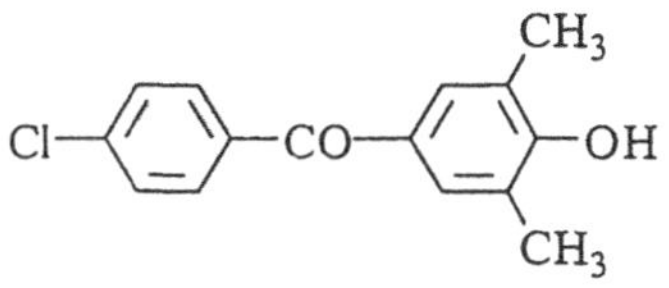

Synthesis

-Preparation by Friedel-Crafts acylation of 2,6-dimethyl-anisole with p-chlorobenzoyl chloride in the presence of aluminium chloride [909].
N.B.: Na and K salts [444].

m.p. 98° [909]; Spectra (NA).

(5-Chloro-2-hydroxy-4-methoxyphenyl)(4-methoxyphenyl)methanone

[136741-44-1] $C_{15}H_{13}ClO_4$ mol.wt. 292.72

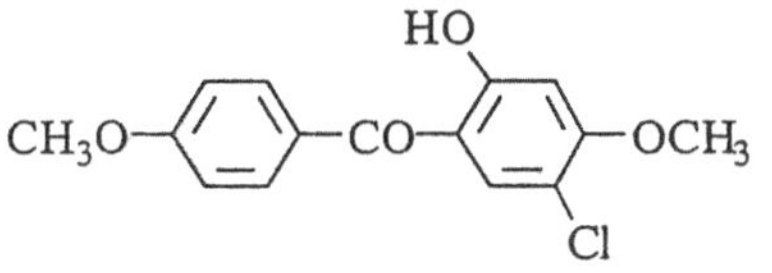

Synthesis

-Preparation by reaction of 4-methoxybenzoyl chloride with 4-chlororesorcinol dimethyl ether in the presence of aluminium chloride in ethylene dichloride (85%) [1193].

m.p. 130-132° [1193]; Spectra (NA).

(2-Chlorophenyl)(2-hydroxy-3,4-dimethoxyphenyl)methanone

[140665-36-7] $C_{15}H_{13}ClO_4$ mol.wt. 292.72

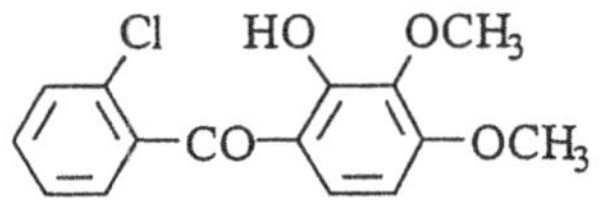

Synthesis

-Preparation by reaction of o-chlorobenzoyl chloride with 1,2,3-trimethoxybenzene in the presence of aluminium chloride in ethylene dichloride between 0 to 5°, then at r.t. for 3 h and at reflux for 1 h (86%) [1194].

m.p. 115-117° [1194]; Spectra (NA).

(2-Chlorophenyl)(2-hydroxy-4,5-dimethoxyphenyl)methanone

[140665-22-1] $C_{15}H_{13}ClO_4$ mol.wt. 292.72

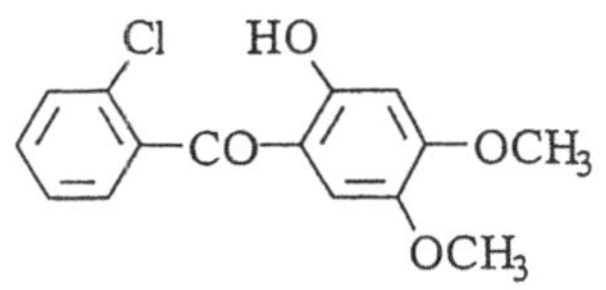

Synthesis

-Preparation by reaction of o-chlorobenzoyl chloride with 1,2,4-trimethoxybenzene in the presence of aluminium chloride in ethylene dichloride between 0 to 5°, then at r.t. for 4 h and at reflux for 2 h (60%) [1194].

m.p. 75-77° [1194]; Spectra (NA).

(4-Chlorophenyl)(2-hydroxy-3,4-dimethoxyphenyl)methanone

[7508-29-4] $C_{15}H_{13}ClO_4$ mol.wt. 292.72

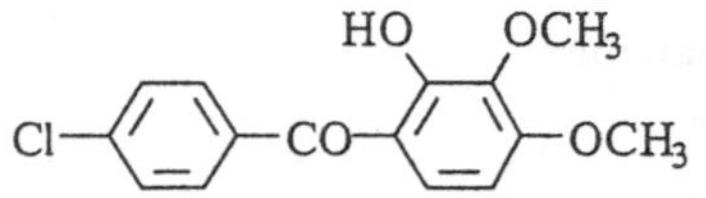

Synthesis

-Preparation by reaction of p-chlorobenzoyl chloride with
 1,2,3-trimethoxybenzene in the presence of aluminium
 chloride in ethylene dichloride between 0 to 5°, then at
 r.t. for 3 h and at reflux for 1 h (54%) [1194].

m.p.   148-149° [1194];   Spectra (NA).

(4-Ethyl-2-hydroxyphenyl)(3-fluorophenyl)methanone

[61466-88-4] $C_{15}H_{13}FO_2$ mol.wt. 244.27

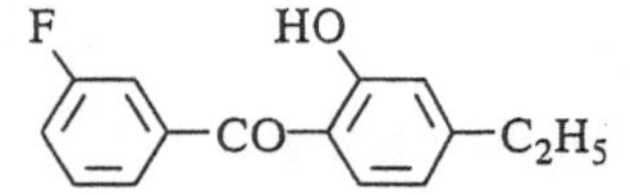

Synthesis

-Obtained (poor yield) by reaction of m-fluorobenzoyl
 chloride with m-ethylphenol in the presence of aluminium
 chloride in tetrachloroethane at 105° for 22 h [1206, 1207],
 (5%) [400].

oil [400];   m.p. <20° [1206, 1207];
n_D^{22} = 1.5962 [1206, 1207];   b.p. and Spectra (NA).

(4-Ethyl-2-hydroxyphenyl)(4-fluorophenyl)methanone

[56394-78-6] $C_{15}H_{13}FO_2$ mol.wt. 244.27

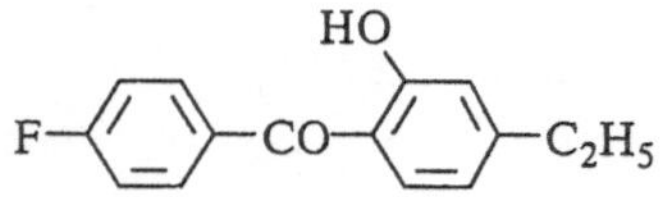

Synthesis

-Obtained by reaction of p-fluorobenzoyl chloride with
 m-ethylphenol in the presence of aluminium chloride in
 tetrachloroethane at 105° for 22 h (64%) [1205], (41%)
 [1206, 1207], (5%) [400, 1208].

-Also refer to: [401].

m.p.   44-48° [400, 1205, 1206, 1207, 1208];
b.p.$_{0.06}$ 129-132° [1205, 1206, 1207];   Spectra (NA).

(5-Ethyl-2-hydroxyphenyl)(4-fluorophenyl)methanone

[108294-75-3] $C_{15}H_{13}FO_2$ mol.wt. 244.27

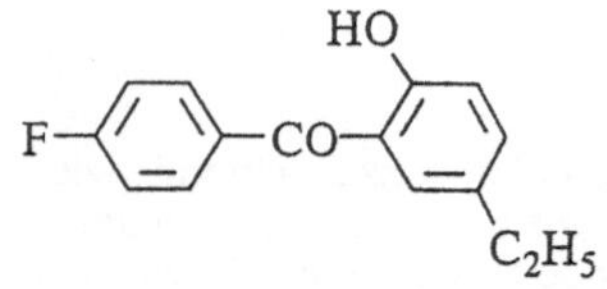

Synthesis

-Preparation by reaction of p-fluorobenzoyl chloride with
 p-ethylphenol in the presence of aluminium chloride in
 tetrachloroethane at 105° for 22 h (38%) [400].

m.p.   44-46° [400];   Spectra (NA).

(4-Fluorophenyl)(4-hydroxy-2,3-dimethylphenyl)methanone

$C_{15}H_{13}FO_2$ mol.wt. 244.27

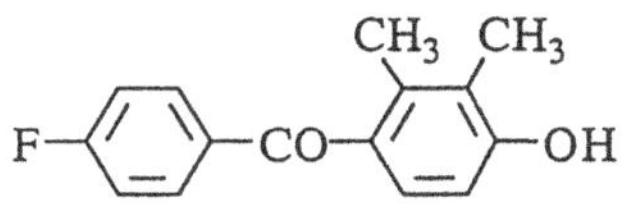

Synthesis

-Preparation by demethylation of 4'-fluoro-4-methoxy-2,3-dimethylbenzophenone (SM) with aluminium chloride in refluxing methylene chloride overnight. SM was obtained by Friedel-Crafts acylation of 2,3-dimethylanisole with p-fluorobenzoyl chloride in the presence of aluminium chloride in refluxing methylene chloride [653].

m.p. and Spectra (NA).

(4-Fluorophenyl)(4-hydroxy-3,5-dimethylphenyl)methanone

[102331-06-6] $C_{15}H_{13}FO_2$ mol.wt. 244.27

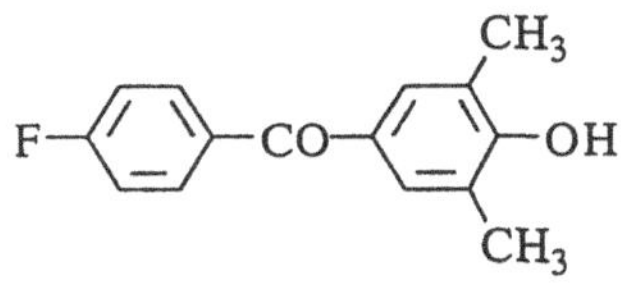

Synthesis

-Preparation by Fries rearrangement of 2,6-dimethylphenyl p-fluorobenzoate in the presence of aluminium chloride in o-dichlorobenzene at 150° (93%) [445].

m.p. and Spectra (NA).

(2-Fluorophenyl)(2-hydroxy-4-methoxy-3-methylphenyl)methanone

[72483-03-5] $C_{15}H_{13}FO_3$ mol.wt. 260.26

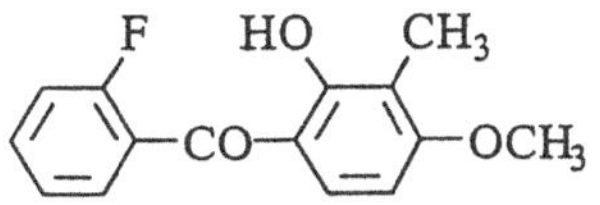

Synthesis

-Preparation by reaction of o-fluorobenzoyl chloride with 2,6-dimethoxytoluene in the presence of aluminium chloride in refluxing ethylene dichloride for 30 min (93%) [1270, 1272].

m.p. 119-120° [1272], 118-120° [1270];   Spectra (NA).

(2-Fluoro-6-methoxyphenyl)(2-hydroxy-6-methoxyphenyl)methanone

[129103-94-2] $C_{15}H_{13}FO_4$ mol.wt. 276.26

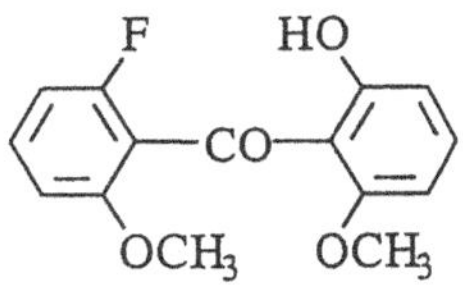

Synthesis

-Preparation from 2-iodo-3-methoxyphenyl 2-fluoro-6-methoxybenzoate by rearrangement on treatment with n-butyllithium in a mixture of ethyl ether, hexane and tetrahydrofuran at -100° followed by heating to -70°, then treatment with saturated aqueous ammonium chloride (88%) [598].

m.p. 79°5-80° [598]; ^{1}H NMR [598], IR [598], MS [598].

(2-Fluorophenyl)(2-hydroxy-3,4-dimethoxyphenyl)methanone

[140665-37-8] $C_{15}H_{13}FO_4$ mol.wt. 276.26

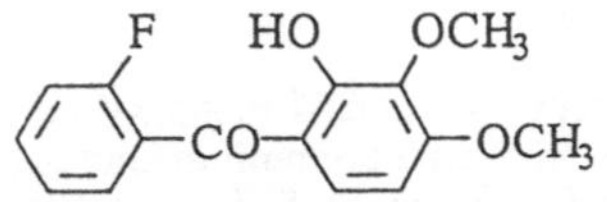

Synthesis

-Preparation by reaction of o-fluorobenzoyl chloride with 1,2,3-trimethoxybenzene in the presence of aluminium chloride in ethylene dichloride between 0 to 5°, then at r.t. for 3 h and at reflux for 1 h (95%) [1194].

m.p. 97-99° [1194];   Spectra (NA).

(2-Fluorophenyl)(2-hydroxy-4,5-dimethoxyphenyl)methanone

[140665-23-2] $C_{15}H_{13}FO_4$ mol.wt. 276.26

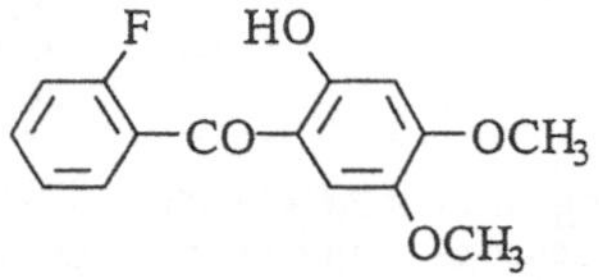

Synthesis

-Preparation by reaction of o-fluorobenzoyl chloride with 1,2,4-trimethoxybenzene in the presence of aluminium chloride in ethylene dichloride between 0 to 5°, then at r.t. for 4 h and at reflux for 2 h (95%) [1194].

m.p. 89-90° [1194];   Spectra (NA).

(3-Fluorophenyl)(2-hydroxy-3,4-dimethoxyphenyl)methanone

[140665-38-9] $C_{15}H_{13}FO_4$ mol.wt. 276.26

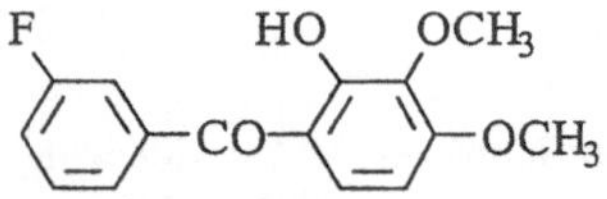

Synthesis

-Preparation by reaction of m-fluorobenzoyl chloride with 1,2,3-trimethoxybenzene in the presence of aluminium chloride in ethylene dichloride between 0 to 5°, then at r.t. for 3 h and at reflux for 1 h (80%) [1194].

m.p. 84-85° [1194];   Spectra (NA).

(4-Fluorophenyl)(2-hydroxy-3,4-dimethoxyphenyl)methanone

[140665-39-0] $C_{15}H_{13}FO_4$ mol.wt. 276.26

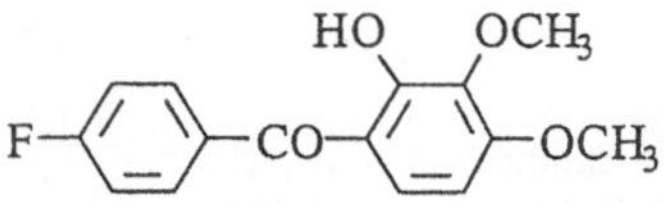

Synthesis

-Preparation by reaction of p-fluorobenzoyl chloride with 1,2,3-trimethoxybenzene in the presence of aluminium chloride in ethylene dichloride between 0 to 5°, then at r.t. for 3 h and at reflux for 1 h (79%) [1194].

m.p. 144-146° [1194];   Spectra (NA).

(4-Fluorophenyl)(2-hydroxy-4,5-dimethoxyphenyl)methanone

$C_{15}H_{13}FO_4$ mol.wt. 276.26

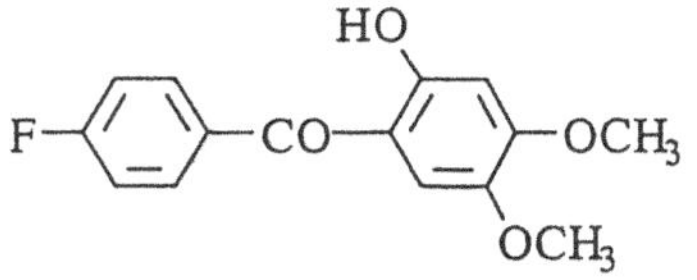

Synthesis

-Preparation by reaction of p-fluorobenzoyl chloride with 3,4-dimethoxyphenol in the presence of boron trichloride in a benzene/methylene chloride mixture at r.t. for 2 h [213].

m.p. (NA); orange solid [213]; ^{1}H NMR [213].

(4-Ethyl-2-hydroxyphenyl)(4-nitrophenyl)methanone

[78473-49-1] $C_{15}H_{13}NO_4$ mol.wt. 271.27

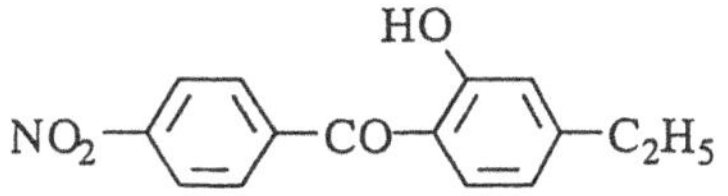

Synthesis

-Refer to: [401].

m.p. and Spectra (NA).

(5-Ethyl-2-hydroxyphenyl)(4-nitrophenyl)methanone

[108294-76-4] $C_{15}H_{13}NO_4$ mol.wt. 271.27

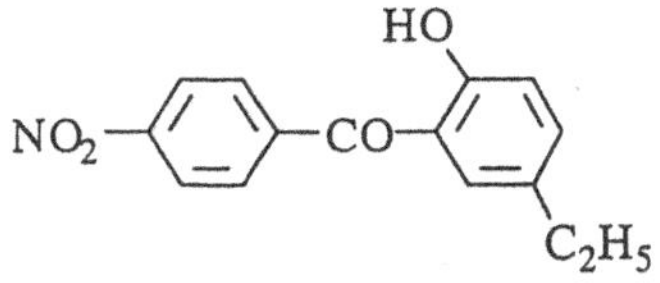

Synthesis

-Preparation by reaction of p-nitrobenzoyl chloride with p-ethylphenol in the presence of aluminium chloride in tetrachloroethane at 105° for 22 h (63%) [400].

m.p. 100-101°5 [400]; Spectra (NA).

(4-Hydroxy-3,5-dimethylphenyl)(4-nitrophenyl)methanone

[85916-09-2] $C_{15}H_{13}NO_4$ mol.wt. 271.27

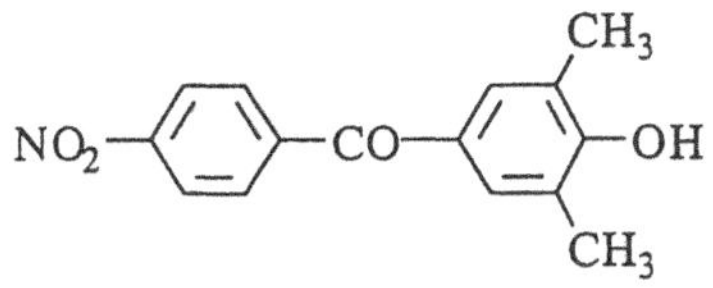

Synthesis

-Refer to: [573].

m.p. and Spectra (NA).

(2-Hydroxy-4-methyl-5-nitrophenyl)(4-methylphenyl)methanone

$C_{15}H_{13}NO_4$ mol.wt. 271.27

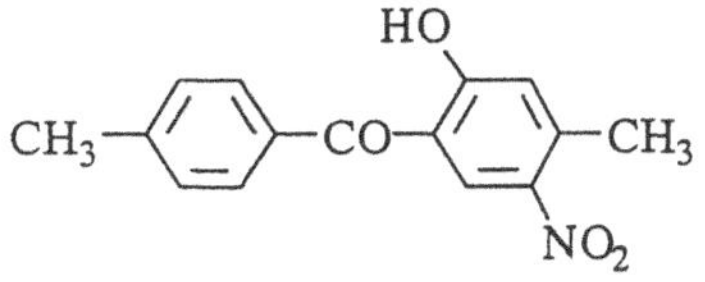

Synthesis

-Obtained by diazotization of 2-amino-4,4'-dimethyl-5-nitrobenzophenone, followed by hydrolysis of the diazonium salt so formed (24%) [264].

m.p. 144° [264]; Spectra (NA).

(4-Hydroxy-3-methyl-5-nitrophenyl)(2-methylphenyl)methanone

[103555-90-4] $C_{15}H_{13}NO_4$ mol.wt. 271.27

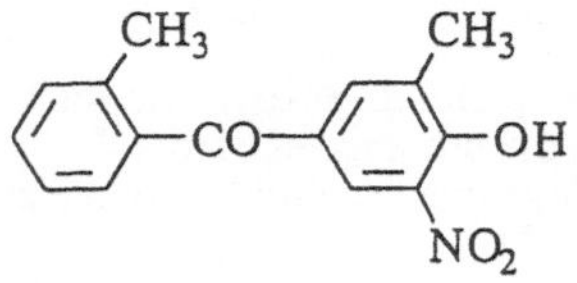

Synthesis

-Preparation from o-methylanisole in three steps: first, acylation of o-methylanisole with o-toluoyl chloride in the presence of ferric chloride during 3 h at 20°. Then, demethylation of 4-methoxy-2',3-dimethylbenzophenone so formed by treatment with refluxing 55% hydriodic acid for 10 h and subsequent nitration of the obtained 4-hydroxy-2',3-dimethylbenzophenone with a concentrated nitric acid/concentrated sulfuric acid mixture under cooling [872].

m.p. and Spectra (NA).

(4-Hydroxy-3-methoxy-5-nitrophenyl)(2-methylphenyl)methanone

[134612-79-6] $C_{15}H_{13}NO_5$ mol.wt. 287.27

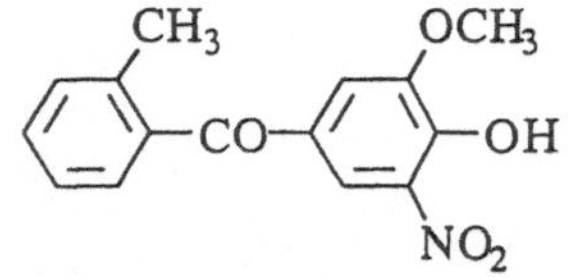

Synthesis

-Preparation by nitration of 4-hydroxy-3-methoxy-2'-methyl-benzophenone with 65% nitric acid in acetic acid at 20° [164].

m.p. 125-127° [164];   Spectra (NA).

(4-Hydroxy-3-methoxy-5-nitrophenyl)(4-methylphenyl)methanone

[134612-80-9] $C_{15}H_{13}NO_5$ mol.wt. 287.27

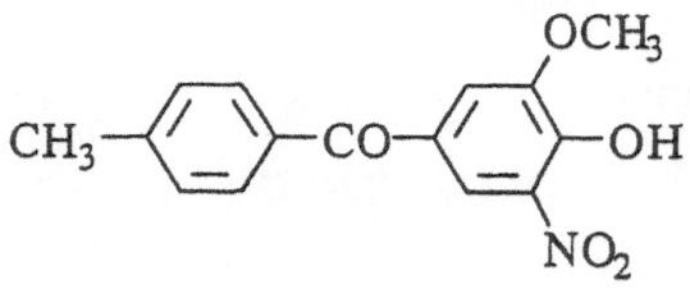

Synthesis

-Preparation by nitration of 4-hydroxy-3-methoxy-4'-methylbenzophenone with 65% nitric acid in acetic acid at 20° [164].

m.p. 137-139° [164];   Spectra (NA).

(2,4-Dimethoxyphenyl)(4-hydroxy-3-nitrophenyl)methanone

[67286-44-6] $C_{15}H_{13}NO_6$ mol.wt. 303.27

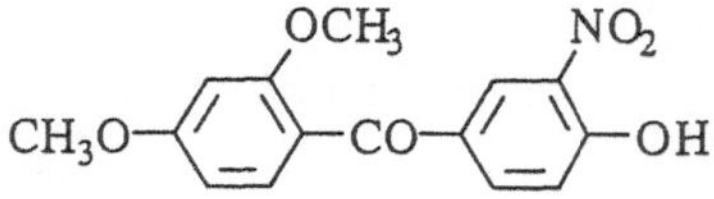

Synthesis

-Obtained by reaction of nitric acid (d = 1.42) with 4-hydroxy-2',4'-dimethoxybenzophenone in acetic acid at 12° (major product) [1083].

m.p. 105° [1083]; ^{1}H NMR [1083].

(3,4-Dimethoxyphenyl)(4-hydroxy-3-nitrophenyl)methanone

[134612-83-2] $C_{15}H_{13}NO_6$ mol.wt. 303.27

Synthesis

-Refer to: [164].

m.p. and Spectra (NA).

(2-Hydroxy-4,6-dimethoxyphenyl)(2-nitrophenyl)methanone

[61736-75-2] $C_{15}H_{13}NO_6$ mol.wt. 303.27

Syntheses

-Obtained by partial demethylation of 2,4,6-trimethoxy-
2'-nitrobenzophenone with boron tribromide in methylene
chloride at 0° for 10 min and at r.t. overnight (97%) [14].
-Also obtained (poor yield) by reaction of o-nitrobenzoyl
chloride with 3,5-dimethoxyphenol in the presence of aluminium chloride in ethyl ether, first at 0°
for 3 h, then at 20° for 3 h (15%) [13].
-Preparation by photo-Fries rearrangement of 3,5-dimethoxyphenyl o-nitrobenzoate in benzene for
1.5 h (quantitative yield) [13].

m.p.   198-199° [13], 133-135° [14]. One of the reported melting points is obviously
wrong.
^{1}H NMR [13, 14], IR [13, 14], UV [13, 14], MS [13, 14].

(4-Hydroxy-2,6-dimethoxyphenyl)(2-nitrophenyl)methanone

[59190-66-8] $C_{15}H_{13}NO_6$ mol.wt. 303.27

Synthesis

-Obtained (poor yield) by reaction of o-nitrobenzoyl
chloride with 3,5-dimethoxyphenol in the presence of
aluminium chloride in ethyl ether, first at 0° for 3 h, then at
20° for 3 h (5%) [13].

m.p.   175-177° [13]; ^{1}H NMR [13], IR [13], UV [13], MS [13].

(2-Hydroxy-4-methoxy-5-nitrophenyl)(2-methoxyphenyl)methanone

[79204-64-1] $C_{15}H_{13}NO_6$ mol.wt. 303.27

Syntheses

-Obtained by oxidation of 6-methoxy-2-(2-methoxy-
4-nitrophenyl)-3-(2-methoxyphenyl)-5-nitrobenzofuran
with chromium trioxide in acetic acid, followed by
saponification of the keto ester so formed (70%) with
potassium hydroxide in boiling ethanol (75%) [579].
-Also obtained by oxidation of 2,3-bis(2-methoxyphenyl)-6-methoxy-5-nitrobenzofuran with
chromium trioxide in acetic acid, followed by saponification of the keto ester so formed (65%) with
potassium hydroxide in boiling ethanol (60%) [579].

m.p. 155° [579];   IR [579], MS [579].

(2-Hydroxy-4-methoxy-5-nitrophenyl)(4-methoxyphenyl)methanone

[66666-07-7] $C_{15}H_{13}NO_6$ mol.wt. 303.27

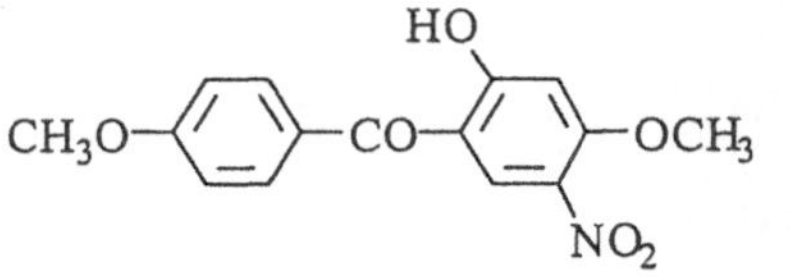

Syntheses

-Obtained by reaction of nitric acid (d = 1.42) with 2-hydroxy-4,4'-dimethoxybenzophenone in acetic acid at 24° (major product) [1083].
-Preparation by saponification of 2-(4-anisoyloxy)-4,4'-dimethoxy-5-nitrobenzophenone (SM) with potassium hydroxide in refluxing ethanol for 1 h (70%). SM was obtained by oxidation of 2,3-bis(4-methoxyphenyl)-6-methoxy-5-nitrobenzofuran with chromium trioxide in refluxing acetic acid for 45 min (75%) [578].

m.p. 162° [1083], 140° [578]. One of the reported melting points is obviously wrong.
Spectra (NA).

(4-Hydroxy-3-methoxy-5-nitrophenyl)(4-methoxyphenyl)methanone

$C_{15}H_{13}NO_6$ mol.wt. 303.27

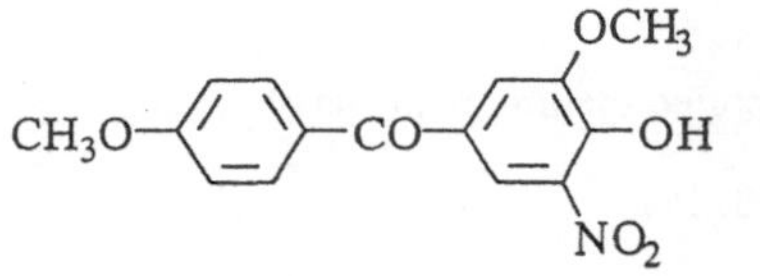

Synthesis

-Obtained by reaction of 65% nitric acid with 3,4'-di-methoxy-4-hydroxybenzophenone in acetic acid at 20° [164].

m.p. 134-136° [164];   Spectra (NA).

(5-Hydroxy-4-methoxy-2-nitrophenyl)(4-methoxyphenyl)methanone

[2898-51-3] $C_{15}H_{13}NO_6$ mol.wt. 303.27

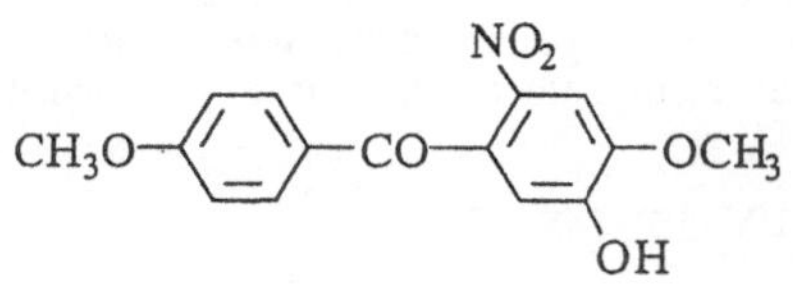

Synthesis

-Obtained by heating 2-nitro-4,4',5-trimethoxy-benzophenone with potassium hydroxide and dilute methanol in an autoclave at 140° for 8 h [10].

m.p. 178° [10];   Spectra (NA).

(2-Hydroxy-3-methylphenyl)(3-methylphenyl)methanone

$C_{15}H_{14}O_2$ mol.wt. 226.27

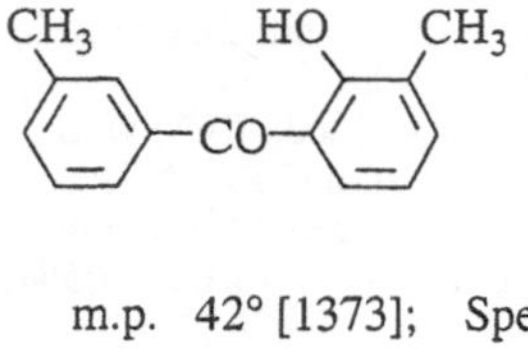

Synthesis

-Preparation (poor yield) by action of m-toluoyl chloride with o-tolyl borate in the presence of aluminium chloride in tetrachloroethane at 100° (8%) [1373].

m.p. 42° [1373];   Spectra (NA).

(2-Hydroxy-4-methylphenyl)(3-methylphenyl)methanone

$C_{15}H_{14}O_2$ mol.wt. 226.27

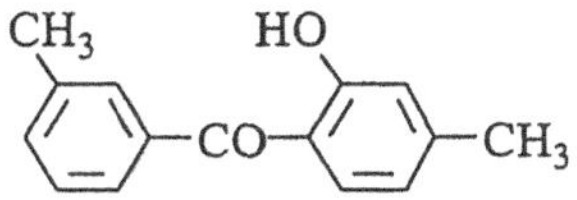

Synthesis

-Preparation by reaction of m-toluoyl chloride with m-tolyl borate in the presence of aluminium chloride in tetrachloroethane at 100° (37%) [1373].

b.p.$_8$ 194-195° [1373];   Spectra (NA).

(2-Hydroxy-4-methylphenyl)(4-methylphenyl)methanone

[81652-53-1] $C_{15}H_{14}O_2$ mol.wt. 226.27

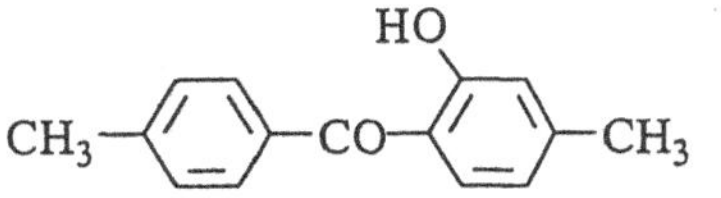

Syntheses

-Obtained (by-product) by Fries rearrangement of m-cresyl p-toluate in the presence of aluminium chloride in nitrobenzene at 62-63° for 18 h (16%) [126].
-Also obtained (poor yield) by diazotization of 2-amino-4,4'-dimethylbenzophenone, followed by hydrolysis of the diazonium salt so obtained (9%) [264]. In this reaction, the major product was 3,6-dimethylfluorenone (70%).
-Also refer to: [17].

m.p.   73-74° [264];   Spectra (NA).

(2-Hydroxy-5-methylphenyl)(2-methylphenyl)methanone

[147029-79-6] $C_{15}H_{14}O_2$ mol.wt. 226.27

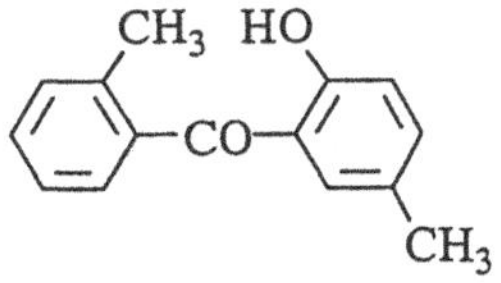

Syntheses

-Preparation by Fries rearrangement of p-tolyl o-toluate with aluminium chloride at 100-150° for 0.5-3 h (73%) [1363].
-Preparation by condensation of the Grignard reagent of p-methylanisole with o-toluoyl chloride, followed by demethylation in acid medium of the resultant methyl ether (excellent yield) [152].

m.p.   95° [152], 94-95° [1363];   Spectra (NA).

(2-Hydroxy-5-methylphenyl)(3-methylphenyl)methanone

[26880-98-8] $C_{15}H_{14}O_2$ mol.wt. 226.27

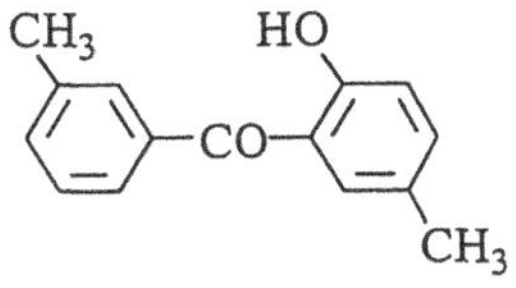

Synthesis

-Preparation by Fries rearrangement of p-tolyl m-toluate with aluminium chloride at 130° for 5 h (54%) [772].

m.p. 54-55° [772];
^{1}H NMR [772], IR [772, 792], UV [772];

pK_a [772, 792];   polarographic study [1223].

(2-Hydroxy-5-methylphenyl)(4-methylphenyl)methanone

[26880-95-5] $C_{15}H_{14}O_2$ mol.wt. 226.27

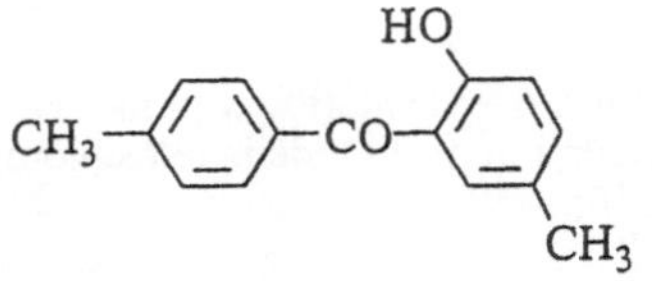

Synthesis

-Preparation by Fries rearrangement of p-tolyl p-toluate in
  the presence of,
  *aluminium chloride [1223], (64%) [754], at 120° (82%)
   [1267] or at 180° for 10 min (85%) [944];
  *Nafion-H, a polymeric perfluorinated resin sulfonic acid,
in refluxing nitrobenzene for 12 h (72%) [1004].
-Also refer to: [435].

  m.p. 89°5-90° [1267], 88-88°5 [772], 88° [944], 63°5-65° [754];
  ^{1}H NMR [754, 772, 1004], ^{13}C NMR [754, 1004],
  IR [754, 772, 792], UV [772];
  pK_a [772, 792];   polarographic study [1223].

(3-Hydroxy-2-methylphenyl)(2-methylphenyl)methanone

[50454-58-5] $C_{15}H_{14}O_2$ mol.wt. 226.27

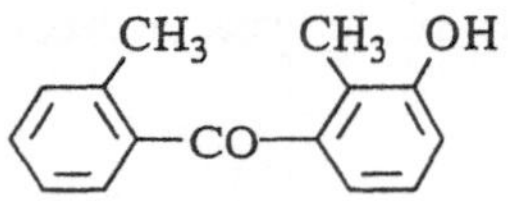

Synthesis

-Refer to: [107] (Japanese patent).

  m.p. and Spectra (NA).

(3-Hydroxy-2-methylphenyl)(4-methylphenyl)methanone

[74167-89-8] $C_{15}H_{14}O_2$ mol.wt. 226.27

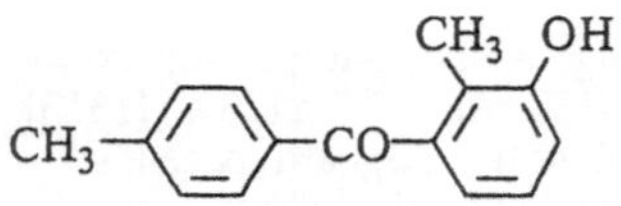

Synthesis

-Preparation by diazotization of 3-amino-2,4'-dimethyl-
  benzophenone, followed by hydrolysis of the diazonium salt
  so obtained (85%) [82], according to [731].
-Also refer to: [436].

  m.p. 102° [82];   Spectra (NA).

(4-Hydroxy-2-methylphenyl)(3-methylphenyl)methanone

                                     $C_{15}H_{14}O_2$ mol.wt. 226.27

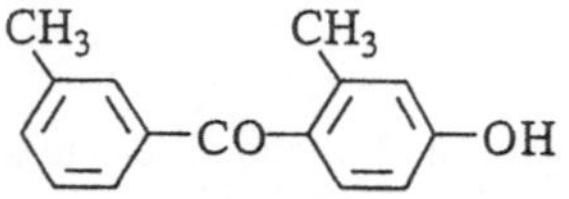

Synthesis

-Obtained by action of m-toluoyl chloride with m-tolyl borate
  in the presence of aluminium chloride in tetrachloroethane at
  100° (30%) [1373].

  m.p. 110° [1373];   Spectra (NA).

(4-Hydroxy-2-methylphenyl)(4-methylphenyl)methanone

$C_{15}H_{14}O_2$ mol.wt. 226.27

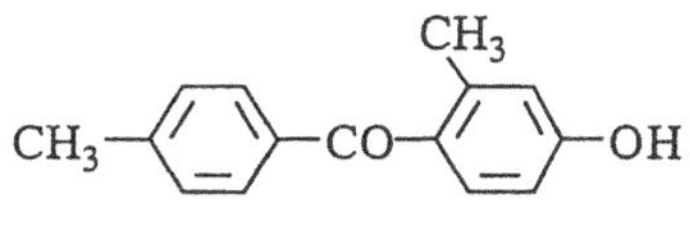

Synthesis

-Preparation by Fries rearrangement of m-cresyl p-toluate with aluminium chloride in nitrobenzene at 62-63° for 18 h (50%) [126].

m.p. 108-109° [126];   Spectra (NA).

(4-Hydroxy-3-methylphenyl)(2-methylphenyl)methanone

[147029-76-3] $C_{15}H_{14}O_2$ mol.wt. 226.27

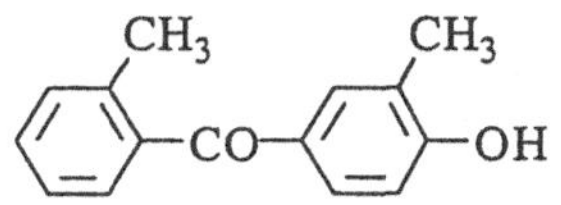

Synthesis

-Preparation by condensation of the Grignard reagent of o-methylanisole with o-toluoyl chloride, followed by demethylation in acid medium of the resultant methyl ether (excellent yield) [152].

m.p. 142° [152];   Spectra (NA).

(4-Hydroxy-3-methylphenyl)(3-methylphenyl)methanone

[62064-85-1] $C_{15}H_{14}O_2$ mol.wt. 226.27

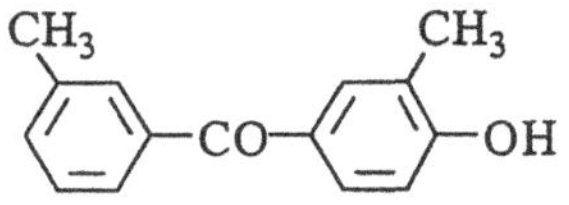

Synthesis

-Obtained by reaction of m-toluoyl chloride with o-cresyl borate in the presence of aluminium chloride in tetrachloroethane at 100° (21%) [1373].

m.p. 158° [1373];   Spectra (NA).

[4-Hydroxy-3-(hydroxymethyl)phenyl](3-methylphenyl)methanone

[62064-89-5] $C_{15}H_{14}O_3$ mol.wt. 242.27

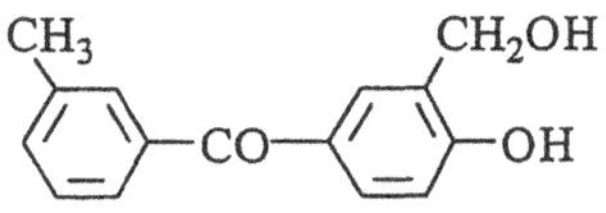

Synthesis

-Not yet described.

N.B.: A metabolite of 3,3'-dimethyl-4-methoxy benzophenone (NK-049) in the rat [439] (Japanese paper).

m.p. and Spectra (NA).

(2-Hydroxy-4-methoxyphenyl)(2-methylphenyl)methanone

[27847-83-2] $C_{15}H_{14}O_3$ mol.wt. 242.27

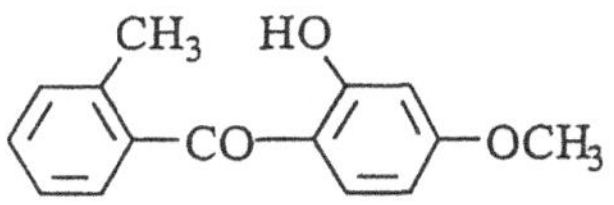

Synthesis

-Refer to: [61].

m.p. and Spectra (NA).

(2-Hydroxy-4-methoxyphenyl)(4-methylphenyl)methanone *(Mexenone)*

[1641-17-4]   C$_{15}$H$_{14}$O$_3$ mol.wt. 242.27

Syntheses

-Preparation by Friedel-Crafts acylation of m-methoxyphenol with p-toluoyl chloride,
*in the presence of boron trichloride in benzene first at
   -10°, then at reflux for 10 h under nitrogen (80%)
   [1064];
*in the presence of titanium tetrachloride in benzene first at -10°, then at reflux for 14 h under
   nitrogen (73%) [1064] or in chlorobenzene for 1 h at 120° (79%) [1140].
-Preparation by reaction of p-toluoyl chloride with 1,3-dimethoxybenzene in the presence of
   aluminium chloride in a chlorobenzene/N,N-dimethylformamide mixture (22:1) at 115° [539,
   1067].
-Also obtained by oxidation of 6-methoxy-3-(4-methylphenyl)-2-phenylbenzofuran with chromium
   trioxide in boiling acetic acid for 30-40 min, followed by saponification of the keto ester so formed
   [1052, 1055].
-Also refer to: [43, 348, 475, 602].

m.p. 98° [1052, 1055], 95-96° [1140], 95° [1064];   IR [1052], UV [474, 475, 1067].

(4-Hydroxy-3-methoxyphenyl)(2-methylphenyl)methanone

[134612-38-7]   C$_{15}$H$_{14}$O$_3$ mol.wt. 242.27

Synthesis

-Preparation by reaction of 33% hydrobromic acid in acetic
   acid with 4-(benzyloxy)-3-methoxy-2'-methylbenzophenone
   in methylene chloride at 20-25° [164].

m.p. 103-105° [164];   Spectra (NA).

(4-Hydroxy-3-methoxyphenyl)(4-methylphenyl)methanone

[134612-39-8]   C$_{15}$H$_{14}$O$_3$ mol.wt. 242.27

Synthesis

-Preparation by reaction of 33% hydrobromic acid in
   acetic acid with 4-(benzyloxy)-3-methoxy-4'-methyl-
   benzophenone in methylene chloride at 20-25° [164].

m.p. 103-105° [164];   Spectra (NA).

(2-Hydroxy-3-methylphenyl)(4-methoxyphenyl)methanone

[65611-79-2]   C$_{15}$H$_{14}$O$_3$ mol.wt. 242.27

Synthesis

-Preparation by reaction between (o-tolyloxy)magnesium
   bromide complexed with HMPT and p-methoxy-
   benzaldehyde in refluxing benzene for 48 h (47%) [250].

m.p.   52° [250]; ^{1}H NMR [250], IR [250], MS [250].

(2-Hydroxy-4-methylphenyl)(4-methoxyphenyl)methanone

[108478-27-9] $C_{15}H_{14}O_3$ mol.wt. 242.27

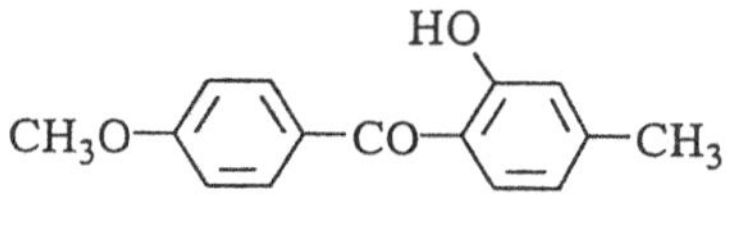

Syntheses

-Preparation by reaction of p-anisic acid with m-cresol
 in the presence of boron trifluoride at 160° for 1 h
 (81%) [730].
-Obtained by Fries rearrangement of m-cresyl p-anisate
with aluminium chloride without solvent at 120° or at 160° [514].

m.p. 96-97° [514, 730];   Spectra (NA).

(2-Hydroxy-5-methylphenyl)(2-methoxyphenyl)methanone

[53271-51-5] $C_{15}H_{14}O_3$ mol.wt. 242.27

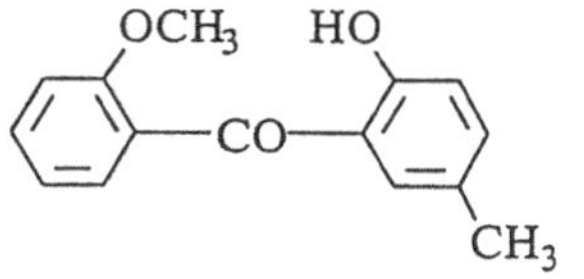

Syntheses

-Obtained (poor yield) by Fries rearrangement of p-cresyl
 o-anisate with polyphosphoric acid at 100° for 20 min (4%)
 [867].
-Also obtained (poor yield) by Friedel-Crafts acylation of
 p-cresol with o-anisic acid in the presence of methanesulfonic acid at 120° for 20 min (5%) [867].
-Also obtained by photo-Fries rearrangement of p-cresyl o-anisate in benzene during 72 h (53%)
 [860].

yellow oil, very viscous [860];   m.p. 60° [867];
^{1}H NMR (Sadtler: standard n° 38494 M) [860],
IR (Sadtler: standard n° 65532 K) [860, 867], UV [867], MS [195, 860].

(2-Hydroxy-5-methylphenyl)(3-methoxyphenyl)methanone

[26880-99-9] $C_{15}H_{14}O_3$ mol.wt. 242.27

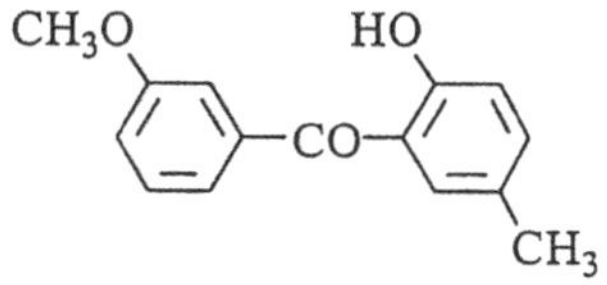

Syntheses

-Preparation by photo-Fries rearrangement of p-tolyl
 m-anisate in methanol at 20° for 10 h (34%) [772].
-Preparation by acylating p-methylanisole with m-anisoyl
 chloride in the presence of aluminium chloride in carbon
 disulfide [1379].

m.p. 130-132° [772];
^{1}H NMR [772], IR [772, 792], UV [772];
pK_a [772, 792];   polarographic study [1223].

(2-Hydroxy-5-methylphenyl)(4-methoxyphenyl)methanone

[26880-96-6] $C_{15}H_{14}O_3$ mol.wt. 242.27

Syntheses

-Preparation by Fries rearrangement of p-tolyl p-anisate (without solvent),
*with aluminium chloride at 120° for 1 h (26%) [867], at 120° or at 160° [514, 772];
*with titanium tetrachloride at 120° for 1 h (89%) [867].
-Preparation by oxidation of 2-phenyl-3-(4-methoxyphenyl)-5-methylbenzofuran with chromium trioxide in boiling acetic acid for 2 h, followed by saponification of the keto ester so formed with 10% sodium hydroxide in boiling ethanol for 15 min [994].
-Obtained (by-product) by reaction of p-anisoyl chloride with p-methylanisole in the presence of aluminium chloride in boiling carbon disulfide [99].

   m.p.   108-109° [99, 514, 772, 994], 107° [867];
   ^{1}H NMR (Sadtler: standard n° 38493 M) [772, 994],
   IR (Sadtler: standard n° 65531 K) [772, 792, 867, 994],
   UV [772, 867], MS [195]; pK_a [772, 792];   polarographic study [1223].

(4-Hydroxy-3-methylphenyl)(2-methoxyphenyl)methanone

[72324-24-4] $C_{15}H_{14}O_3$ mol.wt. 242.27

Syntheses

-Obtained by Fries rearrangement of o-cresyl o-anisate with polyphosphoric acid at 100° for 20 min (20%) [867].
-Preparation by reaction of o-anisic acid with o-cresol in the presence of methanesulfonic acid at 120° for 20 min (48%) [867].

   m.p.   137° [867]; ^{1}H NMR (Sadtler: standard n° 38491 M),
   IR (Sadtler: standard n° 65529 K) [867], UV [867], MS [195].

(4-Hydroxy-3-methylphenyl)(4-methoxyphenyl)methanone

[72324-23-3] $C_{15}H_{14}O_3$ mol.wt. 242.27

Syntheses

-Preparation by Fries rearrangement of o-cresyl p-anisate,
*with titanium tetrachloride in nitromethane at 20° for 170 h (84%) [867];
*with aluminium chloride in nitromethane at 20° for 170 h (42%) [867] or at 75° for 6 h (60%) [867]. The same reaction carried out without solvent at 120° or at 160° gave the same product [514].
-Also obtained (via o-cresyl p-anisate) by heating a mixture of p-anisic acid and o-cresol in the presence of Tonsil [1216].
-Also refer to: [210, 211, 1320].

   m.p.   188° [867], 186° [514], 143° [1216]. One of the reported melting points is obviously wrong.   1H NMR (Sadtler: standard n° 38499 M),
   IR (Sadtler: standard n° 65537 K) [867], UV [867], MS [195].

[3-(Hydroxymethyl)phenyl](4-Hydroxy-3-methylphenyl)methanone

[62064-88-4]   C$_{15}$H$_{14}$O$_3$ mol.wt. 242.27

Synthesis

-Not yet described.

N.B.: A metabolite of 3,3'-dimethyl-4-methoxy-
benzophenone (NK-049) in the rat [439] (Japanese
paper).

m.p. and Spectra (NA).

(2-Hydroxy-3-methoxyphenyl)(4-methoxyphenyl)methanone

[155645-18-4]   C$_{15}$H$_{14}$O$_4$ mol.wt. 258.27

Synthesis

-Preparation by reaction of 2,3-dimethoxybenzoyl
chloride with anisole (74%) [989], according to the
method [1407].

m.p. 96° [989];   Spectra (NA).

(2-Hydroxy-4-methoxyphenyl)(2-methoxyphenyl)methanone

[62495-36-7]   C$_{15}$H$_{14}$O$_4$ mol.wt. 258.27

Syntheses

-Preparation by reaction of o-anisic acid with m-methoxyphenol [1253] in the presence of zinc chloride and
phosphorous oxychloride at 65-70° for 2 h (68%) [1395].
-Preparation by oxidation of 2,3-bis(2-methoxyphenyl)-6-methoxybenzofuran with chromium
trioxide in refluxing acetic acid for 30 min, followed by saponification of the keto ester so formed
(55%) with potassium hydroxide in refluxing ethanol for 1 h (50%) [579].
-Preparation by selective demethylation of 2,2',4-trimethoxybenzophenone with excess beryllium
chloride in refluxing toluene for 3 h (90%) [1250].
-Also obtained, in mixture with 2,4-dimethoxy-2'-hydroxybenzophenone, by reaction of
2-methoxybenzoyl chloride with resorcinol dimethyl ether in ethyl ether in the presence of
aluminium chloride for 8 h at r.t. (total yield: 65%) [825, 826].

m.p. 92-93° [1250], 91° [1253], 88°5-89° [1395], 80° [579];
^{1}H NMR [825, 826, 1250, 1253, 1395],
IR [579, 1250, 1395], UV [1250, 1395], MS [1253];   TLC [1250].

(2-Hydroxy-4-methoxyphenyl)(3-methoxyphenyl)methanone

[62495-37-8]   C$_{15}$H$_{14}$O$_4$ mol.wt. 258.27

Syntheses

-Preparation by reaction of m-anisic acid with
3-methoxyphenol in the presence of zinc chloride and
phosphorous oxychloride at 65-70° for 2 h (54%)
[1395].

-Preparation by partial methylation of 2,4-dihydroxy-3'-methoxybenzophenone with methyl iodide in the presence of potassium carbonate in boiling acetone (69%) [650].
-Preparation by selective demethylation of 2,3',4-trimethoxybenzophenone with excess beryllium chloride in refluxing toluene for 3 h (92%) [1250].

clear oil [1250];   b.p.$_{0.3}$ 165-170° [1395];   m.p. 66° [650];
^{1}H NMR [1250, 1395], IR [1250, 1395], UV [1250, 1395];   TLC [1250].

(2-Hydroxy-4-methoxyphenyl)(4-methoxyphenyl)methanone

[6131-38-0]   C$_{15}$H$_{14}$O$_4$ mol.wt. 258.27

Syntheses

-Preparation by reaction of p-methoxybenzoic acid (p-anisic acid) with resorcinol monomethyl ether in the presence of boron trifluoride [785].
-Preparation by reaction of 2-hydroxy-4-methoxy-benzoic acid with anisole in the presence of stannic chloride at 115-120° for 3-4 h [1112, 1358, 1359].
-Preparation by oxidation of 6-methoxy-3-(4-methoxyphenyl)-2-phenylbenzofuran with chromium trioxide, followed by saponification of the keto ester so formed — the 2-(benzoyloxy)-4,4'-di-methoxybenzophenone — with potassium hydroxide [1052, 1055].
-Preparation by partial methylation,
*of 2,4-dihydroxy-4'-methoxybenzophenone;
  with methyl iodide in the presence of potassium carbonate in boiling acetone (88%) [650];
  with methyl bromide in the presence of potassium carbonate in refluxing acetone for 20 h [150];
*of 2,4'-dihydroxy-4-methoxybenzophenone [1213], with methyl iodide in the presence of potassium carbonate in acetone at r.t. [1083];
*of 2,4,4'-trihydroxybenzophenone with dimethyl sulfate in alkaline solution (40%) [233].
-Preparation by selective demethylation of 2,4,4'-trimethoxybenzophenone with excess beryllium chloride in refluxing toluene for 3 h (90%) [1250].
-Preparation by reaction of p-anisoyl chloride with resorcinol dimethyl ether,
*in chlorobenzene in the presence of titanium tetrachloride for 1 h at 120° (86%) [1140];
*in o-dichlorobenzene or without solvent, in the presence of ferric chloride between 180 to 200° for 6-7 h (by-product) [1467].
-Also refer to: [17, 19, 43, 53, 384, 428, 513, 539, 602, 781, 894, 965, 968, 971, 1023, 1084, 1302].

m.p.   144° [428], 130° [1406], 129-131° [1358, 1359], 118° [650, 1083, 1213],
          117-118° [233, 785], 115° [1052, 1055], 111-112° [1140], 110-112° [150],
          110° [1250]. There is a discrepancy between the various melting points.
^{1}H NMR [267, 1213, 1250], IR [785, 1052, 1213, 1250],
UV [233, 785, 1065, 1250, 1358, 1359, 1406], EPR [267];
TLC [1250];   paper chromatography [911].

(2-Hydroxy-5-methoxyphenyl)(2-methoxyphenyl)methanone

[42833-51-2]   C$_{15}$H$_{14}$O$_4$ mol.wt. 258.27

Syntheses

-Obtained by reaction of o-anisoyl chloride with hydroquinone dimethyl ether in the presence of aluminium chloride in ethyl ether for 19 h [1097].
-Obtained by partial demethylation of 2,2',5-trimethoxy-benzophenone,
*with boron tribromide in methylene chloride at 0° for 1.5 h (50%) [1250];

*with boron trichloride in methylene chloride at 0° for 3 h (50%) [1250];
*with boron trifluoride-etherate in refluxing benzene for 6 h (60%) or in refluxing toluene for 4 h (60%) [1250];
*with beryllium chloride in refluxing benzene for 8-10 h (40%) or in refluxing toluene for 5 h (40%) [1250].
N.B.: In these experiments, only the reactions using boron halides were carried out under nitrogen atmosphere.

m.p. 100-101° [1250]; ^{1}H NMR [1097, 1250], IR [1250];   TLC [1250].

(2-Hydroxy-5-methoxyphenyl)(3-methoxyphenyl)methanone

[183106-25-4] $C_{15}H_{14}O_4$ mol.wt. 258.27

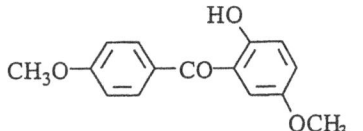

Synthesis

-Preparation by selective demethylation of 2,3',5-tri-methoxybenzophenone with excess beryllium chloride in refluxing toluene for 4 h (90%) [1250].

pale yellow oil [1250];   b.p. (NA);
^{1}H NMR [1250], IR [1250], UV [1250], MS [1250];   TLC [1250].

(2-Hydroxy-5-methoxyphenyl)(4-methoxyphenyl)methanone

[16762-04-2] $C_{15}H_{14}O_4$ mol.wt. 258.27

Syntheses

-Obtained by Fries rearrangement of p-methoxyphenyl p-anisate with titanium tetrachloride without solvent at 120° for 1 h (20 to 35% yields) [863].
-Also obtained by partial demethylation of 2,4',5-tri-methoxybenzophenone (SM),
*with aluminium chloride in nitromethane at 20° for 24 h (27%). SM was prepared by Friedel-Crafts acylation of 1,4-dimethoxybenzene with p-anisoyl chloride in the presence of stannic chloride in nitromethane at 20° for 1 h (78-94%) [863];
*with excess beryllium chloride in refluxing toluene for 3 h (95%) [1250].
-Also refer to: [151, 1055].

m.p. 71° [863], 66-68° [1250];
^{1}H NMR [1250], IR [863, 1250], UV [863, 1250], MS [1250];   TLC [1250].

(2-Hydroxy-6-methoxyphenyl)(2-methoxyphenyl)methanone

[129103-90-8] $C_{15}H_{14}O_4$ mol.wt. 258.27

Synthesis

-Preparation from 2-iodo-3-methoxyphenyl 2-methoxy-benzoate by rearrangement on treatment with n-butyllithium in a mixture of ethyl ether, hexane and tetrahydrofuran at -70° for 2 h, then treatment with saturated aqueous ammonium chloride (81%) [598].

m.p. 83-83°5 [598]; ^{1}H NMR [598], IR [598], MS [598].

(5-Hydroxy-2-methoxyphenyl)(4-methoxyphenyl)methanone

[80427-36-7]   C$_{15}$H$_{14}$O$_4$ mol.wt. 258.27

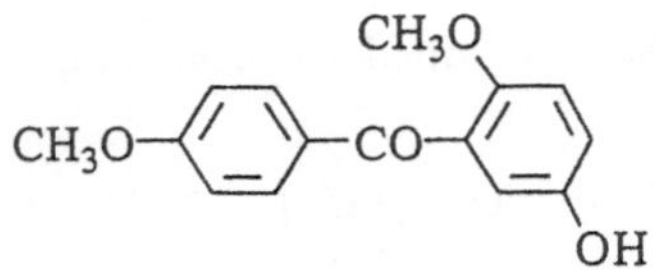

Synthesis

-Preparation by reaction of p-methoxybenzoyl chloride with p-methoxyphenyl benzoate in the presence of stannic chloride in nitromethane at 20° for 2 days, followed by saponification of the m-keto ester so obtained — the 2,4'-dimethoxy-5-(4-methoxybenzoyloxy)benzophenone (73%) — with sodium hydroxide in refluxing methanol for 1 h (quantitative yield) [863].

oil [863];   b.p. (NA); ^{1}H NMR [863], IR [863], UV [863].

[3-Hydroxy-4-(methylamino)phenyl](4-methoxyphenyl)methanone

[136134-37-7]   C$_{15}$H$_{15}$NO$_3$ mol.wt. 257.29

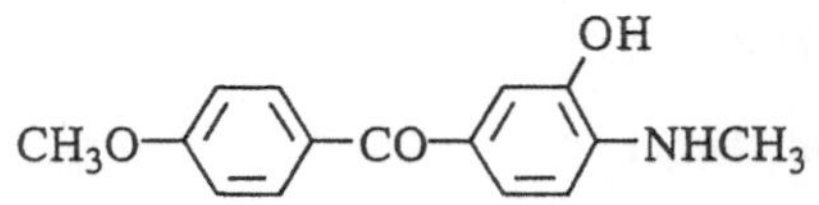

Synthesis

-Preparation from 6-(4-methoxybenzoyl)-3-methyl-benzoxazolinone by hydrolysis with 10% aqueous sodium hydroxide solution [148].

m.p. and Spectra (NA).

(2-Aminophenyl)(2-hydroxy-4,6-dimethoxyphenyl)methanone

[61736-72-9]   C$_{15}$H$_{15}$NO$_4$ mol.wt. 273.29

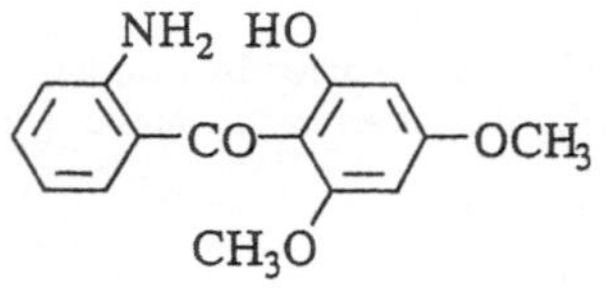

Syntheses

-Preparation by treatment of 2-hydroxy-4,6-dimethoxy-2'-nitrobenzophenone with a mixture of ammonium chloride and zinc moss in dilute ethanol at r.t. for 8 h (quantitative yield) [13], (77%) [14].

-Preparation by photo-Fries rearrangement of 3,5-dimethoxyphenyl 2-aminobenzoate in benzene for 12.5 h (45%) [13].

m.p.   71-75° [13], 61-63° [14];
^{1}H NMR [13, 14], IR [13, 14], UV [13, 14], MS [13, 14].

(2-Bromophenyl)[5-chloro-2-hydroxy-3-(2-propenyl)phenyl]methanone

[93575-78-1]   C$_{16}$H$_{12}$BrClO$_2$ mol.wt. 351.63

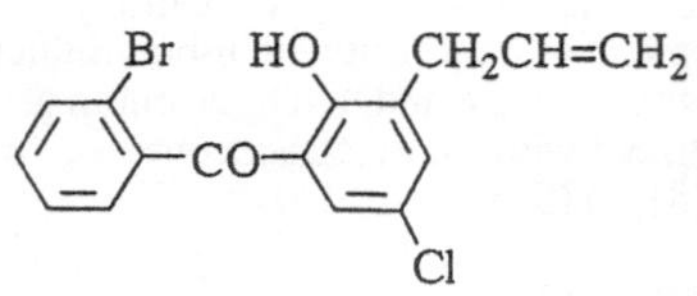

Synthesis

-Preparation by Fries rearrangement of 2-allyl-4-chloro-phenyl o-bromobenzoate with aluminium chloride at 160° for 15 min [698].

oil [698];   b.p. and Spectra (NA).

(4-Chlorophenyl)[5-fluoro-2-hydroxy-3-(2-propenyl)phenyl]methanone

[93575-77-0]   C16H12ClFO2   mol.wt. 290.72

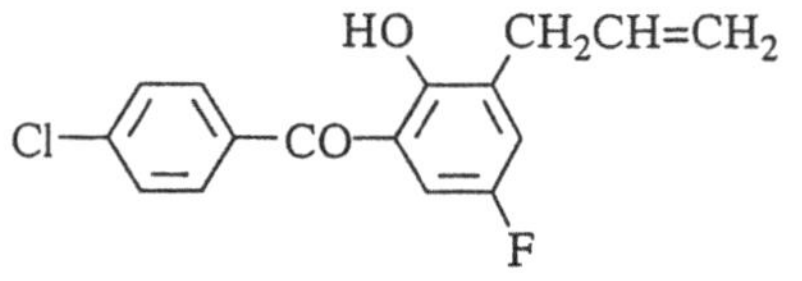

Synthesis

-Preparation by Fries rearrangement of 2-allyl-
4-fluorophenyl 4-chlorobenzoate with aluminium
chloride at 160° for 15 min [698].

m.p. 44-46° [698];   Spectra (NA).

(5-Ethyl-2-hydroxyphenyl)[4-(trifluoromethyl)phenyl]methanone

[61750-29-6]   C16H13F3O2   mol.wt. 294.27

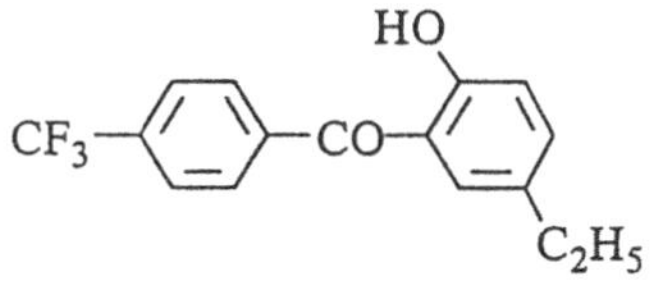

Synthesis

-Preparation by demethylation of 5-ethyl-2-methoxy-
4'-(trifluoromethyl)benzophenone (SM) by heating with
55% hydrobromic acid at 110-120° for 5 h. SM was
obtained by reaction of p-(trifluoromethyl)benzoyl
chloride with p-ethylanisole in the presence of aluminium
chloride in methylene chloride at r.t. overnight (92%) [1207].

m.p. and Spectra (NA).

(2-Hydroxy-3,4-dimethoxyphenyl)[3-(trifluoromethyl)phenyl]methanone

[140665-42-5]   C16H13F3O4   mol.wt. 326.27

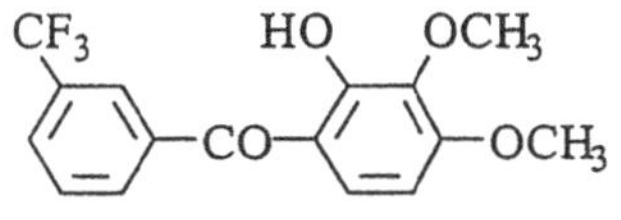

Synthesis

-Obtained by reaction of m-(trifluoromethyl)benzoyl
chloride with 1,2,3-trimethoxybenzene in the presence of
aluminium chloride in ethylene dichloride between 0 to 5°,
then at r.t. for 3 h and at reflux for 1 h (33%) [1194].

m.p. 124-126° [1194];   Spectra (NA).

[5-Chloro-2-hydroxy-3-(1-methylethyl)phenyl](4-chlorophenyl)methanone

[93575-39-4]   C16H14Cl2O2   mol.wt. 309.19

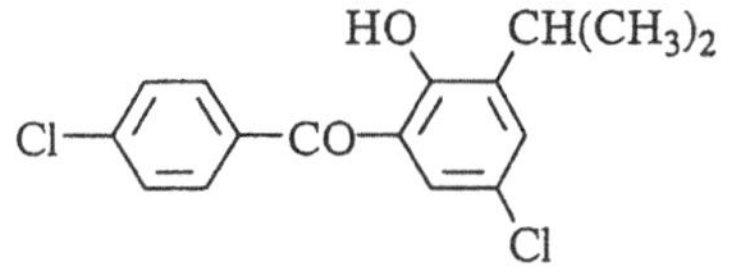

Synthesis

-Preparation by Fries rearrangement of 4-chloro-
2-isopropylphenyl 4-chlorobenzoate with aluminium
chloride at 160° for 15 min [698].

m.p. 54-55° [698];   Spectra (NA).

(5-Chloro-2-hydroxy-3-propylphenyl)(4-chlorophenyl)methanone

[93575-68-9]   C16H14Cl2O2   mol.wt. 309.19

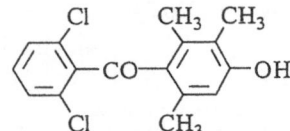

Synthesis

-Preparation by Fries rearrangement of 4-chloro-2-propyl-
 phenyl p-chlorobenzoate with aluminium chloride at 160°
 for 15 min [698].
-Also refer to: [683].

m.p. 55-56° [698];   Spectra (NA).

(3,5-Dichloro-4-hydroxyphenyl)(2,4,6-trimethylphenyl)methanone

C16H14Cl2O2   mol.wt. 309.19

Synthesis

-Preparation by Fries rearrangement of 2,6-dichloro-
 phenyl mesitylenecarboxylate (2,6-dichlorophenyl
 2,4,6-trimethylbenzoate) with aluminium chloride at 155°
 for 1 h (79%) [1357].

m.p. 201°5-203° [1357];   Spectra (NA).

(2,6-Dichlorophenyl)(4-hydroxy-2,3,6-trimethylphenyl)methanone

[183724-89-2]   C16H14Cl2O2   mol.wt. 309.19

Synthesis

-Preparation by demethylation of 2',6'-dichloro-4-methoxy-
 2,3,6-trimethylbenzophenone with hydrobromic acid in
 acetic acid for 1.5 h at 75° [313].

m.p. 122° [313];   Spectra (NA).

(2,6-Dichlorophenyl)(2-hydroxy-3,4-dimethoxy-6-methylphenyl)methanone

[183726-43-4]   C16H14Cl2O4   mol.wt. 341.19

Syntheses

-Preparation by acylation of 3,4,5-trimethoxytoluene with
 2,6-dichlorobenzoyl chloride in methylene chloride in the
 presence of aluminium chloride for 1 h at 0° and for 16 h at
 r.t. (30%) [313].
-Also obtained by partial demethylation of 2',6'-dichloro-
6-methyl-2,3,4-trimethoxybenzophenone with hydrobromic acid in acetic acid for 1.5 h at 75°
[313].
N.B.: K salt [313].

m.p. 161° [313];   Spectra (NA).

(4-Ethenylphenyl)(2-hydroxy-4-methoxyphenyl)methanone

[48177-42-0] $C_{16}H_{14}O_3$ mol.wt. 254.29

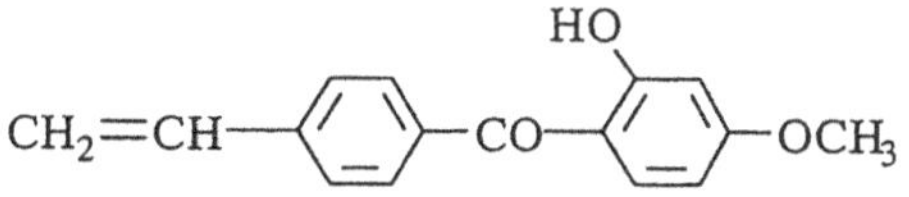

Synthesis

-Preparation by reaction of aqueous potassium hydroxide with 2-hydroxy-4-methoxy-4'-(2-bromoethyl)benzophenone in the presence of hydroquinone in refluxing methanol for 1.5 h under nitrogen bubbling (30%) [675].
-Also refer to: [1068].

m.p. 69-71° [675];
^{1}H NMR [675], IR [675], UV [675], MS [675].

[2-(Acetyloxy)phenyl](2-hydroxy-5-methoxyphenyl)methanone

[83570-58-5] $C_{16}H_{14}O_5$ mol.wt. 286.28

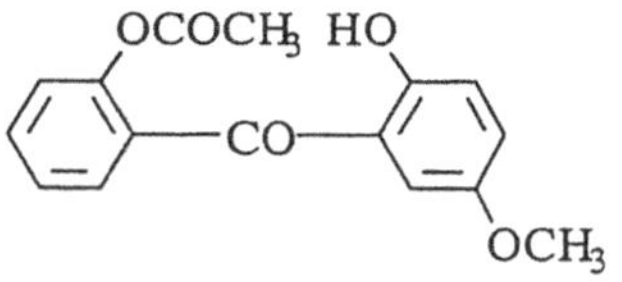

Synthesis

-Obtained by UV light irradiation of p-methoxyphenyl o-acetoxybenzoate (p-methoxyphenyl acetylsalicylate) in benzene [924], for 11 h (40%) [351].

yellow oil [351]; b.p. (NA);
^{1}H NMR [351], IR [351], UV [351], MS [351].

(5-Chloro-3-ethyl-2-hydroxyphenyl)(4-methylphenyl)methanone

[93575-40-7] $C_{16}H_{15}ClO_2$ mol.wt. 274.75

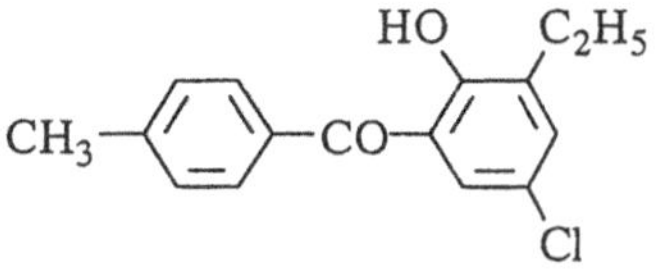

Synthesis

-Preparation by Fries rearrangement of 4-chloro-2-ethyl-phenyl p-toluate with aluminium chloride at 160° for 15 min [698].

m.p. 53-55° [698]; Spectra (NA).

(5-Chloro-2-hydroxy-3-methylphenyl)(3,4-dimethylphenyl)methanone

[86914-89-8] $C_{16}H_{15}ClO_2$ mol.wt. 274.75

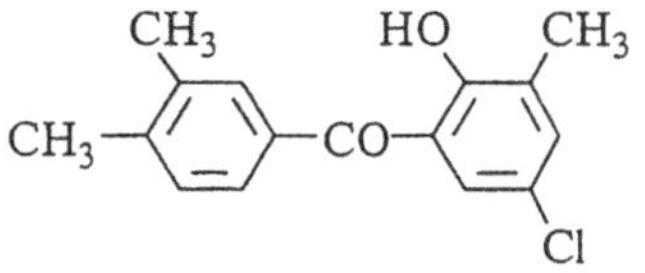

Synthesis

-Preparation by Fries rearrangement of 4-chloro-2-methyl-phenyl 3,4-dimethylbenzoate with aluminium chloride at 180° for 10 min [688].

m.p. 108° [688]; Spectra (NA).

(5-Chloro-2-hydroxy-3-methylphenyl)(4-ethylphenyl)methanone

[86914-74-1] $C_{16}H_{15}ClO_2$ mol.wt. 274.75

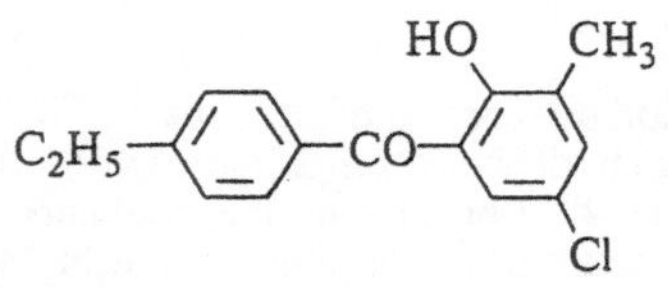

Synthesis

-Preparation by Fries rearrangement of 4-chloro-2-methyl-phenyl 4-ethylbenzoate with aluminium chloride at 180-185° for 10 min [684, 688].

m.p. 152-155° [684, 688];   Spectra (NA).

(4-Chlorophenyl)(2-hydroxy-3-propylphenyl)methanone

[108294-79-7] $C_{16}H_{15}ClO_2$ mol.wt. 274.75

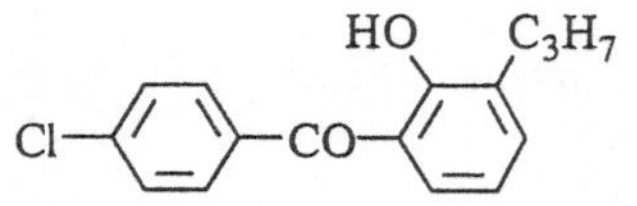

Synthesis

-Preparation by reaction of 2-hydroxy-3-propylbenzoyl chloride with chlorobenzene in the presence of aluminium chloride in refluxing carbon disulfide (35%) [400].

m.p. 68-70° [400];   Spectra (NA).

(4-Chlorophenyl)(2-hydroxy-5-propylphenyl)methanone

[61466-81-7] $C_{16}H_{15}ClO_2$ mol.wt. 274.75

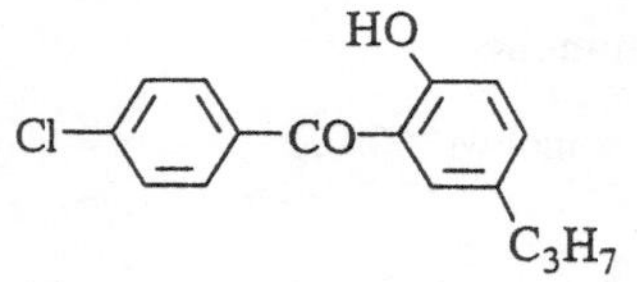

Synthesis

-Preparation by reaction of p-chlorobenzoyl chloride with p-propylphenol in the presence of aluminium chloride in tetrachloroethane at 105° for 22 h (27-28%) [400, 1206].

b.p.$_{0.02}$ 151-153° [400, 1206];   Spectra (NA).

(4-Fluorophenyl)(2-hydroxy-4-propoxyphenyl)methanone

                                      $C_{16}H_{15}FO_3$ mol.wt. 274.29

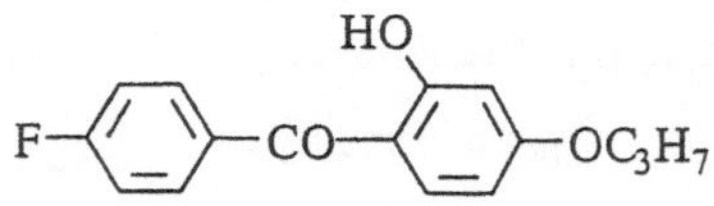

Synthesis

-Preparation by partial alkylation of 4'-fluoro-2,4-di-hydroxybenzophenone with a propyl halide in the presence of an alkali [394].

m.p. and Spectra (NA).

(2,3-Dimethyl-5-nitrophenyl)(2-hydroxy-3-methylphenyl)methanone

[110969-54-5] $C_{16}H_{15}NO_4$ mol.wt. 285.30

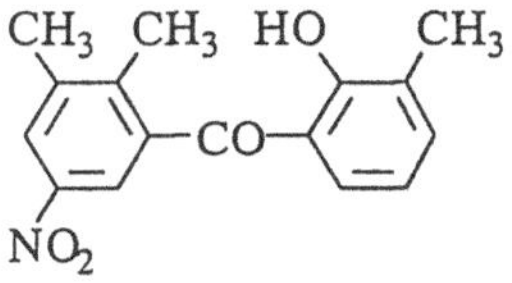

Synthesis

-Obtained (by-product) by Fries rearrangement of o-cresyl 2,3-dimethyl-5-nitrobenzoate with aluminium chloride at 160° for 2 h (<4%) [843].

m.p. 104° [843]; ^{1}H NMR [843], IR [843], UV [843].

(2,3-Dimethyl-5-nitrophenyl)(2-hydroxy-4-methylphenyl)methanone

[110969-57-8] $C_{16}H_{15}NO_4$ mol.wt. 285.30

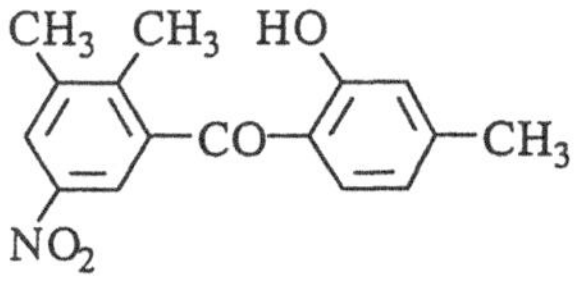

Synthesis

-Preparation by Fries rearrangement of m-cresyl 2,3-dimethyl-5-nitrobenzoate with aluminium chloride at 160° for 2 h (43%) [843].

m.p. 158° [843]; ^{1}H NMR [843], IR [843], UV [843].

(2,3-Dimethyl-5-nitrophenyl)(2-hydroxy-5-methylphenyl)methanone

[110969-60-3] $C_{16}H_{15}NO_4$ mol.wt. 285.30

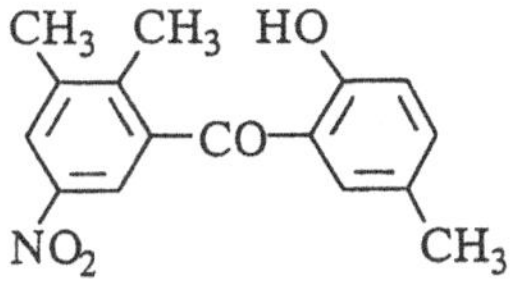

Synthesis

-Preparation by Fries rearrangement of p-cresyl 2,3-dimethyl-5-nitrobenzoate with aluminium chloride at 160° for 2 h (70%) [843].

m.p. 165° [843]; ^{1}H NMR [843], IR [843], UV [843].

(2,3-Dimethyl-5-nitrophenyl)(4-hydroxy-2-methylphenyl)methanone

[110969-58-9] $C_{16}H_{15}NO_4$ mol.wt. 285.30

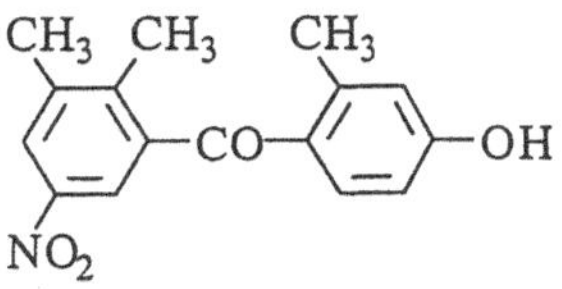

Synthesis

-Obtained (by-product) by Fries rearrangement of m-cresyl 2,3-dimethyl-5-nitrobenzoate with aluminium chloride at 160° for 2 h (12%) [843].

m.p. 222° [843]; ^{1}H NMR [843], IR [843], UV [843].

(2,3-Dimethyl-5-nitrophenyl)(4-hydroxy-3-methylphenyl)methanone

[110969-55-6] $C_{16}H_{15}NO_4$ mol.wt. 285.30

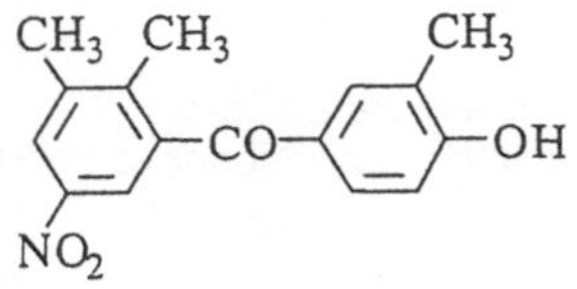

Synthesis

-Preparation by Fries rearrangement of o-cresyl 2,3-di-
methyl-5-nitrobenzoate with aluminium chloride at 160° for
2 h (56%) [843].

m.p. 210° [843]; ^{1}H NMR [843], IR [843], UV [843].

(2-Hydroxy-3-nitrophenyl)[4-(1-methylethyl)phenyl]methanone

[35698-18-1] $C_{16}H_{15}NO_4$ mol.wt. 285.30

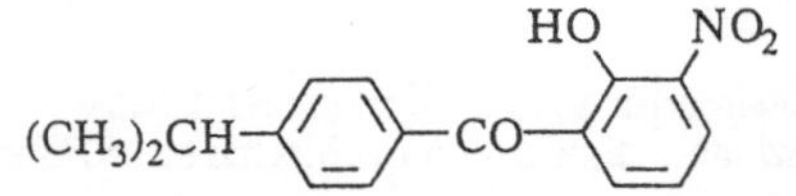

Synthesis

-Obtained by reaction of fuming nitric acid with
2-hydroxy-4'-isopropylbenzophenone in an acetic
acid/acetic anhydride mixture (4:3), first at 10° for
50 min, then at 20° for 40 min (19%) [469, 882].

m.p. and Spectra (NA).

(2-Hydroxy-5-nitrophenyl)[4-(1-methylethyl)phenyl]methanone

[35698-19-2] $C_{16}H_{15}NO_4$ mol.wt. 285.30

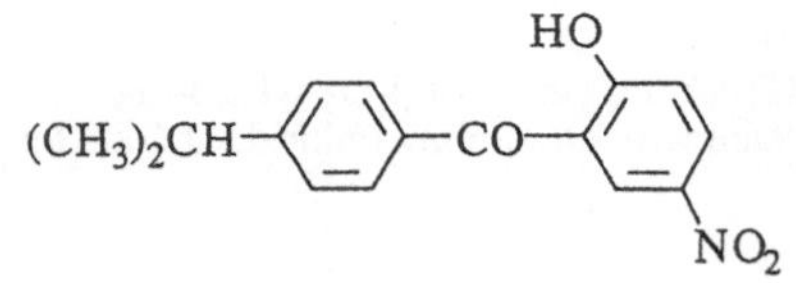

Synthesis

-Obtained by reaction of fuming nitric acid with
2-hydroxy-4'-isopropylbenzophenone in an acetic
acid/acetic anhydride mixture (4:3), first at 10° for
50 min, then at 20° for 40 min (10%) [469, 882].

m.p. and Spectra (NA).

(4,5-Dimethoxy-2-nitrophenyl)(2-hydroxy-4-methoxyphenyl)methanone

                                                     $C_{16}H_{15}NO_7$ mol.wt. 333.30

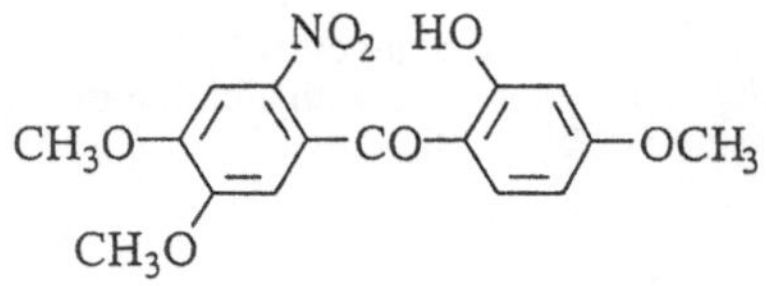

Synthesis

-Obtained by action of a mixture of nitric acid
(d = 1.42)/concentrated sulfuric acid with
2-hydroxy-3',4,4'-trimethoxybenzophenone in acetic
acid at 48-52° for 16 min [1106].

m.p. 211° [1106];   Spectra (NA).

(3,4-Dimethoxyphenyl)(4-hydroxy-3-methoxy-5-nitrophenyl)methanone

[134612-42-3] $C_{16}H_{15}NO_7$ mol.wt. 333.30

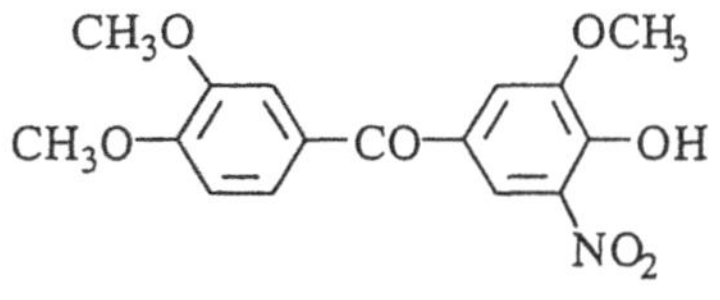

Synthesis

-Preparation by reaction of 65% nitric acid with
  4-hydroxy-3,3',4'-trimethoxybenzophenone in acetic
  acid at 20° [164].

m.p. 170-180° [164];   Spectra (NA).

(2,4-Dimethylphenyl)(2-hydroxy-5-methylphenyl)methanone

[0018-48-0] $C_{16}H_{16}O_2$ mol.wt. 240.30

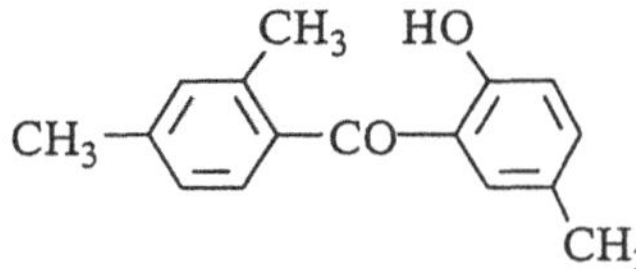

Synthesis

-Preparation by Fries rearrangement of p-cresyl 2,4-di-
  methylbenzoate with aluminium chloride [694].

m.p. and Spectra (NA).

(2,4-Dimethylphenyl)(3-hydroxy-2-methylphenyl)methanone

[76981-57-2] $C_{16}H_{16}O_2$ mol.wt. 240.30

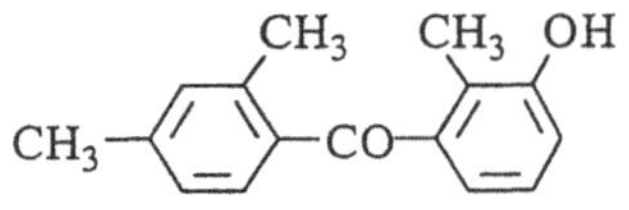

Synthesis

-Preparation by diazotization of 3-amino-2,2',4'-trimethyl-
  benzophenone, followed by hydrolysis of the diazonium salt
  so obtained (54%) [82], according to [731].

m.p.   124° [82];   Spectra (NA).

(2,6-Dimethylphenyl)(2-hydroxy-5-methylphenyl)methanone

                                  $C_{16}H_{16}O_2$ mol.wt. 240.30

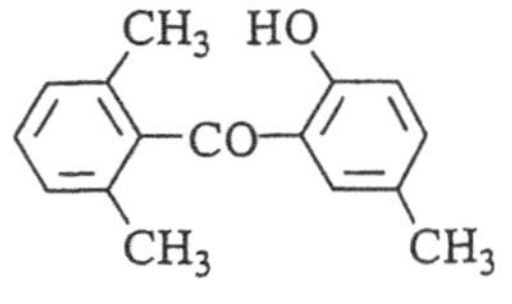

Synthesis

-Obtained by Fries rearrangement of p-cresyl 2,6-dimethyl-
  benzoate in the presence of aluminium chloride, first in
  refluxing carbon disulfide for 2 h, then at 150° for 1 h after
  elimination of solvent (22%) [454].

m.p.   89°7-90°7 [454];   Spectra (NA).

(: Ethyl-2-hydroxyphenyl)(2-methylphenyl)methanone

.170799-15-2] $C_{16}H_{16}O_2$ mol.wt. 240.30

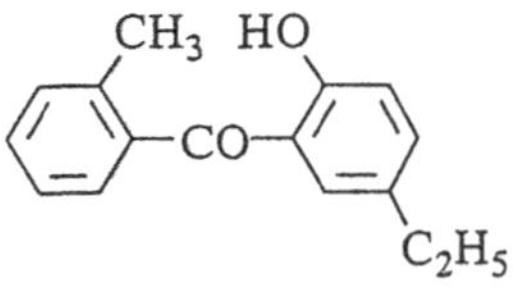

Synthesis

-Preparation by Fries rearrangement of p-ethylphenyl
  o-toluate with aluminium chloride at 100-150° for 0.5-3 h
  (95%) [1363].

oil [1363];   b.p. and Spectra (NA).

(5-Ethyl-2-hydroxyphenyl)(4-methylphenyl)methanone

[61750-25-2] $C_{16}H_{16}O_2$ mol.wt. 240.30

Syntheses

-Preparation by reaction of p-methylbenzoyl chloride with p-ethylphenol in the presence of aluminium chloride in tetrachloroethane at 105° for 22 h (38%) [400].
-Preparation by Fries rearrangement of p-ethylphenyl p-toluate with aluminium chloride in tetrachloroethane at 125° for 6 h [1207].

   m.p. 49-51° [400, 1207];   Spectra (NA).

(2-Hydroxy-3,4-dimethylphenyl)(4-methylphenyl)methanone

[16762-05-3] $C_{16}H_{16}O_2$ mol.wt. 240.30

Syntheses

-Preparation by Fries rearrangement of 2,3-dimethyl-phenyl p-toluate with aluminium chloride at 180° for 10 min (55%) [944].

-Also obtained by degradation of 6,7-dimethyl-3-(4-methylphenyl)-2-phenylbenzofuran with chromium trioxide in boiling acetic acid, followed by saponification of the keto ester so obtained (2-benzoyloxy-3,4-dimethyl-4'-methylbenzophenone) [1052, 1055].

   m.p. 84° [944, 1051, 1052, 1055];   IR [1052].

(2-Hydroxy-3,5-dimethylphenyl)(2-methylphenyl)methanone

[86914-79-6] $C_{16}H_{16}O_2$ mol.wt. 240.30

Synthesis

-Preparation by Fries rearrangement of 2,4-dimethylphenyl o-toluate with aluminium chloride at 180° for 10 min [688].

   oil [688];   b.p. and Spectra (NA).

(2-Hydroxy-3,5-dimethylphenyl)(4-methylphenyl)methanone

[86914-85-4] $C_{16}H_{16}O_2$ mol.wt. 240.30

Syntheses

-Preparation by Fries rearrangement of 2,4-dimethylphenyl p-toluate with aluminium chloride at 180° for 10 min [688], (60%) [944].
-Obtained by degradation of 5,7-dimethyl-3-(4-methyl-phenyl)-2-phenylbenzofuran in boiling acetic acid, followed by saponification of the keto ester so obtained (2-benzoyloxy-3,4',5-trimethyl-benzophenone) [1052].

   oil [1052];   m.p. 55-56° [688], 54° [944]; ^{1}H NMR [944], IR [944, 1052].

(2-Hydroxy-4,5-dimethylphenyl)(2-methylphenyl)methanone

[93433-88-6] $C_{16}H_{16}O_2$ mol.wt. 240.30

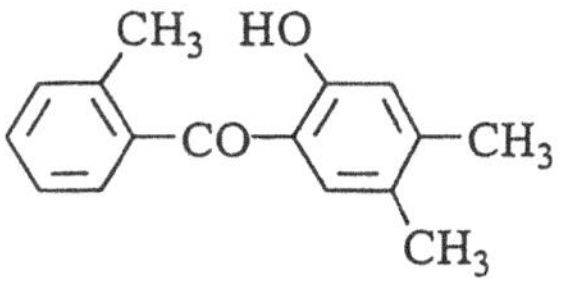

Synthesis

-Preparation by Fries rearrangement of 3,4-dimethylphenyl
o-toluate with aluminium chloride at 100-150° for 0.5-3 h
(80%) [1363].

m.p. 56-57° [1363];   Spectra (NA).

(2-Hydroxy-4,6-dimethylphenyl)(2-methylphenyl)methanone

[62261-96-5] $C_{16}H_{16}O_2$ mol.wt. 240.30

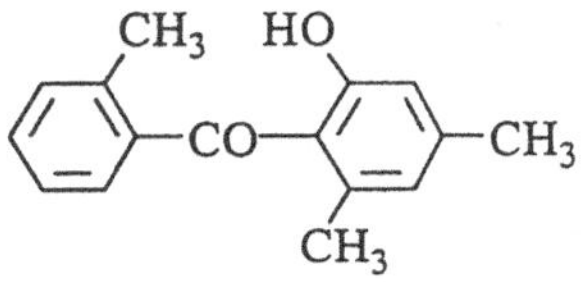

Synthesis

-Refer to: [147].

m.p. and Spectra (NA).

(2-Hydroxy-4,6-dimethylphenyl)(4-methylphenyl)methanone

[62261-95-4] $C_{16}H_{16}O_2$ mol.wt. 240.30

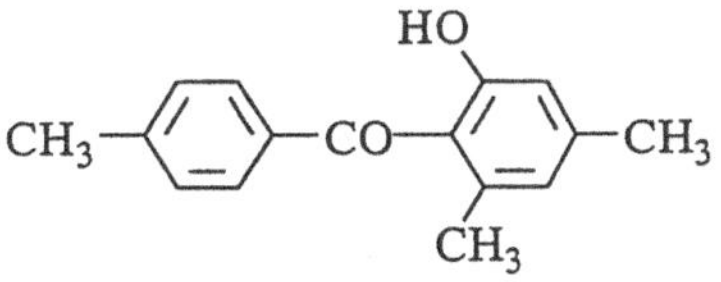

Synthesis

-Refer to: [147].

m.p. and Spectra (NA).

(4-Ethylphenyl)(2-hydroxy-4-methoxyphenyl)methanone

$C_{16}H_{16}O_3$ mol.wt. 256.30

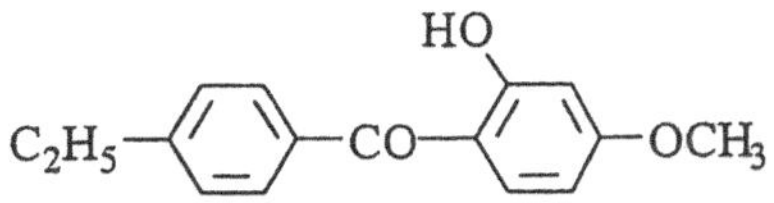

Synthesis

-Preparation by partial demethylation of 2,4-di-
methoxy-4'-ethylbenzophenone with aluminium
chloride in chlorobenzene at 80-100° [602].
-Also refer to: [43].

m.p. and Spectra (NA).

(2-Hydroxy-3,4-dimethylphenyl)(4-methoxyphenyl)methanone

[16762-06-4] $C_{16}H_{16}O_3$ mol.wt. 256.30

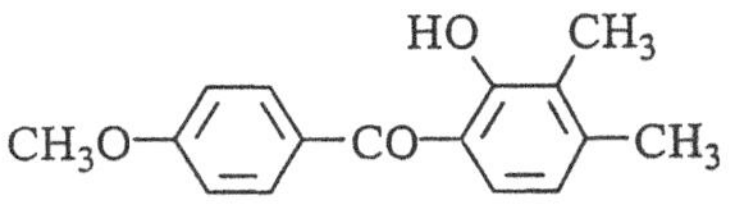

Synthesis

-Preparation by oxidation of 6,7-dimethyl-2-phenyl-
3-(4-methoxyphenyl)benzofuran with chromium
trioxide in boiling acetic acid, followed by
saponification of the keto ester so formed with sodium
hydroxide [1052] or potassium hydroxide [1055] in refluxing ethanol.

m.p.   86° [1052], 85° [1055];   IR [1052, 1055].

(2-Hydroxy-3,5-dimethylphenyl)(2-methoxyphenyl)methanone

[72324-22-2]   C$_{16}$H$_{16}$O$_3$ mol.wt. 256.30

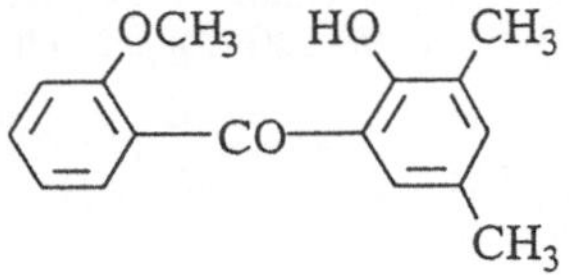

Synthesis

-Obtained (poor yields) by reaction of o-anisic acid with
 2,4-dimethylphenol,
*in the presence of methanesulfonic acid at 100° for 1 h (6%)
 [867];
*in the presence of butanesulfonic acid at 140° for 1 h (4%)
 [867].

m.p.   75° [867]; ^{1}H NMR (Sadtler: standard n° 38495 M),
IR (Sadtler: standard n° 65533 K) [867], UV [867], MS [195].

(2-Hydroxy-3,5-dimethylphenyl)(4-methoxyphenyl)methanone

[72324-20-0]   C$_{16}$H$_{16}$O$_3$ mol.wt. 256.30

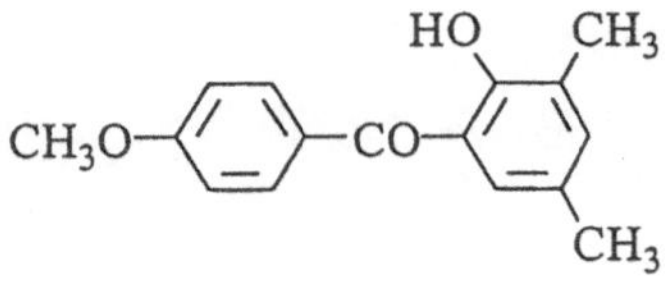

Synthesis

-Preparation by Fries rearrangement of 2,4-dimethylphenyl
 p-anisate with titanium tetrachloride at 120° for 1 h (82%)
 [867].

m.p.   66° [867]; ^{1}H NMR (Sadtler: standard n° 38496 M),
IR (Sadtler: standard n° 65534 K) [867], UV [867], MS [195].

(2-Hydroxy-4,5-dimethylphenyl)(4-methoxyphenyl)methanone

[54921-19-6]   C$_{16}$H$_{16}$O$_3$ mol.wt. 256.30

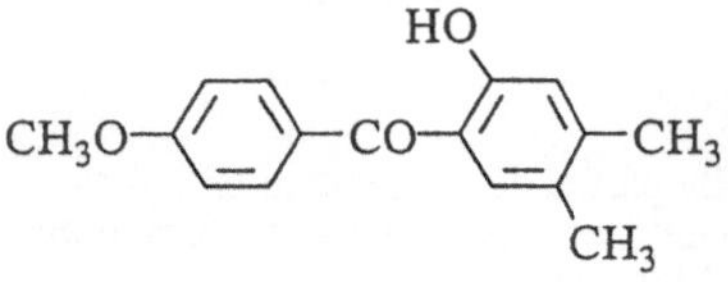

Synthesis

-Preparation by oxidation of 2-phenyl-3-(4-methoxy-
 phenyl)-5,6-dimethylbenzofuran with chromium
 trioxide in boiling acetic acid for 2 h, followed by
 saponification of the keto ester so formed with 10%
 sodium hydroxide in boiling ethanol for 15 min [994].

m.p.   92-93° [994]; ^{1}H NMR [994], IR [994].

(4-Hydroxy-3,5-dimethylphenyl)(2-methoxyphenyl)methanone

[72324-21-1]   C$_{16}$H$_{16}$O$_3$ mol.wt. 256.30

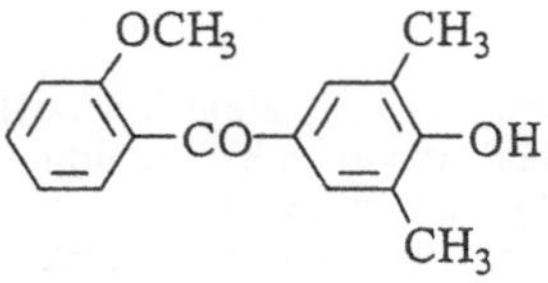

Synthesis

-Preparation by reaction of o-anisic acid with 2,6-dimethyl-
 phenol in the presence of methanesulfonic acid at 120° for
 20 min (46%) [867].

m.p. 145° [867];
^{1}H NMR (Sadtler: standard n° 38492 M),
IR (Sadtler: standard n° 65530 K) [867], UV [867], MS [195].

(4-Hydroxy-3,5-dimethylphenyl)(4-methoxyphenyl)methanone

[72324-19-7] $C_{16}H_{16}O_3$ mol.wt. 256.30

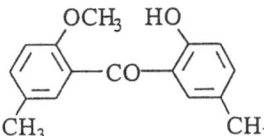

Synthesis

-Preparation by Fries rearrangement of 2,6-dimethyl-
 phenyl p-anisate in nitromethane,
*in the presence of titanium tetrachloride at 20° for
 170 h (82%) [867];
*in the presence of antimony pentachloride at 20° for
 25 h (40%) [867].

-Also refer to: [211].

m.p. 125° [867];
^{1}H NMR (Sadtler: standard n° 38500 M),
IR (Sadtler: standard n° 65538 K) [867], UV [867], MS [195].

(2-Hydroxy-4-methylphenyl)(2-methoxy-4-methylphenyl)methanone

[54468-79-0] $C_{16}H_{16}O_3$ mol.wt. 256.30

Synthesis

-Obtained by oxidation of 6-methyl-2-phenyl-
 3-(2-methoxy-4-methylphenyl)benzofuran with
 chromium trioxide, then saponification of the keto
 ester so formed, the 2-(benzoyloxy)-2'-methoxy-
 4,4'-dimethylbenzophenone [1053].

oil [1053];   b.p. (NA);   IR [1053].

(2-Hydroxy-5-methylphenyl)(2-methoxy-5-methylphenyl)methanone

[32229-35-9] $C_{16}H_{16}O_3$ mol.wt. 256.30

Syntheses

-Obtained by Friedel-Crafts acylation of excess p-methoxytoluene with 2-methoxy-5-methylbenzoyl chloride in the
 presence of aluminium chloride, first at 0° for 2 h, then at r.t.
 overnight and at reflux for 3 h (23%) [671].
-Also obtained (by-product) by Friedel-Crafts acylation of
p-methoxytoluene with 2-methoxy-5-methylbenzoyl chloride in carbon disulfide in the presence of
aluminium chloride, first at 0° for 5 h, then at r.t. overnight and at reflux for 2 h (10%) [671].
-Also refer to: [672].

m.p. 68-69° [671, 672];   IR [672], UV [672].

(4-Hydroxy-3-methylphenyl)(4-methoxy-3-methylphenyl)methanone

[79002-05-4] $C_{16}H_{16}O_3$ mol.wt. 256.30

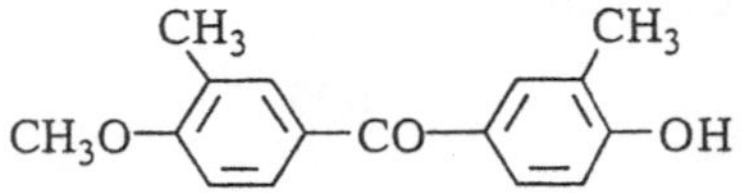

Synthesis

-Obtained by rapid degradation of 4-methoxy-3,3'-di-
methylbenzophenone [41295-28-7] (an herbicide) in
reductive flooded soils (main metabolite) [639].

m.p. and Spectra (NA).

(4-Ethoxy-2-hydroxyphenyl)(4-methoxyphenyl)methanone

$C_{16}H_{16}O_4$ mol.wt. 272.30

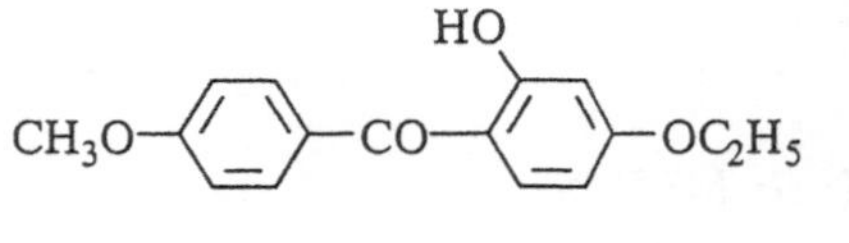

Synthesis

-Preparation by partial demethylation of 2,4'-di-
methoxy-4-ethoxybenzophenone with aluminium
chloride in chlorobenzene at 80-100° [602].
-Also refer to: [43].

m.p. and Spectra (NA).

(4-Ethoxyphenyl)(2-hydroxy-4-methoxyphenyl)methanone

$C_{16}H_{16}O_4$ mol.wt. 272.30

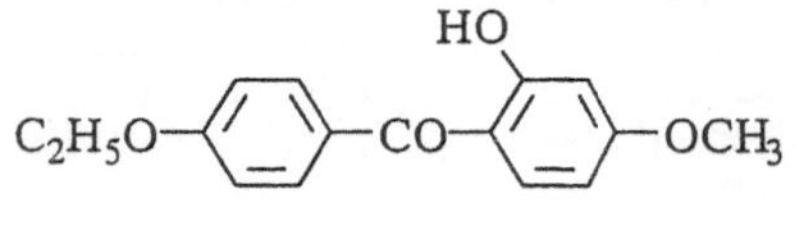

Syntheses

-Preparation by reaction of 2,4-dimethoxybenzoyl
chloride with phenetole in the presence of
aluminium chloride in a chlorobenzene/N,N-di-
methylformamide mixture (22:1) at 115° [539,
1067].
-Preparation by partial demethylation of 4'-ethoxy-2,4-dimethoxybenzophenone with aluminium
chloride or aluminium bromide in chlorobenzene at 90-95° (good yield) [602].

m.p. (NA);   UV [1067].

(2-Hydroxy-3,4-dimethoxyphenyl)(2-methylphenyl)methanone

[140665-41-4] $C_{16}H_{16}O_4$ mol.wt. 272.30

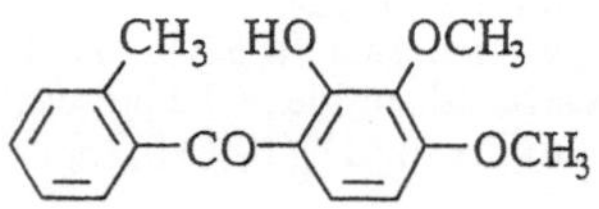

Synthesis

-Preparation by reaction of o-toluoyl chloride with 1,2,3-tri-
methoxybenzene in the presence of aluminium chloride in
ethylene dichloride between 0 to 5°, then at r.t. for 3 h and at
reflux for 1 h (63%) [1194].

m.p. 93-95° [1194];   Spectra (NA).

(2-Hydroxy-4-methoxy-3-methylphenyl)(2-methoxyphenyl)methanone

[115296-09-8] $C_{16}H_{16}O_4$ mol.wt. 272.30

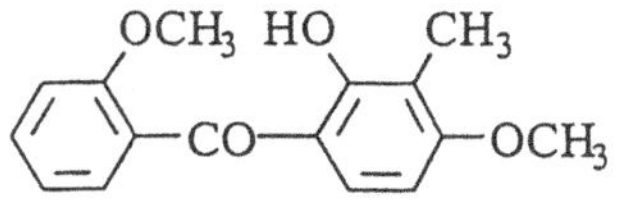

Synthesis

-Preparation by reaction of o-anisoyl chloride with 3-methoxy-2-methylphenol in methylene chloride in the presence of aluminium chloride first at 0° for 2 h and at r.t. for 1 h (31%) [456].

m.p. 111-115° [456];   Spectra (NA).

(2-Hydroxy-4-methoxy-3-methylphenyl)(4-methoxyphenyl)methanone

$C_{16}H_{16}O_4$ mol.wt. 272.30

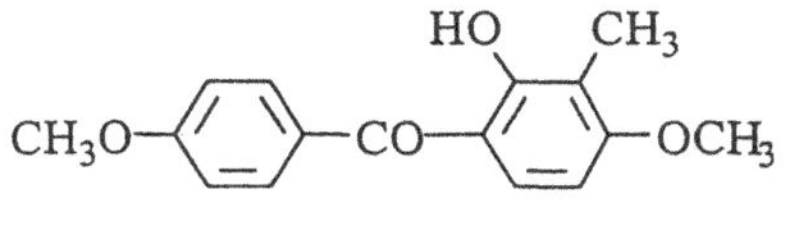

Synthesis

-Preparation by partial demethylation of 2,4,4'-trimethoxy-3-methylbenzophenone with aluminium chloride in chlorobenzene at 80-100° [602].
-Also refer to: [43].

m.p. and Spectra (NA).

(2,3-Dimethoxyphenyl)(2-hydroxy-3-methoxyphenyl)methanone

[35040-42-7] $C_{16}H_{16}O_5$ mol.wt. 288.30

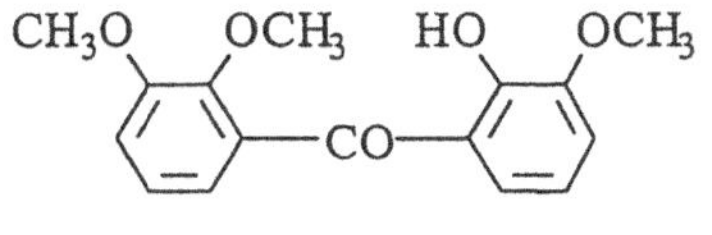

Synthesis

-Obtained by photo-Fries rearrangement of o-methoxyphenyl 2,3-dimethoxybenzoate in ethanol at r.t. for 22 h (16%) [420].

m.p. 110-110°5 [420]; ^{1}H NMR [420], IR [420], UV [420].

(2,3-Dimethoxyphenyl)(2-hydroxy-4-methoxyphenyl)methanone

[147188-10-1] $C_{16}H_{16}O_5$ mol.wt. 288.30

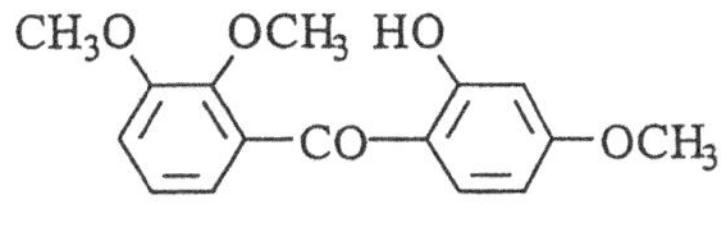

Synthesis

-Obtained, in mixture with 2'-hydroxy-2,3',4-trimethoxybenzophenone, by reaction of 2,3-dimethoxybenzoyl chloride with resorcinol dimethyl ether in ethyl ether in the presence of aluminium chloride for 8 h at r.t. (total yield: 54%) [825].

-Also refer to: [556, 826].

m.p. (NA); ^{1}H NMR [825].

(2,3-Dimethoxyphenyl)(2-hydroxy-5-methoxyphenyl)methanone

[35040-36-9] $C_{16}H_{16}O_5$ mol.wt. 288.30

Synthesis

-Obtained by photo-Fries rearrangement of p-methoxyphenyl 2,3-dimethoxybenzoate in ethanol at 30-40° for 9 h
 (42%) [420].

m.p. 74-76° [420]; ^{1}H NMR [420], IR [420], UV [420].

(2,3-Dimethoxyphenyl)(4-hydroxy-3-methoxyphenyl)methanone

[35042-49-0] $C_{16}H_{16}O_5$ mol.wt. 288.30

Synthesis

-Obtained (poor yield) by photo-Fries rearrangement of
 o-methoxyphenyl 2,3-dimethoxybenzoate in ethanol at r.t.
 for 22 h (<7%) [420].

m.p.   143-143°5 [420]; ^{1}H NMR [420], IR [420], UV [420].

(2,4-Dimethoxyphenyl)(2-hydroxy-3-methoxyphenyl)methanone

$C_{16}H_{16}O_5$ mol.wt. 288.30

Synthesis

-Obtained, in mixture with 2-hydroxy-2',3',4-tri-
 methoxybenzophenone, by reaction of 2,3-di-
 methoxybenzoyl chloride with resorcinol dimethyl
 ether in ethyl ether in the presence of aluminium
 chloride for 8 h at r.t. (total yield: 54%) [825].

-Also refer to: [826].

m.p. (NA); ^{1}H NMR [825].

(2,4-Dimethoxyphenyl)(2-hydroxy-4-methoxyphenyl)methanone

[4142-51-2] $C_{16}H_{16}O_5$ mol.wt. 288.30

Syntheses

-Preparation by reaction of 2,4-dimethoxybenzoic
 acid with m-methoxyphenol in the presence of
 boron trifluoride [785].
-Preparation by partial methylation of 2,2'-di-
hydroxy-4,4'-dimethoxybenzophenone with methyl iodide in the presence of potassium carbonate
in refluxing acetone for 12 h (68%) [542].
-Preparation by treatment of a difluoroboron chelate (SM) in methanol at 50° for 10 min (95%)
[1085, 1086]. SM was obtained by reaction of 2,2',4,4'-tetramethoxybenzophenone with boron
trifluoride-etherate in refluxing toluene (86%, m.p. 160-161°) [1085, 1086]. **N.B.**: A mixture of
products is obtained from aluminium chloride, induced cleavage of 2,2',4,4'-tetramethoxy-
benzophenone [480].
-Also obtained by reaction of 2,4-dimethoxybenzoyl chloride with resorcinol dimethyl ether in ethyl
ether in the presence of aluminium chloride for 8 h at r.t. (52%) [825].

-Also refer to: [43, 826, 1395].

m.p.   110-111°5 [542], 108-109° [1085, 1086], 104-105° [785];
^{1}H NMR [542, 1086], IR [542, 785, 825, 1086], UV [785], MS [542].

(2,4-Dimethoxyphenyl)(2-hydroxy-5-methoxyphenyl)methanone

[169455-12-3] $C_{16}H_{16}O_5$ mol.wt. 288.30

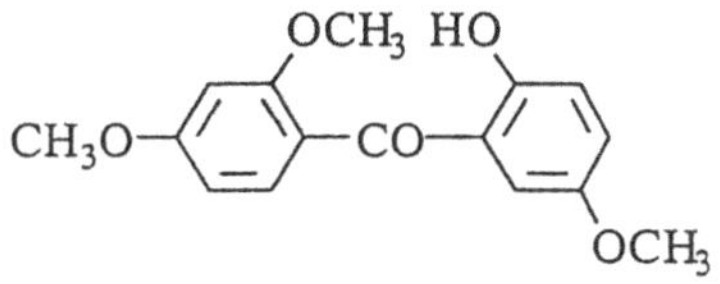

Synthesis

-Obtained, in mixture with 2'-hydroxy-
2,4',5-trimethoxybenzophenone, by reaction of
2,4-dimethoxybenzoyl chloride with hydroquinone
dimethyl ether in ethyl ether in the presence of aluminium
chloride for 8 h at r.t. (total yield: 54%) [825].

-Also refer to: [556, 826].

m.p. (NA);   ^{1}H NMR [825].

(2,4-Dimethoxyphenyl)(2-hydroxy-6-methoxyphenyl)methanone

[147188-11-2] $C_{16}H_{16}O_5$ mol.wt. 288.30

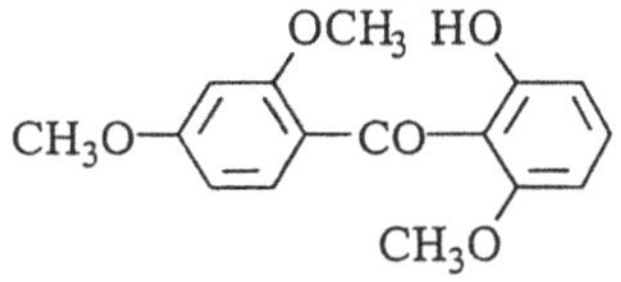

Synthesis

-Obtained, in mixture with 2'-hydroxy-2,4',6-tri-
methoxybenzophenone, by reaction of 2,4-dimethoxy-
benzoyl chloride with resorcinol dimethyl ether in ethyl
ether in the presence of aluminium chloride for 8 h at r.t.
(total yield: 52%) [825].

-Also refer to: [556, 826].

m.p. (NA);   ^{1}H NMR [825].

(2,5-Dimethoxyphenyl)(2-hydroxy-3-methoxyphenyl)methanone

[129168-52-1] $C_{16}H_{16}O_5$ mol.wt. 288.30

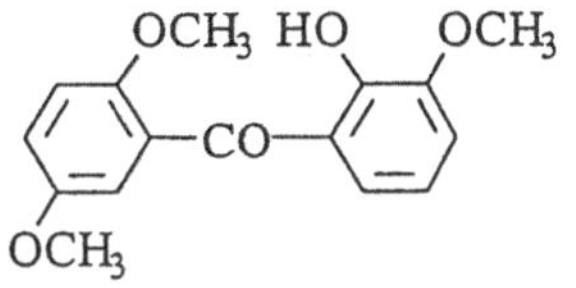

Synthesis

-Preparation by Friedel-Crafts acylation of hydroquinone
dimethyl ether with 2,3-dimethoxybenzoyl chloride in ethyl
ether in the presence of aluminium chloride for 1 h at 45-50°
(72%) [845].

m.p. 100-100°5 [845];   Spectra (NA).

(2,5-Dimethoxyphenyl)(2-hydroxy-4-methoxyphenyl)methanone

[62495-96-9] $C_{16}H_{16}O_5$ mol.wt. 288.30

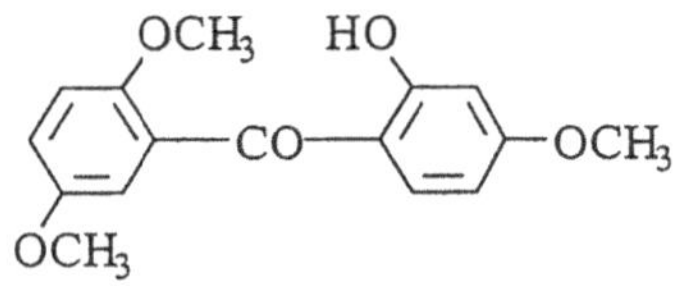

Syntheses

-Preparation by reaction of 2,5-dimethoxybenzoic acid
with 3-methoxyphenol in the presence of zinc chloride
and phosphorous oxychloride at 65-70° for 2 h (21%)
[1395].

-Also obtained, in mixture with 2-hydroxy-2',4',5-trimethoxybenzophenone, by reaction of
  2,4-dimethoxybenzoyl chloride with hydroquinone dimethyl ether in ethyl ether in the presence of
  aluminium chloride for 8 h at r.t. (total yield: 54%) [825].
-Also refer to: [826].

  m.p.   110° [1395];
  ^{1}H NMR [825, 1395], IR [1395], UV [1395].

(2,6-Dimethoxyphenyl)(2-hydroxy-3-methoxyphenyl)methanone

[37570-57-3]   C$_{16}$H$_{16}$O$_5$ mol.wt. 288.30

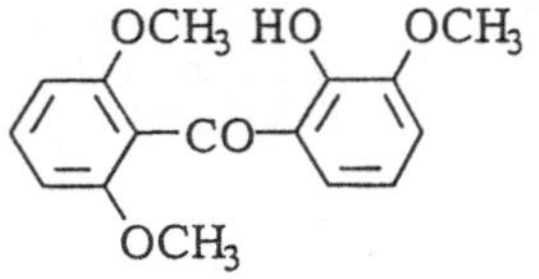

Synthesis

-Obtained by demethylation of 2,2',3,6'-tetramethoxy-
  benzophenone,
  *in the presence of boron trichloride in methylene chloride
   for 30 min (88%) (high selectivity) [1096];
  *in the presence of 40% aqueous hydrobromic acid in
refluxing acetic acid for 2 h (17%) [1096].

  m.p.   146-148° [1096];
  ^{1}H NMR [1096], IR [1096], MS [1096].

(2,6-Dimethoxyphenyl)(2-hydroxy-4-methoxyphenyl)methanone

[147188-12-3]   C$_{16}$H$_{16}$O$_5$ mol.wt. 288.30

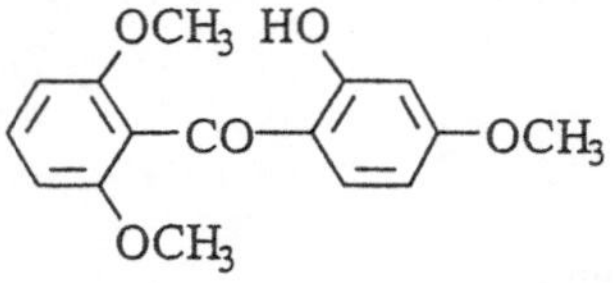

Synthesis

-Obtained, in mixture with 2-hydroxy-2',4',6-trimethoxy-
  benzophenone, by reaction of 2,4-dimethoxybenzoyl
  chloride with resorcinol dimethyl ether in ethyl ether in the
  presence of aluminium chloride for 8 h at r.t. (total yield:
  52%) [825].

-Also refer to: [826].

  m.p. (NA); ^{1}H NMR [825].

(2,6-Dimethoxyphenyl)(2-hydroxy-6-methoxyphenyl)methanone

[129103-93-1]   C$_{16}$H$_{16}$O$_5$ mol.wt. 288.30

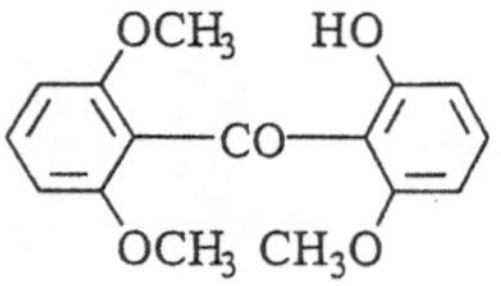

Synthesis

-Obtained from 2-iodo-3-methoxyphenyl 2,6-dimethoxy-
  benzoate by rearrangement on treatment with n-butyllithium
  in a mixture of ethyl ether, hexane and tetrahydrofuran at
  -100° followed by heating to -70°, then treatment with
  saturated aqueous ammonium chloride (9%) [598].

  m.p. and Spectra (NA).

(3,4-Dimethoxyphenyl)(2-hydroxy-4-methoxyphenyl)methanone

[42045-60-3] $C_{16}H_{16}O_5$ mol.wt. 288.30

Syntheses

-Obtained by reaction of methyl iodide with 2,4-di-
hydroxy-3',4'-dimethoxybenzophenone in the
presence of potassium carbonate in refluxing
acetone [1274], (89%) [650].

-Also obtained from β-(5-methoxy-2-veratroylphenoxy)propionic acid by heating with aqueous
  sodium hydroxide solution [1274].
-Preparation by reaction of 3,4-dimethoxybenzoic acid with m-methoxyphenol in the presence of
  boron trifluoride [785].
-Preparation by reaction of veratroyl chloride,
*with resorcinol monomethyl ether in the presence of aluminium chloride in nitrobenzene, first at 0°
  for 3 h, then at 10-16° for 18 h [1106];
*with resorcinol dimethyl ether in the presence of aluminium chloride in refluxing carbon disulfide
  for 1 h (61%) [58].
-Also obtained by condensation of ethyl m-methoxyphenoxypropionate with veratroyl chloride in
  the presence of aluminium chloride in nitrobenzene at 0° for 12 h [1106].
-Also refer to: [43, 982, 1067, 1395].

m.p.   141° [1106, 1274], 140-141° [58], 140° [650], 135-136° [785];
IR [785], UV [785].

(3,4-Dimethoxyphenyl)(4-hydroxy-2-methoxyphenyl)methanone

$C_{16}H_{16}O_5$ mol.wt. 288.30

Syntheses

-Obtained by reaction of veratroyl chloride with
resorcinol monomethyl ether in the presence of
aluminium chloride in nitrobenzene, first at 0° for 3 h,
then at 10-16° for 18 h [1106].

-Also obtained from β-(3-methoxy-4-veratroylphenoxy)propionic acid on heating with an aqueous
  sodium hydroxide solution [1274].

m.p.   179° [1274], 175° [1106];   Spectra (NA).

(2-Hydroxy-3,4-dimethoxyphenyl)(2-methoxyphenyl)methanone

[147188-08-7] $C_{16}H_{16}O_5$ mol.wt. 288.30

Synthesis

-Obtained, in mixture with 2'-hydroxy-2,3,4-trimethoxy-
benzophenone, by reaction of 2-methoxybenzoyl chloride
with pyrogallol trimethyl ether in ethyl ether in the presence
of aluminium chloride for 8 h at r.t. (total yield: 56-59%)
[825, 826].

-Also refer to: [556].

m.p. (NA);   ¹H NMR [825].

(2-Hydroxy-4,5-dimethoxyphenyl)(2-methoxyphenyl)methanone

[42833-48-7]   C16H16O5   mol.wt. 288.30

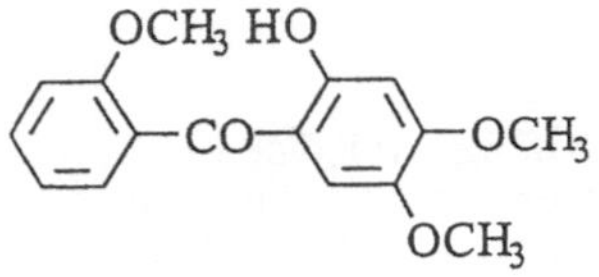

Syntheses

-Obtained, in mixture with 2,4,5-trimethoxy-2'-hydroxy-
benzophenone, by reaction of o-anisoyl chloride with
1,2,4-trimethoxybenzene in ethyl ether in the presence of
aluminium chloride for 20 h at r.t. (total yield: 56%) [825,
1097] or for 48 h at r.t. [372].

-Also obtained from 2,5-dihydroxy-2',4-dimethoxybenzophenone on treatment with ethereal
methanolic diazomethane for 4.5 h [1097].
-Also refer to: [826].

   m.p.   104-105° [1097], 98° [372];
   1H NMR [372, 825, 1097], UV [1097], MS [1097].

(2-Hydroxy-4,5-dimethoxyphenyl)(3-methoxyphenyl)methanone

[51106-90-2]   C16H16O5   mol.wt. 288.30

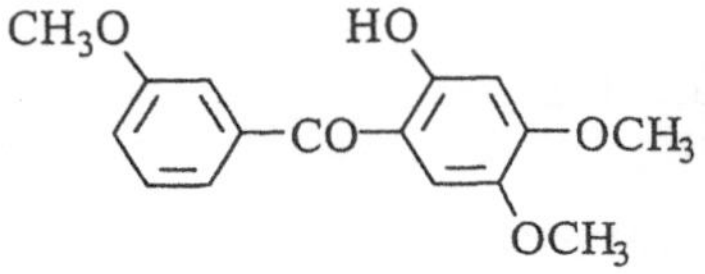

Synthesis

-Preparation by acylation of 1,2,4-trimethoxybenzene with
m-anisoyl chloride in ethyl ether in the presence of
aluminium chloride at r.t. for 48 h (35%) [372].

   m.p. 94° [372];   1H NMR [372].

(2-Hydroxy-4,5-dimethoxyphenyl)(4-methoxyphenyl)methanone

[58115-11-0]   C16H16O5   mol.wt. 288.30

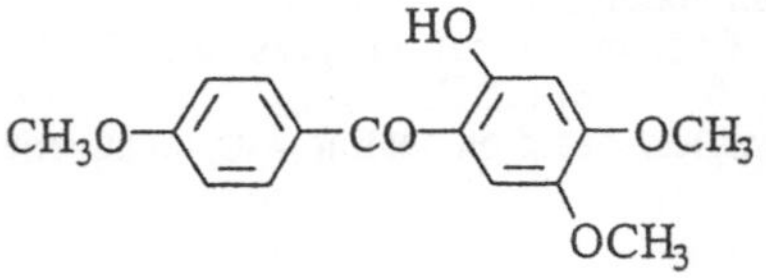

Synthesis

-Obtained (by-product) by condensation of p-anisoyl
chloride with 1,2,4-trimethoxybenzene in the
presence of aluminium chloride in refluxing carbon
disulfide for 6 h [134, 371] or at r.t. several days
[135].

   m.p.   127-128° [134, 135], 124-125° [371];   Spectra (NA).

(2-Hydroxy-4,6-dimethoxyphenyl)(2-methoxyphenyl)methanone

[147188-04-3]   C16H16O5   mol.wt. 288.30

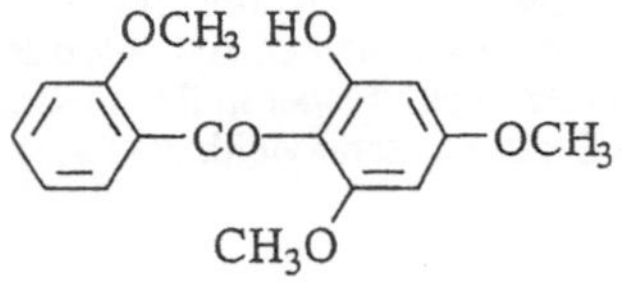

Synthesis

-Obtained, in mixture with 2'-hydroxy-2,4,6-trimethoxy-
benzophenone, by reaction of 2-methoxybenzoyl chloride
with phloroglucinol trimethyl ether in ethyl ether in the
presence of aluminium chloride for 8 h at r.t. (total yield:
59%) [825].

-Also refer to: [556, 826].

   m.p. (NA);   1H NMR [825].

(2-Hydroxy-4,6-dimethoxyphenyl)(3-methoxyphenyl)methanone

[21382-23-0] $C_{16}H_{16}O_5$ mol.wt. 288.30

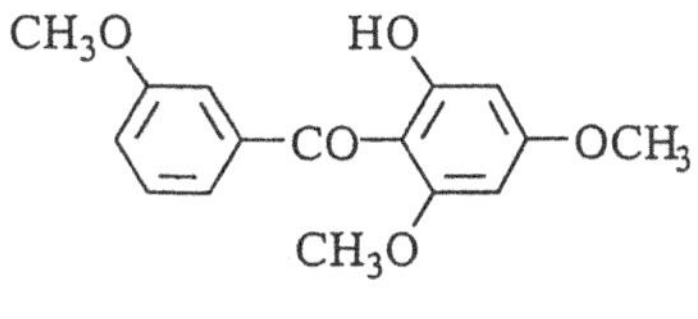

Synthesis

-Preparation by partial methylation of 2,3',4,6-tetra-
hydroxybenzophenone or of 2,4,6-trihydroxy-
3'-methoxybenzophenone with dimethyl sulfate in the
presence of potassium carbonate in refluxing acetone for
6 h (45-50%) [85].

Isolation from natural source

-From fresh *Gentiana lutea* rhizome (Gentianaceae) [84].

m.p. 123-125° [85];   b.p.$_{0.1}$ 180-188° [85];
^{1}H NMR [85], IR [85], UV [85].

(2-Hydroxy-4,6-dimethoxyphenyl)(4-methoxyphenyl)methanone

[97746-14-0] $C_{16}H_{16}O_5$ mol.wt. 288.30

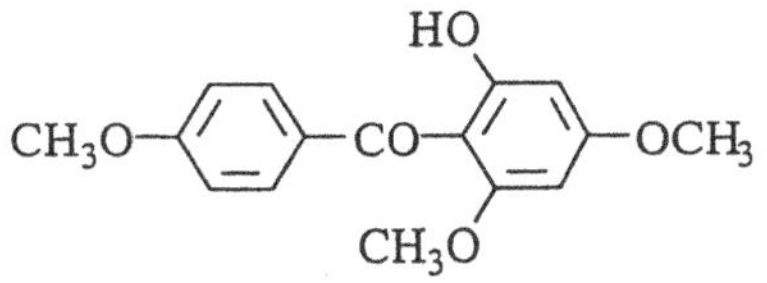

Synthesis

-Preparation by Friedel-Crafts acylation of 1,3,5-tri-
methoxybenzene with p-methoxybenzoyl chloride
in the presence of aluminium chloride in ethyl ether
for 4 h (70%) [29].

m.p. 150-151° [29]; ^{1}H NMR [29].

(4-Hydroxy-3,5-dimethoxyphenyl)(4-methoxyphenyl)methanone

[54808-44-5] $C_{16}H_{16}O_5$ mol.wt. 288.30

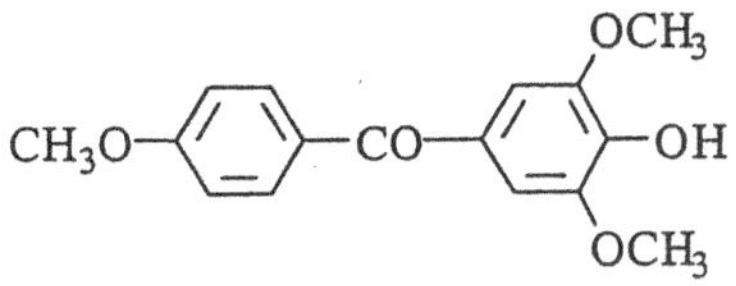

Synthesis

-Refer to: [284].

m.p. and Spectra (NA).

(2-Amino-4,5-dimethoxyphenyl)(2-hydroxy-4-methoxyphenyl)methanone *(Hydrochloride)*

$C_{16}H_{17}NO_5$,HCl   mol.wt. 339.78

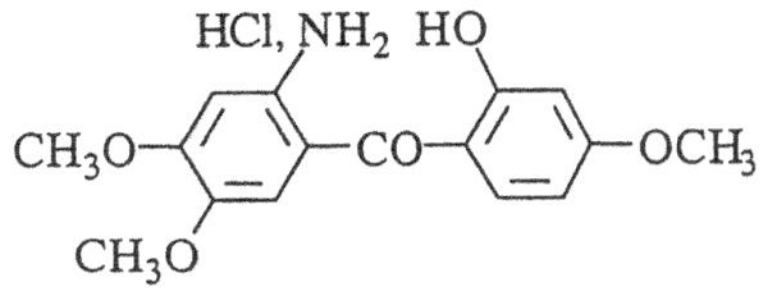

Synthesis

-Preparation by reduction of 2-hydroxy-2'-nitro-
4,4',5'-trimethoxybenzophenone in ethanol with
stannous chloride, tin foil and concentrated
hydrochloric acid on a steam bath for 15 min [1106].

m.p. 240° [1106];   Spectra (NA).

[5-Chloro-2-hydroxy-3-(1-methyl-2-propenyl)phenyl](4-chlorophenyl)methanone

[93575-41-8] $C_{17}H_{14}Cl_2O_2$ mol.wt. 321.20

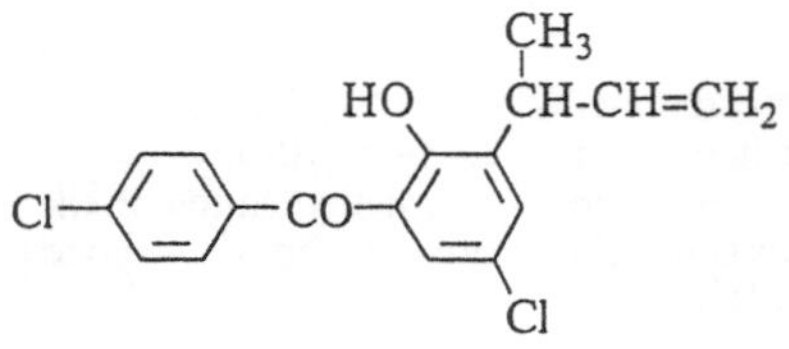

Synthesis

-Preparation by Fries rearrangement of 4-chloro-2-(1-methylallyl)phenyl p-chlorobenzoate with aluminium chloride at 160° for 15 min [698].

m.p. 45-47° [698];   Spectra (NA).

[5-(1,1-Dimethylethyl)-2-hydroxy-3-nitrophenyl](4-chlorophenyl)methanone

$C_{17}H_{16}ClNO_4$ mol.wt. 333.77

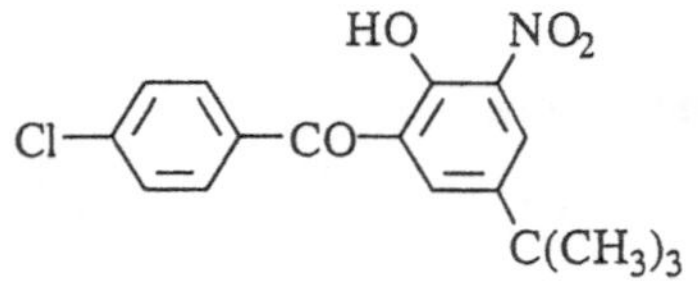

Synthesis

-Preparation by reaction of a concentrated nitric acid/concentrated sulfuric acid mixture with 5-tert-butyl-4'-chloro-2-hydroxybenzophenone in methylene chloride at 5° for 30 min (43%) [932].

m.p.   96-98° [932]; ^{1}H NMR [932];   TLC [932].

(3-Butyl-5-chloro-2-hydroxyphenyl)(4-chlorophenyl)methanone

[93575-74-7] $C_{17}H_{16}Cl_2O_2$ mol.wt. 323.22

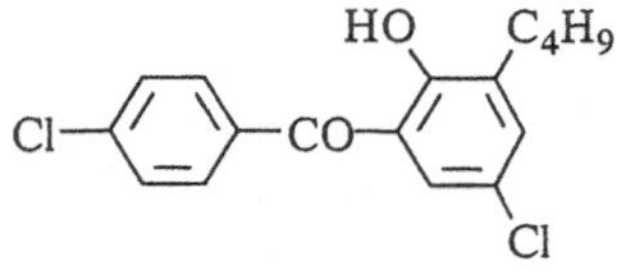

Synthesis

-Preparation by Fries rearrangement of 2-butyl-4-chlorophenyl p-chlorobenzoate with aluminium chloride at 160° for 15 min [698].

oil [698];   b.p.$_{0.04}$ 168-170° [698];   Spectra (NA).

[5-Chloro-3-(1,1-dimethylethyl)-2-hydroxyphenyl](4-chlorophenyl)methanone

[93575-72-5] $C_{17}H_{16}Cl_2O_2$ mol.wt. 323.22

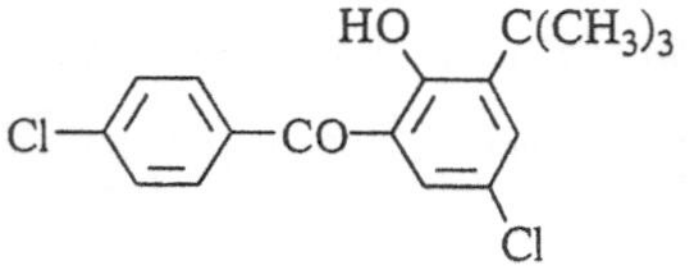

Synthesis

-Preparation by photo-Fries rearrangement of 2-tert-butyl-4-chlorophenyl p-chlorobenzoate in benzene under nitrogen for 32 h [698].

m.p.   86°5-87° [698];   Spectra (NA).

[5-Chloro-2-hydroxy-3-(2-methylpropyl)phenyl](4-chlorophenyl)methanone

[93575-75-8] $C_{17}H_{16}Cl_2O_2$ mol.wt. 323.22

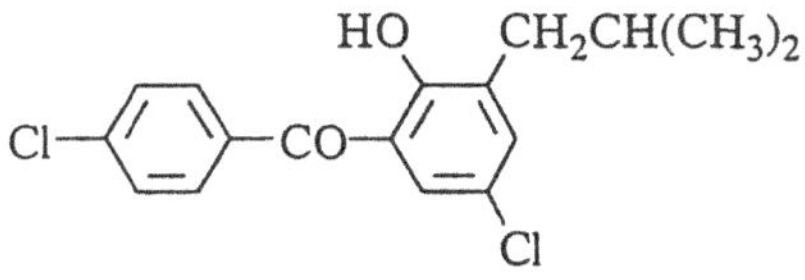

Synthesis

-Preparation by Fries rearrangement of 2-iso-
butyl-4-chlorophenyl 4-chlorobenzoate with
aluminium chloride at 160° for 15 min [698].

m.p. 65-66° [698];   Spectra (NA).

(2,4-Dichlorophenyl)[5-(1,1-dimethylethyl)-2-hydroxyphenyl]methanone

[61709-37-3] $C_{17}H_{16}Cl_2O_2$ mol.wt. 323.22

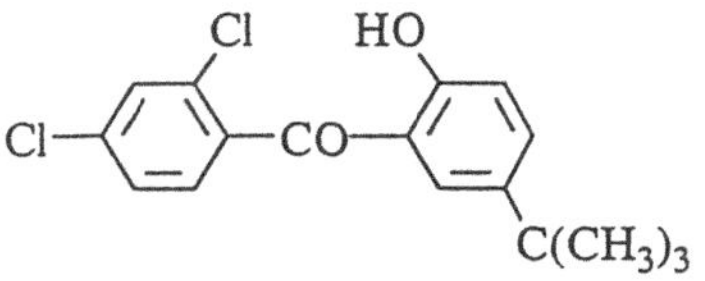

Synthesis

-Preparation by demethylation of 5-tert-butyl-2',4'-di-
chloro-2-methoxybenzophenone (SM) with aluminium
chloride in methylene chloride at 10°. SM was obtained
by Friedel-Crafts acylation of p-tert-butylanisole with
2,4-dichlorobenzoyl chloride in methylene chloride
in the presence of aluminium chloride at 10° [812].
-Also refer to: [1226].

m.p. 42-43° [812];   Spectra (NA).

(2,6-Dichlorophenyl)[3-(1,1-dimethylethyl)-4-hydroxyphenyl]methanone

[124979-18-6] $C_{17}H_{16}Cl_2O_2$ mol.wt. 323.22

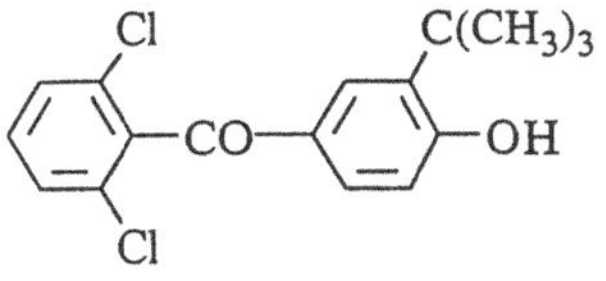

Synthesis

-Preparation by Friedel-Crafts acylation of o-tert-butylphenol with 2,6-dichlorobenzoyl chloride in ethylene
dichloride in the presence of titanium tetrachloride, first at
0°, then at r.t. [733].

m.p. 229-230° [733];   Spectra (NA).

(3,4-Dichlorophenyl)[5-(1,1-dimethylethyl)-2-hydroxyphenyl]methanone

$C_{17}H_{16}Cl_2O_2$ mol.wt. 323.22

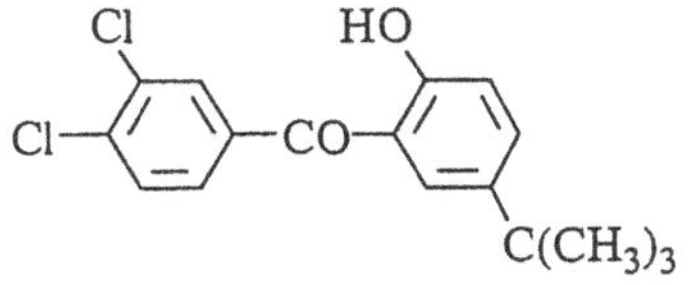

Synthesis

-Obtained by photo-Fries rearrangement of p-tert-
butylphenyl 3,4-dichlorobenzoate in benzene (37%)
[743].

m.p. 110-111° [743];   Spectra (NA).

(2,4-Dichlorophenyl)[4-hydroxy-2-methyl-5-(1-methylethyl)phenyl]methanone

[72236-97-6] $C_{17}H_{16}Cl_2O_2$ mol.wt. 323.22

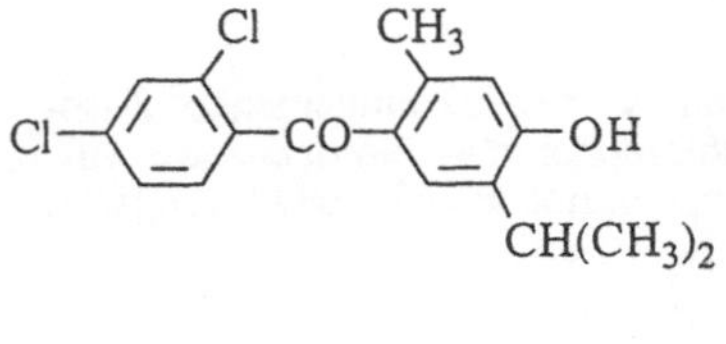

Synthesis

-Preparation by Fries rearrangement of the intermediate 5-methyl-2-isopropylphenyl 2,4-dichlorobenzoate formed *in situ* by reaction of 2,4-dichlorobenzoyl chloride with thymol in the presence of aluminium chloride in nitrobenzene at r.t. for 24 h (70%) [108].

m.p. 132° [108]; ^{1}H NMR [108].

(3,4-Dichlorophenyl)[4-hydroxy-2-methyl-5-(1-methylethyl)phenyl]methanone

[72236-99-8] $C_{17}H_{16}Cl_2O_2$ mol.wt. 323.22

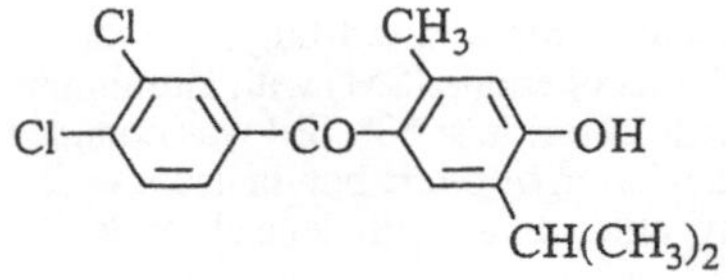

Synthesis

-Preparation by demethylation of 4-(3,4-dichloro-benzoyl)-5-methyl-2-isopropylanisole with boiling pyridinium chloride for 15 min (80%) [108].

m.p. 133° [108]; ^{1}H NMR [108].

(2,4-Dichlorophenyl)[2-hydroxy-5-(1-methylpropyl)phenyl]methanone

[59746-94-0] $C_{17}H_{16}Cl_2O_2$ mol.wt. 323.22

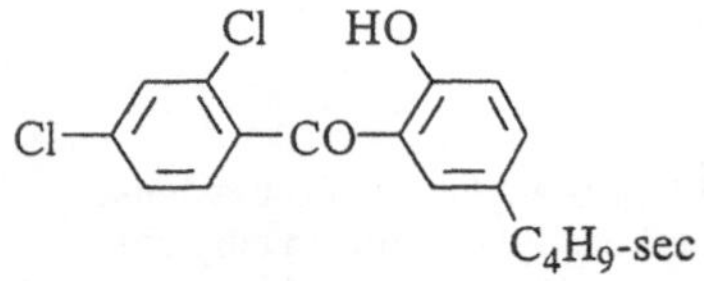

Synthesis

-Preparation by reaction of 2,4-dichlorobenzotrichloride with 4-sec-butylphenol in hydrofluoric acid in the presence of water at -10°, then between 0 to -10° for 2 h, at r.t. for 7 h and at 80° for 30 min into an autoclave (68%) [393].

-Also refer to: [1226].

yellow oil [393];   b.p.$_{0.05}$ 170° [393];   Spectra (NA).

(2,5-Dichlorophenyl)[2-hydroxy-5-(1-methylpropyl)phenyl]methanone

$C_{17}H_{16}Cl_2O_2$ mol.wt. 323.22

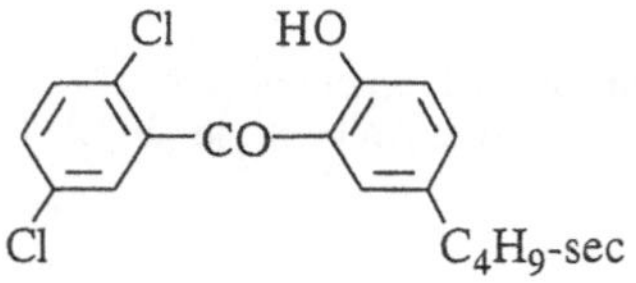

Synthesis

-Preparation by reaction of 2,5-dichlorobenzotrichloride with 4-sec-butylphenol in hydrofluoric acid in the presence of water at -10°, then between 0 to -10° for 2 h, at r.t. for 7 h and at 80° for 30 min into an autoclave [393].

m.p. and Spectra (NA).

(3,4-Dichlorophenyl)[2-hydroxy-5-(1-methylpropyl)phenyl]methanone

$C_{17}H_{16}Cl_2O_2$ mol.wt. 323.22

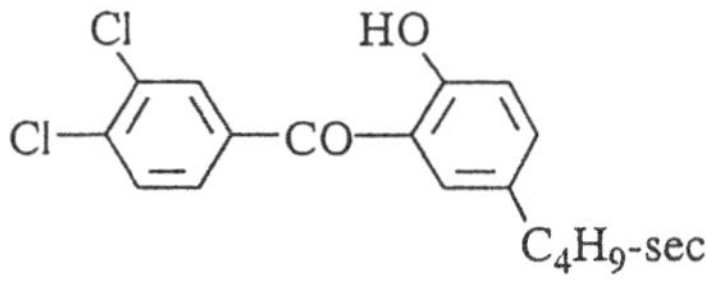

Synthesis

-Preparation by reaction of 3,4-dichlorobenzo-
trichloride with 4-sec-butylphenol in hydrofluoric acid
in the presence of water at -10°, then between 0 to
-10° for 2 h, at r.t. for 7 h and at 80° for 30 min into
an autoclave [393].

m.p. and Spectra (NA).

(3,5-Dinitrophenyl)[5-(1,1-dimethylethyl)-2-hydroxyphenyl]methanone

[93332-04-8] $C_{17}H_{16}N_2O_6$ mol.wt. 344.32

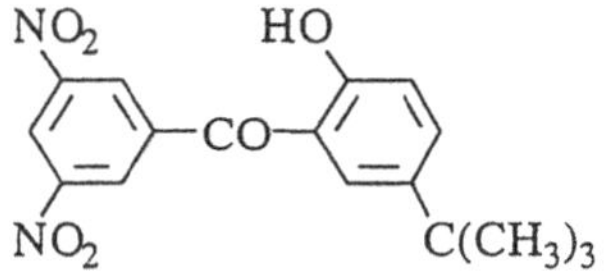

Synthesis

-Obtained (poor yield) by photo-Fries rearrangement of
p-tert-butylphenyl 3,5-dinitrobenzoate in benzene (7%)
[743].

m.p. 133-134° [743];   Spectra (NA).

(4-Ethenylphenyl)(4-ethoxy-2-hydroxyphenyl)methanone

[80167-00-6] $C_{17}H_{16}O_3$ mol.wt. 268.31

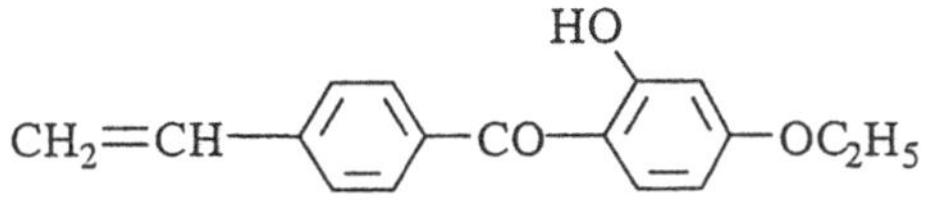

Synthesis

-Preparation by reaction of aqueous
potassium hydroxide with 4'-(2-bromo-
ethyl)-4-ethoxy-2-hydroxybenzophenone in
the presence of hydroquinone in refluxing
methanol for 1.5 h with nitrogen bubbling (51%) [675].

m.p. 76-79° [675];
^{1}H NMR [675], IR [675], UV [675], MS [675].

(2-Bromophenyl)[5-(1,1-dimethylethyl)-2-hydroxyphenyl]methanone

$C_{17}H_{17}BrO_2$ mol.wt. 333.22

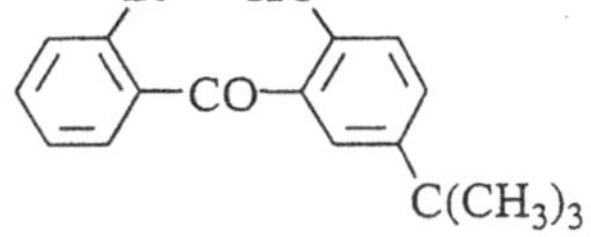

Synthesis

-Preparation by demethylation of 2'-bromo-5-tert-butyl-
2-methoxybenzophenone with aluminium chloride in
benzene at 50-60° for 5 h (80%) [775].

b.p.$_{12}$ 208-212° [775];   Spectra (NA).

(3-Bromophenyl)[5-(1,1-dimethylethyl)-2-hydroxyphenyl]methanone

$C_{17}H_{17}BrO_2$ mol.wt. 333.22

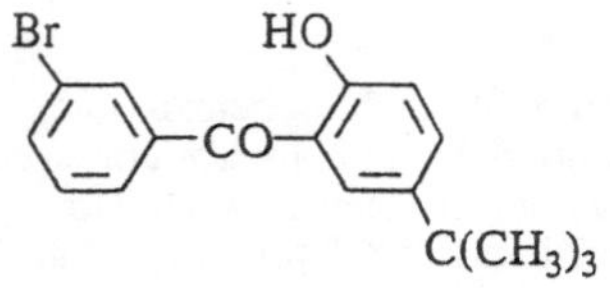

Synthesis

-Preparation by demethylation of 5-tert-butyl-3'-bromo-
2-methoxybenzophenone with aluminium chloride in
benzene at 50-60° for 5 h (75%) [775].

b.p.$_{12}$ 225-227° [775];   Spectra (NA).

(4-Bromophenyl)[5-(1,1-dimethylethyl)-2-hydroxyphenyl]methanone

[75060-50-3] $C_{17}H_{17}BrO_2$ mol.wt. 333.22

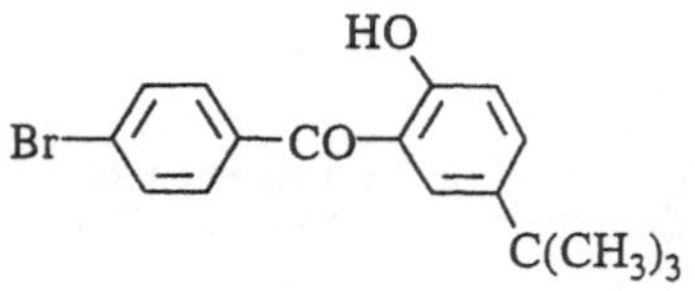

Synthesis

-Preparation by demethylation of 2-(4-bromobenzoyl)-
4-tert-butylanisole with a mixture of 57% hydriodic
acid and 47% hydrobromic acid in refluxing acetic
acid (87%) [636].

m.p. 83-85° [636];   Spectra (NA).

(2-Bromophenyl)[4-hydroxy-2-methyl-5-(1-methylethyl)phenyl]methanone

[72237-01-5] $C_{17}H_{17}BrO_2$ mol.wt. 333.22

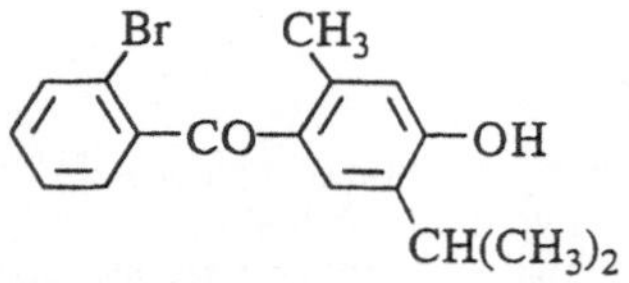

Synthesis

-Preparation by demethylation of 4-(2-bromobenzoyl)-
5-methyl-2-isopropylanisole with boiling pyridinium
chloride for 15 min (85%) [108].

m.p. 126° [108]; ^{1}H NMR [108].

(4-Bromophenyl)[4-hydroxy-2-methyl-5-(1-methylethyl)phenyl]methanone

[72237-03-7] $C_{17}H_{17}BrO_2$ mol.wt. 333.22

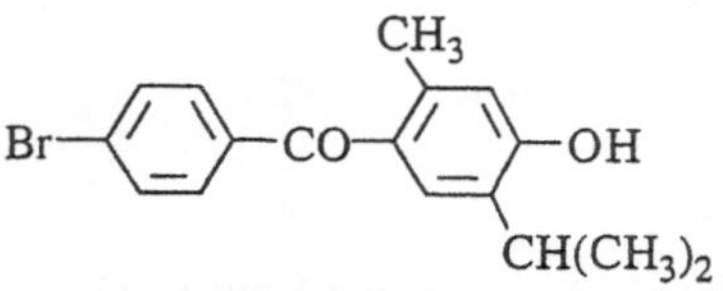

Synthesis

-Preparation by demethylation of 4-(4-bromobenzoyl)-
5-methyl-2-isopropylanisole with boiling pyridinium
chloride for 15 min (85%) [108].

m.p. 150° [108]; ^{1}H NMR [108].

(5-Butyl-2-hydroxyphenyl)(4-chlorophenyl)methanone

[108294-80-0] $C_{17}H_{17}ClO_2$ mol.wt. 288.77

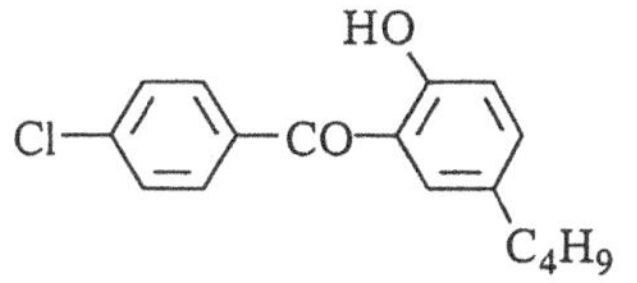

Synthesis

-Preparation by reaction of p-chlorobenzoyl chloride with
p-butylphenol in the presence of aluminium chloride in
tetrachloroethane at 105° for 22 h (54%) [400].

b.p.$_{0.75}$ 174° [400];   Spectra (NA).

(5-Chloro-3-ethyl-2-hydroxyphenyl)(4-ethylphenyl)methanone

[93575-37-2] $C_{17}H_{17}ClO_2$ mol.wt. 288.77

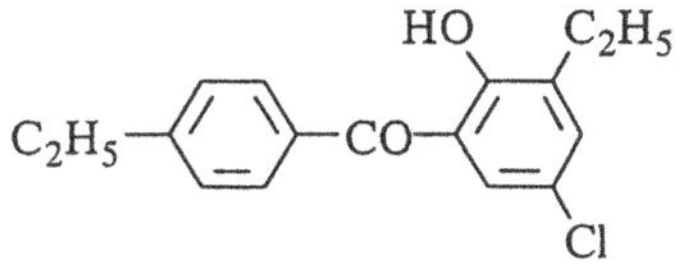

Synthesis

-Preparation by Fries rearrangement of 4-chloro-
2-ethylphenyl 4-ethylbenzoate with aluminium chloride
at 160° for 15 min [698].

oil [698];   b.p. (NA); n_D^{21} = 1.6060 [698];   Spectra (NA).

(5-Chloro-2-hydroxy-3-methylphenyl)[4-(1-methylethyl)phenyl]methanone

[86914-75-2] $C_{17}H_{17}ClO_2$ mol.wt. 288.77

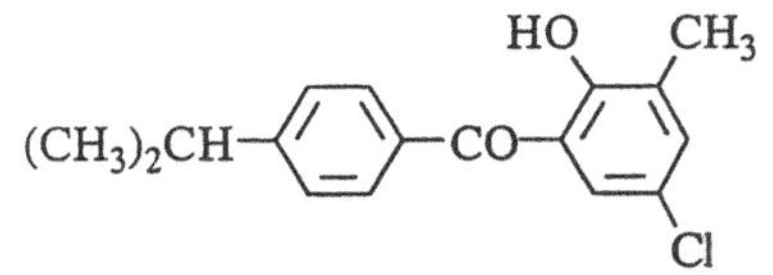

Synthesis

-Preparation by Fries rearrangement of 4-chloro-
2-methylphenyl 4-isopropylbenzoate with aluminium
chloride at 180° for 10 min [684, 688].

yellow oil [684, 688];   b.p. (NA); n_D^{22} = 1.619 [684, 688];
Spectra (NA).

(5-Chloro-2-hydroxy-3-propylphenyl)(4-methylphenyl)methanone

[92739-94-1] $C_{17}H_{17}ClO_2$ mol.wt. 288.77

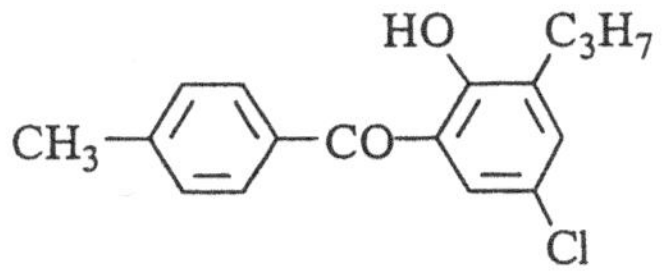

Synthesis

-Preparation by Fries rearrangement of 4-chloro-2-propyl-
phenyl p-toluate with aluminium chloride at 160° for
15 min [698].

m.p. 63-64° [698];   Spectra (NA).

(4-Chlorophenyl)(3,5-diethyl-2-hydroxyphenyl)methanone

[108294-82-2]   C$_{17}$H$_{17}$ClO$_2$ mol.wt. 288.77

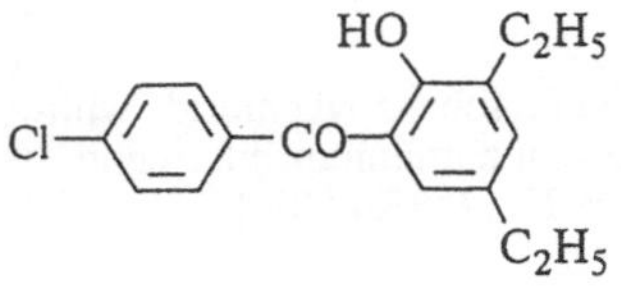

Synthesis

-Preparation by reaction of p-chlorobenzoyl chloride with
 2,4-diethylphenol in the presence of aluminium chloride in
 tetrachloroethane at 105° for 22 h (53%) [400].

b.p.$_{1.4}$ 188° [400];   Spectra (NA).

(4-Chlorophenyl)(4,5-diethyl-2-hydroxyphenyl)methanone

[61750-26-3]   C$_{17}$H$_{17}$ClO$_2$ mol.wt. 288.77

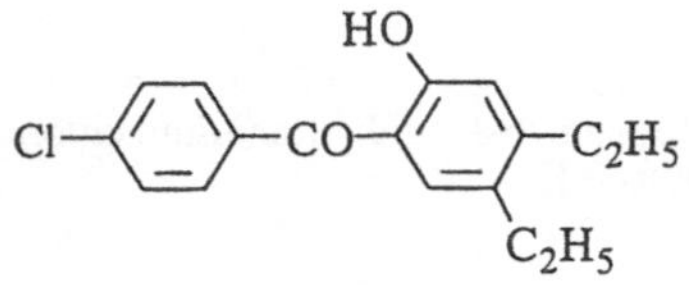

Synthesis

-Preparation by Fries rearrangement of 3,4-diethyl-
 phenyl p-chlorobenzoate with aluminium chloride in
 tetrachloroethane at 125° for 6 h [1207].

b.p.$_{1.4}$ 188° [1207];   Spectra (NA).

(2-Chlorophenyl)[5-(1,1-dimethylethyl)-2-hydroxyphenyl]methanone

C$_{17}$H$_{17}$ClO$_2$ mol.wt. 288.77

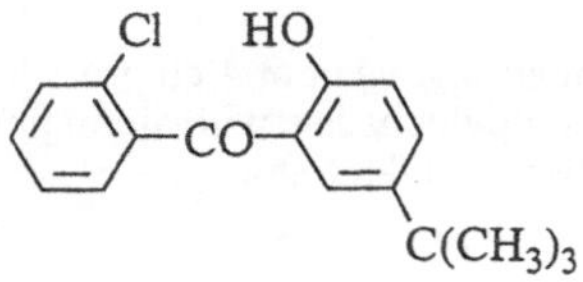

Synthesis

-Preparation by demethylation of 5-tert-butyl-2'-chloro-
 2-methoxybenzophenone with aluminium chloride in
 benzene at 50-60° for 5 h (70%) [775].

b.p.$_{12}$ 200-205° [775];   Spectra (NA).

(3-Chlorophenyl)[3-(1,1-dimethylethyl)-4-hydroxyphenyl]methanone

[124979-06-2]   C$_{17}$H$_{17}$ClO$_2$ mol.wt. 288.77

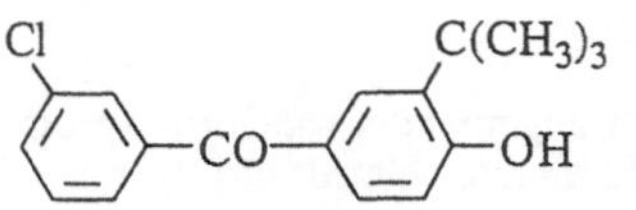

Synthesis

-Preparation by Friedel-Crafts acylation of o-tert-butylphenol with m-chlorobenzoyl chloride in ethylene
 dichloride in the presence of titanium tetrachloride, first at
 0°, then at r.t. [733].

m.p. 173-174° [733];   Spectra (NA).

(4-Chlorophenyl)[3-(1,1-dimethylethyl)-4-hydroxyphenyl]methanone

[124979-05-1] $C_{17}H_{17}ClO_2$ mol.wt. 288.77

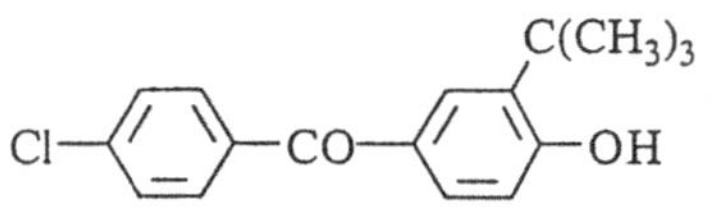

Synthesis

-Preparation by Friedel-Crafts acylation of o-tert-butylphenol with p-chlorobenzoyl chloride in ethylene
dichloride in the presence of titanium tetrachloride, first
at 0°, then at r.t. [733].

m.p. 205-206° [733];   Spectra (NA).

(4-Chlorophenyl)[5-(1,1-dimethylethyl)-2-hydroxyphenyl]methanone

[72083-19-3] $C_{17}H_{17}ClO_2$ mol.wt. 288.77

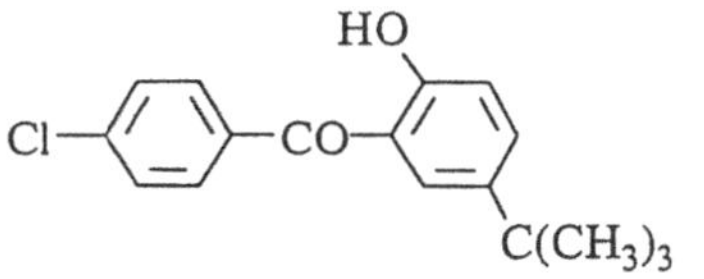

Syntheses

-Preparation by demethylation of 5-tert-butyl-4'-chloro-
2-methoxybenzophenone with aluminium chloride in
benzene at 50-60° for 5 h (68%) [775] or at 70° for
12 h [693].
-Obtained by photo-Fries rearrangement of p-tert-
butylphenyl p-chlorobenzoate in benzene or in ethanol (55 and 48% yields, respectively) [743].
-Also refer to: [932, 1336].

m.p. 94-95° [775], 92-94° [743], 64-65° [693]. One of the reported melting points is
obviously wrong. b.p.$_{12}$ 218-220° [775];   Spectra (NA).

(4-Chlorophenyl)[4-hydroxy-2-methyl-5-(1-methylethyl)phenyl]methanone

                                      $C_{17}H_{17}ClO_2$ mol.wt. 288.77

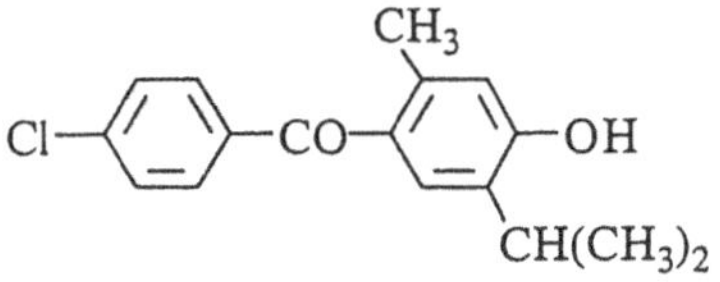

Synthesis

-Preparation by demethylation of 4'-chloro-4-methoxy-
2-methyl-5-isopropylbenzophenone with boiling
pyridinium chloride for 15 min (85%) [108].

m.p. 156° [108];   Spectra (NA).

(2-Chlorophenyl)[2-hydroxy-5-(1-methylpropyl)phenyl]methanone

[59746-95-1] $C_{17}H_{17}ClO_2$ mol.wt. 288.77

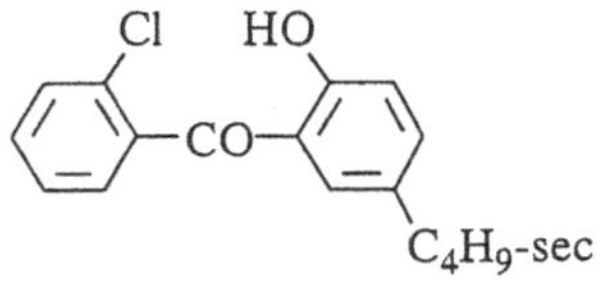

Synthesis

-Preparation by reaction of 2-chlorobenzotrichloride with
4-sec-butylphenol in hydrofluoric acid in the presence of
water at -10°, then between 0 to -10° for 2 h, at r.t. for
7 h and at 80° for 30 min into an autoclave (72%) [393].

-Also refer to: [1226].

yellow oil [393];   b.p.$_{0.2}$ 152° [393];   Spectra (NA).

(4-Chlorophenyl)[2-hydroxy-5-(1-methylpropyl)phenyl]methanone

[59746-96-2] $C_{17}H_{17}ClO_2$ mol.wt. 288.77

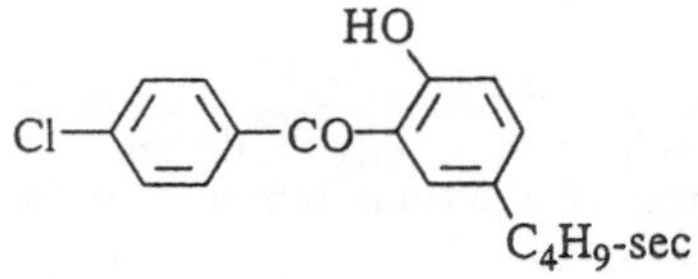

Synthesis

-Preparation by reaction of 4-chlorobenzotrichloride with 4-sec-butylphenol in hydrofluoric acid in the presence of water at -10°, then between 0 to -10° for 2 h, at r.t. for 7 h and at 80° for 30 min into an autoclave (51%) [393].

-Also refer to: [1226].

yellow oil [393];   b.p._1 175-177° [393];   Spectra (NA).

[3-(1,1-Dimethylethyl)-4-hydroxyphenyl](2-fluorophenyl)methanone

[124979-11-9] $C_{17}H_{17}FO_2$ mol.wt. 272.32

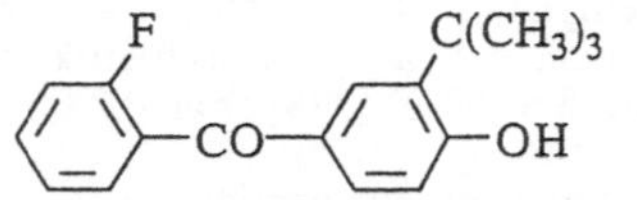

Synthesis

-Preparation by Friedel-Crafts acylation of o-tert-butylphenol with o-fluorobenzoyl chloride in ethylene dichloride in the presence of titanium tetrachloride, first at 0°, then at r.t. [733].

m.p.   152-153° [733];   Spectra (NA).

[3-(1,1-Dimethylethyl)-4-hydroxyphenyl](4-fluorophenyl)methanone

[124979-09-5] $C_{17}H_{17}FO_2$ mol.wt. 272.32

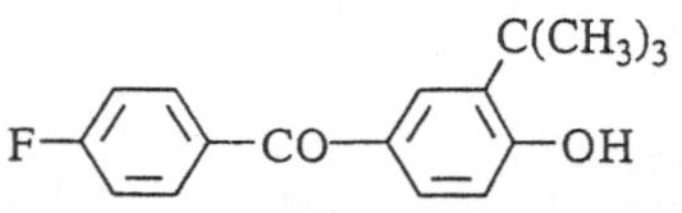

Synthesis

-Preparation by Friedel-Crafts acylation of o-tert-butylphenol with p-fluorobenzoyl chloride in ethylene dichloride in the presence of titanium tetrachloride, first at 0°, then at r.t. [733].

m.p.   203-204° [733];   Spectra (NA).

(4-Butoxy-2-hydroxyphenyl)(4-fluorophenyl)methanone

$C_{17}H_{17}FO_3$ mol.wt. 288.32

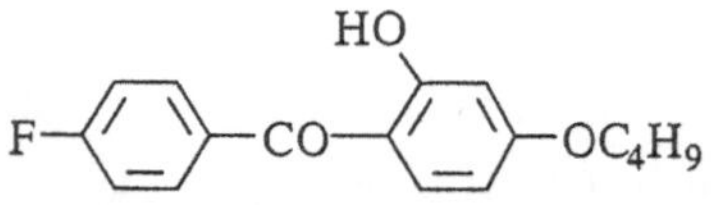

Synthesis

-Preparation by partial alkylation of 4'-fluoro-2,4-dihydroxybenzophenone with a butyl halide in the presence of an alkali [394].

m.p. and Spectra (NA).

(2-Fluoro-4,6-dimethoxyphenyl)(2-hydroxy-4,5-dimethoxyphenyl)methanone

[129103-95-3] $C_{17}H_{17}FO_6$ mol.wt. 336.32

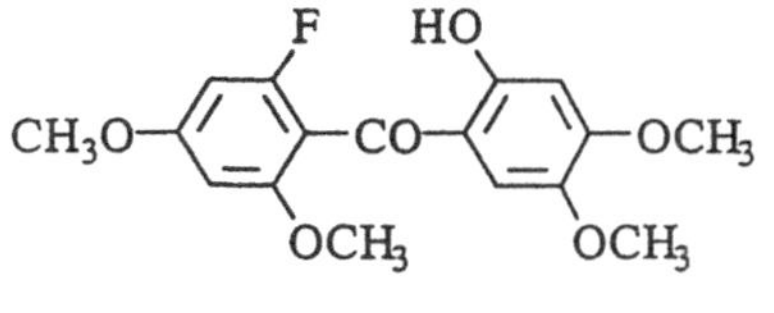

Synthesis

-Preparation from 2-iodo-4,5-dimethoxyphenyl
2-fluoro-4,6-dimethoxybenzoate by rearrangement
on treatment with sec-butyllithium in a mixture of
ethyl ether, hexane and tetrahydrofuran at -100°
followed by heating to -70° for 2 h, then treatment
with saturated aqueous ammonium chloride (89%) [598].

    m.p. 133-134° [598];
    ^{1}H NMR [598], IR [598], MS [598].

(2,3-Dimethyl-5-nitrophenyl)(2-hydroxy-3,4-dimethylphenyl)methanone

[110969-70-5] $C_{17}H_{17}NO_4$ mol.wt. 299.33

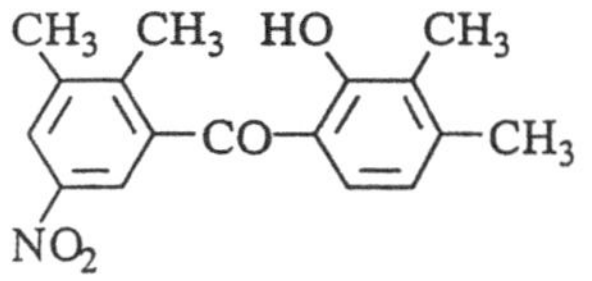

Synthesis

-Preparation by Fries rearrangement of 2,3-dimethylphenyl
2,3-dimethyl-5-nitrobenzoate with aluminium chloride at
160° for 2 h (56%) [843].

    m.p. 165° [843];
    ^{1}H NMR [843], IR [843], UV [843].

(2,3-Dimethyl-5-nitrophenyl)(2-hydroxy-3,5-dimethylphenyl)methanone

[110969-73-8] $C_{17}H_{17}NO_4$ mol.wt. 299.33

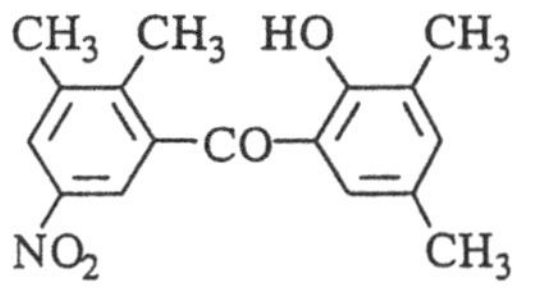

Synthesis

-Preparation by Fries rearrangement of 2,4-dimethylphenyl
2,3-dimethyl-5-nitrobenzoate with aluminium chloride at
160° for 2 h (56%) [843].

    m.p. 133° [843];
    ^{1}H NMR [843], IR [843], UV [843].

(2,3-Dimethyl-5-nitrophenyl)(2-hydroxy-3,6-dimethylphenyl)methanone

[110969-75-0] $C_{17}H_{17}NO_4$ mol.wt. 299.33

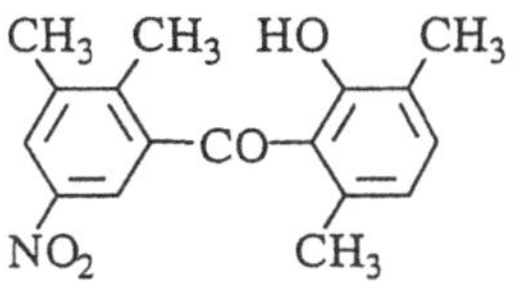

Synthesis

-Obtained by Fries rearrangement of 2,5-dimethylphenyl
2,3-dimethyl-5-nitrobenzoate with aluminium chloride at
160° for 2 h (20%) [843].

    m.p. 120° [843];
    ^{1}H NMR [843], IR [843], UV [843].

(2,3-Dimethyl-5-nitrophenyl)(2-hydroxy-4,5-dimethylphenyl)methanone

[110969-80-7] $C_{17}H_{17}NO_4$ mol.wt. 299.33

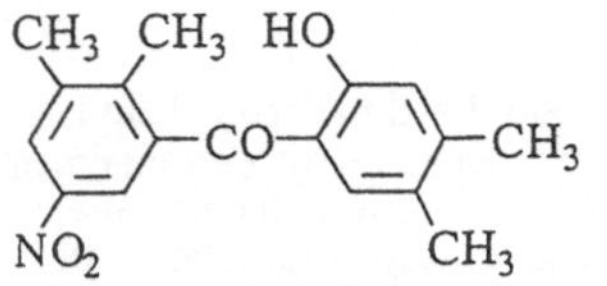

Synthesis

-Preparation by Fries rearrangement of 3,4-dimethylphenyl
2,3-dimethyl-5-nitrobenzoate with aluminium chloride at
160° for 2 h (60%) [843].

m.p.   169° [843];
^{1}H NMR [843], IR [843], UV [843].

(2,3-Dimethyl-5-nitrophenyl)(2-hydroxy-4,6-dimethylphenyl)methanone

[110993-12-9] $C_{17}H_{17}NO_4$ mol.wt. 299.33

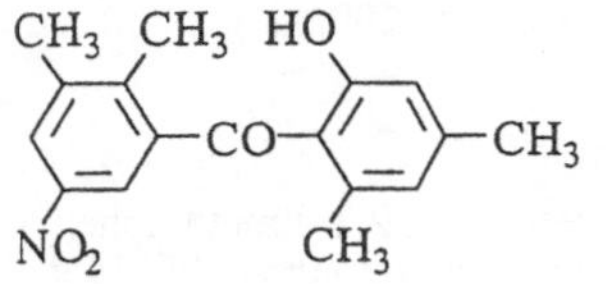

Synthesis

-Preparation by Fries rearrangement of 3,5-dimethylphenyl
2,3-dimethyl-5-nitrobenzoate with aluminium chloride at
160° for 2 h (70%) [843].

m.p.   121° [843];
^{1}H NMR [843], IR [843], UV [843].

(2,3-Dimethyl-5-nitrophenyl)(4-hydroxy-2,3-dimethylphenyl)methanone

[110969-71-6] $C_{17}H_{17}NO_4$ mol.wt. 299.33

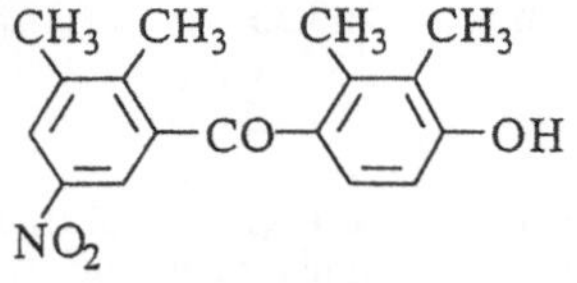

Synthesis

-Obtained by Fries rearrangement of 2,3-dimethylphenyl
2,3-dimethyl-5-nitrobenzoate with aluminium chloride at
160° for 2 h (12%) [843].

m.p.   212° [843];
^{1}H NMR [843], IR [843], UV [843].

(2,3-Dimethyl-5-nitrophenyl)(4-hydroxy-2,5-dimethylphenyl)methanone

[110969-76-1] $C_{17}H_{17}NO_4$ mol.wt. 299.33

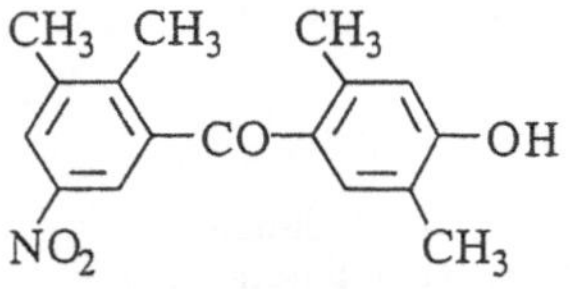

Synthesis

-Preparation by Fries rearrangement of 2,5-dimethylphenyl
2,3-dimethyl-5-nitrobenzoate with aluminium chloride at
160° for 2 h (30%) [843].

m.p.   232° [843];
^{1}H NMR [843], IR [843], UV [843].

(2,3-Dimethyl-5-nitrophenyl)(4-hydroxy-3,5-dimethylphenyl)methanone

[110969-78-3] $C_{17}H_{17}NO_4$ mol.wt. 299.33

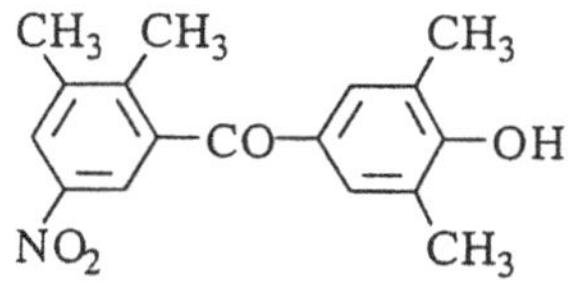

Synthesis

-Preparation by Fries rearrangement of 2,6-dimethylphenyl
2,3-dimethyl-5-nitrobenzoate with aluminium chloride at
160° for 2 h (55%) [843].

m.p. 214° [843];
^{1}H NMR [843], IR [843], UV [843].

[5-(1,1-Dimethylethyl)-2-hydroxyphenyl](4-nitrophenyl)methanone

$C_{17}H_{17}NO_4$ mol.wt. 299.33

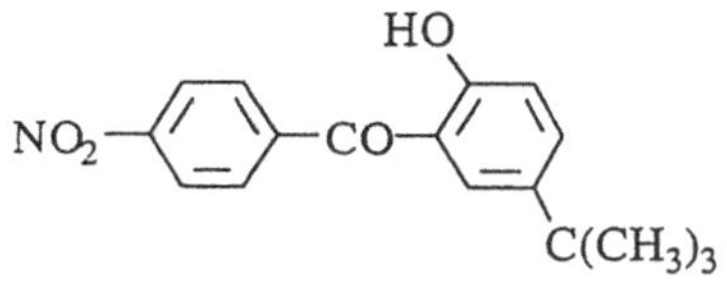

Synthesis

-Obtained (poor yield) by photo-Fries rearrangement of
p-tert-butylphenyl p-nitrobenzoate in benzene (10%)
[743].

m.p. 102-103° [743];   Spectra (NA).

[3-Amino-4-(1,1-dimethylethyl)-2-hydroxyphenyl](4-chlorophenyl)methanone

$C_{17}H_{18}ClNO_2$ mol.wt. 303.79

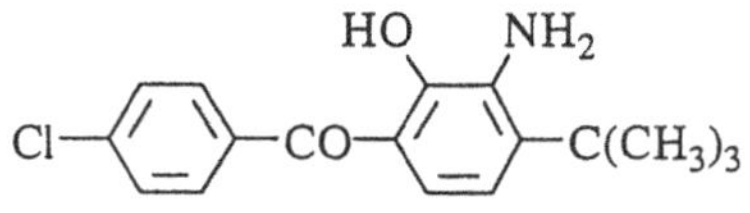

Synthesis

-Preparation by reduction of 4-tert-butyl-4'-chloro-
2-hydroxy-3-nitrobenzophenone with titanium
trichloride in a benzene/tetrahydrofuran mixture for
20 h at r.t. [932].

m.p. (NA);   red oil [932];
^{1}H NMR [932], IR [932], MS [932];   TLC [932].

[3-Amino-4-(1,1-dimethylethyl)-2-hydroxyphenyl](4-chlorophenyl)methanone
(Hydrochloride)

$C_{17}H_{18}ClNO_2,HCl$ mol.wt. 340.25

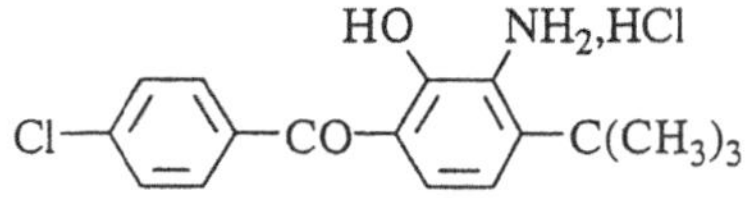

Synthesis

-Obtained by reaction of concentrated hydrochloric
acid with 3-amino-4-tert-butyl-4'-chloro-2-hydroxy-
benzophenone (52%) [932].

m.p. 153-158° [932];   Spectra: see the corresponding amino base.

(2,4-Dimethylphenyl)(2-hydroxy-3,5-dimethylphenyl)methanone

[86914-80-9] $C_{17}H_{18}O_2$ mol.wt. 254.33

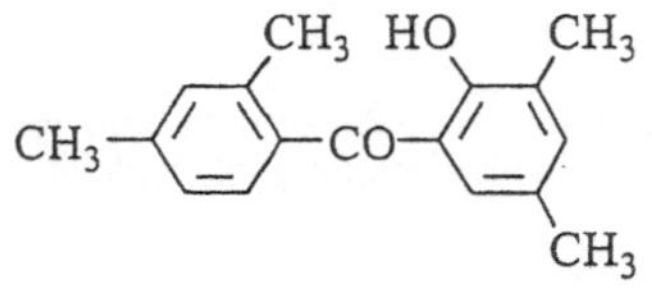

Synthesis

-Preparation by Fries rearrangement of 2,4-dimethylphenyl
2,4-dimethylbenzoate with aluminium chloride at 180° for
10 min [688].

oil [688];   b.p. and Spectra (NA).

(2,5-Dimethylphenyl)(2-hydroxy-3,5-dimethylphenyl)methanone

[86914-78-5] $C_{17}H_{18}O_2$ mol.wt. 254.33

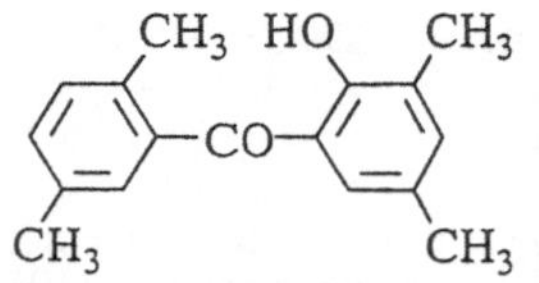

Synthesis

-Preparation by Fries rearrangement of 2,4-dimethylphenyl
2,5-dimethylbenzoate with aluminium chloride at 180° for
10 min [688].

oil [688];   b.p. and Spectra (NA).

(3,4-Dimethylphenyl)(2-hydroxy-3,5-dimethylphenyl)methanone

[86914-88-7] $C_{17}H_{18}O_2$ mol.wt. 254.33

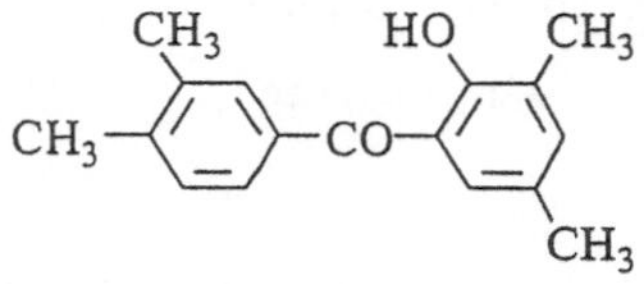

Synthesis

-Preparation by Fries rearrangement of 2,4-dimethyl-
phenyl 3,4-dimethylbenzoate with aluminium chloride at
180° for 10 min [688].

m.p. 68° [688];   Spectra (NA).

(3-Ethyl-2-hydroxy-5-methylphenyl)(4-methylphenyl)methanone

[92739-95-2] $C_{17}H_{18}O_2$ mol.wt. 254.33

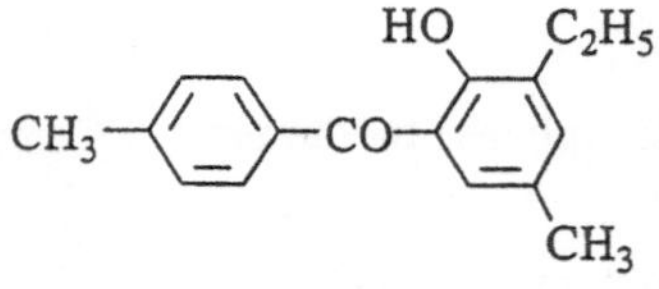

Synthesis

-Preparation by Fries rearrangement of 2-ethyl-4-methyl-
phenyl p-toluate with aluminium chloride at 160° for
15 min [698].
-Also refer to: [689].

m.p. 38-39° [698];   Spectra (NA).

[2-Hydroxy-5-(1-methylethyl)phenyl](2-methylphenyl)methanone

[170799-16-3] $C_{17}H_{18}O_2$ mol.wt. 254.33

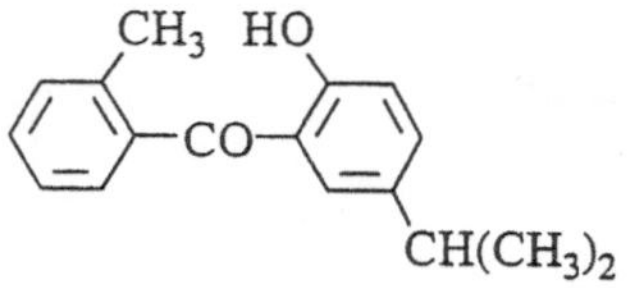

Synthesis

-Preparation by Fries rearrangement of p-isopropylphenyl
o-toluate with aluminium chloride at 100-150° for 0.5-3 h
(98%) [1363].

oil [1363];   b.p. and Spectra (NA).

(2-Hydroxy-5-methylphenyl)(2,4,6-trimethylphenyl)methanone

$C_{17}H_{18}O_2$ mol.wt. 254.33

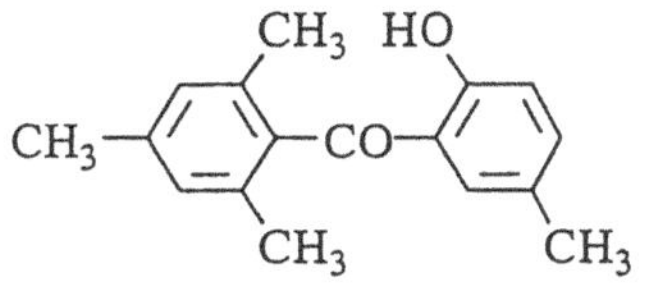

Synthesis

-Preparation by Fries rearrangement of p-cresyl 2,4,6-tri-
methylbenzoate with aluminium chloride at 150° for 2 h
(63%) [454].

m.p. 86° [454]; Spectra (NA).

(2,6-Dimethylphenyl)(5-hydroxy-4-methoxy-2-methylphenyl)methanone

$C_{17}H_{18}O_3$ mol.wt. 270.33

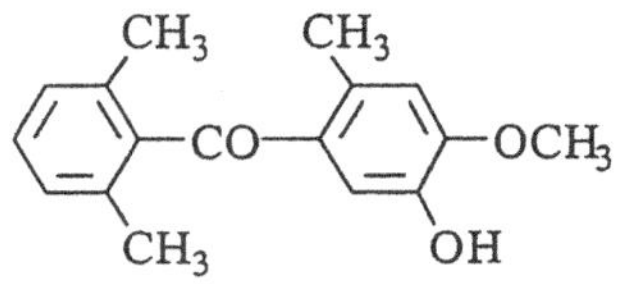

Synthesis

-Obtained by partial demethylation of 4,5-dimethoxy-
2,2',6'-trimethylbenzophenone [313].

m.p. and Spectra (NA).

(2-Hydroxy-4-propoxyphenyl)(2-methylphenyl)methanone

[172479-21-9] $C_{17}H_{18}O_3$ mol.wt. 270.33

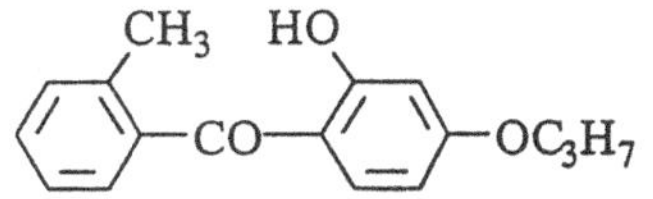

Synthesis

-Prepared by standard techniques [1065].

m.p. (NA); UV [1065].

(2,4-Dimethylphenyl)(4-hydroxy-3,5-dimethoxyphenyl)methanone

$C_{17}H_{18}O_4$ mol.wt. 286.33

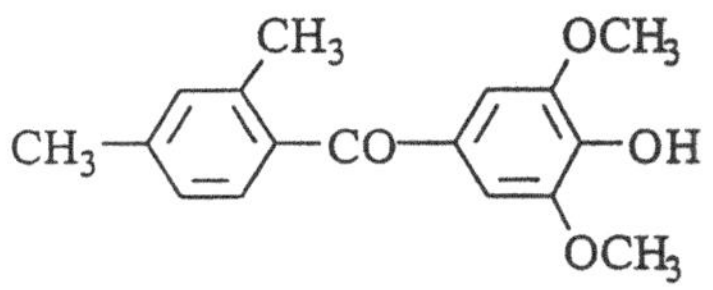

Synthesis

-Preparation by Friedel-Crafts acylation of m-xylene with
3,4,5-trimethoxybenzoyl chloride [593].

m.p. and Spectra (NA).

(5-Ethyl-2-hydroxy-4-methoxyphenyl)(4-methoxyphenyl)methanone

[66666-17-9] $C_{17}H_{18}O_4$ mol.wt. 286.33

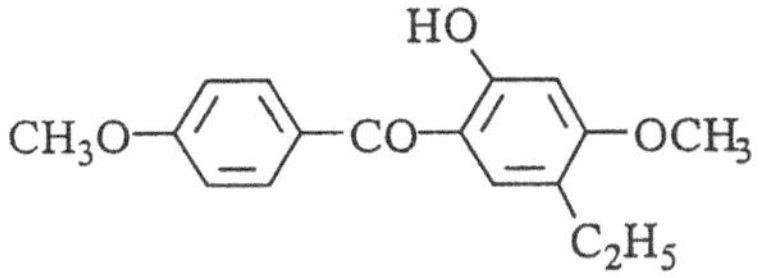

Synthesis

-Preparation by saponification of 2-(p-anisoyloxy)-
4,4'-dimethoxy-5-ethylbenzophenone (SM) with
potassium hydroxide in refluxing ethanol for 1 h
(81%). SM was obtained by oxidation of 5-ethyl-
6-methoxy-2,3-bis(p-methoxyphenyl)benzofuran

with chromium trioxide in refluxing acetic acid for 45 min [578].

oil [578];   b.p. and Spectra (NA).

(2,4-Dimethoxy-6-methylphenyl)(2-hydroxy-5-methoxyphenyl)methanone

[78044-92-5] $C_{17}H_{18}O_5$ mol.wt. 302.33

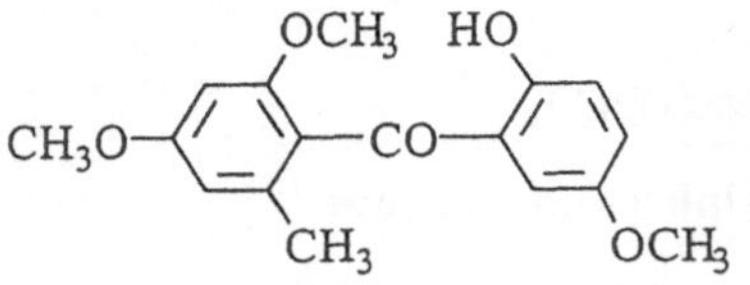

Synthesis

-Obtained by photo-Fries rearrangement of p-methoxyphenyl 2,4-dimethoxy-6-methylbenzoate in benzene
 for 4 h under nitrogen (37%) [816].

m.p.   120-123° [816]; ^{1}H NMR [816], IR [816], UV [816], MS [816].

(4-Ethoxyphenyl)(2-hydroxy-3,4-dimethoxyphenyl)methanone

[69471-31-4] $C_{17}H_{18}O_5$ mol.wt. 302.33

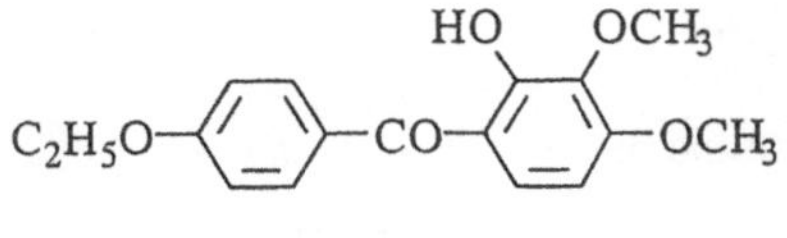

Synthesis

-Preparation by selective demethylation of
 4'-ethoxy-2,3,4-trimethoxybenzophenone with
 aluminium chloride in nitrobenzene at 100° for 1 h
 (70%) [1008].

m.p.   113-114° [1008]; ^{1}H NMR [1008], IR [1008].

(2-Hydroxy-3,4-dimethoxyphenyl)(3-methoxy-4-methylphenyl)methanone

                                      $C_{17}H_{18}O_5$ mol.wt. 302.33

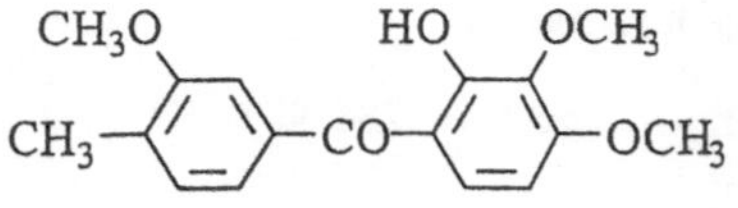

Synthesis

-Obtained by Friedel-Crafts acylation of pyrogallol
 trimethyl ether with 3-methoxy-4-methylbenzoyl
 chloride in the presence of aluminium chloride in
 boiling carbon disulfide for 4 h (22%) [1049, 1050].

m.p.   109° [1049, 1050];   Spectra (NA).

[3-Hydroxy-2-methoxy-6-(methoxymethyl)phenyl](2-methoxyphenyl)methanone

[133386-99-9] $C_{17}H_{18}O_5$ mol.wt. 302.33

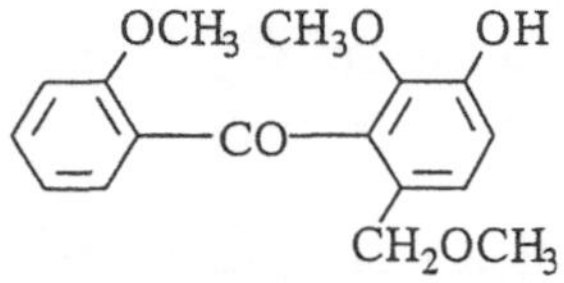

Synthesis

-Preparation from 4-hydroxy-3-methoxybenzyl methyl ether
 (SM) in two steps: first, direct lithiation of SM with
 n-butyl lithium in tetrahydrofuran at r.t., followed by
 quenching with o-anisaldehyde. Then, oxidation of the
 8-(2-methoxyphenyl)-2,7-dimethoxybicyclo[4.2.0]octa-
 1,3,5-triene-3,8-diol so formed (good yield) [286].

m.p. and Spectra (NA).

(2,3-Dimethoxyphenyl)(2-hydroxy-4,6-dimethoxyphenyl)methanone

[151417-67-3] $C_{17}H_{18}O_6$ mol.wt. 318.33

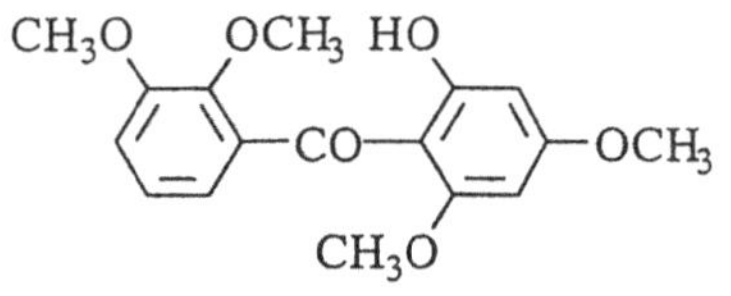

Synthesis

-Obtained, in mixture with 2'-hydroxy-2,3',4,6-tetra-
methoxybenzophenone, by reaction of 2,3-dimethoxy-
benzoyl chloride with 1,3,5-trimethoxybenzene in ethyl
ether in the presence of aluminium chloride for 15 h at
r.t. (total yield: 92%) [826].

-Also refer to: [556].

m.p. (NA); ^1H NMR [826].

(2,5-Dimethoxyphenyl)(2-hydroxy-3,6-dimethoxyphenyl)methanone

[109092-84-4] $C_{17}H_{18}O_6$ mol.wt. 318.33

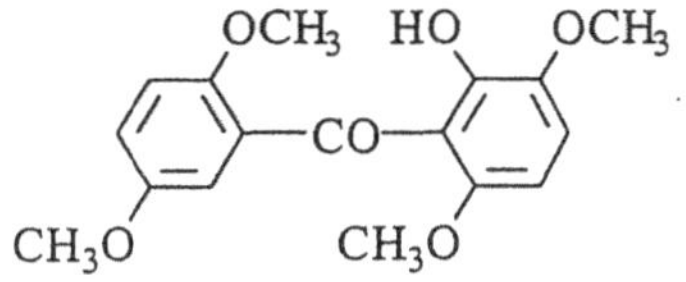

Synthesis

-Obtained by reaction of 2,3,6-trimethoxybenzoyl chloride
with hydroquinone dimethyl ether in the presence of
aluminium chloride in ethyl ether at r.t. for 2 days (17%)
[1062] or for 1 h at 45-50°, then 20 h at r.t. (69%) [845].

m.p. 115-116° [1062], 115° [845];   Spectra (NA).

(2,5-Dimethoxyphenyl)(2-hydroxy-4,5-dimethoxyphenyl)methanone

[88133-95-3] $C_{17}H_{18}O_6$ mol.wt. 318.33

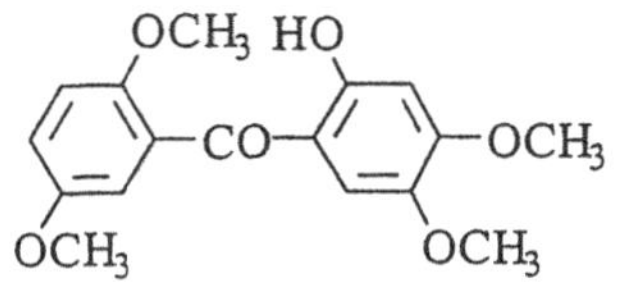

Synthesis

-Obtained by reaction of 2,5-dimethoxybenzoyl chloride with
1,2,4-trimethoxybenzene in the presence of aluminium
chloride in ethyl ether for 30 h [370], or for 1 h at 45-50°,
then 20 h at r.t. (65-70%) [845].

m.p. 97-98° [370], 82°5-83° [845]; ^1H NMR [370].

(2,5-Dimethoxyphenyl)(2-hydroxy-4,6-dimethoxyphenyl)methanone

[42833-59-0] $C_{17}H_{18}O_6$ mol.wt. 318.33

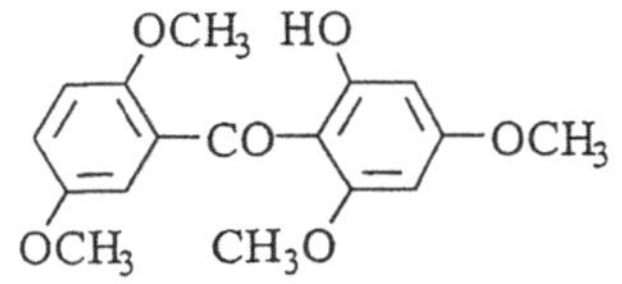

Synthesis

-Preparation by reaction of 2,5-dimethoxybenzoyl chloride
with 1,3,5-trimethoxybenzene in the presence of aluminium
chloride in ethyl ether for 28 h (minor product) [1097],
(52%, estimated, not isolated) [845].

m.p. (NA); ^1H NMR [1097], MS [1097].

(2,6-Dimethoxyphenyl)(2-hydroxy-3,4-dimethoxyphenyl)methanone

[42833-55-6] $C_{17}H_{18}O_6$ mol.wt. 318.33

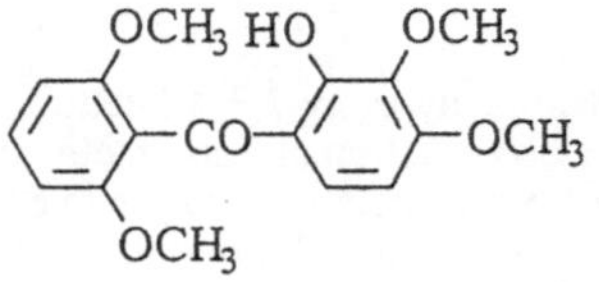

Synthesis

-Obtained by reaction of 2,6-dimethoxybenzoyl chloride with
1,2,3-trimethoxybenzene in the presence of aluminium
chloride in ethyl ether for 20 h [1097].

m.p. (NA); ^{1}H NMR [1097], MS [1097].

(2,6-Dimethoxyphenyl)(2-hydroxy-4,5-dimethoxyphenyl)methanone

[42833-53-4] $C_{17}H_{18}O_6$ mol.wt. 318.33

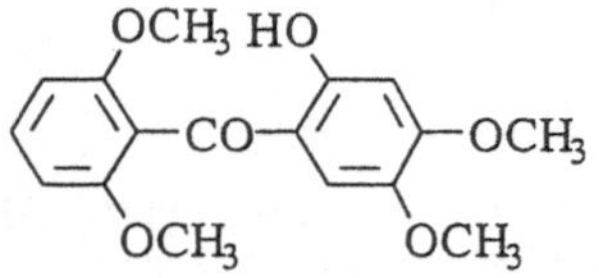

Synthesis

-Obtained by reaction of 2,6-dimethoxybenzoyl chloride with
1,2,4-trimethoxybenzene in the presence of aluminium
chloride in ethyl ether at r.t. for 40 h [1097].

m.p. (NA); ^{1}H NMR [1097], UV [1097], MS [1097].

(3,4-Dimethoxyphenyl)(2-hydroxy-4,5-dimethoxyphenyl)methanone

$C_{17}H_{18}O_6$ mol.wt. 318.33

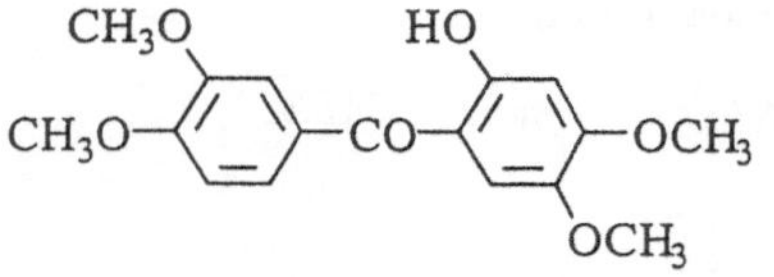

Synthesis

-Preparation by reaction of veratroyl chloride
(3,4-dimethoxybenzoyl chloride) with hydroxyhydroquinone trimethyl ether (1,2,4-trimethoxy-
benzene) in the presence of aluminium chloride,
*in ethyl ether at r.t. for 48 h [1395], (47%) [369];
*in refluxing carbon disulfide for 8 h, then at r.t. for 12 h (37%) [460].

m.p.   148-149° [460], 145-146° [369];   UV [460].

(3,4-Dimethoxyphenyl)(2-hydroxy-4,6-dimethoxyphenyl)methanone

[62495-41-4] $C_{17}H_{18}O_6$ mol.wt. 318.33

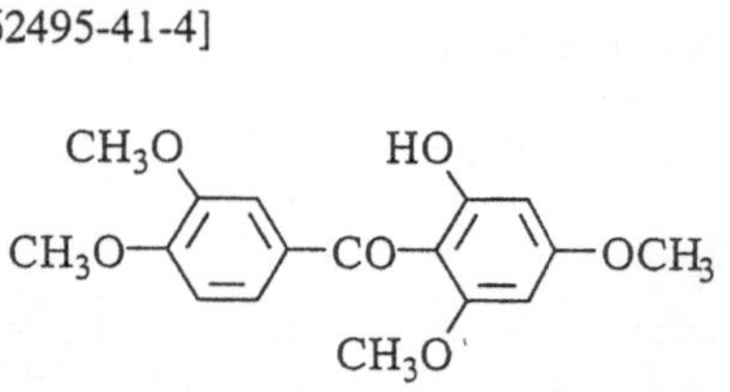

Syntheses

-Preparation by reaction of 3,4-dimethoxybenzoyl
chloride with 1,3,5-trimethoxybenzene in the
presence of aluminium chloride in ethyl ether for
4 h (50%) [29].
-Preparation by reaction of 3,4-dimethoxybenzoyl
chloride with 3,5-dimethoxyphenol in the presence of aluminium chloride in ethyl ether at r.t. for
48 h (49%) [1395].

m.p.   135-136° [29], 134° [1395];
^{1}H NMR [29, 1395], IR [1395], UV [1395].

(3,5-Dimethoxyphenyl)(2-hydroxy-3,5-dimethoxyphenyl)methanone

$C_{17}H_{18}O_6$ mol.wt. 318.33

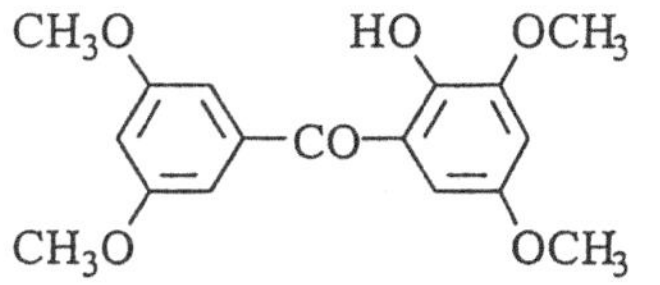

Synthesis

-Preparation by reaction of 3,5-dimethoxybenzoyl chloride with pyrogallol trimethyl ether in the presence of aluminium chloride in boiling carbon disulfide for 2 h (40%) [884].

m.p. 123-124° [884]; Spectra (NA).

(3,5-Dimethoxyphenyl)(2-hydroxy-4,6-dimethoxyphenyl)methanone

[53250-54-7] $C_{17}H_{18}O_6$ mol.wt. 318.33

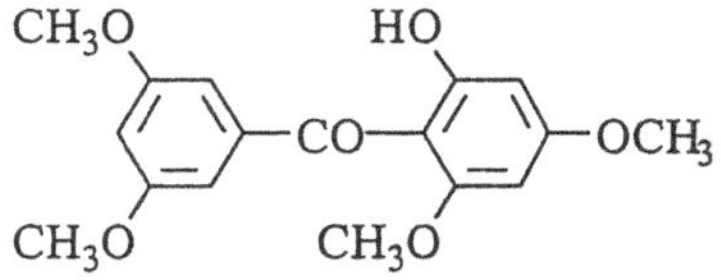

Synthesis

-Obtained by reaction of 3,5-dimethoxybenzoyl chloride with phloroglucinol trimethyl ether in ethyl ether in the presence of aluminium chloride at 25° for 48 h (16%) [1100].

m.p. 118° [1100]; Spectra (NA).

[2-Hydroxy-4-(2-hydroxyethoxy)phenyl][4-(2-hydroxyethoxy)phenyl]methanone

$C_{17}H_{18}O_6$ mol.wt. 318.33

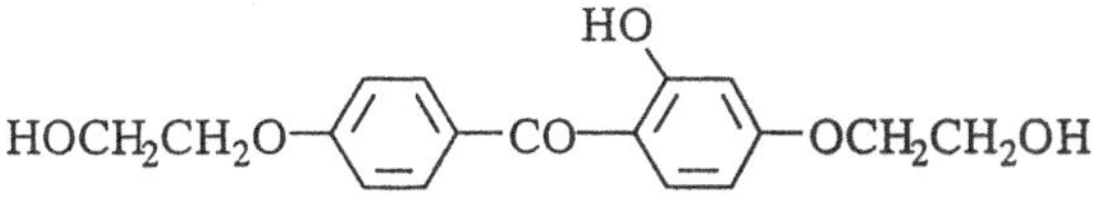

Synthesis

-Refer to: [377, 606].

m.p. 128° [606]; UV [606];
pK_a [606]; TLC [377].

(2-Hydroxy-3-methoxyphenyl)(2,4,6-trimethoxyphenyl)methanone

[6343-00-6] $C_{17}H_{18}O_6$ mol.wt. 318.33

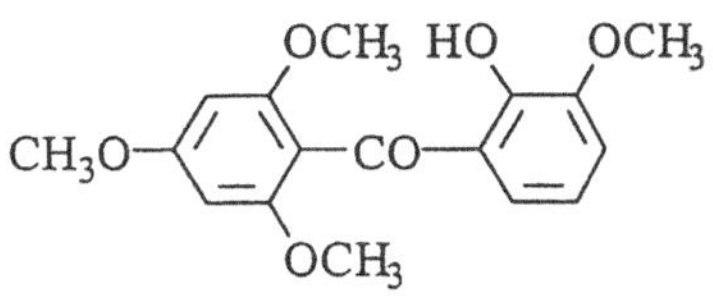

Synthesis

-Obtained, in mixture with 2-hydroxy-2',3',4,6-tetra-methoxybenzophenone, by reaction of 2,3-dimethoxy-benzoyl chloride with 1,3,5-trimethoxybenzene in ethyl ether in the presence of aluminium chloride for 15 h at r.t. (total yield: 92%) [826].

-Also refer to: [556].

m.p. (NA); ^1H NMR [826].

(2-Hydroxy-4-methoxyphenyl)(3,4,5-trimethoxyphenyl)methanone

[62495-39-0] $C_{17}H_{18}O_6$ mol.wt. 318.33

Synthesis

-Preparation by reaction of dimethyl sulfate with
2,4'-dihydroxy-3',4,5'-trimethoxybenzophenone in
the presence of potassium carbonate in refluxing
acetone for 30 min (57%) [1395].

m.p. 106-107° [1395]; ^{1}H NMR [1395], IR [1395], UV [1395].

(2-Hydroxy-5-methoxyphenyl)(2,4,6-trimethoxyphenyl)methanone

[42832-64-4] $C_{17}H_{18}O_6$ mol.wt. 318.33

Synthesis

-Obtained by reaction of 2,5-dimethoxybenzoyl chloride
with 1,3,5-trimethoxybenzene in the presence of
aluminium chloride in ethyl ether for 28 h (major
product) [1097], (30%, estimated, not isolated) [845].
-Also refer to: [313].

m.p. (NA); ^{1}H NMR [1097], MS [1097].

(3-Hydroxy-4-methoxyphenyl)(3,4,5-trimethoxyphenyl)methanone *(Phenstatin)*

[203448-32-2] $C_{17}H_{18}O_6$ mol.wt. 318.33

Synthesis

-Obtained by deprotection of 3-[(tert-butyldimethylsilyl)oxy]-3',4,4',5'-tetramethoxybenzophenone
(SM) in tetrahydrofuran with 1 M tetrabutyl-
ammonium fluoride for 15 min under argon (83%),
but only 30% overall yield [1057].

SM was prepared in two steps: first, formation of N-[3-[(tert-butyldimethylsilyl)oxy]-
4-methoxybenzoyl]morpholine (SM1) by action of 3-[(tert-butyldimethylsilyl)oxy]-4-methoxy-
benzoyl chloride with morpholine in toluene for 4 h at r.t. under argon. Then, the amide SM1 was
allowed to react with the lithium derivative prepared from 3,4,5-trimethoxybromobenzene [659] and
tert-butyllithium in tetrahydrofuran at -78° to give SM [1057].

m.p. 149-150° [1057]; ^{1}H NMR [1057], ^{13}C NMR [1057], IR [1057], MS [1057];
crystal data [1057];   TLC [1057].

(4-Hydroxy-3-methoxyphenyl)(2,4,6-trimethoxyphenyl)methanone

$C_{17}H_{18}O_6$ mol.wt. 318.33

Synthesis

-Obtained by reaction of veratronitrile with
phloroglucinol trimethyl ether (Hoesch reaction)
(15%) [755].

m.p. 242° [755];   Spectra (NA).

(2-Hydroxy-3,4,5-trimethoxyphenyl)(2-methoxyphenyl)methanone

[42833-60-3] $C_{17}H_{18}O_6$ mol.wt. 318.33

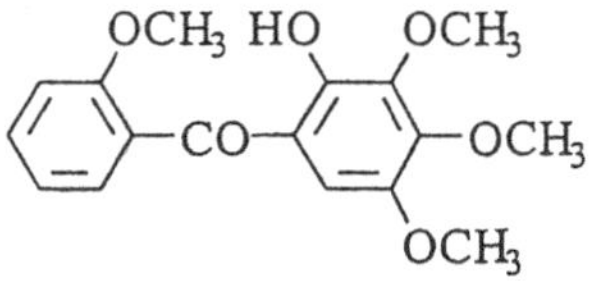

Synthesis

-Obtained by Friedel-Crafts acylation of 1,2,3,4-tetra-
methoxybenzene with o-anisoyl chloride in the presence of
aluminium chloride in ethyl ether for 44 h [1097].

m.p. (NA); ^{1}H NMR [1097], UV [1097].

(4-Aminophenyl)[5-(1,1-dimethylethyl)-2-hydroxyphenyl]methanone

[98031-50-6] $C_{17}H_{19}NO_2$ mol.wt. 269.34

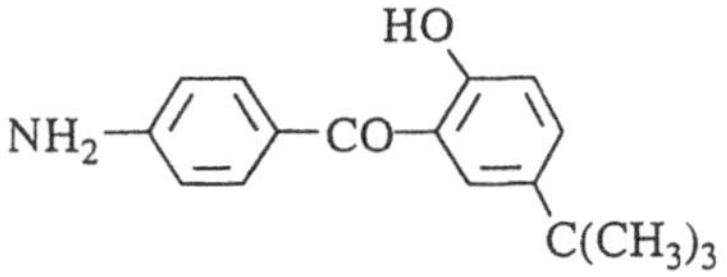

Synthesis

-Obtained by photo-Fries rearrangement of p-tert-
butylphenyl p-aminobenzoate in benzene (12%) [743].

m.p. 98-100° [743];   Spectra (NA).

[3-(1,1-Dimethylethyl)-4-hydroxyphenyl][2-(trifluoromethyl)phenyl]methanone

[124979-17-5] $C_{18}H_{17}F_3O_2$ mol.wt. 322.33

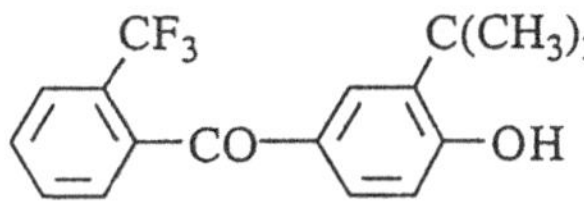

Synthesis

-Preparation by Friedel-Crafts acylation of o-tert-butylphenol
with o-(trifluoromethyl)benzoyl chloride in ethylene
dichloride in the presence of titanium tetrachloride, first at
0°, then at r.t. [733].

m.p. 160-161° [733];   Spectra (NA).

**(3-Chloro-4,6-dimethoxy-2-methylphenyl)(3-chloro-6-hydroxy-2,4-dimethoxyphenyl)
methanone**

[68048-21-5] $C_{18}H_{18}Cl_2O_6$ mol.wt. 401.24

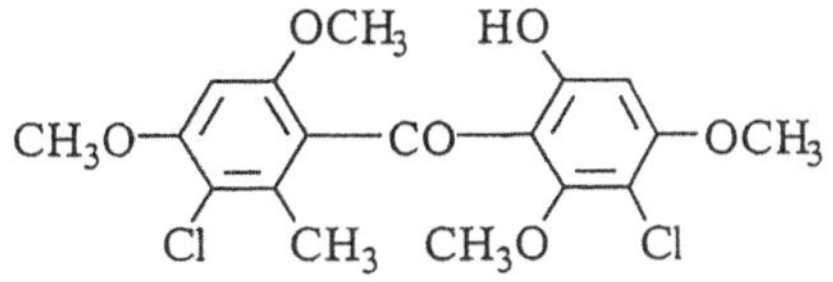

Synthesis

-Preparation by hydrogenolysis of 6-(benzyloxy)-
3,3'-dichloro-2,4,4',6-tetramethoxy-2'-methyl-
benzophenone (SM) under hydrogen in the
presence of 10% Pd/C at 25°. SM was obtained by
condensation of 3-chloro-4,6-dimethoxy-2-methyl-
benzoic acid with 4-chloro-3,5-dimethoxyphenol
benzyl ether in the presence of trifluoroacetic anhydride in methylene under nitrogen for 80 min
(50%) [1328].

m.p. 196-197° [1328];
^{1}H NMR [1328], IR [1328], MS [1328].

[2-Hydroxy-4-methoxy-3-(2-propenyl)phenyl](4-methoxyphenyl)methanone

[74079-07-5] $C_{18}H_{18}O_4$ mol.wt. 298.34

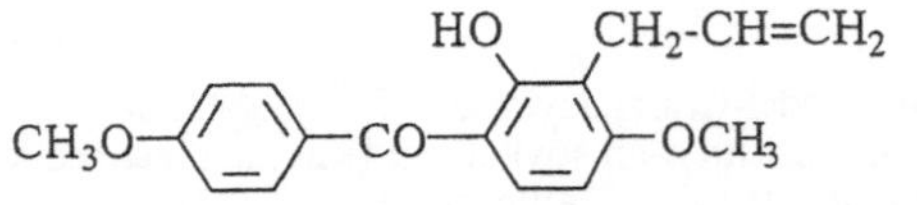

Synthesis

-Obtained by heating 2-(allyloxy)-4,4'-di-methoxybenzophenone (SM) to 240° that which initiates exothermic heating to 290° (60%) (Claisen rearrangement). SM was obtained by reaction of allyl bromide with resbenzophenone (90%) [513].

m.p. and Spectra (NA).

[4-(Acetyloxy)-3,5-dimethoxyphenyl](2-hydroxy-4-methoxyphenyl)methanone

[62495-40-3] $C_{18}H_{18}O_7$ mol.wt. 346.34

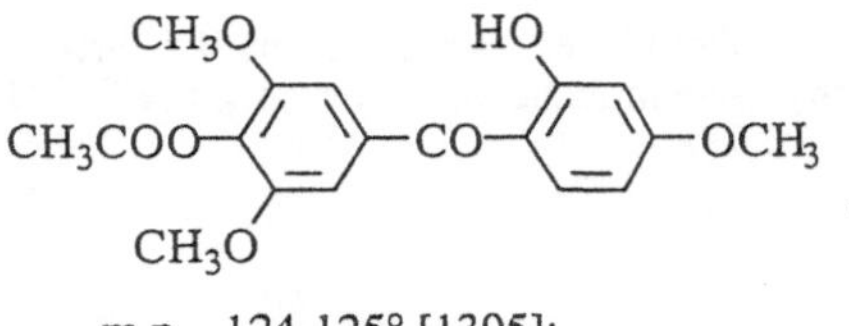

Synthesis

-Preparation by reaction of acetic anhydride with 2,4'-dihydroxy-3',4,5'-trimethoxybenzophenone in the presence of pyridine at r.t. for 24 h (80%) [1395].

m.p. 124-125° [1395];
^{1}H NMR [1395], IR [1395], UV [1395].

(3-Butyl-5-chloro-2-hydroxyphenyl)(4-methylphenyl)methanone

[92739-93-0] $C_{18}H_{19}ClO_2$ mol.wt. 302.80

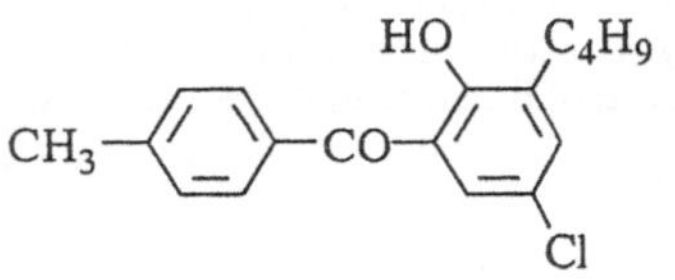

Synthesis

-Preparation by Fries rearrangement of 2-n-butyl-4-chlorophenyl p-toluate with aluminium chloride at 160° for 15 min [698].

oil [698];   b.p.$_{0.108}$ 148-150° [698];   Spectra (NA).

(5-Chloro-3-ethyl-2-hydroxyphenyl)[4-(1-methylethyl)phenyl]methanone

[93575-43-0] $C_{18}H_{19}ClO_2$ mol.wt. 302.80

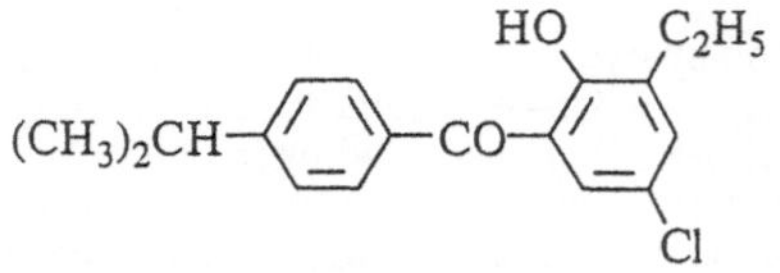

Synthesis

-Preparation by Fries rearrangement of 4-chloro-2-ethylphenyl 4-isopropylbenzoate with aluminium chloride at 160° for 15 min [698].

m.p. 42-43° [698];   Spectra (NA).

[5-Chloro-2-hydroxy-3-(2-methylpropyl)phenyl](4-methylphenyl)methanone

[93575-76-9] $C_{18}H_{19}ClO_2$ mol.wt. 302.80

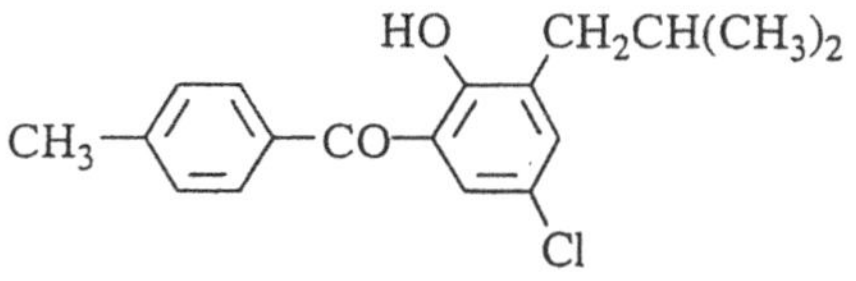

Synthesis

-Preparation by Fries rearrangement of 2-iso-butyl-4-chlorophenyl p-toluate with aluminium chloride at 160° for 15 min [698].

m.p. 71-72° [698];   Spectra (NA).

(3-Chloro-6-hydroxy-2,4-dimethoxyphenyl)(2,4-dimethoxy-6-methylphenyl)methanone

[68048-15-7] $C_{18}H_{19}ClO_6$ mol.wt. 366.80

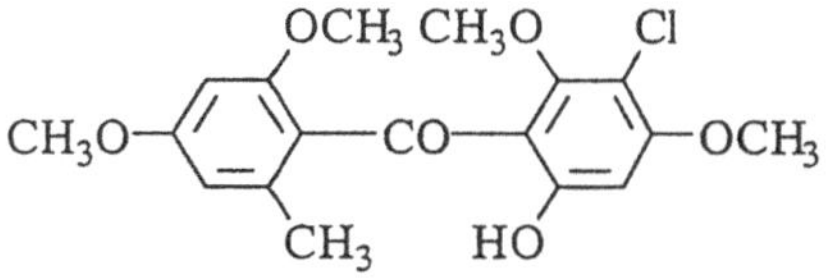

Synthesis

-Preparation by hydrogenolysis of 6-(benzyloxy)-3-chloro-2,2',4,4'-tetramethoxy-6'-methyl-benzophenone (SM) with hydrogen in ethyl acetate/tetrahydrofuran in the presence of 10% Pd/C at 25°. SM was obtained by condensation of 2,4-dimethoxy-6-methylbenzoic anhydride with 4-chloro-3,5-dimethoxyphenol benzyl ether in the presence of trifluoroacetic anhydride in methylene chloride under nitrogen for 10 min (47%) [1328].

m.p. 140°5-142° [1328]; ^{1}H NMR [1328], IR [1328], MS [1328].

[3-(Aminomethyl)-5-(1,1-dimethylethyl)-2-hydroxyphenyl](4-bromophenyl)methanone

[75061-00-6] $C_{18}H_{20}BrNO_2$ mol.wt. 362.27

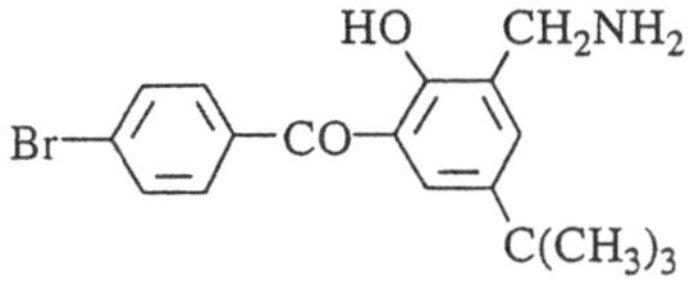

Synthesis

-Refer to: [636]; see the hydrochloride below.

m.p. and Spectra (NA).

[3-(Aminomethyl)-5-(1,1-dimethylethyl)-2-hydroxyphenyl](4-bromophenyl)methanone
(Hydrochloride)

[75060-65-0] $C_{18}H_{20}BrNO_2,HCl$ mol.wt. 398.73

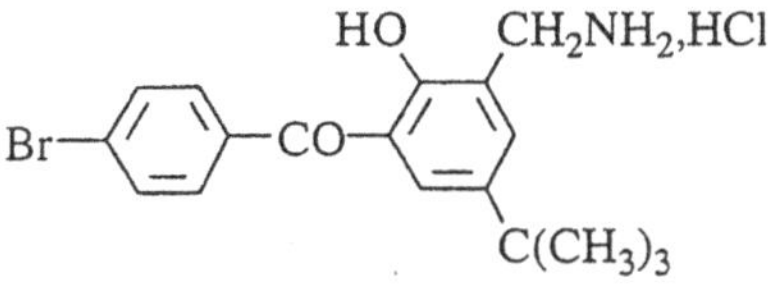

Synthesis

-Preparation by reaction of concentrated hydrochloric acid with 2-(4-bromobenzoyl)-4-tert-butyl-6-(N-chloroacetylaminomethyl)phenol in refluxing ethanol for 20 h (89%) [636].

m.p. 240-245° [636];   Spectra (NA).

[3-(1,1-Dimethylethyl)-4-hydroxyphenyl](4-methylphenyl)methanone

[124979-04-4] $C_{18}H_{20}O_2$ mol.wt. 268.36

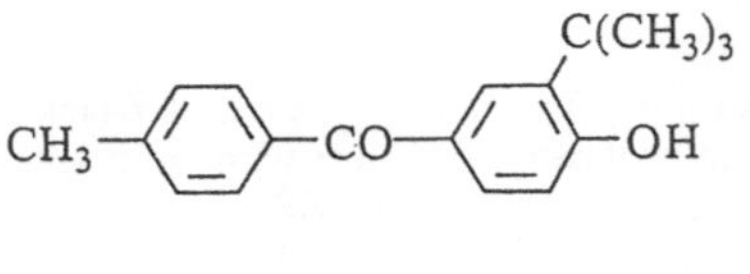

Synthesis

-Preparation by Friedel-Crafts acylation of o-tert-
butylphenol with p-toluoyl chloride in ethylene
dichloride in the presence of titanium tetrachloride,
first at 0°, then at r.t. [733].

m.p. 209-210° [733];   Spectra (NA).

[5-(1,1-Dimethylethyl)-2-hydroxyphenyl](4-methylphenyl)methanone

[75919-94-7] $C_{18}H_{20}O_2$ mol.wt. 268.36

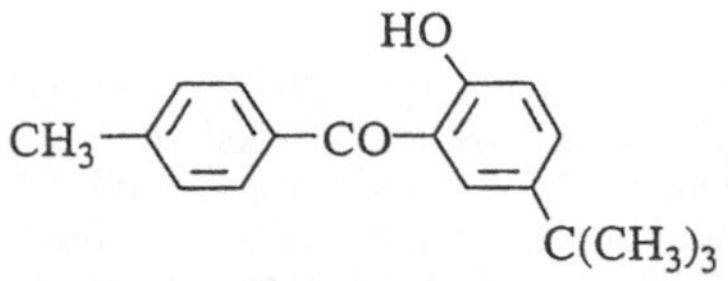

Synthesis

-Preparation by demethylation of 5-tert-butyl-
2-methoxy-4'-methylbenzophenone with aluminium
chloride in benzene between 50 to 60° for 5 h (85%)
[775].

b.p.$_{12}$ 210-212° [775];   m.p. 94-95° [775];
Spectra (NA).

(2-Hydroxy-4,6-dimethylphenyl)(2,4,6-trimethylphenyl)methanone

[100923-74-8] $C_{18}H_{20}O_2$ mol.wt. 268.36

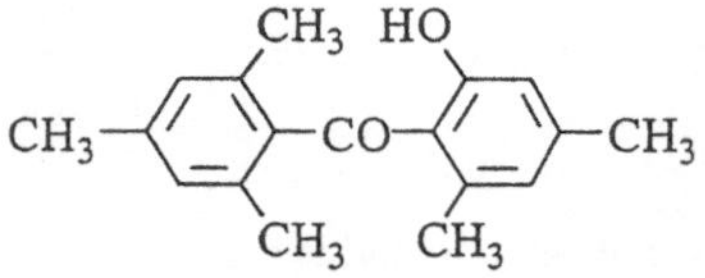

Synthesis

-Preparation by reaction of mesitoyl chloride with
3,5-dimethylphenol in the presence of aluminium
chloride in refluxing nitromethane for 90 min (39%)
[1181].

m.p. 116-117° [1181], 116° [1178, 1179];
^{1}H NMR [1177, 1178, 1179, 1181], ^{13}C NMR [1177],
IR [1178, 1179, 1181];   thermal behaviour [1178, 1179].

(4-Hydroxy-3,5-dimethylphenyl)(2,4,6-trimethylphenyl)methanone

[69795-00-2] $C_{18}H_{20}O_2$ mol.wt. 268.36

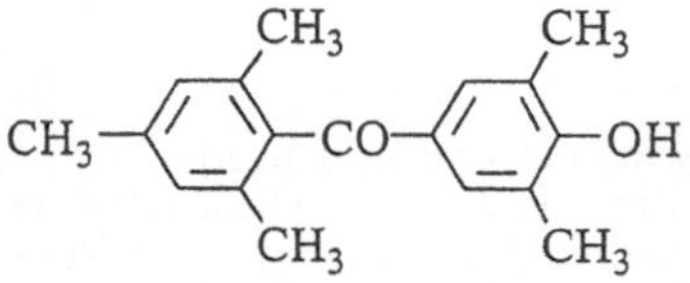

Synthesis

-Obtained by photo-Fries rearrangement of 2,6-di-
methylphenyl mesitoate in pentane (5%), absorbed on
a silica gel-pentane (18%) or on dry silica gel (37%)
[102].

m.p. and Spectra (NA).

[4-Hydroxy-2-methyl-5-(1-methylethyl)phenyl](4-methylphenyl)methanone

[109250-36-4] $C_{18}H_{20}O_2$ mol.wt. 268.36

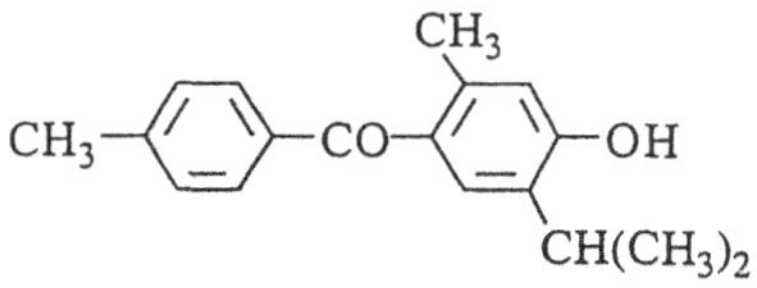

Synthesis

-Obtained by demethylation of 4-methoxy-2,4'-di-
methyl-5-isopropylbenzophenone with refluxing
pyridinium chloride [239].

m.p. 176° [239];   Spectra (NA).

[5-(1,1-Dimethylethyl)-2-hydroxyphenyl][4-(methylthio)phenyl]methanone

[75060-57-0] $C_{18}H_{20}O_2S$ mol.wt. 300.42

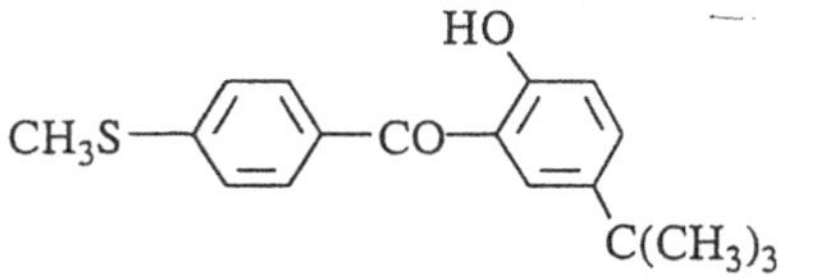

Synthesis

-Preparation by reaction of 30% methyl mercaptan
with 2-(p-bromobenzoyl)-4-tert-butylphenol in the
presence of sodium methoxide in refluxing
methanol for 4 days (34%) [636].

m.p. 130-104° [636]. A typing error probably occurred in the published data.
Spectra (NA).

(4-Butylphenyl)(2-hydroxy-4-methoxyphenyl)methanone

$C_{18}H_{20}O_3$ mol.wt. 284.36

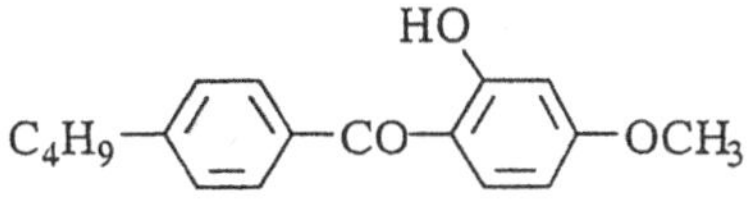

Synthesis

-Preparation by partial demethylation of 4'-butyl-
2,4-dimethoxybenzophenone with aluminium
chloride or aluminium bromide in chlorobenzene at
90-95° (good yield) [602].

m.p. and Spectra (NA).

[5-(1,1-Dimethylethyl)-2-hydroxyphenyl](4-methoxyphenyl)methanone

[116496-22-1] $C_{18}H_{20}O_3$ mol.wt. 284.36

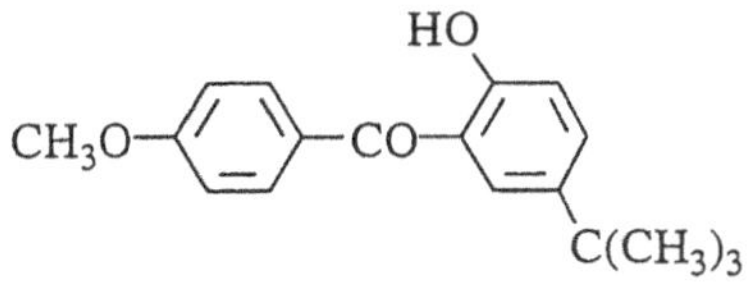

Syntheses

-Preparation by partial demethylation of 5-tert-butyl-
2,4'-dimethoxybenzophenone with aluminium
chloride in benzene at 50-60° for 5 h (76%) [775].
-Preparation by treatment of 2-hydroxy-4'-methoxy-
benzophenone at 120° with a mixture of
isobutylene/nitrogen (1:1) in the presence of a macroreticular acid ion exchanger (Wofatit OK 80)
as catalyst, for 3 h (60%) [111].

b.p.$_{12}$ 235° [775], b.p.$_{0.15}$ 180-190° [111];
m.p. 93-94° [775];   Spectra (NA).

[4-(1,1-Dimethylethyl)phenyl](2-hydroxy-4-methoxyphenyl)methanone

[50739-53-2] $C_{18}H_{20}O_3$ mol.wt. 284.36

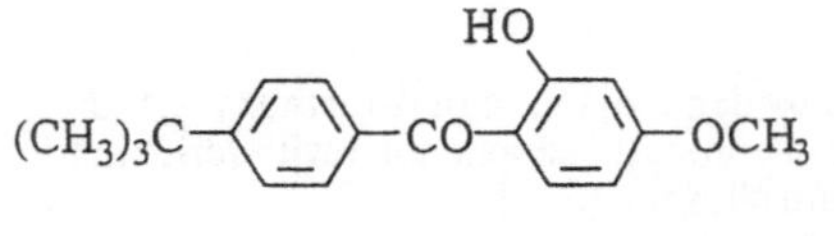

Synthesis

-Preparation by reaction of p-tert-butylbenzoyl
   chloride with resorcinol dimethyl ether,
 *in tetrachloroethane in the presence of
   aluminium chloride at 90° [43];
*in chlorobenzene in the presence of titanium tetrachloride for 1 h at 120° (75%) [1140].
-Also refer to: [384, 541].

m.p.   75-77° [43], 73-75° [1140];   UV [1067].

[4-(1,1-Dimethylethyl)phenyl](2-hydroxy-5-methoxyphenyl)methanone

[162657-94-5] $C_{18}H_{20}O_3$ mol.wt. 284.36

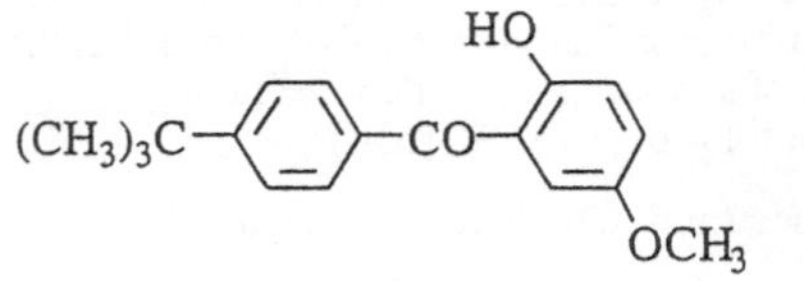

Synthesis

-Preparation by partial demethylation of 4'-tert-butyl-
   2,5-dimethoxybenzophenone with aluminium
   chloride in benzene under nitrogen at 80° for 12 h
   (83%) [1047, 1048].

viscous oil [1047];   b.p. (NA);
^{1}H NMR [1047], ^{13}C NMR [1047], MS [1047].

(4-Ethoxy-2-hydroxyphenyl)(4-propylphenyl)methanone

$C_{18}H_{20}O_3$ mol.wt. 284.36

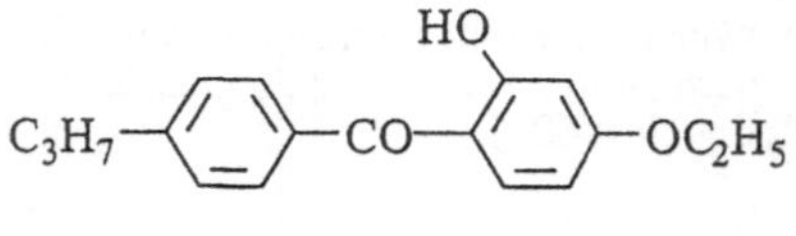

Synthesis

-Preparation by partial deethylation of 2,4-diethoxy-
   4'-propylbenzophenone with aluminium chloride in
   chlorobenzene at 80-100° [602].
-Also refer to: [43, 384].

m.p. (NA);   UV [1067].

(2-Hydroxy-3,4-dimethylphenyl)(2-methoxy-3,4-dimethylphenyl)methanone

[54468-80-3] $C_{18}H_{20}O_3$ mol.wt. 284.36

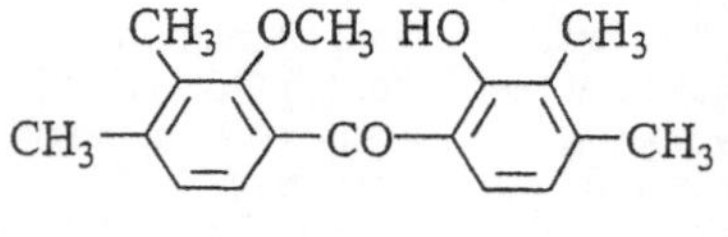

Synthesis

-Obtained by oxidation of 6,7-dimethyl-2-phenyl-
   3-(2-methoxy-3,4-dimethylphenyl)benzofuran with
   chromium trioxide, then saponification of the keto
   ester so formed, the 2-(benzoyloxy)-2'-methoxy-
   3,3',4,4'-tetramethylbenzophenone [1053].

oil [1053];   b.p. (NA);   IR [1053].

(2-Hydroxy-4,5-dimethylphenyl)(2-methoxy-4,5-dimethylphenyl)methanone

[54468-82-5] $C_{18}H_{20}O_3$ mol.wt. 284.36

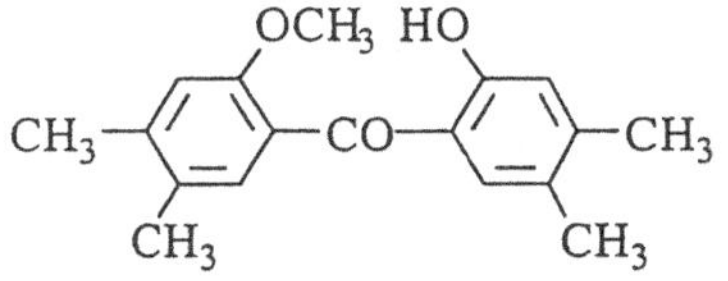

Synthesis

-Obtained by oxidation of 5,6-dimethyl-2-phenyl-3-(2-methoxy-4,5-dimethylphenyl)benzofuran with chromium trioxide, then saponification of the obtained keto ester, the 2-(benzoyloxy)-2'-methoxy-4,4',5,5'-tetramethylbenzophenone [1053].

oil [1053];   b.p. (NA);   IR [1053].

[2-Hydroxy-4-methyl-5-(1-methylethyl)phenyl](4-methoxyphenyl)methanone

[129375-12-8] $C_{18}H_{20}O_3$ mol.wt. 284.36

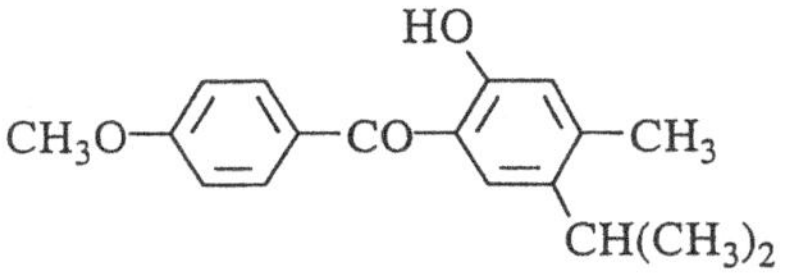

Synthesis

-Preparation by Fries rearrangement of 3-methyl-4-isopropylphenyl p-methoxybenzoate with titanium tetrachloride in nitromethane at 20° for 170 h (66%) [865].

m.p. 100° [865];
^{1}H NMR (Sadtler: standard n° 52709 M) [865],
IR (Sadtler: standard n° 79766 K) [865], UV [865], MS [865].

(2,6-Dimethoxyphenyl)(3-hydroxy-6-methoxy-2,4-dimethylphenyl)methanone

[42594-58-1] $C_{18}H_{20}O_5$ mol.wt. 316.35

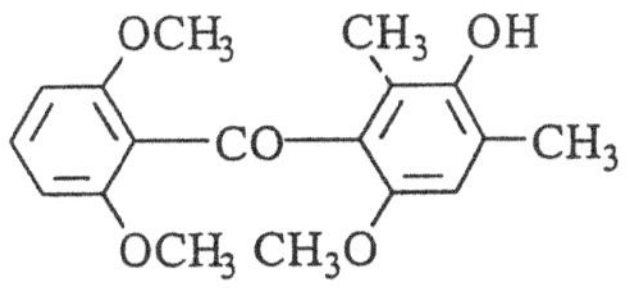

Synthesis

-Preparation by hydrogenolysis of 3-(benzyloxy)-2',6,6'-tri-methoxy-2,4-dimethylbenzophenone with hydrogen in the presence of Pd/C in an ethyl acetate/methanol solution containing 70% perchloric acid (4 drops) (96%) [430].

m.p. 185-186° [430];
^{1}H NMR [430], IR [430], UV [430], MS [430].

(5-Ethoxy-2-hydroxy-4-methoxyphenyl)(3-ethoxyphenyl)methanone

[51106-93-5] $C_{18}H_{20}O_5$ mol.wt. 316.35

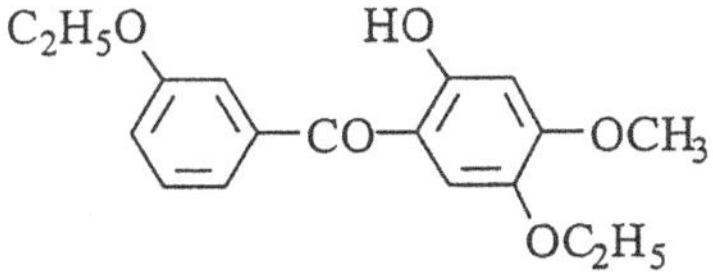

Synthesis

-Obtained (by-product) by acylation of 2,5-diethoxy-anisole with m-ethoxybenzoyl chloride in ethyl ether in the presence of aluminium chloride at r.t. for 48 h (<2%) [372].

m.p. 79-80° [372]; ^{1}H NMR [372].

(2,4-Dimethoxy-6-methylphenyl)(2-hydroxy-4,6-dimethoxyphenyl)methanone

[93904-08-6] $C_{18}H_{20}O_6$ mol.wt. 332.35

Synthesis

-Obtained by partial methylation of griseophenone C,
*with dimethyl sulfate in the presence of potassium
  carbonate in refluxing acetone for 24 h (45%) [1115];
*with diazomethane in an ethyl ether/tetrahydrofuran
  mixture at 0° during 40 h (quantitative yield) [1115].

m.p.   124-125° [1115];   IR [1115], UV [1115].

(4,5-Dimethoxy-2-methylphenyl)(2-hydroxy-4,5-dimethoxyphenyl)methanone

[101744-11-0] $C_{18}H_{20}O_6$ mol.wt. 332.35

Syntheses

-Preparation by reaction of 2,4,5-trimethoxybenzoyl
  chloride with 4-methylveratrole in the presence of
  aluminium chloride in refluxing carbon disulfide for
  8 h, then at r.t. for 12 h [460].
-Preparation by reaction of 4,5-dimethoxy-2-methyl-
benzoyl chloride with hydroxyhydroquinone trimethyl ether in the same conditions that previously
[460].

m.p.   138-139° [460];   UV [460].

(2,3-Dimethoxyphenyl)[3-hydroxy-2-methoxy-6-(methoxymethyl)phenyl]methanone

[133387-00-5] $C_{18}H_{20}O_6$ mol.wt. 332.35

Synthesis

-Preparation from 4-hydroxy-3-methoxybenzyl methyl
  ether (SM) in two steps: first, direct lithiation of SM with
  n-butyl lithium in tetrahydrofuran at r.t., followed by
  quenching with 2,3-dimethoxybenzaldehyde. Then,
  oxidation of the 8-(2,3-dimethoxyphenyl)-2,7-di-
methoxybicyclo[4.2.0]octa-1,3,5-triene-3,8-diol formed (good yield) [286].

m.p. and Spectra (NA).

(2,4-Dimethoxyphenyl)[3-hydroxy-2-methoxy-6-(methoxymethyl)phenyl]methanone

[133386-98-8] $C_{18}H_{20}O_6$ mol.wt. 332.35

Synthesis

-Preparation from 4-hydroxy-3-methoxybenzyl methyl
  ether (SM1) in two steps: first, direct lithiation of SM1
  with n-butyl lithium in tetrahydrofuran at r.t., followed
  by quenching with 2,4-dimethoxybenzaldehyde. Then,
  oxidation of the 8-(2,4-dimethoxyphenyl)-

2,7-dimethoxybicyclo[4.2.0]octa-1,3,5-triene-3,8-diol formed (SM2) (good yield) [286]. Actually, SM2 was quantitatively converted back into the expected ketone by heating a toluene solution to reflux for 14 h [286].

m.p. and Spectra (NA).

(2-Hydroxy-3-methoxy-6-methylphenyl)(2,4,5-trimethoxyphenyl)methanone

[129103-92-0] $C_{18}H_{20}O_6$ mol.wt. 332.35

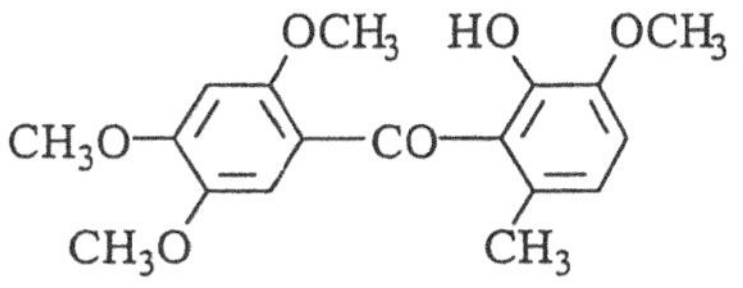

Synthesis

-Preparation from 2-iodo-6-methoxy-3-methylphenyl 2,4,5-trimethoxybenzoate on treatment with n-butyl-lithium in a mixture of ethyl ether, hexane and tetra-hydrofuran at -70° for 2 h, followed by treatment with saturated aqueous ammonium chloride (40%) [598].

m.p. 160-162° [598]; ^{1}H NMR [598], IR [598], MS [598].

(2-Hydroxy-4-methoxy-6-methylphenyl)(2,4,6-trimethoxyphenyl)methanone

[74628-37-8] $C_{18}H_{20}O_6$ mol.wt. 332.35

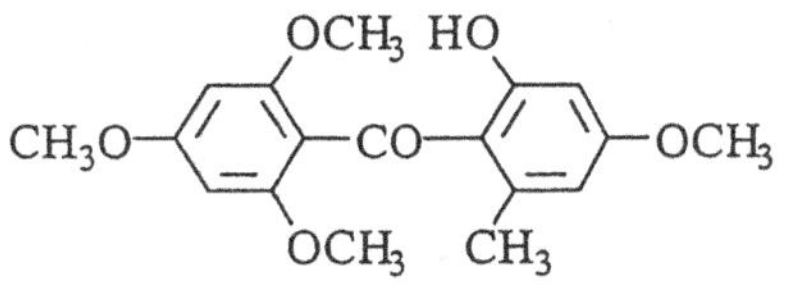

Synthesis

-Preparation by selective methylation of 2,4'-di-hydroxy-2',4,6'-trimethoxy-6-methylbenzo-phenone [1188], with dimethyl sulfate in the presence of potassium carbonate in refluxing acetone for 23 h (85%) [1189].

m.p. 161-162° [1189]; ^{1}H NMR [1189], MS [1189].

(2,3-Dimethoxyphenyl)(2-hydroxy-3,4,5-trimethoxyphenyl)methanone

[42833-83-0] $C_{18}H_{20}O_7$ mol.wt. 348.35

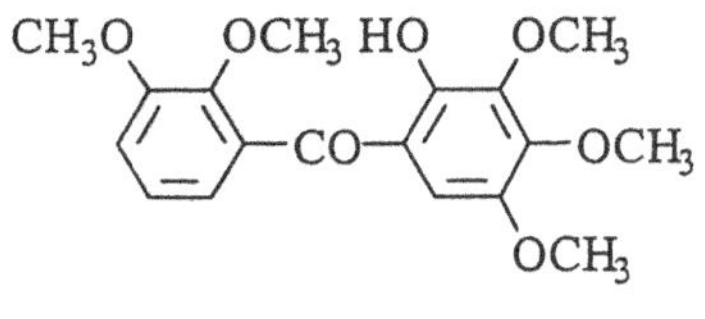

Synthesis

-Obtained by reaction of 2,3-dimethoxybenzoyl chloride with 1,2,3,4-tetramethoxybenzene in the presence of aluminium chloride in ethyl ether for 42 h [1097].

m.p. (NA); ^{1}H NMR [1097], UV [1097], MS [1097].

(2,3-Dimethoxyphenyl)(2-hydroxy-3,4,6-trimethoxyphenyl)methanone

                                  $C_{18}H_{20}O_7$ mol.wt. 348.35

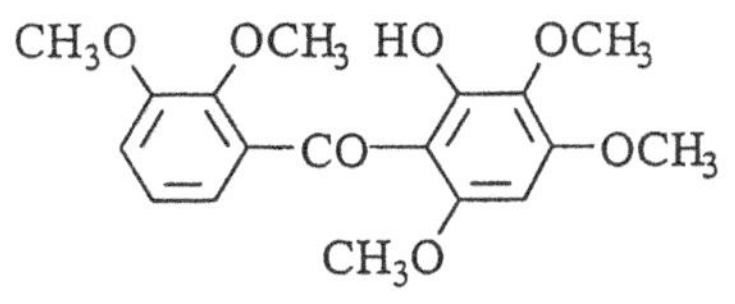

Synthesis

-Obtained by reaction of 2,3-dimethoxybenzoic acid with 1,2,3,5-tetramethoxybenzene in trifluoroacetic anhydride for 23 h at r.t. [1311].

m.p. and Spectra (NA);   TLC [1311].

(2,3-Dimethoxyphenyl)(6-hydroxy-2,3,4-trimethoxyphenyl)methanone

[22804-59-7] $C_{18}H_{20}O_7$ mol.wt. 348.35

Synthesis

-Obtained by condensation of 2,3-dimethoxybenzoic
acid with 1-acetoxy-3,4,5-trimethoxybenzene in
trifluoroacetic anhydride, followed by hydrolysis of the
reaction mixture (26%) [1310].

   m.p. 110°8-111°2 [1310]; ^{1}H NMR [1310], UV [1310].

(2,5-Dimethoxyphenyl)(2-hydroxy-3,4,5-trimethoxyphenyl)methanone

[129168-53-2] $C_{18}H_{20}O_7$ mol.wt. 348.35

Syntheses

-Preparation by reaction of 2,5-dimethoxybenzoyl chloride
with 1,2,3,4-tetramethoxybenzene in benzene in the presence
of aluminium chloride at 50° (60%). The same result was
obtained in ethyl ether at r.t. [845].
-Also obtained (poor yield) by reaction of 2,3,4,5-tetra-
methoxybenzoyl chloride with hydroquinone dimethyl ether in benzene in the presence of
aluminium chloride (11%) [845].

   m.p. 74-74°5 [845];   Spectra (NA).

(2,5-Dimethoxyphenyl)(2-hydroxy-3,4,6-trimethoxyphenyl)methanone

[23251-65-2] $C_{18}H_{20}O_7$ mol.wt. 348.35

Syntheses

-Obtained by reaction of 2,5-dimethoxybenzoyl chloride with
1,2,3,5-tetramethoxybenzene in the presence of aluminium
chloride in ethyl ether at r.t. for 18 h [1097], (75%) [845].
-Also obtained by selective demethylation of

2,2',3,4,5',6-hexamethoxybenzophenone with aluminium chloride in ethyl ether (30%) [1311].

   m.p. 128-130° [845, 1311]; ^{1}H NMR [1097, 1311], UV [1311], MS [1097].

(2,5-Dimethoxyphenyl)(6-hydroxy-2,3,4-trimethoxyphenyl)methanone

[22804-57-5] $C_{18}H_{20}O_7$ mol.wt. 348.35

Synthesis

-Preparation by reaction of 2,5-dimethoxybenzoic acid with
1-acetoxy-3,4,5-trimethoxybenzene in the presence of
trifluoroacetic anhydride for two weeks at r.t., followed by
hydrolysis of the reaction mixture (52%) [1310], (45%)
[845].

   yellow oil [1310];   m.p. 65-66° [845]; ^{1}H NMR [1310], UV [1310].

(2,6-Dimethoxyphenyl)(2-hydroxy-3,4,6-trimethoxyphenyl)methanone

[22804-60-0] $C_{18}H_{20}O_7$ mol.wt. 348.35

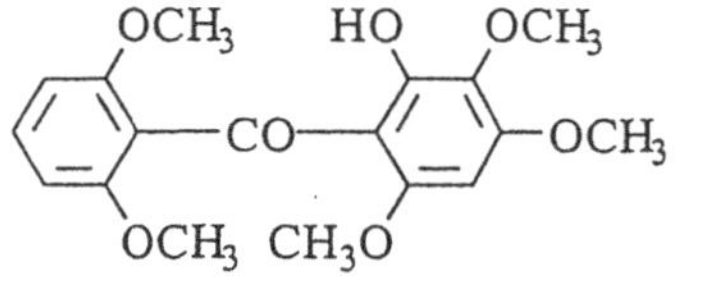

Synthesis

-Preparation by selective demethylation of 2,2',3,4,6,6'-hexamethoxybenzophenone (SM) with aluminium chloride in refluxing ethyl ether (60%). SM was obtained by condensation of 2,6-dimethoxybenzoyl chloride with 1,2,3,5-tetramethoxybenzene in nitrobenzene with aluminium chloride for two days at r.t. (27%, m.p. 136-136°5) [1310].

m.p. 167-168° [1310];
^{1}H NMR [1310], UV [1310].

(2-Hydroxy-3,4-dimethoxyphenyl)(2,4,5-trimethoxyphenyl)methanone

$C_{18}H_{20}O_7$ mol.wt. 348.35

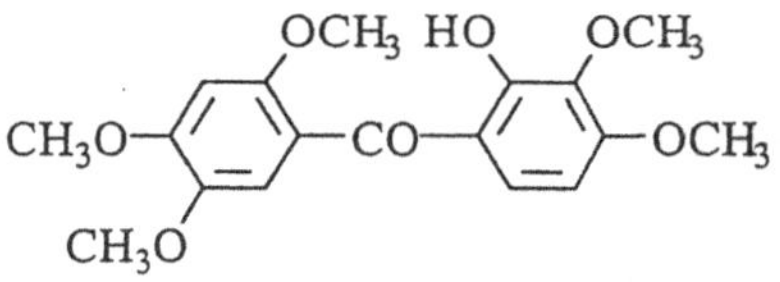

Synthesis

-Refer to: [826] (compound 42).

m.p. and Spectra (NA).

(2-Hydroxy-3,4-dimethoxyphenyl)(2,4,6-trimethoxyphenyl)methanone

[42833-67-0] $C_{18}H_{20}O_7$ mol.wt. 348.35

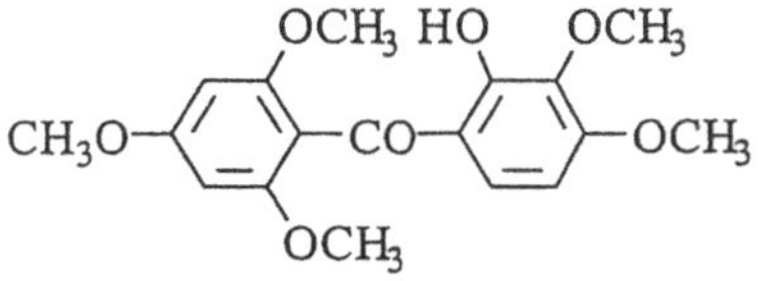

Synthesis

-Obtained by reaction of 2,3,4-trimethoxybenzoyl chloride with 1,3,5-trimethoxybenzene in the presence of aluminium chloride in ethyl ether for 15 h [1097].

m.p. (NA); ^{1}H NMR [1097].

(2-Hydroxy-3,4-dimethoxyphenyl)(3,4,5-trimethoxyphenyl)methanone

$C_{18}H_{20}O_7$ mol.wt. 348.35

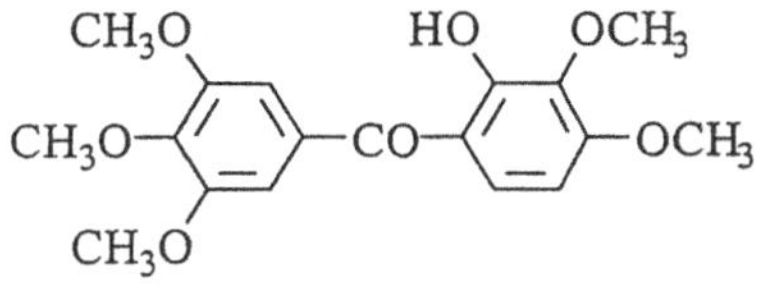

Synthesis

-Preparation by Friedel-Crafts acylation of pyrogallol trimethyl ether with 3,4,5-trimethoxybenzoyl chloride in carbon disulfide in the presence of aluminium chloride in a water bath [1049, 1050].

m.p. 133-134° [1049, 1050];   Spectra (NA).

(2-Hydroxy-4,5-dimethoxyphenyl)(2,3,4-trimethoxyphenyl)methanone

$C_{18}H_{20}O_7$ mol.wt. 348.35

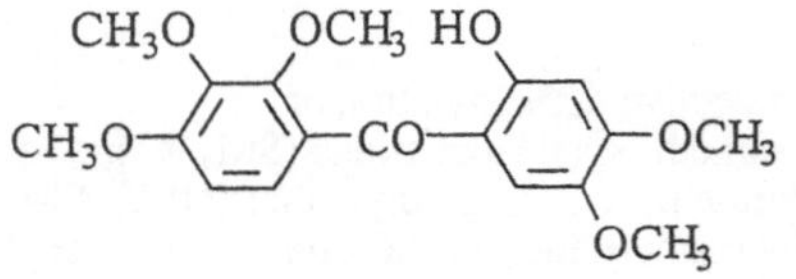

Synthesis

-Refer to: [826] (compound 43).

m.p. and Spectra (NA).

(2-Hydroxy-4,5-dimethoxyphenyl)(2,4,6-trimethoxyphenyl)methanone

[42833-68-1] $C_{18}H_{20}O_7$ mol.wt. 348.35

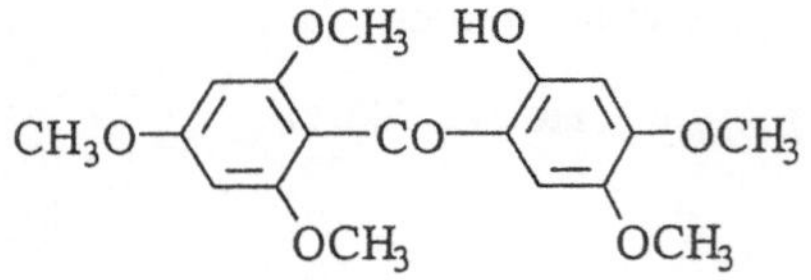

Syntheses

-Obtained (by-product) by partial demethylation of 2,2',4,4',5',6-hexamethoxybenzophenone with boron trichloride in methylene chloride at 18° for 20 min (26%) [300].
-Also obtained by Friedel-Crafts acylation of 1,3,5-trimethoxybenzene with 2,4,5-trimethoxybenzoyl chloride in the presence of aluminium chloride in ethyl ether for 21 h [1097] or in boiling ethyl ether for 48 h (12%) [300].

m.p.   171-173° [1095], 165-166° [300];
^{1}H NMR [300, 1097], IR [300], MS [300].

(2-Hydroxy-4,6-dimethoxyphenyl)(2,4,5-trimethoxyphenyl)methanone

[76013-33-7] $C_{18}H_{20}O_7$ mol.wt. 348.35

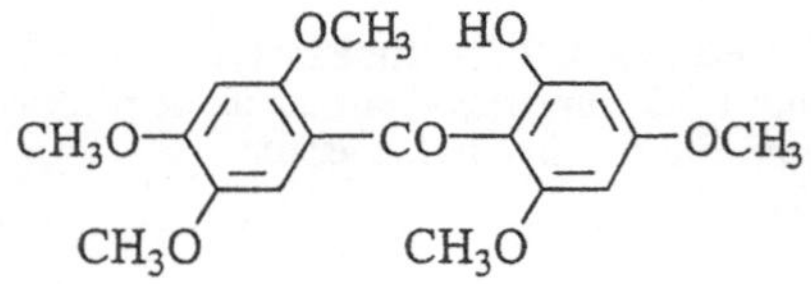

Synthesis

-Preparation by partial demethylation of 2,2',4,4',5',6-hexamethoxybenzophenone with boron trichloride in methylene chloride at 18° for 20 min (57%) [300].

m.p.   129-130° [300];
^{1}H NMR [300], IR [300], MS [300].

(2-Hydroxy-3-methoxyphenyl)(2,3,4,6-tetramethoxyphenyl)methanone

$C_{18}H_{20}O_7$ mol.wt. 348.35

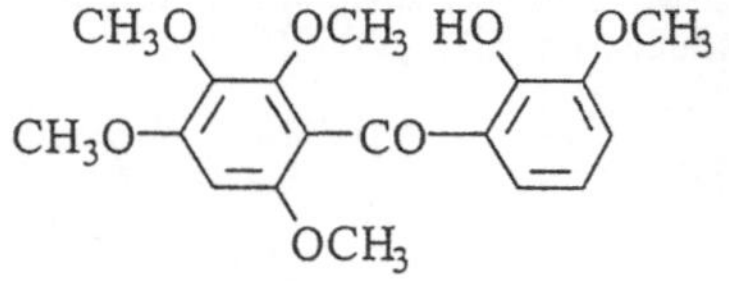

Synthesis

-Obtained by reaction of 2,3-dimethoxybenzoic acid with 1,2,3,5-tetramethoxybenzene in trifluoroacetic acid for 23 days at r.t. [1311].

m.p. and Spectra (NA);   TLC [1311].

[4-(4-Bromophenoxy)-2-hydroxyphenyl](3,4-dichlorophenyl)methanone

[35698-03-4] $C_{19}H_{11}BrCl_2O_3$ mol.wt. 438.10

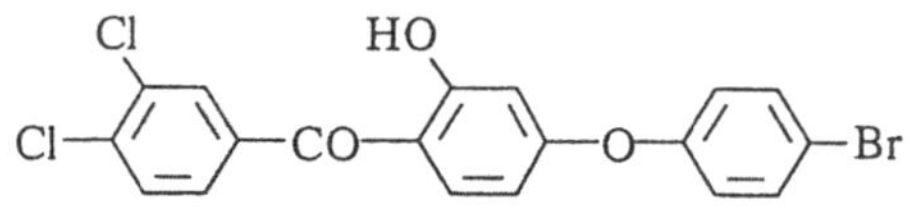

Synthesis

-Refer to: [101] (compound 33).

m.p. (NA);   UV [101].

(4-Chlorophenyl)(2-hydroxy-4-phenoxyphenyl)methanone

[35698-40-9] $C_{19}H_{13}ClO_3$ mol.wt. 324.76

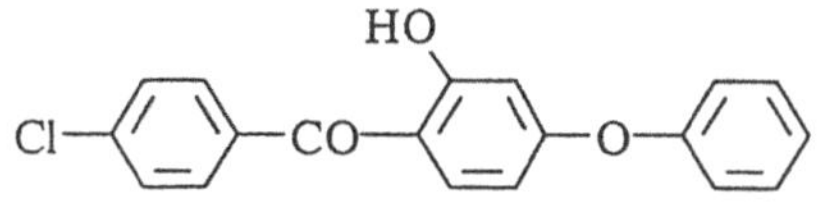

Synthesis

-Preparation by reaction of p-chlorobenzoyl chloride with 3-methoxydiphenyl ether in chlorobenzene in the presence of aluminium chloride: first at r.t., then for 4 h at 90-95° [101].

m.p. 125-127° [101];   UV [101].

[3-(Cyclohexyloxy)-4-hydroxy-5-nitrophenyl](2-nitrophenyl)methanone

[190585-64-9] $C_{19}H_{18}N_2O_7$ mol.wt. 386.36

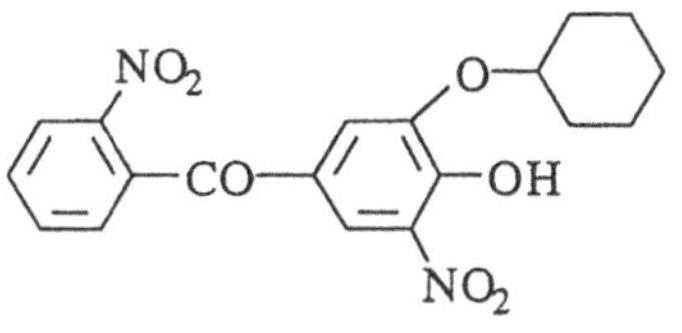

Synthesis

-Preparation by reaction of cyclohexene with 3,4-dihydroxy-2',5-dinitrobenzophenone in the presence of boron trifluoride ethyl ether complex for 12 h at reflux (24%). -Refer to: Chem. Abstr., 127, 17465u (1997).

m.p. and Spectra (NA).

[4-(Acetyloxy)-2-methoxy-6-methylphenyl](3-chloro-2-hydroxy-4,6-dimethoxyphenyl)-methanone

[95276-66-7] $C_{19}H_{19}ClO_7$ mol.wt. 394.81

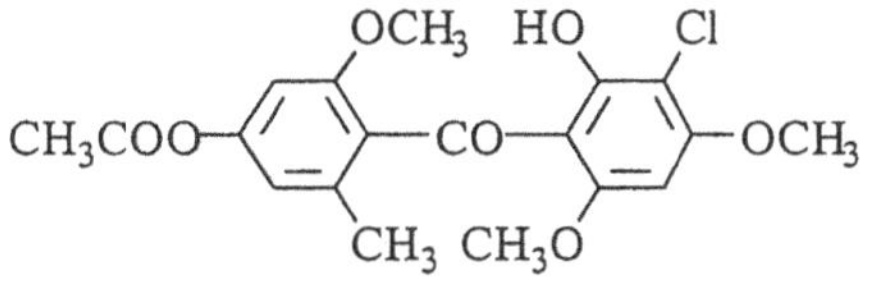

Synthesis

-Obtained by reaction of 4-acetoxy-2-methoxy-6-methylbenzoic acid with 2-chloro-3,5-dimethoxyphenol in the presence of trifluoroacetic anhydride at 20° for 18 h (17%) [1360] or at 25° (50%) [1361, 1362].

-Also refer to: [784].

m.p. 199-203° [1360];   Spectra (NA).

[4-(Acetyloxy)-2-methoxy-6-methylphenyl](3-fluoro-2-hydroxy-4,6-dimethoxyphenyl)-methanone

$C_{19}H_{19}FO_7$ mol.wt. 378.35

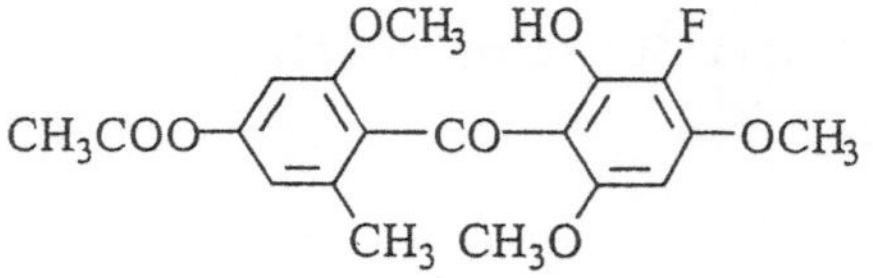

Synthesis

-Preparation by addition of 4-acetoxy-2-methoxy-6-methylbenzoic acid to 2-fluoro-3,5-dimethoxyphenol in the presence of trifluoroacetic anhydride (60%) [1361], first at 0°, then at 20-25° for 20 h (40%) [1362].

m.p. 195-200° [1361, 1362];   IR [1362], UV [1361, 1362].

(5-Chloro-3-hexyl-2-hydroxyphenyl)(4-chlorophenyl)methanone

[92739-91-8] $C_{19}H_{20}Cl_2O_2$ mol.wt. 351.27

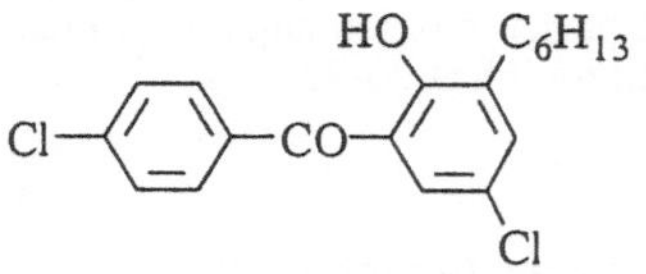

Synthesis

-Preparation by Fries rearrangement of 4-chloro-2-hexyl-phenyl p-chlorobenzoate with aluminium chloride at 160° for 15 min [698].

oil [698];   b.p.$_{0.133}$ 204° [698];   Spectra (NA).

(4-Butoxy-2-hydroxyphenyl)(4-ethenylphenyl)methanone

[80167-01-7] $C_{19}H_{20}O_3$ mol.wt. 296.37

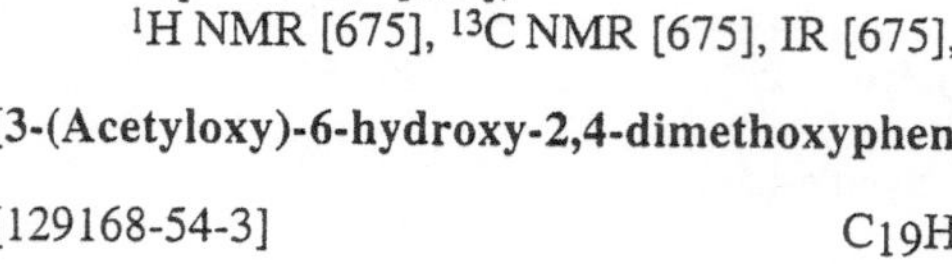

Synthesis

-Obtained by reaction of aqueous potassium hydroxide with 2-hydroxy-4-butoxy-4'-(2-bromoethyl)benzophenone in the presence of hydroquinone in refluxing methanol for 1.5 h with nitrogen bubbling (24%) [675].

m.p. 78-80° [675];
^{1}H NMR [675], ^{13}C NMR [675], IR [675], UV [675], MS [675].

[3-(Acetyloxy)-6-hydroxy-2,4-dimethoxyphenyl](2,5-dimethoxyphenyl)methanone

[129168-54-3] $C_{19}H_{20}O_8$ mol.wt. 376.36

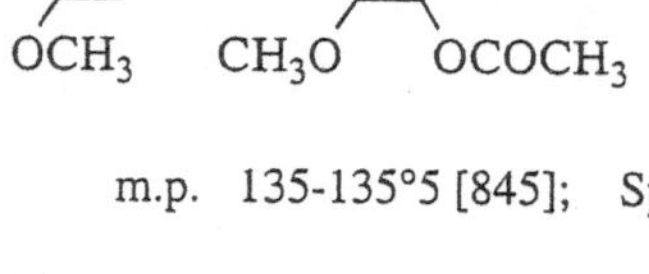

Synthesis

-Obtained (poor yield) by reaction of 2,5-dimethoxybenzoic acid with 2,6-dimethoxy-1,4-hydroquinone diacetate in the presence of trifluoroacetic anhydride for two weeks at r.t. (<5%) [845].

m.p. 135-135°5 [845];   Spectra (NA).

(4-Fluorophenyl)[4-(hexyloxy)-2-hydroxyphenyl]methanone

$C_{19}H_{21}FO_3$ mol.wt. 316.37

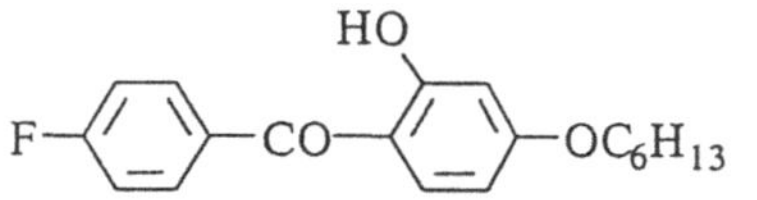

Synthesis

-Preparation by partial alkylation of 4'-fluoro-2,4-di-hydroxybenzophenone with an hexyl halide in the presence of an alkali [394].

m.p. and Spectra (NA).

[3-(1,1-Dimethylethyl)-2-hydroxy-5-methylphenyl](4-methylphenyl)methanone

[93575-42-9] $C_{19}H_{22}O_2$ mol.wt. 282.38

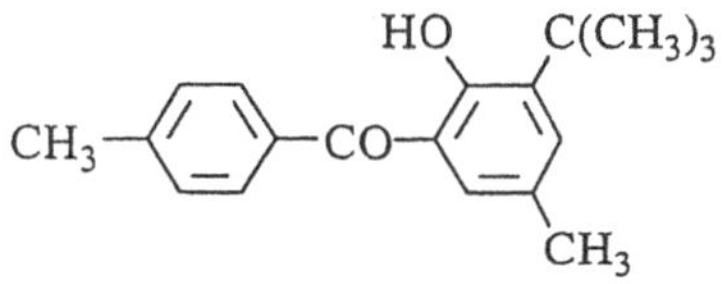

Synthesis

-Preparation by photo-Fries rearrangement of 2-tert-butyl-4-methylphenyl p-toluate in benzene under nitrogen for 32 h [698].

m.p. 110-111° [698];   Spectra (NA).

[3-(1,1-Dimethylethyl)-4-hydroxyphenyl](2,4-dimethylphenyl)methanone

[203786-32-7] $C_{19}H_{22}O_2$ mol.wt. 282.38

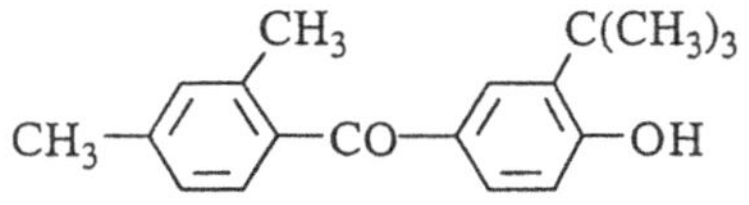

Synthesis

-Refer to: Chem. Abstr., **128**, 210826m (1998) (Japanese patent).

m.p. and Spectra (NA).

(3-Ethyl-2-hydroxy-5-methylphenyl)(4-propylphenyl)methanone

[93575-38-3] $C_{19}H_{22}O_2$ mol.wt. 282.38

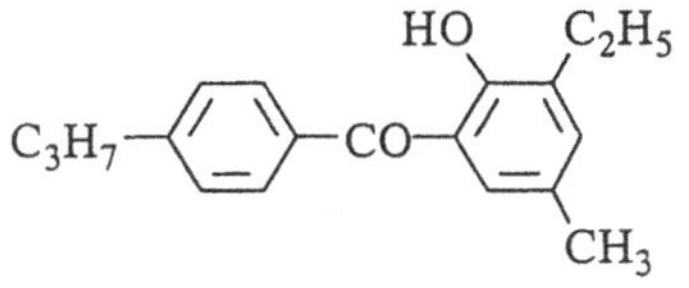

Synthesis

-Preparation by Fries rearrangement of 2-ethyl-4-methylphenyl 4-n-propylbenzoate with aluminium chloride at 160° for 15 min [698].

oil [698]; b.p. (NA); $n_D^{21} = 1.5861$ [698]; Spectra (NA).

(2-Hydroxy-4-propoxyphenyl)(4-propoxyphenyl)methanone

[6131-39-1] $C_{19}H_{22}O_4$ mol.wt. 314.38

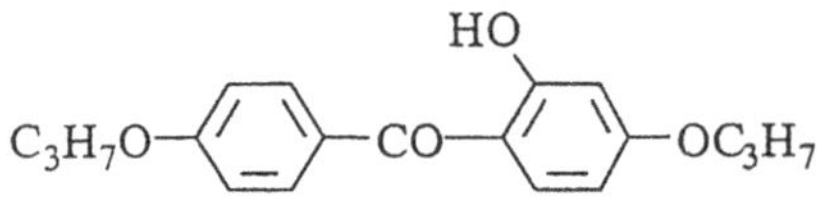

Synthesis

-Preparation by reaction of propyl bromide with 2,4,4'-trihydroxybenzophenone in the presence of potassium carbonate (50%) [233].

m.p. 60-60°5 [233];   IR [991], UV [233, 991].

(3,6-Diethoxy-2-hydroxyphenyl)(2,5-dimethoxyphenyl)methanone

[110047-51-3] $C_{19}H_{22}O_6$ mol.wt. 346.38

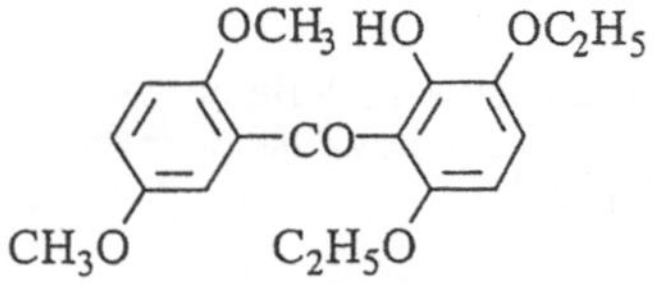

Synthesis

-Obtained (poor yield) by reaction of 3,6-diethoxy-
2-methoxybenzoyl chloride with hydroquinone dimethyl
ether in ethyl ether in the presence of aluminium chloride
at r.t. for 2 days (4%) [1062].

m.p. 102-103° [1062];   Spectra (NA).

(2,5-Diethoxyphenyl)(2-hydroxy-3,6-dimethoxyphenyl)methanone

[110049-41-7] $C_{19}H_{22}O_6$ mol.wt. 346.38

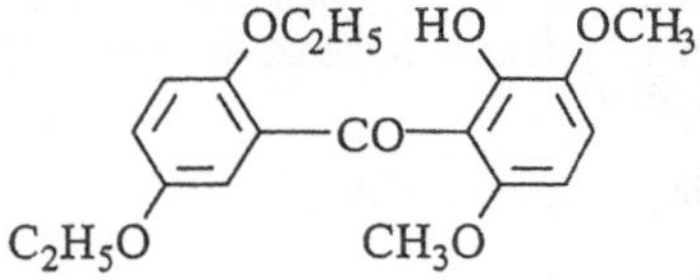

Synthesis

-Obtained (poor yield) by reaction of 2,3,6-trimethoxy-
benzoyl chloride with hydroquinone diethyl ether in
ethyl ether in the presence of aluminium chloride at r.t.
for 2 days (7%) [1062].

m.p. 100-101° [1062];   Spectra (NA).

(3,4-Dimethoxy-2,6-dimethylphenyl)(2-hydroxy-4,5-dimethoxyphenyl)methanone

                                                            $C_{19}H_{22}O_6$ mol.wt. 346.38

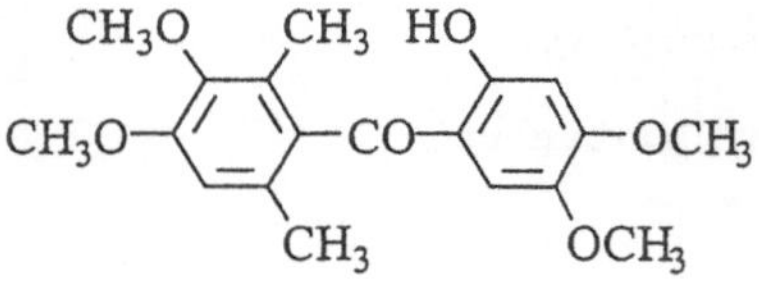

Synthesis

-Preparation by reaction of 2,4,5-trimethoxybenzoyl
chloride with 3,5-dimethylveratrole in the presence
of aluminium chloride in refluxing carbon disulfide
for 8 h, then at r.t. for 12 h [460].

m.p. 138-139° [460];   UV [460].

(5-Ethoxy-2-hydroxy-3,4-dimethoxyphenyl)(4-ethoxyphenyl)methanone

[69471-33-6] $C_{19}H_{22}O_6$ mol.wt. 346.38

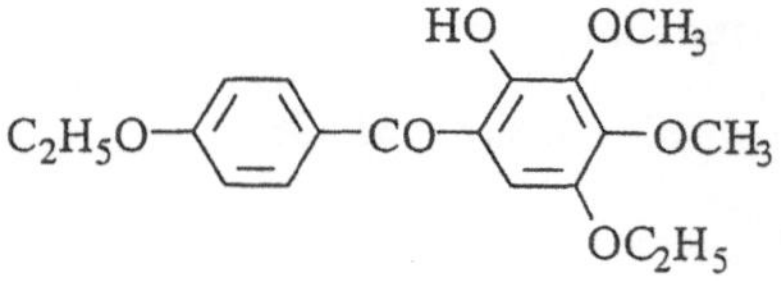

Synthesis

-Preparation by reaction of ethyl iodide with
4'-ethoxy-2,5-dihydroxy-3,4-dimethoxybenzo-
phenone in the presence of potassium carbonate in
refluxing acetone during 16 h (79%) [1008].

m.p. 60-61°5 [1008]; ^{1}H NMR [1008], IR [1008].

(4-Ethoxy-2-hydroxy-5-methoxyphenyl)(3-ethoxy-4-methoxyphenyl)methanone

[18008-38-3] $C_{19}H_{22}O_6$ mol.wt. 346.38

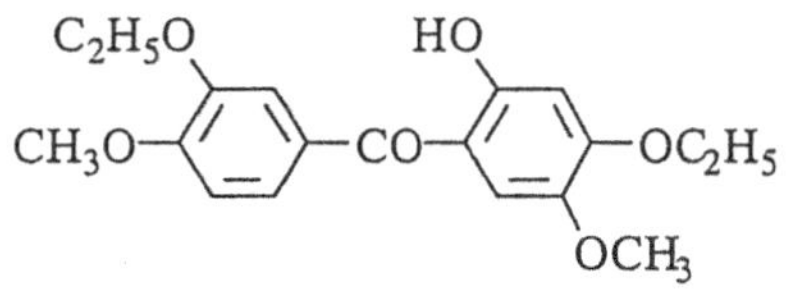

Synthesis

-Preparation by reaction of 3-ethoxy-4-methoxy-
benzoyl chloride with 2,5-dimethoxyphenetole in
the presence of aluminium chloride in ethyl ether at
r.t. for 48 h (23%) [369].

m.p. 132-133° [369]; ^{1}H NMR [369], IR [369], UV[369].

(5-Ethoxy-2-hydroxy-4-methoxyphenyl)(3-ethoxy-4-methoxyphenyl)methanone

[17892-44-3] $C_{19}H_{22}O_6$ mol.wt. 346.38

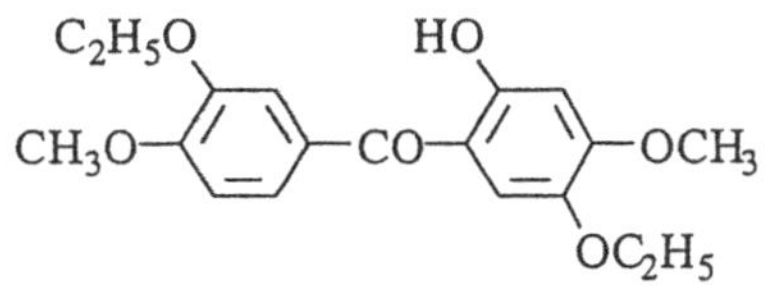

Syntheses

-Obtained by action of diethyl sulfate with 2,3',5-tri-
hydroxy-4,4'-dimethoxybenzophenone in the
presence of potassium carbonate in refluxing acetone
for 6 h (46%) [371].
-Preparation by reaction of 3-ethoxy-4-methoxy-
benzoyl chloride with 2,4-dimethoxyphenetole in the presence of aluminium chloride in ethyl ether
at r.t. for 48 h (26%) [369].

m.p. 121-122° [369], 121° [371]; ^{1}H NMR [369], IR [369], UV [369].

(2-Ethyl-4,5-dimethoxyphenyl)(2-hydroxy-4,5-dimethoxyphenyl)methanone

$C_{19}H_{22}O_6$ mol.wt. 346.38

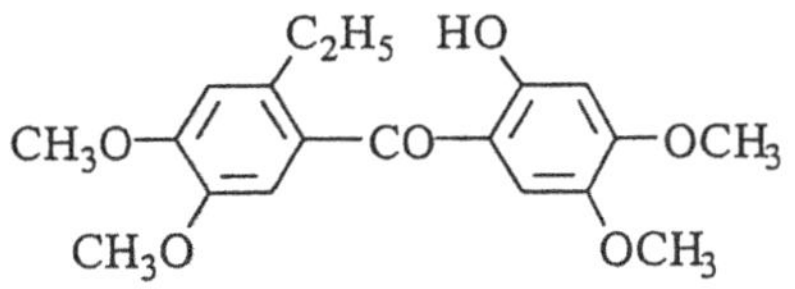

Synthesis

-Preparation by reaction of 2,4,5-trimethoxybenzoyl
chloride with 4-ethylveratrole in the presence of
aluminium chloride in refluxing carbon disulfide for
8 h, then at r.t. for 12 h [460].

m.p. 113-114° [460];   UV [460].

(2,3-Dimethoxyphenyl)(2-hydroxy-3,4,5,6-tetramethoxyphenyl)methanone

[22804-56-4] $C_{19}H_{22}O_8$ mol.wt. 378.38

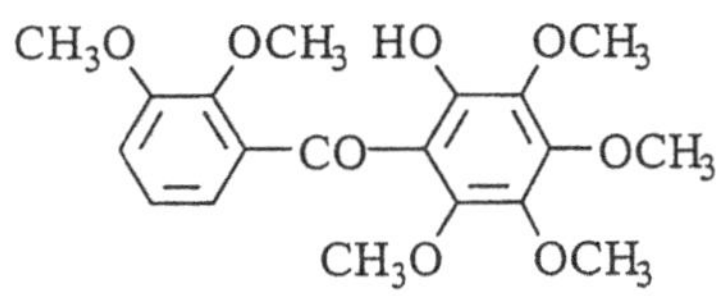

Synthesis

-Obtained by condensation of 2,3-dimethoxybenzoyl
chloride with pentamethoxybenzene in the presence of
aluminium chloride in refluxing ethyl ether for 1 h,
then overnight at r.t. (19%) [1310].

m.p. 116°5-117°5 [1310]; ^{1}H NMR [1310], UV [1310].

(2,5-Dimethoxyphenyl)(2-hydroxy-3,4,5,6-tetramethoxyphenyl)methanone

[22961-80-4] $C_{19}H_{22}O_8$ mol.wt. 378.38

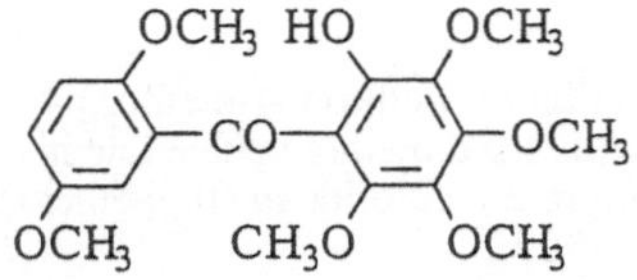

Syntheses

-Preparation by reaction of 2,5-dimethoxybenzoyl chloride with pentamethoxybenzene in nitrobenzene in the presence of aluminium chloride for 6 h at r.t. (38%) [845], (13%) [1310].
-Preparation by selective demethylation of 2,2',3,4,5,5',6-heptamethoxybenzophenone with aluminium chloride in ethyl ether for 1.5 h at r.t. (42%) [1310].

yellow oil [1310];   m.p. 59° [845];
^{1}H NMR [1310], UV [1310].

(2,6-Dimethoxyphenyl)(2-hydroxy-3,4,5,6-tetramethoxyphenyl)methanone

[22804-62-2] $C_{19}H_{22}O_8$ mol.wt. 378.38

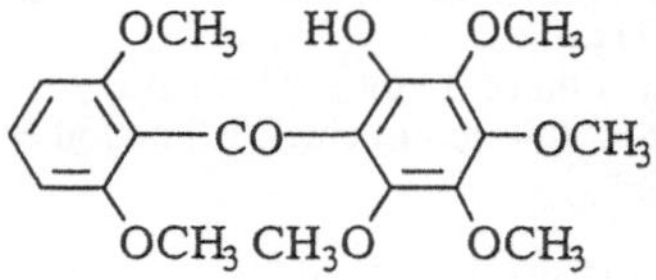

Synthesis

-Obtained (trace) by condensation of 2,6-dimethoxybenzoyl chloride with pentamethoxybenzene in nitrobenzene with aluminium chloride for 4 h at r.t. (<1%) [1310].

oil [1310];   b.p. (NA); ^{1}H NMR [1310], UV [1310].

(2-Hydroxy-3,4,5-trimethoxyphenyl)(2,3,4-trimethoxyphenyl)methanone

[197355-26-3] $C_{19}H_{22}O_8$ mol.wt. 378.38

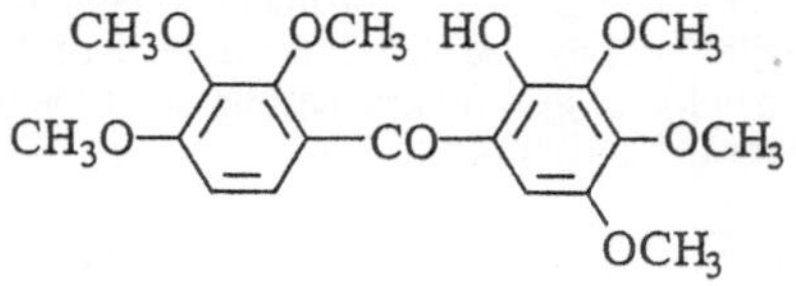

Synthesis

-Refer to: Chem. Abstr., **127**, 303322p (1997).

m.p. and Spectra (NA).

(2-Hydroxy-3,4,5-trimethoxyphenyl)(2,4,6-trimethoxyphenyl)methanone

[42833-85-2] $C_{19}H_{22}O_8$ mol.wt. 378.38

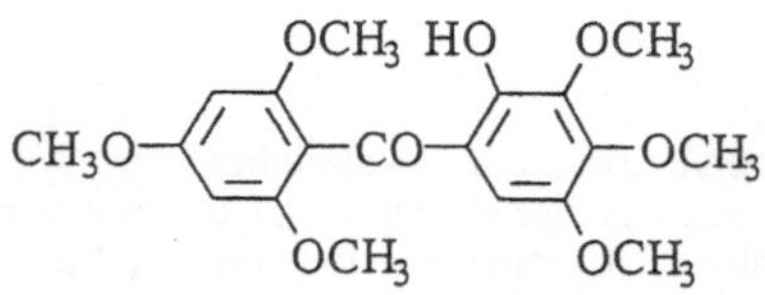

Synthesis

-Obtained by reaction of 2,4,6-trimethoxybenzoyl chloride with 1,2,3,4-tetramethoxybenzene in the presence of aluminium chloride in ethyl ether for 18 h [1097].

m.p. (NA); ^{1}H NMR [1097].

[3-(Aminomethyl)-5-(1,1-dimethylethyl)-2-hydroxyphenyl][4-(methylthio)phenyl]-methanone

[75061-01-7] $C_{19}H_{23}NO_2S$ mol.wt. 329.46

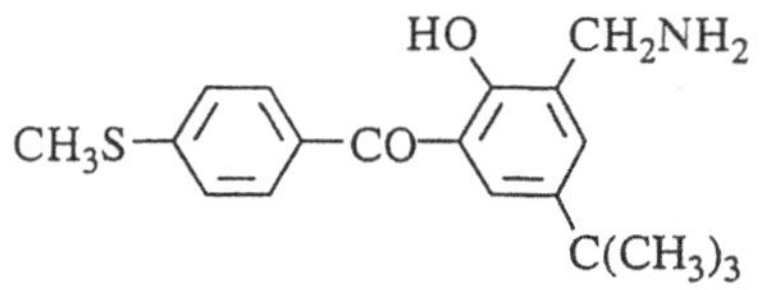

Synthesis

-Refer to: [636]; see the hydrochloride below.

m.p. and Spectra (NA).

[3-(Aminomethyl)-5-(1,1-dimethylethyl)-2-hydroxyphenyl][4-(methylthio)phenyl]-methanone *(Hydrochloride)*

[75060-69-4] $C_{19}H_{23}NO_2S,HCl$ mol.wt. 365.92

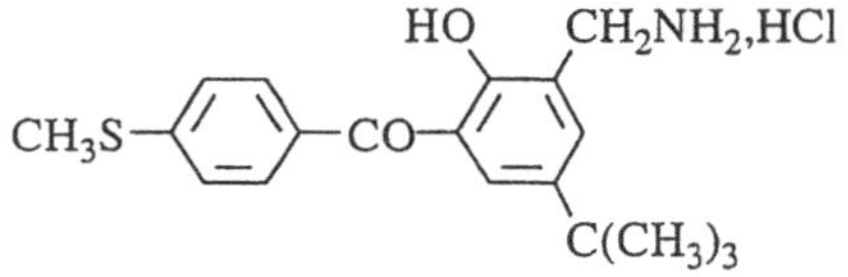

Synthesis

-Preparation by reaction of concentrated hydrochloric acid with 4-tert-butyl-6-(N-chloroacetylaminomethyl)-2-(4-methyl-thiobenzoyl)phenol in refluxing ethanol for 20 h (86%) [636].

m.p.   205-208° [636];   Spectra (NA).

[3-[(2,6-Dichlorophenyl)methoxy]-4-hydroxy-5-nitrophenyl](2-nitrophenyl)methanone

[190585-65-0] $C_{20}H_{12}Cl_2N_2O_7$ mol.wt. 463.03

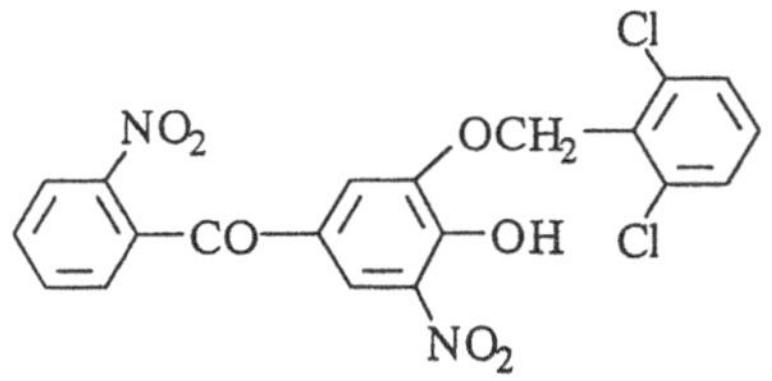

Synthesis

-Obtained by adding 2,6-dichlorobenzyl bromide to a solution of 3,4-dihydroxy-2',5-dinitrobenzo-phenone in N,N-dimethylformamide, first treated with sodium hydride, then stirring for 2 h at 35° (66%). -Refer to: Chem. Abstr., **127**, 17465u (1997).

m.p. and Spectra (NA).

[2,3-Dichloro-4-hydroxy-5-(phenylmethoxy)phenyl](2-fluorophenyl)methanone

[103843-59-0] $C_{20}H_{13}Cl_2FO_3$ mol.wt. 391.23

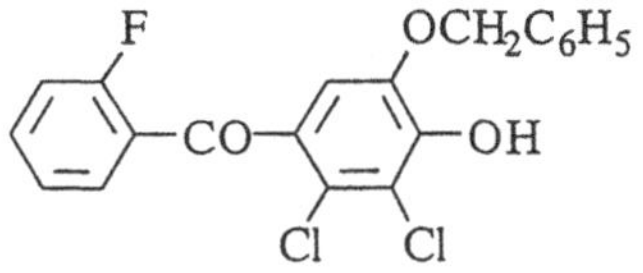

Synthesis

-Preparation by reaction of benzyl bromide with 2,3-dichloro-4,5-dihydroxy-2'-fluorobenzophenone in the presence of sodium hydride in N,N-dimethyl-formamide at r.t. for 15 min (71%) [634, 635].

m.p.   156-157° [634, 635];
^{1}H NMR [634, 635], IR [634, 635];   X-ray analysis [634].

[2,3-Dichloro-5-hydroxy-4-(phenylmethoxy)phenyl](2-fluorophenyl)methanone

[103843-64-7]   C20H13Cl2FO3   mol.wt. 391.23

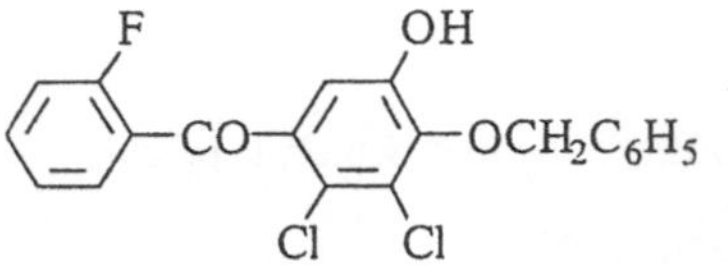

Synthesis

-Preparation by reaction of benzyl bromide with
2,3-dichloro-3,4-dihydroxy-2'-fluorobenzophenone in
the presence of sodium hydride in N,N-dimethyl-
formamide at 100° for 2 h (52%) [634].

m.p.   109-110° [634];   ¹H NMR [634], IR [634].

[4-Hydroxy-3-nitro-5-(phenylmethoxy)phenyl](2-nitrophenyl)methanone

[190585-63-8]   C20H14N2O7   mol.wt. 294.34

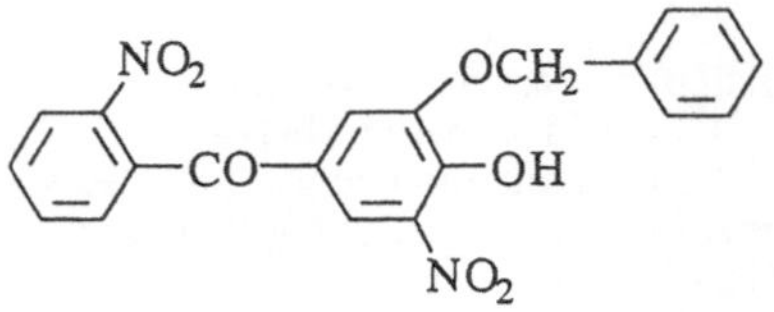

Synthesis

-Obtained by adding benzyl bromide to a solution
of 3,4-dihydroxy-2',5-dinitrobenzophenone in
N,N-dimethylformamide, first treated with sodium
hydride,
*then stirring for 2 h at 35° (66%);
*or then refluxing for 1 h and stirring for 12 h at
25° (75%). -Refer to: Chem. Abstr., **127**, 17465u (1997)*.

m.p. (NA);   ¹H NMR*.

(4-Chlorophenyl)[2-hydroxy-4-(4-methylphenoxy)phenyl]methanone

[35698-48-7]   C20H15ClO3   mol.wt. 338.79

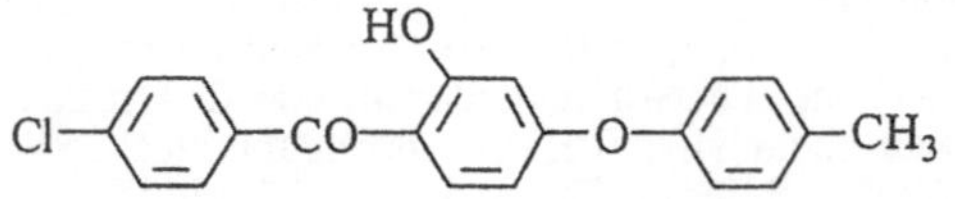

Synthesis

-Preparation by Fries rearrangement of
3-(p-chlorobenzoyloxy)-4'-methyldiphenyl
ether in 1,2,4-trichlorobenzene in the
presence of aluminium chloride at first for
30 min at 140°, then for 2 h at 200° [101].

m.p.   166-169° [101];   UV [101].

(2-Hydroxy-4-methoxyphenyl)(4-phenoxyphenyl)methanone

[127724-93-0]   C20H16O4   mol.wt. 320.34

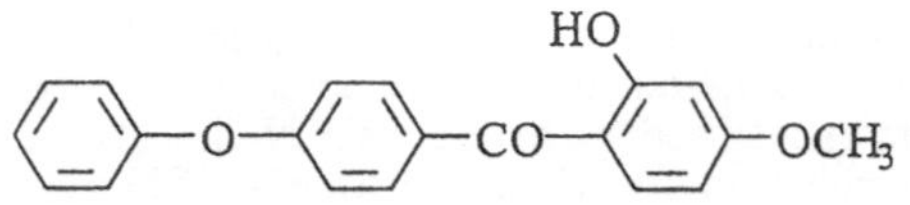

Synthesis

-Refer to: [968].

m.p. and Spectra (NA).

(5-Chloro-3-hexyl-2-hydroxyphenyl)(4-methylphenyl)methanone

[93739-92-9] $C_{20}H_{23}ClO_2$ mol.wt. 330.86

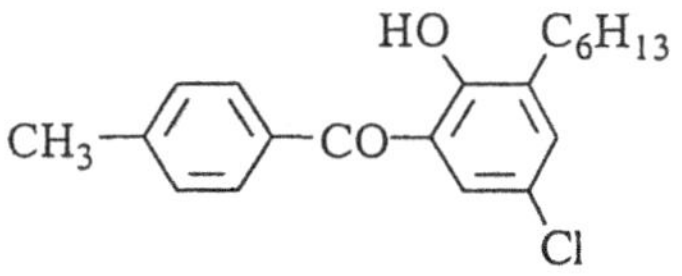

Synthesis

-Preparation by Fries rearrangement of 4-chloro-
2-hexylphenyl p-toluate with aluminium chloride at
160° for 15 min [698].

oil [698]; b.p.$_{0.133}$ 192° [698]; Spectra (NA).

(2,4-Dimethylphenyl)[2-hydroxy-4-(pentyloxy)phenyl]methanone

[36130-60-6] $C_{20}H_{24}O_3$ mol.wt. 312.41

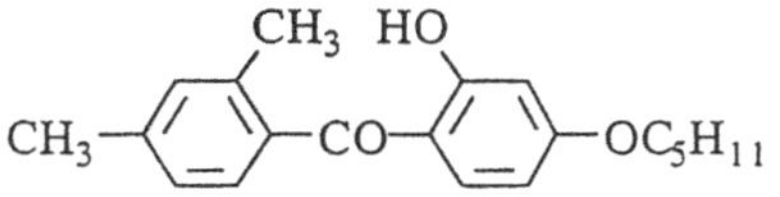

Synthesis

-Preparation by reaction of n-amyl chloride
(1-chloropentane) with 2,4-dihydroxy-2',4'-di-
methylbenzophenone in the presence of potassium
hydroxide and antimony triiodide in diethylene
glycol at 140° for 1.5 h (85%) [387].

b.p.$_{0.4}$ 218-222° [387]; Spectra (NA).

(2-Hydroxy-3,4-dimethoxy-6-methyl)(2,3,5,6-tetramethylphenyl)methanone

[183725-95-3] $C_{20}H_{24}O_4$ mol.wt. 328.41

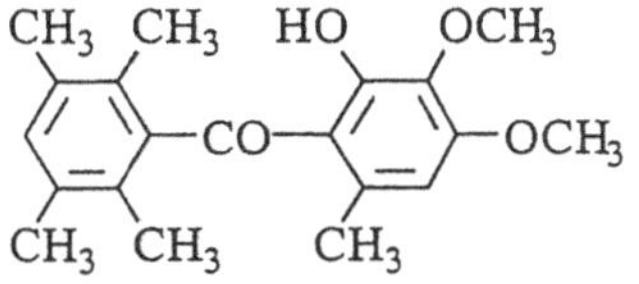

Syntheses

-Preparation by Friedel-Crafts acylation of 3,4,5-trimethoxy-
toluene with 2,3,5,6-tetramethylbenzoyl chloride [313].
-Also obtained by partial demethylation of 2,3,5,6,6'-penta-
methyl-2',3',4'-trimethoxybenzophenone [313].

m.p. 137° [313]; Spectra (NA).

(4,5-Dimethoxy-2-propylphenyl)(2-hydroxy-4,5-dimethoxyphenyl)methanone

$C_{20}H_{24}O_6$ mol.wt. 360.41

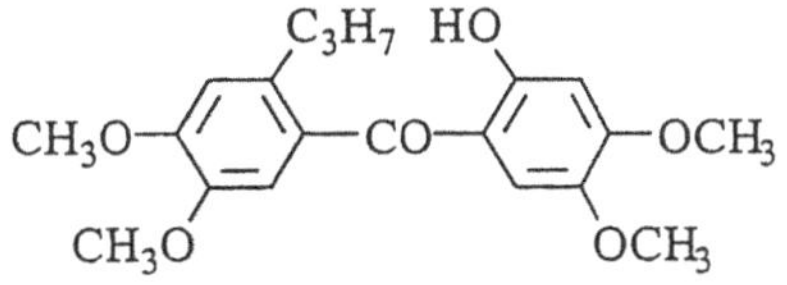

Synthesis

-Preparation by reaction of 2,4,5-trimethoxybenzoyl
chloride with 4-propylveratrole in the presence of
aluminium chloride in refluxing carbon disulfide
for 8 h, then at r.t. for 12 h [460].

m.p. 108-109° [460]; UV [460].

(2,4-Dichlorophenyl)[2-hydroxy-5-(1,1,3,3-tetramethylbutyl)phenyl]methanone

[93885-04-2] $C_{21}H_{24}Cl_2O_2$ mol.wt. 379.36

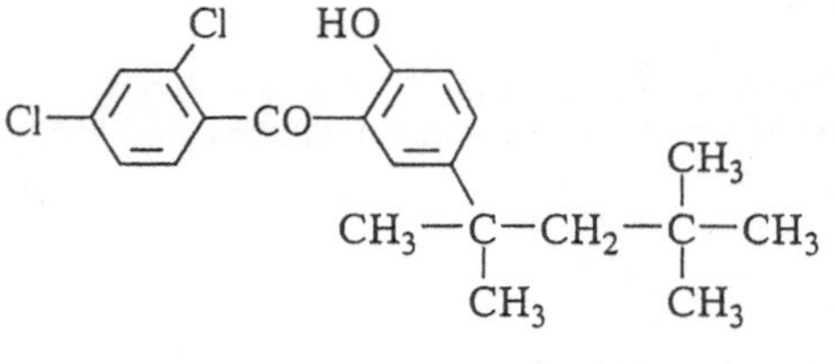

Synthesis

-Preparation by demethylation of 2-(2,4-di-chlorobenzoyl)-4-(1,1,3,3-tetramethyl-butyl)anisole (SM) with aluminium chloride in methylene chloride for 1 h at 10-12° (55%). SM was obtained by Friedel-Crafts acylation of 4-(1,1,3,3-tetramethylbutyl)-anisole with 2,4-dichlorobenzoyl chloride in methylene chloride in the presence of aluminium chloride for 30 min at 10° (74%) [812].

m.p. 89-90° [812];   Spectra (NA);   GC [812].

(3,4-Dichlorophenyl)[2-hydroxy-4-(octyloxy)phenyl]methanone

[36414-88-7] $C_{21}H_{24}Cl_2O_3$ mol.wt. 395.32

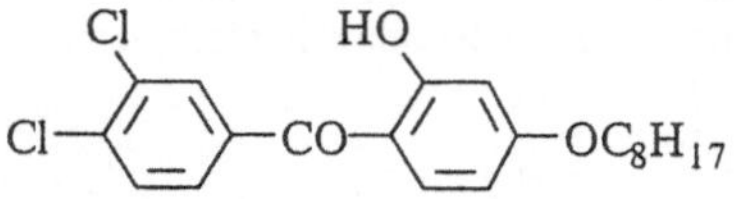

Synthesis

-Refer to: [1067].

m.p. 51° [1067];   UV [1067].

[2-Hydroxy-4,5-dimethoxy-3-(2-propenyl)phenyl](2,4,6-trimethoxyphenyl)methanone

[42833-95-5] $C_{21}H_{24}O_7$ mol.wt. 388.42

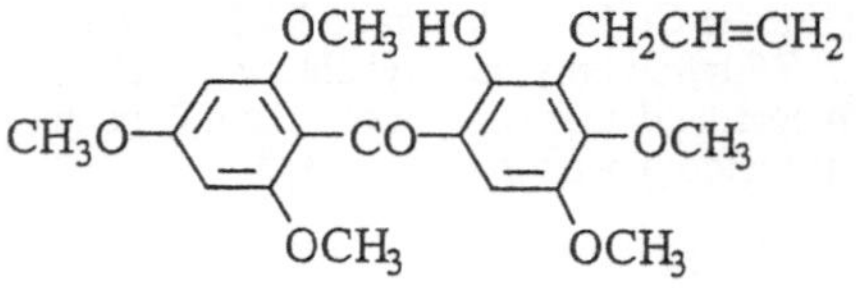

Synthesis

-Obtained by Claisen rearrangement of 2-(allyloxy)-2',4,4',5,6'-pentamethoxy-benzophenone in refluxing N,N-dimethyl-aniline for 4 h (62%) [1097].

pale yellow oil [1097];   b.p. (NA); ^1H NMR [1097].

[3,5-Bis(1,1-dimethylethyl)-4-hydroxyphenyl](2-bromophenyl)methanone

[84700-54-9] $C_{21}H_{25}BrO_2$ mol.wt. 389.33

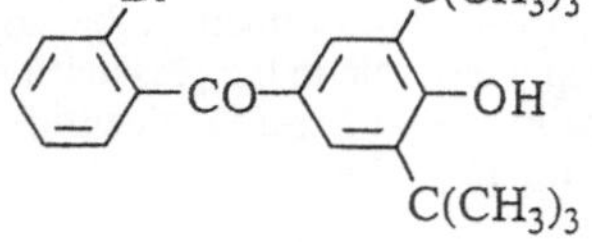

Synthesis

-Preparation by acylation of 2,6-di-tert-butylphenol with o-bromobenzoyl chloride in the presence of aluminium chloride (31%) [1233], according to [1080].

m.p. 74-75° [1233];   IR [1233].

[3,5-Bis(1,1-dimethylethyl)-4-hydroxyphenyl](4-chlorophenyl)methanone

[84700-53-8] $C_{21}H_{25}ClO_2$ mol.wt. 344.88

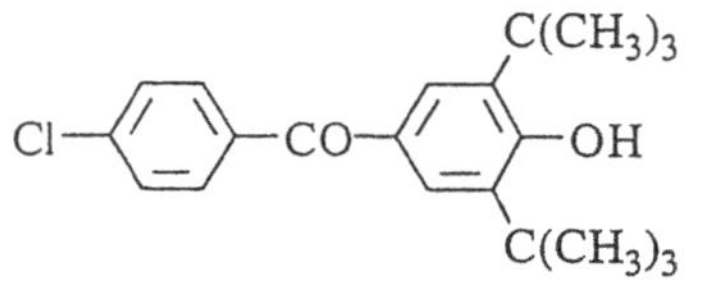

Synthesis

-Preparation by acylation of 2,6-di-tert-butylphenol with p-chlorobenzoyl chloride in the presence of aluminium chloride (41%) [1233], according to [1080].

m.p. 162-163° [1233];   IR [1233].

(4-Chlorophenyl)[2-hydroxy-4-(octyloxy)phenyl]methanone

[18190-30-2] $C_{21}H_{25}ClO_3$ mol.wt. 360.88

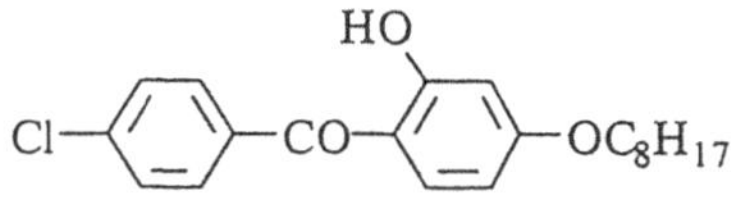

Synthesis

-Refer to: [1067].

m.p. 60° [1067];   UV [1067].

[3,5-Bis(1,1-dimethylethyl)-4-hydroxyphenyl](4-fluorophenyl)methanone

[69451-08-7] $C_{21}H_{25}FO_2$ mol.wt. 328.43

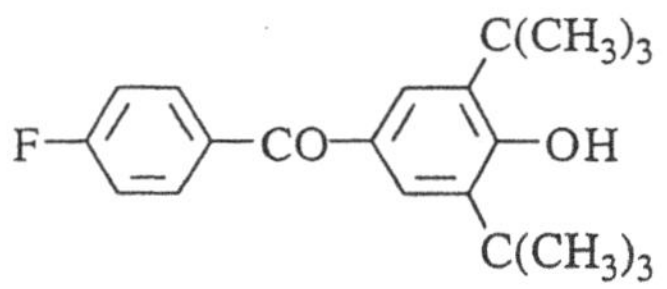

Synthesis

-Preparation by acylation of 2,6-di-tert-butylphenol with p-fluorobenzoyl chloride in the presence of aluminium chloride (35%) [1233], according to [1080].

m.p. 133-134° [1233];   IR [1233].

(4-Fluorophenyl)[2-hydroxy-4-(octyloxy)phenyl]methanone

[84794-99-0] $C_{21}H_{25}FO_3$ mol.wt. 344.43

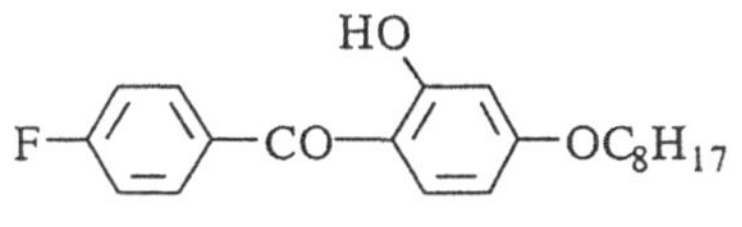

Synthesis

-Preparation by reaction of 1-bromooctane with 4'-fluoro-2,4-dihydroxybenzophenone in the presence of sodium bicarbonate in cyclohexanone for 5 h at 150° (72%) [394].

m.p. 48-49° [394];   Spectra (NA).

[5-(1,1-Dimethylethyl)-2-hydroxyphenyl][4-(1,1-dimethylethyl)phenyl]methanone

$C_{21}H_{26}O_2$ mol.wt. 310.44

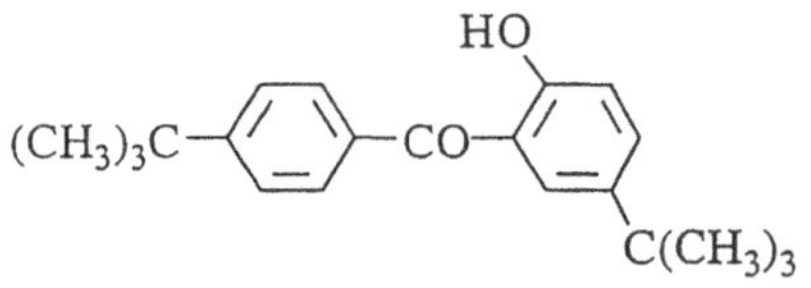

Synthesis

-Obtained by UV light irradiation of p-tert-butyl-phenyl p-tert-butylbenzoate in benzene (48%) [743].

yellow oil [743];   b.p.$_{0.02}$ 150° [743];   Spectra (NA).

(4-Butoxy-2-hydroxyphenyl)(4-butoxyphenyl)methanone

C$_{21}$H$_{26}$O$_4$ mol.wt. 342.44

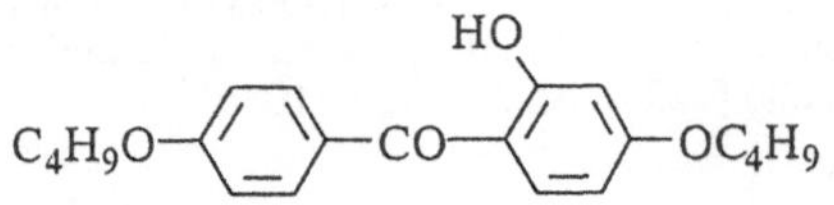

Synthesis

-Preparation by reaction of butyl bromide with
 2,4,4'-trihydroxybenzophenone in the presence of
 potassium carbonate (50%) [233].

m.p. 95-96° [233];   UV [233].

[3,5-Bis(1,1-dimethylethyl)-4-hydroxyphenyl](3-methylphenyl)methanone

[84700-50-5]   C$_{22}$H$_{28}$O$_2$ mol.wt. 324.46

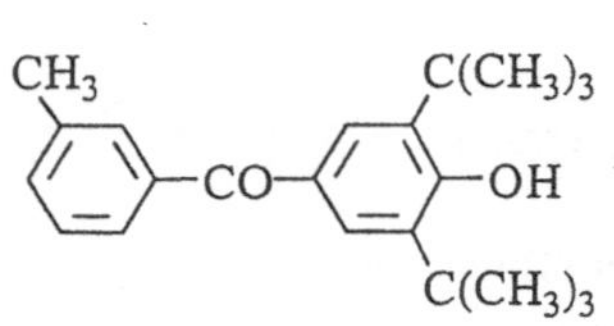

Synthesis

-Preparation by acylation of 2,6-di-tert-butylphenol with
 m-methylbenzoyl chloride in the presence of aluminium
 chloride (38%) [1233], according to [1080].

m.p. 99-100° [1233];   IR [1233].

[3,5-Bis(1,1-dimethylethyl)-4-hydroxyphenyl](4-methylphenyl)methanone

[84700-49-2]   C$_{22}$H$_{28}$O$_2$ mol.wt. 324.46

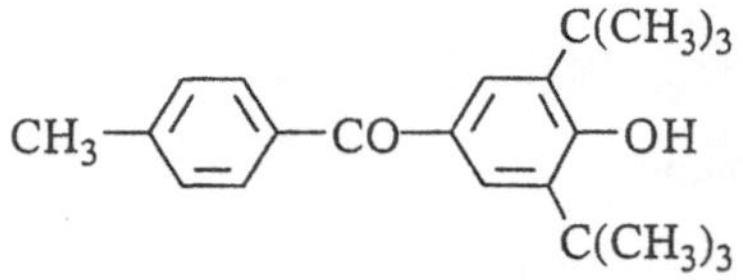

Synthesis

-Preparation by acylation of 2,6-di-tert-butylphenol with
 p-methylbenzoyl chloride in the presence of aluminium
 chloride (40%) [1233], according to [1080].

m.p.   131-132° [1233];   IR [1233].

[3,5-Bis(1,1-dimethylethyl)-2-hydroxyphenyl](4-methoxyphenyl)methanone

[80078-54-2]   C$_{22}$H$_{28}$O$_3$ mol.wt. 340.46

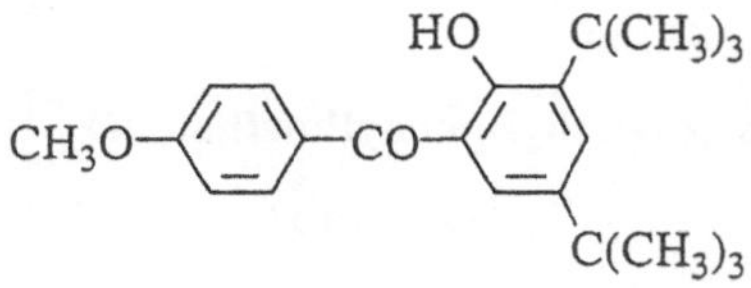

Syntheses

-Preparation by oxidation of 2,4-di-tert-butyl-
 6-(p-methoxybenzyl)phenol with silver oxide in
 boiling acetone for 40 min (24%) [660].
-Preparation by treatment of 2-hydroxy-4'-methoxy-
 benzophenone at 120° with a mixture of
isobutylene/nitrogen (1:1) in the presence of a macroreticular acid ion exchanger (Wofatit OK 80)
as catalyst for 10 h (85%) [111].

b.p.$_{0.15}$ 200-205° [111];   m.p. 116-117° [660];
^{1}H NMR [660], IR [660], UV [660], MS [660].

[3,5-Bis(1,1-dimethylethyl)-4-hydroxyphenyl](2-methoxyphenyl)methanone

[28441-13-6] $C_{22}H_{28}O_3$ mol.wt. 340.46

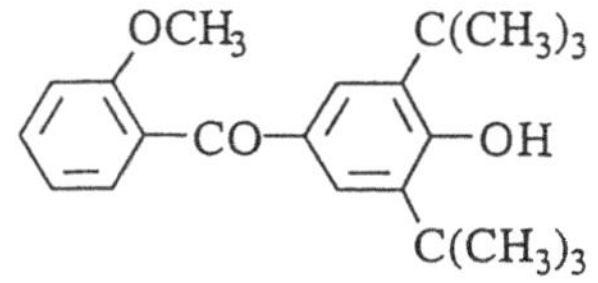

Synthesis

-Preparation by acylation of 2,6-di-tert-butylphenol with
o-methoxybenzoyl chloride in the presence of aluminium
chloride (36%) [1233], according to [1080].

m.p. 87-88° [1233];   IR [1233].

[3,5-Bis(1,1-dimethylethyl)-4-hydroxyphenyl](4-methoxyphenyl)methanone

[28440-99-5] $C_{22}H_{28}O_3$ mol.wt. 340.46

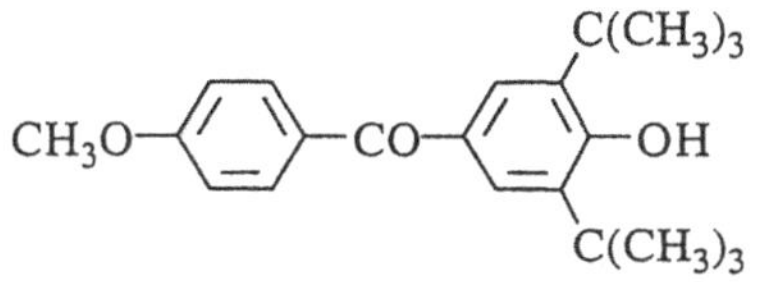

Synthesis

-Preparation by acylation of 2,6-di-tert-butylphenol
with p-methoxybenzoyl chloride in the presence of
aluminium chloride (61%) [1233], according to
[1080].

m.p. 143-144° [1233];   IR [1233].

[2-Hydroxy-4-(octyloxy)phenyl](4-methoxyphenyl)methanone

[36469-90-6] $C_{22}H_{28}O_4$ mol.wt. 356.46

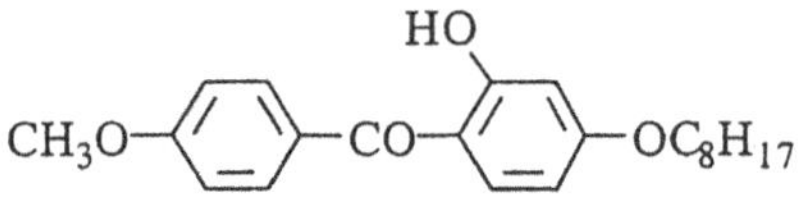

Synthesis

-Refer to: [1067].

m.p. 55° [1067];   UV [1067].

(3,4-Dichlorophenyl)[4-[4-(1,1-dimethylethyl)phenoxy]-2-hydroxyphenyl]methanone

[35698-02-3] $C_{23}H_{20}Cl_2O_3$ mol.wt. 415.32

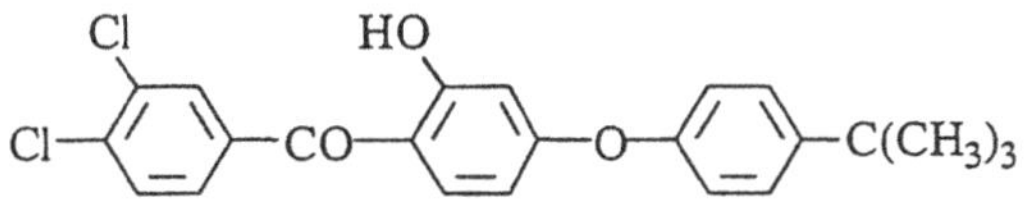

Synthesis

-Refer to: [101] (compound 32).

m.p. (NA);   UV [101].

[4-(1,1-Dimethylethyl)phenyl](2-hydroxy-4-phenoxyphenyl)methanone

[35698-42-1] $C_{23}H_{22}O_3$ mol.wt. 346.94

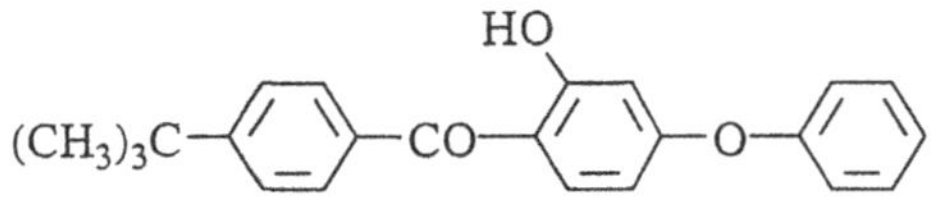

Synthesis

-Preparation by reaction of p-tert-butylbenzoyl chloride with 3-methoxydiphenyl
ether in chlorobenzene in the presence of
aluminium chloride first at r.t., then for 4 h
at 90-95° [101].

m.p. 88-90° [101];   UV [101].

(3-Butoxyphenyl)(2-hydroxy-4-phenoxyphenyl)methanone

[35698-55-6] $C_{23}H_{22}O_4$ mol.wt. 362.43

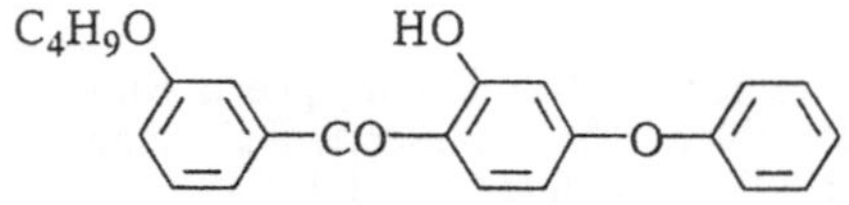

Synthesis

-Refer to: [101] (compound 13).

m.p. (NA); UV [101].

(4-Butoxyphenyl)(2-hydroxy-4-phenoxyphenyl)methanone

[35698-52-3] $C_{23}H_{22}O_4$ mol.wt. 362.43

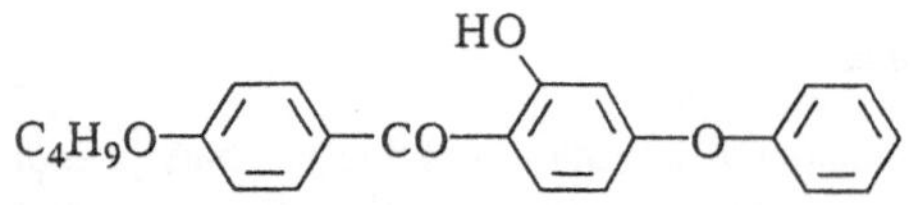

Synthesis

-Refer to: [101] (compound 10).

m.p. (NA); UV [101].

[2-Hydroxy-4-[(3-methyl-2-butenyl)oxy]phenyl][4-[(3-methyl-2-butenyl)oxy]phenyl]-methanone

[63565-01-5] $C_{23}H_{26}O_4$ mol.wt. 366.46

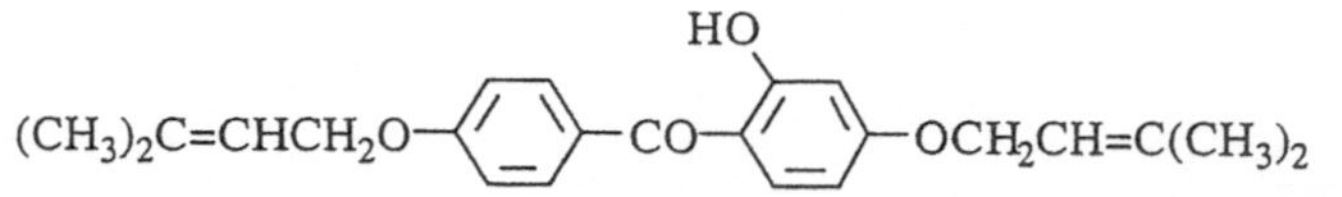

Synthesis

-Obtained (poor yield) by reaction of prenyl bromide with 2,4,4'-trihydroxybenzophenone,
*in the presence of potassium carbonate in refluxing acetone for 3 h (6%) [875];
*in the presence of boron trifluoride-etherate in dioxane at 50-60° for 3 h (<1%) [874].

m.p. 74° [875], 73-74° [874]; ^{1}H NMR [875], IR [875], UV [875].

(4-Ethenylphenyl)[2-hydroxy-4-(octyloxy)phenyl]methanone

[80167-02-3] $C_{23}H_{28}O_3$ mol.wt. 352.47

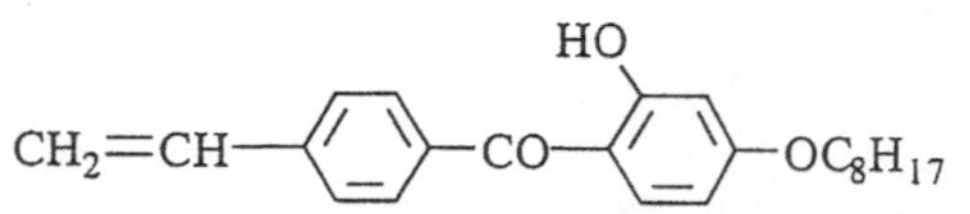

Synthesis

-Preparation by reaction of aqueous potassium hydroxide with 2-hydroxy-4-(octyloxy)-4'-(2-bromoethyl)-benzophenone in the presence of hydroquinone in refluxing methanol for 1.5 h with nitrogen bubbling (41%) [675].

m.p. 57-59° [675]; ^{1}H NMR [675], IR [675], UV [675], MS [675].

[3,5-Bis(1,1-dimethylethyl)-4-hydroxyphenyl](3-ethylphenyl)methanone

[84700-52-7]   C$_{23}$H$_{30}$O$_2$ mol.wt. 338.49

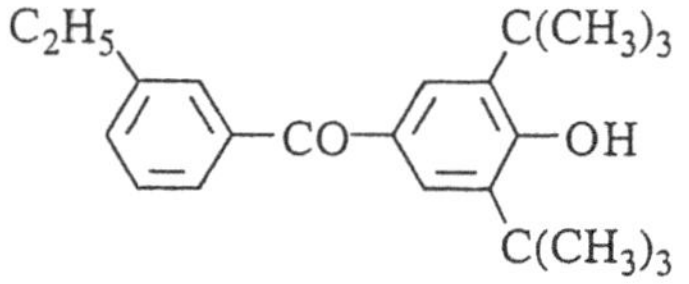

Synthesis

-Preparation by Friedel-Crafts acylation of 2,6-di-tert-butylphenol with 3-ethylbenzoyl chloride in the presence of aluminium chloride at -10° (50%) [1233], according to [1080].

m.p. 69-70° [1233];   Spectra (NA).

[3,5-Bis(1,1-dimethylethyl)-4-hydroxyphenyl](4-ethylphenyl)methanone

[84700-51-6]   C$_{23}$H$_{30}$O$_2$ mol.wt. 338.49

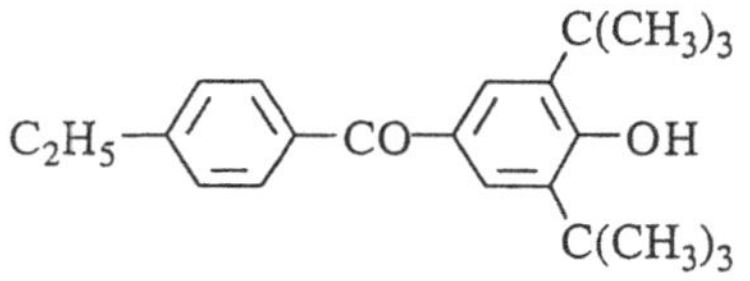

Synthesis

-Preparation by Friedel-Crafts acylation of 2,6-di-tert-butylphenol with p-ethylbenzoyl chloride in the presence of aluminium chloride at -10° (35%) [1233] according to [1080].

m.p. 136-137° [1233];   Spectra (NA).

[3-Chloro-4,6-dimethoxy-2-(phenylmethoxy)phenyl](4-hydroxy-2-methoxy-6-methylphenyl)methanone

C$_{24}$H$_{23}$ClO$_6$ mol.wt. 442.90

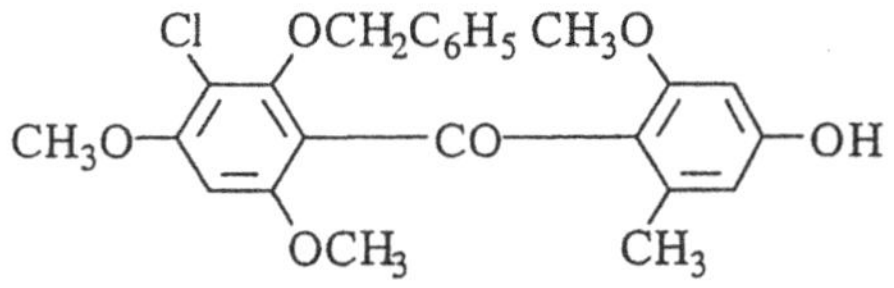

Synthesis

-Preparation by reaction of 2-(benzyloxy)-3-chloro-4,6-dimethoxybenzoyl chloride with mono(trimethylsilyl)derivative of 3-methoxy-5-methylphenol in the presence of stannic chloride (or titanium tetrachloride or aluminium chloride) in refluxing methylene chloride for 2 h [649].

m.p. and Spectra (NA).

[4-(4-Butylphenoxy)-2-hydroxyphenyl](3-methylphenyl)methanone

[35698-58-9]   C$_{24}$H$_{24}$O$_3$ mol.wt. 360.45

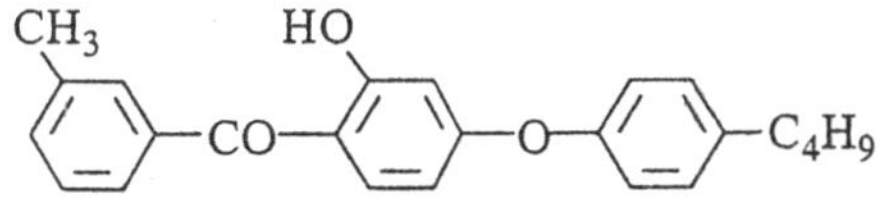

Synthesis

-Refer to: [101] (compound 16).

m.p. (NA);   UV [101].

[4-(4-Butoxyphenoxy)-2-hydroxyphenyl](3-methylphenyl)methanone

[35697-96-2]   C24H24O4   mol.wt. 376.45

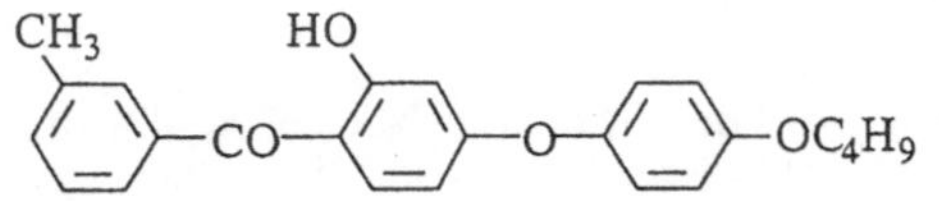

Synthesis

-Refer to: [101] (compound 25).

m.p. (NA);   UV [101].

(4-Butoxyphenyl)[2-hydroxy-4-(3-methylphenoxy)phenyl]methanone

[35698-62-5]   C24H24O4   mol.wt. 376.45

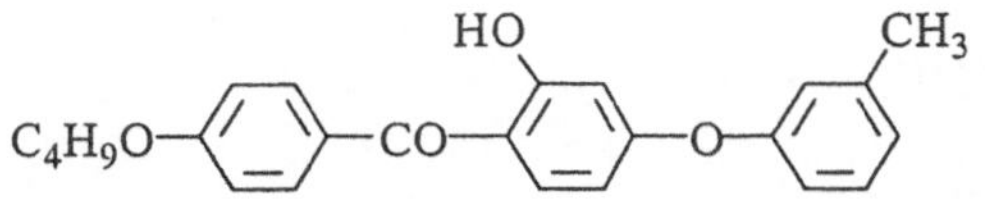

Synthesis

-Refer to: [101] (compound 20).

m.p. (NA);   UV [101].

(4-Butoxyphenyl)[2-hydroxy-4-(4-methylphenoxy)phenyl]methanone

[35698-59-0]   C24H24O4   mol.wt. 376.45

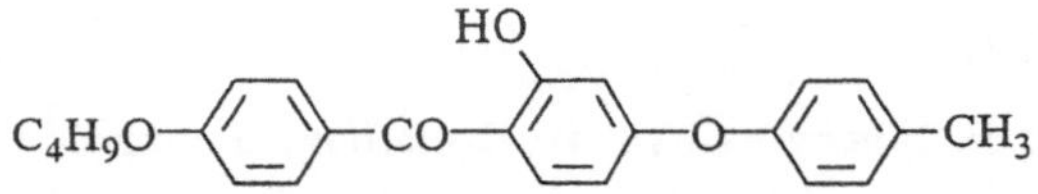

Synthesis

-Refer to: [101] (compound 17).

m.p. (NA);   UV [101].

[3,5-Bis(1,1-dimethylethyl)-4-hydroxyphenyl](3,4,5-trimethoxyphenyl)methanone

[54808-42-3]   C24H32O5   mol.wt. 400.52

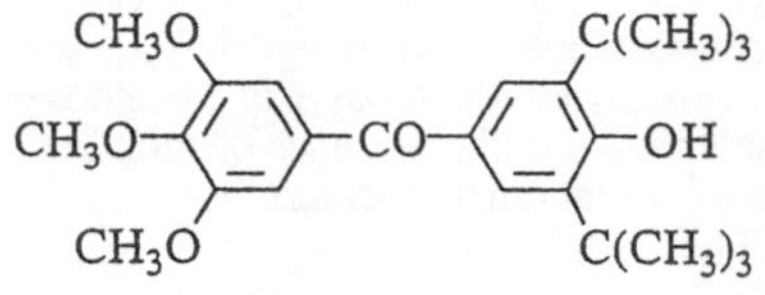

Synthesis

-Preparation by Friedel-Crafts acylation of 2,6-di-tert-butylphenol with 3,4,5-trimethoxybenzoyl chloride in the presence of stannic chloride in refluxing methylene chloride for 19 h (58%) [284].

m.p.   138-138°5 [284]; ^{1}H NMR [284], IR [284].

[3-Hydroxy-5-(phenoxy-d5)phenyl][4-(phenoxy-3,5-d2)phenyl]methanone

[176738-22-0]   C25H11D7O4   mol.wt. 389.46

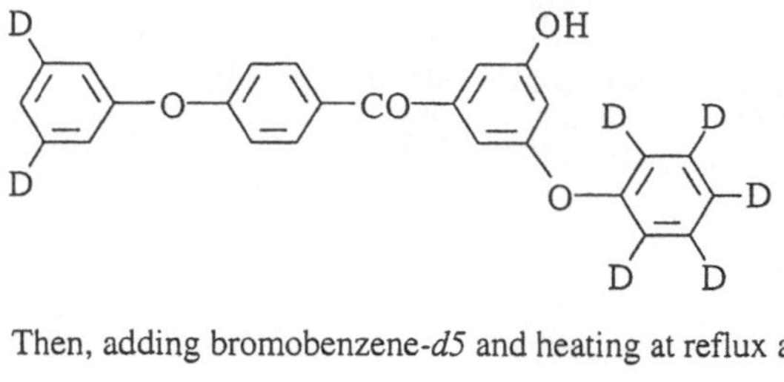

Synthesis

-Preparation by adding 3,5-dihydroxy-4'-(phenoxy-d2)benzophenone, potassium carbonate and copper powder to a solution of N-methyl-pyrrolidone/toluene. The obtained mixture was refluxed for 1 h, with removal of water.

Then, adding bromobenzene-d5 and heating at reflux again for 2 h (22%) (compound 22) [550].

m.p. (NA);
^{1}H NMR [550], ^{13}C NMR [550].

[4-(1,1-Dimethylethyl)phenyl][4-(3,4-dimethylphenoxy)-2-hydroxyphenyl]methanone

[35697-99-5] $C_{25}H_{26}O_3$ mol.wt. 374.48

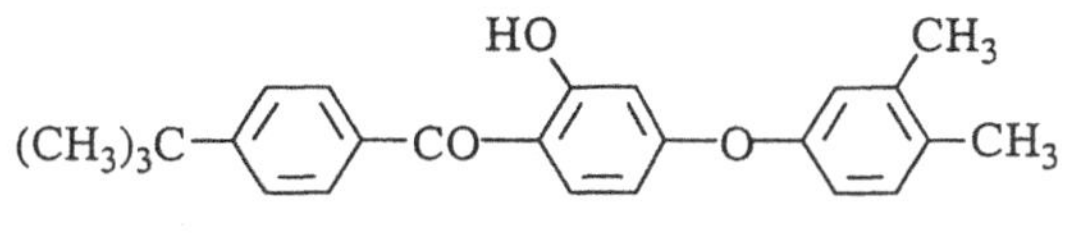

Synthesis

-Refer to: [101] (compound 28).

m.p. (NA);   UV [101].

(4-Dodecylphenyl)(2-hydroxy-3-nitrophenyl)methanone

[35698-24-9] $C_{25}H_{33}NO_4$ mol.wt. 411.54

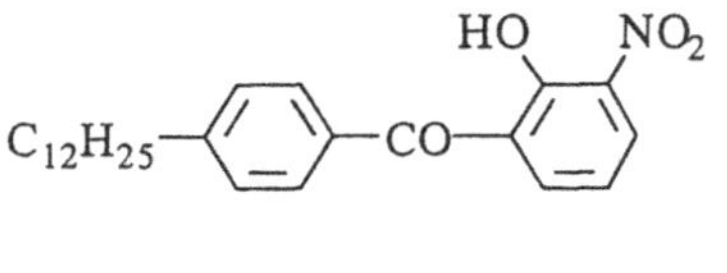

Synthesis

-Obtained by reaction of fuming nitric acid with
2-hydroxy-4'-dodecylbenzophenone in an acetic acid/
acetic anhydride mixture (4:3) at 5-7° for 1 h (27%)
[469, 882].

m.p. and Spectra (NA).

(4-Dodecylphenyl)(2-hydroxy-5-nitrophenyl)methanone

[35698-23-8] $C_{25}H_{33}NO_4$ mol.wt. 411.54

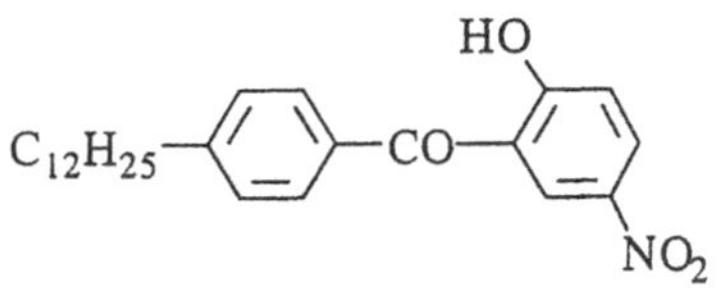

Synthesis

-Obtained by reaction of fuming nitric acid with
2-hydroxy-4'-dodecylbenzophenone in an acetic
acid/acetic anhydride mixture (4:3) at 5-7° for 1 h
(54%) [469, 882].

m.p. and Spectra (NA).

[4-(1,1-Dimethylethyl)phenyl][2-hydroxy-4-(octyloxy)phenyl]methanone

[36419-37-1] $C_{25}H_{34}O_3$ mol.wt. 382.54

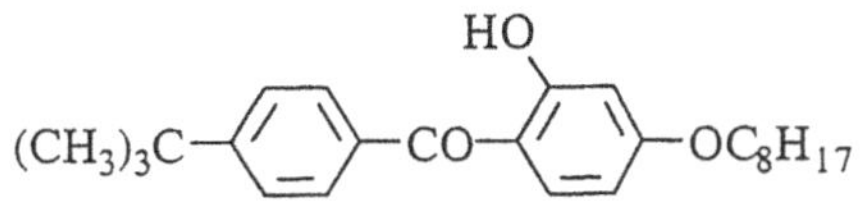

Synthesis

-Refer to: [1067].

oil [1067];   b.p. (NA);   UV [1067].

[4-(Dodecyloxy)-2-hydroxyphenyl](4-methylphenyl)methanone

[36130-67-3] $C_{26}H_{26}O_3$ mol.wt. 386.49

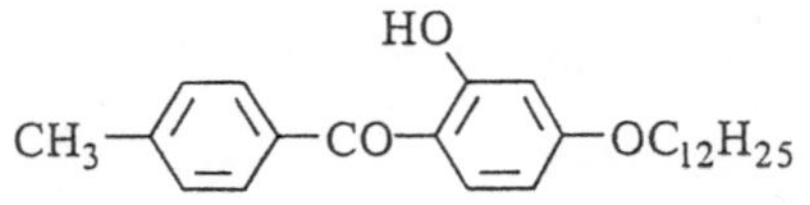

Synthesis

-Preparation by reaction of dodecyl chloride with 2,4-dihydroxy-4'-methylbenzophenone (92%), according to [387].

m.p. 50-51° [387];   Spectra (NA).

(4-Dodecylphenyl)(2-hydroxy-3-methyl-5-nitrophenyl)methanone

[35698-29-4] $C_{26}H_{35}NO_4$ mol.wt. 425.57

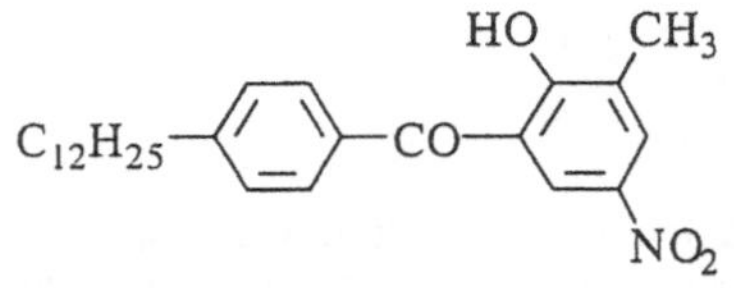

Synthesis

-Preparation by reaction of fuming nitric acid with 4'-dodecyl-2-hydroxy-3-methylbenzophenone in acetic acid/acetic anhydride (95%) [469, 882].

m.p. and Spectra (NA).

(4-Dodecylphenyl)(2-hydroxy-3-methylphenyl)methanone

[35698-28-3] $C_{26}H_{36}O_2$ mol.wt. 380.57

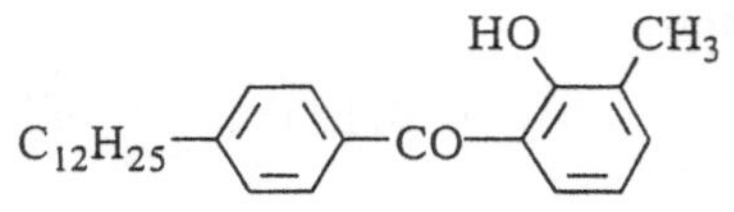

Synthesis

-Obtained (poor yield) by reaction of 2-hydroxy-3-methylbenzoyl chloride with dodecylbenzene in nitrobenzene in the presence of aluminium chloride for 6 h at 40° and at r.t. overnight (8%) [469, 882].

b.p.$_{0.45-0.7}$ 210-217° [469, 882];   Spectra (NA).

[[4-(1,1-Dimethylethyl)phenoxy]-2-hydroxyphenyl][4-(1,1-dimethylethyl)phenyl]-methanone

[35698-44-3] $C_{27}H_{30}O_3$ mol.wt. 402.53

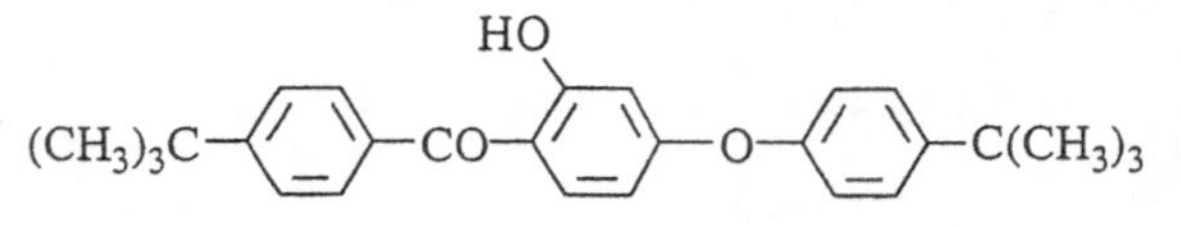

Synthesis

-Preparation by reaction of p-tert-butylbenzoyl chloride with 4-tert-butyl-3'-methoxydiphenyl ether in chlorobenzene in the presence of aluminium chloride first at r.t., then at 90-95° for 4 h [101].

m.p. 134-136° [101];   UV [101].

(2-Hydroxy-4-phenoxyphenyl)[3-(octyloxy)phenyl]methanone

[35698-56-7] $C_{27}H_{30}O_4$ mol.wt. 418.53

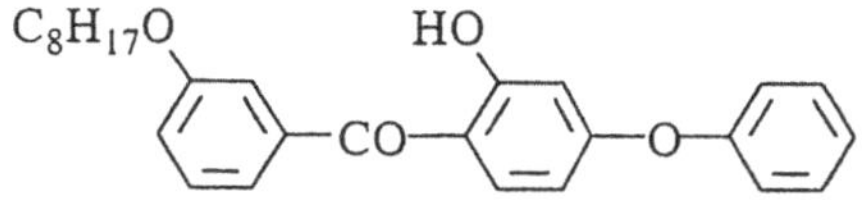

Synthesis

-Refer to: [101] (compound 14).

m.p. (NA);   UV [101].

(2-Hydroxy-4-phenoxyphenyl)[4-(octyloxy)phenyl]methanone

[35698-53-4] $C_{27}H_{30}O_4$ mol.wt. 418.53

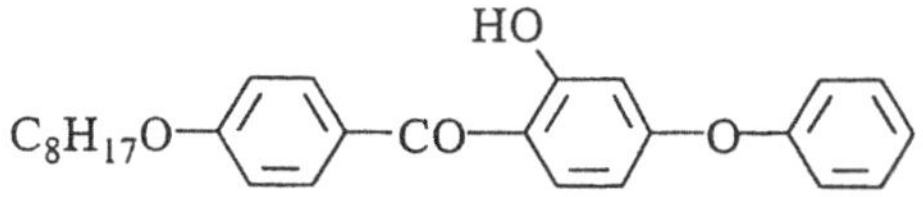

Synthesis

-Refer to: [101] (compound 11).

m.p. (NA);   UV [101].

[4-(4-Butoxyphenoxy)-2-hydroxyphenyl](4-butoxyphenyl)methanone

[35697-93-9] $C_{27}H_{30}O_5$ mol.wt. 434.68

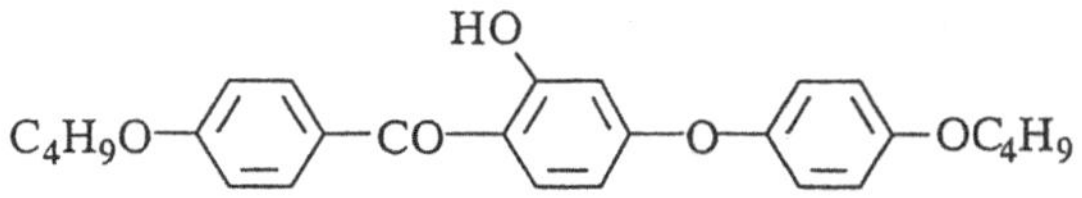

Synthesis

-Refer to: [101] (compound 22).

m.p. (NA);   UV [101].

[4-(Dodecyloxy)-2-hydroxyphenyl](4-ethenylphenyl)methanone

[80167-03-9] $C_{27}H_{36}O_3$ mol.wt. 408.58

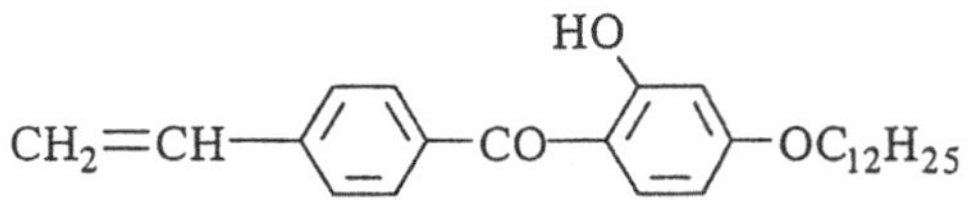

Synthesis

-Obtained by reaction of aqueous potassium hydroxide with 2-hydroxy-4-(dodecyloxy)-4'-(2-bromoethyl)benzophenone in the presence of hydroquinone in refluxing methanol for 1.5 h with nitrogen bubbling (18%) [675].

m.p. 49-50° [675]; ^{1}H NMR [675], IR [675], UV [675], MS [675].

[2-Hydroxy-4-(3-methylphenoxy)phenyl][4-(octyloxy)phenyl]methanone

[35697-92-8] $C_{28}H_{32}O_4$ mol.wt. 432.56

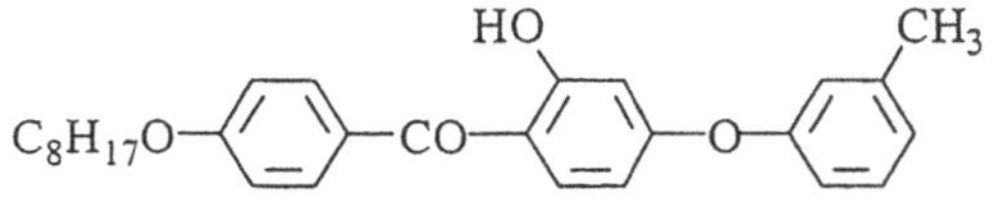

Synthesis

-Refer to: [101] (compound 21).

m.p. (NA);   UV [101].

[2-Hydroxy-4-(4-methylphenoxy)phenyl][4-(octyloxy)phenyl]methanone

[35698-60-3] $C_{28}H_{32}O_4$ mol.wt. 432.56

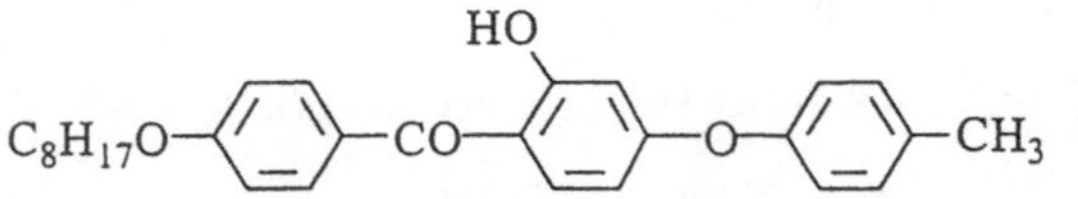

Synthesis

-Refer to: [101] (compound 18).

m.p. (NA);   UV [101].

[2-Hydroxy-4-[4-(octyloxy)phenoxy]phenyl](3-methylphenyl)methanone

[35697-97-3] $C_{28}H_{32}O_4$ mol.wt. 432.56

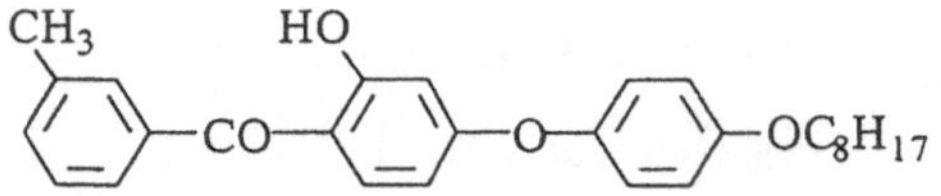

Synthesis

-Refer to: [101] (compound 26).

m.p. (NA);   UV [101].

[2-Hydroxy-4-(octyloxy)phenyl][4-(octyloxy)phenyl]methanone

$C_{29}H_{42}O_4$ mol.wt. 454.65

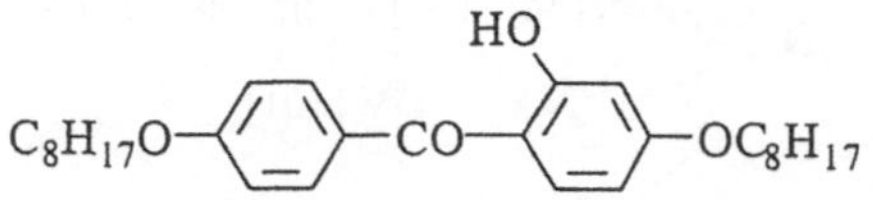

Synthesis

-Preparation by reaction of octyl bromide with 2,4,4'-trihydroxybenzophenone in the presence of potassium carbonate (60%) [233].

m.p.   60°5-61° [233];   UV [233].

(3,4-Dimethylphenyl)[2-hydroxy-4-(4-nonylphenoxy)phenyl]methanone

[35698-01-2] $C_{30}H_{36}O_3$ mol.wt. 444.61

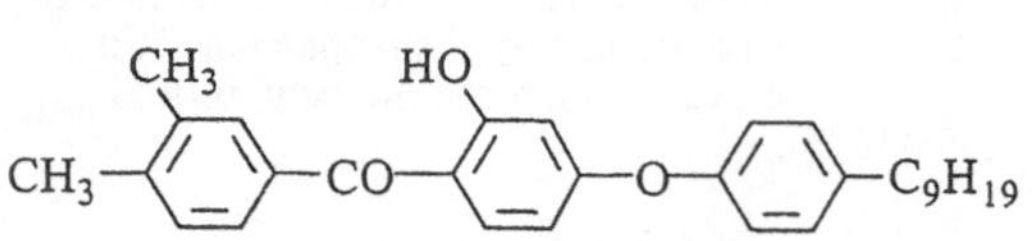

Synthesis

-Refer to: [101] (compound 31).

m.p. (NA);   UV [101].

[3-(Dodecyloxy)phenyl](2-hydroxy-4-phenoxyphenyl)methanone

[35698-57-8] $C_{31}H_{38}O_4$ mol.wt. 474.64

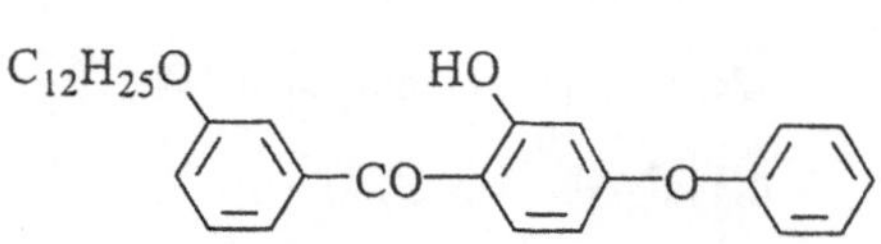

Synthesis

-Refer to: [101] (compound 15).

m.p. (NA);   UV [101].

[4-(Dodecyloxy)phenyl](2-hydroxy-4-phenoxyphenyl)methanone

[35698-54-5] $C_{31}H_{38}O_4$ mol.wt. 474.64

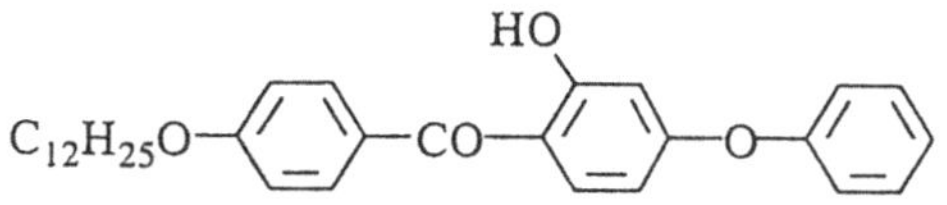

Synthesis

-Refer to: [101] (compound 12).

m.p. (NA);   UV [101].

[2-Hydroxy-4-(2,4,6-trimethylphenoxy)phenyl][4-(isononyloxy)phenyl]methanone

[36118-66-8] $C_{31}H_{38}O_4$ mol.wt. 474.64

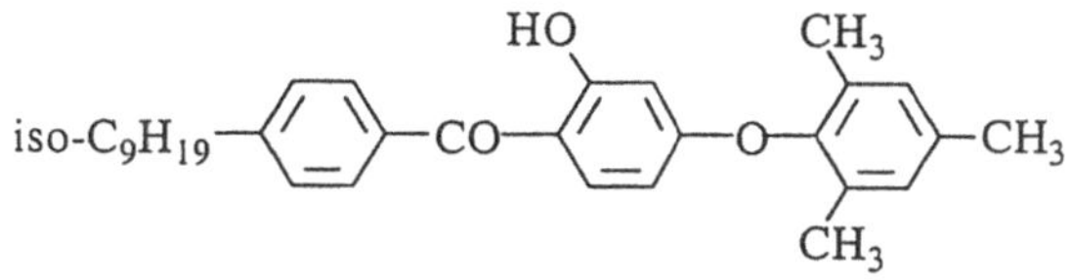

Synthesis

-Refer to: [101] (compound 29).

m.p. (NA);   UV [101].

[4-(4-Dodecylphenoxy)-2-hydroxyphenyl](3-methylphenyl)methanone

[35697-98-4] $C_{32}H_{40}O_3$ mol.wt. 472.67

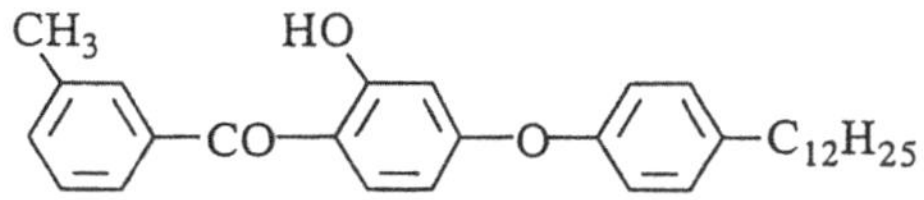

Synthesis

-Refer to: [101] (compound 27).

m.p. (NA);   UV [101].

[4-(Dodecyloxy)phenyl][2-hydroxy-4-(4-methylphenoxy)phenyl]methanone

[35698-61-4] $C_{32}H_{40}O_4$ mol.wt. 488.67

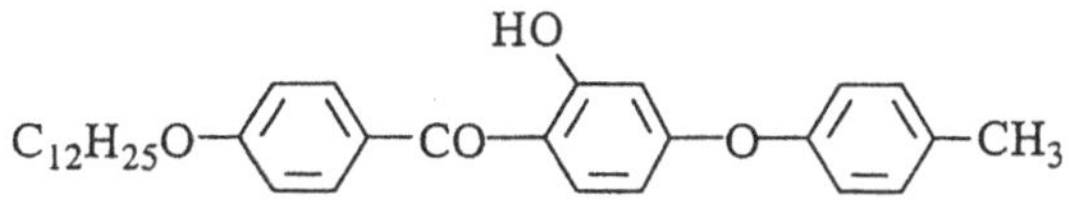

Synthesis

-Refer to: [101] (compound 19).

m.p. (NA);   UV [101].

[2-Hydroxy-4-(octadecyloxy)phenyl](4-methoxyphenyl)methanone

[36130-68-4] $C_{32}H_{48}O_4$ mol.wt. 496.73

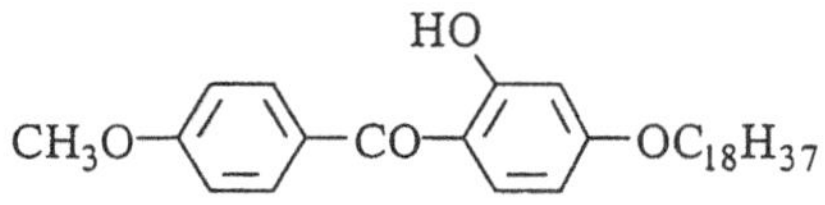

Synthesis

-Preparation by reaction of octadecyl chloride (1-chlorooctadecane) with 2,4-dihydroxy-4'-methoxybenzophenone (81%), according to [387].

m.p. 73-75° [387];   Spectra (NA).

[4-[4-(1,1-Dimethylethyl)-2,6-dimethylphenoxy]-2-hydroxyphenyl][4-(1,1,3,3-tetra-methylbutyl)phenyl]methanone

[35698-00-1] $C_{33}H_{42}O_3$ mol.wt. 486.69

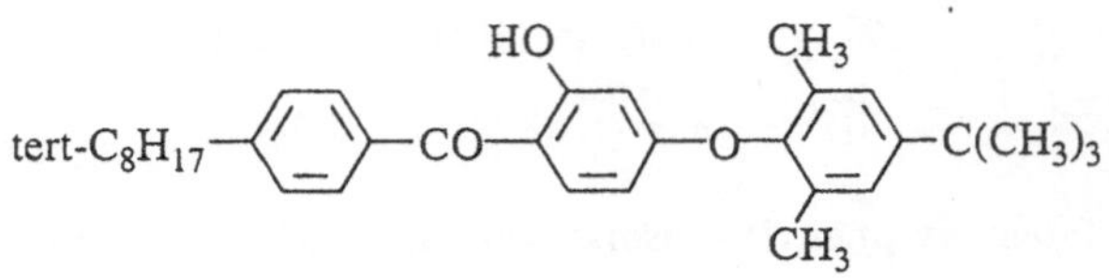

Synthesis

-Refer to: [101] (compound 30).

m.p. (NA); UV [101].

[2-Hydroxy-4-[4-(octyloxy)phenoxy]phenyl][4-(octyloxy)phenyl]methanone

[35697-94-0] $C_{35}H_{46}O_5$ mol.wt. 546.75

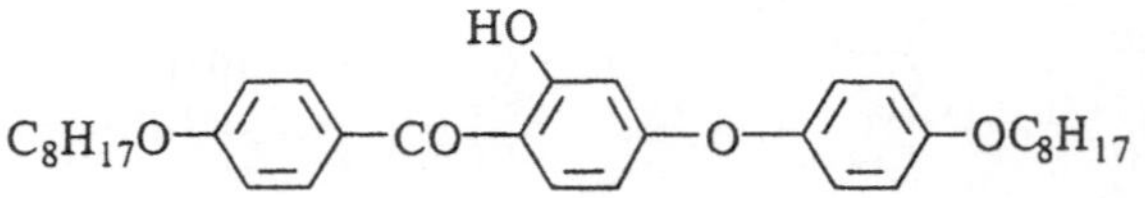

Synthesis

-Refer to: [101] (compound 23).

m.p. (NA); UV [101].

[4-(Dodecyloxy)-2-hydroxyphenyl][4-(dodecyloxy)phenyl]methanone

$C_{37}H_{58}O_4$ mol.wt. 566.87

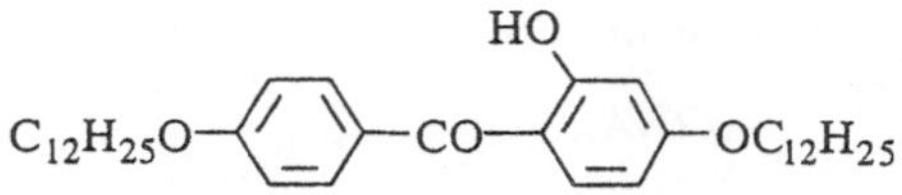

Synthesis

-Refer to: [267].

m.p. (NA); EPR [267].

[4-(4-Dodecylphenoxy)-2-hydroxyphenyl](4-dodecylphenyl)methanone

[35697-95-1] $C_{43}H_{62}O_3$ mol.wt. 626.96

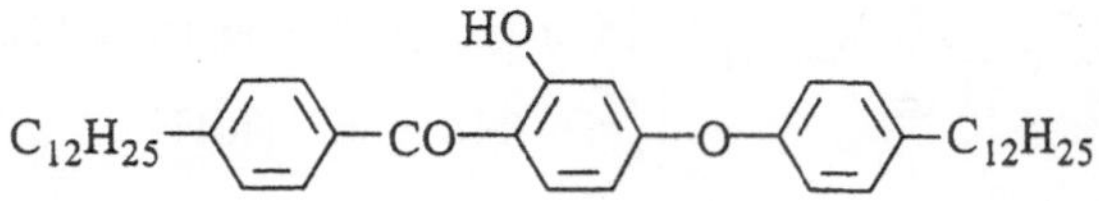

Synthesis

-Refer to: [101] (compound 24).

m.p. (NA); UV [101].

2. DIHYDROXYBENZOPHENONES

2.1. *Hydroxy groups located on the same ring*

2.1.1. *Substituents located on the hydroxylated ring*

Phenyl(2,3,5-trifluoro-4,6-dihydroxyphenyl)methanone

[32541-14-3]   C13H7F3O3   mol.wt. 268.19

Synthesis

-Preparation by demethylation of 2,4-dimethoxy-3,5,6-tri
fluorobenzophenone (SM1) or of 2-hydroxy-4-methoxy-
3,5,6-trifluorobenzophenone (SM2) with aluminium chloride
in methylene chloride at 20° for 6 h (97% and 60% yields,
respectively) [839]. SM1 was obtained in two
steps: first, preparation of 4-methoxy-2,3,5,6-tetrafluorobenzophenone by treatment of
2,3,4,5,6-pentafluorobenzophenone with sodium methoxide in methanol at 20° for 3 days
(91%). Then, this new ketone, by reaction with sodium methoxide in boiling methanol for 15 h
gave SM1 (82%) [839]. The preparation of SM2 was also described in this book (94%) [839].

    m.p. 173-176° [839];   IR [839], UV [839].

(3-Bromo-4-chloro-2,5-dihydroxyphenyl)phenylmethanone

[154700-58-0]   C13H8BrClO3   mol.wt. 327.56

Synthesis

-Obtained by 10% Pd/C-catalyzed hydrogenolysis of *gem*-
diphenylcyclopropane fused with 4-bromo-3,6-dichloro-
benzoquinone in 1,4-dioxane in the presence of water (1%)
for 6 h under atmospheric pressure at r.t. (27%) [1021].

    m.p. (NA); ^{1}H NMR [1021].

(3-Bromo-2,4-dihydroxy-5-nitrophenyl)phenylmethanone

                C13H8BrNO5   mol.wt. 338.11

Syntheses

-Preparation by reaction of bromine with 2,4-dihydroxy-
5-nitrobenzophenone in acetic acid at r.t. for 2 h [320] or in
hot acetic acid for 1 h [321].
-Also obtained by reaction of nitric acid (d = 1.4) with
3,3'-dibenzoyl-5,5'-dibromo-2,2',6,6'-tetrahydroxy-
diphenyl thioether in acetic acid at r.t. for 1 h. The same result was obtained from 3,3'-dibenzoyl-
5,5'-dibromo-4,4',6,6'-tetrahydroxydiphenyl thioether [320].
-Also obtained by action of nitric acid (d = 1.4) with 5-bromo-2,4-dihydroxybenzophenone in acetic
acid, first at low temperature, then at r.t. According to the authors, there is a migration of the
bromine atom during nitration [321].

    m.p. 208-209° [320, 321];   Spectra (NA).

(5-Bromo-2,4-dihydroxy-3-nitrophenyl)phenylmethanone

$C_{13}H_8BrNO_5$ mol.wt. 338.11

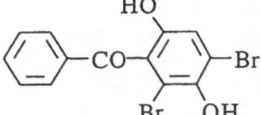

Synthesis

-Obtained by reaction of bromine with 2,4-dihydroxy-
   3-nitrobenzophenone in hot acetic acid solution then at r.t.
   for overnight [321].

    m.p. 110-111° [321];   Spectra (NA).

(2,4-Dibromo-3,6-dihydroxyphenyl)phenylmethanone

[27065-46-9] $C_{13}H_8Br_2O_3$ mol.wt. 372.01

Syntheses

-Preparation by demethylation of 4,6-dibromo-2-hydroxy-
   5-methoxybenzophenone with aluminium chloride in boiling
   benzene for 10 min (72%) [574].
-Also obtained by saponification of 5-acetoxy-2-benzoyloxy-
   4,6-dibromobenzophenone (SM) with potassium

hydroxide in boiling ethanol for 1 h (72%) [576]. SM was prepared in three steps: first,
bromination of 5-hydroxy-2,3-diphenylbenzofuran (bromine/carbon tetrachloride), followed by
acetylation of the 4,6-dibromo-5-hydroxy-2,3-diphenylbenzofuran so obtained (acetic anhydride/
sodium acetate) and oxidation of the resulting 5-acetoxy-4,6-dibromo-2,3-diphenylbenzofuran
(chromium trioxide/acetic acid).

    m.p. 170-171° [576], 170° [574];   Spectra (NA).

(3,4-Dibromo-2,5-dihydroxyphenyl)phenylmethanone

[154700-61-5] $C_{13}H_8Br_2O_3$ mol.wt. 372.01

Synthesis

-Obtained by 10% Pd/C-catalyzed hydrogenolysis of *gem*-
   diphenylcyclopropane fused with 3,4,6-tribromo-
   benzoquinone in 1,4-dioxane in the presence of water (1%)
   for 6 h under atmospheric pressure at r.t. (33%) [1021].

    m.p. (NA); ^{1}H NMR [1021].

(3,5-Dibromo-2,4-dihydroxyphenyl)phenylmethanone

[3286-96-2] $C_{13}H_8Br_2O_3$ mol.wt. 372.01

Syntheses

-Preparation by reaction of bromine with resbenzophenone
   in acetic acid [320, 321], (44%) [1236].
-Also obtained by reaction of bromine with 3,3'-dibenzoyl-
   2,2',6,6'-tetrahydroxydiphenyl thioether in acetic acid in a
   boiling water bath for 4 h. The same result was obtained

from 3,3'-dibenzoyl-4,4',6,6'-tetrahydroxydiphenyl thioether [320].

-Also obtained by saponification of 4-acetoxy-2-benzoyloxy-3,5-dibromobenzophenone (SM) with potassium hydroxide in boiling ethanol for 1 h (73%). SM was prepared in three steps: first, bromination of 6-hydroxy-2,3-diphenylbenzofuran (bromine/carbon tetrachloride), followed by acetylation of the 5,7-dibromo-6-hydroxybenzofuran so obtained (acetic anhydride/sodium acetate) and oxidation of the resulting 6-acetoxy-5,7-dibromo-2,3-diphenylbenzofuran (chromium trioxide/acetic acid) [576].

   m.p. 151-152° [1236], 150-151° [320, 321], 148-149° [576];
   Spectra (NA).

(4,6-Dibromo-2,3-dihydroxyphenyl)phenylmethanone

[65202-42-8] $C_{13}H_8Br_2O_3$ mol.wt. 372.01

Synthesis

-Preparation by saponification of 3-acetoxy-2-(benzoyloxy)-4,6-dibromobenzophenone (SM) with potassium hydroxide in boiling ethanol (72%). SM was obtained by oxidation of 7-acetoxy-4,6-dibromo-2,3-diphenylbenzofuran with chromium trioxide in boiling acetic acid (70%) [1].

   m.p. 205° [1];   Spectra (NA).

(2,3-Dichloro-4,5-dihydroxyphenyl)phenylmethanone

[103843-56-7] $C_{13}H_8Cl_2O_3$ mol.wt. 283.11

Synthesis

-Preparation by Friedel-Crafts acylation of 1,2-dichloro-3,4-dihydroxybenzene with benzoyl chloride in the presence of aluminium chloride in refluxing ethylene dichloride for 24 h (83%) [634, 635].

   m.p. 178-180° [634, 635]; ^{1}H NMR [634], IR [634].

(3,4-Dichloro-2,5-dihydroxyphenyl)phenylmethanone

[21250-79-3] $C_{13}H_8Cl_2O_3$ mol.wt. 283.11

Syntheses

-Obtained by reaction of 2,3-dichloro-1,4-benzoquinone with benzaldehyde, either in the presence of benzoyl peroxide at 80° or in the absence of this one at 155° [745].

-Also obtained by 10% Pd/C-catalyzed hydrogenolysis of gem-diphenylcyclopropane fused with 3,4,6-trichloro-benzoquinone or with 6-bromo-3,4-dichlorobenzoquinone in 1,4-dioxane in the presence of water (2%) for 2 h under atmospheric pressure at r.t. (59% and 75% yields, respectively) [1021].

   m.p. (NA); ^{1}H NMR [1021].

(2,4-Dihydroxy-3,5-diiodophenyl)phenylmethanone

[33427-67-7] $C_{13}H_8I_2O_3$ mol.wt. 466.01

Syntheses

-Preparation by iodination of resbenzophenone,
*with iodine and iodic acid in dilute ethanol for 1 h (62%) [122];
*with iodine and potassium iodide in aqueous ammonia for 15 min [122];
*with iodine monochloride in acetic acid for 2 h at r.t. [122].
-Also obtained from 2,4-dihydroxy-3,5-dinitrobenzophenone by methylation, reduction to 3,5-di-amino-2,4-dimethoxybenzophenone, subsequent diazotization and heating of the diazonium salt with potassium iodide, then demethylation of the 3,5-diiodo-2,4-dimethoxybenzophenone so formed [122].

  m.p. 184-185° [122];   Spectra (NA).

(2,3-Dihydroxy-4,6-dinitrophenyl)phenylmethanone

[65202-37-1] $C_{13}H_8N_2O_7$ mol.wt. 304.22

Synthesis

-Preparation by saponification of 3-acetoxy-2-(benzoyloxy)-4,6-dinitrobenzophenone (SM) with potassium hydroxide in refluxing ethanol (68%). SM was obtained by oxidation of 7-acetoxy-4,6-dinitro-2,3-diphenylbenzofuran with chromium trioxide in boiling acetic acid (81%) [1].

  m.p. 264-265° [1];   Spectra (NA).

(2,4-Dihydroxy-3,5-dinitrophenyl)phenylmethanone

[27065-50-5] $C_{13}H_8N_2O_7$ mol.wt. 304.22

Syntheses

-Obtained by saponification of 4-acetoxy-2-benzoyloxy-3,5-dinitrobenzophenone (SM) with potassium hydroxide in boiling ethanol for 1 h (76%). SM was prepared in three steps: first, nitration of 6-hydroxy-2,3-diphenylbenzofuran (concentrated nitric acid/acetic acid), followed by acetylation of the 6-hydroxy-5,7-dinitro-2,3-diphenylbenzofuran so obtained (acetic anhydride/sodium acetate) and oxidation of the resulting 6-acetoxy-5,7-dinitro-2,3-diphenylbenzofuran (chromium trioxide/acetic acid) [576].
-Preparation by nitration of resbenzophenone with 4 N nitric acid in the presence of sodium nitrite at r.t. for 8 days [122, 576], (60%) [565].

  m.p. 187-188° [565], 185° [576]; ^{1}H NMR [576].

(2-Bromo-4,5-dihydroxyphenyl)phenylmethanone

[91197-10-3] $C_{13}H_9BrO_3$ mol.wt. 293.12

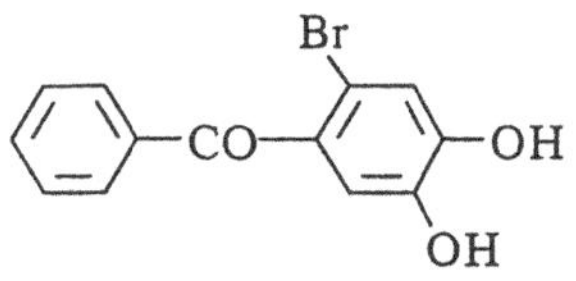

Synthesis

-Preparation by demethylation of 2-bromo-4,5-dimethoxy-
benzophenone (SM) (m.p. 71-72°) with pyridinium chloride
for 2 h at 180-200°. SM was obtained by reaction of benzoic
acid with 1-bromo-3,4-dimethoxybenzene in the presence of
phosphorous pentoxide in methanesulfonic acid for 30 min
at 70° [518].

-Also refer to: [519].

   brown oil [518];   b.p. (NA);   Spectra (NA).

(3-Bromo-2,5-dihydroxyphenyl)phenylmethanone

[112932-43-1] $C_{13}H_9BrO_3$ mol.wt. 293.12

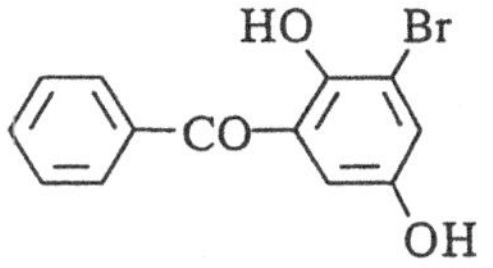

Synthesis

-Refer to: [463, 664].

m.p. and Spectra (NA).

(3-Bromo-2,6-dihydroxyphenyl)phenylmethanone

                               $C_{13}H_9BrO_3$ mol.wt. 293.12

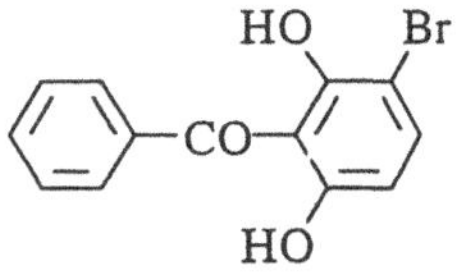

Syntheses

-Preparation by decarboxylation of 3-benzoyl-5-bromo-
2,4-dihydroxybenzoic acid [1239], in the presence of
concentrated hydrochloric acid in refluxing dilute acetic acid
for 24 h (34%) [1240].
-Preparation by hydrolysis of 8-benzoyl-6-bromo-
7-hydroxy-4-methylcoumarin [1240], in refluxing 10% sodium hydroxide for 2.5 h (49%) [1239].

   m.p.   122-123° [1240], 122° [1239];   Spectra (NA).

(5-Bromo-2,4-dihydroxyphenyl)phenylmethanone

[3286-97-3] $C_{13}H_9BrO_3$ mol.wt. 293.12

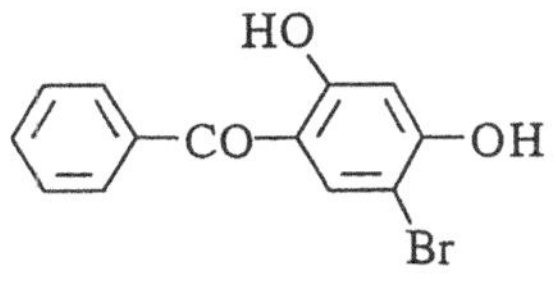

Syntheses

-Obtained by demethylation of 5-bromo-2-hydroxy-
4-methoxybenzophenone with aluminium chloride in boiling
benzene for 10 min (72%) [574].
-Preparation by reaction of bromine with 2,4-dihydroxy-
benzophenone in acetic acid at r.t. overnight [320, 321].

-Also refer to: [1236].

   m.p.   148-149° [320, 321], 148° [574];   Spectra (NA).

(2-Chloro-4,5-dihydroxyphenyl)phenylmethanone

[85525-22-0]   C13H9ClO3   mol.wt. 248.67

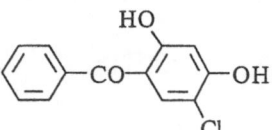

Syntheses

-Preparation by reaction of benzoyl chloride with 4-chloro-
1,2-benzenediol in the presence of aluminium chloride at
140° for 4.5 h [517].
-Preparation by total demethylation of 2-chloro-4,5-di-
methoxybenzophenone (SM) with 48% hydrobromic acid
in refluxing acetic acid for 17 h [517]. SM was obtained by benzoylation of 4-chloro-
1,2-dimethoxybenzene,
*with benzoyl chloride in the presence of iodine;
*with 2-pyridinyl benzoate in the presence of trifluoroacetic acid at 150° for 2 h in a sealed tube.
-Also refer to: [518, 519].

m.p. 130-135° [517];   Spectra (NA).

(3-Chloro-2,6-dihydroxyphenyl)phenylmethanone

                                       C13H9ClO3   mol.wt. 248.67

Syntheses

-Preparation by decarboxylation of 3-benzoyl-5-chloro-
2,4-dihydroxybenzoic acid with concentrated hydrochloric
acid in refluxing dilute acetic acid for 24 h [1239], (30%)
[1240].
-Preparation by hydrolysis of 8-benzoyl-6-chloro-7-hydroxy-
4-methylcoumarin with refluxing 10% aqueous sodium hydroxide or potassium hydroxide
solution for 2.5 h [1240], (40%) [1239].

m.p. 119-120° [1239, 1240];   Spectra (NA).

(5-Chloro-2,4-dihydroxyphenyl)phenylmethanone

[3286-95-1]   C13H9ClO3   mol.wt. 248.67

Syntheses

-Preparation by reaction of benzotrichloride with 4-chloro-
resorcinol in 40% aqueous isopropanol solution at 70-80°
(89%) [1460].
-Preparation in two steps: first, chlorination of resorcinol
dimethyl ether with sulfuryl chloride in chloroform, then
Friedel-Crafts acylation of the 2,4-dimethoxychlorobenzene
so obtained with benzoyl chloride in ethylene dichloride in the presence of aluminium chloride at
r.t. for 3 h (81%) [1195].

m.p. 142-143°5 [1195], 142-143° [1460];
^{1}H NMR [1195], IR [1195], MS [1195].

(2-Fluoro-4,5-dihydroxyphenyl)phenylmethanone

[85525-20-8]   C13H9FO3   mol.wt. 232.21

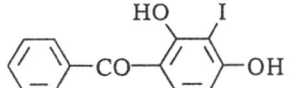

Synthesis

-Preparation by total demethylation of 2-fluoro-4,5-di-
methoxybenzophenone (SM) with 48% hydrobromic acid in
refluxing acetic acid for 17 h. SM was obtained on heating a
mixture of 4-fluoro-1,2-dimethoxybenzene*,
2-pyridinyl benzoate and trifluoroacetic acid at 100° for
4.5 h in a sealed tube [517]. The starting dimethyl ether* was prepared from 3,4-dimethoxyaniline
by a Balz-Schiemann reaction [127].
-Also refer to: [519].

m.p. 169°5-171° [517];   Spectra (NA).

(2,4-Dihydroxy-3-iodophenyl)phenylmethanone

[33427-62-2]   C13H9IO3   mol.wt. 340.12

Synthesis

-Obtained by iodination of resbenzophenone with iodine and
iodic acid in dilute ethanol for 30 min at r.t. (78%) [122].

m.p. 150-151° [122];   Spectra (NA).

(2,4-Dihydroxy-5-iodophenyl)phenylmethanone

[33427-72-4]   C13H9IO3   mol.wt. 340.12

Synthesis

-Obtained from 2,4-dihydroxy-3,5-diiodobenzophenone in
refluxing acetic acid for 8 h. There is an elimination of one
iodine atom [122].

m.p. 151° [122];   Spectra (NA).

(2,4-Dihydroxy-3-nitrophenyl)phenylmethanone

[59746-91-7]   C13H9NO5   mol.wt. 259.22

Syntheses

-Preparation by reaction of benzotrichloride with 2-nitro-
resorcinol in hydrofluoric acid in the presence of water at
-10° for 4 h, then at r.t. overnight (98%) [393].
-Preparation by Fries rearrangement of 2-nitroresorcinol dibenzoate with aluminium chloride [45],
*in nitrobenzene at 100-110° for 3 h or at 25-28° for 70 h (42%);
*without solvent, at 140° for 2 h (23%).
-Also obtained (poor yield) by reaction of concentrated nitric acid with resbenzophenone in acetic
acid at r.t. for 1-2 h (6%) [933].
-Also obtained (by-product) by reaction of benzoyl chloride with 2-nitroresorcinol in the presence
of aluminium chloride in nitrobenzene in a boiling water bath for 1 h [321].

m.p.   145° [45], 144-145° [321], 142° [393];
^{1}H NMR [933], IR [933], MS [933];   TLC [933].

(2,4-Dihydroxy-5-nitrophenyl)phenylmethanone

[40990-79-2]   C$_{13}$H$_9$NO$_5$ mol.wt. 259.22

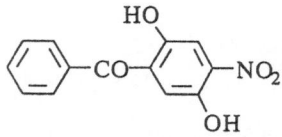

Syntheses

-Obtained by nitration of resbenzophenone,
*with concentrated nitric acid in acetic acid at r.t. for 1-2 h
  (38%) [933] or first at r.t., then at 60° [320];
*with 4 N nitric acid in the presence of sodium nitrite (trace)
  at r.t. for 8 days [576];
*also obtained by adding nitric acid (d = 1.4) to an ice cooled solution of resbenzophenone in acetic
  acid. Then, the ice bath was removed and the reaction stopped when the temperature reached 45°
  [321].
-Preparation by refluxing hydrobromic acid with 2-hydroxy-4-methoxy-5-nitrobenzophenone for
  3 h (61%) [574].
-Also obtained by reaction of nitric acid (d = 1.4) with 3,3'-dibenzoyl-4,4',6,6'-tetrahydroxydiphenyl
  thioether in acetic acid at r.t. for 1 h [320].

m.p.   144-145° [320, 321], 143° [574];
^{1}H NMR [933], IR [933], MS [933];   TLC [933].

(2,5-Dihydroxy-4-nitrophenyl)phenylmethanone

[40990-70-3]   C$_{13}$H$_9$NO$_5$ mol.wt. 259.22

Syntheses

-Preparation by saponification of 5-(acetyloxy)-4-nitro-
  2-(p-nitrobenzoyloxy)benzophenone (SM) with potassium
  hydroxide in ethanol for 1 h (81%). SM was obtained by
  oxidation of 5-(acetyloxy)-6-nitro-2-(p-nitrophenyl)-
  3-phenylbenzofuran with chromium trioxide in refluxing
  acetic acid for 30 min [574].
-Preparation by demethylation of 2-hydroxy-5-methoxy-4-nitrobenzophenone with aluminium
  chloride in boiling benzene (62%) [574].

m.p.   167° [574];   Spectra (NA).

(2,6-Dihydroxy-3-nitrophenyl)phenylmethanone

C$_{13}$H$_9$NO$_5$ mol.wt. 259.22

Syntheses

-Preparation by reaction of benzoic anhydride with 4-nitro-
  resorcinol in the presence of aluminium chloride in nitrobenzene on a steam bath for 3 h [957].
-Preparation by nitration of 2,6-dihydroxybenzophenone with
  nitric acid (d = 1.42) at 0° for 10 min [957].
-Preparation by Fries rearrangement,
*of 4-nitroresorcinol 1-monobenzoate with aluminium chloride without solvent at 130-140° for 2 h
  (16%) or in nitrobenzene at 100° for 2 h (43%) [46];

*of 4-nitroresorcinol 3-monobenzoate that, under the above conditions, afforded the same mono-
ketone [46];
*of 4-nitroresorcinol dibenzoate with aluminium chloride without solvent at 140° for 3 h or in
nitrobenzene at 100° or 140° for 2 h [46].

   m.p.   159-160° [46], 158° [957];   Spectra (NA).

(3,4-Dihydroxy-5-nitrophenyl)phenylmethanone

[125628-96-8] $C_{13}H_9NO_5$ mol.wt. 259.22

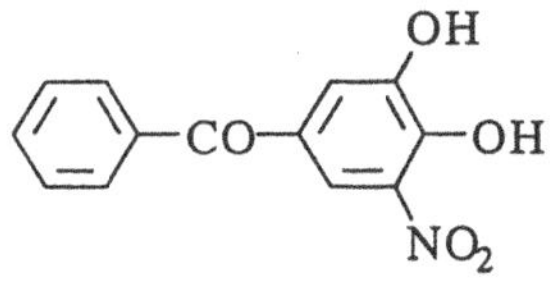

Synthesis

-Preparation by total demethylation of 3,4-dimethoxy-5-nitro-
   benzophenone with 48% hydrobromic acid in acetic acid at
   110° for 30 h [164].
-Also refer to: [162, 163, 197].

   m.p. 132° [164];   Spectra (NA); pK_a [197].

(3,5-Dihydroxy-4-nitrophenyl)phenylmethanone

[51787-06-5] $C_{13}H_9NO_5$ mol.wt. 259.22

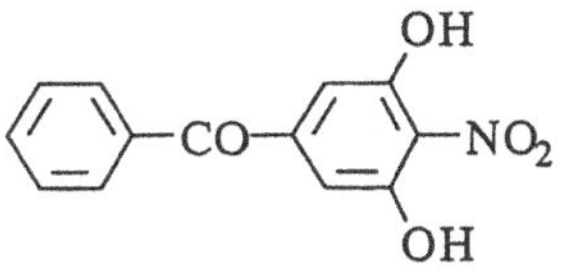

Synthesis

-Preparation from 3,5-dihydroxybenzophenone (and not from
   resorcinol 3-benzoate, as indicated in the paper) by nitration
   with aluminium nitrate in acetic acid for 4 h at r.t. (83%)
   [1098].
N.B.: This ketone was called 2-nitro-5-benzoyl resorcinol in
   the paper (table 1, compound 4) [1098].

   m.p. 121° [1098];   Spectra (NA).

(3-Amino-2,4-dihydroxyphenyl)phenylmethanone

[87119-03-7] $C_{13}H_{11}NO_3$ mol.wt. 229.24

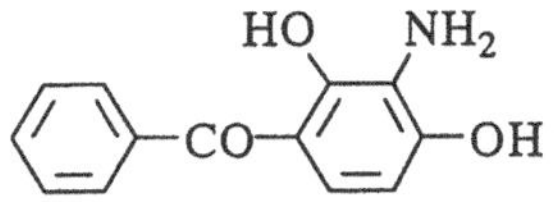

Synthesis

-Obtained from the corresponding hydrochloride described
   below [933].

   m.p. and Spectra (NA).

(3-Amino-2,4-dihydroxyphenyl)phenylmethanone *(Hydrochloride)*

[87119-04-8] $C_{13}H_{11}NO_3,HCl$ mol.wt. 265.70

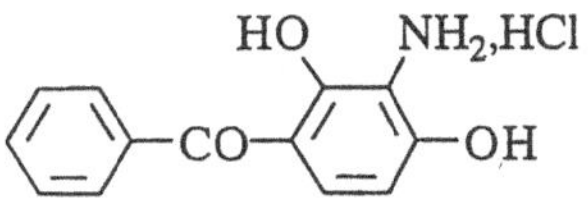

Synthesis

-Preparation by hydrogenation of 2,4-dihydroxy-3-nitro-
   benzophenone in chloroform/ethanol solution in the
   presence of 10% Pd/C, followed by treatment of the amino
compound formed with concentrated hydrochloric acid in ethanol (61%) [933].

m.p.   175-185° (d) [933];
^{1}H NMR [933], IR [933], MS [933];   TLC [933].

(5-Amino-2,4-dihydroxyphenyl)phenylmethanone

[87119-01-5]   C$_{13}$H$_{11}$NO$_3$ mol.wt. 229.24

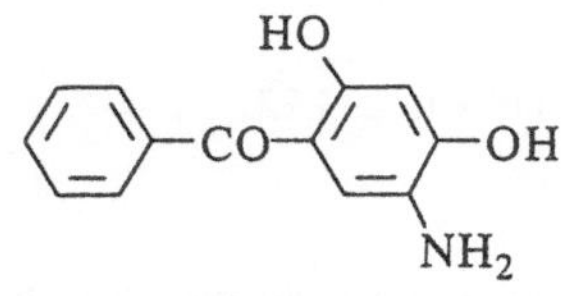

Synthesis

-Obtained from the corresponding hydrochloride described
 below [933].

m.p. and Spectra (NA).

(5-Amino-2,4-dihydroxyphenyl)phenylmethanone   *(Hydrochloride)*

[87119-02-6]   C$_{13}$H$_{11}$NO$_3$,HCl   mol.wt. 265.70

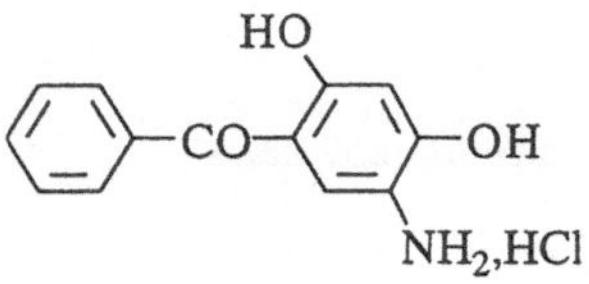

Synthesis

-Preparation by hydrogenation of 2,4-dihydroxy-5-nitro-
 benzophenone in chloroform/ethanol solution in the
 presence of 10% Pd/C, followed by treatment of the amino
 compound formed with concentrated hydrochloric acid in
 ethanol (85%) [933].

m.p.   155-160° (d) [933];   ^{1}H NMR [933], IR [933], MS [933];   TLC [933].

(3,5-Dibromo-2,6-dihydroxy-4-methoxyphenyl)phenylmethanone

C$_{14}$H$_{10}$Br$_2$O$_4$ mol.wt. 402.04

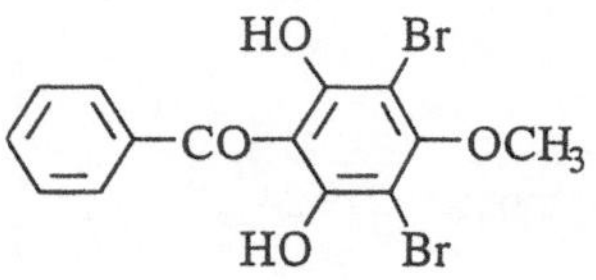

Synthesis

-Obtained by action of bromine with Cotoin in cooled
 chloroform [274].
N.B.: The formula proposed is the more likely.

m.p. 116° [274];   Spectra (NA).

[3-Chloro-2,4 (or 2,5)-dihydroxy-5 (or 4)-methoxyphenyl]phenylmethanone

[140708-51-6]   C$_{14}$H$_{11}$ClO$_4$ mol.wt. 278.69

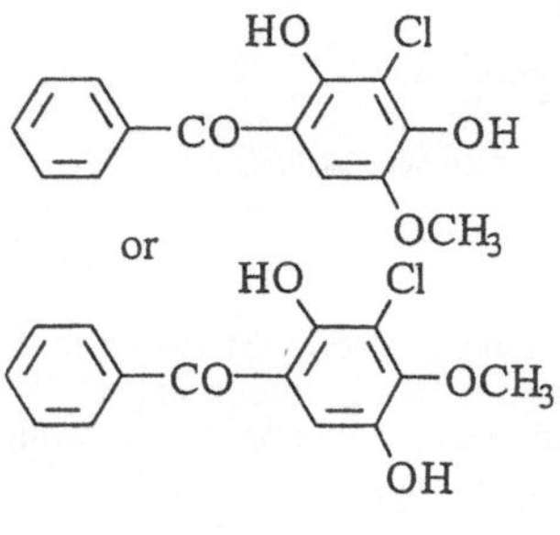

Synthesis

-Obtained by reaction of benzoyl chloride with 3-chloro-
 1,2,4-trimethoxybenzene in the presence of aluminium
 chloride in ethylene dichloride between 0 to 5°, then at r.t.
 for 8 h and at reflux for 1.5 h (58%) [1194].

m.p. 181-182° [1194];
^{1}H NMR [1194], IR [1194], MS [1194].

(2,6-Dihydroxy-4-methoxy-3-nitrosophenyl)phenylmethanone

$C_{14}H_{11}NO_5$ mol.wt. 273.25

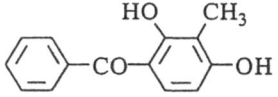

Synthesis

-Preparation by adding an aqueous solution of potassium nitrite to a solution of 2,6-dihydroxy-4-methoxy-benzophenone (so called Cotoin) in an acetic acid/ethanol mixture [1075].

m.p. 153-154° [1075];   Spectra (NA).

(2,4-Dihydroxy-3-methylphenyl)phenylmethanone

[52117-23-4] $C_{14}H_{12}O_3$ mol.wt. 228.25

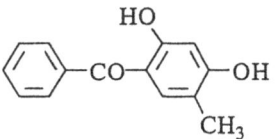

Syntheses

-Preparation by reaction of benzonitrile with 2,6-dihydroxy-toluene in the presence of zinc chloride and hydrochloric acid, followed by hydrolysis of the ketimine hydrochloride so formed with boiling water for 1 h (68%) (Hoesch reaction) [652].

-Preparation by Friedel-Crafts acylation of 3-methoxy-2-methylphenol with benzoyl chloride in the presence of aluminium chloride in boiling carbon disulfide for 1 h [652].

-Preparation by demethylation of 2-hydroxy-4-methoxy-3-methylbenzophenone with hydriodic acid in refluxing acetic anhydride (125-135°) for 2 h (63%) [1242].

-Also obtained by methylation of resbenzophenone with methyl iodide in the presence of potassium hydroxide in refluxing methanol for 6 h [757] according to [1347]. In this case, there is introduction of one methyl group on the benzene nucleus [757].

-Preparation by reaction of benzoic acid with 2-methylresorcinol in tetrachloroethane in the presence of boron trifluoride at 80° for 4 h (70%) [271].

-Also refer to: [22, 23, 27, 521, 889, 917, 929, 955, 1132].

m.p. 177° [652, 1242], 176° [757], 173-174° [271];   ^{1}H NMR [271], UV [271].

(2,4-Dihydroxy-5-methylphenyl)phenylmethanone

[52220-71-0] $C_{14}H_{12}O_3$ mol.wt. 228.25

Syntheses

-Preparation by demethylation of 2-hydroxy-4-methoxy-5-methylbenzophenone with aluminium chloride in boiling benzene (62%) [2].

-Preparation by reaction of benzonitrile with 2,4-dihydroxy-toluene in the presence of zinc chloride and hydrochloric acid in ethyl ether, followed by hydrolysis of the ketimine hydrochloride so formed (65%) (Hoesch reaction) [889].

-Preparation by reaction of benzoyl chloride with 4-methylresorcinol in the presence of aluminium chloride [520].

N.B.: In the paper [340], the formula of the 2,4-dihydroxy-6-methylbenzophenone reported in the discussion (page 392, formula II) is correct since this product is prepared starting from orcinol. It must be pointed out that in the same communication, this compound has been erroneously named as the 2,4-dihydroxy-5-methylbenzophenone in the experimental part (page 394). The obtained compound is actually the 2,4-dihydroxy-6-methylbenzophenone already prepared by an other procedure [585].

m.p.   137°5-138° [889], 137° [2];   Spectra (NA).

(2,4-Dihydroxy-6-methylphenyl)phenylmethanone

[43221-40-5] $C_{14}H_{12}O_3$ mol.wt. 228.25

Syntheses

-Preparation by Hoesch condensation of orcinol with
  benzonitrile (90%) [26], (50%) [585].
-Preparation by reaction of benzoyl chloride with orcinol in
  the presence of aluminium chloride in nitrobenzene at r.t.
  overnight, then heated in a water bath (temperature not
  quoted) for 1 h [340]. N.B.: In the paper [340], the
formula of the 2,4-dihydroxy-6-methylbenzophenone reported in the discussion (page 392,
formula II) is correct since this product is prepared starting from orcinol. It must be pointed out
that in the same communication, this compound has been erroneously named as the 2,4-di-
hydroxy-5-methylbenzophenone in the experimental part (page 394). The obtained compound is
actually the 2,4-dihydroxy-6-methylbenzophenone already prepared by an other procedure [585].
-Preparation from 3-methyl-3'-phenyl-5,5'-diisoxazolylmethane by performing hydrogenolysis and
  subsequent hydrolysis with hydrochloric acid (48%) [90].
-Orcinol by condensation with benzanilide imidochloride in the presence of aluminium chloride in
  ethyl ether gave a keto anil. This one was hydrolyzed by refluxing with ethanolic hydrochloric acid
  yielded the expected ketone (26%) [1061].
-Also refer to: [1368].

m.p.   145-146° [26], 141° [585, 1061], 140° [90], 138° [340];
^1H NMR [26, 90], MS [90].

(2,5-Dihydroxy-4-methylphenyl)phenylmethanone

[59954-93-7] $C_{14}H_{12}O_3$ mol.wt. 228.25

Syntheses

-Preparation by demethylation of,
*2,5-dimethoxy-4-methylbenzophenone on heating with
  pyridinium chloride for 20 min (95%) [1151];
*2-hydroxy-5-methoxy-4-methylbenzophenone with
  aluminium chloride in boiling benzene for 10 min (65%) [2].

m.p.   152°5 [1151], 152° [2];   Spectra (NA).

(2,6-Dihydroxy-4-methylphenyl)phenylmethanone

[68436-77-1] $C_{14}H_{12}O_3$ mol.wt. 228.25

Syntheses

-Obtained by Fries rearrangement of orcinol dibenzoate with
  aluminium chloride at 160-170° for 90 min [338].
-Preparation by heating a mixture of benzoic acid, orcinol,
  zinc chloride and phosphorous oxychloride at 65° for 3 h
  [982].

-Also obtained by partial decarbonylation of 3-benzoyl-2,6-dihydroxy-4-methylbenzophenone by treatment with 85% sulfuric acid at r.t. for 4 h (quantitative yield) [338].
-Also obtained (poor yield) by Hoesch condensation of orcinol with benzonitrile (10%) [26].

m.p.   153-154° [26], 127° [338]; ^{1}H NMR [26].

(3,4-Dihydroxy-5-methylphenyl)phenylmethanone

[108055-13-6] $C_{14}H_{12}O_3$ mol.wt. 228.25

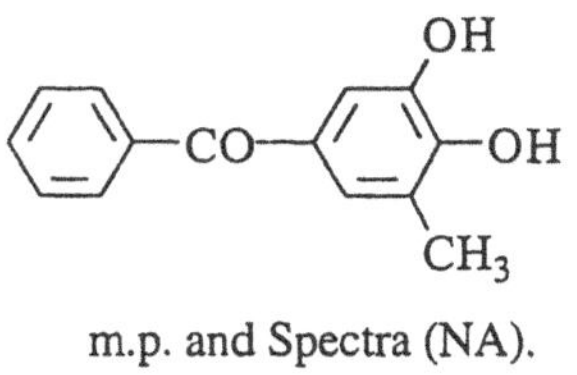

Synthesis

-Refer to: [1354].

m.p. and Spectra (NA).

(2,3-Dihydroxy-4-methoxyphenyl)phenylmethanone
(Alizarine Yellow A, monomethyl ether)

[35836-41-0] $C_{14}H_{12}O_4$ mol.wt. 244.25

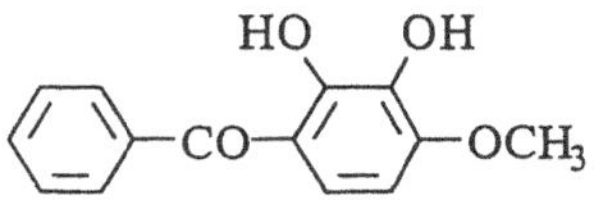

Syntheses

-Obtained by partial methylation of 2,3,4-trihydroxy-
  benzophenone,
*with methyl iodide in the presence of potassium hydroxide
  in methanol at 100° for several hours [495] or in the
presence of lithium carbonate in N,N-dimethylformamide at 30° for 15 h under nitrogen (30%) [802];
*with dimethyl sulfate in the presence of alkali [950].
-Also obtained by reaction of methyl iodide with monosodium salt of 2,3,4-trihydroxy-
  benzophenone at 120° for several hours [495].
-Also obtained by adding a 5% sodium hydrogen carbonate solution to a 2,3-diacetoxy-4-methoxy-
  benzophenone solution in methanol and stirring at 30° for 2 h under nitrogen (25%) [802].
-Also refer to: [414].

m.p.   172-174° [802], 165° [495], 164-165° [950];
^{1}H NMR [802], ^{13}C NMR [802], UV [802], MS [802].

(2,4-Dihydroxy-5-methoxyphenyl)phenylmethanone

                                       $C_{14}H_{12}O_4$ mol.wt. 244.25

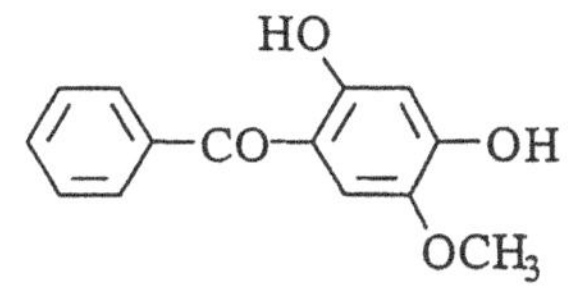

Synthesis

-Obtained by partial demethylation of 2,4,5-trimethoxy-
  benzophenone or 2-hydroxy-4,5-dimethoxybenzophenone
  with hydrobromic acid (d = 1.47) in acetic acid [134, 135].

m.p.   183-185° [134, 135];   Spectra (NA).

(2,4-Dihydroxy-6-methoxyphenyl)phenylmethanone *(Isocotoin)*

[81525-12-4] $C_{14}H_{12}O_4$ mol.wt. 244.25

Syntheses

-Preparation by reaction of benzonitrile with phloroglucinol monomethyl ether in the presence of zinc chloride and hydrochloric acid in ethyl ether, followed by hydrolysis of the ketimine hydrochloride so formed with boiling water for 15 min (good yield) (Hoesch reaction) [702, 704].

-Preparation in one step involving the reaction of phlorobenzophenone with two mol of p-toluene-sulfonyl chloride in acetone in the presence of potassium carbonate followed by methylation with dimethyl sulfate and subsequent detosylation with ethanolic potassium hydroxide [24, 1043].

-Also refer to: [25].

Isolation from natural sources

-From *Helichrysum triplinerve* (Asteraceae) [1099];
-From *genus Leontonyx* [192].

m.p.   162° [702];   Spectra (NA).

(2,5-Dihydroxy-4-methoxyphenyl)phenylmethanone *(Cearoin)*

[52811-37-7] $C_{14}H_{12}O_4$ mol.wt. 244.25

Syntheses

-Obtained (poor yield) by reaction of benzoic acid with 2-methoxyhydroquinone in the presence of boron trifluoride-etherate, on heating at 100° for 30-45 min (5%) [785].

-Also obtained (poor yield) by oxidation of 2-hydroxy-4-methoxybenzophenone (Elbs reaction),

*with lead tetraacetate in acetic acid at 100° for 5 h (3%) [785];
*with potassium persulfate in aqueous potassium hydroxide (22%) [807].

-Preparation by oxidation of 2-hydroxy-4,5-dimethoxybenzophenone with nitric acid (d = 1.2) for 30 min at 15-20°, followed by reduction of the 2-benzoyl-5-methoxy-1,4-benzoquinone formed (93%) with sulfur dioxide in warm ethanol containing a drop of acetic acid for 1 h (79%) [778].

-Also refer to: [503].

Isolation from natural sources

-From *Dalbergia cearensis* (Leguminosae) [512, 1006, 1007, 1235] and *Dalbergia miscolobium* [1006, 1007, 1235];
-From *Dalbergia melanoxylon* Guill. et Perr. heartwood (Leguminosae-Lotoideae) [371];
-From the stem/bark of *Dalbergia volubilis* (Leguminosae) [726];
-From the heartwood of *Dalbergia latifolia* Roxb. [370];
-From the heartwood of *Dalbergia parviflora* Roxb. (Leguminosae) [953];
-From *Dalbergia violacea* (Leguminosae) [1006].

m.p.   188-189° [778], 188° [136, 807], 187-188° [512], 187° [726],
        184°5-185°5 [953], 183-185° [133], 182-184° [370, 371], 182-183° [785];
^{1}H NMR [512, 726, 785, 953], ^{13}C NMR [953],
IR [512, 726, 785, 953], UV [726, 785], MS [512, 726, 953].

(2,6-Dihydroxy-4-methoxyphenyl)phenylmethanone *(Cotoin)*

[479-21-0] $C_{14}H_{12}O_4$ mol.wt. 244.25

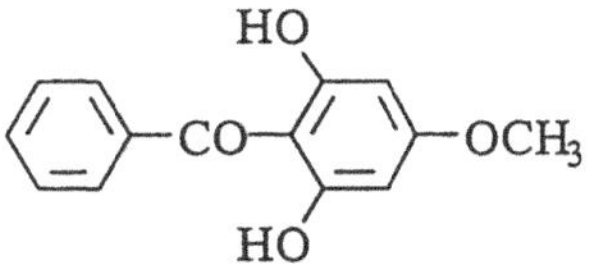

Syntheses

-Obtained by reaction of benzonitrile with phloroglucinol monomethyl ether (Hoesch reaction) [24, 702].
-Also obtained by saponification of 2,6-diacetoxy-4-methoxybenzophenone with boiling aqueous potassium hydroxide solution for 15 min [274].
-Also refer to: [25, 137, 276, 425, 704, 901].

Isolation from natural sources

-From *Coto* bark (Lauraceae) [274, 275, 396, 569, 702, 899, 900, 1075, 1358];
-From the wood of *Aniba dukei* Kostermans (Lauraceae) [494].
N.B.: In 1890 a careful quantitative analysis was kindly undertaken by Julius B. Cohen [396].

m.p.   131-132° [494], 131° [702], 130-131° [274, 900], 130° [396];
^{1}H NMR [403], IR [494, 900], UV [1358].

(3,4-Dihydroxy-2-methoxyphenyl)phenylmethanone

[177703-29-6] $C_{14}H_{12}O_4$ mol.wt. 244.25

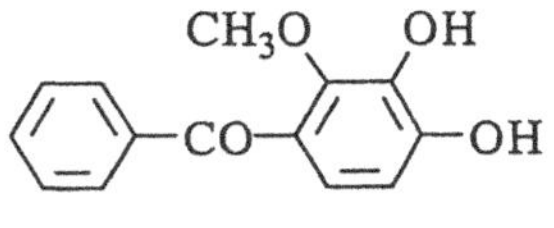

Synthesis

-Preparation by adding a 5% sodium bicarbonate solution to a 3,4-diacetoxy-2-methoxybenzophenone solution in methanol and stirring at 30° for 2 h under nitrogen (50%) [802].

m.p.   130-132° [802]; ^{1}H NMR [802], ^{13}C NMR [802], UV [802], MS [802]; pK_a [802].

(3,4-Dihydroxy-5-methoxyphenyl)phenylmethanone

$C_{14}H_{12}O_4$ mol.wt. 244.25

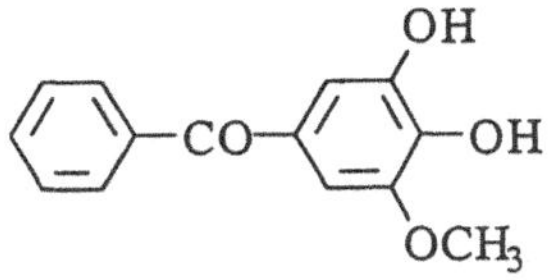

Synthesis

-Obtained by reaction of benzoyl chloride with 2,6-dimethoxyphenol in the presence of aluminium chloride in nitrobenzene, first at 2-3°, then at r.t. for 24 h (6 to 12% yield) [885].

m.p.   168-169° [885];   Spectra (NA).

(3,6-Dihydroxy-2-methoxyphenyl)phenylmethanone

[55137-06-9] $C_{14}H_{12}O_4$ mol.wt. 244.25

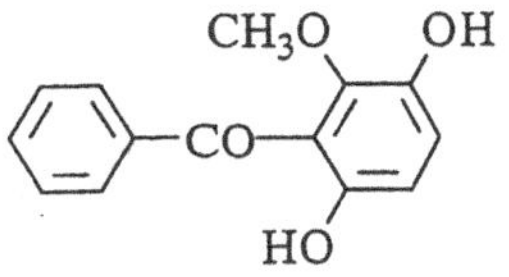

Synthesis

-Obtained from 2-benzoyl-1,4-benzoquinone [1278],
*in refluxing methanol for 48 h without catalyst (26%);
*in refluxing methanol for 12 h in the presence of zinc chloride (23%).

m.p.   141-143° [1278]; ^{1}H NMR [1278], IR [1278], UV [1278].

[2,4-Dihydroxy-3,5-bis[((trifluoromethyl)thio]phenyl]phenylmethanone

[66625-08-9] $C_{15}H_8F_6O_3S_2$ mol.wt. 414.35

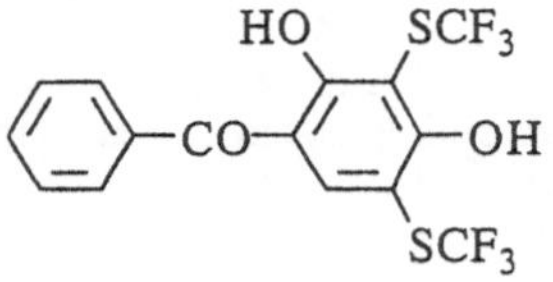

Synthesis

-Preparation by reaction of trifluoromethanesulfenyl chloride
 with 2,4-dihydroxybenzophenone in chloroform in the
 presence of a slight excess of pyridine, first at -40°, then 60°
 for 3 h (80%) [307].

m.p. 91-95° [307]; ^{1}H NMR [307], IR [307].

(2-Ethyl-4,5-dihydroxyphenyl)phenylmethanone

[91197-12-5] $C_{15}H_{14}O_3$ mol.wt. 242.27

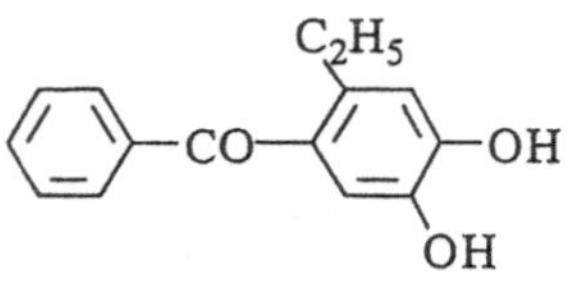

Synthesis

-Preparation by demethylation of 2-ethyl-4,5-dimethoxy-
 benzophenone (SM) with pyridinium chloride for 2 h at
 180-200°. SM was obtained by reaction of benzoic acid with
 1-ethyl-3,4-dimethoxybenzene in the presence of phosphorous
 pentoxide in methanesulfonic acid for 30 min at 70° [518].

m.p. and Spectra (NA).

(3-Ethyl-2,6-dihydroxyphenyl)phenylmethanone

                                            $C_{15}H_{14}O_3$ mol.wt. 242.27

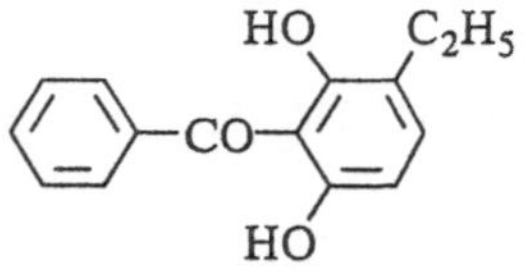

Syntheses

-Preparation from 8-benzoyl-6-ethyl-7-hydroxy-4-methyl-
 coumarin by action of,
*a 20% aqueous sodium hydroxide solution at reflux for
 2-2.5 h (51%) [1369];
*a 10% aqueous sodium hydroxide solution at reflux for 4 h
 (40%) [336].
-Also obtained by decarboxylation of 3-benzoyl-2,4-dihydroxy-5-ethylbenzoic acid with dilute
 hydrochloric acid (1:1) in a sealed tube at 160-170° [1237].

m.p.   128° [336], 125-126° [1369], 125° [1237];   Spectra (NA).

(4-Ethyl-2,5-dihydroxyphenyl)phenylmethanone

[59623-16-4] $C_{15}H_{14}O_3$ mol.wt. 242.27

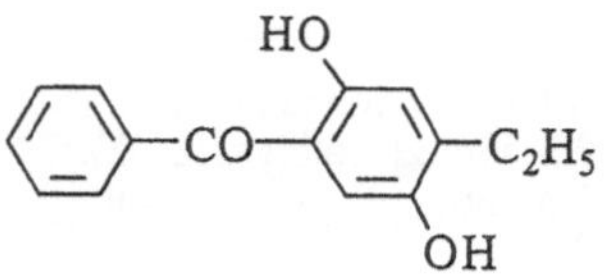

Synthesis

-Preparation by demethylation of 4-ethyl-2-hydroxy-
 5-methoxybenzophenone with aluminium chloride in
 boiling benzene for 10 min (55%) [577].

m.p. 86° [577];   Spectra (NA).

(5-Ethyl-2,4-dihydroxyphenyl)phenylmethanone

[50537-80-9] $C_{15}H_{14}O_3$ mol.wt. 242.27

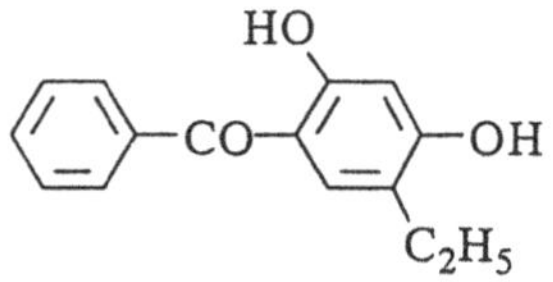

Syntheses

-Preparation by reaction of benzotrichloride with 4-ethyl-
  resorcinol in a 40% aqueous acetic acid solution at 70-80°
  (82%) [1460].
-Preparation by Fries rearrangement of 4-ethylresorcinol
  dibenzoate with aluminium chloride in nitrobenzene at
  50-60° for 3-4 h [1141].
-Preparation by reaction of benzoic acid with 4-ethylresorcinol,
*in hydrofluoric acid at 100° in a stainless steel bomb (48%) [1406];
*in the presence of boron trifluoride in tetrachloroethane on a steam bath for 4 h [1406].
-Preparation by demethylation of 5-ethyl-2-hydroxy-4-methoxybenzophenone with aluminium
  chloride in boiling benzene for 10 min (70%) [577].

  m.p.   109° [577], 104° [1406], 99°5-100°5 [1460], 63-64° [1141]. One of the reported
  melting points is obviously wrong.
  b.p.₁ 240-250° [1406];   Spectra (NA).

(2,6-Dihydroxy-4-methoxy-3-methylphenyl)phenylmethanone

$C_{15}H_{14}O_4$ mol.wt. 258.27

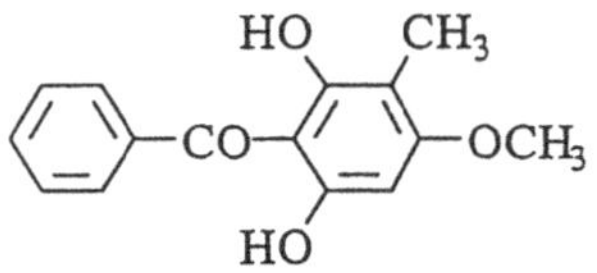

Synthesis

-Preparation by reaction of benzonitrile with 5-methoxy-
  4-methylresorcinol according to Hoesch method (50%)
  [889].

  m.p.   143-144° [889];   Spectra (NA).

(2,3-Dihydroxy-4,5-dimethoxyphenyl)phenylmethanone

$C_{15}H_{14}O_5$ mol.wt. 274.27

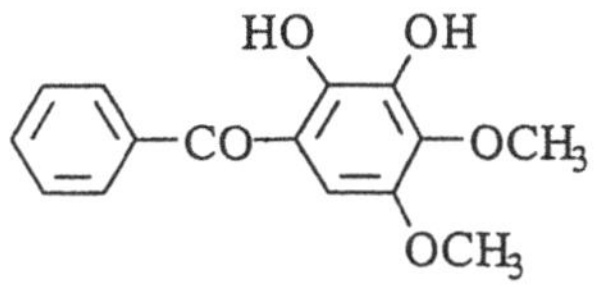

Synthesis

-Refer to: [503].

                          Isolation from natural source

-From *Machaerium scleroxylon* [493].

-Also refer to: [1097].

  m.p. (NA);   ¹H NMR [493], IR [493], UV [493].

(2,4-Dihydroxy-3,5-dimethoxyphenyl)phenylmethanone

[42833-89-6] $C_{15}H_{14}O_5$ mol.wt. 274.27

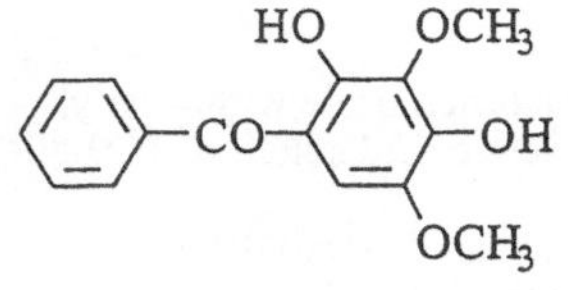

Synthesis

-Obtained by selective demethylation of 2-hydroxy-3,4,5-tri-
methoxybenzophenone with refluxing aqueous piperidine
for 45-65 h [1097].

m.p. 182-184° [1097];   UV [1097], MS [1097].

(2,5-Dihydroxy-3,4-dimethoxyphenyl)phenylmethanone *(Scleroin)*

[4646-78-0] $C_{15}H_{14}O_5$ mol.wt. 274.27

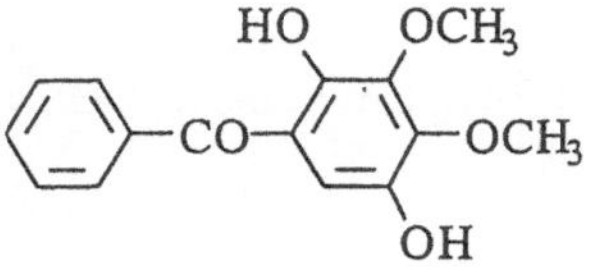

Synthesis

-Obtained (poor yield) by action of potassium persulfate with
2-hydroxy-3,4-dimethoxybenzophenone in the presence of
ferrous sulfate in aqueous potassium hydroxide at r.t. for
3-4 h (10%) (Elbs reaction) [132].

Isolation from natural source

-From *Machaerium scleroxylon* (Leguminosae) [403, 493, 1006, 1235].

-Also refer to: [503, 832, 1097].

m.p.   143°5-144°5 [403], 140-142° [132]; ^{1}H NMR [403], IR [403], UV [403].

[2,4-Dihydroxy-5-(2-propenyl)phenyl]phenylmethanone

$C_{16}H_{14}O_3$ mol.wt. 254.29

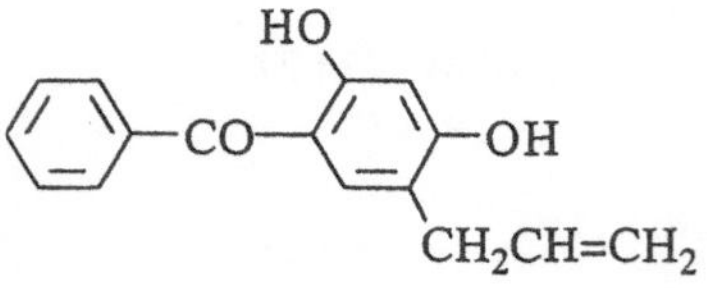

Synthesis

-Preparation by reaction of benzotrichloride with
4-allylresorcinol [1460].

m.p. and Spectra (NA).

[2,6-Dihydroxy-3-(2-propenyl)phenyl]phenylmethanone

$C_{16}H_{14}O_3$ mol.wt. 254.29

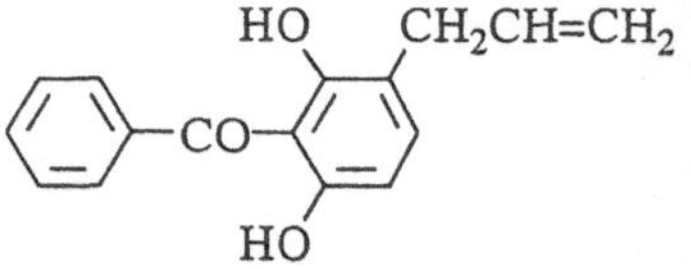

Synthesis

-Preparation from 6-allyl-8-benzoyl-4-methyl-
umbelliferone (6-allyl-8-benzoyl-7-hydroxy-4-methyl-
coumarin), by cleavage with boiling aqueous sodium
hydroxide in the presence of sodium hydrosulfite under
nitrogen for 3 h (90%) [1283].

m.p.   80-81° [1283];   Spectra (NA);   TLC [1283].

(2,4-Dihydroxy-3-propylphenyl)phenylmethanone

[79557-81-6] $C_{16}H_{16}O_3$ mol.wt. 256.30

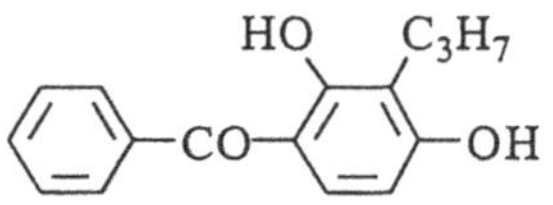

Synthesis

-Preparation by reaction of benzoic acid with 2-propyl-
resorcinol in the presence of zinc chloride at 150° for 2.5 h
(62%) [1029] (Nencki reaction).
-Also refer to: [917, 1028, 1432].

m.p. 152-153° [1029];   Spectra (NA).

(2,4-Dihydroxy-5-propylphenyl)phenylmethanone

$C_{16}H_{16}O_3$ mol.wt. 256.30

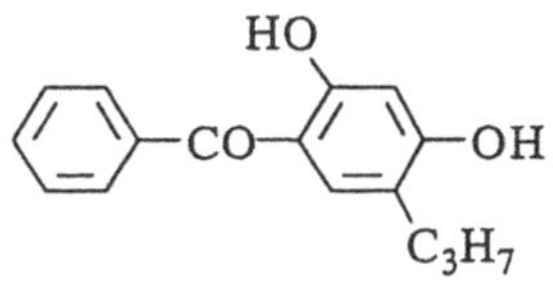

Synthesis

-Obtained by Fries rearrangement of 4-propylresorcinol
dibenzoate with aluminium chloride in nitrobenzene at 50°
for 4 h [1141].

m.p. 138-140° [1141];   b.p.$_9$ 240-245° [1141];   Spectra (NA).

[3-(2-Butenyl)-2,4-dihydroxyphenyl]phenylmethanone

[96836-08-7] $C_{17}H_{16}O_3$ mol.wt. 268.31

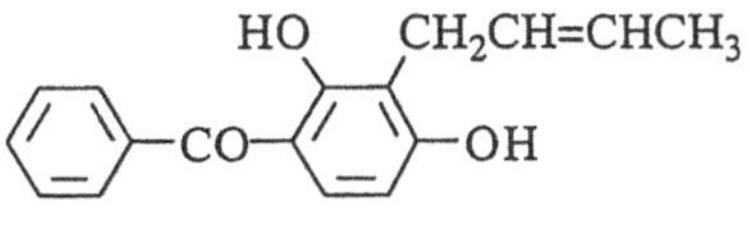

Synthesis

-Preparation by condensation of 2,4-dihydroxy-
benzophenone with 1,3-butadiene in the presence of
orthophosphoric acid in petroleum ether at
30-35° for 24 h (40%) [30].

m.p. 144-146° [30]; ^{1}H NMR [30].

[5-(2-Butenyl)-2,4-dihydroxyphenyl]phenylmethanone

[96859-90-4] $C_{17}H_{16}O_3$ mol.wt. 268.31

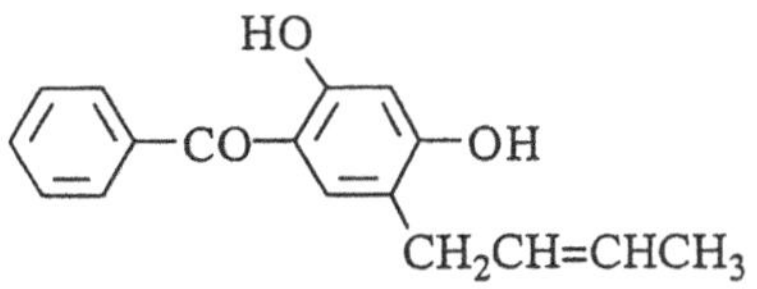

Synthesis

-Preparation by condensation of 2,4-dihydroxy-
benzophenone with 1,3-butadiene in the presence of
orthophosphoric acid in petroleum ether at
30-35° for 24 h (35%) [30].

m.p. 119-121° [30]; ^{1}H NMR [30].

[2,4-Dihydroxy-3-(1-methyl-2-propenyl)phenyl]phenylmethanone

[96836-07-6] $C_{17}H_{16}O_3$ mol.wt. 268.31

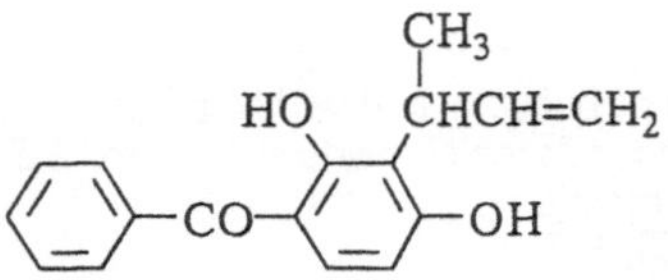

Synthesis

-Obtained (poor yield) by condensation of 2,4-di-
hydroxybenzophenone with 1,3-butadiene in the
presence of orthophosphoric acid in petroleum ether at
30-35° for 24 h (15%) [30].

m.p. 153-155° [30]; ^{1}H NMR [30].

[2,5-Dihydroxy-6-methyl-3-(1-methylethyl)phenyl]phenylmethanone

[101594-97-2] $C_{17}H_{18}O_3$ mol.wt. 270.33

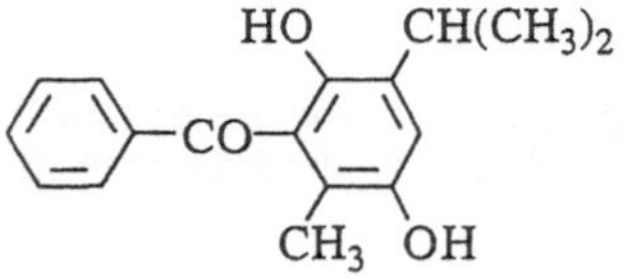

Synthesis

-Obtained by demethylation of 2-hydroxy-5-methoxy-
6-methyl-3-isopropylbenzophenone or 3-hydroxy-
6-methoxy-2-methyl-5-isopropylbenzophenone in
refluxing pyridinium chloride for 20 min [1151].

m.p. 147-147°5 [1151];   Spectra (NA).

[5-(1,1-Dimethylethyl)-2,4-dihydroxyphenyl]phenylmethanone

[4211-67-0] $C_{17}H_{18}O_3$ mol.wt. 270.33

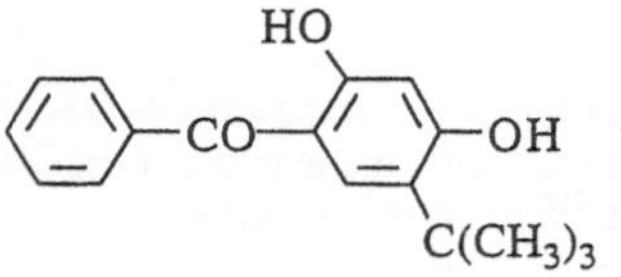

Synthesis

-Preparation by alkylation of resbenzophenone with
isobutylene in benzene in the presence of p-toluenesulfonic
acid [587].
-Also refer to: [378, 379, 380].

m.p. 141° [587, 588];
Spectra (NA);   gel chromatography [1145].

[2,4-Dihydroxy-3-(3-methyl-2-butenyl)phenyl]phenylmethanone

[63565-04-8] $C_{18}H_{18}O_3$ mol.wt. 282.34

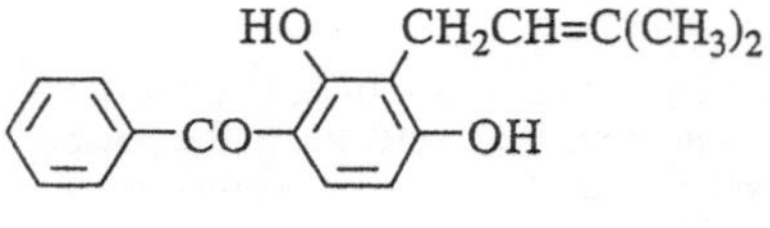

Synthesis

-Obtained (poor yield) by reaction of prenyl
bromide with resbenzophenone,
*in the presence of sodium methoxide in refluxing
methanol for 4 h (10%) [875];
*in the presence of boron trifluoride-etherate in dioxane at 60-70° for 2 h (<3%) [874].

m.p. 121° [875], 120° [874]; ^{1}H NMR [875], IR [875], UV [875].

[3-(2-Butenyl)-2,4-dihydroxy-6-methoxyphenyl]phenylmethanone

[96836-12-3] $C_{18}H_{18}O_4$ mol.wt. 298.34

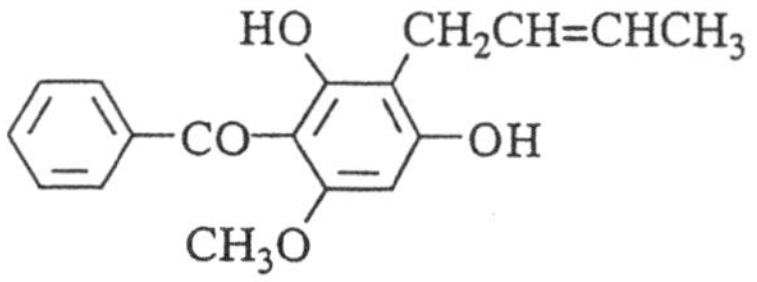

Synthesis

-Obtained by condensation of 2,4-dihydroxy-
6-methoxybenzophenone with 1,3-butadiene in the
presence of orthophosphoric acid in petroleum ether
at 30-35° for 24 h (35%) [30].

m.p. 165-167° [30]; ^{1}H NMR [30].

[3-(2-Butenyl)-4,6-dihydroxy-2-methoxyphenyl]phenylmethanone

[96836-13-4] $C_{18}H_{18}O_4$ mol.wt. 298.34

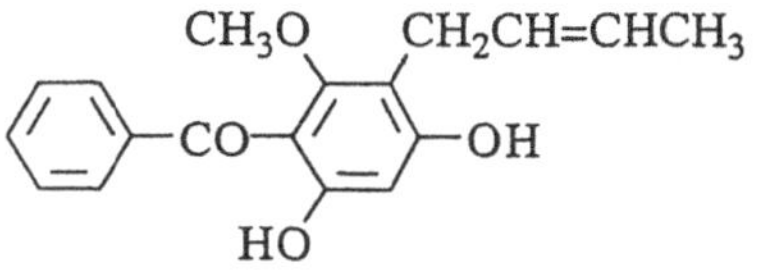

Synthesis

-Obtained by condensation of 2,4-dihydroxy-
6-methoxybenzophenone with 1,3-butadiene in the
presence of orthophosphoric acid in petroleum ether
at 30-35° for 24 h (35%) [30].

m.p. 116-118° [30]; ^{1}H NMR [30].

[2,4-Dihydroxy-6-methoxy-3-(1-methyl-2-propenyl)phenyl]phenylmethanone

[96836-11-2] $C_{18}H_{18}O_4$ mol.wt. 298.34

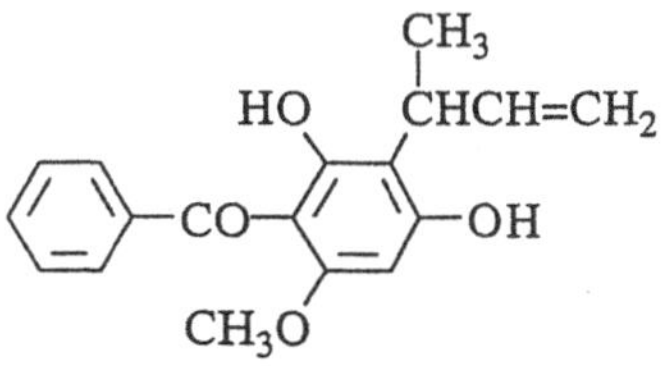

Synthesis

-Obtained (poor yield) by condensation of 2,4-di-
hydroxy-6-methoxybenzophenone with 1,3-butadiene
in the presence of orthophosphoric acid in petroleum
ether at 30-35° for 24 h (15%) [30].

m.p. 117-119° [30]; ^{1}H NMR [30].

[2,6-Dihydroxy-4-[(3-methyl-2-butenyl)oxy]phenyl]phenylmethanone

[70219-83-9] $C_{18}H_{18}O_4$ mol.wt. 298.34

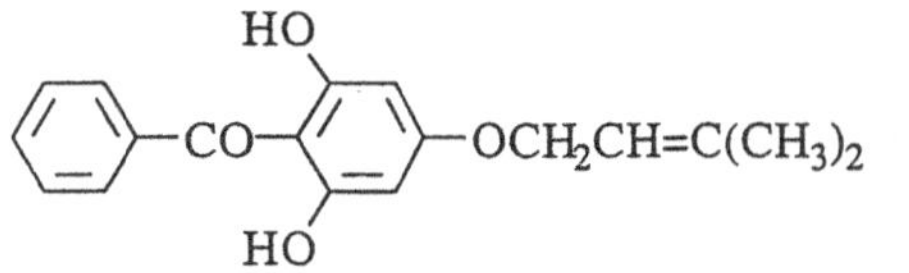

Syntheses

-Obtained (trace) by reaction of 2-methyl-
3-buten-2-ol with 2,4,6-trihydroxy-
benzophenone in the presence of boron
trifluoride-etherate in dioxane at 25-30° (<1%)
[1043].

-Also obtained (trace) by reaction of prenyl bromide with 2,4,6-trihydroxybenzophenone in the
presence of sodium methoxide in refluxing methanol for 3 h (<1%) [1044].

Isolation from natural source

-From *Leontonyx squarrosus* DC (Compositae) [192].

colourless oil [192]. This product is impure or in a metastable state.
m.p.   121° [1043], 120-121° [1044];
^{1}H NMR [192, 1043, 1044], IR [192, 1043, 1044], UV [1043, 1044], MS [192].

[2,4-Dihydroxy-5-(1,1-dimethylpropyl)phenyl]phenylmethanone

$C_{18}H_{20}O_3$ mol.wt. 284.36

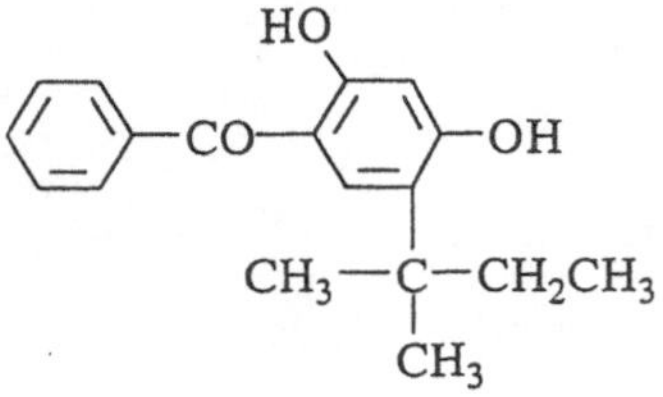

Synthesis

-Preparation by alkylation of resbenzophenone with
2-methylbutene in benzene in the presence of p-toluenesulfonic acid [587].

m.p. 116° [587, 588];   Spectra (NA).

[3,6-Dihydroxy-2-(phenylsulfonyl)phenyl]phenylmethanone

[145746-55-0] $C_{19}H_{14}O_5S$ mol.wt. 354.38

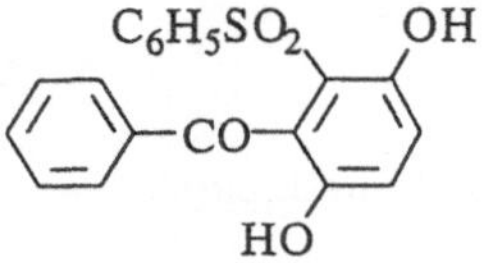

Synthesis

-Preparation by shaking an aqueous solution of sodium
benzenesulfinate with a solution of benzoyl-1,4-benzo-
quinone and trifluoroacetic acid in methylene chloride for
4 h at r.t. (85%) [223].

m.p.   207-210° [223];   ^{1}H NMR [223], IR [223], MS [223].

(5-Cyclohexyl-2,4-dihydroxyphenyl)phenylmethanone

$C_{19}H_{20}O_3$ mol.wt. 296.37

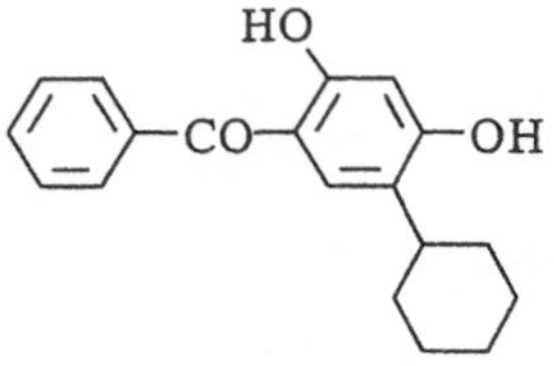

Synthesis

-Preparation by reaction of benzoic acid with 4-cyclohexyl-
resorcinol in hydrofluoric acid at 100° in a stainless steel
bomb [1406].

m.p. 164° [1406];   Spectra (NA).

[2,4-Dihydroxy-6-methoxy-3-(3-methyl-2-butenyl)phenyl]phenylmethanone

[81490-45-1] $C_{19}H_{20}O_4$ mol.wt. 312.37

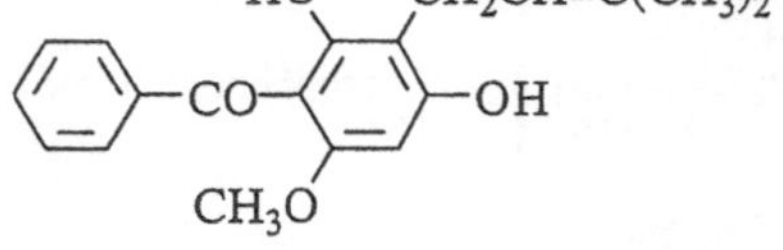

Synthesis

-Obtained by prenylation of 2,4-dihydroxy-
6-methoxybenzophenone (Isocotoin) with
2-methyl-3-buten-2-ol in the presence of boron
trifluoride-etherate [24].

m.p.   160-161° [24];   ^{1}H NMR [24].

[4,6-Dihydroxy-2-methoxy-3-(3-methyl-2-butenyl)phenyl]phenylmethanone

[81490-46-2] $C_{19}H_{20}O_4$ mol.wt. 312.37

CH$_3$O   CH$_2$CH=C(CH$_3$)$_2$

⟨benzene⟩–CO–⟨ring⟩–OH

HO

Synthesis

-Obtained by prenylation of 2,4-dihydroxy-
6-methoxybenzophenone (Isocotoin) with
2-methyl-3-buten-2-ol in the presence of boron
trifluoride-etherate [24].

m.p. 104-105° [24]; ^{1}H NMR [24].

(5-Hexyl-2,4-dihydroxyphenyl)phenylmethanone

[59746-92-8] $C_{19}H_{22}O_3$ mol.wt. 298.38

HO

⟨benzene⟩–CO–⟨ring⟩–OH

C$_6$H$_{13}$

Syntheses

-Preparation by reaction of benzotrichloride with 4-hexyl-
resorcinol in hydrofluoric acid in the presence of water at
-10° for 4 h, then at r.t. overnight (80%) [393].
-Preparation by reaction of benzoic acid with 4-hexyl-
resorcinol in the presence of boron trifluoride in
tetrachloroethane on a steam bath for 4 h [1406].
-Also refer to: [384].

m.p. 81-82° [1406]; Spectra (NA).

[4-(Benzoyloxy)-2,6-dihydroxyphenyl]phenylmethanone

$C_{20}H_{14}O_5$ mol.wt. 334.33

HO

⟨benzene⟩–CO–⟨ring⟩–OCOC$_6$H$_5$

HO

Synthesis

-Obtained by action of benzoyl chloride with phlorobenzophenone (2,4,6-trihydroxybenzophenone) in the
presence of 1% potassium hydroxide aqueous solution
at 0° (14%) [248].

m.p. 186° [248]; Spectra (NA).

[2,4-Dihydroxy-3-(phenylmethyl)phenyl]phenylmethanone

$C_{20}H_{16}O_3$ mol.wt. 304.35

HO   CH$_2$C$_6$H$_5$

⟨benzene⟩–CO–⟨ring⟩–OH

Syntheses

-Obtained (poor yield) by reaction of benzonitrile with
2-benzylresorcinol (8%) (Hoesch reaction) [954].
-Also obtained (poor yield) by hydrolysis of 3-benzyl-
4-(benzyloxy)-2-hydroxybenzophenone (SM) with
concentrated hydrochloric acid in boiling acetic acid for 2 h (2%). SM was obtained by reaction of
benzyl chloride with resbenzophenone in the presence of potassium hydroxide in refluxing
methanol for 5 h [954].
-Preparation by Friedel-Crafts acylation of 2-benzylresorcinol with benzoyl chloride in methylene
chloride in the presence of aluminium chloride [1432].

m.p.   159-160° [954];   Spectra (NA).

[2,4-Dihydroxy-5-(phenylmethyl)phenyl]phenylmethanone

$C_{20}H_{16}O_3$ mol.wt. 304.35

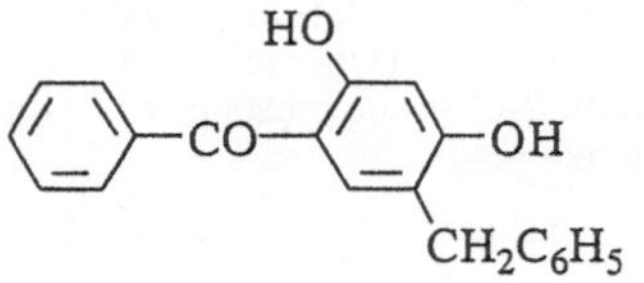

Synthesis

-Preparation by reaction of benzotrichloride with 4-benzyl-resorcinol [1460].

m.p. and Spectra (NA).

[2,4-Dihydroxy-3-(1-phenylethyl)phenyl]phenylmethanone

[65221-07-0] $C_{21}H_{18}O_3$ mol.wt. 318.37

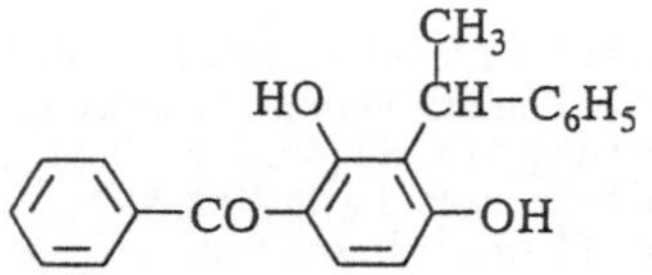

Synthesis

-Refer to: [1411].

m.p. (NA);   UV [1411].

[2,4-Dihydroxy-5-(1-phenylethyl)phenyl]phenylmethanone

[43221-41-6] $C_{21}H_{18}O_3$ mol.wt. 318.37

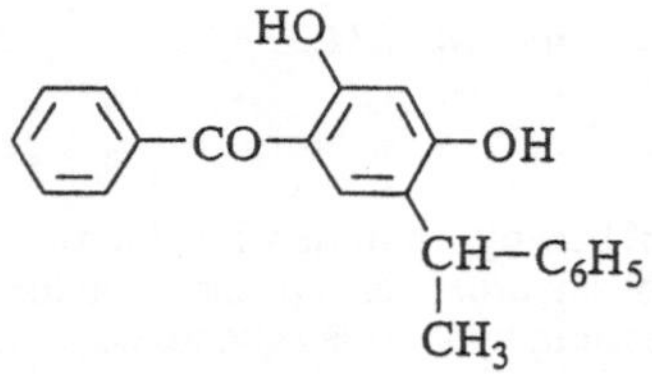

Synthesis

-Refer to: [1411].

m.p. and Spectra (NA).

[4,6-Dihydroxy-3-methyl-2-(phenylmethoxy)phenyl]phenylmethanone

[74627-92-2] $C_{21}H_{18}O_4$ mol.wt. 334.37

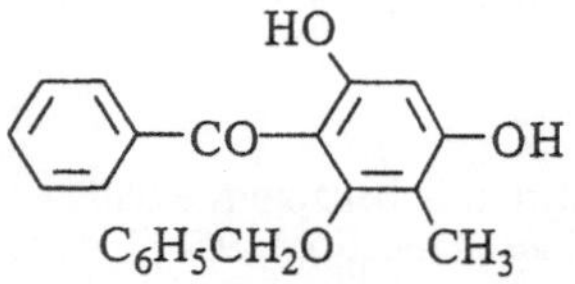

Synthesis

-Preparation from 2-(benzyloxy)-4,6-dimethoxy-3-methyl-benzophenone by treatment with 10% hydrochloric acid in refluxing methanol for 40 min (89%) [57].

m.p. 155° [57]; ^{1}H NMR [57], MS [57].

[5-(2-Ethylhexyl)-2,4-dihydroxyphenyl]phenylmethanone

$C_{21}H_{26}O_3$ mol.wt. 326.44

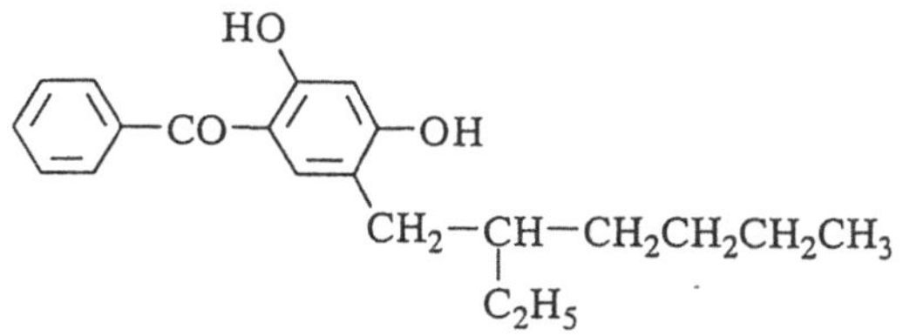

Synthesis

-Preparation by reaction of benzotrichloride with 4-(2-ethylhexyl)-
resorcinol [1460].

m.p. and Spectra (NA).

[2,4-Dihydroxy-3,5-bis(3-methyl-2-butenyl)phenyl]phenylmethanone

[69443-76-1] $C_{23}H_{26}O_3$ mol.wt. 350.46

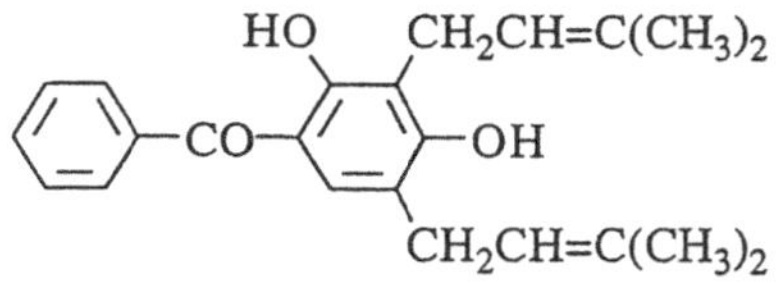

Synthesis

-Obtained (poor yield) by reaction of 2-methyl-
3-buten-2-ol with resbenzophenone in the presence
of boron trifluoride-etherate in dioxane at 60-70°
for 2 h (3%) [874].

m.p. 74-75° [874]; ^{1}H NMR [874], IR [874], UV [874].

[2,6-Dihydroxy-3-(3-methyl-2-butenyl)-4-[(3-methyl-2-butenyl)oxy]phenyl]-phenylmethanone

[83611-01-2] $C_{23}H_{26}O_4$ mol.wt. 366.46

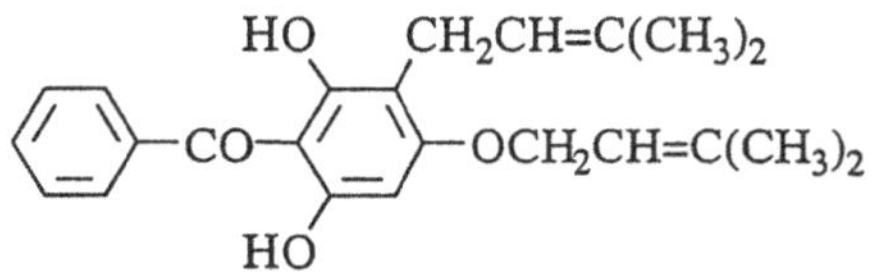

Synthesis

-Obtained (trace) by reaction of 2-methyl-
3-buten-2-ol with 2,4,6-trihydroxy-
benzophenone in the presence of boron
trifluoride-etherate in dioxane at 25-30° (<1%)
[1043].

brown oil [1043];   b.p. (NA); ^{1}H NMR [1043], IR [1043], UV [1043].

[4-[(3,7-Dimethyl-2,6-octadienyl)oxy]-2,6-dihydroxyphenyl]phenylmethanone (E)

[70219-85-1] $C_{23}H_{26}O_4$ mol.wt. 366.46

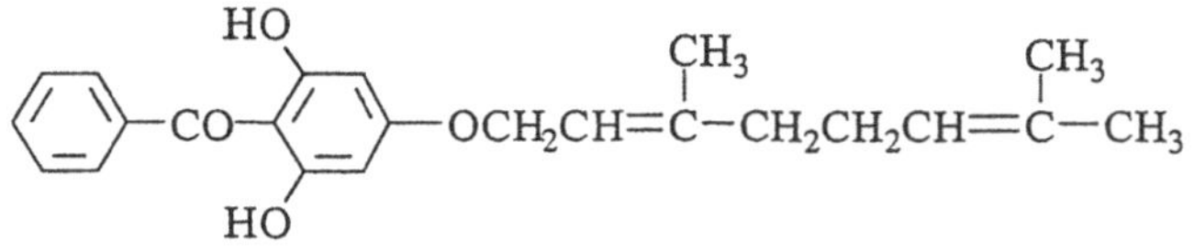

Synthesis

-Not yet described.

Isolation from natural source

-From *Leontonyx spathulatus* Less. (Compositae) [192, 1099].

colourless oil [192];   b.p. (NA); ^{1}H NMR [192], IR [192].

[2,4-Dihydroxy-6-methoxy-3,5-bis(3-methyl-2-butenyl)phenyl]phenylmethanone
(Vismiaphenone A)

[76444-61-6] $C_{24}H_{28}O_4$ mol.wt. 380.48

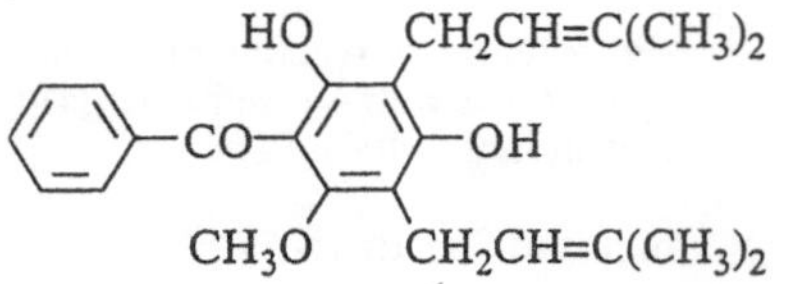

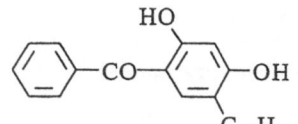

Syntheses

-Obtained by prenylation of 2,4-dihydroxy-
6-methoxybenzophenone (Isocotoin) with
2-methyl-3-buten-2-ol in the presence of boron
trifluoride-etherate [24, 1043].

-Also obtained first by methylation of 6-hydroxy-
3,5-diprenyl-2,4-ditosyloxybenzophenone with dimethyl sulfate in the presence of potassium
carbonate in refluxing acetone for 12 h. Then, the resulting compound was treated with 10%
ethanolic potassium hydroxide for 2 h at 50-55° (17%) [1043].

Isolation from natural source

-From the berries of *Vismia decipiens* Schlecht-Cham. (Guttiferae) [332].

oil [24, 332, 1043];   b.p. (NA);
^{1}H NMR [24, 332, 1043], IR [24, 332, 1043], UV [24, 332, 1043], MS [332].

[2,4-Dihydroxy-5-(dodecyloxy)phenyl]phenylmethanone

$C_{25}H_{34}O_4$ mol.wt. 398.54

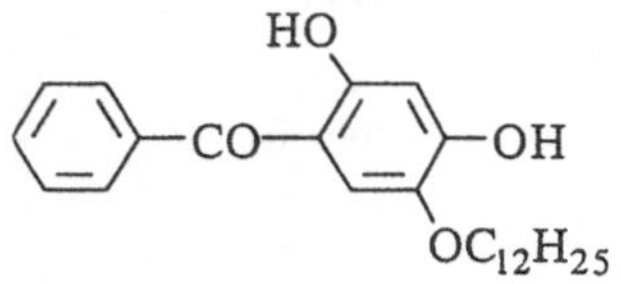

Synthesis

-Refer to: [384].

m.p. and Spectra (NA).

(2,4-Dihydroxy-5-octadecylphenyl)phenylmethanone

$C_{31}H_{46}O_3$ mol.wt. 466.70

Synthesis

-Preparation by reaction of benzotrichloride with 4-stearyl-
resorcinol [1460].

m.p. and Spectra (NA).

2.1.2. *Substituents located on the other ring*

(2,4-Dihydroxyphenyl)(2,4,6-trinitrophenyl)methanone

[188347-38-8] $C_{13}H_7N_3O_9$ mol.wt. 349.21

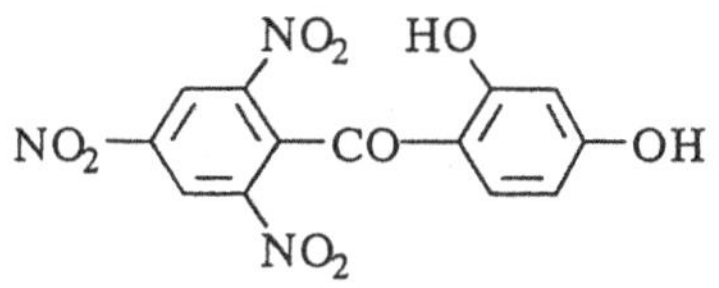

Synthesis

-Refer to: [964] (compound 51) NSC 338104 (no comments).

m.p. and Spectra (NA).

(4-Bromo-3-chlorophenyl)(2,5-dihydroxyphenyl)methanone

[161463-54-3] $C_{13}H_8BrClO_3$ mol.wt. 327.56

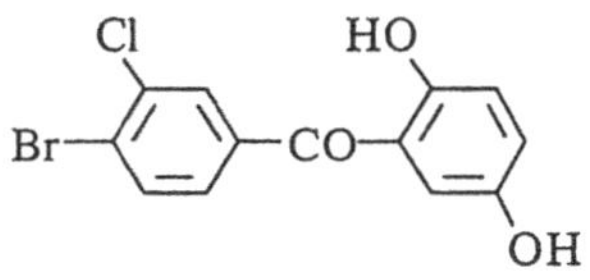

Synthesis

-Preparation by demethylation of 4-bromo-3-chloro-2',5'-dimethoxybenzophenone (SM) with boron tribromide in methylene chloride at 0° for 15-17 h (70-95%). SM was prepared by reaction of 4-bromo-3-chlorobenzoic acid with 1,4-dimethoxybenzene in the presence of polyphosphoric acid at 60-70° for 6-7 h (40-83%) [1279].

m.p. (NA); ^{1}H NMR [1279], IR [1279].

(2-Chloro-4-nitrophenyl)(2,5-dihydroxyphenyl)methanone

[37884-01-8] $C_{13}H_8ClNO_5$ mol.wt. 293.66

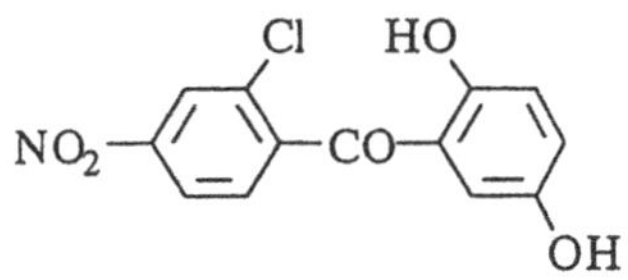

Synthesis

-Preparation by demethylation of 2-chloro-2',5'-dimethoxy-4-nitrobenzophenone with an excess of boiling pyridinium chloride for 6 h (40%) [1153].

m.p. 265° [1153];   Spectra (NA).

(2,4-Dichlorophenyl)(2,4-dihydroxyphenyl)methanone

$C_{13}H_8Cl_2O_3$ mol.wt. 283.11

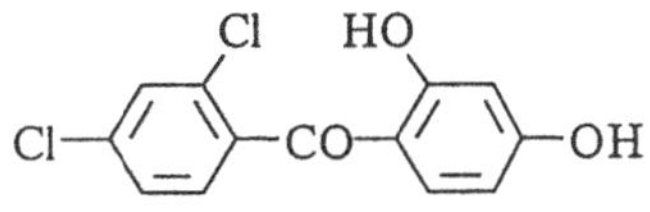

Synthesis

-Preparation by reaction of 2,4-dichlorobenzotrichloride with resorcinol [1460], in hydrofluoric acid in the presence of water at -10°, then at 15° overnight and at 80° for 30 min [393].

m.p. and Spectra (NA).

(2,4-Dichlorophenyl)(2,5-dihydroxyphenyl)methanone

[37884-00-7] $C_{13}H_8Cl_2O_3$ mol.wt. 283.11

Syntheses

-Preparation by reaction of 2,4-dichlorobenzotrichloride with
hydroquinone in hydrofluoric acid in the presence of water
at -10°, then at 15° overnight and at 80° for 30 min [393].
-Also obtained by condensation reaction of hydroquinone
dimethyl ether with excess of 2,4-dichlorobenzoic acid in the
presence of polyphosphoric acid [1280].

-Preparation by demethylation of 2,4-dichloro-2',5'-dimethoxybenzophenone,
*with boiling excess pyridinium chloride for 1 h (95%) [1153];
*with boron tribromide in methylene chloride at 0° (77%) [1280].

m.p.   145° [1153], 126° [1280]; ^{1}H NMR [1280], IR [1280].

(3,4-Dichlorophenyl)(2,4-dihydroxyphenyl)methanone

[36419-34-8] $C_{13}H_8Cl_2O_3$ mol.wt. 283.11

Synthesis

-Preparation by reaction of 3,4-dichlorobenzoic acid with
resorcinol in the presence of boron trifluoride in
tetrachloroethane on a steam bath for 4 h [1406].
-Refer to: [279, 1067].

m.p.   188° [1406], 186° [1067];   UV [279, 1067, 1406].

(2-Bromophenyl)(2,4-dihydroxyphenyl)methanone

                                          $C_{13}H_9BrO_3$ mol.wt. 293.12

Synthesis

-Preparation by reaction of o-bromobenzotrichloride with
resorcinol [1460].

m.p. and Spectra (NA).

(4-Bromophenyl)(2,4-dihydroxyphenyl)methanone

[3286-88-2] $C_{13}H_9BrO_3$ mol.wt. 293.12

Synthesis

-Preparation by reaction of p-bromobenzonitrile with
resorcinol in the presence of zinc chloride and
hydrochloric acid in ethyl ether in an ice bath for
1 or 2 days, followed by hydrolysis of the ketimine
hydrochloride so formed with boiling water (64%) [735], (50%) [755].
-Also refer to: Chem. Abstr., **127**, 108921f (1997).

m.p.   169° [735], 164° [755];   Spectra (NA).

(4-Bromophenyl)(2,5-dihydroxyphenyl)methanone

$C_{13}H_9BrO_3$ mol.wt. 293.12

Synthesis

-Preparation by condensation of p-bromobenzoic acid with hydroquinone in the presence of aluminium chloride and sodium chloride at 180-200° for 2 min (32%) [220].

m.p. 153° [220];   Spectra (NA).

(2-Chlorophenyl)(2,4-dihydroxyphenyl)methanone

[50685-40-0] $C_{13}H_9ClO_3$ mol.wt. 248.67

Syntheses

-Preparation by reaction of o-chlorobenzoyl chloride with O,O-bis(trimethylsilyl)resorcinol in the presence of stannic chloride in refluxing methylene chloride for 2 h (80%) [649].
-Preparation by Fries rearrangement of m-methoxyphenyl o-chlorobenzoate with aluminium chloride without solvent at 180° (40%) [171].
-Preparation by reaction of o-chlorobenzotrichloride with resorcinol in dilute isopropanol at 70-80° (85%) [1460].
-Preparation by condensation of resorcinol and o-chlorobenzoic acid with boron trifluoride-etherate in carbon tetrachloride [1102], according to [1020].

m.p.   138° [171], 135°5-136°5 [1460], 131-132° [649];   Spectra (NA).

(2-Chlorophenyl)(2,5-dihydroxyphenyl)methanone

[37883-99-1] $C_{13}H_9ClO_3$ mol.wt. 248.67

Syntheses

-Preparation by reaction of o-chlorobenzoyl chloride with O,O-bis(trimethylsilyl)hydroquinone in the presence of stannic chloride in refluxing methylene chloride for 2 h (84%) [649].
-Preparation by demethylation of 2'-chloro-2,5-dimethoxy-benzophenone with an excess of boiling pyridinium chloride for 1 h (63%) [1153].
-Also obtained by photochemical addition of o-chlorobenzaldehyde to 1,4-benzoquinone in benzene in the presence of benzophenone for 5 days (78%) [764].

m.p.   160° [1153];   Spectra (NA).

(2-Chlorophenyl)(2,6-dihydroxyphenyl)methanone

[100334-93-8] $C_{13}H_9ClO_3$ mol.wt. 248.67

Synthesis

-Preparation by hydrolysis of 8-(o-chlorobenzoyl)-7-hydroxy-4-methylcoumarin (SM) with sodium hydroxide in refluxing dilute ethanol for 3 h (72%). SM was obtained by Fries rearrangement of 7-(o-chlorobenzoyloxy)-4-methylcoumarin with aluminium chloride, first at 180°, then at 185-190° for 1.5 h (65%, m.p. 165°). -Refer to: Chem. Abstr., **114**, 42490n (1991)*.

m.p. 140°*; IR*.

(3-Chlorophenyl)(2,4-dihydroxyphenyl)methanone

$C_{13}H_9ClO_3$ mol.wt. 248.67

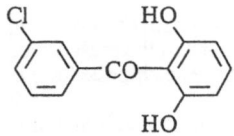

Synthesis

-Preparation by reaction of m-chlorobenzonitrile with resorcinol (52%) (Hoesch reaction) [1017].

m.p. 197-197°5 [1017];   Spectra (NA).

(3-Chlorophenyl)(2,5-dihydroxyphenyl)methanone

[161463-59-8] $C_{13}H_9ClO_3$ mol.wt. 248.67

Synthesis

-Preparation by demethylation of 3'-chloro-2,5-dimethoxy-benzophenone (SM) with boron tribromide in methylene chloride at 0° for 15-17 h (70-95%). SM was prepared by reaction of m-chlorobenzoic acid with 1,4-dimethoxy-benzene in the presence of polyphosphoric acid at 60-70° for 6-7 h (40-83%) [1279].

m.p. (NA);   ¹H NMR [1279], IR [1279].

(3-Chlorophenyl)(2,6-dihydroxyphenyl)methanone

[131425-89-3] $C_{13}H_9ClO_3$ mol.wt. 248.67

Synthesis

-Preparation by hydrolysis of 8-(m-chlorobenzoyl)-7-hydroxy-4-methylcoumarin (SM) with sodium hydroxide in refluxing dilute ethanol for 3 h (67%). SM was obtained by Fries rearrangement of 7-(m-chlorobenzoyloxy)-4-methylcoumarin with aluminium chloride, first at 185°, then at 190-195° for 1 h (64%, m.p. 245°). -Refer to: Chem. Abstr., **114**, 42490n (1991)*.

m.p. 150°*; IR*.

(4-Chlorophenyl)(2,4-dihydroxyphenyl)methanone

[18239-10-6] $C_{13}H_9ClO_3$ mol.wt. 248.67

Syntheses

-Obtained by reaction of p-chlorobenzonitrile with resorcinol in the presence of zinc chloride and hydrochloric acid in ethyl ether (Hoesch reaction) in ice during 24 h and hydrolysis of 4'-chloro-2,4-dihydroxybenzophenone imide hydrochloride so formed (72%) with boiling water for 30 min (46%) [735], (39%) [1017].

-Also obtained by reaction of p-chlorobenzoic acid with resorcinol,
*in hydrofluoric acid at 100° in a stainless steel bomb [1406];
*in the presence of Zeolite-H-beta in refluxing p-chlorotoluene (162°) for 7 h, with stirring and
  azeotropic removal of water (4%) [582].
-Preparation by reaction of 4-chlorobenzotrichloride with resorcinol in 40% aqueous isopropanol
  solution at 70-80° (95%) [1460].
-Also refer to: [279, 524, 919, 1067].

m.p. 155° [735], 152° [1067], 151-152° [1017], 151° [1406], 150° [1079],
      149°5-150°5 [1460];   UV [1067, 1079, 1406].

(4-Chlorophenyl)(2,5-dihydroxyphenyl)methanone

[91290-75-4] $C_{13}H_9ClO_3$ mol.wt. 248.67

Syntheses

-Preparation by demethylation of 4'-chloro-2,5-dimethoxy-
  benzophenone (SM) with boron tribromide in methylene
  chloride,
*at 0° (95%) [1280];
*first at -30° for 30 min, then at 22° for 5 h. SM was
obtained by Friedel-Crafts acylation of hydroquinone dimethyl ether with p-chlorobenzoyl chloride
in methylene chloride in the presence of aluminium chloride at 0° for 8 h (85%) [1046].
-Also obtained by photochemical addition of p-chlorobenzaldehyde to 1,4-benzoquinone in benzene
in the presence of benzophenone for 5 days (73%) [764].

m.p. 132° [1280]; ^{1}H NMR [1280], IR [1280].

(4-Chlorophenyl)(2,6-dihydroxyphenyl)methanone

[29627-01-8] $C_{13}H_9ClO_3$ mol.wt. 248.67

Synthesis

-Preparation by hydrolysis of 8-(p-chlorobenzoyl)-
  7-hydroxy-4-methylcoumarin (SM) with sodium hydroxide
  in refluxing dilute ethanol for 3 h (71%). SM was obtained
  by Fries rearrangement of 7-(p-chlorobenzoyloxy)-
  4-methylcoumarin with aluminium chloride, first at 180°,
then at 185-190° for 1.5 h (60%, m.p. 214°). -Refer to: Chem. Abstr., 114, 42490n (1991)*.

m.p. 130°*;   IR*.

(4-Chlorophenyl)(3,4-dihydroxyphenyl)methanone

[134612-84-3] $C_{13}H_9ClO_3$ mol.wt. 248.67

Syntheses

-Preparation by demethylation of 4'-chloro-4-hydroxy-
  3-methoxybenzophenone with hydrobromic acid in
  refluxing aqueous acetic acid [164].
-Preparation by Friedel-Crafts reaction of p-chlorobenzoyl
  chloride with veratrole [847].

m.p. 190° [847], 174-176° [164];   Spectra (NA).

(2,4-Dihydroxyphenyl)(2-fluorophenyl)methanone

[19390-38-6]   C$_{13}$H$_9$FO$_3$ mol.wt. 232.21

Synthesis

-Preparation by reaction of 2-fluorobenzoyl chloride with resorcinol dimethyl ether in the presence of aluminium chloride in ethylene dichloride at 60° for 90 min [1270], (59%) [1272].

m.p.   112-114° [1272], 109-111° [1270];   Spectra (NA).

(2,4-Dihydroxyphenyl)(4-fluorophenyl)methanone

[84794-97-8]   C$_{13}$H$_9$FO$_3$ mol.wt. 232.21

Synthesis

-Preparation by Friedel-Crafts acylation of resorcinol with p-fluorobenzoyl chloride in the presence of aluminium chloride in ethylene dichloride at 50° for 1.5 h (90%) [394].

-Also refer to: [1386].

m.p.   171-172° [394];   Spectra (NA).

(2,5-Dihydroxyphenyl)(2-fluorophenyl)methanone

[176547-98-1]   C$_{13}$H$_9$FO$_3$ mol.wt. 232.21

Synthesis

-Preparation by total demethylation of 2'-fluoro-2,5-dimethoxybenzophenone (SM) with boron tribromide in methylene chloride, first at -70°, then at 25° for 24 h (86%). SM was obtained by Friedel-Crafts acylation of hydroquinone dimethyl ether with o-fluorobenzoyl chloride in methylene chloride in the presence of aluminium chloride at 0° for 4 h (79%) [1048].

m.p.   118-119° [1048]; ^{1}H NMR [1048].

(2,5-Dihydroxyphenyl)(3-fluorophenyl)methanone

[161463-61-2]   C$_{13}$H$_9$FO$_3$ mol.wt. 232.21

Synthesis

-Preparation by demethylation of 3'-fluoro-2,5-dimethoxybenzophenone (SM) with boron tribromide in methylene chloride at 0° for 15-17 h (70-95%) [1279] or at 22° [1048]. SM was prepared from hydroquinone dimethyl ether,

*by reaction with m-fluorobenzoic acid in the presence of polyphosphoric acid at 60-70° for 6-7 h (40-83%) [1279];
*by reaction with m-fluorobenzoyl chloride in methylene chloride in the presence of aluminium chloride at 0° (78%) [1048].

m.p. (NA); ^{1}H NMR [1279], IR [1279].

(2,5-Dihydroxyphenyl)(4-fluorophenyl)methanone

[83235-21-6] $C_{13}H_9FO_3$ mol.wt. 232.21

Syntheses

-Preparation by total demethylation of 4'-fluoro-2,5-di-
methoxybenzophenone (SM) with boron tribromide in
methylene chloride, first at -30° for 30 min, then at 22° for
5 h (80%). SM was obtained by Friedel-Crafts acylation of
hydroquinone dimethyl ether with p-fluorobenzoyl
chloride in methylene chloride in the presence of aluminium chloride at 0° for 8 h (90%) [1046, 1048].
-Also obtained by photochemical addition of p-fluorobenzaldehyde to 1,4-benzoquinone in benzene
in the presence of benzophenone for 5 days (61%) [764].
-Also obtained by UV light irradiation of α-hydroxy(p-fluorobenzyl)-1,4-benzoquinone in benzene
for 72 h (45%) [4].
-Also refer to: [33].

   m.p. 140-141° [4, 1046]; ^1H NMR [4, 1046], IR [4].

(3,5-Dihydroxyphenyl)(4-fluorophenyl)methanone

[148253-51-4] $C_{13}H_9FO_3$ mol.wt. 232.21

Synthesis

-Preparation by total demethylation of 3,5-dimethoxy-
4'-fluorobenzophenone (SM) with 48% hydrobromic acid in
refluxing acetic acid under nitrogen for 20 h (91%) [272,
550]. SM was obtained by oxidation of 3,5-dimethoxy-
4'-fluorobenzhydrol with pyridinium chlorochromate
in the presence of sodium acetate in methylene chloride at r.t. under nitrogen for 30 min (84%)
[272].

   m.p. 142-143° [272, 550];
   ^1H NMR [272, 550], ^{13}C NMR [272, 550], IR [272, 550], MS [272, 550].

(2,4-Dihydroxyphenyl)(3-nitrophenyl)methanone

                                      $C_{13}H_9NO_5$ mol.wt. 259.22

Syntheses

-Obtained by reaction of m-nitrobenzonitrile with resorcinol
in the presence of zinc chloride and hydrochloric acid in
ethyl ether, followed by hydrolysis of the resulting ketimine
hydrochloride in boiling water (73%) [1079], for
30 min [1453], (11%) [1452] (Hoesch reaction).
-Preparation by reaction of m-nitrobenzotrichloride with resorcinol [1460].

   m.p. 228° [1079, 1452, 1453];   Spectra (NA).

(2,4-Dihydroxyphenyl)(4-nitrophenyl)methanone

[6994-40-7]   C13H9NO5   mol.wt. 259.22

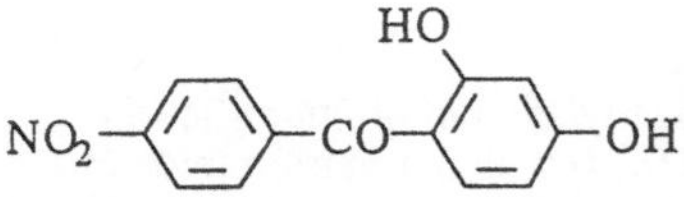

Synthesis

-Preparation by reaction of p-nitrobenzonitrile with resorcinol in the presence of zinc chloride and hydrochloric acid in ethyl ether, followed by hydrolysis of the ketimine hydrochloride so formed with boiling water for 30 min [755, 1344, 1453], (17%) [1452] (Hoesch reaction).

m.p.   203° [1452, 1453], 200° [755];   Spectra (NA).

(3,4-Dihydroxyphenyl)(3-nitrophenyl)methanone

[203060-36-0]   C13H9NO5   mol.wt. 259.22

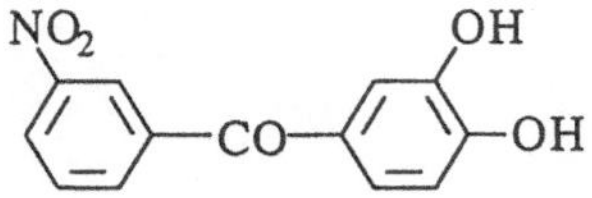

Synthesis

-Refer to: Chem. Abstr., **128**, 181082h (1998).

m.p. and Spectra (NA).

(3,4-Dihydroxyphenyl)(4-nitrophenyl)methanone

[203060-35-9]   C13H9NO5   mol.wt. 259.22

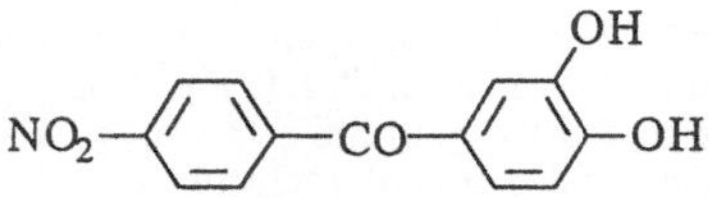

Synthesis

-Refer to: Chem. Abstr., **128**, 181082h (1998).

m.p. and Spectra (NA).

(2,4-Dihydroxyphenyl)[2-(trifluoromethyl)phenyl]methanone

C14H9F3O3   mol.wt. 282.22

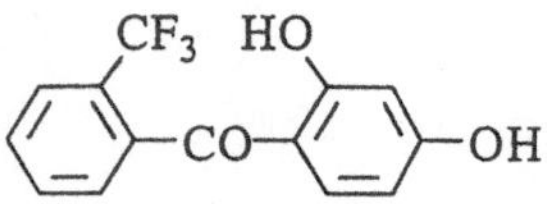

Synthesis

-Preparation by reaction of o-(trifluoromethyl)benzoyl chloride with resorcinol in the presence of aluminium chloride in refluxing carbon disulfide for 20 h [501].

m.p.   168-168°5 [501];   Spectra (NA).

(2,4-Dihydroxyphenyl)[3-(trifluoromethyl)phenyl]methanone

C14H9F3O3   mol.wt. 282.22

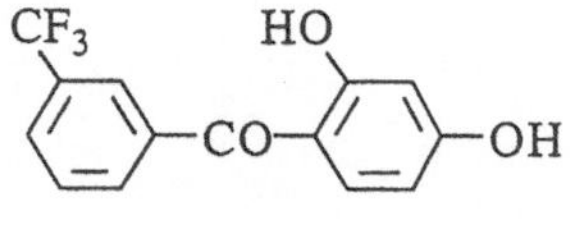

Synthesis

-Preparation by reaction of m-(trifluoromethyl)benzoyl chloride with resorcinol in the presence of aluminium chloride in refluxing carbon disulfide for 20 h [501].

m.p.   175°5-176° [501];   Spectra (NA).

(2,5-Dihydroxyphenyl)[2-(trifluoromethyl)phenyl]methanone

[161463-62-3] $C_{14}H_9F_3O_3$ mol.wt. 282.22

Synthesis

-Preparation by demethylation of 2,5-dimethoxy-2'-(trifluoromethyl)benzophenone (SM) with boron tribromide at
0° for 15-17 h (70-95%). SM was prepared by reaction of
2-(trifluoromethyl)benzoic acid with 1,4-dimethoxybenzene
in the presence of polyphosphoric acid at 60-70° for 6-7 h
(40-83%) [1279].

m.p. (NA); ^{1}H NMR [1279], IR [1279].

(3-Chloro-4-methylphenyl)(2,5-dihydroxyphenyl)methanone

$C_{14}H_{11}ClO_3$ mol.wt. 262.69

Synthesis

-Preparation by demethylation of 3'-chloro-2,5-dimethoxy-
4'-methylbenzophenone (SM) with boron tribromide in
methylene chloride at 0° for 15-17 h (70-95%). SM was
prepared by reaction of 3-chloro-4-methylbenzoic acid with
1,4-dimethoxybenzene in the presence of polyphosphoric
acid at 60-70° for 6-7 h (40-83%) [1279].

m.p. (NA); ^{1}H NMR [1279], IR [1279].

(2,5-Dihydroxyphenyl)(3-fluoro-4-methylphenyl)methanone

[161463-63-4] $C_{14}H_{11}FO_3$ mol.wt. 246.24

Synthesis

-Preparation by demethylation of 2,5-dimethoxy-3'-fluoro-
4'-methylbenzophenone (SM) with boron tribromide in
methylene chloride at 0° for 15-17 h (70-95%). SM was
prepared by reaction of 3-fluoro-4-methylbenzoic acid with
1,4-dimethoxybenzene in the presence of polyphosphoric
acid at 60-70° for 6-7 h (40-83%) [1279].

m.p. (NA); ^{1}H NMR [1279], IR [1279].

(2,5-Dihydroxyphenyl)(2-methyl-3-nitrophenyl)methanone

[161463-58-7] $C_{14}H_{11}NO_5$ mol.wt. 273.25

Synthesis

-Preparation by demethylation of 2,5-dimethoxy-2'-methyl-
3'-nitrobenzophenone (SM) with boron tribromide in
methylene chloride at 0° for 15-17 h (70-95%). SM was
prepared by reaction of 2-methyl-3-nitrobenzoic acid with
1,4-dimethoxybenzene in the presence of polyphosphoric
acid at 60-70° for 6-7 h (40-83%) [1279].

m.p. (NA);   ¹H NMR [1279], IR [1279].

(2,5-Dihydroxyphenyl)(3-methyl-4-nitrophenyl)methanone

[153907-08-5] $C_{14}H_{11}NO_5$ mol.wt. 273.25

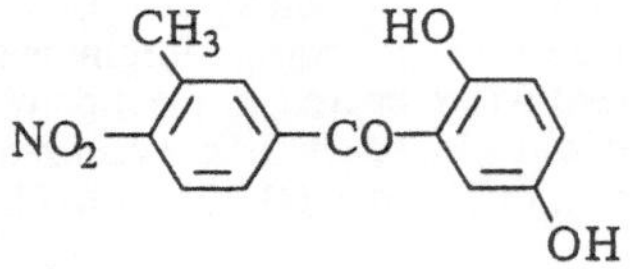

Synthesis

-Preparation by demethylation of 2,5-dimethoxy-
3'-methyl-4'-nitrobenzophenone with boron tribromide in
methylene chloride at 0° (81%) [1280].

m.p. 150° [1280];   ¹H NMR [1280], IR [1280].

(2,4-Dihydroxyphenyl)(2-methylphenyl)methanone

[14446-07-2] $C_{14}H_{12}O_3$ mol.wt. 228.25

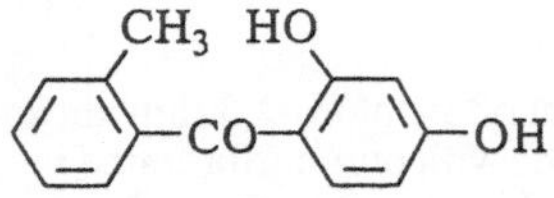

Synthesis

-Preparation by reaction of o-toluic acid (2-methylbenzoic
acid) with resorcinol,
*in the presence of zinc chloride for 20 min at 160° (42%),
(Nencki reaction) [824];
*in the presence of Amberlyst-15 in refluxing 4-chlorotoluene (162°) for 2 h (91%) [582];
*in the presence of Zeolite-H-beta (previously activated at 400°) in refluxing mesitylene or
4-chlorotoluene for 3 h (with removal of water) (76%) [582, 584].
-Also refer to: [583].

m.p.   126-127° [824], 121-123° [582, 584];   ¹H NMR [582, 584].

(2,4-Dihydroxyphenyl)(3-methylphenyl)methanone

$C_{14}H_{12}O_3$ mol.wt. 228.25

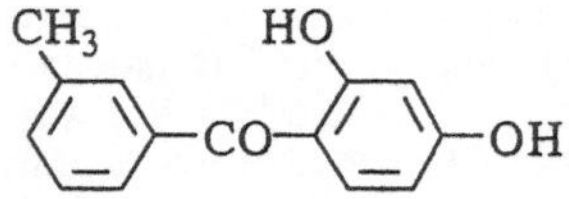

Synthesis

-Preparation by reaction of m-toluic acid with resorcinol,
*in the presence of zinc chloride and phosphorous
oxychloride at 65° for 3 h [982];
*in the presence of zinc chloride at 140° for 20 min (22%)
[167] (Nencki reaction).

m.p.   168° [167];   Spectra (NA).

(2,4-Dihydroxyphenyl)(4-methylphenyl)methanone

[40444-43-7] $C_{14}H_{12}O_3$ mol.wt. 228.25

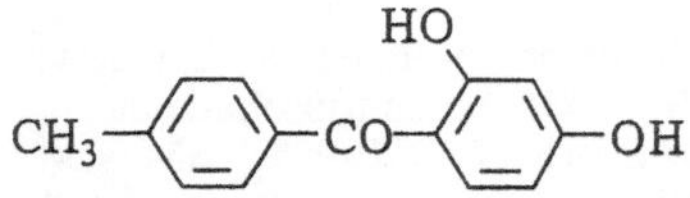

Synthesis

-Preparation by reaction of p-toluic acid (4-methyl-
benzoic acid) with resorcinol,
*in the presence of zinc chloride for 20 min at 165°
(25%) (Nencki reaction) [823];
*in the presence of Zeolite-H-beta in refluxing 4-chlorotoluene (162°) for 16-18 h (70%) [582];

*in the presence of boron trifluoride in tetrachloroethane on a steam bath for 4 h [1406].
-Also refer to: [9, 523, 524, 571, 713, 737, 742, 958, 1101].

    m.p.   139° [1406], 138° [823];   UV [1406].

(2,5-Dihydroxyphenyl)(2-methylphenyl)methanone

[83235-18-1] $C_{14}H_{12}O_3$ mol.wt. 228.25

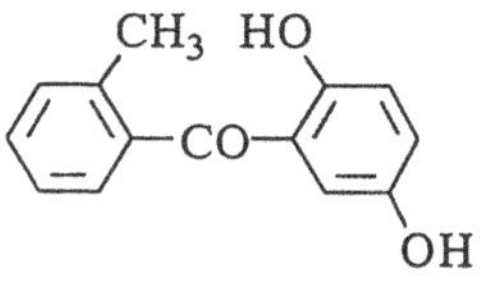

Syntheses

-Obtained by photochemical addition of o-tolualdehyde to
  1,4-benzoquinone in benzene in the presence of
  benzophenone for 5 days (65%) [764].
-Obtained by UV light irradiation of α-hydroxy(o-methyl-
  benzyl)-1,4-benzoquinone in benzene for 72 h (35%) [4].

-Also refer to: [33].

    m.p.   106-108° [4];   ¹H NMR [4], IR [4].

(2,5-Dihydroxyphenyl)(3-methylphenyl)methanone

[83235-19-2] $C_{14}H_{12}O_3$ mol.wt. 228.25

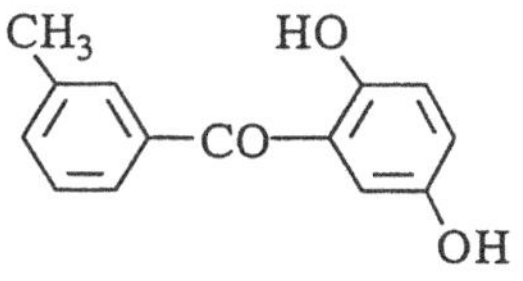

Synthesis

-Obtained by UV light irradiation of α-hydroxy(m-methyl-
  benzyl)-1,4-benzoquinone in benzene for 72 h (37%) [4].
-Also refer to: [33].

    m.p. 114-116° [4];   ¹H NMR [4], IR [4].

(2,5-Dihydroxyphenyl)(4-methylphenyl)methanone

[83235-20-5] $C_{14}H_{12}O_3$ mol.wt. 228.25

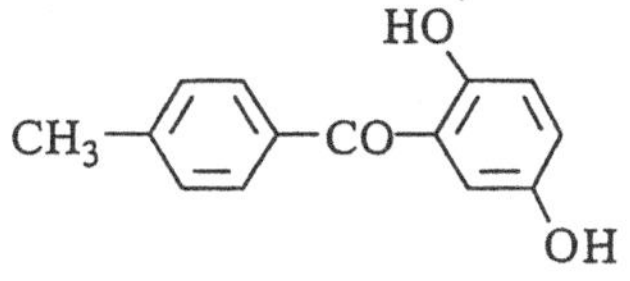

Syntheses

-Obtained by photochemical addition of p-tolualdehyde to
  1,4-benzoquinone in benzene in the presence of
  benzophenone (79%) [764].
-Obtained by UV light irradiation of α-hydroxy(p-methyl-
  benzyl)-1,4-benzoquinone in benzene for 72 h (43%) [4].

    m.p.   137-139° [4];   ¹H NMR [4], IR [4].

(2,6-Dihydroxyphenyl)(2-methylphenyl)methanone

                                          $C_{14}H_{12}O_3$ mol.wt. 228.25

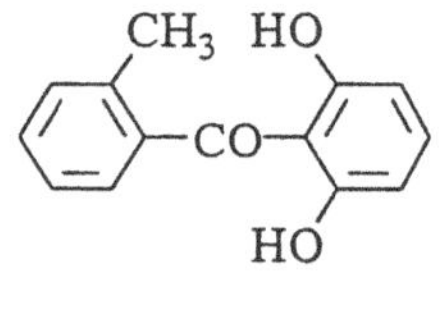

Synthesis

-Obtained from 8-(o-toluoyl)-4-methylumbelliferone by
  refluxing with N aqueous sodium hydroxide for 30 min
  (54%) [824].

    m.p. 140° [824];   Spectra (NA).

(2,6-Dihydroxyphenyl)(3-methylphenyl)methanone

C$_{14}$H$_{12}$O$_3$ mol.wt. 228.25

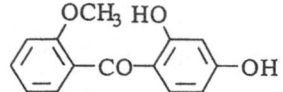

Synthesis

-Obtained from 8-(m-toluoyl)-4-methylumbelliferone by
  refluxing with N aqueous sodium hydroxide for 30 min
  (41%) [167].

m.p. 145° [167];   Spectra (NA).

(2,6-Dihydroxyphenyl)(4-methylphenyl)methanone

C$_{14}$H$_{12}$O$_3$ mol.wt. 228.25

Synthesis

-Obtained from 8-(p-toluoyl)-4-methylumbelliferone by
  refluxing with N aqueous sodium hydroxide for 30 min
  (45%) [823].

m.p. 125° [823];   Spectra (NA).

(2,4-Dihydroxyphenyl)(2-methoxyphenyl)methanone

[79215-32-0]   C$_{14}$H$_{12}$O$_4$ mol.wt. 244.25

Syntheses

-Preparation by oxidation of 6-acetoxy-2,3-bis(2-methoxy-
  phenyl)benzofuran with chromium trioxide in refluxing
  acetic acid for 30 min, followed by saponification of the
  resulting keto ester (65%) with potassium hydroxide in
  refluxing ethanol for 1 h (66%) [579].
-Preparation by condensation of resorcinol and o-anisic acid with boron trifluoride-etherate in
  carbon tetrachloride [1102], according to [1020].
-Also obtained by hydrolysis of various substituted keto anils* with potassium hydroxide in
  refluxing ethanol for 8 h [1402],
*4-(N-phenyl-o-anisimidoyl)resorcinol;
*4-(N-o-tolyl-o-anisimidoyl)resorcinol;
*4-[N-(p-methoxyphenyl)-o-anisimidoyl]resorcinol;
*4-[N-(p-ethoxyphenyl)-o-anisimidoyl]resorcinol.
-Also refer to: [1386].

   m.p. 247-248° [1402], 175° [579]. One of the reported melting points is obviously
   wrong.   IR [579].

(2,4-Dihydroxyphenyl)(3-methoxyphenyl)methanone

C$_{14}$H$_{12}$O$_4$ mol.wt. 244.25

Syntheses

-Preparation by reaction of m-anisoyl chloride with
  resorcinol in the presence of aluminium chloride in
  nitrobenzene at r.t. for 48 h (86%) [650].

-Obtained (poor yield) by condensation of m-anisic acid with resorcinol in the presence of zinc chloride at 160° for 10 min (Nencki reaction) (4%) [650].

m.p. 176° [650]; Spectra (NA).

(2,4-Dihydroxyphenyl)(4-methoxyphenyl)methanone

[5298-27-1]   C14H12O4   mol.wt. 244.25

Syntheses

-Preparation by reaction of p-methoxybenzoic acid (p-anisic acid) with resorcinol,
*in the presence of Zeolite-H-beta in refluxing p-chlorotoluene (162°) for 3 h, with stirring and azeotropic removal of water (55%) [582];
*in the presence of zinc chloride at 160° (Nencki reaction) [748];
*in the presence of zinc chloride and phosphorous oxychloride during 3 h at 65° [982];
*in the presence of zinc chloride and a mixture of polyphosphoric acid/85% phosphoric acid (60:40) at 40°. Then, during 1.5 h, phosphorous trichloride was added and the mixture heated at 60° for 16 h (98%) [1302];
*in the presence of boron trifluoride without solvent at 160° for 2 h in a sealed tube (67%) [993] or in tetrachloroethane on a steam bath for 4 h (81%) [1406];
*in hydrofluoric acid at 100° in a stainless steel bomb [1406].
-Preparation by condensation of p-anisoyl chloride with resorcinol in the presence of aluminium chloride in nitrobenzene during 48 h at r.t. (73%) [650].
-Also obtained by hydrolysis of various substituted keto anils* with potassium hydroxide in refluxing ethanol for 8 h [1402],
*4-(N-phenyl-p-anisimidoyl)resorcinol;
*4-(N-o-tolyl-p-anisimidoyl)resorcinol;
*4-[N-(p-methoxyphenyl)-p-anisimidoyl]resorcinol;
*4-[N-(p-ethoxyphenyl)-p-anisimidoyl]resorcinol.
-Also refer to: [1067, 1104].

m.p. 165° [748, 1406], 164° [993], 163°4-164°8 [1302], 163° [1067],
160° [650], 158-159° [1402];
UV [1067, 1406]; paper chromatography [911].

(2,5-Dihydroxyphenyl)(2-methoxyphenyl)methanone

[140660-43-1]   C14H12O4   mol.wt. 244.25

Syntheses

-Obtained by photochemical addition of o-anisaldehyde to 1,4-benzoquinone in benzene in the presence of benzophenone for 5 days (72%) [764].
-Preparation by irradiation of a solution of 1,4-benzo-quinone and salicylaldehyde in benzene under nitrogen for 5 days (62%) [762].

-Also refer to: [728].

m.p. 145-148° [762];
1H NMR [762], IR [762], MS [762]; TLC [762].

(2,5-Dihydroxyphenyl)(4-methoxyphenyl)methanone

[160720-40-1] $C_{14}H_{12}O_4$ mol.wt. 244.25

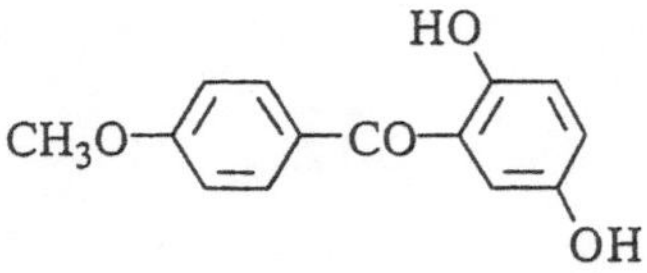

Synthesis

-Obtained by photochemical addition of p-anisaldehyde to 1,4-benzoquinone in benzene in the presence of benzophenone for 5 days (77%) [764].

m.p. and Spectra (NA).

(2,4-Dihydroxyphenyl)[4-(methylsulfonyl)phenyl]methanone

[36419-33-7] $C_{14}H_{12}O_5S$ mol.wt. 292.31

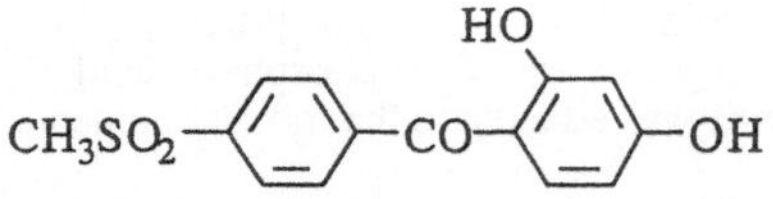

Synthesis

-Refer to: [1067].

m.p. 202° [1067];   UV [1067].

(2,4-Dihydroxyphenyl)(4-ethenylphenyl)methanone

[66787-22-2] $C_{15}H_{12}O_3$ mol.wt. 240.26

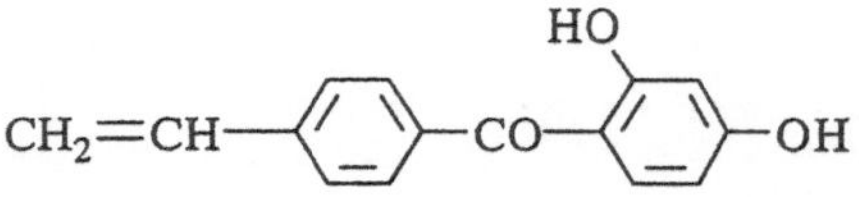

Syntheses

-Preparation by reaction of aqueous potassium hydroxide with 2,4-dihydroxy-4'-(2-bromo-ethyl)benzophenone in the presence of hydroquinone in refluxing methanol for 1.5 h with nitrogen bubbling (40%) [675].
-Preparation, first by dehydrobromination of 2,4-diacetoxy-4'-(1-bromoethyl)benzophenone with tri-n-butylamine in the presence of picric acid as polymerization inhibitor in refluxing N,N-di-methylacetamide (140°) for 80 min, under nitrogen, then hydrolysis of the resulting new keto ester (2,4-diacetoxy-4'-vinylbenzophenone) (42%) with sodium bicarbonate in refluxing aqueous methanol for 1 h (33%) [118].

m.p.   96° [118, 1376], 91-93° [675];
^{1}H NMR [118, 675], IR [118, 675], UV[118, 675], MS [675].

[4-(2-Bromoethyl)phenyl](2,4-dihydroxyphenyl)methanone

[80167-04-0] $C_{15}H_{13}BrO_3$ mol.wt. 321.17

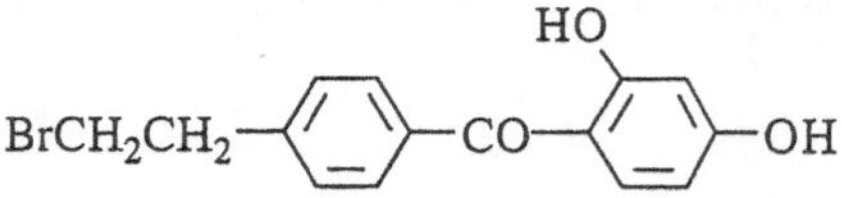

Synthesis

-Preparation by reaction of p-(2-bromoethyl)-benzonitrile with resorcinol (Hoesch reaction) (53%) [675].

m.p.   113-115° [675];   Spectra (NA).

(3-Chloro-4,5-dimethylphenyl)(2,5-dihydroxyphenyl)methanone

[161463-57-6] $C_{15}H_{13}ClO_3$ mol.wt. 276.72

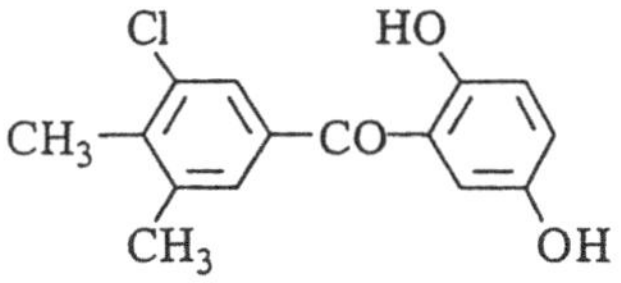

Synthesis

N.B.: This compound is mentioned in [Chem. Abstr., **122**, 187188v (1995)]. Nevertheless, the benzophenone in question is not described in the original paper [1279]. Actually, in this one, all substituted benzophenones without exception have the two *ortho* positions to the carbonyl group occupied.

m.p. and Spectra (NA).

(2,4-Dihydroxyphenyl)(2,4-dimethylphenyl)methanone

[36130-59-3] $C_{15}H_{14}O_3$ mol.wt. 242.27

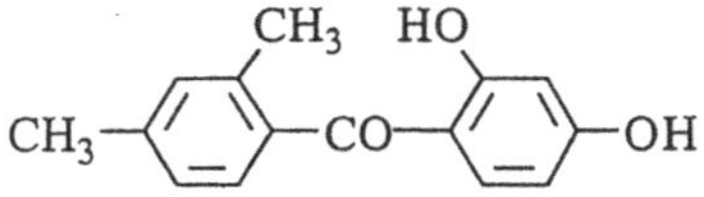

Synthesis

-Refer to: [387].

m.p. and Spectra (NA).

(2,4-Dihydroxyphenyl)(2,6-dimethylphenyl)methanone

[147809-19-6] $C_{15}H_{14}O_3$ mol.wt. 242.27

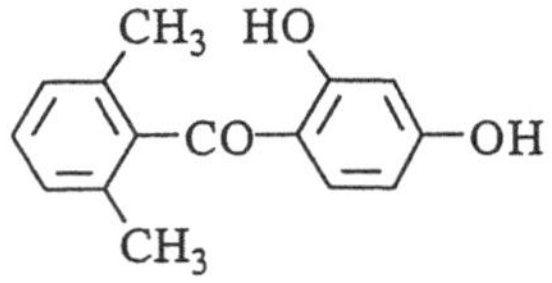

Synthesis

-Obtained by Friedel-Crafts acylation of resorcinol with 2,6-dimethylbenzoic acid in refluxing p-chlorotoluene (162°), with stirring and azeotropic removal of water,
*in the presence of Amberlyst-15 for 2 h (98%) [582];
*in the presence of Zeolite-H-beta for 5 h (32%) [582].

-Also refer to: [583].

m.p. and Spectra (NA).

(2,4-Dihydroxyphenyl)(3,5-dimethylphenyl)methanone

[36419-35-9] $C_{15}H_{14}O_3$ mol.wt. 242.27

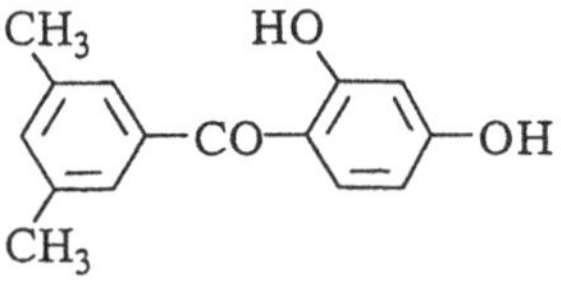

Synthesis

-Refer to: [279] (Japanese patent) and [1067].

m.p. 158° [1067];   UV [1067].

(2,4-Dihydroxyphenyl)(4-ethylphenyl)methanone

[66802-91-3] $C_{15}H_{14}O_3$ mol.wt. 242.27

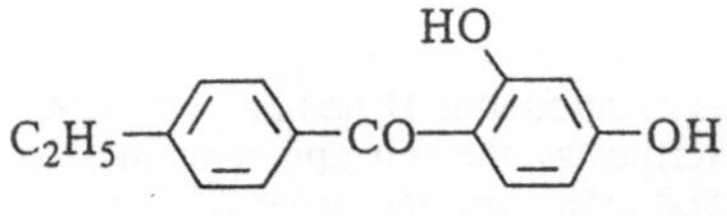

Synthesis

-Preparation by reaction of resorcinol with p-ethyl-
benzoic acid in tetrachloroethane in the presence of
boron trifluoride at 80° for 4 h (59%) [118].
-Also refer to: [1374, 1375].

m.p.   111-113° [118];   b.p.$_{0.1}$ 195-200° [118];
^{1}H NMR [118], IR [118], UV [118].

(2,4-Dihydroxyphenyl)(3,4-dimethoxyphenyl)methanone

[128996-02-1] $C_{15}H_{14}O_5$ mol.wt. 274.27

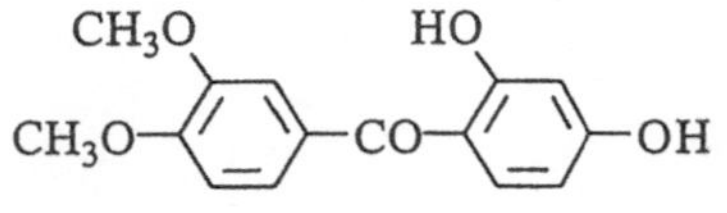

Syntheses

-Preparation by reaction of veratric acid with resorcinol
in the presence of zinc chloride at 160° for 7 min
(24%) (Nencki reaction) [650].
-Preparation by reaction of veratroyl chloride with
resorcinol in the presence of aluminium chloride in nitrobenzene during 3 days (60-64%) [650].
-Also obtained from β-(3-hydroxy-4-veratroylphenoxy)propionic acid with aqueous sodium
hydroxide solution [1274].

m.p.   177° [650], 149° [1274]. One of the reported melting points is obviously wrong.
Spectra (NA).

(2,5-Dihydroxyphenyl)(2,4-dimethoxy-6-methylphenyl)methanone

[78044-94-7] $C_{16}H_{16}O_5$ mol.wt. 288.30

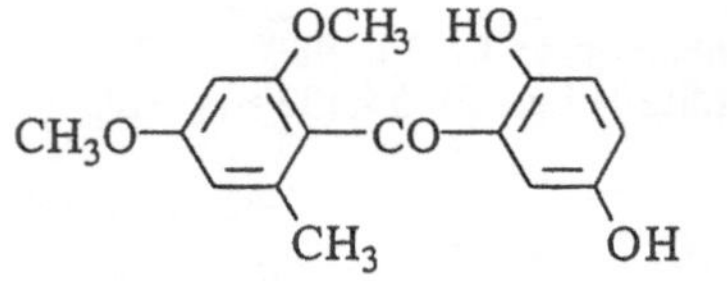

Synthesis

-Obtained by photo-Fries rearrangement of 4-hydroxy-
phenyl 2,4-dimethoxy-6-methylbenzoate in benzene
under nitrogen for 4 h (33%) [816].

m.p.   228-229° [816]; ^{1}H NMR [816], IR [816], UV [816], MS [816].

(3,4-Dihydroxyphenyl)(2,4,6-trimethoxyphenyl)methanone *(Cotogenin)*

$C_{16}H_{16}O_6$ mol.wt. 304.30

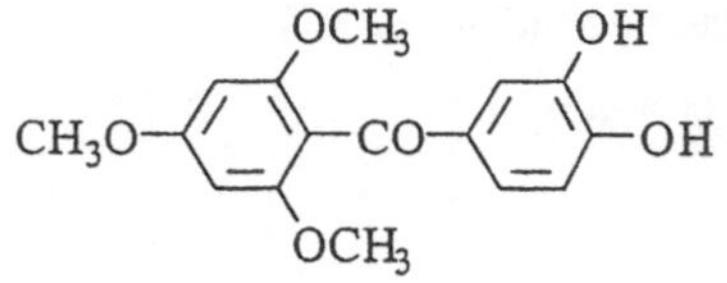

Synthesis

-Obtained by reaction of 3,4-(diacetoxy)benzonitrile
with phloroglucinol trimethyl ether in the presence of
zinc chloride and hydrochloric acid, followed by
hydrolysis of the resulting ketimine hydrochloride
(16%) (Hoesch reaction) [604].

-Also refer to: [275, 277].

m.p.   219-220° [604], 217° [277], 210° [278];   Spectra (NA).

(2,4-Dihydroxyphenyl)[4-(1,1-dimethylethyl)phenyl]methanone

[21332-56-9] $C_{17}H_{18}O_3$ mol.wt. 270.33

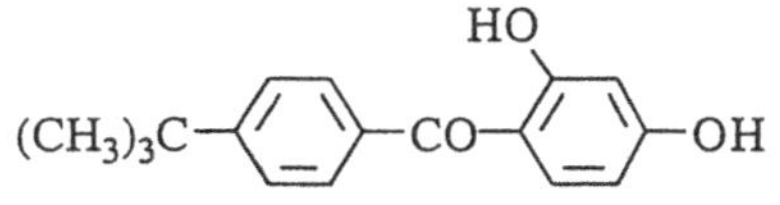

Synthesis

-Refer to: [100, 1067].

m.p. 162° [1067]; UV [1067].

(2,5-Dihydroxyphenyl)[4-(1,1-dimethylethyl)phenyl]methanone

[169696-58-6] $C_{17}H_{18}O_3$ mol.wt. 270.33

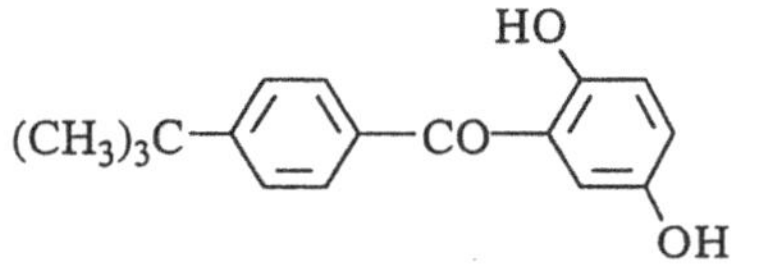

Synthesis

-Preparation by total demethylation of 4'-tert-butyl-
2,5-dimethoxybenzophenone (SM) with boron
tribromide in methylene chloride, first at -30° for
30 min, then at 22° for 5 h (86%). SM was obtained
by Friedel-Crafts acylation of hydroquinone dimethyl
ether with p-tert-butylbenzoyl chloride in methylene chloride in the presence of aluminium chloride
at 0° for 8 h (84%) [1046].

m.p. 115-116° [1046]; ^{1}H NMR [1046].

(2,4-Dihydroxyphenyl)[4-(1-methylbutyl)phenyl]methanone

$C_{18}H_{20}O_3$ mol.wt. 284.36

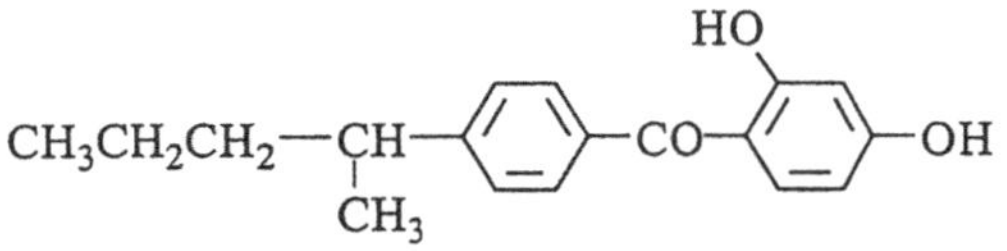

Synthesis

-Preparation by reaction of p-sec-amyl-
benzoic acid with resorcinol in the
presence of boron trifluoride in tetrachloroethane on a steam bath for 4 h
[1406].

b.p.$_{0.75}$ 235-240° [1406]; Spectra (NA).

(3,5-Dihydroxyphenyl)(4-phenoxy-3,5-d_2-phenyl)methanone

[176738-21-9] $C_{19}H_{12}D_2O_4$ mol.wt. 308.33

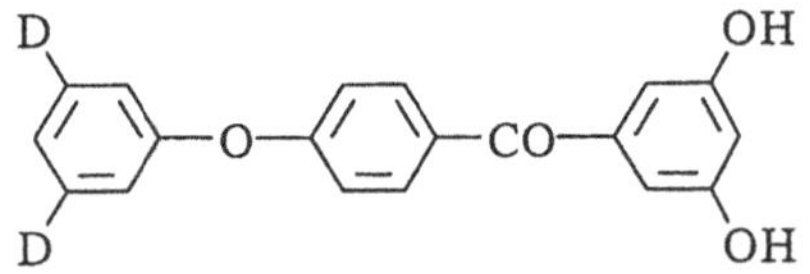

Synthesis

-Preparation by demethylation of 3,5-dimethoxy-
4'-(phenoxy-d_5)benzophenone with 48%
hydrobromic acid in refluxing acetic acid for 16 h
(92%) (compound 21). The acidic treatment
exchanges exclusively the deuterium atoms in the
2", 4" and 6" positions, because of the mesomeric inductive effect of the exocyclic oxygen [550].

m.p. (NA), white solid [550]; ^{1}H NMR [550], ^{13}C NMR [550].

2.1.3. *Substituents located on both rings*

(2,3,4,5,6-Pentafluorophenyl)(2,3,5-trifluoro-4,6-dihydroxyphenyl)methanone

[32541-20-1] $C_{13}H_2F_8O_3$ mol.wt. 358.14

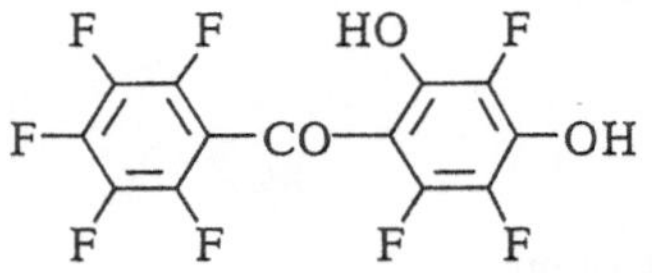

Synthesis

-Preparation by total demethylation of 2,4-dimethoxy-2',3,3',4',5,5',6,6'-octafluorobenzophenone in methylene chloride with aluminium chloride (2 mol) at 20° (73%) [839].

m.p. 118-120° [839];   IR [839].

(2,3-Dichloro-4,5-dihydroxyphenyl)(2-fluorophenyl)methanone

[103843-57-8] $C_{13}H_7Cl_2FO_3$ mol.wt. 301.10

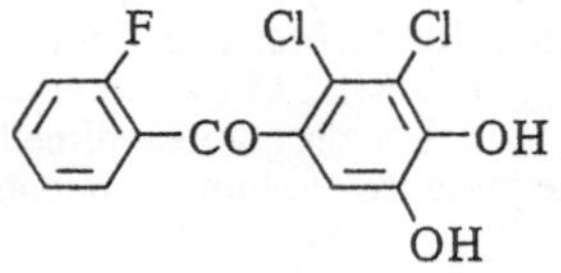

Synthesis

-Preparation by reaction of 2-fluorobenzoyl chloride with 3,4-dichloro-1,2-dihydroxybenzene in the presence of aluminium chloride in refluxing ethylene dichloride during 24 h (75%) [634, 635].

m.p. 164-165° [634, 635];
^{1}H NMR [634, 635], IR [634, 635].

(5-Chloro-2,4-dihydroxy-3-nitrophenyl)(4-chlorophenyl)methanone

$C_{13}H_7Cl_2NO_5$ mol.wt. 328.11

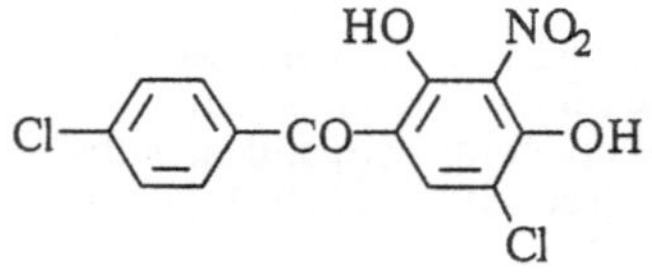

Synthesis

-Obtained by reaction of concentrated nitric acid with 4',5-dichloro-2,4-dihydroxybenzophenone (50%) [933].

m.p. 120° [933];   TLC [933];
^{1}H NMR [933], IR [933], MS [933].

(5-Chloro-2,4-dihydroxyphenyl)(2,4-dichlorophenyl)methanone

$C_{13}H_7Cl_3O_3$ mol.wt. 317.55

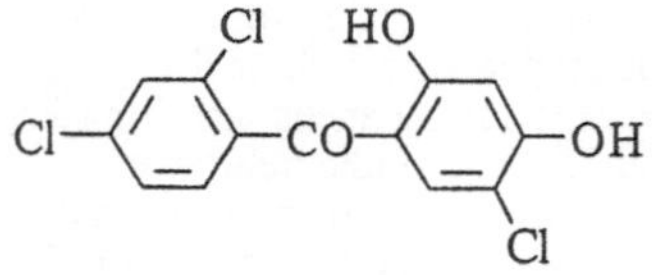

Synthesis

-Preparation by reaction of 2,4-dichlorobenzotrichloride with 4-chlororesorcinol [1460].

m.p. and Spectra (NA).

(2,6-Difluorophenyl)(3,4-dihydroxy-5-nitrophenyl)methanone

[134612-45-6] $C_{13}H_7F_2NO_5$ mol.wt. 295.20

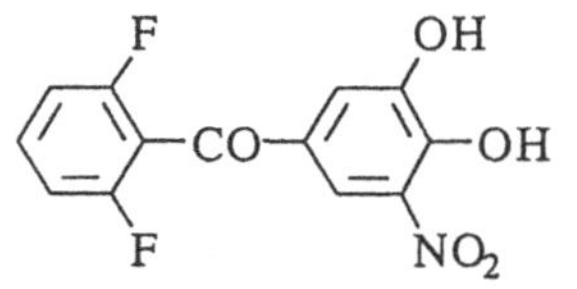

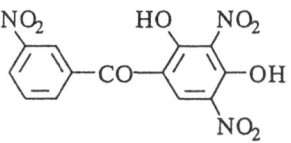

Synthesis

-Preparation by demethylation of 2',6'-difluoro-4-hydroxy-
3-methoxy-5-nitrobenzophenone with hydrobromic acid in
refluxing aqueous acetic acid [164].

m.p. 194-196° [164]; Spectra (NA).

(2,4-Dihydroxy-3,5-dinitrophenyl)(3-nitrophenyl)methanone

$C_{13}H_7N_3O_9$ mol.wt. 349.21

Synthesis

-Preparation by slowly adding (2 h) nitric acid (d = 1.42) to a
solution of resbenzophenone in concentrated sulfuric acid
(d = 1.84) at 5-10°. The mixture was allowed to stand
overnight, then heated at 50° for 30 min (quantitative yield)
[565]. **N.B.:** The first crop from 95% ethanol
contained 12-14% of the 2,4-dihydroxy-3,5-dinitrobenzophenone.

m.p. 178-180° [565]; Spectra (NA).

(3-Bromo-2,5-dihydroxyphenyl)(2-chlorophenyl)methanone

[153907-03-0] $C_{13}H_8BrClO_3$ mol.wt. 327.56

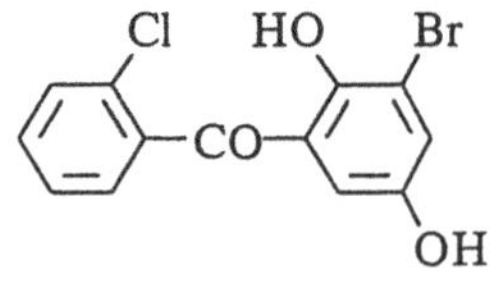

Synthesis

-Preparation by reaction of boron tribromide with 3-bromo-
2'-chloro-2,5-dimethoxybenzophenone in methylene chloride
at 0° (95%) [1280].

m.p. 150° [1280]; ^{1}H NMR [1280], IR [1280].

(3-Bromo-2,5-dihydroxyphenyl)(4-chlorophenyl)methanone

[153907-04-1] $C_{13}H_8BrClO_3$ mol.wt. 327.56

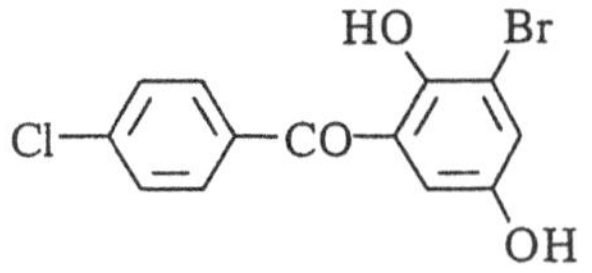

Synthesis

-Preparation by demethylation of 3-bromo-4'-chloro-2,5-di-
methoxybenzophenone with boron tribromide in methylene
chloride at 0° (86%) [1280].

m.p. 90° [1280]; ^{1}H NMR [1280], IR [1280].

(4-Bromophenyl)(4-chloro-2,5-dihydroxyphenyl)methanone

[161463-53-2] $C_{13}H_8BrClO_3$ mol.wt. 327.56

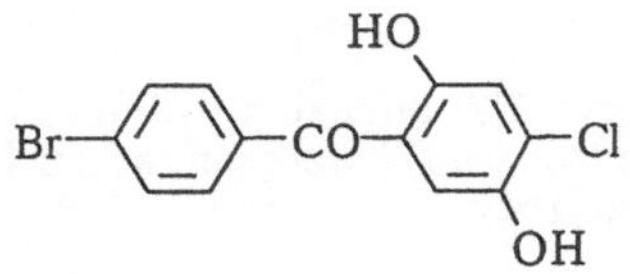

Synthesis

-Preparation by demethylation of 4'-bromo-4-chloro-2,5-di-
methoxybenzophenone (SM) with boron tribromide in
methylene chloride at 0° for 15-17 h (70-95%). SM was
prepared by reaction of 4-bromobenzoic acid with 2-chloro-
1,4-dimethoxybenzene in the presence of polyphosphoric
acid at 60-70° for 6-7 h (40-83%) [1279].

m.p. (NA); ^{1}H NMR [1279], IR [1279].

(4-Chlorophenyl)(5-fluoro-2,3-dihydroxyphenyl)methanone

[92735-05-2] $C_{13}H_8ClFO_3$ mol.wt. 266.65

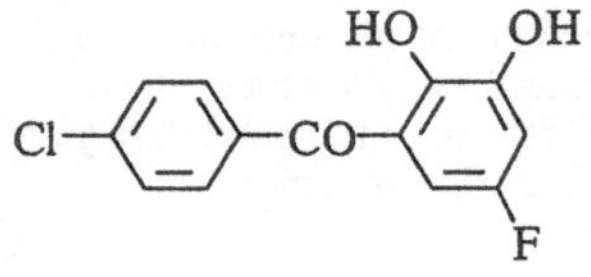

Synthesis

-Obtained by action of dilute hydrogen peroxide with
4'-chloro-5-fluoro-3-formyl-2-hydroxybenzophenone in
refluxing aqueous sodium hydroxide under argon for
30 min [696].

m.p.   124-126° [696]; ^{1}H NMR [696], IR [696].

(2-Chlorophenyl)(3,4-dihydroxy-5-nitrophenyl)methanone

[134612-46-7] $C_{13}H_8ClNO_5$ mol.wt. 293.66

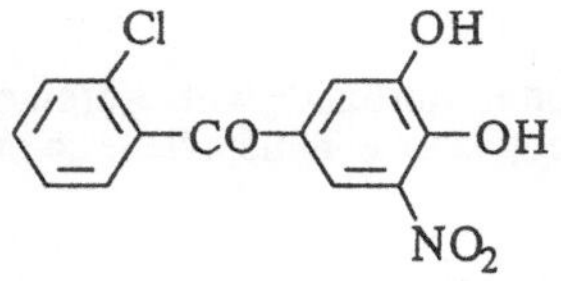

Synthesis

-Preparation by demethylation of 2'-chloro-4-hydroxy-
3-methoxy-5-nitrobenzophenone with hydrobromic acid in
refluxing aqueous acetic acid [164].

m.p.   129-131° [164];   Spectra (NA).

(3-Chlorophenyl)(3,4-dihydroxy-5-nitrophenyl)methanone

[134612-47-8] $C_{13}H_8ClNO_5$ mol.wt. 293.66

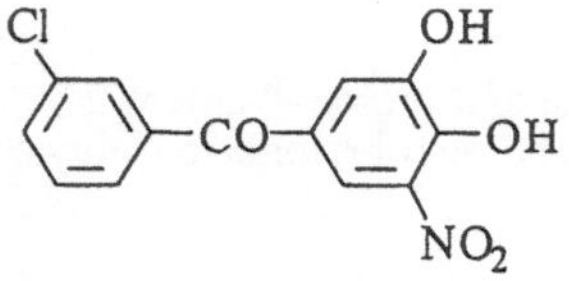

Synthesis

-Preparation by demethylation of 3'-chloro-4-hydroxy-
3-methoxy-5-nitrobenzophenone with hydrobromic acid in
refluxing aqueous acetic acid [164].

m.p.   143-145° [164];   Spectra (NA).

(2-Chloro-4,5-dihydroxyphenyl)(3-chlorophenyl)methanone

[91197-11-4] $C_{13}H_8Cl_2O_3$ mol.wt. 283.11

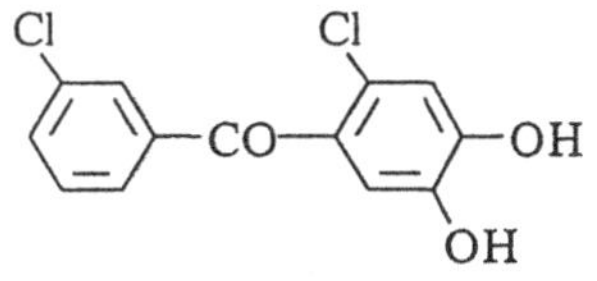

Synthesis

-Preparation by demethylation of 2,3'-dichloro-4,5-di-methoxybenzophenone (SM) with pyridinium chloride for 2 h at 180-200°. SM was obtained by reaction of m-chlorobenzoic acid with 1-chloro-3,4-dimethoxybenzene in the presence of phosphorous pentoxide in methanesulfonic acid for 30 min at 70° [518].

m.p. and Spectra (NA).

(4-Chloro-2,5-dihydroxyphenyl)(3-chlorophenyl)methanone

[161463-60-1] $C_{13}H_8Cl_2O_3$ mol.wt. 283.11

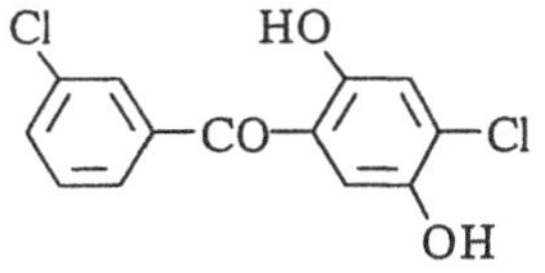

Synthesis

-Preparation by demethylation of 3',4-dichloro-2,5-di-methoxybenzophenone (SM) with boron tribromide in methylene chloride at 0° for 15-17 h (70-95%). SM was prepared by reaction of m-chlorobenzoic acid with 2-chloro-1,4-dimethoxybenzene in the presence of polyphosphoric acid at 60-70° for 6-7 h (40-83%) [1279].

m.p. (NA); ^{1}H NMR [1279], IR [1279].

(5-Chloro-2,3-dihydroxyphenyl)(4-chlorophenyl)methanone

[92735-01-8] $C_{13}H_8Cl_2O_3$ mol.wt. 283.11

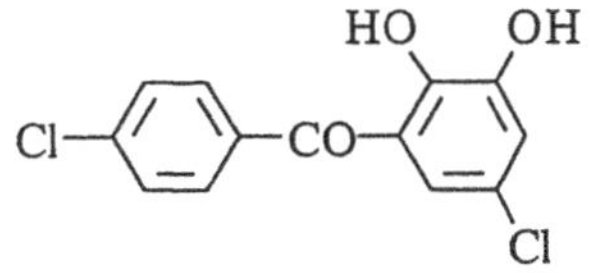

Synthesis

-Obtained by action of dilute hydrogen peroxide with 4',5-dichloro-3-formyl-2-hydroxybenzophenone in refluxing aqueous sodium hydroxide under argon for 30 min [696].

m.p. 133-135° [696]; ^{1}H NMR [696], IR [696].

(5-Chloro-2,4-dihydroxyphenyl)(2-chlorophenyl)methanone

[50685-42-2] $C_{13}H_8Cl_2O_3$ mol.wt. 283.11

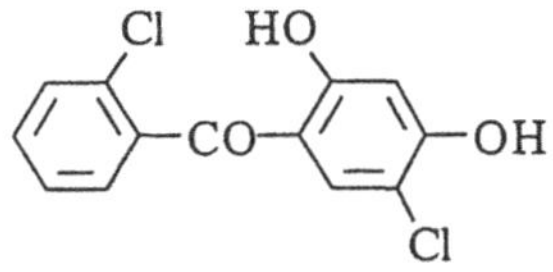

Synthesis

-Preparation by reaction of 2-chlorobenzotrichloride with 4-chlororesorcinol in isopropanol at 70-80° (81%) [1460].

m.p. 187-188° [1460];   Spectra (NA).

(5-Chloro-2,4-dihydroxyphenyl)(4-chlorophenyl)methanone

[50685-41-1] $C_{13}H_8Cl_2O_3$ mol.wt. 283.11

Syntheses

-Preparation by demethylation of 4',5-dichloro-2-hydroxy-
 4-methoxybenzophenone with a 47% aqueous solution of
 hydrobromic acid in refluxing acetic acid for 18 h under
 nitrogen (22%) [933].
-Preparation by reaction of 4-chlorobenzotrichloride with
4-chlororesorcinol in 40% aqueous isopropanol solution at 70-80° (85%) [1460].

m.p. 188°5-189°5 [1460], 170-190° [933]. A typing error probably occurred in the
published data. ^{1}H NMR [933], IR [933], MS [933];   TLC [933].

(3,4-Dihydroxy-5-nitrophenyl)(2-fluorophenyl)methanone

[125628-97-9] $C_{13}H_8FNO_5$ mol.wt. 277.21

Synthesis

-Preparation by demethylation of 2'-fluoro-4-hydroxy-
 3-methoxy-5-nitrobenzophenone with 48% hydrobromic
 acid in refluxing acetic acid for 4 h (86%) [164, 197].
-Also refer to: [161, 835, 854, 1288, 1385].

m.p. 169-171° [164, 197]; ^{1}H NMR [197], MS [197];   LD$_{50}$ [1384].

(3,4-Dihydroxy-5-nitrophenyl)[2-(fluoro-^{18}F)phenyl]methanone

[172546-74-6] $C_{13}H_8FNO_5$ mol.wt. 276.21

Synthesis

-Preparation by demethylation of 2'-[^{18}F] fluoro-3,4-di-
 methoxy-5-nitrobenzophenone (SM) in dimethyl sulfoxide
 with aqueous hydrobromic acid first at r.t. and at 140° for
 30 min. SM was obtained by action of [^{18}F] cesium fluoride
 with 3,4-dimethoxy-2',5-dinitrobenzophenone in
dimethyl sulfoxide for 10 min at 150° in a silicone-coated tube (Vacutainer). -Refer to: Chem.
Abstr., **127**, 17465u (1997).

m.p. and Spectra (NA).

(3,4-Dihydroxy-5-nitrophenyl)(3-fluorophenyl)methanone

[134612-43-4] $C_{13}H_8FNO_5$ mol.wt. 277.21

Synthesis

-Preparation by demethylation of 3'-fluoro-4-hydroxy-
 3-methoxy-5-nitrobenzophenone with hydrobromic acid in
 refluxing aqueous acetic acid [164].

m.p. 124-126° [164];   Spectra (NA).

(3,4-Dihydroxy-5-nitrophenyl)(4-fluorophenyl)methanone

[134612-44-5] $C_{13}H_8FNO_5$ mol.wt. 277.21

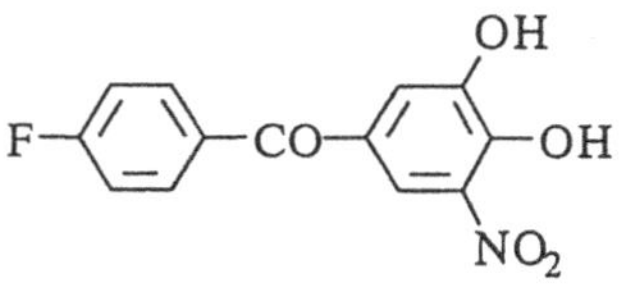

Synthesis

-Preparation by demethylation of 4'-fluoro-4-hydroxy-3-methoxy-5-nitrobenzophenone with hydrobromic acid in refluxing aqueous acetic acid [164].

m.p. 171-173° [164]; Spectra (NA).

(3,4-Dihydroxy-5-nitrophenyl)(2-nitrophenyl)methanone

[190523-00-3] $C_{13}H_8N_2O_7$ mol.wt. 304.22

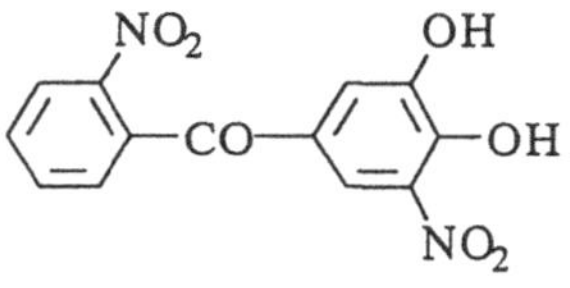

Synthesis

-Preparation by demethylation of 4-hydroxy-3-methoxy-2',5-dinitrobenzophenone with 48% hydrobromic acid in refluxing acetic acid for 6 h (70%). -Refer to: Chem. Abstr., **127**, 17465u (1997)*.

m.p. (NA); ^{1}H NMR*, MS*.

(3-Amino-5-chloro-2,4-dihydroxyphenyl)(4-chlorophenyl)methanone

[87119-05-9] $C_{13}H_9Cl_2NO_3$ mol.wt. 298.12

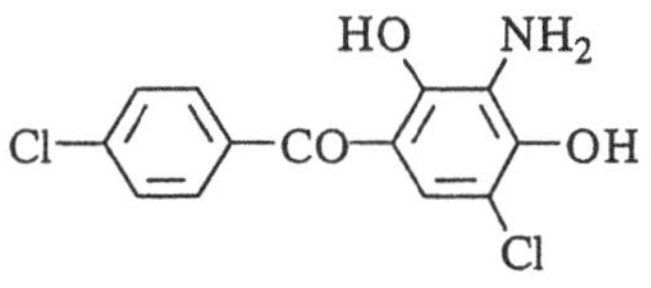

Synthesis

-Obtained from the corresponding hydrochloride described below [933].

m.p. and Spectra (NA).

(3-Amino-5-chloro-2,4-dihydroxyphenyl)(4-chlorophenyl)methanone *(Hydrochloride)*

[87119-06-0] $C_{13}H_9Cl_2NO_3,HCl$ mol.wt. 334.59

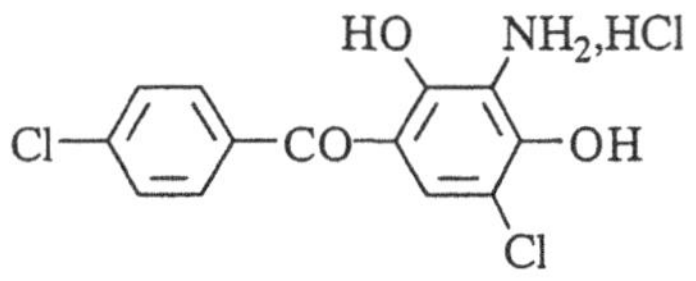

Synthesis

-Preparation by hydrogenation of 4',5-dichloro-2,4-dihydroxy-3-nitrobenzophenone in chloroform/ethanol solution in the presence of 10% Pd/C, followed by treatment of the resulting amino compound with concentrated hydrochloric acid in ethanol (51%) [933].

m.p. 180° [933];
^{1}H NMR [933], IR [933], MS [933]; TLC [933].

(3,4-Dihydroxy-5-nitrophenyl)[2-(trifluoromethyl)phenyl]methanone

[134612-50-3] $C_{14}H_8F_3NO_5$ mol.wt. 327.22

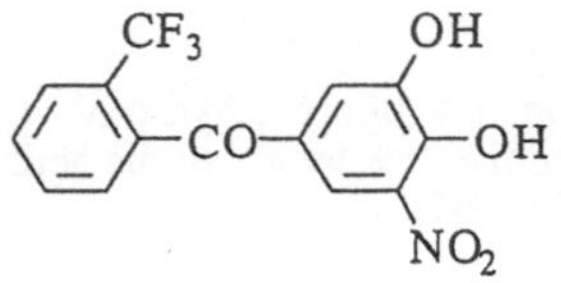

Synthesis

-Preparation by demethylation of 4-hydroxy-3-methoxy-5-nitro-2'-(trifluoromethyl)benzophenone with hydrobromic acid in refluxing aqueous acetic acid [164].

m.p. 146-148° [164];   Spectra (NA).

(3,4-Dihydroxy-5-nitrophenyl)[4-(trifluoromethyl)phenyl]methanone

[134611-76-0] $C_{14}H_8F_3NO_5$ mol.wt. 327.22

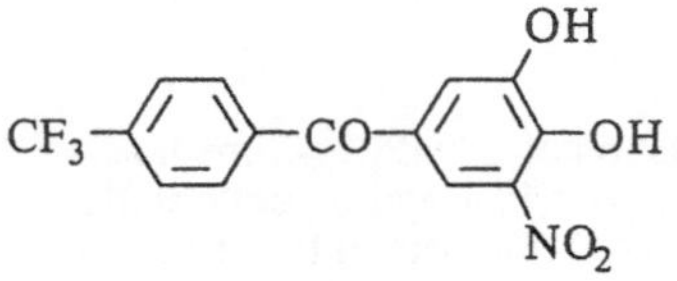

Synthesis

-Preparation by demethylation of 4-hydroxy-3-methoxy-5-nitro-4'-(trifluoromethyl)benzophenone first with 33% hydrobromic acid in acetic acid at 90° for 18 h, then with 48% hydrobromic acid in aqueous acetic acid at 110° for 18 h [164].

m.p. 116-118° [164];   Spectra (NA).

(2,4-Dihydroxy-3,5,6-trinitrophenyl)(4-methoxy-3-nitrophenyl)methanone

[67246-07-5] $C_{14}H_8N_4O_{12}$ mol.wt. 424.24

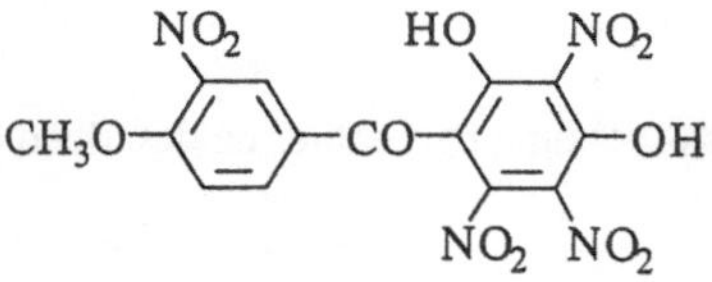

Synthesis

-Obtained by partial methylation of 2,4,4'-trihydroxy-3,5,3',6-tetranitrobenzophenone with dimethyl sulfate in the presence of potassium carbonate in acetone [1083].

m.p. 75° [1083]; ^{1}H NMR [1083].

[3-Chloro-2,4 (or 2,5)-dihydroxy-5 (or 4)-methoxyphenyl](2-fluorophenyl)methanone

[140708-53-8] $C_{14}H_{10}ClFO_4$ mol.wt. 296.68

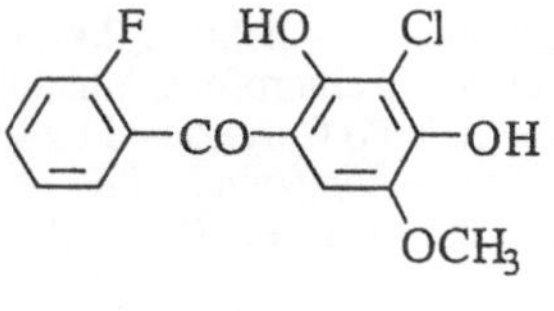

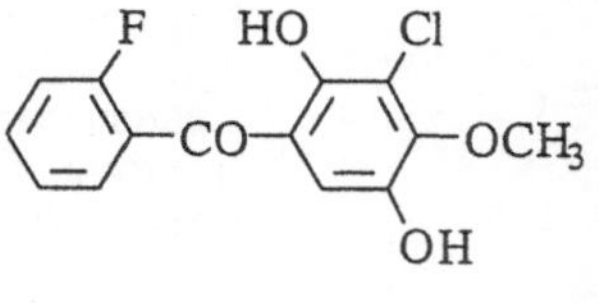

Synthesis

-Obtained by reaction of o-fluorobenzoyl chloride with 3-chloro-1,2,4-trimethoxybenzene

in the presence of aluminium chloride in ethylene dichloride between 0 to 5°, then at r.t. for 8 h and at reflux for 1.5 h (71%) [1194].

m.p. 136-137° [1194];   Spectra (NA).

(4-Chloro-2,5-dihydroxyphenyl)(3-chloro-4-methylphenyl)methanone

[161463-56-5] $C_{14}H_{10}Cl_2O_3$ mol.wt. 297.14

Synthesis

-Preparation by demethylation of 3',4-dichloro-2,5-di-methoxy-4'-methylbenzophenone (SM) with boron tribromide in methylene chloride at 0° for 15-17 h (70-95%). SM was prepared by reaction of 3-chloro-4-methylbenzoic acid with 2-chloro-1,4-dimethoxy-benzene in the presence of polyphosphoric acid at 60-70° for 6-7 h (40-83%) [1279].

m.p. (NA); ^{1}H NMR [1279], IR [1279].

(2,4-Dichlorophenyl)(2,5-dihydroxy-3-methylphenyl)methanone

[153907-06-3] $C_{14}H_{10}Cl_2O_3$ mol.wt. 297.14

Synthesis

-Preparation by demethylation of 2',4'-dichloro-2,5-di-methoxy-3-methylbenzophenone with boron tribromide in methylene chloride at 0° (77%) [1280].

m.p. 150° [1280]; ^{1}H NMR [1280], IR [1280].

(2,6-Dichlorophenyl)(4,5-dihydroxy-2-methylphenyl)methanone

[91197-05-6] $C_{14}H_{10}Cl_2O_3$ mol.wt. 297.14

Syntheses

-Preparation by total demethylation of 2',6'-dichloro-4,5-di-methoxy-2-methylbenzophenone (SM) [313], with pyridinium chloride for 2 h at 180-200° [518]. SM was obtained by reaction of 2,6-dichlorobenzoic acid with 1,2-dimethoxy-4-methylbenzene in the presence of phosphorous pentoxide in methanesulfonic acid for 30 min at 70° [518].
-Preparation by Friedel-Crafts acylation of 4-methylpyrocatechol with 2,6-dichlorobenzoyl chloride [313].
-Also refer to: [519].

m.p. 206-207° [518], 201-203° [313];   Spectra (NA).

(2,4-Dihydroxy-3,5-dinitrophenyl)(2-methoxyphenyl)methanone

[79204-68-5] $C_{14}H_{10}N_2O_8$ mol.wt. 334.24

Synthesis

-Preparation by oxidation of 6-acetoxy-2,3-bis(o-methoxy-phenyl)-5,7-dinitrobenzofuran with chromium trioxide in acetic acid, followed by saponification of the keto ester so formed with potassium hydroxide in ethanol (55%) [579].

m.p. 280° [579];   IR [579].

(2-Bromophenyl)(4,5-dihydroxy-2-methylphenyl)methanone

[91197-04-5] $C_{14}H_{11}BrO_3$ mol.wt. 307.14

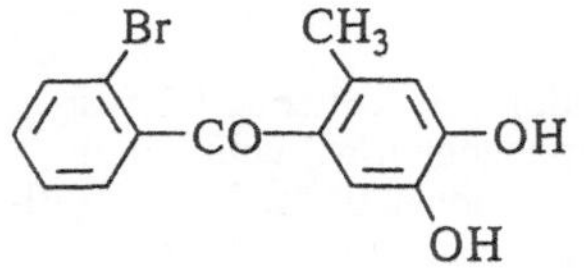

Synthesis

-Preparation by demethylation of 2'-bromo-4,5-dimethoxy-2-methylbenzophenone (SM) with pyridinium chloride for 2 h at 180-200°. SM was obtained by reaction of o-bromobenzoic acid with 1,2-dimethoxy-4-methylbenzene in the presence of phosphorous pentoxide in methanesulfonic acid for 30 min at 70° [518].

m.p. 177-178° [518];   Spectra (NA).

(4-Chloro-2,5-dihydroxyphenyl)(4-methylphenyl)methanone

[161463-55-4] $C_{14}H_{11}ClO_3$ mol.wt. 262.69

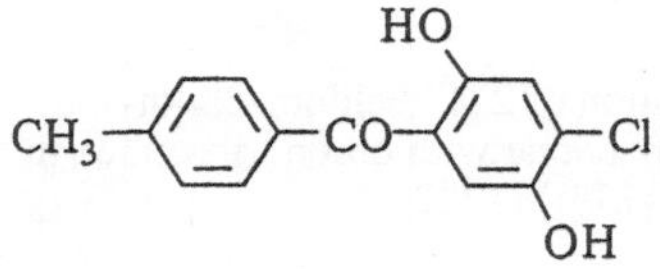

Synthesis

-Preparation by demethylation of 4-chloro-2,5-dimethoxy-4'-methylbenzophenone (SM) with boron tribromide in methylene chloride at 0° for 15-17 h (70-95%). SM was prepared by reaction of p-toluic acid with 2-chloro-1,4-dimethoxybenzene in the presence of polyphosphoric acid at 60-70° for 6-7 h (40-83%) [1279].

m.p. (NA); ^1H NMR [1279], IR [1279].

(2-Chlorophenyl)(2,5-dihydroxy-3-methylphenyl)methanone

[153907-05-2] $C_{14}H_{11}ClO_3$ mol.wt. 262.69

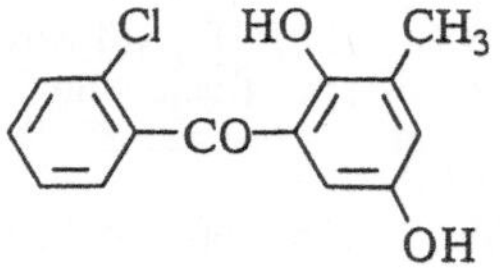

Synthesis

-Preparation by demethylation of 2'-chloro-2,5-dimethoxy-3-methylbenzophenone with boron tribromide in methylene chloride at 0° (76%) [1280].

m.p. 140° [1280]; ^1H NMR [1280], IR [1280].

(4-Chlorophenyl)(2,5-dihydroxy-3-methylphenyl)methanone

[153907-07-4] $C_{14}H_{11}ClO_3$ mol.wt. 262.69

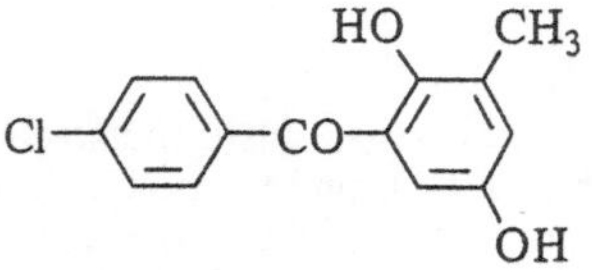

Synthesis

-Preparation by demethylation of 4'-chloro-2,5-dimethoxy-3-methylbenzophenone with boron tribromide in methylene chloride at 0° (70%) [1280].

m.p. 190° [1280]; ^1H NMR [1280], IR [1280].

(3,4-Dihydroxy-5-nitrophenyl)(2-methylphenyl)methanone

[134612-48-9] C$_{14}$H$_{11}$NO$_5$ mol.wt. 273.25

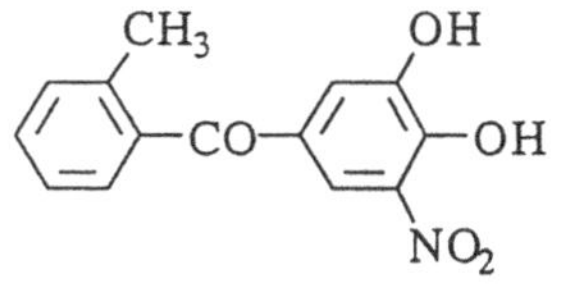

Synthesis

-Preparation by demethylation of 4-hydroxy-3-methoxy-2'-methyl-5-nitrobenzophenone with hydrobromic acid in refluxing aqueous acetic acid [164].

m.p. 164-166° [164]; Spectra (NA).

(3,4-Dihydroxy-5-nitrophenyl)(4-methylphenyl)methanone

[134308-13-7] C$_{14}$H$_{11}$NO$_5$ mol.wt. 273.25

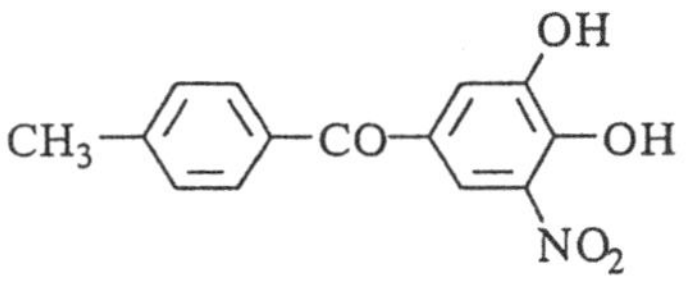

Synthesis

-Preparation by demethylation of 4-hydroxy-3-methoxy-4'-methyl-5-nitrobenzophenone with hydrobromic acid in refluxing aqueous acetic acid [164].
-Also refer to: [316, 846, 1366].

m.p. 146-148° [164]; Spectra (NA).

(2,6-Dichlorophenyl)(2,3-dihydroxy-4-methoxy-6-methylphenyl)methanone

[183725-80-6] C$_{15}$H$_{12}$Cl$_2$O$_4$ mol.wt. 327.16

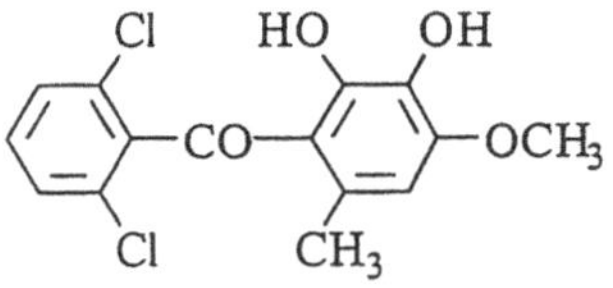

Synthesis

-Preparation by partial demethylation of 2',6'-dichloro-2,3,4-trimethoxy-6-methylbenzophenone in acetic acid in the presence of 30% hydrobromic acid for 2 h at 75° (39%) [313].

m.p. 182° [313]; Spectra (NA).

(2,4-Dihydroxy-3-methylphenyl)(2-methylphenyl)methanone

[147809-15-2] C$_{15}$H$_{14}$O$_3$ mol.wt. 242.27

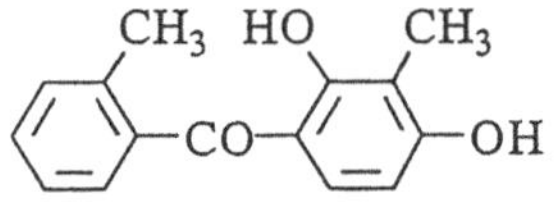

Synthesis

-Preparation by treating o-toluic acid with 2-methylresorcinol in the presence of Zeolite-H-beta (previously activated at 400°), in refluxing p-chlorotoluene or n-decane for 2-3 h (with water removal) (74%) [582, 584].

-Also refer to: [583].

m.p. 187-188°5 [582, 584]; ^{1}H NMR [582, 584].

(4,5-Dihydroxy-2-methylphenyl)(2-methylphenyl)methanone

[91197-07-8] $C_{15}H_{14}O_3$ mol.wt. 242.27

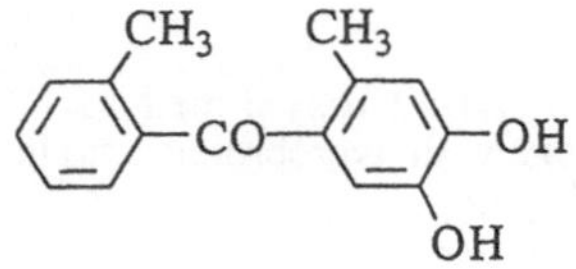

Synthesis

-Preparation by demethylation of 4,5-dimethoxy-2,2'-di-
methylbenzophenone (SM) with pyridinium chloride for 2 h
at 180-200°. SM was obtained by reaction of o-toluic acid
with 1,2-dimethoxy-4-methylbenzene in the presence of
phosphorous pentoxide in methanesulfonic acid for 20 min
at 70° [518].

-Also refer to: [519].

m.p.   149-150° [518];   Spectra (NA).

(2,4-Dihydroxy-3-methylphenyl)(2-methoxyphenyl)methanone

[85636-84-6] $C_{15}H_{14}O_4$ mol.wt. 258.27

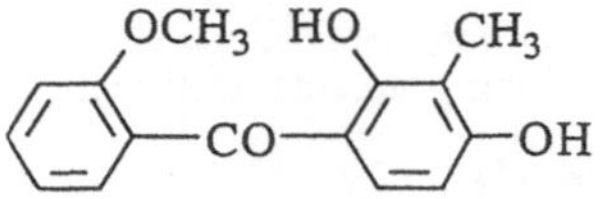

Synthesis

-Refer to: [982].

m.p. and Spectra (NA).

(2,4-Dihydroxy-3-methylphenyl)(4-methoxyphenyl)methanone

[79861-83-9] $C_{15}H_{14}O_4$ mol.wt. 258.27

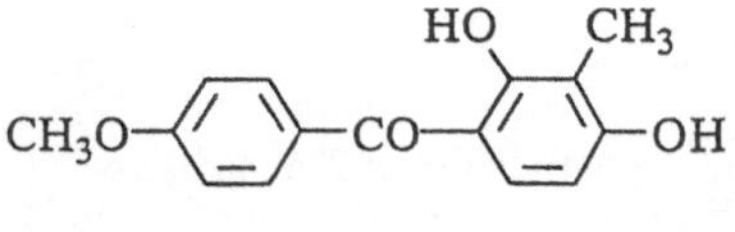

Synthesis

-Preparation by reaction of p-anisic acid with 2-methyl-
resorcinol in tetrachloroethane in the presence of
boron trifluoride at 80° for 4 h (66%) [271].

m.p.   194-195° [271]; ^{1}H NMR [271].

(2,5-Dihydroxy-4-methoxyphenyl)(2-methoxyphenyl)methanone

[42833-90-9] $C_{15}H_{14}O_5$ mol.wt. 274.27

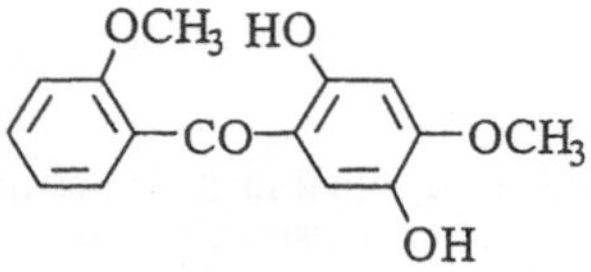

Synthesis

-Preparation by reaction of 2-hydroxy-2',4,5-trimethoxy-
benzophenone with DDQ (2,3-Dichloro-5,6-dicyano-
1,4-benzoquinone) in refluxing benzene for 1.5 h, followed
by complete evaporation of the solvent, then treatment of the
resulting brown oil in boiling methanol (58%) [1097].

m.p.   195-196° [1097];   UV [1097], MS [1097].

(2,5-Dihydroxy-4-methoxyphenyl)(4-methoxyphenyl)methanone

[42045-63-6] $C_{15}H_{14}O_5$ mol.wt. 274.27

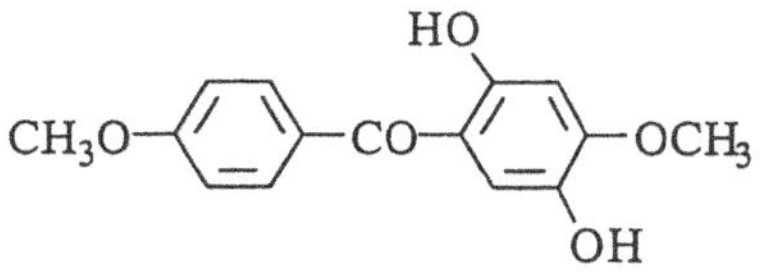

Syntheses

-Preparation by reaction of p-anisic acid with 2-methoxyhydroquinone in the presence of boron trifluoride-etherate, heating at 100° for 30-45 min (26%) [785].
-Also obtained (poor yield) by reaction of 2-hydroxy-4,4'-dimethoxybenzophenone with lead tetraacetate in acetic acid at 100° for 5 h (1%) [785].

m.p. 164-165° [785]; ^{1}H NMR [785], IR [785], UV [785].

(2,6-Dihydroxy-4-methoxyphenyl)(4-methoxyphenyl)methanone

[55051-89-3] $C_{15}H_{14}O_5$ mol.wt. 274.27

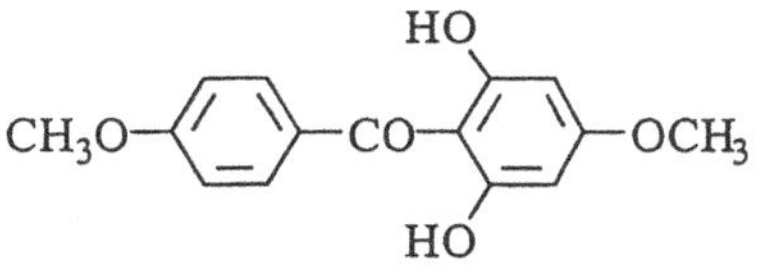

Synthesis

-Obtained by partial methylation of 2,4',6-trihydroxy-4-methoxybenzophenone with diazomethane in ethyl ether [299].

m.p. (NA); ^{1}H NMR [299].

(2-Chlorophenyl)(2,4-dihydroxy-3-propylphenyl)methanone

[115296-10-1] $C_{16}H_{15}ClO_3$ mol.wt. 290.75

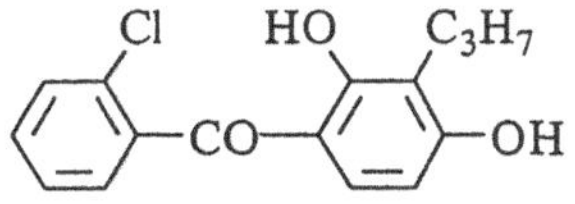

Synthesis

-Obtained (poor yield) by reaction of o-chlorobenzoyl chloride with 1,3-dimethoxy-2-propylbenzene in methylene chloride in the presence of aluminium chloride, first at 0° for 2 h and at r.t. for 1 h (10%) [456].

oil [456];   b.p. and Spectra (NA).

(4,5-Dihydroxy-2-methylphenyl)(2,6-dimethylphenyl)methanone

[91197-06-7] $C_{16}H_{16}O_3$ mol.wt. 256.30

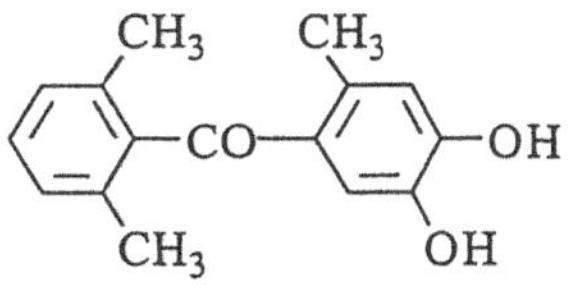

Synthesis

-Preparation by demethylation of 4,5-dimethoxy-2,2',6'-tri-methylbenzophenone (SM) with pyridinium chloride for 2 h at 180-200°. SM was obtained by reaction of 2,6-di-methylbenzoic acid with 1,2-dimethoxy-4-methylbenzene in methanesulfonic acid in the presence of phosphorous pentoxide for 30 min at 70° [518].

-Also refer to: [519].

m.p. 193-195° [518];   Spectra (NA).

(2,5-Dihydroxy-4-methoxyphenyl)(3,4-dimethoxyphenyl)methanone

[62495-45-8] $C_{16}H_{16}O_6$ mol.wt. 304.30

Synthesis

-Obtained (poor yield) by oxidation of 2-hydroxy-
3',4,4',5-tetramethoxybenzophenone with manganese
(III) acetate dihydrate in refluxing acetic acid for 2 h
(9%) [1395].

m.p. 208-209° [1395]; ^{1}H NMR [1395], IR [1395], UV [1395].

(3-Chloro-4,6-dihydroxy-2-methylphenyl)(2,4,6-trimethoxyphenyl)methanone

[68048-19-1] $C_{17}H_{17}ClO_6$ mol.wt. 352.77

Synthesis

-Preparation by hydrogenolysis of 4,6-bis(benzyloxy)-
3-chloro-2',4',6'-trimethoxy-2-methylbenzophenone
(SM) with hydrogen in ethyl acetate/tetrahydrofuran in
the presence of 10% Pd/C at 25°. SM was obtained by
condensation of 4,6-bis(benzyloxy)-3-chloro-
2-methylbenzoic acid with phloroglucinol trimethyl
ether in the presence of trifluoroacetic anhydride in methylene chloride under nitrogen for 3 min
(87%) [1328].

m.p. 132-133° [1328]; ^{1}H NMR [1328], IR [1328], MS [1328].

(2,5-Dihydroxy-3,4-dimethoxyphenyl)(4-ethoxyphenyl)methanone

[69471-32-5] $C_{17}H_{18}O_6$ mol.wt. 318.33

Synthesis

-Preparation by oxidation of 4'-ethoxy-2-hydroxy-
3,4-dimethoxybenzophenone with potassium
persulfate (24%) (Elbs reaction) [1008].

m.p. 140-142° [1008]; ^{1}H NMR [1008], IR [1008].

(2,4-Dihydroxy-6-methylphenyl)(2,4,6-trimethoxyphenyl)methanone

[76631-09-9] $C_{17}H_{18}O_6$ mol.wt. 318.33

Synthesis

-Preparation by reaction of triethylamine with
1-(2,4,6-trimethoxyphenyl)-1,3,5,7-octanetetraone
(SM) in refluxing tetrahydrofuran in a nitrogen
atmosphere for 38 h (quantitative yield) [1184]. SM
can be obtained according to two methods:
*acylation of the trilithium salt of 2,4,6-heptanetrione (prepared with LDA in tetrahydrofuran at 0°)
with methyl 2,4,6-trimethoxybenzoate in tetrahydrofuran for 17 h at r.t. (55%);
*acylation of the dilithium salt of 2,4-pentanedione with methyl 3-(2,4,6-trimethoxyphenyl)-
3-oxopropanoate (sodium salt) (42%).

m.p.   169-173° [1184]; ^{1}H NMR [1184], IR [1184], MS [1184];   TLC [1184].

(2,4-Dihydroxy-3,5-dimethoxyphenyl)(2,5-dimethoxyphenyl)methanone

$C_{17}H_{18}O_7$ mol.wt. 334.33

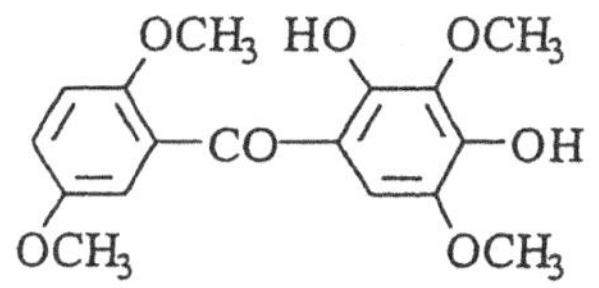

Synthesis

-Obtained (poor yield) by reaction of 2,3,4,5-tetramethoxy-benzoyl chloride with hydroquinone dimethyl ether in benzene in the presence of aluminium chloride (16%) [845].

m.p. and Spectra (NA).

(3,6-Dihydroxy-2,4-dimethoxyphenyl)(2,5-dimethoxyphenyl)methanone

$C_{17}H_{18}O_7$ mol.wt. 334.33

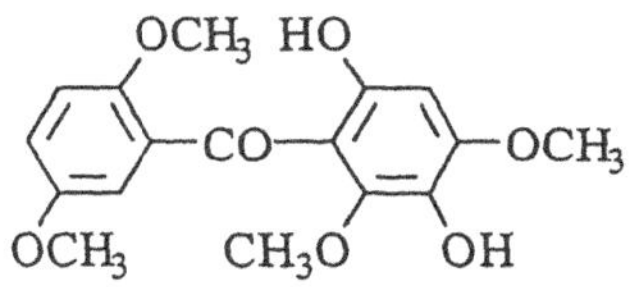

Synthesis

-Preparation by reaction of 2,5-dimethoxybenzoic acid with 2,6-dimethoxyhydroquinone in trifluoroacetic anhydride [1311].

m.p. and Spectra (NA).

(2-Chlorophenyl)(5-hexyl-2,4-dihydroxyphenyl)methanone

[50685-43-3] $C_{19}H_{21}ClO_3$ mol.wt. 332.83

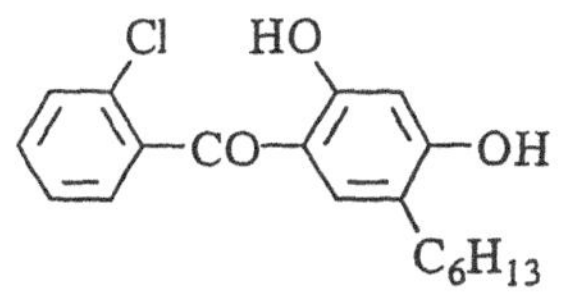

Synthesis

-Preparation by reaction of 2-chlorobenzotrichloride with 4-hexylresorcinol in dilute isobutanol at 70-80° (77%) [1460].

m.p. and Spectra (NA).

(5-Hexyl-2,4-dihydroxyphenyl)(3-nitrophenyl)methanone

$C_{19}H_{21}NO_5$ mol.wt. 343.38

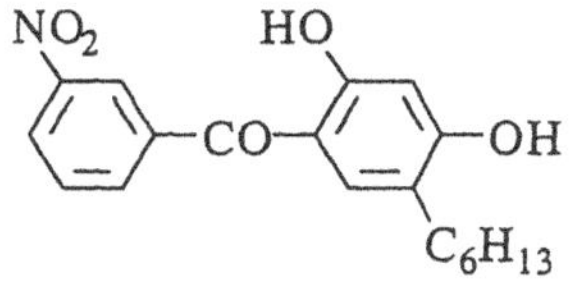

Synthesis

-Refer to: [1406].

m.p. 85° [1406];   Spectra (NA).

2.2. *Hydroxy groups located on both rings*

2.2.1. *Substituents located on one ring*

(2,3-Dichloro-4-hydroxyphenyl)(2-hydroxyphenyl)methanone

[72482-30-5]   C13H8Cl2O3   mol.wt. 283.11

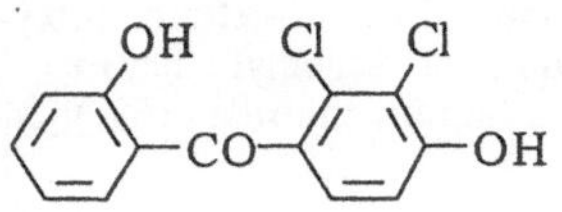

Synthesis

-Preparation by total demethylation of 2,3-dichloro-
2',4-dimethoxybenzophenone (SM) with pyridinium chloride
at 200° for 1 h. SM was obtained by reaction of
o-anisoyl chloride with 2,3-dichloroanisole in ethylene
dichloride in the presence of aluminium chloride first at -10° for 2.5 h, then at +5° [1270].
-Also refer to: [1025].

m.p.   197-201° [1270];   Spectra (NA).

(2,3-Dichloro-4-hydroxyphenyl)(4-hydroxyphenyl)methanone

[78697-41-3]   C13H8Cl2O3   mol.wt. 283.11

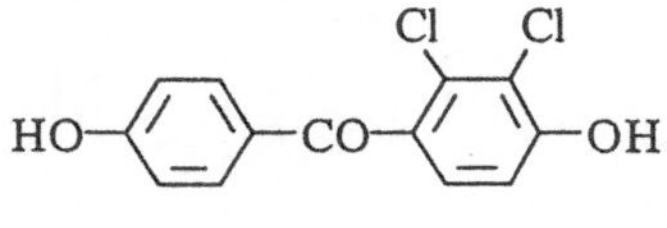

Synthesis

-Obtained by Friedel-Crafts acylation of 2,3-dichloro-
anisole with p-methoxybenzoyl chloride in the presence
of aluminium chloride at r.t. [1071] or in benzene at
75-80° for 1 h (53%) [449].

m.p.   208-210° [449];   Spectra (NA).

(2,5-Dichloro-4-hydroxyphenyl)(4-hydroxyphenyl)methanone

[98155-78-3]   C13H8Cl2O3   mol.wt. 283.11

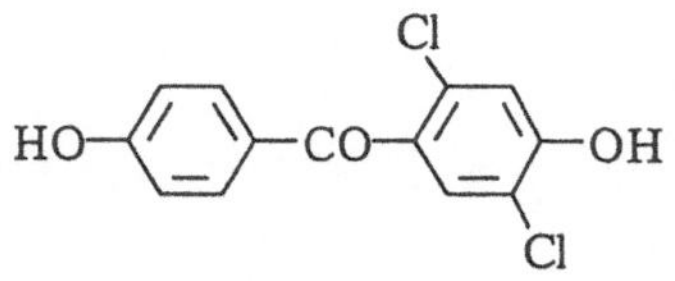

Synthesis

-Refer to: [486] (compound 20).

m.p. and Spectra (NA).

(3,5-Dichloro-4-hydroxyphenyl)(3-hydroxyphenyl)methanone

[92005-28-2]   C13H8Cl2O3   mol.wt. 283.11

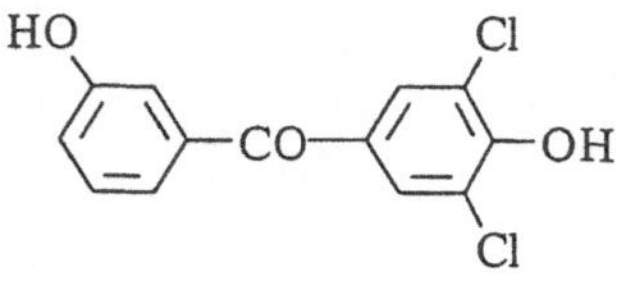

Synthesis

-Refer to: [544].

m.p. and Spectra (NA).

(3,5-Dichloro-4-hydroxyphenyl)(4-hydroxyphenyl)methanone

[92005-19-1] $C_{13}H_8Cl_2O_3$ mol.wt. 283.11

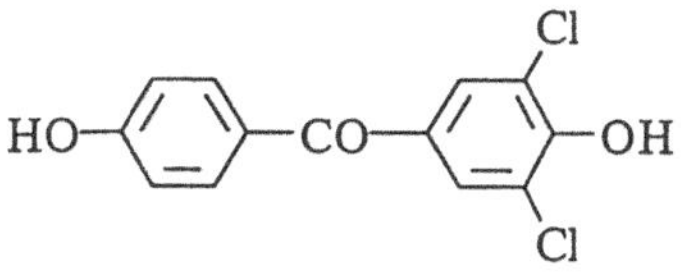

Synthesis

-Preparation by acylation of 2,6-dichlorophenol with
p-(trichloromethyl)phenyl p-(trichloromethyl)benzoate in
methylene chloride in the presence of aluminium chloride
at 0-5° over 1 h, then at r.t. for 1 h, followed by alkaline
hydrolysis of the resulting keto ester [581] (Japanese
patent).

m.p. and Spectra (NA).

(2-Chloro-4-hydroxyphenyl)(2-hydroxyphenyl)methanone

[126165-47-7] $C_{13}H_9ClO_3$ mol.wt. 248.67

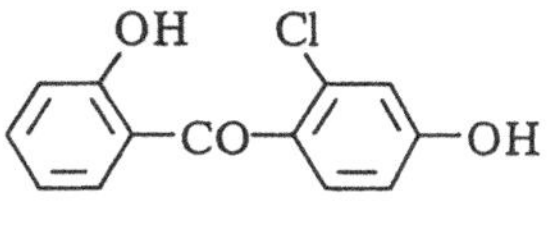

Synthesis

-Preparation by Fries rearrangement of m-chlorophenyl
salicylate with aluminium chloride without solvent at 180-
182° for 3 h (40%) [172].

m.p. 130° [172];
Spectra (NA).

(2-Chloro-4-hydroxyphenyl)(3-hydroxyphenyl)methanone

[126165-56-8] $C_{13}H_9ClO_3$ mol.wt. 248.67

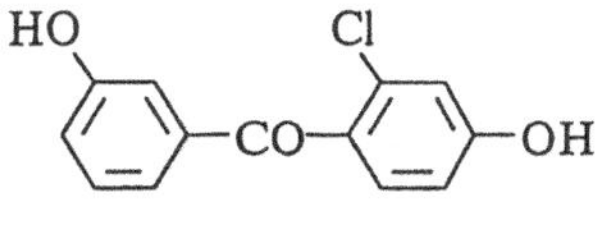

Synthesis

-Obtained (by-product) by Fries rearrangement of m-chlorophenyl m-methoxybenzoate with aluminium chloride at 182-
184° for 3 h (15%) [172].

m.p. 200° [172];
Spectra (NA).

(2-Chloro-4-hydroxyphenyl)(4-hydroxyphenyl)methanone

[98155-77-2] $C_{13}H_9ClO_3$ mol.wt. 248.67

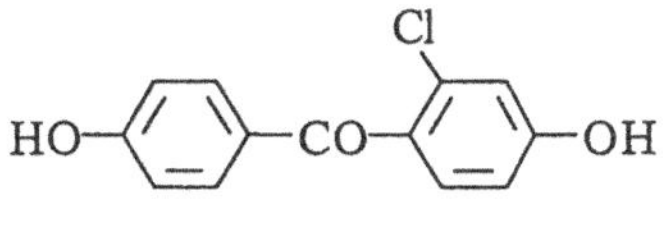

Synthesis

-Preparation by reaction of bis(4-trichloromethylphenoxy)-
ketone with m-chlorophenol in methylene chloride in the
presence of aluminium chloride, first at 0° for 1.5 h, then
at 20° for 3 h, then hydrolysis of the intermediate
product so obtained with 20% sodium hydroxide aqueous solution at 20° for 7 h (64%) [486].

m.p. 176-178° (anhydrous) and 100-108° (hemihydrate) [486];
Spectra (NA).

(3-Chloro-2-hydroxyphenyl)(3-hydroxyphenyl)methanone

[126165-53-5] $C_{13}H_9ClO_3$ mol.wt. 248.67

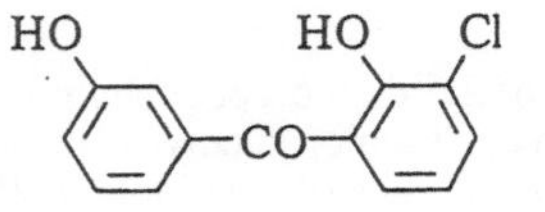

Synthesis

-Obtained (by-product) by Fries rearrangement of o-chlorophenyl m-methoxybenzoate with aluminium chloride at 182-
 184° for 3 h (13%) [172].

m.p. 121° [172];   Spectra (NA).

(3-Chloro-2-hydroxyphenyl)(4-hydroxyphenyl)methanone

[126165-62-6] $C_{13}H_9ClO_3$ mol.wt. 248.67

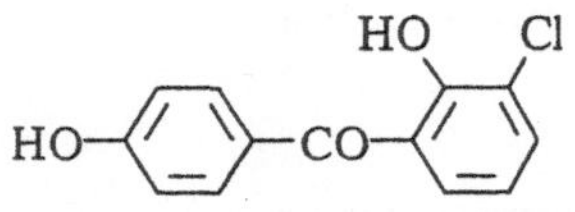

Synthesis

-Obtained (by-product) by Fries rearrangement of o-chlorophenyl p-methoxybenzoate with aluminium chloride at 180°
 for 3 h (10%) [172].

m.p. 152° [172];   Spectra (NA).

(3-Chloro-4-hydroxyphenyl)(2-hydroxyphenyl)methanone

[123861-94-9] $C_{13}H_9ClO_3$ mol.wt. 248.67

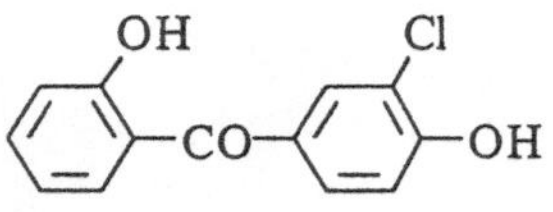

Syntheses

-Preparation by Fries rearrangement of o-chlorophenyl
 salicylate with aluminium chloride at 180-183° for 3 h (35%)
 [172].
-Preparation by hydrolysis of 2-chloro-4-salicyloylphenyl
salicylate [170], a secondary product obtained by Fries rearrangement of o-chlorophenyl salicylate
at 180-183° for 3 h (20%) [172].
-Also obtained (by-product) by Fries rearrangement of o-chlorophenyl 2-(nicotinoyloxy)benzoate
with aluminium chloride at 150-152° for 2 h (4%) [170].

m.p. 113° [170, 172];   Spectra (NA).

(3-Chloro-4-hydroxyphenyl)(3-hydroxyphenyl)methanone

[92005-08-8] $C_{13}H_9ClO_3$ mol.wt. 248.67

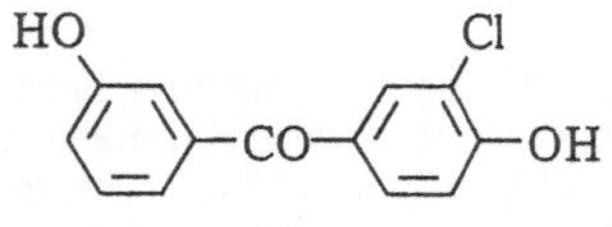

Synthesis

-Preparation by Fries rearrangement of o-chlorophenyl
 m-methoxybenzoate with aluminium chloride at 182-184°
 for 3 h (62%) [172].

m.p. 183° [172];   Spectra (NA).

(3-Chloro-4-hydroxyphenyl)(4-hydroxyphenyl)methanone

[92005-17-9] $C_{13}H_9ClO_3$ mol.wt. 248.67

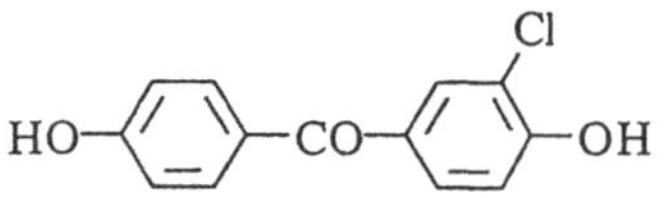

Synthesis

-Preparation by Fries rearrangement of o-chlorophenyl
p-methoxybenzoate with aluminium chloride at 180° for
3 h (57%) [172].

m.p. 204° [172];
Spectra (NA).

(4-Chloro-2-hydroxyphenyl)(3-hydroxyphenyl)methanone

[126165-57-9] $C_{13}H_9ClO_3$ mol.wt. 248.67

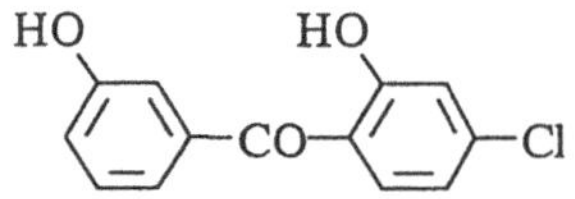

Synthesis

-Preparation by Fries rearrangement of m-chlorophenyl
m-methoxybenzoate with aluminium chloride at 182-184°
for 3 h (55%) [172].

m.p. 132° [172];
Spectra (NA).

(4-Chloro-2-hydroxyphenyl)(4-hydroxyphenyl)methanone

[126165-44-4] $C_{13}H_9ClO_3$ mol.wt. 248.67

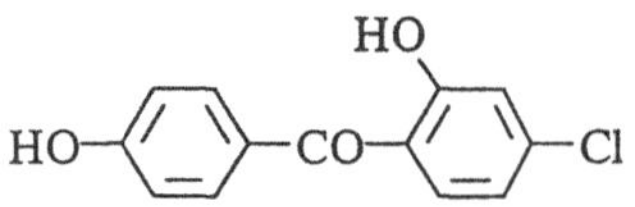

Synthesis

-Preparation by Fries rearrangement of m-chlorophenyl
p-methoxybenzoate with aluminium chloride without
solvent at 182-185° for 3 h (47%) [172].

m.p. 170° [172];
Spectra (NA).

(5-Chloro-2-hydroxyphenyl)(2-hydroxyphenyl)methanone

[76237-02-0] $C_{13}H_9ClO_3$ mol.wt. 248.67

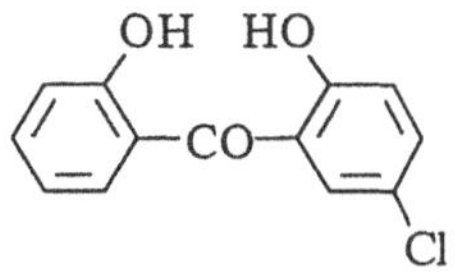

Syntheses

-Obtained by Fries rearrangement of p-chlorophenyl
salicylate with aluminium chloride at 208-210° for 2.5 h
(20%) [172].
-Also obtained by UV light irradiation of p-chlorophenyl
salicylate in methanol for 8 h [269].

m.p. 68-69° [269], 59-60° [172];
^1H NMR [269], ^{13}C NMR [269], IR [269], UV [269], MS [269].

(5-Chloro-2-hydroxyphenyl)(3-hydroxyphenyl)methanone

[126165-59-1]   C$_{13}$H$_9$ClO$_3$ mol.wt. 248.67

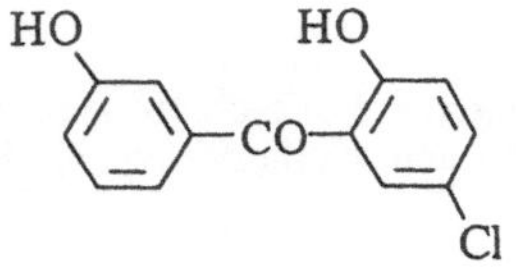

Synthesis

-Preparation by Fries rearrangement of p-chlorophenyl
 m-methoxybenzoate with aluminium chloride at 182-185°
 for 3 h (60%) [172].

m.p. 148° [172];
Spectra (NA).

(5-Chloro-2-hydroxyphenyl)(4-hydroxyphenyl)methanone

[126165-40-0]   C$_{13}$H$_9$ClO$_3$ mol.wt. 248.67

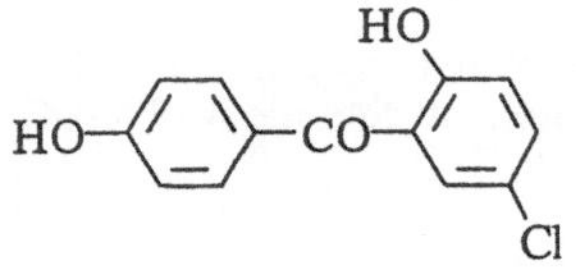

Synthesis

-Preparation by Fries rearrangement of p-chlorophenyl
 p-methoxybenzoate with aluminium chloride without solvent
 at 182-185° for 3 h (67%) [172].

m.p. 160° [172];
Spectra (NA).

(2-Fluoro-4-hydroxyphenyl)(4-hydroxyphenyl)methanone

[98155-81-8]   C$_{13}$H$_9$FO$_3$ mol.wt. 232.21

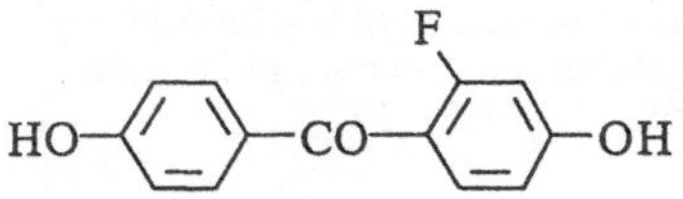

Synthesis

-Refer to: [486] (compound 30).

m.p. and Spectra (NA).

(5-Fluoro-2-hydroxyphenyl)(4-hydroxyphenyl)methanone

[159300-38-6]   C$_{13}$H$_9$FO$_3$ mol.wt. 232.21

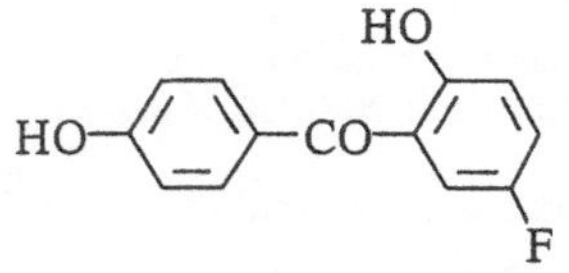

Synthesis

-Obtained by demethylation of 5-fluoro-2-hydroxy-
 4'-methoxybenzophenone with an excess of boron
 tribromide under nitrogen first at -70° for 2 min, then at 0°
 for 1 h (47%) [329].

m.p. 104-105° [329];
^{1}H NMR [328, 329], MS [328, 329];   HPLC [328, 329].

(4-Hydroxy-3-nitrophenyl)(4-hydroxyphenyl)methanone

[94737-85-6] $C_{13}H_9NO_5$ mol.wt. 259.22

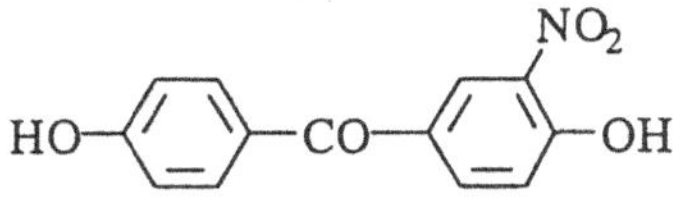

Synthesis

-Refer to: [581] (Japanese patent).

m.p. and Spectra (NA).

(4-Amino-3-hydroxyphenyl)(4-hydroxyphenyl)methanone

[136134-35-5] $C_{13}H_{11}NO_3$ mol.wt. 229.24

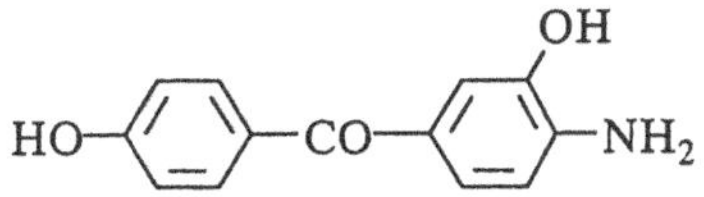

Synthesis

-Preparation by hydrolysis of 6-(4-hydroxybenzoyl)-
benzoxazolinone in aqueous sodium hydroxide [148],
according to [196].

m.p. (NA);   MS [148].

(2-Hydroxy-3-methylphenyl)(3-hydroxyphenyl)methanone

$C_{14}H_{12}O_3$ mol.wt. 228.25

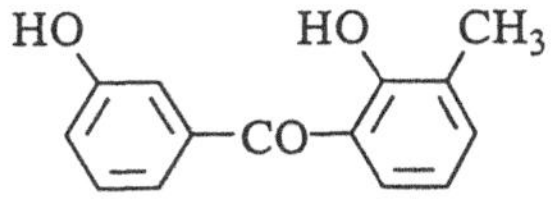

Synthesis

-Obtained by Fries rearrangement of o-cresyl m-anisate with
aluminium chloride without solvent at 120° or 160° for 2 h
[1168].

m.p.   144° [1168];   Spectra (NA).

(2-Hydroxy-4-methylphenyl)(2-hydroxyphenyl)methanone

[86415-67-0] $C_{14}H_{12}O_3$ mol.wt. 228.25

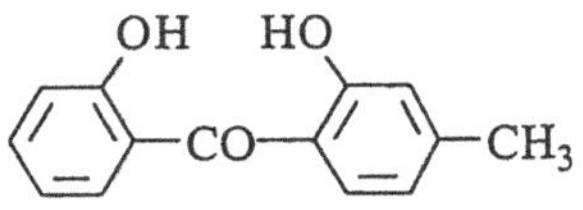

Synthesis

-Refer to: [41].

m.p. and Spectra (NA).

(2-Hydroxy-4-methylphenyl)(3-hydroxyphenyl)methanone

$C_{14}H_{12}O_3$ mol.wt. 228.25

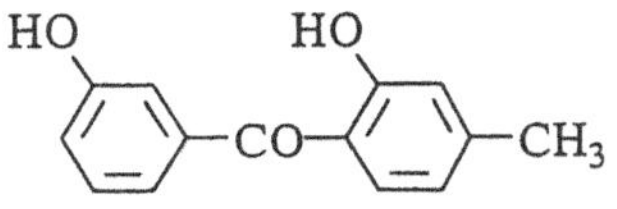

Synthesis

-Obtained by Fries rearrangement of m-cresyl m-anisate with
aluminium chloride without solvent at 120° or 160° for 2 h
[1168].

m.p.   105° [1168];   Spectra (NA).

(2-Hydroxy-4-methylphenyl)(4-hydroxyphenyl)methanone

$C_{14}H_{12}O_3$ mol.wt. 228.25

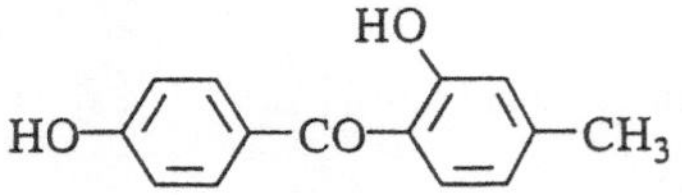

Synthesis

-Preparation by reaction of p-hydroxybenzoic acid with m-cresol in the presence of boron trifluoride at 150° for 3 h (68%) [730].

m.p.   148-149° [730];   Spectra (NA).

(2-Hydroxy-5-methylphenyl)(2-hydroxyphenyl)methanone

[93097-75-7] $C_{14}H_{12}O_3$ mol.wt. 228.25

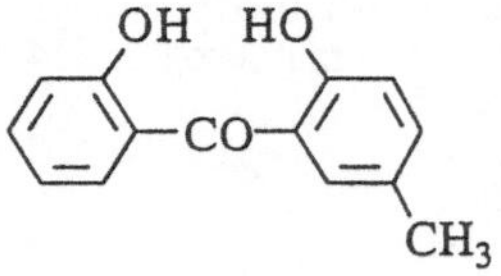

Syntheses

-Preparation by Fries rearrangement of p-tolyl salicylate with aluminium chloride at 140° for 3 h (70%) [50].
-Also obtained by reaction of o-methoxybenzoyl chloride with p-cresol in the presence of aluminium chloride at 100° for 24 h [99].

-Also refer to: [1253].

m.p.   143-144° [50];   Spectra (NA).

(2-Hydroxy-5-methylphenyl)(3-hydroxyphenyl)methanone

[57855-38-6] $C_{14}H_{12}O_3$ mol.wt. 228.25

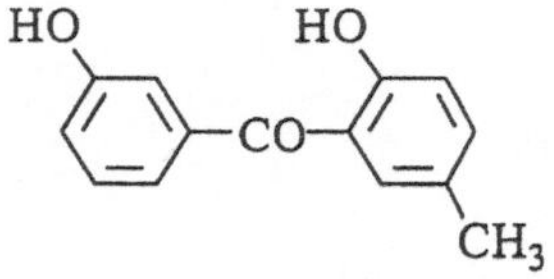

Syntheses

-Preparation by demethylation of 2-hydroxy-3'-methoxy-5-methylbenzophenone with hydrobromic acid in acetic acid [1378, 1379].
-Also obtained by Fries rearrangement of p-tolyl m-anisate with aluminium chloride without solvent at 120° or 160° for 2 h (20 to 30%) [1168].

-Also refer to: [1377].

m.p.   136° [1168];   Spectra (NA).

(2-Hydroxy-5-methylphenyl)(4-hydroxyphenyl)methanone

[25148-21-4] $C_{14}H_{12}O_3$ mol.wt. 228.25

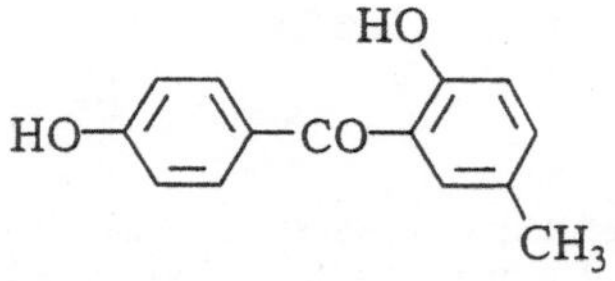

Syntheses

-Preparation by demethylation of 4'-hydroxy-2-methoxy-5-methylbenzophenone (m.p. 160°) with aluminium chloride at 150° [99].
-Preparation by diazotization of 4'-amino-2-hydroxy-5-methylbenzophenone (m.p. 137°), followed by hydrolysis of the resulting diazonium salt [99].

m.p.   150-151° [99];   Spectra (NA).

(3-Hydroxy-4-methylphenyl)(4-hydroxyphenyl)methanone

[75731-48-5] $C_{14}H_{12}O_3$ mol.wt. 228.25

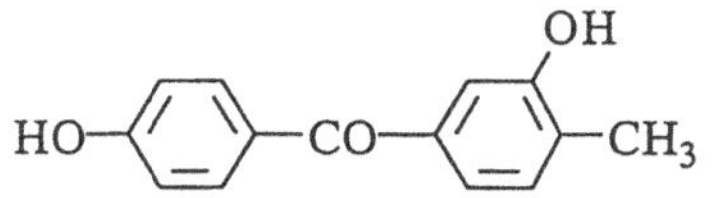

Synthesis

-Refer to: [624, 625].

m.p. and Spectra (NA).

(4-Hydroxy-2-methylphenyl)(2-hydroxyphenyl)methanone

$C_{14}H_{12}O_3$ mol.wt. 228.25

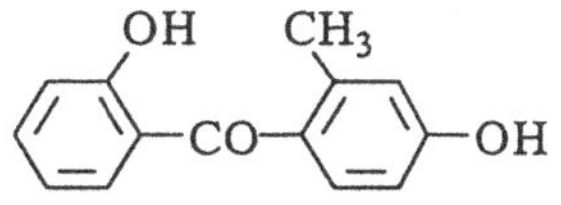

Syntheses

-Obtained by Fries rearrangement,
*of m-cresyl salicylate (1 mol) with aluminium chloride
 (4 mol) at 140° for 3 h (80%) [50];
*of m-cresyl o-methoxybenzoate (1 mol) with aluminium
chloride (2.6 mol) at 120° or 160° for 2 h (low yields) [1168].

m.p. 146° [50, 1168];   Spectra (NA).

(4-Hydroxy-2-methylphenyl)(3-hydroxyphenyl)methanone

$C_{14}H_{12}O_3$ mol.wt. 228.25

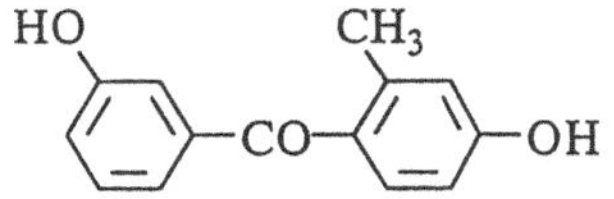

Synthesis

-Obtained by Fries rearrangement of m-cresyl m-anisate with
 aluminium chloride without solvent at 120° or 160° for 2 h
 [1168].

m.p. 173° [1168];   Spectra (NA).

(4-Hydroxy-2-methylphenyl)(4-hydroxyphenyl)methanone

[98155-72-7] $C_{14}H_{12}O_3$ mol.wt. 228.25

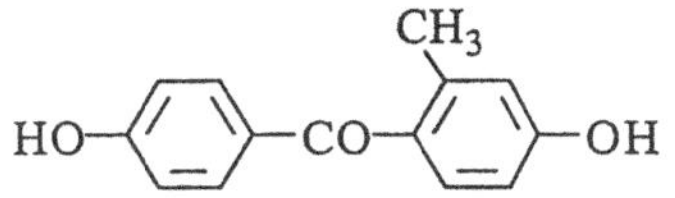

Synthesis

-Preparation by adding m-cresol to a mixture of bis(4-tri-
 chloromethylphenoxy)ketone and aluminium chloride in
 methylene chloride at 0° for 1.5 h, then at 20° for
 3 h. Then, hydrolysis of the intermediate product so
obtained with 20% sodium hydroxide aqueous solution at 20° for 7 h (40%) [486].

m.p. 82-85° [486];   Spectra (NA).

(4-Hydroxy-3-methylphenyl)(2-hydroxyphenyl)methanone

$C_{14}H_{12}O_3$ mol.wt. 228.25

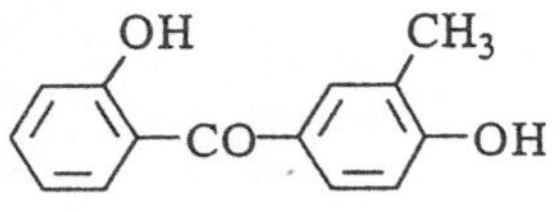

Synthesis

-Preparation by Fries rearrangement of o-cresyl salicylate with aluminium chloride without solvent at 140° for 3 h (80%) [50].

m.p.   112° [50];   Spectra (NA).

(4-Hydroxy-3-methylphenyl)(3-hydroxyphenyl)methanone

[92005-62-4] $C_{14}H_{12}O_3$ mol.wt. 228.25

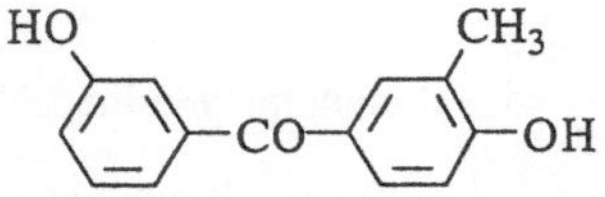

Syntheses

-Obtained by Fries rearrangement of o-cresyl m-methoxy-benzoate with aluminium chloride between 120 to 160° for 2 h (20 to 30%) [1168].

-Preparation by demethylation of 3,4'-dimethoxy-3'-methylbenzophenone with 48% hydrobromic acid in an acetic anhydride/acetic acid mixture (1:1) for 15 h at reflux (78%) [624].

m.p.   174-175° [624], 172° [1168];   Spectra (NA).

(4-Hydroxy-3-methylphenyl)(4-hydroxyphenyl)methanone

[92005-11-3] $C_{14}H_{12}O_3$ mol.wt. 228.25

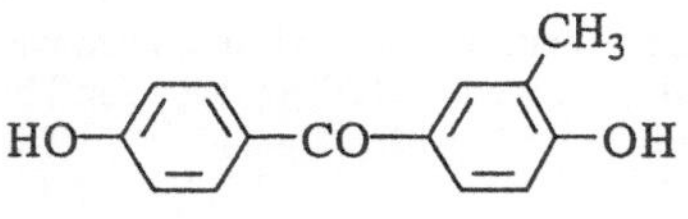

Synthesis

-Preparation by acylation of o-cresol with p-(trichloro-methyl)phenyl p-(trichloromethyl)benzoate in methylene chloride in the presence of aluminium chloride at 0-5° over 1 h, then at r.t. for 1 h, followed by alkaline hydrolysis of the resulting keto ester [581] (Japanese patent).

m.p. and Spectra (NA).

(2-Hydroxy-3-methoxyphenyl)(2-hydroxyphenyl)methanone

[117574-12-6] $C_{14}H_{12}O_4$ mol.wt. 244.25

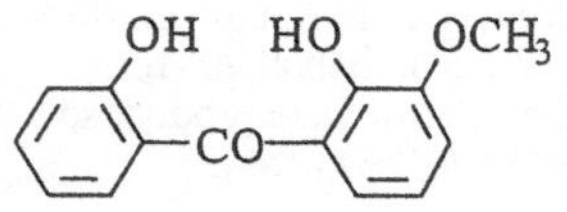

Synthesis

-Refer to: [723].

m.p. and Spectra (NA).

(2-Hydroxy-4-methoxyphenyl)(2-hydroxyphenyl)methanone
(Dioxybenzone, Cyasorb UV 24)

[131-53-3] $C_{14}H_{12}O_4$ mol.wt. 244.25

Syntheses

-Preparation by condensation of salicylic acid and
   m-methoxyphenol [1253].
-Preparation by reaction of o-anisoyl chloride with 1,3-di-
   methoxybenzene in the presence of aluminium chloride in
chlorobenzene, first at 0°, then at 88° (21%) [540, 541].
-Preparation by partial demethylation of 2-hydroxy-2',4'-dimethoxybenzophenone or 2,2',4'-tri-
methoxybenzophenone with aluminium chloride or aluminium bromide in chlorobenzene at 90-95°
(good yield) [602].
-Preparation by partial methylation of 2,2',4-trihydroxybenzophenone with dimethyl sulfate in
alkaline solution (50%) [233].
-Also refer to: [285, 383, 406, 723, 1067].

     b.p.₁ 170-175° [540, 541];
     m.p.   71-72° [1253], 70° [606], 69-70° [233], 69° [1217];   ¹H NMR [177], IR [1253],
     UV [177, 233, 475, 515, 606, 1065, 1067, 1217], MS [1253];
     pK_a [606, 791];   TLC [377, 1396];
     gel permeation chromatography [301, 1145];   vapour pressure [1217];
     paper chromatography [911].

(2-Hydroxy-4-methoxyphenyl)(3-hydroxyphenyl)methanone

[84394-12-7] $C_{14}H_{12}O_4$ mol.wt. 244.25

Synthesis

-Preparation by condensation of m-methoxyphenol with
   m-hydroxybenzoic acid in the presence of stannic chloride
   at reflux for 8 h (26%) [1176].

     m.p. 113-114° [1176];   IR [1176].

(2-Hydroxy-4-methoxyphenyl)(4-hydroxyphenyl)methanone

[33257-86-2] $C_{14}H_{12}O_4$ mol.wt. 244.25

Syntheses

-Obtained by condensation of 2-hydroxy-4-methoxy-
   benzoic acid with phenol,
   *in the presence of stannic chloride at 115-120° for
     3-4 h [1358];
*in the presence of zinc chloride at 115-120° for 3-4 h [1359].
-Preparation by reaction of p-hydroxybenzoic acid with m-methoxyphenol in the presence of
phosphorous oxychloride at 60-70° for 1.5 h (36%) [1103, 1104, 1213].
-Also refer to: [747, 1083, 1084, 1127, 1251, 1252].

     m.p. 200° [1104], 136-138° [1358, 1359]. One of the reported melting points is obviously
     wrong.
     IR [1104], UV [1358, 1359].

(2-Hydroxy-5-methoxyphenyl)(2-hydroxyphenyl)methanone

[83570-57-4] $C_{14}H_{12}O_4$ mol.wt. 244.25

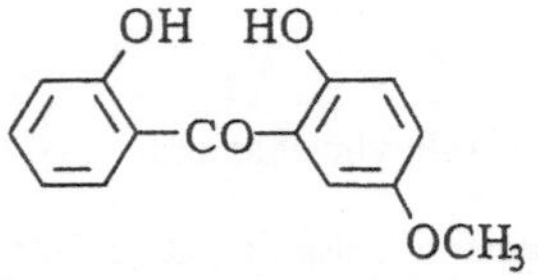

Syntheses

-Preparation by saponification of 2'-(acetylamino)-
  2-hydroxy-5-methoxybenzophenone (SM) with 10%
  sodium hydroxide (37%). SM was obtained by photo-Fries
  rearrangement of p-methoxyphenyl 2-(acetylamino)-
  benzoate in benzene for 2.5 h (40%) [153].
-Also obtained by photo-Fries rearrangement of p-methoxyphenyl salicylate in hexane for 4 h (8%)
  or in methanol for 11 h (23%) [351].
-Also obtained by photo-Fries rearrangement of p-methoxyphenyl o-acetoxybenzoate in benzene,
  followed by saponification of the keto ester so formed [924].
-Preparation by partial demethylation of 2,2',5-trimethoxybenzophenone with aluminium chloride in
  benzene at 50° under nitrogen atmosphere for 12 h (40%) [1250].

   m.p.   93° [351], 88-90° [1250];
        ¹H NMR [267, 351, 1250], IR [351, 1250], UV [1250], MS [1250];   TLC [1250].

(2-Hydroxy-5-methoxyphenyl)(4-hydroxyphenyl)methanone

[80427-40-3] $C_{14}H_{12}O_4$ mol.wt. 244.25

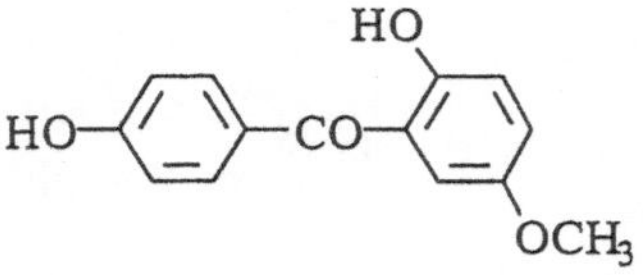

Synthesis

-Obtained by partial demethylation of 2,4',5-trimethoxy-
  benzophenone (SM) with aluminium chloride in nitromethane at 20° for 24 h. SM was obtained by Friedel-
  Crafts acylation of 1,4-dimethoxybenzene with
  p-anisoyl chloride in the presence of stannic chloride in
  nitromethane at 20° for 1 h [863].

-Also refer to: [581] (Japanese patent).

   m.p.   154° [863];   IR [863], UV [863].

(4-Hydroxy-3-methoxyphenyl)(4-hydroxyphenyl)methanone

[147904-63-0] $C_{14}H_{12}O_4$ mol.wt. 244.25

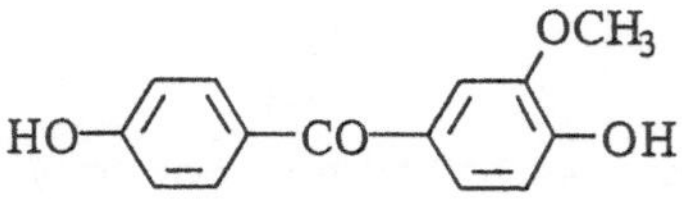

Synthesis

-Refer to: [487].

                              Isolation from natural source

-Identified from lignin dimers [487].

   m.p. (NA);   GC-MS [487].

[3-Hydroxy-4-(methylamino)phenyl](4-hydroxyphenyl)methanone

[136134-36-6] C14H13NO3 mol.wt. 243.26

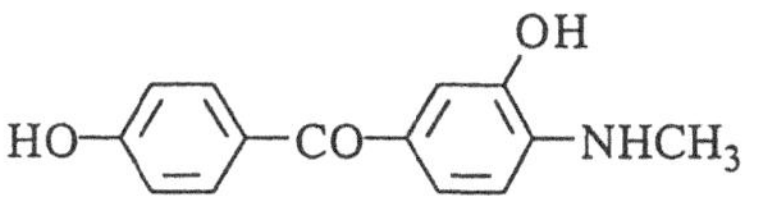

Synthesis

-Preparation from 6-(4-hydroxybenzoyl)-3-methyl-
 benzoxazolinone by alkaline hydrolysis with boiling
 10% aqueous sodium hydroxide solution [148].

m.p. and Spectra (NA).

(4-Hydroxy-2,3-dimethylphenyl)(4-hydroxyphenyl)methanone

[98155-73-8] C15H14O3 mol.wt. 242.27

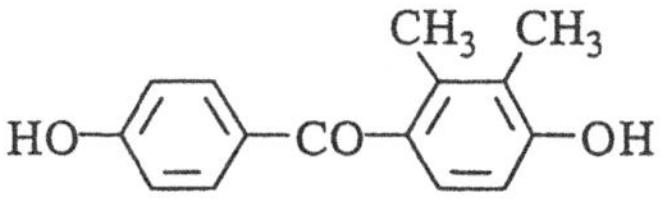

Synthesis

-Refer to: [486] (compound 2).

m.p. and Spectra (NA).

(4-Hydroxy-2,5-dimethylphenyl)(4-hydroxyphenyl)methanone

[98155-75-0] C15H14O3 mol.wt. 242.27

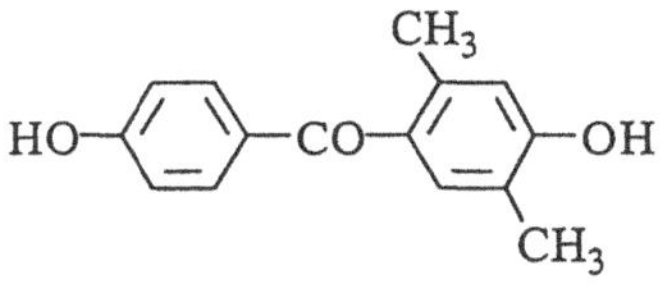

Synthesis

-Refer to: [486] (compound 5).

m.p. and Spectra (NA).

(4-Hydroxy-2,6-dimethylphenyl)(4-hydroxyphenyl)methanone

[93899-05-9] C15H14O3 mol.wt. 242.27

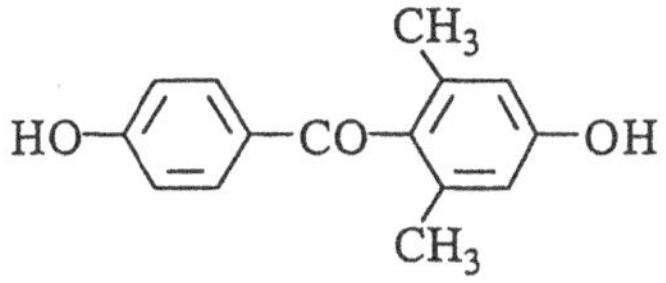

Synthesis

-Preparation by reaction of p-hydroxybenzoic acid with
 3,5-xylenol in the presence of zinc chloride and a mixture
 of polyphosphoric acid/85% phosphoric acid (60:40) at
 40°. Then, during 1.5 h, phosphorous trichloride was
 added and the mixture heated at 70° for 16 h (90%)
 [1302].

-Also refer to: [486] (compound 3).

m.p. and Spectra (NA).

(4-Hydroxy-3,5-dimethylphenyl)(2-hydroxyphenyl)methanone

$C_{15}H_{14}O_3$ mol.wt. 242.27

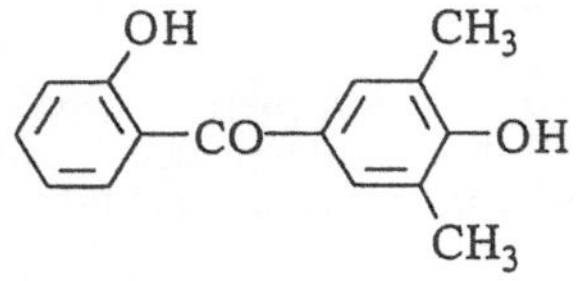

Synthesis

-Obtained by photo-Fries rearrangement of 2,6-dimethyl-
phenyl salicylate (by-product) [970].

m.p. (NA);   UV [970].

(4-Hydroxy-3,5-dimethylphenyl)(3-hydroxyphenyl)methanone

[92005-26-0] $C_{15}H_{14}O_3$ mol.wt. 242.27

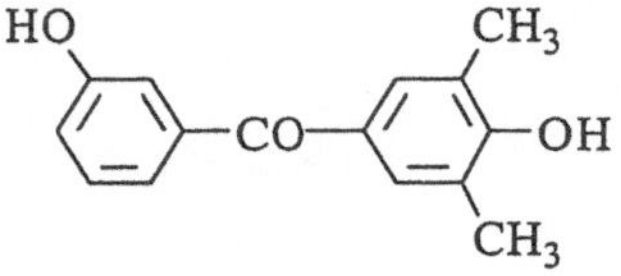

Synthesis

-Preparation by acylation of 2,6-dimethylphenol with m-(trichloromethyl)phenyl m-(trichloromethyl)benzoate in
methylene chloride in the presence of aluminium chloride at
0-5° over 1 h, then at r.t. for 1 h, followed by alkaline
hydrolysis of the resulting keto ester [581] (Japanese
patent).

m.p. and Spectra (NA).

(4-Hydroxy-3,5-dimethylphenyl)(4-hydroxyphenyl)methanone

[92005-13-5] $C_{15}H_{14}O_3$ mol.wt. 242.27

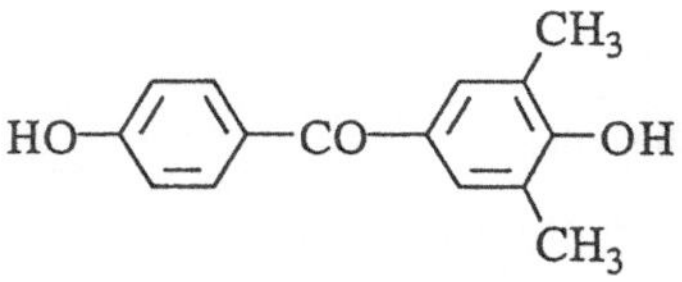

Synthesis

-Preparation by acylation of 2,6-dimethylphenol with
p-(trichloromethyl)phenyl p-(trichloromethyl)benzoate in
methylene chloride in the presence of aluminium
chloride at 0-5° over 1 h, then at r.t. for 1 h, followed by
alkaline hydrolysis of the resulting keto ester [581]
(Japanese patent).

m.p. and Spectra (NA).

(4-Ethoxy-2-hydroxyphenyl)(2-hydroxyphenyl)methanone

$C_{15}H_{14}O_4$ mol.wt. 258.27

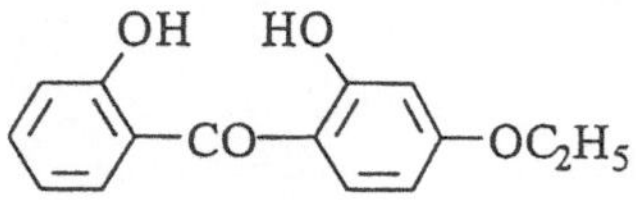

Synthesis

-Preparation by partial deethylation of 2,4-diethoxy-
2'-hydroxybenzophenone with aluminium chloride in
chlorobenzene at 80-100° [602].

m.p. and Spectra (NA).

(2-Hydroxy-3,4-dimethoxyphenyl)(2-hydroxyphenyl)methanone

$C_{15}H_{14}O_5$ mol.wt. 274.27

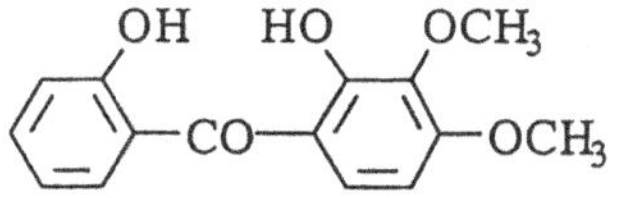

Synthesis

-Preparation by action of o-anisoyl chloride with pyrogallol trimethyl ether in the presence of aluminium chloride.
-Refer to: Chem. Abstr., **9**, 609[5-6] (1915)*.

m.p. 127°*;   Spectra (NA).

(2-Hydroxy-4,6-dimethoxyphenyl)(3-hydroxyphenyl)methanone

[34425-65-5] $C_{15}H_{14}O_5$ mol.wt. 274.27

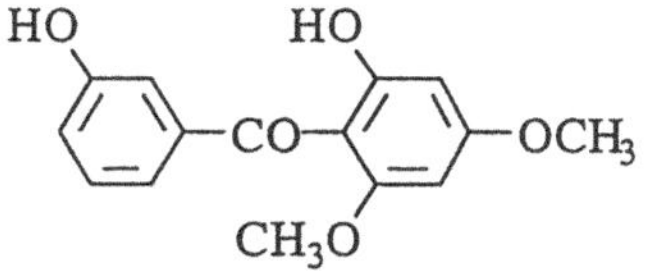

Synthesis

-Not yet described.

                    Isolation from natural source

-From the heartwood of *Allanblackia floribunda* Oliver (Guttiferae) [831].

m.p. 105-107° [831];
^{1}H NMR [831], IR [831], UV [831], MS [831].

[5-(1,1-Dimethylethyl)-2-hydroxyphenyl](2-hydroxyphenyl)methanone

[125182-25-4] $C_{17}H_{18}O_3$ mol.wt. 270.33

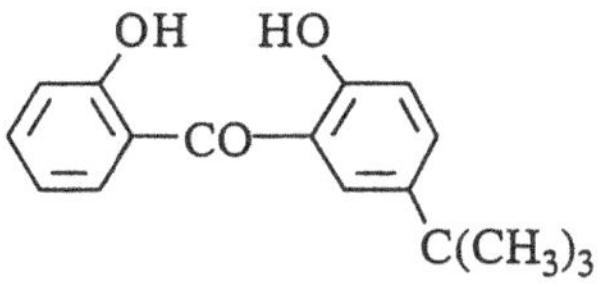

Synthesis

-Obtained by treatment of 2,2'-dihydroxybenzophenone at 120° with a mixture of isobutylene/nitrogen (1:1) in the presence of a macroreticular acid ion exchanger (Wofatit OK 80) as catalyst for 1 h (40%) [111].

b.p.$_{0.15}$ 180-185° [111];   Spectra (NA).

[4-Hydroxy-2-methyl-5-(1-methylethyl)phenyl](2-hydroxyphenyl)methanone

$C_{17}H_{18}O_3$ mol.wt. 270.33

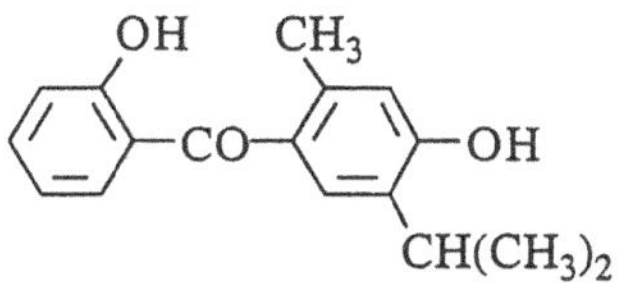

Synthesis

-Preparation by demethylation of 2',4-dimethoxy-2-methyl-5-isopropylbenzophenone with refluxing pyridinium chloride at 205-215° for 2 h (61%) [1148].

b.p.$_{0.6}$ 195-200° [1148];   m.p. 117-118° [1148];   Spectra (NA).

[4-Hydroxy-2-methyl-5-(1-methylethyl)phenyl](4-hydroxyphenyl)methanone

$C_{17}H_{18}O_3$ mol.wt. 270.33

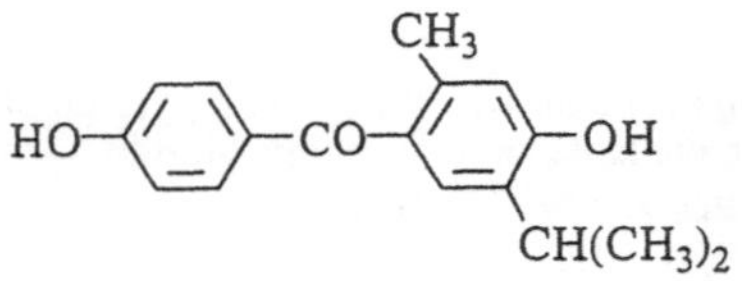

Synthesis

-Obtained by demethylation of 4,4'-dimethoxy-
2-methyl-5-isopropylbenzophenone in refluxing
pyridinium chloride (205-215°) for 2 h (20%) [1148].

b.p._{0.6} 200-210° [1148];   Spectra (NA).

(4-Butoxy-2-hydroxyphenyl)(2-hydroxyphenyl)methanone

$C_{17}H_{18}O_4$ mol.wt. 286.33

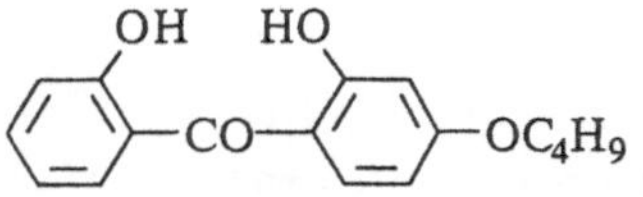

Synthesis

-Refer to: [911].

m.p. and Spectra (NA);   paper chromatography [911].

(2-Hydroxy-4-methoxy-3-propylphenyl)(2-hydroxyphenyl)methanone

[115296-05-4] $C_{17}H_{18}O_4$ mol.wt. 286.33

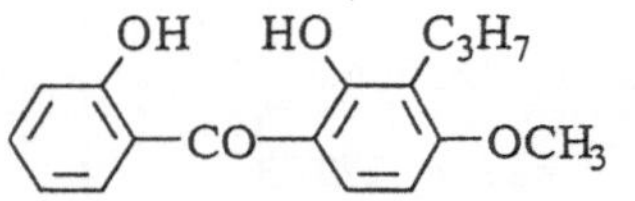

Syntheses

-Preparation by partial methylation of 2,2',4-trihydroxy-
3-propylbenzophenone with dimethyl sulfate in the
presence of potassium carbonate in refluxing 2-butanone
for overnight (93%) [456].
-Also obtained (trace) by reaction of o-anisoyl chloride with 1,3-dimethoxy-2-propylbenzene in
methylene chloride in the presence of aluminium chloride, first at 0° for 2 h and at r.t. for 1 h
[456].

oil [456];   b.p. and Spectra (NA).

[5-(1,1-Dimethylethyl)-2-hydroxy-4-methoxyphenyl](2-hydroxyphenyl)methanone

$C_{18}H_{20}O_4$ mol.wt. 300.35

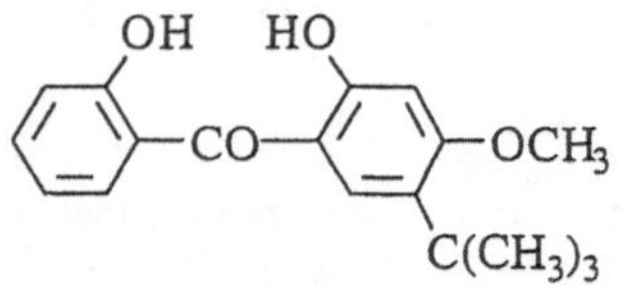

Synthesis

-Preparation by selective methylation of 5-tert-butyl-
2,2',4-trihydroxybenzophenone with dimethyl sulfate in
refluxing methyl ethyl ketone in the presence of potassium
carbonate [587].

m.p. 116°5 [587, 588];   Spectra (NA).

(2-Hydroxyphenyl)[2-hydroxy-4-(2-propenyloxy)-3-propylphenyl]methanone

[115308-88-8] $C_{19}H_{20}O_4$ mol.wt. 312.37

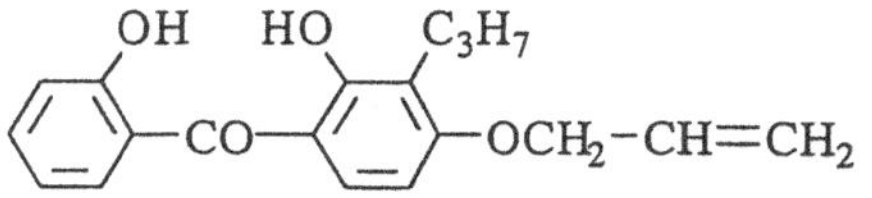

Synthesis

-Obtained by reaction of allyl bromide with 2,2',4-trihydroxy-3-propylbenzophenone in the presence of potassium carbonate and potassium iodide in refluxing methyl ethyl ketone for 3 h (30%) [456].

oil [456];   b.p. and Spectra (NA).

[4-(Hexyloxy)-2-hydroxyphenyl](2-hydroxyphenyl)methanone

[65221-06-9] $C_{19}H_{22}O_4$ mol.wt. 314.38

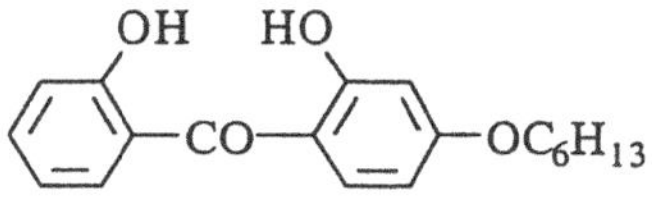

Synthesis

-Refer to: [1411].

m.p. (NA);   UV [1411].

[3,5-Bis(1,1-dimethylethyl)-4-hydroxyphenyl](2-hydroxyphenyl)methanone

$C_{21}H_{26}O_3$ mol.wt. 326.44

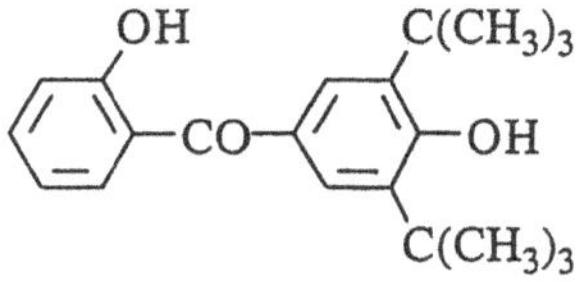

Synthesis

-Obtained by UV light irradiation of phenyl 3,5-di-tert-butyl-4-hydroxybenzoate in isooctane [297].

m.p. 119° [297];   Spectra (NA).

[4-(2-Ethylhexyl)-2-hydroxyphenyl](2-hydroxyphenyl)methanone

[84875-84-3] $C_{21}H_{26}O_3$ mol.wt. 326.44

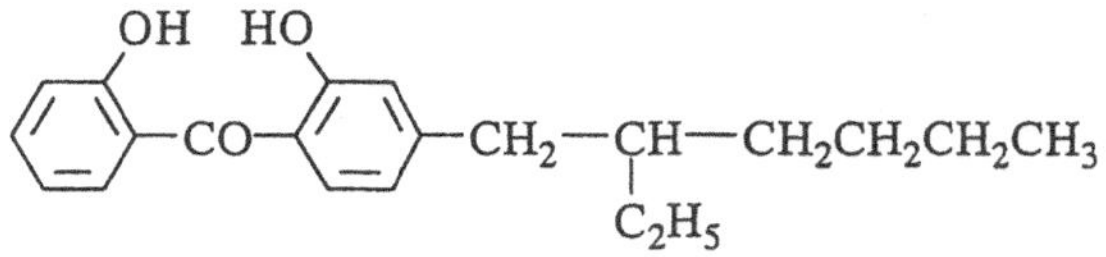

Synthesis

-Refer to: [667].

m.p. and Spectra (NA).

(2-Hydroxy-3-octylphenyl)(2-hydroxyphenyl)methanone

$C_{21}H_{26}O_3$ mol.wt. 326.44

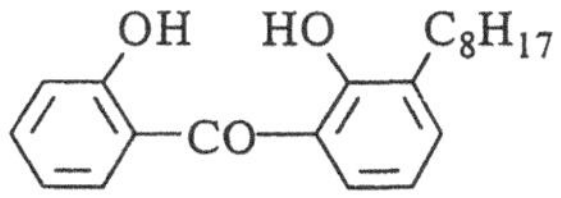

Synthesis

-Obtained by photo-Fries rearrangement of o-octylphenyl salicylate (major product) [970].

m.p. (NA);   UV [970].

(2-Hydroxy-5-octylphenyl)(2-hydroxyphenyl)methanone

$C_{21}H_{26}O_3$ mol.wt. 326.44

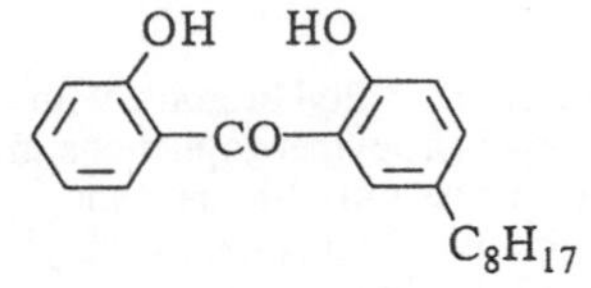

Synthesis

-Obtained by photo-Fries rearrangement of p-octylphenyl
  salicylate (major product) [970].

m.p. (NA);   UV [970].

(4-Hydroxy-3-octylphenyl)(2-hydroxyphenyl)methanone

$C_{21}H_{26}O_3$ mol.wt. 326.44

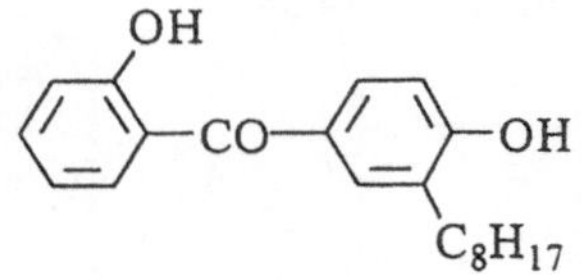

Synthesis

-Obtained by photo-Fries rearrangement of o-octylphenyl
  salicylate (by-product) [970].

m.p. (NA);   UV [970].

[2-Hydroxy-4-(octyloxy)phenyl](2-hydroxyphenyl)methanone

[85-24-5] $C_{21}H_{26}O_4$ mol.wt. 342.44

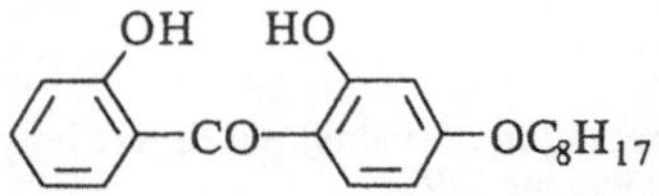

Synthesis

-Preparation by reaction of octyl bromide with 2,2',4-tri-
  hydroxybenzophenone,
  *in the presence of potassium carbonate (50%) [233];
  *in the presence of sodium carbonate in refluxing dilute
ethanol [44].
-Also refer to: [227, 228, 229, 304, 406, 468, 492, 619, 855, 870, 929, 930, 943, 1012, 1218].

m.p.   90°5-91° [233];
UV [233];   paper chromatography [911].

[2-Hydroxy-4-(octyloxy)phenyl](4-hydroxyphenyl)methanone

$C_{21}H_{26}O_4$ mol.wt. 342.44

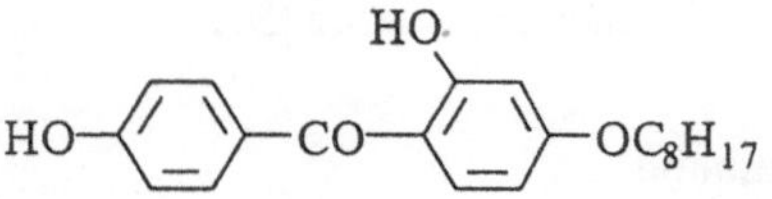

Synthesis

-Refer to: [930] (Japanese patent).

m.p. and Spectra (NA).

**[2-Hydroxy-3-(3-methyl-2-butenyl)-4-[(3-methyl-2-butenyl)oxy]phenyl]-
(4-hydroxyphenyl)methanone**

[63565-06-0] $C_{23}H_{26}O_4$ mol.wt. 366.46

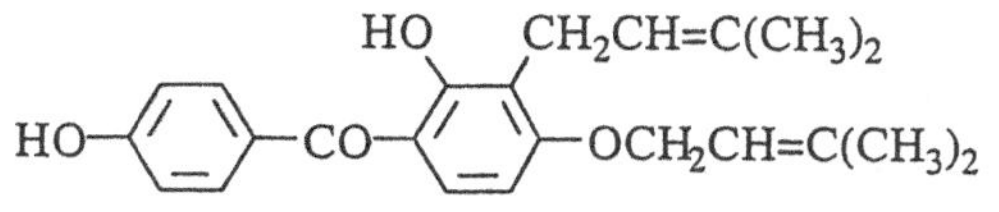

Synthesis

-Obtained (poor yield) by reaction of
prenyl bromide with 2,4,4'-trihydroxy-
benzophenone in the presence of
potassium carbonate in refluxing acetone
for 3 h (7%) [875].

m.p. 80° [875]; ^{1}H NMR [875], IR [875], UV [875].

(3-Dodecyl-2-hydroxy-5-methylphenyl)(2-hydroxyphenyl)methanone

$C_{26}H_{36}O_3$ mol.wt. 396.57

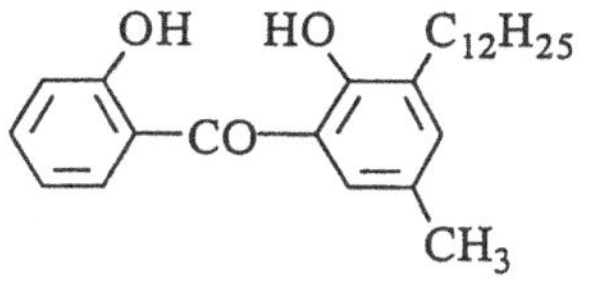

Synthesis

-Obtained by photo-Fries rearrangement of 2-dodecyl-
4-methylphenyl salicylate (major product) [970].

m.p. (NA);   UV [970].

[5-(1,1-Dimethylethyl)-4-(dodecyloxy)-2-hydroxyphenyl](2-hydroxyphenyl)methanone

$C_{29}H_{42}O_4$ mol.wt. 454.65

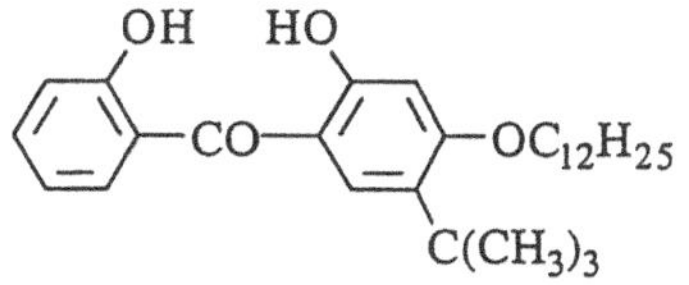

Synthesis

-Preparation by selective alkylation of 5-tert-butyl-
2,2',4-trihydroxybenzophenone with lauryl bromide in
refluxing methyl ethyl ketone in the presence of
potassium carbonate [587].

m.p. 81° [587, 588];   Spectra (NA).

2.2.2. *Substituents located on both rings*

Symmetrical ketones

Bis(2,3,5,6-tetrachloro-4-hydroxyphenyl)methanone

$C_{13}H_2Cl_8O_3$ mol.wt. 489.78

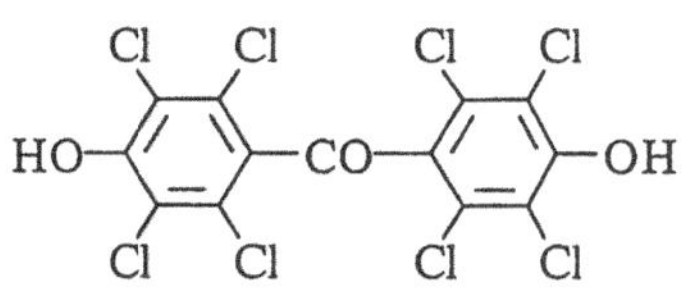

Synthesis

-Refer to: [486] (compound 12).

m.p. and Spectra (NA).

Bis(3-bromo-4-hydroxy-5-nitrophenyl)methanone

$C_{13}H_6Br_2N_2O_7$ mol.wt. 462.01

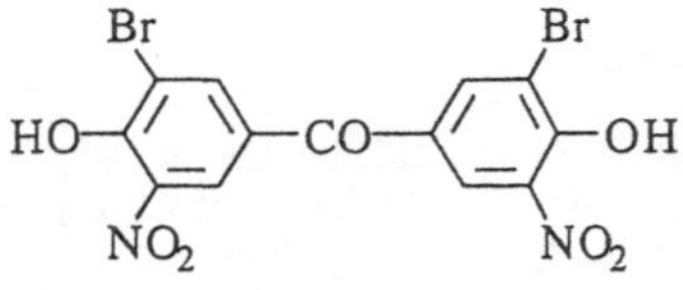

Synthesis

-Preparation by adding sodium nitrite to an hot solution of 3,3',5,5'-tetrabromo-4,4'-dihydroxybenzophenone in acetic acid [1473].

m.p. 246° [1473];   Spectra (NA).

Bis(3,5-dibromo-2-hydroxyphenyl)methanone

$C_{13}H_6Br_4O_3$ mol.wt. 529.80

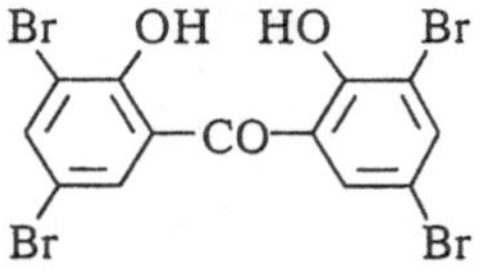

Synthesis

-Obtained by treatment of 2,2'-dihydroxybenzophenone in hot acetic acid with bromine (60%) [551].
-Also refer to: [859].

m.p. 178°5-180°5 [551];   Spectra (NA).

Bis(3,5-dibromo-4-hydroxyphenyl)methanone

[28818-29-3] $C_{13}H_6Br_4O_3$ mol.wt. 529.80

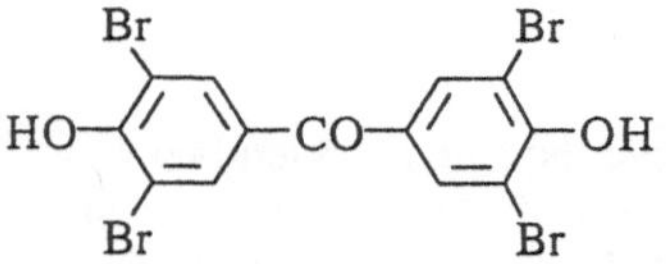

Syntheses

-Obtained by action of bromine with 4,4'-dihydroxy-benzophenone [114, 115], (86%) [551].
-Also obtained by oxidation of α,α-bis(3,5-dibromo-4-hydroxyphenyl)dichloroethylene by heating with chromium trioxide (81%) [962].

m.p. 230-232° [551], 225-226° [962, 1473], 213-214° [114, 115];   Spectra (NA).

Bis(3,5-dichloro-4-hydroxyphenyl)methanone

$C_{13}H_6Cl_4O_3$ mol.wt. 352.00

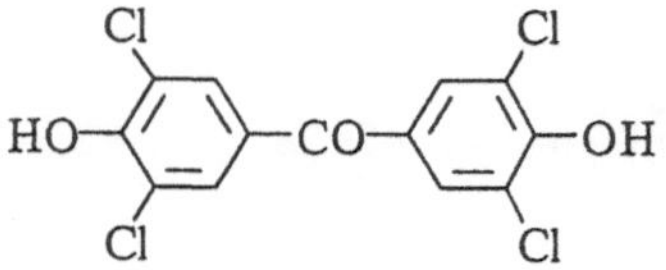

Synthesis

-Preparation by reaction of chlorine with 4,4'-dihydroxy-benzophenone in acetic acid [1473].
-Also refer to: [876, 1003] (Japanese papers).

m.p. 231-232° [1473];   Spectra (NA).

Bis(2-hydroxy-3,5-diiodophenyl)methanone

[33417-57-1] $C_{13}H_6I_4O_3$ mol.wt. 717.80

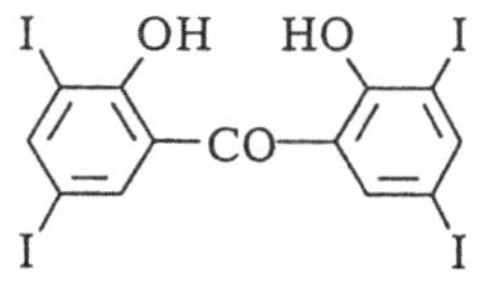

Synthesis

-Preparation by iodination of 2,2'-dihydroxybenzophenone,
*with iodine and iodic acid in dilute ethanol for 20 min (77%)
[122];
*with iodine and potassium iodide in aqueous ammonia
[122].

m.p. 225° [122];   Spectra (NA).

Bis(4-hydroxy-3,5-diiodophenyl)methanone

[15198-16-0] $C_{13}H_6I_4O_3$ mol.wt. 717.80

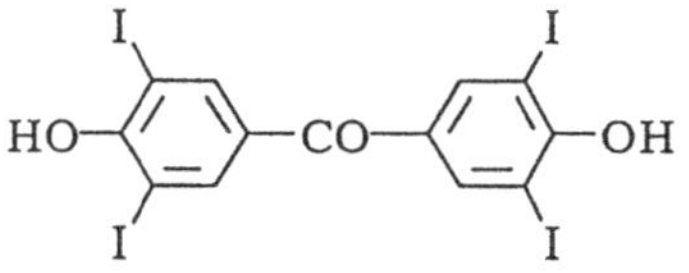

Syntheses

-Preparation by iodination of 4,4'-dihydroxy-
  benzophenone,
*with iodine and potassium iodide in ammonia for 1 h
(65%) [122];
*with iodine and iodic acid in dilute ethanol for 15 min
  [122];
*with iodine monochloride (Wijs'chloride) at r.t. [122].
-Also obtained (by-product) by reaction of 4-hydroxy-3,5-diiodophenylpyruvic acid with
4-hydroxy-3,5-diiodobenzoic acid (6%) [979], according to the methods [898, 1255].

m.p. 255° (d) [979], 254° (d) [122];
IR [979], UV [979].

Bis(2-bromo-4-hydroxyphenyl)methanone

[98155-80-7] $C_{13}H_8Br_2O_3$ mol.wt. 372.01

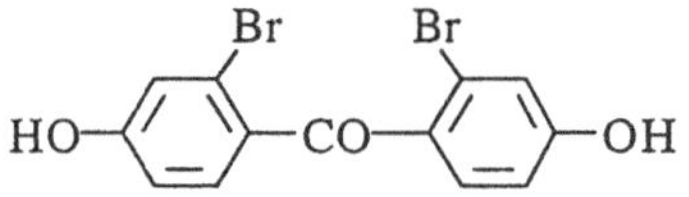

Synthesis

-Refer to: [486] (compound 29).

m.p. and Spectra (NA).

Bis(3-bromo-4-hydroxyphenyl)methanone

[5423-21-2] $C_{13}H_8Br_2O_3$ mol.wt. 372.01

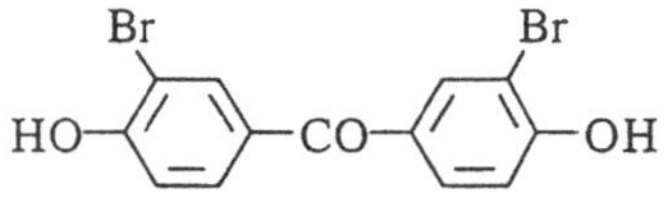

Synthesis

-Preparation by demethylation of 3,3'-dibromo-4,4'-di-
  methoxybenzophenone (SM) with pyridinium chloride at
  210° for 20 min (85%) [813]. SM was obtained by action
  of bromine with 4,4'-dimethoxybenzophenone in acetic
  acid for 1 h in daylight (77%).

m.p. 214-215° [813]; ^{1}H NMR [813], IR [813].

Bis(5-bromo-2-hydroxyphenyl)methanone

$C_{13}H_8Br_2O_3$ mol.wt. 372.01

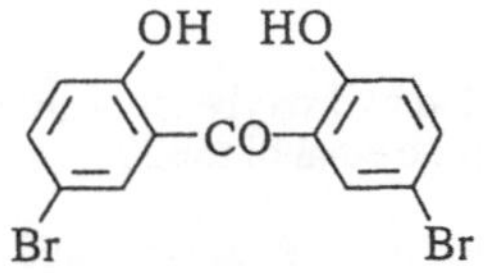

Synthesis

-Refer to: [859].

m.p. and Spectra (NA).

Bis(2-chloro-4-hydroxyphenyl)methanone

[94323-04-3] $C_{13}H_8Cl_2O_3$ mol.wt. 283.11

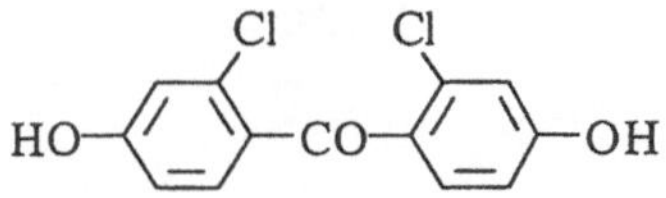

Synthesis

-Refer to: [486] (compound 13).

m.p. and Spectra (NA);   hemihydrate [486].

Bis(3-chloro-4-hydroxyphenyl)methanone

[79616-16-3] $C_{13}H_8Cl_2O_3$ mol.wt. 283.11

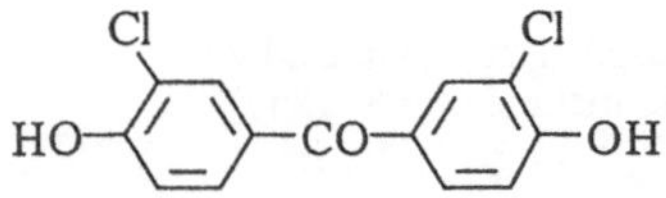

Synthesis

-Preparation by demethylation of 3,3'-dichloro-4,4'-di-
methoxybenzophenone (SM) with pyridinium chloride at
210° for 20 min (84%) [813]. SM was obtained by
chlorination of 4,4'-dimethoxybenzophenone with
sulfuryl chloride in methylene chloride at 45° for 5 h (74%).

m.p.   206-207° [813]; ^{1}H NMR [813], IR [813].

Bis(5-chloro-2-hydroxyphenyl)methanone

[6178-89-8] $C_{13}H_8Cl_2O_3$ mol.wt. 283.11

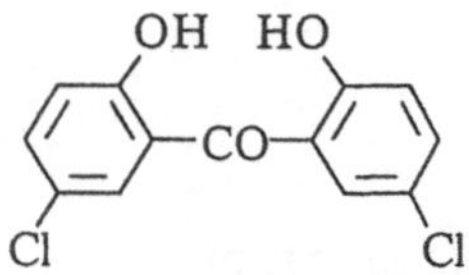

Synthesis

-Preparation by oxidation of 2,2'-methylenebis(4-chloro-
anisole) with chromium trioxide in refluxing acetic acid,
followed by demethylation of the 5,5'-dichloro-2,2'-di-
methoxybenzophenone so obtained with aluminium chloride
in chlorobenzene at 60° for 5.5 h [594], (92%) [405].

-Also refer to: [859, 888, 949, 1091].

m.p.   151-152° [405];   Spectra (NA).

Bis(2-fluoro-4-hydroxyphenyl)methanone

[98155-79-4] $C_{13}H_8F_2O_3$ mol.wt. 250.20

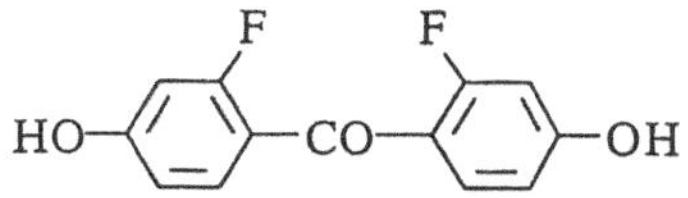

Synthesis

-Refer to: [486] (compound 28).

m.p. and Spectra (NA).

Bis(2-hydroxy-5-iodophenyl)methanone

[33417-58-2] $C_{13}H_8I_2O_3$ mol.wt. 466.01

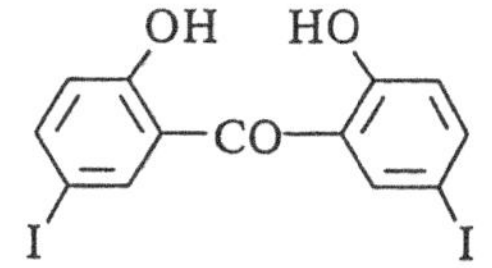

Synthesis

-Obtained by iodination of 2,2'-dihydroxybenzophenone with iodine monochloride in acetic acid for 24 h at r.t. (23%) [122].

m.p. 194-195° [122]; Spectra (NA).

Bis(2-hydroxy-5-nitrophenyl)methanone

$C_{13}H_8N_2O_7$ mol.wt. 304.22

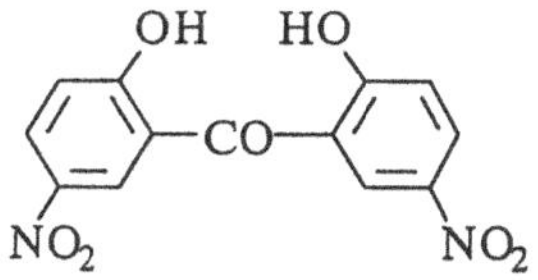

Synthesis

N.B.: The 2,7-dinitroxanthone was obtained on heating 2,2'-dimethoxy-5,5'-dinitrobenzophenone (m.p. 188°) with 75% sulfuric acid at 150° for 1 h *via* 2,2'-dihydroxy-5,5'-dinitrobenzophenone [1072].
-Also refer to: [113, 808].

m.p. and Spectra (NA).

Bis(4-hydroxy-3-nitrophenyl)methanone

[37567-35-4] $C_{13}H_8N_2O_7$ mol.wt. 304.22

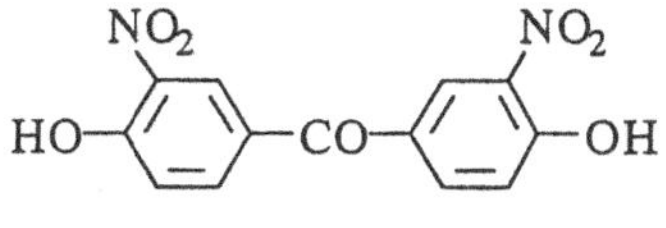

Synthesis

-Preparation by hydrolysis of 4,4'-dichloro-3,3'-dinitrobenzophenone with 5-15% sodium hydroxide at 155-160° for 1-1.5 h (97-98%) [927].

m.p. 195°2-195°5 [927]; Spectra (NA).

Bis(3-amino-4-hydroxyphenyl)methanone

[22445-98-3] $C_{13}H_{12}N_2O_3$ mol.wt. 244.25

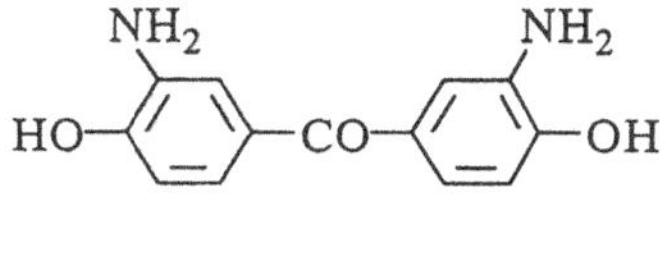

Synthesis

-Preparation by hydrogenation of 4,4'-dihydroxy-3,3'-dinitrobenzophenone in the presence of Raney nickel in water at 85-90° for 1 h under 100 atmospheres (69%) [927].

m.p. >220° (d) [927]; Spectra (NA).

Bis(4-amino-2-hydroxyphenyl)methanone

[107516-91-6] $C_{13}H_{12}N_2O_3$ mol.wt. 244.25

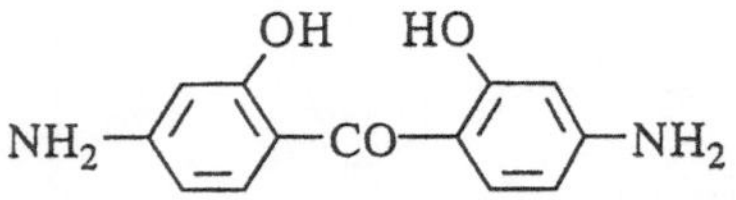

Synthesis

-Preparation from 4-amino-2'-hydroxy-2-methoxy-
4'-nitrobenzophenone by refluxing with freshly
distilled constant-boiling hydriodic acid at 140° for
7 h (81%) [662].

m.p.   247-248° [662];   Spectra (NA).

Bis(4-amino-2-hydroxyphenyl)methanone *(Dihydrochloride)*

$C_{13}H_{12}N_2O_3,2HCl$ mol.wt. 317.17

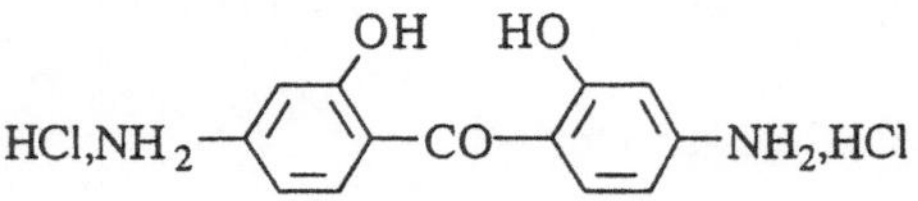

Synthesis

-Preparation from 4,4'-diamino-2,2'-di-
hydroxybenzophenone with hydrochloric
acid in refluxing dilute ethanol (74%) [662].

m.p.   208° (d) [662];   Spectra (NA).

Bis(2-chloro-4-hydroxy-6-methylphenyl)methanone

$C_{15}H_{12}Cl_2O_3$ mol.wt. 311.16

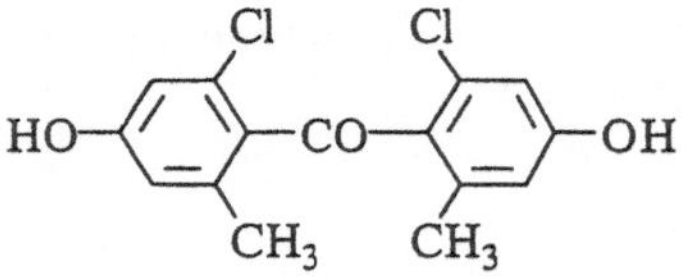

Synthesis

-Refer to: [486] (compound 35).

m.p. and Spectra (NA).

Bis(2-hydroxy-4-methylphenyl)methanone

[24018-76-6] $C_{15}H_{14}O_3$ mol.wt. 242.27

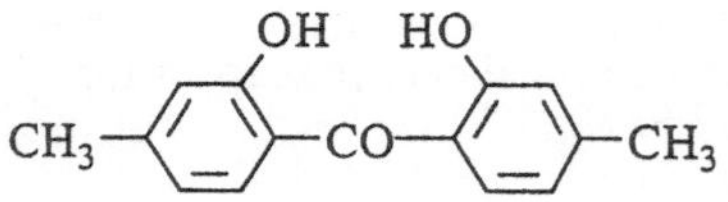

Synthesis

-Refer to: [968] and also [674, 1011, 1261] (Japanese
papers).

m.p. and Spectra (NA).

Bis(2-hydroxy-5-methylphenyl)methanone

[27404-62-2] $C_{15}H_{14}O_3$ mol.wt. 242.27

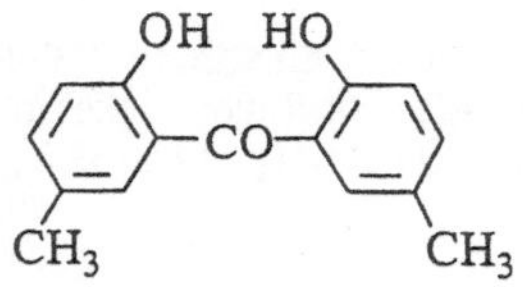

Syntheses

-Preparation by Friedel-Crafts acylation of p-methoxytoluene
with 2-methoxy-5-methylbenzoyl chloride in carbon
disulfide in the presence of aluminium chloride, first at 0° for
5 h, then at r.t. overnight and at reflux for 2 h (74%) [671].

-Also obtained by total demethylation of 2,2'-dimethoxy-5,5'-dimethylbenzophenone (m.p. 84°) with hydriodic acid in refluxing acetic acid for 7 h (83%) [671].
-Also obtained (poor yield) by Fries rearrangement of p-cresyl 2-methoxy-5-methylbenzoate in nitrobenzene with aluminium chloride for 20 min at 130° (9%) [671].
-Also obtained (poor yield) by alkaline melting of p-cresolphthalein at 200° with potassium hydroxide [116, 671].
-Also obtained by fusion of 2,7-dimethylfluoran with potassium hydroxide at 220-240° [1110].
-Also refer to: [391, 672, 673].

m.p. 107° [672], 106-107° [391, 671], 104-106° [1110], 104-105° [116];
IR [672], UV [672], MS [391].

Bis(4-hydroxy-2-methylphenyl)methanone

[98155-74-9] $C_{15}H_{14}O_3$ mol.wt. 242.27

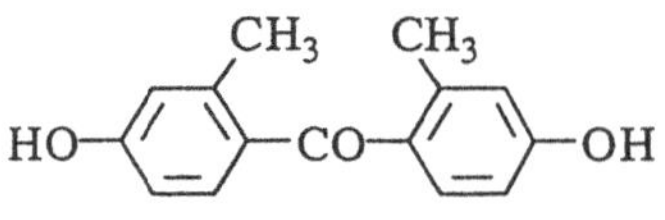

Synthesis

-Refer to: [486] (compound 4).

m.p. and Spectra (NA).

Bis(4-hydroxy-3-methylphenyl)methanone

[94323-02-1] $C_{15}H_{14}O_3$ mol.wt. 242.27

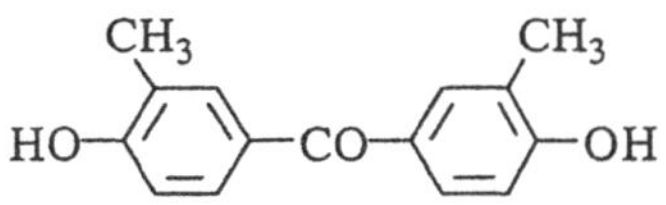

Syntheses

-Obtained by condensation of o-cresol with carbon tetrachloride in the presence of metallic halides (zinc chloride, aluminium chloride or stannic chloride) at 100-130° in an autoclave [484].
-Obtained by oxidation of o-cresaurin in 5% sodium hydroxide solution by passing a very slow stream of air during several days (41%) [484].

m.p. 240° [484]; Spectra (NA).

Bis[4-hydroxy-3-(hydroxymethyl)phenyl]methanone

[74697-55-5] $C_{15}H_{14}O_5$ mol.wt. 274.27

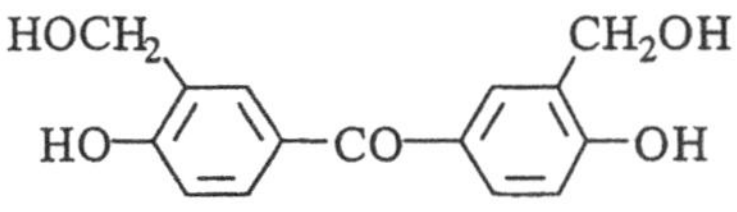

Synthesis

-Refer to: [752].

m.p. and Spectra (NA).

Bis(2-hydroxy-4-methoxyphenyl)methanone
(UV 12, Uvinul D-49, Uvinul 490, Uvinul 3049)

[131-54-4] $C_{15}H_{14}O_5$ mol.wt. 274.27

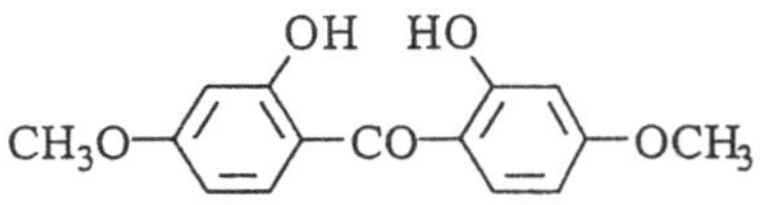

Syntheses

-Preparation by reaction of 2-hydroxy-4-methoxybenzoic acid with resorcinol monomethyl ether in the presence of zinc chloride and phosphorous oxychloride at 70-75° for 2 h (27%) [507].

-Preparation by reaction of 2,4-dimethoxybenzoyl chloride with 1,3-dimethoxybenzene in the presence of aluminium chloride,
*in a chlorobenzene/N,N-dimethylformamide mixture (22:1) at 115° (72%) [539, 1067];
*in chlorobenzene at 90° [43].
-Preparation by partial demethylation of 2-hydroxy-2',4,4'-trimethoxybenzophenone or 2,2',4,4'-tetramethoxybenzophenone with aluminium chloride or aluminium bromide in chlorobenzene at 90-95° (good yield). The same result is obtained using ethylene dichloride or nitrobenzene as the solvent [602].
-Preparation by reaction of phosgene with resorcinol dimethyl ether in the presence of aluminium chloride [1079].
-Also refer to: [43, 285, 383, 384, 475, 541, 565, 615, 810, 958, 968].

      m.p.   139-140° [507], 137° [1079], 136-137° [785], 135° [1217],
             133-136° [606], 133-135° [43];
      EPR [267], IR [785], UV [474, 475, 541, 606, 785, 1065, 1067, 1079, 1217];
      TLC [377];   vapour pressure [1217];   pK_a [606];
      gel permeation chromatography [1145].

Bis(4-hydroxy-3-methoxyphenyl)methanone

[5623-44-9] $C_{15}H_{14}O_5$ mol.wt. 274.27

Syntheses

-Obtained from 4-hydroxy-3,3',4'-trimethoxy-1,2-di-phenyl-1,2-ethanediol (Z configuration) after 3 h radiolysis of aqueous sodium hydroxide solution [473].

-Also obtained by heating a mixture of the 4,4'-dihydroxy-3,3'-dimethoxydiphenylmethane, 2 N sodium hydroxide and nitrobenzene for 2 h at 170° (21%) [255, 1420].
-Also obtained by mild oxidation of various materials* with cupric hydroxide in aqueous sodium hydroxide at 170° for 5 h in a stainless steal bomb,
*three polymers prepared from isoeugenol by oxidative coupling with cuprous chloride in pyridine at 100° under oxygen [402];
*dehydrodiisoeugenol [402];
*milled wood lignin [402];
*lignin and related products [1045].
-Also refer to: [621, 1413, 1414].
N. B.: Na salt [1079].

      m.p. (NA);   IR [1079];
      TLC [402];   GLC [402, 1045];   GC and GC-MS [255, 1420].

Bis(4-hydroxy-3,5-dimethylphenyl)methanone

                                          $C_{17}H_{18}O_3$ mol.wt. 270.33

Synthesis

-Refer to: Chem. Abstr., **127**, 109367s (1997).

m.p. and Spectra (NA).

Bis(4-ethoxy-2-hydroxyphenyl)methanone

[15889-67-5] $C_{17}H_{18}O_5$ mol.wt. 302.33

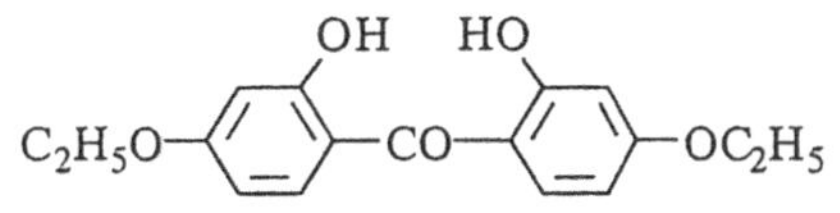

C_2H_5O —⟨ ⟩— CO —⟨ ⟩— OC_2H_5
(OH HO)

Synthesis

-Preparation by partial dealkylation of 2-hydroxy-2'-methoxy-4,4'-diethoxybenzophenone, 2-hydroxy-2',4,4'-triethoxybenzophenone or 2,2',4,4'-tetraethoxybenzophenone with aluminium chloride in chlorobenzene at 80-100° [602].

m.p. and Spectra (NA).

Bis(2-hydroxy-3-methoxy-5-methylphenyl)methanone

[17772-33-7] $C_{17}H_{18}O_5$ mol.wt. 302.33

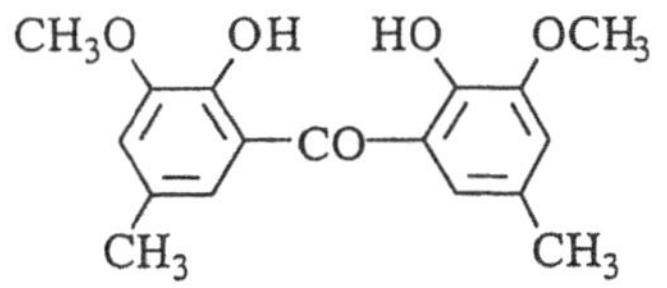

(CH$_3$O OH HO OCH$_3$) —CO— (CH$_3$ CH$_3$)

Synthesis

-Obtained by heating a mixture of the 2,2'-dihydroxy-3,3'-dimethoxy-5,5'-dimethyldiphenylmethane, 2 N sodium hydroxide and nitrobenzene for 2 h at 170° (24%) [255, 1420].

m.p. and Spectra (NA); GC and GC-MS [255, 1420].

Bis(2-hydroxy-4-methoxy-6-methylphenyl)methanone

[78135-60-1] $C_{17}H_{18}O_5$ mol.wt. 302.33

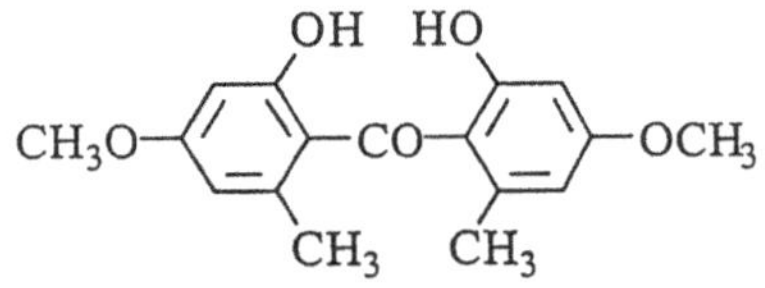

CH_3O —⟨ ⟩— CO —⟨ ⟩— OCH_3
(OH HO)
(CH$_3$ CH$_3$)

Synthesis

-Preparation by reduction of 4',6-dimethoxy-4,6'-dimethylspiro[benzofuran-2,1'-cyclohexa-3',5'-diene]-2',3 (2H) dione with zinc dust in acetic acid at r.t. for 1 h (94%) [1174].

m.p. 159-160° [1174]; ^{1}H NMR [1174], MS [1174].

Bis(4-hydroxy-3,5-dimethoxyphenyl)methanone

[34007-64-2] $C_{17}H_{18}O_7$ mol.wt. 334.33

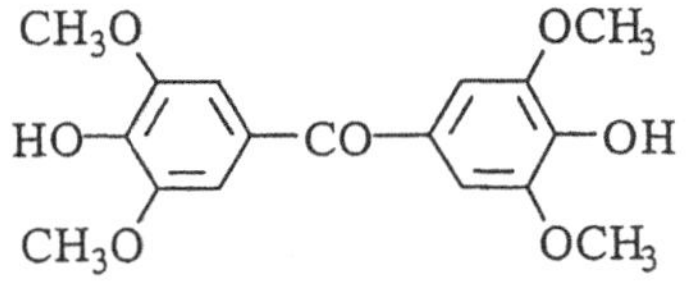

(CH$_3$O OCH$_3$)
HO—⟨ ⟩— CO —⟨ ⟩—OH
(CH$_3$O OCH$_3$)

Synthesis

-Obtained by heating a mixture of the 4,4'-dihydroxy-3,3',5,5'-tetramethoxydiphenylmethane (disyringyl-methane), 2 N sodium hydroxide and nitrobenzene for 2 h at 170° (18%) [255, 1420].

m.p. and Spectra (NA); GC and GC-MS [255, 1420].

Bis[2-hydroxy-4-(2-hydroxyethoxy)phenyl]methanone

[15577-13-6] $C_{17}H_{18}O_7$ mol.wt. 334.33

HOCH$_2$CH$_2$O—(OH)—CO—(HO)—OCH$_2$CH$_2$OH

Synthesis

-Preparation by bubbling ethylene oxide into a hot mixture of 2,2',4,4'-tetrahydroxy-benzophenone and aqueous sodium hydroxide during 1 h at 55° (98%) [985].

-Also refer to: [377, 606].

   m.p. 152° [606];   UV [606];   pK_a [606];   TLC [377].

Bis[5-(1,1-dimethylethyl)-2-hydroxyphenyl]methanone

[25446-98-4] $C_{21}H_{26}O_3$ mol.wt. 326.44

(CH$_3$)$_3$C—(OH)—CO—(HO)—C(CH$_3$)$_3$

Syntheses

-Obtained by photo-Fries rearrangement of p-tert-butylphenyl carbonate,
*in ethanol for 24 h (26%) [600];
*in benzene for 24 h (36%) [600];
*in ethylene dichloride [614], where the rearrangement proceeds *via* the formation of an intermediate ester, the p-tert-butylphenyl 5-tert-butylsalicylate [614].

-Also obtained by treatment of 2,2'-dihydroxybenzophenone at 120° with a mixture of isobutylene/nitrogen (1:1) in the presence of a macroreticular acid ion exchanger (Wofatit OK 80) as catalyst for 1 h (30%) [111].

   b.p.$_{0.15}$ 200-205° [111];   m.p. 104-106° [600];   IR [600].

Bis[4-hydroxy-2-methyl-5-(1-methylethyl)phenyl]methanone

                                      $C_{21}H_{26}O_3$ mol.wt. 326.44

HO—(CH$_3$)—CO—(CH$_3$)—OH
(CH$_3$)$_2$CH   CH(CH$_3$)$_2$

Synthesis

-Obtained by total demethylation of 5,5'-di-isopropyl-4,4'-dimethoxy-2,2'-dimethyl-benzophenone with refluxing pyridinium chloride (205-215°) for 3 h (26%) [1148].

   b.p.$_{0.9}$ 215-225° [1148];   Spectra (NA).

Bis[3,5-bis(1,1-dimethylethyl)-2-hydroxyphenyl]methanone

[30381-72-7] $C_{29}H_{42}O_3$ mol.wt. 438.65

(CH$_3$)$_3$C—(OH)—CO—(HO)—C(CH$_3$)$_3$
(CH$_3$)$_3$C   C(CH$_3$)$_3$

Syntheses

-Preparation by oxidation of bis(2-hydroxy-3,5-di-tert-butylphenyl)methane (SM) with chromium trioxide in acetic anhydride. SM was obtained by condensation of 2,4-di-tert-butylphenol with formaldehyde under acidic condition [640].

-Preparation by treatment of 2,2'-dihydroxybenzophenone at 120° with a mixture of isobutylene/nitrogen (1:1) in the presence of a macroreticular acid ion exchanger (Wofatit OK 80) as catalyst for 10 h (75%) [111].

b.p._{0.19} 230-240° [111]; m.p. 202-204° [640]; MS [640].

Bis[3,5-bis(1,1-dimethylethyl)-4-hydroxyphenyl]methanone

[29372-72-3]   C29H42O3   mol.wt. 438.65

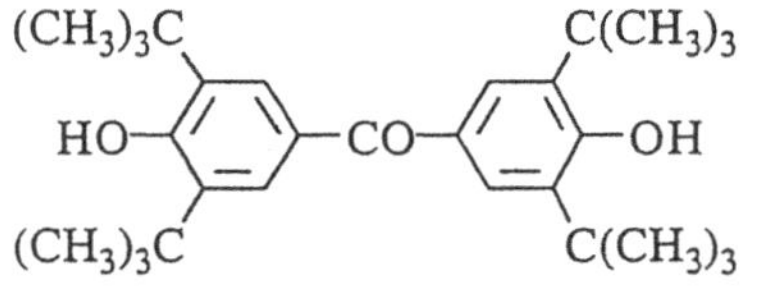

Synthesis

-Preparation by demethylation of 3,3',5,5'-tetra-tert-butyl-4,4'-dimethoxybenzophenone (SM) by means of sodium thioethoxide, under nitrogen, in refluxing N,N-dimethylformamide for 15 h (95%). SM was obtained by a two-step synthesis: at first, total methylation of 3,3',5,5'-tetra-tert-butyl-4,4'-dihydroxydiphenylmethane with methyl iodide in the presence of sodium hydride, under nitrogen, in refluxing tetrahydrofuran for 2 h. Then, by adding a solution of chromium trioxide in dilute sulfuric acid to an acetonic solution of the dimethyl ether previously formed (82%) and stirring at r.t. for 70 h, one obtains the expected ketone SM (86%) [284].

m.p. 226°5-227° [284]; ¹H NMR [284], IR [284].

Bis(2-hydroxy-5-octylphenyl)methanone

C29H42O3   mol.wt. 438.65

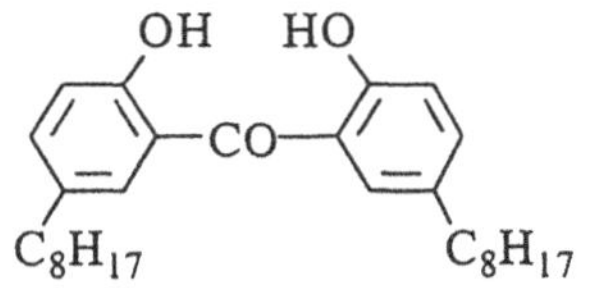

Synthesis

-Obtained by photo-Fries rearrangement of p-octylphenyl 5-octylsalicylate (major product) [970].

m.p. (NA); UV [970].

Asymmetric ketones

(3,5-Dibromo-2-hydroxyphenyl)(3,5-dibromo-4-hydroxyphenyl)methanone

C13H6Br4O3   mol.wt. 529.80

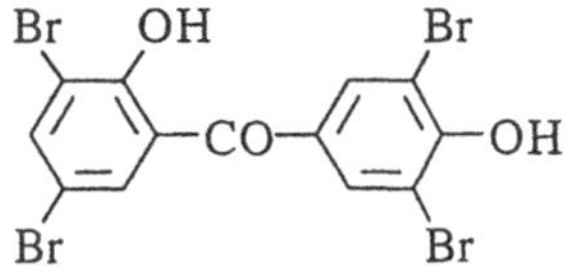

Synthesis

-Preparation by reaction of bromine with 2,4'-dihydroxybenzophenone in acetic acid (84%) [551].

m.p. 193-195°5 [551]; Spectra (NA).

(2-Chloro-4-hydroxyphenyl)(2-fluoro-4-hydroxyphenyl)methanone

C13H8ClFO3   mol.wt. 266.65

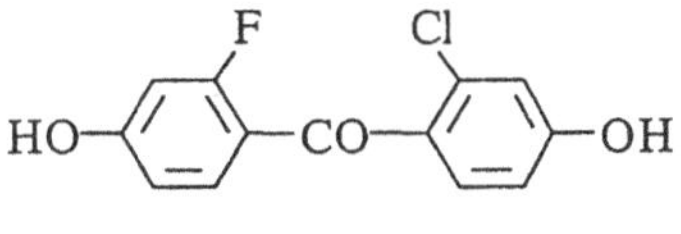

Synthesis

-Refer to: [486] (compound 32).

m.p. and Spectra (NA).

(2-Chloro-3-hydroxyphenyl)(4-chloro-3-hydroxyphenyl)methanone

$C_{13}H_8Cl_2O_3$ mol.wt. 283.11

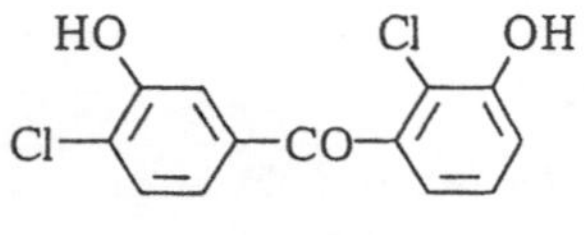

Synthesis

-Preparation by demethylation of 2,4'-dichloro-3,3'-di-
methoxybenzophenone with aluminium chloride in
chlorobenzene at 60° for 5.5 h (88%) [405].

m.p. 157-158°5 [405];   Spectra (NA).

(2-Chloro-5-hydroxyphenyl)(4-chloro-3-hydroxyphenyl)methanone

$C_{13}H_8Cl_2O_3$ mol.wt. 283.11

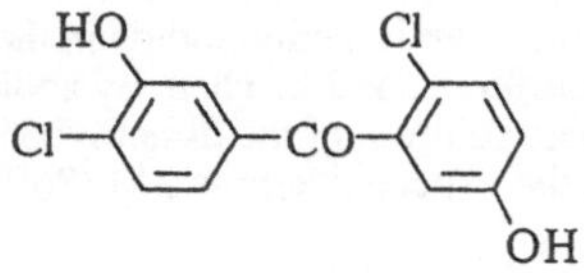

Synthesis

-Preparation by diazotization of 3',5-diamino-2,4'-dichloro-
benzophenone, followed by treatment of the diazonium salt
obtained with boiling 70% sulfuric acid (160°) for 10 min
(32%) [405].

m.p. 161° [405];   Spectra (NA).

(4-Amino-2-hydroxyphenyl)(2-hydroxy-4-nitrophenyl)methanone

$C_{13}H_{10}N_2O_5$ mol.wt. 274.23

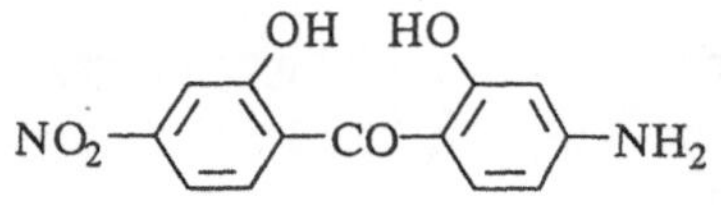

Syntheses

-Preparation by hydrolysis of 4-acetamido-2,2'-di-
hydroxy-4'-nitrobenzophenone with refluxing 20%
hydrochloric acid for 3 h [662].
-Preparation by demethylation of 4-amino-2'-hydroxy-
2-methoxy-4'-nitrobenzophenone with 50% hydrobromic acid in refluxing acetic acid for 6 h
(78%) [662].

m.p. 224-226° [662];   Spectra (NA).

(2-Amino-4-hydroxyphenyl)(4-amino-2-hydroxyphenyl)methanone

[107518-30-9] $C_{13}H_{12}N_2O_3$ mol.wt. 244.25

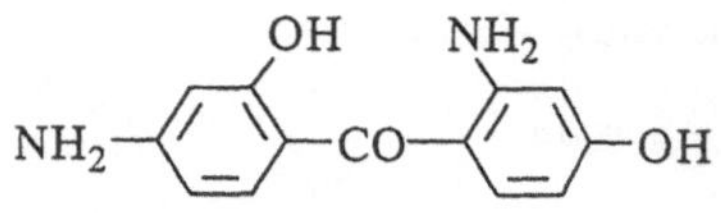

Synthesis

-Preparation from 2-amino-2'-hydroxy-4-methoxy-
4'-nitrobenzophenone by refluxing with freshly distilled
constant-boiling hydriodic acid at 140° for 7 h (63%)
[662].

m.p. 190-191° [662];   Spectra (NA).

(2-Amino-4-hydroxyphenyl)(4-amino-2-hydroxyphenyl)methanone *(Dihydrochloride)*

$C_{13}H_{12}N_2O_3,2HCl$ mol.wt. 317.17

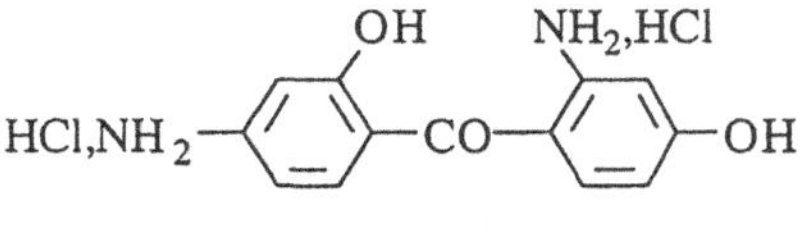

OH   NH$_2$,HCl

HCl,NH$_2$—⟨⟩—CO—⟨⟩—OH

Synthesis

-Preparation from 2,4'-diamino-2',4-dihydroxy-
benzophenone with hydrochloric acid in refluxing
dilute ethanol (72%) [662].

m.p.   218° (d) [662];   Spectra (NA).

(3-Bromo-2-hydroxy-5-methylphenyl)(3,5-dibromo-2-hydroxyphenyl)methanone

$C_{14}H_9Br_3O_3$ mol.wt. 464.94

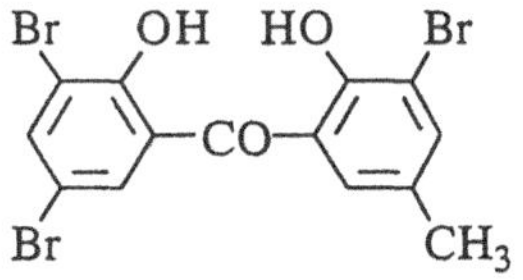

Br   OH   HO   Br

⟨⟩—CO—⟨⟩

Br   CH$_3$

Synthesis

-Obtained by reaction of excess bromine with 2,2'-di-
hydroxy-5-methylbenzophenone in acetic acid [99].

m.p. 190° [99];   Spectra (NA).

(3-Bromo-2-hydroxy-5-methylphenyl)(3,5-dibromo-4-hydroxyphenyl)methanone

$C_{14}H_9Br_3O_3$ mol.wt. 464.94

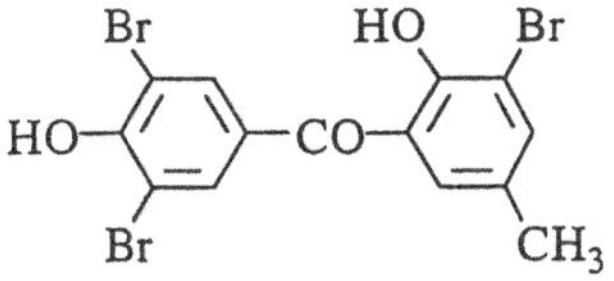

Br   HO   Br

HO—⟨⟩—CO—⟨⟩

Br   CH$_3$

Synthesis

-Obtained by reaction of bromine with 2,4'-dihydroxy-
5-methylbenzophenone in acetic acid [99].

m.p. 211°5-212°5 [99];   Spectra (NA).

(2-Hydroxy-4-methoxy-5-nitrophenyl)(4-hydroxy-3-nitrophenyl)methanone

[67246-05-3] $C_{14}H_{10}N_2O_8$ mol.wt. 334.24

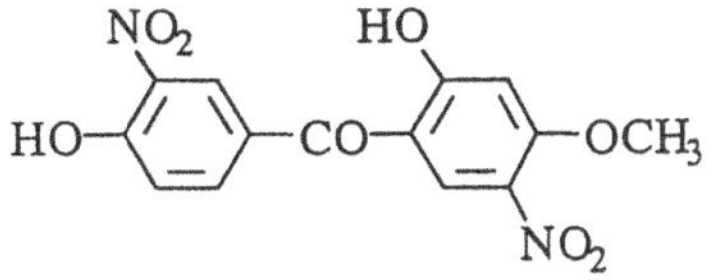

NO$_2$ HO

HO—⟨⟩—CO—⟨⟩—OCH$_3$

                         NO$_2$

Synthesis

-Obtained by reaction of nitric acid (d = 1.42) with
2,4'-dihydroxy-4-methoxybenzophenone in acetic acid
at 32° [1083].

m.p. 185° [1083]; ^{1}H NMR [1083].

(2-Chloro-4-hydroxyphenyl)(4-hydroxy-2-methylphenyl)methanone

[98155-83-0] $C_{14}H_{11}ClO_3$ mol.wt. 262.69

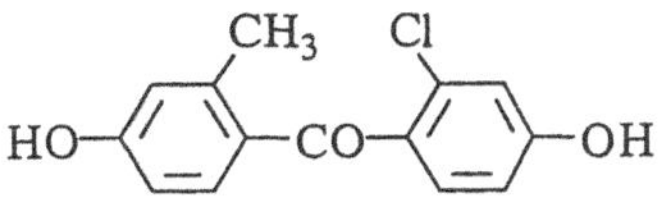

CH$_3$ Cl

HO—⟨⟩—CO—⟨⟩—OH

Synthesis

-Refer to: [486] (compound 34).

m.p. and Spectra (NA).

(2-Chloro-4-hydroxyphenyl)(4-hydroxy-2-methoxyphenyl)methanone

$C_{14}H_{11}ClO_4$ mol.wt. 278.69

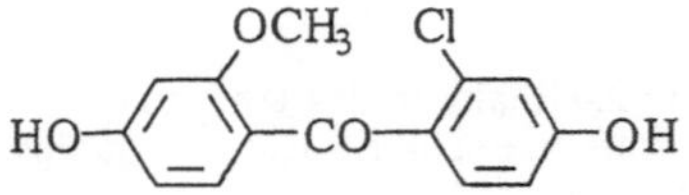

Synthesis

-Refer to: [486] (compound 36).

m.p. and Spectra (NA).

(4-Chloro-2-hydroxyphenyl)(2-hydroxy-4-methoxyphenyl)methanone

$C_{14}H_{11}ClO_4$ mol.wt. 278.69

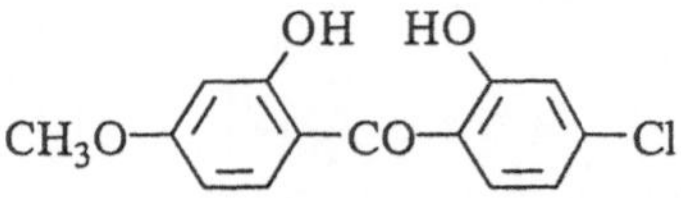

Synthesis

-Preparation by reaction of 2,4-dimethoxybenzoyl chloride with m-chloroanisole in chlorobenzene in the presence of aluminium chloride at 90° [43].

m.p. and Spectra (NA).

(2-Hydroxy-5-methylphenyl)(2-hydroxy-5-nitrophenyl)methanone

[145804-70-2] $C_{14}H_{11}NO_5$ mol.wt. 273.25

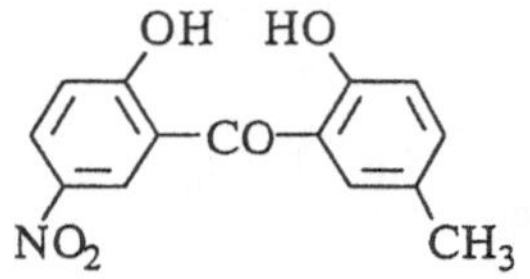

Synthesis

-Obtained by cleavage of 2-methyl-7-nitroxanthone with 10% potassium hydroxide in refluxing methanol for 12 h [895].

m.p. 146-148° [895];   Spectra (NA).

(3-Chloro-2-hydroxy-5-methylphenyl)(2-hydroxy-5-methylphenyl)methanone

[27404-63-3] $C_{15}H_{13}ClO_3$ mol.wt. 276.72

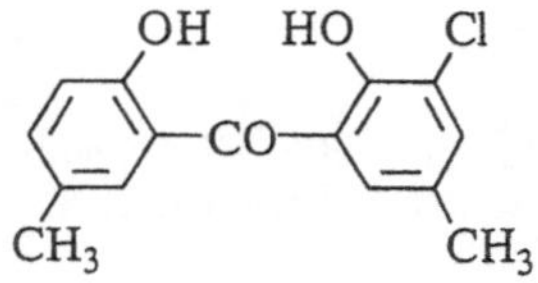

Synthesis

-Refer to: [391] (compound IV).

m.p. 146-147° [391];   MS [391].

(2-Hydroxy-4-methoxyphenyl)(2-hydroxy-4-methylphenyl)methanone

[105515-30-8] $C_{15}H_{14}O_4$ mol.wt. 258.27

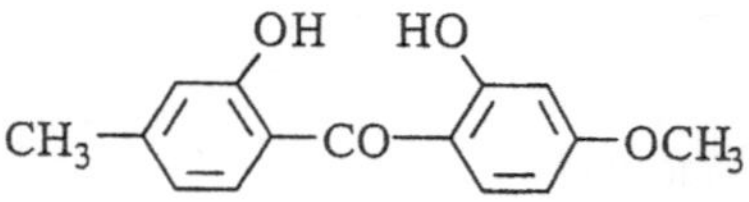

Synthesis

-Refer to: [1196] (Japanese patent).

m.p. and Spectra (NA).

(2-Ethyl-4-hydroxyphenyl)(4-hydroxy-2-methylphenyl)methanone

$C_{16}H_{16}O_3$ mol.wt. 256.30

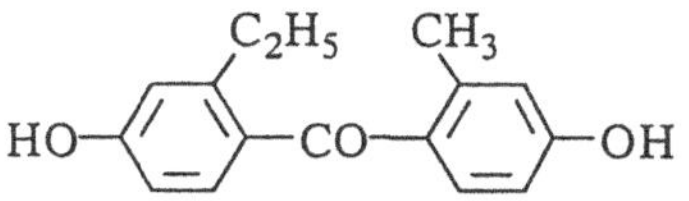

Synthesis

-Refer to: [486] (compound 6).

m.p. and Spectra (NA).

(4-Hydroxy-2,5-dimethylphenyl)(4-hydroxy-3-methylphenyl)methanone

$C_{16}H_{16}O_3$ mol.wt. 256.30

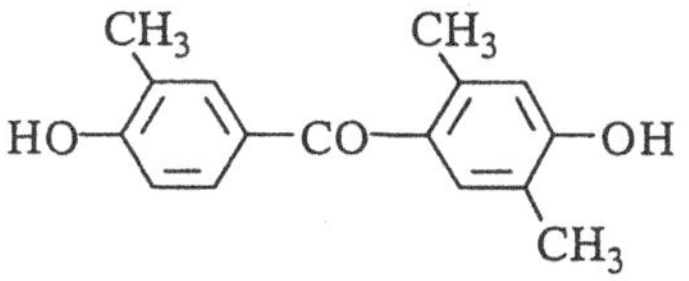

Synthesis

-Refer to: [486] (compound 8).

m.p. and Spectra (NA).

(4-Ethoxy-2-hydroxyphenyl)(2-hydroxy-4-methoxyphenyl)methanone

$C_{16}H_{16}O_5$ mol.wt. 288.30

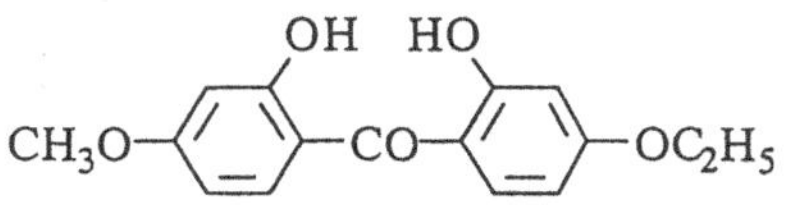

Syntheses

-Preparation by reaction of 2,4-dimethoxybenzoyl chloride with 1,3-diethoxybenzene in the presence of aluminium chloride,
*in a chlorobenzene/N,N-dimethylformamide mixture (22:1) at 115° [539];
*in chlorobenzene at 90° [43].
-Preparation by partial demethylation of 4-ethoxy-2-hydroxy-2',4'-dimethoxybenzophenone or 4-ethoxy-2,2',4'-trimethoxybenzophenone with aluminium chloride or aluminium bromide in chlorobenzene at 90-95° (good yield) [602].
-Also refer to: [1067].

m.p. (NA);   UV [1067].

(2-Hydroxy-4,6-dimethoxyphenyl)(4-hydroxy-2-methylphenyl)methanone

[4650-75-3]

$C_{16}H_{16}O_5$ mol.wt. 288.30

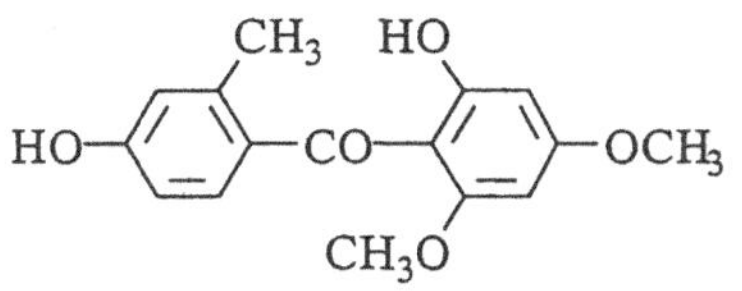

Synthesis

-Refer to: [125].

m.p. (NA);   MS [125].

(2-Hydroxy-3-methoxy-5-methylphenyl)(4-hydroxy-3-methoxyphenyl)methanone

[199735-29-0] $C_{16}H_{16}O_5$ mol.wt. 288.30

Synthesis

-Obtained by heating a mixture of the 2,4'-dihydroxy-3,3'-dimethoxy-5-methyldiphenylmethane, 2 N sodium hydroxide and nitrobenzene for 2 h at 170° (10%) [255, 1420].

m.p. and Spectra (NA);   GC and GC-MS [255, 1420].

(2-Hydroxy-4,5-dimethoxyphenyl)(2-hydroxy-4-methoxyphenyl)methanone

$C_{16}H_{16}O_6$ mol.wt. 304.30

Synthesis

-Preparation by reaction of 2-hydroxy-4-methoxybenzoic acid with 3,4-dimethoxyphenol in the presence of zinc chloride and phosphorous oxychloride on heating at 65-70° for 1 h [785].
-Also refer to: [757].

m.p.   147-148° [785];   IR [785], UV [785].

(4-Hydroxy-3,5-dimethoxyphenyl)(2-hydroxy-4-methoxyphenyl)methanone

[62495-38-9] $C_{16}H_{16}O_6$ mol.wt. 304.30

Synthesis

-Preparation by reaction of 3,4,5-trimethoxybenzoic acid with 3-methoxyphenol in the presence of phosphorous oxychloride and zinc chloride at 65-70° for 2 h (32%) [1395].

m.p.   142-143° [1395]; ^{1}H NMR [1395], IR [1395], UV [1395].

(3-Chloro-6-hydroxy-4-methoxy-2-methylphenyl)(3,5-dichloro-2-hydroxy-4-methoxy-6-methylphenyl)methanone

[69709-92-8] $C_{17}H_{15}Cl_3O_5$ mol.wt. 405.66

Synthesis

-Preparation by hydrogenolysis of 2,2'-bis (benzyloxy)-3,5,5'-trichloro-4,4'-dimethoxy-6,6'-dimethylbenzophenone (SM) with hydrogen in the presence of 10% Pd/C in ethyl acetate containing concentrated hydrochloric acid (5 drops) (>90%) [1174]. SM was obtained by Friedel-Crafts acylation of 5-(benzyloxy)-2-chloro-3-methoxy-toluene with 2-(benzyloxy)-3,5-dichloro-4-methoxy-6-methylbenzoic acid in the presence of trifluoroacetic anhydride in refluxing ethylene dichloride for 5 h (28%).
-Also refer to: [1173].

m.p.   137-138° [1174]; ^{1}H NMR [1174], MS [1174].

(3-Chloro-2-hydroxy-6-methoxy-4-methylphenyl)(4-hydroxy-2-methoxy-6-methyl-phenyl)methanone

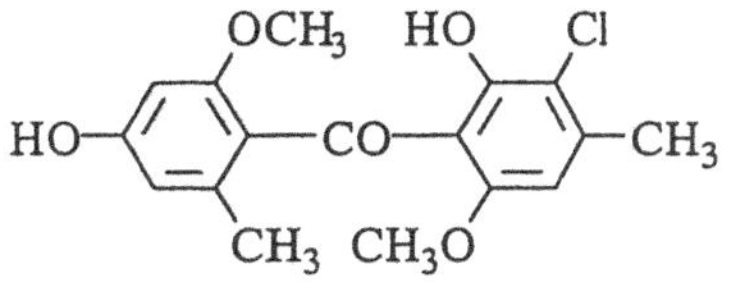

$C_{17}H_{17}ClO_5$ mol.wt. 336.77

Synthesis

-Not yet described.

Isolation from natural source

-From cultures of *Penicillium patulum* [890].

m.p. 181-182° [890];
IR [890], UV [890], MS [890].

(3-Chloro-2-hydroxy-4,6-dimethoxyphenyl)(4-hydroxy-2-methoxy-6-methylphenyl)-methanone *(Griseophenone A)*

[2151-17-9] $C_{17}H_{17}ClO_6$ mol.wt. 352.77

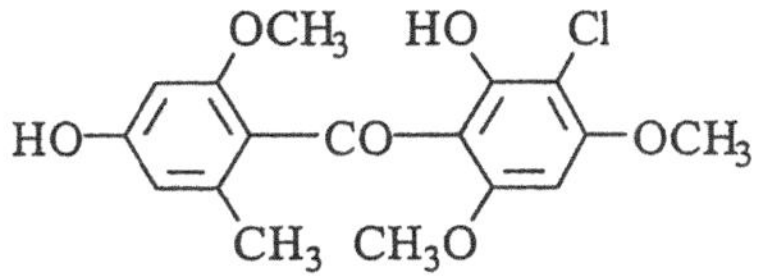

Syntheses

-Obtained by acylation of 2-chloro-3,5-dimethoxy-
   phenol,
*with 4-acetoxy-2-methoxy-6-methylbenzoyl chloride
   in the presence of aluminium chloride in nitrobenzene
   at r.t. for 20 h (35%) [1360] (Friedel-Crafts
   reaction);
*with 4-acetoxy-2-methoxy-6-methylbenzoic acid in the presence of trifluoroacetic anhydride at 20°
   for 18 h (50%) [1360];
*with 2-methoxy-4-methoxycarbonyl-6-methylbenzoyl chloride in the presence of aluminium
   chloride in nitrobenzene at r.t. [326].
-Also obtained by hydrogenation of dehydrogriseofulvin,
*with hydrogen in the presence of Pd/C in ethanol [1227];
*over Rh/C containing 3% of selenium in ethanol (70%) [326].
-Also obtained by saponification of two intermediate esters* (formed in the above reactions) with
   2.5% sodium hydroxide in dilute methanol at 25° (quantitative yields) [1360],
*of 4'-acetoxy-3-chloro-2-hydroxy-2',4,6-trimethoxy-6'-methylbenzophenone;
*of 4'-acetoxy-3-chloro-2-(trifluoroacetoxy)-2',4,6-trimethoxybenzophenone.
-Also obtained from dehydrogriseofulvin by reductive scission with chromous chloride or with zinc
   in acetic acid [784].
-Also obtained from 2-chloro-3,5-dimethoxyphenyl 4-hydroxy-2-methoxy-6-methylbenzoate,
*by Fries rearrangement in the presence of titanium tetrachloride in nitrobenzene at 20° for 18 h
   (65%) [1360];
*by light-catalyzed Fries rearrangement in ethanol (10-15%) [784], at 40° for 66 h (8%) [1360].
-Also refer to: [181, 543].

Isolation from natural source

-From cultures of *Penicillium patulum* [326, 890, 1115, 1116].

m.p. 213-214° [1115, 1116], 212°5-215° [1360], 212-214° [890], 210-212° [784];
IR [326, 890, 1115], UV [326, 784, 890], MS [890];
paper chromatography [1115].

(3-Fluoro-2-hydroxy-4,6-dimethoxyphenyl)(4-hydroxy-2-methoxy-6-methylphenyl)-methanone

$C_{17}H_{17}FO_6$ mol.wt. 336.32

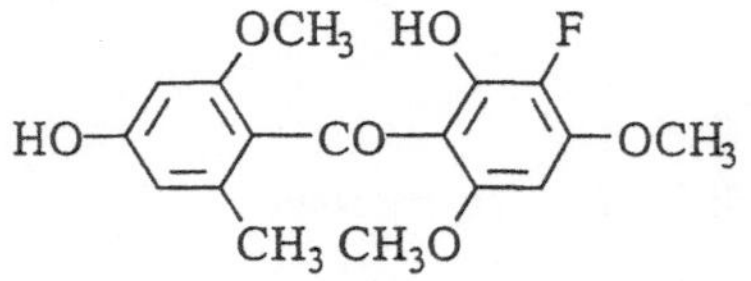

Syntheses

-Obtained by addition of 4-acetoxy-2-methoxy-6-methylbenzoic acid to 2-fluoro-3,5-dimethoxy-phenol in the presence of trifluoroacetic anhydride, first at 0°, then at 20-25° for 20 h (21%) [1362].
-Preparation by saponification of the corresponding acetate [1361], with 5% aqueous sodium hydroxide at 20° under nitrogen for 2 h (quantitative yield) [1362].

    m.p. 200-203° [1361], double melting point: 186-190°, then 200-203° [1362];
    IR [1362], UV [1361, 1362].

(2-Hydroxy-4-methoxy-6-methylphenyl)(4-hydroxy-2-methoxy-6-methylphenyl)-methanone

[81574-67-6] $C_{17}H_{18}O_5$ mol.wt. 302.33

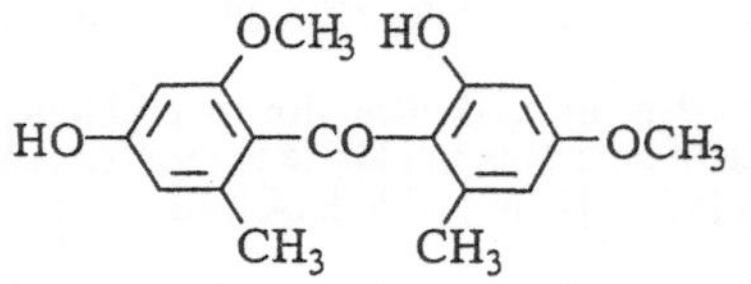

Synthesis

-Preparation by reduction of the 2',6-dimethoxy-4,6'-dimethylspiro[benzofuran-2,1'-cyclohexa-2',5'-diene]-3(2H),4'-dione (SM) with zinc dust in acetic acid for 1 h (83%). SM was obtained from methyl 4-methoxy-2-(2,4-dimethoxy-6-methyl-phenoxy)-6-methylbenzoate by treatment with titanium tetrachloride and hydrogen chloride for 40 h (65%, m.p. 190-192°) [1189].

    m.p. 176-177° [1189];
    ^{1}H NMR [1189], MS [1189].

(4-Hydroxy-2,6-dimethoxyphenyl)(2-hydroxy-4-methoxy-6-methylphenyl)methanone

[74628-36-7] $C_{17}H_{18}O_6$ mol.wt. 318.33

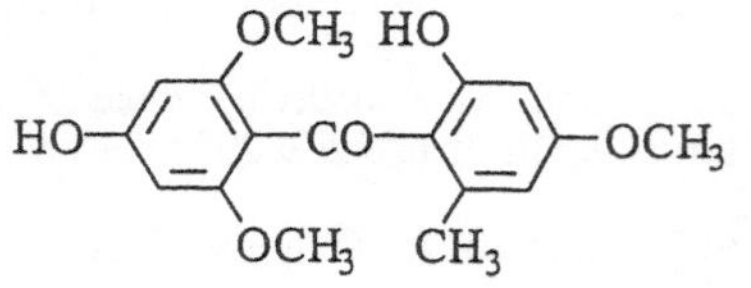

Synthesis

-Preparation by reduction of the 2',6,6'-trimethoxy-4-methylspiro[benzofuran-2,1'-cyclohexa-2',5'-diene]-3(2H),4'-dione (SM) with zinc dust in acetic acid [1188], for 1 h (82%). SM was obtained from methyl 4-methoxy-2-(2,4,6-trimethoxy-phenoxy)-6-methylbenzoate by treatment with titanium tetrachloride and hydrogen chloride for 12 h (90%, m.p. 273-274°) [1189].

    m.p. 192-193° [1189];
    ^{1}H NMR [1189], MS [1189].

(4-Hydroxy-3,5-dimethoxyphenyl)(2-hydroxy-3-methoxy-5-methylphenyl)methanone

[25138-53-8] $C_{17}H_{18}O_6$ mol.wt. 318.33

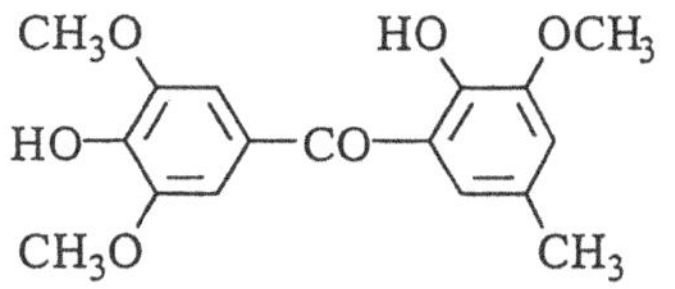

Synthesis

-Obtained (poor yield) by heating a mixture of 2,4'-di-hydroxy-3,3',5'-trimethoxy-5-methyldiphenylmethane, 2 N sodium hydroxide and nitrobenzene for 2 h at 170° (5%) [255, 1420].

m.p. and Spectra (NA);   GC and GC-MS [255, 1420].

(4-Hydroxy-3,5-dimethoxyphenyl)(5-hydroxy-4-methoxy-2-methylphenyl)methanone

[199735-38-1] $C_{17}H_{18}O_6$ mol.wt. 318.33

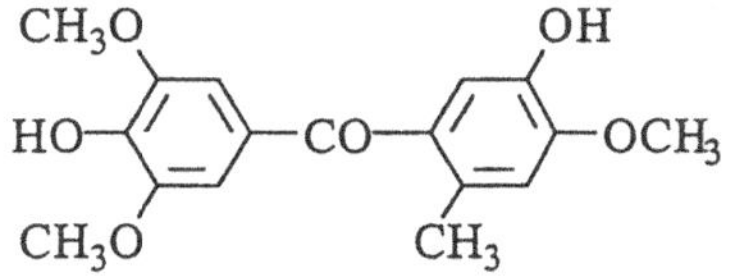

Synthesis

-Obtained (poor yield) by heating a mixture of 4,5'-dihydroxy-3,4',5-trimethoxy-2-methyldiphenyl-methane, 2 N sodium hydroxide and nitrobenzene for 2 h at 170° (4%) [255, 1420].

m.p. and Spectra (NA);   GC and GC-MS [255, 1420].

[4-Hydroxy-2-methyl-5-(1-methylethyl)phenyl](2-hydroxy-5-methylphenyl)methanone

$C_{18}H_{20}O_3$ mol.wt. 284.36

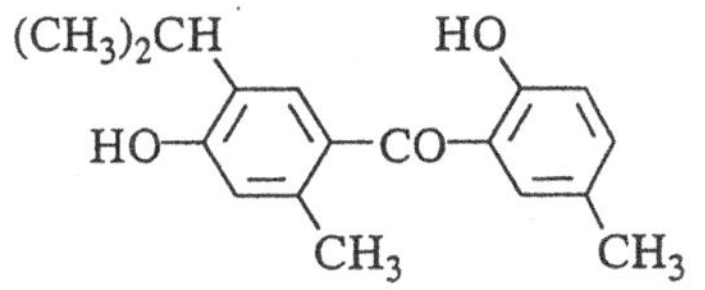

Synthesis

-Obtained by demethylation of 2',4-dimethoxy-2,5'-di-methyl-5-isopropylbenzophenone with pyridinium chloride at reflux (205-215°) for 1.5 h (30%) [1148].

b.p.$_{0.7}$ 200-210° [1148];   Spectra (NA).

(4-Butoxy-2-hydroxyphenyl)(2-hydroxy-4-methoxyphenyl)methanone

$C_{18}H_{20}O_5$ mol.wt. 316.35

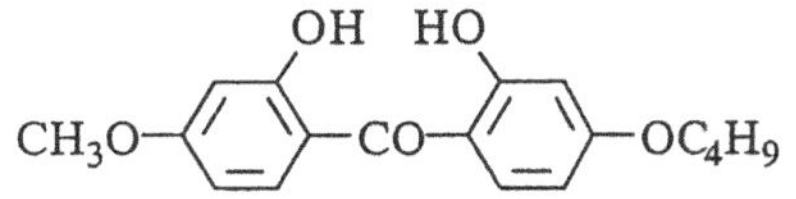

Syntheses

-Preparation by partial demethylation of 4-butoxy-2-hydroxy-2',4'-dimethoxybenzophenone with aluminium chloride or aluminium bromide in chlorobenzene at 90-95° (good yield) [602].
-Preparation by reaction of 2,4-dimethoxybenzoyl chloride with 1,3-dibutoxybenzene in the presence of aluminium chloride,
*in a chlorobenzene/N,N-dimethylformamide mixture (22:1) at 115° [539];
*in chlorobenzene at 90° [43].
-Also refer to: [1067].

m.p. (NA);   UV [1067].

(6-Hydroxy-2,4-dimethoxy-3-methylphenyl)(4-hydroxy-2,6-dimethoxyphenyl)methanone

[81574-66-5] $C_{18}H_{20}O_7$ mol.wt. 348.35

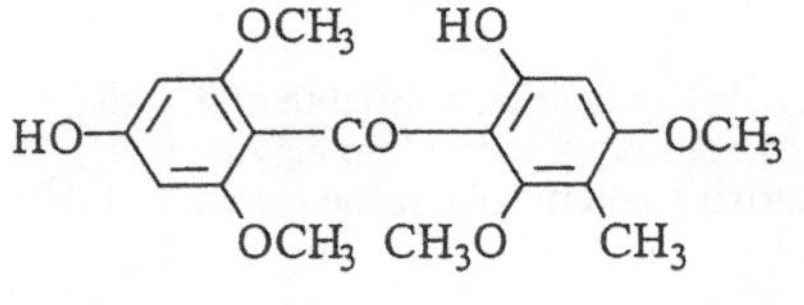

Synthesis

-Preparation by reduction of 2',4,6,6'-tetramethoxy-
5-methylspiro[benzofuran-2,1'-cyclohexa-
2',5'-diene]-3(2H),4'-dione (SM) with zinc dust in
acetic acid for 1 h (85%). SM was prepared from
methyl 2,4-dimethoxy-6-(2,4,6-trimethoxy-
phenoxy)-3-methylbenzoate by treatment with titanium tetrachloride and hydrogen chloride for
65 h (78%, m.p. 265-267°) [1189].

m.p. 239-241° [1189];
^{1}H NMR [1189], MS [1189].

[5-(1,1-Dimethylethyl)-2-hydroxyphenyl](4-hydroxy-3,5-dimethylphenyl)methanone

$C_{19}H_{22}O_3$ mol.wt. 298.38

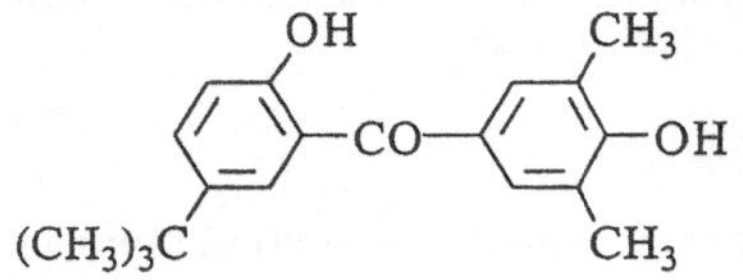

Synthesis

-Obtained by photo-Fries rearrangement of 2,6-di-
methylphenyl 5-tert-butylsalicylate (by-product)
[970].

m.p. and Spectra (NA).

[3,5-Bis(1,1-dimethylethyl)-4-hydroxyphenyl](4-chloro-2-hydroxyphenyl)methanone

$C_{21}H_{25}ClO_3$ mol.wt. 360.88

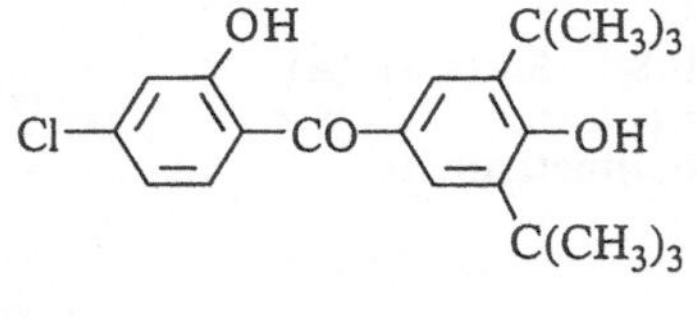

Synthesis

-Obtained by UV light irradiation of m-chlorophenyl
3,5-di-tert-butyl-4-hydroxybenzoate in isooctane [297].

m.p. 162°3 [297];   Spectra (NA).

[3,5-Bis(1,1-dimethylethyl)-4-hydroxyphenyl](5-chloro-2-hydroxyphenyl)methanone

$C_{21}H_{25}ClO_3$ mol.wt. 360.88

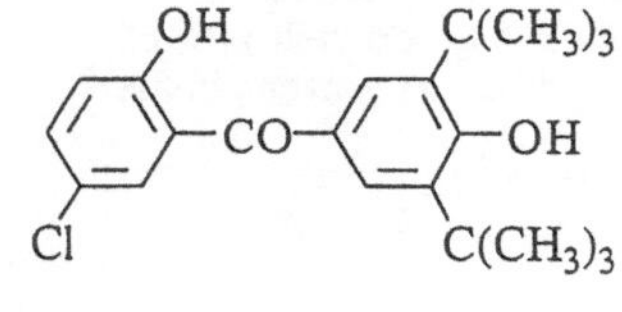

Synthesis

-Obtained by UV light irradiation of p-chlorophenyl 3,5-di-
tert-butyl-4-hydroxybenzoate in isooctane [297].

m.p. 164-165° [297];   Spectra (NA).

[3,5-Bis(1,1-dimethylethyl)-4-hydroxyphenyl](2-hydroxy-4-methylphenyl)methanone

$C_{22}H_{28}O_3$ mol.wt. 340.46

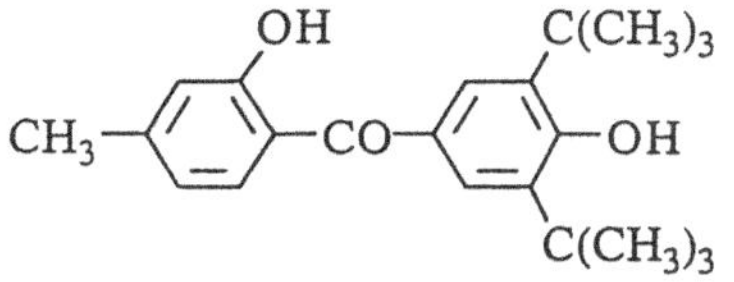

Synthesis

-Obtained by UV light irradiation of m-cresyl 3,5-di-tert-butyl-4-hydroxybenzoate in isooctane [297].

m.p. 126° [297];   Spectra (NA).

[3,5-Bis(1,1-dimethylethyl)-4-hydroxyphenyl](2-hydroxy-5-methylphenyl)methanone

$C_{22}H_{28}O_3$ mol.wt. 340.46

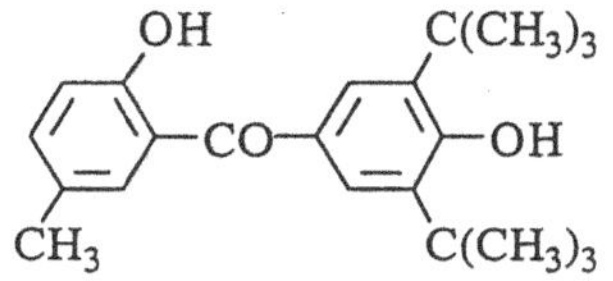

Synthesis

-Obtained by UV light irradiation of p-cresyl 3,5-di-tert-butyl-4-hydroxybenzoate in isooctane [297].

m.p. 148-149° [297];   Spectra (NA).

[3,5-Bis(1,1-dimethylethyl)-4-hydroxyphenyl](2-hydroxy-4-methoxyphenyl)methanone

$C_{22}H_{28}O_4$ mol.wt. 356.46

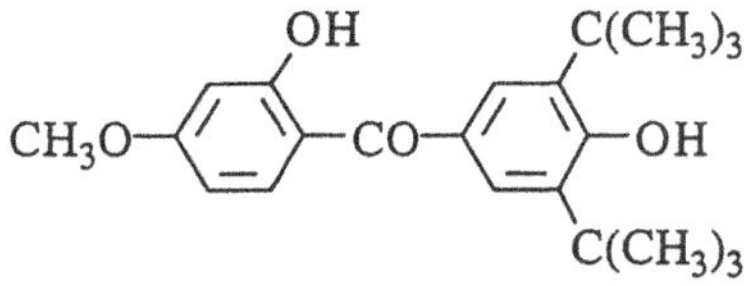

Synthesis

-Obtained by UV light irradiation of m-methoxyphenyl 3,5-di-tert-butyl-4-hydroxybenzoate in isooctane [297].

m.p. 137-138° [297];   Spectra (NA).

[3,5-Bis(1,1-dimethylethyl)-4-hydroxyphenyl](2-hydroxy-5-methoxyphenyl)methanone

$C_{22}H_{28}O_4$ mol.wt. 356.46

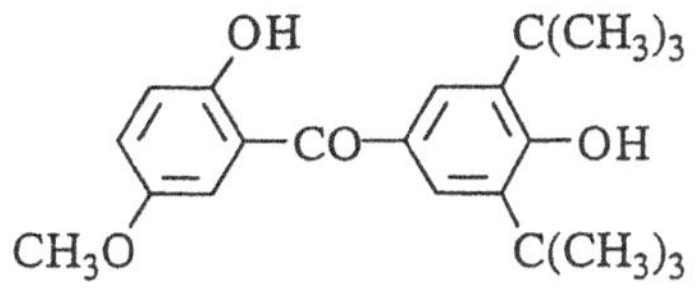

Synthesis

-Obtained by UV light irradiation of p-methoxyphenyl 3,5-di-tert-butyl-4-hydroxybenzoate in isooctane (60%) [297].

m.p. 142-143° [297];   Spectra (NA).

[3,5-Bis(1,1-dimethylethyl)-4-hydroxyphenyl](4-hydroxy-3,5-dimethoxyphenyl)-methanone

[54808-43-4] $C_{23}H_{30}O_5$ mol.wt. 386.49

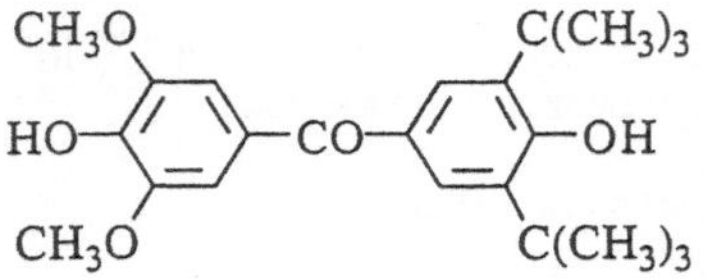

Synthesis

-Preparation by selective demethylation of 3,5-di-tert-butyl-4-hydroxy-3',4',5'-trimethoxybenzophenone (SM),
*with 45% anhydrous hydrobromic acid in acetic acid at r.t. for 48 h (49%) [284];

*by adding a freshly prepared ethereal magnesium iodide solution to a solution of SM in toluene, elimination of ethyl ether by distillation, toluene being added to maintain the original volume. Then, heating at reflux for 10 h (68%) [284].

     m.p. 181°5-182°5 [284]; ^{1}H NMR [284], IR [284].

[3,5-Bis(1,1-dimethylethyl)-2-hydroxyphenyl][5-(1,1-dimethylethyl)-2-hydroxy-phenyl]methanone

[125182-26-5] $C_{25}H_{34}O_3$ mol.wt. 382.54

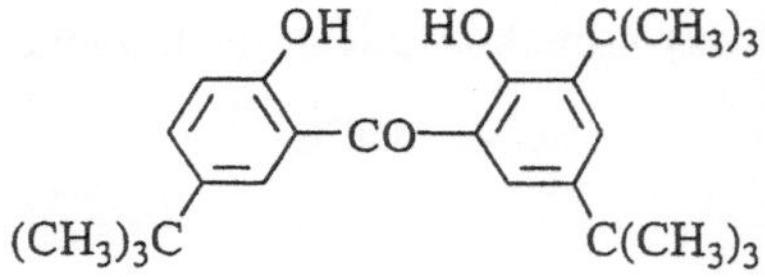

Synthesis

-Obtained by treatment of 2,2'-dihydroxy-benzophenone at 120° with a mixture of isobutylene/nitrogen (1:1) in the presence of a macroreticular acid ion exchanger (Wofatit OK 80) as catalyst for 5 h (40%) [111].

     b.p.$_{0.15}$ 215-220° [111];   Spectra (NA).

[4-(Dodecyloxy)-2-hydroxyphenyl](4-hydroxy-3,5-dimethylphenyl)methanone

                                               $C_{27}H_{38}O_4$ mol.wt. 426.60

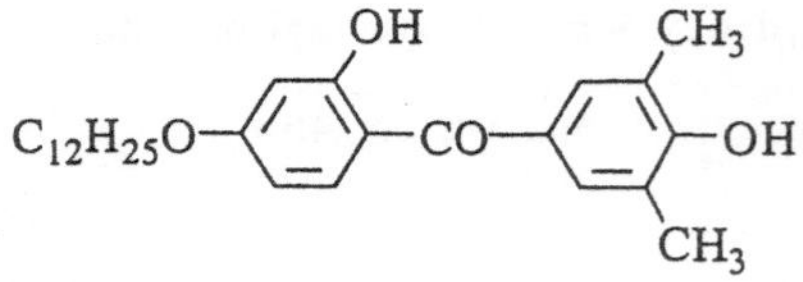

Synthesis

-Obtained by photo-Fries rearrangement of 2,6-dimethylphenyl 4-dodecyloxysalicylate (by-product) [970].

     m.p. and Spectra (NA).

[4-(Dodecyloxy)-2-hydroxyphenyl][5-(1,1-dimethylethyl)-2-hydroxyphenyl]methanone

                                               $C_{29}H_{42}O_4$ mol.wt. 454.65

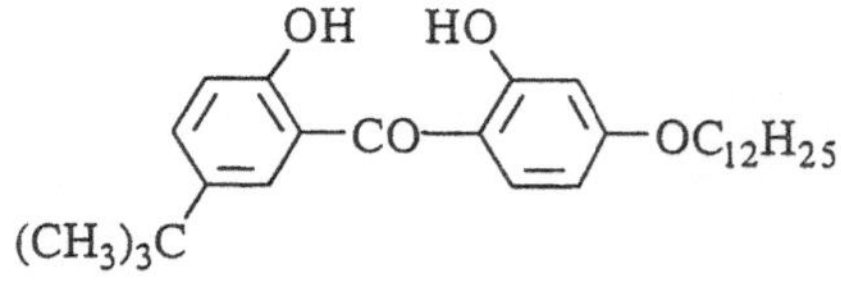

Synthesis

-Obtained by photo-Fries rearrangement of p-tert-butylphenyl 4-dodecyloxysalicylate (major product) [970].

     m.p. (NA);   UV [970].

(2-Hydroxy-4-methoxy-6-methylphenyl)[2-hydroxy-4-(octadecyloxy)phenyl]methanone

[128464-15-3] $C_{33}H_{50}O_5$ mol.wt. 526.76

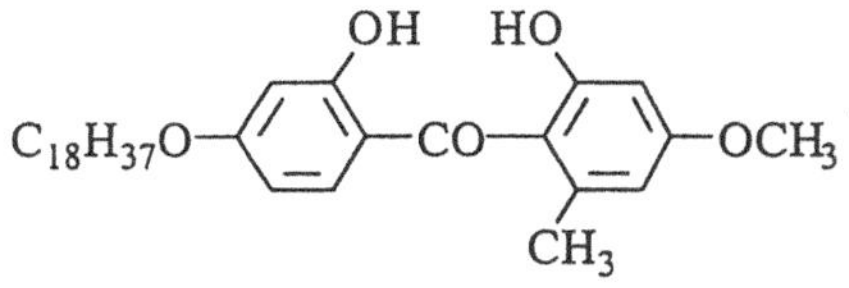

Synthesis

-Refer to: [701] and [780] (Japanese patent).

m.p. and Spectra (NA).

3. TRIHYDROXYBENZOPHENONES

3.1. *Hydroxy groups located on the same ring*

3.1.1. *Substituents located on the hydroxylated ring*

Phenyl(2,4,6-trihydroxy-3-methylphenyl)methanone

[68223-56-3] $C_{14}H_{12}O_4$ mol.wt. 244.25

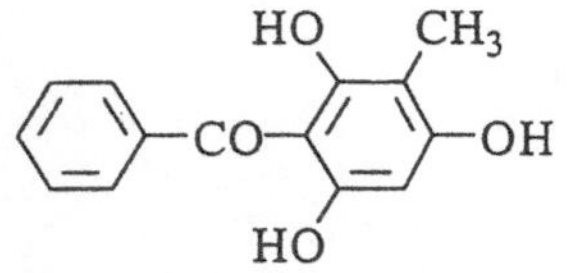

Synthesis

-Preparation by reaction of benzonitrile with 2-methyl-
  phloroglucinol in the presence of zinc chloride and
  hydrochloric acid in ethyl ether, followed by hydrolysis of
  the ketimine hydrochloride so formed (51%) [889, 1081].
-Also refer to: [663, 818] (Chinese papers).

m.p.   146-147° [1081], 139-140° [889];   Spectra (NA).

Phenyl[2,3,4-trihydroxy-5-(hydroxymethyl)phenyl]methanone

[138250-28-9] $C_{14}H_{12}O_5$ mol.wt. 260.25

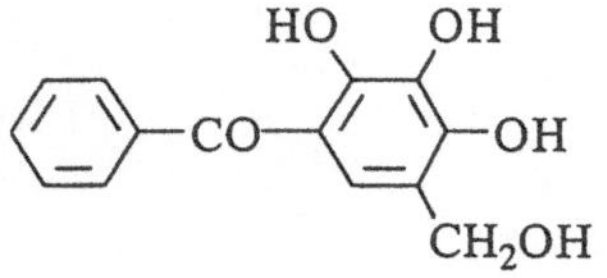

Synthesis

-Refer to: [1001, 1387, 1388].

m.p. and Spectra (NA).

Phenyl(2,4,6-trihydroxy-3,5-dimethylphenyl)methanone

[22744-25-8] $C_{15}H_{14}O_4$ mol.wt. 258.27

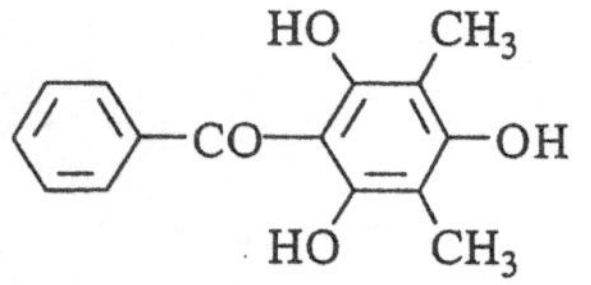

Synthesis

-Preparation by reaction of benzonitrile with 2,4-dimethyl-
  phloroglucinol (42%) (Hoesch reaction) [951].

m.p. 134° [951];   Spectra (NA).

[3-(3,7-Dimethyl-2,6-octadienyl)-2,4,6-trihydroxyphenyl]phenylmethanone

[70219-87-3] *(E)*
[76015-48-0] *(Z)* $C_{23}H_{26}O_4$ mol.wt. 366.46

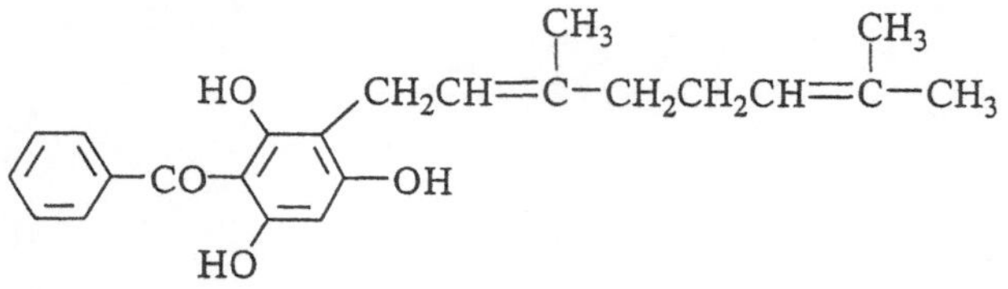

Synthesis

-Not yet described.

Isolation from natural sources

-From *Leontonyx spathulatus* Less. (Compositae) [192];
-From *Leontonyx Squarrosus* DC (Compositae) [192];
-From *Helichrysum crispum* Less. (Compositae) [193];
-From *Helichrysum monticola* Hilliard (Compositae) [194].

colourless oil [192, 194];   b.p. (NA);
^{1}H NMR [192, 194], IR [192, 194], MS [192, 194].

Phenyl[2,4,6-trihydroxy-3,5-bis(3-methyl-2-butenyl)phenyl]methanone

[70219-84-0]   C$_{23}$H$_{26}$O$_4$ mol.wt. 366.46

Syntheses

-Obtained (poor yield) by reaction of 2-methyl-3-buten-2-ol with 2,4,6-trihydroxybenzophenone in the presence of boron trifluoride-etherate in dioxane at 25-30° (6%) [1043].
-Also obtained (poor yield) by reaction of prenyl bromide with 2,4,6 trihydroxybenzophenone in the presence of sodium methoxide in refluxing methanol for 3 h [1043], (<1%) [1044].

Isolation from natural sources

-From *Leontonyx spathulatus Less* and from *Leontonyx sqarrosus* DC (Compositae) [192];
-From *Helichrysum crispum* Less (Compositae) [193].

Colourless oil [192]. This product is impure or in a metastable state.
m.p. 94-95° [1043, 1044];
^{1}H NMR [192, 1043, 1044], IR [192, 1043, 1044], UV [1043, 1044], MS [192].

3.1.2. *Substituents located on the other ring*

(2,4-Dichlorophenyl)(2,4,6-trihydroxyphenyl)methanone

[61101-87-9]   C$_{13}$H$_8$Cl$_2$O$_4$ mol.wt. 299.11

Synthesis

-Preparation by reaction of 2,4-dichlorobenzoyl chloride with O,O,O-tris(trimethylsilyl)phloroglucinol in the presence of stannic chloride in refluxing methylene chloride for 2 h (76%) [649].

oily product, not distillable [649];   b.p. and Spectra (NA).

(2-Chlorophenyl)(2,4,6-trihydroxyphenyl)methanone

[61101-84-6]   C$_{13}$H$_9$ClO$_4$ mol.wt. 264.66

Synthesis

-Preparation by adding O,O,O-tris(trimethylsilyl)phloroglucinol to a solution of o-chlorobenzoyl chloride and stannic chloride in methylene chloride and the resulting solution stirred overnight at r.t. (80%) [649].

m.p.   141-143° [649];   Spectra (NA).

(3-Chlorophenyl)(2,4,6-trihydroxyphenyl)methanone

$C_{13}H_9ClO_4$ mol.wt. 264.66

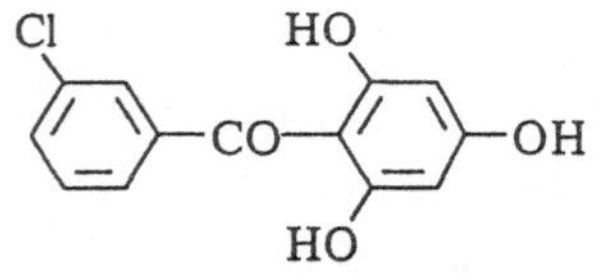

Synthesis

-Preparation by reaction of m-chlorobenzonitrile with
 phloroglucinol (67%) (Hoesch reaction) [1017].

m.p.   169°5-170° [1017];   Spectra (NA).

(4-Chlorophenyl)(2,3,4-trihydroxyphenyl)methanone

$C_{13}H_9ClO_4$ mol.wt. 264.66

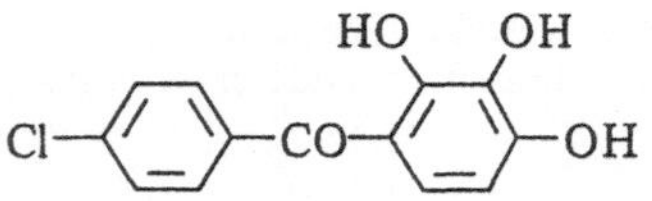

Synthesis

-Preparation by reaction of p-chlorobenzonitrile with
 pyrogallol in the presence of zinc chloride and hydrochloric
 acid in ethyl ether, followed by hydrolysis of the resulting
 ketimine hydrochloride with boiling water for 10 min under
carbon dioxide (25%) [755].

m.p.   157-158° [755];   Spectra (NA).

(4-Chlorophenyl)(2,4,5-trihydroxyphenyl)methanone

$C_{13}H_9ClO_4$ mol.wt. 264.66

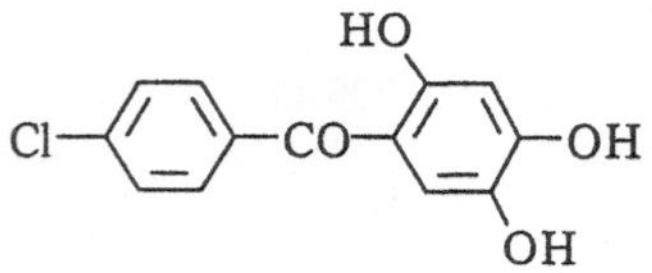

Synthesis

-Preparation by reaction of p-chlorobenzonitrile with
 hydroxyhydroquinone in the presence of zinc chloride and
 hydrochloric acid in ethyl ether, followed by hydrolysis of
 the resulting ketimine hydrochloride with boiling water for
 30 min under carbon dioxide (Hoesch reaction) (55%)
 [755].

m.p.   260° [755];   Spectra (NA).

(4-Chlorophenyl)(2,4,6-trihydroxyphenyl)methanone

$C_{13}H_9ClO_4$ mol.wt. 264.66

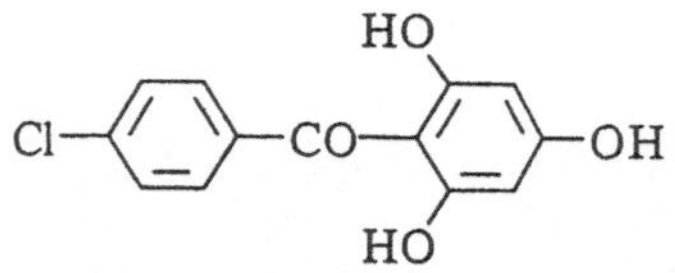

Synthesis

-Preparation by reaction of p-chlorobenzonitrile with
 phloroglucinol (43%) (Hoesch reaction) [1017].

m.p.   169-169°5 [1017];   Spectra (NA).

(4-Fluorophenyl)(2,3,4-trihydroxyphenyl)methanone

[84795-00-6] $C_{13}H_9FO_4$ mol.wt. 248.21

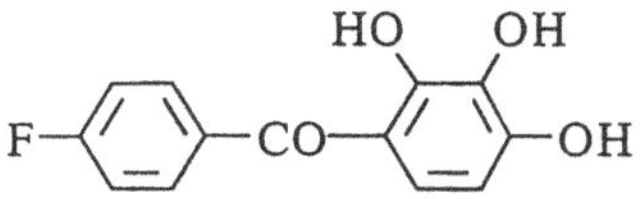

Synthesis

-Preparation by reaction of p-fluorobenzotrichloride with pyrogallol in ethanol at 65° for 30 min (80%) [394].

m.p. 149-150° [394]; Spectra (NA).

(2-Nitrophenyl)(2,4,6-trihydroxyphenyl)methanone

[61736-69-4] $C_{13}H_9NO_6$ mol.wt. 275.22

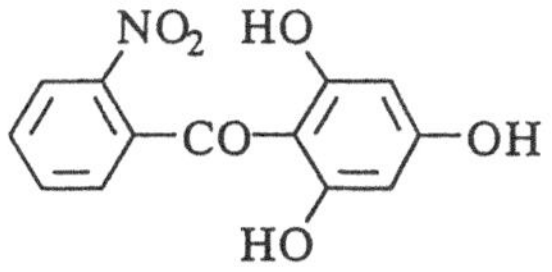

Synthesis

-Obtained by reaction of o-nitrobenzoyl chloride with phloroglucinol in the presence of aluminium chloride in ethyl ether at 0° for 3 h (20%) [14].
-Also refer to: [204].

m.p. 182-184° [14];
^{1}H NMR [14], IR [14], UV [14], MS [14].

(3-Nitrophenyl)(2,4,6-trihydroxyphenyl)methanone

$C_{13}H_9NO_6$ mol.wt. 275.22

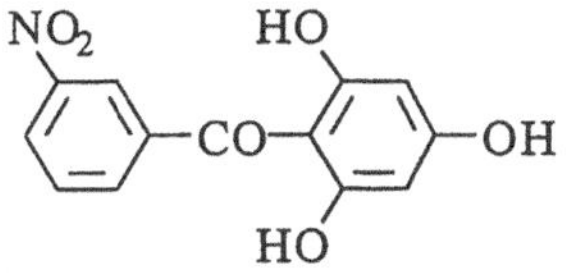

Synthesis

-Preparation by reaction of m-nitrobenzonitrile with phloroglucinol in the presence of zinc chloride and hydrochloric acid in ethyl ether, followed by hydrolysis of the resulting ketimine hydrochloride with boiling water for 30 min (Hoesch reaction) [1452, 1453].

m.p. 194° [1452, 1453]; Spectra (NA).

(4-Nitrophenyl)(2,4,6-trihydroxyphenyl)methanone

$C_{13}H_9NO_6$ mol.wt. 275.22

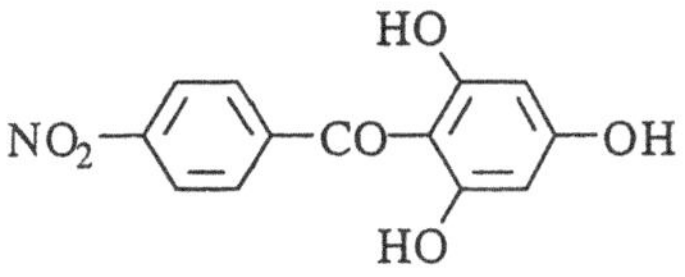

Synthesis

-Preparation by reaction of p-nitrobenzonitrile with phloroglucinol in the presence of zinc chloride and hydrochloric acid, followed by hydrolysis of the resulting ketimine hydrochloride with boiling water for 10-30 min [1452, 1453], (46%) [755] (Hoesch reaction).

m.p. 246-247° [1452, 1453], 244-245° [755];
monohydrate [755, 1452]; Spectra (NA).

(2-Aminophenyl)(2,3,4-trihydroxyphenyl)methanone

$C_{13}H_{11}NO_4$ mol.wt. 245.23

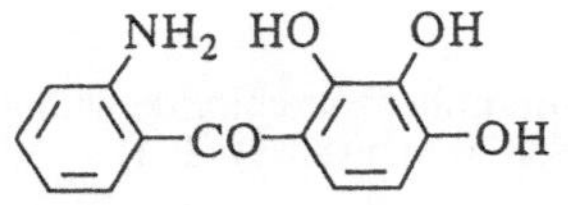

Synthesis

-Refer to: [204].

m.p. and Spectra (NA).

(2-Aminophenyl)(2,4,6-trihydroxyphenyl)methanone

$C_{13}H_{11}NO_4$ mol.wt. 245.23

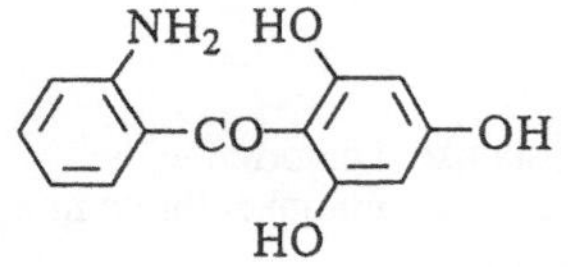

Synthesis

-Presumably obtained by treatment of 2-nitro-2',4',6'-tri-
   hydroxybenzophenone in the presence of tin and hydrochloric
   acid at r.t., before cyclodehydration and quantitative
   conversion into 1,3-dihydroxyacridan-9-one [204].
-Also refer to: [14, 203].

m.p. and Spectra (NA).

(2-Methylphenyl)(2,3,4-trihydroxyphenyl)methanone

[120506-54-9] $C_{14}H_{12}O_4$ mol.wt. 244.25

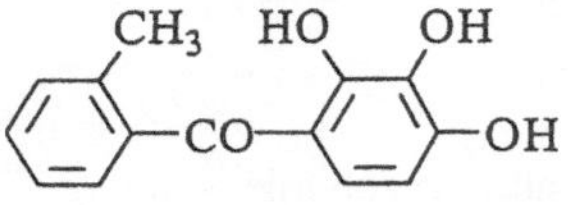

Synthesis

-Preparation by acylation of pyrogallol with o-toluic acid in
   the presence of Amberlyst-15 in refluxing toluene with
   azeotropical water removal [1400].
-Also refer to: [739, 740, 1401] (Japanese patents).

m.p. and Spectra (NA).

(4-Methylphenyl)(2,3,4-trihydroxyphenyl)methanone

[120506-55-0] $C_{14}H_{12}O_4$ mol.wt. 244.25

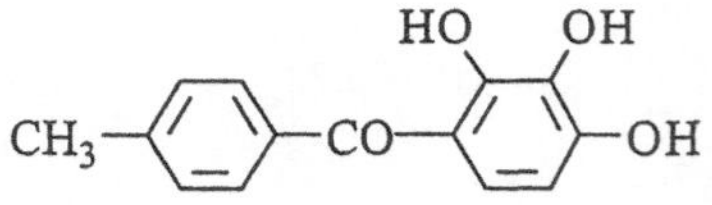

Synthesis

-Preparation by acylation of pyrogallol with p-toluic acid
   in the presence of Amberlyst-15 in refluxing toluene
   with azeotropical water removal [1400].

m.p. and Spectra (NA).

(2-Methoxyphenyl)(2,3,4-trihydroxyphenyl)methanone

[156333-16-3] $C_{14}H_{12}O_5$ mol.wt. 260.25

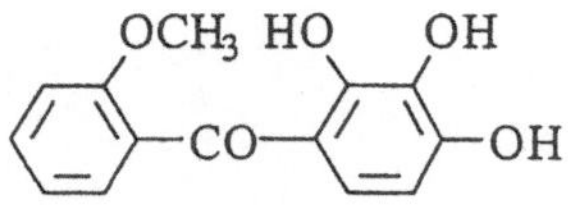

Synthesis

-Preparation by condensation of pyrogallol and o-anisic acid
   with boron trifluoride-etherate in carbon tetrachloride [1102],
   according to [1020].

m.p. and Spectra (NA).

(3-Methoxyphenyl)(2,4,6-trihydroxyphenyl)methanone

[21554-79-0] $C_{14}H_{12}O_5$ mol.wt. 260.25

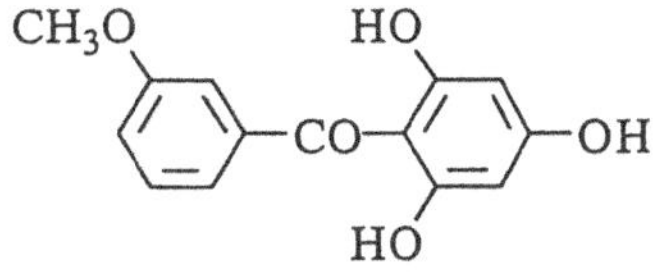

Synthesis

-Obtained (poor yield) by reaction of m-anisoyl chloride
with phloroglucinol in the presence of aluminium chloride
in ethyl ether at r.t. for 60 h [415], (11%) [85, 87].

m.p. 168-171° [87]; ^{1}H NMR [87], IR [87], UV [87].

(4-Methoxyphenyl)(2,3,4-trihydroxyphenyl)methanone

[105443-50-3] $C_{14}H_{12}O_5$ mol.wt. 260.25

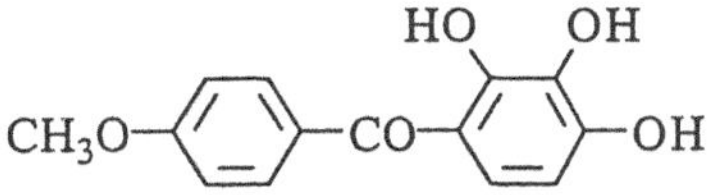

Synthesis

-Preparation by reaction of p-anisic acid with
  pyrogallol,
*in tetrachloroethane bubbling boron trifluoride at 110°
  for 1 h (84%) [1443];
*in the presence of Amberlyst-15 in refluxing toluene with azeotropical water removal [1400].
-Also refer to: [668].

m.p. and Spectra (NA).

(4-Ethoxyphenyl)(2,3,4-trihydroxyphenyl)methanone

[69471-29-0] $C_{15}H_{14}O_5$ mol.wt. 274.27

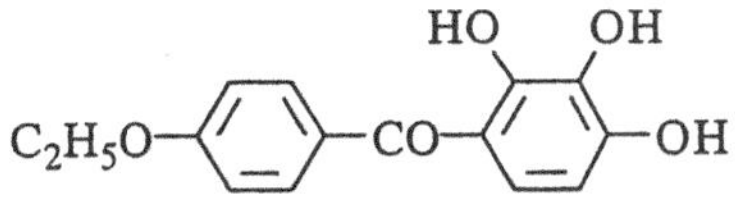

Synthesis

-Preparation by reaction of p-ethoxybenzoic acid with
pyrogallol in the presence of boron trifluoride-ethyl
ether complex at 100° for 2 h (73%) [1008].

m.p. 104-105° [1008]; ^{1}H NMR [1008], IR [1008], MS [1008].

(2,6-Dimethoxyphenyl)(2,4,6-trihydroxyphenyl)methanone

[61101-86-8] $C_{15}H_{14}O_6$ mol.wt. 290.27

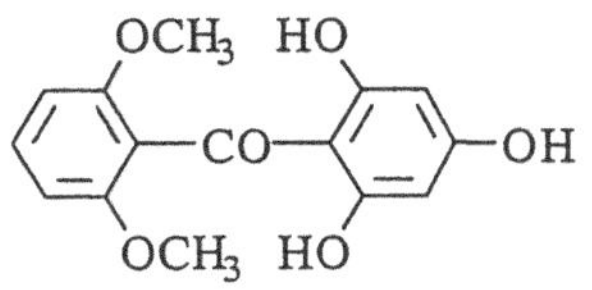

Syntheses

-Preparation by reaction of 2,6-dimethoxybenzoyl chloride
with O,O,O-tris(trimethylsilyl)phloroglucinol in the presence
of stannic chloride in refluxing methylene chloride for 2 h
(64%) [649].
-Also obtained (poor yield) by reaction of 2,6-dimethoxy-
benzonitrile with phloroglucinol in the presence of zinc chloride and hydrochloric acid in ethyl
ether at 0° for 5 days, followed by hydrolysis of the resulting ketimine with boiling water for 2 h
(7%) (Hoesch reaction) [755].

m.p. 216-218° [755], 195-202° (d) [649];   Spectra (NA).

(2,4-Dimethoxy-6-methylphenyl)(2,4,6-trihydroxyphenyl)methanone

[38071-50-0] $C_{16}H_{16}O_6$ mol.wt. 304.30

Synthesis

-Obtained by treatment of methyl 7-(4-orcinyl)-
3,5,7-trioxoheptanoate dimethyl ether with aqueous
potassium hydroxide (25%) [552].

m.p. and Spectra (NA).

3.2. *Hydroxy groups located on both rings*

3.2.1. *Substituents located on one ring*

(2-Chloro-3-hydroxyphenyl)(2,6-dihydroxyphenyl)methanone

$C_{13}H_9ClO_4$ mol.wt. 264.66

Synthesis

-Preparation by reaction of 2-chloro-3-hydroxybenzoyl
chloride with bis(trimethylsilyl) derivative of resorcinol in
the presence of stannic chloride (or titanium tetrachloride or
aluminium chloride) in refluxing methylene chloride for 2 h
[649].

m.p. and Spectra (NA).

(4-Chloro-2-hydroxyphenyl)(2,4-dihydroxyphenyl)methanone

[95481-60-0] $C_{13}H_9ClO_4$ mol.wt. 264.66

Synthesis

-Preparation by reaction of 4-chlorosalicylic acid with
resorcinol in the presence of zinc chloride and a mixture of
polyphosphoric acid/85% phosphoric acid (60:40) at 27°.
Then, during 2 h, phosphorous trichloride was added
between 27 to 37° and the mixture heated at 60° for 16 h [1302].
-Also refer to: [159].

pale yellow crystals [1302];   m.p. and Spectra (NA).

(2,4-Dihydroxy-3-nitrophenyl)(2-hydroxyphenyl)methanone

[69169-87-5] $C_{13}H_9NO_6$ mol.wt. 275.22

Synthesis

-Obtained (poor yield) by action of salicylic acid with
2-nitroresorcinol in the presence of zinc chloride and
phosphorous oxychloride at 60-65° for 3 h (10%) [455].

m.p. 118-120° [455];   Spectra (NA).

(3,4-Dihydroxy-5-nitrophenyl)(4-hydroxyphenyl)methanone

[134612-51-4] $C_{13}H_9NO_6$ mol.wt. 275.22

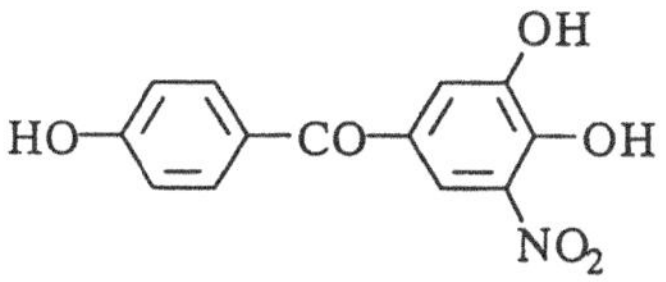

Synthesis

-Preparation by demethylation of 3,4'-dimethoxy-4-hydroxy-5-nitrobenzophenone with hydrobromic acid in refluxing acetic acid [164].

m.p. 212-214° [164];   Spectra (NA).

(2,6-Dihydroxyphenyl)(3-hydroxy-2-nitrophenyl)methanone

$C_{13}H_9NO_6$ mol.wt. 275.22

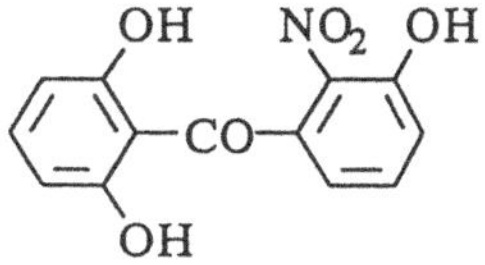

Synthesis

-Preparation by reaction of 3-hydroxy-2-nitrobenzoyl chloride with bis(trimethylsilyl) derivative of resorcinol in the presence of stannic chloride (or titanium tetrachloride or aluminium chloride) in refluxing methylene chloride for 2 h [649].

m.p. and Spectra (NA).

(2,4-Dihydroxy-3-methylphenyl)(2-hydroxyphenyl)methanone

[57654-18-9] $C_{14}H_{12}O_4$ mol.wt. 244.25

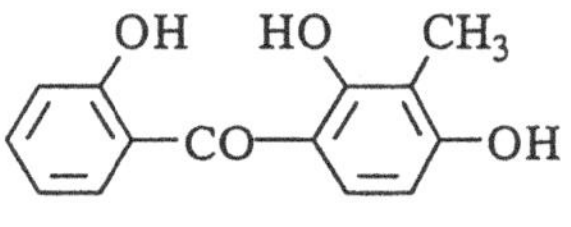

Synthesis

-Preparation by reaction of salicylic acid with 2-methyl-resorcinol in the presence of zinc chloride and phosphorous oxychloride for 4 h at 60-65° (55%) [177].

m.p. 145°5-147° [177];
^{1}H NMR [177], IR [177], UV [177], MS [177].

(2,4-Dihydroxy-3-methylphenyl)(3-hydroxyphenyl)methanone

[61227-13-2] $C_{14}H_{12}O_4$ mol.wt. 244.25

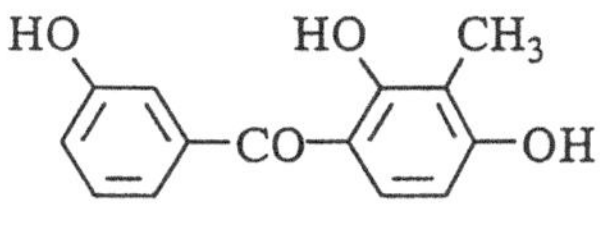

Syntheses

-Preparation by condensation of m-acetoxybenzonitrile with 2-methylresorcinol in the presence of zinc chloride in ethyl ether (Hoesch reaction) (42%) [398].
-Also obtained by [397] according to the method [1402].

m.p. 181° [398];   Spectra (NA).

(2,4-Dihydroxy-3-methylphenyl)(4-hydroxyphenyl)methanone

[79861-84-0] $C_{14}H_{12}O_4$ mol.wt. 244.25

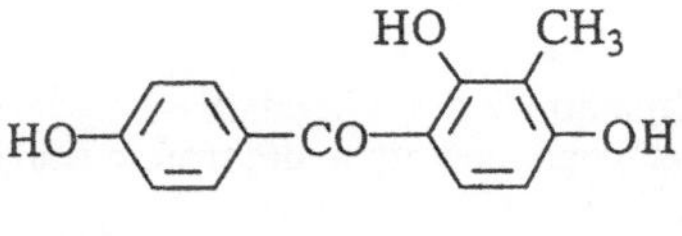

Synthesis

-Preparation by reaction of p-hydroxybenzoic acid with
2-methylresorcinol in tetrachloroethane in the presence
of boron trifluoride at 80° for 4 h (79%) [271].

m.p. 227-228° [271]; ^{1}H NMR [271].

(2,4-Dihydroxy-5-methylphenyl)(3-hydroxyphenyl)methanone

[61227-14-3] $C_{14}H_{12}O_4$ mol.wt. 244.25

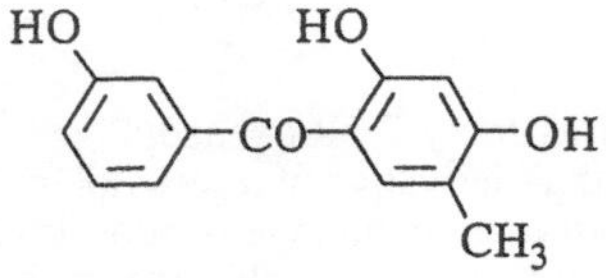

Syntheses

-Preparation by condensation of m-acetoxybenzonitrile with
4-methylresorcinol (Hoesch reaction) (70%) [398].
-Also obtained by [397] according to the method [1402].

m.p. 180° [398];   Spectra (NA).

(2,6-Dihydroxy-4-methylphenyl)(4-hydroxyphenyl)methanone

[190728-23-5] $C_{14}H_{12}O_4$ mol.wt. 244.25

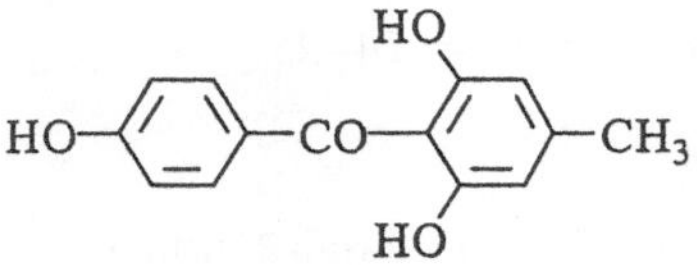

Synthesis

-Refer to: Chem. Abstr., **127**, 34137f (1997).

m.p. and Spectra (NA).

(2,4-Dihydroxyphenyl)(2-hydroxy-3-methylphenyl)methanone

[107412-87-3] $C_{14}H_{12}O_4$ mol.wt. 244.25

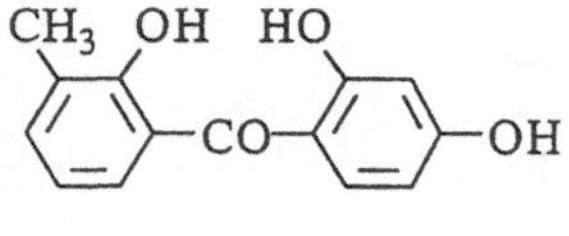

Synthesis

-Preparation by heating o-cresotic acid (3-methylsalicylic acid
or 2-hydroxy-3-methylbenzoic acid) and resorcinol with zinc
chloride and phosphorous oxychloride [679].

m.p. 116-117° [679];   Spectra (NA).

(2,4-Dihydroxyphenyl)(2-hydroxy-4-methylphenyl)methanone

[92254-59-6] $C_{14}H_{12}O_4$ mol.wt. 244.25

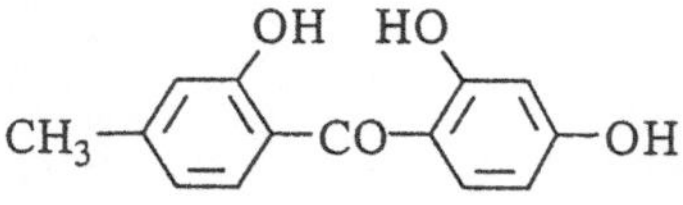

Synthesis

-Preparation by reaction of 4-methylsalicylic acid
(m-cresotic acid or 2-hydroxy-4-methylbenzoic acid)
with resorcinol,
*in the presence of zinc chloride and a mixture of

polyphosphoric acid/85% phosphoric acid (60:40) at 27°. Then, during 2 h, phosphorous trichloride was added between 27 to 37° and the mixture heated at 60° for 16 h [1302]; *in the presence of zinc chloride and phosphorous oxychloride [679]. -Also refer to: [38].

pale yellow crystals [1302];   m.p. 153-154° [679];   Spectra (NA).

(2,4-Dihydroxyphenyl)(4-hydroxy-2-methylphenyl)methanone

[4520-99-4] $C_{14}H_{12}O_4$ mol.wt. 244.25

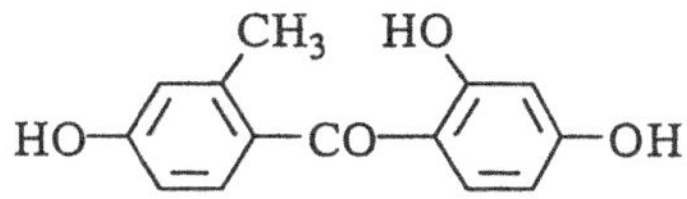

Synthesis

-Refer to: [125].

m.p. (NA);   MS [125].

(2,4-Dihydroxyphenyl)(5-hydroxy-2-methylphenyl)methanone

[42470-88-2] $C_{14}H_{12}O_4$ mol.wt. 244.25

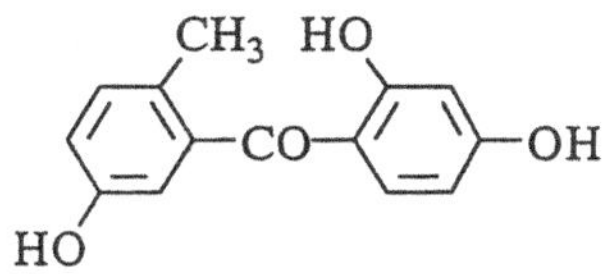

Synthesis

-Preparation by condensation of 5-hydroxy-2-methylbenzoic acid with resorcinol in the presence of zinc chloride and phosphorous oxychloride at 75° for 1 h (78%) [431].

m.p. 162° [431]; ^{1}H NMR [431], IR [431], UV [431], MS [431].

(2,4-Dihydroxy-6-methoxyphenyl)(3-hydroxyphenyl)methanone

[61227-12-1] $C_{14}H_{12}O_5$ mol.wt. 260.25

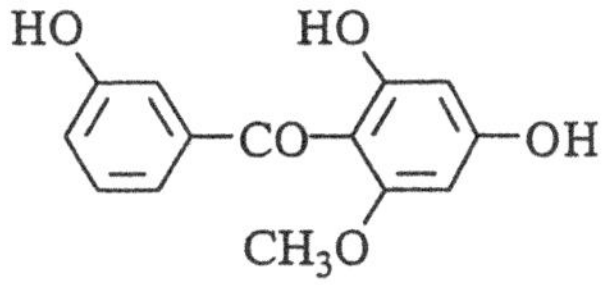

Syntheses

-Preparation by condensation of m-acetoxybenzonitrile with phloroglucinol monomethyl ether (Hoesch reaction) (33%) [398]. -Also obtained by [397] according to the method [1402].

m.p. 178° [398];   Spectra (NA).

(2,4-Dihydroxy-6-methoxyphenyl)(4-hydroxyphenyl)methanone

[56308-11-3] $C_{14}H_{12}O_5$ mol.wt. 260.25

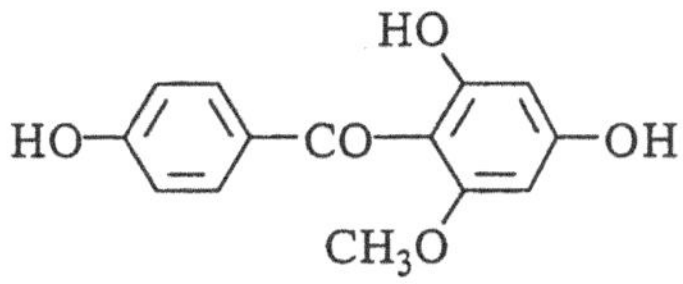

Synthesis

-Refer to: [1337].

m.p. and Spectra (NA).

(2,5-Dihydroxy-4-methoxyphenyl)(4-hydroxyphenyl)methanone

[58115-06-3] $C_{14}H_{12}O_5$ mol.wt. 260.25

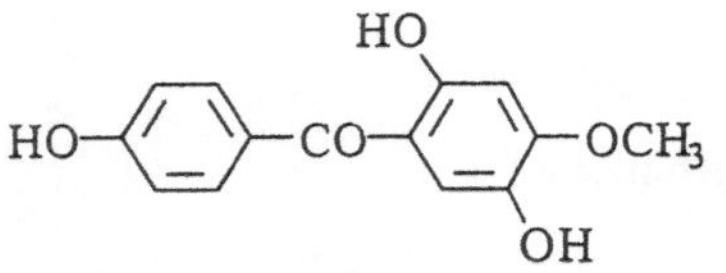

Synthesis

-Obtained by partial demethylation of 2-hydroxy-
4,4',5-trimethoxybenzophenone with hydriodic acid in
acetic anhydride [371].

Isolation from natural source

-From *Dalbergia melanoxylon* Guill. et Perr. heartwood (Leguminosae-Lotoideae) [371].

m.p.   228-229°5 [371]; ^{1}H NMR [371].

(2,6-Dihydroxy-4-methoxyphenyl)(4-hydroxyphenyl)methanone

[55051-85-9] $C_{14}H_{12}O_5$ mol.wt. 260.25

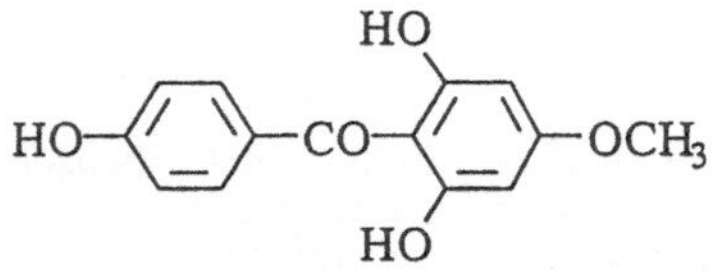

Synthesis

-Refer to: [1170].

Isolation from natural sources

-From *Aniba duckei* Kostern (Lauraceae) [299];
-From rhizome of *Anemarrhena asphodeloides* [977], according to [1170] or from *Anemarrhena
asphodeloides Bge* [176].

m.p.   179-181° [299], 146-150° [176]. There is discrepancy between the two melting points.
^{1}H NMR [176, 299], ^{13}C NMR [176], IR [299],
UV [299], MS [176, 299];   TLC [977].

(2,3-Dihydroxyphenyl)(2-hydroxy-6-methoxyphenyl)methanone

[25577-00-8] $C_{14}H_{12}O_5$ mol.wt. 260.25

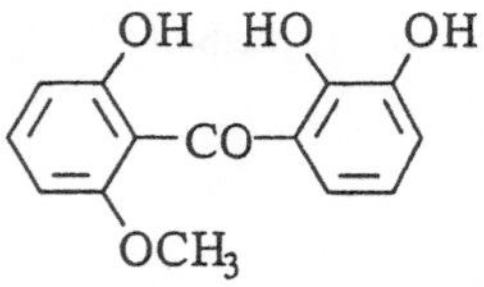

Synthesis

-Preparation by partial demethylation of 2,2',3',6-tetra-
methoxybenzophenone with boron tribromide in benzene at
r.t. for 5 h (76%) [830].

m.p.   147-150° [830]; ^{1}H NMR [830], IR [830], UV [830], MS [830].

(2,4-Dihydroxyphenyl)(2-hydroxy-4-methoxyphenyl)methanone

[7392-62-3] $C_{14}H_{12}O_5$ mol.wt. 260.25

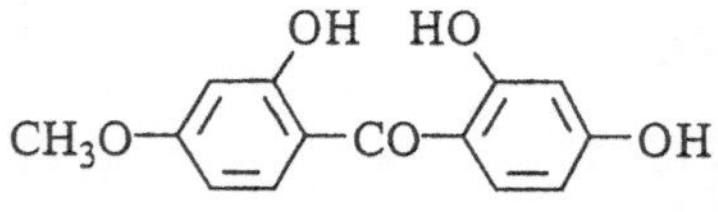

Synthesis

-Obtained by action of 2-hydroxy-4-methoxybenzoic
acid with resorcinol in the presence of zinc chloride
and phosphorous oxychloride at 60-65° for 3 h (30%)
[455].

-Also refer to: [61, 1253, 1301].

m.p. 101-102° [455]; Spectra (NA).

(2,4-Dihydroxyphenyl)(4-hydroxy-2-methoxyphenyl)methanone

[71655-03-3] $C_{14}H_{12}O_5$ mol.wt. 260.25

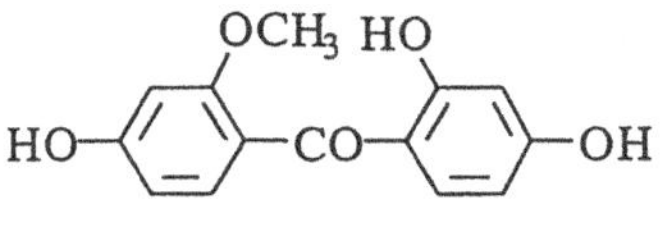

Synthesis

-Preparation by reaction of β-resorcylic acid with resorcinol monomethyl ether in the presence of zinc chloride and a mixture of polyphosphoric acid/85% phosphoric acid (60:40) at 27°. Then, during 2 h, phosphorous trichloride was added between 27 to 37° and the mixture heated at 60° for 16 h [1302].
-Also refer to: [938] (Japanese patent).

yellow crystals [1302]; m.p. and Spectra (NA).

(2,4-Dihydroxyphenyl)(4-hydroxy-3-methoxyphenyl)methanone

$C_{14}H_{12}O_5$ mol.wt. 260.25

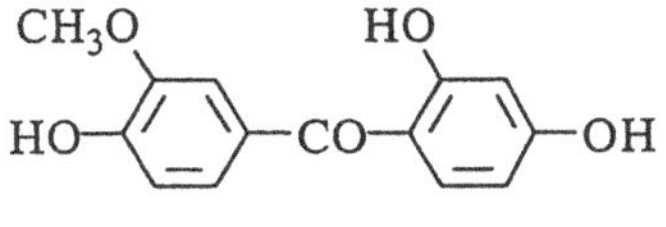

Synthesis

-Preparation by reaction of vanillonitrile (4-hydroxy-3-methoxybenzonitrile) with resorcinol (Hoesch reaction) (20%) [586].

m.p. 210° [586]; Spectra (NA).

(2,4-Dihydroxyphenyl)(5-hydroxy-2-methoxyphenyl)methanone

[61227-15-4] $C_{14}H_{12}O_5$ mol.wt. 260.25

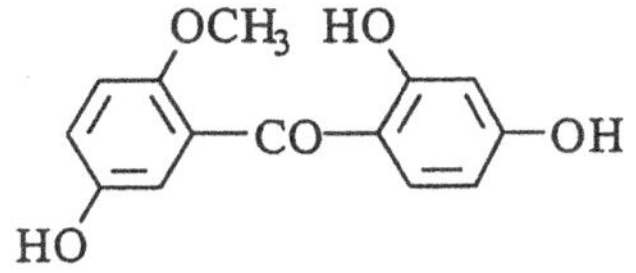

Synthesis

-Preparation by treatment of N-[5-acetoxy-α-(2,4-dihydroxyphenyl)-2-methoxybenzylidene]aniline with 20% sulfuric acid at reflux for 5 h under nitrogen (92%) [398].

m.p. 194° [398]; IR [398].

(2,5-Dihydroxyphenyl)(2-hydroxy-4-methoxy-6-methylphenyl)methanone

[78044-96-9] $C_{15}H_{14}O_5$ mol.wt. 274.27

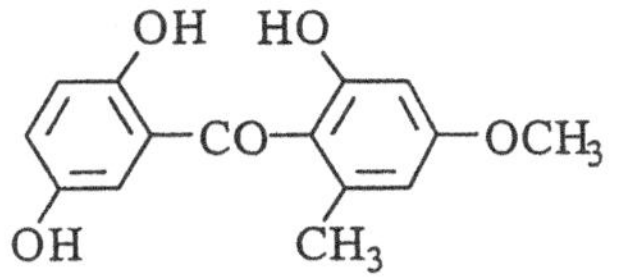

Synthesis

-Preparation by partial demethylation of 2',5'-dihydroxy-2,4-dimethoxy-6-methylbenzophenone with boron trichloride in methylene chloride at r.t. for 5 days (82%) [816].

m.p. 176-179° [816]; ^{1}H NMR [816], IR [816], UV [816].

(2,4-Dihydroxy-3-propylphenyl)(2-hydroxyphenyl)methanone

[115296-04-3] $C_{16}H_{16}O_4$ mol.wt. 272.30

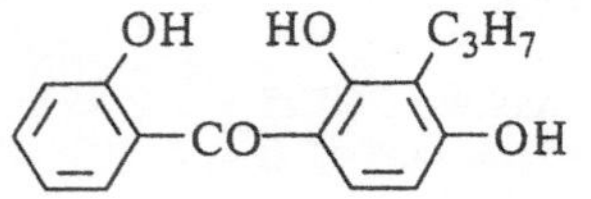

Synthesis

-Obtained by total demethylation of 2,2',4-trimethoxy-
 3-propylbenzophenone or of 2,4-dimethoxy-2'-hydroxy-
 3-propylbenzophenone with pyridinium chloride at 180° for
 3 h [456].
-Also refer to: [1432].

m.p. 110-113° [456];   Spectra (NA).

[2,4-Dihydroxy-5-(1,1-dimethylethyl)phenyl](2-hydroxyphenyl)methanone

$C_{17}H_{18}O_4$ mol.wt. 286.33

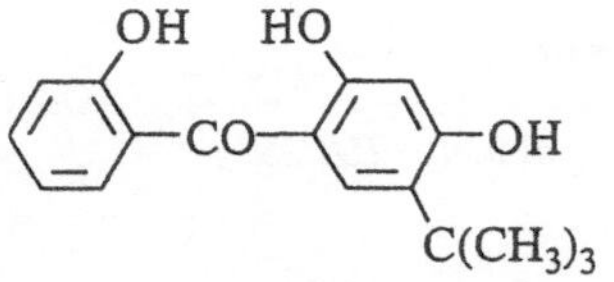

Synthesis

-Preparation by alkylation of 2,2',4-trihydroxy-
 benzophenone with isobutylene in benzene in the presence
 of p-toluenesulfonic acid for 2 h at 65-75° (70%) [587,
 588].

m.p. 194-196° [587, 588];   UV [588].

(2,4-Dihydroxyphenyl)[4-hydroxy-2-methyl-5-(1-methylethyl)phenyl]methanone

$C_{17}H_{18}O_4$ mol.wt. 286.33

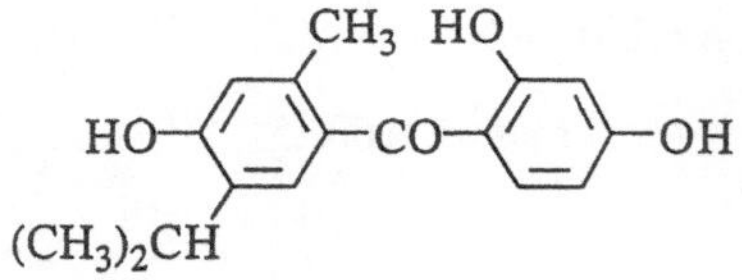

Synthesis

-Obtained by demethylation of 2',4,4'-trimethoxy-
 2-methyl-5-isopropylbenzophenone in refluxing
 pyridinium chloride (205-215°) for 1.3 h (34%)
 [1148].

b.p.$_{0.9}$ 220-230° [1148];   Spectra (NA).

[2,4-Dihydroxy-5-(1,1-dimethylpropyl)phenyl](2-hydroxyphenyl)methanone

$C_{18}H_{20}O_4$ mol.wt. 300.35

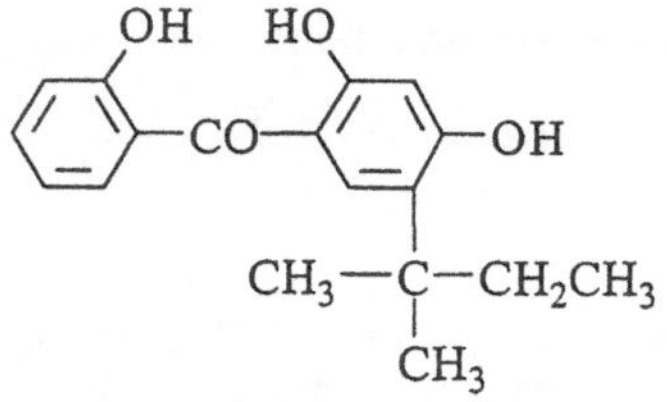

Synthesis

-Preparation by alkylation of 2,2',4-trihydroxy-
 benzophenone with 2-methylbutene in benzene in the
 presence of concentrated sulfuric acid (82%) [587].

m.p. 142-144° [587, 588];   Spectra (NA).

[2,4-Dihydroxy-5-(2-propenyl)-3-propylphenyl](2-hydroxyphenyl)methanone

$C_{19}H_{20}O_4$ mol.wt. 312.37

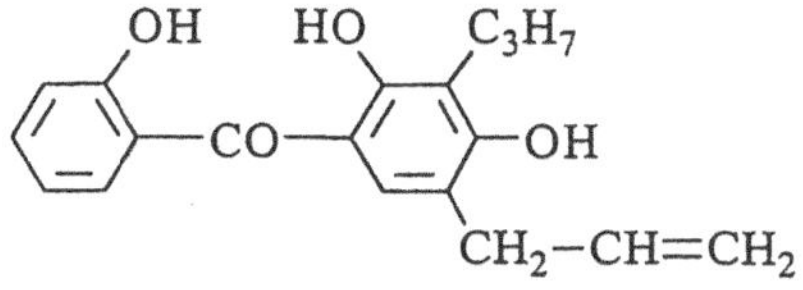

Synthesis

-Obtained (poor yield) by heating 4-allyloxy-
2,2'-dihydroxy-3-propylbenzophenone at 180° for
6 h (19%) [456] (Claisen rearrangement).

m.p. and Spectra (NA).

[5-(1,1-Dimethylbutyl)-2,4-dihydroxyphenyl](2-hydroxyphenyl)methanone

$C_{19}H_{22}O_4$ mol.wt. 314.38

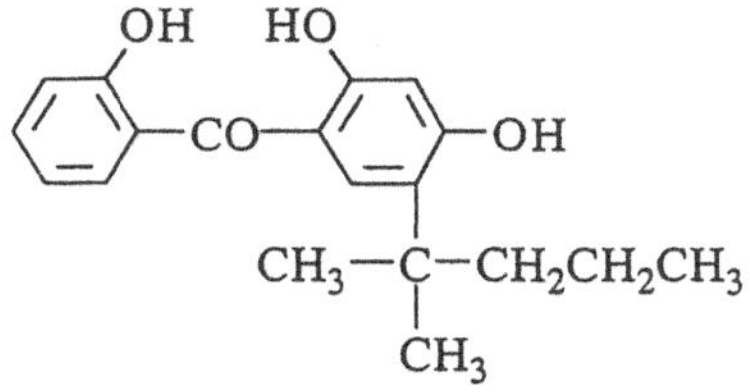

Synthesis

-Preparation by alkylation of 2,2',4-trihydroxy-
benzophenone with 2-methylpentene in benzene in
the presence of concentrated sulfuric acid [587].

m.p. 125° [587, 588];   Spectra (NA).

(2,4-Dihydroxy-3,5-dipropylphenyl)(2-hydroxyphenyl)methanone

$C_{19}H_{22}O_4$ mol.wt. 314.38

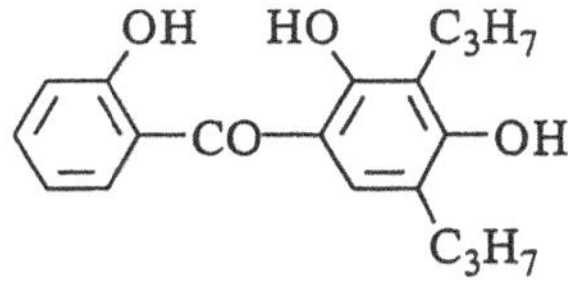

Synthesis

-Preparation by catalytic hydrogenation of 5-allyl-3-propyl-
2,2',4-trihydroxybenzophenone in ethyl acetate in the
presence of 5% Pd/C (94%) [456].

m.p. 110-112° [456];   Spectra (NA).

[2,4-Dihydroxy-5-(1,1,3,3-tetramethylbutyl)phenyl](2-hydroxyphenyl)methanone

$C_{21}H_{26}O_4$ mol.wt. 342.44

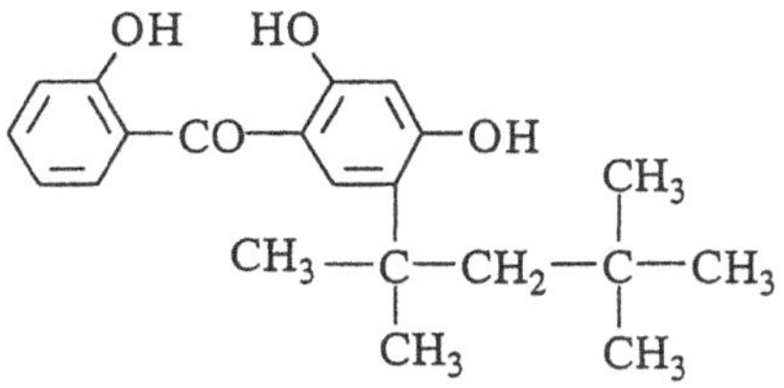

Synthesis

-Preparation by alkylation of 2,2',4-trihydroxy-
benzophenone with 2,2,4-trimethylpentene in
benzene in the presence of concentrated sulfuric
acid [587].

m.p. 165-168° [587, 588];   Spectra (NA).

(2,4-Dihydroxyphenyl)(2-hydroxy-5-nonylphenyl)methanone

$C_{22}H_{28}O_4$ mol.wt. 356.46

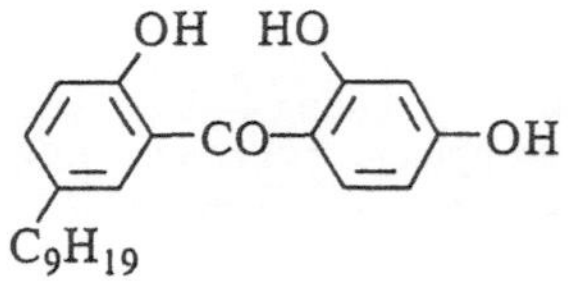

Synthesis

-Preparation by reaction of β-resorcylic acid with p-nonyl-phenol in the presence of a mixture of polyphosphoric acid/85% phosphoric acid (60:40) at 40-45°. Then, during 2 h, phosphorous trichloride was added and the mixture heated at 60° for 16 h (73%) [1302].

m.p. and Spectra (NA).

3.2.2. Substituents located on both rings

(2,4-Dihydroxy-3,5,6-trinitrophenyl)(4-hydroxy-3-nitrophenyl)methanone

[67246-03-1] $C_{13}H_6N_4O_{12}$ mol.wt. 410.21

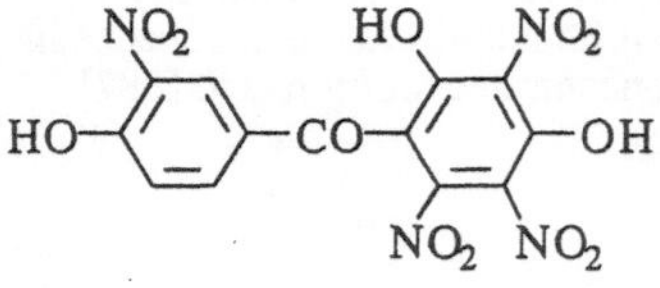

Synthesis

-Obtained by reaction of nitric acid (d = 1.42) with 4-hydroxy-2',4'-dimethoxybenzophenone in acetic acid at 32° [1083].

m.p. 110° [1083]; ^1H NMR [1083].

(2,4-Dihydroxy-3-methylphenyl)(5-hydroxy-2-methoxyphenyl)methanone

[61227-17-6] $C_{15}H_{14}O_5$ mol.wt. 274.27

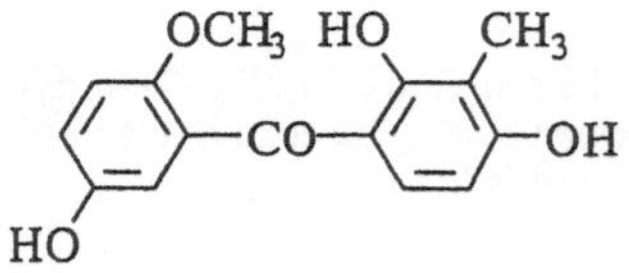

Synthesis

-Preparation by hydrolysis of N-[5-acetoxy-α-(2,4-di-hydroxyphenyl)-2-methoxy-3-methylbenzylidene]aniline with refluxing 20% sulfuric acid under nitrogen during 2 h (90%) [398].

m.p. 210° [398];   Spectra (NA).

(2,4-Dihydroxy-5-methylphenyl)(5-hydroxy-2-methoxyphenyl)methanone

$C_{15}H_{14}O_5$ mol.wt. 274.27

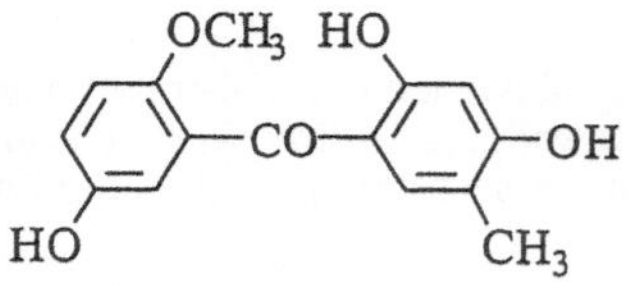

Syntheses

-Preparation by hydrolysis of N-[5-acetoxy-α-(2,4-di-hydroxyphenyl)-2-methoxy-5-methylbenzylidene]-aniline with refluxing 20% sulfuric acid during 5 h under nitrogen (90%) [398].
-Also obtained by [397] according to the method [1402].

m.p. 199° [398];   Spectra (NA).

(2,4-Dihydroxy-6-methylphenyl)(5-hydroxy-2-methoxyphenyl)methanone

[61227-16-5] $C_{15}H_{14}O_5$ mol.wt. 274.27

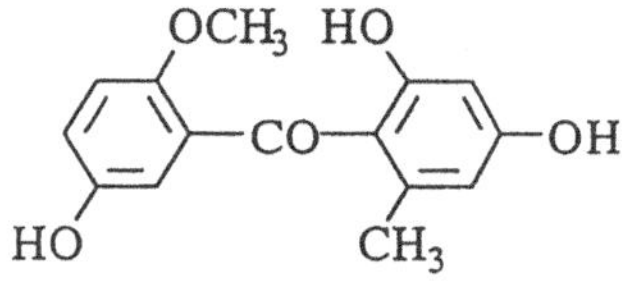

Synthesis

N.B.: This compound, mentioned in [Chem. Abstr., **86**, 16504d (1977)], is not described in the original paper [398], where the sole corresponding ketone indicated is the 2,4,5'-trihydroxy-2'-methoxy-5-methylbenzophenone or (2,4-dihydroxy-5-methylphenyl)(5-hydroxy-2-methoxyphenyl)methanone.

m.p. and Spectra (NA).

(2,5-Dihydroxy-4-methoxyphenyl)(3-hydroxy-4-methoxyphenyl)methanone
(Melanoxoin)

[58115-05-2] $C_{15}H_{14}O_6$ mol.wt. 290.27

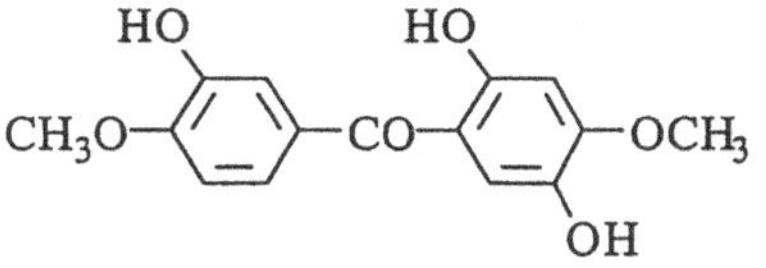

Synthesis

-Not yet described.

Isolation from natural source

-From *Dalbergia melanoxylon* Guill. et Perr. heartwood (Leguminosae-Lotoideae) [371].

m.p. 232-234° [371]; IR [371], UV [371].

(2,6-Dihydroxy-4-methoxyphenyl)(2-hydroxy-3-methoxyphenyl)methanone

$C_{15}H_{14}O_6$ mol.wt. 290.27

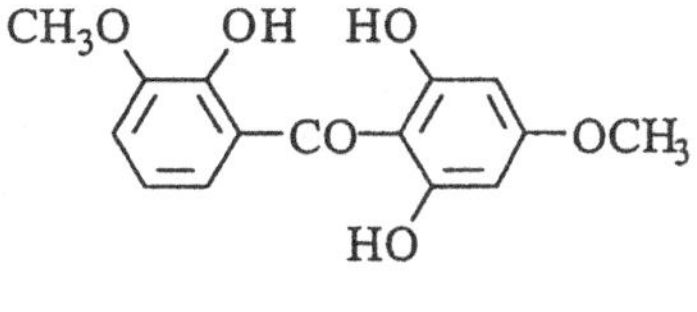

Synthesis

-Obtained by condensation of 2,3-dimethoxybenzoic acid and 1,3,5-trimethoxybenzene in the presence of aluminium chloride, zinc chloride and phosphorous oxychloride, prior to cyclisation into 1-hydroxy-3,5-dimethoxyxanthone [832].

m.p. and Spectra (NA).

(3-Chloro-4,6-dihydroxy-2-methylphenyl)(3,5-dichloro-2-hydroxy-4-methoxy-6-methylphenyl)methanone

[69709-89-3] $C_{16}H_{13}Cl_3O_5$ mol.wt. 391.63

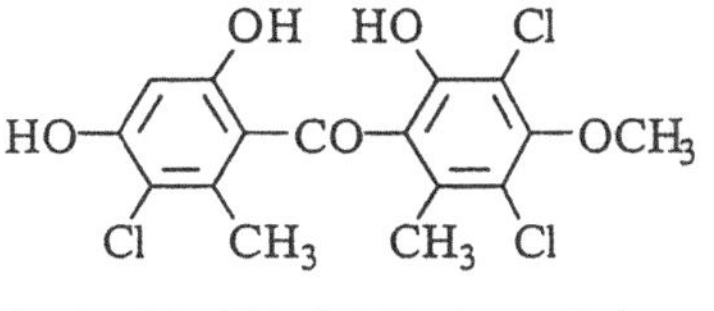

Synthesis

-Preparation by hydrogenolysis of 2,2',4'-tris-(benzyloxy)-3,5,5'-trichloro-4-methoxy-6,6'-dimethylbenzophenone (SM) in the presence of 10% Pd/C in ethyl acetate containing concentrated hydrochloric acid (5 drops) (>90%) [1174]. SM was obtained by Friedel-Crafts acylation of 3,5-bis(benzyloxy)-2-chlorotoluene with 2-(benzyloxy)-3,5-dichloro-4-methoxy-6-methylbenzoic acid in the presence of trifluoroacetic anhydride in

refluxing ethylene dichloride for 5 h (42%).
-Also refer to: [1173].

m.p.   168°5-169°5 [1174]; ^{1}H NMR [1174], MS [1174].

(3-Chloro-6-hydroxy-4-methoxy-2-methylphenyl)(3,5-dichloro-2,4-dihydroxy-6-methylphenyl)methanone

C$_{16}$H$_{13}$Cl$_3$O$_5$ mol.wt. 391.63

Synthesis

-Preparation by hydrogenolysis of 2,2',4-tris-(benzyloxy)-3,5,5'-trichloro-4'-methoxy-6,6'-di-methylbenzophenone (SM) in the presence of 10% Pd/C in ethyl acetate containing concentrated hydrochloric acid (5 drops) (>90%) [1174]. SM was obtained by Friedel-Crafts acylation of 5-(benzyloxy)-2-chloro-3-methoxytoluene with 2,4-bis-(benzyloxy)-3,5-dichloro-6-methylbenzoic acid in the presence of trifluoroacetic anhydride in refluxing ethylene dichloride for 5 h (42%).

m.p.   174-175°5 [1174]; ^{1}H NMR [1174], MS [1174].

(3,5-Dibromo-2-hydroxy-4-methoxy-6-methylphenyl)(2,4-dihydroxy-6-methylphenyl)-methanone

C$_{16}$H$_{14}$Br$_2$O$_5$ mol.wt. 446.09

Synthesis

-Preparation by hydrogenolysis of 2,2',4'-tris-(benzyloxy)-3,5-dibromo-4-methoxy-6,6'-dimethyl-benzophenone (SM) in the presence of 10% Pd/C in ethanol/ethyl acetate mixture at r.t. and atmospheric pressure. SM was obtained by condensation of orcinol dibenzyl ether with 3,5-dibromoeverninic acid benzyl ether in the presence of a trifluoroacetic anhydride/trifluoroacetic acid mixture under nitrogen in refluxing chloroform for 5 h [564].

m.p. and Spectra (NA).

(3,5-Dibromo-2-hydroxy-6-methoxy-4-methylphenyl)(2,4-dihydroxy-6-methylphenyl)-methanone

[39803-81-1]   C$_{16}$H$_{14}$Br$_2$O$_5$ mol.wt. 446.09

Synthesis

-Preparation by hydrogenolysis of 2,2',4'-tris-(benzyloxy)-3,5-dibromo-6-methoxy-4,6'-dimethyl-benzophenone (SM) in ethanol/trifluoroacetic acid mixture in the presence of 10% Pd/C at atmospheric pressure (quantitative yield). SM was obtained by condensation of O-benzyl-O-methyl-3,5-dibromo-γ-orsellinic acid — 2-(benzyloxy)-3,5-dibromo-6-methoxy-4-methylbenzoic acid — with orcinol dibenzyl ether in the presence of trifluoroacetic anhydride/trifluoroacetic acid mixture under nitrogen in refluxing chloroform for 6.25 h [564].

m.p. and Spectra (NA).

(3-Chloro-4,6-dihydroxy-2-methylphenyl)(3-chloro-6-hydroxy-4-methoxy-2-methylphenyl)methanone

[78135-54-3] $C_{16}H_{14}Cl_2O_5$ mol.wt. 357.19

Synthesis

-Preparation by hydrogenolysis (4 bars) of 4,6,6'-tris(benzyloxy)-3,3'-dichloro-4'-methoxy-2,2'-dimethylbenzophenone (m.p. 168-169°) in ethyl acetate in the presence of 10% Pd/C and 5 drops of concentrated hydrochloric acid (75%) [1092].

m.p. 180-182° [1092]; ^{1}H NMR [1092], IR [1092].

(3,5-Dichloro-2,6-dihydroxy-4-methylphenyl)(2-hydroxy-4-methoxy-6-methylphenyl)methanone

[23565-77-7] $C_{16}H_{14}Cl_2O_5$ mol.wt. 357.19

Synthesis

-Obtained (poor yield) by photo-Fries rearrangement of 2,4-dichloro-5-hydroxy-3-methylphenyl 2-hydroxy-4-methoxy-6-methylbenzoate in ethanol at 20° for 75 h (3%) [16].

m.p. 142-144° [16]; ^{1}H NMR [16], IR [16], UV [16].

(3,5-Dichloro-2-hydroxy-4-methoxy-6-methylphenyl)(2,4-dihydroxy-6-methylphenyl)methanone

[39803-63-9] $C_{16}H_{14}Cl_2O_5$ mol.wt. 357.19

Synthesis

-Preparation by hydrogenolysis of 2,2',4'-tris-(benzyloxy)-3,5-dichloro-4-methoxy-6,6'-dimethylbenzophenone (SM) in the presence of 10% Pd/C in ethanol containing a few drops of trifluoroacetic acid for 3-4 h. SM was obtained by condensation of orcinol dibenzyl ether with 3,5-dichloroeverninic acid benzyl ether in the presence of a trifluoroacetic anhydride/trifluoroacetic acid mixture in refluxing chloroform for 6 h [564].

m.p. and Spectra (NA).

(3-Chloro-4,6-dihydroxy-2-methylphenyl)(3-chloro-6-hydroxy-2,4-dimethoxyphenyl)methanone

[68048-23-7] $C_{16}H_{14}Cl_2O_6$ mol.wt. 373.19

Synthesis

-Preparation by hydrogenolysis of 4,6,6'-tris-(benzyloxy)-3,3'-dichloro-2',4'-dimethoxy-2-methylbenzophenone (SM) in ethyl acetate/tetrahydrofuran in the presence of 10% Pd/C at 25° (29%). SM was obtained by condensation of 4,6-bis(benzyloxy)-

3-chloro-2-methylbenzoic acid with 4-chloro-3,5-dimethoxyphenol benzyl ether in the presence of trifluoroacetic anhydride in methylene chloride under nitrogen for 20 min [1328].

m.p.   217-219° [1328];
^{1}H NMR [1328], IR [1328], MS [1328].

(3-Chloro-2,6-dihydroxy-4-methoxyphenyl)(4-hydroxy-2-methoxy-6-methylphenyl)-methanone *(Griseophenone B)*

[3811-00-5] $C_{16}H_{15}ClO_6$ mol.wt. 338.74

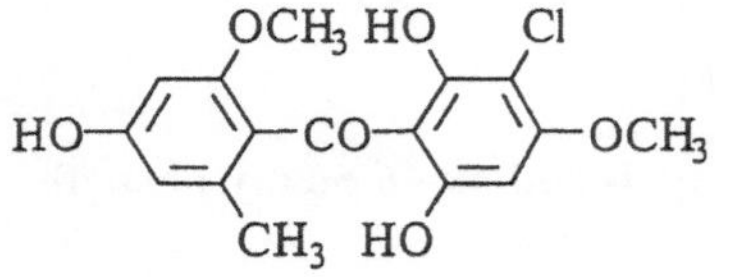

Synthesis

-Not yet described.

Isolation from natural source

-From cultures of *Penicillium patulum* [1115, 1116].

-Also refer to: [181, 543, 1199, 1200, 1201, 1202, 1203].

m.p.   204°5-205°5 [1115];   IR [1115], UV [1115].

(3-Chloro-6-hydroxy-2,4-dimethoxyphenyl)(2,4-dihydroxy-6-methylphenyl)methanone

[68048-17-9] $C_{16}H_{15}ClO_6$ mol.wt. 338.74

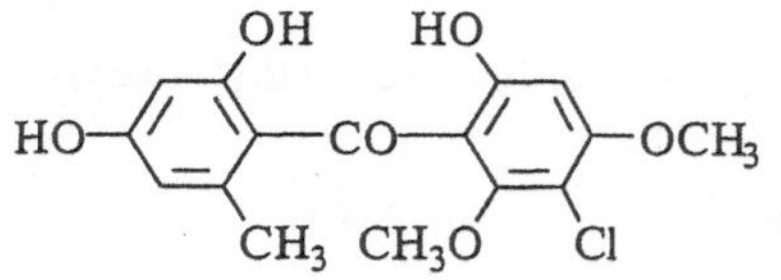

Synthesis

-Preparation by hydrogenolysis of 2',4',6-tris-(benzyloxy)-3-chloro-2,4-dimethoxy-6'-methyl-benzophenone (SM) in ethyl acetate/tetrahydrofuran in the presence of 10% Pd/C at 25°. SM was obtained by condensation of 2,4-bis(benzyloxy)-6-methylbenzoic acid with 4-chloro-3,5-dimethoxyphenol benzyl ether in the presence of trifluoroacetic anhydride in methylene chloride under nitrogen for 15 min (61%) [1328].

m.p.   181-182° [1328];
^{1}H NMR [1328], IR [1328], MS [1328].

(3,6-Dihydroxy-2,4-dimethylphenyl)(2-hydroxy-6-methoxyphenyl)methanone

[42594-59-2] $C_{16}H_{16}O_5$ mol.wt. 288.30

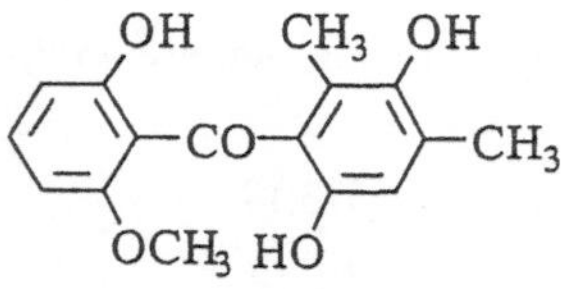

Synthesis

-Preparation by partial demethylation of 3-hydroxy-2',6,6'-trimethoxy-2,4-dimethylbenzophenone with boron tribromide in methylene chloride (quantitative yield) [430].

m.p.   188-191° [430];
^{1}H NMR [430], IR [430], UV [430], MS [430].

(2,4-Dihydroxy-6-methylphenyl)(2-hydroxy-4-methoxy-6-methylphenyl)methanone

[21147-34-2] $C_{16}H_{16}O_5$ mol.wt. 288.30

Syntheses

-Preparation by hydrogenolysis of 2,2',4'-tris-
(benzyloxy)-4-methoxy-6,6'-dimethylbenzophenone
(SM) in the presence of 10% Pd/C in ethyl acetate
containing concentrated hydrochloric acid (5 drops)
[1174]. SM was obtained by Friedel-Crafts
acylation of 3,5-bis(benzyloxy)toluene with 2-(benzyloxy)-4-methoxy-6-methylbenzoic acid in the
presence of trifluoroacetic anhydride in refluxing ethylene dichloride for 5 h (95%).
-Also obtained by photo-Fries rearrangement of 5-hydroxy-3-methylphenyl 2-hydroxy-4-methoxy-
6-methylbenzoate in ethanol at r.t. for 80 h (21%) [16].

    m.p. 158-160° [1174], 158-159° [16];
    ^{1}H NMR [16, 1174], IR [16], UV [16], MS [125, 1174].

(2,4-Dihydroxy-6-methylphenyl)(4-hydroxy-2-methoxy-6-methylphenyl)methanone

[23573-47-9] $C_{16}H_{16}O_5$ mol.wt. 288.30

Synthesis

-Obtained by reaction of orcinol with 2-methoxy-
4-O-methoxycarbonyl-6-methylbenzoic acid in
trifluoroacetic anhydride, first at 0° for 30 min, then at
20° for 15 h (30%) [16].

    m.p. 118-120° [16];
    ^{1}H NMR [16], IR [16], UV [16].

(2,6-Dihydroxy-4-methylphenyl)(2-hydroxy-4-methoxy-6-methylphenyl)methanone

[21147-33-1] $C_{16}H_{16}O_5$ mol.wt. 288.30

Syntheses

-Obtained (poor yields) by photo-Fries rearrangement
of two esters in ethanol at r.t.,
*from 3-hydroxy-5-methylphenyl 2-hydroxy-
4-methoxy-6-methylbenzoate during 80 h (5%) [16];
*from 3-methoxy-5-methylphenyl 2,6-dihydroxy-
4-methylbenzoate during 60 h (<2%) [16].
-Preparation by hydrogenolysis of 3',5'-dichloro-2,2',6'-trihydroxy-4-methoxy-4',6-dimethyl-
benzophenone in the presence of Pd/C in 1 N sodium hydroxide [16].

    m.p. 56-58° [16];
    ^{1}H NMR [16], IR [16], UV [16], MS [125].

(2,6-Dihydroxy-4-methylphenyl)(4-hydroxy-2-methoxy-6-methylphenyl)methanone

[23565-89-1] $C_{16}H_{16}O_5$ mol.wt. 288.30

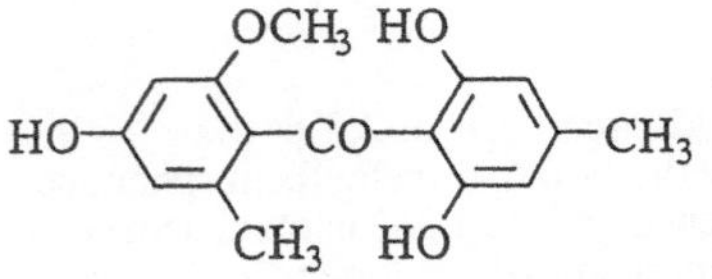

Synthesis

-Obtained (by-product) by reaction of orcinol with
2-methoxy-4-O-methoxycarbonyl-6-methylbenzoic
acid in trifluoroacetic anhydride, first at 0° for 30 min,
then at 20° for 15 h (11%) [16].

m.p. 139-140° [16]; ^{1}H NMR [16], IR [16], UV [16].

(2,6-Dihydroxy-4-methoxyphenyl)(4-hydroxy-2-methoxy-6-methylphenyl)methanone
(Griseophenone C)

[3733-72-0] $C_{16}H_{16}O_6$ mol.wt. 304.30

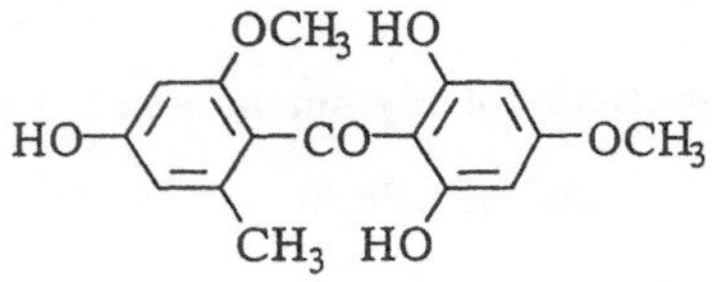

Synthesis

-Not yet described.

Isolation from natural source

-From cultures of *Penicillium patulum* [215, 1115, 1116].
-Also refer to: [543, 552, 1199, 1200, 1203].

m.p. 183-185° [1116], 183-184° [1115], 175-178° [215];
^{1}H NMR [215], IR [1115], UV [1115].

(3-Chloro-6-hydroxy-4-methoxy-2,5-dimethylphenyl)(2,4-dihydroxy-6-methylphenyl)-methanone

[60138-98-9] $C_{17}H_{17}ClO_5$ mol.wt. 336.77

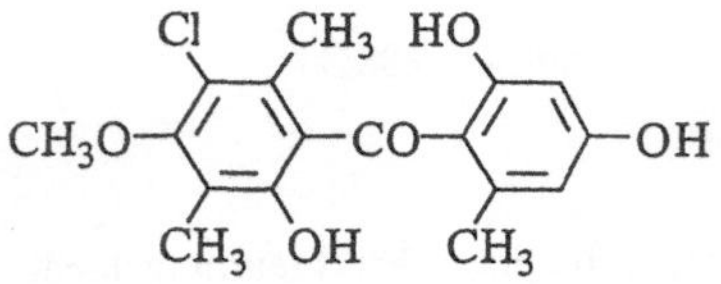

Synthesis

-Preparation by hydrogenolysis of 2,2',4-tris-
(benzyloxy)-5'-chloro-4'-methoxy-3',6,6'-trimethyl-
benzophenone in the presence of 10% Pd/C in ethyl
acetate containing a small quantity of concentrated
hydrochloric acid (89%) [1190].

m.p. 157-158° [1190]; ^{1}H NMR [1190].

(3,5-Dichloro-2,6-dihydroxy-4-methoxyphenyl)(4-hydroxy-2-methyl-6-propoxyphenyl)-methanone

[72614-88-1] $C_{18}H_{18}Cl_2O_6$ mol.wt. 401.24

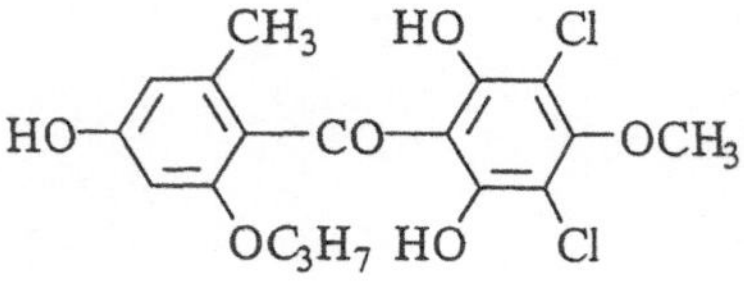

Synthesis

-Obtained by transformation of the 2-propoxy analog
of griseophenone B by *Penicillium urticae* [1200],
(10%) [1199].

m.p. (NA);   MS [1199, 1200].

(3-Chloro-6-hydroxy-4-methoxy-2,5-dimethylphenyl)(2,4-dihydroxy-3,6-dimethyl-phenyl)methanone

[61852-15-1] $C_{18}H_{19}ClO_5$ mol.wt. 350,80

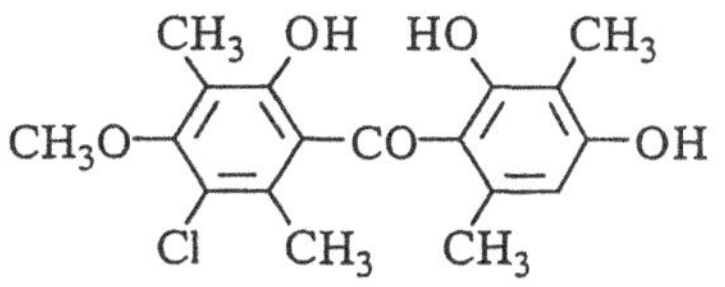

Synthesis

-Preparation by hydrogenolysis of 2,2',4'-tris-(benzyloxy)-5-chloro-4-methoxy-3,3',6,6'-tetra-methylbenzophenone (SM) in the presence of 10% Pd/C in ethyl acetate containing concentrated hydrochloric acid (3 drops) (quantitative yield) [1191], (>90%) [1174]. SM was obtained by Friedel-Crafts acylation of 1,3-bis(benzyloxy)-2,5-dimethylbenzene with 2-(benzyloxy)-5-chloro-4-methoxy-3,6-dimethylbenzoic acid in the presence of trifluoroacetic anhydride in methylene chloride at r.t. for 5.5 h (40%) [1174].
-Also refer to: [1173].

m.p. 176-177° [1174, 1191];
^1H NMR [1191], IR [1191], MS [1191].

(3-Chloro-2,6-dihydroxy-4-methoxyphenyl)(4-hydroxy-2-methyl-6-propoxyphenyl)-methanone

[69218-66-2] $C_{18}H_{19}ClO_6$ mol.wt. 366.80

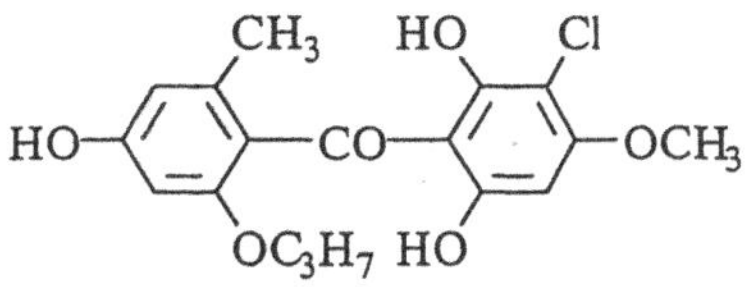

Synthesis

-Refer to: [1199, 1200] (this compound is a 2-propoxy analog of griseophenone B).

m.p. and Spectra (NA).

[3-Bromo-6-hydroxy-4-methoxy-5-methyl-2-(1-methylpropyl)phenyl](2,4-dihydroxy-6-methylphenyl)methanone

[67097-17-0] $C_{20}H_{23}BrO_5$ mol.wt. 423.30

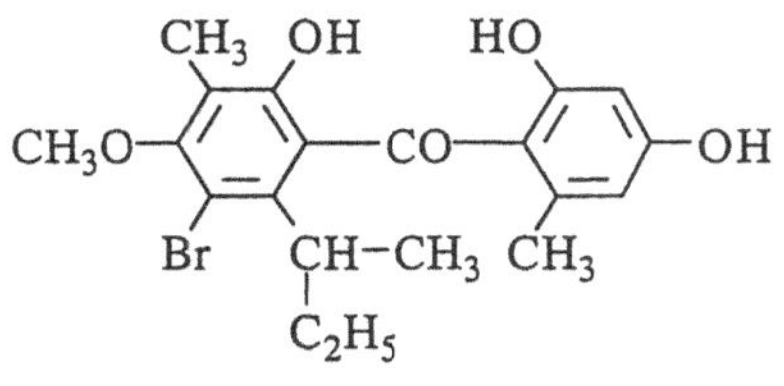

Synthesis

-Preparation by hydrogenolysis of 2,2',4'-tris-(benzyloxy)-5-bromo-4-methoxy-3,6-dimethyl-6-sec-butylbenzophenone in the presence of 10% Pd/C in ethyl acetate containing a small quantity of concentrated hydrochloric acid (89%) [361].
-Also refer to: [1173, 1174].

m.p. 185-188° [361], hydrate: 97-101° [361];
^1H NMR [361], MS [361].

[3-Chloro-6-hydroxy-4-methoxy-5-methyl-2-(1-methylpropyl)phenyl](2,4-dihydroxy-6-methylphenyl)methanone

[78023-64-0] $C_{20}H_{23}ClO_5$ mol.wt. 378.85

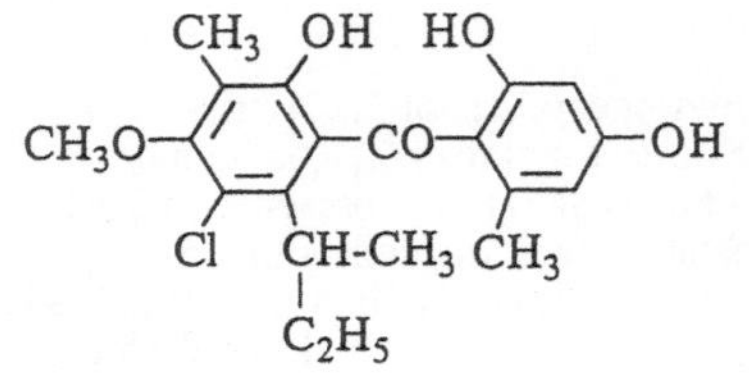

Synthesis

-Preparation by hydrogenolysis of 2,2',4'-tris-(benzyloxy)-5-chloro-4-methoxy-3,6'-dimethyl-6-sec-butylbenzophenone in the presence of 10% Pd/C in ethyl acetate containing concentrated hydrochloric acid (10 drops) (94%) [416].

m.p.   177-179° [416]; ^{1}H NMR [416], MS [416].

(2,4-Dihydroxy-3-methylphenyl)[2-hydroxy-4-(octyloxy)phenyl]methanone

$C_{22}H_{28}O_4$ mol.wt. 356.46

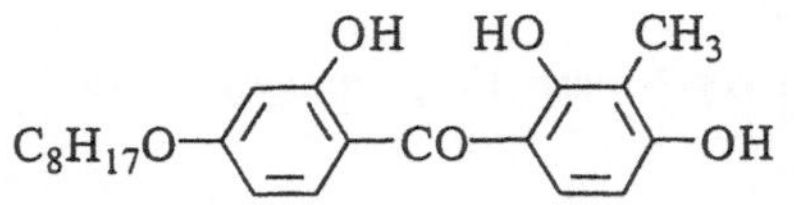

Synthesis

-Refer to: [271].

m.p. and Spectra (NA).

4. TETRAHYDROXYBENZOPHENONES

4.1. *Hydroxy groups located on one ring* ***Not described till December 1999.***

4.2. *Hydroxy groups located on both rings*

4.2.1. *Substituents located on one ring*

(3,5-Dichloro-4-hydroxyphenyl)(2,3,4-trihydroxyphenyl)methanone

[105443-52-5] $C_{13}H_8Cl_2O_5$ mol.wt. 315.11

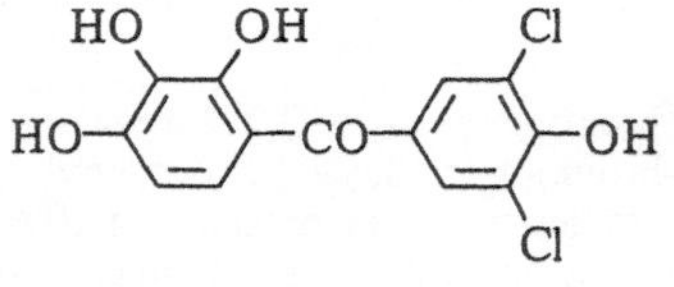

Synthesis

-Preparation by action of 3,5-dichloro-4-hydroxybenzoic acid on pyrogallol with boron trifluoride or its complexes [1443].

m.p. and Spectra (NA).

(3,4-Dihydroxy-5-nitrophenyl)(3,4-dihydroxyphenyl)methanone

[134612-52-5] $C_{13}H_9NO_7$ mol.wt. 291.22

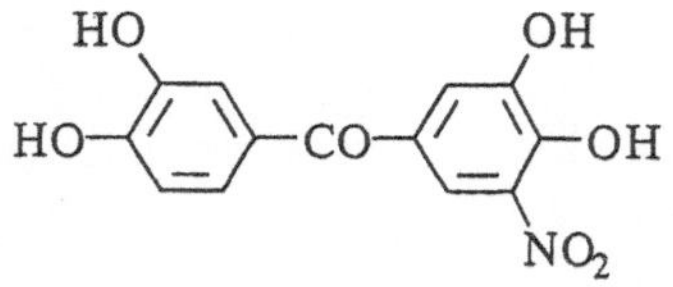

Synthesis

-Preparation by total demethylation of 4-hydroxy-5-nitro-3,3',4'-trimethoxybenzophenone with hydrobromic acid in refluxing dilute acetic acid [164].

m.p. 222-224° [164];   Spectra (NA).

(5-Hydroxy-2-nitrophenyl)(2,4,6-trihydroxyphenyl)methanone

$C_{13}H_9NO_7$ mol.wt. 291.22

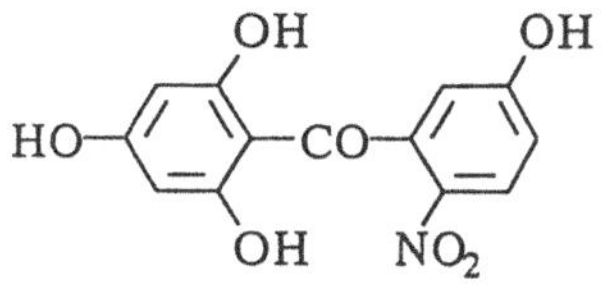

Synthesis

-Preparation by reaction of 5-hydroxy-2-nitrobenzoyl chloride with tris(trimethylsilyl) derivative of phloroglucinol in the presence of stannic chloride (or titanium tetrachloride or aluminium chloride) in refluxing methylene chloride for 2 h [649].

m.p. and Spectra (NA).

(2-Chloro-5-hydroxy-4-methoxyphenyl)(2,4,6-trihydroxyphenyl)methanone

$C_{14}H_{11}ClO_6$ mol.wt. 310.69

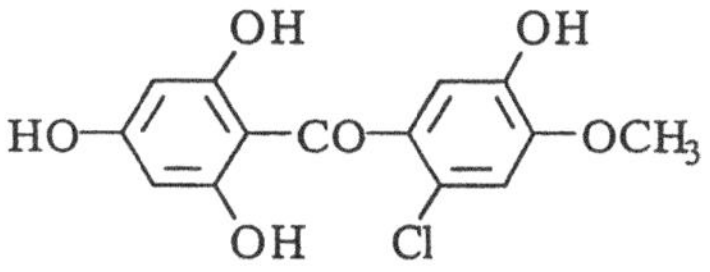

Synthesis

-Preparation by reaction of 2-chloro-5-hydroxy-4-methoxybenzoyl chloride with tris-(trimethylsilyl) derivative of phloroglucinol in the presence of stannic chloride (or titanium tetrachloride or aluminium chloride) in refluxing methylene chloride for 2 h [649].

m.p. and Spectra (NA).

(3-Hydroxy-6-methoxy-2-nitrophenyl)(2,4,6-trihydroxyphenyl)methanone

$C_{14}H_{11}NO_8$ mol.wt. 321.24

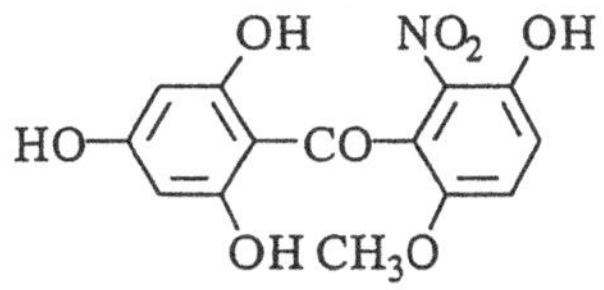

Synthesis

-Preparation by reaction of 3-hydroxy-6-methoxy-2-nitro-benzoyl chloride with tris-(trimethylsilyl) derivative of phloroglucinol in the presence of stannic chloride (or titanium tetrachloride or aluminium chloride) in refluxing methylene chloride for 2 h [649].

m.p. and Spectra (NA).

(2,4-Dihydroxy-3-methylphenyl)(2,5-dihydroxyphenyl)methanone

[61234-46-6] $C_{14}H_{12}O_5$ mol.wt. 260.25

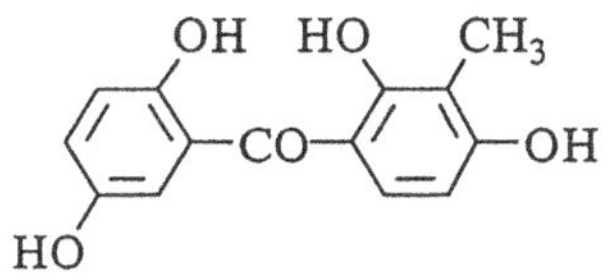

Synthesis

-Preparation by demethylation of 2,4,5'-trihydroxy-2'-methoxy-3-methylbenzophenone with hydrobromic acid in refluxing acetic acid during 2 h (79%) [398].

m.p. 238-240° [398]; Spectra (NA).

(2,4-Dihydroxy-5-methylphenyl)(2,5-dihydroxyphenyl)methanone

[61234-45-5]   C$_{14}$H$_{12}$O$_5$ mol.wt. 260.25

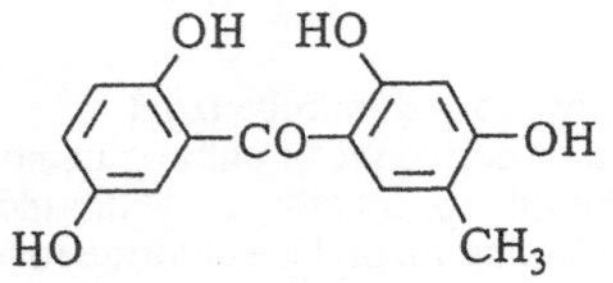

Synthesis

-Preparation by demethylation of 2,4,5'-trihydroxy-
2'-methoxy-5-methylbenzophenone with hydrobromic acid
in refluxing acetic acid during 2 h (77%) [398].

m.p. 244° [398];   Spectra (NA).

(2,4-Dihydroxy-5-methylphenyl)(3,5-dihydroxyphenyl)methanone

[61234-68-2]   C$_{14}$H$_{12}$O$_5$ mol.wt. 260.25

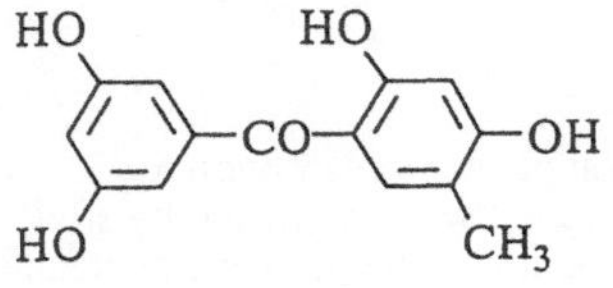

Synthesis

-Preparation by condensation of 3,5-diacetoxybenzonitrile
with 4-methylresorcinol (80%) (Hoesch reaction) [398].

m.p. 263-265° [398];   Spectra (NA).

(2-Hydroxy-3-methylphenyl)(2,3,4-trihydroxyphenyl)methanone

[107412-94-2]   C$_{14}$H$_{12}$O$_5$ mol.wt. 260.25

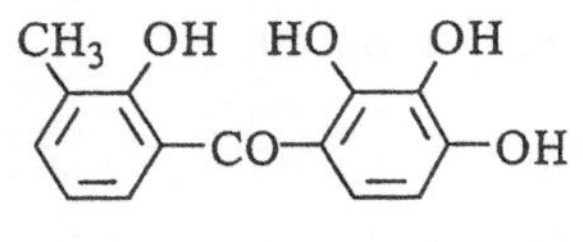

Synthesis

-Preparation by heating o-cresotic acid and pyrogallol with
zinc chloride and phosphorous oxychloride (40%) [679].

m.p. 137-138° [679];   Spectra (NA).

(2-Hydroxy-4-methylphenyl)(2,3,4-trihydroxyphenyl)methanone

[109067-41-6]   C$_{14}$H$_{12}$O$_5$ mol.wt. 260.25

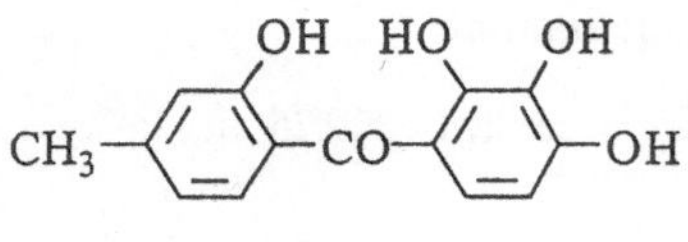

Synthesis

-Preparation by heating m-cresotic acid and pyrogallol
with zinc chloride and phosphorous oxychloride (40%)
[679].

m.p. 122-123° [679];   Spectra (NA).

(2-Hydroxy-5-methylphenyl)(2,3,4-trihydroxyphenyl)methanone

[105443-51-4] $C_{14}H_{12}O_5$ mol.wt. 260.25

Synthesis

-Preparation by treatment of pyrogallol with 5-methylsalicylic acid,
*in tetrachloroethane bubbling boron trifluoride at 110° for 1 h [1443];
*in the presence of Amberlyst-15 in refluxing toluene under azeotropical water removal for 21 h [1400].
-Also refer to: [1257].

   m.p. and Spectra (NA).

(4-Hydroxy-3-methoxyphenyl)(2,4,6-trihydroxyphenyl)methanone

                    $C_{14}H_{12}O_6$ mol.wt. 276.25

Synthesis

-Preparation by reaction of vanillonitrile with phloroglucinol in the presence of zinc chloride and hydrochloric acid in ethyl ether, first at r.t. for 1 h, then at 50° for 4 h, followed by hydrolysis of the ketimine hydrochloride so obtained with boiling water for 1-2 h (38%) [586].

   monohydrate [586];   m.p. >200° (d) [586];   Spectra (NA).

(5-Hydroxy-2-methoxyphenyl)(2,4,6-trihydroxyphenyl)methanone

                    $C_{14}H_{12}O_6$ mol.wt. 276.25

Synthesis

-Preparation by reaction of 5-hydroxy-2-methoxybenzoyl chloride with tris(trimethylsilyl) derivative of phloroglucinol in the presence of stannic chloride (or titanium tetrachloride or aluminium chloride) in refluxing methylene chloride for 2 h [649].

   m.p. and Spectra (NA).

(3,6-Dihydroxy-2,4-dimethylphenyl)(2,4-dihydroxyphenyl)methanone

[42470-91-7] $C_{15}H_{14}O_5$ mol.wt. 274.27

Synthesis

-Preparation by condensation of 3,6-dihydroxy-2,4-dimethylbenzoic acid with resorcinol in the presence of zinc chloride and phosphorous oxychloride at 40° for 1 h (69%) [431].

   m.p. 226° [431];
   1H NMR [431], IR [431], UV [431], MS [431].

(3,6-Dihydroxy-2,4-dimethylphenyl)(2,6-dihydroxyphenyl)methanone

[42594-60-5] $C_{15}H_{14}O_5$ mol.wt. 274.27

Synthesis

-Obtained by demethylation of 2',3,6-trihydroxy-
6'-methoxy-2,4-dimethylbenzophenone with boron
tribromide in refluxing methylene chloride for 90 h (58%)
[430].

m.p. 171-174° [430]; ^{1}H NMR [430], IR [430], UV [430], MS [430].

(2-Hydroxy-4-methoxy-6-methylphenyl)(2,4,6-trihydroxyphenyl)methanone

$C_{15}H_{14}O_6$ mol.wt. 290.27

Synthesis

-Preparation by hydrogenolysis of 2,2',4,6-tetrakis-
(benzyloxy)-4'-methoxy-6'-methylbenzophenone
(SM) in ethyl acetate/tetrahydrofuran mixture in the
presence of 10% Pd/C at 25°. SM was obtained by
condensation of 2-(benzyloxy)-4-methoxy-6-methyl-
benzoic acid with phloroglucinol tribenzyl ether in the presence of trifluoroacetic anhydride in
methylene chloride under nitrogen for 15 min (81%) [1328].

m.p. (NA); ^{1}H NMR [1328].

(4-Hydroxy-2-methoxy-6-methylphenyl)(2,4,6-trihydroxyphenyl)methanone

[60556-49-2] $C_{15}H_{14}O_6$ mol.wt. 290.27

Synthesis

-Preparation by catalytic hydrogenolysis of 2,4,4',6-tetra-
kis(benzyloxy)-2'-methoxy-6'-methylbenzophenone (SM)
in the presence of 10% Pd/C in ethanol (72%). SM was
obtained by condensation of 4-(benzyloxy)-6-methoxy-
o-toluic acid with phloroglucinol tribenzyl ether in the
presence of trifluoroacetic anhydride in methylene chloride (88%) [543].
-Also refer to: [181].

m.p. 177-178° [543]; ^{1}H NMR [543], IR [543], UV [543], MS [543].

(3-Hydroxy-2,6-dimethoxyphenyl)(2,4,6-trihydroxyphenyl)methanone

$C_{15}H_{14}O_7$ mol.wt. 306.27

Synthesis

-Preparation by reaction of 3-hydroxy-2,6-dimethoxy-
benzoyl chloride with tris(trimethylsilyl) derivative of
phloroglucinol in the presence of stannic chloride (or
titanium tetrachloride or aluminium chloride) in refluxing
methylene chloride for 2 h [649].

m.p. and Spectra (NA).

4.2.2. *Substituents located on both rings*

Symmetrical ketones

Bis(2,4-Dihydroxy-6-methylphenyl)methanone

[39803-53-7] $C_{15}H_{14}O_5$ mol.wt. 274.27

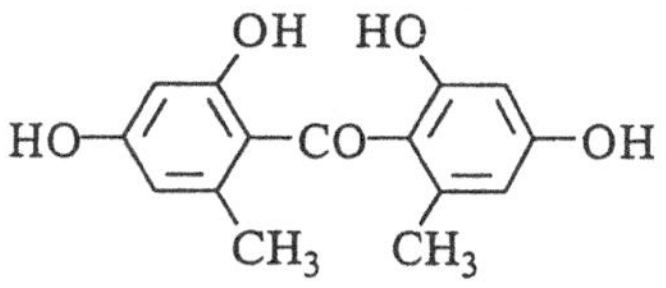

Synthesis

-Preparation by hydrogenolysis of 2,2',4,4'-tetrakis-
(benzyloxy)-6,6'-dimethylbenzophenone (SM) in the
presence of 10% Pd/C in ethyl acetate (90%) [564]. SM
was obtained by condensation of orcinol dibenzyl ether
with orsellinic acid dibenzyl ether in methylene chloride
in the presence of trifluoroacetic anhydride for 5 min (75%).
-Also refer to: [1327].

m.p. 195-198° [564];
^{1}H NMR [564], IR [564], UV [564].

Bis(2,6-Dihydroxy-4-methoxyphenyl)methanone

$C_{15}H_{14}O_7$ mol.wt. 306.27

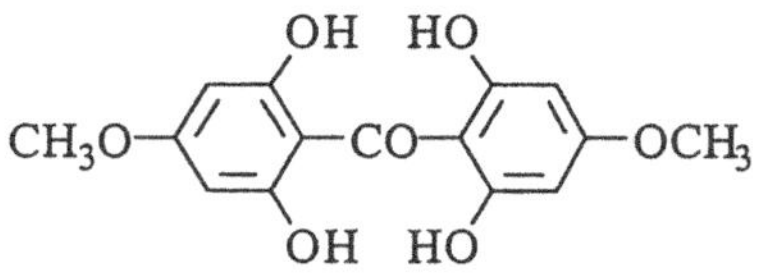

Synthesis

-Refer to: [1228].

m.p. and Spectra (NA).

Asymmetric ketones

(3-Chloro-4,6-dihydroxy-2-methylphenyl)(3,5-dichloro-2,4-dihydroxy-6-methylphenyl)-methanone

[69709-91-7] $C_{15}H_{11}Cl_3O_5$ mol.wt. 377.61

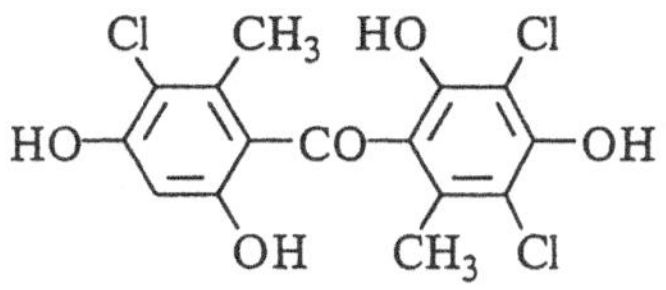

Synthesis

-Preparation by hydrogenolysis of 2,2',4,4'-tetrakis-
(benzyloxy)-3,5,5'-trichloro-6,6'-dimethylbenzophenone
(SM) in the presence of 10% Pd/C in ethyl acetate
containing concentrated hydrochloric acid (5 drops)
(>90%) [1174]. SM was obtained by Friedel-Crafts
acylation of 3,5-bis(benzyloxy)-2-chlorotoluene with 2,4-bis(benzyloxy)-3,5-dichloro-
6-methylbenzoic acid in the presence of trifluoroacetic anhydride in refluxing ethylene dichloride
for 5 h (33%).
-Also refer to: [1173].

m.p. 201-203° [1174]; ^{1}H NMR [1174], MS [1174].

(3,5-Dichloro-2,4-dihydroxy-6-methylphenyl)(2,4-dihydroxy-6-methylphenyl)methanone

[39803-58-2] $C_{15}H_{12}Cl_2O_5$ mol.wt. 343.16

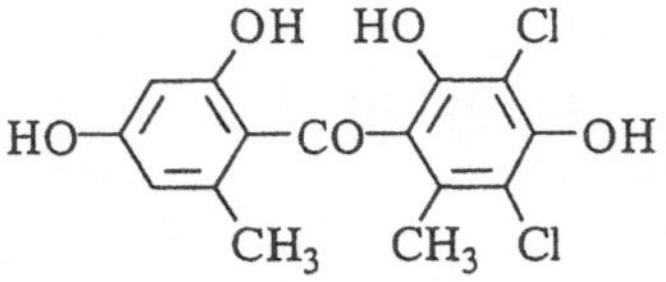

Synthesis

-Preparation by hydrogenolysis of 2,2',4,4'-tetrakis-
(benzyloxy)-3,5-dichloro-6,6'-dimethylbenzophenone
(SM) in ethanol in the presence of 10% Pd/C for 3 h
(51%). SM was obtained by condensation of 3,5-di-
chloroorsellinic acid dibenzyl ether with orcinol
dibenzyl ether in refluxing chloroform in the presence of trifluoroacetic anhydride/trifluoroacetic
acid for 3 h [564].

m.p. 213-216° [564];   ¹H NMR [564], UV [564].

(2,6-Dihydroxy-4-methoxyphenyl)(2,4-dihydroxy-6-methylphenyl)methanone

[60556-46-9] $C_{15}H_{14}O_6$ mol.wt. 290.27

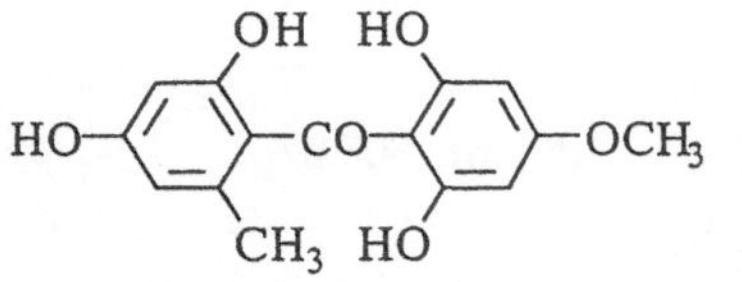

Synthesis

-Preparation by catalytic hydrogenolysis of
2,2',4',6-tetrakis(benzyloxy)-4-methoxy-6'-methyl-
benzophenone (SM) in the presence of 10% Pd/C in
ethyl acetate at r.t. (71%). SM was obtained by
condensation of 2,6-bis(benzyloxy)-4-methoxy-
benzoic acid with 3,5-bis(benzyloxy)toluene in the presence of trifluoroacetic anhydride in
methylene chloride for 5 min at r.t. (69%) [543].
-Also refer to: [1199, 1200].

m.p. 250-251° (first melting point at 115°) [543];
¹H NMR [543], IR [543], UV [543], MS [543].

(2,3-Dihydroxy-4-methoxyphenyl)(3,5-dihydroxy-4-methoxyphenyl)methanone

                                                $C_{15}H_{14}O_7$ mol.wt. 306.27

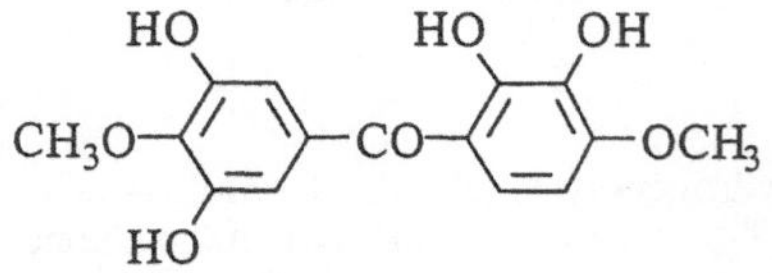

Synthesis

-Obtained by reaction of methyl iodide with exifone
(2,3,3',4,4',5'-hexahydroxybenzophenone) in the
presence of lithium carbonate in N,N-dimethyl-
formamide at 30° for 15 h under nitrogen (30%)
[802].

m.p. 216-218° [802];
¹H NMR [802], ¹³C NMR [802], MS [802];   pK_a [802].

[4-(Benzoylmethoxy)-3,5-dihydroxyphenyl](2,3-dihydroxy-4-methoxyphenyl)methanone
2-[4-(2,3-Dihydroxy-4-methoxybenzoyl)-2,6-dihydroxyphenoxy]-1-phenylethanone

$C_{22}H_{18}O_8$ mol.wt. 410.38

Synthesis

-Obtained by action of methyl iodide with 4'-phenacyl ether of exifone (2,3,3',4,4',5'-hexahydroxybenzophenone) in the presence of lithium carbonate in N,N-dimethylformamide at 30° for 15 h (30%) [802].

m.p. (NA); ^{1}H NMR [802], ^{13}C NMR [802].

5. PENTAHYDROXYBENZOPHENONES

5.1. *Hydroxy groups located on one ring* ***Not described till December 1999.***

5.2. *Hydroxy groups located on both rings*

5.2.1. *Substituents located on one ring*

(3-Chloro-4,6-dihydroxy-2-methylphenyl)(2,4,6-trihydroxyphenyl)methanone

[68048-30-6] $C_{14}H_{11}ClO_6$ mol.wt. 310.69

Synthesis

-Preparation by hydrogenolysis of 2',4,4',6,6'-pentakis-(benzyloxy)-3-chloro-2-methylbenzophenone (SM) in ethyl acetate/tetrahydrofuran in the presence of 10% Pd/C at 25°. SM was obtained by condensation of 4,6-bis-(benzyloxy)-3-chloro-2-methylbenzoic acid with phloroglucinol tribenzyl ether in the presence of trifluoroacetic anhydride in methylene chloride under nitrogen for 2 min (95%) [1328].

m.p. (NA); ^{1}H NMR [1328].

(2,4-Dihydroxy-6-methylphenyl)(2,4,6-trihydroxyphenyl)methanone

[55018-96-7] $C_{14}H_{12}O_6$ mol.wt. 276.25

Syntheses

-Preparation by hydrogenolysis of 2,2',4,4',6-pentakis-(benzyloxy)-6'-methylbenzophenone (SM), in the presence of 10% Pd/C in ethyl acetate or in a ethanol/ethyl acetate mixture at atmospheric pressure and r.t. (>95%) [543, 1328, 1329]. SM was obtained,
*by condensation of 4,6-bis(benzyloxy)-o-toluic acid with phloroglucinol tribenzyl ether in the presence of trifluoroacetic anhydride (TFAA) in a chloroform/methylene chloride mixture at r.t. for 5 min (61%, colourless oil) [543];

*by condensation of 2,4,6-tris(benzyloxy)benzoic acid with orcinol dibenzyl ether in the presence of trifluoroacetic anhydride (TFAA) in methylene chloride for 10 min [1328], (83%, m.p. 33°) [1329].

-Also obtained (poor yield) by demethylation of 2,2',4,4',6-pentamethoxy-6'-methylbenzophenone (m.p. 126°5-127°) in methylene chloride in the presence of boron tribromide at r.t. for overnight (15%) [1329].

-Also obtained by Friedel-Crafts acylation of phloroglucinol with o-orsellinic acid [1187].

-Also obtained by Fries rearrangement of phloroglucinyl o-orsellinate (m.p. 185-188°) [1187].

-Also refer to: [890, 1186, 1199, 1200, 1228].

N.B.: Discussion on hypothetical formation from various polyketones [1184]. This ketone is the biosynthetic precursor of the antibiotic griseofulvin and the various fungal and lichen xanthones [1184]. It was very unstable and underwent facile cyclization to norlichexanthone (1,3,6-tri-hydroxy-8-methylxanthen-9-one) [1328].

> viscous, pale yellow oil [543];   b.p. (NA);
> pale yellow powder [1329];   m.p. (NA);
> light yellow compound [1328];
> N.B.: The description of the physical state of this ketone was imprecise.
> ^{1}H NMR [543, 1329], IR [543, 1329], UV [543, 1329],
> MS [543];   TLC [1329].

(2,3-Dihydroxy-4-methoxyphenyl)(3,4,5-trihydroxyphenyl)methanone

[177703-30-9] $C_{14}H_{12}O_7$ mol.wt. 292.25

Synthesis

-Preparation from exifone (2,3,3',4,4',5'-hexahydroxy-benzophenone) in three steps: at first, phenacylation of exifone at the 4'-position, according to [801]. After, methylation of the 4'-phenacyl ether of exifone at the 4-position. Then, removal of the phenacyl protecting group of 4-methoxy-4'-phenacyl exifone derivative obtained in the presence of zinc dust in an acetic acid/methanol mixture at r.t. for 2 min (50%) [802].

> m.p. 260-262° [802];
> ^{1}H NMR [802], ^{13}C NMR [802], MS [802];   pK_a [802].

(3,5-Dihydroxy-4-methoxyphenyl)(2,3,4-trihydroxyphenyl)methanone

[170630-11-2] $C_{14}H_{12}O_7$ mol.wt. 292.25

Synthesis

-Obtained by reaction of methyl iodide with exifone (2,3,3',4,4',5'-hexahydroxybenzophenone) in the presence of lithium carbonate in N,N-dimethyl-formamide at 30° for 15 h under nitrogen (15%) [802].

-Also refer to: [427].

> m.p. 202-204° [802];
> ^{1}H NMR [802], ^{13}C NMR [802], MS [802];   pK_a [802].

[4-(Benzoylmethoxy)-3,5-dihydroxyphenyl](2,3,4-trihydroxyphenyl)methanone
2-[2,6-Dihydroxy-4-(2,3,4-trihydroxybenzoyl)phenoxy]-1-phenylethanone

$C_{21}H_{16}O_8$ mol.wt. 396.35

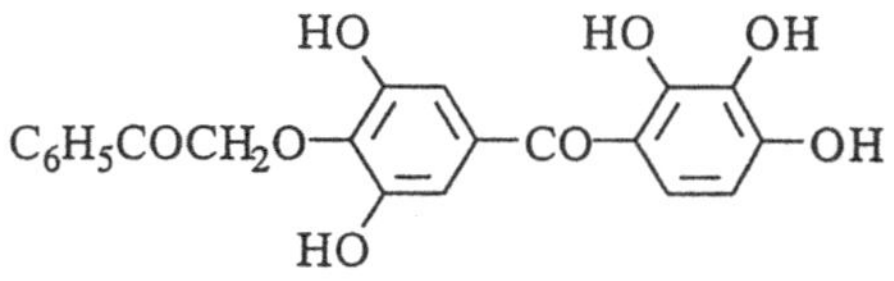

Synthesis

-Obtained by reaction of α-bromoacetophenone with exifone (2,3,3',4,4',5'-hexahydroxy-benzophenone) in the presence of lithium carbonate in N,N-dimethylformamide at r.t. for 15 h under nitrogen (40%) [801].

m.p. decomposes above 180° [801];
^{1}H NMR [801], ^{13}C NMR [801], MS [801].

5.2.2. Substituents located on both rings

(3-Bromo-4,5-dihydroxyphenyl)(3,5-dibromo-2,4,6-trihydroxyphenyl)methanone

$C_{13}H_7Br_3O_6$ mol.wt. 498.91

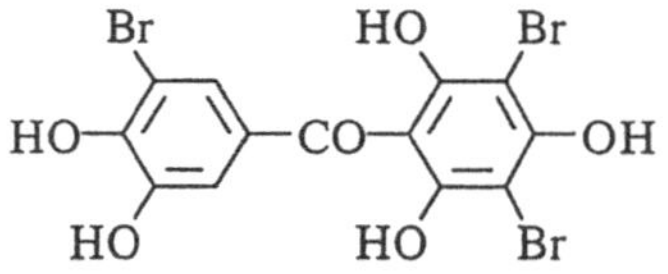

Synthesis

-Obtained by reaction of maclurin with bromine in boiling water [157].
N.B.: The formula proposed is the more likely.

monohydrate [157];   m.p. and Spectra (NA).

(3-Chloro-4,6-dihydroxy-2-methylphenyl)(3-chloro-2,4,6-trihydroxyphenyl)methanone

[68048-32-8] $C_{14}H_{10}Cl_2O_6$ mol.wt. 345.14

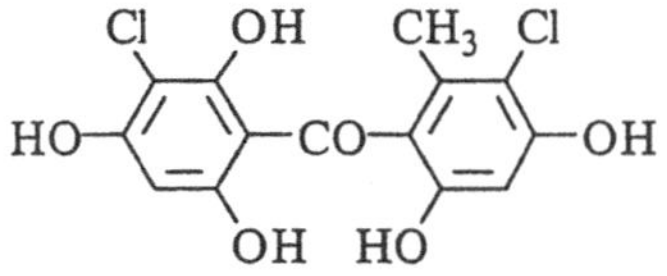

Synthesis

-Preparation by hydrogenolysis of 2,4,4',6,6'-pentakis-(benzyloxy)-3,3'-dichloro-2'-methylbenzophenone (SM) in ethyl acetate/tetrahydrofuran mixture in the presence of 10% Pd/C at 25°. SM was obtained by condensation of 2,4,6-tris(benzyloxy)-3-chlorobenzoic acid with 3,5-bis (benzyloxy)-2-chlorotoluene in the presence of trifluoroacetic anhydride in methylene chloride under nitrogen for 20 min (18%) [1328].

m.p. (NA); ^{1}H NMR [1328].

(3-Chloro-2,4,6-trihydroxyphenyl)(2,4-dihydroxy-6-methylphenyl)methanone

[68048-31-7] $C_{14}H_{11}ClO_6$ mol.wt. 310.69

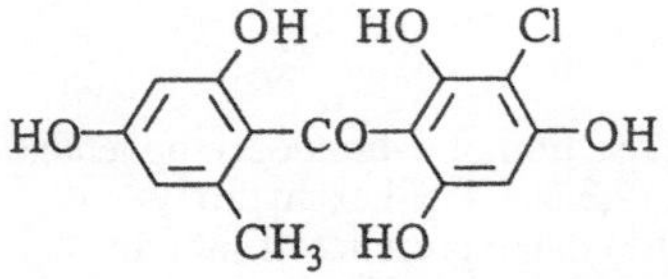

Synthesis

-Preparation by hydrogenolysis of 2,2',4,4',6-pentakis-
(benzyloxy)-3-chloro-6'-methylbenzophenone (SM) in
ethyl acetate/tetrahydrofuran mixture in the presence of
10% Pd/C at 25°. SM was obtained by condensation of
3-chloro-2,4,6-tris(benzyloxy)benzoic acid with orcinol
dibenzyl ether in the presence of trifluoroacetic anhydride in methylene chloride under nitrogen for
15 min [1328].

m.p. (NA); ^{1}H NMR [1328].

6. HEXAHYDROXYBENZOPHENONE

(2,3,4-Trihydroxy-5-methylphenyl)(3,4,5-trihydroxyphenyl)methanone

[112232-18-5] $C_{14}H_{12}O_7$ mol.wt. 292.25

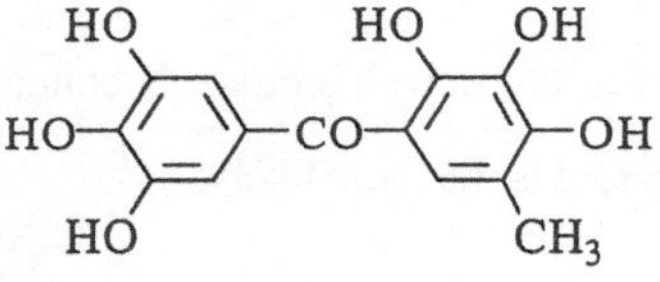

Synthesis

-Refer to: [601] (Japanese patent).

m.p. and Spectra (NA).

Chapter 3. Polyphenyl phenyl methanones *(Class of METHANONES)*

1. BIPHENYL PHENYL METHANONES

1.1. *Monohydroxylated ketones*

(6-Bromo-5-hydroxy[1,1'-biphenyl]-2-yl)phenylmethanone

[133721-72-9] $C_{19}H_{13}BrO_2$ mol.wt. 353.22

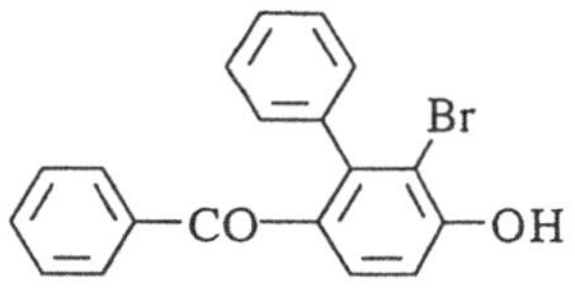

Synthesis

-Preparation by adding trimethylsilyl trifluoromethane-
  sulfonate to a solution of 5-bromo-2,3-dihydro-2,2-di-
  methyl-6-phenyl-4*H*-pyran-4-one, di-(tert-butyl)pyridine
  and 1-benzoylacetylene in chloroform at -30°. Then, the
  mixture was stirred at 20° for 24 h (59%) [992].

m.p. 155-156° [992]; ¹H NMR [992], IR [992], MS [992].

(4-Chlorophenyl)(4'-hydroxy[1,1'-biphenyl]-4-yl)methanone

[38304-24-4] $C_{19}H_{13}ClO_2$ mol.wt. 308.76

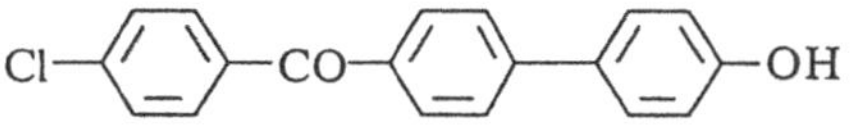

Syntheses

-Preparation by reaction of p-chlorobenzoic acid
  with 4-hydroxybiphenyl in the presence of
  trifluoromethanesulfonic acid, first at 50° for
23 h, then at 70° for 4 h (97%) [1138].
-Preparation by Friedel-Crafts acylation of p-acetoxybiphenyl with p-chlorobenzoyl chloride,
  followed by hydrolysis of the keto ester so obtained (74%) [1356].

m.p. 196-197° [1138, 1356]; ¹H NMR [1138, 1356], IR [1138, 1356].

(4-Fluorophenyl)(4'-hydroxy[1,1'-biphenyl]-4-yl)methanone

[112782-46-4] $C_{19}H_{13}FO_2$ mol.wt. 292.31

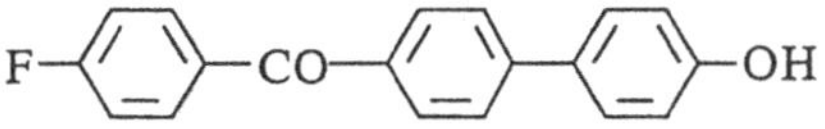

Synthesis

-Preparation by Fries rearrangement of
  4-(4-fluorobenzoyloxy)biphenyl with aluminium
  chloride in o-dichlorobenzene at 120° for 8 h (good yield) [288].

m.p. and Spectra (NA).

[1,1'-Biphenyl]-4-yl(4-hydroxyphenyl)methanone

[3558-83-6] $C_{19}H_{14}O_2$ mol.wt. 274.32

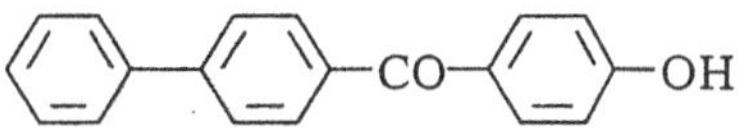

Syntheses

-Preparation by demethylation of p-anisoylbiphenyl,
  *with pyridinium chloride at reflux a few min [238];
*with aluminium chloride in benzene [781].

-Also obtained by reaction of EKONOL(RM), an aromatic polyester as Friedel-Crafts reagent, with biphenyl in triflic acid at 25° for 18 h (95%) [287]. Similar results can be obtained using hydrofluoric acid/boron trifluoride or aluminium chloride in place of triflic acid [287].
-Also refer to: Chem. Abstr., **127**, 34137f (1997).

m.p.   186° [238], 185-187° [781]; ^{13}C NMR [287], MS [287];   HPLC [287].

(2-Hydroxy[1,1'-biphenyl]-3-yl)phenylmethanone

$C_{19}H_{14}O_2$ mol.wt. 274.32

Synthesis

-Obtained by irradiation of 2-biphenylyl benzoate with 254 nm light in benzene (23%) [741].

m.p. and Spectra (NA).

(4-Hydroxy[1,1'-biphenyl]-3-yl)phenylmethanone

$C_{19}H_{14}O_2$ mol.wt. 274.32

Synthesis

-Obtained by irradiation of 4-biphenylyl benzoate with 254 nm light in benzene (54%) [741].

m.p. and Spectra (NA).

(4'-Hydroxy[1,1'-biphenyl]-4-yl)phenylmethanone

[5623-46-1] $C_{19}H_{14}O_2$ mol.wt. 274.32

Syntheses

-Obtained by Fries rearrangement of 4-benzoyloxy-biphenyl in the presence of aluminium chloride without solvent for 30 min at 160° (22%) [413] or in tetrachloroethane for 1 h at 140° [187]. There is an heteronuclear migration of the benzoyl group, since the Fries reaction catalysed by Lewis acids is not a true rearrangement, but more probably an intermolecular acylation [861, 868].
-Preparation by Friedel-Crafts acylation of p-acetoxybiphenyl with benzoyl chloride, followed by hydrolysis of the resulting keto ester (94%) [1356].

m.p.   194-195°5 [1356], 193-195° [187, 413]; ^{1}H NMR [1356], IR [1356].

(5-Hydroxy[1,1'-biphenyl]-2-yl)phenylmethanone

[133721-67-2] $C_{19}H_{14}O_2$ mol.wt. 274.32

Syntheses

-Obtained by heating a solution of 5-[(triisopropylsilyl)oxy]-biphenyl-2-yl phenyl ketone in methanol with 2 N aqueous hydrochloric acid for 3 h at 80° in a sealed tube (93%) [992].
-Preparation by adding trimethylsilyl trifluoromethane-

sulfonate to a solution of 2,3-dihydro-2,2-dimethyl-6-phenyl-*4H*-pyran-4-one, di-(tert-butyl)pyridine and 1-benzoylacetylene in chloroform at -30°. Then, the mixture was stirred at 0° for 30 min and at 20° for 24 h (84%) [992].

m.p. 179-180° [992]; ¹H NMR [992], IR [992], MS [992].

(6-Hydroxy[1,1'-biphenyl]-3-yl)phenylmethanone

[84627-07-6] $C_{19}H_{14}O_2$ mol.wt. 274.32

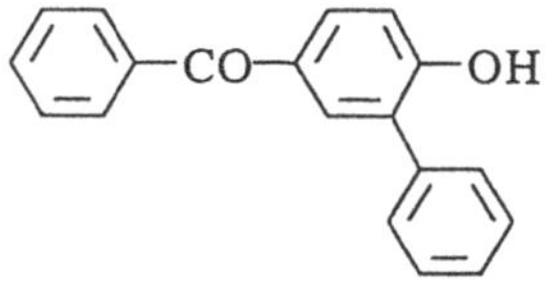

Syntheses

-Preparation by decarboxylation of 2-[(2'-hydroxy-5'-biphenylyl)carbonyl]benzoic acid in the presence of cupric acetate in refluxing quinoline for 30 min (96%) [1163].
-Also obtained by irradiation of 2-biphenylyl benzoate with 254 nm light in benzene (27%) [741].

m.p. 195°2-197° [1163];
¹H NMR [1163], IR [1163], UV [1163]; TLC [1163].

[1,1'-Biphenyl]-4-yl(2-Hydroxy-5-methylphenyl)methanone

$C_{20}H_{16}O_2$ mol.wt. 288.35

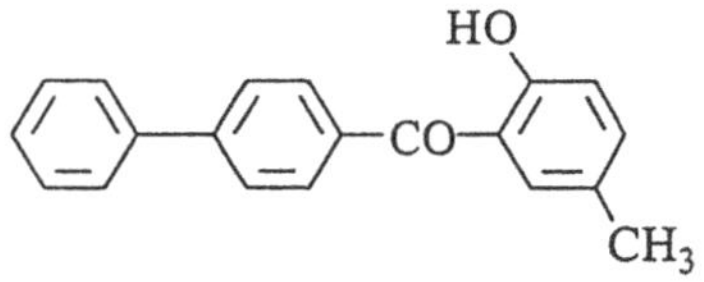

Syntheses

-Preparation by reaction of 4-biphenylcarbonyl chloride with p-cresol in the presence of aluminium chloride in trichlorobenzene at 190-200° for 2 h (64%) [1273].
-Preparation by Fries rearrangement of p-cresyl 4-biphenylcarboxylate with aluminium chloride in trichlorobenzene, first at 140° for 30 min, then between 140-200° for 30 min and at 200° for 3 h (82%) [1273].
-Also refer to: [1067].

m.p. 79-80° [1273]; UV [1273].

(2'-Hydroxy-5'-methyl[1,1'-biphenyl]-3-yl)phenylmethanone

[132555-32-9] $C_{20}H_{16}O_2$ mol.wt. 288.35

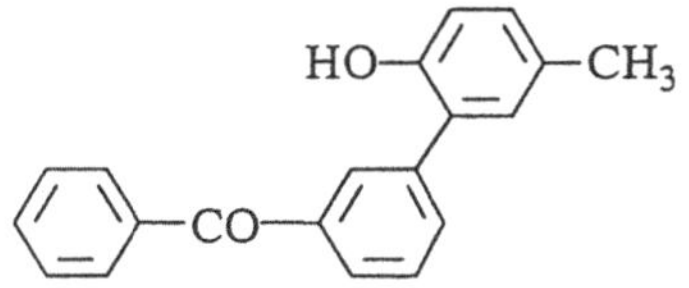

Synthesis

-Obtained by irradiation of a (Z)-(3-benzoylphenyl)azo tert-butyl sulfide and p-cresol mixture in the presence of potassium tert-butoxide in dimethyl sulfoxide for 3.5 h (53%) [1056].

glassy oil [1056]; b.p. (NA); ¹H NMR [1056].

(2'-Hydroxy-5'-methyl[1,1'-biphenyl]-4-yl)phenylmethanone

[132555-33-0] $C_{20}H_{16}O_2$ mol.wt. 288.35

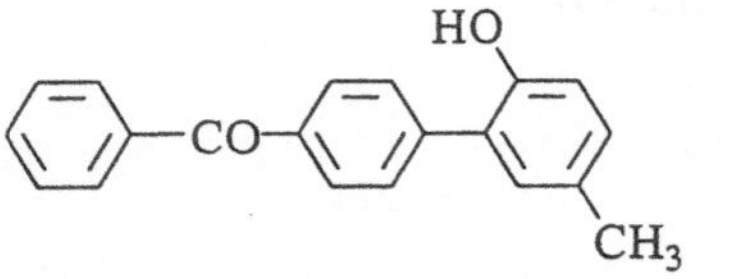

Synthesis

-Obtained by irradiation of a (Z)-(4-benzoylphenyl)azo tert-butyl sulfide and p-cresol mixture in the presence of potassium tert-butoxide in dimethyl sulfoxide for 2.5 h (54%) [1056].

m.p. 144°5-145°9 [1056]; ^{1}H NMR [1056], IR [1056].

(5-Hydroxy-3-methyl[1,1'-biphenyl]-2-yl)phenylmethanone

[133721-68-3] $C_{20}H_{16}O_2$ mol.wt. 288.35

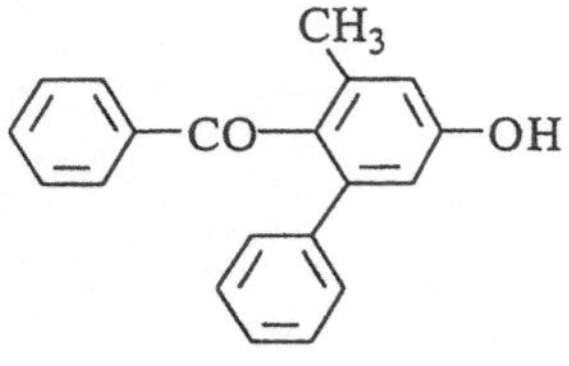

Synthesis

-Preparation by heating 3-methyl-5-[(triisopropylsilyl)oxy]-biphenyl-2-yl phenyl ketone in ethanol with 2 N aqueous hydrochloric acid for 4.5 h at 80° in a sealed tube (93%) [992].

m.p. 148-148°5 [992]; ^{1}H NMR [992], IR [992], MS [992].

[1,1'-Biphenyl]-4-yl(2-hydroxy-4-methoxyphenyl)methanone

[90986-69-9] $C_{20}H_{16}O_3$ mol.wt. 304.35

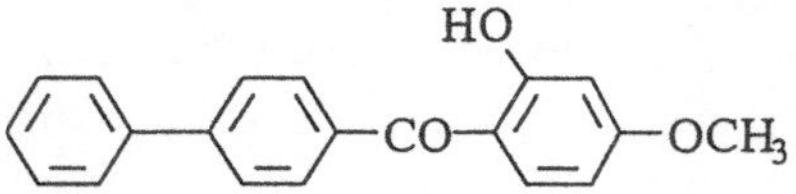

Synthesis

-Preparation by reaction of 4-biphenylcarbonyl chloride with resorcinol dimethyl ether in the presence of aluminium chloride [1212],
*in nitrobenzene, first between 25 to 30° for 2 h,
then at 80° for 4 h (56%) [1273];
*in refluxing carbon disulfide for 4 h (20-30%) [781].
-Also refer to: [1067].

m.p. 105-106° [1273], 104-105° [781];
^{1}H NMR [781], IR [781], UV [1273], MS [781].

(5-Hydroxy-2'-methoxy[1,1'-biphenyl]-2-yl)phenylmethanone

[133721-75-2] $C_{20}H_{16}O_3$ mol.wt. 304.35

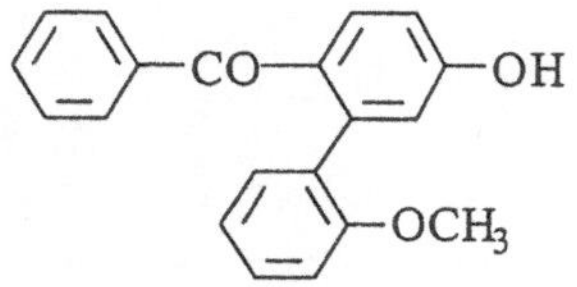

Synthesis

-Obtained by heating a solution of 5-[[dimethyl(1,1,2-tri-methylpropyl)silyl]oxy]-2'-methoxybiphenyl-2-yl phenyl ketone in ethanol with 2 N aqueous hydrochloric acid for 2.5 h at 80° in a sealed tube (93%) [992].

m.p. 177°5-178°5 [992]; ^{1}H NMR [992], IR [992], MS [992].

[4-(Acetyloxy)-2-hydroxyphenyl][1,1'-biphenyl]-4-ylmethanone

[36415-12-0] $C_{21}H_{16}O_4$ mol.wt. 332.36

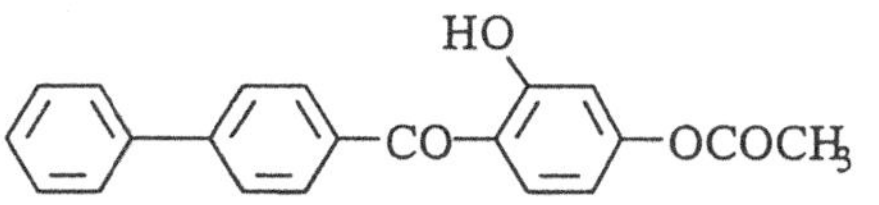

Synthesis

-Refer to: [1067].

m.p. 108° [1067];   UV [1067].

[1,1'-Biphenyl]-4-yl(4-ethoxy-2-hydroxyphenyl)methanone

$C_{21}H_{18}O_3$ mol.wt. 318.37

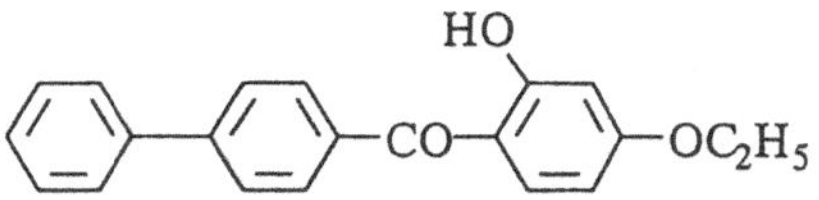

Synthesis

-Preparation by reaction of 4-biphenylcarbonyl chloride with resorcinol diethyl ether in the presence of aluminium chloride in nitrobenzene, first between 25 to 30° for 2 h, then at 80° for 4 h [1273].
-Also refer to: [1067].

m.p. 114-115° [1273];   UV [1273].

(2-Hydroxy-4-methoxyphenyl)(5-methyl[1,1'-biphenyl]-2-yl)methanone

[80988-17-6] $C_{21}H_{18}O_3$ mol.wt. 318.37

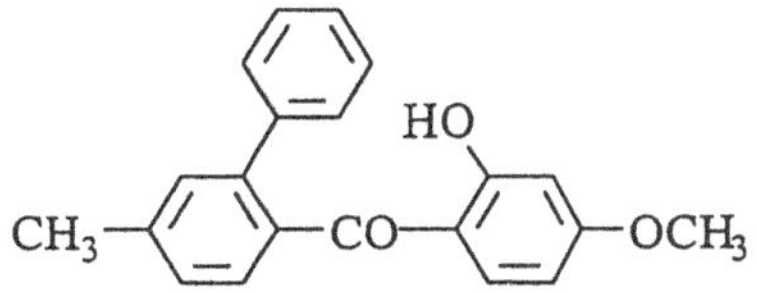

Synthesis

-Refer to: [1093, 1094].

m.p. and Spectra (NA).

[1,1'-Biphenyl]-4-yl[4-(2-butenyloxy)-2-hydroxyphenyl]methanone

[36414-90-1] $C_{23}H_{20}O_3$ mol.wt. 344.41

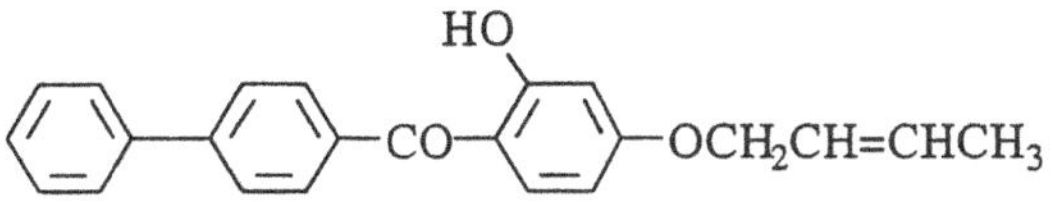

Synthesis

-Refer to: [1067].

m.p. 117° [1067];   UV [1067].

[1,1'-Biphenyl]-4-yl[4-(1,1-dimethylethoxy)-2-hydroxyphenyl]methanone

[36488-90-1] $C_{23}H_{22}O_3$ mol.wt. 346.94

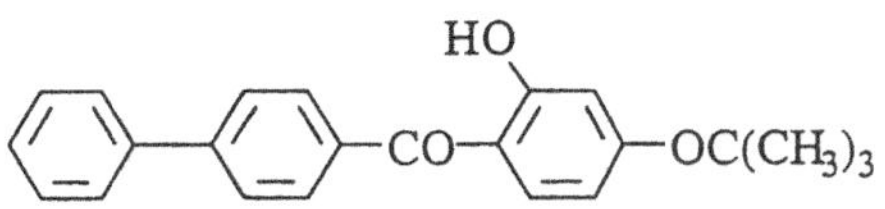

Synthesis

-Refer to: [1067].

m.p. 80° [1067];   UV [1067].

[1,1'-Biphenyl]-4-yl[2-hydroxy-4-(4-nitrophenoxy)phenyl]methanone

[36469-48-4]   C$_{25}$H$_{17}$NO$_5$ mol.wt. 411.41

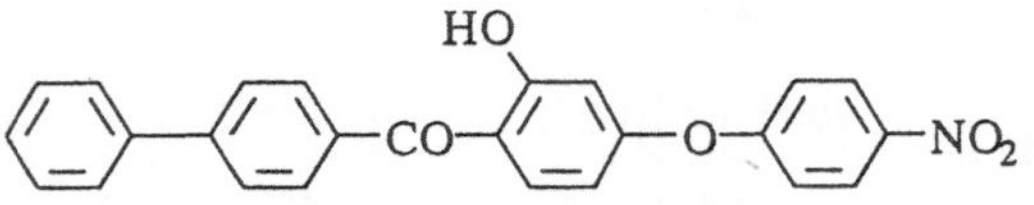

Synthesis

-Refer to: [1067].

m.p. 234° [1067];   UV [1067].

[1,1'-Biphenyl]-4-yl(4-hydroxy[1,1'-biphenyl]-3-yl)methanone
4',5-Diphenyl-2-hydroxybenzophenone

[95818-93-2]   C$_{25}$H$_{18}$O$_2$ mol.wt. 350.42

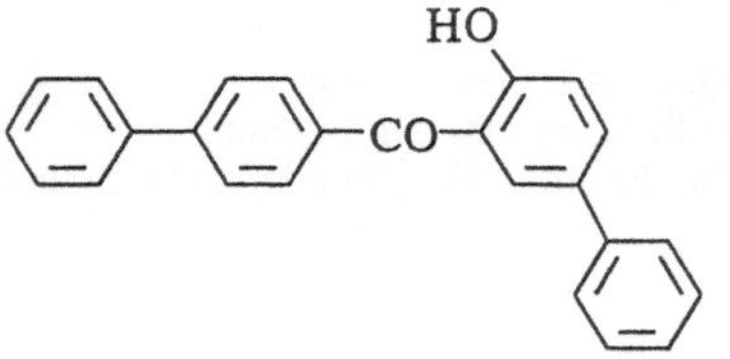

Syntheses

-Preparation by reaction of 4-biphenylcarbonyl
 chloride with 4-phenylphenol in the presence of
 aluminium chloride in trichlorobenzene at 190-200°
 for 2 h [1273].
-Also obtained by Fries rearrangement of 4-biphenylyl
 biphenyl-4-carboxylate, (4-biphenylyl 4-phenyl-
 benzoate), with aluminium bromide in chlorobenzene
 at 110° for 6 h [347].

m.p.   135-136° [1273], 135° [347];   IR [347], UV [347, 1273];   GLC [347].

[4-(4-Aminophenoxy)-2-hydroxyphenyl][1,1'-biphenyl]-4-ylmethanone

[36414-91-2]   C$_{25}$H$_{19}$NO$_3$ mol.wt. 381.43

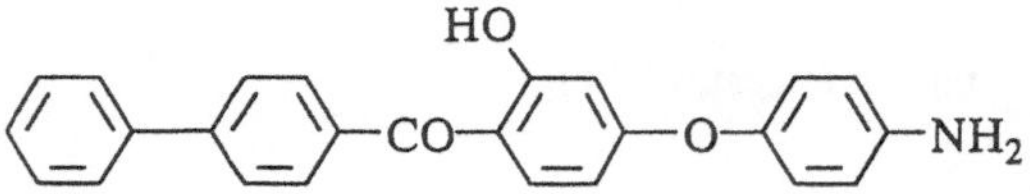

Synthesis

-Refer to: [1067].

m.p. 163° [1067];   UV [1067].

[1,1'-Biphenyl]-4-yl[4-(cyclohexyloxy)-2-hydroxyphenyl]methanone

[36414-95-6]   C$_{25}$H$_{24}$O$_3$ mol.wt. 372.46

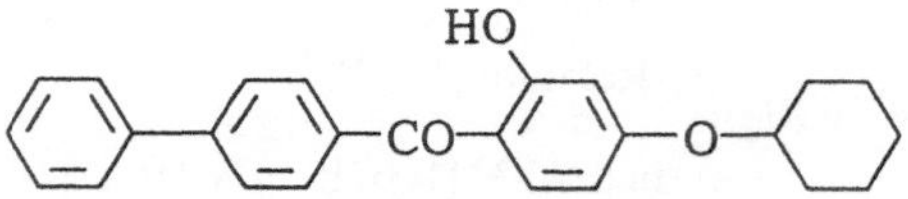

Synthesis

-Refer to: [1067].

m.p. 123° [1067];   UV [1067].

[1,1'-Biphenyl]-4-yl[4-(hexyloxy)-2-hydroxyphenyl]methanone

[36419-22-4]   C$_{25}$H$_{26}$O$_3$ mol.wt. 374.48

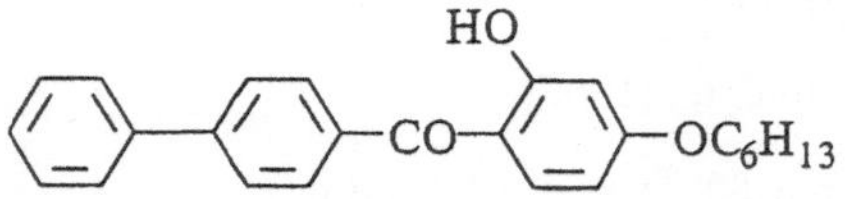

Synthesis

-Refer to: [1067].

m.p. 89° [1067];   UV [1067].

[1,1'-Biphenyl]-4-yl[2-hydroxy-4-(1-propylbutoxy)phenyl]methanone

[36414-89-8] $C_{26}H_{28}O_3$ mol.wt. 388.51

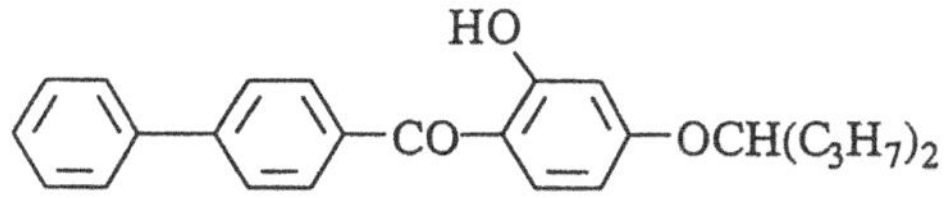

HO

—CO—   —OCH(C₃H₇)₂ *(—OCH(C_3H_7)₂)*

Synthesis

-Refer to: [1067].

oil [1067];   b.p. (NA);   UV [1067].

[1,1'-Biphenyl]-4-yl[2-hydroxy-4-(octyloxy)phenyl]methanone

[36130-58-2] $C_{27}H_{30}O_3$ mol.wt. 402.53

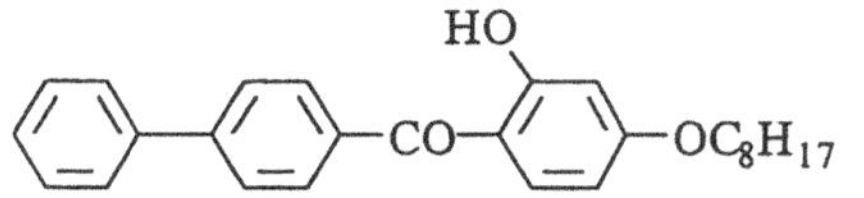

HO

—CO—   —OC₈H₁₇ *(—OC_8H_{17})*

Synthesis

-Preparation by reaction of n-octyl chloride with 2,4-dihydroxy-4'-phenylbenzophenone in the presence of sodium hydroxide and antimony triiodide in diethylene glycol (diglycol) at 140° for 1 h (90%) [387].

-Also refer to: [1067].

m.p.   69-70° [387], 67° [1067];   UV [1067].

[1,1'-Biphenyl]-4-yl[4-(decyloxy)-2-hydroxyphenyl]methanone

[36419-24-6] $C_{29}H_{34}O_3$ mol.wt. 430.59

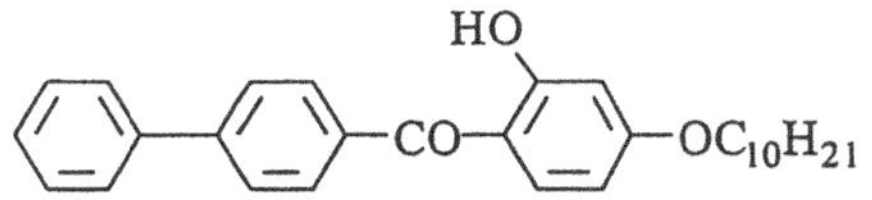

HO

—CO—   —OC₁₀H₂₁ *(—O$C_{10}H_{21}$)*

Synthesis

-Refer to: [1067].

m.p. 82° [1067];   UV [1067].

[1,1'-Biphenyl]-4-yl[4-(hexadecyloxy)-2-hydroxyphenyl]methanone

[36419-25-7] $C_{35}H_{46}O_3$ mol.wt. 514.75

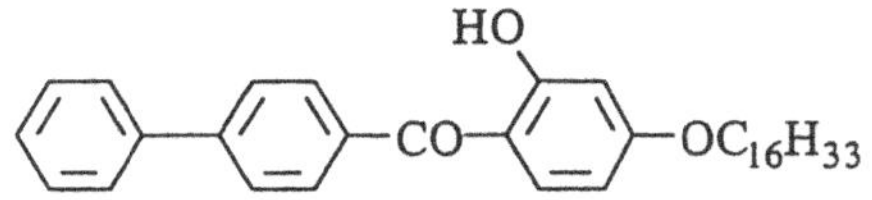

HO

—CO—   —OC₁₆H₃₃ *(—O$C_{16}H_{33}$)*

Synthesis

Refer to: [1067].

m.p. 88° [1067];   UV [1067].

1.2. *Dihydroxylated ketones*

(2,4-Dihydroxyphenyl)(4'-nitro[1,1'-biphenyl]-4-yl)methanone

[36414-94-5] $C_{19}H_{13}NO_5$ mol.wt. 335.32

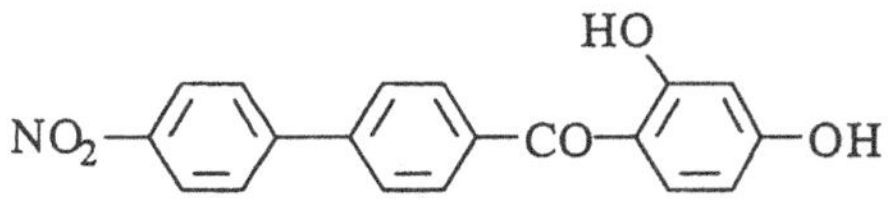

NO₂—   —CO—   —OH

HO

Synthesis

-Refer to: [1067].

m.p. 299° [1067];   UV [1067].

[1,1'-Biphenyl]-4-yl(2,4-dihydroxyphenyl)methanone

[36130-57-1]   C₁₉H₁₄O₃   mol.wt. 290.32

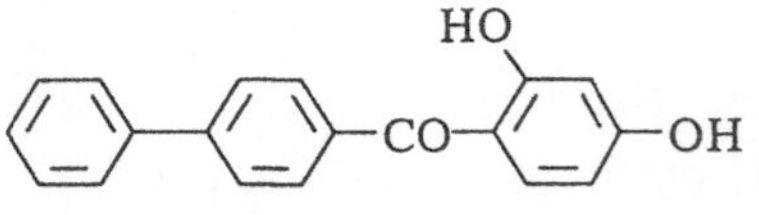

Syntheses

-Preparation by reaction of 4-biphenylcarboxylic acid
 with resorcinol in the presence of boron trifluoride in
 tetrachloroethane at <45° for 5 h (67%) [1273].
-Preparation by reaction of 4-biphenylcarbonyl
chloride with resorcinol in the presence of aluminium chloride in nitrobenzene, first between 25 to
30° for 1 h, then at 55-60° for 4 h (91%) [1273].
-Also refer to: [387, 1067].

     m.p.   188° [1067], 183-184° [1273];   UV [1067, 1273].

(3,5-Dihydroxy[1,1'-biphenyl]-2-yl)phenylmethanone

[54439-82-6]   C₁₉H₁₄O₃   mol.wt. 290.32

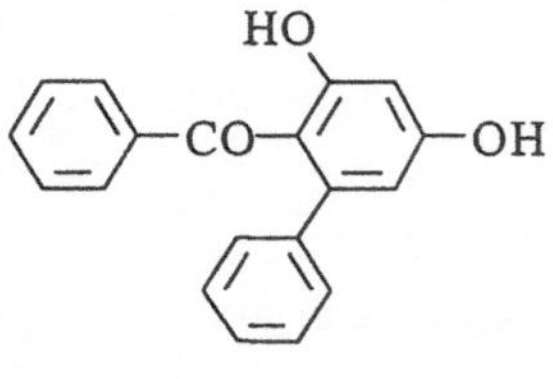

Synthesis

-Preparation from 3,3'-diphenyl-5,5'-diisoxazolylmethane by
 performing hydrogenolysis and subsequent hydrolysis with
 hydrochloric acid (55%) [90].

m.p. 148° [90];   ¹H NMR [90], MS [90].

(4,4'-Dihydroxy[1,1'-biphenyl]-3-yl)phenylmethanone

[52189-86-3]   C₁₉H₁₄O₃   mol.wt. 290.32

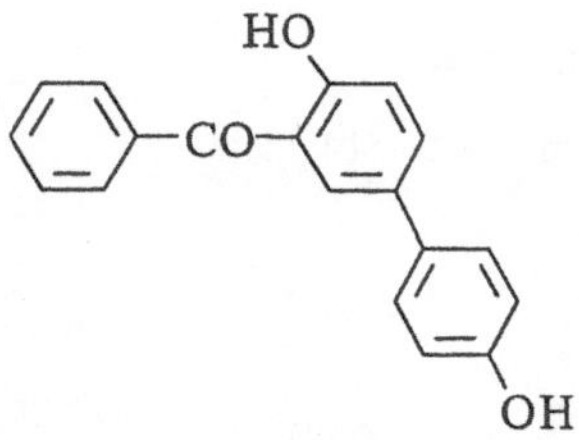

Syntheses

-Obtained by irradiation of [bi(cyclohexa-2,5-dienylidene)-
 4,4'-dione], so called 4,4'-diphenoquinone, in benzaldehyde
 for 2 days (21%) [221].
-Also obtained by Fries rearrangement of 4,4'-biphenyldiyl
 dibenzoate with aluminium chloride in o-dichlorobenzene
 during 3 h at 180° (19%) [1451].

m.p. 142-144° [1451], 130° [221];
¹H NMR [221, 1451], IR [221, 1451], UV [1451], MS [221].

(2-Hydroxy[1,1'-biphenyl]-3-yl)(2-hydroxyphenyl)methanone

C₁₉H₁₄O₃   mol.wt. 290.32

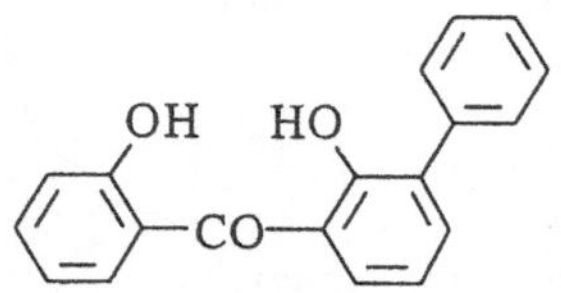

Synthesis

-Obtained by photo-Fries rearrangement of phenyl m-phenyl-
 salicylate (major product) [970].

m.p. and Spectra (NA).

(4'-Hydroxy[1,1'-biphenyl]-4-yl)(3-hydroxyphenyl)methanone

[75731-50-9]   C$_{19}$H$_{14}$O$_3$ mol.wt. 290.32

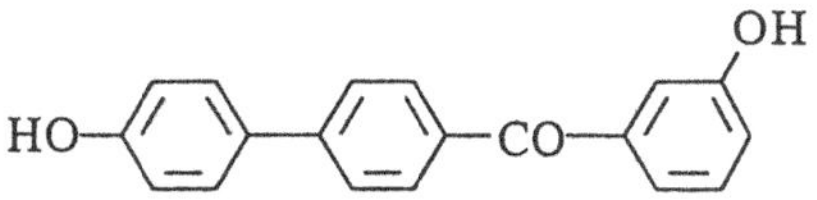

Synthesis

-Preparation by acylation of 4-hydroxybiphenyl with 3-hydroxybenzoic acid in hydrofluoric acid in the presence of boron trifluoride (pressure: 207 KPa) in an autoclave for 6 h at 0° (94%) [624].

m.p. 213-219° [624];   Spectra (NA).

(4'-Hydroxy[1,1'-biphenyl]-4-yl)(4-hydroxyphenyl)methanone

[86432-13-5]   C$_{19}$H$_{14}$O$_3$ mol.wt. 290.32

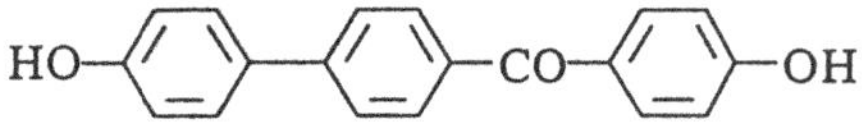

Syntheses

-Preparation by reaction of 4-hydroxybenzoic acid with 4-hydroxybiphenyl in the presence of trifluoromethanesulfonic acid at r.t. for
overnight, then at 50° for 3.5 h and at 65° for 1.5 h (95%) [1138].
-Preparation by Friedel-Crafts acylation of p-acetoxybiphenyl with p-hydroxybenzoyl chloride, followed by hydrolysis of the keto ester so obtained (65%) [1356].
-Preparation by heating a mixture of 4-hydroxybiphenyl-4'-carboxylic acid and phenol in nitrobenzene in the presence of boron trifluoride at 80° for 30 min (84%) [390].

m.p. 230-232° [1356], 190-198° [1138]. There is discrepancy between the two melting points.
^{1}H NMR [1138, 1356], IR [1356].

[1,1'-Biphenyl]-4-yl[2-hydroxy-4-(2-hydroxyethoxy)phenyl]methanone

C$_{21}$H$_{18}$O$_4$ mol.wt. 334.37

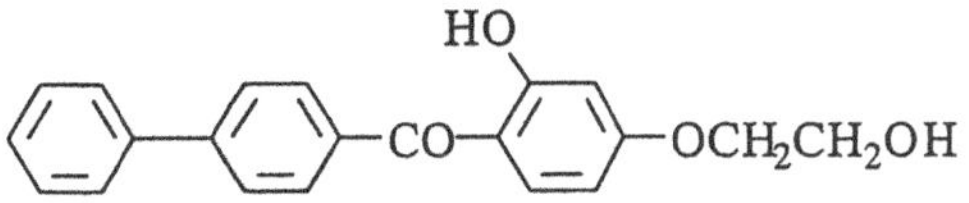

Synthesis

-Preparation by reaction of ethylene carbonate with 2,4-dihydroxy-4'-phenylbenzophenone in the presence of sodium methoxide in diisobutyl ketone for 2 h at 130° (84%) [388].

m.p. 157-158° [388];   Spectra (NA).

[1,1'-Biphenyl]-4-yl[2-hydroxy-4-(3-hydroxypropoxy)phenyl]methanone

[36469-47-3]   C$_{22}$H$_{20}$O$_4$ mol.wt. 348.40

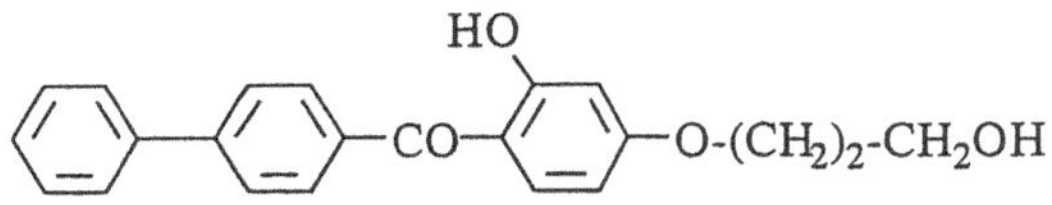

Synthesis

-Refer to: [1067].

m.p. 145° [1067];   UV [1067].

2. TERPHENYL PHENYL METHANONES

(2,4-Dihydroxyphenyl)[1,1';4',1'']terphenyl-4''-ylmethanone

$C_{25}H_{18}O_3$ mol.wt. 366.42

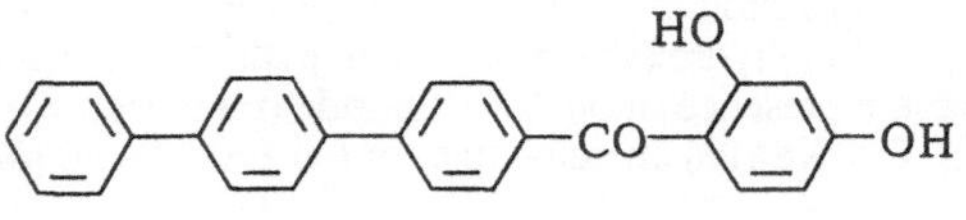

Synthesis

-Preparation by Friedel-Crafts acylation of p-terphenyl with 2,4-dihydroxybenzoic acid or its derivatives [711].

m.p. and Spectra (NA).

(2-Hydroxy-4-methoxyphenyl)[1,1',4',1'']terphenyl-4''-ylmethanone

$C_{26}H_{20}O_3$ mol.wt. 380.44

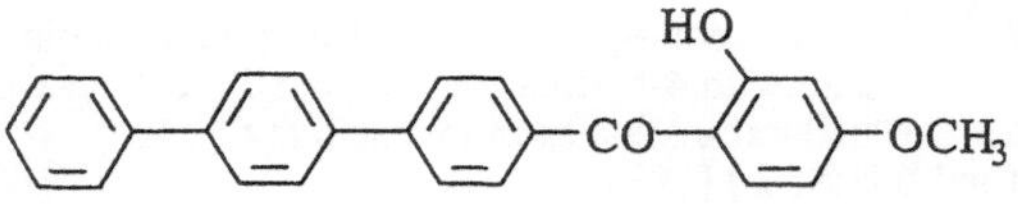

Synthesis

-Preparation by Friedel-Crafts acylation of p-terphenyl with 2-hydroxy-4-methoxybenzoic acid or its derivatives [711].

m.p. (NA);   UV [711].

Chapter 4. Cyclohexyl phenyl methanones *(Class of METHANONES)*

1. MONOHYDROXYLATED KETONES

(3-Bromo-2-hydroxyphenyl)cyclohexylmethanone

[81066-14-0] $C_{13}H_{15}BrO_2$ mol.wt. 283.16

Syntheses

-Preparation by reaction of cyclohexanecarbonyl chloride with o-bromophenol in the presence of aluminium chloride without solvent at 140° for 1 h (48%) [984].
-Preparation by Fries rearrangement of o-bromophenyl cyclohexanecarboxylate with aluminium chloride without solvent at 100° for 1 h (36%) [1461].

b.p._1 140-145° [984];   m.p. 51-52° [984];   Spectra (NA).

(3-Bromo-4-hydroxyphenyl)cyclohexylmethanone

[81066-15-1] $C_{13}H_{15}BrO_2$ mol.wt. 283.16

Syntheses

-Obtained by reaction of cyclohexanecarbonyl chloride with o-bromophenol in the presence of aluminium chloride without solvent at 120° for 2 h (16%) [984].
-Also obtained by Fries rearrangement of o-bromophenyl cyclohexanecarboxylate with aluminium chloride without solvent at 120° for 1 h (19%) [1461].

b.p._1 160-170° [984];   m.p. 138-139° [984];   Spectra (NA).

(4-Bromo-2-hydroxyphenyl)cyclohexylmethanone

[81066-16-2] $C_{13}H_{15}BrO_2$ mol.wt. 283.16

Syntheses

-Preparation by reaction of cyclohexanecarbonyl chloride with m-bromophenol in the presence of aluminium chloride without solvent at 120° for 1-2 h (57-64%) [984, 1461].
-Preparation by Fries rearrangement of m-bromophenyl cyclohexanecarboxylate with aluminium chloride without solvent at 100° for 1 h (71%) [1461].

b.p._1 165-175° [984];   m.p. 56°5-57°5 [984];   Spectra (NA).

(5-Bromo-2-hydroxyphenyl)cyclohexylmethanone

[81066-17-3] $C_{13}H_{15}BrO_2$ mol.wt. 283.16

Syntheses

-Preparation by Fries rearrangement of p-bromophenyl cyclohexanecarboxylate with aluminium chloride without solvent at 120° for 1 h (55%) [1461].
-Also obtained by reaction of cyclohexanecarbonyl chloride with p-bromophenol in the presence of aluminium chloride

without solvent at 120° for 2 h (32%) [984] or between 100 to 140° for 1 h (12 to 22% yields) [1461].

b.p.$_1$ 150-170° [984];   m.p. 98°5-99°5 [984];   Spectra (NA).

Cyclohexyl(5-fluoro-2-hydroxyphenyl)methanone

[183280-18-4]   C$_{13}$H$_{15}$FO$_2$ mol.wt. 222.26

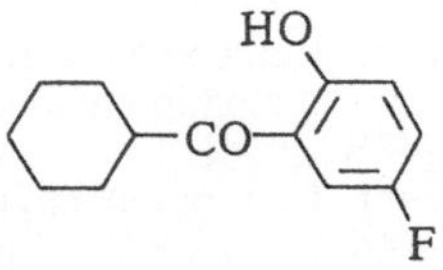

Synthesis

-Preparation by Fries rearrangement of p-fluorophenyl
cyclohexanecarboxylate with aluminium chloride at 150-
180° for 20 min (47%) [765].

m.p. 61°5 [765]; ^{1}H NMR [765], MS [765].

Cyclohexyl(2-hydroxyphenyl)methanone

[18066-52-9]   C$_{13}$H$_{16}$O$_2$ mol.wt. 204.27

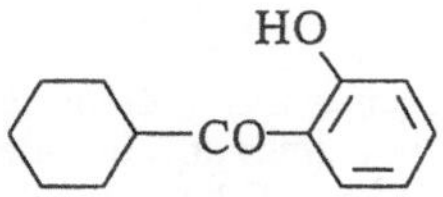

Syntheses

-Preparation by Fries rearrangement of phenyl cyclohexanecarboxylate with aluminium chloride,
*without solvent at 120° or at 160° for 2 h (45% and 27%
yields, respectively) [168] or at 150° for 3 h (70%) [925];
*in nitrobenzene at 70° for 2 h [632].
-Preparation by oxidative cleavage of 3-cyclohexylbenzofuran in methylene chloride with ozone at
-78° for 15 min, then at r.t. for 1 h and saponification of the cyclohexyl 2-(formyloxy)phenyl
ketone so obtained in methanol with 1 N sodium hydroxide at r.t. for 12 h [1224].

m.p. 104-105° [168];
b.p.$_2$ 189-190° [168], b.p.$_4$ 127-129° [632]. There is discrepancy between the two boiling
points.
^{1}H NMR [1224], IR [1224], MS [1224].

Cyclohexyl(3-hydroxyphenyl)methanone

[148493-08-7]   C$_{13}$H$_{16}$O$_2$ mol.wt. 204.27

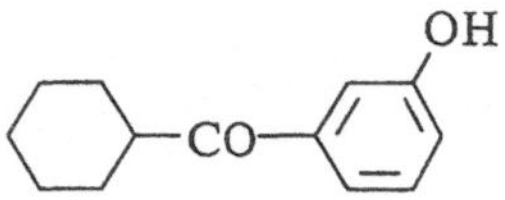

Synthesis

-Preparation by treatment of m-bromophenol in tetrahydrofuran with tert-butyllithium in pentane for 15 min at
-78° under argon, after which a solution of cyclohexyl-
N,O-dimethylhydroxamide in tetrahydrofuran was slowly
added (60%). -Refer to: Chem. Abstr., 119, 49010t (1993).

m.p. 70-72°;   Spectra (NA).

Cyclohexyl(4-hydroxyphenyl)methanone

[38459-58-4] $C_{13}H_{16}O_2$ mol.wt. 204.27

Syntheses

-Obtained (by-product) by Fries rearrangement of phenyl cyclohexanecarboxylate with aluminium chloride at 120° for 2 h (1%) [168].

-Preparation by demethylation of 4-methoxyphenyl cyclohexyl ketone (SM) by treatment with boron tribromide in methylene chloride (76%). SM was obtained by Friedel-Crafts acylation of anisole with cyclohexanecarboxylic acid chloride [883].

m.p. and Spectra (NA).

Cyclohexyl(4-hydroxy-3-methoxy-5-nitrophenyl)methanone

[125629-27-8] $C_{14}H_{17}NO_5$ mol.wt. 279.29

Synthesis

-Preparation by reaction of 65% nitric acid with cyclohexyl 4-hydroxy-3-methoxyphenyl ketone in acetic acid at r.t. (75%) [197].

m.p. 135-141° [197]; ^1H NMR [197], MS [197].

Cyclohexyl(2-hydroxy-3-methylphenyl)methanone

$C_{14}H_{18}O_2$ mol.wt. 218.30

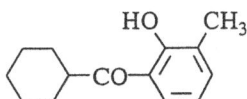

Synthesis

-Preparation by Fries rearrangement of o-cresyl cyclohexane-carboxylate with aluminium chloride at 120° for 2 h (56%) or at 160° for 2 h [168].

m.p. 130° [168];   Spectra (NA).

Cyclohexyl(2-hydroxy-4-methylphenyl)methanone

$C_{14}H_{18}O_2$ mol.wt. 218.30

Syntheses

-Obtained by Fries rearrangement of m-cresyl cyclohexane-carboxylate,
*with alumina in methanesulfonic acid for 25 min at 160° (86%). -Refer to: Chem. Abstr., **130**, 81248q (1999);
*with aluminium chloride without solvent at 120° or 160° for 2 h (poor yield) [168].
-Preparation by Friedel-Crafts acylation of m-cresol with cyclohexanecarboxylic acid in the presence of alumina in methanesulfonic acid for 5 min at 120° (87%). -Refer to: Chem. Abstr., **130**, 81248q (1999).

b.p.$_2$ 115° [168];   Spectra (NA).

Cyclohexyl(2-hydroxy-5-methylphenyl)methanone

$C_{14}H_{18}O_2$ mol.wt. 218.30

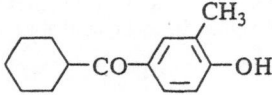

Syntheses

-Preparation by reaction of cyclohexanecarboxylic acid with
 p-cresol in the presence of aluminium chloride and
 phosphorous trichloride at 160° for 2 h (77%) [983].
-Also obtained (poor yield) by Fries rearrangement of
 p-cresyl cyclohexanecarboxylate with aluminium chloride
without solvent at 120° for 2 h (9%) or at 160° [168].

   m.p.   75-76° [983];
   b.p.$_2$ 120° [168], b.p.$_5$ 160-170° [983]. There is discrepancy between the two boiling
   points.   Spectra (NA).

Cyclohexyl(4-hydroxy-3-methylphenyl)methanone

$C_{14}H_{18}O_2$ mol.wt. 218.30

Synthesis

-Obtained (trace) by Fries rearrangement of o-cresyl
 cyclohexanecarboxylate with aluminium chloride at 120° or
 at 160° for 2 h (<1%) [168].

   m.p.   118-119° [168];   Spectra (NA).

Cyclohexyl(2-hydroxy-4-methoxyphenyl)methanone

[69210-88-4] $C_{14}H_{18}O_3$ mol.wt. 234.30

Syntheses

-Preparation by Fries rearrangement of m-methoxyphenyl
 cyclohexanecarboxylate in a nitromethane/toluene mixture
 in the presence of hafnium triflate/lithium perchlorate for
 6 h at 50°.
-Preparation by Friedel-Crafts acylation of m-methoxyphenol with cyclohexanecarbonyl chloride,
cyclohexanecarboxylic acid or its anhydride in the same conditions. -Refer to: Chem. Abstr., **127**,
278063v (1997).

   m.p. and Spectra (NA).

Cyclohexyl(4-hydroxy-3-methoxyphenyl)methanone

[125629-26-7] $C_{14}H_{18}O_3$ mol.wt. 234.30

Synthesis

-Preparation by adding a solution of 30% hydrobromic acid
 in acetic acid to a solution of 4-(benzyloxy)-3-methoxy-
 phenyl cyclohexyl ketone in methylene chloride with stirring
 2 h at r.t. (77%) [197].

   m.p.   113-115° [197]; ^{1}H NMR [197], MS [197].

Cyclohexyl(2-hydroxy-3,4-dimethoxyphenyl)methanone

[121638-96-8] $C_{15}H_{20}O_4$ mol.wt. 264.32

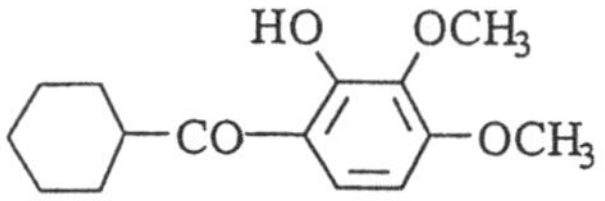

Synthesis

-Obtained (poor yield) by Friedel-Crafts acylation of pyrogallol trimethyl ether with cyclohexanecarbonyl chloride in the presence of aluminium chloride in refluxing methylene chloride for 2 h (7%) [1407].

m.p. 85° [1407]; Spectra (NA).

Cyclohexyl[3-(1,1-dimethylethyl)-4-hydroxyphenyl]methanone

[124979-10-8] $C_{17}H_{24}O_2$ mol.wt. 260.38

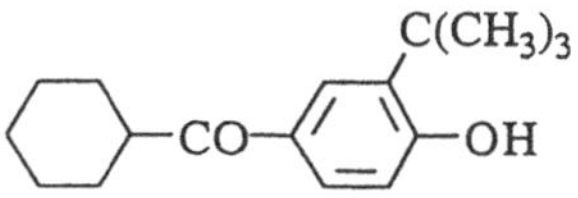

Synthesis

-Preparation by Friedel-Crafts acylation of o-tert-butyl-phenol with cyclohexanecarbonyl chloride in ethylene dichloride in the presence of titanium tetrachloride, first at 0°, then at r.t. [733].

m.p. 142-143° [733]; Spectra (NA).

Cyclohexyl[5-(1,1-dimethylethyl)-2-hydroxyphenyl]methanone

[18738-74-4] $C_{17}H_{24}O_2$ mol.wt. 260.38

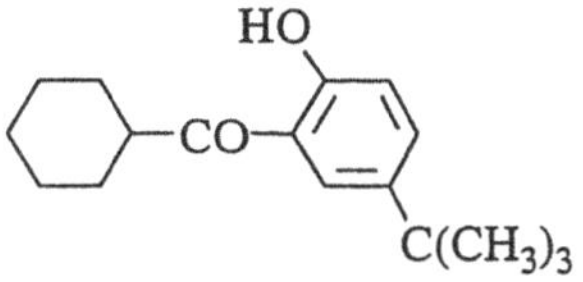

Syntheses

-Preparation by demethylation of 2-(cyclohexylcarbonyl)-4-tert-butylanisole with a mixture of 47% hydrobromic acid and 57% hydriodic acid in refluxing acetic acid for 2 h (95%) [636].
-Also obtained (poor yields) by Fries rearrangement of p-tert-butylphenyl cyclohexanecarboxylate with aluminium chloride at 120° for 2 h (5%) or at 160° for 2 h (8%) [168].

b.p.$_6$ 219-220° [168]; Spectra (NA).

Cyclohexyl[2-hydroxy-5-(1,1-dimethylpropyl)phenyl]methanone

$C_{18}H_{26}O_2$ mol.wt. 274.40

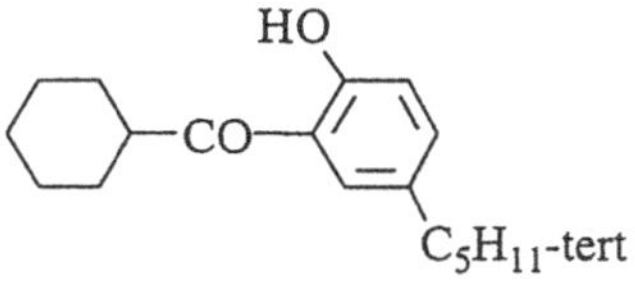

Synthesis

-Obtained (poor yields) by Fries rearrangement of p-tert-pentylphenyl cyclohexanecarboxylate with aluminium chloride at 120° for 2 h (16%) or at 160° for 2 h (7%) [168].

oil [168]; b.p.$_5$ 219-220° [168]; Spectra (NA).

[3-(Aminomethyl)-5-(1,1-dimethylethyl)-2-hydroxyphenyl]cyclohexylmethanone

[75060-98-9] $C_{18}H_{27}NO_2$ mol.wt. 289.42

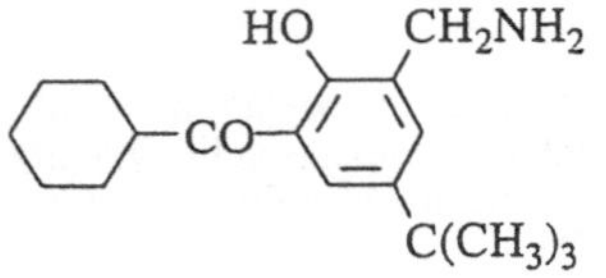

HO   CH$_2$NH$_2$

Synthesis

-Refer to: [636]; see the hydrochloride below.

m.p. and Spectra (NA).

C(CH$_3$)$_3$

[3-(Aminomethyl)-5-(1,1-dimethylethyl)-2-hydroxyphenyl]cyclohexylmethanone
(Hydrochloride)

[75060-63-8] $C_{18}H_{27}NO_2,HCl$ mol.wt. 325.89

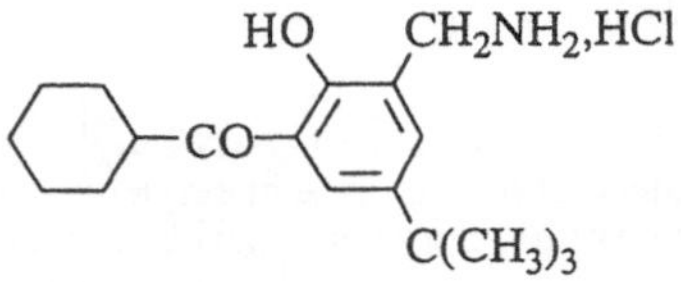

HO   CH$_2$NH$_2$,HCl

Synthesis

-Preparation by reaction of concentrated hydrochloric acid with 2-(cyclohexylcarbonyl)-4-tert-butyl-6-(N-chloroacetylaminomethyl)phenol in refluxing ethanol for 30 h (67%) [636].

C(CH$_3$)$_3$

m.p. 205-209° [636];   Spectra (NA).

[3,5-Bis(1,1-Dimethylethyl)-4-hydroxyphenyl]cyclohexylmethanone

[28440-98-4] $C_{21}H_{32}O_2$ mol.wt. 316.48

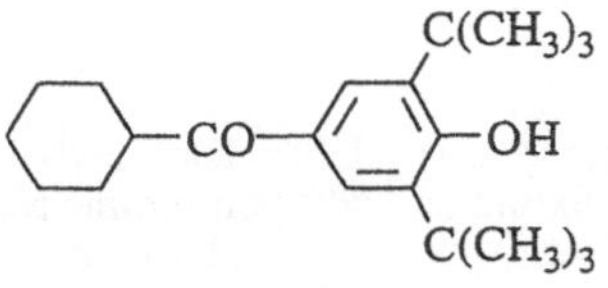

C(CH$_3$)$_3$

Synthesis

-Preparation by action of cyclohexanecarbonyl chloride with 2,6-di-tert-butylphenol in ethylene dichloride in the presence of titanium tetrachloride for 15-30 min at r.t..
-Refer to: Chem. Abstr., **90**, 121219v (1979)*.

C(CH$_3$)$_3$

m.p. 125-127°*;   Spectra (NA).

2. DIHYDROXYLATED KETONES

Cyclohexyl(3,4-dihydroxy-5-nitrophenyl)methanone

[125628-95-7] $C_{13}H_{15}NO_5$ mol.wt. 265.27

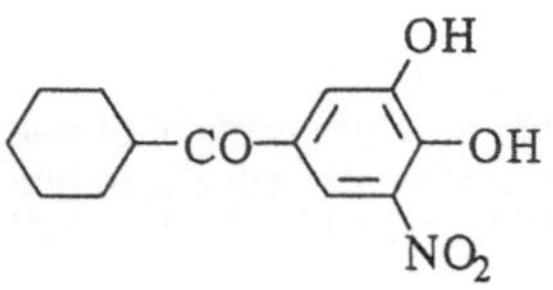

OH

Synthesis

-Preparation by demethylation of cyclohexyl (4-hydroxy-3-methoxy-5-nitrophenyl) ketone with 48% hydrobromic acid in refluxing acetic acid for 20 h (55%) [197].

NO$_2$

m.p. 145-147° [197]; ^{1}H NMR [197], MS [197].

Cyclohexyl(2,4-dihydroxyphenyl)methanone

[97231-21-5] $C_{13}H_{16}O_3$ mol.wt. 220.27

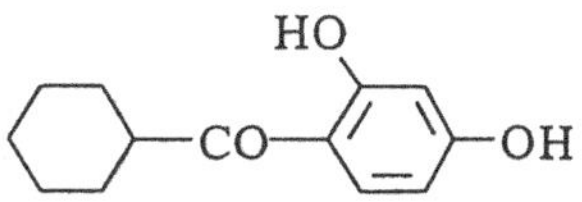

Synthesis

-Preparation by reaction of cyclohexanecarboxylic acid with resorcinol in the presence of zinc chloride at 125-135° (54%) (Nencki reaction) [1355].
-Also refer to: [958].

m.p. 115°5-116° [1355];   b.p.$_4$ 200-202° [1355];
Spectra (NA).

(1-Hydroxycyclohexyl)(4-hydroxyphenyl)methanone

[200420-24-2] $C_{13}H_{16}O_3$ mol.wt. 220.27

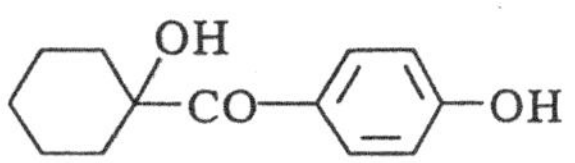

Synthesis

-Refer to: Chem. Abstr., **128**, 68498k (1998).

m.p. and Spectra (NA).

(2-Hydroxycyclohexyl)(2-hydroxyphenyl)methanone

[148077-95-6] $C_{13}H_{16}O_3$ mol.wt. 220.27

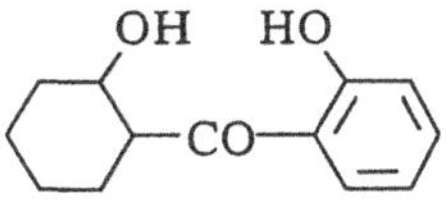

Synthesis

-Refer to: Chem. Abstr., **119**, 8806p (1993).

m.p. and Spectra (NA).

3. TRIHYDROXYLATED KETONES

Cyclohexyl(2,4,6-trihydroxyphenyl)methanone

[85602-45-5] $C_{13}H_{16}O_4$ mol.wt. 236.27

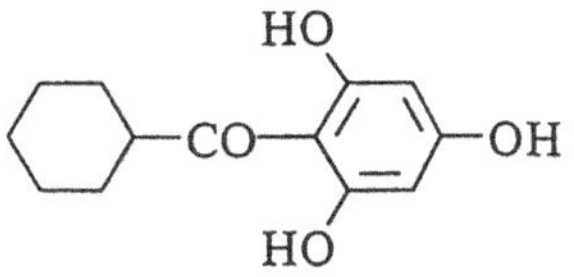

Synthesis

-Preparation by action of cyclohexanecarbonyl chloride with phloroglucinol in the presence of aluminium chloride in refluxing carbon disulfide/nitrobenzene mixture for 3 h.
-Refer to: Chem. Abstr., **99**, 5339w (1983)*.
-Also refer to: Chem. Abstr., **101**, 151650s (1984).

m.p. 110-113°*   Spectra (NA).

Cyclohexyl[2,4,6-trihydroxy-3-(3-methyl-2-butenyl)phenyl]methanone

[85602-18-2]   C18H24O4   mol.wt. 304.39

Synthesis

-Preparation by reaction of prenyl chloride with cyclohexyl 2,4,6-trihydroxyphenyl ketone in ethyl ether in the presence of a saturated aqueous sodium carbonate solution and a catalytic amount of cuprous chloride for 3 h at r.t. (48%). -Refer to: Chem. Abstr., **101**, 151650s (1984)*.

-Also refer to: Chem. Abstr., **101**, 221168s (1984)**.

m.p.   144-147°*; ^{13}C NMR*, **, IR*, MS*.

PART 2 - DIAROYLPHENOLS AND POLYAROYLPHENOLS

Chapter 5. Phenols with one benzoyl group and one or several acetyl groups
(Class of ETHANONES)

1. MONOHYDROXYLATED KETONES

1-[3-(4-Chlorobenzoyl)-4-hydroxyphenyl]ethanone

[108294-81-1] $C_{15}H_{11}ClO_3$ mol.wt. 274.70

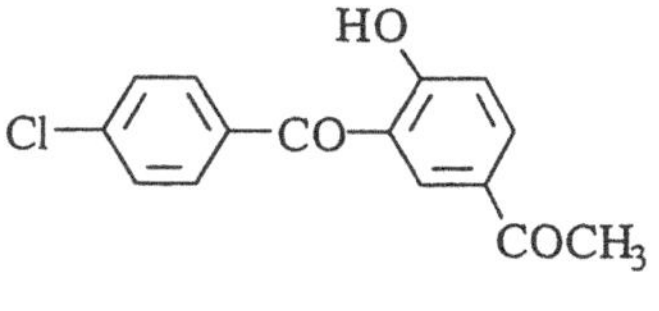

Synthesis

-Obtained (poor yield) by reaction of 5-acetyl-2-hydroxy-benzoyl chloride with chlorobenzene in the presence of aluminium chloride at 100° for 24 h (7%) [400].

m.p. 127-130° [400];   Spectra (NA).

1-(3-Benzoyl-2-hydroxyphenyl)ethanone

[85558-60-7] $C_{15}H_{12}O_3$ mol.wt. 240.26

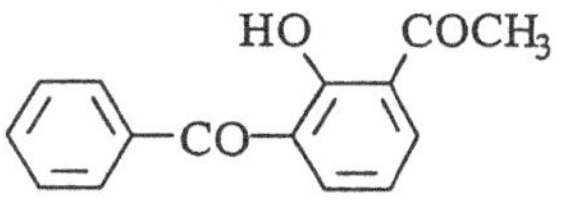

Synthesis

-Obtained by hydrolysis of 2-acetoxy-3-benzoyl-acetophenone [458].

m.p. 108-109° [458]; ^{1}H NMR [458], IR [458], MS [458].

1-(3-Benzoyl-4-hydroxyphenyl)ethanone

[13043-37-3] $C_{15}H_{12}O_3$ mol.wt. 240.26

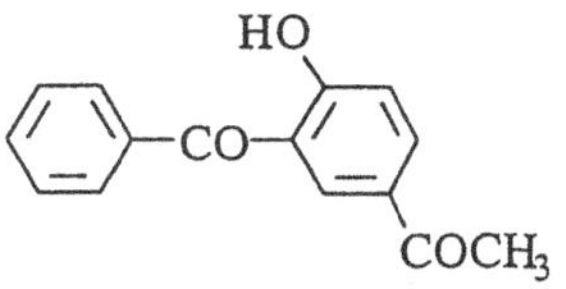

Synthesis

-Preparation by Fries rearrangement of p-acetylphenyl benzoate with aluminium chloride at 153° for 8 h (24%) [608].

m.p. 95° [608]; ^{1}H NMR [610], UV [607, 610, 793, 794];
pK_a [608, 791];   polarographic study [1223];   TLC [377].

1-(5-Benzoyl-2-hydroxyphenyl)ethanone

[2589-80-2] $C_{15}H_{12}O_3$ mol.wt. 240.26

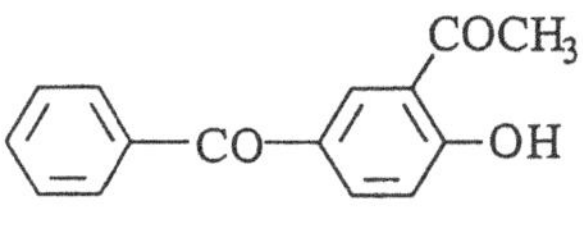

Synthesis

-Refer to: Chem. Abstr., **113**, 97396w (1990).

m.p. and Spectra (NA).

1-[2-(2-Hydroxybenzoyl)phenyl]ethanone

[17526-21-5] $C_{15}H_{12}O_3$ mol.wt. 240.26

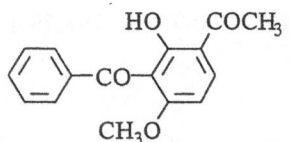

Syntheses

-Obtained by rearrangement of 10-methylanthronyl
  hydroperoxide (SM) with a mixture of acetic acid/4 N
  sulfuric acid (4:3). SM was obtained either by self-oxidation
  of 10-methylanthrone or by action of methylmagnesium
iodide on 10-methylanthryl acetate [429].
-Also obtained (poor yield) from meso-photoxide of 9-methylanthracene in refluxing o-dichlorobenzene for 3 h under nitrogen (3%) [1125].

   m.p.   108° [429], 107-108° [1125];
   IR [429], UV [429], MS [429].

(3-Benzoyl-2-hydroxy-5-methylphenyl)ethanone

[79877-07-9] $C_{16}H_{14}O_3$ mol.wt. 254.29

Syntheses

-Preparation by Fries rearrangement of 2-(benzoyloxy)-
  5-methylacetophenone with aluminium chloride at 130-140°
  for 2 h (60-65%) [758].
-Preparation by Fries rearrangement of 2-(acetyloxy)-
  5-methylbenzophenone with aluminium chloride at 140° for
  1 h [1440].
-Preparation by reaction of benzoyl chloride with 2-hydroxy-5-methylacetophenone in the presence
  of aluminium chloride in nitrobenzene at 80° for 8 h (40%) [852].
-Also refer to: [974].

   m.p.   102-103° [1440], 101-102° [852];
   ^{1}H NMR [852], IR [852], UV [852];
   structural data [699, 1105].

1-(3-Benzoyl-2-hydroxy-4-methoxyphenyl)ethanone

[64857-84-7] $C_{16}H_{14}O_4$ mol.wt. 270.28

Syntheses

-Obtained by Friedel-Crafts acylation of paeonol with
  benzoyl chloride in the presence of aluminium chloride in
  nitrobenzene at r.t. or at 100° [644].
-Also obtained by Friedel-Crafts acylation of 2-hydroxy-
  6-methoxybenzophenone with acetic anhydride in the
presence of aluminium chloride in nitrobenzene in a boiling water bath for 2 h [644].

   m.p.   169° [644];   Spectra (NA).

2. DIHYDROXYLATED KETONES

Symmetrical ketones

Bis(5-Acetyl-2-hydroxyphenyl)methanone

[33427-60-0] $C_{17}H_{14}O_5$ mol.wt. 298.30

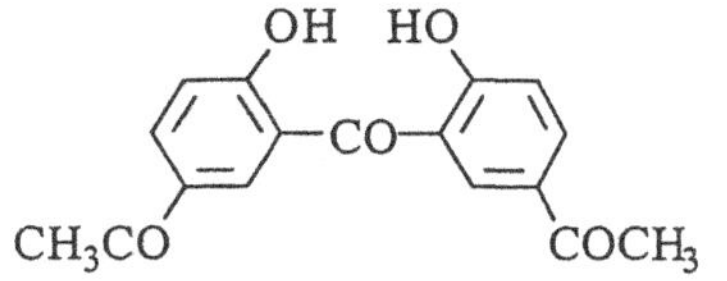

-Also refer to: [638].

Syntheses

-Obtained (poor yield) by Fries rearrangement of 2,2'-diacetoxybenzophenone with aluminium chloride without solvent at 160-180° for 20 min (10%) [949].
-Also obtained by reaction of sodium hydroxide with 2,7-diacetylxanthone (94%) [467].

m.p. >300° [949], 175-177° [467]. There is discrepancy between the two melting points.
^{1}H NMR [467, 949], IR [949], MS [467, 949].

Bis(3-Acetyl-2-hydroxy-5-methylphenyl)methanone

$C_{19}H_{18}O_5$ mol.wt. 326.35

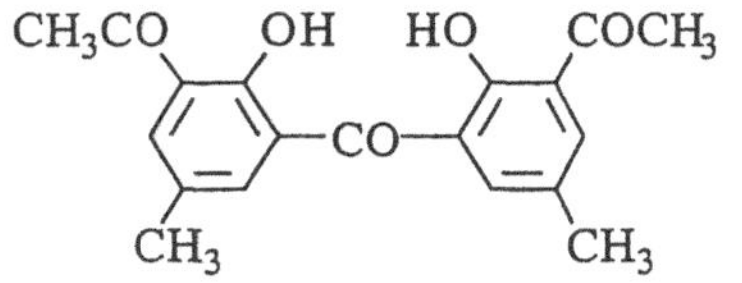

m.p. 207-208° [671];   Spectra (NA).

Synthesis

-Preparation by Fries rearrangement of 2,2'-diacetoxy-5,5'-dimethylbenzophenone with aluminium chloride without solvent, first at 120°, then at 170° for 20 min (80%) [671].

Asymmetric ketones

1-(3-Benzoyl-2,4-dihydroxyphenyl)ethanone

$C_{15}H_{12}O_4$ mol.wt. 256.26

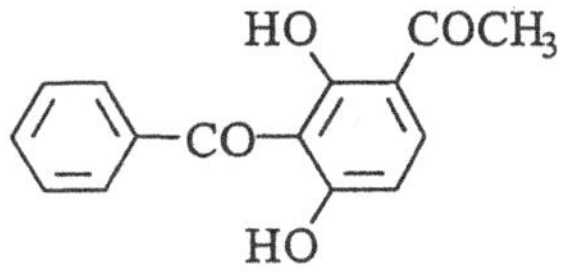

Syntheses

-Resacetophenone, by condensation with benzanilide imidochloride in the presence of aluminium chloride in nitrobenzene gave a keto anil. This one was hydrolyzed with hydrochloric acid in refluxing ethanol and yielded the expected ketone (18%) [1061].
-Also obtained by Fries rearrangement of 4-(benzoyloxy)-2-hydroxyacetophenone in the presence of aluminium chloride without solvent at 140° for 1.5 h (15%) [340].
-Also obtained (poor yield) by condensation of benzoyl chloride with resacetophenone in the presence of aluminium chloride in nitrobenzene in a water bath for 2 h (2%) [340].
-Also obtained by decarboxylation of 5-acetyl-3-benzoyl-2,4-dihydroxybenzoic acid on heating in a test-tube at 220-225° for 1 h [340].

m.p. 167-168° [1061], 165° [340];   Spectra (NA).

1-(3-Benzoyl-2,6-dihydroxyphenyl)ethanone

$C_{15}H_{12}O_4$ mol.wt. 256.26

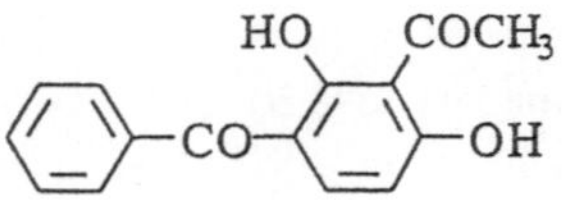

Syntheses

-Obtained by Friedel-Crafts acylation of resbenzophenone,
*with acetic anhydride in the presence of aluminium chloride
  in nitrobenzene in a water bath for 6 h (30%) [340];
*with acetyl chloride in the presence of aluminium chloride
  in o-dichlorobenzene [280].
-Also obtained by Friedel-Crafts acylation of 2,6-dihydroxyacetophenone with benzoyl chloride in
the presence of aluminium chloride in nitrobenzene, first at r.t., then in a water bath for 1 h [340].

   m.p.   108° [340], 99°5-100° [280];   Spectra (NA).

1-(5-Benzoyl-2,4-dihydroxyphenyl)ethanone

[64857-83-6] $C_{15}H_{12}O_4$ mol.wt. 256.26

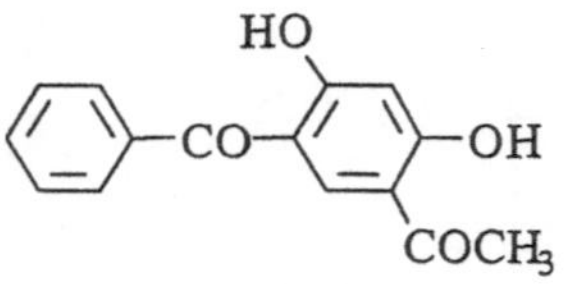

Synthesis

-Obtained by Friedel-Crafts acylation of paeonol with benzoyl
  chloride in the presence of aluminium chloride in nitrobenzene at r.t. or at 100° [644].

   m.p.   138° [644];   Spectra (NA).

1-[2-(2,4-Dihydroxybenzoyl)phenyl]ethanone

[36414-93-4] $C_{15}H_{12}O_4$ mol.wt. 256.26

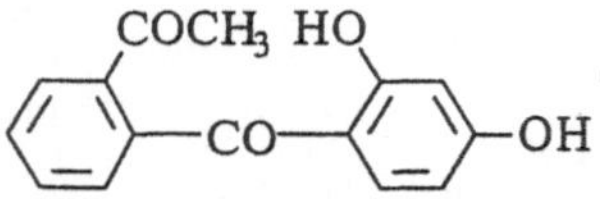

Synthesis

-Refer to: [1067].

   m.p.   184° [1067];   UV [1067].

1-[2-Hydroxy-5-(2-hydroxybenzoyl)phenyl]ethanone

[124208-69-1] $C_{15}H_{12}O_4$ mol.wt. 256.26

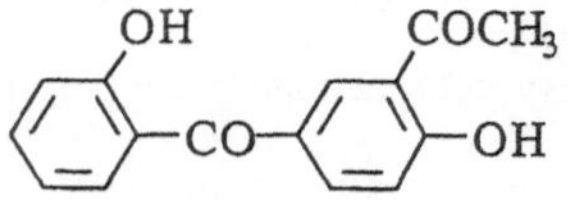

Syntheses

-Obtained by Fries rearrangement,
*of 2,4'-diacetoxybenzophenone with aluminium chloride at
  158-160° for 2 h (15%) [171];
*of 4'-acetoxy-2-methoxybenzophenone with aluminium
chloride at 153-155° for 2 h (60%) [171].

   m.p.   130-131° [171];   Spectra (NA).

1-[4-Hydroxy-3-(4-hydroxybenzoyl)phenyl]ethanone

[124208-64-6] $C_{15}H_{12}O_4$ mol.wt. 256.26

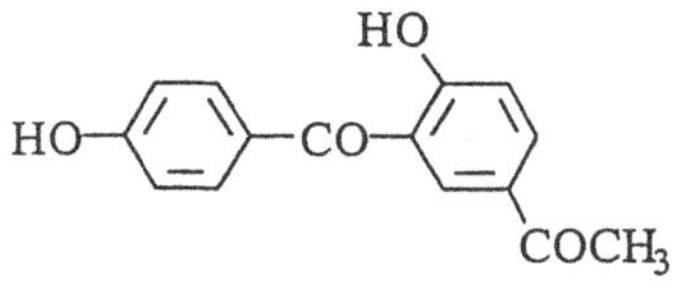

Syntheses

-Obtained by Fries rearrangement,
*of phenyl 2-acetoxybenzoate with aluminium chloride at 180° for 3 h (38%) [171];
*of phenyl 5-acetyl-2-hydroxybenzoate with aluminium chloride at 180° for 3 h (38%) [171].

m.p. 188° [171];   Spectra (NA).

1-(3-Acetyl-5-benzoyl-2,4-dihydroxyphenyl)ethanone
3,5-Diacetyl-2,4-dihydroxybenzophenone

[16832-72-7] $C_{17}H_{14}O_5$ mol.wt. 298.30

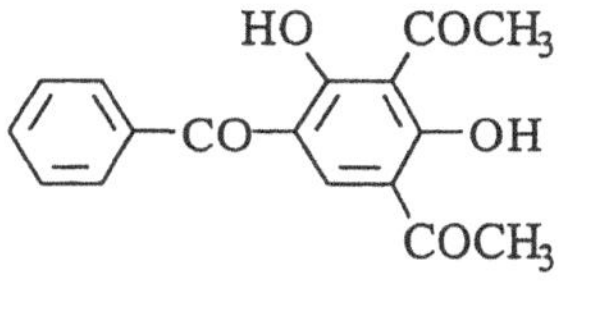

Syntheses

-Obtained by acetylation of 2,4-dihydroxybenzophenone,
*with acetyl chloride in the presence of aluminium chloride in o-dichlorobenzene [280];
*with acetic anhydride in the presence of aluminium chloride in nitrobenzene in a water bath for 6 h (5%) [340].

-Also obtained by condensation of 2-acetyl-4-benzoylresorcinol with acetic anhydride according to Friedel-Crafts [340].

m.p. 151-151°5 [280], 151° [340];   Spectra (NA).

3. TRIHYDROXYLATED KETONE

1-(3-Benzoyl-2,4,6-trihydroxyphenyl)ethanone

[31188-65-5] $C_{15}H_{12}O_5$ mol.wt. 272.26

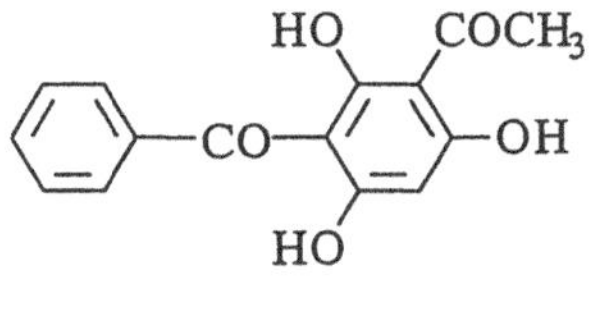

Synthesis

-Preparation by two successive Friedel-Crafts acylations of phloroglucinol, first with acetic acid, then with benzoic acid, always in the presence of boron trifluoride-etherate (compound 17). -Refer to: Chem. Abstr., **114**, 94606j (1991).

m.p. and Spectra (NA).

4. TETRAHYDROXALATED KETONE

1-[3-(3,6-Dihydroxy-2-methylbenzoyl)-2,4-dihydroxyphenyl]ethanone

[115834-34-9] $C_{16}H_{14}O_6$ mol.wt. 302.28

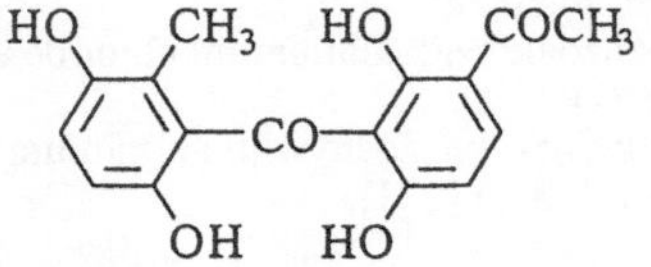

HO   CH₃   HO   COCH₃

Synthesis

-Not yet described.

Isolation from natural source

-This ketone, named baishouwubenzophenone, is one of constituents of Baishouwu. The botanical source of this one is chiefly the tuber of *Cynanchum auriculatum* Royle ex Wight (Asclepiadaceae). -Refer to: Chem. Abstr., **109**, 79560h (1988)*.

m.p. (NA);
^{1}H NMR*, ^{13}C NMR*, IR*, UV*, MS*.

Chapter 6. Phenols with two or several benzoyl groups
(Class of METHANONES)

1. MONOHYDROXYLATED KETONES

Symmetrical ketones

(5-Hydroxy-1,3-phenylene)bis[phenylmethanone

$C_{20}H_{14}O_3$ mol.wt. 302.33

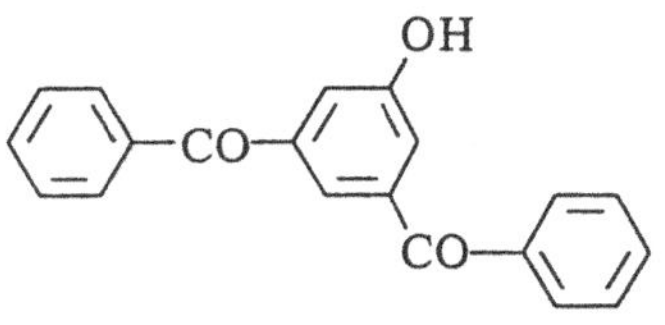

Synthesis

-Preparation by diazotization of 3,5-dibenzoylaniline, followed by hydrolysis of the resulting diazonium salt [359].

m.p. 135° [359];   Spectra (NA).

(2-Hydroxy-5-methyl-1,3-phenylene)bis[phenylmethanone

[77347-19-4] $C_{21}H_{16}O_3$ mol.wt. 316.36

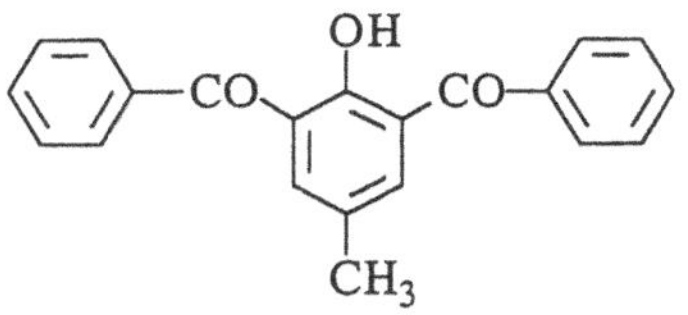

Syntheses

-Preparation by Fries rearrangement of 2-(benzoyl-oxy)-5-methylbenzophenone with aluminium chloride at 130-140° for 2 h (60-65%) [758].
-Preparation by reaction of benzoyl chloride with p-cresol in the presence of aluminium chloride in nitrobenzene at 80° for 8 h (30%) [852].

-Preparation by reaction of benzotrichloride,
*with 2-hydroxy-5-methylbenzophenone in the presence of aluminium chloride in nitrobenzene at 70° for 2 h [915], (59%) [974], (54%) [538];
*with p-cresol in the presence of aluminium chloride in carbon disulfide at 0° for 2 h (4%) [974].
-Also refer to: [537, 853, 972, 973].

m.p. 166-166°5 [974], 164°2-165°6 [538], 163-164° [915], 160° [852];
^{1}H NMR [852], IR [852], UV[852].

(2-Hydroxy-4,6-dimethyl-5-nitro-1,3-phenylene)bis[phenylmethanone

[85450-69-7] $C_{22}H_{17}NO_5$ mol.wt. 375.38

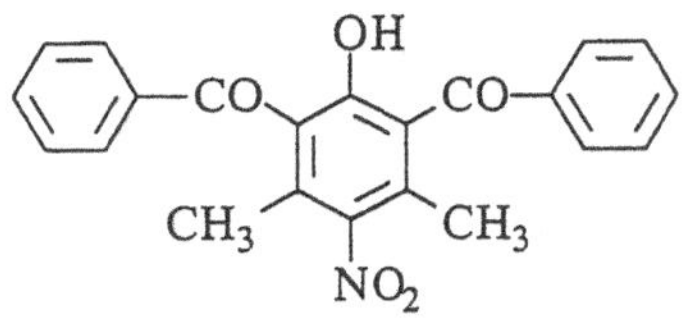

Synthesis

-Preparation by reaction of nitromethane with 1,3-dibenzoyl-4,6-dimethylpyrone in the presence of potassium tert-butoxide in tert-butanol at 60° for 4 h (77%) [392].

m.p. 208-210° [392];   Spectra (NA).

(5-Ethyl-2-hydroxy-1,3-phenylene)bis[phenylmethanone

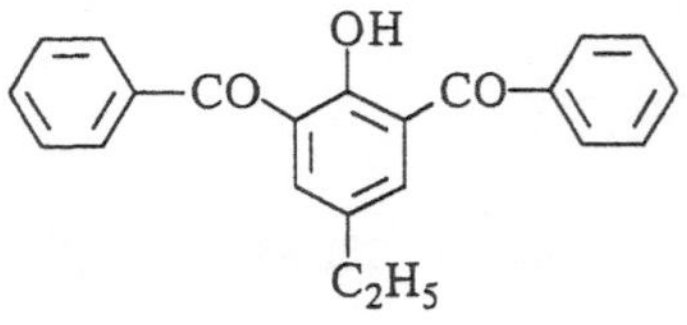

$C_{22}H_{18}O_3$ mol.wt. 330.38

Synthesis

-Preparation by reaction of benzotrichloride with 5-ethyl-2-hydroxybenzophenone in the presence of aluminium chloride in nitrobenzene at 70° for 2 h [538].
-Also refer to: [537].

m.p. and Spectra (NA).

(2-Hydroxy-4,6-dimethoxy-1,3-phenylene)bis[phenylmethanone

[197169-08-7] $C_{22}H_{18}O_5$ mol.wt. 362.38

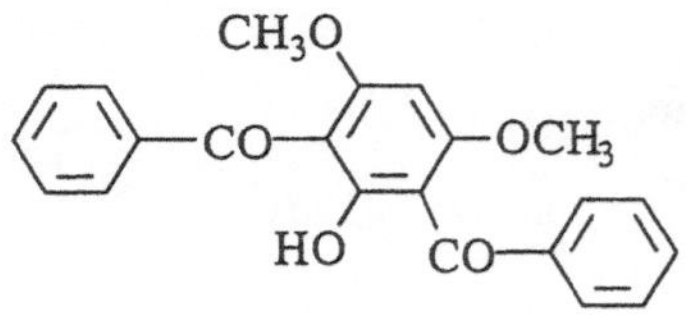

Synthesis

-Preparation from 1,3,5-trimethoxybenzene and benzoyl chloride by Friedel-Crafts acylation reaction.
-Refer to: Chem. Abstr., **127**, 292966y (1997).

m.p. and Spectra (NA).

(2-Hydroxy-4,6-dimethoxy-1,3-phenylene)bis[phenylmethanone
or
(6-Hydroxy-2,4-dimethoxy-1,3-phenylene)bis[phenylmethanone

$C_{22}H_{18}O_5$ mol.wt. 362.38

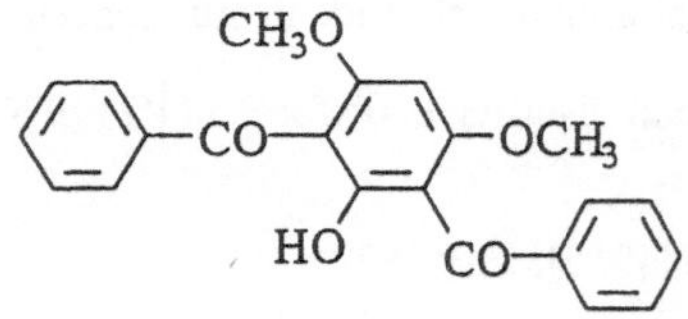

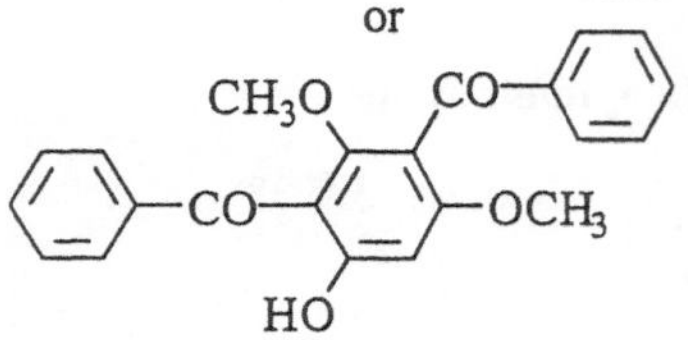

Synthesis

-Preparation by saponification of 2,4- or 2,6-di-benzoyl- 3,5-dimethoxyphenyl benzoate (SM) with potassium hydroxide in ethanol for 3 h at 100° in a sealed tube (90%) [425]. SM was obtained by Friedel-Crafts acylation of 3,5-dimethoxyphenyl benzoate with benzoyl chloride in the presence of zinc chloride in refluxing benzene for 45 min (70%, m.p. 194°). SM can also be obtained by Friedel-Crafts acylation of hydrocotoin (2-hydroxy-4,6-dimethoxy-benzophenone) with benzoyl chloride in the presence of zinc chloride in refluxing benzene for 90 min (72%, m.p. 194°).
N.B.: K salt : [425].

m.p. 170° [425];   Spectra (NA).

(5-Amino-2-hydroxy-4,6-dimethyl-1,3-phenylene)bis[phenylmethanone

[85450-78-8] $C_{22}H_{19}NO_3$ mol.wt. 345.40

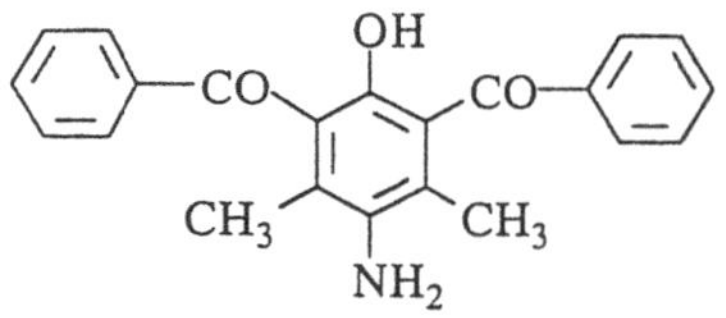

Synthesis

-Obtained (poor yield) by catalytic hydrogenation of
3-benzoyl-2-hydroxy-4,6-dimethyl-5-nitro-
benzophenone in the presence of 10% Pd/C at 40°
for 3 days (10%) [392].

m.p.   208-211° [392];   Spectra (NA).

(5-Butyl-2-hydroxy-1,3-phenylene)bis[phenylmethanone

$C_{23}H_{20}O_3$ mol.wt. 344.41

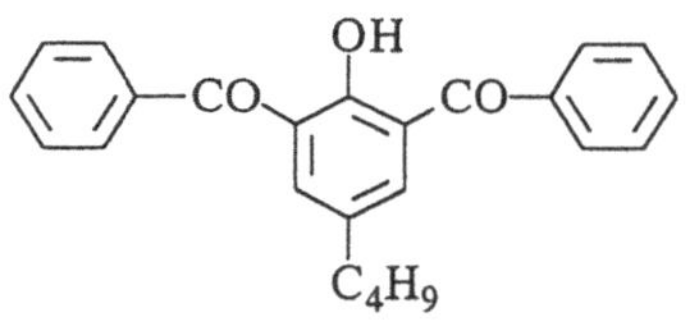

Synthesis

-Preparation by reaction of benzotrichloride with
5-butyl-2-hydroxybenzophenone in the presence of
aluminium chloride in nitrobenzene at 70° for 2 h
[538].

m.p. and Spectra (NA).

(2,4-Diethoxy-6-hydroxy-1,3-phenylene)bis[phenylmethanone
or
(4,6-Diethoxy-2-hydroxy-1,3-phenylene)bis[phenylmethanone

$C_{24}H_{22}O_5$ mol.wt. 390.44

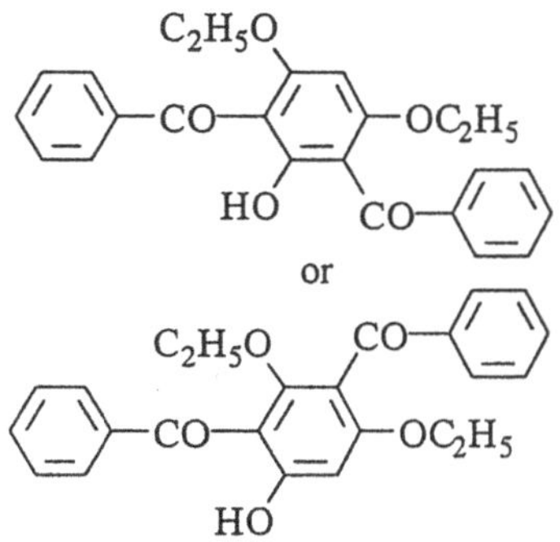

Synthesis

-Preparation by saponification of 2,4- or 2,6-di-
benzoyl-3,5-diethoxyphenyl benzoate (SM) with
potassium hydroxide in ethanol for 3 h at 100° in a
sealed tube (90%). SM was obtained by Friedel-Crafts
acylation of 3,5-diethoxyphenyl benzoate (m.p. 84°)
with benzoyl chloride in the presence of zinc chloride
in refluxing benzene for 45 min (72-81%, m.p. 163-
164°) [425].

m.p. 156° [425];   Spectra (NA).

[5-(Acetyloxy)-2-hydroxy-4,6-dimethoxy-1,3-phenylene]bis[(2,5-dimethoxyphenyl)-methanone

[129168-55-4] $C_{28}H_{28}O_{11}$ mol.wt. 492.53

Synthesis

-Obtained (poor yield) by reaction of
2,5-dimethoxybenzoic acid with 2,6-di-
methoxy-1,4-hydroquinone diacetate in
the presence of trifluoroacetic anhydride
for two weeks at r.t. (3%) [845].

m.p. 114-115° [845];   MS [845].

Asymmetric ketones

(2-Benzoylphenyl)(2-hydroxyphenyl)methanone

[14596-74-8] $C_{20}H_{14}O_3$ mol.wt. 302.33

Syntheses

-From photoxide of 9-phenylanthracen (SM1):
*SM1 in dilute acetic acid under nitrogen during 3 h
isomerises into 1-(2-hydroxyphenyl)-3-phenylisobenzofuran
(SM2). In this mixture, SM2, by self-oxidation in the
presence of air and sunlight during 3 days gave the
2-(2-hydroxybenzoyl)benzophenone (22%) [1124].

*Also obtained (poor yield) by action of sulfuric acid with SM1 in acetic acid/acetic anhydride at r.t.
for 5 min (8%) [1124].
*Also obtained (poor yield) from SM1 in refluxing toluene (at 110°) or in o-dichlorobenzene (at
180°) for 3 h under nitrogen (2 to 5%, yields respectively). The reaction carried out in the presence
of N-maleimide in refluxing toluene in the same conditions gave a 3% yield [1125].
-From 1-(2-hydroxyphenyl)-3-phenylisobenzofuran (SM2):
*Obtained by self-oxidation of SM2 in chloroformic solution in the presence of air and sunlight
during 3 days (quantitative yield) [1124].
*Also obtained by action of sulfuric acid with SM2 in acetic acid/acetic anhydride for 5 min (51%)
[1124].
*Also obtained by action of N-methylmaleimide with SM2 in methylene chloride at 0° (20%)
[1124].
-From photoxide of 9-methoxy-10-phenylanthracene (SM3): Obtained by action of concentrated
sulfuric acid (d = 1.83) with SM3 in acetone for 24 h at r.t. (25%) [1126].
-From 3,3-diphenyl-1,2-indandione (SM4): Obtained by irradiation of SM4 in the presence of air
in benzene during 70 h at r.t. (10%) [1122, 1123].
-From 10-phenylanthrone (SM5): Obtained by irradiation of SM5 in the presence of air in benzene
for 2.5 h at r.t. (20%) [1122], (30%) [1123].
-From 10-phenylanthronyl hydroperoxide (SM6): Obtained by photochemical degradation of SM6
in benzene for 5.5 h (29%) [1122]. Also obtained by chemical degradation of SM6 with sulfuric
acid in acetone for 1 to 1.5 h (21%) [1122].
-From photoxide of 1,3-diphenylisobenzofuran (SM7): Obtained by isomerization of SM7 in acetic
acid solution [1130].

m.p.   139-140° [1122, 1123, 1124, 1125], 138-139° [1126];
^{1}H NMR [1126], IR [1122, 1123, 1126], UV [1122].

(2-Benzoylphenyl)(3-hydroxyphenyl)methanone

[57436-75-6] $C_{20}H_{14}O_3$ mol.wt. 302.33

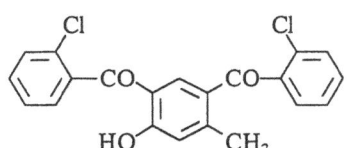

Synthesis

-Obtained by isomerization of photoxide of 1,3-diphenyl-
isobenzofuran, *via* an arene oxide formation, in acetic acid
solution [1130].

m.p. 135-137° [1130];   Spectra (NA).

(4-Hydroxy-1,3-phenylene)bis[phenylmethanone

[2589-81-3] $C_{20}H_{14}O_3$ mol.wt. 302.33

Syntheses

-Preparation by Friedel-Crafts acylation,
*of phenol with benzoyl chloride in the presence of
aluminium chloride at 200° for 15 min (69%) [864];
*of 4-hydroxybenzophenone with benzoyl chloride in the
presence of aluminium chloride from 100° up to 150°
[1066].
-Also obtained by Fries rearrangement of 4-benzoyloxybenzophenone with aluminium chloride at
160-165° for 1 h [169].

m.p. 105-106° [169], 105° [1066], 101° [864], 98° [608];
1H NMR (Sadtler: standard n° 20032 M) [610],
IR (Sadtler: standard n° 47041) [864], UV [607, 610, 864];
pK_a [608];   TLC [377].

(4-Hydroxy-6-methyl-1,3-phenylene)bis[(2-chlorophenyl)methanone

[147167-72-4] $C_{21}H_{14}Cl_2O_3$ mol.wt. 385.25

Syntheses

-Preparation by reaction of o-chlorobenzoic with
2'-chloro-2-hydroxy-4-methylbenzophenone in the
presence of phosphoric acid and phosphorous
pentoxide at 70° for 20 h (76%) [1249].
-Also obtained by reaction of o-chlorobenzoic acid
with m-cresol in the presence of phosphoric acid and phosphorous pentoxide at 130° for 2 h
(15%) [1249].
-Also obtained by Fries rearrangement of m-cresyl o-chlorobenzoate with phosphoric acid and
phosphorous pentoxide at 130° for 1 h (16%) [1249].
-Also obtained by isomerization of 2'-chloro-4-hydroxy-2-methylbenzophenone in the presence of
phosphoric acid and phosphorous pentoxide at 130° for 4 h (15%) [1249].

m.p. 105-107° [1249];   1H NMR [1249], IR [1249].

2. DI- AND POLYHYDROXYLATED KETONES

Symmetrical ketones

(4,6-Dihydroxy-1,3-phenylene)bis[(2,4-dichlorophenyl)methanone]

[13340-61-9] $C_{20}H_{10}Cl_4O_4$ mol.wt. 456.11

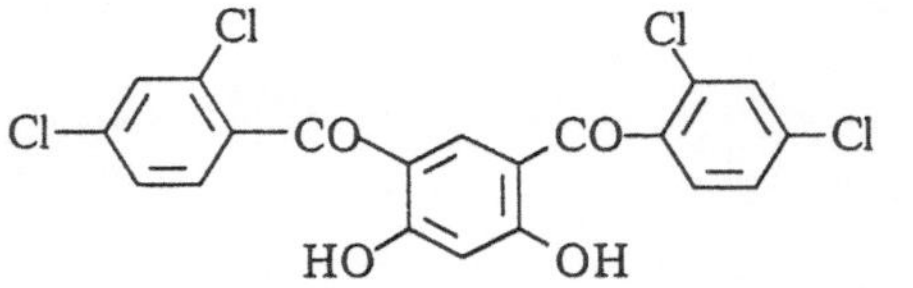

Synthesis

-Preparation by Friedel-Crafts acylation of resorcinol with 2,4-dichlorobenzoyl chloride in the presence of trifluoromethanesulfonic acid and methanesulfonic acid, *via* a Fries rearrangement,

*without solvent, at 180° for 3 h (92%) [1434];
*in o-dichlorobenzene at 100° for 12 h (76%) [1434].
-Also refer to: [1435].

m.p.   209° [1434]; ^{13}C NMR [1434].

(5-Fluoro-2-hydroxy-1,3-phenylene)bis[(5-fluoro-2-hydroxyphenyl)methanone]

[78563-16-3] $C_{20}H_{11}F_3O_5$ mol.wt. 388.30

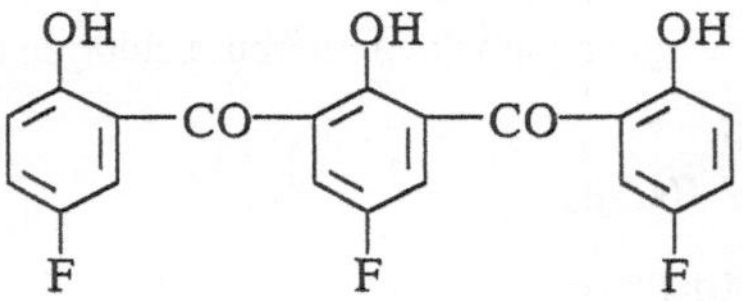

Synthesis

-Preparation by treatment of the corresponding triacetate (SM) with 15% aqueous sodium hydroxide solution (90%). SM was obtained by oxidation of the methylene groups in 1-acetoxy-2,6-bis (2'-acetoxy-5'-fluoro-α-tolyl)-4-fluorobenzene to carbonyl groups by means of chromium trioxide in acetic anhydride (97%) [526].

m.p.   154-155° [526];   IR [526], MS [526].

(4,6-Dihydroxy-1,3-phenylene)bis[(2-chlorophenyl)methanone]

[152383-56-7] $C_{20}H_{12}Cl_2O_4$ mol.wt. 387.22

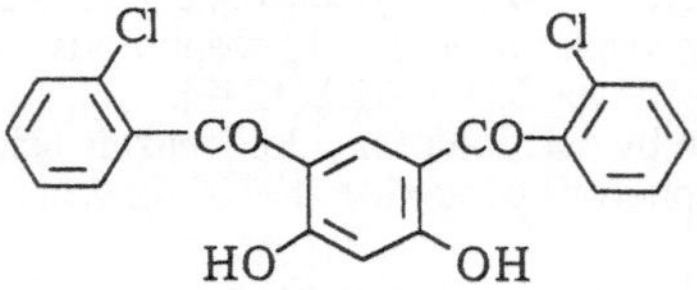

Synthesis

-Preparation by reaction of o-chlorobenzoic acid with resorcinol diacetate in the presence of hexafluoro-isopropylsulfonic acid in methanesulfonic acid first at 160° for 1 h, then at 180° for 12 h [1434].
-Also refer to: [1435].

m.p. (NA); ^{13}C NMR [1434].

(4,6-Dihydroxy-5-nitro-1,3-phenylene)bis[phenylmethanone

[102160-16-7] $C_{20}H_{13}NO_6$ mol.wt. 363.33

Syntheses

-Obtained by condensation of benzoyl chloride with
2-nitroresorcinol in the presence of aluminium chloride
in nitrobenzene at 100° [45, 321].
-Preparation by Fries rearrangement of 2-nitro-
resorcinol dibenzoate with aluminium chloride in
nitrobenzene at 100-110° for 3 h (40%) [45].
-Preparation by nitration of 4,6-dibenzoylresorcinol with nitric acid in a sulfuric acid and acetic acid
solution at 0° for 1 h [45].
-Also refer to: [356].

m.p.   215-216° [321], 215° [45];   Spectra (NA).

(2,3-Dihydroxy-1,4-phenylene)bis[phenylmethanone

[31709-42-9] $C_{20}H_{14}O_4$ mol.wt. 318.33

Syntheses

-Obtained by hydrogenolysis of 3,6-diphenyl-
benzo[1,2-d:4,3-d']diisoxazole (SM) (m.p. 206-207°)
in the presence of Raney nickel at r.t., and treating
the intermediate with 20% sulfuric acid (21%) [166].
SM was prepared through a four-
steps synthesis: first, reaction of benzonitrile oxide with 1,3-cyclohexadiene in boiling ethyl ether.
Second, bromination of the 5,10-diphenyl-3,12-dioxa-4,11-diazatricyclo[7,3,0,0 ^{2,6}]dodeca-
4,10-diene (m.p. 186-187°) obtained with N-bromosuccinimide in carbon tetrachloride.
After which, dehydrohalogenation of the intermediate compound in boiling triethylamine and
aromatization with N-bromosuccinimide in carbon tetrachloride.
-Preparation by hydrolysis of 2,3-dibenzoyloxy-1,4-dibenzoylbenzene (SM) in concentrated
sulfuric acid at r.t. for 15 min (75%) [358]. SM (m.p. 144°) was obtained by oxidation of
2,3,6,7-tetraphenylbenzo[2,1-b:3,4-b']difuran (o-benzotetraphenyldifurfuran) with chromium
trioxide in boiling acetic acid for 75 min.

m.p.   159°5 [358], 156-158° [166];   IR [166].

(4,6-Dihydroxy-1,3-phenylene)bis[phenylmethanone

[3088-15-1] $C_{20}H_{14}O_4$ mol.wt. 318.33

Syntheses

-Preparation by saponification of 4,6-di-(benzoyloxy)-
1,3-dibenzoylbenzene (SM) with potassium hydroxide
in refluxing ethanol [356, 367]. The keto ester SM was
obtained by acylation of resorcinol dibenzoate with
benzoyl chloride in the presence of zinc chloride
at 100-120° during several days [367].
-Preparation by Fries rearrangement of resorcinol dibenzoate,
*with aluminium chloride [45];
*with zinc chloride (25%) [798].

-Also obtained by photo-Fries rearrangement of resorcinol dibenzoate in benzene for 8 h under
nitrogen (15%) [1254].
-Also refer to: [483, 732, 1248].

    oily liquid [1254]; b.p. (NA). This product is impure or in a metastable state.
    m.p.   149-150° [356], 149° [367], 145° [1406]; ^{1}H NMR [1254], UV [1406].

1,2-Phenylenebis[(2-hydroxyphenyl)methanone

[119838-11-8]   C$_{20}$H$_{14}$O$_4$ mol.wt. 318.33

Synthesis

-Preparation by demethylation of bis(2-methoxybenzoyl)-
  benzene with 47% hydrobromic acid at reflux for 5 h (95%)
  [811].

m.p. 160-161° [811];
^{1}H NMR [811], IR [811], MS [811].

1,3-Phenylenebis[(4-hydroxyphenyl)methanone

[5436-05-5]   C$_{20}$H$_{14}$O$_4$ mol.wt. 318.33

Syntheses

-Preparation by demethylation of 1,3-di-(p-methoxybenzoyl)-
  benzene (SM) with hydrobromic acid (d = 1.49) in refluxing
  acetic acid for 6 h (68%) [1428] or for 10 h [1364]. SM was
  obtained by Friedel-Crafts acylation of anisole with
  isophthaloyl chloride [1364].
-Preparation by Friedel-Crafts acylation of anisole with
  isophthaloyl chloride in the presence of aluminium chloride
in boiling carbon disulfide (major product) [1428], (<75%) [186].
-Preparation by Fries rearrangement of diphenyl isophthalate with aluminium chloride at 185-195°
for 25 min [186].
-Obtained by action of isophthalic acid with phenol in the presence of trifluoromethanesulfonic acid,
first at 50° for 23 h, then at 70° for 4 h (50%) [1138].
-Also refer to: [323, 869, 912, 1175, 1421].

    m.p.   215° [1428], 207-209° [186], 194° [1138]. There is discrepancy between the melting
    points. ^{1}H NMR [1138], IR [1138].

1,4-Phenylenebis[(2-hydroxyphenyl)methanone

[66832-95-9]   C$_{20}$H$_{14}$O$_4$ mol.wt. 318.33

Synthesis

-Preparation by total demethylation of 1,4-bis
  (2-anisoyl)-benzene. This one was obtained by
  condensation of 2-anisoyl magnesium bromide
  with terephthalonitrile [892].

-Also refer to: [806].

    m.p. and Spectra (NA).

1,4-Phenylenebis[(4-hydroxyphenyl)methanone]

[15517-46-1] $C_{20}H_{14}O_4$ mol.wt. 318.33

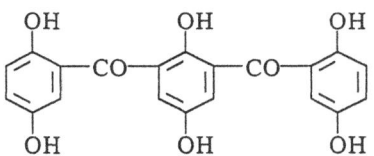

Syntheses

-Obtained by Fries rearrangement of phenyl terephthalate (diester) with aluminium chloride at 185-195° for 25 min [186].
-Preparation by Friedel-Crafts acylation of anisole (2 mol) with terephthaloyl chloride (1 mol) in the presence of aluminium chloride in refluxing carbon disulfide for 1 h [186].
-Preparation by reaction of terephthalic acid with phenol in the presence of trifluoromethanesulfonic acid at 50° for 23 h, then at 70° for 4 h (34%) [1138].
-Also refer to: Chem. Abstr., **101**, 15450z (1984).

m.p. 298-299° [186], 292° [1138];
^{1}H NMR [1138], IR [1138].

(2,5-Dihydroxy-1,3-phenylene)bis[(2,5-dihydroxyphenyl)methanone]

[78563-18-5] $C_{20}H_{14}O_8$ mol.wt. 382.33

Synthesis

-Preparation by saponification of the corresponding hexaacetate (SM) with 6 N sodium hydroxide in methanol or with 1% sodium hydroxide in aqueous solution (15% yield). SM was obtained by oxidation of the methylene groups of 1,4-diacetoxy-2,6-bis (2',5'-diacetoxy-α-tolyl)benzene to carbonyl groups by means of chromium trioxide in acetic anhydride (80% yield) [526].

m.p. >300° [526];   Spectra (NA).

(4,6-Dihydroxy-2-methyl-1,3-phenylene)bis[(2,6-dichlorophenyl)methanone]

[152383-58-9] $C_{21}H_{12}Cl_4O_4$ mol.wt. 470.13

Synthesis

-Refer to: [1435] (compound 6).

m.p. and Spectra (NA).

(4,6-Dihydroxy-5-methyl-1,3-phenylene)bis[(2,4-dichlorophenyl)methanone]

[153167-54-5] $C_{21}H_{12}Cl_4O_4$ mol.wt. 470.13

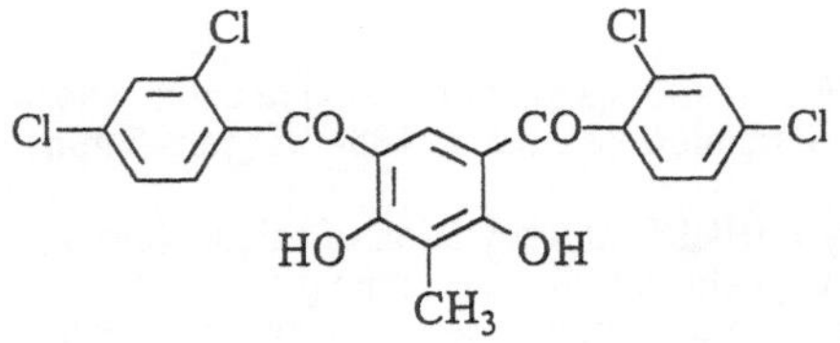

Synthesis

-Preparation by Fries rearrangement of
2-methylresorcinol 2,4-dichlorobenzoate
(diester) with fluoroalcanesulfonic acid and
alcanesulfonic acid mixture [1434].

m.p. (NA); ^{13}C NMR [1434].

(4,6-Dihydroxy-5-methyl-1,3-phenylene)bis[(2,6-dichlorophenyl)methanone]

[153167-55-6] $C_{21}H_{12}Cl_4O_4$ mol.wt. 470.13

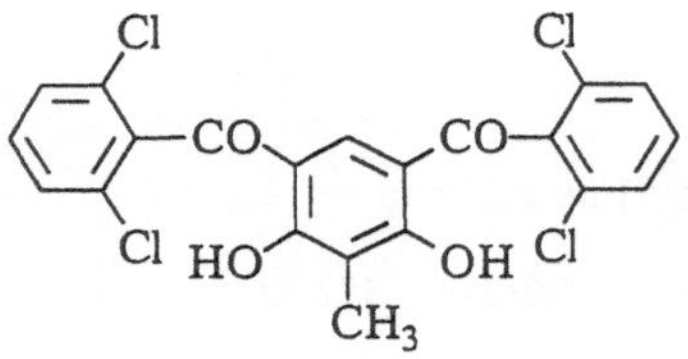

Synthesis

-Preparation by Fries rearrangement of 2-methyl-
resorcinol 2,6-dichlorobenzoate (diester) with
trifluoromethanesulfonic acid or hexafluoro-
isopropanesulfonic acid in methanesulfonic acid
[1434].

m.p. (NA); ^{13}C NMR [1434].

(4,6-Dihydroxy-5-methyl-1,3-phenylene)bis[(2-fluorophenyl)methanone]

$C_{21}H_{14}F_2O_4$ mol.wt. 368.34

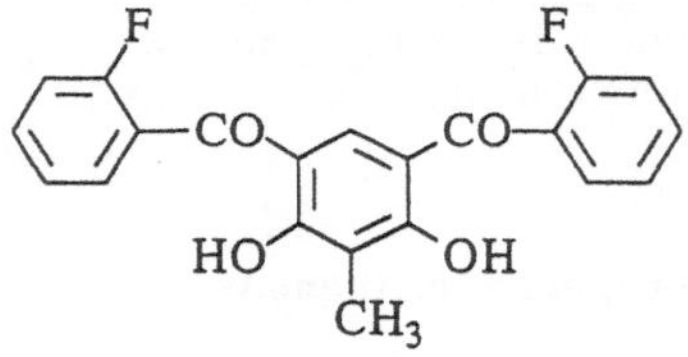

Synthesis

-Refer to: [1435] (compound 3).

m.p. and Spectra (NA).

1,4-Phenylenebis[(2-hydroxy-5-methylphenyl)methanone]

[4084-62-2] $C_{22}H_{18}O_4$ mol.wt. 346.38

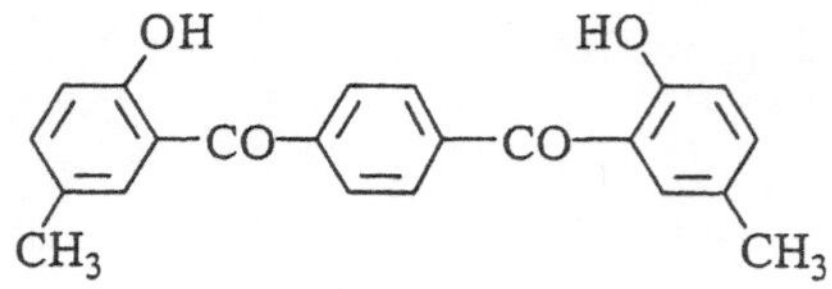

Syntheses

-Preparation by Fries rearrangement of p-tolyl
terephthalate (diester) in the presence of
aluminium chloride,
*without solvent
at 210° (90%) [1322] or at 220-230° for 15 min,
then a short time at 250° (27%) [1002],

*with solvent
in chlorobenzene at 132° for 5 h (31%) [1192] or at reflux for 4 h (68%) [806] or for 38 h (25%)
[1372]. The diester was obtained by heating p-cresol and terephthaloyl chloride at 200° for 90 min
[806].

in nitrobenzene at 100° for 3 h [936].

-Also obtained by photo-Fries rearrangement of p-tolyl terephthalate (diester) in benzene for 20 h under nitrogen (10%) [1002].

-Preparation by reaction of 1,4-bis(trichloromethyl)benzene with p-cresol in the presence of 20% or 30% sodium hydroxide, first at 50°, then 95° for 1 h (69% and 77% yields, respectively) [806].

    m.p. 190-190°5 [1372], 187-189° [1002], 165-168° [806]. There is discrepancy between the melting points. IR [806, 1372], UV [806].

1,4-Phenylenebis[(2-hydroxy-4-methoxyphenyl)methanone]

$$C_{22}H_{18}O_6 \qquad mol.wt.\ 378.38$$

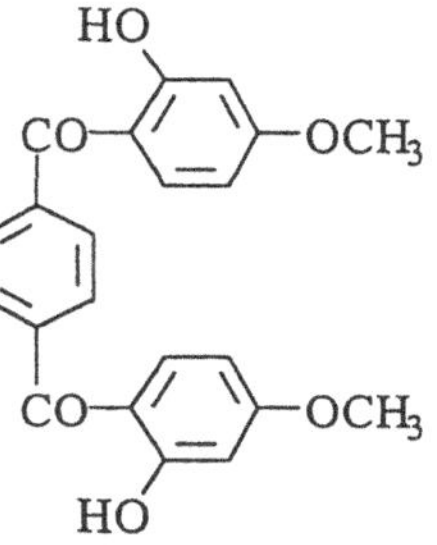

Synthesis

-Preparation by reaction of 1,4-bis(trichloromethyl)benzene with resorcinol monomethyl ether in the presence of 30% sodium hydroxide, first at 70°, then at 95° for 1 h (71%) [806].

glassy brown compound [806];
m.p. (NA);
IR [806], UV [806].

(2-Hydroxy-5-methyl-1,3-phenylene)bis[(2-hydroxy-5-methylphenyl)methanone]

[27404-61-1] $C_{23}H_{20}O_5$ mol.wt. 376.41

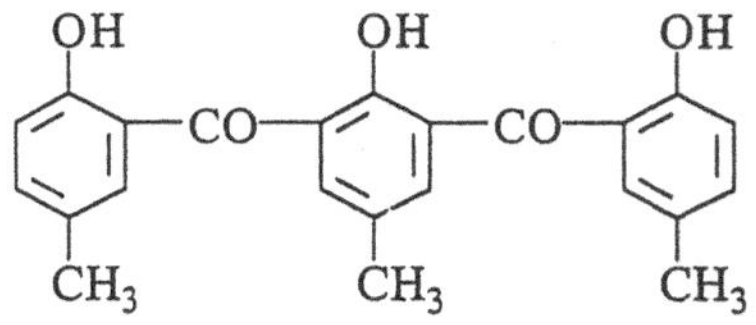

Synthesis

-Preparation by Friedel-Crafts acylation of 4-methoxytoluene with 2-methoxy-5-methyl-benzene-1,3-dicarbonyl dichloride in carbon disulfide in the presence of aluminium chloride, first at r.t. for overnight, then at reflux for 4 h (40%). The same reaction, carried out in nitrobenzene at 120° for 2 h, gave a 37% yield [671].

-Also refer to: [1294] (compound VII).

    m.p. 149° [671], 122-123° [391]. There is discrepancy between the two melting points. MS [391].

Bis[5-Chloro-3-(5-chloro-2-hydroxybenzoyl)-2-hydroxyphenyl]methanone

[78563-33-4] $C_{27}H_{14}Cl_4O_7$ mol.wt. 592.21

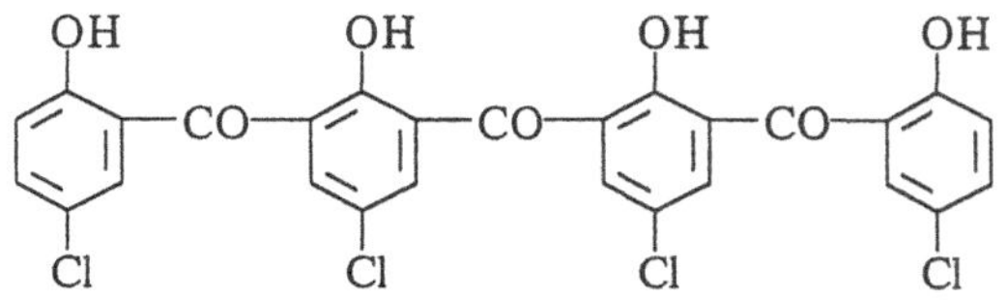

Synthesis

-Preparation by saponification of the corresponding tetraacetate with 15% aqueous sodium hydroxide solution at 40° for 1 h (90%) [526].

    m.p. 224° [526]; MS [526].

Bis[5-Fluoro-3-(5-fluoro-2-hydroxybenzoyl)-2-hydroxyphenyl]methanone

[78563-35-6] $C_{27}H_{14}F_4O_7$ mol.wt. 526.40

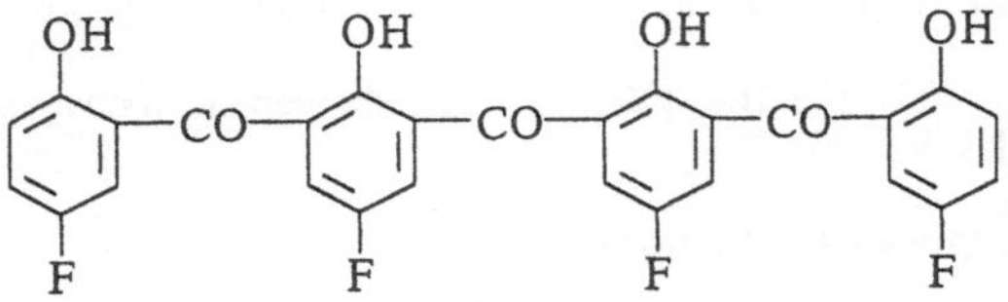

Synthesis

-Preparation by saponification of the corresponding tetraacetate with 15% aqueous sodium hydroxide solution at 40° for 1 h (95%) [526].

m.p. 165-167° [526];   IR [526], UV [526], MS [526].

(2,4-Dihydroxy-1,3,5-benzenetriyl)tris[phenylmethanone

[82-67-7] $C_{27}H_{18}O_5$ mol.wt. 422.44

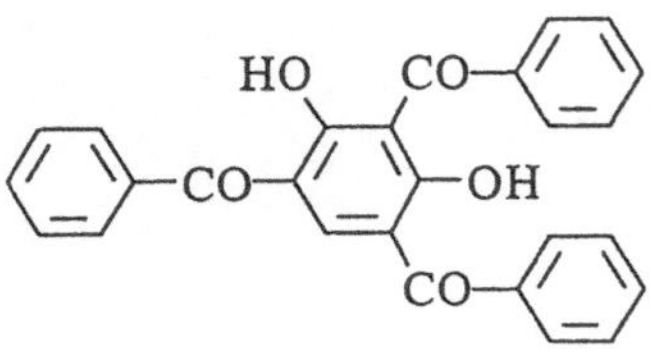

Synthesis

-Obtained by action of benzoyl chloride with resorcinol dibenzoate in the presence of aluminium chloride at 100° for 8 h, then at 150° [1066].

m.p. 185° [1066, 1406]; UV [1406].

(2,4,6-Trihydroxy-1,3,5-benzenetriyl)tris[phenylmethanone

[1818-24-2] $C_{27}H_{18}O_6$ mol.wt. 438.44

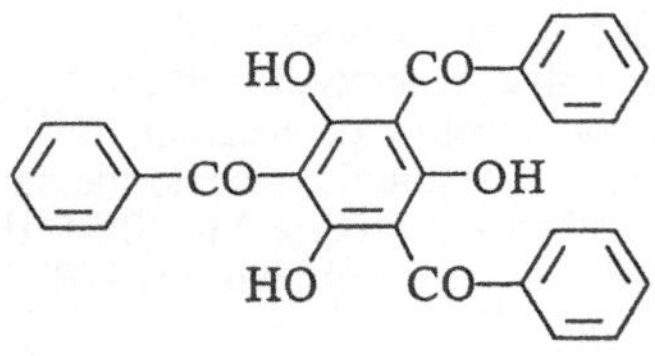

Syntheses

-Obtained by Fries rearrangement of phloroglucinol tribenzoate with aluminium chloride at 130-140° for 30 min (30%) [1142].
-Also obtained by action of benzoyl chloride with phloroglucinol in the presence of boron trifluoride at 50° (72-78%) [1454].

m.p. 185° [1142], 134-136° [1454]. There is discrepancy between the two melting points. Spectra (NA).

Bis[3-(2-Bromobenzoyl)-2-hydroxy-5-methylphenyl]methanone

$C_{29}H_{20}Br_2O_5$ mol.wt. 608.28

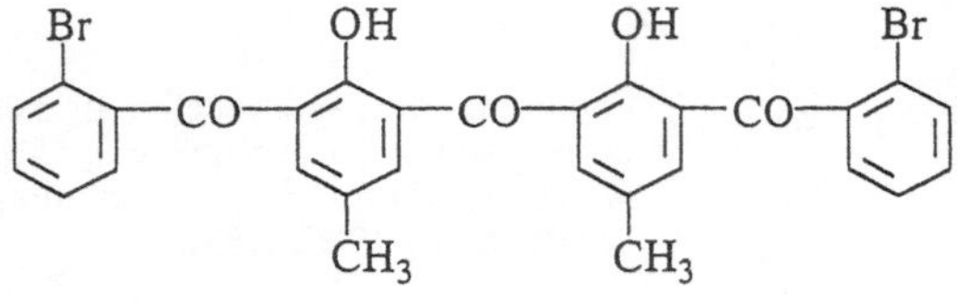

Synthesis

-Preparation by Fries rearrangement of 2,2'-di-(o-bromobenzoyloxy)-5,5'-dimethylbenzophenone with aluminium chloride without solvent for 30 min at 160° (50%) [671].

m.p. 170° [671];   Spectra (NA).

Bis(3-Benzoyl-2-hydroxy-5-methylphenyl)methanone

$C_{29}H_{22}O_5$ mol.wt. 450.49

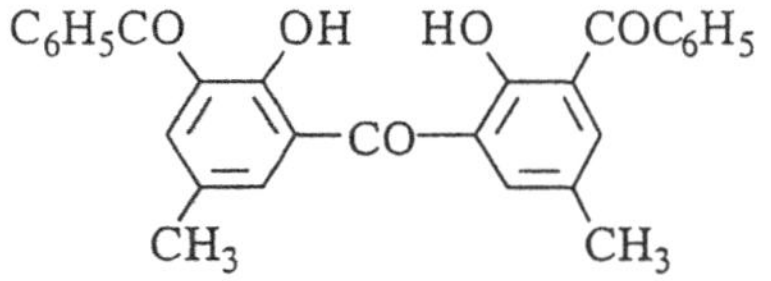

Synthesis

-Preparation by Fries rearrangement of 2,2'-di-(benzoyloxy)-5,5'-dimethylbenzophenone with aluminium chloride without solvent at 160° for 30 min (50%) [671].

m.p. 204° [671];   Spectra (NA).

Bis[3-(4-Methylbenzoyl)-2-hydroxy-5-methylphenyl]methanone

$C_{31}H_{26}O_5$ mol.wt. 478.54

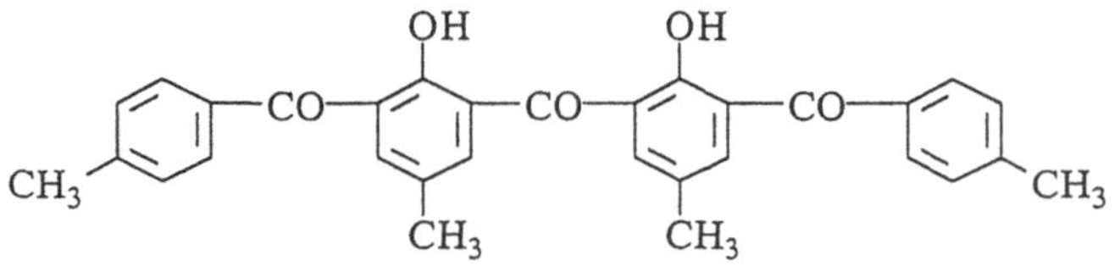

Synthesis

-Preparation by Fries rearrangement of 2,2'-di-(p-toluoyloxy)-5,5'-dimethyl-benzophenone with aluminium chloride

without solvent for 30 min at 160° (81%) [671].

m.p. 184-185° [671];   Spectra (NA).

1,4-Phenylenebis[2-hydroxy-5-(1,1,3,3-tetramethylbutyl)phenyl]methanone

$C_{36}H_{46}O_4$ mol.wt. 542.76

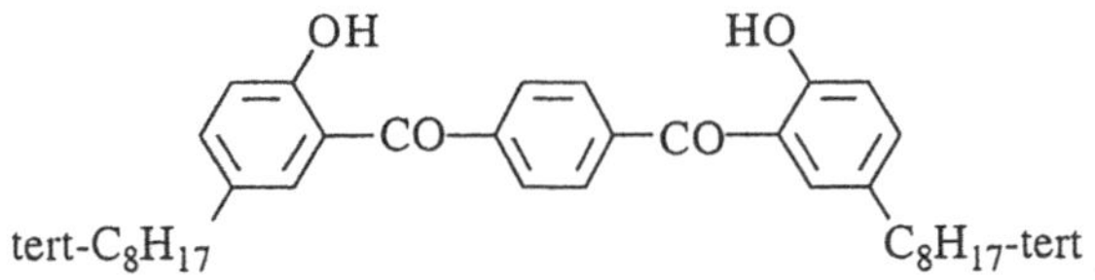

Syntheses

-Preparation by reaction of 1,4-bis-(trichloromethyl)benzene with p-(1,1,3,3-tetramethylbutyl)phenol in the presence of 30% sodium hydroxide, first at 70°,

then at 95° for 1 h (71%) [806].
-Preparation by reaction of terephthaloyl chloride with p-(1,1,3,3-tetramethylbutyl)phenol in ethylene dichloride in the presence of aluminium chloride, first at 0°, then between 5 to 10° for 7 h (58%) [806].

glassy brown compound [806];   m.p. (NA);
IR [806], UV [806].

1,4-Phenylenebis[2-hydroxy-4-(octyloxy)phenyl]methanone

$C_{36}H_{46}O_6$ mol.wt. 574.76

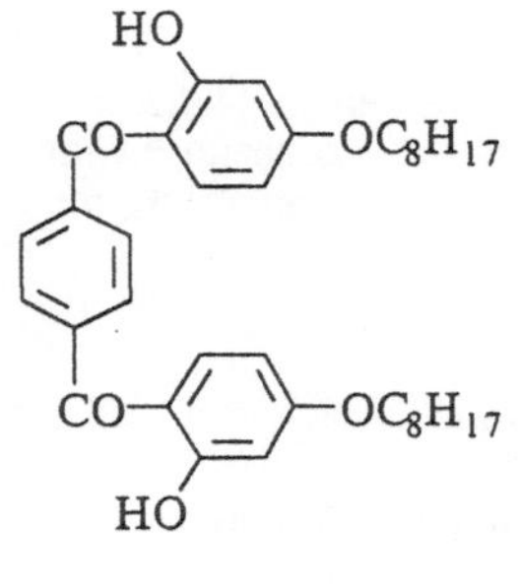

Synthesis

-Preparation by reaction of 1,4-bis(trichloromethyl)benzene with resorcinol monooctyl ether in the presence of 30% sodium hydroxide, first at 70°, then at 95° for 1 h (74%) [806].

glassy light brown compound [806];   m.p. (NA); IR [806], UV [806].

1,4-Phenylenebis[(2-hydroxy-5-dodecylphenyl)methanone

$C_{44}H_{62}O_4$ mol.wt. 654.97

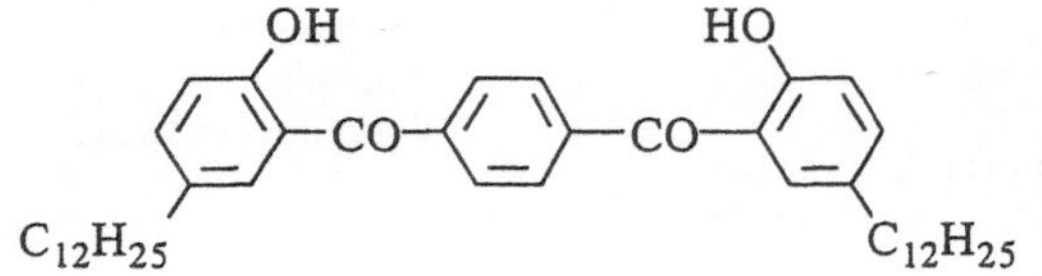

Synthesis

-Preparation by reaction of 1,4-bis-(trichloromethyl)benzene with 4-do-decylphenol in the presence of 30% sodium hydroxide, first at 70°, then at 95° for 1 h (68%) [806].

glassy light brown compound [806];   m.p. (NA); IR [806], UV [806].

1,4-Phenylenebis[2-hydroxy-4-(dodecyloxy)phenyl]methanone

$C_{44}H_{62}O_6$ mol.wt. 686.97

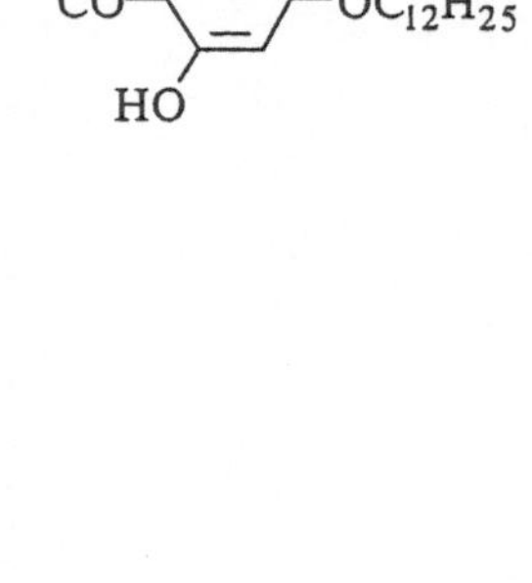

Synthesis

-Preparation by reaction of 1,4-bis(trichloromethyl)benzene with resorcinol monododecyl ether in the presence of 30% sodium hydroxide, first at 70°, then at 95° for 1 h (69%) [806].

glassy dark brown compound [806]; m.p. (NA); IR [806], UV [806].

Asymmetric ketones

(2,4-Dihydroxy-1,3-phenylene)bis[(2,4-dichlorophenyl)methanone

[153167-57-8] $C_{20}H_{10}Cl_4O_4$ mol.wt. 456.11

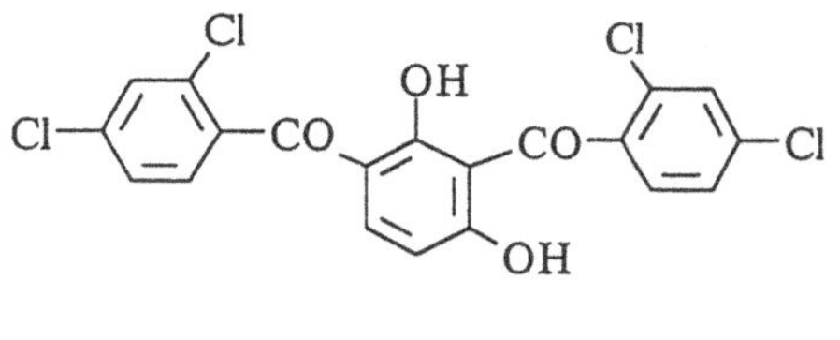

Synthesis

-Preparation by reaction of 2,4-dichloro-
benzoyl chloride with resorcinol in the
presence of trifluoromethanesulfonic acid
in methanesulfonic acid, at 180° for 3 h
(by-product) or in o-dichlorobenzene at
100° for 12 h (24%) [1434].

-Also refer to: [1435] (compound 4).

m.p. (NA); ^{13}C NMR [1434].

(3-Benzoyl-4-hydroxyphenyl)(2-hydroxyphenyl)methanone

$C_{20}H_{14}O_4$ mol.wt. 318.33

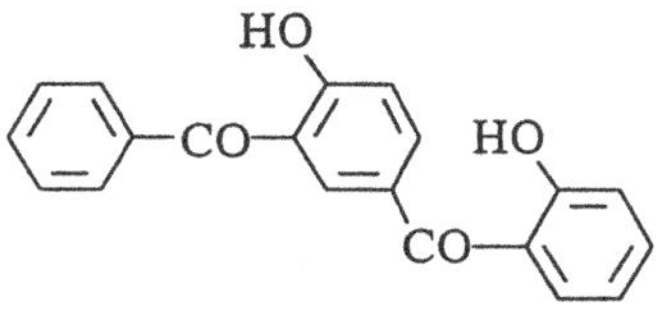

Synthesis

-Obtained (by-product) by heating a mixture of o-anisoyl
chloride, benzene and aluminium chloride between 79 to
90° for 2.25 h (14%) [1425].

m.p. 127°9-130°1 [1425]; Spectra (NA).

(5-Benzoyl-2-hydroxyphenyl)(4-hydroxyphenyl)methanone

[124208-66-8] $C_{20}H_{14}O_4$ mol.wt. 318.33

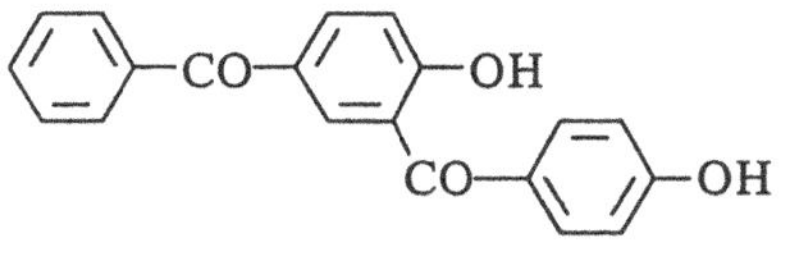

Syntheses

-Obtained by Fries rearrangement of phenyl
5-benzoyl-2-hydroxybenzoate with aluminium
chloride at 180° for 3 h (20%); the same reaction, in
the presence of phenol, gave a 36% yield [171].
-Also obtained by Fries rearrangement of phenyl
2-benzoyloxybenzoate with aluminium chloride at 180° for 3 h (10%). The same reaction, in the
presence of phenol, gave a 21% yield [171].

m.p. 162° [171]; Spectra (NA).

(2,4-Dihydroxy-1,3-phenylene)bis[phenylmethanone

[82-64-4] $C_{20}H_{14}O_4$ mol.wt. 318.33

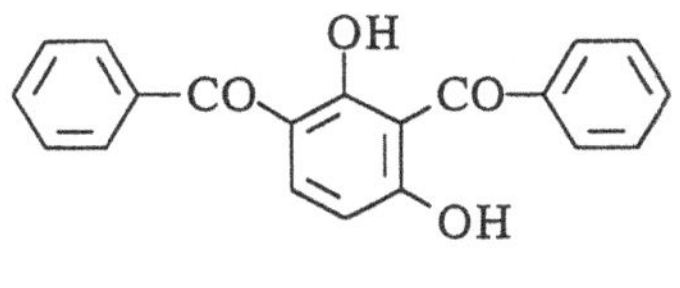

Syntheses

-Obtained by reaction of chromium trioxide with
angular meta benzotetraphenyldifurfuran in boiling
acetic acid for 1 h, followed by saponification of the
non-isolated keto ester formed with 10% sodium
hydroxide in refluxing ethanol for 1.5 h [356].

-Also obtained on decarboxylation of 2,4-dihydroxy-3,5-dibenzoylbenzoic acid [339].
-Also obtained by photo-Fries rearrangement of resorcinol dibenzoate in benzene for 8 h under
 nitrogen (10%) [1254].
-Also refer to: [629, 987] (Japanese patents).

   oil [1254]. This product is impure or in a metastable state.
   m.p.   105° [1406], 103-104° [356], 102° [339];
   ^{1}H NMR [267, 1254], UV [475, 515, 1406].

(2,5-Dihydroxy-1,4-phenylene)bis[phenylmethanone]

[97971-75-0] $C_{20}H_{14}O_4$ mol.wt. 318.33

Syntheses

-Preparation from lin-parabenzotetraphenyl-
   difurfuran by oxidation with chromium trioxide in
   refluxing acetic anhydride for >1 h, followed by
   saponification of the 2,5-dibenzoylhydroquinone
   dibenzoate so formed with potassium hydroxide in
ethanol in a water bath for 30 min (43%) [357].
-Also obtained by reaction of benzoyl chloride with hydroquinone in the presence of aluminium
 chloride at 190-200° [364, 368, 534] or at 200-205° [191] for 2 days, followed by saponification of
 the keto ester so formed (2,5-dibenzoylhydroquinone dibenzoate) with potassium hydroxide in
 ethanol, (2%) [368], (12-15%) [191].
-Also obtained by photo-Fries rearrangement of hydroquinone dibenzoate in benzene for 8 h under
 nitrogen (10%) [1254].

   m.p.   210°5-211°1 [191], 207° [364, 368], 203° [357], 172-173° [1254]. There is
   discrepancy between the melting points.
   ^{1}H NMR [1254].

(3,4-Dihydroxy-1,2-phenylene)bis[phenylmethanone]

[97971-73-8] $C_{20}H_{14}O_4$ mol.wt. 318.33

Synthesis

-Obtained by photo-Fries rearrangement of 1,2-di-
   (benzoyloxy)benzene in benzene for 8 h under nitrogen
   (28%) [1254].

   m.p. 134-136° [1254]; ^{1}H NMR [1254].

(2,4-Dihydroxy-5-hexyl-1,3-phenylene)bis[phenylmethanone]

                                              $C_{26}H_{26}O_4$ mol.wt. 402.49

Synthesis

-Preparation by reaction of benzoyl chloride with
   4-hexylresorcinol in the presence of aluminium chloride
   in nitrobenzene at 80-90° for 3 h (52%) [1406].

   m.p. 68-69° [1406];   Spectra (NA).

PART 3 - MISCELLANEOUS RELATED COMPOUNDS
(Class of METHANONES)

1. DIPHENYL DERIVATIVES

(4,4'-Dihydroxy[1,1'-biphenyl]-3,3'-diyl)bis[(2,4-dichorophenyl)methanone

[152383-57-8] $C_{26}H_{14}Cl_4O_4$ mol.wt. 532.21

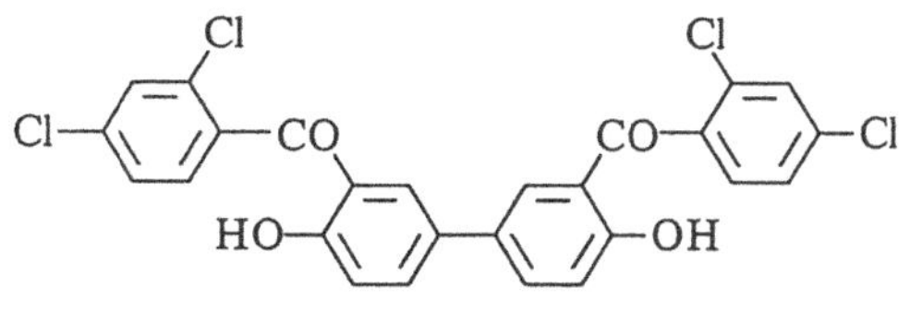

Synthesis

-Preparation by Fries rearrangement of 4,4'-dihydroxybiphenyl di-(2,4-di-chlorobenzoate) in the presence of tri-fluoromethanesulfonic acid and ethane-sulfonic acid in refluxing dichlorobenzene for 16 h (94%) [1434].

N.B.: Di-Na salt [1435].

m.p. (NA); ^{13}C NMR [1434].

(4,4'-Dihydroxy[1,1'-biphenyl]-2,2'-diyl)bis[(3-fluorophenyl)methanone

[176548-03-1] $C_{26}H_{16}F_2O_4$ mol.wt. 430.41

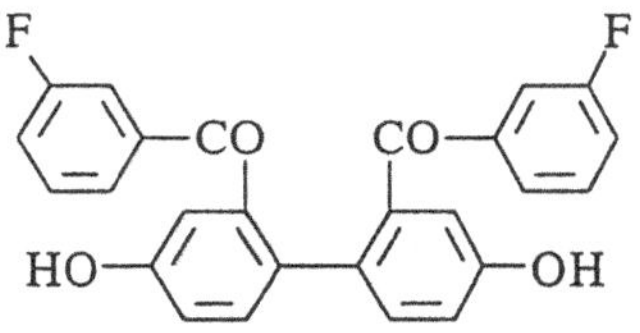

Synthesis

-Preparation by demethylation of 2,2'-bis-(m-fluoro-benzoyl)-4,4'-dimethoxybiphenyl with boron tribromide in methylene chloride, first at -70°, then at 25° for 8 h (80%) [1048].

m.p. 239-241° [1048]; ^1H NMR [1048].

(4,4'-Dihydroxy[1,1'-biphenyl]-2,2'-diyl)bis[(4-fluorophenyl)methanone

[162658-02-8] $C_{26}H_{16}F_2O_4$ mol.wt. 430.41

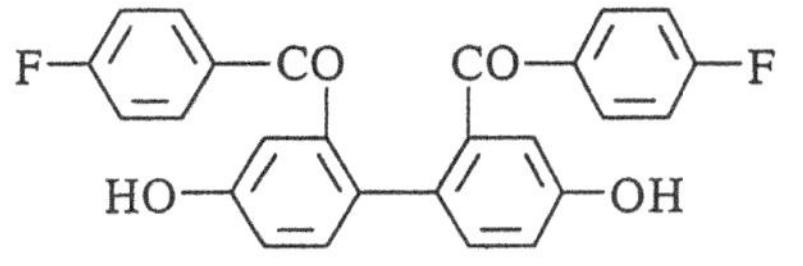

Synthesis

-Preparation by total demethylation of 2,2'-bis(p-fluorobenzoyl)-4,4'-di-methoxybiphenyl with boron tribromide in methylene chloride under nitrogen first at -70°, then at 25° for 8 h (94%) [1047, 1048].

m.p. 256-258° [1047];
^1H NMR [1047], ^{13}C NMR [1047], MS [1047].

[1,1'-Biphenyl]-4,4'-diylbis[(4-hydroxyphenyl)methanone

[106647-50-1]   C$_{26}$H$_{18}$O$_4$ mol.wt. 394.43

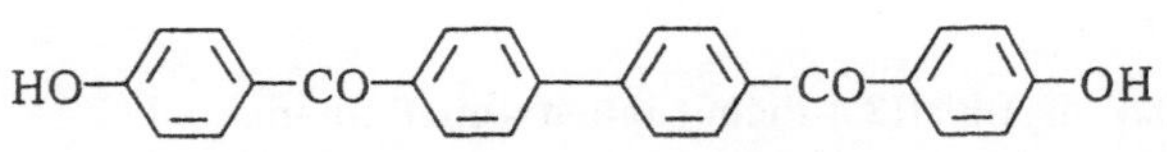

Synthesis

HO—⟨ ⟩—CO—⟨ ⟩—⟨ ⟩—CO—⟨ ⟩—OH   -Preparation by Fries
                                                                                    rearrangement of 4,4'-bis-
(phenoxycarbonyl)diphenyl in hydrofluoric acid between -10 to 0° for 20 h (96%) [789].

m.p. and Spectra (NA).

(4,4'-Dihydroxy[1,1'-biphenyl]-2,2'-diyl)bis[phenylmethanone

[162658-01-7]   C$_{26}$H$_{18}$O$_4$ mol.wt. 394.43

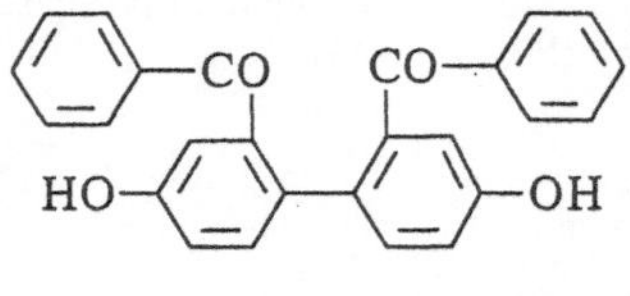

Synthesis

-Preparation by total demethylation of 2,2'-dibenzoyl-
4,4'-dimethoxybiphenyl with boron tribromide in
methylene chloride under nitrogen first at -70°, then at
25° for 8 h (83%) [1047, 1048].

m.p. (NA); ^{1}H NMR [1047].

(4,4'-Dihydroxy[1,1'-biphenyl]-3,3'-diyl)bis[phenylmethanone

[71182-85-9]   C$_{26}$H$_{18}$O$_4$ mol.wt. 394.43

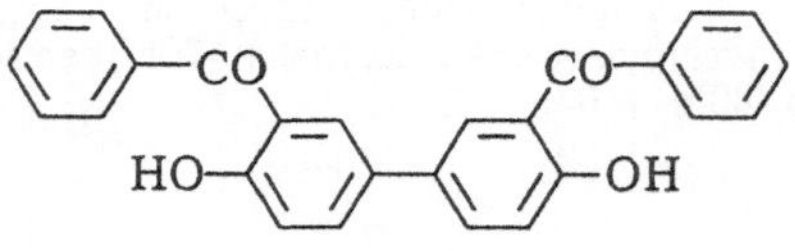

Synthesis

-Preparation by Fries rearrangement of 4,4'-bi-
phenyldiyl dibenzoate,
*with aluminium chloride in o-dichlorobenzene
during 3 h at 180° (70%) [1451] or in refluxing
chlorobenzene for 3 days (30%) [1032];
*with aluminium chloride and sodium chloride mixture, first at 140°, then at 200° for 20 min (51%)
[928].

m.p.   187°5-189°5 [1451], 184-185° [1032];
^{1}H NMR [1451], IR [1451], UV [1451].

(4,4'-Dihydroxy-6,6'-dimethoxy[1,1'-biphenyl]-3,3'-diyl)bis[phenylmethanone

[42045-62-5]   C$_{28}$H$_{22}$O$_6$ mol.wt. 454.48

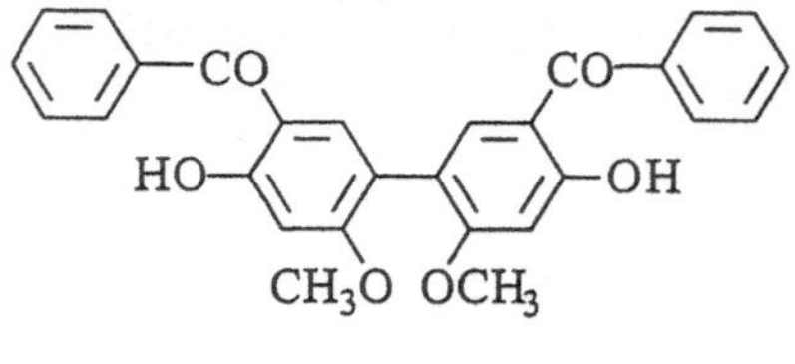

Synthesis

-Obtained by reaction of 2-hydroxy-4-methoxy-
benzophenone with lead tetraacetate in acetic acid
at 100° for 5 h (15%). The same reaction using
manganic acetate gave 7% yield [785].

m.p.   199-200° [785];
^{1}H NMR [785], IR [785], UV [785].

(4,4'-Dihydroxy-6,6'-dimethoxy[1,1'-biphenyl]-3,3'-diyl)bis[(4-methoxyphenyl)-methanone

[42045-61-4] $C_{30}H_{26}O_8$ mol.wt. 514.53

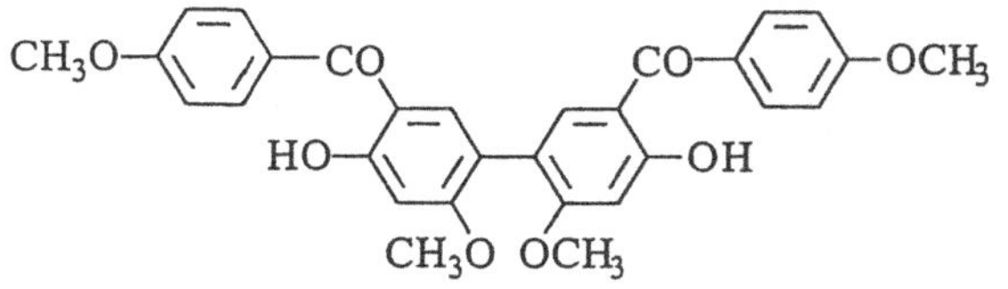

Synthesis

-Obtained by reaction of
2-hydroxy-4,4'-dimethoxybenzo-
phenone with lead tetraacetate in
acetic acid at 100° for 5 h (21%)
[933].

m.p. 232-234° [933];
^1H NMR [933], IR [933], UV [933].

(4,4'-Dihydroxy[1,1'-biphenyl]-2,2'-diyl)bis[4-(1,1-dimethylethyl)phenyl]methanone

[162658-03-9] $C_{34}H_{34}O_4$ mol.wt. 506.64

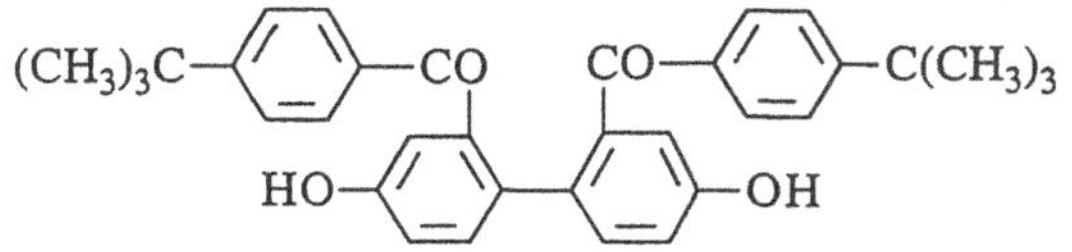

Synthesis

-Preparation by total demethylation of
2,2'-bis(p-tert-butylbenzoyl)-
4,4'-dimethoxybiphenyl with boron
tribromide in methylene chloride
under nitrogen first at -70°, then at
25° for 8 h (77%) [1047, 1048].

m.p. 236-238° [1047];
^1H NMR [1047], ^{13}C NMR [1047], MS [1047].

2. DIPHENYLMETHANE DERIVATIVES

Phenyl[2,3,4-trihydroxy-5-[(2,4,6-trihydroxyphenyl)methyl]phenyl]methanone

[138250-29-0] $C_{20}H_{16}O_7$ mol.wt. 368.34

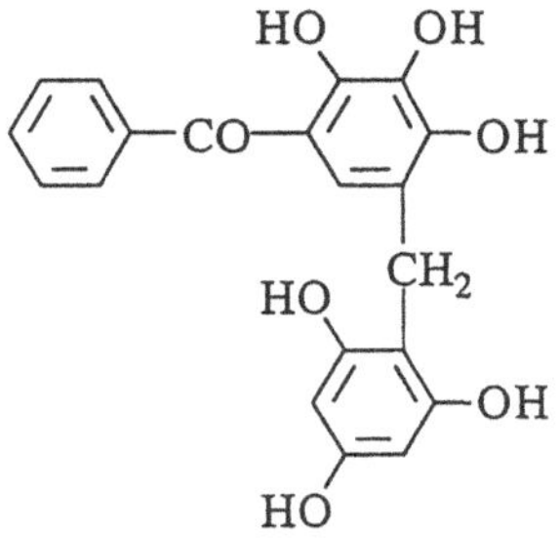

Synthesis

-Refer to: [1001, 1387, 1388].

m.p. and Spectra (NA).

Phenyl[2,3,4-trihydroxy-5-[(2-hydroxy-4-methylphenyl)methyl]phenyl]methanone

$C_{21}H_{18}O_5$ mol.wt. 350.37

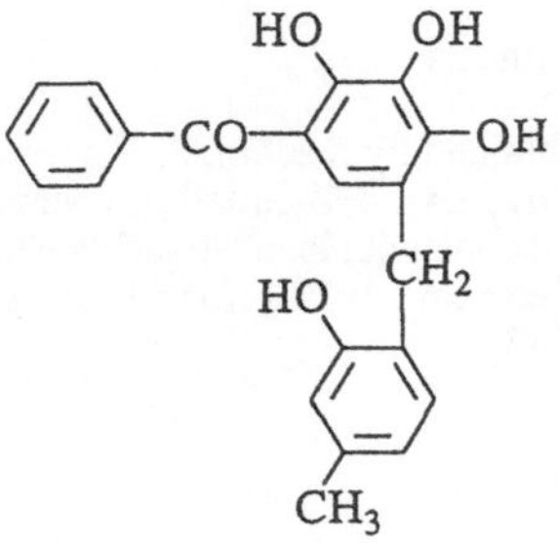

Synthesis

-Refer to: [1387].

m.p. and Spectra (NA).

Phenyl[2,3,4-trihydroxy-5-[(4-hydroxy-2-methylphenyl)methyl]phenyl]methanone

$C_{21}H_{18}O_5$ mol.wt. 350.37

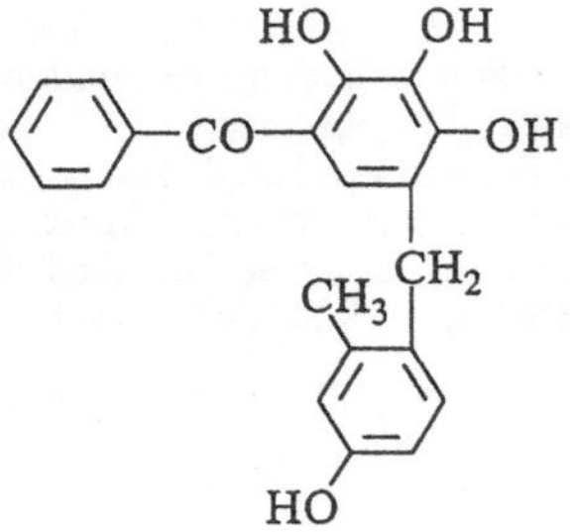

Synthesis

-Refer to: [1387].

m.p. and Spectra (NA).

[Methylenebis(2,4-dihydroxy-5-methyl-3,1-phenylene)]bis[phenylmethanone

$C_{29}H_{24}O_6$ mol.wt. 468.51

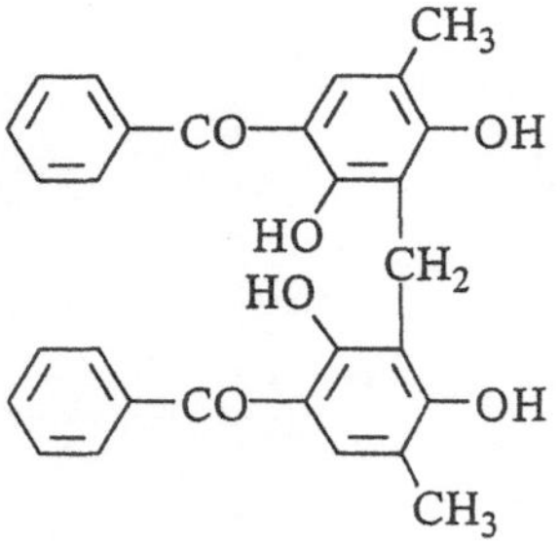

Synthesis

-Preparation by reaction of 40% aqueous formaldehyde solution with 2,4-dihydroxy-5-methylbenzophenone in ethanol in the presence of concentrated sulfuric acid for 24 h at r.t. (73%) [889].

m.p. 240° [889];   Spectra (NA).

[Methylenebis(4,6-dihydroxy-5-methyl-3,1-phenylene)]bis[phenylmethanone

$C_{29}H_{24}O_6$ mol.wt. 468.51

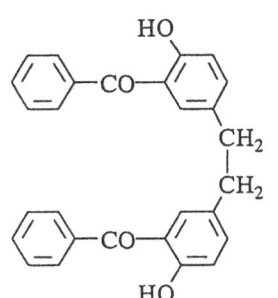

Synthesis

-Preparation by reaction of 40% aqueous formaldehyde
 solution with 2,4-dihydroxy-3-methylbenzophenone in
 ethanol in the presence of concentrated sulfuric acid for
 3 days at r.t. (68%) [889].

m.p. 207-208° [889];   Spectra (NA).

[Methylenebis(2,4,6-trihydroxy-5,1,3-benzenetriyl)]tetrakis[phenylmethanone

[98149-22-5] $C_{41}H_{28}O_{10}$ mol.wt. 680.67

Synthesis

-Preparation by condensation of 2,4-di-
 benzoylphloroglucinol with 40%
 aqueous formaldehyde solution (58 to
 64%) [1454].

m.p. 215-217° [1454];
Spectra (NA).

3. DIPHENYLETHANE DERIVATIVE

[1,2-Ethanediylbis(6-hydroxy-3,1-phenylene)]bis[phenylmethanone

[76346-16-2] $C_{28}H_{22}O_4$ mol.wt. 422.48

Synthesis

-Obtained (poor yield) by reaction of benzoyl peroxide with
 2-hydroxy-5-methylbenzophenone in refluxing chloroform
 for 6 h (<4%) [1211].

m.p. 172-173° [1211];
^{1}H NMR [1211], IR [1211], UV [1211].

4. DIPHENYLPROPANE DERIVATIVES

3-Benzoyldiphenylolpropane 4'-monobenzoate

$C_{29}H_{24}O_4$ mol.wt. 436.51

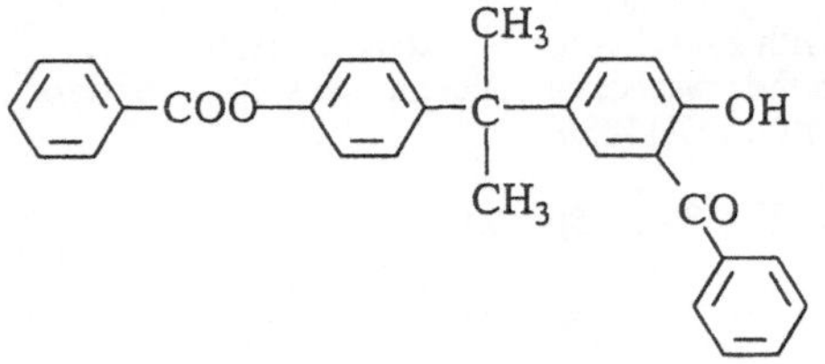

Synthesis

-Obtained by photo-Fries rearrangement of diphenylolpropane dibenzoate (Bisphenol A, dibenzoate) in benzene during 96 h under nitrogen (22%) [842].

b.p.$_{0.15}$ 220° [842];
IR [842], UV [842].

2,2-Bis(3-Benzoyl-4-hydroxyphenyl)propane

$C_{29}H_{24}O_4$ mol.wt. 436.51

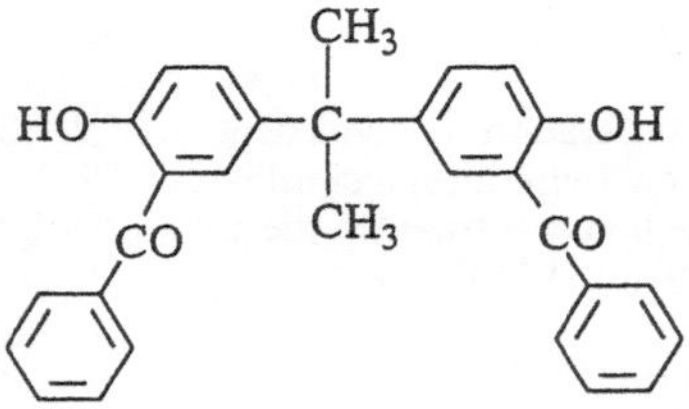

Synthesis

-Obtained by reaction of benzoyl chloride with diphenylolpropane dimethyl ether in the presence of aluminium chloride in ethylene dichloride, first at 0° for 45 min, then at r.t. overnight and at 45° for 90 min (14%) [842].

m.p. 159°5-160°5 [842];   IR [842], UV [842].

5. DIPHENYL OXIDE DERIVATIVES

(4-Chlorophenyl)[4-(4-hydroxyphenoxy)phenyl]methanone

[86405-16-5] $C_{19}H_{13}ClO_3$ mol.wt. 324.76

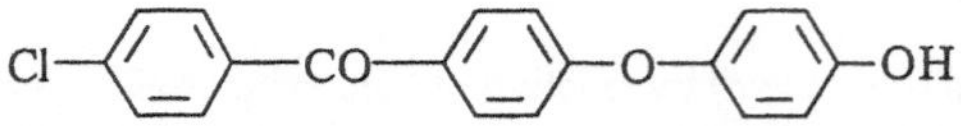

Synthesis

-Refer to: Chem. Abstr., **127**, 278690x (1997).

m.p. and Spectra (NA).

[4-(4-Hydroxyphenoxy)phenyl]phenylmethanone

$C_{19}H_{14}O_3$ mol.wt. 290.22

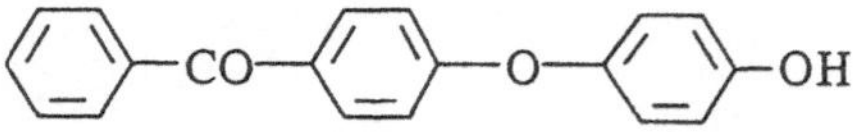

Synthesis

-Preparation by demethylation of 4-(4-methoxyphenoxy)benzophenone (m.p. 104-105°) with aluminium chloride in boiling benzene during some hours [355].

m.p.   109° [355];   Spectra (NA).

2-(4-Benzoyl-3-hydroxyphenoxy)cyclohexanone

[125426-75-7] $C_{19}H_{18}O_4$ mol.wt. 310.35

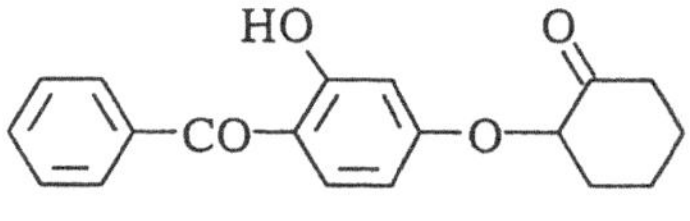

Synthesis

-Preparation by reaction of α-bromocyclohexanone with resbenzophenone in the presence of potassium carbonate in refluxing acetone for 6 h (70%) [23].

m.p. 132-133° [23]; ^{1}H NMR [23].

2-(4-Benzoyl-3-hydroxy-2-methylphenoxy)cyclohexanone

[125426-85-9] $C_{20}H_{20}O_4$ mol.wt. 324.38

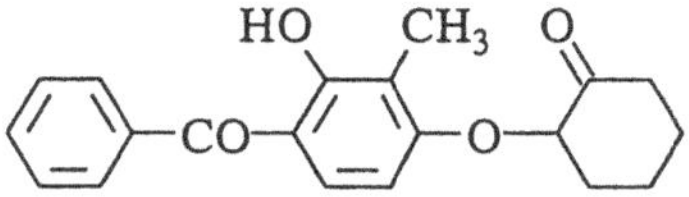

Synthesis

-Preparation by reaction of α-bromocyclohexanone with 2,4-dihydroxy-3-methylbenzophenone in the presence of potassium carbonate in refluxing acetone for 6 h [23].

m.p. 121-122° [23]; ^{1}H NMR [23].

Bis[4-(4-Hydroxyphenoxy)phenyl]methanone

$C_{25}H_{18}O_5$ mol.wt. 398.41

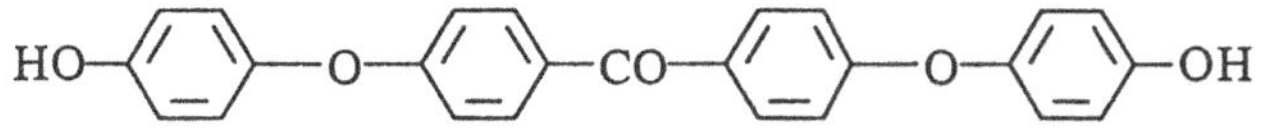

Synthesis

-Preparation by demethylation of 4,4'-(4-methoxyphenoxy)

benzophenone (m.p. 198-199°) with excess aluminium chloride in boiling benzene (30%) [355].

m.p. 214° [355];   Spectra (NA).

(Oxydi-4,1-phenylene)bis(4-hydroxyphenyl)methanone

[86432-12-4] $C_{26}H_{18}O_5$ mol.wt. 258.23

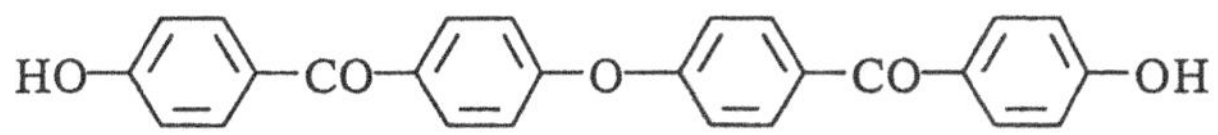

Syntheses

-Obtained by reaction of 4,4'-Oxybis(benzoic acid) with phenol in the presence

of trifluoromethanesulfonic acid at r.t. for 6 days [1138].
-Preparation by Fries rearrangement of 4,4'-bis(phenoxycarbonyl)diphenyl ether,
*with methanesulfonic acid at 120° for 2 h (70%) [787];
*with trifluoromethanesulfonic acid (high yield) [287];
*with hydrofluoric acid at -10 to 0° for 20 h (96%) [788].
-Also obtained by reaction of EKONOL(TM), an aromatic polyester as Friedel-Crafts reagent, with diphenyl ether in triflic acid solution at 25° for 18 h (98%) [287]. Similar results can be obtained using hydrofluoric acid/boron trifluoride or aluminium chloride in place of triflic acid [287].

m.p. (NA); ^{1}H NMR [1138], ^{13}C NMR [287], MS [287];
HPLC [287].

6. DIPHENYL SULFOXIDE DERIVATIVES

[Sulfinylbis(6-hydroxy-4-methoxy-3,1-phenylene)]bis[phenylmethanone]

[35839-46-4] $C_{28}H_{22}O_7S$ mol.wt. 502.54

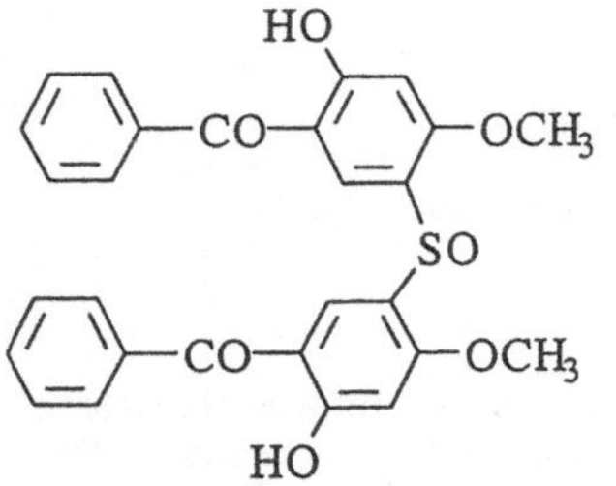

Synthesis

-Obtained by action of thionyl chloride with 2-hydroxy-
4-methoxybenzophenone in the presence of aluminium
chloride in nitrobenzene at r.t. for 24 h [914].

m.p. 202-203° [914];   Spectra (NA).

[Sulfinylbis[4-(cyclohexyloxy)-6-hydroxy-3,1-phenylene]]bis[phenylmethanone]

$C_{38}H_{38}O_7S$ mol.wt. 638.79

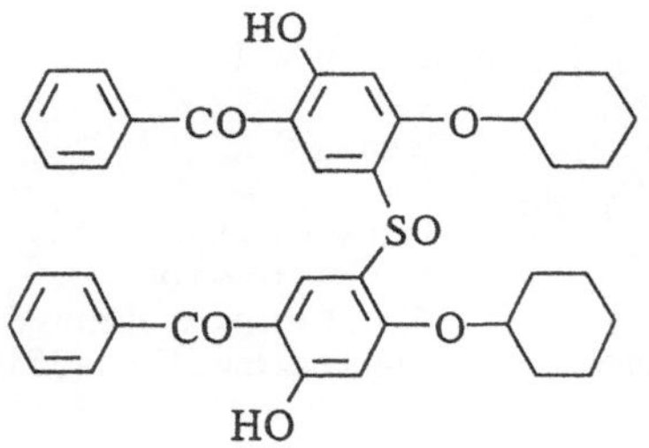

Synthesis

-Obtained by action of thionyl chloride with
2-hydroxy-4-(cyclohexyloxy)benzophenone in the
presence of aluminium chloride in nitrobenzene at r.t.
for 24 h [914].

m.p. and Spectra (NA).

[Sulfinylbis[4-(benzyloxy)-6-hydroxy-3,1-phenylene]]bis[phenylmethanone]

$C_{40}H_{30}O_7S$ mol.wt. 654.74

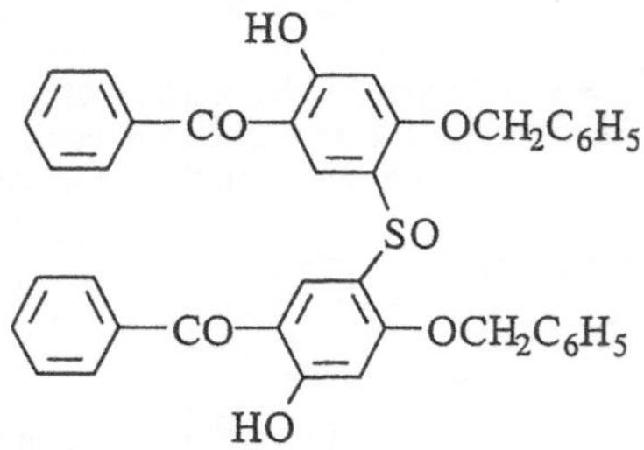

Synthesis

-Obtained by action of thionyl chloride with
2-hydroxy-4-(benzyloxy)benzophenone in the
presence of aluminium chloride in nitrobenzene at r.t.
for 24 h [914].

m.p. and Spectra (NA).

[Sulfinylbis[6-hydroxy-4-(octyloxy)-3,1-phenylene]]bis[phenylmethanone

[35839-47-5] $C_{42}H_{50}O_7S$ mol.wt. 698.92

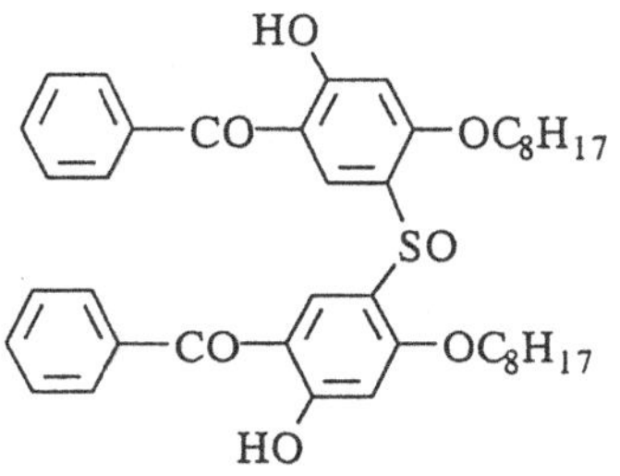

Synthesis

-Obtained by action of thionyl chloride with 2-hydroxy-4-(octyloxy)benzophenone in the presence of aluminium chloride in nitrobenzene at r.t. for 24 h [914].

m.p. 131°5-132° [914];   Spectra (NA).

7. DIPHENYL SULFONE DERIVATIVES

[Sulfonylbis(6-hydroxy-3,1-phenylene)]bis[(2,4-dichlorophenyl)methanone

[153167-56-7] $C_{26}H_{14}Cl_4O_6S$ mol.wt. 596.27

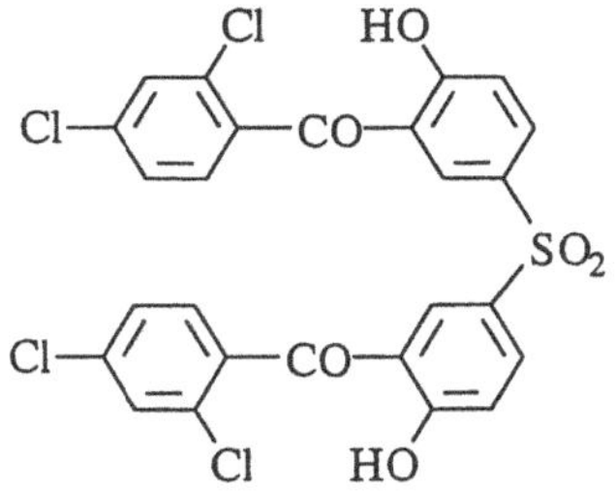

Synthesis

-Preparation by Fries rearrangement of bis(4-hydroxyphenyl)sulfone di-(2,4-dichlorobenzoate) with a mixture of fluoroalcanesulfonic acid and alcanesulfonic acid [1434].

m.p. (NA); ^{13}C NMR [1434].

[Sulfonylbis(4,6-dihydroxy-3,1-phenylene)]bis[phenylmethanone

[35839-48-6] $C_{26}H_{18}O_8S$ mol.wt. 490.49

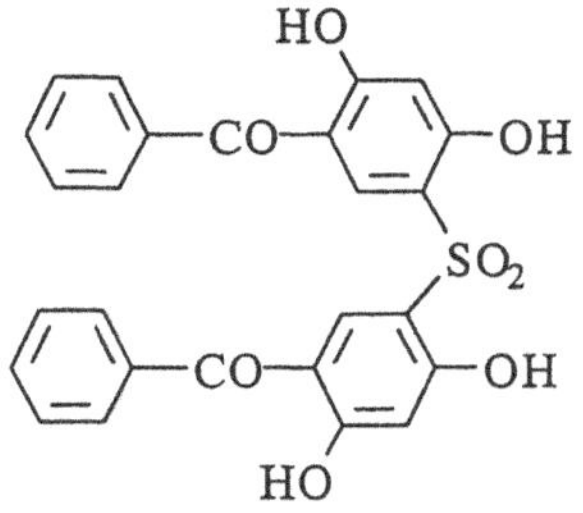

Synthesis

-Obtained by reaction of a 30% aqueous solution of hydrogen peroxide with 5,5'-thiobis(2,4-dihydroxybenzophenone) in acetic acid. The mixture was then heated on a steam bath for 6 h [914].

m.p. 217-219° [914];   Spectra (NA).

[Sulfonylbis(6-hydroxy-4-methoxy-3,1-phenylene)]bis[phenylmethanone

[35698-04-5] $C_{28}H_{22}O_8S$ mol.wt. 518.54

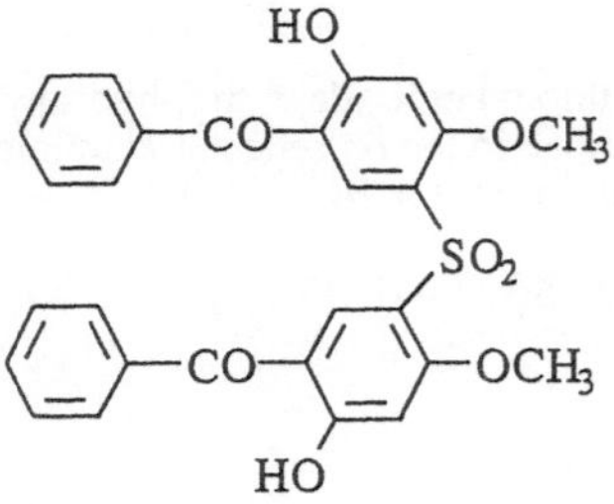

Synthesis

-Obtained by reaction of a 30% aqueous solution of
hydrogen peroxide with 5,5'-thiobis(2-hydroxy-
4-methoxybenzophenone) in acetic acid. The mixture was
then heated on a steam bath for 4 h [914].

m.p. 264-267° [914];   Spectra (NA).

[Sulfonylbis(6-hydroxy-4-methoxy-3,1-phenylene)]bis[(2-methylphenyl)methanone

[35698-06-7] $C_{30}H_{26}O_8S$ mol.wt. 546.60

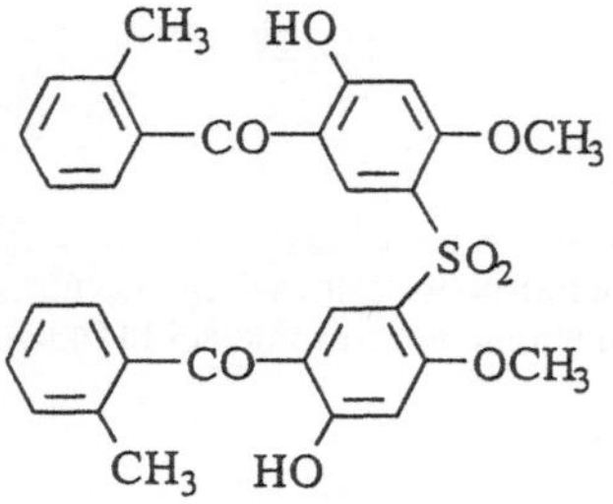

Synthesis

-Obtained by reaction of a 30% aqueous solution of
hydrogen peroxide with 5,5'-thiobis(2-hydroxy-
4-methoxy-2'-methylbenzophenone) in acetic acid. The
mixture was then heated on a steam bath for 6 h [914].

m.p. 268-269° [914];   Spectra (NA).

[Sulfonylbis[4-(cyclopentyloxy)-6-hydroxy-3,1-phenylene]]bis[phenylmethanone

$C_{36}H_{34}O_8S$ mol.wt. 626.73

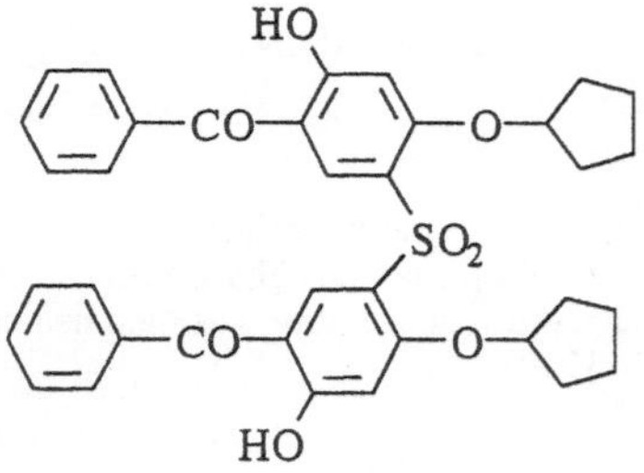

Synthesis

-Obtained by reaction of a 30% aqueous solution of
hydrogen peroxide with 5,5'-thiobis[2-hydroxy-
4-(cyclopentyloxy)benzophenone] in acetic acid. The
mixture was then heated on a steam bath for 6 h [914].

m.p. and Spectra (NA).

[Sulfonylbis[4-(benzyloxy)-6-hydroxy-3,1-phenylene]]bis[phenylmethanone

$C_{40}H_{30}O_8S$ mol.wt. 670.74

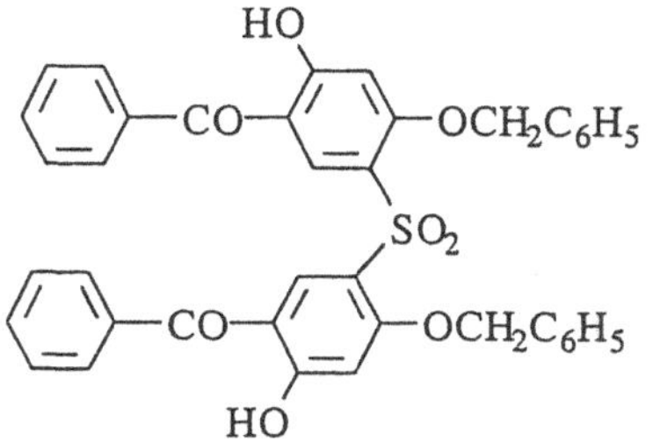

Synthesis

-Obtained by reaction of a 30% solution of hydrogen peroxide with 5,5'-sulfinylbis(2-hydroxy-4-benzyloxybenzophenone) in acetic acid. The mixture was then heated on a water bath for 6 h [914].

m.p. and Spectra (NA).

[Sulfonylbis[6-hydroxy-4-(octyloxy)-3,1-phenylene]]bis[phenylmethanone

[35698-05-6]

$C_{42}H_{50}O_8S$ mol.wt. 714.92

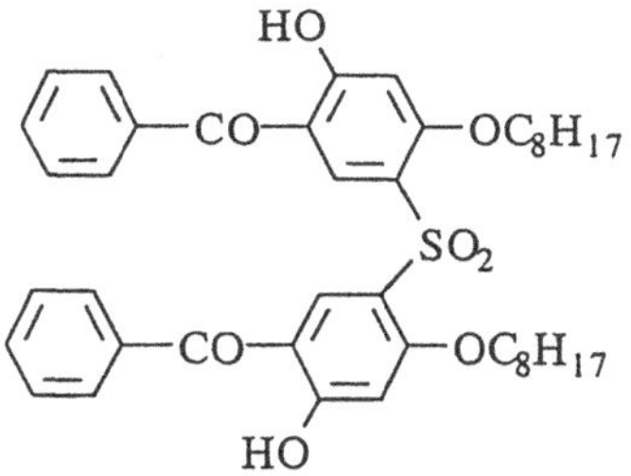

Synthesis

-Obtained by reaction of a 30% solution of hydrogen peroxide with 5,5'-sulfinylbis(2-hydroxy-4-octyloxy-benzophenone) in acetic acid. The mixture was then heated on a steam bath for 6 h [914].

m.p. and Spectra (NA).

8. OTHER ACYLATED COMPOUNDS

2-Chloro-4-(2-hydroxybenzoyl)phenyl 2-hydroxybenzoate
2-Hydroxybenzoic acid, 2-chloro-4-(2-hydroxybenzoyl)phenyl ester

[123861-93-8]

$C_{20}H_{13}ClO_5$ mol.wt. 368.77

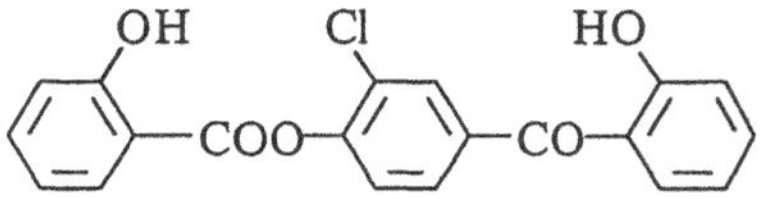

Syntheses

-Obtained by Fries rearrangement,
*of o-chlorophenyl 2-(nicotinoyloxy)benzoate in the presence of aluminium chloride without solvent at 150-152° for 2 h (26%) [170];
*of o-chlorophenyl salicylate in the presence of aluminium chloride without solvent at 180-183° for 3 h (20%) [172] according to [342].

m.p. 151-152° [170, 172]; Spectra (NA).

4-(2-Hydroxybenzoyl)phenyl 2-hydroxybenzoate
2-Hydroxybenzoic acid 4-(2-hydroxybenzoyl)phenyl ester

[124011-55-8]   C$_{20}$H$_{14}$O$_5$ mol.wt. 334.33

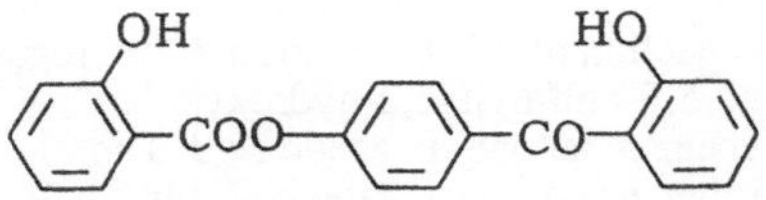

Syntheses

-Preparation by total demethylation of
  4-(2-methoxybenzoyl)phenyl 2-methoxybenzoate
  with aluminium chloride in refluxing benzene for
  1 h (50%) [170].
-Also obtained by Fries rearrangement of phenyl salicylate (salol) with aluminium chloride [342],
  without solvent at 180-182° for 3 h (10%) [171].
-Also obtained (by-product) by Fries rearrangement of phenyl 2-(nicotinoyloxy)benzoate with
  aluminium chloride without solvent at 140-145° for 2 h (16%) [170].

  m.p.   110° [171], 109-110° [170];   Spectra (NA).

REFERENCES

1   Abd el Rahman, A. H. and Basha, R. M.: Z. Naturforsch., B: Anorg. Chem., Org. Chem., **32B** (9) 1084-1088 (1977).
2   Abd el Rahman, A. H. and Ismail, E. M.: Arzneim.-Forsch. (Drug Res.) **26** (5) 756-759 (1976).
3   Abdel-Nabi, Ismail M.; Kadry, Abdelrazak M.; Davis, Richard A. and Abdel-Rahman, Mohamed S.: J. Appl. Toxicol., **12** (4) 255-259 (1992); Chem. Abstr., **117**, 144887b (1992).
4   Abdulla, Khalid A.; Abdul-Rahman, Azhar L.; Al-Hamdany, Ra'Ad and Al-Saigh, Zeki Y.: J. Prakt. Chem., **324** (3) 498-504 (1982).
5   Abe, Taku; Kimura, Takeshi; Ayabe, Yoshimoto and Shiomi, Taiichi: Jpn. Kokai Tokkyo Koho JP 61,282,335 [86,282,335] (1986); Chem. Abstr., **107**, 39399q (1987).
6   Abe, Toshiyuki; Gonda, Michihiro; Otomo, Kikuo and Yoshikawa, Katsumasa: Jpn. Kokai Tokkyo Koho JP 62 59,086 (1987); Chem. Abstr., **107**, 124719p (1987).
7   Abe, Toyohiko; Mishina, Makoto and Kohtoh, Noriaki: Polym. Adv. Technol., **4** (4) 188-293 (1993); Chem. Abstr., **120**, 120514p (1994).
8   Abe, Yoshio and Okawa, Katsuaki: Jpn. Kokai Tokkyo Koho JP 03, 167,151 [91,167,151] (1991); Chem. Abstr., **116**, 6240p (1992).
9   Abe, Yoshio; Tanaka, Naoki and Ookawa, Katsuaki: Jpn. Kokai Tokkyo Koho JP 05,311,151 [93, 311,151] (1993); Chem. Abstr., **120**, 165988n (1994).
10 Aboulezz, A. F. and Quelet, R.: J. Chem. U.A.R., **5** (2) 137-146 (1962); Chem. Abstr., **63**, 1411h (1965).
11 Abramovitch, R. A.; Hey, D. H. and Long, R. A. J.: J. Chem. Soc., 1781-1788 (1957).
12 Adam, W. and Schulz, M. H.: Chem. Ber., **125**, 2455-2461 (1992).
13 Adams, J. H.; Brown, P.M.; Gupta, P.; Khan, M. S. and Lewis, J. R.: Tetrahedron, **37** (1) 209-217 (1981).
14 Adams, J. H.; Gupta, P.; Khan, M. S.; Lewis, J. R. and Watt, R. A.: J. Chem. Soc., Perkin Trans. 1, **19**, 2089-2093 (1976).
15 Aebi, J.; Guerry, P.; Jolidon, S. and Morand, O.: Eur. Pat. Appl. EP 636,367 (1995); Chem. Abstr., **122**, 178398m (1995).
16 Afzal, M.; Davies, J. S. and Hassal, C. H.: J. Chem. Soc. C, **13**, 1721-1727 (1969).
17 Agarwal, S. K. and Murray, R. W.: Photochem. Photobiol., **35** (1) 31-35 (1982).
18 Aggarwal, S. C. and Saharia, G. S.: J. Indian Chem. Soc., **37** (5) 295-298 (1960).
19 Aggarwal, S. K.; Grover, S. K. and Seshadri, T. R.: Indian J. Chem., **10** (9) 911-913 (1972).
20 Agrawal, V. K. and Sharma, S.: Indian J. Chem., Sect. B, **23B** (9) 839-843 (1984).
21 Agzatova, K. O.; Yuldashev, Kh. Yu and Sidorova, N. G. (USSR) : Deposited Doc., SPSTL 366Khp-D80, 12 pp. (1980); Chem. Abstr., **97**, 55405u (1982).
22 Ahluwalia, V. K.; Gupta, R.; Grover, M.; Mukherjee, I. and Khanduri, C. H.: Indian J. Chem., Sect. B, **27B** (12) 1138-1139 (1988).
23 Ahluwalia, V. K. and Khanduri, C. H.: Indian J. Chem., Sect. B, **28B** (7) 599-601 (1989).
24 Ahluwalia, V. K.; Khanna, M. and Singh, R. P.: Indian J. Chem., Sect. B, **20B** (11) 990-991 (1981).
25 Ahluwalia, V. K.; Khanna, M. and Singh, R. P.: Synthesis, **5**, 404-406 (1983).
26 Ahluwalia, V. K.; Kumar, D. and Gupta, M. C.: Indian J. Chem., Sect. B, **16B** (7) 574-578 (1978).
27 Ahluwalia, V. K.; Mann, R. R. and Singh, S. B.: J. Indian Chem. Soc., **65** (11) 768-770 (1988).
28 Ahluwalia, V. K. and Mehta, B.: Indian J. Chem., Sect. B, **25B** (11) 1171 (1986).
29 Ahluwalia, V. K.; Singh, D. and Singh, R. P.: Monatsh. Chem., **116** (6-7) 869-872 (1985).
30 Ahluwalia, V. K.; Singh, R. and Singh, R. P.: Gazz. Chim. Ital., **114** (11-12) 501-503 (1984).
31 Aichaoui, H.; Lesieur, I. and Henichart, J. P.: Synthesis, **8**, 679-680 (1990).
32 Al-Hamdany, R. and Ali, B.: J. Chem. Soc., Chem. Commun., **9**, 397 (1978).
33 Al-Hamdany, R.; Al-Rawi, J. M.; Ahmed, B. A. and Al-Shahiry, K. F.: J. Prakt. Chem., **329** (2) 337-342 (1987).

34 Al-Ka'bi, J.; Farooqi, J. A.; Gore, P. H.; Moonga, B. S. and Waters, D. N.: J. Chem. Res., Synop., **3**, 80-81 (1989).
35 Al-Ka'bi, J.; Gore, P. H.; Moonga, B.; Al-Shiebani, I. S.; Shibaldain, N. L. and Kamounah, F. S.: J. Chem. Res. (M), **7**, 2201-2226 (1986).
36 Al-Malaika, S.; Goonetileka, M. D. R. J. and Scott, G.: Polym. Degrad. Stab., 32 (2) 231-247 (1991); Chem. Abstr., **114**, 208492s (1991).
37 Alberola, A. and Fernandez, M. I.: An. Quim., **65** (12) 1121-1124 (1969).
38 Albert, B.; Eilingsfeld, H. and Neumann, P.: Eur. Pat.Appl. EP 194,526 (1986); Chem. Abstr., **106**, 205256s (1987).
39 Allen, J. and Giffard, D.: J. Labelled Compd. Radiopharm., **19** (2) 301-307 (1982).
40 Allen, N. S.; Binkley, J. P.; Parsons, B. J.; Phillips, G. O. and Tennent, N. H.: Dyes Pigm., **4** (1) 11-24 (1983); Chem. Abstr., **98**, 73805z (1983).
41 Allen, N. S.; Luc-Gardette, J. and Lemaire, J.: Polym. Photochem., **3** (4) 251-265 (1983); Chem. Abstr., **99**, 39247e (1983).
42 Ambrovic, P. and Mikovic, J.: Eur. Polym. J.-Supplement, 371-377 (1969).
43 American Cyanamid Co.: French Patent 1,098,344 (1955).
44 American Cyanamid Co.: Brit. 890,476 (1962); Chem. Abstr., **57**, 736c (1962).
45 Amin, G. C.; Chaughuley, A. S. U. and Jadhav, G. V.: J. Indian Chem. Soc., **36** (9) 617-621 (1959).
46 Amin, G. C.; Chaughuley, A. S. U. and Jadhav, G. V.: J. Indian Chem. Soc., **36** (12) 833-837 (1959).
47 Amin, G. C. and Shah, N. M.: J. Indian Chem. Soc., **25** (8) 377-384 (1948).
48 Amin, G. C. and Shah, N. M.: J. Indian Chem. Soc., **27** (10) 531-534 (1950).
49 Amin, G. C. and Shah, N. M.: J. Indian Chem. Soc., **29** (5) 351-356 (1952).
50 Amin, J. H. and Desai, R. D.: J. Sci. Ind. Res., Sect. B, **13B**, 178-180 (1954).
51 Amin, K. C. and Amin, G. C.: J. Indian Chem. Soc., **37** (8) 469-472 (1960).
52 Anderson, J. C. and Reese, C. B.: J. Chem. Soc., 1781-1784 (1963).
53 Anon.: Res. Discl., **155**, 38 (1977); Chem. Abstr., **86**, 141475w (1977).
54 Anschütz, R.: Justus Liebigs Ann. Chem., **346**, 381-382 (1906).
55 Anthony, Blair T.: Can. CA 1,175,056 (1984); Chem. Abstr., **102**, 63082y (1985).
56 Anthony, Blair T.: U.S. US 4,366,207 (1982); Chem. Abstr., **98**, 90555k (1983).
57 Antus, S.; Schindlbeck, E.; Ahmad, S.; Seligmann, O.; Chari, V. M. and Wagner, H.: Tetrahedron, **38** (1) 133-137 (1982).
58 Appel, H.; Baker, W.; Hagenbach, H. and Robinson, R.: J. Chem. Soc., 738-744 (1937).
59 Arai, Naoto; Tuji, Takuji and Yoneda, Hiroshi: Ger. Offen. DE 3,219,278 (1982); Chem. Abstr., **98**, 98842x (1983).
60 Arcadi, A.; Cacchi, S.; Rosario, M. D.; Fabrizi, G. and Marinelli, F.: J. Org. Chem., **61** (26) 9280-9288 (1996).
61 Aries, R.: Fr. 1,559,837 (1969); Chem. Abstr., **72**, 24515j (1970).
62 Arisawa, M.; Morita, N.; Kondo, Y. and Takemoto, T.: Chem. Pharm. Bull., **21** (10) 2323-2328 (1973).
63 Armitage, J. B.; Dessauer, R. and Hyson, A. M.: U.S. 3,098,842 (1963); Chem. Abstr., **59**, 10309f (1963).
64 Arventi, B. I.: Bull. Soc. Chim. Fr., 598-603 (1936).
65 Arventi, B. I.: Bull. Soc. Chim. Fr., 999-1007 (1937).
66 Arventiev, B.; Gabe, I. and Cascaval, A.: An. Stiint. Univ. "Al. I. Cuza" Iasi, Sect. 1c, **13** (1) 53-58 (1967).
67 Arventiev, B.; Gabe, I.; Offenberg, H. and Nicolaescu, T.: An. Stiint Univ. "Al I Cuza" Iasi, Sect. 1c, **10**, 173-182 (1964).
68 Arventiev, B.; Gabe, I.; Offenberg, H.; Nicolaescu, T.; Harnagea, F. and Badilescu, S.: An. Stiint Univ. "Al. I. Cuza", Sect. 1c, **12** (2) 181-183 (1966).
69 Arventiev, B. and Offenberg, H.: Acad. Repub. Pop. Rom., Fil. Iasi, Stud. cercet. stiint., Chim., **11**, 305-310 (1960); Chem. Abstr., **56**, 11554c (1962).
70 Arventiev, B. and Offenberg, H.: An. Stiint Univ. "Al. I. Cuza", Sect. 1c, **8** (1) 217-224 (1962).
71 Arventiev, B.; Offenberg, H. and Nicolaescu, T.: An. Stiint Univ. "Al I. Cuza" Iasi, Sect. 1c, **10**, 65-70 (1964).

72 Arventiev, B.; Singurel, L.; Ofenberg, H. L. and Nicolaescu, T.: An. Stiint. Univ. "Al. I. Cuza" Iasi, Sect. 1c, **20** (1) 41-45 (1974).

73 Arventiev, B.; Singurel, L.; Offenberg, G. and Nicolaescu, T.: An. Stiint. Univ. "Al. I. Cuza", Sect. 1c, **13** (2) 135-138 (1967).

74 Arventiev, B.; Strul, M. and Wexler, H.: Acad. Repub. Pop. Rom., Fil. Iasi, Stud. cercet. stiint., Chim., **11**, 53-62 (1960); Chem. Abstr., **55**, 15452h (1961).

75 Arventiev, B. and Wexler, H.: An. Stiint. Univ. "Al. I. Cuza" Iasi, Sect. 1c, **12** (1) 51-55 (1966); Chem. Abstr., **67**, 53981 (1967).

76 Arventiev, B. and Wexler, H.: An. Stiint. Univ. "Al. I. Cuza" Iasi, Sect. 1c, **17** (1) 61-65 (1971).

77 Arventiev, B.; Wexler, H.; Gabe, I.; Harnagea, F. and Badilescu, S.: An. Stiint. Univ. "Al. I. Cuza" Iasi, Sect. 1c, **13** (1) 59-60 (1967).

78 Arventiev, B.; Wexler, H. and Strul, M.: Acad. Repub. Pop. Rom., Fil. Iasi, Stud. cercet. stiint., Chim., **11**, 63-73 (1960); Chem. Abstr., **55**, 15453a (1961).

79 Asahi-Dow Ltd.: Jpn. Kokai Tokkyo Koho JP 57,105,452 [82,105,452] (1982); Chem. Abstr., **98**, 5082m (1983).

80 Ashbrook, A. W.; Itzkovitch, I. J. and Sowa, W.: CIM Spec. Vol., **21** (2, Proc. Int. Solvent Extr. Conf., 1977), 741-749 (1979); Chem. Abstr., **92**, 87593s (1980).

81 Aslam, M. and Aguilar, D. A.: U.S. US 5,130,448 (1992); Chem. Abstr., **118**, 38609y (1993).

82 Astoin, J.; Lepage, F.; Fromantin, J. P. and Poisson, M.: Eur. J. Med. Chem. - Chim. Ther., **15** (5) 457-462 (1980).

83 Atkinson, H. and Heilbron, I. M.: J. Chem. Soc., 2688-2691 (1926).

84 Atkinson, J. E.; Gupta, P. and Lewis, J. R.: Chem. Commun., 1386-1387 (1968).

85 Atkinson, J. E.; Gupta, P. and Lewis, J. R.: Tetrahedron, **25**, 1507-1511 (1969).

86 Atkinson, J. E. and Lewis, J. R.: Chem. Commun., **16**, 803 (1967).

87 Atkinson, J. E. and Lewis, J. R.: J. Chem. Soc. C, 281-287 (1969).

88 Attwood, T. E.; Dawson, P. C.; Freeman, J. L.; Hoy, L. R. J.; Rose, J. B. and Staniland, P. A.: Polymer, **22** (8) 1096-1103 (1981).

89 Aumueller, A. and Goetze, W.: Eur. Pat. Appl. EP 425,974 (1991); Chem. Abstr., **115**, 49103t (1991).

90 Auricchio, S.; Morrocchi, S. and Ricca, A.: Tetrahedron Lett., **33**, 2793-2796 (1974).

91 Auwers, K.: Ber. Dtsch. Chem. Ges., **36**, 3890-3892 (1903).

92 Auwers, K.: Ber. Dtsch. Chem. Ges., **36**, 3893-3902 (1903).

93 Auwers, K. and Betteridge, F. H.: Z. Phys. Chem., Stoechiom. Verwandtschaftsl., **32**, 39-45 (1900).

94 Auwers, K. and Czerny, H.: Ber. Dtsch. Chem. Ges., **31**, 2692-2698 (1898).

95 Auwers, K. and Janssen, E.: Justus Liebigs Ann. Chem., **483**, 44-65 (1930).

96 Auwers, K. and Markovits, T.: Ber. Dtsch. Chem. Ges., **41**, 2332-2340 (1908).

97 Auwers, K. and Mauss, W.: Ber. Dtsch. Chem. Ges., **61**, 1495-1507 (1928).

98 Auwers, K. and Mauss, W.: Justus Liebigs Ann. Chem., **464**, 293-311 (1928).

99 Auwers, K. and Rietz, E.: Ber. Dtsch. Chem. Ges., **40**, 3514-3521 (1907).

100 Avar, Lajos and Hess, Erwin: Ger. Offen. DE 3,320,615 (1983); Chem. Abstr., **100**, 193628a (1984).

101 Avar, Lajos and Hofer, Kurt: Ger. Offen. 2,146,075 (1972); Chem. Abstr., **76**, 153345k (1972).

102 Avnir, D.; De Mayo, P. and Ono, Isao: J. Chem. Soc., Chem. Commun., **24**, 1109-1110 (1978).

103 Azzolina, O.; Vercesi, D. and Ghislandi, V.: Farmaco, Ed. Sci., **43** (5) 469-478 (1988).

104 B.A.S.F.: D.R.P. 49,149 (1889).

105 B.A.S.F.: D.R.P. 50451 (1889).

106 B.A.S.F.: D.R.P. 54661 (1890).

107 Baba, Y. and Kodama, H.: Japan 72 38,062 (1972); Chem. Abstr., **79**, 147160s (1973).

108 Bachelet, J. P.; Cavier, R.; Lemoine, J.; Rigothier, M. C.; Gayral, P. and Royer, R.: Eur. J. Med. Chem., **14** (4) 321-324 (1979).

109 Baddar, F. G.; El-Assal, L. S. and Baghos, V. B.: J. Chem. Soc., 1714-1718 (1955).

110 Baddeley, G.: J. Chem. Soc., 273-274 (1943).

111 Baeseler, M.; Seiffarth, K.; Dahlmann, J.; Hoeft, E. and Woydowski, K.: Ger. (East) DD
     268,930 (1989); Chem. Abstr., **112**, 76607e (1990).
112 Baeyer, A.: Justus Liebigs Ann. Chem., **353**, 152-204 (1907).
113 Baeyer, A.: Justus Liebigs Ann. Chem., **372**, 80-151 (1910).
114 Baeyer, A. and Burkhardt, J. B.: Ber. Dtsch. Chem. Ges., **11**, 1299-1301 (1878).
115 Baeyer, A. and Burkhardt, J. B.: Justus Liebigs Ann. Chem., **202**, 36-140 (1880).
116 Baeyer, A. and Drewsen, V.: Liebigs Ann. Chem., **212**, 340-347 (1882).
117 Bagolini, C. A.; Pacifici, L. and Quaresima, E. T.: Eur. Pat. Appl. EP 323,416 (1989); Chem.
     Abstr., **112**, 20792u (1990).
118 Bailey, D.; Tirrell, D.; Pinazzi, C. and Vogl, O.: Macromolecules, **11** (2) 312-320 (1978).
119 Bailey, D.; Tirrell, D. and Vogl, O.: J. Macromol. Sci., Chem., **A12** (5) 661-699 (1978).
120 Baker, W. and Smith, A. R.: J. Chem. Soc., 346-348 (1936).
121 Balani, R. A. and Sethna, S.: J. Indian Chem. Soc., **44** (1) 52-56 (1967).
122 Balani, R. A. and Sethna, S.: J. Indian Chem. Soc., **48** (5) 417-422 (1971).
123 Balazs, E.; Toth, A.; Bogsch, E.; Stefko, B. Gebhardt, I. and Mathe, D.: Hung. Teljes HU
     26,079 (1983); Chem. Abstr., **100**, 47089j (1984).
124 Balazs, E.; Toth, A.; Bogsch, E.; Stefko, B.; Gebhardt, I. and Mathe, D.: Pat. Specif. (Aust.)
     AU 514,160 (1981), Chem. Abstr., **96**, 195119h (1982).
125 Ballantine, J. A. and Pillinger, C. T.: Org. Mass. Spectrom., **1** (3) 425-445 (1968); Chem.
     Abstr., **69**, 111296v (1968).
126 Baltzly, R.; Ide, W. S. and Phillips, A. P.: J. Am. Chem. Soc., **77**, 2522-2533 (1955).
127 Balz, G. and Schiemann, G.: Ber. Dtsch. Chem. Ges., **60**, 1186-1190 (1927).
128 Barabas, E. S.; Mallya, P. and Gromelski, S. J., Jr.: U.S. US 4,301,267 (1981); Chem. Abstr.,
     96, 86464w (1982).
129 Barabas, E. S.; Mallya, P. and Gromelski, S. J., Jr.: U.S. US 4,302,606 (1981); Chem. Abstr.,
     96, 104914z (1982).
130 Barabas, E. S.; Mallya, P. and Gromelski, S. J., Jr.: U.S. US 4,310,687 (1982); Chem. Abstr.,
     96, 200337j (1982).
131 Barabas, E. S.; Mallya, P. and Gromelski, S. J., Jr.: U.S. US 4,312,995 (1982); Chem. Abstr.,
     96, 218388m (1982).
132 Bargellini, G.: Gazz. Chim. Ital., **46**, 249-255 (1916).
133 Bargellini, G. and Martegiani, E.: Atti. R. Accad. Naz. Lincei, **20**, 118-124 (1911).
134 Bargellini, G. and Martegiani, E.: Atti. R. Accad. Naz. Lincei, **20**, 183-190 (1911).
135 Bargellini, G. and Martegiani, E.: Gazz. Chim. Ital., **41**, 603-612 (1911).
136 Barnes, H. C.; Ollis, W. D.; Sutherland, I. O.; Gottlieb, O. R. and Taveira Magalhaes, M.:
     Tetrahedron, **21**, 2707-2715 (1965).
137 Bartolotti, P.: Gazz. Chim. Ital., **26**, 433-441 (1896).
138 Bartolotti, P.: Gazz. Chim. Ital., **27**, 280-288 (1897).
139 Bartolotti, P.: Gazz. Chim. Ital., **28**, 283-290 (1898).
140 Bartolotti, P.: Gazz. Chim. Ital., **30**, 56-98 (1900).
141 Bartolotti, P.: Gazz. Chim. Ital., **30**, 229-234 (1900).
142 Bartolotti, P. and Linari, A.: Gazz. Chim. Ital., **32**, 271-276 (1902).
143 Bartolotti, P. and Linari, A.: Gazz. Chim. Ital., **32**, 494-503 (1902).
144 Barton, H.: J. Pharmacol. Pharm., **31** (2) 169-174 (1979).
145 Basha, A.; Ahmed, S. S. and Farooqui, T. A.: Tetrahedron Lett., **36**, 3217-3220 (1976).
146 Basil, B.; Coffee, E. C. J.; Gell, D. L.; Maxwell, D. R.; Sheffield, D. J. and Wooldridge, K.
     R. H.: J. Med. Chem., **13** (3) 403-406 (1970).
147 Basil, B. and Wooldridge, K. R. H.: Ger. Offen. 2,627,210 (1976); Chem. Abstr., **86**,
     139626c (1977).
148 Bastide, M.; Chabard, J. L.; Lartigue, C.; Bargnoux, H.; Petit, J.; Berger, J. A.; Ait Mansour,
     H.; Lesieur, D. and Busch, N.: Biol. Mass Spectrom., **20** (8) 484-492 (1991); Chem. Abstr.,
     115, 149652q (1991).
149 Becker, H. D.; Björk, A. and Adler, E.: J. Org. Chem., **45**, 1596-1600 (1980).
150 Beger, J.; Binte, H. J.; Brunne, L. and Neumann, R.: J. Prakt. Chem./Chem.-Ztg., **334** (3)
     269-277 (1992); Chem. Abstr., **117**, 130886p (1992).

151 Beger, J.; Neumann, R.; Vogel, T.; Luecke, L.; Kaestner, G.; Runge, H. J.; Schewe, T.; Schewe, C.; Ludwig, P. and Slapke, J.: Ger. (East) DD 297,155 (1992); Chem. Abstr., 116, 214145p (1992).
152 Bellamy, F.; Horton, D.; Millet, J.; Picart, F.; Samreth, S. and Chazan, J. B.: J. Med. Chem., 36 (7) 898-903 (1993).
153 Belled, C.; Miranda, M. A. and Simon-Fuentes, A.: An. Quim., 86, 431-435 (1990).
154 Belled, C.; Miranda, M. A. and Simon-Fuentes, A.: An. Quim., Ser. C, 85 (1) 39-47 (1989).
155 Belokon, Yu. N.; Pritula, L. K.; Tararov, V. I.; Bakhmutov, V. I.; Struchkov, Yu. T.; Timofeeva, T. V. and Belikov, V. M.: J. Chem. Soc., Dalton Trans., 6, 1867-1872 (1990).
156 Bencze, W. L.: US 3,007, 935 (1961); Chem. Abstr., 56, 4681e (1962).
157 Benedikt, R.: Justus Liebigs Ann. Chem., 185, 114-119 (1877).
158 Bennetau, B.; Rajarison, F.; Dunogues, J. and Babin, P.: Tetrahedron, 50 (4) 1179-1188 (1994).
159 Bennett, B. and Clough, D.: Ger. Offen., DE 3,417,782 (1984); Chem. Abstr., 102, 133487k (1985).
160 Bennett, O. F.; Sister, M. J. B.; Malloy, R.; Dervin, P. and Saluti, G.: J. Org. Chem., 37 (9) 1356-1359 (1972).
161 Bergstroem, M.; Lu, L.; Marouez, M.; Moulder, R.; Jacobsson, G.; Oegren, M.; Eriksson, B.; Watanabe, Y. and Langstroem, B.: Scand. J. Gastroenterol., 31 (12) 1216-1222 (1996); Chem. Abstr., 126, 98810g (1997).
162 Bernauer, K.; Borgulya, J.; Bruderer, H.; Da Prada, M. and Zürcher, G.: Eur. Pat. Appl. 1,034,690 (1987).
163 Bernauer, K.; Borgulya, J.; Bruderer, H.; Da Prada, M. and Zürcher, G.: Swiss Pat. Appl. CH 980/86 (1986).
164 Bernauer, K.; Borgulya, J.; Bruderer, H.; Da Prada, M. and Zürcher, G.: Pat. Specif. (Aust.) AU 603,788 (1990); Chem. Abstr., 115, 49134d (1991).
165 Berthier, C.; Allaigre, J. P. and Desbois, J.: Fr. Demande FR 2,553,763 (1985); Chem. Abstr., 103, 141480p (1985).
166 Bettinetti, G. F. and Gamba, A.: Gazz. Chim. Ital., 100 (12) 1144-1159 (1970).
167 Bhagwat, V. K. and Shahane, R. Y.: Rasayanam, 1, 191-194 (1939); Chem. Abstr., 34, 5070⁹ (1940).
168 Bhargava, S. S.; Jain, S. K. and Saharia, G. S.: Indian J. Chem., 5 (11) 543-544 (1967).
169 Bhatt, M. R. and Shah, N. M.: J. Indian Chem. Soc., 33 (5) 318-320 (1956).
170 Bhavsar, M. D. and Desai, V. B.: Man-Made Text. India, 31 (10) 431, 433, 435, 437-439 (1988); Chem. Abstr., 111, 232222n (1989).
171 Bhavsar, M. D. and Desai, V. B.: Man-Made Text. India, 31 (12) 529-535, 556 (1988); Chem. Abstr., 112, 7121e (1990).
172 Bhavsar, M. D. and Desai, V. B.: Man-Made Text. India, 32 (1) 8-11, 15 (1989); Chem. Abstr., 113, 213772s (1990).
173 Bhavsar, M. D. and Saraiya, P. N.: Man-Made Text. India, 29 (5) 224-230 (1986); Chem. Abstr., 107, 77603j (1987).
174 Bhavsar, M. D. and Shah, B. M.: Man-Made Text. India, 31 (7) 287-288, 291, 303 (1988); Chem. Abstr., 110, 136814r (1989).
175 Bhavsar, M. D. and Surendranath, V.: Man-Made Text. India, 28 (11) 425, 427-429, 431 (1985); Chem. Abstr., 106, 49685y (1987).
176 Bian, Ji; Xu, Suisu; Huang, Song and Wang, Zexing: Shenyang Yaoke Daxue Xuebao, 13 (1) 34-40 (1996); Chem. Abstr., 125 (7) 81806s (1996).
177 Bichan, D. J. and Yates, P.: Can. J. Chem., 53 (14) 2054-2063 (1975).
178 Bier, G. and Kricheldorf, H. R.: Ger. Offen. DE 3,211,421 (1983); Chem. Abstr., 100, 7431h (1984).
179 Billon, P.: Ann. Chim. (Paris), 7, 336-384 (1927).
180 Binev, I.; Kolev, Ts. and Yukhnovski, I.: Izv. Khim., 14 (3) 341-354 (1982); Chem. Abstr., 97, 215337d (1982).
181 Birch, A. J.: Proc. Chem. Soc., London, 3-13 (1962).
182 Blakey, W.; Jones, W. I. and Scarborough, H. A.: J. Chem. Soc., 2865-2872 (1927).
183 Blatt, A. H.: J. Org. Chem., 20, 591-602 (1955).
184 Blatt, A. H.: Org. React., 1, 342-369 (1942).

185 Bleuler, H. and Perkin, A. G.: J. Chem. Soc., **109**, 529-543 (1916).
186 Blicke, F. F. and Patelski, R. A.: J. Am. Chem. Soc., **60**, 2283-2285 (1938).
187 Blicke, F. F. and Weinkauff, O. J.: J. Am. Chem. Soc., **54**, 330-334 (1932).
188 Blicke, F. F. and Weinkauff, O. J.: J. Am. Chem. Soc., **54**, 1446-1453 (1932).
189 Boboli, E.; Malasnicki, W. L. and Kowalski, M.: Pol. 83,600 (1977); Chem. Abstr., **90**, 137491p (1979).
190 Boehme, W. R. and Scharpf, W. G.: J. Org. Chem., **26**, 1692-1695 (1961).
191 Bogert, M. T. and Howells, H. P.: J. Am. Chem. Soc., **52**, 837-850 (1930).
192 Bohlmann, F. and Suwita, A.: Phytochemistry, **17** (11) 1929-1934 (1978).
193 Bohlmann, F. and Suwita, A.: Phytochemistry, **18** (12) 2046-2049 (1979).
194 Bohlmann, F. and Zdero, C.: Phytochemistry, **19** (4) 683-684 (1980).
195 Böhmer, V.; Lüderwald, I. and Martin, R.: Fresenius' Z. Anal. Chem., **297** (5) 365-369 (1979).
196 Bonte, J. P.; Lesieur, D.; Lespagnol, C.; Cazin, J. C. and Cazin, M.: Eur. J. Med. Chem., **9** (5) 497-500 (1974).
197 Borgulya, J.; Bruderer, H.; Bernauer, K.; Zürcher, G. and Da Prada, M.: Helv. Chim. Acta, **72** (5) 952-968 (1989).
198 Borsche, W.; Löwenstein, H. and Quast, R.: Ber. Dtsch. Chem. Ges., **50**, 1339-1355 (1917).
199 Borsche, W. and Hahn-Weinheimer, P.: Justus Liebigs Ann. Chem., **570**, 155-164 (1950).
200 Boulton, A. J.; Tsoungas, P. G. and Tsiamis, C.: J. Chem. Soc., Perkin Trans. 1, **9**, 1665-1667 (1986).
201 Bourgeois, Y.; Devaux, J.; Legras, R. and Parsons, I. W.: Polymer, **37** (14) 3171-3176 (1996); Chem. Abstr., **125** (12) 143421y (1996).
202 Bouzard, D.; Weber, A. and Le Henaff, P.: Bull. Soc. Chim. Fr., 3375-3384 (1972).
203 Bowen, I. H.: Thesis, University of Aberdeen, UK (1967).
204 Bowen, I. H.; Gupta, P. and Lewis, J. R.: Chem. Commun., 1625-1626 (1970).
205 Bowen, I. H. and Lewis, J. R.: J. Chem. Soc., Perkin Trans. 1, **5**, 683-685 (1972).
206 Bradshaw, J. S., Loveridge, E. L. and White, L.: J. Org. Chem., **33**, 4127-4132 (1968).
207 Bragole, R. A. and Shepard, R. A.: J. Org. Chem., **25**, 1230-1232 (1960).
208 Bredereck, H.; Lehmann, G.; Fritzsche, E. and Schönfeld, C.: Angew. Chem., **26**, 445-446 (1939).
209 Bredereck, H.; Lehmann, G.; Schönfeld, C. and Fritzsche, E.: Ber. Dtsch. Chem. Ges., **72B**, 1414-1429 (1939).
210 Breipohl, G.; Knolle, J. and Stueber, W.: Ger. Offen. DE 3,711,866 (1988); Chem. Abstr., **110**, 154897z (1989).
211 Breipohl, G.; Knolle, J. and Stueber, W.: Tetrahedron Lett., **28** (46) 5651-5654 (1987).
212 Brevet Polonia 105,666 (1980).
213 Brieaddy, L. E.: PCT Int. Appl. WO 96 05,188 (1996); Chem. Abstr., **125**, 114724u (1996).
214 Britton, E. C.: U. S. 1,961,630 (1934); Chem. Abstr., **28**, 4744[5] (1934).
215 Broadbent, D.; Mabelis, R. P. and Spencer, H.: Phytochemistry, **14** (9) 2082-2083 (1975).
216 Broadhurst, M. J.; Hassall, C. H. and Thomas, G. J.: J. Chem. Soc., Perkin Trans. 1, **22**, 2502-2512 (1977).
217 Brose, T.; Holzscheiter, F.; Mattersteig, G.; Pritzkow, W. and Voerckel, V.: J. Prakt. Chem./Chem.-Ztg., **334** (6) 497-504 (1992).
218 Brossi, A. and Teitel, S.: Tetrahedron Lett., **6**, 417-419 (1970).
219 Brown, R. E. and Shavel, J., Jr.: U.S. 3,927,023 (1975); Chem. Abstr., **84**, 90154q (1976).
220 Bruce, D. B.; Sorrie, A. J. S. and Thomson, R. H.: J. Chem. Soc., 2403-2406 (1953).
221 Bruce, J. M. and Chaudhry, A.-u.-h.: J. Chem. Soc., Perkin Trans. 1, **2**, 295-297 (1974).
222 Bruce, J. M.; Fitzjohn, S. and Pardasani, R. T.: J. Chem. Res., Synop., **8**, 252-253 (1981).
223 Bruce, J. M. and Lloyd-Williams, P.: J. Chem. Soc., Perkin Trans. 1, **21**, 2877-2884 (1992).
224 Bruna, R. F.; Fergus, J. H.; Coughenour, L. L.; Courtland, G. G.; Pugsley, T. A.; Dodd, J. H. and Tinney, F. J.: Mol. Pharmacol., 38 (6) 950-958 (1990); Chem. Abstr., **114**, 220744h (1991).
225 Brunelle, D. J.: U.S. US 4,452,932 (1984); Chem. Abstr., **101**, 73299t (1984).
226 Buchanan, R. L.; Partyka, R. A. and Standridge, R. T.: U.S. 4,056,540 (1977); Chem. Abstr., **88**, 22931y (1978).
227 Burgess, R. H.: Brit. 992,312 (1965); Chem. Abstr., **63**, 5841g (1965).

228 Burgess, R. H.: Brit. 1,001,062 (1965); Chem. Abstr., **63**, 11801h (1965).
229 Burgess, R. H.: Brit. 1,001,471 (1965); Chem. Abstr., **63**, 15061d (1965).
230 Burke, H. M. and Joullie, M. M.: J. Med. Chem., **21** (10) 1084-1086 (1978).
231 Burke, J. T.; Durand, A.; Ferrandes, B. and Morselli, P. L.: Biopharm. Pharmacokinet., Eur. Congr., 2nd, **2**, 493-500 (1984); Chem. Abstr., **102**, 197455f (1985).
232 Burkhardt, J. B.: Justus Liebigs Ann. Chem., **202**, 126-135 (1880).
233 Burmistrova, R. S.; Gushchina, N. A.; Florentseva, L. I. and Yanovskii, D. M.: Zh. Prikl. Khim., **38** (10) 2383-2386 (1965); Chem. Abstr., **64**, 3770d (1966).
234 Bursey, M. M. and Twine, C. E., Jr.: J. Org. Chem., **36** (1) 137-140 (1971).
235 Buu-Hoi, N. P.; Jacquignon, P. and Périn, F.: Bull. Soc. Chim. Fr., 1622-1624 (1962).
236 Buu-Hoi, N. P.; Lavit, D. and Xuong, N. D.: J. Org. Chem., **19**, 1617-1621 (1954).
237 Buu-Hoi, N. P.; Royer, R. and Eckert, B.: J. Org. Chem., **17**, 1463-1465 (1952).
238 Buu-Hoi, N. P.; Royer, R. and Hubert-Habart, M.: Recl. Trav. Chim. Pays-Bas, **70** (9-10) 825-832 (1951).
239 Buu-Hoi, N. P.; Royer, R.; Xuong, N. D. and Thang, K. V.: Bull. Soc. Chim. Fr., 1204-1207 (1955).
240 Buu-Hoi, N. P.; Xuong, N. D.; Binon, F. and Nam, N. H.: C. R. Acad. Sci., **235**, 329-331 (1952).
241 Buu-Hoi, N. P.; Xuong, N. D. and Lavit, D.: J. Chem. Soc., 1034-1038 (1954).
242 Bymark, R. M.; Kirk, A. R.; Griggs, A. L. and Martin, S. J.: Eur. Pat. Appl. EP 342,035 (1989); Chem. Abstr., **112**, 218894x (1990).
243 Cadogan, J. I. G.; Hutchison, H. S. and McNab, H.: J. Chem. Soc., Perkin Trans. 1, **2**, 385-393 (1991).
244 Campbell, N. and Thomson, A.: Proc. Roy. Soc. Edinburgh, Sect. A, **68** (3) 245-256 (1970); Chem. Abstr., **72**, 110958n (1970).
245 Campbell, T. W. and Coppinger, G. M.: U.S. 2,686,123 (1954); Chem. Abstr., **49**, 4203h (1955).
246 Campbell, T. W. and Coppinger, G. M.; J. Am. Chem. Soc., **73**, 2708-2712 (1951).
247 Canon, K. K.: Jpn. Kokai Tokkyo Koho 81 05,567 (1981); Chem. Abstr., **95**, 124004k (1981).
248 Canter, F. W.; Curd, F. H. and Robertson, A.: J. Chem. Soc., 1245-1255 (1931).
249 Caro, H. and Graebe, C.: Ber. Dtsch. Chem. Ges., **11**, 1348-1351 (1878).
250 Casnati, G.; Colli, M.; Pochini, A. and Ungaro, R.: Chim. Ind. (Milan), **59** (11) 764-765 (1977).
251 Cassebaum, H.: J. Prakt. Chem., **13**, 141-151 (1961).
252 Cassebaum, H. and Drux, R.: Ger. 1,134,084 (1962); Chem. Abstr., **58**, 3358b (1963).
253 Ceccato, G.; Geri, S. and Colombo, L.: Ger. Offen., 2,350,293 (1974); Chem. Abstr., **81**, 137428n (1974).
254 Chakravarti, D. and Bera, B. C.: J. Indian Chem. Soc., **21**, 109-111 (1944).
255 Chan, F. D.; Nguyen, Kien Loi and Wallis, A. F. A.: J. Wood Chem. Technol., **15** (3) 329-347 (1995).
256 Chandrakumar, N. S.; Chen, B. B.; Clare, M.; Desai, B. N.; Djurie, S. W.; Docter, S. H.; Gasiecki, A. F.; Haack, R. A.; Liang, Chi-Dean; Miyashiro, J. M.; Penning, T. D.; Russell, M. A. and Yu, S. Siu-tzyy: PCT Int. Appl. WO 96 10, 999 (1996); Chem. Abstr., **125** (11) 142725p (1996).
257 Chandrakumar, N. S.; Chen, B. B.; Clare, M.; Desai, B. N.; Djurie, S. W.; Docter, S. H.; Gasiecki, A. F.; Haack, R. A.; Liang, Chi-Dean; Miyashiro, J. M.; Penning, T. D.; Russell, M. A. and Yu, S. Siu-tzyy: PCT Int. Appl. WO 96 11,192 (1996); Chem. Abstr., **125** (11) 142545e (1996).
258 Changani, V. S.; Kalavadia, A. V.; Manvar, U. V. and Joshi, G. K.: J. Indian Chem. Soc., **66** (1) 63-64 (1989).
259 Chao, T. S; Hutchison, D. A. and Kjonaas, M.: Ind. Eng. Chem. Prod. Res. Dev., **23** (1) 21-27 (1984); Chem. Abstr., **100**, 70883x (1984).
260 Chao, T. S. and Kjonaas, M.: Prepr.- Am. Chem. Soc., Div. Pet. Chem., **27** (2) 362-379 (1982); Chem. Abstr., **99**, 160972r (1983).
261 Chao, T. S.; Kjonaas, M. and DeJovine, J.: Prepr., Div. Pet. Chem., Am. Chem. Soc., **24** (3) 836-846 (1979); Chem. Abstr., **95**, 9579q (1981).

262 Chardonneus, L.; Laroche, B. and Gamba, G.: Helv. Chim. Acta, **48**, 1800-1803 (1965).
263 Chardonneus, L. and Schlapbach, W.: Helv. Chim. Acta, **29**, 1413-1424 (1946).
264 Chardonneus, L. and Würmli, A.: Helv. Chim. Acta, **29**, 922-928 (1946).
265 Charlesworth, E. H. and Charleson, P.: Can. J. Chem., **46**, 1843-1847 (1968).
266 Chatterjea, J. N.; Gupta, S. N. P. and Mehrotra, V. N.: J. Indian Chem. Soc., **42** (4) 205-210 (1965).
267 Chaudet, J. H. and Tamblyn, J. W.: SPE (Soc. Plastics Engrs.) Trans., **1**, 57-62 (1961); Chem. Abstr., **56**, 4936c (1962).
268 Chênevert, R. and Voyer, N.: Tetrahedron Lett., **25** (44) 5007-5008 (1984).
269 Chiang, Hung-Cheh and Chien, Chung-Hsin: Hua Hsueh, **1**, 7-9 (1979); Chem. Abstr., **94**, 46445t (1981).
270 Chiavari, G.; Conciali, V. and Vitali, P.: J. Chromatogr., **249** (2) 385-392 (1982).
271 Ching, Ta Yen: U.S. 4,288,631 (1981); Chem. Abstr., **95**, 204987y (1981).
272 Chu, F. and Hawker, C. J.: Polym. Bull. (Berlin), **30** (3) 265-272 (1993); Chem. Abstr., **119**, 28729u (1993).
273 Ciamician, G. and Silber, P.: Ber. Dtsch. Chem. Ges., **24**, 299-301 (1891).
274 Ciamician, G. and Silber, P.: Ber. Dtsch. Chem. Ges., **27**, 409-426 (1894).
275 Ciamician, G. and Silber, P.: Ber. Dtsch. Chem. Ges., **27**, 1497-1501 (1894).
276 Ciamician, G. and Silber, P.: Ber. Dtsch. Chem. Ges., **27**, 1627-1633 (1894).
277 Ciamician, G. and Silber, P.: Ber. Dtsch. Chem. Ges., **28**, 1393-1398 (1895).
278 Ciamician, G. and Silber, P.: Gazz. Chim. Ital., **22** (1) 461-492 (1892).
279 Ciba-Geigy A.-G., Jpn. Tokkyo Koho 80, 12,586 (1980); Chem. Abstr., **93**, 195471e (1980).
280 Clark, G. A. and Havens, C. B.: U.S. 2,891,996 (1959); Chem. Abstr., **54**, 432e (1960).
281 Clark, G. A. and Havens, C. B.: U.S. 3,072,602 (1963); Chem. Abstr., **58**, 6992c (1963).
282 Coffield, T. H.; Filbey, A. H.; Ecke, G. G. and Kolka, A. J.: J. Am. Chem. Soc., **79**, 5019-5023 (1957).
283 Cohn, P.: Monatsh. Chem., **17**, 102-109 (1896).
284 Colegate, S. M.; Hewgill, F. R. and Howie, G. B.: Aust. J. Chem., **28** (2) 343-353 (1975).
285 Coleman, R. A. and Weicksel, J. A.: Modern Plastics, **36** (12) 117, 119, 121, 198, 200 (1959); Chem. Abstr., **53**, 19435h (1959).
286 Coll, G.; Costa, A.; Deya, P. M. and Saa, J. M.: Tetrahedron Lett., **32** (2) 263-266 (1991).
287 Colquhoun, H. M.: Polym. Commun., **29** (6) 154-155 (1988); Chem. Abstr., **109**, 93743b (1988).
288 Colquhoun, H. M.; Daniels, J. A. and Lewis, D. F.: Eur. Pat. Appl. EP 232,992 (1987); Chem. Abstr., **109**, 149073c (1988).
289 Comarmond, J.; Purcell, T. and Zard, L.: Eur. Pat. Appl. EP 239,461 (1987); Chem. Abstr., **108**, 150318x (1988).
290 Comarmond, J.; Purcell, T. and Zard, L.: Fr. Demande FR 2,575,469 (1986); Chem. Abstr., **107**, 134303n (1987).
291 Comarmond, J.; Purcell, T. and Zard, L.: Fr. Demande FR 2,575,470 (1986); Chem. Abstr., **106**, 102283u (1987).
292 Compagnie de Saint-Gobain, Fr. Addn. 2,064,557 (1971); Chem. Abstr., **76**, 153344j (1972).
293 Cook, C. D. and Gilmour, N. D.: J. Org. Chem., **25**, 1429-1431 (1960).
294 Cooper, D. G.; King, R. J. and Brown, T. H.: PCT Int. Appl. WO 95 11,238 (1995); Chem. Abstr., **123**, 55712x (1995).
295 Cooper, D. G.; King, R. J. and Brown, T. H.: PCT Int. Appl. WO 95 11,240 (1995); Chem. Abstr., **123**, 55711w (1995).
296 Copping, L. G.; Kerry, J. C.; Watkins, T. I.; Willis, R. J. and Palmer, B. H.: U.S. US 4,344,893 (1982); Chem. Abstr., **98**, 106973h (1983).
297 Coppinger, G. M. and Bell, E. R.: J. Phys. Chem., **70** (11) 3479-3489 (1966).
298 Cornu-Chagnon, M.-C.; Dupont, H. and Edgar, A.: Fundam. Appl. Toxicol., **26** (1) 63-74 (1995); Chem. Abstr., **123**, 47328m (1995).
299 Correa, Dirceu de B. and Gottlieb, O. R.: Phytochemistry, **14** (1) 271-272 (1975).
300 Cotterill, P. J. and Scheinmann, F.: J. Chem. Soc., Perkin Trans. 1, 2353-2357 (1980).
301 Coupek, J.; Pokorny, S.; Protivova, J.; Holcik, J.; Karvas, M. and Pospisil, J.: J. Chromatogr., **65** (1) 279-286 (1972).
302 Cox, E. H.: J. Am. Chem. Soc., **49**, 1028-1030 (1927).

303 Cox, E.: J. Am. Chem. Soc., **52**, 352-358 (1930).
304 Craven, W. J.: U.S. 3,000,857 (1959); Chem. Abstr., **56**, 1626g (1962).
305 Creveling, C. R.; Morris, N.; Shimizu, H.; Ong, H. and Daly, J.: Mol. Pharmacol., **8** (4) 398-409 (1972); Chem. Abstr., **77**, 71863j (1972).
306 Criodain, T. O.; O'Sullivan, M.; Meegan, M. J. and Donnelly, D. M. X.: Phytochemistry, **20** (5) 1089-1092 (1981).
307 Croft, T. S.: Phosphorous Sulfur, **2** (1-2-3) 129-132 (1976).
308 Cullinane, N. M. and Edwards, B. F. R.: J. Appl. Chem., 133-136 (1959).
309 Cullinane, N. M. and Edwards, B. F. R.: J. Chem. Soc., 434-438 (1958).
310 Cullinane, N. M. and Edwards, B. F. R.: J. Chem. Soc., 1311-1312 (1958).
311 Cullinane, N. M. and Edwards, B. F. R.: J. Chem. Soc., 2926-2929 (1958).
312 Cullinane, N. M.; Morgan, N. M. E. and Plummer, C. A. J.: Recl. Trav. Chim. Pays-Bas, **56**, 627-631 (1937).
313 Curtz, J.; Rudolph, C. H. G.; Schroeder, L.; Albert, G.; Rehnig, A. E. E. and Sieverding, E. G.: Can. Pat. Appl. CA 2,167,550 (1996); Chem. Abstr., **126** (1) 7819c (1997).
314 Czech, Z.: Ger. Offen. DE 19,501,025 (1996); Chem. Abstr., **125** (16) 196681q (1996).
315 D.R.P. 295,495
316 Da Prada, M.; Zuercher, G.; Kettler, R. and Colzi, A.: Adv. Behav. Biol., **39** (Basal Ganglia 3) 723-732 (1991); Chem. Abstr., **115**, 198214g (1991).
317 Daglish, A. F. and Faulkner, D.: U.S. 2,659,709 (1953); Chem. Abstr., **48**, 4253f (1954).
318 Dahl, K. J. and Jansons, V.: Eur. Pat. Appl. EP 69,598 (1983); Chem. Abstr., **99**, 38197b (1983).
319 Dale, R. S. and Schorlemmer, C.: Justus Liebigs Ann. Chem., **217**, 387-388 (1883).
320 Dalvi, V. J. and Jadhav, G. V.: J. Indian Chem. Soc., **33** (11) 807-811 (1956).
321 Dalvi, V. J. and Jadhav, G. V.: J. Univ. Bombay, **25A**, Pt.3, 19-22 (1957); Chem. Abstr., **52**, 1967g (1968).
322 Dann, O. and Mylius, G.: Justus Liebigs Ann. Chem., **587**, 1-15 (1954).
323 Darms, R. and Monnier, C. E.: Eur. Pat. Appl. EP 157,740 (1985); Chem. Abstr., **104**, 225761q (1986).
324 Davis, B.: J. Labelled Compd. Radiopharm., **24** (10) 1221-1227 (1987); Chem. Abstr., **109**, 54439n (1988).
325 Dawson, I. M.; Hart, L. S. and Littler, J. S.: J. Chem. Soc., Perkin Trans. 2, **2** (9) 1601-1606 (1985).
326 Day, A. C.; Nabney, J. and Scott, A. I.: Proc. Chem. Soc. (London), 284-285 (1960).
327 De Cointet, P.; Loppinet, V.; Sornay, R.; Morinere, J. L.; Boucherle, A.; Renson, F. J.; Voegelin, H. and Dumont, C.: Chim. Ther., **8** (5) 574-587 (1973).
328 De Vos, F. and Slegers, G.: J. Chromatogr., A **692** (1 + 2) 97-102 (1995).
329 De Vos, F. and Slegers, G.: J. Labelled Compd. Radiopharm., **34** (7) 643-652 (1994).
330 Dean, F. M.; Goodchild, J.; Houghton, L. E.; Martin, J. A.; Morton, R. B.; Parton, B.; Price, A. W. and Somvichien, N.: Tetrahedron Lett., **35**, 4153-4159 (1966).
331 Dean, F. M.; Herbin, G. A.; Matkin, D. A.; Price, A. W. and Robinson, M. L.: J. Chem. Soc., Perkin Trans. 1, **9**, 1986-1993 (1980).
332 Delle Monache, G.; Gonzalez, J. G.; Delle Monache, F. and Marini Bettolo, G. B.: Phytochemistry, **19**, 2025-2028 (1980).
333 Desai, R. D. and Ekhlas, M.: Proc. Indian Acad. Sci., **8A**, 567-577 (1938).
334 Desai, R. D. and Mavani, C. K.: Curr. Sci., **10**, 524 (1941).
335 Desai, R. D. and Mavani, C. K.: J. Sci. Ind. Res., Sect. B, **12B**, 236-239 (1953).
336 Desai, R. D. and Mavani, C. K.: Proc. Indian Acad. Sci., **25A**, 341-344 (1947).
337 Desai, R. D. and Mavani, C. K.: Proc. Indian Acad. Sci., **25A**, 353-358 (1947).
338 Desai, R. D. and Mavani, C. K.: Proc. Indian Acad. Sci., **29A**, 269-273 (1949).
339 Desai, R. D. and Radha, K. S.: Proc. Indian Acad. Sci., **12A**, 46-49 (1940).
340 Desai, R. D. and Vakil, V. M.: Proc. Indian Acad. Sci., **12A**, 391-398 (1940).
341 Desai, S. M. and Trivedi, K. N.: Indian J. Chem., Sect. B, **24 B** (1) 47-50 (1985).
342 Desai, V. B.: M. Sc. Thesis, University of Bombay, Bombay (India) (1974).
343 Desbois, M.: Eur. Pat. Appl. EP 84,742 (1983); Chem. Abstr., **100**, 6085m (1984).
344 Destrade, C.; Nguyen Huu Tinh and Gasparoux, H.: Mol. Cryst. Liq. Cryst., **59** (3-4) 273-288 (1980); Chem. Abstr., **93**, 58557y (1980).

345 DeTar, D. F. and Relyea, D. I.: J. Am. Chem. Soc., 76, 1680-1685 (1954).
346 Dewar, M. J. S. and Hart, L. S.: Tetrahedron, 26, 973-1000 (1970).
347 Dewar, M. J. S. and Hart, L. S.: Tetrahedron, 26, 1001-1008 (1970).
348 Dewhirst, F. E.: U.S. 4,244,970 (1981); Chem. Abstr., 94, 156549n (1981).
349 Dey, B. B. and Raman, M. V. S., Laboratory manual of organic chemistry; S. Vishwanathan, ed., Madras, (India), 1957, page 279.
350 Dhar, K. L. and Kalla, Ashok K.: Phytochemistry, 13 (12) 2894 (1974).
351 Diaz-Mondejar, M. R. and Miranda, M. A.: Tetrahedron, 38 (10) 1523-1526 (1982).
352 Dillard, R. D.: Eur. Pat. Appl. EP 132,366 (1985); Chem. Abstr., 103, 141966b (1985).
353 Dillard, R. D.; Carr, F. P.; McCullough, D.; Haisch, K. D.; Rinkema, L. E. and Fleisch; J. H.: J. Med. Chem., 30 (5) 911-918 (1987).
354 Dilling, W. L.: J. Org. Chem., 31, 1045-1050 (1966).
355 Dilthey, W. and Harenberg, F.: J. Prakt. Chem., 136, 49-74 (1933).
356 Dischendorfer, O.: Monatsh. Chem., 62, 263-283 (1933).
357 Dischendorfer, O.: Monatsh. Chem., 66, 201-217 (1935).
358 Dischendorfer, O. and Limontschew, W.: Monatsh. Chem., 80, 741-748 (1949).
359 Dischendorfer, O. and Verdino, A.: Monatsh. Chem., 66, 255-284 (1935).
360 Dischendorfer, O. and Verdino, A.: Monatsh. Chem., 68, 10-20 (1936).
361 Djura, P. and Sargent, M. V.: J. Chem. Soc., Perkin Trans. 1, 4, 395-400 (1978).
362 DoAmaral, J. R.; Blanz, E. J. and French, F. A.: J. Med. Chem., 12 (1) 21-25 (1969).
363 Dobratz, E. H. and Kolka, A. J.: US 3,403,183 (1968); Chem. Abstr., 69, 106263c (1968).
364 Doebner, O.: Justus Liebigs Ann. Chem., 210, 246-284 (1881).
365 Doebner, O. and Stackmann, W.: Ber. Dtsch. Chem. Ges., 9, 1918-1920 (1876).
366 Doebner, O. and Stackmann, W.: Ber. Dtsch. Chem. Ges., 10, 1968-1972 (1877).
367 Doebner, O. and Stackmann, W.: Ber. Dtsch. Chem. Ges., 11, 2268-2274 (1878).
368 Doebner, O. and Wolff, W.: Ber. Dtsch. Chem. Ges., 12, 661-663 (1879).
369 Donnelly, B. J.; Donnelly, D. M. X. and O'Sullivan, A. M.: Tetrahedron, 24 (5) 2617-2622 (1968).
370 Donnelly, D. M. X.; Criodain, T. O. and O'Sullivan, M.: Proc. R. Ir. Acad., Sect. B, 83B (1-16) 39-48 (1983); Chem. Abstr., 100, 20402u (1984).
371 Donnelly, D. M. X.; O'Reilly, J. and Whalley, W. B.: Phytochemistry, 14 (10) 2287-2290 (1975).
372 Donnelly, D. M. X.; Thompson, J. C.; Whalley, W. B. and Ahmad, S.: J. Chem. Soc., Perkin Trans., 16, 1737-1745 (1973).
373 Downey, P. M. and Zerbe, R. O.: US 2,670,382 (1954); Chem. Abstr., 49, 4020b (1955).
374 Dubini, M.; Cicchetti, O.; Vicario, G. P. and Bua, E.: Eur. Polym. J., 3 (3) 473-479 (1967); Chem. Abstr., 67, 91187p (1967).
375 Duennenberger, M. and Schellenbaum, M.: Ger. Offen. 2,033,720 (1971); Chem. Abstr., 75, 117482y (1971).
376 Dumitrescu, V. and Dumitrescu, N.: Bul. Inst. Politeh. "Gheorghe Gheorghiu-Dej" Bucuresti, Ser. Chim., 45, 72-78 (1983); Chem. Abstr., 101, 239502f (1984).
377 Durisinova, L. and Bellus, D.: J. Chromatogr., 32 (3) 584-587 (1968).
378 Durmis, J.; Balogh, A.; Caucik, P. and Karvas, M.: Brit. 1,396,240 (1975); Chem. Abstr., 83, 165164h (1975).
379 Durmis, J.; Caucik, P.; Balogh, A.; Karvas, M. and Holcik, J.: Czech. 157,337 (1975); Chem. Abstr., 83, 96742f (1975).
380 Durmis, J.; Karvas, M.; Balogh, A. and Caucik, P.: Ger. Offen. 2,350,180 (1975); Chem. Abstr., 83, 58438d (1975).
381 Durmis, J.; Karvas, M.; Caucik, P. and Holcik, J.: Eur. Polym. J., 11 (3) 219-222 (1975).
382 Dziomko, V. M.; Markovich, I. S. and Kruglova, N. V.: Khim. Geterotsikl. Soedin., 3, 536-537 (1968); Chem. Abstr., 69, 96653p (1968).
383 E. I. du Pont de Nemours & Co. Brit. 835,841 (1960); Chem. Abstr., 54, 21850h (1960).
384 Eastman Kodak Co., French Patent 1,256,112 (1961).
385 Effenberger, F. and Gutmann, R.: Chem. Ber., 115, 1089-1102 (1982).
386 Effenberger, F.; Klenk, H. and Reiter, P. L.: Angew. Chem., 85 (18) 819-820 (1973).
387 Eggensperger, H.; Diehl, K. H. and Kloss, W.: Ger. 1,768,599 (1971); Chem. Abstr., 76, 85557d (1972).

388 Eggensperger, H.; Franzen, V. and Kloss, W.: Brit. 1,186,818 (1970); Chem. Abstr., 72, 132325f (1970).
389 Egyed, J.; Furka, A. and Szell, T.: Acta Phys. Chem. Szeged, 11, 51-54 (1965).
390 Eichenauer, U. and Neumann, P.: Ger. Offen. DE 3,831,092 (1990); Chem. Abstr., 113, 114819f (1990).
391 Eichhoff, H. J.; Kämmerer, H. and Weller, D.: Makromol. Chem., 132, 163-177 (1970).
392 Eiden, F.; Leister, H. P. and Mayer, D.: Arzneim. Forsch./ Drug Res., 33 (1) 101-105 (1983).
393 Eiglmeier, K.: Ger. Offen., 2,451,037 (1976); Chem. Abstr., 85, 46202a (1976).
394 Eiglmeier, K. and Schulz, J.: Ger. Offen. DE 3,206,129 (1982); Chem. Abstr., 98, 106978p (1983).
395 Eisai Co., Ltd.: Jpn. Kokai Tokkyo Koho JP 82,114,509 (1982); Chem. Abstr., 97, 188282y (1982).
396 Elborne, W. : Pharm. J., 24 (3) 168-171 (1893).
397 Ellis, R. C.; Whalley, W. B. and Ball, K.: Chem. Commun., 16, 803-804 (1967).
398 Ellis, R. C.; Whalley, W. B. and Ball, K.: J. Chem. Soc., Perkin Trans. 1, 13, 1377-1382 (1976).
399 Endo, T.; Takada, T. and Komatsu, S.: Jpn. Kokai Tokkyo Koho JP 05,125,180 [93,125,180] (1993); Chem. Abstr., 120, 9190r (1994).
400 Evans, D.; Cracknell, M. E.; Saunders, J. C.; Smith, C. E.; Williamson, W. R. N.; Dawson, W. and Sweatman, W. J.: J. Med. Chem., 30 (8) 1321-1327 (1987).
401 Evans, D.; Saunders, J. C. and Williamson, W. R. N.: Brit. 1,586,466 (1981); Chem. Abstr., 95, 61748h (1981).
402 Evliya, H. and Olcay, A.: Holzforschung, 28 (4) 130-135 (1974); Chem. Abstr., 82, 60215b (1975).
403 Eyton, W. B.; Ollis, W. D.; Fineberg, M.; Gottlieb, O. R.; Guimaraes, I. S. de S. and Magalhaes, M. T.: Tetrahedron, 21, 2697-2705 (1965).
404 Fabre, J. L.; James, C. and Lave, D.: Eur. Pat. Appl. EP 253,711 (1988); Chem. Abstr., 109, 128993n (1988).
405 Faith, H. E.; Bahler, M. E. and Florestano, H. J.: J. Am. Chem. Soc., 77, 543-547 (1955).
406 Farbenfabriken Bayer A.-G.: Neth. Appl. 6,413,419 (1965); Chem. Abstr., 63, 16557f (1965).
407 Farooqui, M Y. H. and Metcalf, R. L.: Pestic. Biochem. Physiol., 19 (2) 210-220 (1983); Chem. Abstr., 98, 193319r (1983).
408 Farraj, N. F.; Davis, S. S.; Parr, G. D. and Stevens, H. N. E.: Int. J. Pharm., 43 (1-2) 93-100 (1988); Chem. Abstr., 108, 192701x (1988).
409 Farraj, N. F.; Davis, S. S.; Parr, G. D. and Stevens, H. N. E.: Int. J. Pharm., 46 (3) 231-239 (1988); Chem. Abstr., 112, 221862d (1988).
410 Farraj, N. F.; Davis, S. S.; Parr, G. D. and Stevens, H. N. E.: Pharm. Res., 4 (1) 28-32 (1987).
411 Farraj, N. F.; Davis, S. S.; Parr, G. D. and Stevens, H. N. E.: Pharm. Res., 5 (4) 226-231 (1988).
412 Ferrandes, B.; Durand, A.; Fraisse-Andre, J. and Morselli, P. L.: Metab. Antiepileptic Drugs, [Proc. Workshop], 183-190 (1982) (Pub. 1984); Chem. Abstr., 101, 103519j (1984).
413 Fieser, L. F. and Bradsher, C. K.: J. Am. Chem. Soc., 58, 2337-2338 (1936).
414 Findlay, G. H. and Nel, S. J.: Brit. J. Dermatol., Suppl., 7, 44-49 (1971); Chem. Abstr., 76, 121420q (1972).
415 Findlay, J. W. A.; Gupta, P. and Lewis, J. R.: J. Chem. Soc. C, 19, 2761-2762 (1969).
416 Finlay-Jones, P. F.; Sala, T. and Sargent, M. V.: J. Chem. Soc., Perkin Trans. 1, 3, 874-876 (1981).
417 Finnegan, R. A. and Knutson, D.: Chem. Ind. (London), 1837-1838 (1965).
418 Finnegan, R. A. and Knutson, D.: Tetrahedron Lett., 30, 3429-3432 (1968).
419 Finnegan, R. A. and Mattice, J. J.: Tetrahedron, 21, 1015-1026 (1965).
420 Finnegan, R. A. and Merkel, K. E.: J. Org. Chem., 37 (19) 2986-2989 (1972).
421 Finnegan, R. A.; Merkel, K. E. and Patel, J. K.: J. Pharm. Sci., 62 (3) 483-485 (1973).
422 Finnegan, R. A. and Patel, J. K.: J. Chem. Soc., Perkin Trans. 1, 15, 1896-1901 (1972).
423 Fischer, A.; Greig, C. C. and Roederer, R.: Can. J. Chem., 53 (11) 1570-1578 (1975).

424 Fischer, E.: Ber. Dtsch. Chem. Ges., **42**, 1015-1022 (1909).
425 Fischer, E.: Justus Liebigs Ann. Chem., **371**, 303-318 (1910).
426 Fischer, E. and Rapaport, M.: Ber. Dtsch. Chem. Ges., **46**, 2389-2401(1913).
427 Fleury, M. B.; Maurette, J. M. and Largeron, M.: Fr. Demande FR 2,714,381 (1995); Chem. Abstr., **123**, 340160x (1995).
428 Fomenko, A. S.; Abramova, T. M.; Platonova, E. P. and Furman, E. G.: Vysokomol. Soedin., Ser. B, **11** (5) 387-390 (1969); Chem. Abstr., **71**, 39618y (1969).
429 Franck, B.; Radtke, V. and Zeidler, U.: Angew. Chem., Int. Ed. Engl., **6** (11) 952-953 (1967).
430 Franck, B.; Stöckigt, J.; Zeidler, U. and Gerhard, F.: Chem. Ber., **106** (4) 1198-1220 (1973).
431 Franck, B. and Zeidler, U.: Chem. Ber., **106** (4) 1182-1197 (1973).
432 Freeman, H. S.: PCT Int. Appl. WO 89 10,384 (1989); Chem. Abstr., **112**, 160482t (1990).
433 Freeman, H. S.; Hao, Z.; McIntosh, S. A.; Posey, J. C., Jr. and Hsu, W. N.: Dyes Pigm., **12** (3) 233-242 (1990); Chem. Abstr., **112**, 141216t (1990).
434 Freeman, H. S. and Posey, J. C., Jr.: Dyes Pigm., **20** (3) 171-195 (1992); Chem. Abstr., **118**, 82822m (1993).
435 Freyberg, D. P. and Mockler, G. M. and Sinn, E.: Inorg. Chem., **18** (3) 808-815 (1979).
436 Fromantin, J. P. M. J.: Brit. UK Pat. Appl. 1,016,460 (1979); Chem. Abstr., **93**, 46189s (1980).
437 Fromantin, J. P. M. J.: U.S. Patent 4,277,497 (1981).
438 Fujii, S.; Yaegashi, T.; Nakayama, T.; Sakurai, Y.; Nunomura, S. and Okutome, T.: Fr. Demande FR 2,500,825 (1982); Chem. Abstr., **98**, 88995k (1983).
439 Fujii, Y.; Kurokawa, T.; Ishida, S.; Yamaguchi, I. and Misato, T.: Nippon Noyaku Gakkaishi, 1 (4) 313-320 (1976); Chem. Abstr., **86**, 115837w (1977).
440 Fujio, M.; Mishima, M.; Tsuno, Y.; Yukawa, Y. and Takai, Y.: Bull. Chem. Soc. Jpn., **48** (7) 2127-2133 (1975).
441 Fujita, T.; Ishiguro, M.; Takahata, K. and Saeki, K.: Jpn. Kokai Tokkyo Koho JP 60,252,444 [85,252,444] (1985); Chem. Abstr., **104**, 168144q (1986).
442 Fukuoka, N. and Suzuki, M.: Jpn. Kokai Tokkyo Koho JP 01 71,835 [89 71,835] (1989); Chem. Abstr., **111**,173764r (1989).
443 Fukuoka, N. and Yasuda, H.: Jpn. Kokai Tokkyo Koho JP 01,153,651 [89,153,651] (1989); Chem. Abstr., **112**, 37420u (1990).
444 Fukuoka, S. and Matsuda, H.: Jpn. Kokai Tokkyo Koho JP 61 12,643 [86 12,643] (1986); Chem. Abstr., **105**, 42483b (1986).
445 Fukuoka, S. and Matsuda, H.: Jpn. Kokai Tokkyo Koho, JP 61 07,230 [86 07,230] (1986); Chem. Abstr., **104**, 206905y (1986).
446 Fukuoka, S.; Shimizu, A. and Yamataka, K.: Jpn. Kokai Tokkyo Koho JP 62,114,927 [87,114,927] (1987); Chem. Abstr., **108**, 21502d (1988).
447 Fukuoka, S. and Tojo, M.: Jpn. Kokai Tokkyo Koho JP 61,221,147 [86,221,147] (1986); Chem. Abstr., **107**, 84169e (1987).
448 Fukuoka, S. and Tojo, M.: Jpn. Kokai Tokkyo Koho JP 63 57,546 [88 57,546] (1988); Chem. Abstr., **109**, 92474c (1988).
449 Fung, A. K. L.; Morrison, D. E. and Pernet, A. G.: U.S. 4,268,691 (1981); Chem. Abstr., **95**, 97357u (1981).
450 Furka, A. and Szell, T.: Acta Phys. Chem. Szeged, **6**, 113-115 (1960).
451 Furka, A. and Szell, T.: Acta Phys. Chem. Szeged, **7**, 70-72 (1961).
452 Fusco, R. and Sannicolo, F.: J. Org. Chem., **46** (1) 90-92 (1981).
453 Fuson, R. C. and Hornberger, C., Jr., J. Org. Chem., **16**, 637-642 (1951).
454 Fuson, R. C.; Scott, S. L. and Speck, S. B.: J. Am. Chem. Soc., **63**, 2845-2846 (1941).
455 Gaekwad, Y. G. and Sethna, S.: J. Indian Chem. Soc., **55** (8) 794-800 (1978).
456 Gapinski, D. M.: Eur. Pat. Appl. EP 242,989 (1987); Chem. Abstr., **109**, 54488c (1988).
457 Garcia, E. E.; Benjamin, L. E.; Fryer, R. I.; Sternbach, L. H. and Archer, G.: J. Med. Chem., **15** (9) 986-987 (1972).
458 Garcia, H.; Martinez-Utrilla, R.; Miranda, M. A. and Roquet-Jalmar, M. F.: J. Chem. Res., Synop., **12**, 350-351 (1982).
459 Gardner, P. D.: J. Am. Chem. Soc., **77**, 4674-4675 (1955).
460 Garofano, T. and Werber, G.: Ann. Chim. (Rome), **50**, 245-276 (1960).
461 Gattermann, L.: Ber. Dtsch. Chem. Ges., **29**, 3034-3037 (1896).

462 Gattermann, L. and Rudt, H.: Ber. Dtsch. Chem. Ges., **27**, 2293-2297 (1894).
463 Gaur, V. B.; Shah, V. H. and Parikh, A. R.: Indian J. Heterocycl. Chem., **1** (3) 141-146 (1991).
464 Gay, M. and Lavault, S.: Eur. Pat. Appl. EP 401,134 (1990); Chem. Abstr., **114**, 248540n (1991).
465 Gazave, J. M.; Hayaux du Tilly-Achard, M. and Parrot, J. L.: Conv. Int. Polifenoli, [Relaz. Comun.], 135-141 (1975); Chem. Abstr., **86**, 83512t (1977).
466 Gazave, J. M.; Rancurel, A. and Grenier, G.: Ger. Offen. 2,501,443 (1975); Chem. Abstr., **83**, 188522n (1975).
467 Ge, Dalun; Liang, Xiaotian; Lu, Yuhua and Qi, Jianxin: Yaoxue Xuebao, **22** (11) 822-826 (1987); Chem. Abstr., **109**, 110207z (1988).
468 Geigy, J. R.; A.-G.: Belg. 652,265 (1964); Chem. Abstr., **64**, 14364g (1966).
469 General Mills, Inc., Fr. 2,093,173 (1972); Chem. Abstr., **77**, 114047r (1972).
470 Gerecs, A.; Szell, T. and Windholz, M.: Acta Chim. (Budapest), **3**, 459-467 (1953).
471 Ghosh, C. K. and Mukhopadhyay, K. K.: I. Indian Chem. Soc., **55** (1) 52-55 (1978).
472 Gibson, M. S.; Vines, S. M. and Walthew, J. W.: J. Chem. Soc., Perkin Trans. 1, **2**, 155-160 (1975).
473 Gierer, J.; Yang, E. and Reitberger, T.: Holzforschung, **50** (4) 353-359 (1996); Chem. Abstr., **125** (20) 250692j (1996).
474 Giesen, M.: Chim. peintures, **23**, 69-81 (1960); Chem. Abstr., **54**, 17905h (1960).
475 Giesen, M.: Fette, Seifen, Anstrichm., **61** (12) 1245-1251 (1959)
476 Giles, D. P.; Kerry, J. C.; Kozlik, A.; Palmer, B. H.; Shutler, S. W. and Willis, R. J.: Eur. Pat. Appl. 26,040 (1981); Chem. Abstr., **95**, 115059e (1981).
477 Gillet, G.; Fraisse-Andre, J.; Lee, C. R.; Dring, L. G. and Morselli, P. L.: J. Chromatogr., **230** (1) 154-161 (1982).
478 Giovannini, E.; Rosales, J. and de Souza, B.: Helv. Chim. Acta, **54** (7) 2111-2113 (1971).
479 Glahn, W. H. and Stanley, L. N.: U.S. 2,861,104 (1958); Chem. Abstr., **53**, 8081a (1959).
480 Glahn, W. H. and Stanley, U.S. Pat. 2,789,140 (1957); Chem. Abstr., **51**, 13927d (1957).
481 Glanz, K. D.: Eur. Pat. Appl. EP 100,196 (1984); Chem. Abstr., **101**, 31195a (1984).
482 Glanz, K. D.: U.S. US 4,470,057 (1984); Chem. Abstr., **102**, 15251f (1985).
483 Gol'denberg, V. I.; Shlyapintokh, V. Ya.; Postnikov, L. M. and Sukhanov, G. A.: Izv. Akad. Nauk SSSR, Ser. Khim., **10**, 2232-2236 (1968).
484 Gomberg, M. and Anderson, L. C.: J. Am. Chem. Soc., **47**, 2022-2033 (1925).
485 Gomberg, M. and Snow, H. R.: J. Am. Chem. Soc., **47**, 198-211 (1925).
486 Gonda, M.; Kitsukawa, K.; Suzuka, S.; Kanasugi, M.; Sato, Y. and Otomo, K.: Eur. Pat. Appl. EP 140,050 (1985); Chem. Abstr., **103**, 132484r (1985).
487 Goni, M. A. and Hedges, J. I.: Geochim. Cosmochim. Acta, **56** (11) 4025-4043 (1992).
488 Gordon, D. A.: U.S. Patent 3,086,988.
489 Gordon, D. A. and Burgert, B. E.: U.S. Patent 2,967,186 (1961).
490 Gore, P. H.; Smith, G. H. and Thorburn, S.: J. Chem. Soc. C, **1**, 650-652 (1971).
491 Goszczynski, S.; Kopczynski, T.; Szymanowski, J.; Borowiak, A. and Blaszczak, J.: Pol. 105,987 (1980); Chem. Abstr., **93**, 135805g (1980).
492 Gottfried, C. and Dutzer, M. J.: J. Appl. Polymer Sci., **5**, 612-619 (1961).
493 Gottlieb, O. R.; Fineberg, M.; Guimaraes, I. S. de S.; Magalhaes, M. T.; Ollis, W. D. and Eyton, W. B.: An. Acad. Brasil. Cienc., **36** (1) 33-34 (1964); Chem. Abstr., **62**, 5491h (1965).
494 Gottlieb, O. R. and Mors, W. B.: J. Am. Chem. Soc., **80**, 2263-2265 (1958).
495 Graebe, C. and Eichengrün, A.: Justus Liebigs Ann. Chem., **269**, 295-325 (1892).
496 Graebe, C. and Feer, A.: Ber. Dtsch. Chem. Ges., **19**, 2607-2614 (1886).
497 Graebe, C. and Ullmann, F.: Ber. Dtsch. Chem. Ges., **27**, 3483-3484 (1894).
498 Graebe, C. and Ullmann, F.: Ber. Dtsch. Chem. Ges., **29**, 824-825 (1896).
499 Graebe, C. and Ullmann, F.: Justus Liebigs Ann. Chem., **291**, 8-17 (1896).
500 Gray, D. N. and Burton, G.: J. Chem. Eng. Data, **11** (1) 59-60 (1966); Chem. Abstr., **65**, 62h (1966).
501 Gray, D. N. and Knight, R. D.: U.S. 3,387,035 (1968); Chem. Abstr., **69**, 35730q (1968).
502 Green, P. N. and Green, W. A.: Eur. Pat. Appl. EP 279,475 (1988); Chem. Abstr., **111**, 39004q (1989).

503 Gregson, M.: Ph. D Thesis, University of Sheffield, UK (1965).
504 Grenier, G. and Pacheco, H.: Chim. Ther., **7**, 408-414 (1966).
505 Gronowska, J.; Pilat, E. and Ruminski, J.: Rocz. Chem., **47**, 1949-1955 (1973).
506 Gronowska, J. and Ruminski, J.: Rocz. Chem., **45** (11) 1957-1965 (1971).
507 Grover, P. K.; Shah, G. D. and Shah, R. C.: J. Chem. Soc., 3982-3985 (1955).
508 Grucarevic, S. and Merz, V.: Ber. Dtsch. Chem. Ges., **6**, 1238-1246 (1873).
509 Grzywa, E. and Zdrojek, T.: Pol. PL 114,150 (1982); Chem. Abstr., **98**, 144821q (1983).
510 Gu, Caixian and Chen, Fangfang: Yiyao Gongye, **6**, 5-6 (1983); Chem. Abstr., **99**, 157938d (1983).
511 Guerry, P.; Jolidon, S. and Zurflueh, R.: Eur. Pat. Appl. EP 410,359 (1991); Chem. Abstr., **115**, 71122v (1991).
512 Guimaraes, I. S. de S.; Gottlieb, O. R.; Andrade, C. H. S. and Magalhaes, M. T.: Phytochemistry, **14** (5/6) 1452-1453 (1975).
513 Gupta, A. and Yavrouian, A. H.: U.S. US 4,310,650 (1982); Chem. Abstr., **96**,143553a (1982).
514 Gupta, A. R. and Saharia, G. S.: J. Indian Chem. Soc., **35** (2) 133-135 (1958).
515 Gysling, H. and Heller, H. J.: Kunststoffe, **51**, 13-17 (1961); Chem. Abstr., **55**, 11906f (1961).
516 Haas, G.; Neisius, K. H. and Stein, A.: Ger. Offen. DE 3,220,816 (1983); Chem. Abstr., **100**, 219047n (1984).
517 Habicht, E. and Zbinden, P.: Eur. Pat. Appl. EP 64,027 (1982); Chem. Abstr., **98**, 179352q (1983).
518 Habicht, E. and Zbinden, P.: Eur. Pat. Appl. EP 107,623 (1984); Chem. Abstr., **101**, 72715p (1984).
519 Habicht, E. and Zbinden, P.: U.S. US 4,517,184 (1985); Chem. Abstr., **103**, 87860x (1985).
520 Haga, T.; Fukutani, H.; Nagasaka, H.; Nishimura, A. and Imai, Y.: Japan Kokai 74 27,497 (1974); Chem. Abstr., **81**, 137030b (1974).
521 Haga, T.; Fukutani, H.; Nagasawa, H. and Nishimura, A.: Japan Kokai 74 01,545 (1974); Chem. Abstr., **81**, 3596g (1974).
522 Hagihara, T.: Shikizai Kyokaishi, **44** (10) 449-464 (1971); Chem. Abstr., **76**, 47424a (1972).
523 Hahn, E. and Neumann, P.: Ger. Offen. DE 3,814,781 (1989); Chem. Abstr., **112**, 216426x (1990).
524 Hahn, E. and Neumann, P.: U.S. US 4,885,396 (1989); Chem. Abstr., **113**, 23370m (1990).
525 Haigh, D. B. and Hindley, R. M.: Eur. Pat. Appl. EP 299,620 (1989); Chem. Abstr., **111**, 7394a (1989).
526 Hakimelahi, G. H. and Moshfegh, A. A.: Helv. Chim. Acta, **64** (2) 599-609 (1981).
527 Hamada, Chiomatsu: Science Repts. Tôhoku Imp. Univ., First Ser. **22**, 55-60 (1933); Chem. Abstr., **27**, 3928[5] (1933).
528 Hamana, R.; Saito, M. and Mori, S.: Jpn. Kokai Tokkyo Koho JP 61,200,941 [86,200,941] (1986); Chem. Abstr., **106**, 84165a (1987).
529 Hamann, P. R.; Hinman, L.; Hollander, I.; Holcomb, R.; Tsou, H. R.; Hallett, W. and Weiss, M. J.: Eur. Pat. Appl. EP 689,845 (1996); Chem. Abstr., **124**, 261609t (1996).
530 Hamazaki, Y.; Kawabata, S.; Yamamoto, T.: Shiraishi, Y.; Ueno, A. Amemiya, K. and Saga, K.: U.S. 4,124,726 (1978); Chem. Abstr., **90**, 168289h (1979).
531 Hamazaki, Y.; Kawabata, S.; Yamamoto, T.; Shiraishi, Y.; Ueno, A.; Amemiya, K. and Saga, K.: Ger. Offen. 2,659,580 (1977); Chem. Abstr., **87**, 167732y (1977).
532 Hamazaki, Y.; Yamamoto, T.; Seri, K.; Sakasai, M.; Sato, R. and Ishiyama, N.: Ger. Offen. 2,757,459 (1978); Chem. Abstr., **89**, 146626t (1978).
533 Hanabusa, K.; Shirai, H.; Hojo, N.; Kondo, K. and Takemoto, K.: Makromol. Chem., **183** (5) 1101-1111 (1982).
534 Hantzsch, A. and Blackler, M. B.: Ber. Dtsch. Chem. Ges., **39**, 3080-3102 (1906).
535 Harada, H.; Matsushita, Y.; Yodo, M.; Nakamura, M. and Yonetani, Y.: Chem. Pharm. Bull., **35** (8) 3215-3226 (1987).
536 Hardegger, E.; Widmer, E.; Steiner, K. and Pfiffner, A.: Helv. Chim. Acta, **47**, (7) 2027-2030 (1964).
537 Hardy, W. B.: U.S. 2,890,201 (1959); Chem. Abstr., **54**, 4045b (1960).
538 Hardy, W. B.: U.S. 2,890,193 (1959); Chem. Abstr., **54**, 4044h (1960).

539 Hardy, W. B. and Forster, W. S. : U.S. 2,773,903 (1956); Chem. Abstr., **51**, 16552d (1957).
540 Hardy, W. B.; Forster, W. S. and Coleman, R. A.: U.S. 2,853,521 (1958); Chem. Abstr., **53**, 5206b (1959).
541 Hardy, W. B.; Forster, W. S. and Coleman, R. A.: Ger. Offen., 1,093,374 (1956).
542 Harper, S. D. and Arduengo, A. J.: J. Am. Chem. Soc., **104** (9) 2497-2501 (1982).
543 Harris, C. M.; Roberson, J. S. and Harris, T. M.: J. Am. Chem. Soc., **98** (17) 5380-5386 (1976).
544 Harris, J. F., Jr.: U.S. US 4,447,592 (1984); Chem. Abstr., **101**, 132352e (1984).
545 Harrison, C. R.; Hodge, P. and Khan, N.: J. Chem. Soc., Perkin Trans. 1, 1592-1594 (1980).
546 Hart, L. S.: Ph. D. (London) (1967).
547 Hartmann, C. and Gattermann, L.: Ber. Dtsch. Chem. Ges., **25**, 3531-3534 (1892).
548 Hauser, C. R. and Man, E. H.: J. Org. Chem., **17**, 390-396 (1952).
549 Havens, C. B. and Clark, G. A.: U.S.P. 2,964,554.
550 Hawker, C. J. and Chu, Fengkui: Macromolecules, **29** (12) 4370-4380 (1996).
551 Hawkins, E. G. E.: J. Appl. Chem., **6**, 131-136 (1956).
552 Hay, J. V. and Harris, T. M.: J. Chem. Soc., Chem. Commun., **16**, 953-955 (1972).
553 Hayashi, I.; Ogihara, K.; Itikawa, T. and Shimizu, K.: Eur. Pat. Appl. EP 127,342 (1984); Chem. Abstr., **102**, 185507m (1985).
554 Hayashi, I.; Ogihara, K. and Shimizu, K.: Bull. Chem. Soc. Jpn., **56** (8) 2432-2437 (1983).
555 Hayashi, M.: J. Prakt. Chem., **123**, 289-312 (1929).
556 Hayashi, T.; Okina, G.; Ryu, K. and Ryu, S.: Jpn. Kokai Tokkyo Koho JP 07 82, 263 [95 82,263] (1995); Chem. Abstr., **123**, 285780x (1996).
557 Head, F. S. H.: Brit. 1,088,755 (1967); Chem. Abstr., **69**, 10258d (1968).
558 Head, F. S. H.: J. Chem. Soc. C, 34-37 (1969).
559 Hechenbleikner, I.: Ger. Offen., 2,061,019 (1971); Chem. Abstr., **76**, 4503h (1972).
560 Heeres, J.: U.S. 4,101,665 (1978); Chem. Abstr., **90**, 87466m (1979).
561 Heiber, F.: Ber. Dtsch. Chem. Ges., **24**, 3677-3687 (1891).
562 Heller, G.: Ber. Dtsch. Chem. Ges., **46**, 1497-1504 (1913).
563 Hellwinkel, D.; Laemmerzahl, F. and Hofmann, G.: Chem. Ber., **116** (10) 3375-3405 (1983).
564 Hendrickson, J. B.; Ramsay, M. V. J. and Kelly, T. R.: J. Am. Chem. Soc., **94** (19) 6834-6843 (1972).
565 Henry, R. A. and Tait, C. W.: U.S. 3,070,473 (1962); Chem. Abstr., **58**, 6639f (1963).
566 Hensel, W. and Hoyer, H.: Z. Physik Chem. (Frankfurt), **36** (5/6) 387-391 (1963).
567 Hercouet, A. and Le Corre, M.: Tetrahedron, **37** (16) 2867-2873 (1981).
568 Herzig, J. and Hofmann, B.: Ber. Dtsch. Chem. Ges., **41**, 143-145 (1908).
569 Hesse, O.: Ber. Dtsch. Chem. Ges., **26**, 2790-2795 (1893).
570 Hibbert, D. B.; Sandall, J. P. B.; Lovering, J. R.; Ridd, J. H. and Yousaf, T. I.: J. Chem. Soc., Perkin Trans. 2, **9**, 1739-1742 (1988).
571 Hiller, Gary Lynn: Ger. Offen. 2,212,302 (1972); Chem. Abstr., **78**, 91034p (1973).
572 Hinnen, A.: Ger. Offen. 1,949,867 (1970); Chem. Abstr., **73**, 77083u (1970).
573 Hinshaw, J. C.; Toner, J. L. and Reynolds, G. A.: Eur. Pat. Appl. EP 68,875 (1983); Chem. Abstr., **98**, 218134s (1983).
574 Hishmat, O. H. and Abd El Rahman, A. H.: J. Prakt. Chem., **315** (2) 227-234 (1973).
575 Hishmat, O. H. and Abd El Rahman, A. H.: Z. Naturforsch., B: Anorg. Chem., Org. Chem., **31B** (8) 1138-1141 (1976).
576 Hishmat, O. H. and Abd el Rahman, A. H.: Justus Liebigs Ann. Chem., **733**, 120-124 (1970).
577 Hishmat, O. H.; Abd El-Rahman, A. H. and Kandeel, Ez Eldin M.: Indian J. Chem., Sect. B, **14B** (1) 41-42 (1976).
578 Hishmat, O. H.; Abd-el-Rahman, A. H. and Wahba, N. F.: Pol. J. Chem., **52** (1) 87-95 (1978).
579 Hishmat, O. M.; Abd El Rahman, A. H.; Khalil, K. M. A. and Atta, S. M. S.: J. Indian Chem. Soc., **58** (7) 697-700 (1981).
580 Hlasiwetz, H. and Pfaundler, L.: Justus Liebigs Ann. Chem., **127**, 351-361 (1863).
581 Hodogaya Chemical Co., Ltd. : Jpn. Kokai Tokkyo Koho JP 59,170,033 [84,170,033] (1984); Chem. Abstr., **102**, 78568k (1985).
582 Hoefnagel, A. J. and van Bekkum, H.: Appl. Catal., **97** (2) 87-102 (1993).

583   Hoefnagel, A. J. and van Bekkum, H.: Appl. Catal., **102** (1) N16 (1993); Chem. Abstr., **119**, 273835y (1993).
584   Hoefnagel, A. J. and van Bekkum, H.: PCT Int. Appl. WO 93 22,268 (1993); Chem. Abstr., **120**, 191347x (1994).
585   Hoesch, K.: Ber. Dtsch. Chem. Ges., **48**, 1122-1133 (1915).
586   Hoesch, K. and Zarzecki, T.: Ber. Dtsch. Chem. Ges., **50**, 462-468 (1917).
587   Hoeschele, G. K. and Verbanc, J. J.: U.S. 3,008,995 (1961); Chem. Abstr., **57**, 736f (1962).
588   Hoeschele, G. K. and Verbanc, J. J.: U.S. 3,113,880 (1963); Chem. Abstr., **60**, 6995g (1964).
589   Hoffmann-La Roche, F. & Co., A.-G.: Brit. 929,254 (1963); Chem. Abstr., **60**, 2827g (1964).
590   Hoffmann-La Roche, F. & Co., A.-G.: Ger. 1,145,626 (1963); Chem. Abstr., **60**, 12033h (1964).
591   Hoffmann-La Roche, F., & Co., A.-G.: Fr. 1,531,765 (1968); Chem. Abstr., **71**, 112635h (1969).
592   Hoffmann-La Roche, F., & Co., A.-G.: Fr. 1489246 (1967); Chem. Abstr., **69**, 10254z (1968).
593   Hogberg, Bertil; Fex, Hans; Bracke, Bo Fredholm; Perklev, Torsten and Veige, Sten: Ger. Offen. 2,240,229 (1973); Chem. Abstr., **78**, 147556e (1973).
594   Hohlweg, R.; Joergensen, T. K.; Andersen, K. E.; Olsen, U. B.; Madsen, P.; Polivka, Z.; Koenigova, O.; Miksik, F.; Kovandova, M.; Silhankova, A. and Sindelar, K.: PCT Int. Appl. WO 97 11,071 (1997); Chem. Abstr., **126** (23) 305593f (1997).
595   Hokko Chemical Industry Co., Ltd.: Jpn. Kokai Tokkyo Koho JP 57,175,109 [82,175,109] (1982); Chem. Abstr., **98**, 139011w (1983).
596   Horace A. De Wald: U.S. 3,272,841 (1966); Chem. Abstr., **66**, 2349h (1967).
597   Horace A. De Wald: U.S. 3,288,806.(1966); Chem. Abstr., **66**, 37765s (1967).
598   Horne, S. and Rodrigo, R.: J. Org. Chem., **55** (15) 4520-4522 (1990).
599   Horner, L. and Baston, D. W.: Justus Liebigs Ann. Chem., **5-6**, 910-935 (1973).
600   Horspool, W. M. and Pauson, P. L.: J. Chem. Soc., 5162-5166 (1965).
601   Hosaka, Y.; Nozue, I.; Takatori, M. and Harita, Y.: Jpn. Kokai Tokkyo Koho JP 62,150,245 [87,150,245] (1987); Chem. Abstr., **108**, 46855j (1988).
602   Hosler, J. F. and Storfer, S. J.: U.S. 2,928,878 (1960); Chem. Abstr., **54**, 14195f (1960).
603   Hotta, H. and Akasaka, M.: Jpn. Kokai Tokkyo Koho JP 62,138,422 [87,138,422] (1987); Chem. Abstr., **107**, 183352m (1987).
604   Houben, J. and Fischer, W.: J. Prakt. Chem., **123**, 89-109 (1929).
605   Houtman, T., Jr.: U.S. 2,419,553 (1947); Chem. Abstr., **41**, 5150d (1947).
606   Hrdlovic, P. and Bellus, D.: Chem. Zvesti, **21** (6) 410-416 (1967).
607   Hrdlovic, P. and Bellus, D.: Chem. Zvesti, **22** (7) 508-513 (1968).
608   Hrdlovic, P.; Bellus, D. and Lazar, M.: Collect. Czech. Chem. Commun., **33**, 59-67 (1968).
609   Hrdlovic, P.; Schubertova, N.; Arventiev, B. and Wexler, H.: Collect. Czech. Chem. Commun., **36** (5) 1948-1954 (1971).
610   Hrdlovic, T.; Schubertova, N. and Pavlovcik, R.: Collect. Czech. Chem. Commun., **36** (5) 1942-1947 (1971).
611   Huber, H. and Brunner, K.: Monatsh. Chem., **56**, 322-330 (1930).
612   Huls, R. and Hubert, A.: Bull. Soc. Chim. Belg., **65**, 596-602 (1956).
613   Humbert, D.; Dagnaux, M.; Cohen, N. C.; Fournex, R. and Clemence, F.: Eur. J. Med. Chem. - Chim. Ther., **18** (1) 67-78 (1983).
614   Humphrey, J. S.: Polym. Prepr., Amer. Chem. Soc., Div. Polym. Chem., **9** (1) 453-460 (1968); Chem. Abstr., **71**, 125332g (1969).
615   Hung, W. M. and Su, Kai C.: Eur. Pat. Appl. EP 388,356 (1990); Chem. Abstr., **114**, 150261x (1991).
616   Huston, R. C. and Robinson, K. R.: J. Am. Chem. Soc., **73**, 2483-2486 (1951).
617   Il'in, S. N.; Biryukov, V. P.; Levin, P. I. and Zimin, Yu. B.: U.S.S.R. 307,085 (1971); Chem. Abstr., **76**, 4510h (1972).
618   Imaki, K.; Arai, Y. and Okegawa, T.: Eur. Pat. Appl. EP 210,772 (1987); Chem. Abstr., **106**, 175951v (1987).
619   Imperial Chemical Industries Ltd.: Neth. Appl. 6,401,477 (1964); Chem. Abstr., **64**, 3781a (1966).

620 Inoue, Y.; Hata, K. and Oishi, T.: Jpn. Kokai Tokkyo Koho JP 61,293,946 [86,293,946] (1986); Chem. Abstr., **107**, 39398p (1987).
621 Ioffé, I. S.: Zh. Obshch. Khim., **17** (7) 1359-1369 (1947).
622 Ioffé, I. S.: Zh. Obshch. Khim., **20** (2) 346-355 (1950).
623 Irwin, R. S.: Eur. Pat. Appl. 26,991 (1981); Chem. Abstr., **95**, 26546u (1981).
624 Irwin, R. S.: Ger. Offen., 2,932,178 (1980); Chem. Abstr., **94**, 4880a (1981).
625 Irwin, R. S.: U.S. 4,245,082 (1981); Chem. Abstr., **94**, 123029q (1981).
626 Irwin, R. S.; Sweeny, W.; Gardner, K. H.; Gochanour, C. R. and Weinberg, M.: Macromolecules, **22** (3) 1065-1074 (1989).
627 Ishida, S.; Hashida, Y.; Shizuka, H. and Matsui, K.: Bull. Chem. Soc. Jpn., **52** (4) 1135-1138 (1979).
628 Ishigami, M.; Arimoto, K. and Hamada, M.: Japan 74 48,315 (1974); Chem. Abstr., **82**, 155806j (1975).
629 Ishikura, S.; Kanda, K. and Mizuguchi, R.: Jpn. Kokai Tokkyo Koho JP 61 23,663 [86 23,663] (1986); Chem. Abstr., **105**, 116678t (1986).
630 Ishikura, Y.; Kino, M. and Kanehisa, S.: Jpn. Kokai Tokkyo Koho 03,170,414 [91,170,414] (1991); Chem. Abstr., **116**, 136247g (1992).JP
631 Ishizuka Glass Co., Ltd.: Jpn. Kokai Tokkyo Koho JP 57,165,466 [82,165,466]; Chem. Abstr., **98**, 181234w (1983).
632 Ismailov, A. G. and Salimova, B. A.: Zh. Org. Khim., **4** (1) 85-88 (1968).
633 Israelstam, S. S. and Stephen, H.: J. S. African Chem. Inst., **26**, 41-48 (1943); Chem. Abstr., **38**, 7315 (1944).
634 Itazaki, H.; Hayashi, K.; Matsuura, M.; Yonetani, Y. and Nakamura, M.: Chem. Pharm. Bull., **36** (9) 3404-3432 (1988).
635 Itazaki, H.; Hayashi, K.; Matsuura, M.; Yonetani, Y. and Nakamura, M.: Eur. Pat. Appl. EP 182,302 (1986); Chem. Abstr., **105**, 97483g (1986).
636 Itoh, H.; Konno, M.; Tokuhiro, T.; Iguchi, S. and Hayashi, M.: Brit. UK Pat. Appl. 2,026,480 (1980); Chem. Abstr., **93**, 167893a (1980).
637 Itoh, Y.; Ohne, K.; Tanaka, H.; Goto, S. and Ieda, S.: PCT Int. Appl. WO 94 26,738 (1994); Chem. Abstr., **123**, 143640c (1995).
638 Iwagaki, M.; Kono, J. and Kaguchi, H.: Jpn. Kokai Tokkyo Koho JP 62,136,641 [87,136,641] (1987); Chem. Abstr., **108**, 29489a (1988).
639 Izawa, T.; Fujii, Y. and Asaka, Y.: Nippon Noyaku Gakkaishi, 6 (2) 223-226 (1981); Chem. Abstr., **95**, 145173e (1981).
640 Izuoka, A.; Miya, S. and Sugawara, T.: Tetrahedron Lett., **29** (44) 5673-5676 (1988).
641 Jackson, L. B. and Waring, A. J.: J. Chem. Soc., Perkin Trans. 2, **11**, 1893-1898 (1990).
642 Jasching, W.: Kunststoffe, **52** (8) 458-464 (1962).
643 Jaunin, R.: Ger. Offen. 2,733,886 (1978); Chem. Abstr., **88**, 152424w (1978).
644 Jhaveri, D. B.; Thakor, V. M. and Naik, H. B.: Vidya, B, **19** (2) 149-152 (1976); Chem. Abstr., **87**, 201001w (1977).
645 Jobst, J. and Hesse, O.: Justus Liebigs Ann. Chem., **199**, 17-96 (1879).
646 Jochanan, Ramat: Israeli 39,037 (1975); Chem. Abstr., **83**, 80381b (1975).
647 John, H. and Beetz, P.: J. Prakt. Chem., **143**, 342-346 (1935).
648 John, H. and Beetz, P.: J. Prakt. Chem., **149**, 164-170 (1937).
649 Johnson, F. and Cricchio, R.: U.S. 3,985,783 (1976); Chem. Abstr., **85**, 192377c (1976).
650 Johnson, P. C. and Robertson, A.: J. Chem. Soc., 2381-2389 (1950).
651 Jolad, S. D.; Badiger, V. V. and Nargund, K. S.: J. Karnatak Univ., **5**, 1-9 (1960); Chem. Abstr., **58**, 3341g (1963).
652 Jones, E. T. and Robertson, A.: J. Chem. Soc., 1689-1693 (1932).
653 Jones, P. H.; Bariana, D. S.; Fung, A. K. L.; Martin, Y. C.; Kyncl, J. and Lall, A.: Ger. Offen. 2,642, 879 (1977); Chem. Abstr., **87**, 22795m (1977).
654 Jones, T. G. H. and White, M.: J. Proc. Roy. Soc. Qd., **43** (4) 24-27 (1931).
655 Joshi, G. G. and Shah, N. M.: J. Indian Chem. Soc., **31**, 220-222 (1954).
656 Joshi, K. C. and Bahel, S. C.: J. Indian Chem. Soc., **37** (11) 687-689 (1960).
657 Joshi, U. K.; Kelkar, R. M. and Paradkar, M. V.: Indian J. Chem., **23B** (5) 456-457 (1984).
658 Julia, M. and Baillargé, M.: Bull. Soc. Chim. Fr., 639-642 (1952).

659 Jung, M. E.; Lam, P. Yuk-Sun; Mansuri, M. M. and Speltz, L. M.: J. Org. Chem., **50**, 1087-1105 (1985).
660 Jurd, L. and Wong, R. Y.: Aust. J. Chem., **34** (8) 1633-1644 (1981).
661 Kad, G. L.; Trahan, I. R.; Kaur, J.; Nayyar, S.; Arora, A. and Brar, J. S.: Indian J. Chem., Sect. B: Org. Chem. Incl. Med. Chem., **35B** (7) 734-736 (1996).
662 Kahn, A. M. M.; Linnell, W. H. and Sharp, L. K.: J. Chem. Soc., 1618-1621 (1960).
663 Kai, Yuan-Chu; Li, Ying; Yu, Pei-Lin; Cheng, Ya-Ping; Wang, Te-Sheng; Chen, I-Sin and Li, Liang-Chuan: Hua Hsueh Hsueh Pao, **36** (2) 143-148 (1978); Chem. Abstr., **89**, 197115h (1978).
664 Kalaiya, S. B. and Parikh, A. R.: J. Indian Chem. Soc., **64** (3) 172-175 (1987).
665 Kamal, A. and Sattur, P. B.: Synth. Commun., **12** (2) 157-162 (1982).
666 Kamal, A.; Rao, A. B. and Sattur, P. B.: J. Org. Chem., **53** (17) 4112-4114 (1988).
667 Kamas, F.; Svoboda, P.; Kralicek, J. and Kral, I.: Czech. CS 200,061 (1982); Chem. Abstr., **98**, 108434g (1983).
668 Kamisaka, T.; Watanuki, T.; Tsukamoto, K. and Ogata, M.: Jpn. Kokai Tokkyo Koho JP 01 63,965 [89 63,965] (1989); Chem. Abstr., **111**, 144071g (1989).
669 Kamiya, T.; Tanaka, K.; Tsutomu, K.; Kishimoto, T.; Hemmi, K/; Sakane, K. and Gotoh, J.: Ger. Offen. 2,724,073 (1977); Chem. Abstr., **88**, 190856w (1978).
670 Kamiya, T. and Teraji, T.: U.S. US 4,350,692 (1982); Chem. Abstr., **98**, 71808d (1983).
671 Kämmerer, H.; Büsing, G. and Haub, H. G.: Makromol. Chem., **66**, 82-90 (1963).
672 Kämmerer, H.; Gros, G. and Schweikert, H.: Makromol. Chem., **143**, 135-152 (1971).
673 Kämmerer, H.; Hegemann, G.; Lotz, W.; Ritz, J.; Pachta, K. and Mueller, O.: Makromol. Chem., **180** (7) 1635-1650 (1979).
674 Kamogawa, H.: Kogyo Gijutsuin Sen'i Kogyo Shikensho Kenkyu Hokoku, **86**, 95-96 (1969); Chem. Abstr., **72**, 3996b (1970).
675 Kamogawa, H.; Takayanagi, Y. and Nanasawa, M.: J. Polym. Sci., Polym. Chem. Ed., **19** (11) 2947-2953 (1981).
676 Kamounah, Fadhil S.; Al-Sheibani, Ikbal S.; Shibaldin, Nazar I. and Salman, Salman R.: Magn. Reson. Chem., **23** (7) 521-523 (1985); Chem. Abstr., **104**, 129370r (1986).
677 Kandaswamy, P. and Reddy, S. J.: Indian J. Environ. Prot., **9** (8) 607-608 (1989); Chem. Abstr., **113**, 243982c (1990).
678 Kane, M.; Dean, J. R.; Hitchen, S. M.; Tomlinson, W. R.; Tranter, R. I. and Dowle, C. J.: Analyst (Cambridge, U. K.), **118** (10) 1261-1264 (1993); Chem. Abstr., **120**, 216617a (1994).
679 Kane, V. V.; Kulkarni, A. B. and Shah, R. C.: J. Sci. Ind. Res., Sect. B, **18B**, 28-32 (1959).
680 Kankare, J. J. and Haapakka, K. E.: Ger. Offen. DE 3,908,918 (1989); Chem. Abstr., **113**, 207868m (1990).
681 Kankare, J.; Haapakka, K.; Kulmala, S.; Nanto, V.; Eskola, J. and Takalo, H.: Anal. Chim. Acta, **266** (2) 205-212 (1992).
682 Kanzaki Paper Mfg. Co.: Jpn. Kokai Tokkyo Koho JP 58,224,787 [83,224,787] (1983); Chem. Abstr., **101**, 46392m (1984).
683 Kaplan, J. P.: Eur. Pat. Appl. EP 125975 (1984); Chem. Abstr., **102**, 95399y (1985).
684 Kaplan, J. P.: Fr. Demande FR 2,535,318 (1984); Chem. Abstr., **102**, 5705t (1985).
685 Kaplan, J. P.: Fr. Demande FR 2,535,319 (1984); Chem. Abstr., **101**, 191151m (1984).
686 Kaplan, J. P.: Fr. Demande FR 2,548,183 (1985); Chem. Abstr., **103**, 123154f (1985).
687 Kaplan, J. P.: Ger. Offen. 2,914,801 (1979); Chem. Abstr., **92**, 146448y (1980).
688 Kaplan, J. P.: Ger. Offen. DE 3,242,442 (1983); Chem. Abstr., **99**, 104980e (1983).
689 Kaplan, J. P.: Ger. Offen. DE 3,342,999 (1984); Chem. Abstr., **101**, 191323u (1984).
690 Kaplan, J. P.: Ger. Offen. DE 3,414,050 (1984); Chem. Abstr., **102**, 113038m (1985).
691 Kaplan, J. P.: Ger. Offen. DE 3,414,051 (1984); Chem. Abstr., **102**, 113052m (1985).
692 Kaplan, J. P.; Jalfre, M. and Giudicelli, Don P. R. L.: Ger. Offen. 2,634,288 (1977); Chem. Abstr., **86**, 189530n (1977).
693 Kaplan, J. P. and Raizon, B. M.: Ger. Offen. 2,907,379 (1979); Chem. Abstr., **92**, 6279u (1980).
694 Kaplan, J. P. and Raizon, B.: Fr. Demande FR 2,475,543 (1981); Chem. Abstr., **96**, 6370z (1982).

695 Kaplan, J. P. and Raizon, B.: Fr. Demande FR 2,501,201 (1982); Chem. Abstr., **98**, 71653z (1983).
696 Kaplan, J. P. and Raizon, B.: Ger. Offen. DE 3,341,198 (1984); Chem. Abstr., **101**, 191321s (1984).
697 Kaplan, J. P.; Raizon, B. M.; Desarmenien, M.; Feltz, P.; Headley, P. M.; Worms, P.; Lloyd, K. G. and Bartholini, G.: J. Med. Chem., **23** (6) 702-704 (1980).
698 Kaplan, J. P.; Raizon, B.; Peynot, M. and Mangane, M.: Ger. Offen. DE 3,343,000 (1984); Chem. Abstr., **102**, 5913j (1985).
699 Kar, Tanusree. and Sen Gupta, S. P.: Indian J. Phys., A, **66A** (5) 645-651 (1992); Chem. Abstr., **118**, 264010c (1993).
700 Karanjgoakar, C. G.; Rao, A. V. R.; Venkataraman, K.; Yemul, S. S. and Palmer, K. J.: Tetrahedron Lett., **50**, 4977-4980 (1973).
701 Karasawa, H.; Kometani, K. and Nakamura, K.: Eur. Pat. Appl. EP 351,732 (1990); Chem. Abstr., **113**, 60647c (1990).
702 Karrer, P.: Helv. Chim. Acta, **2**, 486-489 (1919).
703 Karrer, P.: Helv. Chim. Acta, **4**, 992-993 (1921).
704 Karrer, P. and Lichtenstein, N.: Helv. Chim. Acta, **11**, 789-795 (1928).
705 Karrer, W.: Konstitution und Vorkommen Organischen Pflanzenstoffe, pp. 187-188 (1958), Bükhauser (Basel).
706 Karvas, M.; Jexova, E.; Holcik, J. and Balogh, A.: Chem. Prum., **18** (7-8) 427-429 (1968); Chem. Abstr., **69**, 86549v (1968).
707 Kashiwai, K.; Yoshida, T.; Suga, A.; Ikeda, Y. and Kumagai, S.: Jpn. Kokai Tokkyo Koho JP 03 31,235 [91 31,235] (1991); Chem. Abstr., **115**, 49102s (1991).
708 Katakami, T.; Yokoyama, T.; Miyamoto, M.; Mori, H.; Kawauchi, N.; Nobori, T.; Kamiya, J. and Ishii, M.: Eur. Pat. Appl. EP 454,498 (1991); Chem. Abstr., **116**, 128950r (1992).
709 Katakami, T.; Yokoyama, T.; Miyamoto, M.; Mori, H.; Kawauchi, N.; Nobori, T.; Sannohe, K.; Kamiya, J.; Ishii, M. and Yoshihara, K.: Eur. Pat. Appl. EP 369,627 (1990); Chem. Abstr., **115**, 29363z (1991).
710 Kataoka, M.; Ando, T. and Nakagawa, M.: Bull. Chem. Soc. Jpn., **44** (1) 177-184 (1971).
711 Kath, J.; Baron, H. and Doeller, W.: Ger. Offen. 1,468,202 (1969).
712 Kato, T.; Takahashi, K.; Obara, H. and Nakayama, T.: Jpn. Kokai Tokkyo Koho JP 05,289,332 [93,289,332] (1993); Chem. Abstr., **120**, 204680c (1994).
713 Kato, Taizo: Juzen Igakkai Zasshi, **93** (2) 275-290 (1984); Chem. Abstr., **102**, 73168v (1985).
714 Katz, R. and Jacobson, A. E.: Mol. Pharmacol., **8** (5) 594-599 (1972); Chem. Abstr., **77**, 161447f (1972).
715 Katzenellenbogen, J. A.; Tatee, T. and Robertson, D. W.: J. Labelled Compd. Radiopharm., **18** (6) 865-879 (1981).
716 Kauffmann, H. and Grombach, A.: Ber. Dtsch. Chem. Ges., **38**, 794-801 (1905).
717 Kauffmann, H. and Grombach, A.: Justus Liebigs Ann. Chem., **344**, 30-77 (1906).
718 Kauffmann, H. and Pannwitz, P.: Ber. Dtsch. Chem. Ges., **43**, 1205-1213 (1910).
719 Kaukeinen, J. Y. and Rockafellow, D. A.: Ger. Offen. 2,533,688 (1976); Chem. Abstr., **85**, 169687v (1976).
720 Kaukeinen, J. Y. and Rockafellow, D. A.: U.S. 4,234,670 (1980); Chem. Abstr., **94**, 165665z (1981).
721 Kawai, T.; Shimizu, T. and Chiba, H.: J. Pharm. Soc. Jpn., **72**, 660-665 (1956).
722 Kawai, Y.; Yamazaki, H.; Kayakiri, N.; Yoshihara, K.; Yatabe, T. and Oku, T.: PCT Int. Appl. WO 95 29,907 (1995); Chem. Abstr., **124** (15) 202000q (1996).
723 Kawatsuki, N.; Fujisawa, K.; Matsuzaki, I. and Uetsuki, M.: Eur. Pat. Appl. EP 271,002 (1988); Chem. Abstr., **109**, 219622g (1988).
724 Keller, R. T.: US 3,366,691 (1968); Chem. Abstr., **69**, 35747a (1968).
725 Khanna, R. N.; Singh, K. P. and Sharma, J.: Org. Prep. Proced. Int., **24** (6) 687-690 (1992).
726 Khera, U. and Chibber, S. S.: Indian J. Chem., Sect. B, **16B** (1) 78-79 (1978).
727 Kim, Bongsub and Kaack, H.: U.S. US 4,978,797 (1990); Chem. Abstr., **114**, 184976z (1991).
728 Kim, Hakwon: J. Korean Chem. Soc., **40** (8) 549-556 (1996); Chem. Abstr., **125**, 275477f (1996).

729 Kindler, H. and Oelschlager, H.: Chem. Ber., **87**, 194-202 (1954).
730 Kindler, K., Oelschlager, H. and Henrich, P.: Arch. Pharm. (Weinheim, Ger.), **287**, 210-223 (1954).
731 King, F. E.; King, T. J. and Muir, I. H. M.: J. Chem. Soc., 5-10 (1946).
732 Kirkpatrick, A. and Maclaren, J. A.: Text. Res. J., **37** (6) 510-511 (1967); Chem. Abstr., **67**, 82891m (1967).
733 Kise, M.; Yoshimoto, Y.; Fujisawa, H.; Sasaki, Y. and Yasufuku, S.: Eur. Pat. Appl. EP 331,195 (1989); Chem. Abstr., **112**, 76603a (1990).
734 Kishimura, S.; Yamaguchi, A.; Yamada, Y. and Nagata, H.: Polym. Eng. Sci., **32** (20) 1550-1555 (1992); Chem. Abstr., **118**, 29738n (1993).
735 Klarmann, E. and Wowern, J. V.: J. Am. Chem. Soc., **51**, 605-610 (1929).
736 Klinger, H. and Standke, O.: Ber. Dtsch. Chem. Ges., **24**, 1340-1346 (1891).
737 Klopman, G. and Buyukbingol, E.: Mol. Pharmacol., **34** (6) 852-862 (1988); Chem. Abstr., **110**, 150323m (1989).
738 Knapp, Laszlo.; Wein, Tibor; L. Tarjanyi, Eva; Varfalvi, Ferenc; Ivadi, Laszlo and Ersek, Laszlo: Hung. Teljes HU 39,708 (1986); Chem. Abstr., **107**, 39400h (1987).
739 Ko, M. and Tachiki, S.: Jpn. Kokai Tokkyo Koho JP 04,209,667 [92,209,667] (1992); Chem. Abstr., **118**, 222903r (1993).
740 Ko, M. and Tachiki, S.: Jpn. Kokai Tokkyo Koho JP 04,209,666 [92,209,666] (1992); Chem. Abstr., **118**, 104897u (1993).
741 Ko, Seung Hye and Chae, Woo Ki: Bull. Korean Chem. Soc., **19** (5) 513-514 (1998); Chem. Abstr., **129**, 95171a (1998).
742 Kobayashi, Hajime; Yano, Yasuhiro and Endo, Ichiro: Ger. Offen., 2,702,919 (1977); Chem. Abstr., **89**, 34144c (1978).
743 Kobsa, H.: J. Org. Chem., **27**, 2293-2298 (1962).
744 Koga, H. and Sato, H.: Jpn. Kokai Tokkyo Koho JP 01,131,180 [89,131,180] (1989); Chem. Abstr., **111**, 214474s (1989).
745 Koike, Denzo: Gunma Daigaku Kyoyobu Kiyo, **2**, 13-28 (1968); Chem. Abstr., **70**, 57356v (1969).
746 Kolbach, D.; Blazevic, N.; Hannoun, M.; Kajfez, F.; Kovac, T.; Rendic, S. and Sunjic, V.: Helv. Chim. Acta, **60** (1) 265-283 (1977).
747 Kole, P. L. and Ray, S.: J. Labelled Compd. Radiopharm., **16** (2) 373-375 (1979); Chem. Abstr., **91**, 140669d (1979).
748 Komarowsky, A. and von Kostanecki, S.: Ber. Dtsch. Chem. Ges., **27**, 1997-2000 (1894).
749 Kometani, T. and Shiotani, S.: J. Med. Chem., **21** (11) 1105-1110 (1978).
750 König, E. and Kostanecki, S.: Ber. Dtsch. Chem. Ges., **27**, 1994-1997 (1894).
751 König, E. and Kostanecki, S.: Ber. Dtsch. Chem. Ges., **39**, 4027-4031 (1906).
752 Konishi, Jinemon: Eur. Pat. Appl. 6,407 (1980); Chem. Abstr., **93**, 114516y (1980).
753 Konwar, D.; Boruah, R. C. and Sandhu, J. S.: Chem. Ind. (London), 6, 191 (1989).
754 Kopczynski, T.; Krzyzanowska, E. and Olszanowski, A.: J. Prakt. Chem., **331** (3) 486-492 (1989).
755 Korczynski, A. and Nowakowski, A.: Bull. Soc. Soc. Chim. Fr., 329-337 (1928).
756 Kostanecki, S.; Lampe, V. and Marschalk, C.: Ber. Dtsch. Chem. Ges., **40**, 3660-3669 (1907).
757 Kostanecki, S. and Tambor, J.: Ber. Dtsch. Chem. Ges., **28**, 2302-2309 (1895).
758 Kotali, A.; Glaveri, U.; Pavlidou, E. and Tsoungas, P. G.: Synthesis, **12**, 1172-1173 (1990).
759 Kouskov, V. K. and Naoumov, I. A.: Zh. Obshch. Khim., **31** (1) 54-59 (1961).
760 Koutek, B.; Musil, L.; Velek, J.; Lycka, A.; Snobl, D. Synackova, M. and Soucek, M.: Collect. Czech. Chem. Commun., **46** (10) 2540-2556 (1981).
761 Koyama, S.; Horikawa, Y. and Kimura, O.: Jpn. Kokai Tokkyo Koho JP O5 70,397 [93 70,397] (1993); Chem. Abstr., **119**, 138880u (1993).
762 Kraus, G. A. and Kirihara, M.: J. Org. Chem., **57** (11) 3256-3257 (1992).
763 Kraus, G. A.; Kirihara, M. and Wu, Y.: ACS Symp. Ser. 1994, 577 (Benign by Design), 76-83; Chem. Abstr., **123**, 21803n (1995).
764 Kraus, G. A. and Liu, P.: Tetrahedron Lett., **35** (42) 7723-7726 (1994).
765 Krause, M.; Rouleau, A.; Stark, H.; Garbarg, M.; Schwartz, J. C. and Schunack, W.: Pharmazie, **51** (10) 720-726 (1996).

766 Krause, M.; Rouleau, A.; Stark, H.; Luger, P.; Lipp, R.; Garbarg, M.; Schwartz, J. C. and
     Schunack, W.: J. Med. Chem., **38** (20) 4070-4079 (1995).
767 Kreilick, R. W.: J. Am. Chem. Soc., **88**, 5284-5288 (1966).
768 Kreshkov, A. P.; Gurvich, Ya. A. and Gel'pern, G. M.: Zh. Anal. Khim., **28** (12) 2440-2445
     (1973); Chem. Abstr., **80**, 90983b (1974).
769 Kricheldorf, H. B.; Chen, Xiangdong and Masri, Majdi Al.: Macromolecules, **28** (7) 2112-
     2117 (1995).
770 Kruber, O.: Ber. Dtsch. Chem. Ges., **65**, 1382-1396 (1932).
771 Kubodera, S. and Morigaki, M.: Jpn. Kokai Tokkyo Koho JP 01,171,887 [89,171,887]
     (1989); Chem. Abstr., **112**, 149106c (1990).
772 Kulickova, M.; Slama, P. and Hrdlovic, P.: Chem. Zvesti, **33** (5) 630-635 (1979).
773 Kuliev, A. M.; Sardarova, S. A. and Agamalieva, M. M.: Prisadki Smaz. Maslam, **7**, 3-5
     (1981); Chem. Abstr., **97**, 55410s (1982).
774 Kulka, M.: Can. 560,324 (1958); Chem. Abstr., **53**, 10130h (1959).
775 Kulka, M.: J. Am. Chem. Soc., **76**, 5469-5471 (1954).
776 Kulkarni, Y. D.; Sharma, R.; Sharma, V. L. and Dua, P. R.: Biol. Mem., **13** (2) 183-187
     (1987); Chem. Abstr., **108**, 198210h (1988).
777 Kulkarni, Y. D.; Sharma, R.; Sharma, V. L. and Dua, P. R.: J. Indian Chem. Soc., **64** (1) 46-
     48 (1987).
778 Kulshresth, S. K.; Mukerjee, S. K. and Seshadri, T. R.: Indian J. Chem., **12** (1) 10-14
     (1974).
779 Kumagai, S.; Kashiwai, K. and Suga, A.: Jpn. Kokai Tokkyo Koho JP 02,180,909
     [90,180,909] (1990); Chem. Abstr., **113**, 232242k (1990).
780 Kumaki, J.; Nakamura, S. and Yonetani, K.: Jpn. Kokai Tokkyo Koho JP 02,292,356
     [90,292,356]; Chem. Abstr., **115**, 51130t (1991).
781 Kumar, S.; Seth, M.; Bhaduri, A. P.; Agnihotri, A. and Srivastava, A. K.: Indian J. Chem.,
     Sect. B, **23B** (2) 154-157 (1984).
782 Kumiai Chemical Industry Co., Ltd.: Jpn. Kokai Tokkyo Koho JP 46,904 (1982); Chem.
     Abstr., **97**, 34709f (1982).
783 Kumiai Chemical Industry Co., Ltd.: Jpn. Kokai Tokkyo Koho JP 82 46,905 (1982); Chem.
     Abstr., **97**, 19054v (1982).
784 Kuo, C. H.; Hoffsommer, R. D.; Slates, H. L.; Taub, D. and Wendler, N. L.: Chem. Ind.
     (London), 1627-1628 (1960).
785 Kurosawa, K.; Sasaki, Y. and Ikeda, M.: Bull. Chem. Soc. Jpn., **46**, 1498-1501 (1973).
786 Kurosu, Yasuhisa; Kanasugi, Haruki; Sakuraba, Yasuya and Imamura, Shosuke: Japan Kokai
     78 59,026 (1978); Chem. Abstr., **89**, 101972m (1978).
787 Kusuda, M. and Matsuoka, Y.: Jpn. Kokai Tokkyo Koho JP 62,234,041 [87,234,041]
     (1987); Chem. Abstr., **109**, 170034a (1988).
788 Kusuda, M. and Matsuoka, Y.: Jpn. Kokai Tokkyo Koho JP 62,234,042 [87,234,042]
     (1987); Chem. Abstr., **109**, 149056z (1988).
789 Kusuda, M. and Matsuoka, Y.: Jpn. Kokai Tokkyo Koho JP 63 99,033 [88 99,033] (1988);
     Chem. Abstr., **109**, 210704c (1988).
790 Kysel, O.: Kinet. Mech. Polyreactions, Int. Symp. Macromol. Chem., Prep., **5**, 263-270
     (1969); Chem. Abstr., **75**, 68099t (1971).
791 Kysel, O.: Collect. Czech. Chem. Commun., **39** (11) 3256-3267 (1974).
792 Kysel, O. and Jany, I.: Chem. Zvesti, **28** (1) 70-76 (1974).
793 Kysel, O. and Zahradnik, R.: Collect. Czech. Chem. Commun., **35** (10) 3030-3044 (1970).
794 Kysel, O.; Zahradnik, R. and Pakula, B.: Collect. Czech. Chem. Commun., **35** (10) 3020-
     3029 (1970).
795 Lacey, H. T.: Ph. D. Thesis, Cornell University Library (1926).
796 Lacey, R. N.: Brit. 951,435 (1964); Chem. Abstr., **60**, 15779b (1964).
797 Laidlaw, R. A. and Smith, G. A.: Chem. Ind. (London), 1604-1605 (1959).
798 Lakshmi, C., Mrs; Raj, N. Giridhar; Srinivasan, K. V. S. Chandra, Mrs. and Kumar, K.
     Akshaya: J. Inst. Chem. (India), **60** (3) 114 (1988); Chem. Abstr., **110**, 153838u (1989).
799 Lamchen, M. and Wicken, A. J.: J. Chem. Soc., 2779-2782 (1959).
800 Lappin, G. L. and Tamblyn, J. W.: USP 2,861,053 (1958); Chem. Abstr., **53**, 4818i (1959).
801 Largeron, M.; Dupuy, H. and Fleury, M. B.: Tetrahedron, **51** (17) 4953-4968 (1995).

802 Largeron, M.; Langevin-Bermond, D. and Fleury, M. B.: J. Chem. Soc., Perkin Trans. 2, **5**, 893-899 (1996).
803 Larine, N. A., Matveeva, E. N. and Kajoutkina, L. V.: Zh. Obshch. Khim., **32** (2) 367-369 (1962).
804 Larine, N. A.; Matveeva, E. N. and Petrova, T. G.: Zh. Obshch. Khim., **34** (3) 864-866 (1964).
805 Larine, N. A.; Matveeva, E. N. and Smirnova, V. S.: Zh. Obshch. Khim., **30** (7) 2377-2379 (1960).
806 Laskorin, B. N.; Yakshin, V. V.; Ul'yanov, V. S. and Mirokhin, A. M.: U.S.S.R. 591,452 (1978); Chem. Abstr., **89**, 108668a (1978).
807 Launas, K. R.; Neelakantan, S. and Seshadri, T. R.: Proc. Indian Acad. Sci., **46A**, 343-348 (1957).
808 Le Fèvre, R. W.: J. Chem. Soc., 3249-3252 (1928).
809 Leary, G. and Oliver, J. A.: Tetrahedron Lett., **3**, 299-302 (1968).
810 Lebedeva, E. P.; Matyushin, G. A.; Mezentseva, G. A.; Tavrizova, M. A. and Fain, V. Ya.: Zh. Prikl. Spektrosk., **31** (1) 104-108 (1979); Chem. Abstr., **91**, 174466a (1979).
811 Lee, W. Y.; Moon, B. G.; Park, C. H.; Bang, S. H. and Lee, J. H.: Bull. Korean Chem. Soc., **9** (5) 325-328 (1988); Chem. Abstr., **110**, 153839v (1989).
812 Lemoine, J.; Amgar, A. and Rasquin, B.: Fr. Demande FR 2,540,103 (1984); Chem. Abstr., **102**, 45603q (1985).
813 Lesiak, T. and Nowakowski, J.: J. Prakt. Chem., **323** (4) 684-690 (1981).
814 Lewis, J. R.: J. Chem. Soc. C, **4**, 629-631 (1971).
815 Lewis, J. R.: Proc. Chem. Soc., London, 373 (1963).
816 Lewis, J. R. and Paul, J. G.: J. Chem. Soc., Perkin Trans. 1, **3**, 770-775 (1981).
817 Lewis, J. R. and Warrington, B. H.: J. Chem. Soc., 5074-5077 (1964).
818 Li, Liangquan and Gai, Yuanzhu: Huaxue Shiji, **4** (4) 226-227 (1982); Chem. Abstr., **97**, 197917m (1982).
819 Liebermann, C.: Ber. Dtsch. Chem. Ges., **6**, 951-953 (1873).
820 Liebermann, C.: Ber. Dtsch. Chem. Ges., **11**, 1434-1438 (1878).
821 Limaye, D. B.: Ber. Dtsch. Chem. Ges., **67B**,12-15 (1934).
822 Limaye, D. B. and Munje, R. H.: Rasayanam, **1**, 80-86 (1937); Chem. Abstr., **32**, 2096[1] (1938).
823 Limaye, D. B. and Shenolikar, G. S.: Rasayanam, **1**, 93-100 (1937); Chem. Abstr., **32**, 2096[6] (1938).
824 Limaye, D. B. and Talwalkar, S. S.: Rasayanam, **1**, 141-146 (1938); Chem. Abstr., **33**, 1698[3] (1939).
825 Lin, Chun-Nan; Liou, Shorong Shii; Ko, Feng Nien and Teng, Che Ming: J. Pharm. Sci., **82** (1) 11-16 (1993).
826 Lin, Chun-Nan; Won, Shen-Jeu; Lieu, Hsiao-Sheng and Liou, Shorong-Shii: U.S. US 5,741,813 (1998); Chem. Abstr., **128**, 317250 (1998).
827 Linari, A.: Gazz. Chim. Ital., **33**, 60-65 (1903).
828 Lo, Young S.: Can. CA 1,262,353 (1989); Chem. Abstr., **113**, 114820z (1990).
829 Locksley, H. D.; Moore, I. and Scheinmann, F.: Tetrahedron, **23**, 2229-2234 (1967).
830 Locksley, H. D. and Murray, I. G.: J. Chem. Soc. C, 392-398 (1970).
831 Locksley, H. D. and Murray, I. G.: J. Chem. Soc. C, **7**, 1332-1340 (1971).
832 Locksley, H. D.; Quillinan, A. J. and Scheinmann, F.: J. Chem. Soc. (C), 3804-3814 (1971).
833 Loewe, H.; Urbanietz, J.; Duewel, D. and Kirsch, R.: Ger. Offen. 2,443,297 (1976); Chem. Abstr., **85**, 78128x (1976).
834 Lothrop, W. C. and Goodwin, P. A.: J. Am. Chem. Soc., **65**, 363-367 (1943).
835 Lotta, T.; Taskinen, J.; Backstrom, R. and Nissinen, E.: J. Comput.-Aided Mol. Des., **6** (3) 253-272 (1992); Chem. Abstr., **117**, 142869y (1992).
836 Löwe, J.: Z. Anal. Chem., **14**, 117-130 (1875).
837 Löwenberg, E.: Dissertation, Bonn (Germany), 39-46 (1904).
838 Löwenberg, E.: Justus Liebigs Ann. Chem., **346**, 386-389 (1906).
839 Lubenets, E. G.; Gerasimova, T. N. and Fokin, E. P.: Zh. Org. Khim., **7** (4) 805-812 (1971).
840 Luston, J.; Gunis, J. and Manasek, Z.: J. Macromol. Sci.-Chem. **A7** (3) 587-599 (1973)
841 Luu Duc Cuong: Arzneim.-Forsch., **26** (5) 894-895 (1976).

842 Maerov, S. B.: J. Polymer Sci., part A, **3**, 487-499 (1965).
843 Maguer, P.: Bull. Soc. Chim. Fr., 199-204 (1987).
844 Mahajan, A. R.; Dutta, D. K.; Boruah, R. C. and Sandhu, J. S.: Tetrahedron Lett., **31** (27) 3943-3944 (1990).
845 Mahfouz, N. M. A.; Hambloch, H.; Omar, N. M. and Frahm, A. W.: Arch. Pharm. (Weinheim, Ger.), **323** (3) 163-169 (1990).
846 Maj, J.; Rogoz, Z.; Skuza, G.; Sowinska, H. and Superata, J.: J. Neural Transm.: Parkinson's Dis. Dementia Sect., **2** (2) 101-112 (1990); Chem. Abstr., **115**, 126794y (1991).
847 Majoie, B.: Ger. Offen. 2,637,098 (1977); Chem. Abstr., **86**, 189537v (1977).
848 Majoie, B.: U.S. Patent 4,146,385 (1979).
849 Makishima, H. and Nakano, K.: Jpn. Kokai Tokkyo Koho JP 04,214,563 [92,214,563] (1992); Chem. Abstr., **118**, 90911e (1993).
850 Malik, V. P. and Saharia, G. S.: J. Sci. Ind. Res., Sect. B, **15B**, 633-635 (1956).
851 Malin, G.: Justus Liebigs Ann. Chem., **138**, 76-83 (1866).
852 Mandal, S. K. and Nag, K.: J. Chem. Soc., Dalton Trans., **11**, 2429-2434 (1983).
853 Mandal, S. K.; Thompson, L. K.; Newlands, M. J.; Biswas, A. K.; Adhikary, B.; Nag, K.; Gabe, E. and Lee, F. L.: Can. J. Chem., **67** (4) 662-670 (1989).
854 Mannisto, P. T. and Tuomainen, P.: Naunyn-Schmiedeberg's Arch. Pharmacol., **344** (4) 412-418 (1991); Chem. Abstr., **115**, 270584x (1991).
855 Marcantonatos, M.; Marcantonatos, A. and Monnier, D.: Helv. Chim. Acta, **48** (1) 194-201 (1965).
856 Marcincin, A. and Pikler, A.: Zb. Pr. Chemickotechnol. Fak. SVST (Slov. Vys. Sk. Tech.) 197-202 (1969-1970) (Pub. 1971); Chem. Abstr., **76**, 114573p (1972).
857 Marini-Bettolo, G. B.; Ballio, A. and Baroni, G.: Gazz. Chim. Ital., **78**, 301-303 (1948).
858 Marini-Bettolo, G. B. and Paolini, L.: Ital 435,779 (1948); Chem. Abstr., **44**, 8375d (1950).
859 Marsh, P. B. and Butler, M. L.: Ind. Eng. Chem., **38**, 701-705 (1946).
860 Martens, J. and Praefcke, K.: Chem. Ber., **107**, 2319-2325 (1974).
861 Martin, R.: Bull. Soc. Chim. Fr., 983-988 (1974).
862 Martin, R.: Bull. Soc. Chim. Fr., 901-905 (1977).
863 Martin, R.: Monatsh. Chem., **112**, 1155-1163 (1981).
864 Martin, R. and Coton, G.: Bull. Soc. Chim. Fr., 1442-1445 (1973).
865 Martin, R. and Demerseman, P.: Monatsh. Chem., **121**, 227-236 (1990).
866 Martin, R. and Demerseman, P.: Synthesis, **8**, 738-740 (1992).
867 Martin, R.; Gros, N.; Böhmer, V. and Kämmerer, H.: Monatsh. Chem., **110**, 1057-1066 (1979).
868 Martin, R.; Lafrance, J. R. and Demerseman, P.: Bull. Soc. Chim. Belg., **100** (7) 539-548 (1991).
869 Martinez Nunez, F.; De Abajo, J.; Mercier, R. and Sillion, B.: Polymer, **33** (15) 3286-3291 (1992); Chem. Abstr., **117**, 151431b (1992).
870 Martinovich, R. J.: U.S. 3,445,424 (1969); Chem. Abstr., **71**, 39845v (1969).
871 Maruyuma, K. and Miyagi, Y.: Bull. Chem. Soc. Jpn., **47** (5) 1303-1304 (1974).
872 Masutani, T.; Itami, Y.; Nishioka, T.; Maeda, M. and Kitahara, K.: Jpn. Kokai Tokkyo Koho JP 61 22,052 [86 22,052] (1986); Chem. Abstr., **105**, 78656z (1986).
873 Mather, J.: Brit. 1,060,855 (1967); Chem. Abstr., **67**, 99858d (1967).
874 Mathur, A. K.: Curr. Sci., **47** (23) 889-890 (1978).
875 Mathur, A. K.; Mathur, K. B. L. and Seshadri, T. R.: Indian J. Chem., Sect. B, **15B** (1) 54-57 (1977);
876 Matsui, K.: J. Soc. Org. Synthetic. Chem. (Japan), **9**, 92-96 (1951); Chem. Abstr., **47**, 815c (1953).
877 Matsui, K. and Motoi, M.: Bull. Chem. Soc. Jpn., **46**, 565-569 (1973).
878 Matsumura, S. and Inada, H.: Jpn. Kokai Tokkyo Koho JP 03 41,047 [91 41,047] (1991); Chem. Abstr., **115**, 8309w (1991).
879 Matsushita, H.; Yamahata, T. and Kakimoto, H.: U.S. 3,983,279 (1976); Chem. Abstr., **86**, 113700x (1977).
880 Matsuura, T. and Kitaura, Y.: Tetrahedron, **25** (18) 4501-4514 (1969).
881 Mattison, P. L. and Swanson, R. R.: S. African 70 08,071 (1970); Chem. Abstr., **76**, 153340e (1972).

882 Mattison, P. L. and Swanson, R. R.: U.S. 3,939,203 (1976); Chem. Abstr., **84**, 164446k (1976).
883 Matzke, M.; Mohrs, K. H.; Raddatz, S.; Fruchtmann, R.; Hatzelmann, A.; Kohlsdorfer, C.; Mueller-Peddinghau, R. and Theisen,-Popp, P.: Eur. Pat. Appl. EP 525,571 (1993); Chem. Abstr., **119**, 95356h (1993).
884 Mauthner, F.: J. Prakt. Chem., **87**, 403-409 (1913).
885 Mauthner, N.: Mat. naturw. Anz. ungar. Akad. Wiss., **50**, 484-486 (1933); Chem. Abstr., **28**, 3392^9 (1934).
886 McDonald, P. D.: Ph. D. Thesis, Pennsylvania State University, U.S.A. (1970).
887 McDonald, P. D. and Hamilton, G. A.: J. Am. Chem. Soc., **95** (23) 7752-7758 (1973).
888 McGarry, E. J.; Forsyth, B. A. and Wilshire, C.: Ger. Offen., 2,848,493 (1979); Chem. Abstr., **91**, 157428x (1979).
889 McGookin, A.; Robertson, A. and Simpson, T. H.: J. Chem. Soc., 2021-2029 (1951).
890 McMaster, B. J.; Scott, A. I. and Trippett, S.: J. Chem. Soc., 4628-4631 (1960).
891 McOmie, J. F. W.; Watts, M. L. and West, D. E.: Tetrahedron, **24**, 2289-2292 (1968).
892 Meador, M. A.; Abdulaziz, M. and Meador, M. A. B.: ACS Symp. Ser., 1990, 417 (Radiat. Curing Polym. Mater.), 220-237; Chem. Abstr., **113**, 24612d (1990).
893 Meguro, K.; Tawada, H. and Ikeda, H. PCT Int. Appl. WO 91 12,249 (1991); Chem. Abstr., **115**, 279815f (1991).
894 Mehrotra, P. K.; Prashad, M.; Seth, M.; Bhaduri, A. P. and Kamboj, V. P.: Indian J. Exp. Biol., **17** (12) 1317-1319 (1979); Chem. Abstr., **92**, 104744m (1980).
895 Meisenheimer, J.; Hanssen, R. and Wachterowitz, A.: J. Prakt. Chem., **119**, 315-367 (1928).
896 Melchore, J. A.: Ind. Eng. Chem., Proc. Res. Develop., **1**, 232-235 (1962); Chem. Abstr., **59**, 3553g (1963).
897 Melton, J. W. and Henze, H. R.: J. Am. Chem. Soc., **69**, 2018-2020 (1947).
898 Meltzer, R. I. and Stanaback, R. J.: J. Org. Chem., **26**, 1977-1979 (1961).
899 Messner, J.: Pharm. Zentralhalle Dtschl., **67**, 625-627 (1926).
900 Messner, J.: Pharm. Zentralhalle Dtschl., **67**, 680-688 (1926).
901 Messner, J.: Pharm. Zentralhalle Dtschl., **67**, 696-699 (1926).
902 Meussdoerffer, J. N. and Niederpruem, H.: Ger. Offen. 2,653,601 (1978); Chem. Abstr., **89**, 129265g (1978).
903 Meyer, R. and Conzetti, A.: Ber. Dtsch. Chem. Ges., **30**, 969-973 (1897).
904 Meyer, R. and Conzetti, A.: Ber. Dtsch. Chem. Ges., **32**, 2103-2108 (1899).
905 Michael, A.: Amer. Chem. J., **5**, 81-97 (1883).
906 Michael, A.: Ber. Dtsch. Chem. Ges., **14**, 656-658 (1881).
907 Miertus, S. and Kysel, O.: J. Mol. Struct., **26** (2) 163-173 (1975); Chem. Abstr., **83**, 113408w (1975).
908 Mieville, A.: Ger. Offen. 2,250,327 (1973); Chem. Abstr., **79**, 53029d (1973).
909 Mieville, A.: Ger. Offen. 2,605,382 (1976); Chem. Abstr., **85**, 192382a (1976).
910 Mikhailov, N. V.; Tokareva, L. G. and Popov, A. G.: Vysokomolekul. Soedin., 5 (2) 188-194 (1963); Chem. Abstr., **59**, 2980c (1963).
911 Mikhailova, N. N. and Vorozheeva, V. P.: Zavodsk. Lab., **30** (7) 802-803 (1964); Chem. Abstr., **61**, 10048g (1964).
912 Mikitaev, A. K.: Acta Polym., **32** (8) 453-460 (1981); Chem. Abstr., **95**, 220388j (1981).
913 Mikovic, J.; Ambrovic, P.; Manasek, Z. and Karvas, M.: Chem. Zvesti, **27** (2) 255-262 (1973).
914 Milionis, J. P.: U.S. 3,649,695 (1972); Chem. Abstr., **76**, 153346m (1972).
915 Miller, G. J. and Quackenbush, F. W.: J. Am. Oil Chem. Soc., **34**, 404-407 (1957).
916 Miller, J. A.: J. Org. Chem., **52**, 322-323 (1987).
917 Milligan, B. and Holt, L. A.: Polym. Degrad. Stab., **10** (4) 335-352 (1985); Chem. Abstr., **102**, 205341h (1985).
918 Minafuji, M. and Mori, S.: Jpn. Kokai Tokkyo Koho JP 63,253,050 [88,253,050] (1988); Chem. Abstr., **110**, 172879z (1989).
919 Minagawa, M. and Kubota, N.: Japan Kokai 74 78,692 (1974); Chem. Abstr., **83**, 60596x (1975).
920 Minajew, W.: Chem. Zentralbl., **1**, 84 (1927).

921 Minajew, W.: Zh. Russ. Fiz. Khim. O-va., Chast Khim., **58**, 307-317 (1926); Chem. Zentralbl., **1**, 84 (1927).
922 Mindl, J. and Vecera, M.: Collect. Czech. Chem. Commun., **38** (11) 3496-3505 (1973).
923 Miquel, J. F.; Muller, P. and Buu-Hoi, N. P.: Bull. Soc. Chim. Fr., 633-636 (1956).
924 Miranda Alonso, M. A; Diaz Mondejar, M. R. and Corma Canos, A.: Span. ES 508,568 (1982); Chem. Abstr., **99**, 38199d (1983).
925 Miranda, M. A. and Tormos, R.: J. Org. Chem., **58** (12) 3304-3307 (1993).
926 Mironov, G. S.; Budnij, I. V.; Cerniakovskaja, K. A. and Farberov, M. I.: Zh. Org. Khim., **8**, 597-600 (1972).
927 Mironov, G. S.; Oustinov, V. A. and Farberov, M. I.: Zh. Org. Khim., **8** (7) 1509-1515 (1972).
928 Misra, G. C.; Pande, L. M.; Joshi, G. C. and Misra, A. K.: Aust. J. Chem., **25**, 1579-1581 (1972).
929 Mitsubishi Chemical Industries Co., Ltd., Neth. Appl. 74 01,240 (1975); Chem. Abstr., **85**, 20855r (1976).
930 Mitsubishi Rayon Co., Ltd. Japan 9624('65) (1961); Chem. Abstr., **63**, 18376c (1965).
931 Mitter, B. and Saharia, G. S.: Vikram, J. Vikram Univ., **2**, 143-147 (1958); Chem. Abstr., **54**, 12062f (1960).
932 Miyamoto, T.; Mohri, T.; Shigeoka, S.; Itoh, H. and Hayashi, M.: Eur. Pat. Appl. EP 65,874 (1982); Chem. Abstr., **98**, 143114z (1983).
933 Miyamoto, T.; Watsuka, H.; Hashimoto, S.; Itoh, H.; Mohri, T. and Hayashi, M.: Eur. Appl. EP 79,141 (1983); Chem. Abstr., **99**, 121998j (1983).
934 Miyano, M. and Deason, J. R.: U.S. US 4,683,241 (1987); Chem. Abstr., **108**, 37399m (1988).
935 Miyano, M.; Deason, J. R.; Nakao, A.; Stealey, M. A.; Villamil, C. I.; Sohn, D. D. and Mueller, R. A.: J. Med. Chem., **31** (5) 1052-1061 (1988).
936 Miyata, S.; Sugawara, K.; Sasada, K.; Mori, S.; Yonezawa, N. and Murao, A.: Jpn. Kokai Tokkyo Koho JP 03,141,240 [91,141,240] (1991); Chem. Abstr., **115**, 233094s (1991).
937 Miyazaki, N. and Takayanagi, T.: Jpn. Kokai Tokkyo Koho JP 02 40,268 [90 40,268] (1990); Chem. Abstr., **113**, 80631w (1990).
938 Miyazawa, S. and Enomoto, K.: Jpn. Kokai Tokkyo Koho 79 65,535 (1979); Chem. Abstr., **91**, 166380c (1979).
939 Mompon, B.; Loyaux, D.; Kauffmann, E. and Krstulovic, Ante L.: J. Chromatogr., **363** (2) 372-391 (1986).
940 Montagne, P. J.: Ber. Dtsch. Chem. Ges., **49**, 2243-2262 (1916).
941 Montagne, P. J.: Recl. Trav. Chim. Pays-Bas, **39**, 339-349 (1920).
942 Montagne, P.: Chem. Weekblad, **14**, 526-529 (1917).
943 "Montecatini" Societa Generale per l'Industria Mineraria e Chimica: Brit. 923,407 (1963); Chem. Abstr., **59**, 5309d (1963).
944 Montfort, B.; Laude, B.; Vebrel, J. and Cerutti, E.: Bull. Soc. Chim. Fr., 848-854 (1987).
945 Morand, O. H.; Aebi, J. D.; Dehmlow, H.; Ji, Yu-Hua; Gains, N.; Lengsfeld, H. and Himber, J.: J. Lipid Res., **38** (2) 373-390 (1997); Chem. Abstr., **126** (22) 287791n (1997).
946 Moriconi, E. J.; O'Connor, W. F. and Forbes, W. F.: J. Am. Chem. Soc., **82**, 5454-5459 (1960).
947 Morselli, P. L.; Burke, J. T.; Ferrandes, B.; Padovani, P.; Bianchetti, G.; Gomeni, R.; Thenot, J. P. and Thiercelin, J. P.: Metab. Antiepileptic Drugs, [Proc. Workshop], 191-197 (1982) (Pub. **1984**); Chem. Abstr., **101**, 143454r (1984).
948 Morton, A. and Erlam, T.: J. Chem. Soc., 159-169 (1941).
949 Moshfegh, A. A.; Badri, R.; Hojjatie, M.; Kaviani, M.; Naderi, B.; Nazmi, A. H.; Ramezanian, M.; Roozpeikar, B. and Hakimelahi, G. H.: Helv. Chim. Acta, **65** (4) 1221-1228 (1982).
950 Motylewski, S.: Ber. Dtsch. Chem. Ges., **42**, 3148-3152 (1909).
951 Moussa, G. E. M.: Acta Chem. Scand., **22** (10) 3329-3330 (1968).
952 Moussavi, Z.; Lesieur, D.; Lespagnol, C.; Sauzieres, J. and Olivier, P.: Eur. J. Med. Chem., **24** (1) 55-60 (1989).
953 Muangnoicharoen, N. and Frahm, A. W.: Phytochemistry, **21** (3) 767-772 (1982).
954 Mullaji, B. Z. and Shah,, R. C.: Proc. Indian Acad. Sci., **34A**, 88-96 (1951).

955   Munakata, H. and Imaki, N.: Ger. Offen. 2,338,823 (1974); Chem. Abstr., **80**, 145793x (1974).
956   Nachbaur, C.: Justus Liebigs Ann. Chem., **107**, 243-248 (1858).
957   Naik, R. M.; Thakor, V. M. and Shah, R. C.: Proc. Indian Acad. Sci., **37A**, 765-773 (1953).
958   Nakai, Noboru; Fujii, Yutaka; Kobashi, Kyoichi and Nomura, Keiichi: Arch. Biochem. Biophys., **239** (2) 491-496 (1985); Chem. Abstr., **103**, 34008u (1985).
959   Nakazawa, K. and Baba, S.: J. Pharm. Soc. Jpn., **75**, 378-381 (1955); Chem. Abstr., **50**, 2510b (1956).
960   Nakazawa, K. and Kusuda, K.: J. Pharm. Soc. Jpn., **75**, 257-260 (1955).
961   Nakazawa, K.; Matsuura, S. and Baba, S.: J. Pharm. Soc. Jpn., **74**, 498-501 (1954).
962   Nandi, M.; Verma, S. D.; Mehrotra, J. K. and Sircar, A. K.: J. Indian Chem. Soc., **40** (4) 296-298 (1963).
963   Natsugari, H.; Ikeda, H.; Ishimaru, T. and Doi, T.: Eur. Pat. Appl. EP 585,913 (1994); Chem. Abstr., **122**, 56051x (1995).
964   Neamati, N.; Hong, H.; Mazumder, A.; Wang, S.; Sunder, S.; Nicklaus, M. C.; Milne, G. W. A.; Proksa, B. and Pommier, Y.: J. Med. Chem., **40** (6) 942-951 (1997).
965   Negi, A. S.; Dwivedi, I.; Setty, B. S. and Ray, S.: Indian J. Pharm. Sci., **56** (3) 105-108 (1994); Chem. Abstr., **121**, 293002u (1994).
966   Nemes, A.; Baciu, D.; Muresian, F. O.; Glatz, A. M. and Stanescu, L.: Rom. RO 86,745 (1985); Chem. Abstr., **105**, 155055g (1986).
967   Nencki, M. and Stoeber, E.: Ber. Dtsch. Chem. Ges., **30B**, 1768-1772 (1897).
968   Neumann, P.; Eilingsfeld, H. and Aumueller, A.: Eur. Pat. Appl. EP 351,615 (1990); Chem. Abstr., **113**, 23371n (1990).
969   Neumann, S. M. and Henzel, R. P.: Eur. Pat. Appl. EP 727,320 (1996); Chem. Abstr., **125** (20) 261448k (1996).
970   Newland, G. C. and Tamblyn, J. W.: J. Appl. Polymer Sci., **8** (5) 1949-1956 (1964); Chem. Abstr., **62**, 693c (1965).
971   Newland, G. C. and Tamblyn, J. W.: U.S. 3;003,996 (1959); Chem. Abstr., **56**, 4953b (1962).
972   Newman, M. S. and Pinkus, A. G.: J. Org. Chem., **19**, 978-984 (1954).
973   Newman, M. S. and Pinkus, A. G.: J. Org. Chem., **19**, 985-991 (1954).
974   Newman, M. S. and Pinkus, A. G.: J. Org. Chem., **19**, 992-995 (1954).
975   Newman, M. S. and Pinkus, A. G.: J. Org. Chem., **19**, 996-1002 (1954).
976   Nierenstein, M.: J. Indian Chem. Soc., **8**, 143-145 (1931).
977   Nikaido, T.; Ohmoto, T.; Noguchi, H.; Kinoshita, T.; Saitoh, H. and Sankawa, U.: Planta Med., **43** (1) 18-23 (1981).
978   Nishikawa, H. and Robinson, R.: J. Chem. Soc., Trans. **121**, 839-843 (1922).
979   Nishinaga, A. and Matsuura, T.: J. Org. Chem., **29**, 1812-1817 (1964).
980   Nishinaga, A., Shimizu, T. and Matsuura, T.: Tetrahedron Lett., **22** (52) 5293-5296 (1981).
981   Nishinaga, A.; Shimizu, T.; Toyoda, Y.; Matsuuda, T. and Hirotsu, K.: J. Org. Chem., **47** (12) 2278-2285 (1982).
982   Nishino, H. and Kurosawa, K.: Bull. Chem. Soc. Jpn., **56** (2) 474-480 (1983).
983   Niyazov, A. N. and Atlyev, K.: Izv. Akad. Nauk Turkm. SSR, Ser. Fiz.-Tekhn., Khim. i Geol. Nauk, **4**, 38-41 (1966); Chem. Abstr., **66**, 104785p (1967).
984   Niyazov, A. N.; Yurchenko, N. N. and Atlyev, K. A.: Izv. Akad. Nauk Turkm. SSR, Ser. Fiz.-Tekh., Khim. Geol. Nauk, **6**, 69-75 (1981); Chem. Abstr., **96**, 122354w (1982).
985   Nogi, T.; Yasuhara, Y.; Nagai, M.; Hanada, T. and Tanaka, T.: Japan 69 08,656 (1969); Chem. Abstr., **71**, 21890w (1969).
986   Norell, J. R.: J. Org. Chem., **38**, 1924-1928 (1973).
987   Nozaki, S.; Kanno, T. and Watanabe, H.: Jpn. Kokai Tokkyo Koho JP 01 47,590 [89 47, 590] (1989); Chem. Abstr., **111**, 244475k (1989).
988   Nufer, H. L.: US 3,407,235 (1968); Chem. Abstr., **70**, 19805h (1969).
989   Nuhrich, A.; Varache-Lembege, M.; Renard, P. and Devaux, G.: Eur. J. Med. Chem., **29** (1) 75-83 (1994).
990   Nunes, R. C.; Diaz-Calleja, R.; Pinto, M.; Saiz, E. and Riande, E.: J. Phys. Chem., **99** (34) 12962-12970 (1995).

991 Nurmukhametov, R. N.; Betin, O. I. and Shigorin, D. N.: Dokl. Akad. Nauk SSSR, **234** (5) 1128-1131 (1977); Chem. Abstr., **87**, 85677v (1977).
992 Obrecht, D.: Helv. Chim. Acta., **74** (1) 27-46 (1991).
993 Oelschlager, H.: Arch. Pharm. (Weinheim, Ger.), **288**, 102-113 (1955).
994 Ofenberg, H.; Pham Xuan Hoi and Arventiev, B.: An. Stiint. Univ. "Al. I. Cuza", Iasi, Sect. 1c, **20** (2) 177-181 (1974).
995 Offenberg, H. and Arventiev, B.: An. Stiint. Univ. "Al. I. Cuza" Iasi, Sect. 1c, **11** (2) 155-163 (1965).
996 Offenberg, H. L. and Arventiev, B.: An. Stiint. Univ. "Al. I. Cuza" Iasi, Sect. 1c, **17** (1) 45-48 (1971).
997 Offenberg, H. L.; Ludatser, F. and Arventiev, B.: An. Stiint. Univ. "Al. I. Cuza" Iasi, Sect. 1c, **17** (1) 73-78 (1971).
998 Offenberg, H.; Harnagea, F.; Badilescu, S. and Arventiev, B.: An. Stiint Univ. "Al. I. Cuza" Iasi, Sect. 1c, **12** (1) 67-72 (1966); Chem. Abstr., **67**, 11034y (1967).
999 Ogata, M.; Matsumoto, H.; Kida, S. and Shimizu, S.: Fukusokan Kagaku Toronkai Koen Yoshishu, 12th, 71-75 (1979); Chem. Abstr., **93**, 71640q (1980).
1000 Ogata, Naoya: Kino Zairyo, **7** (12) 43-46 (1987); Chem. Abstr., **108**, 229441r (1988).
1001 Oie, M.; Kawata, S. and Yamada, T.: Eur. Pat. Appl. EP 419,147 (1991); Chem. Abstr., **116**, 13368q (1992).
1002 Okawara, M.; Tani, S. and Imoto, E.: Kogyo Kagaku Zasshi, **68** (1) 223-228 (1965); Chem. Abstr., **63**, 3068g (1965).
1003 Okazaki, K.; Kawaguchi, T. and Matsui, K.: J. Pharm. Soc. Jpn., **72**, 1403-1404 (1952).
1004 Olah, G. A.; Arvanaghi, M. and Krishnamurthy, V. V.: J. Org. Chem., **48** (19) 3359-3360 (1983).
1005 Olin, J.: U.S. 3,736,343 (1973); Chem. Abstr., **79**, 42219q (1973).
1006 Ollis, W. D.: Experientia, **22**, 777-783 (1966).
1007 Ollis, W. D.: Recent Advances in Phytochemistry (edited by J. T. Mabry), Vol. I, pp. 329-374, North Holland Publication, Amsterdam, 1968.
1008 Ollis, W. D.; Redman, B. T.; Roberts, R. J.; Sutherland, I. O.; Gottlieb, O. R. and Magalhaes, M. T.: Phytochemistry, **17**, 1383-1388 (1978).
1009 Olszanowski, A.: J. Prakt. Chem., **332** (6) 1093-1098 (1990).
1010 Olszanowski, A.; Wisniewski, M. and Szymanowski, J.: Chem. Stosow., **30** (3) 439-446 (1986); Chem. Abstr., **108**, 186261n (1988).
1011 Omori, A.; Tomihashi, N.; Shimizu, Y. and Nakai, K.: Jpn. Kokai Tokkyo Koho JP 61 12,740 [86 12,740] (1986); Chem. Abstr., **105**, 44828s (1986).
1012 O'Neill, W. A.; Stimpson, J. W. and Mather, J.: Brit. 954,387 (1964); Chem. Abstr., **61**, 7190c (1964).
1013 Ono Pharmaceutical Co., Ltd.: Jpn. Kokai Tokkyo Koho JP 58,140,016 [83,140,016] (1983); Chem. Abstr., **99**, 200517e (1983).
1014 Ono, H.; Masai, T. and Suzuki, H.: Japan 72 29,199 (1972); Chem. Abstr., **78**, 98936t (1973).
1015 Oota, T.; Kurokawa, M.; Yamachika, M.; Tsuji, A. and Hayase, R.: Jpn. Kokai Tokkyo Koho JP 06,202,320 [94,202,320] (1994); Chem. Abstr., **121**, 289741s (1994).
1016 Orban, S.: Plaste Kautsch., **23** (4) 260-263 (1976); Chem. Abstr., **85**, 33886b (1976).
1017 Orito, I.: Science Repts. Tohoku Imp. Univ., 1st Ser., **18**, 121-128 (1929); Chem. Abstr., **24**, 98 (1930).
1018 Orndorff, W. R. and Lacey, H. T.: J. Am. Chem. Soc., **49**, 818-826 (1927).
1019 Orndorff, W. R. and McNulty, S. A.: J. Am. Chem. Soc., **49**, 992-997 (1927).
1020 Orndorff, W. R. and Wang, C.: J. Am. Chem. Soc., **49**, 1284-1289 (1927).
1021 Oshima, T.; Nakajima, Y. and Nagai, T.: Chem. Lett., **11**, 1977-1980 (1993).
1022 Ost, H.: J. Prakt. Chem., **20**, 208 (1879).
1023 Osuga, S.; Washitsuka, S.; Kaida, K.; Otsuki, H. and Morioka, T.: Jpn. Kokai Tokkyo Koho JP °4 80,029 [92 80,029] (1992); Chem. Abstr., **117**, 152867k (1992).
1024 Ours, C. W. and Lee, C. M.: Ger. Offen. 3,038,011 (1981); Chem. Abstr., **95**, 42695x (1981).
1025 Ours, C. W. and Lee, C. M.: U.S. US 4,323,691 (1982); Chem. Abstr., **97**, 23475a (1982).

1026 Oustinov, V. A.; Mironov, G. S.; Kopeikin, V. V.; Lapteva, N. E.; Pachomov, V. I. and
      Koloskova, G. N.: Otkrytiya, Izobret., Prom. Obraztsy, Tovarnye Znaki, **53** (35) 56-57
      (1978); Chem. Abstr., **85**, 192378d (1976).
1027 Oustinov, V. A.; Mironov, G. S.; Kopeikin, V. V.; Lapteva, N. E.; Pachomov, V. I. and
      Koloskova, G. N.: U.S.S.R. 529,149 (1976); Chem. Abstr., **85**, 192378d (1976).
1028 Oxford, A. W. and Ellis, F.: Brit. UK Pat. Appl. GB 2,058,785 (1981); Chem. Abstr., **96**,
      51977p (1982).
1029 Oxford, A. W. and Ellis, F.: Eur. Pat. Appl. EP 28,063 (1981); Chem. Abstr., **96**, 34874k
      (1982).
1030 Padovani, P.; Deves, C.; Bianchetti, G.; Thenot, J. P. and Morselli, P. L.: J. Chromatogr., **308**,
      229-239 (1984).
1031 Padovani, P.; Thenot, J. P.; Warrington, S.; Hermann, P.; Fraisse-Andre, F.; Thiercelin, J. F.;
      Larribaud, J. and Morselli, P. L.: Adv. Epileptol., **15th**, 169-175 (1984); Chem. Abstr., **102**,
      72314c (1985).
1032 Pakkal, R.; Thomas, II, F. D. and Fernelius, W. C.: J. Org. Chem., **25**, 282-283 (1960).
1033 Pallares, E. S. and Garza, H. M.: Arch. inst. cardiol. Mex., **17**, 833-849 (1947); Chem.
      Abstr., **42**, 2730i (1948).
1034 Pande, C. D.; Tripathi, B. N. and Venkataramani, B.: Indian J. Chem., **6** (9) 542-543 (1968).
1035 Parke, Davis & Co.: Fr. 1,464,253 (1966); Chem. Abstr., **67**, 64241v (1967).
1036 Parmar, V. S.; Khanduri, C. H.; Tyagi, O. D.; Prasad, A. K.; Gupta, S.; Bisht, K. S.; Pati, H.
      N. and Sharma, N. K.: Indian J. Chem., Sect. B, **31B** (12) 925-929 (1992).
1037 Parrish, J. R.: J. S. Afr. Chem. Inst., **23** (3) 129-135 (1970); Chem. Abstr., **74**, 42077p
      (1971).
1038 Patel, J. K.: M.S. thesis, State University of New-York at Buffalo, Buffalo, N.Y. (1967).
1039 Patel, R. B. and Keemtilal: Acta Cienc. Indica, [Ser.] Chem., **9** (1-4) 49-51 (1983); Chem.
      Abstr., **101**, 16246z (1984).
1040 Patel, S. V.; Bhadani, G. V. and Joshi, G. B.: J. Indian Chem. Soc., **61** (2) 169-171 (1984).
1041 Patel, S. V.; Bhadani, G. V. and Joshi, G. B.: J. Indian Chem. Soc., **61** (4) 372-374 (1984).
1042 Patel, S. V.; Nagar, N. Y. and Joshi, G. B.: J. Indian Chem. Soc., **60** (3) 304-306 (1983).
1043 Pathak, V. P. and Khanna, R. N.: Bull. Chem. Soc. Jpn., **55** (7) 2264-2268 (1982).
1044 Pathak, V. P. and Khanna, R. N.: Indian J. Chem., Sect. B, **21B** (3) 253-254 (1982).
1045 Pearl, I. A. and Olcay, A.: Tappi, **54** (10) 1656-1658 (1971); Chem. Abstr., **76**, 47541m
      (1972).
1046 Percec, V.; Bae, J. Y.; Zhao, M. and Hill, D. H.: Macromolecules, **28** (20) 6726-6734 (1995).
1047 Percec, V.; Bae, J.-Y.; Zhao, M. and Hill, D. H.: J. Org. Chem., **60** (4) 1066-1069 (1995).
1048 Percec, V.; Zhao, M.; Bae, J. Y. and Hill, D. H.: Macromolecules, **29** (11) 3727-3735 (1996).
1049 Perkin Jr., W. H. and Weizmann, C.: Chem. Zentralbl., **1**, 406-408 (1907).
1050 Perkin Jr., W. H. and Weizmann, C.: J. Chem. Soc., **89**, 1649-1665 (1906).
1051 Perrot, C.: Thèse d'Etat, Besançon, France, (1972).
1052 Perrot, C. and Cerutti, E.: Bull. Soc. Chim. Fr., 2225-2232 (1974).
1053 Perrot, C. and Cerutti, E.: Bull. Soc. Chim. Fr., 2591-2595 (1974).
1054 Perrot, C. and Cerutti, E.: C. R. Acad. Sci., Ser. C, **264** (15) 1301-1303 (1967).
1055 Perrot, C. and Cerutti, E.: C. R. Acad. Sci., Ser. C, **265**, 320-323 (1967).
1056 Petrillo, G.; Novi, M.; Dell'Erba, C.; Tavani, C. and Berta, G.: Tetrahedron, **46** (23) 7977-
      7990 (1990).
1057 Pettit, G. H.; Toki, B.; Herald, D. L.; Verdier-Pinard, P.; Boyd, M. R.; Hamel, E. and Pettit,
      R. K.: J. Med. Chem., **41** (10) 1688-1695 (1998).
1058 Pfeiffer, P.: Justus Liebigs Ann. Chem., **398**, 137-196 (1913).
1059 Pfeiffer, P. and Loewe, W.: J. Prakt. Chem., **147**, 293-310 (1937).
1060 Pfeiffer, P. and Wang, Liu: Angew. Chem., **40**, 983-991 (1927).
1061 Phadke, R. and Shah, R. C.: J. Indian Chem. Soc., **27** (7), 349-356 (1950).
1062 Philbin, E. M.; Swirski, J. and Wheeler, T. S.: J. Chem. Soc., 4455-4458 (1956).
1063 Picart, F.: Eur. Pat. Appl. EP 51,023 (1982); Chem. Abstr., **97**, 198512n (1982).
1064 Piccolo, O.; Filippini, L.; Tinucci, L.; Valoti, E. and Citterio, A.: Tetrahedron, **42** (3) 885-891
      (1986).
1065 Pickett, J. E. and Moore, J. R.: Angew. Makromol. Chem., **232**, 229-238 (1995); Chem.
      Abstr., **124**, 57753f (1996).

1066 Pieroni, A. and Longhini, S.: Gazz. Chim. Ital., **62**, 387-393 (1932).
1067 Piller, B.; Meier, M.; Puennenberger, M.; Biland, R. and Luethi, C.: Ger. Offen. 2,111,766 (1971); Chem. Abstr., **76**, 29514a (1972).
1068 Pinazzi, C. P. and Fernandez, A.: Makromol. Chem., **168**, 19-26 (1973).
1069 Pitchumani, K.; Warrier, M. and Ramamurthy, V.: J. Am. Chem. Soc., **118** (39) 9428-9429 (1996).
1070 Plank, A.: Tetrahedron Lett., **52**, 5423-5426 (1968).
1071 Plattner, J. J.; Fung, A. K. L.; Smital, J. R.; Lee, C. M.; Crowley, S. R.; Pernet, A. G.; Bunnell, P. R.; Buckner, S. A. and Sennello, L. T.: J. Med. Chem., **27** (12) 1587-1596 (1984).
1072 Pohlmann, C. W.: Recl. Trav. Chim. Pays-Bas, **55**, 737-752 (1936).
1073 Pola Chemical Industries, Inc.: Jpn. Kokai Tokkyo Koho JP 58,110,535 [83,110,535] (1983); Chem. Abstr., **99**, 212282n (1983).
1074 Pollak, J.: Monatsh. Chem., **18**, 736-748 (1897).
1075 Pollak, J.: Monatsh. Chem., **22**, 996-1001 (1901).
1076 Polonsky, J. and Auguste, F.: Bull. Soc. Chim. Fr., 541-549 (1955).
1077 Pong, S. F. and Huang, C. L.: J. Pharm. Sci., **63** (10) 1527-1532 (1974).
1078 Popa, G.; Dumitrescu, V.; Dumitrescu, N. and Vegh, B.: Rev. Chim. (Bucharest), **26** (2) 160-163 (1975); Chem. Abstr., **83**, 107785a (1975).
1079 Popova, Z. V.; Ianovskii, D. M.; Zilberman, E. N.; Reibakova, N. A. and Ganina, V. I.: Zh. Prikl. Khim.(Leningrad), **34**, 874-881 (1961).
1080 Portnykh, N. V.; Volodd'kin, A. A. and Ershov, V. V.: Izv. Akad. Nauk SSSR, Ser. Khim., **12**, 2243-2244 (1966).
1081 Powell, V. H. and Sutherland, M. D.: Aust. J. Chem., **16** (2) 282-284 (1963).
1082 Power, M. B.; Bott, S. G.; Bishop, E. J.; Tierce, K. D.; Atwood, J. L. and Barron, A. R.: J. Chem. Soc., Dalton Trans., **2**, 241-247 (1991).
1083 Prashad, M.; Ray, S. and Bhaduri, A. P.: Indian J. Chem., Sect. B, **16B** (2) 142-143 (1978).
1084 Prashad, M.; Seth, M.; Bhaduri, A. P. and Srimal, R. C.: Indian J. Chem., Sect. B, **17B** (5) 496-498 (1979).
1085 Preston, P. N.; Winwick, T. and Morley, J. O.: J. Chem. Soc., Chem. Commun., **2**, 89-90 (1983).
1086 Preston, P. N.; Winwick, T. and Morley, J. O.: J. Chem. Soc., Perkin Trans. 1, **7**, 1439-1441 (1983).
1087 Price, D. N. and Wain, R. L.: Ann. Appl. Biol., **83** (1) 115-124 (1976); Chem. Abstr., **85**, 15221b (1976).
1088 Price, P. and Israelstam, S. S.: J. Org. Chem., **29** (9) 2800-2802 (1964).
1089 Prichard, J. H.: U.S. 3,219,621 (1965); Chem. Abstr., **64**, 3780d (1966).
1090 Priestley, H. M. and Moness, E.: J. Org. Chem., **5**, 355-361 (1940).
1091 Prows, B. L. and McIlhenny, W. F.: U. S. Environ. Prot. Agency, Off. Res. Dev., [Rep.] EPA, EPA-660/3-73-006, 126 pp. (1973); Chem. Abstr., **84**, 100689j (1976).
1092 Pulgarin, C.; Gunzinger, J. and Tabacchi, R.: Helv. Chim. Acta, **68** (4) 945-948 (1985).
1093 Qualitz, M. and Krupp, V. A.: Ger. Offen. DE 3,009,754 (1981); Chem. Abstr., **96**, 208444n (1982).
1094 Qualitz, M. and Krupp, V. A.: Ger. Offen. DE 3,111,904 (1982); Chem. Abstr., **99**, 46047t (1983).
1095 Quillinan, A. J.: Ph. D. thesis, University of Salford, UK (1971).
1096 Quillinan, A. J. and Scheinmann, F.: J. Chem. Soc., Perkin Trans. 1, **11**, 1382-1387 (1972).
1097 Quillinan, A. J. and Scheinmann, F.: J. Chem. Soc., Perkin Trans. 1, **13**, 1329-1337 (1973).
1098 Rajamohan, K. and Rao, N. V. S.: Indian J. Chem., **11** (10) 1076-1077 (1973).
1099 Randriaminahy, M.; Proksch, P.; Witte, L. and Wray, V.: Z. Naturforsch., C : Biosci., **47** (1-2) 10-16 (1992).
1100 Rao, A. V. R.; Sarma, M. R.; Venkataraman, K. and Yemul, S. S.: Phytochemistry, **13** (7) 1241-1244 (1974).
1101 Rao, E. V. S. Bhushana; Kumari, S. Subhadra and Rao, K. S. R. Krishna Mohan: Indian J. Chem., Sect. B, **26B** (7) 620-623 (1987).
1102 Ravi, P.; Vathani, P. and Reddy, G. C.: Indian J. Heterocycl. Chem., **3** (3) 209-210 (1994).

1103   Ray, S.; Grover, P. K. and Anand, N.: Indian 129,188 (1974); Chem. Abstr., **81**, 169437f
       (1974).
1104   Ray, S.; Grover, P. K. and Anand, N.: Indian J. Chem., **9** (7) 619-623 (1971).
1105   Ray, Tanusree and Gupta, S. P. Sen.: Cryst. Struct. Commun., **10** (3) 1123-1128 (1981);
       Chem. Abstr., **95**, 213415n (1981).
1106   Rây, J. N.; Silooja, S. S. and Wadha, P. R.: J. Indian Chem. Soc., **10**, 617-620 (1933).
1107   Raychaudhuri, S.; Seshadri, T. R. and Mukerjee, S. K.: Indian J. Chem., **10** (1) 56-58
       (1972).
1108   Razus, A. C.; Arvay, Z.; Bartha, E.; Condeiu, C. and Glatz, A. M.: Rev. Chim. (Bucharest), **36**
       (4) 299-302 (1985); Chem. Abstr., **103**, 123109v (1985).
1109   Reich, D. A. and Nightingale, D. V.: J. Org. Chem., **21**, 825-826 (1956).
1110   Reilly, J. and Drumm, P. J.: J. Chem. Soc., 2814-2819 (1927).
1111   Reiter, P. L.: Ph. D., University of Stuttgart, Stuttgart (Germany) (1974).
1112   René, L.; Buisson, J. P. and Royer, R.: Bull. Soc. Chim. Fr., 475-476 (1974).
1113   Rey, M. E. and Waters, W. A.: J. Chem. Soc., 2753-2755 (1955).
1114   Reyes, J.; Greco, F.; Motais, R. and Latorre, R.: J. Membr. Biol., **72** (1-2) 93-103 (1983);
       Chem. Abstr., **98**, 175236g (1983).
1115   Rhodes, A.; Boothryod, B.; Gonagle, M. P. and Somerfield, G. A.: Biochem. J., **81** (1) 28-37
       (1961).
1116   Rhodes, A.; Somerfield, G. A. and Gonagle, M. P.: Biochem. J., **88** (1) 349-357 (1963).
1117   Richter, P.; Besch, A.; Wunderlich, I. and Hagen, A.: Ger. (East) DD 290,184 (1991); Chem.
       Abstr., **115**, 255827k (1991).
1118   Richter, R.: J. Prakt. Chem., **28**, 273-309 (1883).
1119   Richtzenhain, H. and Alfredsson, B.: Acta Chem. Scand., **8** (9) 1519-1529 (1954).
1120   Ricoh Co., Ltd.: Jpn. Kokai Tokkyo Koho JP 57,146,689 [82,146,689] (1982); Chem.
       Abstr., **99**, 222478v (1983).
1121   Ridd, J. H.; Yousaf, T. I. and Rose, J. B.: J. Chem. Soc., Perkin Trans. 2, **9**, 1729-1734
       (1988).
1122   Rigaudy, J. and Paillous, N.: Bull. Soc. Chim. Fr., 585-591 (1971).
1123   Rigaudy, J. and Paillous, N.: Tetrahedron Lett., **40**, 4825-4831 (1966).
1124   Rigaudy, J.; Perlat, M. C. and Nguyen Kim Cuong: Bull. Soc. Chim. Fr., 2521-2526 (1974).
1125   Rigaudy, J.; Perlat, M. C.; Simon, D. and Nguyen Kim Cuong: Bull. Soc. Chim. Fr., 493-
       500 (1976).
1126   Rigaudy, J. and Sparfel, D.: Bull. Soc. Chim. Fr., 3441-3446 (1972).
1127   Rink, H.: Eur. Pat. Appl. EP 285,562 (1988); Chem. Abstr., **111**, 97721g (1989).
1128   Rink, H.: Tetrahedron Lett., **28** (33) 3787-3790 (1987).
1129   Rio, G. and Berthelot, J.: Bull. Soc. Chim. Fr., 1705-1707 (1971).
1130   Rio, G. and Scholl, M. J.: J. Chem. Soc., Chem. Commun., **12**, 474 (1975).
1131   Ritchie, E.: J. Proc. Roy. Soc. N. S. Wales, **80**, 33-40 (1946); Chem. Abstr., **41**, 3094i
       (1947).
1132   Robertson, A.; Waters, R. B. and Jones, E. T.: J. Chem. Soc., 1681-1688 (1932).
1133   Roche, M. and Cerutti, E.: C. R. Acad. Sci., Ser. C, **279** (15) 663-666 (1974).
1134   Rodrigo, R.: Personal communication (05.08.1993).
1135   Rohr, T. M. and Kuhn, J.: Eur. Pat. Appl. EP 169,808 (1986); Chem. Abstr., **104**, 216418f
       (1986).
1136   Rojdestvenskii, N.: Zh. Russ. Fiz. Khim. O-va, Chast. Khim., **46**, 1075-1077 (1914); Chem.
       Abstr., **9**, 1899 (1915).
1137   Rose, J. B.: Ger. Offen., 2,730,991 (1978); Chem. Abstr., **88**, 104911f (1978).
1138   Rose, J. B. and Cinderey, M. B.: Eur. Pat. Appl. EP 75,390 (1983); Chem. Abstr., **99**,
       70383v (1983).
1139   Rosenblatt, D. H.; Epstein, J. and Levitch, M.: J. Am. Chem. Soc., **75**, 3277-3278 (1953).
1140   Rosenkranz, H. J. and Lachmann, B.: Ger. Offen. 2,209,527 (1973); Chem. Abstr., **80**, 3265c
       (1974).
1141   Rosenmund, K. W.; Buchwald, R. and Deligiannis, T.: Arch. Pharm. Ber. Dtsch. Pharm.
       Ges., **271**, 342-352 (1933).
1142   Rosenmund, K. and Lohfert, H.: Ber. Dtsch. Chem. Ges., **61**, 2601-2607 (1928).
1143   Rosenmund, K. and Schnurr, W.: Justus Liebigs Ann. Chem., **460**, 56-98 (1928).

1144   Rosenmund, K. and Schulz, H.: Arch. Pharm. Ber. Dtsch. Pharm. Ges., **265**, 308-319
       (1927).
1145   Rotschova-Protivova, J.; Pospisil, J.; Holcik, J. and Durmis, J.: J. Chromatogr., **106** (2) 343-
       348 (1975).
1146   Roussel-UCLAF: Fr. Demande 2,269,330 (1975); Chem. Abstr., **85**, 5378u (1976).
1147   Roussel-UCLAF: Fr. Demande 2,405,252 (1979); Chem. Abstr., **92**, 41964k (1980).
1148   Royer, R. and Bisagni, E.: Bull. Soc. Chim. Fr., 486-492 (1954).
1149   Royer, R. and Demerseman, P.: Bull. Soc. Chim. Fr., 1682-1686 (1959).
1150   Royer, R.; Demerseman, P. and Cheutin, A.: Bull. Soc. Chim. Fr., 275-277 (1960).
1151   Royer, R.; Demerseman, P.; Cheutin, A.; Allegrini, E. and Michelet, R.: Bull. Soc. Chim. Fr.,
       1379-1388 (1957).
1152   Royer, R.; Demerseman, P.; Michelet, R. and Cheutin, A.: Bull. Soc. Chim. Fr., 1148-1157
       (1959).
1153   Royer, R.; Lechartier, J. P. and Demerseman, P.: Bull. Soc. Chim. Fr., 2948-2951 (1972).
1154   Royer, R.; René, L. and Demerseman, P.: Chim. Ther., **8** (2) 139-142 (1973).
1155   Royer, R.; René, L.; Demerseman, P. and Cavier, R.: Fr. Demande 2,198,735 (1974); Chem.
       Abstr., **82**, 31247h (1975).
1156   Rubtsova, T. A. and Volkotrub, M. N. (USSR): Deposited Doc. 1974, VINITI 1703-74, 12
       pp.; Chem. Abstr., **88**, 158779n (1978).
1157   Ruegg, R.; Ryser, G. and Schwieter, U.: Brit. 1,115,158 (1968); Chem. Abstr., **69**, 76941c
       (1968).
1158   Ruenitz, P. C.: PCT Int. Appl. WO 92 04,310 (1992); Chem. Abstr., **117**, 7654j (1992).
1159   Ruenitz, P. C.; Bourne, C. S.; Sullivan, K. J. and Moore, S. A.: J. Med. Chem., **39** (24) 4853-
       4859 (1996).
1160   Ruminski, J. K.: Chem. Ber., **116**, 970-979 (1983).
1161   Ruminski, J. K.: Chem. Anal. (Warsaw), **37** (2) 171-175 (1992); Chem. Abstr., **118**,
       212388n (1993).
1162   Ruminski, J. K.; Dabkowska, H. and Gronowska, J.: Pol. J. Chem., **52** (3) 629-635 (1978).
1163   Ruminski, J. K. and Mokhtar, H. M.: Pol. J. Chem., **55**, 995-1005 (1981).
1164   Ruminski, J. K. and Przewoska, K. D.: Chem. Ber., **115**, 3436-3443 (1982).
1165   Sadrmohaghegh, C.; Scott, G. and Setoudeh, E.: Eur. Polym. J., **18** (11) 1007-1010 (1982);
       Chem. Abstr., **98**, 108235t (1983).
1166   Saharia, G. S. and Sharma, B. R.: J. Indian Chem. Soc., **33**, 788-790 (1956).
1167   Saharia, G. S. and Sharma, B. R.: J. Sci. Ind. Res., Sect. B, **14B**, 263-267 (1955).
1168   Saharia, G. S. and Sharma, B. R.: J. Sci. Ind. Res., Sect. B, **16B**, 125-128 (1957).
1169   Saharia, G. S. and Sharma, H. R.: Sci. Cult., **45** (4) 139-144 (1979); Chem. Abstr., **92**, 6202p
       (1980).
1170   Saito, T.; Noguchi, H. and Shibata, S.: Proceedings of Annual Meeting of the Japanese
       Society of Pharmacognosy (Chiba, 1975), p. 33.
1171   Sakai, J.; Ikeda, K. and Ishida, Y.: Iyakuhin Kenkyu, **15** (6) 1078-1090 (1984); Chem.
       Abstr., **102**, 119508m (1985).
1172   Sakai, M. and Takahashi, O.: Jpn. Kokai Tokkyo Koho JP 63,264,748 [88,264,748] (1988);
       Chem. Abstr., **114**, 72224w (1991).
1173   Sala, T. and Sargent, M. V.: J. Chem. Soc., Chem. Commun., **23**, 1043-1044 (1978).
1174   Sala, T. and Sargent, M. V.: J. Chem. Soc., Perkin Trans. 1, **3**, 855-869 (1981).
1175   Salaskin, S. N.; Kalachev, A. I.; Korshak, V. V. and Vinogradova, S. V.: Deposited Doc.
       1975, VINITI 1064-1075, 23 pp.; Chem. Abstr., **87**, 68688g (1977).
1176   Salman, M.; Ray, S.; Agarwal, A. K.; Durani, S.; Setty, B. S.; Kamboj, V. P. and Anand, N.:
       J. Med. Chem., **26** (4) 592-595 (1983).
1177   Salman, S. R.: Can. J. Spectrosc., **34** (2) 50-51 (1989).
1178   Salman, S. R. and Abas, K. F.: Thermochim. Acta, **152** (2) 381-386 (1989); Chem. Abstr.,
       112, 157512x (1990).
1179   Salman, S. R. and Abas, K. F.: Thermochim. Acta, **149**, 381-386 (1989); Chem. Abstr., **112**,
       157570q (1990).
1180   Salman, S. R. and Kamounah, F. S.: Magn. Reson. Chem., **25** (11) 966-969 (1987).
1181   Salman, S. R.; Kamounah, F. S. and Tameesh, A. H. H.: J. Chromatogr., **483**, 390-393
       (1989).

1182 Samreth, S.; Bellamy, F. and Millet, J.: Eur. Pat. Appl. EP 290,321 (1988); Chem. Abstr.,
        110, 232029f (1989).
1183 Samreth, S.; Bellamy, F. and Millet, J.: Eur. Pat. Appl. EP 367,671(1990); Chem. Abstr., **114**,
        24485k (1991).
1184 Sandifer, R. M.; Bhattacharya, A. K. and Harris, T. M.: J. Org. Chem., **46** (11) 2260-2267
        (1981).
1185 Sandner, M. R.; Hedaya, E. and Trecker, D. J.: J. Am. Chem. Soc., **90** (26) 7249-7254
        (1968).
1186 Santesson, J.: Acta Chem. Scand., **22**, 1698-1699 (1968).
1187 Santesson, J. and Sundholm, G.: Arkiv. Für Kem., **30**, 427-431 (1969).
1188 Sargent, M. V.: J. Chem. Soc., Chem. Commun., **6**, 285 (1980).
1189 Sargent, M. V.: J. Chem. Soc., Perkin Trans. 1, **2**, 403-411 (1982).
1190 Sargent, M. V. and Vogel, P.: Aust. J. Chem., **29** (4) 907-914 (1976).
1191 Sargent, M. V.; Vogel, P.; Elix, J. A. and Ferguson, B. A.: Aust. J. Chem., **29** (10) 2263-
        2269 (1976).
1192 Sasada, K.; Sugawara, K.; Miyata, S.; Mori, S.; Yonezawa, N. and Murao, A.: Jpn. Kokai
        Tokkyo Koho JP 03,141,239 [91,141,239] (1991); Chem. Abstr., **115**, 256876n (1991).
1193 Sato, H.; Dan, T.; Onuma, E.; Tanaka, H.; Aoki, B. and Koga, H.: Chem. Pharm. Bull., **39** (7)
        1760-1772 (1991).
1194 Sato, H.; Dan, T.; Onuma, E.; Tanaka, H.; Aoki, B. and Koga, H.: Chem. Pharm. Bull., **40** (1)
        109-116 (1992).
1195 Sato, H.; Kuromaru, K.; Ishizawa, T.; Aoki, B. and Koga, H.: Chem. Pharm. Bull., **40** (10)
        2597-2601 (1992).
1196 Sato, K.: Jpn. Kokai Tokkyo Koho JP 61 89,092 [86 89,092] (1986); Chem. Abstr., **106**,
        25850x (1987).
1197 Sato, T.: Jpn. Kokai Tokkyo Koho JP 02 43,540 [90 43,540] (1990); Chem. Abstr., **114**,
        217938u (1991).
1198 Sato, T.; Kato, T.; Kajiwara, S.; Miyamori, C. and Takata, I.: Int. Congr. Ser.- Excerpta Med.,
        605 (Curr. Probl. Thyroid Res.), 147-150 (1983); Chem. Abstr., **99**, 188304y (1983).
1199 Sato, Y.; Ajiro, Y. and Oda, T.: Tennen Yuki Kagobutsu Toronkai Koen Yoshishu, 21st, 152-
        158 (1978); Chem. Abstr., **90**, 83393n (1979).
1200 Sato, Y.; Ajiro, Y.; and Oda, T.: Symp. Pap. IUPAC Int. Symp. Chem. Nat. Prod., 11th, **1**,
        175-178 (1978); Chem. Abstr., **92**, 144892q (1980).
1201 Sato, Y.; Machida, T. and Oda, T.: Tetrahedron Lett., **51**, 4571-4574 (1975).
1202 Sato, Y. and Oda, T.: J. Chem. Soc., Chem. Commun., **3**, 135-136 (1978).
1203 Sato, Y. and Oda, T.: Tetrahedron Lett., **44**, 3971-3974 (1976).
1204 Satoh, T.; Itaya, T.; Miura, M. and Nomura, M.: Chem. Lett., **9**, 823-824 (1996).
1205 Saunders, J. C. and Williamson, W. R. N.: Ger. Offen. 2,450,053 (1975); Chem. Abstr., **83**,
        97263n (1975).
1206 Saunders, J. C. and Williamson, W. R. N.: Ger. Offen. 2,546,738 (1976); Chem. Abstr., **86**,
        43409h (1977).
1207 Saunders, J. C. and Williamson, W. R. N.: Ger. Offen. 2,615,487 (1976); Chem. Abstr., **86**,
        72208x (1977).
1208 Saunders, J. C. and Williamson, W. R. N.: J. Med. Chem., **22** (12) 1554-1558 (1979).
1209 Sawada, H.; Hara, A.; Asano, S. and Matsumoto, Y.: Clin. Chem. (Winston-Salem, N. C.), **22**
        (10) 1596-1603 (1976); Chem. Abstr., **86**, 11657w (1977).
1210 Sawada, S.; Yasui, K. and Takahashi, S.: Biosci. Biotechnol. Biochem., **56** (9) 1506-1507
        (1992).
1211 Sawhney, K. N. and Mathur, K. B. L.: Indian J. Chem., Sect. B, **19B** (7) 590-592 (1980).
1212 Saxena, S. K.; Sahib, M. K.; Kumar, S.; Seth, M. and Bhaduri, A. P.: Indian J. Biochem.
        Biophys., **21** (2) 139-141 (1984); Chem. Abstr., **101**, 48792r (1984).
1213 Saxena, S. K.; Seth, M.; Bhaduri, A. P. and Sahib, M. K.: J. Steroid Biochem., **18** (3) 303-
        308 (1983); Chem. Abstr., **98**, 173374b (1983).
1214 Schellenbaum, M. and Duennenberger, M.: Ger. Offen. 2,033,749 (1971); Chem. Abstr., **75**,
        151537e (1971).
1215 Schering A.-G.: Fr. Demande 2,267,101 (1975); Chem. Abstr., **86**, 29806d (1977).
1216 Schering-Kahlbaum A.-G.: E. P. 397 505 (1933); Chem. Zentralbl., **105**, 129 (1934).

1217 Schmitt, R. G. and Hirt, R. C.: J. Polym. Sci., **45**, 35-47 (1960).
1218 Schmitt, R. G. and Hirt, R. C.: J. Appl. Polymer Sci., **7** (5) 1565-1580 (1963).
1219 Schmitz, F.: Dissertation, Bonn (Germany), pp. 40-43 (1904).
1220 Schmitz, F.: Justus Liebigs Ann. Chem., **346**, 389-391 (1906).
1221 Schroetter, E.; Hoegel, E. and Jeschke, H. J.: Ger. (East) 106,635 (1974); Chem. Abstr., **82**, 155759w (1975).
1222 Schroetter, E.; Weuffen, W. and Herudek, D.: Pharmazie, **29** (6) 374-382 (1974).
1223 Schubertova, N. and Hrdlovic, P.: Chem. Zvesti, **23** (7) 495-500 (1969).
1224 Schultz, A. G.; Napier, J. J. and Ravichandran, R.: J. Org. Chem., **48**, 3408-3412 (1983).
1225 Schwartz, J. C.; Arrang, J. M.; Garbarg, M.; Quemener, A.; Lecomte, J. M.; Ligneau, X.; Schunack, W. G.; Stark, H.; Purand, K. and al.: PCT Int. Appl. WO 96 29,315 (1996); Chem. Abstr., **126** (2) 18872r (1997).
1226 Schwenk, U.;Eiglmeier, K. and Pfueller, P.: Ger. Offen. 2,518,631 (1976); Chem. Abstr., **86**, 72209y (1977).
1227 Scott, A. I.: Proc. Chem. Soc. (London), 195 (1958).
1228 Scott, A. I.; Pike, D. G.; Ryan, J. J. and Guilford, H.: Tetrahedron, **27** (14) 3051-3063 (1971).
1229 Scott, G.: Food Addit. Contam., **5**(Suppl. 1) 421-432 (1988); Chem. Abstr., **110**, 113323r (1989).
1230 Sekiguchi, T. and Tanaka, M.: Jpn. Kokai Tokkyo Koho JP 60,185,743 [85,185,743] (1985); Chem. Abstr., **104**, 68598x (1986).
1231 Sekiguchi, T. and Tanaka, M.: Nippon Kagaku Kaishi, **4**, 742-746 (1985); Chem. Abstr., **104**, 109139u (1986).
1232 Selwood, D. L.; Livingston, D. J.; Comley, J. C. W.; O'Dowd, A. B.; Hudson, A. T.; Kackson, P.; Jandu, K. S.; Rose, V. S. and Stables, J. N.: J. Med. Chem., **33** (1) 136-142 (1990).
1233 Sergovskaya, N. L.; Kornienko, N. I.; Shekhter, O. V. and Tsizin, Yu. S.: Zh. Org. Khim., **18** (10) 2167-2170 (1982).
1234 Seshadri, T. R.: J. Indian Chem. Soc., **40** (7) 497-504 (1963).
1235 Seshadri, T. R.: Phytochemistry, **11**, 881-898 (1972).
1236 Seshadri, T. R.; Subramani, P. E. and Varadarajan, S.: J. Sci. Ind. Res., Sect. B, **11B**, 56-62 (1952).
1237 Setalvad, J. I.; Amin, G. C. and Shah, N. M.: J. Indian Chem. Soc., **29** (12) 915-920 (1952).
1238 Setalvad, J. I.; Amin, G. C. and Shah, N. M.: J. Indian Chem. Soc., **33** (4) 249-252 (1956).
1239 Setalvad, J. I. and Shah, N. M.: J. Indian Chem. Soc., **31**, 600-604 (1954).
1240 Setalvad, J. I. and Shah, N. M.: J. Indian Chem. Soc., **32**, 529-532 (1955).
1241 Seymour, R. B. and Tsang, Hing Shya: Tex. J. Sci., **23** (2) 187-200 (1971); Chem. Abstr., **76**, 58639a (1972).
1242 Shah, H. A. and Shah, R. C.: J. Indian Chem. Soc., **17**, 32-36 (1940).
1243 Shah, R. C.; Deshpande, R. K. and Chaubal, J. S.: J. Chem. Soc., 642-650 (1932).
1244 Shah, R. C. and Mehta, P. R.: J. Indian Chem. Soc., **13**, 368-371 (1936).
1245 Shah, R. R.; Mehta, R. D. and Parikh, A. R.: J. Indian Chem. Soc., **58** (11) 1113-1115 (1981).
1246 Shah, V. H.; Gaur, V. B.; Patel, H. H. and Parikh, A. R.: Acta Cienc. Indica, [Ser.] Chem., **8** (4) 212-217 (1982); Chem. Abstr., **98**, 160620d (1983).
1247 Shahane, R. Y.: Curr. Sci., **10**, 523-524 (1941).
1248 Shapiro, Anna; Nathan, H. C.; Hutner, S. H.; Garofalo, Joanne; McLaughlin, Susan, Dittus; Rescigno, Diane and Bacchi, C. J.: J. Protozool., **29** (1) 85-90 (1982); Chem. Abstr., **97**, 103749h (1982).
1249 Sharghi, H. and Eshghi, H.: Bull. Chem. Soc. Jpn., **66**, 135-139 (1993).
1250 Sharghi, H. and Tamaddon, F.: Tetrahedron, **52** (43) 13623-13640 (1996).
1251 Sharma, I. and Ray, S.: Indian J. Chem., Sect. B, **24B** (1) 59-61 (1985).
1252 Sharma, I. and Ray, S.: Indian J. Chem., Sect. B; **27B** (4) 374-375 (1988).
1253 Sharma, I.; Salman, M.; Koley, P. L. and Ray, S.: Indian J. Chem., Sect. B, **23B** (6) 567-570 (1984).
1254 Sharma, P. K. and Khanna, R. N.: Monatsh. Chem., **116**, 353-356 (1985).
1255 Shiba, T. and Cahnmann, H. J.: J. Org. Chem., **27**, 1773-1778 (1962).

1256 Shibaldain, N. L.: M.Sc. Thesis, Basrah, IRAQ (1984).
1257 Shimizu, Y.; Nakanishi, H.; Kuwana, K. and Ninomiya, T.: Jpn. Kokai Tokkyo Koho JP
      63,208,840 [88,208,840] (1988); Chem. Abstr., **110**, 182980q (1989).
1258 Shinagawa, Yoshiya and Shinagawa, Yasuko: Int. J. Quantum Chem., Quantum Biol. Symp.,
      5, 269-279 (1978); Chem. Abstr., **90**, 117100g (1979).
1259 Shioda, Hirohisa; Namiki, Isamu; Hori, Hisako and Katsuyama, Yoshihisa: U.S. 3,639,483
      (1972); Chem. Abstr., **76**, 153339m (1972).
1260 Shioda, Hirohisa; Namiki, Isamu; Hori, Hisako and Katsuyma, Yoshihisa: Brit. 1,246,958
      (1971); Chem. Abstr., **76**, 26022c (1972).
1261 Shiraishi, Shohei: Japan Kokai 75 23,219 (1975); Chem. Abstr., **83**, 88707h (1975).
1262 Shlyapintokh, V. Ya.: Pure Appl. Chem., 55 (10) 1661-1668 (1983); Chem. Abstr., **100**,
      122112s (1984).
1263 Shoesmith, J. B. and Haldane, J.: J. Chem. Soc., **125**, 113-115 (1924).
1264 Shoji, R.; Watanabe, T.; Tashiro, S. and Shi, S.: Iyakuhin Kenkyu, **26** (6) 386-397 (1995);
      Chem. Abstr., **123**, 296397d (1995).
1265 Shores, J. H.: Dissertation, Bonn (Germany), 9-16 (1898).
1266 Shores, J. H.: Justus Liebigs Ann. Chem., **346**, 382-386 (1906).
1267 Shriner, R. L. and Moffett, R. B.: J. Am. Chem. Soc., **63**, 1694-1698 (1941).
1268 Shulgin, A. T. and Kerlinger, H. O.: Chem. Commun., **9**, 249-250 (1966).
1269 Shultz, A. R.; Young, A. L.; Alessi, S. and Stewart, M.: J. Appl. Polym. Sci., **28** (5) 1685-
      1700 (1983); Chem. Abstr., **98**, 216421j (1983).
1270 Shutske, G. M.; Setescak, L. L. and Allen, R. C.: Eur. Pat. Appl. 2,666 (1979); Chem. Abstr.,
      92, 94378d (1980).
1271 Shutske, G. M.; Setescak, L. L. and Allen, R. C.: Eur. Pat. Appl. EP 45,078 (1982); Chem.
      Abstr., **96**, 217825q (1982).
1272 Shutske, G. M.; Setescak, L. L.; Allen, R. C.; Davis, L.; Effland, R. C.; Ranbom, K.; Kitzen,
      J. M.; Wilker, J. C. and Novick, W. J., Jr.: J. Med. Chem., **25** (1) 36-44 (1982).
1273 Siegrist, A. E.; Maeder, E. and Dünnenberger, M.: Ger. Offen. 1,093,373 (1958); Chem.
      Abstr., **57**, 9746c (1962).
1274 Simpson, E. M. D.; Tomlinson, M. L. and Taylor, H. V.: J. Chem. Soc., 2239-2243 (1951).
1275 Simunkova, D. and Marcek, O.: Plaste Kautsch., **30** (1)18-20 (1983); Chem. Abstr., **98**,
      144348r (1983).
1276 Singh, A. K. and Sonar, S. M.: Synth. Commun., **15** (12) 1113-1121 (1985).
1277 Singh, J. M. and Turner, A. B.: J. Chem. Soc., Perkin Trans. 1, **18**, 2294-2296 (1972).
1278 Singh, J. M. and Turner, A. B.: J. Chem. Soc., Perkin Trans. 1, **22**, 2556-2559 (1974).
1279 Singh, P.; Pardasani, R. T.; Prashant, A.; Pokharna, C. P. and Chaudhary, B.: J. Indian
      Chem. Soc., **71** (6-8) 409-414 (1994).
1280 Singh, P.; Pardasani, R. T.; Prashant, A.; Pokharna, C. P. and Chaudhary, B.: Pharmazie, **48**
      (9) 699-700 (1993).
1281 Sinitsyna, T. A.; Alekseeva, I. A. and Voronina, N. M.: Zh. Prikl. Spektrosk., **32** (4) 648-651
      (1980); Chem. Abstr., **93**, 185247x (1980).
1282 Sinitsyna, T. A.; Sidorov, E. O.; Voronina, N. M.; Yalovets, I. A.; Latosh, N. I. and Alekseeva,
      I. A.: Zh. Obshch. Khim., **50** (5) 1174-1177 (1980).
1283 Sinyavskaya, L. P. and Shamshurin, A. A.: Zh. Org. Khim., **4** (7) 1267-1270 (1968).
1284 Sivakumar, A. and Reddy, S. J.: Trans. SAEST, **18** (4) 331-333 (1983); Chem. Abstr., **100**,
      182126j (1984).
1285 Sivakumar, A. and Reddy, S. J.: Trans. SAEST, **19** (4) 295-297 (1984); Chem. Abstr., **103**,
      122895m (1985).
1286 Sivakumar, A. and Reddy, S. J.: Trans. SAEST, **29** (1) 27-30 (1985); Chem. Abstr., **103**,
      149432v (1985).
1287 Sivakumar, A. and Reddy, S. J.: Indian J. Chem., Sect. A, **23A** (9) 732-735 (1984).
1288 Smit, N. P. M.; Latter, A. J. M.; Naish-Byfield, S.; Westerhof, W.; Pavel, S. and Riley, P. A.:
      Biochem. Pharmacol., **48** (4) 743-752 (1994); Chem. Abstr., **121**, 195222w (1994).
1289 Smith, A. W.: Ber. Dtsch. Chem. Ges., **24**, 4025-4058 (1891).
1290 Smith, M. G.; Renga, J. M.; Riley, B. K.; Ray, P. G. and Marlowe, C.: U.S. US 4,938, 790
      (1990); Chem. Abstr., **113**, 191411u (1990).
1291 Snyder, H. R. and Elston, C. T. : J. Am. Chem. Soc., **77**, 364-366 (1955).

1292 SORI, French Patent 2,321,276 (1976).
1293 Sornay, R.; Gurrieri, J.; Tourne, C.; Renson, F. J.; Majoie, B. and Wulfert, E.: Arzneim.-Forsch., **26** (5) 885-889 (1976).
1294 Spada, A.; Cameroni, R. and Bernabei, M. T.: Gazz. Chim. Ital., **86**, 46-55 (1956).
1295 Srivastava, A. K. and Bahel, S. C.: J. Indian Chem. Soc., **53** (8) 841-845 (1976).
1296 Staedel, W.: Justus Liebigs Ann. Chem., **283**, 149-151 (1894).
1297 Staedel, W.: Justus Liebigs Ann. Chem., **283**, 164-181 (1894).
1298 Städel, W. and Gail, F.: Justus Liebigs Ann. Chem., **194**, 307-372 (1878).
1299 Staedel, W. and Gail, F.: Ber. Dtsch. Chem. Ges., **11**, 744-746 (1878).
1300 Stahlhofen, P.: Eur. Pat. Appl. EP 68,346 (1983); Chem. Abstr., **99**, 30749n (1983).
1301 Stanley, L. N.: U.S. 2;921,962 (1960); Chem. Abstr., **54**, 16431c (1960).
1302 Stanley, L. N.: U.S. 3,073,866 (1963); Chem. Abstr., **59**, 11348d (1963).
1303 Stanley, T.: U.S. 3,330,884 (1967); Chem. Abstr., **68**, 96546k (1968).
1304 Stark, P. A.; Stewart, E. G. and Everaerts, A. I.: PCT Int. Appl. WO 97 07,161 (1997); Chem. Abstr., **126** (19) 252329r (1997).
1305 Starkov, S. P.; Panasenko, A. I., Volkotrub, M. N. and Zhidkova, L. A.: Zh. Obshch. Khim., **63** (2) 406-408 (1993).
1306 Staudinger, H. and Kon, N.: Justus Liebigs Ann. Chem., **384**, 38-135 (1911).
1307 Sternbach, L. H.; Fryer, R. I.; Metlesics, W.; Sach, G. and Stempel, A.: J. Org. Chem., **27**, 3781-3788 (1962).
1308 Stewart, R.; Granger, M. R.; Moodie, R. B. and Muenster, L. J.: Can. J. Chem., **41**, 1065-1070 (1963).
1309 Stoermer, R.; Friderici, E. and Altgelt, H.: Ber. Dtsch. Chem. Ges., **41**, 321-324 (1908).
1310 Stout, G. H. and Balkenhol, W. J.: Tetrahedron, **25** (7) 1947-1960 (1969).
1311 Stout, G. H.; Christensen, E. N.; Balkenhol, W. J. and Stevens, K. L.: Tetrahedron, **25** (7) 1961-1973 (1969).
1312 Strat, G.: An. Stiint. Univ. "Al. I. Cuza" Iasi, Sect. 1b, **28**, 13-18 (1982).
1313 Strat, M.; Singurel, L. and Strat, G.: An. Stiint. Univ. "Al. I. Cuza" Iasi, Sect. 1b, **20** (1) 67-72 (1974).
1314 Strat, M.; Singurel, L. and Strat, G.: An. Stiint. Univ. "Al. I. Cuza" Iasi, Sect. 1b, **20** (2) 157-162 (1974).
1315 Strat, M.; Umreiko, D. S. and Khovratovich, N. N.: Zh. Prikl. Spektrosk., **19** (1) 103-108 (1973); Chem. Abstr., **79**, 77607s (1973).
1316 Strat, M.; Umreiko, D. S. and Khovratovich, N. N.: Zh. Prikl. Spektrosk., **19** (2) 288-293 (1973); Chem. Abstr., **79**, 114652a (1973).
1317 Strubell, W. and Baumgartel, H.: J. Prakt. Chem., **9**, 213-216 (1959).
1318 Strupczewski, J. T.; Helsley, G. C.; Chiang, Y. and Bordeau, K. J.: Eur. Pat. Appl. EP 402,644 (1990); Chem. Abstr., **114**, 185553w (1991).
1319 Strupczewski, J. T.; Helsley, G. C.; Chiang, Y.; Bordeau, K. J. and Glamkowski, E. J.: Eur. Pat. Appl. EP 542,136 (1993); Chem. Abstr., **120**, 54553x (1994).
1320 Stueber, W.; Knolle, J. and Breipohl, G.: Int. J. Pept. Protein Res., **34** (3) 215-221 (1989); Chem. Abstr., **112**, 199053d (1990).
1321 Suda, H.; Kanoh, S.; Hasegawa, H. and Motoi, M.: Kanazawa Daigaku Kogakubu Kiyo, **15** (1) 71-74 (1982); Chem. Abstr., **97**, 197910d (1982).
1322 Sugawara, K.; Sasada, K.; Miyata, S.; Mori, S.; Yonezawa, N. and Murao, A.: Jpn. Kokai Tokkyo Koho JP 03,145,438 [91,145,438] (1991); Chem. Abstr., **115**, 231871u (1991).
1323 Sumathi, T. and Balasubramanian, K. K.: Tetrahedron Lett., **33** (16) 2213-2216 (1992).
1324 Sumathi, T. and Balasubramanian, K. K.: Tetrahedron Lett., **34** (24) 3915-3918 (1993).
1325 Sumitomo Chemical Co. Ltd., Jpn. Kokai Tokkyo Koho JP 57,119,941 [82,119,941] (1982); Chem. Abstr., **98**, 90464e (1983).
1326 Sumitomo Chemical Co., Ltd.: Jpn. Kokai Tokkyo Koho JP 58 38,229 [83 38,229] (1983); Chem. Abstr., **99**, 38201y (1983).
1327 Sundholm, E. G.: Acta Chem. Scand., Ser. B, **B32** (3) 177-181 (1978).
1328 Sundholm, E. G.: Tetrahedron, **34** (5) 577-586 (1978).
1329 Sundholm, G.: Acta Chem. Scand., **B28** (9) 1102-1103 (1974).
1330 Sundstrom, G.: J. Agric. Food Chem., **25** (1) 18-21 (1977); Chem. Abstr., **86**, 38307u (1977).

1331   Surendranath, V.: Sasmira Tech. Dig., **3**, 9-11 (1980); Chem. Abstr., **96**, 51894g (1982).
1332   Suzuki, K.; Kajigaeshi, S. and Sano, M.: Yûki Gôsei Kagaku Kyôkaishi, **16**, 82-87 (1958); Chem. Abstr., **52**, 10019f (1958).
1333   Suzuki, T.; Hibino, K.; Murai, M. and Fujita, T.: Eur. Pat. Appl. EP 93,194 (1983); Chem. Abstr., **100**, 28796w (1984).
1334   Suzuki, T.; Hibino, K.; Murai, M. and Fujita, T.: U.S. U.S. 4,425,404 (1984); Chem. Abstr., **100**, 113786n (1984).
1335   Syamala, M. S.; Rao, B. N. and Ramamurthy, V.: Tetrahedron, **44** (23) 7234-7242 (1988).
1336   Synthelabo S. A.: Israeli IL 56,737 (1982); Chem. Abstr., **100**, 34271n (1984).
1337   Szabo, V.; Borbely, S.; Farkas, E. and Tolnai, S.: Magy. Kem. Foly., 81 (5) 220-224 (1975); Chem. Abstr., **83**, 79033h (1975).
1338   Szell, T.; Hajas, E. and Sipos, S.: Acta Phys. Chem. Szeged, **11**, 47-50 (1965).
1339   Szymanowski, J. and Blaszczak, J.: Chem. Stosow., **26** (1) 99-109 (1982); Chem. Abstr., **98**, 215259a (1983).
1340   Szymanowski, J. and Prochaska, K.: J. Colloid Interface Sci., **123** (2) 456-465 (1988); Chem. Abstr., **109**, 116566n (1988).
1341   Szymanowski, J. and Prochaska, K.: J. Colloid Interface Sci., **125** (2) 649-666 (1988); Chem. Abstr., **110**, 45408v (1989).
1342   Szymanowski, J.; Voelkel, A.; Beger, J. and Binte, H. J.: J. Prakt. Chem., **327** (3) 353-361 (1985).
1343   Szymanowski, J.; Voelkel, A. and Rashid, Z. A.: J. Chromatogr., **402**, 55-64 (1987).
1344   Tadkod, R. S.; Kulkarni, S. N. and Nargund, K. S.: J. Karnatak Univ., **3**, 78-80 (1958); Chem. Abstr., **54**, 8717 (1960).
1345   Tadkod, R. S.; Sattur, P. B.; Kulkarni, S. N. and Nargund, K. S.: J. Karnatak Univ., 3 (1) 29-32 (1957); Chem. Abstr., **53**, 8062f (1959).
1346   Tadros, W. and Latif, A.: J. Chem. Soc., 3337-3340 (1949).
1347   Tahara, Y.: Ber. Dtsch. Chem. Ges., **24**, 2459-2462 (1891).
1348   Taher, B.; Schleusener, A. and Baltes, W.: Dtsch. Lebensm.-Rundsch., **90** (2) 35-38 (1994); Chem. Abstr., **121**, 238039m (1994).
1349   Takahashi, A.; Inada, H. and Matsumura, S.: Jpn. Kokai Tokkyo Koho JP 03 81,245 [91 81,245] (1991); Chem. Abstr., **115**, 160067y (1991).
1350   Takahata, K.; Kaya, H.; Taniguchi, K. and Takai, T.: PCT Int. Appl. WO 90 00,087 (1990); Chem. Abstr., **113**, 8410f (1990).
1351   Takai, T. and Taniguchi, K.: Jpn. Kokai Tokkyo Koho JP 01,224,345 [89,224,345] (1989); Chem. Abstr., **112**, 118455m (1990).
1352   Takayanagi, T.; Miyazaki, N. and Sasao, Y.: Jpn. Kokai Tokkyo Koho JP 01,287,160 [89,287,160] (1989); Chem. Abstr., **112**, 200807m (1990).
1353   Takematsu, T.; Konnai, M.; Konno, K.; Hayashi, Y.; Ikeda, K.; Go, A. and Ishigaki, A.: Jpn. Kokai Tokkyo Koho, 79 02,323 (1979); Chem. Abstr., **90**, 181589a (1979).
1354   Takuwa, A.; Iwamoto, H.; Soga, O. and Maruyama, K.: J. Chem. Soc., Perkin Trans. 1, **9**, 1627-1631 (1986).
1355   Talbot, R. H. and Adams, R.: J. Am. Chem. Soc., **49**, 2040-2042 (1927).
1356   Tanigaki, Teiichi: Ger. Offen. DE 3,402,831 (1984); Chem. Abstr., **101**, 230145j (1984).
1357   Tarbell, D. S. and Fanta, P. E.: J. Am. Chem. Soc., **65**, 2169-2174 (1943).
1358   Tasaki, T.: Acta Phytochim., **2**, 49-73 (1925).
1359   Tasaki, T.: Chem. Zentralbl., **2**, 1354-1356 (1925).
1360   Taub, D.; Kuo, C. H.; Slates, H. L. and Wendler, N. L.: Tetrahedron, **19** (1) 1-17 (1963).
1361   Taub, D.; Kuo, C. H. and Wendler, N. L.: Chem. Ind. (London), 557-558 (1962).
1362   Taub, D.; Kuo, C. H. and Wendler, N. L.: J. Org. Chem., **28**, 2752-2755 (1963).
1363   Tawada, H.; Natsugari, H.; Ishikawa, E.; Sugiyama, Y.; Ikeda, H. and Meguro, K.: Chem. Pharm. Bull., **43** (4) 616-625 (1995).
1364   Tayama, T.: Jpn. Kokai Tokkyo Koho JP 63 96,148 [88 96,148] (1988); Chem. Abstr., **109**, 231700p (1988).
1365   Teasley, M. F. and Hsiao, B. S.: Macromolecules, **29** (20) 6432-6441 (1996).
1366   Tedroff, J.; Hartvig, P.; Bjurling, P.; Andersson, Y.; Antoni, G. and Laangstroem; B.: J. Neural Transm.: Gen. Sect., **85** (1) 11-17 (1991); Chem. Abstr., **115**, 247522s (1991).

1367 Teijin Chemicals, Ltd.: Jpn. Kokai Tokkyo Koho JP 57,149,335 [82,149,335] (1982); Chem. Abstr., **98**, 108960a (1983).
1368 Temchin, Yu. I.; Burmistrov, E. F.; Medvadev, A. I.; Egidis, F. M. and Rozantsev, E. G.: Sint. Issled. Eff. Khim. Polim. Mater., Mater. Vses. Nauchno-Tekh. Konf., 4th, 81-86 (1972) (Pub. **1974**); Chem. Abstr., **82**, 140955h (1975).
1369 Thakor, V. M. and Shah, N. M.: J. Indian Chem. Soc., **23**, 423-424 (1946).
1370 The Merck Index, Merck & CO., Inc. publishers, 11th edition, (1989).
1371 Thiele, K.; Ahmed, Q.; Jahn, U. and Adrian, R. W.: Arzneim.-Forsch. 29 (5) 711-720 (1979).
1372 Thomas II, F. D.; Shamma, M. and Fernelius, W. C.: J. Am. Chem. Soc., **80**, 5864-5867 (1958).
1373 Thomas, L. H. and Vlismas, T.: J. Chem. Soc., 612-615 (1963).
1374 Tirrell, D. A.: ACS Symp. Ser. 1981, 151(Photodegradation Photostab. Coat.), 43-49; Chem. Abstr., **95**, 8090m (1981).
1375 Tirrell, D.; Bailey, D. and Vogl, O.: Polym. Drugs, [Proc. Int. Symp.] 1977 (Pub. 1978), 77-101; Chem. Abstr., **91**, 145926v (1979).
1376 Tirrell, D. and Vogl, O.: Makromol. Chem., **181**, 2097-2109 (1980).
1377 Tomita, M.; Nakata, M. and Nakano, J.: Japan Kokai 75,151,874 (1975); Chem. Abstr., **85**, 21103f (1976).
1378 Tomita, M.; Nakata, M. and Nakano, J.: Japan Kokai 75,154,259 (1975); Chem. Abstr., **85**, 78001a (1976).
1379 Tomita, M.; Nakata, M. and Nakano, J.: Japan Kokai 75 53,370 (1975); Chem. Abstr., **84**, 43842c (1976).
1380 Toray Industries, Inc.: Jpn. Kokai Tokkyo Koho JP 60 15,455 [85 15,455] (1985); Chem. Abstr., **102**, 205262h (1985).
1381 Toray Industries, Inc.: Jpn. Kokai Tokkyo Koho JP 60 49,060 [85 49,060] (1985); Chem. Abstr., **103**, 38525h (1985).
1382 Toray Industries, Inc.: Jpn. Kokai Tokkyo Koho JP 60 49,061 [85 49,061] (1985); Chem. Abstr., **103**, 143216f (1985).
1383 Toray Industries, Inc.: Jpn. Kokai Tokkyo Koho JP 60 53,558 [85 53,558] (1985); Chem. Abstr., **103**, 143217g (1985).
1384 Tornwall, M. and Mannisto, P. T.: Pharmacol. Toxicol. (Copenhagen), **69** (1) 64-70 (1991); Chem. Abstr., **115**, 105480c (1991).
1385 Tornwall, M.; Tuomainen, P. and Mannisto, P. T.: Arch. Int. Pharmacodyn. Ther., **320**, 5-20 (1992); Chem. Abstr., **118**, 224969r (1993).
1386 Toth, E.; Torley, J.; Fekete, G.; Szporny, L.; Vereczkey, L.; Palosi, E.; Klebovich, I.; Vittay, P.; Gorog, S. and Hajdu, I.: Eur. Pat. Appl. EP 112,587 (1984); Chem. Abstr., **102**, 5915m (1985).
1387 Toukhy, M. A. and Beauchemin, B. T., Jr., Polym. Microelectron. Proc. Int. Symp. 1989 (Pub. **1990**), 363-374; Chem. Abstr., **117**, 201687n (1992).
1388 Toukhy, M. A. and Jeffries, A. T.: U.S. US 5,019,478 (1991); Chem. Abstr., **116**, 117237n (1992).
1389 Tozzi, A.: Polym. Age, **5** (10) 272-274 (1974); Chem. Abstr., **83**, 132481f (1975).
1390 Tsai, S. C. and Liu, S. T.: J. Chin. Chem. Soc. (Taipei), **38** (1) 39-45 (1991).
1391 Tsekhanskii, R. S.: Izv. Vyssh. Ucheb. Zaved., Khim. Khim. Tekhnol., **17** (1) 59-61 (1974); Chem. Abstr., **80**, 94804s (1974).
1392 Tsekhanskii, R. S.: Zh. Prikl. Spektrosk., **5** (3) 316-322 (1966); Chem. Abstr., **66**, 33262c (1967).
1393 Tsekhanskii, R. S. and Fedorov, Yu. A.: Uch. Zap. Chuvashsk. Gos. Ped. Inst., **14**, 116-117 (1962); Chem. Abstr., **60**, 15759a (1964).
1394 Tsekhanskii, R. S.; Zobova, N. N. and Ushenina, V. F.: Izv. Vysshikh Uchebn. Zavedenii, Khim. i Khim. Tekhnol., **4**, 985-987 (1961); Chem. Abstr., **57**, 16449h (1962).
1395 Ueda, S. and Kurosawa, K.: Bull. Chem. Soc. Jpn., 50 (1) 193-196 (1977).
1396 Uhde, W. J. and Zydek, G.: Fresenius' Z. analyt. Chem., **239** (1) 25-26 (1968); Chem. Abstr., **69**, 49051f (1968).
1397 Ullmann, F. and Goldberg, I.: Ber. Dtsch. Chem. Ges., **35**, 2811-2814 (1902).
1398 Ullmann, F. and Mallet, E.: Ber. Dtsch. Chem. Ges., **31**, 1694-1696 (1898).

1399 Ullmann, F.; Engi, G.; Wosnessensky, N.; Kuhn, E. and Herre, E.: Justus Liebigs Ann.
       Chem., **366**, 78-118 (1909).
1400 Urano, H. and Kikuchi, H.: Jpn. Kokai Tokkyo Koho JP 63,264,543 [88,264,543]; Chem.
       Abstr., **111**, 7061q (1989).
1401 Urano, H.; Sugiura, H.; Kikuchi, H. and Yamazawa, Y.: Jpn. Kokai Tokkyo Koho
       JP 01 17,049 [89 17,049] (1989); Chem. Abstr., **111**, 31356k (1989).
1402 Usgaonkar, U. R. and Jadhav, G. V.: J. Indian Chem. Soc., **40** (1) 27-30 (1963).
1403 Valette, M.: Bull. Soc. Chim. Fr., 289-300 (1930).
1404 Van Daele, G.: Eur. Pat. Appl. EP 68,544 (1983); Chem. Abstr., **99**, 22493j (1983).
1405 Van Daele, G.: Eur. Pat. Appl. EP 76,530 (1983); Chem. Abstr., **99**, 194812d (1983).
1406 VanAllan, J. and Tinker, J. F.: J. Org. Chem., **19**, 1243-1251 (1954).
1407 Varache-Beranger, M.; Nuhrich, A. and Devaux, G.: Eur. J. Med. Chem., **23** (6) 501-510
       (1988).
1408 Varadi, G.; Toth, G. K. and Penke, B.: Int. J. Pept. Protein Res., **43** (1) 29-30 (1994); Chem.
       Abstr., **120**, 164889u (1994).
1409 Venkatachalapathy, C. and Pitchumani, K.: Tetrahedron, **53** (50) 17171-17176 (1997).
1410 Verkman, A. S. and Solomon, A. K.: J. Gen. Physiol., **80** (4) 557-581 (1982); Chem. Abstr.,
       97, 194667u (1982).
1411 Volkotrub, M. N.; Rubtsova, T. A. and Lukovnikov, A. F.: Vysokomol. Soedin., Ser. B, 19
       (10) 762-765 (1977); Chem. Abstr., **88**, 38473m (1978).
1412 Vollenweider, J. K. and Fischer, H.: Chem. Phys., **124** (3) 333-345 (1988).
1413 Voronina, N. M.; Ivakin, A. A.; Podgornaya, I. V. and Klyachina, K. N.: Otkrytiya, Izobret.,
       Prom. Obraztsy, Tovarnye Znaki, **51** (25) 79 (1974); Chem. Abstr., **81**, 105020g (1974).
1414 Voronina, N. M.; Ivakin, A. A.; Podgornaya, I. V. and Klyachina, K. N.: U.S.S.R. 435,229
       (1974); Chem. Abstr., **81**, 105020g (1974).
1415 Wagle, D. R.; Usgaonkar, U. and Usgaonkar, R. N.: Indian J. Chem., Sect. B, **16B** (5) 378-
       380 (1978).
1416 Wagner, H.; Seligmann, O.; Chari, M. V.; Wollenweber, E.; Dietz, V. H.; Donnelly, D. M.;
       Meegan, M. J. and O'Donnell, B.: Tetrahedron Lett., **44**, 4269-4272 (1979).
1417 Wagner, R.: J. Prakt. Chem., [1] **51** (1) 82-106 (1850).
1418 Waldemar, A. and Arce de Sanabia, J.: J. Org. Chem., **38** (14) 2571-2572 (1973).
1419 Walker, G. N.: J. Org. Chem., **27**, 1929-1930 (1962).
1420 Wallis, A. F. A.; Chan, F. D. and Nguyen, Kien Loi: Int. Symp. Wood. Pulping Chem., 8th,
       1, 549-556 (1995); Chem. Abstr., **128**, 36144a (1998).
1421 Wan, I. Y.; Priddy, D. B.; Lyle, G. D. and McGrath, J. E.: Polym. Prepr. (Am. Chem. Soc.,
       Div. Polym. Chem.), **34** (1) 806-807 (1993); Chem. Abstr., **120**, 192415e (1994).
1422 Wang, R. H. S. and Gott, S. L.: PCT Int. Appl. WO 89 02,906 '1989); Chem. Abstr., **111**,
       196060x (1989).
1423 Wang, R. H. S. and Irick, G., Jr.: U.S. 4,043,973 (1977); Chem. Abstr., **87**, 168757x (1977).
1424 Wang, R. H. S. and Zannucci, J. S.: U.S. 4,115,348 (1978); Chem. Abstr., **90**, 72866t
       (1979).
1425 Warren S. Forster: U.S. 2,818,400 (1957); Chem. Abstr., **52**, 5036d (1958).
1426 Watanabe, H.; Kenji, M. and Tsuboi, A.: Jpn. Kokai Tokkyo Koho JP 03,127,753
       [91,127,753] (1991), Chem. Abstr., **115**, 233090n (1991).
1427 Watanabe, Y.; Yoshiwara, H. and Kanao, M.: J. Heterocycl. Chem., **30** (2) 445-451 (1993).
1428 Weiss, R. and Chledowski, L.: Monatsh. Chem., **65**, 357-366 (1935).
1429 Wexler, H. and Arventiev, B.: An. Stiint. Univ. "Al. I. Cuza" Iasi, Sect. 1c, **17** (1) 67-71
       1971); Chem. Abstr., **75**, 118198x (1971).
1430 Whelen Myron S.: US 3,371,119 (1968); Chem. Abstr., **69**, 35748b (1968).
1431 White, W. N.; Gwynn, D.; Schlitt, R.; Girard, C. and Fife, W.: J. Am. Chem. Soc., **80**, 3271-
       3277 (1958).
1432 Whitesitt, C. A.; Simon, R. L.; Reel, J. K.; Sigmund, S. K.; Phillips, M. L.; Shadle, J. K.;
       Heinz, L. J.; Koppel, G. A.; Hunden, D. C.; Lifer, S. L.; Berry, D.; Ray, J.; Little, S. P.; Liu
       Xiadong; Marshall, W. S. and Panetta, J. A.: Bioorg. Med. Chem. Lett., **6** (18) 2157-2162
       (1996).
1433 Widdowson, K. L.; Veber, D. F.; Jurewicz, A. J.; Rutledge, M. C. Jr. and Hertzberg, R. P.:
       PCT Int. Appl. WO 96 25,157 (1996); Chem. Abstr., **125** (21) 275430k (1996).

1434 Wilharm, P.: Eur. Pat. Appl. EP 564,981 (1993); Chem. Abstr., 120, 163696s (1994).
1435 Wilharm, P.: Ger. Offen. DE 4,211,420 (1993); Chem. Abstr., 120, 134450b (1994).
1436 Williamson, W. R. N. and Saunders, J. C.: Brit. 1,488,004 (1977); Chem. Abstr., 88, 120799f (1978).
1437 Winkler, L. W.: Arch. Pharm. Ber. Dtsch. Pharm. Ges., 266, 45-62 (1928).
1438 Wittig, G.; Baugert, F. and Richter, H. E.: Justus Liebigs Ann. Chem., 446, 155-204 (1926).
1439 Wittig, G.; Oppermann, A. and Faber, K.: J. Prakt. Chem., 158, 61-71 (1941).
1440 Wittig, G. and Schulze, W.: J. Prakt. Chem., 130, 81-91 (1931).
1441 Yaegashi, T.; Nunomura, S.; Okutome, T.; Nakayama, T.; Kurumi, M.; Sakurai, Y.; Aoyama, T. and Fujii, S.: Chem. Pharm. Bull., 32 (11) 4466-4477 (1984).
1442 Yakshin, V. V.; Mirokhin, A. M. and Ignat'ev, M. M.: Kompleksn. Ispol'z. Miner. Syr'ya, 4, 60-64 (1984); Chem. Abstr., 102, 95339d (1985).
1443 Yamada, H.; Takao, M. and Fukuhara, C.: Jpn. Kokai Tokkyo Koho JP 61 97,240 [86 97,240] (1986); Chem. Abstr., 105, 226048v (1986).
1444 Yamada, O.; Ishida, H.; Nitani, F.; Ito, K. and Yamamoto, H.: Japan 74 11,410 (1974); Chem. Abstr., 81, 115925v (1974).
1445 Yamaguchi, S.; Ohhira, Y.; Yamada, M.; Michitani, H. and Kawase, Y.: Bull. Chem. Soc. Jpn., 63 (3) 952-954 (1990).
1446 Yamamoto, J.: Ikeda, Y.; Inohara, T.; Nakata, H. and Umezu, M.: Nippon Kagaku Kaishi, 12, 1911-1915 (1981).
1447 Yamamoto, J.; Ishikawa, T. and Okamoto, Y.: Nippon Kagaku Kaishi, 11, 1870-1875 (1989).
1448 Yamamoto, J.; Kisida, M.; Takenaka, Y. and Okamoto, Y.: Nippon Kagaku Kaishi, 3, 288-293 (1988).
1449 Yamamoto, J.; Kurokawa, H. and Sugita, K.: Nippon Kagaku Kaishi, 11, 2107-2110 (1985).
1450 Yamamoto, J.; Nakane, I.; Nakashima, M.; Asano, M.; Akamatsu, H.; Okamoto, Y. and Sugita, K.: Nippon Kagaku Kaishi, 9, 1587-1592 (1989).
1451 Yamamoto, J.; Yamana, H.; Haraguchi, Y. and Sasaki, H.: Nippon Kagaku Kaishi, 8, 747-750 (1996).
1452 Yamashita, M.: Bull. Chem. Soc. Jpn., 3, 180-182 (1928).
1453 Yamashita, M.: Science Repts. Tohoku Imp. Univ. 1st Ser., 18, 129-133 (1929); Chem. Abstr., 24, 98 (1930).
1454 Yao, Run-hua; Ma, Rong-sheng; Chen, Yao-qing and Huang, Lan-sun: Yaoxue Xuebao, 19 (3) 228-231 (1984); Chem. Abstr., 103, 123103p (1985).
1455 Yardley, J. P. and Fletcher, H., 3rd, Synthesis, 4, 244 (1976).
1456 Yazaki, T.; Makino, M.; Yamamoto, T.; Tsuji, K.; Zenda, H. and Kosuge, T.: Yakugaku Zasshi, 98 (7) 914-922 (1978); Chem. Abstr., 89, 197912r (1978).
1457 Yoshikawa, T.; Kimura, K. and Fujimura, S.: J. Appl. Polym. Sci., 15 (10) 2513-2519 (1971); Chem. Abstr., 76, 34680f (1972).
1458 Yousaf, T. I.: Ph. D. Thesis, London, UK (1983).
1459 Yuan, Mu; Zhong, Yuguo; Pang, Qijie and Li, Zhenghua: Huaxi Yike Daxue Xuebao, 21 (3) 310-314 (1990); Chem. Abstr., 114, 206687k (1991).
1460 Yukutomi, M.; Tanaka, Y.; Genda, S. and Kitaura, M.: U.S. 3,769,349 (1973); Chem. Abstr., 79, 146221a (1973).
1461 Yurchenko, N. N.; Niyazov, A. N. and Vakhabova, K. D.: Izv. Akad. Nauk Turkm. SSR, Ser. Fiz.-Tekh., Khim. Geol. Nauk, 3, 48-54 (1987); Chem. Abstr., 108, 111914u (1988).
1462 Zaitseva, N. P.; Koval, E. K.; Kobel'chuk, Yu. M.; Zherebtsova, L. P. and Moshchinskaya, N. K.: Vopr. Khim. Khim. Tekhnol., 72, 37-39 (1983); Chem. Abstr., 102, 24231b (1985).
1463 Zakett, D.; Flynn, R. G. A. and Cooks, R. G.: J. Phys. Chem., 82 (22) 2359-2362 (1978).
1464 Zakhs, E. R.; Zvenigorodskaya, L. A. and Efros, L. S.: Khim. Geterotsikl. Soedin., 12, 1618-1623 (1973); Chem. Abstr., 81, 51080f (1974).
1465 Zannucci, J. S.: U.S. US 4,355,080 (1982); Chem. Abstr., 98, 17801h (1983).
1466 Zannucci, J. S. and Pruett, W. P.: U.S. US 4,418,000 (1983); Chem. Abstr., 100, 52643q (1984).
1467 Zemzina, I. N. and Inagamova, M. I.: Nauch. Tr. Tashkent. Un-T, 462, 45-48 (1974); Chem. Abstr., 84, 4593c (1976).
1468 Zemzina, I. N. and Karaul'uykh, L. V.: Zh. Org. Khim., 9 (10) 2163-2167 (1973).

1469 Zemzina, I. N. and Sidorova, N. G.: Kratkie Tezisy-Vsesoyuznoe Soveshchanie, probleme
      Mekhanizmy Geteroliticheskikh reaktsü, Leningrad, 184-185 (1974); Chem. Abstr., 85,
      62765s (1976).
1470 Zil'berman, E. N. and Rybakova, N. A.: Zh. Obshch. Khim., 30 (6) 1992-1996 (1960).
1471 Zil'berman, E. N. and Rybakova, N. A.: Zh. Obshch. Khim., 32 (2) 591-596 (1962).
1472 Zimmerman, H. E. and Swenton, J. S.: J. Am. Chem. Soc., 89 (4) 906-912 (1967).
1473 Zincke, T. and Birschel: Justus Liebigs Ann. Chem., 362, 221-241 (1908).
1474 Zincke, T. and Siebert, K.: Ber. Dtsch. Chem. Ges., 39, 1930-1938 (1906).
1475 Zonis, L. S.; Khaletskii, A. M. and Pesin, V. G.: Zh. Obshch. Khim., 33 (1) 3141-3142
      (1963).

MOLECULAR FORMULA INDEX

$C_{13}H_6N_4O_{12}$

(2,4-Dihydroxy-3,5,6-trinitrophenyl)(4-hydroxy-3-nitrophenyl)methanone, 450

$C_{13}H_7BrCl_2O_2$

(4-Bromophenyl)(2,4-dichloro-6-hydroxyphenyl)methanone, 173
(4-Bromophenyl)(2,6-dichloro-4-hydroxyphenyl)methanone, 174
(4-Bromophenyl)(3,5-dichloro-4-hydroxyphenyl)methanone, 174

$C_{13}H_7BrF_2O_2$

(4-Bromophenyl)(2,5-difluoro-4-hydroxyphenyl)methanone, 174

$C_{13}H_7Br_3O_6$

(3-Bromo-4,5-dihydroxyphenyl)(3,5-dibromo-2,4,6-trihydroxyphenyl)methanone, 467

$C_{13}H_7ClFNO_4$

(4-Chlorophenyl)(5-fluoro-2-hydroxy-3-nitrophenyl)methanone, 174

$C_{13}H_7Cl_2FO_2$

(2,3-Dichloro-4-hydroxyphenyl)(2-fluorophenyl)methanone, 175
(2,3-Dichloro-4-hydroxyphenyl)(3-fluorophenyl)methanone, 175
(2,3-Dichloro-4-hydroxyphenyl)(4-fluorophenyl)methanone, 175

$C_{13}H_7Cl_2FO_3$

(2,3-Dichloro-4,5-dihydroxyphenyl)(2-fluorophenyl)methanone, 382

$C_{13}H_7Cl_2NO_4$

(5-Chloro-2-hydroxy-3-nitrophenyl)(4-chlorophenyl)methanone, 176
(2,3-Dichloro-4-hydroxyphenyl)(4-nitrophenyl)methanone, 176

$C_{13}H_7Cl_2NO_5$

(5-Chloro-2,4-dihydroxy-3-nitrophenyl)(4-chlorophenyl)methanone, 382

$C_{13}H_7Cl_3O_2$

(2-Chloro-4-hydroxyphenyl)(2,4-dichlorophenyl)methanone, 176
(3-Chloro-4-hydroxyphenyl)(2,4-dichlorophenyl)methanone, 176
(5-Chloro-2-hydroxyphenyl)(2,4-dichlorophenyl)methanone, 177
(5-Chloro-2-hydroxyphenyl)(3,4-dichlorophenyl)methanone, 177
(2-Chlorophenyl)(2,3-dichloro-4-hydroxyphenyl)methanone, 177
(2-Chlorophenyl)(2,4-dichloro-6-hydroxyphenyl)methanone, 177
(2-Chlorophenyl)(3,5-dichloro-2-hydroxyphenyl)methanone, 178
(2-Chlorophenyl)(3,5-dichloro-4-hydroxyphenyl)methanone, 178
(3-Chlorophenyl)(2,3-dichloro-4-hydroxyphenyl)methanone, 178
(4-Chlorophenyl)(2,3-dichloro-4-hydroxyphenyl)methanone, 178
(4-Chlorophenyl)(2,4-dichloro-6-hydroxyphenyl)methanone, 179
(4-Chlorophenyl)(2,6-dichloro-4-hydroxyphenyl)methanone, 179
(4-Chlorophenyl)(3,5-dichloro-4-hydroxyphenyl)methanone, 179
Phenyl(2,3,5-trichloro-6-hydroxyphenyl)methanone, 35

$C_{13}H_7Cl_3O_3$

(5-Chloro-2,4-dihydroxyphenyl)(2,4-dichlorophenyl)methanone, 382

C₁₃H₇F₂NO₅

(2,6-Difluorophenyl)(3,4-dihydroxy-5-nitrophenyl)methanone, 383

C₁₃H₇F₃O₂

(2,6-Difluorophenyl)(3-fluoro-4-hydroxyphenyl)methanone, 179

C₁₃H₇F₃O₃

Phenyl(2,3,5-trifluoro-4,6-dihydroxyphenyl)methanone, 339

C₁₃H₇I₃O₂

(3-Hydroxy-2,4,6-triiodophenyl)phenylmethanone, 35

C₁₃H₇N₃O₉

(2,4-Dihydroxy-3,5-dinitrophenyl)(3-nitrophenyl)methanone, 383
(2,4-Dihydroxyphenyl)(2,4,6-trinitrophenyl)methanone, 365

C₁₃H₈BrClO₂

(3-Bromo-4-chlorophenyl)(4-hydroxyphenyl)methanone, 126
(2-Bromophenyl)(5-chloro-2-hydroxyphenyl)methanone, 180
(4-Bromophenyl)(3-chloro-4-hydroxyphenyl)methanone, 180

C₁₃H₈BrClO₃

(3-Bromo-4-chloro-2,5-dihydroxyphenyl)phenylmethanone, 339
(4-Bromo-3-chlorophenyl)(2,5-dihydroxyphenyl)methanone, 365
(3-Bromo-2,5-dihydroxyphenyl)(2-chlorophenyl)methanone, 383
(3-Bromo-2,5-dihydroxyphenyl)(4-chlorophenyl)methanone, 383
(4-Bromophenyl)(4-chloro-2,5-dihydroxyphenyl)methanone, 384

C₁₃H₈BrFO₂

(3-Bromo-5-fluoro-4-hydroxyphenyl)phenylmethanone, 35
(4-Bromo-2-fluorophenyl)(4-hydroxyphenyl)methanone, 126
(3-Bromophenyl)(5-fluoro-2-hydroxyphenyl)methanone, 180
(4-Bromophenyl)(2-fluoro-4-hydroxyphenyl)methanone, 180
(4-Bromophenyl)(3-fluoro-4-hydroxyphenyl)methanone, 180

C₁₃H₈BrNO₅

(3-Bromo-2,4-dihydroxy-5-nitrophenyl)phenylmethanone, 339
(5-Bromo-2,4-dihydroxy-3-nitrophenyl)phenylmethanone, 340

C₁₃H₈Br₂O₂

(3-Bromo-4-hydroxyphenyl)(4-bromophenyl)methanone, 181
(3,5-Dibromo-2-hydroxyphenyl)phenylmethanone, 35
(3,5-Dibromo-4-hydroxyphenyl)phenylmethanone, 36
(2,4-Dibromophenyl)(2-hydroxyphenyl)methanone, 126

C₁₃H₈Br₂O₃

Bis(2-bromo-4-hydroxyphenyl)methanone, 415
Bis(3-bromo-4-hydroxyphenyl)methanone, 415
Bis(5-bromo-2-hydroxyphenyl)methanone, 416
(2,4-Dibromo-3,6-dihydroxyphenyl)phenylmethanone, 340
(3,4-Dibromo-2,5-dihydroxyphenyl)phenylmethanone, 340

$C_{13}H_8Cl_2O_3$

$C_{13}H_8Cl_2O_4$

$C_{13}H_8Cl_2O_5$

$C_{13}H_8FNO_4$

$C_{13}H_8FNO_5$

$C_{13}H_9ClO_4$

$C_{13}H_9Cl_2NO_2$

$C_{13}H_9Cl_2NO_3$

$C_{13}H_9FO_2$

$C_{13}H_9FO_3$

$C_{13}H_9FO_4$

$C_{13}H_9F_2NO_2$

$C_{13}H_9IO_2$

C₁₃H₉IO₃

C₁₃H₉NO₄

C₁₃H₉NO₅

C₁₃H₉NO₆

C₁₃H₉NO₇

C₁₃H₁₀ClNO₂

C13H10FNO2

(3-Amino-5-fluoro-2-hydroxyphenyl)phenylmethanone,   50

C13H10N2O4

(2-Amino-3-hydroxy-5-nitrophenyl)phenylmethanone,   51
(4-Amino-2-hydroxyphenyl)(4-nitrophenyl)methanone,   192
(4-Amino-3-nitrophenyl)(4-hydroxyphenyl)methanone,   142

C13H10N2O5

(4-Amino-2-hydroxyphenyl)(2-hydroxy-4-nitrophenyl)methanone,   424

C13H10O2

(2-Hydroxyphenyl)phenylmethanone,   3
(3-Hydroxyphenyl)phenylmethanone,   5
(4-Hydroxyphenyl)phenylmethanone,   6

C13H10O3

Bis(2-hydroxyphenyl)methanone,   13
Bis(3-hydroxyphenyl)methanone,   13
Bis(4-hydroxyphenyl)methanone,   13
(2,3-Dihydroxyphenyl)phenylmethanone,   8
(2,4-Dihydroxyphenyl)phenylmethanone,   9
(2,5-Dihydroxyphenyl)phenylmethanone,   11
(2,6-Dihydroxyphenyl)phenylmethanone,   11
(3,4-Dihydroxyphenyl)phenylmethanone,   12
(3,5-Dihydroxyphenyl)phenylmethanone,   12
(2-Hydroxyphenyl)(3-hydroxyphenyl)methanone,   15
(2-Hydroxyphenyl)(4-hydroxyphenyl)methanone,   15
(3-Hydroxyphenyl)(4-hydroxyphenyl)methanone,   16

C13H10O4

(2,3-Dihydroxyphenyl)(4-hydroxyphenyl)methanone,   19
(2,4-Dihydroxyphenyl)(2-hydroxyphenyl)methanone,   19
(2,4-Dihydroxyphenyl)(3-hydroxyphenyl)methanone,   19
(2,4-Dihydroxyphenyl)(4-hydroxyphenyl)methanone,   20
(2,5-Dihydroxyphenyl)(2-hydroxyphenyl)methanone,   20
(2,5-Dihydroxyphenyl)(4-hydroxyphenyl)methanone,   20
(2,6-Dihydroxyphenyl)(2-hydroxyphenyl)methanone,   21
(2,6-Dihydroxyphenyl)(3-hydroxyphenyl)methanone,   21
(2,6-Dihydroxyphenyl)(4-hydroxyphenyl)methanone,   21
(3,4-Dihydroxyphenyl)(4-hydroxyphenyl)methanone,   21
(3,5-Dihydroxyphenyl)(4-hydroxyphenyl)methanone,   22
Phenyl(2,3,4-trihydroxyphenyl)methanone,   17
Phenyl(2,3,5-trihydroxyphenyl)methanone,   17
Phenyl(2,4,5-trihydroxyphenyl)methanone,   18
Phenyl(2,4,6-trihydroxyphenyl)methanone,   18
Phenyl(3,4,5-trihydroxyphenyl)methanone,   18

C13H10O5

Bis(2,3-dihydroxyphenyl)methanone,   22
Bis(2,4-dihydroxyphenyl)methanone,   22
Bis(2,5-dihydroxyphenyl)methanone,   23
Bis(3,4-dihydroxyphenyl)methanone,   23
(2,3-Dihydroxyphenyl)(2,4-dihydroxyphenyl)methanone,   24
(2,3-Dihydroxyphenyl)(2,5-dihydroxyphenyl)methanone,   24

$C_{13}H_{10}O_6$

$C_{13}H_{10}O_7$

$C_{13}H_{11}NO_2$

C$_{13}$H$_{11}$NO$_3$

(3-Amino-2,4-dihydroxyphenyl)phenylmethanone, 347
(3-Amino-2,4-dihydroxyphenyl)phenylmethanone *(Hydrochloride)*, 347
(5-Amino-2,4-dihydroxyphenyl)phenylmethanone, 348
(5-Amino-2,4-dihydroxyphenyl)phenylmethanone *(Hydrochloride)*, 348
(4-Amino-3-hydroxyphenyl)(4-hydroxyphenyl)methanone, 401

C$_{13}$H$_{11}$NO$_4$

(2-Aminophenyl)(2,3,4-trihydroxyphenyl)methanone, 440
(2-Aminophenyl)(2,4,6-trihydroxyphenyl)methanone, 440

C$_{13}$H$_{12}$N$_2$O$_2$

(3-Amino-4-hydroxyphenyl)(3-aminophenyl)methanone, 193
(3-Amino-4-hydroxyphenyl)(4-aminophenyl)methanone, 193
(3,4-Diaminophenyl)(4-hydroxyphenyl)methanone, 144

C$_{13}$H$_{12}$N$_2$O$_3$

(2-Amino-4-hydroxyphenyl)(4-amino-2-hydroxyphenyl)methanone, 424
(2-Amino-4-hydroxyphenyl)(4-amino-2-hydroxyphenyl)methanone *(Dihydrochloride)*, 425
Bis(3-amino-4-hydroxyphenyl)methanone, 417
Bis(4-amino-2-hydroxyphenyl)methanone, 418
Bis(4-amino-2-hydroxyphenyl)methanone *(Dihydrochloride)*, 418

C$_{13}$H$_{15}$BrO$_2$

(3-Bromo-2-hydroxyphenyl)cyclohexylmethanone, 479
(3-Bromo-4-hydroxyphenyl)cyclohexylmethanone, 479
(4-Bromo-2-hydroxyphenyl)cyclohexylmethanone, 479
(5-Bromo-2-hydroxyphenyl)cyclohexylmethanone, 479

C$_{13}$H$_{15}$FO$_2$

Cyclohexyl(5-fluoro-2-hydroxyphenyl)methanone, 480

C$_{13}$H$_{15}$NO$_5$

Cyclohexyl(3,4-dihydroxy-5-nitrophenyl)methanone, 484

C$_{13}$H$_{16}$O$_2$

Cyclohexyl(2-hydroxyphenyl)methanone, 480
Cyclohexyl(3-hydroxyphenyl)methanone, 480
Cyclohexyl(4-hydroxyphenyl)methanone, 481

C$_{13}$H$_{16}$O$_3$

Cyclohexyl(2,4-dihydroxyphenyl)methanone, 485
(1-Hydroxycyclohexyl)(4-hydroxyphenyl)methanone, 485
(2-Hydroxycyclohexyl)(2-hydroxyphenyl)methanone, 485

C$_{13}$H$_{16}$O$_4$

Cyclohexyl(2,4,6-trihydroxyphenyl)methanone, 485

C$_{14}$H$_4$F$_8$O$_3$

(2,3,4,5,6-Pentafluorophenyl)(2,3,5-trifluoro-6-hydroxy-4-methoxyphenyl)methanone, 193

C$_{14}$H$_7$F$_5$O$_3$

(2-Hydroxy-4-methoxyphenyl)(2,3,4,5,6-pentafluorophenyl)methanone, 193

C$_{14}$H$_8$ClF$_3$O$_2$

(5-Chloro-2-hydroxyphenyl)[2-(trifluoromethyl)phenyl]methanone, 194

C$_{14}$H$_8$F$_3$NO$_5$

(3,4-Dihydroxy-5-nitrophenyl)[2-(trifluoromethyl)phenyl]methanone, 388
(3,4-Dihydroxy-5-nitrophenyl)[4-(trifluoromethyl)phenyl]methanone, 388

C$_{14}$H$_8$F$_4$O$_2$

(5-Fluoro-2-hydroxyphenyl)[4-(trifluoromethyl)phenyl]methanone, 194

C$_{14}$H$_8$N$_4$O$_{12}$

(2,4-Dihydroxy-3,5,6-trinitrophenyl)(4-methoxy-3-nitrophenyl)methanone, 388

C$_{14}$H$_9$Br$_3$O$_2$

(4-Bromophenyl)[2-(dibromomethyl)-4-hydroxyphenyl]methanone, 194

C$_{14}$H$_9$Br$_3$O$_3$

(3-Bromo-2-hydroxy-5-methylphenyl)(3,5-dibromo-2-hydroxyphenyl)methanone, 425
(3-Bromo-2-hydroxy-5-methylphenyl)(3,5-dibromo-4-hydroxyphenyl)methanone, 425

C$_{14}$H$_9$ClF$_2$O$_3$

(3-Chloro-2-hydroxy-4-methoxyphenyl)(2,3-difluorophenyl)methanone, 194
(3-Chloro-2-hydroxy-4-methoxyphenyl)(2,5-difluorophenyl)methanone, 195
(5-Chloro-2-hydroxy-4-methoxyphenyl)(2,4-difluorophenyl)methanone, 195
(5-Chloro-2-hydroxy-4-methoxyphenyl)(2,6-difluorophenyl)methanone, 195

C$_{14}$H$_9$Cl$_2$NO$_4$

(2,6-Dichlorophenyl)(4-hydroxy-2-methyl-5-nitrophenyl)methanone, 195

C$_{14}$H$_9$Cl$_3$O$_2$

(3-Chloro-4-hydroxy-5-methylphenyl)(2,4-dichlorophenyl)methanone, 196
(3-Chloro-6-hydroxy-2-methylphenyl)(2,5-dichlorophenyl)methanone, 196
(2-Methylphenyl)(2,3,5-trichloro-6-hydroxyphenyl)methanone, 196
(4-Methylphenyl)(2,3,5-trichloro-6-hydroxyphenyl)methanone, 196

C$_{14}$H$_9$F$_2$NO$_5$

(2,6-Difluorophenyl)(4-hydroxy-3-methoxy-5-nitrophenyl)methanone, 197

C$_{14}$H$_9$F$_3$O$_2$

(2-Hydroxyphenyl)[2-(trifluoromethyl)phenyl]methanone, 144
(3-Hydroxyphenyl)[3-(trifluoromethyl)phenyl]methanone, 144
(3-Hydroxyphenyl)[4-(trifluoromethyl)phenyl]methanone, 144
(4-Hydroxyphenyl)[2-(trifluoromethyl)phenyl]methanone, 145
(4-Hydroxyphenyl)[3-(trifluoromethyl)phenyl]methanone, 145
(4-Hydroxyphenyl)[4-(trifluoromethyl)phenyl]methanone, 145
[2-Hydroxy-5-(trifluoromethyl)phenyl]phenylmethanone, 53

$C_{14}H_9F_3O_3$

(2,4-Dihydroxyphenyl)[2-(trifluoromethyl)phenyl]methanone,   372
(2,4-Dihydroxyphenyl)[3-(trifluoromethyl)phenyl]methanone,   372
(2,5-Dihydroxyphenyl)[2-(trifluoromethyl)phenyl]methanone,   373
Phenyl(3,5,6-trifluoro-2-hydroxy-4-methoxyphenyl)methanone,   54

$C_{14}H_{10}BrFO_2$

(3-Bromo-2-hydroxy-5-methylphenyl)(3-fluorophenyl)methanone,   197
(3-Bromophenyl)(3-fluoro-2-hydroxy-5-methylphenyl)methanone,   197

$C_{14}H_{10}BrNO_4$

(2-Bromo-5-nitrophenyl)(2-hydroxy-5-methylphenyl)methanone,   197

$C_{14}H_{10}BrNO_5$

(2-Bromo-6-hydroxy-3-methoxy-4-nitrophenyl)phenylmethanone,   54

$C_{14}H_{10}Br_2O_2$

(3,5-Dibromo-2-hydroxyphenyl)(4-methylphenyl)methanone,   198

$C_{14}H_{10}Br_2O_3$

(2,4-Dibromo-6-hydroxy-3-methoxyphenyl)phenylmethanone,   54

$C_{14}H_{10}Br_2O_4$

(3,5-Dibromo-2,6-dihydroxy-4-methoxyphenyl)phenylmethanone,   348

$C_{14}H_{10}ClFO_3$

(2-Chloro-6-hydroxy-4-methoxyphenyl)(2-fluorophenyl)methanone,   198
(3-Chloro-2-hydroxy-4-methoxyphenyl)(2-fluorophenyl)methanone,   198
(5-Chloro-2-hydroxy-4-methoxyphenyl)(2-fluorophenyl)methanone,   198

$C_{14}H_{10}ClFO_4$

[3-Chloro-2,4 (or 2,5)-dihydroxy-5 (or 4)-methoxyphenyl](2-fluorophenyl)methanone,   388

$C_{14}H_{10}ClNO_4$

(2-Chloro-4-nitrophenyl)(2-hydroxy-5-methylphenyl)methanone,   199
(2-Chloro-5-nitrophenyl)(2-hydroxy-5-methylphenyl)methanone,   199

$C_{14}H_{10}ClNO_5$

(5-Chloro-2-hydroxy-3-nitrophenyl)(4-methoxyphenyl)methanone,   199
(2-Chlorophenyl)(4-hydroxy-3-methoxy-5-nitrophenyl)methanone,   199
(3-Chlorophenyl)(4-hydroxy-3-methoxy-5-nitrophenyl)methanone,   199
(4-Chlorophenyl)(4-hydroxy-3-methoxy-5-nitrophenyl)methanone,   200

$C_{14}H_{10}Cl_2O_2$

(3-Chloro-4-hydroxy-5-methylphenyl)(2-chlorophenyl)methanone,   200
(3-Chloro-4-hydroxy-5-methylphenyl)(4-chlorophenyl)methanone,   200
(5-Chloro-2-hydroxy-3-methylphenyl)(2-chlorophenyl)methanone,   200
(5-Chloro-2-hydroxy-3-methylphenyl)(3-chlorophenyl)methanone,   201
(5-Chloro-2-hydroxy-3-methylphenyl)(4-chlorophenyl)methanone,   201

$C_{14}H_{10}N_2O_7$

(4-Hydroxy-3-methoxy-5-nitrophenyl)(2-nitrophenyl)methanone,   209

$C_{14}H_{10}N_2O_8$

(2,4-Dihydroxy-3,5-dinitrophenyl)(2-methoxyphenyl)methanone,   389
(2-Hydroxy-4-methoxy-5-nitrophenyl)(4-hydroxy-3-nitrophenyl)methanone,   425

$C_{14}H_{11}BrO_2$

(3-Bromo-2-hydroxy-5-methylphenyl)phenylmethanone,   55
(3-Bromo-4-hydroxy-5-methylphenyl)phenylmethanone,   55
(4-Bromo-2-hydroxy-3-methylphenyl)phenylmethanone,   55
(4-Bromo-2-hydroxy-5-methylphenyl)phenylmethanone,   55
(5-Bromo-2-hydroxy-3-methylphenyl)phenylmethanone,   56
(2-Bromophenyl)(2-hydroxy-5-methylphenyl)methanone,   210
(3-Bromophenyl)(2-hydroxy-4-methylphenyl)methanone,   210
(3-Bromophenyl)(2-hydroxy-5-methylphenyl)methanone,   210
(3-Bromophenyl)(4-hydroxy-2-methylphenyl)methanone,   210
(4-Bromophenyl)(2-hydroxy-4-methylphenyl)methanone,   211
(4-Bromophenyl)(2-hydroxy-5-methylphenyl)methanone,   211

$C_{14}H_{11}BrO_3$

(4-Bromo-2-hydroxy-3-methoxyphenyl)phenylmethanone,   56
(5-Bromo-2-hydroxy-4-methoxyphenyl)phenylmethanone,   56
(2-Bromophenyl)(4,5-dihydroxy-2-methylphenyl)methanone,   390
(2-Bromophenyl)(2-hydroxy-4-methoxyphenyl)methanone,   211
(2-Bromophenyl)(2-hydroxy-5-methoxyphenyl)methanone,   211
(4-Bromophenyl)(2-hydroxy-4-methoxyphenyl)methanone,   212

$C_{14}H_{11}ClO_2$

(3-Chloro-2-hydroxy-5-methylphenyl)phenylmethanone,   56
(3-Chloro-4-hydroxy-5-methylphenyl)phenylmethanone,   57
(4-Chloro-2-hydroxy-5-methylphenyl)phenylmethanone,   57
(5-Chloro-2-hydroxy-3-methylphenyl)phenylmethanone,   57
(5-Chloro-2-hydroxy-4-methylphenyl)phenylmethanone,   57
(3-Chloro-4-hydroxyphenyl)(2-methylphenyl)methanone,   212
(3-Chloro-4-hydroxyphenyl)(3-methylphenyl)methanone,   212
(5-Chloro-2-hydroxyphenyl)(2-methylphenyl)methanone,   212
(5-Chloro-2-hydroxyphenyl)(3-methylphenyl)methanone,   213
(5-Chloro-2-hydroxyphenyl)(4-methylphenyl)methanone,   213
(2-Chloro-4-methylphenyl)(4-hydroxyphenyl)methanone,   146
(3-Chloro-4-methylphenyl)(4-hydroxyphenyl)methanone,   146
(4-Chloro-2-methylphenyl)(4-hydroxyphenyl)methanone,   147
(4-Chloro-3-methylphenyl)(4-hydroxyphenyl)methanone,   147
[3-(Chloromethyl)-4-hydroxyphenyl]phenylmethanone,   58
[5-(Chloromethyl)-2-hydroxyphenyl]phenylmethanone,   58
(2-Chlorophenyl)(2-hydroxy-3-methylphenyl)methanone,   213
(2-Chlorophenyl)(2-hydroxy-4-methylphenyl)methanone,   213
(2-Chlorophenyl)(2-hydroxy-5-methylphenyl)methanone,   214
(2-Chlorophenyl)(4-hydroxy-2-methylphenyl)methanone,   214
(2-Chlorophenyl)(4-hydroxy-3-methylphenyl)methanone,   214
(3-Chlorophenyl)(2-hydroxy-3-methylphenyl)methanone,   215
(3-Chlorophenyl)(2-hydroxy-4-methylphenyl)methanone,   215
(3-Chlorophenyl)(2-hydroxy-5-methylphenyl)methanone,   215
(3-Chlorophenyl)(4-hydroxy-2-methylphenyl)methanone,   215
(3-Chlorophenyl)(4-hydroxy-3-methylphenyl)methanone,   216
(4-Chlorophenyl)(2-hydroxy-3-methylphenyl)methanone,   216
(4-Chlorophenyl)(2-hydroxy-4-methylphenyl)methanone,   216

(4-Chlorophenyl)(2-hydroxy-5-methylphenyl)methanone, 216
(4-Chlorophenyl)(3-hydroxy-2-methylphenyl)methanone, 217
(4-Chlorophenyl)(3-hydroxy-4-methylphenyl)methanone, 217
(4-Chlorophenyl)(4-hydroxy-2-methylphenyl)methanone, 217
(4-Chlorophenyl)(4-hydroxy-3-methylphenyl)methanone, 218
(4-Chlorophenyl)(5-hydroxy-2-methylphenyl)methanone, 218

$C_{14}H_{11}ClO_3$

(4-Chloro-2,5-dihydroxyphenyl)(4-methylphenyl)methanone, 390
(2-Chloro-6-hydroxy-4-methoxyphenyl)phenylmethanone, 58
(3-Chloro-2-hydroxy-4-methoxyphenyl)phenylmethanone, 58
(5-Chloro-2-hydroxy-4-methoxyphenyl)phenylmethanone, 59
(2-Chloro-4-hydroxyphenyl)(4-hydroxy-2-methylphenyl)methanone, 425
(5-Chloro-2-hydroxyphenyl)(2-methoxyphenyl)methanone, 218
(5-Chloro-2-hydroxyphenyl)(4-methoxyphenyl)methanone, 218
(3-Chloro-4-methoxyphenyl)(4-hydroxyphenyl)methanone, 147
(3-Chloro-4-methylphenyl)(2,5-dihydroxyphenyl)methanone, 373
(2-Chlorophenyl)(2,5-dihydroxy-3-methylphenyl)methanone, 390
(4-Chlorophenyl)(2,5-dihydroxy-3-methylphenyl)methanone, 390
(2-Chlorophenyl)(2-hydroxy-4-methoxyphenyl)methanone, 219
(2-Chlorophenyl)(2-hydroxy-5-methoxyphenyl)methanone, 219
(2-Chlorophenyl)(4-hydroxy-3-methoxyphenyl)methanone, 219
(3-Chlorophenyl)(2-hydroxy-4-methoxyphenyl)methanone, 219
(3-Chlorophenyl)(4-hydroxy-3-methoxyphenyl)methanone, 220
(4-Chlorophenyl)(2-hydroxy-4-methoxyphenyl)methanone, 220
(4-Chlorophenyl)(4-hydroxy-3-methoxyphenyl)methanone, 220

$C_{14}H_{11}ClO_4$

[3-Chloro-2,4 (or 2,5)-dihydroxy-5 (or 4)-methoxyphenyl]phenylmethanone, 348
(2-Chloro-4-hydroxyphenyl)(4-hydroxy-2-methoxyphenyl)methanone, 426
(4-Chloro-2-hydroxyphenyl)(2-hydroxy-4-methoxyphenyl)methanone, 426

$C_{14}H_{11}ClO_6$

(3-Chloro-4,6-dihydroxy-2-methylphenyl)(2,4,6-trihydroxyphenyl)methanone, 465
(2-Chloro-5-hydroxy-4-methoxyphenyl)(2,4,6-trihydroxyphenyl)methanone, 459
(3-Chloro-2,4,6-trihydroxyphenyl)(2,4-dihydroxy-6-methylphenyl)methanone, 468

$C_{14}H_{11}FO_2$

(5-Fluoro-2-hydroxyphenyl)(2-methylphenyl)methanone, 221
(5-Fluoro-2-hydroxyphenyl)(3-methylphenyl)methanone, 221
(5-Fluoro-2-hydroxyphenyl)(4-methylphenyl)methanone, 221
(3-Fluoro-4-methylphenyl)(3-hydroxyphenyl)methanone, 147
(2-Fluorophenyl)(2-hydroxy-5-methylphenyl)methanone, 221
(3-Fluorophenyl)(2-hydroxy-5-methylphenyl)methanone, 221
(4-Fluorophenyl)(2-hydroxy-4-methylphenyl)methanone, 222
(4-Fluorophenyl)(4-hydroxy-2-methylphenyl)methanone, 222

$C_{14}H_{11}FO_3$

(2,5-Dihydroxyphenyl)(3-fluoro-4-methylphenyl)methanone, 373
(5-Fluoro-2-hydroxyphenyl)(4-methoxyphenyl)methanone, 222
(5-Fluoro-2-hydroxyphenyl)[4-(methoxy-^{11}C)phenyl]methanone, 222
(2-Fluorophenyl)(2-hydroxy-4-methoxyphenyl)methanone, 223
(2-Fluorophenyl)(2-hydroxy-5-methoxyphenyl)methanone, 223
(2-Fluorophenyl)(4-hydroxy-3-methoxyphenyl)methanone, 223
(3-Fluorophenyl)(2-hydroxy-4-methoxyphenyl)methanone, 223
(3-Fluorophenyl)(2-hydroxy-5-methoxyphenyl)methanone, 224
(3-Fluorophenyl)(4-hydroxy-3-methoxyphenyl)methanone, 224

(4-Fluorophenyl)(2-hydroxy-4-methoxyphenyl)methanone,   224
(4-Fluorophenyl)(2-hydroxy-5-methoxyphenyl)methanone,   224
(4-Fluorophenyl)(4-hydroxy-3-methoxyphenyl)methanone,   225

$C_{14}H_{11}NO_4$

(2-Hydroxy-3-methyl-4-nitrophenyl)phenylmethanone,   59
(2-Hydroxy-3-methyl-5-nitrophenyl)phenylmethanone,   59
(2-Hydroxy-4-methyl-5-nitrophenyl)phenylmethanone,   59
(2-Hydroxy-5-methyl-3-nitrophenyl)phenylmethanone,   60
(2-Hydroxy-5-methyl-4-nitrophenyl)phenylmethanone,   60
(4-Hydroxy-3-methyl-5-nitrophenyl)phenylmethanone,   60
(2-Hydroxy-3-methylphenyl)(3-nitrophenyl)methanone,   225
(2-Hydroxy-3-methylphenyl)(4-nitrophenyl)methanone,   225
(2-Hydroxy-4-methylphenyl)(3-nitrophenyl)methanone,   225
(2-Hydroxy-4-methylphenyl)(4-nitrophenyl)methanone,   226
(2-Hydroxy-5-methylphenyl)(3-nitrophenyl)methanone,   226
(2-Hydroxy-5-methylphenyl)(4-nitrophenyl)methanone,   226
(2-Hydroxy-6-methylphenyl)(4-nitrophenyl)methanone,   227
(4-Hydroxy-2-methylphenyl)(3-nitrophenyl)methanone,   227
(4-Hydroxy-2-methylphenyl)(4-nitrophenyl)methanone,   227
(4-Hydroxy-3-methylphenyl)(3-nitrophenyl)methanone,   227
(4-Hydroxy-3-methylphenyl)(4-nitrophenyl)methanone,   228

$C_{14}H_{11}NO_5$

(2,6-Dihydroxy-4-methoxy-3-nitrosophenyl)phenylmethanone,   349
(3,4-Dihydroxy-5-nitrophenyl)(2-methylphenyl)methanone,   391
(3,4-Dihydroxy-5-nitrophenyl)(4-methylphenyl)methanone,   391
(2,5-Dihydroxyphenyl)(2-methyl-3-nitrophenyl)methanone,   373
(2,5-Dihydroxyphenyl)(3-methyl-4-nitrophenyl)methanone,   374
(2-Hydroxy-3-methoxy-6-nitrophenyl)phenylmethanone,   60
(2-Hydroxy-4-methoxy-5-nitrophenyl)phenylmethanone,   61
(2-Hydroxy-5-methoxy-4-nitrophenyl)phenylmethanone,   61
(2-Hydroxy-4-methoxyphenyl)(3-nitrophenyl)methanone,   228
(2-Hydroxy-4-methoxyphenyl)(4-nitrophenyl)methanone,   228
(2-Hydroxy-5-methoxyphenyl)(4-nitrophenyl)methanone,   228
(4-Hydroxy-3-methoxyphenyl)(2-nitrophenyl)methanone,   229
(5-Hydroxy-2-methoxyphenyl)(4-nitrophenyl)methanone,   229
(2-Hydroxy-5-methylphenyl)(2-hydroxy-5-nitrophenyl)methanone,   426

$C_{14}H_{11}NO_8$

(3-Hydroxy-6-methoxy-2-nitrophenyl)(2,4,6-trihydroxyphenyl)methanone,   459

$C_{14}H_{12}ClNO_2$

(3-Amino-5-chloro-2-hydroxyphenyl)(4-methylphenyl)methanone,   229
(2-Amino-5-chlorophenyl)(2-hydroxy-5-methylphenyl)methanone,   229
(3-Amino-2-hydroxy-5-methylphenyl)(4-chlorophenyl)methanone,   230
(4-Chlorophenyl)[3-hydroxy-4-(methylamino)phenyl]methanone,   230

$C_{14}H_{12}ClNO_3$

(3-Amino-5-chloro-2-hydroxyphenyl)(4-methoxyphenyl)methanone,   230
(3-Amino-5-chloro-2-hydroxyphenyl)(4-methoxyphenyl)methanone *(Hydrochloride)*,   230

$C_{14}H_{12}FNO_2$

(3-Amino-5-fluoro-2-hydroxyphenyl)(4-methylphenyl)methanone,   231
(3-Amino-2-hydroxy-5-methylphenyl)(4-fluorophenyl)methanone,   231

C$_{14}$H$_{12}$FNO$_3$

(3-Amino-5-fluoro-2-hydroxyphenyl)(4-methoxyphenyl)methanone,   231

C$_{14}$H$_{12}$N$_2$O$_5$

(2-Amino-4-methoxyphenyl)(2-hydroxy-4-nitrophenyl)methanone,   231
(4-Amino-2-methoxyphenyl)(2-hydroxy-4-nitrophenyl)methanone,   231

C$_{14}$H$_{12}$O$_2$

(2-Hydroxy-3-methylphenyl)phenylmethanone,   61
(2-Hydroxy-4-methylphenyl)phenylmethanone,   62
(2-Hydroxy-5-methylphenyl)phenylmethanone,   63
(2-Hydroxy-6-methylphenyl)phenylmethanone,   64
(3-Hydroxy-2-methylphenyl)phenylmethanone,   65
(3-Hydroxy-4-methylphenyl)phenylmethanone,   65
(4-Hydroxy-2-methylphenyl)phenylmethanone,   65
(4-Hydroxy-3-methylphenyl)phenylmethanone,   66
(2-Hydroxyphenyl)(2-methylphenyl)methanone,   147
(2-Hydroxyphenyl)(3-methylphenyl)methanone,   148
(2-Hydroxyphenyl)(4-methylphenyl)methanone,   148
(3-Hydroxyphenyl)(2-methylphenyl)methanone,   149
(3-Hydroxyphenyl)(4-methylphenyl)methanone,   149
(4-Hydroxyphenyl)(2-methylphenyl)methanone,   149
(4-Hydroxyphenyl)(3-methylphenyl)methanone,   149
(4-Hydroxyphenyl)(4-methylphenyl)methanone,   150

C$_{14}$H$_{12}$O$_2$S

(2-Hydroxy-5-methylphenyl)(2-mercaptophenyl)methanone,   232
(4-Hydroxyphenyl)[4-(methylthio)phenyl]methanone,   150

C$_{14}$H$_{12}$O$_3$

(2,4-Dihydroxy-3-methylphenyl)phenylmethanone,   349
(2,4-Dihydroxy-5-methylphenyl)phenylmethanone,   349
(2,4-Dihydroxy-6-methylphenyl)phenylmethanone,   350
(2,5-Dihydroxy-4-methylphenyl)phenylmethanone,   350
(2,6-Dihydroxy-4-methylphenyl)phenylmethanone,   350
(3,4-Dihydroxy-5-methylphenyl)phenylmethanone,   351
(2,4-Dihydroxyphenyl)(2-methylphenyl)methanone,   374
(2,4-Dihydroxyphenyl)(3-methylphenyl)methanone,   374
(2,4-Dihydroxyphenyl)(4-methylphenyl)methanone,   374
(2,5-Dihydroxyphenyl)(2-methylphenyl)methanone,   375
(2,5-Dihydroxyphenyl)(3-methylphenyl)methanone,   375
(2,5-Dihydroxyphenyl)(4-methylphenyl)methanone,   375
(2,6-Dihydroxyphenyl)(2-methylphenyl)methanone,   375
(2,6-Dihydroxyphenyl)(3-methylphenyl)methanone,   376
(2,6-Dihydroxyphenyl)(4-methylphenyl)methanone,   376
(2-Hydroxy-3-methoxyphenyl)phenylmethanone,   67
(2-Hydroxy-4-methoxyphenyl)phenylmethanone,   67
(2-Hydroxy-4-methoxyphenyl)phenylmethanone-^{14}C,   68
(2-Hydroxy-5-methoxyphenyl)phenylmethanone,   68
(2-Hydroxy-6-methoxyphenyl)phenylmethanone,   69
(3-Hydroxy-4-methoxyphenyl)phenylmethanone,   70
(4-Hydroxy-2-methoxyphenyl)phenylmethanone,   70
(4-Hydroxy-3-methoxyphenyl)phenylmethanone,   70
(5-Hydroxy-2-methoxyphenyl)phenylmethanone,   71
(2-Hydroxy-3-methylphenyl)(3-hydroxyphenyl)methanone,   401
(2-Hydroxy-4-methylphenyl)(2-hydroxyphenyl)methanone,   401
(2-Hydroxy-4-methylphenyl)(3-hydroxyphenyl)methanone,   401

$C_{14}H_{12}O_3S$

$C_{14}H_{12}O_4$

$C_{14}H_{12}O_5$

$C_{14}H_{12}O_5S$

$C_{14}H_{12}O_6$

$C_{14}H_{12}O_7$

$C_{14}H_{13}NO_2$

$C_{14}H_{13}NO_3$

$C_{14}H_{14}N_2O_2$

$C_{14}H_{17}NO_5$

$C_{14}H_{18}O_2$

$C_{14}H_{18}O_3$

$C_{15}H_8F_6O_3S_2$

$C_{15}H_{10}ClF_3O_3$

$C_{15}H_{10}Cl_4O_2$

$C_{15}H_{10}F_3NO_5$

$C_{15}H_{10}O_2$

$C_{15}H_{11}ClO_3$

$C_{15}H_{11}Cl_3O_2$

$C_{15}H_{11}Cl_3O_5$

$C_{15}H_{11}F_3O_2$

$C_{15}H_{11}F_3O_3$

C₁₅H₁₂BrIO₃

[4-(2-Bromoethoxy)phenyl](4-hydroxy-3-iodophenyl)methanone,   238

C₁₅H₁₂Br₂O₄

(3,5-Dibromo-2-hydroxy-4,6-dimethoxyphenyl)phenylmethanone,   72
(3,5-Dibromo-2-hydroxy-4-methoxyphenyl)(4-methoxyphenyl)methanone,   239

C₁₅H₁₂ClFO₄

(5-Chloro-2-hydroxy-3,4-dimethoxyphenyl)(4-fluorophenyl)methanone,   239

C₁₅H₁₂ClNO₄

(2-Chloro-4-hydroxyphenyl)(2,3-dimethyl-5-nitrophenyl)methanone,   239
(3-Chloro-2-hydroxyphenyl)(2,3-dimethyl-5-nitrophenyl)methanone,   239
(3-Chloro-4-hydroxyphenyl)(2,3-dimethyl-5-nitrophenyl)methanone,   240
(4-Chloro-2-hydroxyphenyl)(2,3-dimethyl-5-nitrophenyl)methanone,   240
(5-Chloro-2-hydroxyphenyl)(2,3-dimethyl-5-nitrophenyl)methanone,   240

C₁₅H₁₂Cl₂O₂

(5-Chloro-3-ethyl-2-hydroxyphenyl)(4-chlorophenyl)methanone,   240
(3-Chloro-6-hydroxy-2,4-dimethylphenyl)(2-chlorophenyl)methanone,   240
(5-Chloro-2-hydroxy-3-methylphenyl)(4-chloro-2-methylphenyl)methanone,   241
(2,3-Dichloro-4-hydroxyphenyl)(2,3-dimethylphenyl)methanone,   241
(2,4-Dichlorophenyl)(3-ethyl-2-hydroxyphenyl)methanone,   241
(2,4-Dichlorophenyl)(5-ethyl-2-hydroxyphenyl)methanone,   241
(3,4-Dichlorophenyl)(5-ethyl-2-hydroxyphenyl)methanone,   242
(2,4-Dichlorophenyl)(2-hydroxy-4,6-dimethylphenyl)methanone,   242
(3,4-Dichlorophenyl)(4-hydroxy-2,6-dimethylphenyl)methanone,   242

C₁₅H₁₂Cl₂O₃

Bis(2-chloro-4-hydroxy-6-methylphenyl)methanone,   418
(2,6-Dichlorophenyl)(2-hydroxy-4-methoxy-6-methylphenyl)methanone,   242
(2,6-Dichlorophenyl)(5-hydroxy-4-methoxy-2-methylphenyl)methanone,   243

C₁₅H₁₂Cl₂O₄

(2,6-Dichlorophenyl)(2,3-dihydroxy-4-methoxy-6-methylphenyl)methanone,   391
(2,6-Dichlorophenyl)(2-hydroxy-4,5-dimethoxyphenyl)methanone,   243

C₁₅H₁₂Cl₂O₅

(3,5-Dichloro-2,4-dihydroxy-6-methylphenyl)(2,4-dihydroxy-6-methylphenyl)methanone,   464

C₁₅H₁₂N₂O₇

[4-Hydroxy-3-(methoxymethyl)-5-nitrophenyl](2-nitrophenyl)methanone,   243

C₁₅H₁₂N₂O₈

(2,4-Dimethoxyphenyl)(4-hydroxy-3,5-dinitrophenyl)methanone,   243
(2-Hydroxy-4-methoxy-3,5-dinitrophenyl)(2-methoxyphenyl)methanone,   244
(2-Hydroxy-4-methoxy-3,5-dinitrophenyl)(4-methoxyphenyl)methanone,   244

C₁₅H₁₂O₃

1-(3-Benzoyl-2-hydroxyphenyl)ethanone,   487

$C_{15}H_{12}O_4$

$C_{15}H_{12}O_5$

$C_{15}H_{13}BrO_2$

$C_{15}H_{13}BrO_3$

$C_{15}H_{13}BrO_4$

$C_{15}H_{13}ClO_2$

$C_{15}H_{13}ClO_3$

$C_{15}H_{13}ClO_4$

$C_{15}H_{13}FO_2$

$C_{15}H_{13}FO_3$

$C_{15}H_{13}FO_4$

$C_{15}H_{13}NO_4$

$C_{15}H_{13}NO_5$

$C_{15}H_{13}NO_6$

$C_{15}H_{14}O_2$

$C_{15}H_{14}O_3$

C$_{15}$H$_{14}$O$_4$

(2-Hydroxy-5-methoxyphenyl)(3-methoxyphenyl)methanone,   265
(2-Hydroxy-5-methoxyphenyl)(4-methoxyphenyl)methanone,   265
(2-Hydroxy-6-methoxyphenyl)(2-methoxyphenyl)methanone,   265
(5-Hydroxy-2-methoxyphenyl)(4-methoxyphenyl)methanone,   266
(2-Hydroxyphenyl)[2-(methoxymethoxy)phenyl]methanone,   160
(2-Hydroxyphenyl)[4-(methoxymethoxy)phenyl]methanone,   160
Phenyl(2,4,6-trihydroxy-3,5-dimethylphenyl)methanone,   436

$C_{15}H_{14}O_5$

Bis(2,4-dihydroxy-6-methylphenyl)methanone,   463
Bis[4-hydroxy-3-(hydroxymethyl)phenyl]methanone,   419
Bis(2-hydroxy-4-methoxyphenyl)methanone,   419
Bis(4-hydroxy-3-methoxyphenyl)methanone,   420
(2,3-Dihydroxy-4,5-dimethoxyphenyl)phenylmethanone,   355
(2,4-Dihydroxy-3,5-dimethoxyphenyl)phenylmethanone,   356
(2,5-Dihydroxy-3,4-dimethoxyphenyl)phenylmethanone,   356
(3,6-Dihydroxy-2,4-dimethylphenyl)(2,4-dihydroxyphenyl)methanone,   461
(3,6-Dihydroxy-2,4-dimethylphenyl)(2,6-dihydroxyphenyl)methanone,   462
(2,5-Dihydroxy-4-methoxyphenyl)(2-methoxyphenyl)methanone,   392
(2,5-Dihydroxy-4-methoxyphenyl)(4-methoxyphenyl)methanone,   393
(2,6-Dihydroxy-4-methoxyphenyl)(4-methoxyphenyl)methanone,   393
(2,4-Dihydroxy-3-methylphenyl)(5-hydroxy-2-methoxyphenyl)methanone,   450
(2,4-Dihydroxy-5-methylphenyl)(5-hydroxy-2-methoxyphenyl)methanone,   450
(2,4-Dihydroxy-6-methylphenyl)(5-hydroxy-2-methoxyphenyl)methanone,   451
(2,4-Dihydroxyphenyl)(3,4-dimethoxyphenyl)methanone,   380
(2,5-Dihydroxyphenyl)(2-hydroxy-4-methoxy-6-methylphenyl)methanone,   447
(4-Ethoxyphenyl)(2,3,4-trihydroxyphenyl)methanone,   441
(2-Hydroxy-3,4-dimethoxyphenyl)(2-hydroxyphenyl)methanone,   409
(2-Hydroxy-4,6-dimethoxyphenyl)(3-hydroxyphenyl)methanone,   409

$C_{15}H_{14}O_6$

(2,6-Dihydroxy-4-methoxyphenyl)(2,4-dihydroxy-6-methylphenyl)methanone,   464
(2,5-Dihydroxy-4-methoxyphenyl)(3-hydroxy-4-methoxyphenyl)methanone,   451
(2,6-Dihydroxy-4-methoxyphenyl)(2-hydroxy-3-methoxyphenyl)methanone,   451
(2,6-Dimethoxyphenyl)(2,4,6-trihydroxyphenyl)methanone,   441
(2-Hydroxy-4-methoxy-6-methylphenyl)(2,4,6-trihydroxyphenyl)methanone,   462
(4-Hydroxy-2-methoxy-6-methylphenyl)(2,4,6-trihydroxyphenyl)methanone,   462

$C_{15}H_{14}O_7$

Bis(2,6-dihydroxy-4-methoxyphenyl)methanone,   463
(2,3-Dihydroxy-4-methoxyphenyl)(3,5-dihydroxy-4-methoxyphenyl)methanone,   464
(3-Hydroxy-2,6-dimethoxyphenyl)(2,4,6-trihydroxyphenyl)methanone,   462

$C_{15}H_{15}NO_2$

(2-Amino-3-hydroxy-4,6-dimethylphenyl)phenylmethanone,   88
[4-(Dimethylamino)phenyl](3-hydroxyphenyl)methanone,   160
[4-(Dimethylamino)phenyl](4-hydroxyphenyl)methanone,   161

$C_{15}H_{15}NO_3$

[3-Hydroxy-4-(methylamino)phenyl](4-methoxyphenyl)methanone,   266

$C_{15}H_{15}NO_4$

(2-Aminophenyl)(2-hydroxy-4,6-dimethoxyphenyl)methanone,   266

C$_{16}$H$_{14}$O$_2$

[4-Hydroxy-3-(2-propenyl)phenyl]phenylmethanone,   89

C$_{16}$H$_{14}$O$_3$

(3-Benzoyl-2-hydroxy-5-methylphenyl)ethanone,   488
[2,4-Dihydroxy-5-(2-propenyl)phenyl]phenylmethanone,   356
[2,6-Dihydroxy-3-(2-propenyl)phenyl]phenylmethanone,   356
(4-Ethenylphenyl)(2-hydroxy-4-methoxyphenyl)methanone,   269
[2-Hydroxy-4-(1-propenyloxy)phenyl]phenylmethanone,   89
[2-Hydroxy-4-(2-propenyloxy)phenyl]phenylmethanone,   89

C$_{16}$H$_{14}$O$_4$

1-(3-Benzoyl-2-hydroxy-4-methoxyphenyl)ethanone,   488
[2-Hydroxy-4-(oxiranylmethoxy)phenyl]phenylmethanone,   89

C$_{16}$H$_{14}$O$_5$

[2-(Acetyloxy)-5-methoxyphenyl](2-hydroxyphenyl)methanone,   161
[2-(Acetyloxy)phenyl](2-hydroxy-5-methoxyphenyl)methanone,   269

C$_{16}$H$_{14}$O$_6$

1-[3-(3,6-Dihydroxy-2-methylbenzoyl)-2,4-dihydroxyphenyl]ethanone,   492

C$_{16}$H$_{15}$ClO$_2$

(5-Chloro-3-ethyl-2-hydroxyphenyl)(4-methylphenyl)methanone,   269
(5-Chloro-2-hydroxy-3-methylphenyl)(3,4-dimethylphenyl)methanone,   269
(5-Chloro-2-hydroxy-3-methylphenyl)(4-ethylphenyl)methanone,   270
(4-Chlorophenyl)(2-hydroxy-3-propylphenyl)methanone,   270
(4-Chlorophenyl)(2-hydroxy-5-propylphenyl)methanone,   270

C$_{16}$H$_{15}$ClO$_3$

(2-Chlorophenyl)(2,4-dihydroxy-3-propylphenyl)methanone,   393

C$_{16}$H$_{15}$ClO$_6$

(3-Chloro-2,6-dihydroxy-4-methoxyphenyl)(4-hydroxy-2-methoxy-6-methylphenyl)-
methanone,   454
(3-Chloro-6-hydroxy-2,4-dimethoxyphenyl)(2,4-dihydroxy-6-methylphenyl)methanone,   454

C$_{16}$H$_{15}$FO$_3$

(4-Fluorophenyl)(2-hydroxy-4-propoxyphenyl)methanone,   270

C$_{16}$H$_{15}$NO$_4$

(2,3-Dimethyl-5-nitrophenyl)(2-hydroxy-3-methylphenyl)methanone,   271
(2,3-Dimethyl-5-nitrophenyl)(2-hydroxy-4-methylphenyl)methanone,   271
(2,3-Dimethyl-5-nitrophenyl)(2-hydroxy-5-methylphenyl)methanone,   271
(2,3-Dimethyl-5-nitrophenyl)(4-hydroxy-2-methylphenyl)methanone,   271
(2,3-Dimethyl-5-nitrophenyl)(4-hydroxy-3-methylphenyl)methanone,   272
(2-Hydroxy-3-nitrophenyl)[4-(1-methylethyl)phenyl]methanone,   272
(2-Hydroxy-5-nitrophenyl)[4-(1-methylethyl)phenyl]methanone,   272

(2-Hydroxy-3,4-dimethoxyphenyl)(2-methylphenyl)methanone, 278
[2-Hydroxy-4-(2-hydroxypropoxy)phenyl]phenylmethanone, 94
(2-Hydroxy-4-methoxy-3-methylphenyl)(2-methoxyphenyl)methanone, 279
(2-Hydroxy-4-methoxy-3-methylphenyl)(4-methoxyphenyl)methanone, 279
[2-Hydroxy-4-(2-methoxyethoxy)phenyl]phenylmethanone, 94

$C_{16}H_{16}O_5$

(3,6-Dihydroxy-2,4-dimethylphenyl)(2-hydroxy-6-methoxyphenyl)methanone, 454
(2,4-Dihydroxy-6-methylphenyl)(2-hydroxy-4-methoxy-6-methylphenyl)methanone, 455
(2,4-Dihydroxy-6-methylphenyl)(4-hydroxy-2-methoxy-6-methylphenyl)methanone, 455
(2,6-Dihydroxy-4-methylphenyl)(2-hydroxy-4-methoxy-6-methylphenyl)methanone, 455
(2,6-Dihydroxy-4-methylphenyl)(4-hydroxy-2-methoxy-6-methylphenyl)methanone, 456
(2,5-Dihydroxyphenyl)(2,4-dimethoxy-6-methylphenyl)methanone, 380
(2,3-Dimethoxyphenyl)(2-hydroxy-3-methoxyphenyl)methanone, 279
(2,3-Dimethoxyphenyl)(2-hydroxy-4-methoxyphenyl)methanone, 279
(2,3-Dimethoxyphenyl)(2-hydroxy-5-methoxyphenyl)methanone, 280
(2,3-Dimethoxyphenyl)(4-hydroxy-3-methoxyphenyl)methanone, 280
(2,4-Dimethoxyphenyl)(2-hydroxy-3-methoxyphenyl)methanone, 280
(2,4-Dimethoxyphenyl)(2-hydroxy-4-methoxyphenyl)methanone, 280
(2,4-Dimethoxyphenyl)(2-hydroxy-5-methoxyphenyl)methanone, 281
(2,4-Dimethoxyphenyl)(2-hydroxy-6-methoxyphenyl)methanone, 281
(2,5-Dimethoxyphenyl)(2-hydroxy-3-methoxyphenyl)methanone, 281
(2,5-Dimethoxyphenyl)(2-hydroxy-4-methoxyphenyl)methanone, 281
(2,6-Dimethoxyphenyl)(2-hydroxy-3-methoxyphenyl)methanone, 282
(2,6-Dimethoxyphenyl)(2-hydroxy-4-methoxyphenyl)methanone, 282
(2,6-Dimethoxyphenyl)(2-hydroxy-6-methoxyphenyl)methanone, 282
(3,4-Dimethoxyphenyl)(2-hydroxy-4-methoxyphenyl)methanone, 283
(3,4-Dimethoxyphenyl)(4-hydroxy-2-methoxyphenyl)methanone, 283
(4-Ethoxy-2-hydroxyphenyl)(2-hydroxy-4-methoxyphenyl)methanone, 427
[2-Hydroxy-3,5-di(hydroxymethyl)-4-methoxyphenyl]phenylmethanone, 94
(2-Hydroxy-4,6-dimethoxyphenyl)(4-hydroxy-2-methylphenyl)methanone, 427
(2-Hydroxy-3,4-dimethoxyphenyl)(2-methoxyphenyl)methanone, 283
(2-Hydroxy-4,5-dimethoxyphenyl)(2-methoxyphenyl)methanone, 284
(2-Hydroxy-4,5-dimethoxyphenyl)(3-methoxyphenyl)methanone, 284
(2-Hydroxy-4,5-dimethoxyphenyl)(4-methoxyphenyl)methanone, 284
(2-Hydroxy-4,6-dimethoxyphenyl)(2-methoxyphenyl)methanone, 284
(2-Hydroxy-4,6-dimethoxyphenyl)(3-methoxyphenyl)methanone, 285
(2-Hydroxy-4,6-dimethoxyphenyl)(4-methoxyphenyl)methanone, 285
(4-Hydroxy-3,5-dimethoxyphenyl)(4-methoxyphenyl)methanone, 285
(2-Hydroxy-3-methoxy-5-methylphenyl)(4-hydroxy-3-methoxyphenyl)methanone, 428
(2-Hydroxyphenyl)(2,3,4-trimethoxyphenyl)methanone, 163
(2-Hydroxyphenyl)(2,4,5-trimethoxyphenyl)methanone, 163
(2-Hydroxyphenyl)(2,4,6-trimethoxyphenyl)methanone, 163
(3-Hydroxyphenyl)(2,4,6-trimethoxyphenyl)methanone, 163
(4-Hydroxyphenyl)(2,4,6-trimethoxyphenyl)methanone, 164
(4-Hydroxyphenyl)(3,4,5-trimethoxyphenyl)methanone, 164
(2-Hydroxy-3,4,5-trimethoxyphenyl)phenylmethanone, 94

$C_{16}H_{16}O_6$

(2,5-Dihydroxy-4-methoxyphenyl)(3,4-dimethoxyphenyl)methanone, 394
(2,6-Dihydroxy-4-methoxyphenyl)(4-hydroxy-2-methoxy-6-methylphenyl)methanone, 456
(3,4-Dihydroxyphenyl)(2,4,6-trimethoxyphenyl)methanone, 380
(2,4-Dimethoxy-6-methylphenyl)(2,4,6-trihydroxyphenyl)methanone, 442
(2-Hydroxy-4,5-dimethoxyphenyl)(2-hydroxy-4-methoxyphenyl)methanone, 428
(4-Hydroxy-3,5-dimethoxyphenyl)(2-hydroxy-4-methoxyphenyl)methanone, 428

$C_{16}H_{17}NO_2$,

[3-[(Dimethylamino)methyl]-4-hydroxyphenyl]phenylmethanone *(Hydrochloride)*, 95

C$_{16}$H$_{17}$NO$_5$

(2-Amino-4,5-dimethoxyphenyl)(2-hydroxy-4-methoxyphenyl)methanone *(Hydrochloride)*, 285

C$_{17}$H$_{14}$Cl$_2$O$_2$

[5-Chloro-2-hydroxy-3-(1-methyl-2-propenyl)phenyl](4-chlorophenyl)methanone, 286

C$_{17}$H$_{14}$O$_5$

1-(3-Acetyl-5-benzoyl-2,4-dihydroxyphenyl)ethanone, 491
Bis(5-acetyl-2-hydroxyphenyl)methanone, 489

C$_{17}$H$_{14}$O$_6$

[2,3-Bis(acetyloxy)-4-hydroxyphenyl]phenylmethanone, 95
[3,4-Bis(acetyloxy)-2-hydroxyphenyl]phenylmethanone, 95

C$_{17}$H$_{15}$Cl$_3$O$_5$

(3-Chloro-6-hydroxy-4-methoxy-2-methylphenyl)(3,5-dichloro-2-hydroxy-4-methoxy-6-methylphenyl)methanone, 428

C$_{17}$H$_{16}$ClNO$_4$

[5-(1,1-Dimethylethyl)-2-hydroxy-3-nitrophenyl](4-chlorophenyl)methanone, 286

C$_{17}$H$_{16}$Cl$_2$O$_2$

(3-Butyl-5-chloro-2-hydroxyphenyl)(4-chlorophenyl)methanone, 286
[5-Chloro-3-(1,1-dimethylethyl)-2-hydroxyphenyl](4-chlorophenyl)methanone, 286
[5-Chloro-2-hydroxy-3-(2-methylpropyl)phenyl](4-chlorophenyl)methanone, 287
(2,4-Dichlorophenyl)[5-(1,1-dimethylethyl)-2-hydroxyphenyl]methanone, 287
(2,6-Dichlorophenyl)[3-(1,1-dimethylethyl)-4-hydroxyphenyl]methanone, 287
(3,4-Dichlorophenyl)[5-(1,1-dimethylethyl)-2-hydroxyphenyl]methanone, 287
(2,4-Dichlorophenyl)[4-hydroxy-2-methyl-5-(1-methylethyl)phenyl]methanone, 288
(3,4-Dichlorophenyl)[4-hydroxy-2-methyl-5-(1-methylethyl)phenyl]methanone, 288
(2,4-Dichlorophenyl)[2-hydroxy-5-(1-methylpropyl)phenyl]methanone, 288
(2,5-Dichlorophenyl)[2-hydroxy-5-(1-methylpropyl)phenyl]methanone, 288
(3,4-Dichlorophenyl)[2-hydroxy-5-(1-methylpropyl)phenyl]methanone, 289

C$_{17}$H$_{16}$N$_2$O$_6$

(3,5-Dinitrophenyl)[5-(1,1-dimethylethyl)-2-hydroxyphenyl]methanone, 289

C$_{17}$H$_{16}$O$_2$

[3-(2-Butenyl)-4-hydroxyphenyl]phenylmethanone, 95
[4-Hydroxy-3-(1-methyl-2-propenyl)phenyl]phenylmethanone, 96
[4-Hydroxy-3-(2-methyl-2-propenyl)phenyl]phenylmethanone, 96

C$_{17}$H$_{16}$O$_3$

[3-(2-Butenyl)-2,4-dihydroxyphenyl]phenylmethanone, 357
[5-(2-Butenyl)-2,4-dihydroxyphenyl]phenylmethanone, 357
[2,4-Dihydroxy-3-(1-methyl-2-propenyl)phenyl]phenylmethanone, 358
(4-Ethenylphenyl)(4-ethoxy-2-hydroxyphenyl)methanone, 289
[2-Hydroxy-6-methoxy-3-(2-propenyl)phenyl]phenylmethanone, 96
[6-Hydroxy-2-methoxy-3-(2-propenyl)phenyl]phenylmethanone, 96

$C_{17}H_{17}BrO_2$

[3-Bromo-2-hydroxy-6-methyl-5-(1-methylethyl)phenyl]phenylmethanone,   97
[3-Bromo-6-hydroxy-2-methyl-5-(1-methylethyl)phenyl]phenylmethanone,   97
[4-Bromo-2-hydroxy-6-methyl-3-(1-methylethyl)phenyl]phenylmethanone,   97
(2-Bromophenyl)[5-(1,1-dimethylethyl)-2-hydroxyphenyl]methanone,   289
(3-Bromophenyl)[5-(1,1-dimethylethyl)-2-hydroxyphenyl]methanone,   290
(4-Bromophenyl)[5-(1,1-dimethylethyl)-2-hydroxyphenyl]methanone,   290
(2-Bromophenyl)[4-hydroxy-2-methyl-5-(1-methylethyl)phenyl]methanone,   290
(4-Bromophenyl)[4-hydroxy-2-methyl-5-(1-methylethyl)phenyl]methanone,   290

$C_{17}H_{17}ClO_2$

(5-Butyl-2-hydroxyphenyl)(4-chlorophenyl)methanone,   291
[3-Chloro-5-(1,1-dimethylethyl)-2-hydroxyphenyl]phenylmethanone,   97
[5-Chloro-3-(1,1-dimethylethyl)-2-hydroxyphenyl]phenylmethanone,   98
(5-Chloro-3-ethyl-2-hydroxyphenyl)(4-ethylphenyl)methanone,   291
(5-Chloro-2-hydroxy-3-methylphenyl)[4-(1-methylethyl)phenyl]methanone,   291
(5-Chloro-2-hydroxy-3-propylphenyl)(4-methylphenyl)methanone,   291
[4-(Chloromethyl)-2-hydroxy-3-propylphenyl]phenylmethanone,   98
(4-Chlorophenyl)(3,5-diethyl-2-hydroxyphenyl)methanone,   292
(4-Chlorophenyl)(4,5-diethyl-2-hydroxyphenyl)methanone,   292
(2-Chlorophenyl)[5-(1,1-dimethylethyl)-2-hydroxyphenyl]methanone,   292
(3-Chlorophenyl)[3-(1,1-dimethylethyl)-4-hydroxyphenyl]methanone,   292
(4-Chlorophenyl)[3-(1,1-dimethylethyl)-4-hydroxyphenyl]methanone,   293
(4-Chlorophenyl)[5-(1,1-dimethylethyl)-2-hydroxyphenyl]methanone,   293
(4-Chlorophenyl)[4-hydroxy-2-methyl-5-(1-methylethyl)phenyl]methanone,   293
(2-Chlorophenyl)[2-hydroxy-5-(1-methylpropyl)phenyl]methanone,   293
(4-Chlorophenyl)[2-hydroxy-5-(1-methylpropyl)phenyl]methanone,   294

$C_{17}H_{17}ClO_5$

(3-Chloro-6-hydroxy-4-methoxy-2,5-dimethylphenyl)(2,4-dihydroxy-6-methylphenyl)-
methanone,   456
(3-Chloro-2-hydroxy-6-methoxy-4-methylphenyl)(4-hydroxy-2-methoxy-6-methylphenyl)-
methanone,   429

$C_{17}H_{17}ClO_6$

(3-Chloro-4,6-dihydroxy-2-methylphenyl)(2,4,6-trimethoxyphenyl)methanone,   394
(3-Chloro-2-hydroxy-4,6-dimethoxyphenyl)(4-hydroxy-2-methoxy-6-methylphenyl)-
methanone,   429

$C_{17}H_{17}FO_2$

[3-(1,1-Dimethylethyl)-4-hydroxyphenyl](2-fluorophenyl)methanone,   294
[3-(1,1-Dimethylethyl)-4-hydroxyphenyl](4-fluorophenyl)methanone,   294

$C_{17}H_{17}FO_3$

(4-Butoxy-2-hydroxyphenyl)(4-fluorophenyl)methanone,   294

$C_{17}H_{17}FO_6$

(2-Fluoro-4,6-dimethoxyphenyl)(2-hydroxy-4,5-dimethoxyphenyl)methanone,   295
(3-Fluoro-2-hydroxy-4,6-dimethoxyphenyl)(4-hydroxy-2-methoxy-6-methylphenyl)-
methanone,   430

$C_{17}H_{17}NO_4$

(2,3-Dimethyl-5-nitrophenyl)(2-hydroxy-3,4-dimethylphenyl)methanone,   295

C$_{17}$H$_{18}$ClNO$_2$

C$_{17}$H$_{18}$O$_2$

C$_{17}$H$_{18}$O$_3$

$C_{17}H_{18}O_4$

(4-Butoxy-2-hydroxyphenyl)(2-hydroxyphenyl)methanone,   410
[2,4-Dihydroxy-5-(1,1-dimethylethyl)phenyl](2-hydroxyphenyl)methanone,   448
(2,4-Dihydroxyphenyl)[4-hydroxy-2-methyl-5-(1-methylethyl)phenyl]methanone,   448
(2,4-Dimethylphenyl)(4-hydroxy-3,5-dimethoxyphenyl)methanone,   299
(5-Ethyl-2-hydroxy-4-methoxyphenyl)(4-methoxyphenyl)methanone,   299
(2-Hydroxy-4-methoxy-3-propylphenyl)(2-hydroxyphenyl)methanone,   410

$C_{17}H_{18}O_5$

Bis(4-ethoxy-2-hydroxyphenyl)methanone,   421
Bis(2-hydroxy-3-methoxy-5-methylphenyl)methanone,   421
Bis(2-hydroxy-4-methoxy-6-methylphenyl)methanone,   421
(2,4-Dimethoxy-6-methylphenyl)(2-hydroxy-5-methoxyphenyl)methanone,   300
(4-Ethoxyphenyl)(2-hydroxy-3,4-dimethoxyphenyl)methanone,   300
(2-Hydroxy-3,4-dimethoxyphenyl)(3-methoxy-4-methylphenyl)methanone,   300
[3-Hydroxy-2-methoxy-6-(methoxymethyl)phenyl](2-methoxyphenyl)methanone,   300
(2-Hydroxy-4-methoxy-6-methylphenyl)(4-hydroxy-2-methoxy-6-methylphenyl)methanone,   430

$C_{17}H_{18}O_6$

(2,5-Dihydroxy-3,4-dimethoxyphenyl)(4-ethoxyphenyl)methanone,   394
(2,4-Dihydroxy-6-methylphenyl)(2,4,6-trimethoxyphenyl)methanone,   394
(2,3-Dimethoxyphenyl)(2-hydroxy-4,6-dimethoxyphenyl)methanone,   301
(2,5-Dimethoxyphenyl)(2-hydroxy-3,6-dimethoxyphenyl)methanone,   301
(2,5-Dimethoxyphenyl)(2-hydroxy-4,5-dimethoxyphenyl)methanone,   301
(2,5-Dimethoxyphenyl)(2-hydroxy-4,6-dimethoxyphenyl)methanone,   301
(2,6-Dimethoxyphenyl)(2-hydroxy-3,4-dimethoxyphenyl)methanone,   302
(2,6-Dimethoxyphenyl)(2-hydroxy-4,5-dimethoxyphenyl)methanone,   302
(3,4-Dimethoxyphenyl)(2-hydroxy-4,5-dimethoxyphenyl)methanone,   302
(3,4-Dimethoxyphenyl)(2-hydroxy-4,6-dimethoxyphenyl)methanone,   302
(3,5-Dimethoxyphenyl)(2-hydroxy-3,5-dimethoxyphenyl)methanone,   303
(3,5-Dimethoxyphenyl)(2-hydroxy-4,6-dimethoxyphenyl)methanone,   303
(4-Hydroxy-2,6-dimethoxyphenyl)(2-hydroxy-4-methoxy-6-methylphenyl)methanone,   430
(4-Hydroxy-3,5-dimethoxyphenyl)(2-hydroxy-3-methoxy-5-methylphenyl)methanone,   431
(4-Hydroxy-3,5-dimethoxyphenyl)(5-hydroxy-4-methoxy-2-methylphenyl)methanone,   431
[2-Hydroxy-4-(2-hydroxyethoxy)phenyl][4-(2-hydroxyethoxy)phenyl]methanone,   303
(2-Hydroxy-3-methoxyphenyl)(2,4,6-trimethoxyphenyl)methanone,   303
(2-Hydroxy-4-methoxyphenyl)(3,4,5-trimethoxyphenyl)methanone,   304
(2-Hydroxy-5-methoxyphenyl)(2,4,6-trimethoxyphenyl)methanone,   304
(3-Hydroxy-4-methoxyphenyl)(3,4,5-trimethoxyphenyl)methanone,   304
(4-Hydroxy-3-methoxyphenyl)(2,4,6-trimethoxyphenyl)methanone,   304
(2-Hydroxy-3,4,5-trimethoxyphenyl)(2-methoxyphenyl)methanone,   305

$C_{17}H_{18}O_7$

Bis(4-hydroxy-3,5-dimethoxyphenyl)methanone,   421
Bis[2-hydroxy-4-(2-hydroxyethoxy)phenyl]methanone,   422
(2,4-Dihydroxy-3,5-dimethoxyphenyl)(2,5-dimethoxyphenyl)methanone,   395
(3,6-Dihydroxy-2,4-dimethoxyphenyl)(2,5-dimethoxyphenyl)methanone,   395

$C_{17}H_{19}NO_2$

[3-Amino-5-(1,1-dimethylethyl)-2-hydroxyphenyl]phenylmethanone,   105
[3-Amino-5-(1,1-dimethylethyl)-2-hydroxyphenyl]phenylmethanone *(Hydrochloride)*,   105
(4-Aminophenyl)[5-(1,1-dimethylethyl)-2-hydroxyphenyl]methanone,   305

$C_{17}H_{24}O_2$

Cyclohexyl[3-(1,1-dimethylethyl)-4-hydroxyphenyl]methanone,   483
Cyclohexyl[5-(1,1-dimethylethyl)-2-hydroxyphenyl]methanone,   483

C₁₈H₁₇F₃O₂

[3-(1,1-Dimethylethyl)-4-hydroxyphenyl][2-(trifluoromethyl)phenyl]methanone, 305

C₁₈H₁₈Cl₂O₆

(3-Chloro-4,6-dimethoxy-2-methylphenyl)(3-chloro-6-hydroxy-2,4-dimethoxyphenyl)-methanone, 305
(3,5-Dichloro-2,6-dihydroxy-4-methoxyphenyl)(4-hydroxy-2-methyl-6-propoxyphenyl)-methanone, 456

C₁₈H₁₈O₂

[4-Hydroxy-3-(3-methyl-2-butenyl)phenyl]phenylmethanone, 105

C₁₈H₁₈O₃

[2,4-Dihydroxy-3-(3-methyl-2-butenyl)phenyl]phenylmethanone, 358
[2-Hydroxy-4-[(3-methyl-2-butenyl)oxy]phenyl]phenylmethanone, 105

C₁₈H₁₈O₄

[3-(2-Butenyl)-2,4-dihydroxy-6-methoxyphenyl]phenylmethanone, 359
[3-(2-Butenyl)-4,6-dihydroxy-2-methoxyphenyl]phenylmethanone, 359
[2,4-Dihydroxy-6-methoxy-3-(1-methyl-2-propenyl)phenyl]phenylmethanone, 359
[2,6-Dihydroxy-4-[(3-methyl-2-butenyl)oxy]phenyl]phenylmethanone, 359
[2-Hydroxy-4-methoxy-3-(2-propenyl)phenyl](4-methoxyphenyl)methanone, 306

C₁₈H₁₈O₇

[4-(Acetyloxy)-3,5-dimethoxyphenyl](2-hydroxy-4-methoxyphenyl)methanone, 306

C₁₈H₁₉ClO₂

(3-Butyl-5-chloro-2-hydroxyphenyl)(4-methylphenyl)methanone, 306
(5-Chloro-3-ethyl-2-hydroxyphenyl)[4-(1-methylethyl)phenyl]methanone, 306
[5-Chloro-2-hydroxy-3-(2-methylpropyl)phenyl](4-methylphenyl)methanone, 307

C₁₈H₁₉ClO₅

(3-Chloro-6-hydroxy-4-methoxy-2,5-dimethylphenyl)(2,4-dihydroxy-3,6-dimethylphenyl)-methanone, 457

C₁₈H₁₉ClO₆

(3-Chloro-2,6-dihydroxy-4-methoxyphenyl)(4-hydroxy-2-methyl-6-propoxyphenyl)-methanone, 457
(3-Chloro-6-hydroxy-2,4-dimethoxyphenyl)(2,4-dimethoxy-6-methylphenyl)methanone, 307

C₁₈H₂₀BrNO₂

[3-(Aminomethyl)-5-(1,1-dimethylethyl)-2-hydroxyphenyl](4-bromophenyl)methanone, 307
[3-(Aminomethyl)-5-(1,1-dimethylethyl)-2-hydroxyphenyl](4-bromophenyl)methanone *(Hydrochloride)*, 307

C₁₈H₂₀O₂

[2-(1,1-Dimethylethyl)-4-hydroxy-6-methylphenyl]phenylmethanone, 106
[3-(1,1-Dimethylethyl)-2-hydroxy-5-methylphenyl]phenylmethanone, 106
[3-(1,1-Dimethylethyl)-2-hydroxy-6-methylphenyl]phenylmethanone, 106
[3-(1,1-Dimethylethyl)-4-hydroxyphenyl](4-methylphenyl)methanone, 308

[5-(1,1-Dimethylethyl)-2-hydroxyphenyl](4-methylphenyl)methanone,   308
[5-(1,1-Dimethylpropyl)-2-hydroxyphenyl]phenylmethanone,   106
[2-Hydroxy-5,6-dimethyl-3-(1-methylethyl)phenyl]phenylmethanone,   107
(2-Hydroxy-4,6-dimethylphenyl)(2,4,6-trimethylphenyl)methanone,   308
(4-Hydroxy-3,5-dimethylphenyl)(2,4,6-trimethylphenyl)methanone,   308
[4-Hydroxy-2-methyl-5-(1-methylethyl)phenyl](4-methylphenyl)methanone,   309
(4-Hydroxyphenyl)(4-pentylphenyl)methanone,   165

C$_{18}$H$_{20}$O$_2$S

[5-(1,1-Dimethylethyl)-2-hydroxyphenyl][4-(methylthio)phenyl]methanone,   309

C$_{18}$H$_{20}$O$_3$

(4-Butylphenyl)(2-hydroxy-4-methoxyphenyl)methanone,   309
[2,4-Dihydroxy-5-(1,1-dimethylpropyl)phenyl]phenylmethanone,   360
(2,4-Dihydroxyphenyl)[4-(1-methylbutyl)phenyl]methanone,   381
[5-(1,1-Dimethylethyl)-2-hydroxy-4-methoxyphenyl]phenylmethanone,   107
[5-(1,1-Dimethylethyl)-2-hydroxyphenyl](4-methoxyphenyl)methanone,   309
[4-(1,1-Dimethylethyl)phenyl](2-hydroxy-4-methoxyphenyl)methanone,   310
[4-(1,1-Dimethylethyl)phenyl](2-hydroxy-5-methoxyphenyl)methanone,   310
(4-Ethoxy-2-hydroxyphenyl)(4-propylphenyl)methanone,   310
(2-Hydroxy-3,4-dimethylphenyl)(2-methoxy-3,4-dimethylphenyl)methanone,   310
(2-Hydroxy-4,5-dimethylphenyl)(2-methoxy-4,5-dimethylphenyl)methanone,   311
[2-Hydroxy-4-(1-methylbutoxy)phenyl]phenylmethanone,   107
[2-Hydroxy-4-(3-methylbutoxy)phenyl]phenylmethanone,   107
[2-Hydroxy-4-methyl-5-(1-methylethyl)phenyl](4-methoxyphenyl)methanone,   311
[4-Hydroxy-2-methyl-5-(1-methylethyl)phenyl](2-hydroxy-5-methylphenyl)methanone,   431
[2-Hydroxy-4-(pentyloxy)phenyl]phenylmethanone,   108
(3-Hydroxyphenyl)[4-methoxy-2-methyl-5-(1-methylethyl)phenyl]methanone,   165
(4-Hydroxyphenyl)[4-methoxy-2-methyl-5-(1-methylethyl)phenyl]methanone,   165
(4-Hydroxyphenyl)[6-methoxy-2-methyl-3-(1-methylethyl)phenyl]methanone,   166

C$_{18}$H$_{20}$O$_4$

[2,4-Dihydroxy-5-(1,1-dimethylpropyl)phenyl](2-hydroxyphenyl)methanone,   448
(2,4-Dimethoxy-3-propylphenyl)(2-hydroxyphenyl)methanone,   166
[5-(1,1-Dimethylethyl)-2-hydroxy-4-methoxyphenyl](2-hydroxyphenyl)methanone,   410

C$_{18}$H$_{20}$O$_5$

(4-Butoxy-2-hydroxyphenyl)(2-hydroxy-4-methoxyphenyl)methanone,   431
(2,6-Dimethoxyphenyl)(3-hydroxy-6-methoxy-2,4-dimethylphenyl)methanone,   311
(5-Ethoxy-2-hydroxy-4-methoxyphenyl)(3-ethoxyphenyl)methanone,   311

C$_{18}$H$_{20}$O$_6$

(2,4-Dimethoxy-6-methylphenyl)(2-hydroxy-4,6-dimethoxyphenyl)methanone,   312
(4,5-Dimethoxy-2-methylphenyl)(2-hydroxy-4,5-dimethoxyphenyl)methanone,   312
(2,3-Dimethoxyphenyl)[3-hydroxy-2-methoxy-6-(methoxymethyl)phenyl]methanone,   312
(2,4-Dimethoxyphenyl)[3-hydroxy-2-methoxy-6-(methoxymethyl)phenyl]methanone,   312
[2-Hydroxy-4,6-bis(methoxymethyl)-3-methylphenyl]phenylmethanone,   108
(2-Hydroxy-3-methoxy-6-methylphenyl)(2,4,5-trimethoxyphenyl)methanone,   313
(2-Hydroxy-4-methoxy-6-methylphenyl)(2,4,6-trimethoxyphenyl)methanone,   313

C$_{18}$H$_{20}$O$_7$

(2,3-Dimethoxyphenyl)(2-hydroxy-3,4,5-trimethoxyphenyl)methanone,   313
(2,3-Dimethoxyphenyl)(2-hydroxy-3,4,6-trimethoxyphenyl)methanone,   313
(2,3-Dimethoxyphenyl)(6-hydroxy-2,3,4-trimethoxyphenyl)methanone,   314
(2,5-Dimethoxyphenyl)(2-hydroxy-3,4,5-trimethoxyphenyl)methanone,   314
(2,5-Dimethoxyphenyl)(2-hydroxy-3,4,6-trimethoxyphenyl)methanone,   314

$C_{18}H_{21}NO_2$

$C_{18}H_{24}O_4$

$C_{18}H_{26}O_2$

$C_{18}H_{27}NO_2$

$C_{19}H_{11}BrCl_2O_3$

$C_{19}H_{12}D_2O_4$

$C_{19}H_{13}BrO_2$

$C_{19}H_{13}BrO_3$

$C_{19}H_{13}ClO_2$

$C_{19}H_{13}ClO_3$

$C_{19}H_{13}FO_2$

C$_{19}$H$_{13}$NO$_5$

(2,4-Dihydroxyphenyl)(4'-nitro[1,1'-biphenyl]-4-yl)methanone,   475

C$_{19}$H$_{14}$O$_2$

[1,1'-Biphenyl]-4-yl(4-hydroxyphenyl)methanone,   469
(2-Hydroxy[1,1'-biphenyl]-3-yl)phenylmethanone,   470
(4-Hydroxy[1,1'-biphenyl]-3-yl)phenylmethanone,   470
(4'-Hydroxy[1,1'-biphenyl]-4-yl)phenylmethanone,   470
(5-Hydroxy[1,1'-biphenyl]-2-yl)phenylmethanone,   470
(6-Hydroxy[1,1'-biphenyl]-3-yl)phenylmethanone,   471

C$_{19}$H$_{14}$O$_3$

[1,1'-Biphenyl]-4-yl(2,4-dihydroxyphenyl)methanone,   476
(3,5-Dihydroxy[1,1'-biphenyl]-2-yl)phenylmethanone,   476
(4,4'-Dihydroxy[1,1'-biphenyl]-3-yl)phenylmethanone,   476
(2-Hydroxy[1,1'-biphenyl]-3-yl)(2-hydroxyphenyl)methanone,   476
(2-Hydroxy-4-phenoxyphenyl)phenylmethanone,   109
[4-(4-Hydroxyphenoxy)phenyl]phenylmethanone,   514
(2-Hydroxyphenyl)(2-phenoxyphenyl)methanone,   166
(3-Hydroxyphenyl)(4-phenoxyphenyl)methanone,   166
(4-Hydroxyphenyl)(4-phenoxyphenyl)methanone,   167
(4'-Hydroxy[1,1'-biphenyl]-4-yl)(3-hydroxyphenyl)methanone,   477
(4'-Hydroxy[1,1'-biphenyl]-4-yl)(4-hydroxyphenyl)methanone,   477

C$_{19}$H$_{14}$O$_5$S

[3,6-Dihydroxy-2-(phenylsulfonyl)phenyl]phenylmethanone,   360

C$_{19}$H$_{18}$N$_2$O$_7$

[3-(Cyclohexyloxy)-4-hydroxy-5-nitrophenyl](2-nitrophenyl)methanone,   317

C$_{19}$H$_{18}$O$_2$

[3-(1-Hexynyl)-4-hydroxyphenyl]phenylmethanone,   109

C$_{19}$H$_{18}$O$_4$

2-(4-Benzoyl-3-hydroxyphenoxy)cyclohexanone,   515

C$_{19}$H$_{18}$O$_5$

Bis(3-acetyl-2-hydroxy-5-methylphenyl)methanone,   489

C$_{19}$H$_{19}$ClO$_7$

[4-(Acetyloxy)-2-methoxy-6-methylphenyl](3-chloro-2-hydroxy-4,6-dimethoxyphenyl)-
methanone,   317

C$_{19}$H$_{19}$FO$_7$

[4-(Acetyloxy)-2-methoxy-6-methylphenyl](3-fluoro-2-hydroxy-4,6-dimethoxyphenyl)-
methanone,   318

C$_{19}$H$_{20}$Cl$_2$O$_2$

(5-Chloro-3-hexyl-2-hydroxyphenyl)(4-chlorophenyl)methanone,   318

C$_{19}$H$_{20}$O$_2$

C$_{19}$H$_{20}$O$_3$

C$_{19}$H$_{20}$O$_4$

C$_{19}$H$_{20}$O$_8$

C$_{19}$H$_{21}$BrO$_3$

C$_{19}$H$_{21}$ClO$_3$

C$_{19}$H$_{21}$FO$_3$

C$_{19}$H$_{21}$NO$_5$

C$_{19}$H$_{22}$O$_2$

C$_{19}$H$_{22}$O$_3$

C$_{19}$H$_{22}$O$_4$

$C_{19}H_{22}O_6$

$C_{19}H_{22}O_8$

$C_{19}H_{23}NO_2S$

$C_{20}H_{10}Cl_4O_4$

$C_{20}H_{11}F_3O_5$

$C_{20}H_{12}Cl_2N_2O_7$

$C_{20}H_{12}Cl_2O_4$

$C_{20}H_{13}ClO_5$

$C_{20}H_{13}Cl_2FO_3$

$C_{20}H_{13}NO_6$

$C_{20}H_{14}Cl_2O_3$

C$_{20}$H$_{14}$N$_2$O$_7$

[4-Hydroxy-3-nitro-5-(phenylmethoxy)phenyl](2-nitrophenyl)methanone, 324

C$_{20}$H$_{14}$O$_3$

(2-Benzoylphenyl)(2-hydroxyphenyl)methanone, 496
(2-Benzoylphenyl)(3-hydroxyphenyl)methanone, 497
(4-Hydroxy-1,3-phenylene)bis[phenylmethanone, 497
(5-Hydroxy-1,3-phenylene)bis[phenylmethanone, 493

C$_{20}$H$_{14}$O$_4$

(3-Benzoyl-4-hydroxyphenyl)(2-hydroxyphenyl)methanone, 507
(5-Benzoyl-2-hydroxyphenyl)(4-hydroxyphenyl)methanone, 507
[2-(Benzoyloxy)-4-hydroxyphenyl]phenylmethanone, 112
[3-(Benzoyloxy)-2-hydroxyphenyl]phenylmethanone, 112
[3-(Benzoyloxy)-4-hydroxyphenyl]phenylmethanone, 112
[4-(Benzoyloxy)-2-hydroxyphenyl]phenylmethanone, 113
[5-(Benzoyloxy)-2-hydroxyphenyl]phenylmethanone, 113
(2,3-Dihydroxy-1,4-phenylene)bis[phenylmethanone, 499
(2,4-Dihydroxy-1,3-phenylene)bis[phenylmethanone, 507
(2,5-Dihydroxy-1,4-phenylene)bis[phenylmethanone, 508
(3,4-Dihydroxy-1,2-phenylene)bis[phenylmethanone, 508
(4,6-Dihydroxy-1,3-phenylene)bis[phenylmethanone, 499
Phenyl 5-benzoyl-2-hydroxybenzoate, 113
5-Benzoyl-2-hydroxybenzoic acid phenyl ester, 113
1,2-Phenylenebis[(2-hydroxyphenyl)methanone, 500
1,3-Phenylenebis[(4-hydroxyphenyl)methanone, 500
1,4-Phenylenebis[(2-hydroxyphenyl)methanone, 500
1,4-Phenylenebis[(4-hydroxyphenyl)methanone, 501

C$_{20}$H$_{14}$O$_5$

[4-(Benzoyloxy)-2,6-dihydroxyphenyl]phenylmethanone, 361
4-(2-Hydroxybenzoyl)phenyl 2-hydroxybenzoate, 520
2-Hydroxybenzoic acid 4-(2-hydroxybenzoyl)phenyl ester, 520

C$_{20}$H$_{14}$O$_8$

(2,5-Dihydroxy-1,3-phenylene)bis[(2,5-dihydroxyphenyl)methanone, 501

C$_{20}$H$_{15}$ClO$_3$

(4-Chlorophenyl)[2-hydroxy-4-(4-methylphenoxy)phenyl]methanone, 324

C$_{20}$H$_{15}$NO$_5$

[2-Hydroxy-4-[(4-nitrophenyl)methoxy]phenyl]phenylmethanone, 114

C$_{20}$H$_{16}$O$_2$

[1,1'-Biphenyl]-4-yl(2-Hydroxy-5-methylphenyl)methanone, 471
(2'-Hydroxy-5'-methyl[1,1'-biphenyl]-3-yl)phenylmethanone, 471
(2'-Hydroxy-5'-methyl[1,1'-biphenyl]-4-yl)phenylmethanone, 472
(5-Hydroxy-3-methyl[1,1'-biphenyl]-2-yl)phenylmethanone, 472

C$_{20}$H$_{16}$O$_3$

[1,1'-Biphenyl]-4-yl(2-hydroxy-4-methoxyphenyl)methanone, 472
[2,4-Dihydroxy-3-(phenylmethyl)phenyl]phenylmethanone, 361

$C_{20}H_{16}O_4$

$C_{20}H_{16}O_7$

$C_{20}H_{20}O_4$

$C_{20}H_{23}BrO_3$

$C_{20}H_{23}BrO_5$

$C_{20}H_{23}ClO_2$

$C_{20}H_{23}ClO_5$

$C_{20}H_{24}O_2$

$C_{20}H_{24}O_3$

$C_{20}H_{24}O_4$

$C_{20}H_{24}O_6$

$C_{21}H_{12}Cl_4O_4$

C₂₁H₁₄Cl₂O₃

$C_{21}H_{14}Cl_2O_3$

(4-Hydroxy-6-methyl-1,3-phenylene)bis[(2-chlorophenyl)methanone,   497

$C_{21}H_{14}F_2O_4$

(4,6-Dihydroxy-5-methyl-1,3-phenylene)bis[(2-fluorophenyl)methanone,   502

$C_{21}H_{14}O_2$

[4-Hydroxy-3-(phenylethynyl)phenyl]phenylmethanone,   115

$C_{21}H_{16}O_3$

(2-Hydroxy-5-methyl-1,3-phenylene)bis[phenylmethanone,   493

$C_{21}H_{16}O_4$

[4-(Acetyloxy)-2-hydroxyphenyl][1,1'-biphenyl]-4-ylmethanone,   473

$C_{21}H_{16}O_8$

[4-(Benzoylmethoxy)-3,5-dihydroxyphenyl](2,3,4-trihydroxyphenyl)methanone,   467

$C_{21}H_{18}O_2$

[2-Hydroxy-5-(1-phenylethyl)phenyl]phenylmethanone,   115

$C_{21}H_{18}O_3$

[1,1'-Biphenyl]-4-yl(4-ethoxy-2-hydroxyphenyl)methanone,   473
[2,4-Dihydroxy-3-(1-phenylethyl)phenyl]phenylmethanone,   362
[2,4-Dihydroxy-5-(1-phenylethyl)phenyl]phenylmethanone,   362
(2-Hydroxy-4-methoxyphenyl)(5-methyl[1,1'-biphenyl]-2-yl)methanone,   473

$C_{21}H_{18}O_4$

[1,1'-Biphenyl]-4-yl[2-hydroxy-4-(2-hydroxyethoxy)phenyl]methanone,   477
[4,6-Dihydroxy-3-methyl-2-(phenylmethoxy)phenyl]phenylmethanone,   362

$C_{21}H_{18}O_5$

Phenyl[2,3,4-trihydroxy-5-[(2-hydroxy-4-methylphenyl)methyl]phenyl]methanone,   512
Phenyl[2,3,4-trihydroxy-5-[(4-hydroxy-2-methylphenyl)methyl]phenyl]methanone,   512

$C_{21}H_{24}Cl_2O_2$

(2,4-Dichlorophenyl)[2-hydroxy-5-(1,1,3,3-tetramethylbutyl)phenyl]methanone,   326

$C_{21}H_{24}Cl_2O_3$

(3,4-Dichlorophenyl)[2-hydroxy-4-(octyloxy)phenyl]methanone,   326

$C_{21}H_{24}O_7$

[2-Hydroxy-4,5-dimethoxy-3-(2-propenyl)phenyl](2,4,6-trimethoxyphenyl)methanone,   326

$C_{21}H_{25}BrO_2$

[3,5-Bis-(1,1-dimethylethyl)-4-hydroxyphenyl](2-bromophenyl)methanone,   326

[3,5-Bis(1,1-dimethylethyl)-4-hydroxyphenyl](2-hydroxy-5-methoxyphenyl)methanone,   433
(2,4-Dihydroxy-3-methylphenyl)[2-hydroxy-4-(octyloxy)phenyl]methanone,   458
(2,4-Dihydroxyphenyl)(2-hydroxy-5-nonylphenyl)methanone,   450
[2-Hydroxy-4-(octyloxy)phenyl](4-methoxyphenyl)methanone,   329

C$_{23}$H$_{20}$Cl$_2$O$_3$

(3,4-Dichlorophenyl)[4-[4-(1,1-dimethylethyl)phenoxy]-2-hydroxyphenyl]methanone,   329

C$_{23}$H$_{20}$O$_3$

[1,1'-Biphenyl]-4-yl[4-(2-butenyloxy)-2-hydroxyphenyl]methanone,   473
(5-Butyl-2-hydroxy-1,3-phenylene)bis[phenylmethanone,   495

C$_{23}$H$_{20}$O$_5$

(2-Hydroxy-5-methyl-1,3-phenylene)bis[(2-hydroxy-5-methylphenyl)methanone,   503

C$_{23}$H$_{22}$O$_3$

[1,1'-Biphenyl]-4-yl[4-(1,1-dimethylethoxy)-2-hydroxyphenyl]methanone,   473
[4-(4-Butylphenoxy)-2-hydroxyphenyl]phenylmethanone,   120
[4-(1,1-Dimethylethyl)phenyl](2-hydroxy-4-phenoxyphenyl)methanone,   329

C$_{23}$H$_{22}$O$_4$

(3-Butoxyphenyl)(2-hydroxy-4-phenoxyphenyl)methanone,   330
(4-Butoxyphenyl)(2-hydroxy-4-phenoxyphenyl)methanone,   330

C$_{23}$H$_{26}$O$_3$

[2,4-Dihydroxy-3,5-bis(3-methyl-2-butenyl)phenyl]phenylmethanone,   363

C$_{23}$H$_{26}$O$_4$

[2,6-Dihydroxy-3-(3-methyl-2-butenyl)-4-[(3-methyl-2-butenyl)oxy]phenyl]phenylmethanone,   363
[3-(3,7-Dimethyl-2,6-octadienyl)-2,4,6-trihydroxyphenyl]phenylmethanone *(E) (Z)*,   436
[4-[(3,7-Dimethyl-2,6-octadienyl)oxy]-2,6-dihydroxyphenyl]phenylmethanone *(E)*,   363
[2-Hydroxy-3-(3-methyl-2-butenyl)-4-[(3-methyl-2-butenyl)oxy]phenyl](4-hydroxyphenyl)-
methanone,   413
[2-Hydroxy-4-[(3-methyl-2-butenyl)oxy]phenyl][4-[(3-methyl-2-butenyl)oxy]phenyl]-
methanone,   330
Phenyl[2,4,6-trihydroxy-3,5-bis(3-methyl-2-butenyl)phenyl]methanone,   437

C$_{23}$H$_{28}$O$_3$

(4-Ethenylphenyl)[2-hydroxy-4-(octyloxy)phenyl]methanone,   330

C$_{23}$H$_{29}$BrO$_3$

[4-[(10-Bromodecyl)oxy]phenyl](2-hydroxyphenyl)methanone,   168

C$_{23}$H$_{30}$O$_2$

[3,5-Bis(1,1-dimethylethyl)-4-hydroxyphenyl](3-ethylphenyl)methanone,   331
[3,5-Bis(1,1-dimethylethyl)-4-hydroxyphenyl](4-ethylphenyl)methanone,   331

C$_{23}$H$_{30}$O$_3$

[4-(Decyloxy)-2-hydroxyphenyl]phenylmethanone,   121
[2-Hydroxy-4-(isodecyloxy)phenyl]phenylmethanone,   121

C_{23}H_{30}O_3S

[2-Hydroxy-4-[2-(octylthio)ethoxy]phenyl]phenylmethanone, 121

C_{23}H_{30}O_5

[3,5-Bis(1,1-dimethylethyl)-4-hydroxyphenyl](4-hydroxy-3,5-dimethoxyphenyl)methanone, 434

C_{24}H_{22}O_5

(2,4-Diethoxy-6-hydroxy-1,3-phenylene)bis[phenylmethanone, 495
(4,6-Diethoxy-2-hydroxy-1,3-phenylene)bis[phenylmethanone, 495

C_{24}H_{23}ClO_6

[3-Chloro-4,6-dimethoxy-2-(phenylmethoxy)phenyl](4-hydroxy-2-methoxy-6-methylphenyl)-methanone, 331

C_{24}H_{24}O_3

[4-(4-Butylphenoxy)-2-hydroxyphenyl](3-methylphenyl)methanone, 331

C_{24}H_{24}O_4

[4-(4-Butoxyphenoxy)-2-hydroxyphenyl](3-methylphenyl)methanone, 332
(4-Butoxyphenyl)[2-hydroxy-4-(3-methylphenoxy)phenyl]methanone, 332
(4-Butoxyphenyl)[2-hydroxy-4-(4-methylphenoxy)phenyl]methanone, 332

C_{24}H_{28}O_4

[2,4-Dihydroxy-6-methoxy-3,5-bis(3-methyl-2-butenyl)phenyl]phenylmethanone, 364
[4-[(3,7-Dimethyl-2,6-octadienyl)oxy]-2-hydroxy-6-methoxyphenyl]phenylmethanone *(E)*, 121

C_{24}H_{31}BrO_3

[4-[(11-Bromoundecyl)oxy]phenyl](2-hydroxyphenyl)methanone, 168

C_{24}H_{32}O_3

[2-Hydroxy-4-(undecyloxy)phenyl]phenylmethanone, 122

C_{24}H_{32}O_5

[3,5-Bis(1,1-dimethylethyl)-4-hydroxyphenyl](3,4,5-trimethoxyphenyl)methanone, 332

C_{25}H_{11}D_7O_4

[3-Hydroxy-5-(phenoxy-*d5*)phenyl][4-(phenoxy-3,5-*d2*)phenyl]methanone, 332

C_{25}H_{17}NO_5

[1,1'-Biphenyl]-4-yl[2-hydroxy-4-(4-nitrophenoxy)phenyl]methanone, 474

C_{25}H_{18}O_2

[1,1'-Biphenyl]-4-yl(4-hydroxy[1,1'-biphenyl]-3-yl)methanone, 474

C_{25}H_{18}O_3

(2,4-Dihydroxyphenyl)[1,1';4',1"]terphenyl-4"-yl-methanone, 478

C$_{25}$H$_{18}$O$_5$

Bis[4-(4-hydroxyphenoxy)phenyl]methanone,   515

C$_{25}$H$_{19}$NO$_3$

[4-(4-Aminophenoxy)-2-hydroxyphenyl][1,1'-biphenyl]-4-ylmethanone,   474

C$_{25}$H$_{24}$O$_3$

[1,1'-Biphenyl]-4-yl[4-(cyclohexyloxy)-2-hydroxyphenyl]methanone,   474

C$_{25}$H$_{26}$O$_3$

[1,1'-Biphenyl]-4-yl[4-(hexyloxy)-2-hydroxyphenyl]methanone,   474
[4-(1,1-Dimethylethyl)phenyl][4-(3,4-dimethylphenoxy)-2-hydroxyphenyl]methanone,   333

C$_{25}$H$_{33}$BrO$_3$

[4-[(12-Bromododecyl)oxy]phenyl](2-hydroxyphenyl)methanone,   169

C$_{25}$H$_{33}$NO$_4$

(5-Dodecyl-2-hydroxy-3-nitrophenyl)phenylmethanone,   122
(4-Dodecylphenyl)(2-hydroxy-3-nitrophenyl)methanone,   333
(4-Dodecylphenyl)(2-hydroxy-5-nitrophenyl)methanone,   333

C$_{25}$H$_{34}$O$_2$

(5-Dodecyl-2-hydroxyphenyl)phenylmethanone,   122
(4-Dodecylphenyl)(2-hydroxyphenyl)methanone,   169

C$_{25}$H$_{34}$O$_3$

[3,5-Bis(1,1-dimethylethyl)-2-hydroxyphenyl][5-(1,1-dimethylethyl)-2-hydroxyphenyl]-
methanone,   434
[5-(1,1-Dimethylethyl)-2-hydroxy-4-(octyloxy)phenyl]phenylmethanone,   122
[4-(1,1-Dimethylethyl)phenyl][2-hydroxy-4-(octyloxy)phenyl]methanone,   333
[4-(Dodecyloxy)-2-hydroxyphenyl]phenylmethanone,   122

C$_{25}$H$_{34}$O$_4$

[2,4-Dihydroxy-5-(dodecyloxy)phenyl]methanone,   364

C$_{26}$H$_{14}$Cl$_4$O$_4$

(4,4'-Dihydroxy[1,1'-biphenyl]-3,3'-diyl)bis[(2,4-dichlorophenyl)methanone,   509

C$_{26}$H$_{14}$Cl$_4$O$_6$S

[Sulfonylbis(6-hydroxy-3,1-phenylene)]bis[(2,4-dichlorophenyl)methanone,   517

C$_{26}$H$_{16}$F$_2$O$_4$

(4,4'-Dihydroxy[1,1'-biphenyl]-2,2'-diyl)bis[(3-fluorophenyl)methanone,   509
(4,4'-Dihydroxy[1,1'-biphenyl]-2,2'-diyl)bis[(4-fluorophenyl)methanone,   509

C$_{26}$H$_{18}$O$_4$

[1,1'-Biphenyl]-4,4'-diylbis[(4-hydroxyphenyl)methanone,   510
(4,4'-Dihydroxy[1,1'-biphenyl]-2,2'-diyl)bis[phenylmethanone,   510

(4,4'-Dihydroxy[1,1'-biphenyl]-3,3'-diyl)bis[phenylmethanone,   510

C₂₆H₁₈O₅

(Oxydi-4,1-phenylene)bis(4-hydroxyphenyl)methanone,   515

C₂₆H₁₈O₈S

[Sulfonylbis(4,6-dihydroxy-3,1-phenylene)]bis[phenylmethanone,   517

C₂₆H₂₀O₃

(2-Hydroxy-4-methoxyphenyl)[1,1',4',1'']terphenyl-4''-yl-methanone,   478

C₂₆H₂₆O₃

[4-(Dodecyloxy)-2-hydroxyphenyl](4-methylphenyl)methanone,   334

C₂₆H₂₆O₄

(2,4-Dihydroxy-5-hexyl-1,3-phenylene)bis[phenylmethanone,   508

C₂₆H₂₈O₃

[1,1'-Biphenyl]-4-yl[2-hydroxy-4-(1-propylbutoxy)phenyl]methanone,   475

C₂₆H₃₅NO₄

(4-Dodecylphenyl)(2-hydroxy-3-methyl-5-nitrophenyl)methanone,   334

C₂₆H₃₆O₂

(4-Dodecylphenyl)(2-hydroxy-3-methylphenyl)methanone,   334

C₂₆H₃₆O₃

(3-Dodecyl-2-hydroxy-5-methylphenyl)(2-hydroxyphenyl)methanone,   413

C₂₇H₁₄Cl₄O₇

Bis[5-chloro-3-(5-chloro-2-hydroxybenzoyl)-2-hydroxyphenyl]methanone,   503

C₂₇H₁₄F₄O₇

Bis[5-fluoro-3-(5-fluoro-2-hydroxybenzoyl)-2-hydroxyphenyl]methanone,   504

C₂₇H₁₈O₅

(2,4-Dihydroxy-1,3,5-benzenetriyl)tris[phenylmethanone,   504

C₂₇H₁₈O₆

(2,4,6-Trihydroxy-1,3,5-benzenetriyl)tris[phenylmethanone,   504

C₂₇H₂₂O₃

[2-Hydroxy-4-(phenylmethoxy)-3-(phenylmethyl)phenyl]phenylmethanone,   123

C₂₇H₃₀O₃

[1,1'-Biphenyl]-4-yl[2-hydroxy-4-(octyloxy)phenyl]methanone,   475

[[4-(1,1-Dimethylethyl)phenoxy]-2-hydroxyphenyl][4-(1,1-dimethylethyl)phenyl]methanone,   334

C27H30O4

(2-Hydroxy-4-phenoxyphenyl)[3-(octyloxy)phenyl]methanone,   335
(2-Hydroxy-4-phenoxyphenyl)[4-(octyloxy)phenyl]methanone,   335

C27H30O5

[4-(4-Butoxyphenoxy)-2-hydroxyphenyl](4-butoxyphenyl)methanone,   335

C27H36O3

[4-(Dodecyloxy)-2-hydroxyphenyl](4-ethenylphenyl)methanone,   335

C27H38O4

[4-(Dodecyloxy)-2-hydroxyphenyl](4-hydroxy-3,5-dimethylphenyl)methanone,   434

C28H22O4

[1,2-Ethanediylbis(6-hydroxy-3,1-phenylene)]bis[phenylmethanone,   513

C28H22O6

(4,4'-Dihydroxy-6,6'-dimethoxy[1,1'-biphenyl]-3,3'-diyl)bis[phenylmethanone,   510

C28H22O7S

[Sulfinylbis(6-hydroxy-4-methoxy-3,1-phenylene)]bis[phenylmethanone,   516

C28H22O8S

[Sulfonylbis(6-hydroxy-4-methoxy-3,1-phenylene)]bis[phenylmethanone,   518

C28H28O11

[5-(Acetyloxy)-2-hydroxy-4,6-dimethoxy-1,3-phenylene]bis[(2,5-dimethoxyphenyl)-
methanone,   496

C28H32O3

[2-Hydroxy-4-(4-nonylphenoxy)phenyl]phenylmethanone,   123

C28H32O4

[2-Hydroxy-4-(3-methylphenoxy)phenyl][4-(octyloxy)phenyl]methanone,   335
[2-Hydroxy-4-(4-methylphenoxy)phenyl][4-(octyloxy)phenyl]methanone,   336
[2-Hydroxy-4-[4-(octyloxy)phenoxy]phenyl](3-methylphenyl)methanone,   336

C28H34O3

[2-Hydroxy-3,5-bis(3-methyl-2-butenyl)-4-[(3-methyl-2-butenyl)oxy]phenyl]phenyl-
methanone,   123

C28H34O4

[2-Hydroxy-3-(3-methyl-2-butenyl)-4,6-bis[(3-methyl-2-butenyl)oxy]phenyl]phenyl-
methanone,   123

C$_{29}$H$_{20}$Br$_2$O$_5$

C$_{29}$H$_{22}$O$_5$

C$_{29}$H$_{24}$O$_4$

C$_{29}$H$_{24}$O$_6$

C$_{29}$H$_{26}$O$_2$

C$_{29}$H$_{34}$O$_3$

C$_{29}$H$_{42}$O$_3$

C$_{29}$H$_{42}$O$_4$

C$_{30}$H$_{26}$O$_8$

C$_{30}$H$_{26}$O$_8$S

C$_{30}$H$_{36}$O$_3$

C$_{31}$H$_{26}$O$_5$

C$_{31}$H$_{38}$O$_4$

[4-(Dodecyloxy)phenyl](2-hydroxy-4-phenoxyphenyl)methanone,   337
[2-Hydroxy-4-(2,4,6-trimethylphenoxy)phenyl][4-(isononyloxy)phenyl]methanone,   337

$C_{31}H_{44}O_4$

[2-Hydroxy-4-(octadecanoyloxy)phenyl]phenylmethanone,   125

$C_{31}H_{46}O_3$

(2,4-Dihydroxy-5-octadecylphenyl)phenylmethanone,   364
[2-Hydroxy-4-(octadecyloxy)phenyl]phenylmethanone,   125

$C_{32}H_{40}O_3$

[4-(4-Dodecylphenoxy)-2-hydroxyphenyl](3-methylphenyl)methanone,   337

$C_{32}H_{40}O_4$

[4-(Dodecyloxy)phenyl][2-hydroxy-4-(4-methylphenoxy)phenyl]methanone,   337

$C_{32}H_{48}O_3$

[2-Hydroxy-4-(nonadecyloxy)phenyl]phenylmethanone,   125

$C_{32}H_{48}O_4$

[2-Hydroxy-4-(octadecyloxy)phenyl](4-methoxyphenyl)methanone,   337

$C_{33}H_{42}O_3$

[4-[4-(1,1-Dimethylethyl)-2,6-dimethylphenoxy]-2-hydroxyphenyl][4-(1,1,3,3-tetramethylbutyl)-phenyl]methanone,   338

$C_{33}H_{50}O_5$

(2-Hydroxy-4-methoxy-6-methylphenyl)[2-hydroxy-4-(octadecyloxy)phenyl]methanone,   435

$C_{34}H_{34}O_4$

(4,4'-Dihydroxy[1,1'-biphenyl]-2,2'-diyl)bis[[4-(1,1-dimethylethyl)phenyl]methanone,   511

$C_{35}H_{46}O_3$

[1,1'-Biphenyl]-4-yl[4-(hexadecyloxy)-2-hydroxyphenyl]methanone,   475

$C_{35}H_{46}O_5$

[2-Hydroxy-4-[4-(octyloxy)phenoxy]phenyl][4-(octyloxy)phenyl]methanone,   338

$C_{36}H_{34}O_8S$

[Sulfonylbis[4-(cyclopentyloxy)-6-hydroxy-3,1-phenylene]]bis[phenylmethanone,   518

$C_{36}H_{46}O_4$

1,4-Phenylenebis[2-hydroxy-5-(1,1,3,3-tetramethylbutyl)phenyl]methanone,   505

$C_{36}H_{46}O_6$

1,4-Phenylenebis[2-hydroxy-4-(octyloxy)phenyl]methanone,   506

CHEMICAL ABSTRACTS REGISTRY NUMBERS

[82-64-4]   (2,4-Dihydroxy-1,3-phenylene)bis[phenylmethanone, 507
[82-67-7]   (2,4-Dihydroxy-1,3,5-benzenetriyl)tris[phenylmethanone, 504
[82-69-9]   (2,6-Dihydroxyphenyl)(2-hydroxyphenyl)methanone, 21
[85-19-8]   (5-Chloro-2-hydroxyphenyl)phenylmethanone, 44
[85-24-5]   [2-Hydroxy-4-(octyloxy)phenyl](2-hydroxyphenyl)methanone, 412
[85-28-9]   (4-Chlorophenyl)(2-hydroxy-4-methoxyphenyl)methanone, 220
[117-99-7] (2-Hydroxyphenyl)phenylmethanone, 3
[131-53-3] (2-Hydroxy-4-methoxyphenyl)(2-hydroxyphenyl)methanone, 405
[131-54-4] Bis(2-hydroxy-4-methoxyphenyl)methanone, 419
[131-55-5] Bis(2,4-dihydroxyphenyl)methanone, 22
[131-56-6] (2,4-Dihydroxyphenyl)phenylmethanone, 9
[131-57-7] (2-Hydroxy-4-methoxyphenyl)phenylmethanone, 67
[134-92-9] (4-Hydroxyphenyl)(4-methylphenyl)methanone, 150
[342-18-7] (5-Fluoro-2-hydroxyphenyl)(3-methylphenyl)methanone, 221
[362-47-0] (5-Fluoro-2-hydroxyphenyl)phenylmethanone, 46
[365-14-0] (3-Fluoro-4-hydroxyphenyl)phenylmethanone, 46
[479-21-0] (2,6-Dihydroxy-4-methoxyphenyl)phenylmethanone, 353
[519-34-6] (3,4-Dihydroxyphenyl)(2,4,6-trihydroxyphenyl)methanone, 31
[579-15-7] (3-Bromo-5-fluoro-4-hydroxyphenyl)phenylmethanone, 35
[606-12-2] (2-Hydroxyphenyl)(4-hydroxyphenyl)methanone, 15
[611-80-3] Bis(3-hydroxyphenyl)methanone, 13
[611-81-4] (3-Hydroxyphenyl)(4-hydroxyphenyl)methanone, 16
[611-99-4] Bis(4-hydroxyphenyl)methanone, 13
[727-93-5] (5-Fluoro-2-hydroxyphenyl)(4-methoxyphenyl)methanone, 222
[732-55-8] (4-Hydroxyphenyl)[3-(trifluoromethyl)phenyl]methanone, 145
[738-15-8] [4-Hydroxy-3,5-bis(1-methylethyl)phenyl]phenylmethanone, 111
[784-41-8] (2-Amino-5-chlorophenyl)(4-hydroxyphenyl)methanone, 142
[835-11-0] Bis(2-hydroxyphenyl)methanone, 13
[837-60-5] (2,4-Dihydroxyphenyl)(3-hydroxyphenyl)methanone, 19
[1137-42-4] (4-Hydroxyphenyl)phenylmethanone, 6
[1143-72-2] Phenyl(2,3,4-trihydroxyphenyl)methanone, 17
[1470-57-1] (2-Hydroxy-5-methylphenyl)phenylmethanone, 63
[1470-79-7] (2,4-Dihydroxyphenyl)(4-hydroxyphenyl)methanone, 20
[1641-17-4] (2-Hydroxy-4-methoxyphenyl)(4-methylphenyl)methanone, 260
[1818-24-2] (2,4,6-Trihydroxy-1,3,5-benzenetriyl)tris[phenylmethanone, 504
[1834-88-4] (2-Hydroxy-4-nitrophenyl)phenylmethanone, 47
[1834-89-5] (2-Hydroxy-4-nitrophenyl)(3-nitrophenyl)methanone, 190
[1843-05-6] [2-Hydroxy-4-(octyloxy)phenyl]phenylmethanone, 117
[2004-55-9] (4-Hydroxyphenyl)(2,4,6-trimethylphenyl)methanone, 162
[2038-92-8] (4-Bromophenyl)(2-hydroxyphenyl)methanone, 132
[2050-37-5] (2,5-Dihydroxyphenyl)phenylmethanone, 11
[2151-17-9] (3-Chloro-2-hydroxy-4,6-dimethoxyphenyl)(4-hydroxy-2-methoxy-6-methyl-
            phenyl)methanone, 429
[2162-63-2] [4-(Decyloxy)-2-hydroxyphenyl]phenylmethanone, 121
[2341-94-8] (2-Chlorophenyl)(5-fluoro-2-hydroxyphenyl)methanone, 182
[2549-87-3] [2-Hydroxy-4-(2-propenyloxy)phenyl]phenylmethanone, 89
[2549-90-8] [4-[(2-Ethylhexyl)oxy]-2-hydroxyphenyl]phenylmethanone, 118
[2559-64-0] (5-Fluoro-2-hydroxyphenyl)(4-fluorophenyl)methanone, 189
[2589-80-2] 1-(5-Benzoyl-2-hydroxyphenyl)ethanone, 487
[2589-81-3] (4-Hydroxy-1,3-phenylene)bis[phenylmethanone, 497
[2898-51-3] (5-Hydroxy-4-methoxy-2-nitrophenyl)(4-methoxyphenyl)methanone, 256
[2929-45-5] (2-Hydroxy-4,6-dimethylphenyl)phenylmethanone, 80
[2985-59-3] [4-(Dodecyloxy)-2-hydroxyphenyl]phenylmethanone, 122
[2985-79-7] (4-Chlorophenyl)(2-hydroxyphenyl)methanone, 135
[2985-80-0] (4-Chloro-2-hydroxyphenyl)phenylmethanone, 44
[3088-11-7] (2-Hydroxy-4-propoxyphenyl)phenylmethanone, 92
[3088-15-1] (4,6-Dihydroxy-1,3-phenylene)bis[phenylmethanone, 499
[3097-56-1] (5-Cyclohexyl-2-hydroxyphenyl)phenylmethanone, 110

[3098-18-8] (2-Hydroxy-4-methylphenyl)phenylmethanone, 62
[3119-86-6] (2-Hydroxy-4-methoxyphenyl)[2-(trifluoromethyl)phenyl]methanone, 237
[3119-88-8] (2-Fluorophenyl)(2-hydroxy-4-methoxyphenyl)methanone, 223
[3132-42-1] (5-Ethyl-2-hydroxyphenyl)phenylmethanone, 77
[3286-88-2] (4-Bromophenyl)(2,4-dihydroxyphenyl)methanone, 366
[3286-91-7] (5-Chloro-2-hydroxy-4-methoxyphenyl)phenylmethanone, 59
[3286-93-9] (5-Bromo-2-hydroxy-4-methoxyphenyl)phenylmethanone, 56
[3286-95-1] (5-Chloro-2,4-dihydroxyphenyl)phenylmethanone, 344
[3286-96-2] (3,5-Dibromo-2,4-dihydroxyphenyl)phenylmethanone, 340
[3286-97-3] (5-Bromo-2,4-dihydroxyphenyl)phenylmethanone, 343
[3293-97-8] [4-(Hexyloxy)-2-hydroxyphenyl]phenylmethanone, 111
[3333-96-8] (4-Amino-2-hydroxyphenyl)phenylmethanone, 52
[3457-13-4] [2-Hydroxy-4-(octadecyloxy)phenyl]phenylmethanone, 125
[3457-17-8] [4-(Hexadecyloxy)-2-hydroxyphenyl]phenylmethanone, 124
[3550-43-4] [4-(Heptyloxy)-2-hydroxyphenyl]phenylmethanone, 115
[3555-86-0] Phenyl(2,4,6-trihydroxyphenyl)methanone, 18
[3558-83-6] [1,1'-Biphenyl]-4-yl(4-hydroxyphenyl)methanone, 469
[3602-47-9] (4-Fluorophenyl)(2-hydroxy-4-methoxyphenyl)methanone, 224
[3733-72-0] (2,6-Dihydroxy-4-methoxyphenyl)(4-hydroxy-2-methoxy-6-methylphenyl)-
             methanone, 456
[3811-00-5] (3-Chloro-2,6-dihydroxy-4-methoxyphenyl)(4-hydroxy-2-methoxy-6-methyl-
             phenyl)methanone, 454
[4072-08-6] (2-Hydroxy-3-methylphenyl)phenylmethanone, 61
[4072-14-4] (2-Hydroxy-4,5-dimethylphenyl)phenylmethanone, 79
[4072-16-6] [2-Hydroxy-6-methyl-3-(1-methylethyl)phenyl]phenylmethanone, 101
[4072-17-7] (2-Hydroxy-3,6-dimethylphenyl)phenylmethanone, 78
[4072-22-4] (2-Hydroxy-3-methyl-4-nitrophenyl)phenylmethanone, 59
[4072-24-6] (2-Hydroxy-5-methyl-4-nitrophenyl)phenylmethanone, 60
[4072-26-8] (2-Hydroxy-5-methyl-3-nitrophenyl)phenylmethanone, 60
[4084-62-2] 1,4-Phenylenebis[(2-hydroxy-5-methylphenyl)methanone, 502
[4090-99-7] [2-Hydroxy-5-(1,1,3,3-tetramethylbutyl)phenyl]phenylmethanone, 117
[4142-51-2] (2,4-Dimethoxyphenyl)(2-hydroxy-4-methoxyphenyl)methanone, 280
[4211-67-0] [5-(1,1-Dimethylethyl)-2,4-dihydroxyphenyl]phenylmethanone, 358
[4369-50-0] (4-Bromophenyl)(4-hydroxyphenyl)methanone, 132
[4520-99-4] (2,4-Dihydroxyphenyl)(4-hydroxy-2-methylphenyl)methanone, 445
[4646-78-0] (2,5-Dihydroxy-3,4-dimethoxyphenyl)phenylmethanone, 356
[4650-75-3] (2-Hydroxy-4,6-dimethoxyphenyl)(4-hydroxy-2-methylphenyl)methanone, 427
[4998-51-0] [2-Hydroxy-5-(octyloxy)phenyl]phenylmethanone, 118
[5298-27-1] (2,4-Dihydroxyphenyl)(4-methoxyphenyl)methanone, 377
[5326-42-1] (4-Hydroxy-3-methylphenyl)phenylmethanone, 66
[5336-56-1] (4-Hydroxy-3,5-dimethylphenyl)phenylmethanone, 81
[5423-21-2] Bis(3-bromo-4-hydroxyphenyl)methanone, 415
[5436-05-5] 1,3-Phenylenebis[(4-hydroxyphenyl)methanone, 500
[5464-98-2] (4-Hydroxy-3-nitrophenyl)phenylmethanone, 49
[5623-44-9] Bis(4-hydroxy-3-methoxyphenyl)methanone, 420
[5623-46-1] (4'-Hydroxy[1,1'-biphenyl]-4-yl)phenylmethanone, 470
[6079-76-1] [2-Hydroxy-4-(phenylmethoxy)phenyl]phenylmethanone, 114
[6131-38-0] (2-Hydroxy-4-methoxyphenyl)(4-methoxyphenyl)methanone, 264
[6131-39-1] (2-Hydroxy-4-propoxyphenyl)(4-propoxyphenyl)methanone, 319
[6178-89-8] Bis(5-chloro-2-hydroxyphenyl)methanone, 416
[6279-04-5] (4-Chlorophenyl)(2-hydroxy-3-methylphenyl)methanone, 216
[6279-05-6] (4-Chlorophenyl)(2-hydroxy-5-methylphenyl)methanone, 216
[6279-06-7] (4-Chlorophenyl)(4-hydroxy-3-methylphenyl)methanone, 218
[6280-52-0] (2-Chlorophenyl)(2-hydroxy-5-methylphenyl)methanone, 214
[6280-54-2] (3-Chlorophenyl)(2-hydroxy-5-methylphenyl)methanone, 215
[6343-00-6] (2-Hydroxy-3-methoxyphenyl)(2,4,6-trimethoxyphenyl)methanone, 303
[6721-06-8] (4-Bromo-2-hydroxy-3,6-dimethylphenyl)phenylmethanone, 74
[6723-04-2] (4-Bromo-2-hydroxyphenyl)phenylmethanone, 42
[6723-07-5] (4-Bromo-2-hydroxy-5-methylphenyl)phenylmethanone, 55
[6723-09-7] (3-Bromo-2-hydroxy-5-methylphenyl)phenylmethanone, 55
[6723-13-3] (5-Bromo-2-hydroxy-3-methylphenyl)phenylmethanone, 56
[6758-89-0] (4-Bromo-2-hydroxy-3-methylphenyl)phenylmethanone, 55

[35042-49-0]   (2,3-Dimethoxyphenyl)(4-hydroxy-3-methoxyphenyl)methanone, 280
[35042-50-3]   Bis(2,3-dihydroxyphenyl)methanone, 22
[35486-63-6]   (3-Aminophenyl)(3-hydroxy-5-methylphenyl)methanone, 233
[35486-64-7]   (3-Aminophenyl)(2-hydroxyphenyl)methanone, 143
[35582-86-6]   (3-Chloro-2-hydroxyphenyl)phenylmethanone, 43
[35697-92-8]   [2-Hydroxy-4-(3-methylphenoxy)phenyl][4-(octyloxy)phenyl]methanone, 335
[35697-93-9]   [4-(4-Butoxyphenoxy)-2-hydroxyphenyl](4-butoxyphenyl)methanone, 335
[35697-94-0]   [2-Hydroxy-4-[4-(octyloxy)phenoxy]phenyl][4-(octyloxy)phenyl]methanone, 338
[35697-95-1]   [4-(4-Dodecylphenoxy)-2-hydroxyphenyl](4-dodecylphenyl)methanone, 338
[35697-96-2]   [4-(4-Butoxyphenoxy)-2-hydroxyphenyl](3-methylphenyl)methanone, 332
[35697-97-3]   [2-Hydroxy-4-[4-(octyloxy)phenoxy]phenyl](3-methylphenyl)methanone, 336
[35697-98-4]   [4-(4-Dodecylphenoxy)-2-hydroxyphenyl](3-methylphenyl)methanone, 337
[35697-99-5]   [4-(1,1-Dimethylethyl)phenyl][4-(3,4-dimethylphenoxy)-2-hydroxyphenyl]-
               methanone, 333
[35698-00-1]   [4-[4-(1,1-Dimethylethyl)-2,6-dimethylphenoxy]-2-hydroxyphenyl]-
               [4-(1,1,3,3-tetramethylbutyl)phenyl]methanone, 338
[35698-01-2]   (3,4-Dimethylphenyl)[2-hydroxy-4-(4-nonylphenoxy)phenyl]methanone, 336
[35698-02-3]   (3,4-Dichlorophenyl)[4-[4-(1,1-dimethylethyl)phenoxy]-2-hydroxyphenyl]-
               methanone, 329
[35698-03-4]   [4-(4-Bromophenoxy)-2-hydroxyphenyl](3,4-dichlorophenyl)methanone, 317
[35698-04-5]   [Sulfonylbis(6-hydroxy-4-methoxy-3,1-phenylene)]bis[phenylmethanone, 518
[35698-05-6]   [Sulfonylbis[6-hydroxy-4-(octyloxy)-3,1-phenylene]]bis[phenylmethanone, 519
[35698-06-7]   [Sulfonylbis(6-hydroxy-4-methoxy-3,1-phenylene)]bis[(2-methylphenyl)-
               methanone, 518
[35698-16-9]   (5-Dodecyl-2-hydroxyphenyl)phenylmethanone, 122
[35698-17-0]   (5-Dodecyl-2-hydroxy-3-nitrophenyl)phenylmethanone, 122
[35698-18-1]   (2-Hydroxy-3-nitrophenyl)[4-(1-methylethyl)phenyl]methanone, 272
[35698-19-2]   (2-Hydroxy-5-nitrophenyl)[4-(1-methylethyl)phenyl]methanone, 272
[35698-22-7]   (4-Dodecylphenyl)(2-hydroxyphenyl)methanone, 169
[35698-23-8]   (4-Dodecylphenyl)(2-hydroxy-5-nitrophenyl)methanone, 333
[35698-24-9]   (4-Dodecylphenyl)(2-hydroxy-3-nitrophenyl)methanone, 333
[35698-28-3]   (4-Dodecylphenyl)(2-hydroxy-3-methylphenyl)methanone, 334
[35698-29-4]   (4-Dodecylphenyl)(2-hydroxy-3-methyl-5-nitrophenyl)methanone, 334
[35698-39-6]   (2-Hydroxy-4-phenoxyphenyl)phenylmethanone, 109
[35698-40-9]   (4-Chlorophenyl)(2-hydroxy-4-phenoxyphenyl)methanone, 317
[35698-42-1]   [4-(1,1-Dimethylethyl)phenyl](2-hydroxy-4-phenoxyphenyl)methanone, 329
[35698-44-3]   [[4-(1,1-Dimethylethyl)phenoxy]-2-hydroxyphenyl][4-(1,1-dimethylethyl)-
               phenyl]methanone, 334
[35698-46-5]   [2-Hydroxy-4-(methylphenoxy)phenyl]phenylmethanone, 114
[35698-48-7]   (4-Chlorophenyl)[2-hydroxy-4-(4-methylphenoxy)phenyl]methanone, 324
[35698-49-8]   [4-(4-Butylphenoxy)-2-hydroxyphenyl]phenylmethanone, 120
[35698-50-1]   [2-Hydroxy-4-(4-nonylphenoxy)phenyl]phenylmethanone, 123
[35698-51-2]   [4-(3-Bromophenoxy)-2-hydroxyphenyl]phenylmethanone, 109
[35698-52-3]   (4-Butoxyphenyl)(2-hydroxy-4-phenoxyphenyl)methanone, 330
[35698-53-4]   (2-Hydroxy-4-phenoxyphenyl)[4-(octyloxy)phenyl]methanone, 335
[35698-54-5]   [4-(Dodecyloxy)phenyl](2-hydroxy-4-phenoxyphenyl)methanone, 337
[35698-55-6]   (3-Butoxyphenyl)(2-hydroxy-4-phenoxyphenyl)methanone, 330
[35698-56-7]   (2-Hydroxy-4-phenoxyphenyl)[3-(octyloxy)phenyl]methanone, 335
[35698-57-8]   [3-(Dodecyloxy)phenyl](2-hydroxy-4-phenoxyphenyl)methanone, 336
[35698-58-9]   [4-(4-Butylphenoxy)-2-hydroxyphenyl](3-methylphenyl)methanone, 331
[35698-59-0]   (4-Butoxyphenyl)[2-hydroxy-4-(4-methylphenoxy)phenyl]methanone, 332
[35698-60-3]   [2-Hydroxy-4-(4-methylphenoxy)phenyl][4-(octyloxy)phenyl]methanone, 336
[35698-61-4]   [4-(Dodecyloxy)phenyl][2-hydroxy-4-(4-methylphenoxy)phenyl]methanone, 337
[35698-62-5]   (4-Butoxyphenyl)[2-hydroxy-4-(3-methylphenoxy)phenyl]methanone, 332
[35836-41-0]   (2,3-Dihydroxy-4-methoxyphenyl)phenylmethanone, 351
[35839-45-3]   (2-Hydroxyphenyl)[4-(1-methylethyl)phenyl]methanone, 161
[35839-46-4]   [Sulfinylbis(6-hydroxy-4-methoxy-3,1-phenylene)]bis[phenylmethanone, 516
[35839-47-5]   [Sulfinylbis[6-hydroxy-4-(octyloxy)-3,1-phenylene]]bis[phenylmethanone, 517
[35839-48-6]   [Sulfonylbis(4,6-dihydroxy-3,1-phenylene)]bis[phenylmethanone, 517
[36118-66-8]   [2-Hydroxy-4-(2,4,6-trimethylphenoxy)phenyl][4-(isononyloxy)phenyl]-
               methanone, 337
[36130-57-1]   [1,1'-Biphenyl]-4-yl(2,4-dihydroxyphenyl)methanone, 476

[41796-26-3] (3-Chloro-2-hydroxyphenyl)(3-chlorophenyl)methanone, 186
[42019-78-3] (4-Chlorophenyl)(4-hydroxyphenyl)methanone, 135
[42045-60-3] (3,4-Dimethoxyphenyl)(2-hydroxy-4-methoxyphenyl)methanone, 283
[42045-61-4] (4,4'-Dihydroxy-6,6'-dimethoxy[1,1'-biphenyl]-3,3'-diyl)bis[(4-methoxyphenyl)-
              methanone, 511
[42045-62-5] (4,4'-Dihydroxy-6,6'-dimethoxy[1,1'-biphenyl]-3,3'-diyl)bis[phenylmethanone, 510
[42045-63-6] (2,5-Dihydroxy-4-methoxyphenyl)(4-methoxyphenyl)methanone, 393
[42204-63-7] (2-Hydroxyphenyl)(2,3,4-trihydroxyphenyl)methanone, 26
[42404-41-1] (3-Amino-4-hydroxyphenyl)phenylmethanone, 52
[42470-88-2] (2,4-Dihydroxyphenyl)(5-hydroxy-2-methylphenyl)methanone, 445
[42470-91-7] (3,6-Dihydroxy-2,4-dimethylphenyl)(2,4-dihydroxyphenyl)methanone, 461
[42594-58-1] (2,6-Dimethoxyphenyl)(3-hydroxy-6-methoxy-2,4-dimethylphenyl)
              methanone, 311
[42594-59-2] (3,6-Dihydroxy-2,4-dimethylphenyl)(2-hydroxy-6-methoxyphenyl)
              methanone, 454
[42594-60-5] (3,6-Dihydroxy-2,4-dimethylphenyl)(2,6-dihydroxyphenyl)methanone, 462
[42832-64-4] (2-Hydroxy-5-methoxyphenyl)(2,4,6-trimethoxyphenyl)methanone, 304
[42833-48-7] (2-Hydroxy-4,5-dimethoxyphenyl)(2-methoxyphenyl)methanone, 284
[42833-51-2] (2-Hydroxy-5-methoxyphenyl)(2-methoxyphenyl)methanone, 264
[42833-53-4] (2,6-Dimethoxyphenyl)(2-hydroxy-4,5-dimethoxyphenyl)methanone, 302
[42833-55-6] (2,6-Dimethoxyphenyl)(2-hydroxy-3,4-dimethoxyphenyl)methanone, 302
[42833-59-0] (2,5-Dimethoxyphenyl)(2-hydroxy-4,6-dimethoxyphenyl)methanone, 301
[42833-60-3] (2-Hydroxy-3,4,5-trimethoxyphenyl)(2-methoxyphenyl)methanone, 305
[42833-67-0] (2-Hydroxy-3,4-dimethoxyphenyl)(2,4,6-trimethoxyphenyl)methanone, 315
[42833-68-1] (2-Hydroxy-4,5-dimethoxyphenyl)(2,4,6-trimethoxyphenyl)methanone, 316
[42833-83-0] (2,3-Dimethoxyphenyl)(2-hydroxy-3,4,5-trimethoxyphenyl)methanone, 313
[42833-85-2] (2-Hydroxy-3,4,5-trimethoxyphenyl)(2,4,6-trimethoxyphenyl)methanone, 322
[42833-88-5] (2-Hydroxy-3,4,5-trimethoxyphenyl)phenylmethanone, 94
[42833-89-6] (2,4-Dihydroxy-3,5-dimethoxyphenyl)phenylmethanone, 356
[42833-90-9] (2,5-Dihydroxy-4-methoxyphenyl)(2-methoxyphenyl)methanone, 392
[42833-95-5] [2-Hydroxy-4,5-dimethoxy-3-(2-propenyl)phenyl](2,4,6-trimethoxyphenyl)-
              methanone, 326
[43221-40-5] (2,4-Dihydroxy-6-methylphenyl)phenylmethanone, 350
[43221-41-6] [2,4-Dihydroxy-5-(1-phenylethyl)phenyl]phenylmethanone, 362
[46795-43-1] (2,4-Dichlorophenyl)(2-hydroxyphenyl)methanone, 128
[46795-44-2] (2,4-Difluorophenyl)(2-hydroxyphenyl)methanone, 130
[46863-20-1] (2-Hydroxyphenyl)(2,4,6-trimethylphenyl)methanone, 162
[48177-42-0] (4-Ethenylphenyl)(2-hydroxy-4-methoxyphenyl)methanone, 269
[50454-58-5] (3-Hydroxy-2-methylphenyl)(2-methylphenyl)methanone, 258
[50537-80-9] (5-Ethyl-2,4-dihydroxyphenyl)phenylmethanone, 355
[50597-28-9] (2-Hydroxy-6-methylphenyl)phenylmethanone, 64
[50685-40-0] (2-Chlorophenyl)(2,4-dihydroxyphenyl)methanone, 367
[50685-41-1] (5-Chloro-2,4-dihydroxyphenyl)(4-chlorophenyl)methanone, 386
[50685-42-2] (5-Chloro-2,4-dihydroxyphenyl)(2-chlorophenyl)methanone, 385
[50685-43-3] (2-Chlorophenyl)(5-hexyl-2,4-dihydroxyphenyl)methanone, 395
[50739-53-2] [4-(1,1-Dimethylethyl)phenyl](2-hydroxy-4-methoxyphenyl)methanone, 310
[51106-90-2] (2-Hydroxy-4,5-dimethoxyphenyl)(3-methoxyphenyl)methanone, 284
[51106-93-5] (5-Ethoxy-2-hydroxy-4-methoxyphenyl)(3-ethoxyphenyl)methanone, 311
[51339-44-7] (3,5-Dinitrophenyl)(4-hydroxyphenyl)methanone, 130
[51439-89-5] (4-Hydroxy-3-methoxyphenyl)phenylmethanone, 70
[51787-06-5] (3,5-Dihydroxy-4-nitrophenyl)phenylmethanone, 347
[51974-19-7] (2-Hydroxyphenyl)(2-methylphenyl)methanone, 147
[52117-23-4] (2,4-Dihydroxy-3-methylphenyl)phenylmethanone, 349
[52189-86-3] (4,4'-Dihydroxy[1,1'-biphenyl]-3-yl)phenylmethanone, 476
[52196-46-0] [3-(1,1-Dimethylethyl)-2-hydroxy-5-methylphenyl]phenylmethanone, 106
[52196-47-1] [5-Chloro-3-(1,1-dimethylethyl)-2-hydroxyphenyl]phenylmethanone, 98
[52220-71-0] (2,4-Dihydroxy-5-methylphenyl)phenylmethanone, 349
[52220-72-1] [2-Hydroxy-3-methyl-4-(octyloxy)phenyl]phenylmethanone, 120
[52220-73-2] [2-Hydroxy-5-methyl-4-(octyloxy)phenyl]phenylmethanone, 120
[52479-85-3] (2,3,4-Trihydroxyphenyl)(3,4,5-trihydroxyphenyl)methanone, 33
[52591-10-3] (4-Hydroxyphenyl)(2,4,6-trihydroxyphenyl)methanone, 28
[52811-37-7] (2,5-Dihydroxy-4-methoxyphenyl)phenylmethanone, 352

[59746-91-7] (2,4-Dihydroxy-3-nitrophenyl)phenylmethanone, 345
[59746-92-8] (5-Hexyl-2,4-dihydroxyphenyl)phenylmethanone, 361
[59746-93-9] (2,4-Dichlorophenyl)(2-hydroxy-4-methylphenyl)methanone, 203
[59746-94-0] (2,4-Dichlorophenyl)[2-hydroxy-5-(1-methylpropyl)phenyl]methanone, 288
[59746-95-1] (2-Chlorophenyl)[2-hydroxy-5-(1-methylpropyl)phenyl]methanone, 293
[59746-96-2] (4-Chlorophenyl)[2-hydroxy-5-(1-methylpropyl)phenyl]methanone, 294
[59746-97-3] [2-Hydroxy-5-(1-methylpropyl)phenyl]phenylmethanone, 103
[59802-03-8] (2-Hydroxy-5-isononylphenyl)phenylmethanone, 119
[59954-92-6] (2-Hydroxy-5-methoxy-4-methylphenyl)phenylmethanone, 84
[59954-93-7] (2,5-Dihydroxy-4-methylphenyl)phenylmethanone, 350
[59954-97-1] (2-Hydroxy-4-methoxy-5-methylphenyl)phenylmethanone, 84
[60013-02-7] (3,4-Dichlorophenyl)(4-hydroxyphenyl)methanone, 129
[60014-09-7] (4-Amino-3-nitrophenyl)(4-hydroxyphenyl)methanone, 142
[60044-21-5] (4-Chlorophenyl)(4-hydroxyphenyl)methanone-^{14}C, 136
[60138-98-9] (3-Chloro-6-hydroxy-4-methoxy-2,5-dimethylphenyl)(2,4-dihydroxy-6-methyl-phenyl)methanone, 456
[60302-91-2] (2-Amino-3-hydroxy-5-nitrophenyl)phenylmethanone, 51
[60487-86-7] Phenyl(3,4,5-trihydroxyphenyl)methanone, 18
[60556-46-9] (2,6-Dihydroxy-4-methoxyphenyl)(2,4-dihydroxy-6-methylphenyl)-methanone, 464
[60556-49-2] (4-Hydroxy-2-methoxy-6-methylphenyl)(2,4,6-trihydroxyphenyl)methanone, 462
[60805-30-3] (4-Chloro-3-hydroxyphenyl)(4-chlorophenyl)methanone, 187
[60805-31-4] (4-Chloro-2-hydroxyphenyl)(4-chlorophenyl)methanone, 186
[61002-51-5] (4-Chlorophenyl)(4-hydroxy-2-methylphenyl)methanone, 217
[61002-52-6] (3-Chlorophenyl)(4-hydroxyphenyl)methanone, 134
[61002-53-7] (2,6-Dichlorophenyl)(4-hydroxyphenyl)methanone, 129
[61002-54-8] (4-Hydroxyphenyl)(4-methoxyphenyl)methanone, 153
[61002-55-9] (2,6-Dimethylphenyl)(4-hydroxyphenyl)methanone, 155
[61002-59-3] (4-Chlorophenyl)(4-hydroxy-3,5-dimethylphenyl)methanone, 249
[61101-84-6] (2-Chlorophenyl)(2,4,6-trihydroxyphenyl)methanone, 437
[61101-86-8] (2,6-Dimethoxyphenyl)(2,4,6-trihydroxyphenyl)methanone, 441
[61101-87-9] (2,4-Dichlorophenyl)(2,4,6-trihydroxyphenyl)methanone, 437
[61101-88-0] (4-Hydroxyphenyl)(2-nitrophenyl)methanone, 140
[61227-12-1] (2,4-Dihydroxy-6-methoxyphenyl)(3-hydroxyphenyl)methanone, 445
[61227-13-2] (2,4-Dihydroxy-3-methylphenyl)(3-hydroxyphenyl)methanone, 443
[61227-14-3] (2,4-Dihydroxy-5-methylphenyl)(3-hydroxyphenyl)methanone, 444
[61227-15-4] (2,4-Dihydroxyphenyl)(5-hydroxy-2-methoxyphenyl)methanone, 447
[61227-16-5] (2,4-Dihydroxy-6-methylphenyl)(5-hydroxy-2-methoxyphenyl)methanone, 451
[61227-17-6] (2,4-Dihydroxy-3-methylphenyl)(5-hydroxy-2-methoxyphenyl)methanone, 450
[61234-44-4] (2,4-Dihydroxyphenyl)(2,5-dihydroxyphenyl)methanone, 25
[61234-45-5] (2,4-Dihydroxy-5-methylphenyl)(2,5-dihydroxyphenyl)methanone, 460
[61234-46-6] (2,4-Dihydroxy-3-methylphenyl)(2,5-dihydroxyphenyl)methanone, 459
[61234-68-2] (2,4-Dihydroxy-5-methylphenyl)(3,5-dihydroxyphenyl)methanone, 460
[61445-49-6] Bis(3,4-dihydroxyphenyl)methanone, 23
[61445-50-9] (2,4-Dihydroxyphenyl)(3,4-dihydroxyphenyl)methanone, 25
[61445-51-0] (3,4-Dihydroxyphenyl)(2,3,4-trihydroxyphenyl)methanone, 30
[61445-52-1] (3,4-Dihydroxyphenyl)(2,4,5-trihydroxyphenyl)methanone, 30
[61466-73-7] (4-Chlorophenyl)(2-ethyl-4-hydroxyphenyl)methanone, 247
[61466-78-2] (2,4-Dichlorophenyl)(3-ethyl-2-hydroxyphenyl)methanone, 241
[61466-80-6] (4-Chlorophenyl)(3-ethyl-2-hydroxyphenyl)methanone, 247
[61466-81-7] (4-Chlorophenyl)(2-hydroxy-5-propylphenyl)methanone, 270
[61466-83-9] (2,4-Dichlorophenyl)(5-ethyl-2-hydroxyphenyl)methanone, 241
[61466-85-1] (3-Chlorophenyl)(5-ethyl-2-hydroxyphenyl)methanone, 246
[61466-87-3] (3,4-Dichlorophenyl)(5-ethyl-2-hydroxyphenyl)methanone, 242
[61466-88-4] (4-Ethyl-2-hydroxyphenyl)(3-fluorophenyl)methanone, 250
[61709-37-3] (2,4-Dichlorophenyl)[5-(1,1-dimethylethyl)-2-hydroxyphenyl]methanone, 287
[61736-69-4] (2-Nitrophenyl)(2,4,6-trihydroxyphenyl)methanone, 439
[61736-72-9] (2-Aminophenyl)(2-hydroxy-4,6-dimethoxyphenyl)methanone, 266
[61736-75-2] (2-Hydroxy-4,6-dimethoxyphenyl)(2-nitrophenyl)methanone, 255
[61750-25-2] (5-Ethyl-2-hydroxyphenyl)(4-methylphenyl)methanone, 274
[61750-26-3] (4-Chlorophenyl)(4,5-diethyl-2-hydroxyphenyl)methanone, 292
[61750-29-6] (5-Ethyl-2-hydroxyphenyl)[4-(trifluoromethyl)phenyl]methanone, 267

[72614-88-1] (3,5-Dichloro-2,6-dihydroxy-4-methoxyphenyl)(4-hydroxy-2-methyl-6-propoxy-phenyl)methanone, 456
[73720-57-7] [4-Hydroxy-3-(1-methyl-2-propenyl)phenyl]phenylmethanone, 96
[73720-75-9] [4-Hydroxy-3-(2-propenyl)phenyl]phenylmethanone, 89
[74079-07-5] [2-Hydroxy-4-methoxy-3-(2-propenyl)phenyl](4-methoxyphenyl)methanone, 306
[74167-86-5] (4-Chlorophenyl)(3-hydroxy-2-methylphenyl)methanone, 217
[74167-87-6] (3-Hydroxy-2-methylphenyl)phenylmethanone, 65
[74167-88-7] (2,4-Dichlorophenyl)(3-hydroxy-2-methylphenyl)methanone, 204
[74167-89-8] (3-Hydroxy-2-methylphenyl)(4-methylphenyl)methanone, 258
[74167-90-1] (2,4-Dimethylphenyl)(3-hydroxyphenyl)methanone, 155
[74177-55-2] (4-Chlorophenyl)(3-hydroxy-4-methylphenyl)methanone, 217
[74627-90-0] [2-Hydroxy-4,6-bis(methoxymethoxy)-3-methylphenyl]phenylmethanone, 108
[74627-92-2] [4,6-Dihydroxy-3-methyl-2-(phenylmethoxy)phenyl]phenylmethanone, 362
[74627-93-3] [6-Hydroxy-4-methoxy-3-methyl-2-(phenylmethoxy)phenyl]phenyl-methanone, 119
[74628-36-7] (4-Hydroxy-2,6-dimethoxyphenyl)(2-hydroxy-4-methoxy-6-methylphenyl)-methanone, 430
[74628-37-8] (2-Hydroxy-4-methoxy-6-methylphenyl)(2,4,6-trimethoxyphenyl)methanone, 313
[74697-55-5] Bis[4-hydroxy-3-(hydroxymethyl)phenyl]methanone, 419
[75060-50-3] (4-Bromophenyl)[5-(1,1-dimethylethyl)-2-hydroxyphenyl]methanone, 290
[75060-57-0] [5-(1,1-Dimethylethyl)-2-hydroxyphenyl][4-(methylthio)phenyl]methanone, 309
[75060-63-8] [3-(Aminomethyl)-5-(1,1-dimethylethyl)-2-hydroxyphenyl]cyclohexylmethanone *(Hydrochloride)*, 484
[75060-64-9] [3-(Aminomethyl)-5-(1,1-dimethylethyl)-2-hydroxyphenyl]phenylmethanone *(Hydrochloride)*, 108
[75060-65-0] [3-(Aminomethyl)-5-(1,1-dimethylethyl)-2-hydroxyphenyl](4-bromophenyl)-methanone *(Hydrochloride)*, 307
[75060-69-4] [3-(Aminomethyl)-5-(1,1-dimethylethyl)-2-hydroxyphenyl][4-(methylthio)-phenyl]methanone *(Hydrochloride)*, 323
[75060-98-9] [3-(Aminomethyl)-5-(1,1-dimethylethyl)-2-hydroxyphenyl]cyclohexyl-methanone, 484
[75060-99-0] [3-(Aminomethyl)-5-(1,1-dimethylethyl)-2-hydroxyphenyl]phenyl-methanone, 108
[75061-00-6] [3-(Aminomethyl)-5-(1,1-dimethylethyl)-2-hydroxyphenyl](4-bromo-phenyl)methanone, 307
[75061-01-7] [3-(Aminomethyl)-5-(1,1-dimethylethyl)-2-hydroxyphenyl][4-(methylthio)-phenyl]methanone, 323
[75440-84-5] Bis(2,3,4-trihydroxyphenyl)methanone, 32
[75629-21-9] (3,4-Dihydroxyphenyl)(2,4,6-trihydroxyphenyl-1,3,5-$^{14}C_3$)methanone, 31
[75731-48-5] (3-Hydroxy-4-methylphenyl)(4-hydroxyphenyl)methanone, 403
[75731-50-9] (4'-Hydroxy[1,1'-biphenyl]-4-yl)(3-hydroxyphenyl)methanone, 477
[75919-94-7] [5-(1,1-Dimethylethyl)-2-hydroxyphenyl](4-methylphenyl)methanone, 308
[76013-33-7] (2-Hydroxy-4,6-dimethoxyphenyl)(2,4,5-trimethoxyphenyl)methanone, 316
[76015-48-0] [3-(3,7-Dimethyl-2,6-octadienyl)-2,4,6-trihydroxyphenyl]phenyl-methanone *(Z)*, 436
[76237-02-0] (5-Chloro-2-hydroxyphenyl)(2-hydroxyphenyl)methanone, 399
[76346-15-1] [3-(Benzoyloxy)-4-hydroxyphenyl]phenylmethanone, 112
[76346-16-2] [1,2-Ethanediylbis(6-hydroxy-3,1-phenylene)]bis[phenylmethanone, 513
[76444-61-6] [2,4-Dihydroxy-6-methoxy-3,5-bis(3-methyl-2-butenyl)phenyl]phenyl-methanone, 364
[76631-09-9] (2,4-Dihydroxy-6-methylphenyl)(2,4,6-trimethoxyphenyl)methanone, 394
[76981-50-5] (3-Hydroxyphenyl)(2,4,6-trimethylphenyl)methanone, 162
[76981-53-8] (3-Hydroxyphenyl)(4-phenoxyphenyl)methanone, 166
[76981-57-2] (2,4-Dimethylphenyl)(3-hydroxy-2-methylphenyl)methanone, 273
[76981-65-2] (3,4-Dichlorophenyl)(3-hydroxy-2-methylphenyl)methanone, 205
[77151-84-9] (2,3-Dichlorophenyl)(2-hydroxy-5-methylphenyl)methanone, 203
[77156-44-6] (2,6-Dichlorophenyl)(2-hydroxy-4-methoxyphenyl)methanone, 208
[77347-19-4] (2-Hydroxy-5-methyl-1,3-phenylene)bis[phenylmethanone, 493
[78023-64-0] [3-Chloro-6-hydroxy-4-methoxy-5-methyl-2-(1-methylpropyl)phenyl]-(2,4-dihydroxy-6-methylphenyl)methanone, 458
[78044-92-5] (2,4-Dimethoxy-6-methylphenyl)(2-hydroxy-5-methoxyphenyl)methanone, 300
[78044-94-7] (2,5-Dihydroxyphenyl)(2,4-dimethoxy-6-methylphenyl)methanone, 380

[134612-52-5] (3,4-Dihydroxy-5-nitrophenyl)(3,4-dihydroxyphenyl)methanone, 458
[134612-73-0] (3-Fluorophenyl)(4-hydroxy-3-methoxy-5-nitrophenyl)methanone, 208
[134612-74-1] (4-Fluorophenyl)(4-hydroxy-3-methoxy-5-nitrophenyl)methanone, 209
[134612-75-2] (2,6-Difluorophenyl)(4-hydroxy-3-methoxy-5-nitrophenyl)methanone, 197
[134612-76-3] (2-Chlorophenyl)(4-hydroxy-3-methoxy-5-nitrophenyl)methanone, 199
[134612-77-4] (3-Chlorophenyl)(4-hydroxy-3-methoxy-5-nitrophenyl)methanone, 199
[134612-78-5] (4-Chlorophenyl)(4-hydroxy-3-methoxy-5-nitrophenyl)methanone, 200
[134612-79-6] (4-Hydroxy-3-methoxy-5-nitrophenyl)(2-methylphenyl)methanone, 254
[134612-80-9] (4-Hydroxy-3-methoxy-5-nitrophenyl)(4-methylphenyl)methanone, 254
[134612-82-1] (4-Hydroxy-3-methoxy-5-nitrophenyl)[2-(trifluoromethyl)phenyl]methanone, 236
[134612-83-2] (3,4-Dimethoxyphenyl)(4-hydroxy-3-nitrophenyl)methanone, 255
[134612-84-3] (4-Chlorophenyl)(3,4-dihydroxyphenyl)methanone, 369
[134994-27-7] (2,3-Dimethylphenyl)(4-hydroxyphenyl)methanone, 154
[136134-35-5] (4-Amino-3-hydroxyphenyl)(4-hydroxyphenyl)methanone, 401
[136134-36-6] [3-Hydroxy-4-(methylamino)phenyl](4-hydroxyphenyl)methanone, 407
[136134-37-7] [3-Hydroxy-4-(methylamino)phenyl](4-methoxyphenyl)methanone, 266
[136741-43-0] (5-Chloro-2-hydroxy-4-methoxyphenyl)(2-chlorophenyl)methanone, 206
[136741-44-1] (5-Chloro-2-hydroxy-4-methoxyphenyl)(4-methoxyphenyl)methanone, 249
[136741-45-2] (5-Chloro-2-hydroxy-4-methoxyphenyl)(2,6-difluorophenyl)methanone, 195
[136741-46-3] (5-Chloro-2-hydroxy-4-methoxyphenyl)(2,4-difluorophenyl)methanone, 195
[136741-50-9] (2-Chloro-6-hydroxy-4-methoxyphenyl)phenylmethanone, 58
[138250-28-9] Phenyl[2,3,4-trihydroxy-5-(hydroxymethyl)phenyl]methanone, 436
[138250-29-0] Phenyl[2,3,4-trihydroxy-5-[(2,4,6-trihydroxyphenyl)methyl]phenyl]-
               methanone, 511
[140158-57-2] [4-[(3,7-Dimethyl-2,6-octadienyl)oxy]-2-hydroxy-6-methoxyphenyl]phenyl-
               methanone *(E)*, 121
[140660-43-1] (2,5-Dihydroxyphenyl)(2-methoxyphenyl)methanone, 377
[140665-22-1] (2-Chlorophenyl)(2-hydroxy-4,5-dimethoxyphenyl)methanone, 249
[140665-23-2] (2-Fluorophenyl)(2-hydroxy-4,5-dimethoxyphenyl)methanone, 252
[140665-35-6] (5-Chloro-2-hydroxy-3,4-dimethoxyphenyl)phenylmethanone, 75
[140665-36-7] (2-Chlorophenyl)(2-hydroxy-3,4-dimethoxyphenyl)methanone, 249
[140665-37-8] (2-Fluorophenyl)(2-hydroxy-3,4-dimethoxyphenyl)methanone, 252
[140665-38-9] (3-Fluorophenyl)(2-hydroxy-3,4-dimethoxyphenyl)methanone, 252
[140665-39-0] (4-Fluorophenyl)(2-hydroxy-3,4-dimethoxyphenyl)methanone, 252
[140665-40-3] (5-Chloro-2-hydroxy-3,4-dimethoxyphenyl)(4-fluorophenyl)methanone, 239
[140665-41-4] (2-Hydroxy-3,4-dimethoxyphenyl)(2-methylphenyl)methanone, 278
[140665-42-5] (2-Hydroxy-3,4-dimethoxyphenyl)[3-(trifluoromethyl)phenyl]methanone, 267
[140708-51-6] [3-Chloro-2,4 (or 2,5)-dihydroxy-5 (or 4)-methoxyphenyl]phenylmethanone, 348
[140708-53-8] [3-Chloro-2,4 (or 2,5)-dihydroxy-5 (or 4)-methoxyphenyl](2-fluorophenyl)-
               methanone, 388
[143815-11-6] (3-Bromo-2-hydroxy-5,6-dimethylphenyl)phenylmethanone, 73
[143815-12-7] (3-Bromo-2-hydroxy-4,5-dimethylphenyl)phenylmethanone, 73
[143815-13-8] (4-Bromo-6-hydroxy-2,3-dimethylphenyl)phenylmethanone, 74
[143815-17-2] [3-Bromo-2-hydroxy-6-methyl-5-(1-methylethyl)phenyl]phenylmethanone, 97
[143824-87-7] (2,4-Dimethylphenyl)(2-hydroxyphenyl)methanone, 154
[145300-05-6] (2-Fluoro-5-hydroxyphenyl)phenylmethanone, 45
[145723-29-1] [2-(Acetyloxy)phenyl](4-hydroxyphenyl)methanone, 153
[145746-55-0] [3,6-Dihydroxy-2-(phenylsulfonyl)phenyl]phenylmethanone, 360
[145747-24-6] [2-(Acetyloxy)-4-hydroxyphenyl]phenylmethanone, 72
[145804-70-2] (2-Hydroxy-5-methylphenyl)(2-hydroxy-5-nitrophenyl)methanone, 426
[147029-76-3] (4-Hydroxy-3-methylphenyl)(2-methylphenyl)methanone, 259
[147029-77-4] (3-Hydroxyphenyl)(4-nitrophenyl)methanone, 140
[147029-78-5] (3-Hydroxyphenyl)(2-methylphenyl)methanone, 149
[147029-79-6] (2-Hydroxy-5-methylphenyl)(2-methylphenyl)methanone, 257
[147167-72-4] (4-Hydroxy-6-methyl-1,3-phenylene)bis[(2-chlorophenyl)methanone, 497
[147188-04-3] (2-Hydroxy-4,6-dimethoxyphenyl)(2-methoxyphenyl)methanone, 284
[147188-05-4] (2-Hydroxyphenyl)(2,4,6-trimethoxyphenyl)methanone, 163
[147188-07-6] (2-Hydroxyphenyl)(2,4,5-trimethoxyphenyl)methanone, 163
[147188-08-7] (2-Hydroxy-3,4-dimethoxyphenyl)(2-methoxyphenyl)methanone, 283
[147188-09-8] (2-Hydroxyphenyl)(2,3,4-trimethoxyphenyl)methanone, 163
[147188-10-1] (2,3-Dimethoxyphenyl)(2-hydroxy-4-methoxyphenyl)methanone, 279
[147188-11-2] (2,4-Dimethoxyphenyl)(2-hydroxy-6-methoxyphenyl)methanone, 281

USUAL NAMES INDEX

COMMON ABBREVIATIONS

Common abbreviations used in the dictionary for organic chemistry

Å	Angström units
b.p.	Boiling point (for example, b.p.$_{0.1}$ 100° means boils at 100° if the pressure is 0.1 mm Hg)
(d)	Decomposition
20°	20 degrees Celsius
d	Density (for example, d^{20} specific gravity at 20°C referred to water at 4°C)
DDQ	2,3-Dichloro-5,6-dicyanobenzoquinone
EPR	Electron paramagnetic resonance
HMPT	Hexamethylphosphoric Triamide
^{13}C NMR	Nuclear magnetic resonance relative to carbon 13
(E)	Geometric stereodescriptor used for compounds having achiral elements resulting from double bonds where the groups of highest priority are on the opposite sides of the vertical reference plane
^{19}F NMR	Nuclear magnetic resonance relative to fluorine 19
GC	Gas chromatography
GC-MS	Gas chromatography-mass spectrometry
GLC	Gas-liquid chromatography
h	Hour
^{1}H NMR	Nuclear magnetic resonance relative to proton
HPLC	High pressure liquid chromatography
IR	Infrared (spectra)
iso-	Aliphatic hydrocarbon having two methyl groups on the terminal carbon atom of the chain (for example, isoamyl $(CH_3)_2CH-CH_2-CH_2-$)
LDA	lithium diisopropylamide
m-	Meta-
M	Molar (concentration)
min	Minute

mol	Molecule
mol.wt.	Molecular weight
m.p.	Melting point
MS	Mass spectra
n-	Normal, as n-butyl
N	Normal (equivalents per liter, as applied to concentration)
NA	Not available
N.B.:	Nota bene
n_D^{20}	Index of refraction (n_D^{20} for 20°C and sodium light)
o-	Ortho-
p-	Para-
Pa	pascal
Pd/C	Palladium on charcoal
pK_a	Log of the reciprocal of the dissociation constant, $1/\log K_a$
Rh/C	Rhodium on charcoal
r.t.	Room temperature
Sadtler	Sadtler Research Laboratories, Philadelphia (USA)
SDS	Sodium dodecyl sulfate
sec-	Secondary
SM	Starting material
tert-	Tertiary-
TFAA	Trifluoroacetic anhydride
TFMS	Trifluoromethanesulfonic acid
TLC	Thin layer chromatography
UV	Ultraviolet (spectra)
Vol.	Volume
(Z)	Opposite of (E)